30 *Years*

Previous Solved Papers

GATE 2025

Computer Science & Information Technology

- Fully solved with explanations
- Analysis of previous papers
- Topicwise presentation
- Thoroughly revised & updated

MADE EASY PUBLICATIONS

MADE EASY Publications Pvt. Ltd.

Corporate Office: 44-A/4, Kalu Sarai (Near Hauz Khas Metro Station), New Delhi-110016 | **Ph. :** 9021300500, 8860378007

Email : infomep@madeeasy.in | **Web :** www.madeeasypublications.org

GATE - 2025

Computer Science & IT

Topicwise Previous GATE Solved Papers (1995-2024)

Editions

1st Edition	:	2007
2nd Edition	:	2008
3rd Edition	:	2009
4th Edition	:	2010
5th Edition	:	2011
6th Edition	:	2012
7th Edition	:	2013
8th Edition	:	2014
9th Edition	:	2015
10th Edition	:	2016
11th Edition	:	2017
12th Edition	:	2018
13th Edition	:	2019
14th Edition	:	2020
15th Edition	:	2021
16th Edition	:	2022
17th Edition	:	2023
18th Edition	**:**	**2024**

Preface

B. Singh (Ex. IES)

Over the period of time the GATE examination has become more challenging due to increasing number of candidates. Though every candidate has ability to succeed but competitive environment, in-depth knowledge, quality guidance and good source of study is required to achieve high level goals.

The new edition of **GATE 2025 Solved Papers : Computer Science & Information Technology** has been fully revised, updated and edited. The whole book has been divided into topicwise sections.

At the beginning of each subject, analysis of previous papers are given to improve the understanding of subject.

I have true desire to serve student community by way of providing good source of study and quality guidance. I hope this book will be proved an important tool to succeed in GATE examination. Any suggestions from the readers for the improvement of this book are most welcome.

B. Singh (Ex. IES)
Chairman and Managing Director
MADE EASY Group

GATE-2025

Computer Science & IT

CONTENTS

UNIT

I

Discrete & Engineering Mathematics

CONTENTS

Discrete & Engineering Mathematics

Syllabus

Mathematical Logic: Propositional and first order logic.

Set Theory & Algebra: Sets, relations, functions, partial orders and lattices. Groups.

Combinatorics: Counting, recurrence relations, generating functions.

Graph Theory: Connectivity, matching, coloring.

Probability: Random variables. Uniform, normal, exponential, poisson and binomial distributions. Mean, median, mode and standard deviation. Conditional probability and Bayes theorem.

Linear Algebra: Matrices, determinants, system of linear equations, eigenvalues and eigenvectors, LU decomposition.

Calculus: Limits, continuity and differentiability. Maxima and minima. Mean value theorem. Integration.

Analysis of Previous GATE Papers

Exam Year	1 Mark Ques.	2 Marks Ques.	Total Marks
1995	4	4	**12**
1996	7	6	**19**
1997	3	2	**7**
1998	6	7	**20**
1999	2	3	**8**
2000	2	5	**12**
2001	4	5	**14**
2002	5	4	**13**
2003	5	15	**35**
2004	5	11	**27**
2005	5	10	**25**
2006	3	10	**23**
2007	4	9	**22**
2008	4	10	**24**
2009	4	6	**16**
2010	6	8	**24**
2011	–	5	**10**
2012	6	5	**16**
2013	6	3	**12**

Exam Year	1 Mark Ques.	2 Marks Ques.	Total Marks
2014 Set-1	6	9	**24**
2014 Set-2	5	8	**21**
2014 Set-3	7	8	**23**
2015 Set-1	5	8	**21**
2015 Set-2	5	8	**21**
2015 Set-3	7	6	**19**
2016 Set-1	5	4	**13**
2016 Set-2	6	3	**12**
2017 Set-1	2	5	**12**
2017 Set-2	5	5	**15**
2018	6	5	**16**
2019	6	3	**12**
2020	3	5	**13**
2021 Set-1	5	6	**17**
2021 Set-2	5	4	**13**
2022	5	7	**19**
2023	6	5	**16**
2024 Set-1	5	4	**13**
2024 Set-2	5	4	**13**

1 Mathematical Logic

1.1 If the proposition $\neg p \Rightarrow q$ is true, then the truth value of the proposition $\neg p \vee (p \Rightarrow q)$, where $\neg$ is negation, '$\vee$' is inclusive or and '$\Rightarrow$' is implication, is

(a) True
(b) Multiple valued
(c) False
(d) Cannot be determined

[1995 : 2 M]

1.2 Which one of the following is false? Read $\wedge$ as AND, $\vee$ as OR, $\sim$ as NOT, $\rightarrow$ as one way implication and $\leftrightarrow$ as two way implication.

(a) $((x \rightarrow y) \wedge x) \rightarrow y$
(b) $((\sim x \rightarrow y) \wedge (\sim x \rightarrow \sim y)) \rightarrow x$
(c) $(x \rightarrow (x \vee y))$
(d) $((x \vee y) \leftrightarrow (\sim x \rightarrow \sim y))$

[1996 : 2 M]

1.3 Let a, b, c, d be propositions. Assume that the equivalence $a \leftrightarrow (b \vee \neg b)$ and $b \leftrightarrow c$ hold. Then the truth-value of the formula $(a \wedge b) \rightarrow (a \wedge c) \vee d$ is always

(a) True
(b) False
(c) Same as the truth-value of b
(d) Same as the truth-value of d

[2000 : 2 M]

1.4 What is the converse of the following assertion?
I stay only if you go

(a) I stay if you go
(b) If I stay then you go
(c) If you do not go then I do not stay
(d) If I do not stay then you go

[2001 : 1 M]

1.5 Consider two well-formed formulas in propositional logic:

$F_1 : P \Rightarrow \neg P$
$F_2 : (P \Rightarrow \neg P) \vee (\neg P \Rightarrow P)$

Which of the following statements is correct?

(a) F_1 is satisfiable, F_2 is valid
(b) F_1 is unsatisfiable, F_2 is satisfiable
(c) F_1 is unsatisfiable, F_2 is valid
(d) F_1 and F_2 are both satisfiable

[2001 : 1 M]

1.6 "If X then Y unless Z" is represented by which of the following formulas in propostional logic? ("$\neg$") is negation, "$\wedge$" is conjunction, and "$\rightarrow$" is implication)

(a) $(X \wedge \neg Z) \rightarrow Y$
(b) $(X \wedge Y) \rightarrow \neg Z$
(c) $X \rightarrow (Y \wedge \neg Z)$
(d) $(X \rightarrow Y) \wedge \neg Z$

[2002 : 1 M]

1.7 Which of the following is a valid first order formula? (Here α and β are first order formulae with x as their only free variable)

(a) $((\forall x)\ [\alpha] \Rightarrow (\forall x)[\beta]) \Rightarrow (\forall x)\ [\alpha \Rightarrow \beta]$
(b) $(\forall x)\ [\alpha] \Rightarrow (\exists x)\ [\alpha \wedge \beta]$
(c) $(\forall x)\ [\alpha \vee \beta\ ] \Rightarrow (\exists x)\ [\alpha] \Rightarrow (\forall x)\ [\alpha]$
(d) $(\forall x)\ [\alpha \Rightarrow \beta\ ] \Rightarrow ((\forall x)\ [\alpha] \Rightarrow (\forall x)\ [\beta])$

[2003 : 2 M]

1.8 Consider the following formula α and its two interpretations I_1 and I_2.

$\alpha : (\forall x)\ [P_x \Leftrightarrow (\forall y)\ [Q_{xy} \Leftrightarrow \neg Q_{yy}]] \Rightarrow (\forall x)\ [\neg P_x]$

I_1 : Domain : the set of natural numbers
$P_x \equiv$ 'x is a prime number'
$Q_{xy} \equiv$ 'y divides x'

I_2 : Same as I_1 except that $P_x =$ 'x is a composite number.'

Which of the following statements is true?

(a) I_1 satisfies α, I_2 does not
(b) I_2 satisfies α, I_1 does not
(c) Neither I_2 nor I_1 satisfies α
(d) Both I_1 and I_2 satisfy α

[2003 : 2 M]

1.9 The following resolution rule is used in logic programming: Derive clause $(P \vee Q)$ from clauses $(P \vee R)$, $(Q \vee \neg R)$
Which of the following statemnets related to this rule is FALSE?

(a) $(P \vee R) \wedge (Q \vee \neg R) \Rightarrow (P \vee Q)$ is logically valid
(b) $(P \vee Q) \Rightarrow (P \vee R) \wedge (Q \vee \neg R)$ is logically valid
(c) $(P \vee Q)$ is satisfiable if and only if $(P \vee R) \wedge (Q \vee \neg R)$ is satisfiable
(d) $(P \vee Q) \Rightarrow$ FALSE if and only if both P and Q are unsatisfiable

[2003 : 2 M]

1.10 Identify the correct translation into logical notation of the following assertion. Some boys in the class are taller than all the girls

Note: Taller (x, y) is true if x is taller than y.

(a) $(\exists x)$ (boy$(x) \rightarrow (\forall y)$ (girl$(y) \wedge$ taller (x, y)))

(b) $(\exists x)$ (boy$(x) \wedge (\forall y)$ (girl$(y) \wedge$ taller (x, y)))

(c) $(\exists x)$ (boy$(x) \rightarrow (\forall y)$ (girl$(y) \rightarrow$ taller (x, y)))

(d) $(\exists x)$ (boy$(x) \wedge (\forall y)$ (girl$(y) \rightarrow$ taller (x, y)))

[2004 : 1 M]

1.11 Let $a(x, y)$, $b(x, y,)$ and $c(x, y)$ be three statements with variables x and y chosen from some universe. Consider the following statement:

$$(\exists x)(\forall y)[(a(x,y) \wedge b(x,y)) \wedge \neg c(x,y)]$$

Which one of the following is its equivalent?

(a) $(\forall x)(\exists y)[(a(x,y) \vee b(x,y)) \rightarrow c(x,y)]$

(b) $(\exists x)(\forall y)[(a(x,y) \vee b(x,y)) \wedge \neg c(x,y)]$

(c) $\neg(\forall x)(\exists y)[(a(x,y) \wedge b(x,y)) \rightarrow c(x,y)]$

(d) $\neg(\forall x)(\exists y)[(a(x,y) \vee b(x,y)) \rightarrow c(x,y)]$

[2004 : 1 M]

1.12 Let p, q, r and s be four primitive statements. Consider the following arguments:

P: $[(\neg p \vee q) \wedge (r \rightarrow s) \wedge (p \vee r)] \rightarrow (\neg s \rightarrow q)$

Q: $[(\neg p \wedge q) \wedge [q \rightarrow (p \rightarrow r)]] \rightarrow \neg r$

R: $[[(q \wedge r) \rightarrow p] \wedge (\neg q \vee p)] \rightarrow r$

S: $[p \wedge (p \rightarrow r) \wedge (q \vee \neg r)] \rightarrow q$

Which of the above arguments are valid?

(a) P and Q only (b) P and R only

(c) P and S only (d) P, Q, R and S

[2004 : 2 M]

1.13 The following propositional statement is

$$(P \rightarrow (Q \vee R)) \rightarrow ((P \wedge Q) \rightarrow R)$$

(a) Satisfiable but not valid

(b) Valid

(c) A contradiction

(d) None of the above **[2004 : 2 M]**

1.14 Let P, Q and R be three atomic propositional assertions. Let X denote $(P \vee Q) \rightarrow R$ and Y denote $(P \rightarrow R) \vee (Q \rightarrow R)$. Which one of the following is a tautology?

(a) $X \equiv Y$ (b) $X \rightarrow Y$

(c) $Y \rightarrow X$ (d) $\neg Y \rightarrow X$ **[2005 : 2 M]**

1.15 What is the first order predicate calculus statement equivalent to the following? Every teacher is liked by some student

(a) $\forall(x)$ [teacher $(x) \rightarrow \exists (y)$ [student $(y) \rightarrow$ likes (y, x)]]

(b) $\forall(x)$ [teacher $(x) \rightarrow \exists (y)$ [student $(y) \wedge$ likes (y, x)]]

(c) $\exists(y)\ \forall(x)$ [teacher $(x) \rightarrow$ [student $(y) \wedge$ likes (y, x)]]

(d) $\forall(x)$ [teacher $(x) \wedge \exists (y)$ [student $(y) \rightarrow$ likes (y, x)]]

[2005 : 2 M]

1.16 Let $P(x)$ and $Q(x)$ be arbitrary predicates. Which of the following statements is always TRUE?

(a) $(\forall x(P(x) \vee Q(x))) \Rightarrow ((\forall x P(x)) \vee (\forall x Q(x)))$

(b) $(\forall x(P(x) \Rightarrow Q(x))) \Rightarrow ((\forall x P(x)) \Rightarrow (\forall x Q(x)))$

(c) $(\forall x(P(x) \Rightarrow (\forall x Q(x)))) \Rightarrow (\forall x(P(x) \Rightarrow Q(x)))$

(d) $((\forall x(P(x)) \Leftrightarrow (\forall x Q(x))) \Rightarrow (\forall x(P(x) \Leftrightarrow Q(x)))$

[2005 : 2 M]

1.17 Consider the following first order logic formula in which R is a binary relation symbol.

$$\forall x \forall y\ (R(x,y) \Rightarrow R(y,x))$$

The formula is

(a) Satisfiable and valid

(b) Satisfiable and so is its negation

(c) Unsatisfiable but its negation is valid

(d) Satisfiable but its negation is unsatisfiable

[2006 : 2 M]

1.18 Which one of the first order predicate calculus statements given below correctly expresses the following English statement?

Tigers and lions attack if they are hungry or threatened.

(a) $\forall x$ [(tiger $(x) \wedge$ lion $(x)) \rightarrow$ {(hungry $(x) \vee$ threatened $(x)) \rightarrow$ attacks (x)}]

(b) $\forall x$ [(tiger $(x) \vee$ lion $(x)) \rightarrow$ {(hungry $(x) \vee$ threatened $(x)) \wedge$ attacks (x)}]

(c) $\forall x$ [(tiger $(x) \vee$ lion $(x)) \rightarrow$ {(attacks $(x) \rightarrow$ (hungry $(x) \vee$ Threatened (x))}]

(d) $\forall x$ [(tiger $(x) \vee$ lion $(x)) \rightarrow$ {(hungry $(x) \vee$ threatened $(x)) \rightarrow$ attacks (x)}]

[2006 : 2 M]

1.19 Consider the following propositional statements:

$P_1: ((A \wedge B) \rightarrow C)) \equiv ((A \rightarrow C) \wedge (B \rightarrow C))$

$P_2: ((A \vee B) \rightarrow C)) \equiv ((A \rightarrow C) \vee (B \rightarrow C))$

Which one of the following is true?

(a) P_1 is a tautology, but not P_2

(b) P_2 is a tautology, but not P_1

(c) P_1 and P_2 are both tautologies

(d) Both P_1 and P_2 are not tautologies

[2006 : 2 M]

1.20 A logical binary relation $\odot$, is defined as follows:

A	B	A $\odot$ B
True	True	True
True	False	True
False	True	False
False	False	True

Let ~ be the unary negation (NOT) operator, with higher precedence, than ⊙. Which one of the following is equivalent to $A \wedge B$?

(a) $(\sim A \odot B)$ (b) $\sim (A \odot \sim B)$

(c) $\sim (\sim A \odot \sim B)$ (d) $\sim (\sim A \odot B)$

[2006 : 2 M]

1.21 Let Graph (x) be a predicate which denotes that x is a graph. Let Connected (x) be a predicate which denotes that x is connected. Which of the following first order logic sentences DOES NOT represent the statement; "Not every graph is connected"?

(a) $\neg \forall x\,(\text{Graph}(x) \Rightarrow \text{Connected}(x))$

(b) $\exists x\,(\text{Graph}(x) \wedge \neg \text{Connected}(x))$

(c) $\neg \forall x\,(\neg \text{Graph}(x) \vee \text{Connected}(x))$

(d) $\forall x\,(\text{Graph}(x) \Rightarrow \neg \text{Connected}(x))$

[2007 : 2 M]

1.22 Which of the following is TRUE about formulae in Conjunctive Normal Form?

(a) For any formula, there is a truth assignment for which at least half the clauses evaluate to true.

(b) For any formula, there is a truth assignment for a which all the clauses evaluate to true.

(c) There is a formula such that for each truth assignment at most one-fourth of the clauses evaluate to true.

(d) None of the above **[2007 : 2 M]**

1.23 Which one of these first-order logic formulae is valid?

(a) $\forall x(P(x) \Rightarrow Q(x)) \Rightarrow ((\forall x P(x)) \Rightarrow (\forall x Q(x)))$

(b) $\exists x(P(x) \vee Q(x)) \Rightarrow ((\exists x P(x)) \Rightarrow (\exists x Q(x)))$

(c) $\exists x(P(x) \wedge Q(x)) \Leftrightarrow ((\exists x P(x)) \wedge (\exists x Q(x)))$

(d) $\forall x \exists y\, P(x, y) \Rightarrow \exists y \forall x P(x, y)$

[2007 : 2 M]

1.24 Let *fsa* and *pda* be two predicates such that $fsa(x)$ means x is a finite state automaton, and $pda(y)$ means, that y is a pushdown automaton. Let equivalent be another predicate such that equivalent (a, b) means a and b are equivalent. Which of the following first order logic statement represents the following:

Each finite state automaton has an equivalent pushdown automaton.

(a) $\forall x\,(fsa(x) \Rightarrow \exists y\,(pda(y) \wedge \text{equivalent}\,(x, y)))$

(b) $\sim\forall y(\exists x\, fsa(x) \Rightarrow pda(y) \wedge \text{equivalent}\,(x, y))$

(c) $\forall x\, \exists y\,(fsa(x) \wedge pda(y) \wedge \text{equivalent}\,(x, y))$

(d) $\forall x\, \exists y\,(fsa(y) \wedge pda(x) \wedge \text{equivalent}\,(x, y))$

[2008 : 1 M]

1.25 Which of the following first order formulae is logically valid? Here $\alpha(x)$ is a first order formula with x as a free variable, and β is a first order formula with no free variable.

(a) $[\beta \rightarrow (\exists x, \alpha(x))] \rightarrow [\forall x, \beta \rightarrow \alpha(x)]$

(b) $[\exists x, \beta \rightarrow \alpha(x)] \rightarrow [\beta \rightarrow (\forall x, \alpha(x))]$

(c) $[(\exists x, \alpha(x)) \rightarrow \beta] \rightarrow [\forall x, \alpha(x) \rightarrow \beta]$

(d) $[(\forall x, \alpha(x)) \rightarrow \beta] \rightarrow [\forall x, \alpha(x) \rightarrow \beta]$

[2008 : 2 M]

1.26 Which of the following is the negation of $[(\forall x, \alpha \rightarrow (\exists y, \beta \rightarrow (\forall u, \exists v, \gamma))]$

(a) $[\exists x, \alpha \rightarrow (\forall y, \beta \rightarrow (\exists u, \forall v, \gamma))]$

(b) $[\exists x, \alpha \rightarrow (\forall y, \beta \rightarrow (\exists u, \forall v, \neg\gamma))]$

(c) $[\forall x, \neg\alpha \rightarrow (\exists y, \neg\beta \rightarrow (\forall u, \exists v, \neg\gamma))]$

(d) $[\forall x, \alpha \wedge (\exists y, \beta \wedge (\exists u, \forall v, \neg\gamma))]$

[2008 : 2 M]

1.27 P and Q are two propositions. which of the following logical expressions are equivalent?

1. $P \vee \sim Q$
2. $\sim (\sim P \wedge Q)$
3. $(P \wedge Q) \vee (P \wedge \sim Q) \vee (\sim P \wedge \sim Q)$
4. $(P \wedge Q) \vee (P \wedge \sim Q) \vee (\sim P \wedge Q)$

(a) Only 1 and 2 (b) Only 1, 2 and 3

(c) Only 1, 2 and 4 (d) All of these

[2008 : 2 M]

1.28 Which one of the following is the most appropriate logical formula to represent the statement: "Gold and silver ornaments are precious"

The following notations are used:

$G(x)$: x is a gold ornament

$S(x)$: x is a silver ornament

$P(x)$: x is precious

(a) $\forall x(P(x) \rightarrow (G(x) \wedge S(x)))$

(b) $\forall x(G(x) \wedge (S(x) \rightarrow P(x)))$

(c) $\exists x((G(x) \wedge S(x)) \rightarrow P(x))$

(d) $\forall x((G(x) \vee S(x)) \rightarrow P(x))$ **[2009 : 2 M]**

1.29 The binary operation □ is defined as follows:

P	Q	P□Q
T	T	T
T	F	T
F	T	F
F	F	T

Which one of the following is equivalent to $P \vee Q$?

(a) $\neg Q \,\square\, \neg P$ (b) $P \,\square\, \neg Q$

(c) $\neg P \,\square\, Q$ (d) $\neg P \,\square\, \neg Q$

[2009 : 2 M]

1.30 Consider the following well-formed formulae:

I. $\neg\forall x(P(x))$

II. $\neg\exists x(P(x))$

III. $\neg\exists x(\neg P(x))$

IV. $\exists x(\neg P(x))$

Which of the above are equivalent?

(a) I and III (b) I and IV

(c) II and III (d) II and IV

[2009 : 2 M]

1.31 Suppose the predicate $F(x, y, t)$ is used to represent the statement that person x can fool person y at time t. Which one of the statements below expresses best the meaning of the formula $\forall x\exists y\exists t(\neg F(x, y, t))$?

(a) Everyone can fool some person at some time

(b) No one can fool everyone all the time

(c) Everyone cannot fool some person all the time

(d) No one can fool some person at some time

[2010 : 2 M]

1.32 Which one of the following options is CORRECT given three positive integers x, y and z, and a predicate

$$P(x) = \neg(x=1) \wedge \forall y(\exists z(x = y * z) \Rightarrow (y = x) \vee (y = 1)$$

(a) $P(x)$ being true means that x is a prime number

(b) $P(x)$ being true means that x is a number other than 1

(c) $P(x)$ is always true irrespective of the value of x

(d) $P(x)$ being true means that x has exactly two factors other than 1 and x

[2011 : 2 M]

1.33 Consider the following logical inferences.

I_1 : If it rains then the cricket match will not be played.
The cricket match was played.
Inference: There was no rain.

I_2 : If it rains then the cricket match will not be played.
It did not rain.
Inference: The cricket match was played.

Which of the following is TRUE?

(a) Both I_1 and I_2 are correct inferences

(b) I_1 is correct but I_2 is not a correct inference

(c) I_1 is not correct but I_2 is a correct inference

(d) Both I_1 and I_2 are not correct inferences

[2012 : 1 M]

1.34 What is the correct translation of the following statement into mathematical logic?

"Some real numbers are rational"

(a) $\exists x$ (real $(x) \vee$ rational (x))

(b) $\forall x$ (real $(x) \rightarrow$ rational (x))

(c) $\exists x$ (real $(x) \wedge$ rational (x))

(d) $\exists x$ (rational $(x) \rightarrow$ real (x))

[2012 : 1 M]

1.35 What is the logical translation of the following statements?

"None of my friends are perfect"

(a) $\exists x(F(x) \wedge \neg P(x))$

(b) $\exists x(\neg F(x) \wedge P(x))$

(c) $\exists x(\neg F(x) \wedge \neg P(x))$

(d) $\neg\exists x(F(x) \wedge P(x))$

[2013 : 2 M]

1.36 Which one of the following is NOT logically equivalent to $\neg\exists x(\forall y(\alpha) \wedge \forall z(\beta))$

(a) $\forall x(\exists z(\neg\beta) \rightarrow \forall y(\alpha))$

(b) $\forall x(\forall z(\beta) \rightarrow \exists y(\neg\alpha))$

(c) $\forall x(\forall y(\alpha) \rightarrow \exists z(\neg\beta))$

(d) $\forall x(\exists y(\neg\alpha) \vee \exists z(\neg\beta))$

[2013 : 2 M]

1.37 Consider the statement:

"Not all that glitters is gold"

Predicate glitters(x) is true if x glitters and predicate gold(x) is true if x is gold. Which one of the following logical formulae represents the above statement?

(a) $\forall x$: glitters$(x) \Rightarrow \neg$gold(x)

(b) $\forall x$: gold$(x) \Rightarrow$ glitters(x)

(c) $\exists x$: gold$(x) \wedge \neg$glitters(x)

(d) $\exists x$: glitters$(x) \wedge \neg$gold(x)

[2014 (Set-1) : 1 M]

1.38 Which one of the following propositional logic formulas is TRUE when exactly two of p, q, and r are **TRUE**?

(a) $((p \leftrightarrow q) \wedge r) \vee (p \wedge q \wedge \sim r)$

(b) $(\sim(p \leftrightarrow q) \wedge r) \vee (p \wedge q \wedge \sim r)$

(c) $((p \rightarrow q) \wedge r) \vee (p \wedge q \wedge \sim r)$

(d) $(\sim(p \leftrightarrow q) \wedge r) \wedge (p \wedge q \wedge \sim r)$

[2014 (Set-1) : 2 M]

1.39 Which one of the following Boolean expressions is NOT a tautology?

(a) $((a \rightarrow b) \wedge (b \rightarrow c)) \rightarrow (a \rightarrow c)$

(b) $(a \leftrightarrow c) \rightarrow (\sim b \rightarrow (a \wedge c))$

(c) $(a \wedge b \wedge c) \rightarrow (c \vee a)$

(d) $a \rightarrow (b \rightarrow a)$

[2014 (Set-2) : 2 M]

1.40 Consider the following statements:

P: Good mobile phones are not cheap

Q: Cheap mobile phones are not good

L: P implies Q

M: Q implies P

N: P is equivalent to Q

Which one of the following about L, M, and N is **CORRECT**?

(a) Only L is TRUE (b) Only M is TRUE

(c) Only N is TRUE (d) L, M and N are TRUE.

[2014 (Set-3) : 1 M]

1.41 The CORRECT formula for the sentence, "not all rainy days are cold" is

(a) $\forall d$(Rainy $(d) \wedge \sim$Cold (d))

(b) $\forall d$($\sim$Rainy $(d) \rightarrow$ Cold (d))

(c) $\exists d$($\sim$Rainy $(d) \rightarrow$ Cold (d))

(d) $\exists d$(Rainy $(d) \wedge \sim$ Cold (d))

[2014 (Set-3) : 2 M]

1.42 Which one of the following is Not equivalent to $p \leftrightarrow q$?

(a) $(\neg p \vee q) \wedge (p \vee \neg q)$

(b) $(\neg p \vee q) \wedge (q \rightarrow p)$

(c) $(\neg p \wedge q) \vee (p \wedge \neg q)$

(d) $(\neg p \wedge \neg q) \vee (p \wedge q)$ **[2015 (Set-1) : 1 M]**

1.43 The binary operator $\neq$ is defined by the following truth table.

p	q	$p \neq q$
0	0	0
0	1	1
1	0	1
1	1	0

Which one of the following is true about the binary operator $\neq$?

(a) Both commutative and associative

(b) Commutative but not associative

(c) Not commutative but associative

(d) Neither commutative nor associative

[2015 (Set-1) : 2 M]

1.44 Consider the following two statements:

S_1: If a candidate is known to be corrupt, then he will not be elected.

S_2: If a candidate is kind, he will be elected.

Which one of the following statements follows from S_1 and S_2 as per sound inference rules of logic?

(a) If a person is known to be corrupt, he is kind

(b) If a person is not known to be corrupt, he is not kind

(c) If a person is kind, he is not known to be corrupt

(d) If a person is not kind, he is not known to be corrupt **[2015 (Set-2) : 1 M]**

1.45 Which one of the following well formed formulae is a tautology?

(a) $\forall x \, \exists y \, R(x, y) \leftrightarrow \exists y \, \forall x \, R(x, y)$

(b) $(\forall x \, [\exists y \, R(x, y) \rightarrow S(x, y)]) \rightarrow \forall x \, \exists y \, S(x, y)$

(c) $[\forall x \, \exists y \, (P(x, y) \rightarrow R(x, y))] \leftrightarrow [\forall x \, \exists y \, (\neg P(x, y) \vee R(x, y))]$

(d) $\forall x \, \forall y \, P(x, y) \rightarrow \forall x \, \forall y \, P(y, x)$

[2015 (Set-2) : 2 M]

1.46 In a room there are only two types of people, namely Type 1 and Type 2. Type 1 people always tell the truth and Type 2 people always lie. You give a fair coin to a person in that room, without knowing which type he is from and tell him to toss it and hide the result from you till you ask for it. Upon asking, the person replies the following:

"The result of the toss is head if and only if I am telling the truth."

Which of the following options is correct?

(a) The result is head

(b) The result is tail

(c) If the person is of Type 2, then the result is tail

(d) If the person is of Type 1, then the result is tail **[2015 (Set-3) : 1 M]**

1.47 Let p, q, r, s represent the following propositions.

p : $x \in \{8, 9, 10, 11, 12\}$

q : x is a composite number

r : x is a perfect square

s : x is a prime number

The integer $x \geq 2$ which satisfies $\neg((p \Rightarrow q) \wedge (\neg r \vee \neg s))$ is ______.

[2016 (Set-1) : 1 M]

1.48 Consider the following expressions:

(i) False (ii) Q

(iii) True (iv) $P \vee Q$

(v) $\neg Q \vee P$

The number of expressions given above that are logically implied by $P \wedge (P \Rightarrow Q)$ is ______.

[2016 (Set-2) : 1 M]

1.49 Which one of the following well-formed formulae in predicate calculus is **NOT** valid?

(a) $(\forall x \, p(x) \Rightarrow \forall x \, q(x)) \Rightarrow (\exists x \neg p(x) \vee \forall x \, q(x))$

(b) $(\exists x \, p(x) \vee \exists x \, q(x)) \Rightarrow \exists x (p(x) \vee q(x))$

(c) $\exists x (p(x) \wedge q(x)) \Rightarrow (\exists x \, p(x) \wedge \exists x \, q(x))$

(d) $\forall x (p(x) \vee q(x)) \Rightarrow (\forall x \, p(x) \vee \forall x \, q(x))$

[2016 (Set-2) : 2 M]

1.50 Consider the first-order logic sentence $F: \forall x(\exists y R(x,y))$. Assuming non-empty logical domains, which of the sentences below are implied by F?

I. $\exists y(\exists x\, R(x,y))$ II. $\exists y(\forall x\, R(x,y))$
III. $\forall y(\exists x\, R(x,y))$ IV. $\neg\exists x(\forall y\, \neg R(x,y))$

(a) IV only (b) I and IV only
(c) II only (d) II and III only

[2017 (Set-1) : 1 M]

1.51 The statement $(\neg p) \Rightarrow (\neg q)$ is logically equivalent to which of the statements below?

I. $p \Rightarrow q$ II. $q \Rightarrow p$
III. $(\neg q) \vee p$ IV. $(\neg p) \vee q$

(a) I only (b) I and IV only
(c) II only (d) II and III only

[2017 (Set-1) : 1 M]

1.52 Let p, q and r be propositions and the expression $(p \to q) \to r$ be a contradiction. Then, the expression $(r \to p) \to q$ is

(a) A tautology
(b) A contradiction
(c) Always TRUE when p is FALSE
(d) Always TRUE when q is TRUE

[2017 (Set-1) : 2 M]

1.53 Let p, q, r denote the statements "*It is raining*", "*It is cold*", and "*It is pleasant*", respectively. Then the statement "*It is not raining and it is pleasant, and it is not pleasant only if it is raining and it is cold*" is represented by

(a) $(\neg p \wedge r) \wedge (\neg r \to (p \wedge q))$
(b) $(\neg p \wedge r) \wedge ((p \wedge q) \to \neg r)$
(c) $(\neg p \wedge r) \vee ((p \wedge q) \to \neg r)$
(d) $(\neg p \wedge r) \vee (r \to (p \wedge q))$

[2017 (Set-2) : 1 M]

1.54 Consider the first order predicate formula φ:

$\forall x[(\forall z\, z \mid x \Rightarrow ((z = x) \vee (z = 1))) \Rightarrow \exists w\, (w > x) \wedge (\forall z\, z \mid w \Rightarrow ((w = z) \vee (z = 1)))]$

Here '$a \mid b$' denotes that 'a divides b', where a and b are integers. Consider the following sets:

S_1 : {1, 2, 3, ..., 100}
S_2 : Set of all positive integers
S_3 : Set of all integers

Which of the above sets satisfy φ?

(a) S_1 and S_3 (b) S_2 and S_3
(c) S_1, S_2 and S_3 (d) S_1 and S_2

[2019 : 2 M]

1.55 Which one of the following predicate formulae is NOT logically valid?

Note that W is a predicate formula without any free occurrence of x.

(a) $\exists x(p(x) \wedge W) \equiv \exists x\, p(x) \wedge W$
(b) $\forall x(p(x) \to W) \equiv \forall x\, p(x) \to W$
(c) $\exists x(p(x) \to W) \equiv \forall x\, p(x) \to W$
(d) $\forall x(p(x) \vee W) \equiv \forall x\, p(x) \vee W$ **[2020 : 2 M]**

1.56 Let p and q be two propositions. Consider the following two formulae in prepositional logic.

S_1 : $(\neg p \wedge (p \vee q)) \to q$
S_2 : $q \to (\neg p \wedge (p \vee q))$

Which one of the following choices is correct?

(a) Neither S_1 nor S_2 is a tautology.
(b) Both S_1 and S_2 are tautologies.
(c) S_1 is a tautology but S_2 is not a tautology.
(d) S_1 is not a tautology but S_2 is a tautology.

[2021 (Set-1) : 1 M]

1.57 Choose the correct choice(s) regarding the following propositional logic assertion S:

$$S : ((P \wedge Q) \to R) \to ((P \wedge Q) \to (Q \to R))$$

(a) S is a tautology.
(b) S is neither a tautology nor a contradiction.
(c) The antecedent of S is logically equivalent to the consequent of S.
(d) S is a contradiction. **[2021 (Set-2) : 1 M]**

1.58 Geetha has a conjecture about integers, which is of the form $\forall x(P(x) \Rightarrow \exists y\, Q(x,y))$, where P is a statement about integers, and Q is a statement about pairs of integers.

Which of the following (one or more) option(s) would imply Geetha's conjecture?

(a) $\exists x(P(x) \wedge \forall y\, Q(x,y))$
(b) $\forall x \forall y\, Q(x,y)$
(c) $\exists y\, \forall x(P(x) \Rightarrow Q(x,y))$
(d) $\exists x(P(x) \wedge \exists y\, Q(x,y))$ **[2023 : 1 M]**

1.59 Let p and q be the following propositions:

p : Fail grade can be given.
q : Student scores more than 50% marks.

Consider the statement: "*Fail grade cannot be given when student scores more than 50%marks.*"

Which one of the following is the CORRECT representation of the above statement in propositional logic?

(a) $q \to p$ (b) $q \to \neg p$
(c) $p \to q$ (d) $\neg p \to q$

[2024 (Set-2) : 1 M]

■■■■

Answers Mathematical Logic

1.1 (d)	**1.2** (d)	**1.3** (a)	**1.4** (a)	**1.5** (a)	**1.6** (a)	**1.7** (d)	**1.8** (d)	**1.9** (b)
1.10 (d)	**1.11** (c)	**1.12** (c)	**1.13** (a)	**1.14** (b)	**1.15** (b)	**1.16** (b)	**1.17** (b)	**1.18** (d)
1.19 (d)	**1.20** (d)	**1.21** (d)	**1.22** (a)	**1.23** (a)	**1.24** (a)	**1.25** (c)	**1.26** (d)	**1.27** (b)
1.28 (d)	**1.29** (b)	**1.30** (b)	**1.31** (b)	**1.32** (a)	**1.33** (b)	**1.34** (c)	**1.35** (d)	**1.36** (a)
1.37 (d)	**1.38** (b)	**1.39** (b)	**1.40** (d)	**1.41** (d)	**1.42** (c)	**1.43** (a)	**1.44** (c)	**1.45** (c)
1.46 (a)	**1.47** (11)	**1.48** (4)	**1.49** (d)	**1.50** (b)	**1.51** (d)	**1.52** (d)	**1.53** (a)	**1.54** (b)
1.55 (b)	**1.56** (c)	**1.57** (a, c)	**1.58** (b, c)	**1.59** (b)				

Explanations Mathematical Logic

1.1 (d)

p	q	$\neg p \rightarrow q$
0	0	0
0	1	1
1	0	1
1	1	1

Now since $\neg p \rightarrow q$ is given true, we reduce the truth table as follows:

p	q	$\neg p \rightarrow q$
0	1	1
1	0	1
1	1	1

In the reduced truth table we need to find the truth value of $\neg p \vee (p \rightarrow q) \equiv p' + (p \rightarrow q)$

$\equiv p' + p' + q \equiv p' + q$

The truth value of p′ + q in the reduced truth table is given below:

p	q	$p' + q$
0	1	1
1	0	0
1	1	1

Since in the reduced truth table also, the given expression is sometimes true and sometimes false, therefore the truth value of proposition $\neg p \vee (p \rightarrow q)$ can not be determined.

1.2 (d)

(a) $((x \rightarrow y) \wedge x) \rightarrow y$

$= \sim((\sim x \vee y) \wedge x) \vee y$

$= \sim((\sim x \wedge x) \vee (y \wedge x)) \vee y$

$= \sim(F \vee (y \wedge x) \vee y$

$= \sim(y \wedge x) \vee y = \sim y \vee \sim x \vee y$

$= (\sim y \vee y) \vee \sim x$

$= T \vee \sim x = T$

(b) $((\sim x \rightarrow y) \wedge (\sim x \rightarrow \sim y)) \rightarrow x$

$= \sim(\sim(\sim x) \vee y) \wedge (\sim(\sim x) \vee \sim y)) \vee x$

$= \sim((x \vee y) \wedge (x \vee \sim y)) \vee x$

$= \sim(x \vee (y \wedge \sim y)) \vee x = \sim(x \vee F) \vee x$

$= \sim x \vee x = T$

(c) $(x \rightarrow (x \vee y))$

$= (\sim x \vee (x \vee y)) = ((\sim x \vee x) \vee y)$

$= (T \vee y) = T$

(d) $((x \vee y) \leftrightarrow (\sim x \rightarrow \sim y))$

$= (x \vee y) \leftrightarrow (\sim (\sim x) \vee \sim y)$

$= (x \vee y) \leftrightarrow (x \vee \sim y)$

$= ((x \vee y) \wedge (x \vee \sim y)) \vee (\sim(x \vee y) \wedge \sim(x \vee \sim y))$

$= ((x \vee (y \wedge \sim y)) \vee ((\sim x \wedge \sim y) \wedge (\sim x \wedge y))$

$= (x \vee F) \vee ((\sim x \wedge \sim y) \wedge (\sim x \wedge y))$

$= x \vee (\sim x \wedge (y \wedge \sim y)) = x \vee (\sim x \wedge F)$

$= x \vee F = x$

1.3 (a)

- $a \leftrightarrow (b \vee \neg b)$
  $a \leftrightarrow$ True
  So a is true, i.e. $a = 1$
- $b \leftrightarrow c$ holds. So $b = c$
  Now the given expression is
  $(a \wedge b) \rightarrow ((a \wedge c) \vee d) \equiv (a \cdot b) \rightarrow ((a \cdot c) + d)$
  Putting a = 1 in above expression we get
  $1 \cdot b \rightarrow ((1 \cdot c) + d)$
  $\equiv b \rightarrow c + d \equiv b' + c + d$

Now putting $b = c$ in above expression we get

$\equiv c' + c + d \equiv 1 + d \equiv 1$

So the expression is always true.

1.4 (a)

Let, p : I stay, q : you go

I stay only if you go $p \rightarrow q$

Converse of $p \rightarrow q$ is $q \rightarrow p$

Now convert the answers one-by-one into boolean form. Only option (a) i.e. "*I stay if you go*" converts to $q \rightarrow p$.

1.5 (a)

$F_1 : P \to \sim P \equiv p \to p' \equiv p' + p' \equiv p'$

So F_1 is contingency. Hence, F_1 is satisfiable but not valid.

$F_2 : (P \to \sim P) \vee (\sim P \to P)$

$\equiv (p \to p') + (p' \to p) \equiv (p' + p') + (p + p)$

$\equiv p' + p \equiv 1$

So F_2 is tautology and therefore valid.

1.6 (a)

If X then Y unless Z is represented by

$$X \to Y \text{ unless } Z \equiv (X \to Y) + Z \equiv X' + Y + Z$$

Now convert the answers one-by-one into boolean form only choice (a) converts to $X' + Y + Z$ as can be seen below:

$$(X \wedge \neg Z) \to Y \equiv XZ' \to Y$$
$$\equiv (XZ')' + Y \equiv X' + Y + Z$$

1.7 (d)

This is valid LHS is saying that if α is holding for any x, then β also holds for that x. RHS is saying if x is holding for all x, then β also holds for all x.

Clearly LHS $\Rightarrow$ RHS (but RHS does not imply LHS)

$$[(\alpha_1 \to \beta_1) \wedge (\alpha_2 \to \beta_2)] \to [(\alpha_1 \wedge \alpha_2) \to (\beta_1 \wedge \beta_2)]$$

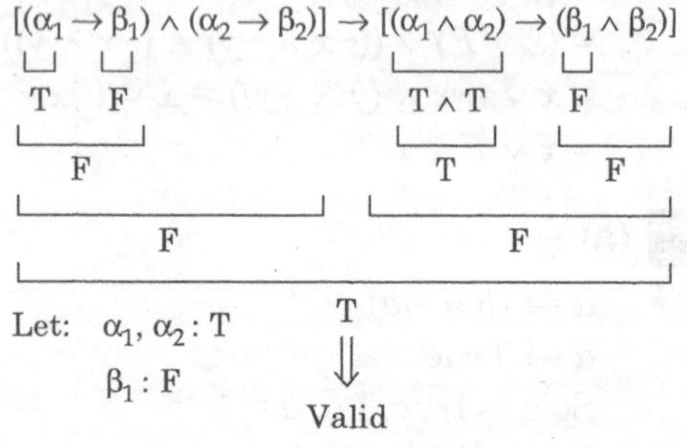

1.8 (d)

$Q_{yy} \equiv$ "y divides y" is always true

$\therefore Q_{xy} \Leftrightarrow \neg Q_{yy}$ is same as $Q_{xy} \Leftrightarrow$ False

Now α becomes

$(\forall x)[P(x) \Leftrightarrow (\forall y)(Q_{xy} \Leftrightarrow \text{false})]$

$\Rightarrow (\forall x)\,[\neg P(x)]$

Now consider $I_1 : P(x) \equiv$ "x is a prime number".

α becomes ($\forall x$ x is a prime number if and only if $\forall y$ (y does not divide x)) $\Rightarrow (\forall x)$ x is not prime.

which means that x is a prime number if and only if no number divides x implies that no number is prime.

Since x always divides x, the above sentence is true.

Now consider $I_2 : P(x) \equiv$ "x is a composite number".

Now α becomes

($\forall x$ x is a composite number if and only if $\forall y$ (y does not divide x)) $\Rightarrow (\forall x)$ x is not composite.

Which means that x is a composite number if and only if no number divides x implies that no number is composite.

Since x always divides x, the above sentence is true.

$\therefore$ Both I_1 and I_2 satisfy α.

1.9 (b)

Derive clause $P \vee Q$ from clauses $P \vee R$, $Q \vee \neg R$ means that $(P \vee R) \wedge (Q \vee \neg R) \Rightarrow P \vee Q$

$\therefore$ (a) is true

Since, $x \Rightarrow y$ does not imply that $y \Rightarrow x$

$\therefore$ $P \vee Q \Rightarrow (P \vee R) \wedge (Q \vee \neg R)$

$\therefore$ may or may not be true.

Hence (b) is false.

1.10 (d)

The statement is "some boys in the class are taller than all the girls".

So the notation for the given statement is

$(\exists x)$ (boy$(x) \wedge (\forall y)$ (girl$(y) \to$ taller (x, y)))

1.11 (c)

Choice (c) is

$\neg(\forall x)(\exists y)[(a(x,y) \wedge b(x,y)) \to c(x,y)]$

$\equiv \neg(\forall x)(\exists y)[a \wedge b \to c]$

$\equiv \neg(\forall x)(\exists y)[(ab)' + c]$

$\equiv \exists x\, \forall y[(ab)' + c]'$

$\equiv \exists x\, \forall y[abc'] \equiv \exists x\, \forall y[a \wedge b \wedge \neg c]$

which is same as the given expression.

$(\exists x)(\forall y)[(a(x,y) \wedge b(x,y)) \wedge \neg c(x,y)]$

1.12 (c)

P: $[(\neg p \vee q) \wedge (r \to s) \wedge (p \vee r)] \to (\neg s \to q)$

$\equiv [(p \to q) \wedge (r \to s) \wedge (p \vee r)] \to (q \vee s)$

which is a rule of inference called constructive dilemma and therefore valid.

S: $[p \wedge (p \to r) \wedge (q \vee \neg r)] \to q$

$\equiv p(p' + r)(q + r') \to q$

$\equiv pr(q + r') \to q \equiv prq \to q$

$\equiv (prq)' + q \equiv p' + r' + q' + q$

$\equiv p' + r' + 1 \equiv 1$

Therefore S is valid.

Q and R can be similarly simplified in boolean algebra to show that they are both not equivalent to 1.

So only P and S are valid.

1.13 (a)

$(P \to (Q \vee R)) \to ((P \wedge Q) \to R)$

$\equiv (P \to Q + R) \to (PQ \to R)$

$\equiv [P' + Q + R] \to [(PQ)' + R]$

$\equiv [P' + Q + R] \to [P' + Q' + R]$

$\equiv (P' + Q + R)' + P' + Q' + R$

$\equiv P\,Q'\,R' + P' + Q' + R$

$\equiv Q' + Q'\,P\,R' + P' + R$

$\equiv Q' + P' + R$ (by absorption law)

Which is a contingency (i.e. satisfiable but not valid).

1.14 (b)

X: $(P \vee Q) \to R$

Y: $(P \to R) \vee (Q \to R)$

X: $P + Q \to R \equiv (P + Q)' + R \equiv P'\,Q' + R$

Y: $(P' + R) + (Q' + R) \equiv P' + Q' + R$

Clearly $X \neq Y$

Consider $X \to Y$

$\equiv (P'Q' + R) \to (P' + Q' + R)$

$\equiv (P'Q' + R)' + P' + Q' + R$

$\equiv (P'Q')' \cdot R' + P' + Q' + R$

$\equiv (P + Q) \cdot R' + P' + Q' + R$

$\equiv PR' + QR' + P' + Q' + R$

$\equiv (PR' + R) + (QR' + Q') + P'$

$\equiv (P + R)\,(R' + R) + (Q + Q') \times (R' + Q') + P'$

$\equiv (P + R) + (R' + Q') + P'$

$\equiv P + P' + R + R' + Q' \equiv 1 + 1 + Q' \equiv 1$

$\therefore$ $X \to Y$ is a tautology.

1.15 (b)

Every teacher is liked by some student: then the logical expression is $\forall(x)$ [teacher$(x) \to \exists(y)$ [student $(y) \wedge$ likes (y, x)]]

Where likes (y, x) means y likes x, such that y represent the student and x represents the teacher.

1.16 (b)

Consider choice (b)

$(\forall x(P(x) \Rightarrow Q(x))) \Rightarrow ((\forall x P(x)) \Rightarrow (\forall x Q(x)))$

Let the LHS of this implication be true

This means that

$P_1 \to Q_1$

$P_2 \to Q_2$

$\vdots$

$P_n \to Q_n$

Now we need to check if the RHS is also true.

The RHS is $((\forall x P(x)) \Rightarrow (\forall x Q(x)))$

To check this let us take the LHS of this as true i.e. take $\forall x\, P(x)$ to be true. This means that $(P_1, P_2, \ldots P_n)$ is taken to be true. Now P_1 along with $P_1 \to Q_1$ will imply that Q_1 is true. Similarly P_2 along with $P_2 \to Q_2$ will imply that Q_2 is true. And so on...

Therefore $(Q_1, Q_2, \ldots Q_n)$ all true.

i.e. $\forall x\, Q(x)$ is true. Therefore the statement (b) is a valid predicate statement.

1.17 (b)

Since a relation may or may not be symmetric, the given predicate is satisfiable but not valid. So (a) is clearly false.

Whenever a predicate is satisfiable its negation also is satisfiable. So option (b) is the correct answer.

1.18 (d)

The given statement should be read as "If an animal is a tiger or a lion, then (if the animal is hungry or threatened, then it will attack). Therefore the correct translation is $\forall x$ [(tiger $(x) \vee$ lion $(x)) \to$ {(hungry $(x) \vee$ threatened $(x)) \to$ attacks (x)}]

which is choice (d).

1.19 (d)

$P_1 : ((A \wedge B) \to C)) \equiv ((A \to C) \wedge (B \to C))$

LHS :

$(A \wedge B) \to C$

$\equiv AB \to C \equiv (A\,B)' + C$

$\equiv A' + B' + C$

RHS :

$(A \to C) \wedge (B \to C)$

$\equiv (A' + C)\,(B' + C) \equiv A'B' + C$

Clearly, LHS $\neq$ RHS

P_1 is not a tautology

$P_2 : ((A \vee B \to C)) \equiv ((A \to C) \vee (B \to C))$

LHS $\equiv (A + B \to C)$

$\equiv (A + B)' + C$

$\equiv A'B' + C$

RHS $\equiv (A \to C) \vee (B \to C)$

$\equiv (A' + C) + (B' + C)$

$\equiv A' + B' + C$

Clearly, LHS $\neq$ RHS $\Rightarrow P_2$ is also not a tautology.

Therefore, both P_1 and P_2 are not tautologies.

Correct choice is (d).

1.20 (d)

By using min terms we can define

$A \odot B = AB + AB' + A'B' = A + A'B'$

$= (A + A') \cdot (A + B') = A + B'$

(a) $\sim A \odot B = A' \odot B = A' + B'$

(b) $\sim (A \odot \sim B) = (A \odot B')'$

$= (A + (B')')'$

$= (A + B)' = A'B'$

(c) $\sim (\sim A \odot \sim B) = (A' \odot B')' = (A' + (B')')'$

$= (A' + B)' = AB'$

(d) $\sim (\sim A \odot B) = (A' \odot B)' = (A' + B')'$

$= A \cdot B = A \wedge B$

$\therefore$ Only, choice (d) $\equiv A \wedge B$

Note: This problem can also be done by constructing truth table for each choice and comparing with truth table for $A \wedge B$.

1.21 (d)

The statement "Not every graph is connected" is same as "There exists some graph which is not connected" which is same as

$\exists x$ {graph $(x) \wedge \neg$ connected (x)}

Which is choice (b)

By boolean algebra we can see that option (a) and (c) are same as (b). Only option (d) is not the same as (b).

Infact option (d) means that "all graphs are not connected".

Alternate solution:

We can translate the given statement "NOT (every graph is connected)" as $\neg(\forall x$ graph (x) $\rightarrow$ connected $(x))$

$\equiv \exists x \neg$(graph $(x) \rightarrow$ connected (x))

$\equiv \exists x \neg(\neg$graph $(x) \vee$ connected (x))

$\equiv \exists x$ (graph $(x) \wedge \neg$ connected (x))

By boolean algebra we can see that option (a) and (c) are same as (b). Only option (d) is not the same as (b).

Infact option (d) means that "all graphs are not connected".

1.22 (a)

In conjunction normal form, for any particular assignment of truth values, all except one clause, will always evaluate to true. So, the proportion of clauses which evaluate to true to the total number of clauses is equal to $\frac{2^n - 1}{2^n}$.

Now putting $n = 1, 2, \ldots$, we get $\frac{1}{2}, \frac{3}{4}, \frac{7}{8} \ldots$

All of these proportions are $\geq \frac{1}{2}$ and so choice (a) atleast half of the clauses evaluate to true, is the correct answer.

1.23 (a)

Option (a) is a standard one way distributive property of predicates.

1.24 (a)

"For x which is an *fsa*, there exists a y which is a *pda* and which is equivalent to x."

$\forall x\ (fsa(x) \Rightarrow \exists y\ (pda(y) \wedge \text{equivalent } (x, y)))$ is the logical representation.

1.25 (c)

Option (c) is $[(\exists x, \alpha(x)) \rightarrow \beta] \rightarrow [\forall x, \alpha(x) \rightarrow \beta]$

Let us check the validity of this predicate.

Let the LHS of this predicate be true.

This means that some $\alpha \rightarrow \beta$.

Let $\alpha_5 \rightarrow \beta$

Now we will check if the RHS is true. The RHS is $[\forall x, \alpha(x) \rightarrow \beta]$ to check this implication let us take $\forall x, \alpha(x)$ to be true.

This means that all the α are true. It means that α_5 is also true.

But $\alpha_5 \rightarrow \beta$. Therefore β is true.

So the RHS $[\forall x, \alpha(x) \rightarrow \beta]$ is true.

Whenever the LHS $[(\exists x, \alpha(x)) \rightarrow \beta]$ is true. So option (c) is valid.

1.26 (d)

The given predicate is

$[\forall x, \alpha \rightarrow (\exists y, \beta \rightarrow (\forall u, \exists v, \gamma))]$

The negation is of this predicate is

$\neg[\forall x, \alpha \rightarrow (\exists y, \beta \rightarrow \forall u, \exists v, \gamma)]$

$\neg[\forall x, \alpha \rightarrow (\forall y, \neg\beta \vee \forall u, \exists v, \gamma)]$

$\neg[\exists x, \neg\alpha \vee (\forall y, \neg\beta \vee \forall u, \exists v, \gamma)]$

$[\forall x, \alpha \wedge (\exists y, \beta \wedge (\exists u, \forall v, \neg\gamma))]$

Which is option (d).

1.27 (b)

1. $P \vee \sim Q \equiv P + Q'$
2. $\sim(\sim P \wedge Q) \equiv (P'\, Q)' \equiv P + Q'$
3. $(P \wedge Q) \vee (P \wedge \sim Q) \vee (\sim P \wedge \sim Q)$

$\equiv PQ + PQ' + P'Q'$

$\equiv P(Q + Q') + P'Q'$

$\equiv P + P'Q'$

$\equiv (P + P')\,(P + Q') \equiv P + Q'$

4. $(P \wedge Q) \vee (P \wedge \sim Q) \vee (\sim P \wedge Q)$
$\equiv PQ + PQ' + P'Q$
$\equiv P(Q + Q') + P'Q \equiv P + P'Q$
$\equiv (P + P')(P + Q) = P + Q$

Clearly (i), (ii) and (iii) are equivalent. Correct choice is (b).

1.28 (d)

The correct translation of "Gold and silver ornaments are precious" is choice (d)

$$\forall x ((G(x) \vee S(x)) \rightarrow P(x))$$

which is read as "if an ornament is gold or silver, then it is precious".
Now since a given ornament cannot be both gold and silver at the same time.
Choice (b) $\forall x ((G(x) \wedge S(x)) \rightarrow P(x))$ is incorrect.

1.29 (b)

The given table can be converted into boolean function by adding minterms corresponding to true rows.

∴

P	Q	P□Q
T	T	T
T	F	T
F	T	F
F	F	T

Since there is only one false in the above truth table, we can represent the function P□Q more efficiently, in conjunctive normal form.
Translates P□Q = P + Q′ (the max-term corresponding to the third row, where the function is false).
Now, we can easily translate the choices into boolean algebra as follows:
Choice (a) ¬Q □¬P ≡ Q′ □P′ ≡ Q′ + P
Choice (b) P □¬Q ≡ P □Q′ ≡ P + Q
Choice (c) ¬P□Q ≡ P′ □Q ≡ P′ + Q′
Choice (d) ¬P□¬Q ≡ P′ □Q′ ≡ P′ + Q
As we can clearly see only choice (b) P □ ¬ Q is equivalent to P + Q.

1.30 (b)

I $\quad \neg \forall x P(x) \equiv \exists x \neg P(x)$
and IV $\quad \exists x \neg P(x)$
Clearly, choices I and IV are equivalent.
II $\quad \neg \exists x P(x) \equiv \forall x \neg P(x)$
and III $\quad \neg \exists x [\neg P(x)] \equiv \forall x P(x)$
Clearly II and III are not equivalent to each other or to I and IV.

1.31 (b)

$\forall x\ \exists y\ \exists t\ \neg F(x, y, t)$
$\equiv \neg\{\exists x\ \forall y\ \forall t\ F(x, y, t)\}$
≡ it is not true that (someone can fool all people at all time)
≡ no one can fool everyone all the time

1.32 (a)

If $P(x)$ is true, then
$x \neq 1$ and also
x is broken into two factors, only if, one of the factors is x itself and the other factor is 1, which is exactly the definition of a prime number.
So $P(x)$ is true means x is a prime number.

1.33 (b)

Let p : It rains
q : cricket match will not be played.

$I_1: \quad p \Rightarrow q$
$\quad\quad \underline{\sim q}$
$\therefore \quad \sim p$

Clearly I_1 is correct since it is in the form of Modus Tollens (rule of contrapositive)

$I_2: \quad p \Rightarrow q$
$\quad\quad \underline{\sim p}$
$\therefore \quad \sim q$

which corresponds $[p \Rightarrow q \wedge \sim p] \Rightarrow \sim q$
$\equiv [(p' + q)\, p'] \Rightarrow q'$
$\equiv [p' + qp'] \Rightarrow q' \equiv p' \Rightarrow q'$
$\equiv (p')' + q' \equiv p + q'$
which is not a tautology.
So I_2 is incorrect inference.

1.34 (c)

Some real numbers are rational
≡ ∃x [real (x) ∧ rational (x)]

1.35 (d)

None of my friends are perfect i.e., all of my friends are not perfect
$\forall x((F(x) \rightarrow \neg P(x))$
$\forall x(\neg F(x) \vee \neg P(x))$
$\neg \exists x\ (F(x) \wedge P(x))$

Alternatively Method:

$\exists x\ (F(x) \wedge P(x))$ gives
there exist some of my friends who are perfect.
$\neg \exists x\ (F(x) \wedge P(x))$
there does not exits any friend who is perfect i.e., none of my friends are perfect.
So (d) is correct option.

1.36 (a)

Let, $\forall y(\alpha) = P,\ \forall z(\beta) = Q$

Then, $\exists y(\neg\alpha) = \neg P$ and $\exists z(\neg\beta) = \neg Q$

Given, $\neg\exists x(\forall y(\alpha) \wedge \forall z(\beta))$

$$= \neg\exists x(P \wedge Q) = \neg\forall x(\neg P \vee \neg Q)$$

$$= \forall x(P \to \neg Q)$$

(a) $\forall x(\exists z(\neg\beta) \to \forall y(\alpha)) = \forall x(\neg Q \to P)$

(b) $\forall x(\forall z(\beta) \to \exists y(\neg\alpha)) = \forall x(Q \to \neg P)$

$= \forall x(P \to \neg Q)$

(c) $\forall x(\forall y(\alpha) \to \exists z(\neg\beta)) = \forall x(P \to \neg Q)$

(d) $\forall x(\exists y(\neg\alpha) \vee \exists z(\neg\beta)) = \forall x(\neg P \vee \neg Q)$

$= \forall x(P \to \neg Q)$

$\therefore$ Only (a) is not logically equivalent to $\forall x(P \to \neg Q)$.

1.37 (d)

(a) $\forall x$ glitters$(x) \Rightarrow \neg$gold(x)

All glitters are not gold

(b) $\forall x$ gold$(x) \Rightarrow$ glitters(x)

All golds are glitters

(c) $\exists x$ gold$(x) \wedge \neg$ glitters(x)

There exist gold which is not glitter i.e. not all golds are glitters.

(d) $\exists x$ glitters$(x) \wedge \neg$ gold(x)

Not all that glitters is gold i.e., there exist some which glitters and which is not gold.

1.38 (b)

Option (b) is $(\sim(p \leftrightarrow q) \wedge r) \vee (p \wedge q \wedge \sim r)$

$\equiv ((p \oplus q)\ r) + pqr' \equiv (pq' + p'q)\ r + pqr'$

$\equiv pq'r + p'qr + pqr' \equiv pqr' + pq'r + p'qr$

This is exactly the min-term form of a logical formula which is true when exactly two variables are true (only p, q true or only p, r true or only q, r true).

1.39 (b)

$(a \leftrightarrow c) \to (\sim b \to (a \wedge c))$

$\equiv (a \leftrightarrow c)' + (b' \to ac) \equiv (a \oplus c)' + (b' \to ac)$

$\equiv ac' + a'c + b + ac \equiv a(c' + c) + a'c + b$

$\equiv a + a'c + b \equiv a + c + b$

which is not a tautology.

1.40 (d)

g : mobile is good

c : mobile is cheap

P: Good mobile phones are not cheap

$$g \to c' \equiv (g' + c')$$

Q: Cheap mobile phones are not good

$$c \to g' \equiv (c' + g')$$

$\therefore$ Both P and Q are equivalent.

L: P → Q

M: Q → P

N: P ≡ Q

Since both P and Q are equivalent, all three of L, M, N are true. So correct option is (d).

1.41 (d)

Not (all rainy days are cold):

$\sim(\forall d\ (\text{Rainy}\ (d) \to \text{Cold}\ (d)))$

$\equiv\sim(\forall d\ (\sim \text{Rainy}\ (d) \vee \text{Cold}\ (d)))$

$\equiv \exists d\ (\text{Rainy}\ (d) \wedge \sim \text{Cold}\ (d))$

Alternate Method:

Not all rainy days are cold is same as some rainy days are not cold which is same as

$\exists d\ (\text{Rainy}\ (d) \wedge \sim \text{Cold}\ (d))$

1.42 (c)

Here, option (a) and (b) can be reduced to $(p \to q) \wedge (q \to p)$ and hence $\equiv p \leftrightarrow q$

Option (d) is $p'q' + pq \equiv p \leftrightarrow q$.

Option (c) is $p'q + pq' = p \oplus q$ which is not equivalent to $p \leftrightarrow q$.

1.43 (a)

The given truth table corresponds to $p'q + pq' = p \oplus q$.

$\oplus$ is known to be both commutative and associative.

1.44 (c)

C : Person is corrupt

K : Person is kind.

E : Person is elected

$S_1 : C \to \bar{E}$

$S_2 : K \to E$

$S_2 \equiv \bar{E} \to \bar{K}$

So from S_1 and S_2: $(C \to \bar{E}) \wedge (\bar{E} \to \bar{K}) \equiv C \to \bar{K}$

We can conclude $C \to \bar{K}$ which is same as $K \to \bar{C}$, which is same as option (c).

1.45 (c)

Since $P \to R \equiv \neg P \vee R$, and the quantifiers on both the sides are same $(\forall x \exists y)$.

Option (c) is clearly a tautology.

1.46 (a)

Type 1 always tells truth.
Type 2 always tells lies.
The statement is "toss is head if and only if I am telling the truth".
There are two cases possible.
Case-1: Let the person be type 1.
Case-2: Let the person be type 2.
In each case we shall prove that result is head.
Case-1: Let the person be type 1. Type 1 always tells truth.
So the statement "toss is head if and only if I am telling the truth" is true.
So toss head $\Leftrightarrow$ telling truth. Since type 1 is telling truth so toss head is also true. So in case 1 result is that toss is head.
Case-2: Let the person be type 2. Type 2 always tells lies.
So the statement "toss is head if and only if I am telling thc truth" is false.
So toss head $\Leftrightarrow$ telling truth is false. So toss head $\oplus$ telling truth. So toss head and telling truth have opposite truth values. Now, since type 2 telling truth is false, so toss head has to have opposite truth value which is true.
So toss head is true. So in case 2 also, result is head.
So in both cases we have proved that the result is head. So option (a) is correct.

1.47 (11)

We wish to make
$\neg((p \Rightarrow q) \wedge (\neg r \vee \neg s)) = 1$
$\Rightarrow (p \Rightarrow q) \wedge (\neg r \vee \neg s) = 0$
$\Rightarrow (p \Rightarrow q) = 0$ ...(1)
or $\neg r \vee \neg s = 0$ ...(2)
Now (1) is satisfies only when $p = 1$ and $q = 0$.
Equation (2) $\neg r \vee \neg s = 0$, iff $r \wedge s = 1$
i.e. $r = 1$ and $s = 1$
i.e. x is a perfect square and x is a prime number. Which is not possible so condition (2) cannot be satisfied by any x.
So condition (1) must be satisfies which is $p = 1$ and $q = 0$ i.e. $x \in \{8, 9, 10, 11, 12,\}$ and x is not a composite.
Now the only value of x which satisfies this is $x = 11$. So correct answer is $x = 11$.

1.48 (4)

$p \wedge (p \Rightarrow q) = p(p' + q) \equiv pq$
Take (i) false
$pq \Rightarrow \text{false} \equiv pq \Rightarrow 0$
$\equiv (pq)' + 0 \equiv p' + q' + 0$
$\equiv$ not valid
Take (ii)
$pq \Rightarrow q \equiv (pq)' + q \equiv (pq)' + q$
$\equiv p' + 1 \equiv 1 \equiv$ valid
Take (iii)
$pq \Rightarrow \text{true} \equiv pq \Rightarrow 1$
$\equiv (pq)' + 1 \equiv 1 \equiv$ valid
Take (iv)
$pq \Rightarrow p + q \equiv (pq)' + p + q$
$\equiv p' + q' + p + q$
$\equiv 1 + 1 \equiv 1 \equiv$ valid
Take (v)
$pq \Rightarrow q' + p \equiv (pq)' + q' + p$
$\equiv p' + q' + q' + p \equiv 1 \equiv$ valid
So the number of expressions that are logically implies by $p \wedge (p \Rightarrow q)$ is 4.

1.49 (d)

$\forall x$ is only one way distribution over "$\vee$" so let us check (d)
(d) $\forall x\, (p(x) \vee q(x)) \Rightarrow (\forall x\, p(x) \vee \forall x\, q(x))$
Let LHS be true, so we have,
$p_1 \vee q_1$ (true)
$p_2 \vee q_2$ (true)
$p_3 \vee q_3$ (true)
$\vdots$
Now take p_1 is true, q_1 is false and p_2 is false, q_2 is true.
Now LHS is true, but RHS, $\forall x\, p(x)$ is false (since p_2 is false) and $\forall x\, q(x)$ is also false (since q_1 is false)
So, LHS $\not\Rightarrow$ RHS.

1.50 (b)

I. $\forall x \exists y\, R(x,y) \to \exists y (\exists x\, R(x,y))$ is true, since $\exists y\, (\exists x\, R(x,y)) \equiv \exists x\, (\exists y\, R(x,y))$

II. $\forall x \exists y\, R(x,y) \to \exists y (\forall x\, R(x,y))$ is false
Since $\exists y$ when it is outside is stronger then when it is inside.

III. $\forall x \exists y\, R(x,y) \to \forall y \exists x\, R(x,y)$ is false
Since R(x, y) may not be symmetric in x and y.

IV. $\forall x \exists y\, R(x,y) \to \neg(\exists x\, \forall y\, \neg R(x,y))$ is true
Since $\neg(\exists x \forall y \neg\, R(x,y)) \equiv \forall x \exists y\, R(x,y)$

So, IV will reduce to $\forall x \exists y\, R(x,y) \to \forall x \exists y\, R(x,y)$ which is trivially true.
So correct answer is I and IV only which is option (b).

1.51 (d)

Statement $p' \Rightarrow q' \equiv p + q'$

I. $p \rightarrow q \equiv p' + q$

II. $q \rightarrow p \equiv q' + p$

III. $(\neg q \vee p) \equiv q' + p$

IV. $(\neg p \vee q) \equiv p' + q$

So, clearly the given statement is same as II and III only.

1.52 (d)

$$(p \rightarrow q) \rightarrow r = 0$$

i.e., $(p' + q)' + r = 0$

i.e., $pq' + r = 0$

This is possible only if $pq' = 0$ and $r = 0$

Now, $pq' = 0$ iff $p = 0$ or $q' = 0$

So, final solution is $r = 0$ and ($p = 0$ or $q' = 0$)

Now let us see

$$(r \rightarrow p) \rightarrow q$$
$$\equiv (r' + p)' + q \equiv rp' + q$$

but we have previously shown that $r = 0$

So, $rp' = 0$

So, $(r \rightarrow p) \rightarrow q \equiv rp' + q \equiv 0 + q \equiv q$

So, truth value of $(r \rightarrow p) \rightarrow q$ is same as that of q. So, it is true whenever q is true.

So option (d) is correct.

Note: Since we have shown that ($p = 0$ (or) $q' = 0$) is true,

Option (c) is false because when p is false i.e., when $p = 0$, q is free to be 0 or 1.

1.53 (a)

p : *"It is raining"*, **q** : *"It is cold"*, and **r** : *"It is pleasant"*

So the correct representation of ***"It is not raining and it is pleasant, and it is not pleasant only if it is raining and it is cold"*** is

$(\neg p \wedge r) \wedge (\neg r$ only if $p \wedge q) \equiv (\neg p \wedge r) \wedge (\neg r \rightarrow (p \wedge q))$

1.54 (b)

$\forall x[\forall z | x \Rightarrow ((z = x) \vee (z = 1)) \Rightarrow \exists w\ (w > x) \wedge (\forall z\ z | w \Rightarrow ((w = z) \vee (z = 1)))]$

The predicate φ simply says that if z is a prime number in the set then there exists another prime number is the set which is larger.

Clearly φ is true in S_2 and S_3 since in set of all integers as well as all positive integers, there is a prime number greater than any given prime number.

However, in $S_1 : \{1, 2, 3, ...100\}$ φ is false since for prime number $97 \in S_1$ there exists no prime number in the set which is greater.

1.55 (b)

$\forall x\ (p(x) \rightarrow W) \Rightarrow (\forall x\ p(x)) \rightarrow W$ is true.

But $(\forall x\ p(x)) \rightarrow W \Rightarrow \forall x\ (p(x) \rightarrow W)$ is false.

Since if LHS is true then $(P_1 \wedge P_2 \wedge P_3 \ldots \wedge P_n) \rightarrow W$ will be true and RHS will be false since $P_1 \rightarrow W$, itself will be false, since only P_1 true cannot make W true (we need all the $P_1, P_2 \ldots, P_n$ to be true to make W true)

So LHS $\Rightarrow$ RHS true and RHS $\Rightarrow$ LHS is false so LHS $\neq$ RHS.

So option (b) is invalid.

Note that option (c) is true because by Boolean algebra,

$$\text{LHS} = \exists x\ (p(x) \rightarrow W) \equiv \exists x\ (\sim p(x) \vee W)$$
$$\equiv \exists x\ (\sim p(x)) \vee W$$
$$\text{RHS} = \forall x\ p(x) \rightarrow W \equiv \sim(\forall x\ p(x)) \vee W$$
$$\equiv \exists x \sim p(x) \vee W$$

So LHS = RHS.

1.56 (c)

$$S_1 : (\neg p \wedge (p \vee q)) \rightarrow p$$
$$S_2 : q \rightarrow (\neg p \wedge (p \vee q))$$
$$S_1 : [p'\ (p + q)] \rightarrow q$$
$$\equiv (p'p + p'q) \rightarrow q$$
$$\equiv p'q \rightarrow q$$
$$\equiv (p'q)' + q$$
$$\equiv p + q' + q$$
$$\equiv p + 1$$
$$\equiv 1 \text{ (tautology)}$$
$$S_2 : q \rightarrow (p'\ (p + q))$$
$$\equiv q \rightarrow (p'p + p'q)$$
$$\equiv q \rightarrow p'q$$
$$\equiv q' + p'q$$
$$\equiv (q' + p')\ (q' + q)$$
$$\equiv q' + p' \text{ (contigency) (not a tautology)}$$

So, option (c) S_1 is a tautology and S_2 is not a tautology is correct.

1.57 (a, c)

$$S : ((P \wedge Q) \rightarrow R) \rightarrow ((P \wedge Q) \rightarrow (Q \rightarrow R))$$
$$\equiv (pq \rightarrow r) \rightarrow (pq \rightarrow (q \rightarrow r))$$
$$\equiv [(pq)' + r] \rightarrow [(pq)' + (q' + r)]$$
$$\equiv [(pq)' + r]' + [(pq)' + q' + r]$$
$$\equiv [pq \cdot r'] + [p' + q' + q' + r]$$
$$\equiv pqr' + p' + q' + r$$
$$\equiv (p + p')(qr' + p') + q' + r$$
$$\equiv qr' + p' + q' + r$$
$$\equiv (q + q')(r' + q') + p' + r$$

$\equiv r' + q' + p' + r \equiv r' + r + q' + p'$

$\equiv 1 + q' + p' \equiv 1$ (Tautology)

So, S is a tautology.

So, option (a) is true.

Option (b) and (d) are false.

Option (c) antecedent of S is

$pq \to r \equiv (pq)' + r$

$\equiv p' + q' + r$

The consequent of S is $pq \to (q \to r)$

$\equiv (pq)' + q' + r$

$\equiv p' + q' + q' + r$

$\equiv p' + q' + r$

So,Antecedent of $S \equiv$ Consequent of S

So, option (c) is also true.

1.58 (b, c)

(a) $\exists x(P(x) \wedge \forall y\, Q(x,y)) \to (\forall x(P(x) \to \exists y\, Q(x,y)))$ is false whenever

$\exists x(P(x) \wedge \forall y\, Q(x,y))$ is true it does not guarantee that $\forall x(P(x) \to \exists y\, Q(x,y))$ is true.

(b) $\forall x \forall y\, Q(x,y) \to (\forall x(P(x) \to \exists y\, Q(x,y))$ is true whenever $\forall x \forall y\, Q(x,y)$ is true then

$\forall x(P(x) \to \exists y\, Q(x,y))$ is true irrespective of $P(x)$.

(c) $\exists y\, \forall x(P(x) \Rightarrow Q(x,y)) \to \forall x(P(x) \to \exists y\, Q(x,y))$ is true whenever

$\exists y\, \forall x(P(x) \to Q(x,y))$ is true for some fixed 'y' then $\forall x(P(x) \to \exists y\, Q(x,y))$ is also true.

(d) $\exists x(P(x) \wedge \exists y\, Q(x,y)) \to \forall x(P(x) \to \exists y\, Q(x,y))$ is false for some 'x' if

$\exists x(P(x) \wedge \exists y\, Q(x,y))$ is true it does not guarantee that $\forall x(P(x) \to \exists y\, Q(x,y))$ is true.

1.59 (b)

A when B

If B then A

$B \to A$

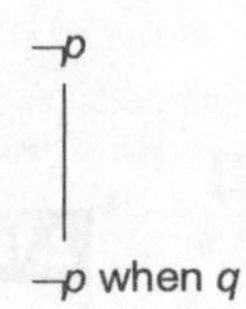

$\Rightarrow q \to \neg px$

2 Set Theory and Algebra

2.1 Let R be a symmetric and transitive relation on a set A. Then

(a) R is reflexive and hence an equivalence relation

(b) R is reflexive and hence a partial order

(c) R is reflexive and hence not an equivalence relation

(d) None of the above

[1995 : 1 M]

2.2 The number of elements in the power set P(S) of the set S = {(ϕ), 1, (2, 3)} is:

(a) 2 (b) 4

(c) 8 (d) None of these

[1995 : 1 M]

2.3 Let A be the set of all nonsingular matrices over real numbers and let $*$ be the matrix multiplication operator. Then

(a) A is closed under $*$ but $\langle A, *\rangle$ is not a semigroup.

(b) $\langle A, *\rangle$ is a semigroup but not a monoid.

(c) $\langle A, *\rangle$ is a monoid but not a group.

(d) $\langle A, *\rangle$ is a group but not an abelian group

[1995 : 2 M]

2.4 Let A and B be sets and let A^c and B^c denote the complements of the sets A and B. The set $(A - B) \cup (B - A) \cup (A \cap B)$ is equal to

(a) $A \cup B$ (b) $A^c \cup B^c$

(c) $A \cap B$ (d) $A^c \cap B^c$

[1996 : 1 M]

2.5 Let X = {2, 3, 6, 12, 24}. Let ≤ be the partial order defined by $x \le y$ if x divides y. The number of edges in the Hasse diagram of (X, ≤) is

(a) 3 (b) 4

(c) 9 (d) None of these

[1996 : 1 M]

2.6 Suppose X and Y are sets and $|X|$ and $|Y|$ are their respective cardinalities. It is given that there are exactly 97 functions from X to Y. From this one can conclude that

(a) $|X| = 1, |Y| = 97$

(b) $|X| = 97, |Y| = 1$

(c) $|X| = 97, |Y| = 97$

(d) None of the above

[1996 : 1 M]

2.7 Which of the following statements is false?

(a) The set of rational numbers is an abelian group under addition

(b) The set of integers in an abelian group under addition

(c) The set of rational numbers form an abelian group under multiplication

(d) The set of real numbers excluding zero in an abelian group under multiplication

[1996 : 1 M]

2.8 Let R denote the set of real numbers. Let $f : R \times R \to R \times R$ be a bijective function defined by $f(x, y) = (x + y, x - y)$. The inverse function of f is given by

(a) $f^{-1}(x, y) = \left(\dfrac{1}{x+y}, \dfrac{1}{x-y}\right)$

(b) $f^{-1}(x, y) = (x - y, x + y)$

(c) $f^{-1}(x, y) = \left(\dfrac{x+y}{2}, \dfrac{x-y}{2}\right)$

(d) $f^{-1}(x, y) = (2(x - y), 2(x + y))$ **[1996 : 2 M]**

2.9 Let R be a non-empty relation on a collection of sets defined by ARB if and only if $A \cap B = \phi$. Then, (pick the true statement)

(a) R is reflexive and transitive

(b) R is symmetric and not transitive

(c) R is an equivalence relation

(d) R is not reflexive and not symmetric

[1996 : 2 M]

2.10 Which one of the following is false?

(a) The set of all bijective functions on a finite set forms a group under function composition.

(b) The set {1, 2, ..., p – 1} forms a group under multiplication mod p where p is a prime number.

(c) The set of all strings over a finite alphabet Σ forms a group under concatenation.

(d) A subset $s \neq \phi$ of G is a subgroup of the group $\langle G, *\rangle$ if and only if for any pair of elements $a, b \in s$, $a * b^{-1} \in s$. **[1996 : 2 M]**

2.11 The number of equivalence relations on the set {1, 2, 3, 4} is

(a) 15 (b) 16

(c) 24 (d) 4 **[1997 : 1 M]**

2.12 Suppose A is a finite set with n elements. The number of elements in the Largest equivalence relation of A is
(a) n (b) n^2
(c) 1 (d) n + 1 **[1998 : 1 M]**

2.13 Let R_1 and R_2 be two equivalence relations on a set. Consider the following assertions:
(i) $R_1 \cup R_2$ is an equivalence relation
(ii) $R_1 \cap R_2$ is an equivalence relation
Which of the following is correct?
(a) both assertions are true
(b) assertion (i) is true but assertion (ii) is not true
(c) assertion (ii) is true but assertion (i) is not true
(d) neither (i) nor (ii) is true **[1998 : 1 M]**

2.14 The number of functions from an m element set to an n element set is
(a) m + n (b) m^n
(c) n^m (d) m * n **[1998 : 1 M]**

2.15 The binary relation R = {(1, 1), (2, 1), (2, 2), (2, 3), (2, 4), (3, 1), (3, 2), (3, 3), (3, 4)} on the set A = {1, 2, 3, 4} is
(a) Reflexive, symmetric and transitive
(b) Neither reflexive, nor irreflexive but transitive
(c) Irreflexive, symmetric and transitive
(d) Irreflexive and antisymmetric
[1998 : 2 M]

2.16 Suppose A = {a, b, c, d} and Π_1 is the following partition of A
$$\Pi_1 = \{\{a, b, c\}, \{d\}\}$$
(a) List the ordered pairs of the equivalence relations induced by Π_1
(b) Draw the graph of the above equivalence relation **[1998 : 2 M]**

2.17 Let (A, *) be a semigroup. Furthermore, for every a and b in A, if a ≠ b, then a * b ≠ b * a.
(a) Show that for every a in A
a * a = a
(b) Show that for every a, b in A
a * b * a = a
(c) Show that for every a, b, c in A
a * b * c = a * c **[1998 : 2 M]**

2.18 The number of binary relations on a set with n elements is
(a) n^2 (b) 2n
(c) 2^{n^2} (d) None of these
[1999 : 1 M]

2.19 (a) Mr. X claims the following:
If a relation R is both symmetric and transitive, then R is reflexive. For this, Mr. X offers the following proof.
"From xRy, using symmetry we get yRx. Now because R is transitive, xRy and yRx together imply xRx. Therefore, R is reflexive."
Briefly point out the flaw in Mr. X's proof
(b) Give an example of a relation R which is symmetric and transitive but not reflexive.
[1999 : 2 M]

2.20 A relation R is defined on the set of integers as xRy iff (x + y) is even. Which of the following statements is true?
(a) R is not an equivalence relation
(b) R is an equivalence relation having 1 equivalence class
(c) R is an equivalence relation having 2 equivalence classes
(d) R is an equivalence relation having 3 equivalence classes **[2000 : 2 M]**

2.21 Let P(S) denote the power set of a set S. Which of the following is always true?
(a) P(P(S)) = P(S)
(b) P(S) ∩ P(P(S)) = {∅}
(c) P(S) ∩ S = P(S)
(d) S ∉ P(S)
[2000 : 2 M]

2.22 Consider the following relations:
R_1(a, b) iff (a + b) is even over the set of integers
R_2(a, b) iff (a + b) is odd over the set of integers
R_3(a, b) iff a.b > 0 over the set of non-zero rational numbers.
R_4(a, b) iff |a – b| ≤ 2 over the set of natural numbers
Which of the following statements is correct?
(a) R_1 and R_2 are equivalence relations, R_3 and R_4 are not
(b) R_1 and R_3 are equivalence relations, R_2 and R_4 are not
(c) R_1 and R_4 are equivalence relations, R_2 and R_3 are not
(d) R_1, R_2 R_3 and R_4 are all equivalence relations
[2001 : 1 M]

2.23 Consider the following statements.
S_1: There exist infinite sets A, B, C such that A ∩ (B ∩ C) is finite.
S_2: There exist two irrational numbers x and y such that (x + y) is rational.
Which of the following is true about S_1 and S_2?

(a) Only S_1 is correct
(b) Only S_2 is correct
(c) Both S_1 and S_2 are correct
(d) None of S_1 and S_2 is correct **[2001 : 2 M]**

2.24 Let $f : A \rightarrow B$ be a function, and let E and F be subsets of A. Consider the following statements about images.
$S_1 : f(E \cup F) = f(E) \cup f(F)$
$S_2 : f(E \cap F) = f(E) \cap f(F)$
(a) Only S_1 is correct
(b) Only S_2 is correct
(c) Both S_1 and S_2 are correct
(d) None of S_1 and S_2 is correct **[2001 : 2 M]**

2.25 The binary relation S = f(empty set) on set A = {1, 2, 3} is
(a) Neither reflexive nor symmetric
(b) Symmetric and reflexive
(c) Transitive and reflexive
(d) Transitive and symmetric **[2002 : 2 M]**

2.26 Consider the set Σ^* of all strings over the alphabet $\Sigma = \{0, 1\}$. Σ^* with the concatenation operator for strings
(a) does not form a group
(b) forms a non-commutative group
(c) does not have a right identity element
(d) forms a group if the empty string is removed from Σ^* **[2003 : 1 M]**

2.27 Let $(S, \leq)$ be a partial order with two minimal elements a and b, and a maximum element c. Let P : S → {True, False} be a predicate defined on S. Suppose that P(a) = True, P(b) = False and $P(x) \Rightarrow P(y)$ for all $x, y \in S$ satisfying $x \leq y$, where ⇒ stands for logical implication. Which of the following statements CANNOT be true?
(a) P(x) = True for all $x \in S$ such that $x \neq b$
(b) P(x) = False for all $x \in S$ such that $x \neq a$ and $x \neq c$
(c) P(x) = False for all $x \in S$ such that $b \leq x$ such that $x \neq c$
(d) P(x) = False for all $x \in S$ such that $a \leq x$ and $b \leq x$ **[2003 : 2 M]**

2.28 Consider the set {a, b, c} with binary operators + and × defined as follows:

+	a	b	c
a	b	a	c
b	a	b	c
c	a	c	b

x	a	b	c
a	a	b	c
b	b	c	a
c	c	c	b

For example, a + c = c, c + a = a, c × b = c and b × c = a.

Given the following set of equations:
$(a \times x) + (a \times y) = c$
$(b \times x) + (c \times y) = c$
The number of solution (s) (i.e., pair (s) (x, y) that satisfies the equations) is
(a) 0 (b) 1
(c) 2 (d) 3 **[2003 : 2 M]**

2.29 Consider the binary relation:
$S = \{(x, y) \mid y = x + 1 \text{ and } x, y \in \{0, 1, 2, ...\}\}$
The reflexive transitive closure of S is
(a) $\{(x, y) \mid y > x \text{ and } x, y \in \{0, 1, 2, ...\}\}$
(b) $\{(x, y) \mid y \geq x \text{ and } x, y \in \{0, 1, 2, ...\}\}$
(c) $\{(x, y) \mid y < x \text{ and } x, y \in \{0, 1, 2, ...\}\}$
(d) $\{(x, y) \mid y \leq x \text{ and } x, y \in \{0, 1, 2, ...\}\}$
[2004 : 1 M]

2.30 The number of different $n \times n$ symmetric matrices with each element being either 0 or 1 is: (Note : power (2, x) is same as 2^x)
(a) power(2, n) (b) power(2, n^2)
(c) power(2, $(n^2 + n)/2$) (d) power (2, $(n^2 - n)/2$)
[2004 : 1 M]

2.31 In a class of 200 students, 125 students have taken Programming Language course, 85 students have taken Data Structures course, 65 students have taken Computer Organization course; 50 students have taken both Programming Language and Data Structures, 35 students have taken both Data Structures and Computer Organization; 30 students have taken both Programming Language and Computer Organization, 15 students have taken all the three courses. How many students have not taken any of the three courses?
(a) 15 (b) 20
(c) 25 (d) 30 **[2004 : 1 M]**

2.32 Let R_1 be a relation from A = {1, 3, 5, 7} to B = {2, 4, 6, 8} and R_2 be another relation from B to C = {1, 2, 3, 4} as defined below:
(i) An element x in A is related to an element y in B (under R_1) if x + y is divisible by 3.
(ii) An element x in B is related to an element y in C (under R_2) if x + y is even but not divisible by 3.

Which is the composite relation R_1R_2 from A to C?
(a) R_1R_2={(1, 2), (1, 4), (3, 3), (5, 4), (7, 3)}
(b) R_1R_2={(1, 2), (1, 3), (3, 2), (5, 2), (7, 3)}
(c) R_1R_2={(1, 2), (3, 2), (3, 4), (5, 4), (7, 2)}
(d) R_1R_2={(3, 2), (3, 4), (5, 1), (5, 3), (7, 1)}
[2004 : 1 M]

2.33 The following is the incomplete operation table of a 4-element group.

*	e	a	b	c
e	e	a	b	c
a	a	b	c	e
b				
c				

The last row of the table is
(a) c a e b (b) c b a e
(c) c b e a (d) c e a b **[2004 : 2 M]**

2.34 The inclusion of which of the following sets into S = {{1, 2}, {1,2,3}, {1, 3, 5}, {1, 2, 4}, {1, 2, 3, 4, 5}} is necessary and sufficient to make S a complete lattice under the partial order defined by set containment?
(a) {1}
(b) {1}, {2, 3}
(c) {1}, {1, 3}
(d) {1}, {1, 3}, {1, 2, 3, 4}, {1, 2, 3, 5}
[2004 : 2 M]

2.35 Let A, B and C be non-empty sets and let X = (A – B) – C and Y = (A – C) – (B – C)
Which one of the following is TRUE?
(a) X = Y (b) $X \subset Y$
(c) $Y \subset X$ (d) None of these
[2005 : 1 M]

2.36 The following is the Hasse diagram of the poset [{a, b, c, d,e}, ≤]

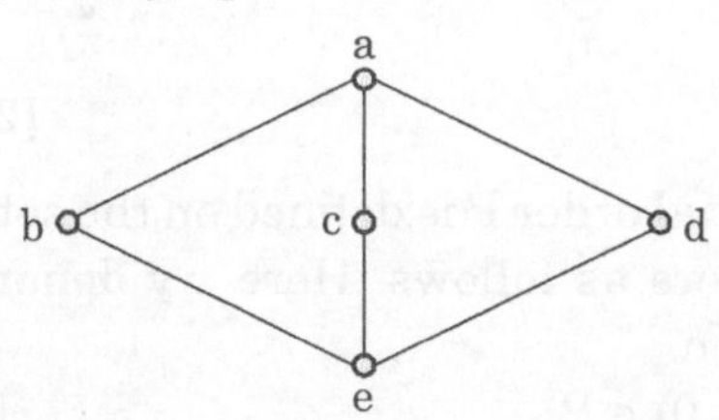

The poset is
(a) not a lattice
(b) a lattice but not a distributive lattice
(c) a distributive lattice but not a Boolean algebra
(d) a Boolean algebra **[2005 : 1 M]**

2.37 The set {1, 2, 4, 7, 8, 11, 13, 14} is a group under multiplication modulo 15. The inverses of 4 and 7 are respectively
(a) 3 and 13 (b) 2 and 11
(c) 4 and 13 (d) 8 and 14
[2005 : 1 M]

2.38 Let R and S be any two equivalence relations on a non-empty set A. Which one of the following statements is TRUE?
(a) $R \cap S$, $R \cup S$ are both equivalence relations
(b) $R \cup S$ is an equivalence relation
(c) $R \cap S$ is an equivalence relation
(d) Neither $R \cup S$ nor $R \cap S$ is an equivalence relation **[2005 : 2 M]**

2.39 Let f: B → C and g: A → B be two functions and let h = f o g. Given that h is an onto function. Which one of the following is TRUE?
(a) f and g should both be onto functions
(b) f should be onto but g need not be onto
(c) g should be onto but f not be onto
(d) both f and g need not be onto **[2005 : 2 M]**

2.40 Consider the set H of all 3 × 3 matrices of the type

$$\begin{bmatrix} a & f & e \\ 0 & b & d \\ 0 & 0 & c \end{bmatrix}$$

where a, b, c, d, e and f are real numbers and abc ≠ 0. Under the matrix multiplication operation, the set H is
(a) a group
(b) a monoid but not group
(c) a semigroup but not a monoid
(d) neither a group nor a semigroup
[2005 : 2 M]

2.41 Let f be a function from a set A to a set B, g a function from B to C, and h a function from A to C, such that h(a) = g(f(a)) for all a∈ A. Which of the following statements is always true for all such functions f and g?
(a) g is onto ⇒ h is onto
(b) h is onto ⇒ f is onto
(c) h is onto ⇒ g is onto
(d) h is onto ⇒ f and g are onto
[2005 : 2 M]

2.42 Let A be a set with n elements. Let C be a collection of distinct subsets of A such that for any two subsets S_1 and S_2 in C, either $S_1 \subset S_2$ or $S_2 \subset S_1$. What is the maximum cardinality of C?
(a) n (b) n+1
(c) $2^{n-1}+1$ (d) n! **[2005: 2 M]**

2.43 For the set N of natural numbers and a binary operation f : N × N → N, an element z ∈ N is called an identity for f, if f(a, z) = a = f(z, a), for all a ∈ N. Which of the following binary operations have a identity?
I. f(x, y) = x + y – 3
II. f(x, y) = max(x, y)
III. f(x, y) = x^y

(a) I and II only (b) II and III only
(c) I and III only (d) None of these
[2006 : 1 M]

2.44 Let, X, Y, Z be sets of sizes x, y and z respectively. Let W = X × Y and E be the set of all subsets of W. The number of functions from Z to E is
(a) z (b) $z \times 2^{xy}$
(c) 2^z (d) 2^{xyz} **[2006 : 1 M]**

2.45 The set {1, 2, 3, 5, 7, 8, 9} under multiplication modulo 10 is not a group. Given below are four possible reasons. Which one of them is false?
(a) It is not closed
(b) 2 does not have an inverse
(c) 3 does not have an inverse
(d) 8 does not have an inverse **[2006 : 1 M]**

2.46 A relation R is defined on ordered pairs of integers as follows: (x, y)R(u,v) if x < u and y > v. Then R is
(a) Neither a Partial Order nor an Equivalence Relation
(b) A Partial Order but not a Total Order
(c) A Total Order
(d) An Equivalence Relation **[2006 : 1 M]**

2.47 Let E, F and G be finite sets:
Let $X = (E \cap F) - (F \cap G)$ and
$Y = (E - (E \cap G)) - (E - F)$.
Which one of the following is true?
(a) $X \subset Y$
(b) $X \supset Y$
(c) $X = Y$
(d) $X - Y \neq \varnothing$ and $Y - X \neq \varnothing$ **[2006 : 2 M]**

2.48 Let S = {1, 2, 3,..., m}, m > 3. Let X_1, X_2, ..., X_n be subsets of S each of size 3. Define a function f from S to the set of natural numbers as, f(i) is the number of sets X_j that contains the element i. That is $f(i) = |\{j \mid i \in X_j\}|$.
Then $\sum_{i=1}^{m} f(i)$ is
(a) 3m (b) 3n
(c) 2m + 1 (d) 2n + 1 **[2006 : 2 M]**

2.49 Let P, Q and R be sets let Δ denote the symmetric difference operator defined as $P \Delta Q = (P \cup Q) - (P \cap Q)$. Using Venn diagrams, determine which of the following is/are TRUE?
I. $P \Delta (Q \cap R) = (P \Delta Q) \cap (P \Delta R)$
II. $P \cap (Q \Delta R) = (P \cap Q) \Delta (P \cap R)$
(a) I only (b) II only
(c) Neither I nor II (d) Both I and II
[2006 : 2 M]

2.50 What is the cardinality of the set of integers X defined below? X = {n | 1 ≤ n ≤ 123, n is not divisible by 2, 3 or 5}
(a) 28 (b) 33
(c) 37 (d) 44 **[2006 : 2 M]**

2.51 Let S be a set of n elements. The number of ordered pairs in the largest and the smallest equivalence relations on S are
(a) n and n (b) n^2 and n
(c) n^2 and 0 (d) n and 1 **[2007 : 1 M]**

2.52 Consider the set S = {a, b, c, d}. Consider the following 4 partitions $\pi_1, \pi_2, \pi_3, \pi_4$ on
S: π_1 = {abcd}, π_2{ab,cd}, π_3 = {abc, d},
π_4 = {a,b,c,d}
Let ≺ be the partial order on the set of partitions $S' = (\pi_1, \pi_2, \pi_3, \pi_4)$ defined as follows: $\pi_i \prec \pi_j$ if and only if π_i refines π_j. The poset diagram for (S′, ≺) is

(a)
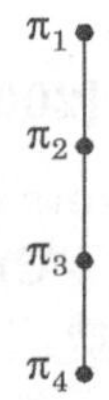

(b)
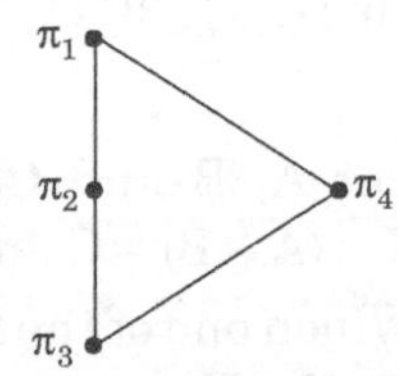

(c)
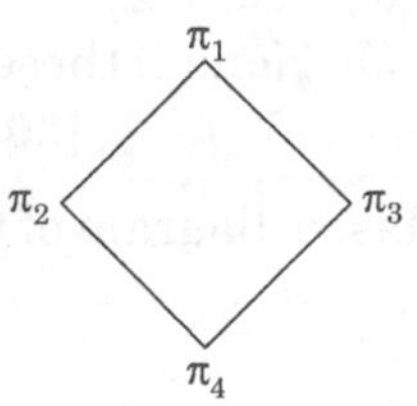

(d)
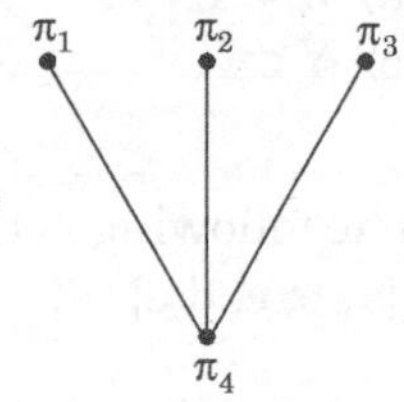

[2007 : 2 M]

2.53 A partial order P is defined on the set of natural numbers as follows. Here x/y denotes integer division.
(i) (0, 0) ∈ P.
(ii) (a, b) ∈ P if and only if a %10 ≤ b %10 and (a/10, b/10) ∈ P.
Consider the following ordered pairs:
(i) (101, 22) (ii) (22, 101)
(iii) (145, 265) (iv) (0, 153)
Which of these ordered pairs of natural numbers are contained in P?
(a) (i) and (iii) (b) (ii) and (iv)
(c) (i) and (iv) (d) (iii) and (iv)
[2007 : 2 M]

2.54 If P, Q, R are subsets of the universal set U, then $(P \cap Q \cap R) \cup (P^c \cap Q \cap R) \cup Q^c \cup R^c$ is
(a) $Q^c \cup R^c$ (b) $P \cup Q^c \cup R^c$
(c) $P^c \cup Q^c \cup R^c$ (d) U **[2008 : 1 M]**

2.55 Consider the following Hasse diagrams:

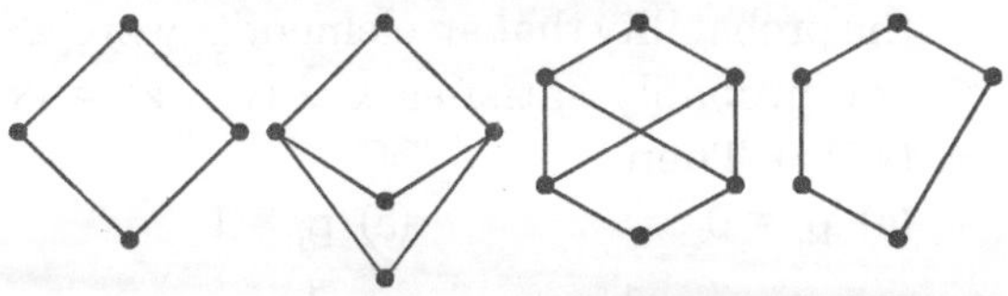

(i) (ii) (iii) (iv)

Which all of the above represent a lattice?
(a) (i) and (iv) only (b) (ii) and (iii) only
(c) (iii) only (d) (i), (ii) and (iv) only
[2008 : 2 M]

2.56 Which one of the following is NOT necessarily a property of a Group?
(a) Commutativity
(b) Associativity
(c) Existence of inverse for every element
(d) Existence of identity **[2009 : 1 M]**

2.57 Consider the binary relation: R = {(x, y), (x, z), (z, x), (z, y)} on the set {x, y, z}. Which one of the following is TRUE?
(a) R is symmetric but NOT antisymmetric
(b) R is NOT symmetric but antisymmetric
(c) R is both symmetric and antisymmetric
(d) R is neither symmetric nor antisymmetric
[2009 : 1 M]

2.58 For the composition table of a cyclic group shown below:

*	a	b	c	d
a	a	b	c	d
b	b	a	d	c
c	c	d	b	a
d	d	c	a	b

Which one of the following choices is correct?
(a) a, b are generators
(b) b, c are generators
(c) c, d are generators
(d) d, a are generators **[2009 : 2 M]**

2.59 What is the possible number of reflexive relations on a set of 5 elements?
(a) 2^{10} (b) 2^{15}
(c) 2^{20} (d) 2^{25} **[2010 : 1 M]**

2.60 Consider the set S = {1, ω, ω²}, where ω and ω² are cube roots of unity. If * denotes the multiplication operation, the structure {S, *} forms
(a) a group (b) a ring
(c) an integral domain (d) a field
[2010 : 1 M]

2.61 How many onto (or surjective) functions are there from an n-element ($n \geq 2$) set to a 2-element set?
(a) 2^n (b) $2^n - 1$
(c) $2^n - 2$ (d) $2(2^n - 2)$ **[2012 : 2 M]**

2.62 A binary operation ⊕ on a set of integers is defined as $x \oplus y = x^2 + y^2$. Which one of the following statements is TRUE about ⊕?
(a) Commutative but not associative
(b) Both commutative and associative
(c) Associative but not commutative
(d) Neither commutative nor associative
[2013 : 1 M]

2.63 Let S denote the set of all functions $f:\{0,1\}^4 \rightarrow \{0,1\}$. Denote by N the number of functions from S to the set {0, 1}. The value of $\log_2 \log_2 N$ is ______. **[2014 (Set-1) : 2 M]**

2.64 Consider the following relation on subsets of the set S of integers between 1 and 2014. For two distinct subsets *U* and *V* of S we say *U* < *V* if the minimum element in the symmetric difference of the two sets is in *U*.
Consider the following two statements:
S_1: There is a subset of S that is larger than every other subset.
S_2: There is a subset of S that is smaller than every other subset.
Which one of the following is CORRECT?
(a) Both S_1 and S_2 are true
(b) S_1 is true and S_2 is false
(c) S_2 is true and S_1 is false
(d) Neither S_1 nor S_2 is true
[2014 (Set-2) : 2 M]

2.65 Let *X* and *Y* be finite sets and $f: X \rightarrow Y$ be a function. Which one of the following statements is TRUE?
(a) For any subsets *A* and *B* of *X*,
$|f(A \cup B)| = |f(A)| + |f(B)|$
(b) For any subsets *A* and *B* of *X*,
$f(A \cap B) = f(A) \cap f(B)$
(c) For any subsets *A* and *B* of *X*,
$|f(A \cap B)| = \min\{|f(A)|, |f(B)|\}$
(d) For any subsets *S* and *T* of *Y*,
$f^{-1}(S \cap T) = f^{-1}(S) \cap f^{-1}(T)$
[2014 (Set-3) : 1 M]

2.66 Let *G* be a group with 15 elements. Let *L* be a subgroup of *G*. It is known that $L \neq G$ and that the size of *L* is at least 4. The size of *L* is ______.
[2014 (Set-3) : 1 M]

2.67 Consider the set of all functions $f: \{0,1, ..., 2014\} \to \{0,1, ...,2014\}$ such that $f(f(i)) = i$, for all $0 \le i \le 2014$. Consider the following statements:

P: For each such function it must be the case that for every i, $f(i) = i$.

Q: For each such function it must be the case that for some i, $f(i) = i$.

R: Each such function must be onto.

Which one of the following is CORRECT?

(a) P, Q and R are true
(b) Only Q and R are true
(c) Only P and Q are true
(d) Only R is true **[2014 (Set-3) : 2 M]**

2.68 There are two elements x, y in a group $(G, *)$ such that every element in the group can be written as a product of some number of x's and y's in some order. It is known that

$$x*x = y*y = x*y*x*y = y*x*y*x = e$$

where e is the identity element. The maximum number of elements in such a group is ______.

[2014 (Set-3) : 2 M]

2.69 If $g(x) = 1 - x$ and $h(x) = \dfrac{x}{x-1}$, then $\dfrac{g(h(x))}{h(g(x))}$ is

(a) $\dfrac{h(x)}{g(x)}$ (b) $\dfrac{-1}{x}$

(c) $\dfrac{g(x)}{h(x)}$ (d) $\dfrac{x}{(1-x)^2}$

[2015 (Set-1) : 1 M]

2.70 For a set A, the power set of A is denoted by 2^A. If $A = \{5, \{6\}, \{7\}\}$, which of the following options are True?

1. $\varnothing \in 2^A$ **2.** $\varnothing \subseteq 2^A$
3. $\{5,\{6\}\} \in 2^A$ **4.** $\{5,\{6\}\} \subseteq 2^A$

(a) 1 and 3 only (b) 2 and 3 only
(c) 1, 2 and 3 only (d) 1, 2 and 4 only

[2015 (Set-1) : 1 M]

2.71 Suppose $\mathcal{L} = \{p, q, r, s, t\}$ is a lattice represented by the following Hasse diagram:

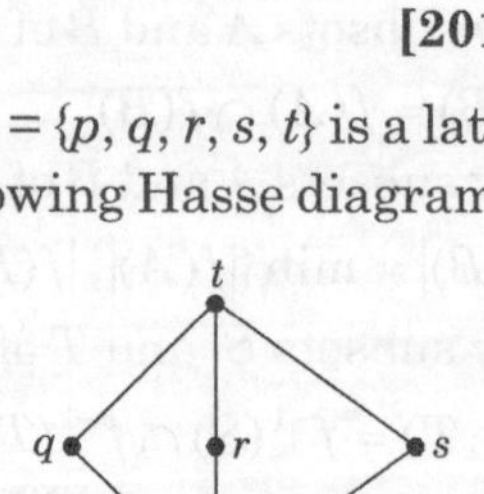

For any $x, y \in \mathcal{L}$, not necessarily distinct, $x \vee y$ and $x \wedge y$ are join and meet of x, y respectively. Let $\mathcal{L}^3 = \{(x,y,z): x, y, z \in \mathcal{L}\}$ be the set of all ordered triplets of the elements of $\mathcal{L}$. Let p_r be the probability that an element $(x,y,z) \in \mathcal{L}^3$ chosen equiprobably satisfies $x \vee (y \wedge z) = (x \vee y) \wedge (x \vee z)$. Then

(a) $p_r = 0$ (b) $p_r = 1$

(c) $0 < p_r \le \dfrac{1}{5}$ (d) $\dfrac{1}{5} < p_r < 1$

[2015 (Set-1) : 2 M]

2.72 Let R be the relation on the set of positive integers such that aRb if and only if a and b are distinct and have a common divisor other than 1. Which one of the following statements about R is True?

(a) R is symmetric and reflexive but not transitive
(b) R is reflexive but not symmetric and not transitive
(c) R is transitive but not reflexive and not symmetric
(d) R is symmetric but not reflexive and not transitive **[2015 (Set-2) : 1 M]**

2.73 The cardinality of the power set of {0, 1, 2..., 10} is ______. **[2015 (Set-2) : 1 M]**

2.74 The number of onto functions (surjective functions) from set X = {1, 2, 3, 4} to set Y = {a, b, c} is ______. **[2015 (Set-2) : 2 M]**

2.75 Let X and Y denote the sets containing 2 and 20 distinct objects respectively and F denote the set of all possible functions defined from X and Y. Let f be randomly chosen from F. The probability of f being one-to-one is ______.

[2015 (Set-2) : 2 M]

2.76 Let # be a binary operator defined as X # Y = X′ + Y′ where X and Y are Boolean variables. Consider the following two statements:

S_1: (P # Q) # R = P # (Q # R)

S_2: Q # R = R # Q

Which of the following is/are true for the Boolean variables P, Q and R?

(a) Only S_1 is True
(b) Only S_2 is True
(c) Both S_1 and S_2 are True
(d) Neither S_1 nor S_2 are True

[2015 (Set-3) : 1 M]

2.77 Suppose U is the power set of the set $S = \{1,2,3,4,5,6\}$. For any $T \in U$, let $|T|$ denote the number of elements in T and T' denote the complement of T. For any $T, R \in U$, let $T \setminus R$ be the set of all elements in T which are not in R. Which one of the following is true?

(a) $\forall X \in U(|X| = |X'|)$
(b) $\exists X \in U \exists Y \in U(|X| = 5, |Y| = 5 \text{ and } X \cap Y = \emptyset)$
(c) $\forall X \in U \forall Y \in U(|X| = 2, |Y| = 3 \text{ and } X \setminus Y = \emptyset)$
(d) $\forall X \in U \forall Y \in U(X \setminus Y = Y' \setminus X')$

[2015 (Set-3) : 1 M]

2.78 Let R be a relation on the set of ordered pairs of positive integers such that $((p, q), (r, s)) \in R$ if and only if $p - s = q - r$. Which one of the following is true about R?
(a) Both reflexive and symmetric
(b) Reflexive but not symmetric
(c) Not reflexive but symmetric
(d) Neither reflexive nor symmetric

[2015 (Set-3) : 2 M]

2.79 A function $f : \mathbb{N}^+ \to \mathbb{N}^+$, defined on the set of positive integers $\mathbb{N}^+$, satisfies the following properties:

$f(n) = f(n/2)$ if n is even

$f(n) = f(n + 5)$ if n is odd

Let $R = \{i \mid \exists j : f(j) = i\}$ be the set of distinct values that f takes. The maximum possible size of R is ______. **[2016 (Set-1) : 2 M]**

2.80 A binary relation R on $N \times N$ is defined as follows: $(a, b) R (c, d)$ if $a \le c$ or $b \le d$. Consider the following propositions:
P : R is reflexive
Q : R is transitive
Which one of the following statements is TRUE?
(a) Both P and Q are true
(b) P is true and Q are false
(c) P is false and Q are true
(a) Both P and Q are false

[2016 (Set-2) : 2 M]

2.81 Consider a set U of 23 different compound in a Chemistry lab. There is a subset S of U of 9 compounds, each of which reacts with exactly 3 compounds of U. Consider the following statements:
I. Each compound in $U \setminus S$ reacts with an odd number of compounds.
II. At least one compound in $U \setminus S$ reacts with an odd number of compounds.
III. Each compound in $U \setminus S$ reacts with an even number of compounds
Which one of the above statements is ALWAYS TRUE?
(a) Only I (b) Only II
(c) Only III (d) None of these

[2016 (Set-2) : 2 M]

2.82 Consider the set $X = \{a, b, c, d, e\}$ under the partial ordering:
$R = \{(a, a), (a, b), (a, c), (a, d), (a, e), (b, b), (b, c), (b, e), (c, c), (c, e), (d, d), (d, e), (e, e)\}$.
The Hasse diagram of the partial order (X, R) is shown below:

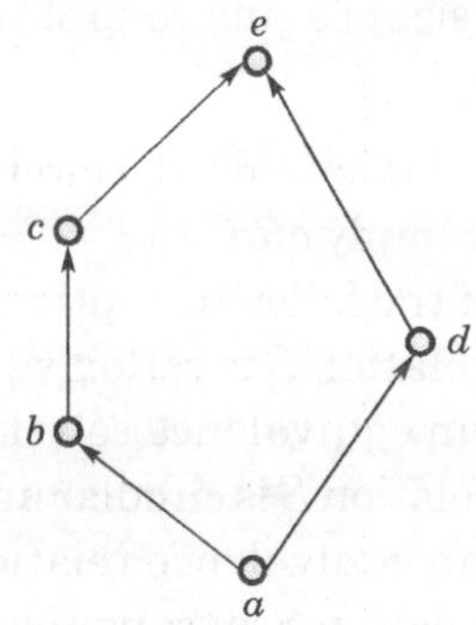

The minimum number of ordered pairs that need to be added to R to make (X, R) a lattice is ______.

[2017 (Set-2) : 1 M]

2.83 The number of integers between 1 and 500 (both inclusive) that are divisible by 3 or 5 or 7 is ______. **[2017 (Set-1) : 1 M]**

2.84 Let G be a finite group on 84 elements. The size of a largest possible proper subgroup of G is ______. **[2018 : 1 M]**

2.85 Let N be the set of natural numbers. Consider the following sets:
P. Set of Rational numbers (positive and negative).
Q. Set of functions from {0, 1} to N.
R. Set of functions from N to {0, 1}.
S. Set of finite subsets of N.
Which of the sets above are countable?
(a) Q and S only (b) P and S only
(c) P and R only (d) P, Q and S only

[2018 : 2 M]

2.86 Let G be an arbitrary group. Consider the following relations on G:
R_1 : $\forall a, b \in G, a\ R_1\ b$ if and only if $\exists g \in G$ such that $a = g^{-1}bg$
R_2 : $\forall a, b \in G, a\ R_2\ b$ if and only if $a = b^{-1}$
Which of the above is/are equivalence relation/relations?
(a) R_1 and R_2
(b) R_1 only
(c) R_2 only
(d) Neither R_1 nor R_2

[2019 : 1 M]

2.87 Let R be the set of all binary relations on the set {1, 2, 3}. Suppose a relation is chosen from R at random. The probability that the chosen relation is reflexive (round off to 3 decimal places) is ________. **[2020 : 1 M]**

2.88 Let G be a group of 35 elements. Then the largest possible size of a subgroup of G other than G itself is ________. **[2020 : 1 M]**

2.89 A relation R is said to be circular if aRb and bRc together imply cRa.
Which of the following options is/are correct?
(a) If a relation S is reflexive and circular, then S is an equivalence relation.
(b) If a relation S is circular and symmetric, then S is an equivalence relation.
(c) If a relation S is transitive and circular, then S is an equivalence relation.
(d) If a relation S is reflexive and symmetric, then S is an equivalence relation.
[2021 (Set-1) : 2 M]

2.90 Let G be a group of order 6, and H be a subgroup of G such that $1 < |H| < 6$. Which one of the following options is correct?
(a) Both G and H are always cyclic.
(b) G is always cyclic, but H may not be cyclic.
(c) G may not be cyclic, but H is always cyclic.
(d) Both G and H may not be cyclic.
[2021 (Set-1) : 2 M]

2.91 Consider the following sets, where $n \geq 2$:
S_1 : Set of all $n \times n$ matrices with entries from the set $\{a, b, c\}$
S_2 : Set of all functions from the set $\{0, 1, 2, ..., n^2 - 1\}$ to the set $\{0, 1, 2\}$
Which of the following choice(s) is/are correct?
(a) There exists a surjection from S_1 to S_2.
(b) There does not exist a bijection from S_1 to S_2.
(c) There does not exist an injection from S_1 to S_2.
(d) There exists a bijection from S_1 to S_2.
[2021 (Set-2) : 1 M]

2.92 Let S be a set consisting of 10 elements. The number of tuples of the form (A, B) such that A and B are subsets of S and $A \subseteq B$ is ________. **[2021 (Set-2) : 2 M]**

2.93 For two n-dimensional real vectors P and Q, the operation $s(P, Q)$ is defined as follows:
$$s(P, Q) = \sum_{i=1}^{n} (P[i].Q[i])$$
Let $\mathcal{L}$ be a set of 10-dimensional non-zero real vectors such that for every pair of distinct vectors $P, Q \in \mathcal{L}$, $s(P, Q) = 0$. What is the maximum cardinality possible for the set $\mathcal{L}$?
(a) 10 (b) 9
(c) 100 (d) 11
[2021 (Set-2) : 2 M]

2.94 Which of the following statements is/are TRUE for a group G?
(a) If for all $x, y \in G$, $(xy)^2 = x^2y^2$, then G is commutative.
(b) If for all $x \in G$, $x^2 = 1$, then G is commutative. Here, 1 is the identity element of G.
(c) If the order of G is 2, then G is commutative.
(d) If G is commutative, then a subgroup of G need not be commutative. **[2022 : 1 M]**

2.95 Let $f: A \to B$ be an onto (or surjective) function, where A and B are non-empty sets. Define an equivalence relation $\sim$ on the set A as
$$a_1 \sim a_2 \text{ if } f(a_1) = f(a_2),$$
where $a_1, a_2 \in A$. Let $\in = \{[x] : x \in A\}$ be the set of all the equivalence classes under $\sim$. Define a new mapping $F : \in \to B$ as
$F([x]) = f(x)$, for all the equivalence classes $[x]$ in $\in$.
Which of the following statements is/are TRUE?
(a) F is NOT well-defined.
(b) F is an onto (or surjective) function.
(c) F is a one-to-one (or injective) function.
(d) F is a bijective function.
[2023 : 2 M]

2.96 Let X be a set and 2^X denote the powerset of X. Define a binary operation Δ on 2^X as follows:
$$A \Delta B = (A - B) \cup (B - A)$$
Let $H = (2^X, \Delta)$. Which of the following statements about H is/are correct?
(a) H is a group.
(b) Every element in H has an inverse, but H is NOT a group.
(c) For every $A \in 2^X$, the inverse of A is the complement of A.
(d) For every $A \in 2^X$, the inverse of A is A.
[2023 : 2 M]

2.97 Let A and B be non-empty finite sets such that there exist one-to-one and onto functions (i) from A to B and (ii) from $A \times A$ to $A \cup B$. The number of possible values of $|A|$ is ________.
[2024 (Set-1) : 1 M]

2.98 Consider the operators ◊ and □ defined by a ◊ $b = a + 2b$, a □ $b = ab$, for positive integers. Which of the following statements is/are TRUE?

(a) Operator ◊ obeys the associative law.

(b) Operator ◊ over the operator □ obeys the distributive law.

(c) Operator □ obeys the associative law.

(d) Operator □ over the operator ◊ obeys the distributive law.

[2024 (Set-1) : 2 M]

2.99 Let P be the partial order defined on the set {1, 2, 3, 4} as follows:

$$P = \{(x, x) \mid x \in \{1, 2, 3, 4\}\} \cup \{(1, 2), (3, 2), (3, 4)\}$$

The number of total orders on {1, 2, 3, 4} that contain P is ________.

[2024 (Set-2) : 1 M]

2.100 Let Z_n be the group of integers {0,1,2, ..., $n-1$} with addition modulo n as the group operation. The number of elements in the group $Z_2 \times Z_3 \times Z_4$ that are their own inverses is ________.

[2024 (Set-2) : 2 M]

■■■■

Answers Set Theory and Algebra

2.1	(d)	2.2	(c)	2.3	(d)	2.4	(a)	2.5	(b)	2.6	(a)	2.7	(c)	2.8	(c)	2.9	(b)
2.10	(c)	2.11	(a)	2.12	(b)	2.13	(c)	2.14	(c)	2.15	(b)	2.16	Sol.	2.17	Sol.	2.18	(c)
2.19	Sol.	2.20	(c)	2.21	(b)	2.22	(b)	2.23	(c)	2.24	(a)	2.25	(d)	2.26	(a)	2.27	(d)
2.28	(c)	2.29	(b)	2.30	(c)	2.31	(c)	2.32	(c)	2.33	(d)	2.34	(a)	2.35	(a)	2.36	(b)
2.37	(c)	2.38	(c)	2.39	(b)	2.40	(a)	2.41	(c)	2.42	(b)	2.43	(a)	2.44	(d)	2.45	(c)
2.46	(a)	2.47	(c)	2.48	(b)	2.49	(b)	2.50	(b)	2.51	(b)	2.52	(c)	2.53	(d)	2.54	(d)
2.55	(a)	2.56	(a)	2.57	(d)	2.58	(c)	2.59	(c)	2.60	(a)	2.61	(c)	2.62	(a)	2.63	(16)
2.64	(a)	2.65	(d)	2.66	(5)	2.67	(b)	2.68	(4)	2.69	(a)	2.70	(c)	2.71	(d)	2.72	(d)
2.73	(2048)	2.74	(36)	2.75	(0.95)	2.76	(b)	2.77	(d)	2.78	(c)	2.79	(2)	2.80	(b)	2.81	(b)
2.82	(0)	2.83	(271)	2.84	(42)	2.85	(d)	2.86	(b)	2.87	(0.125)	2.88	(7)	2.89	(a)	2.90	(c)
2.91	(a, d)	2.92	(3^{10})	2.93	(a)	2.94	(a, b, c)	2.95	(b, c, d)	2.96	(a, d)	2.97	(2)	2.98	(c, d)	2.99	(5)
2.100	(4)																

Explanations Set Theory and Algebra

2.1 (d)

A relation which is symmetric and transitive, need not be reflexive relation.

(i) R = {}; on the set A = {a, b}. The relation R is symmetric and transitive but not reflexive.

(ii) R = {(a, a), (b, b)}; on the set A = {a, b}. The relation R is symmetric, transitive and also reflexive.

∴ A relation is transitive and symmetric relation but need not be reflexive relation.

2.2 (c)

If a set has n elements then its powers set has 2^n elements.

Given set S = {(ϕ), 1, (2, 3)}

Number of elements in S = 3

∴ The number of elements in powerset (S) = $2^3 = 8$.

2.3 (d)

(i) Closure Property: Multiplication of two non-singular matrices is also non-singular matrix. Matrix multiplication over non-singular matrices follows closure properties.

(ii) Associative Property: Multi-plication over any set of matrices is associative.

$$(AB)C = A(BC)$$

Where A, B and C are non-singular matrices

(iii) Identity Element: Identity matrix I is the identity element for matrix multiplication over matrices and I is non-singular.

(iv) Inverse Element: For every non-singular matrix its inverse exists. So, for non-singular matrices inverse element exist.

(v) Commutative: Matrix multi-plication is not commutative

$$AB \neq BA$$

Where A, B are non-singular matrices.

Matrices multiplication is not commutative.

So <A, *> is a group but not an abelian group.

2.4 (a)

$$(A - B) \cup (B - A) \cup (A \cap B)$$

Representing above set using Venn diagram as follows.

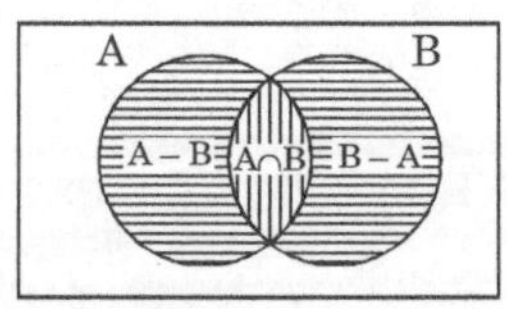

∴ By Venn diagram,

$$(A - B) \cup (B - A) \cup (A \cap B) \equiv A \cup B$$

2.5 (b)

x = {2, 3, 6, 12, 24}

The Hasse diagram of (x, ≤) is

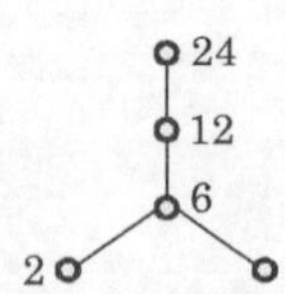

Therefore, the number of edges in the Hasse diagram = 4

2.6 (a)

X and Y are sets. The cardinalities of X and Y are |X| and |Y| respectively.

The number of functions from X to Y = $(|Y|)^{|X|}$

Given that number of functions from X to Y = 97

$$\therefore \quad 97 = (|Y|)^{|X|}$$

So above implies that |X| = 1 and |Y| = 97

2.7 (c)

0 is also a rational number and for 0, inverse doesn't exist under multiplication.
Therefore, the set of rational numbers doesn't form an abelian group under multiplication.

2.8 (c)

$$f(x, y) = ((x + y), (x - y))$$

So let $z_1 = x + y$ ...(i)

$z_2 = x - y$ ...(ii)

$$f(x, y) = (z_1, z_2)$$

So $f^{-1}(z_1, z_2) = (x, y)$

Adding (i) and (ii)

$$z_1 + z_2 = 2x$$

$$x = \frac{z_1 + z_2}{2}$$

Subtracting (i) and (ii)

$$z_1 - z_2 = 2y$$

$$y = \frac{z_1 - z_2}{2}$$

So the inverse function of F is given by

$$f^{-1}(z_1, z_2) = (x, y) = \left(\frac{z_1 + z_2}{2}, \frac{z_1 - z_2}{2}\right)$$

Since z_1 and z_2 are just dummy variables so replacing z_1 and z_2 by x and y respectively

$$f^{-1}(x, y) = \left(\frac{x + y}{2}, \frac{x - y}{2}\right)$$

Alternate Method:

Given $f(x, y) = ((x + y), (x - y))$

Take some random ordered pair say (2, 3)

$f(2, 3) = ((2 + 3), (2 - 3)) = (5, -1)$

Now the correct inverse must map (5, –1) back to (2, 3).

Trying the options one-by-one we find that only option (c) maps (5, –1) to (2, 3).

2.9 (b)

(i) Reflexive

$A \cap A = A \neq \phi$

So; (A, A) doesn't belongs to relation R,

∴ Relation R is not reflexive.

(ii) Symmetric

If $A \cap B = \phi$ then $B \cap A = \phi$ is also true

∴ relation R is symmetric relation.

(iii) Transitive

If $A \cap B = \phi$ and $B \cap C = \phi$, it need not be true that $A \cap C = \phi$.

For example:

$A = \{1, 2\}, B = \{3, 4\}, C = \{1, 5, 6\}$

$A \cap B = \phi$ and $B \cap C = \phi$ but

$A \cap C = \{1\} \neq \phi$

∴ Relation R is not transitive relation.

2.10 (c)

To form a group it has to satisfy following property:

- Closure
- Associativity
- Identity
- Inverse
- Let finite alphabet $\Sigma = \{0, 1\}$ any string $S_1 \cdot S_2$ will belong to given alphabet so closure is satisfied.
- $(S_1 \cdot S_2) \cdot S_3 = S_1 \cdot (S_2 \cdot S_3)$
  ∴ associative.
- $S_1 \cdot \in = S_1 = \in \cdot S_1$
  ∴ $\in$ is identity.
- Inverse of S_1 doesn't exist.
  Let, $S_1 = 001010$
  $S_1 \cdot$ anything will not give $\in$.

2.11 (a)

Corresponding to every partition of the set {1, 2, 3, 4}, there exists a unique equivalence relation. So we count every type of unordered partitions of the set of 4 elements into one block, two block, three block and four block partitions, as shown below:

4 → 4: $= \frac{4!}{4!} = 1$

4 → 1, 3 / 2, 2: $= \frac{4!}{3!} + \frac{4!}{2!\,2!\,2!} = 4 + 3 = 7$

4 → 2, 1, 1: $= \frac{4!}{2!\,2!} = 6$

4 → 1, 1, 1, 1: $= \frac{4!}{4!} = 1$

Total = 1 + 7 + 6 + 1 = 15

So the number of equivalence relations on the set {1, 2, 3, 4} is 15.

2.12 (b)

A is a finite set with n elements. The largest equivalence relation of A is the cross-product $A \times A$ and the number of elements in the cross product $A \times A$ is n^2.

2.13 (c)

A relation is said to be equivalence relation if relation is

(i) Reflexive
(ii) Symmetric
(iii) Transitive

Reflexive and symmetric properties are both closed under $\cup$ & $\cap$.

Transitive property is closed under $\cap$ but not $\cup$.

So equivalence relations are closed under $\cap$ but not $\cup$.

Therefore $R_1 \cap R_2$ is an equivalence relation but $R_1 \cup R_2$ is not necessarily an equivalence relation.

2.14 (c)

The number of functions from an m element set to an n element set is n^m.

2.15 (b)

The relation R doesn't contain (4, 4), so R is not reflexive relation.

Since relation R contains (1, 1), (2, 2) and (3, 3). Therefore, relation R is also not irreflexive.

That R is transitive, can be checked by systematically checking for all (a, b) and (b, c) in R, whether (a, c) also exists in R.

So option (b) is correct.

2.16 Sol.

Given partition of A is

$$\pi_1 = \{(a, b, c), (d)\}$$

(a) The ordered pairs of the equivalence relations induced by π_1 is

$$R = (a, b, c) \times (a, b, c) \cup (d) \times (d)$$
$$R = \{(a, a), (a, b), (a, c), (b, a), (b, b), (b, c), (c, a), (c, b), (c, c), (d, d)\}$$

(b) The diagraph of the above equivalence relation is as follows:

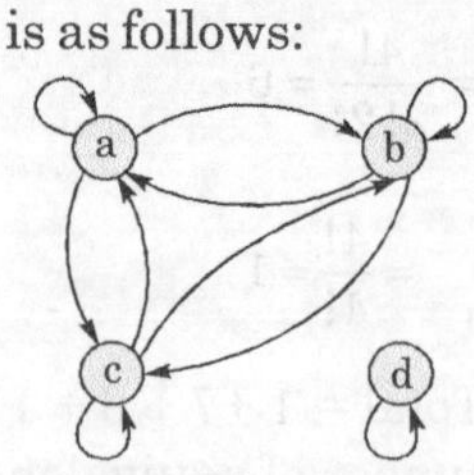

Diagraph of Relation R

2.17 Sol.

Given $a \neq b \Rightarrow a * b \neq b * a$

The contrapositive version of this is

$a*b = b*a \Rightarrow a = b$

(a) Now since (A, *) is given to be a semigroup, * is associative

i.e. $a * (a * a) = (a * a) * a$

Hence $a = a * a$

(b) $(a * b * a) * a = a * b * (a * a)$
$= a * b * a$
$\Rightarrow a * (a * b * a) = (a * a) * b * a = a * b * a$
as $a * a = a$ [proved in part a]
So $(a * b * a) * a = a * (a * b * a)$
Hence $a * b * a = a$

(c) $(a * b * c) * (a * c) = a * b * (c * a * c)$
$= a * b * c$
$(a * c) * (a * b * c) = (a * c * a) * b * c$
$= a * b * c$
So $(a * b * c) * (a * c) = (a * c) * (a * b * c)$
Hence $a * b * c = a * c$

2.18 (c)

The maximum number of elements in a binary relation on a set A with n elements = Number of elements in $A \times A = n^2$

Each element has two choices, either to appear on a binary relation or doesn't appear on a binary relation.

$\therefore$ Number of binary relations $= 2^{n^2}$.

2.19 Sol.

(a) According to Mr. X if xRy is in relation then yRx also in relation because of symmetricity. Now because R is transitive, xRy and yRx implies xRx.

Flaw in Mr. X claim is that if xRy present then only it implies yRx and hence xRx. However, if xRy is not present then according to symmetricity yRx is also not present. So xRx need not be present in relation. So it need not be reflexive.

For Example: Empty relation ϕ is both symmetric and transitive. However, ϕ is not reflexive relation.

(b) Examples of a relations which are symmetric and transitive but not reflexive.

(i) Empty relation ϕ is symmetric and transitive but not reflexive.

(ii) Relation R = {(a, b), (b, a), (a, a), (b, b)} over A = {a, b, c} is both symmetric and transitive but not reflexive.

2.20 (c)

A relation R is defined as xRy iff (x + y) is even over set of integers.

(x + y) is even iff

(i) both x and y are even
(ii) both x and y are odd

Therefore, relation R is equivalence relation because relation is

(i) Reflexive
$x + x = 2x$ = even
So (x, x) belongs to R. So relation is reflexive.

(ii) Symmetric
If $x + y$ = even then $y + x$ is also even
So relation is symmetric.

(iii) Transitive
If $x + y$ = even and $y + z$ = even
Then $x + y + y + z$ = even + even
$\Rightarrow x + z + 2y$ = even
$\Rightarrow x + z$ = even $- 2y \Rightarrow x + z$ = even
$\therefore$ Relation R is transitive.

So relation R is an equivalence relation which divides the set of integer into two equivalence classes: One is of all even and other is of odd integer.

Equivalence classes of R are
$[0] = \{..., -6, -4, -2, 0, 2, 4, 6, ...\}$
$[1] = \{..., -7, -5, -3, -1, 1, 3, 5, 7, ...\}$

2.21 (b)

ϕ always present in any powerset of a set and ϕ is the only common element between P(S) and P(P(S))
$\therefore P(S) \cap P(P(S)) = \{\phi\}$

2.22 (b)

(I) Relation R_1(a, b) iff (a + b) is even over the set of integers.
(i) $a + a = 2a$ which is even
So (a, a) belongs to R_1
$\therefore R_1$ is reflexive relation
(ii) If (a + b) is even, then (b + a) is also even
$\therefore R_1$ is symmetric relation
(iii) If (a + b) and (b + c) are even then
$a + c = (a + b) + (b + c) - 2b$
= even + even – even = even
$\therefore R_1$ is transitive relation.
Since R_1 is reflexive, symmetric and transitive so R_1 is an equivalence relation.

(II) R_2(a, b) iff (a +b) is odd over set of integers.
(i) $a + a = 2a$ which is not odd
So (a, a) doesn't belong to R_2
$\therefore R_2$ is not reflexive relation
Since R_2 is not reflexive, it is not an equivalence relation.

(III) R_3(a, b) iff a.b > 0 over set of non-zero relational numbers.
(i) a. a. > 0 for every non-zero rational number.
$\therefore R_3$ is reflexive relation.
(ii) If a.b > 0 then b.a > 0
$\therefore R_3$ is symmetric relation.
(iii) a.b > 0 and b.c > 0 $\Rightarrow$ All a,b,c are positive or all a,b,c are negative.
So a.c > 0
$\therefore R_3$ is transitive relation.
So R_3 is an equivalence relation.

(IV) R_4 (a, b) iff $|a - b| \le 2$ over the set of natural number
(i) $|a - a| \le 2$
$0 \le 2$
$\therefore R_4$ is reflexive relation.
(ii) If $|a - b| \le 2$ then also $|b - a| \le 2$
$\therefore R_4$ is symmetric relation
(iii) If $|a - b| \le 2$ and $|b - c| \le 2$ then it is not necessary that $|a - c| \le 2$

Example:
$|3 - 5| \le 2$ and $|5 - 7| \le 2$ but
$|3 - 7| \not\le 2$
$\therefore R_4$ is not transitive.

Since R_4 is reflexive and symmetric not transitive, so R_4 is not an equivalence relation.

2.23 (c)

S_1: Let A = set of integers
B = set of odd integers
C = set of even integers
$A \cap (B \cap C) = \phi$ and ϕ is finite set.
Therefore, S_1 is true.

S_2: Let two irrational number x and y are respectively $(1+\sqrt{2})$ and $(1-\sqrt{2})$.
So $x + y = 1+\sqrt{2}+1-\sqrt{2}$
= 2 which is rational number

Therefore, S_2 is true. Since both S_1 and S_2 are true, option (c) is true.

2.24 (a)

Given a function f: x → y and subsets E and F of A then we have
$f(E \cup F) = f(E) \cup f(F)$ and
$f(E \cap F) \subseteq f(E) \cap f(F)$
Therefore S_1 is correct and S_2 is false.

2.25 (d)

The empty relation on any set is always transitive and symmetric but not reflexive.

2.26 (a)

$$\Sigma = \{0, 1\}$$
$$\Sigma^* = \{0, 1\}^* = \{\varepsilon, 0, 1, 01, 10, 11, 000, ...\}$$

So $(\Sigma^*, \cdot)$ is an algebraic system, where $\cdot$ (concatenation) is a binary operation.

So $(\Sigma^*, \cdot)$ is a group if and only if the following conditions are satisfied.

1. $\cdot$ (Concatenation) is a closed operation.
2. $\cdot$ is an associative operation.
3. There is an identity.
4. Every element of Σ^* has a inverse

Condition 1: * is a closed operation because for any $\omega_1 \in \Sigma^*$ and $\omega_2 \in \Sigma^*$, $\omega_1 \cdot \omega_2 \in \Sigma^*$

Condition 2: For any string $x, y, z \in \Sigma^*$,

$$x \cdot (y \cdot z) = (x \cdot y) \cdot z$$

So it is associative for example let

$x = 01, y = 11, z = 00$ then

$$\text{L.H.S.} = x{\cdot}(y{\cdot}z) = 01{\cdot}(11{\cdot}00) = 01{\cdot}(1100) = 011100$$
$$\text{R.H.S.} = (x{\cdot}y){\cdot}z = (01{\cdot}11){\cdot}00 = (0111){\cdot}00 = 011100$$

Condition 3: The Identity is ε or empty string because for any string $\omega \in \Sigma^*$,

$$\varepsilon\omega = \omega\varepsilon = \omega.$$

Now, since $\varepsilon \in \Sigma^*$, identity exists.

Condition 4: There is no inverse exist for Σ^* because any string $\omega \in \Sigma^*$, there is no string ω^{-1} such that $\omega \cdot \omega^{-1} = \varepsilon = \omega^{-1}\omega$.

So Σ^* with the concatenation operator for strings doesn't form a group but it does form a monoid.

2.27 (d)

Maximum element is c : So c is of higher order than any other element in S.

Minimal element are a and b : No other element in S is of lower order than either a or b.

We are given P(a) = true. So for all x such that $a \le x$, P(x) must be true. So, for all x such that $a \le x$, P(x) must be true. We do have at least one such x, which is c as it is maximum element, so (d) is false.

2.28 (c)

The possible solution pairs are (a, a), (a, b), (a, c), (b, a), (b, b), (b, c), (c, a,), (c, b) and (c, c). Substitute them one by one in both equations and see which of them satisfies both the equations.

The given equations are:

$(a \times x) + (a \times y) = c$ ...(i)

$(b \times x) + (c \times y) = c$ ...(ii)

Substitute first (x, y) = (a, a)

LHS of equation (i) becomes

$(a \times a) + (a \times a) = a + a = b$

Now RHS of equation (i) = c

Therefore LHS ≠ RHS. This means that (a, a) is not a solution pair.

Similarly try each of the remaining seven possible solution pairs.

It will be found that only two pairs (b, c) and (c, b) will satisfy both equation (i) & (ii) simultaneously.

Therefore choice (c) is correct.

2.29 (b)

$$S = \{(x, y) \mid y = x + 1, \text{ and } x, y \in \{0, 1, 2, ...\}\} = \{(0, 1), (1, 2), (2, 3), (3, 4), ...\}$$

Now let T_1 be the reflexive closure of S.

$$T_1 = \{(0, 0), (1, 1), (2, 2), (3, 3) ...\} \cup \{(0, 1), (1, 2), (2, 3), (3, 4) ...\}$$
$$= \{(0, 0), (0, 1), (1, 1), (1, 2), (2, 2), (2, 3), (3, 3), (3, 4) ...\}$$

Let T_2 be the transitive closure of S.

$(0, 1), (1, 2) \in S \Rightarrow (0, 2) \in T_2$

$(0, 2), (2, 3) \in S \Rightarrow (0, 3) \in T_2$

$(0, 3), (3, 4) \in S \Rightarrow (0, 4) \in T_2$

and so on ...

Also $(1, 2), (2, 3) \in S \Rightarrow (1, 3) \in T_2$

$(1, 3), (3, 4) \in S \Rightarrow (1, 4) \in T_2$

$(1, 4), (4, 5) \in S \Rightarrow (1, 5) \in T_2$

and so on ...

$$\therefore \; T_2 = \{(0, 1), (0, 2), (0, 3), ..., (1, 2), (1, 3), (1, 4), ...\}$$

Now the reflexive, transitive closure of S will be $T_3 = T_1 \cup T_2 = \{(0, 0), (0, 1), (0, 2), ..., (1, 1), (1, 2), (1, 3), ..., (2, 2) (2, 3), (2, 4), ...\}$. Option (b) is correct.

2.30 (c)

In a symmetric matrix, the lower triangle must be the mirror image of upper triangle using the diagonal as mirror. Diagonal elements may be anything. Therefore, when we are counting symmetric matrices we count how many ways are there to fill the upper triangle and diagonal elements. Since the first row has n elements, second (n – 1) elements, third row (n – 2) elements and so on upto last row, one element.

Total number of elements in diagonal + upper triangle = n + (n – 1) + (n – 2) + ... + 1

$$= \frac{n(n+1)}{2}$$

Now, each one of these elements can be either 0 or 1. So total number of ways we can fill these elements is

$$2^{\frac{n(n+1)}{2}} = \text{power}\left(2, \frac{(n^2+n)}{2}\right)$$

Since there is no choice for lower triangle elements the answer is power $\left(2, \frac{(n^2+n)}{2}\right)$ which is choice (c).

2.31 (c)

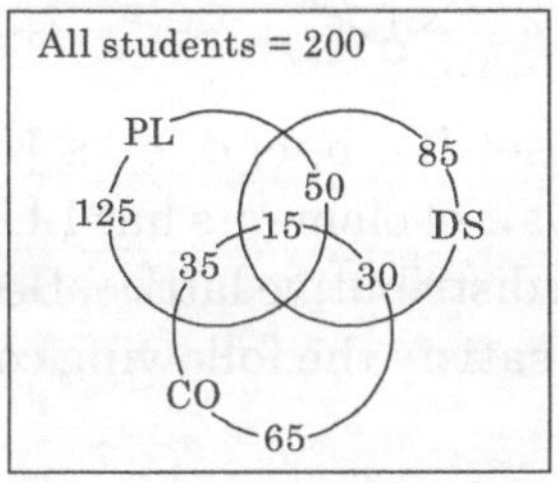

According to inclusion - exclusion formula:

$|PL \cup DS \cup CO| = |PL| + |DS| + |CO| - |PL \cap DS| - |PL - CO| - |DS \cap CO| + |PL \cap DS \cap CO|$

PL = students who have taken programming.
DS = Students who have taken Data structures.
CO = Students who have taken Computer Organization.

So, the number of students who have taken atleast 1 of the 3 courses is given by:

$= 125 + 85 + 65 - 50 - 35 - 30 + 15 = 175$

Therefore, the number of students who have not taken any of the 3 courses is

= Total students – students taken atleast 1 course

$= 200 - 175 = 25$

2.32 (c)

R_1 = {(An element x in A, An element y in B) if x + y divisible by 3}

$R_1 = \{(1, 2), (1, 8), (3, 6), (5, 4), (7, 2), (7, 8)\}$

R_2 = {(An element x in B, An element y in C) if x + y is even but not divisible by 3}

$R_2 = \{(2, 2), (4, 4), (6, 2), (6, 4), (8, 2)\}$

So, $R_1 \circ R_2 = \{(1, 2), (3, 2), (3, 4), (5, 4), (7, 2)\}$.

2.33 (d)

Step 1: By looking at the row for e, we see that it is a copy of the column headers.

So e must be the identity element. Since right identity and left identity element must both be same. The column corresponding to e must be a copy of the row headers.

We can now say that the operation table is:

*	e	a	b	c
e	e	a	b	c
a	a	b	c	e
b	b	–	–	–
c	c	–	–	–

Step 2: From table above we see that a*c = e

∴ c*a must also be = e (if a is the inverse of c, then c is the inverse of a)

Now the operation table looks like

*	e	a	b	c
e	e	a	b	c
a	a	b	c	e
b	b	–	–	–
c	c	e	–	–

Step 3: The blank in second column must be c (since in a cayley table, every row and every column is a unique permutation of the row and column headers).

Now the operation table looks like

*	e	a	b	c
e	e	a	b	c
a	a	b	c	e
b	b	c	–	–
c	c	e	–	–

Step 4: Now the blanks in third row can be filled as a, e or e, a. Let us try each one in turn.

If we fill a, e in third row the operation table will look like

*	e	a	b	c
e	e	a	b	c
a	a	b	c	e
b	b	c	a	e
c	c	e	–	–

Now the blank in fourth row and third column must be filled by e.

However this is not possible since e is already entered in fourth row and second column. Therefore filling a, e in third row blanks is wrong. So let us try filling the third row blanks with e, a

Now the operation table looks like

*	e	a	b	c
e	e	a	b	c
a	a	b	c	e
b	b	c	e	a
c	c	e	–	–

Now the blanks in fourth row has to be filled with a, b

The final operation table looks like

*	e	a	b	c
e	e	a	b	c
a	a	b	c	e
b	b	c	e	a
c	c	e	a	b

which is consistent with all the rules of a Cayley table.
The last row of this table is c, e, a, b
Therefore the correct answer is (d).

2.34 (a)

The hasse diagram of the given poset is:

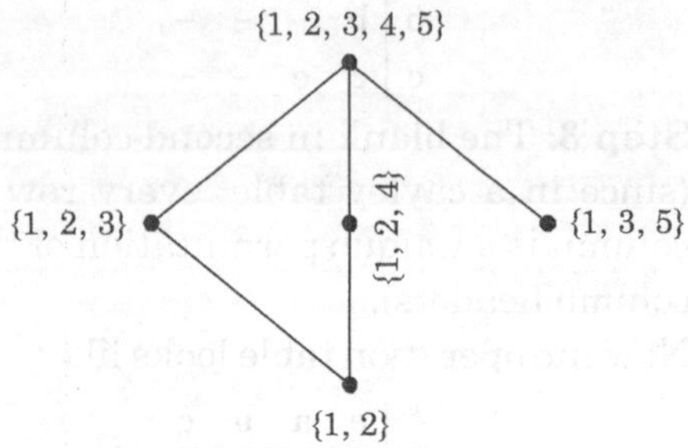

In a complete lattice L, every non empty subset of L, has both LUB and GLB.
Now it is necessary to add {1} since GLB of {1, 2} and {1, 3, 5} is {1}.
The hasse diagram now becomes:

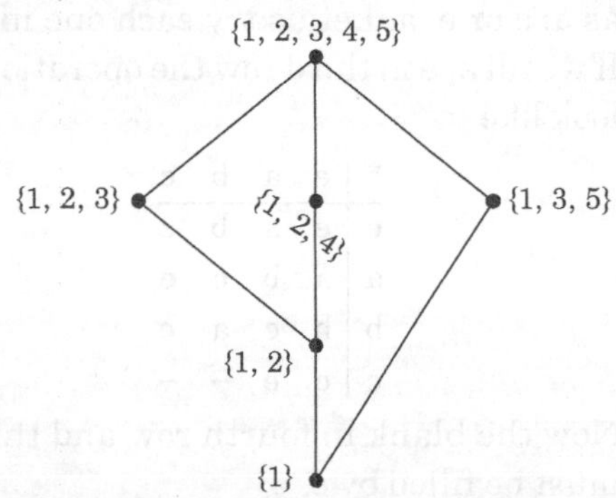

Now the above hasse diagram represents a complete lattice, since every non empty subset has both LUB and GLB. Therefore adding {1} is not only necessary but it is also sufficient to make the given lattice a complete lattice.
Therefore the correct choice is (a).

2.35 (a)

$X = (A - B) - C$
$= (A \cap B') - C = (A \cap B') \cap C' = AB'C'$
$Y = (A - C) - (B - C)$
$= (A \cap C') - (B \cap C')$
$= (AC') - (BC') = (AC') \cap (BC')'$
$= (AC') \cap (B' + C) = (AC') \cdot (B' + C)$
$= AC'B' + AC'C$
$= AC'B'$ (Since $C'C = 0$)
$= AB'C'$ (Commutative property)
$\therefore \quad X = Y$

2.36 (b)

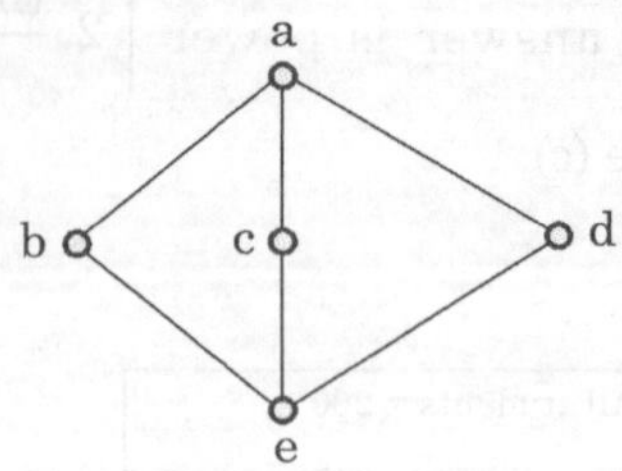

$[\{a, b, c, d, e\}, \leq]$

The poset $[\{a, b, c, d, e\}, \leq]$ is a lattice (since every pair of elements has LUB and GLB) but it is not a distributive lattice. Because distributive lattice satisfy the following conditions. For any x, y, z,

$x \wedge (y \vee z) = (x \wedge y) \vee (x \wedge z)$
$x \vee (y \wedge z) = (x \vee y) \wedge (x \vee z)$

Where $\wedge$ and $\vee$ are meet and join operations but for given poset $[\{a, b, c, d, e\} \leq]$

$b \wedge (c \vee d) = b \wedge a = b$
$(b \wedge c) \vee (b \wedge d) = e \vee e = e$

So it is not distributive. (Also, element b has 2 complements c and d, which is not possible in a distributive lattice, since, in a distributive lattice, complement if it exists, is always unique).

2.37 (c)

The set S = {1, 2, 4, 7, 8, 11, 13, 14} is a group under multiplication modulo 15.
The identity element for this group is e = 1 since, $\forall x \in S$, $1 \cdot x \bmod 15 = x$
Now let the inverse of 4 be 4^{-1}.
Now $(4 \cdot 4^{-1}) \bmod 15 = e = 1$
Since $(4 \cdot 4) \bmod 15 = 1$
$\therefore \quad 4^{-1} = 4$ (This inverse is unique)
Similarly let the inverse of 7 be 7^{-1}.
$(7 \cdot 7^{-1}) \bmod 15 = 1$
putting each element of set as 7^{-1} by trial and error we get
$(7 \cdot 13) \bmod 15 = 91 \bmod 15 = 1$
$\therefore \quad 7^{-1} = 13$
So, 4^{-1} and 7^{-1} are respectively 4 and 13.
Correct choice is (c).

2.38 (c)

$R \cap S$ is an equivalence relation as can be seen from proof given below.
Let $\forall x \in A$ $(x, x) \in R$ and $(x, x) \in S$ (since R and S are reflexive)
$\therefore (x, x) \in R \cap S$ also $\therefore R \cap S$ is reflexive.

Now, $(x, y) \in R \cap S$

$\Rightarrow (x, y) \in R$ and $(x, y) \in S$

$\Rightarrow (y, x) \in R$ and $(y, x) \in S$

(Since R and S are symmetric)

$\Rightarrow$ $(y, x) \in R \cap S$

$\therefore$ $(x, y) \in R \cap S$

$\Rightarrow$ $(y, x) \in R \cap S$

$R \cap S$ is therefore symmetric

Now consider

(x, y) and $(y, z) \in R \cap S$

$\Rightarrow$ (x, y) and $(y, z) \in R$

and (x, y) and $(y, z) \in S$

$\Rightarrow$ $(x, z) \in R$ and $(x, z) \in S$

(Since R and S are transitive)

$\Rightarrow$ $(x, z) \in R \cap S$

$\therefore$ $R \cap S$ is transitive also. Since $R \cap S$ is reflexive, symmetric and transitive.

$\therefore$ $R \cap S$ is equivalence relation.

Note: A similar argument cannot be made from $R \cup S$.

2.39 (b)

Consider the arrow diagram shown below:

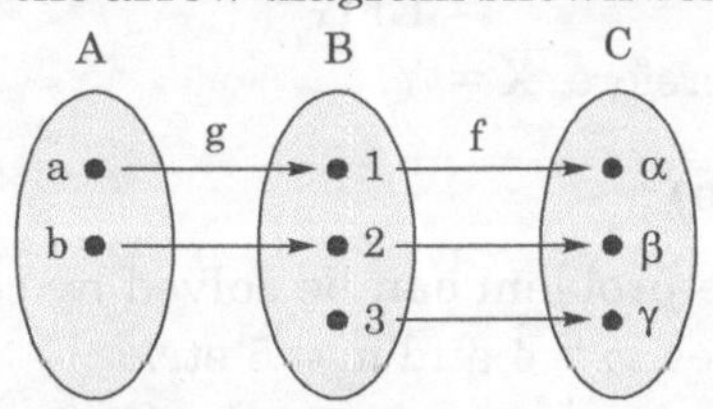

$h(a) = f \cdot g(a) = \alpha$

$h(b) = f \cdot g(b) = \beta$

Here f is onto but g is not onto, yet h is onto.

As can be seen from diagram if f is not onto, h cannot be onto.

$\therefore$ f should be onto, but g need not be onto.

$\therefore$ Answer is (b).

2.40 (a)

(i) The set H is closed, since multiplication of upper triangular matrices will result only in upper triangular matrix.

(ii) Matrix multiplication is associative, i.e. $A * (B * C) = (A * B) * C$.

(iii) Identity element is

$$I = \begin{bmatrix} 1 & 0 & 0 \\ 0 & 1 & 0 \\ 0 & 0 & 1 \end{bmatrix}$$

and this belongs to H as I is an upper triangular as will as lower triangular matrix.

(iv) If $A \in H$, then $|A| = abc$. Since it is given that $abc \neq 0$, this means that $|A| \neq 0$ i.e. every matrix belonging to H is non-singular and has a unique inverse.

$\therefore$ The set H along with matrix multiplication is a group.

2.41 (c)

Given, $h = g(f(x)) = g.f$

Consider the following arrow diagram:

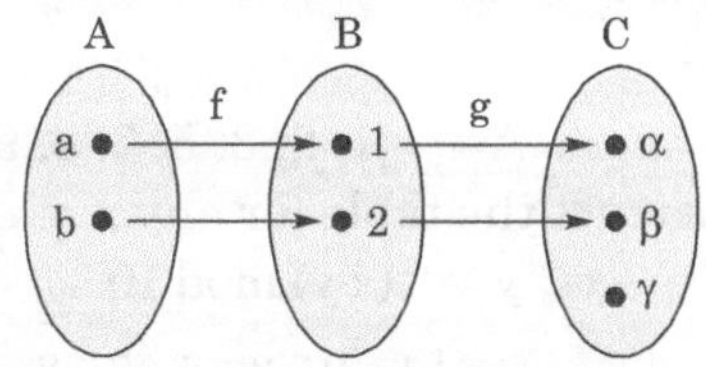

From above diagram it is clear that

g is not onto $\Rightarrow$ h = g.f is also not onto, since the co-domain of g is same as the co-domain of g.f.

The contrapositive version of the above implication is h is onto $\Rightarrow$ g is onto

which also has to be true since direct $\equiv$ contrapositive.

So option (c) is true.

2.42 (b)

The way C is defined in the question, it contains only comparable subsets of A.

i.e. the set C is the set of all comparable subsets of the set A. Such a set is called a chain.

Consider an example with set $A = \{1, 2\}$

Subsets are ϕ, $\{1\}$, $\{2\}$ and $\{1, 2\}$

Maximum cardinality of collection of distinct subsets is $\{\phi, \{1\}$ and $\{1, 2\}\}$ i.e. 3.

So with n elements maximum cardinality is $n + 1$.

2.43 (a)

I: $f(x) = x + y - 3$

$x + a - 3 = x = a + x - 3$

so $a = 3$

Now 3 is unique, and $3 \in N$

So I has identity.

II: $f(x) = \max(x, y)$

$\max(x, a) = x = \max(a, x)$

In N, the only value of a which will satisfy above equation is $a = 1$.

Since 3 is unique, and $3 \in N$

So II has identity.

III: $f(x) = x^y$

$x^a = x = a^x$

Now $x^a = x \Rightarrow a = 1$, but $x = a^x$ has no solution for a in the set N.

So III has no identity.

So only I and II has identity.

2.44 (d)

Given $|X| = x$, $|Y| = y$ and $|Z| = z$

$W = X \times Y$

So $|W| = xy$

$|E| = 2^{|W|} = 2^{xy}$

So the number of functions for Z to E $= |E|^{|Z|}$

$= (2^{xy})^z = 2^{xyz}$

2.45 (c)

Let $A = \{1, 2, 3, 5, 7, 8, 9\}$

Construct the table for any $x, y \in A$ such that

$x * y = (x \cdot y) \bmod 10$

•	1	2	3	5	7	8	9
1	1	2	3	5	7	8	9
2	2	4	6	0	4	6	8
3	3	6	9	5	1	4	7
5	5	0	5	5	5	0	5
7	7	4	1	5	9	6	3
8	8	6	4	0	6	4	2
9	9	8	7	5	3	2	1

We know that $0 \notin A$. So it is not closed. Therefore, (a) is true.

The identity element = 1

$\therefore\ (2 \cdot 2^{-1}) \bmod 10 = 1$

From the table we see that 2^{-1} does not exist.

Since, $(3 \cdot 7) \bmod 10 = 1$

$\therefore$ 7 is the inverse of 3 and $7 \in A$.

$\therefore$ (c) is false.

(d) is true since 8 does not have inverse.

2.46 (a)

$(x, y)\ R\ (u, v)$ iff $x < u$ and $y > v$

$(x, x) \not R\ (x, x)$ since $x \not< x$ and $x \not> x$

So R is not reflexive,

$\therefore$ R is neither a partial order, nor an equivalence relation.

2.47 (c)

Consider the following Venn diagram for

$X = (E \cap F) - (F \cap G)$

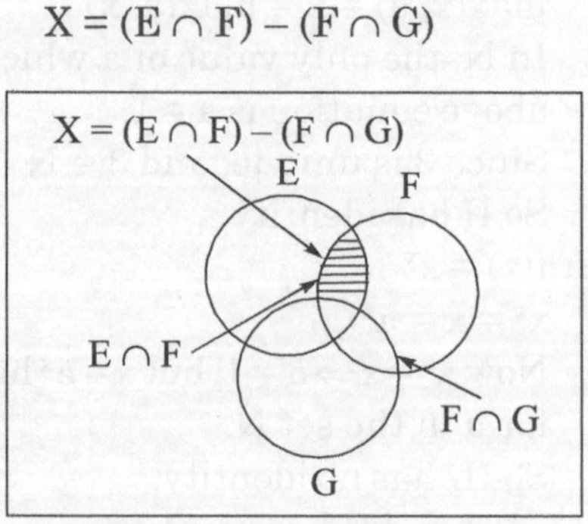

$Y = (E - (E \cap G)) - (E - F)$

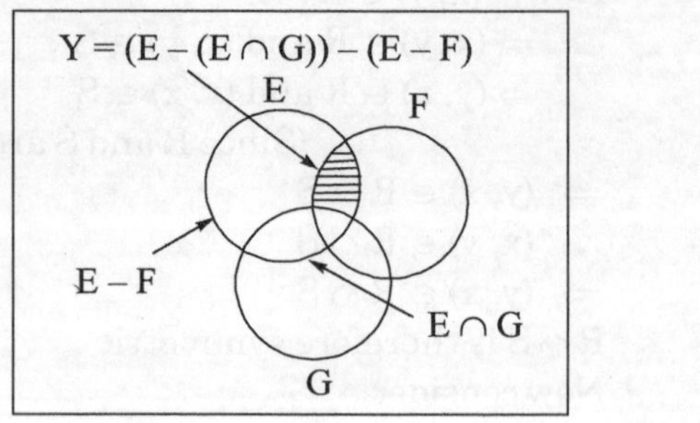

So, $X = Y$

or alternatively the solution can be obtained from boolean algebra as follows:

$X = (E \cap F) - (F \cap G)$

$= EF - FG = EF \cap (FG)'$

$= EF \cdot (F' + G') = EFF' + EFG'$

$= EFG'$

Similarly, $Y = (E - (E \cap G)) - (E - F)$

$= (E - EG) - (E \cdot F')$

$= E \cdot (EG)' - EF'$

$= E \cdot (E' + G') - EF'$

$= EG' - EF' = EG' \cdot (EF')'$

$= EG' \cdot (E' + F) = EE'G' + EFG'$

$= EFG'$

Therefore, $X = Y$

2.48 (b)

The problem can be solved by considering the cases m = 4 and m = 5 etc.

Let, $m = 4$, $S = \{1, 2, 3, 4\}$

n = number of 3 element subsets $= {}^4C_3 = {}^4C_1 = 4$

$\therefore\ n = 4$

The 4 subsets are {1, 2, 3}, {1, 2, 4}, {1, 3, 4} and {2, 3, 4}

f(1) = number of subsets having 1 as an element = 3

f(2) = number of subsets having 2 as an element = 3

f(3) = 3 and f(4) = 3

$\therefore\ \sum_{i=1}^{4} f(i) = 3 + 3 + 3 + 3 = 12$

Both choice (a) and choice (b) are matching the answer since

$3m = 3n = 12$

Now let us try m = 5

$S = \{1, 2, 3, 4, 5\}$

n = number of 3 element subsets $= {}^5C_3 = 10$

$\therefore\ n = 10$

The 10 subsets are {1, 2, 3}, {1, 2, 4}, {1, 2, 5}, {1, 3, 4}, {1, 3, 5}, {1, 4, 5}, {2, 3, 4}, {2, 3, 5}, {2, 4, 5}, {3, 4, 5}

$f(1) = f(2) = f(3) = f(4) = f(5) = 6$

$$\sum_{i=1}^{5} f(i) = 6+6+6+6+6 = 30$$

Clearly 3 m = 3 × 5 = 15 {is not matching Σf(i)}
But 3n = 3 × 10 = 30 {is matching Σf(i) = 30}
∴ 3 n is the only correct answer.
Correct choice is (b).
The problem can also be solved in a more general way as follows:

$$\sum_{i=1}^{m} f(i) = f(1) + f(2) + ... + f(m)$$

Since, $f(1) = f(2) = ... = f(m) = {}^{m-1}C_2$

Therefore, $\sum_{i=1}^{m} f(i) = m \times {}^{(m-1)}C_2$

$$= \frac{m \times (m-1) \times (m-2)}{2}$$

$$= \frac{3 \times m \times (m-1) \times (m-2)}{1 \times 2 \times 3}$$

$$= 3 \times {}^{m}C_3$$

Since n = Number of three elements subsets of a set of m elements = ${}^{m}C_3$

Therefore, $\sum_{i=1}^{m} f(i) = 3 \times {}^{m}C_3 = 3n$

2.49 (b)

P Δ Q = PQ′ + P′Q

where Δ is symmetric difference between P and Q.

I. LHS = P Δ Q ∩ R = PΔ(QR) = P(QR)′ + P′(QR)
= P(Q′ + R′) + P′QR
RHS = (P Δ Q) ∩ (P Δ R) = (PQ′ + P′Q) (PR′ + P′R) = PQ′R′ + P′QR
LHS ≠ RHS
So statement I is false.

II. LHS = P ∩ Q Δ R = P(QR′ + Q′R)
= PQR′ + PQ′R
RHS = (P ∩ Q) Δ (P ∩ R) = PQ Δ PR = PQ(PR)′ + (PQ)′ PR = PQ(P′ + R′) + (P′+Q′)PR
= PQR′ + PQ′R
LHS = RHS
So statement II is true.

2.50 (b)

We know that, number of elements divisible by number 'x' in set of integer (from 1 to X) = $\left\lfloor \frac{X}{x} \right\rfloor$

So, according to inclusion exclusion formula:
n(2 or 3 or 5) = n(2) + n(3) + n(5) – n(2, 3) – n (2, 5) – n(3, 5) – n(2, 3, 5))

No's divisible by 2 in X = $\left\lfloor \frac{123}{2} \right\rfloor = 61$

No's divisible by 3 in X = $\left\lfloor \frac{123}{3} \right\rfloor = 41$

No's divisible by 5 in X = $\left\lfloor \frac{123}{5} \right\rfloor = 24$

No's divisible by 2 and 3 = $\left\lfloor \frac{123}{2\times3} \right\rfloor = \left\lfloor \frac{123}{6} \right\rfloor = 20$

No's divisible by 3 and 5 = $\left\lfloor \frac{123}{3\times5} \right\rfloor = \left\lfloor \frac{123}{15} \right\rfloor = 8$

No's divisible by 2 and 5 = $\left\lfloor \frac{123}{2\times5} \right\rfloor = \left\lfloor \frac{123}{10} \right\rfloor = 12$

No's divisible by 2 and 3 and 5

$$= \left\lfloor \frac{123}{2\times3\times5} \right\rfloor = \left\lfloor \frac{123}{30} \right\rfloor = 4$$

So, n(2 or 3 or 5) = 61 + 41 + 24 – 20 – 8 – 12 + 4 = 90

$\overline{n(2 \text{ or } 3 \text{ or } 5)}$ = Total number – n(2 or 3 or 5)
= 123 – 90 = 33

2.51 (b)

Let S be a set of n elements say {1, 2, 3,..., n}. Now the smallest equivalence relation on S must contain all the reflexive elements {(1, 1), (2, 2), (3, 3),..., (n, n)} and its cardinality is therefore n. The largest equivalence relation on S is S × S, which has cardinality of n × n = n^2.

∴ The largest and smallest equivalence relations on S have cardinalities of n^2 and n respectively.
Correct choice is (b).

2.52 (c)

A partition P_1 is called a refinement of the partition P_2 if every set in P_1, is a subset of one of the sets in P_2.

π_4 is a refinement of π_2, π_3 and π_1
π_2 and π_3 are refinements of π_1
π_2 and π_3 are not comparable since neither is a refinement of the other.

So the poset diagram for (S′, <) would

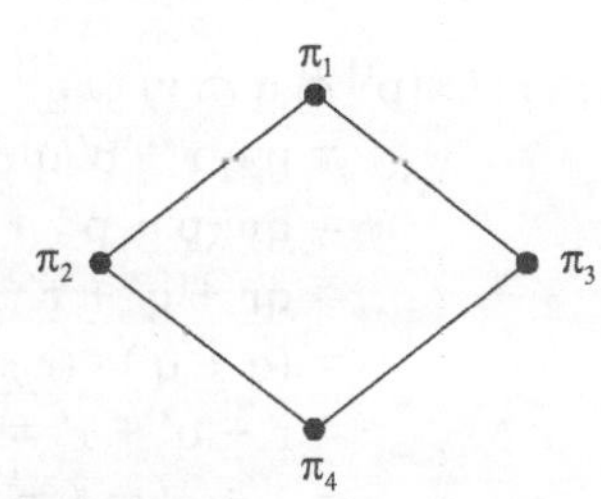

Which is choice (c).

2.53 (d)

In this problem

a % b means the mod function (i.e. residue when a is divided by b).

a/b means integer division (i.e. quotient when a is divided by b)

(i) (101,22):

101% 10 ≤ 22% 10 ⇒ 1 ≤ 2 which is true.

(101/10, 22/10) = (10, 2) ∈ P need to check IS (10, 2) ∈ P?

10% 10 ≤ 2% 10 ⇒ 0 ≤ 2 which is True.

Then (10/10, 2/10) = (1, 0) fails since $1 \nleq 0$.

∴ (101, 22) ∉ P.

(ii) (22, 101)

22% 10 ≤ 101% 10 ⇒ 2 ≤ 1 is False.

∴ (22, 101)∉P

(iii) (145, 265)

145% 10 ≤ 265% 10 ⇒ 5 ≤ 5 is true and (145/10, 265/10) = (14, 26) ∈ P has to be checked.

Now consider (14, 26).

14% 10 ≤ 26% 10 ⇒ 4 ≤ 6 is true and (14/10, 26/10) = (1, 2) ∈ P has to be checked.

Now consider (1, 2)

1% 10 ≤ 2% 10 ⇒ 1 ≤ 2 is true and (1/10, 2/10) = (0, 0) ∈ P which is given to be true.

Therefore (145, 265) ∈ P.

(iv) (0, 153):

0% 10 ≤ 153% 10 ⇒ 0 ≤ 3 is true.

Then (0/10, 153/10) = (0, 15) should be in P.

(0, 15):

0% 10 ≤ 15% 10 ⇒ 0 ≤ 5 is true.

Then (0/10, 15/10) = (0, 1) should be in P.

(0, 1):

0% 10 ≤ 1% 10 ⇒ 0 ≤ 1 is true.

Then (0/10, 1/10) = (0, 0) should be in P.

It is given that (0, 0) ∈ P

Therefore (0, 153) ∈ P.

So (iii) and (iv) are contained in P.

2.54 (d)

The given set theory expression can be converted into equivalent boolean algebra expression as follows:

$$(p \cap q \cap r) \cup (p^C \cap q \cap r) \cup q^C \cup r^C$$
$$= p \cdot q \cdot r + p' \cdot q \cdot r + q' + r'$$
$$= qr(p + p') + q' + r'$$
$$= qr + q' + r'$$
$$= (q + q') \cdot (r + q') + r'$$
$$= r + q' + r' = r + r' + q'$$
$$= 1 + q' = 1 = U$$

2.55 (a)

In lattice every pair in Hasse diagram must have Least Upper Bound (LUB) and Greatest Lower Bound (GLB). In figure

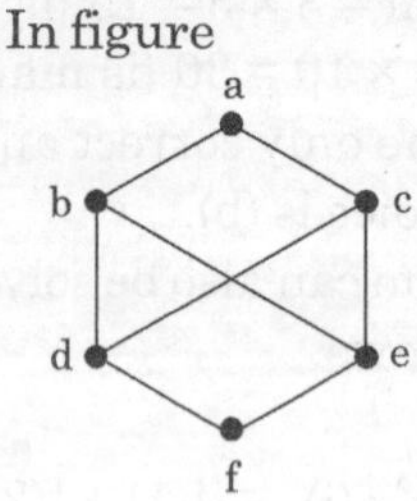

For element d, e

LUB (d, e) = {b, c} but LUB must be unique i.e. only 1 element present in LUB. Same for GLB (b, c) = {d, e}

Hence not lattice.

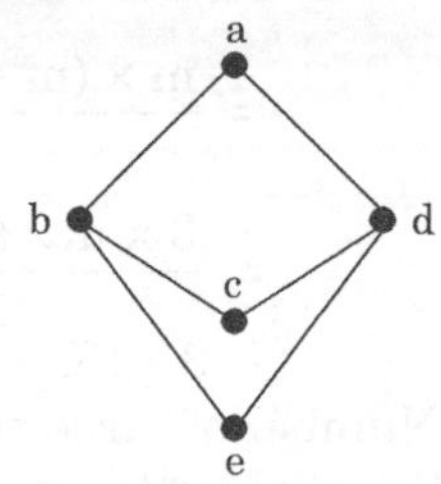

For element b, d

GLB (b, d) = {c, e} but GLB must be unique i.e. only 1 element present in GLB.

Hence not lattice.

So, (ii) and (iii) are not lattice

2.56 (a)

Group properties are closure, associativity existence of identity and existence of inverse for every element. Commutativity is not required for a mathematical structure to become a group.

2.57 (d)

Given, R = {(x, y), (x, z), (z, x), (z, y)} on set {x, y, z}, here (x, y) ∈ R and (y, x) ∉ R.

∴ R is not symmetric

Also (x, z)∈R and (z, x)∈R.

∴ R is not antisymmetric.

R is neither symmetric nor anti-symmetric.

2.58 (c)

If an element is a generator, all elements must be obtained as powers of that element.

Try a, b, c, d one by one to see which are the generators.

$$a = a$$
$$a^2 = a.a = a$$
$$a^3 = a.a^2 = a.a = a \text{ and so on.}$$

∴ a is not the generator.

$$b = b$$
$$b^2 = b.b = a$$
$$b^3 = b.b^2 = b.a = b$$
$$b^4 = b.b^3 = b.b = a \text{ and so on}$$

$\therefore$ b is not the generator

$$c = c$$
$$c^2 = c.c = b$$
$$c^3 = c.c^2 = c.b = d$$
$$c^4 = c.c^3 = c.d = a$$

Since all of a, b, c, d have been generated as powers of c

$\therefore$ c is a generator of this group.

Similarly,

$$d = d$$
$$d^2 = d.d = b$$
$$d^3 = d.d^2 = d.b = c$$
$$d^4 = d.d^3 = d.c = a$$

$\therefore$ d is the other generator.

2.59 (c)

Number of reflexive relations on a set with n elements $= 2^{n^2-n}$.

Here $n = 5$. So, answer is $2^{5^2-5} = 2^{20}$.

2.60 (a)

The structure ({n, n^{th} roots of unity}, $\times$) always forms a group. When n = 3 we get the set ({1, w, w^2}, $\times$) which must also form a group.

or

The fact that ({1, w, w^2}, $\times$) forms a group can also be seen by the fact that it satisfies the four group properties.

1. Closure:

*	1	w	w^2
1	1	w	w^2
w	w	w^2	1
w^2	w^2	1	w

From above operation table, we can see that the given operation is closed, on the given set.

2. Associative: "*" is an associative operation.
3. Identity: From operation table, we can see that the identity element, is "1".
4. Inverse: From operation table we can see that inverse of 1 is 1, inverse of w is w^2 and inverse of w^2 is w.

So, ({1, w, w^2}, *) is a group.

To be ring, integral domain or field, we need two binary operations to be specified, whereas here we have only one operation given. So, choice (a) is correct.

2.61 (c)

Let n = 2. There are only 2 onto functions as shown below:

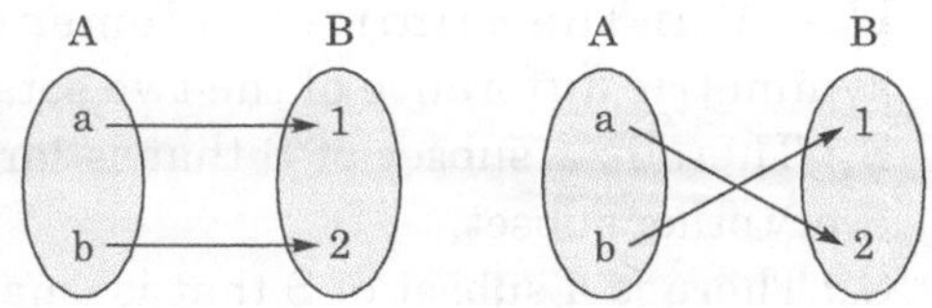

For, n = 2

Option (a) $2^n = 2^2 = 4$

Option (b) $2^n - 1 = 2^2 - 1 = 3$

Option (c) $2^n - 2 = 2^2 - 2 = 2$

Option (d) $2(2^n - 2) = 2(2^2 - 2) = 4$

So only option (c) gives correct answer.

Alternate method:

The number of onto functions from a set A with m elements to set B with n elements where n < m is given by

$$n^m - {}^nC_1 (n-1)^m + {}^nC_2 (n-2)^m \ldots + {}^nC_{n-1} 1^m$$

Here, $m = n = 2$

So number of onto functions $= 2^n - {}^2C_1 1^m$
$$= 2^n - 2$$

which is choice (c).

2.62 (a)

$$x \oplus y = x^2 + y^2$$
$$y \oplus x = y^2 + x^2$$

As '+' sign in commutative so $x^2 + y^2$ is equal to $y^2 + x^2$ so $x \oplus y$ is commutative.

Now check associativity

$$x \oplus (y \oplus z) = x \oplus (y^2 + z^2)$$
$$= x^2 + (y^2 + z^2)^2$$
$$= x^2 + y^4 + z^4 + 2y^2z^2$$
$$(x \oplus y) \oplus z = (x^2 + y^2) \oplus z$$
$$= (x^2 + y^2)^2 + z^2$$
$$= x^4 + y^4 + 2x^2y^2 + z^2$$
$$x \oplus (y \oplus z) \neq (x \oplus y) \oplus z$$

So not associative

Option (a) is correct.

2.63 (16)

$f : \{0,1\}^4 \rightarrow \{0,1\} \Rightarrow$ S is the set of all functions from a 16 element set to a 2 element set.

$|S| = 2^{16}$

N = Number of functions from S to 2 element set $\{0, 1\} = 2^{2^{16}}$]

$$N = 2^{2^{16}}$$

$$\therefore \quad \log \log N = \log \log 2^{2^{16}}$$
$$= \log 2^{16} = 16$$

2.64 (a)

Given the following details:
$S = \{1,2,3,4,...2014\}$
$U < V$ if the minimum element in the symmetric difference of the two sets is in U.
S_1: There is a subset of S that is larger than every other subset.
S_2: There is a subset of S that is smaller than every other subset.
S_1 is true since ϕ (which is subset of S) is larger then every other subset.
i.e. $\forall v \ v < \phi$ is true since the minimum element in $v \Delta \phi = v$, is in v.
(**Note:** $v \Delta \phi = v\phi' + v'\phi = v$)
S_2 is also true since S (which is subset of S) is smaller then every other subset.
i.e. $\forall v \ S < v$ is true since the minimum element in $S \Delta v = v'$, is in S.
(**Note:** $S \Delta v = Sv' + S'v = 1v' + 0v = v'$)
$\therefore$ Both S_1 and S_2 are correct.

2.65 (d)

Example:

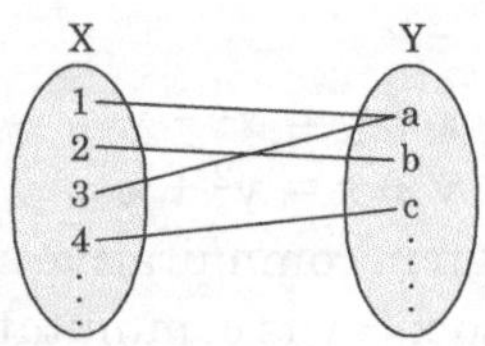

Let $A = \{1, 2\}$, $B = \{3, 4\}$
(a) $|f(A \cup B)| = |f(A)| + |f(B)|$
$\Rightarrow 3 = 2 + 2$ is false
(b) $f(A \cap B) = f(A) \cap f(B)$
$\Rightarrow \phi = \{a\}$ is false
(c) $|f(A \cap B)| = \min\{|f(A)|, |f(B)|\}$
$\Rightarrow 0 = \min(2, 2) = 2$ is false
(d) Let $S = \{a, b\}$, $T = \{a, c\}$
$f^{-1}(S \cap T) = f^{-1}(S) \cap f^{-1}(T)$
$\Rightarrow \{1, 3\} = \{1, 2, 3\} \cap \{1, 3, 4\}$
$\Rightarrow \{1, 3\} = \{1, 3\}$ is true

2.66 (5)

Order of subgroup divides order of group (Lagrange's theorem). So order of L has to be 1, 3, 5 or 15. As it is given that the subgroup has at least 4 elements and it is not equal to the given group, therefore the order of subgroup can't be 1, 3 or 15.
Hence it is 5.

2.67 (b)

Since it is given that $\forall i \ f(f(i)) = i$
It means $f \cdot f = I$ i.e. $f = f^{-1}$
That is f is a symmetric function.
Statement P means that every symmetric function is a identity function which is false, since there are many symmetric functions other than identity function.
Example: {(0, 1), (1, 0), (2, 3), (3, 2),...(2012, 2013), (2013, 2012), (2014, 2014)} is a symmetric function but not the identity function.
Statement Q means that in every such symmetric function at least one element is mapped to itself and this is true, since there are odd number of elements in the set {0, 1, 2, ... 2014}.
Statement R means that every symmetric function is onto which is true, since it is impossible to make an into function symmetric. See below diagram for a into function shown with only 3 elements:

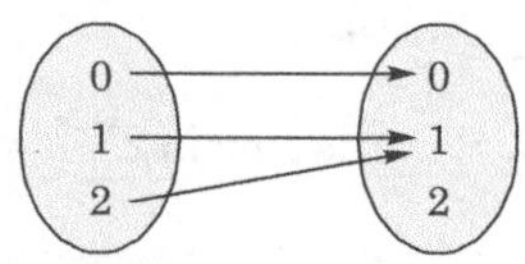

In above function note that (2, 1) exists in the function but not (1, 2) and so it is not symmetric.

2.68 (4)

(i) e is identity element
(ii) $x*x = e$, so $x = x^{-1}$.
(iii) $y*y = e$, so $y = y^{-1}$.
(iv) $(x*y)*(x*y) = e$, so $x*y = (x*y)^{-1}$ ...(i)
and $(y*x)*(y*x) = e$, so $y*x = (y*x)^{-1}$
Now $(x*y)*(y*x) = x*(y*y)*x = x*e*x$
$= x*x = e$
So, $(x*y)^{-1} = (y*x)$ ...(ii)
From (i) and (ii) we get $x*y* = y*x$
There are only 4 distinct elements possible in this group:
1. e 2. x
3. y 4. xy
All other combinations are equal to one of these four as can be seen below:
yx = xy (already proved)
xxx = xe = x
xyy = xe = x
xxy = ey = y
xyx = xxy = y
and so on....
So the group is G = {e, x, y, x*y}
$\Rightarrow$ |G| = 4

2.69 (a)

$$g(x) = 1 - x,\ h(x) = \frac{x}{x-1}$$

$$\frac{g(h(x))}{h(g(x))} = \frac{g\left(\frac{x}{x-1}\right)}{h(1-x)} = \frac{1-\frac{x}{x-1}}{\frac{1-x}{1-x-1}}$$

$$= \frac{\frac{x-1-x}{x-1}}{\frac{1-x}{-x}} = \frac{\frac{-1}{x-1}}{\frac{1-x}{-x}}$$

$$= \frac{x}{(x-1)(1-x)} = \frac{-x}{(1-x)^2}$$

$$\frac{h(x)}{g(x)} = \frac{\frac{x}{x-1}}{1-x} = \frac{x}{(1-x)(x-1)} = \frac{-x}{(1-x)^2}$$

$$\therefore \frac{g(h(x))}{h(g(x))} = \frac{h(x)}{g(x)}$$

2.70 (c)

A = {5, {6}, {7}}

1. $\phi \in 2^A$ is true. Any power set contains an empty set.
2. $\phi \subseteq 2^A$ is true. Empty set is subset of any set.
3. $\{5, \{6\}\} \in 2^A$ is true. power set of A contain {5, {6}} as a 2 element subset of A.
4. $\{5, \{6\}\} \subseteq 2^A$ is false. Power set of A not contain 5 and {6} as elements.

So {5, {6}} can not be subset of 2^A.

2.71 (d)

$\mathcal{L}$ has 5 elements

$\mathcal{L}^3$ has all ordered triplets of elements of $\mathcal{L}$

$\Rightarrow \mathcal{L}^3$ contain $5 \times 5 \times 5 = 5^3 = 125$ elements.

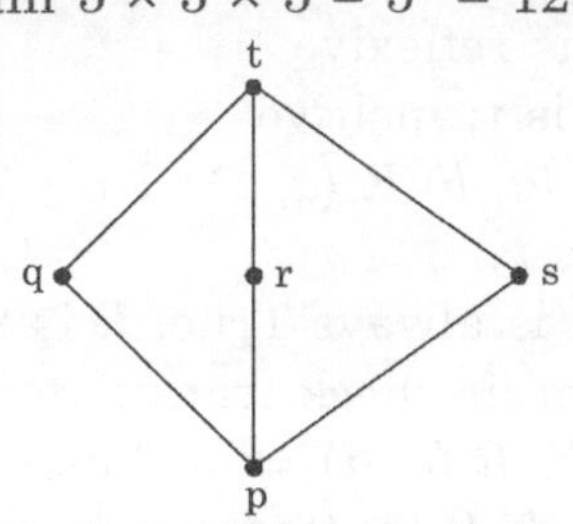

If q, r, s are chosen, then only it will violate the distributive property.

Number of ways to choose q, r, s in any triplet order = 3! = 3 × 2 × 1 = 6.

∴ p(satisfying distributive property)

= 1 – p(violate the distributive property)

$= 1 - \frac{6}{5^3} = 0.952$ which is between $\frac{1}{5}$ and 1.

2.72 (d)

aRb iff a and b are distinct a and b have a common divisor other than 1.

(i) R is not reflexive since a and b are distinct i.e, (a, a) ∉ R

(ii) R is symmetric

If a and b are distinct and have a common divisor other than 1, then b and a also are distinct and have a common divisor other than 1.

(iii) R is not transitive

If (a, b)∈ R and (b, c) ∈ R then (a, c) need not be in R.

Example: (2, 6) ∈ R and (6, 2) ∈ R, but (2, 2) ∉ R

∴ R is symmetric but not reflexive and not transitive.

2.73 (2048)

Let, X = {0, 1, 2, ...10}

|X| = 11

$|P(X)| = 2^{11} = 2048$

2.74 (36)

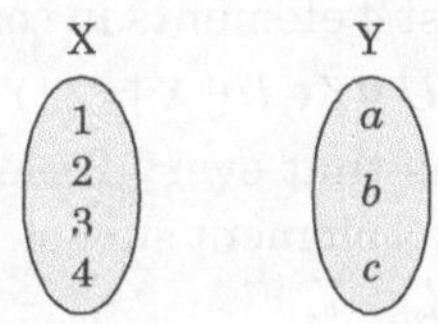

The number of onto functions from A → B where |A| = m and |B| = n is given by the formula

$$n^m - {}^nC_1(n-1)^m + {}^nC_2(n-2)^m + \ldots + (-1)^{n-1}\,{}^nC_{n-1}1^m$$

Here, m = 4 and n = 3

Substituting in above formula we get,

$$3^4 - {}^3C_1(3-1)^4 + {}^3C_2(3-2)^4$$
$$= 81 - 48 + 3 = 36$$

2.75 (0.95)

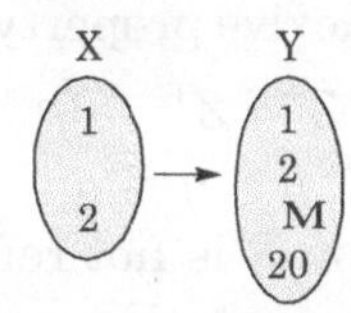

Total possible functions = $20^2 = 400$

Number of one-to-one functions

$= 20P_2 = 20 \times 19$

$$\therefore \text{Required probability} = \frac{20 \times 19}{20 \times 20} = 0.95$$

2.76 (b)

$X \# Y = X' + Y'$

This is the NAND operation. NAND is known to be commutative but not associative.

So, only S_2 is true.

2.77 (d)

$S = \{1, 2, 3, 4, 5, 6\}$

U is the power set of $S \Rightarrow U = P(S)$

$U = \{\{\ \} \{1\}, \{2\}, ...\{1, 2\}, \{1, 3\}, ...\{1, 2, 3\}, ... \{1, 2, 3, 4\}, ...\{1, 2, 3, 4, 5\} ...\{1, 2, 3, 4, 5, 6\}\}$

$|U| = 2^6 = 64$ elements.

$T \in U \Rightarrow T' \in U$

$R \in U, T \setminus R = T - R$

(a) $\forall X \in U(|X| = |X'|)$ means that every subset of S has same size as its complement. Clearly this is False.
(For example, $\{1\} \in U$, and complement of $\{1\} = \{2, 3, 4, 5, 6\}$)

(b) $\exists X \in U \exists Y \in U(|X| = 5, |Y| = 5 \text{ and } X \cap Y = \phi)$
means that there are two 5 element subsets of S which have nothing in common. This is clearly False.
Since any two 5 element subsets will have atleast 4 elements in common.

(c) $\forall X \in U \forall Y \in U(|X| = 2, |Y| = 3 \text{ and } X \setminus Y = \phi)$
means that every 2 element subset X and every 3 element subset Y will have $X - Y = \phi$ i.e., $X \subseteq Y$.
This is clearly False as can be seen from the example $X = \{1, 2\}$, $Y = \{3, 4, 5\}$ where $X \not\subseteq Y$.

(d) $\forall X \in U \forall Y \in U(X \setminus Y = Y' \setminus X')$ means that for any two subsets X and Y, $X \setminus Y = Y' \setminus X'$ i.e. $X - Y = Y' - X'$.
This is clearly True since by boolean algebra LHS $= X - Y = XY'$.

RHS $= Y' - X' = Y'X$ and therefore LHS = RHS.

2.78 (c)

$R = \{(p, q), (r, s) \mid p - s = q - r\}$

(i) Check reflexive property

$\forall (p, q) \in Z^+ \times Z^+ \quad ((p, q), (p, q)) \in R$

is true iff $p - q = q - p$ which is false.

So relation R is not reflexive.

(ii) Check symmetry property

If $((p, q), (r, s)) \in R$ then $((r, s), (p, q)) \in R$

$((p, q), (r, s)) \in R \Rightarrow p - s = q - r$

$((r, s), (p, q)) \in R \Rightarrow r - q = s - p$

If $p - s = q - r$ is true, then $r - q = s - p$ is also true by rearranging the equation.

$\therefore$ R is symmetric.

2.79 (2)

$f(n) = f\left(\frac{n}{2}\right)$ if n is even

$f(n) = f(n + 5)$ if n is odd

$f : N^+ \rightarrow N^+$

Now $f(2) = f\left(\frac{2}{2}\right) = f(1)$

$f(3) = f(3 + 5) = f(8) = f\left(\frac{8}{2}\right) = f(4)$

$= f\left(\frac{4}{2}\right) = f(2) = (1)$

So, $f(1) = f(2) = f(3) = f(4) = f(8)$

Now let us find $f(5) = f(5 + 5) = f(10) = f\left(\frac{10}{2}\right)$

$= f(5)$ so $f(5) = f(10)$

Now let us find $f(9)$

$f(9) = f(9 + 5) = f(14) = f\left(\frac{14}{2}\right) = f(7)$

$= f(7 + 5) = f(12) = f\left(\frac{12}{2}\right) = f(6)$

$= f\left(\frac{6}{2}\right) = f(3)$

So $f(9) = f(7) = f(6) = f(3) = f(1) = f(2) = f(4) = f(8)$

For $n > 10$, the function will be equal to one of $f(1), f(2)....f(10)$

So the maximum no. of distinct values f takes is only 2.

First is $f(1) = f(2) = f(3) = f(4) = f(8) = f(9) = f(7) = f(6)$

Second is $f(5) = f(10)$

All other n values will give only one of these two values.

2.80 (b)

$(a, b)\ R\ (c, d)$ if $a \le c$ or $b \le d$

P : R is reflexive

Q : R is transitive

Since, (a, b) R $(a, b) \Rightarrow a \le a$ or $b \le b$

$\Rightarrow t$ or t

Which is always True, R is reflexive.

Now let us check transitive property

Let$(a,b)\ R\ (c, d) \Rightarrow a \le c$ or $b \le d$

and$(c, d)\ R\ (e, f) \Rightarrow c \le e$ or $d \le f$

Now let us take a situation

$a \le c$ (True) or $b \le d$ (false)

and $c \le e$ (False) or $d \le f$ (True)

Now we can get neither $a \le e$ nor $b \le f$

So, $(a, b)\ R\ (c, d)$ and $(c, d)\ R\ (e, f) \not\Rightarrow (a, b)\ R\ (e, f)$. So clearly R is not transitive.

So P is true and Q is false. Choice (b) is correct.

2.81 (b)

This is actually a graph theory Question.
Let the components be represented by vertices and if component x reacts with component y then let there be an undirected edge between x and y.
Now, 9 of the 23 vertices will have 3 degrees each.
Note: $U \setminus S$ means $U - S$, which is the remaining $23 - 9 = 14$ vertices.
By Handshaking theorem,
9×3 + degree of 14 vertices $= 2e =$ Even.
i.e., 27 + degree of 14 vertices $= 2e =$ Even.
So, the total degree of 14 vertices must be odd.
Statement I means that the degree of each of the 14 vertices is odd, which means that the total degree of 14 vertices will be even (since even number of odd numbers is always even).
But we need the total degree of 14 vertices to be odd so, statement I is false.
Statement III means that everyone of the 14 vertices has even degree, which guarantees that the total degree of 14 vertices will be even. So, statement III is false.
Since **statement II** is the negation of statement III, it must be true.

2.82 (0)

Since the given hasse diagram is already a lattice, three is no need to add any ordered pair. So answer is 0.

2.83 (271)

Let, A → ÷ by 3, B → ÷ by 5, C → ÷ by 7

$$n(A \cup B \cup C) = n(A) + n(B) + n(C) - n(A \cap B) - n(A \cap C) - n(B \cap C) + n(A \cap B \cap C)$$

$$= \left\lfloor \frac{500}{3} \right\rfloor + \left\lfloor \frac{500}{5} \right\rfloor + \left\lfloor \frac{500}{7} \right\rfloor - \left\lfloor \frac{500}{15} \right\rfloor - \left\lfloor \frac{500}{21} \right\rfloor - \left\lfloor \frac{500}{35} \right\rfloor + \left\lfloor \frac{500}{105} \right\rfloor$$

$$= 166 + 100 + 71 - 33 - 23 - 14 + 4 = 271$$

2.84 (42)

Given $|G| = 84$
By Lagrang's theorem any subgroup size is a division of 84.
But a proper subgroup cannot have same size as group.
So largest divisor of 84, other than 84 is 42.
So, largest proper subgroup can have in size of 42.

2.85 (d)

P : Set of rational number → countable
Q : Set of functions from {0, 1} to N → N

0	1

N × N

0 can be assigned in N ways
1 can be assigned in N ways
There are N × N functions, cross product of countable set in countable.
R : Set of functions from N to {0, 1}

1	2	3	4	5	6				

Each of thus boxes can be assigned to 0 or 1 so each such function is a binary number with infinite number of bits.
Example: 0000 ..... is the binary number corresponding to 0 is assigned to all boxes and so on.
Since each such binary number represents a subset of N (the set of natural numbers) by characteristic function method, therefore, the set of such function is same as power set of N which is uncountable due to Cantor's theorem, which says that power set of a countably infinite set is always uncountably infinite.
S : Set of finite subsets of N → countably infinite since we are counting only finite subsets.
So P, Q and S are countable.

2.86 (b)

R_1: $\forall a, b \in G, a\ R_1\ b$ if and only if $\exists g \in G$ such that $a = g^{-1}bg$

Reflexive: $a = g^{-1}ag$ can be satisfied by putting $g = e$, identity "e" always exists in a group. So reflexive

Symmetric: $aRb \Rightarrow a = g^{-1}bg$ for some g
$\Rightarrow b = gag^{-1} = (g^{-1})^{-1}ag^{-1}$
g^{-1} always exists for every $g \in G$.
So symmetric

Transitive: aRb and $bRc \Rightarrow a = g_1^{-1}bg_1$ and $b = g_2^{-1}cg_2$ for some $g_1 g_2 \in G$.
Now $a = g_1^{-1}g_2^{-1}cg_2g_1 = (g_2g_1)^{-1}cg_2g_1$
$g_1 \in G$ and $g_2 \in G \Rightarrow g_2g_1 \in G$ since group is closed so aRb and $aRb \Rightarrow aRc$
hence transitive
Clearly R_1 is equivalence relation.

R_2 is not equivalence it need not even be reflexive, since $aR_2\,a \Rightarrow a = a^{-1}\ \forall a$ which not be true in a group.
R_1 is equivalence relation is the correct answer.

2.87 (0.125)

$$A = \{1, 2, 3\},\ n = |A| = 3$$

Number of relations on A $= 2^{n^2} = 2^{3^2} = 2^9$

Number of reflexive relations on A

$$= 2^{n^2-n} = 2^{3^2-3} = 2^6$$

$$\text{P(reflexive relation)} = \frac{2^6}{2^9} = \frac{1}{8} = 0.125$$

2.88 (7)

Size of group = O(G) = 35

Lagranges theorem says if H is subgroup of G then O(H) | O(G).

So O(H) can be 1, 5, 7 or 35.

So largest subgroup size other than G itself is 7.

2.89 (a)

Let S be reflexive and circular,

Let us checking symmetry:

Symmetry:

Let xSy

Now since S is reflexive ySy true.

So xSy and ySy is true

Now by circular property we get, ySx

So $\quad xSy \Rightarrow ySx$

So S is symmetric.

Transitive:

Let $\quad xSy$ and ySz

Now by circular property we get zSx and by symmetry property proved above, we get

$$zSx \Rightarrow xSz$$

SoxSy and $ySz \Rightarrow xSz$

So S is transitive.

So S is reflexive, symmetric and transitive and hence an equivalence relation.

So option (a) is true.

Option (b): Let S be circular and symmetric.

Let S be defined on set {1, 2, 3}

Now empty relation is circular and symmetric but not reflexive. So S need not be an equivalence relation.

So option (b) is false.

Option (c): Let S be transitive and circular.

Let S be defined on the set {1, 2, 3}

Now empty relation again satisfies transitive and circular but is not reflexive. So S need not be an equivalence relation. So option (c) is false.

Option (d): Reflexive and symmetric need not be transitive for example on {1, 2, 3}.

$$S = \{(1, 1), (2, 2), (3, 3), (1, 2), (2, 1), (2, 3), (3, 2)\}$$

is reflexive and symmetric. But it is not transitive because (1, 2) and (2, 3) belong to S but (1, 3) does not. So option (d) is false.

2.90 (c)

$$|G| = 6$$

H is subgroup, so by Lagrange's theorem

$$|H| = 1, 2, 3 \text{ or } 6 \text{ (Divisor's of 6)}$$

Now it is given that $1 < |H| < 6$

or $\quad |H| = 2$ or 3

Since 2 and 3 are both prime and since every group of prime order is cyclic, H is surely cyclic.

But order of $|G| = 6$ which is not prime.

So G may or may not be cyclic.

So G may not be cyclic, but H is always cyclic.

Option (c) is correct.

2.91 (a, d)

$$|S_1| = 3^{n^2}$$

Since each of the n^2 entries in $n \times n$ matrix can be fills in 3 ways.

$$|S_2| = 3^{n^2}$$

Since $|\{0, 1, 2\}| = 3$ and $|\{0, 1, 2, \ldots.. n^2 - 1\}| = n^2$)

Now the theorem says A bijection $f_{A \to B}$ exists iff $|A| = |B|$.

Here, $\quad |S_1| = |S_2|$

So, there has to be a bijection from S_1 to S_2.

So, option (d) is correct.

If bijection exists surely surjection also exists.

So, option (a) is also correct.

Option (b) and (c) are incorrect.

2.92 (3^{10})

The Venn diagram for this is

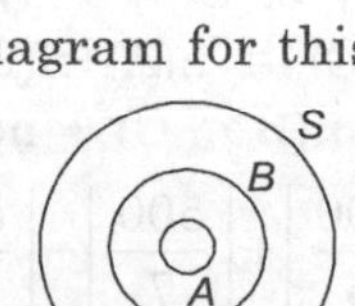

Now every element x in S has only 3 options. It can be $x \in A$ or $x \in B - A$ or $x \in S - B$. So the number of ways to choose A and B such that $A \subseteq B \subseteq S$ is 3^{10}.

2.93 (a)

S(P, Q) is nothing but the dot product of two vectors. The dot product of two vector is zero when they are perpendicular, as $\mathcal{L}$ is a set of 10 dimensional vectors.

So there can be maximum 10 mutually-perpendicular vectors. Which are non-zero.

$$\begin{bmatrix}1\\0\\0\\0\\0\\0\\0\\0\\0\\0\end{bmatrix}\begin{bmatrix}0\\1\\0\\0\\0\\0\\0\\0\\0\\0\end{bmatrix}\begin{bmatrix}0\\0\\1\\0\\0\\0\\0\\0\\0\\0\end{bmatrix}\cdots\cdots\begin{bmatrix}0\\0\\0\\0\\0\\0\\0\\0\\0\\1\end{bmatrix}$$

$\therefore$ 10 is answer.

2.94 (a, b, c)

(a) Given that, $(xy)^2 = x^2y^2$

$$xyxy = xxyy$$
$$yx = xy$$

($\because$ By applying cancelation laws in group)

$\therefore$ $\forall x, y \in G, yx = xy$

$\therefore$ G is commutative.

(b) $\forall x \in G, x^2 = 1$

$\Rightarrow$ $x = x^{-1}$

($\because$ $x^2 = 1, xx = 1, x^{-1}xx = x^{-1}, ex = x^{-1}, x = x^{-1}$)

In a group if every element has its own inverse then group is commutative.

(c) Every group of prime order is commutative so of O(G) = 2, the group 'G' is commutative.

(d) If G is commutative then a subgroup of 'G' is also commutative.

Let H is subgroup of group commutative group 'G'

$\forall a, b, \in H$, we have $a, b, \in G$ and $ab = ba$

($\because$ 'G' is commutative)

$\therefore$ H is commutative.

2.95 (b, c, d)

Every equivalence class of 'x' under F is uniquely mapped with some element 'x' hence F is a function and every function is well defined so option (a) is false.

Distinct equivalence $[x]$, $[y]$ are having distinct images under F hence it is one-one.

Every element of co-domain 'B' is associated with some element of domain of $\in$ under F so F is onto hence F is bijective.

2.96 (a, d)

Let X be a set and 2^X is powerset of X

$$A \Delta B = (A - B) \cup (B - A) \text{ for } A, B \in 2^X$$
$$H = (2^X, \Delta)$$

H satisfies the following properties:

(i) H satisfies closure property under Δ.

(ii) H satisfies associative property under Δ.

(iii) H satisfies identity property under 'ϕ' is identity.

(iv) H is satisfying inverse property for $A \in 2^X$ we have inverse of $A = A$.

(v) H is satisfying commutative property $A \Delta B = B \Delta A$ for all $A, B \in 2^X$

$\therefore$ $H = (2^X, \Delta)$ is abelian group.

$\therefore$ Options (a), (d) are true.

2.97 (2)

$\because$ $f : A \to B$ is one-one and onto

$\Rightarrow$ $|A| = |B| = n$

We check: $f : A \times A \to A \cup B$

$$n^2 \begin{cases} n, \text{ if } (A = B) \\ 2n, \text{ if } (A \neq B, A \cap B = \phi) \end{cases}$$

$\because$ f is one-one

$$|A \times B| = |A \cup B|$$

So either $n^2 = n$ (or) $n^2 = 2n$

$$\Downarrow \qquad\qquad \Downarrow$$

$$n = 0, 1 \qquad n = 0, 2$$

$\therefore$ $n = 1$ or 2

$\therefore$ Number of possible values of cardinality of $A = 2$.

2.98 (c, d)

Check with options:

(a) Consider $a \lozenge (b \lozenge c) = a \lozenge (b + 2c)$

$$= a + 2(b + 2c)$$
$$= a + 2b + 4c$$

Consider $(a \lozenge b) \lozenge c = (a + 2b) \lozenge c$

$$= a + 2b + 2c$$

$\therefore$ $(a \lozenge b) \lozenge c \neq a \lozenge (b \lozenge c)$

Option (a) is false.

(b) Distributive

$$a \lozenge (b \square c) = (a \lozenge b) \square (a \lozenge c)$$
$$a \Delta (bc) \neq (a + 2b) \square (a + 2c)$$
$$a + 2bc \neq (a + 2b)(a + 2c)$$
$$a(b + 2c) = ab + 2ac$$

(d) Distributive law

$$a \square (b \lozenge c) = (a \square b) \lozenge (a \square c)$$
$$a \square (b \lozenge c) = (ab) \lozenge (ac)$$
$$a \square (b + 2c) = ab + 2ac$$
$$a(b + 2c) = ab + 2ac$$

$\therefore$ Option (d) is true.

(c) Associativity

Consider $a \square (b \square c) = a \square (bc) = abc$

and consider $(a \square b) \square c) = (ab) \square c$

$$= abc$$

$\therefore$ $\square$ is associative.

■■■■

3 Combinatorics

3.1 In a room containing 28 people, there are 18 people who speak English, 15 people who speak Hindi and 22 people who speak Kannada. 9 persons speak both English and Hindi, 11 person speak both Hindi and Kannada whereas 13 persons speak both Kannada and English. How many people speak all three languages?

(a) 9 (b) 8
(c) 7 (d) 6 **[1998 : 2 M]**

3.2 Solve the following recurrence relation

$x_n = 2x_{n-1} - 1, n > 1$

$x_1 = 2$ **[1998 : 2 M]**

3.3 Two girls have picked 10 roses, 15 sunflowers and 14 daffodils. What is the number of ways they can divide the flowers among themselves?

(a) 1638 (b) 2100
(c) 2640 (d) None of these **[1999 : 2 M]**

3.4 The minimum number of cards to be dealt from an arbitrarily shuffled deck of 52 cards to guarantee that three cards are from some same suit is

(a) 3 (b) 8
(c) 9 (d) 12 **[2000 : 1 M]**

3.5 How many 4 digit even numbers have all 4 digits distinct?

(a) 2240 (b) 2296
(c) 2620 (d) 4536 **[2001 : 2 M]**

3.6 The solution to the recurrence equation $T(2^k) = 3T(2^{k-1}) + 1$, $T(1) = 1$ is

(a) 2^k (b) $\frac{(3^{k+1} - 1)}{2}$

(c) $3^{\log_2 k}$ (d) $2^{\log_3 k}$ **[2002 : 1 M]**

3.7 The minimum number of colours required to colour the vertices of a cycle with n nodes in such a way that no two adjacent nodes have the same colour is

(a) 2 (b) 3

(c) 4 (d) $n - 2\left\lfloor \frac{n}{2} \right\rfloor + 2$ **[2002 : 1 M]**

3.8 Let A be a sequence of 8 distinct integers sorted in ascending order. How many distinct pairs of sequences, B and C are there such that (i) each is sorted in ascending order, (ii) B has 5 and C has 3 elements, and (iii) the result of merging B and C gives A?

(a) 2 (b) 30
(c) 56 (d) 256 **[2003 : 1 M]**

3.9 n couples are invited to a party with the condition that every husband should be accompanied by his wife. However, a wife need not be accompanied by her husband. The number of different gatherings possible at the party is

(a) $\binom{2n}{n} * 2^n$ (b) 3^n

(c) $\frac{(2n)!}{2^n}$ (d) $\binom{2n}{n}$ **[2003 : 1 M]**

3.10 Mala has a colouring book in which each English letter is drawn two times. She wants to paint each of these 52 prints with one of k colours, such that the colour-pairs used to colour any two letters are different. Both prints of a letter can also be coloured with the same colour. What is the minimum value of k that satisfies this requirement?

(a) 9 (b) 8
(c) 7 (d) 6 **[2004 : 2 M]**

3.11 In how many ways can we distribute 5 distinct balls, $B_1, B_2, \ldots, B_5$ in 5 distinct cells, $C_1, C_2, \ldots, C_5$ such that Ball B_i is not in cell C_i, $\forall_i = 1, 2, \ldots, 5$ and each cell contains exactly one ball?

(a) 44 (b) 96
(c) 120 (d) 3125 **[2004 : 2 M]**

3.12 Let $n = p^2q$, where p and q are distinct prime numbers. How many numbers m satisfy $1 \le m \le n$ and gcd (m, n) = 1? Note that gcd (m, n) is the greatest common divisor of m and n.

(a) $p(q - 1)$
(b) pq
(c) $(p^2 - 1)(q - 1)$
(d) $p(p - 1)(q - 1)$

[2005 : 2 M]

3.13 What is the minimum number of ordered pairs of non-negative numbers that should be chosen to ensure that there are two pairs (a, b) and (c, d) in the chosen set such that $a \equiv c \bmod 3$ and $b \equiv d \bmod 5$

(a) 4 (b) 6
(c) 16 (d) 24 **[2005 : 2 M]**

3.14 Let $G(x) = \dfrac{1}{(1-x)^2} = \sum_{i=0}^{\infty} g(i)x^i$ where $|x| < 1$. What is g(i)?

(a) i (b) i + 1
(c) 2i (d) 2^i **[2005 : 2 M]**

3.15 For each element in a set of size 2n, an unbiased coin is tossed. The 2n coin tossed are independent. An element is chosen if the corresponding coin toss were head. The probability that exactly n elements are chosen is

(a) $\binom{2n}{n}\Big/4^n$ (b) $\binom{2n}{n}\Big/2^n$
(c) $1\Big/\binom{2n}{n}$ (d) $\dfrac{1}{2}$ **[2006 : 2 M]**

Common Data for Q.3.16 & Q.3.17:

Suppose that a robot is placed on the Cartesian plane. At each step it is allowed to move either one unit up or one unit right, i.e., if it is at (i, j) then it can move to either (i + 1, j) or (i, j + 1).

3.16 How many distinct paths are there for the robot to reach the point (10, 10) starting from the initial position (0, 0)?

(a) $\binom{20}{10}$ (b) 2^{20}
(c) 2^{10} (d) None of these
[2007 : 2 M]

3.17 Suppose that the robot is not allowed to traverse the line segment from (4, 4) to (5, 4). With this constraint, how many distinct paths are there for the robot to reach (10, 10) starting from (0, 0)?

(a) 2^9 (b) 2^{19}
(c) $\binom{8}{4}\times\binom{11}{5}$ (d) $\binom{20}{10}-\binom{8}{4}\times\binom{11}{5}$
[2007 : 2 M]

3.18 Let, $P = \sum\limits_{\substack{1\le i\le 2k \\ i\ \text{odd}}} i$ and $Q = \sum\limits_{\substack{1\le i\le 2k \\ i\ \text{even}}} i$ where k is positive integer. Then

(a) P = Q – k (b) P = Q + k
(c) P = Q (d) P = Q + 2k
[2008 : 2 M]

Common Data for Q.3.19 & Q.3.20:

Let x_n denote the number of binary strings of length n that conatin no consecutives 0s.

3.19 Which of the following recurrences does x_n satisfy?

(a) $x_n = 2x_{n-1}$ (b) $x_n = x_{\lfloor n/2\rfloor} + 1$
(c) $x_n = x_{\lfloor n/2\rfloor} + n$ (d) $x_n = x_{n-1} + x_{n-2}$
[2008 : 2 M]

3.20 The value of x_5 is

(a) 5 (b) 7
(c) 8 (d) 13 **[2008 : 2 M]**

3.21 The exponent of 11 in the prime factorization of 300! is

(a) 27 (b) 28
(c) 29 (d) 30 **[2008 : 2 M]**

3.22 In how many ways can *b* blue balls and *r* red balls be distributed in *n* distinct boxes?

(a) $\dfrac{(n+b-1)!(n+r-1)!}{(n-1)!\,b!(n-1)!\,r!}$

(b) $\dfrac{(n+(b+r)-1)!}{(n-1)!(n-1)!\,(b+r)!}$

(c) $\dfrac{n!}{b!\,r!}$

(d) $(n + (b + r) - 1)!\ /n!(b + r - 1)$
[2008 : 2 M]

3.23 A pennant is a sequence of numbers, each number being 1 or 2. An n-pennant is a sequence of numbers with sum equal to n. For example, (1,1,2) is a 4-pennants. The set of all possible 1-pennant is {(1)}, the set of all possible 2-pennants is {(2), (1,1)}and the set of all 3-pennants is {(2,1), (1,1,1), (1,2)}. Note that the pennant (1,2) is not the same as the pennant (2,1). The number of 10-pennants is ______. **[2014 (Set-1) : 2 M]**

3.24 Each of the nine words in the sentence **"The quick brown fox jumps over the lazy dog"** is written on a separate piece of paper. These nine pieces of paper are kept in a box. One of the pieces is drawn at random from the box. The expected length of the word drawn is ______. (The answer should be rounded to one decimal place).
[2014 (Set-2) : 1 M]

3.25 The number of distinct positive integral factors of 2014 is ______. **[2014 (Set-2) : 2 M]**

3.26 $\sum\limits_{x=1}^{99} \dfrac{1}{x(x+1)} =$ ______. **[2015 (Set-1) : 2 M]**

3.27 Let a_n represent the number of bit strings of length n containing two consecutive 1s. What is the recurrence relation for a_n?

(a) $a_{n-2} + a_{n-1} + 2^{n-2}$
(b) $a_{n-2} + 2a_{n-1} + 2^{n-2}$
(c) $2a_{n-2} + a_{n-1} + 2^{n-2}$
(d) $2a_{n-2} + 2a_{n-1} + 2^{n-2}$ **[2015 (Set-1) : 2 M]**

3.28 The number of divisors of 2100 is ________. **[2015 (Set-2) : 1 M]**

3.29 The number of 4 digit numbers having their digits in non-decreasing order (from left to right) constructed by using the digits belonging to the set {1, 2, 3} is ________. **[2015 (Set-3) : 1 M]**

3.30 Let a_n be the number of n-bit strings that do NOT contain two consecutive 1's. Which one of the following is the recurrence relation for a_n?

(a) $a_n = a_{n-1} + 2a_{n-2}$
(b) $a_n = a_{n-1} + a_{n-2}$
(c) $a_n = 2a_{n-1} + a_{n-2}$
(d) $a_n = 2a_{n-1} + 2a_{n-2}$ **[2016 (Set-1) : 1 M]**

3.31 The coefficient of x^{12} in $(x^3 + x^4 + x^5 + x^6 + ...)^3$ is ________. **[2016 (Set-1) : 2 M]**

3.32 Consider the recurrence relation $a_1 = 8$, $a_n = 6n^2 + 2n + a_{n-1}$. Let $a_{99} = K \times 10^4$. The value of K is ________. **[2016 (Set-1) : 2 M]**

3.33 If the ordinary generating function of a sequence $\{a_n\}_{n=0}^{\infty}$ is $\frac{1+z}{(1-z)^3}$, then $a_3 - a_0$ is equal to ________. **[2017 (Set-2) : 2 M]**

3.34 Which one of the following is a closed form expression for the generating function of the sequence $\{a_n\}$, where $a_n = 2n + 3$ for all $n = 0, 1, 2, ...$?

(a) $\frac{3}{(1-x)^2}$
(b) $\frac{3x}{(1-x)^2}$
(c) $\frac{2-x}{(1-x)^2}$
(d) $\frac{3-x}{(1-x)^2}$

[2018 : 2 M]

3.35 Let U = {1, 2, ..., n} and $A = \{(x, X), x \in X$ and $X \subseteq U\}$. Consider the following two statements on $|A|$.

I. $|A| = n.2^{n-1}$

II. $|A| = \sum_{k=1}^{n} k \cdot \binom{n}{k}$

Which of the following is correct?

(a) Both I and II
(b) Neither I nor II
(c) Only II
(d) Only I **[2019 : 1 M]**

3.36 The number of permutations of the characters in LILAC so that no character appears in its original position, if the two L's are indistinguishable, is ________. **[2020 : 2 M]**

3.37 There are 6 jobs with distinct difficulty levels, and 3 computers with distinct processing speeds. Each job is assigned to a computer such that:

- The fastest computer gets the toughest job and the slowest computer gets the easiest job.
- Every computer gets at least one job.

The number of ways in which this can be done is ________. **[2021 (Set-1) : 1 M]**

3.38 The number of arrangements of six identical balls in three identical bins is ________. **[2022 : 1 M]**

3.39 Which one of the following is the closed form for the generating function of the sequence $\{a_n\}_{n \geq 0}$ defined below?

$$a_n = \begin{cases} n+1, & n \text{ is odd} \\ 1, & \text{otherwise} \end{cases}$$

(a) $\frac{x(1+x^2)}{(1-x^2)^2} + \frac{1}{1-x}$

(b) $\frac{x(3-x^2)}{(1-x^2)^2} + \frac{1}{1-x}$

(c) $\frac{2x}{(1-x^2)^2} + \frac{1}{1-x}$

(d) $\frac{x}{(1-x^2)^2} + \frac{1}{1-x}$ **[2022 : 2 M]**

3.40 Consider the following recurrence:

$$f(1) = 1;$$
$$f(2n) = 2f(n) - 1, \text{ for } n \geq 1;$$
$$f(2n+1) = 2f(n) + 1, \text{ for } n \geq 1;$$

Then, which of the following statements is/are TRUE?

(a) $f(2^n - 1) = 2^n - 1$
(b) $f(2^n) = 1$
(c) $f(5.2^n) = 2^{n+1} + 1$
(d) $f(2^n + 1) = 2^n + 1$ **[2022 : 2 M]**

3.41 The Lucas sequence L_n is defined by the recurrence relation:

$$L_n = L_{n-1} + L_{n-2}, \text{ for } n \geq 3,$$

with $L_1 = 1$ and $L_2 = 3$.

Which one of the options given is TRUE?

(a) $L_n = \left(\frac{1+\sqrt{5}}{2}\right)^n + \left(\frac{1-\sqrt{5}}{2}\right)^n$

(b) $L_n = \left(\frac{1+\sqrt{5}}{2}\right)^n - \left(\frac{1-\sqrt{5}}{3}\right)^n$

(c) $L_n = \left(\frac{1+\sqrt{5}}{2}\right)^n + \left(\frac{1-\sqrt{5}}{3}\right)^n$

(d) $L_n = \left(\frac{1+\sqrt{5}}{2}\right)^n - \left(\frac{1-\sqrt{5}}{2}\right)^n$ **[2023 : 1 M]**

3.42 Let $U = \{1, 2, ..., n\}$, where n is a large positive integer greater than 1000. Let k be a positive integer less than n. Let A, B be subsets of U with $|A| = |B| = k$ and $A \cap B = \phi$. We say that a permutation of U separates A from B if one of the following is true.

(a) $n!$

(b) $\binom{n}{2k}(n-2k)!$

(c) $\binom{n}{2k}(n-2k)!\,(k!)^2$

(d) $2\binom{n}{2k}(n-2k)!\,(k!)^2$ **[2023 : 2 M]**

■■■■

Answers Combinatiorics

3.1	(d)	**3.2**	Sol.	**3.3**	(c)	**3.4**	(c)	**3.5**	(b)	**3.6**	(b)	**3.7**	(d)	**3.8**	(c)	**3.9**	(b)
3.10	(d)	**3.11**	(a)	**3.12**	(d)	**3.13**	(c)	**3.14**	(b)	**3.15**	(a)	**3.16**	(a)	**3.17**	(d)	**3.18**	(a)
3.19	(d)	**3.20**	(d)	**3.21**	(c)	**3.22**	(a)	**3.23**	(89)	**3.24**	(3.9)	**3.25**	(8)	**3.26**	(0.99)	**3.27**	(a)
3.28	(36)	**3.29**	(15)	**3.30**	(b)	**3.31**	(10)	**3.32**	(198)	**3.33**	(15)	**3.34**	(d)	**3.35**	(a)	**3.36**	(12)
3.37	(65)	**3.38**	(7)	**3.39**	(a)	**3.40**	(a, b, c)	**3.41**	(a)	**3.42**	(d)						

Explanations Combinatiorics

3.1 (d)

E : Persons speaks English
H : Persons speaks Hindi
K : Persons speaks Kannada

Assume that everyone in the room speaks at least one of the languages.

Given, $n(E \cup H \cup K) = 28$

$n(E) = 18$

$n(H) = 15$

$n(K) = 22$

$n(E \cap H) = 9$

$n(H \cap K) = 11$

$n(K \cap E) = 13$

$n(K \cap E \cap H) = ?$

By principle of inclusion and exclusion

$n(E \cup H \cup K) = n(E) + n(H) + n(K) - n(E \cap H) - n(H \cap E) - n(K \cap E) + n(K \cap E \cap H)$

$28 = 18 + 15 + 22 - 9 - 11 - 13 + n(K \cap E \cap H)$

$n(k \cap E \cap H) = 28 - 55 + 33 = 61 - 55 = 6$

Therefore, the number of people who speak all three languages are 6.

3.2 Sol.

$$x_n = 2x_{n-1} - 1$$

For Homogeneous solution:

$x_n - 2x_{n-1} = 0$

$t - 2 = 0$

$t = 2$

So homogeneous solution is

$x_n = C(2)^n$

For Particular solution:

Let particular solution be d_0.

$d_0 = 2d_0 - 1 = 1$

So particular solution is 1

$\therefore$ Complete solution = Homogeneous solution + Particular solution

Complete solution = $C(2)^n + 1$

$\Rightarrow \quad x_n = C(2)^n + 1$

Given initial condition is $x_1 = 2$

$2 = C(2)^1 + 1, \quad 1 = 2C$

$C = \frac{1}{2}$

$x_n = \frac{1}{2}(2)^n + 1$

$\therefore \quad x_n = 2^{n-1} + 1$

3.3 (c)

Number of ways for distributing r similar things among n different things = ${}^{(n-1+r)}C_r$

The number of ways for distributing 10 roses among the two girls = ${}^{(2-1+10)}C_{10} = 11$.

Similarly number of ways for distributing 15 sunflowers among two girls = ${}^{(2-1+15)}C_{15}$

$= {}^{16}C_{15} = {}^{16}C_1 = 16$

Number of ways for distributing 14 daffodils among the two girls

$= {}^{(2-1+14)}C_{14} = {}^{15}C_{14} = {}^{15}C_1 = 15$

$\therefore$ Total number of ways = $11 \times 16 \times 15 = 2640$

3.4 (c)

Let the number of cards to be dealt from an arbitratily shuffled deck of 52 cards be n.

Number of suits = 4

Required number of cards from the same suit = 3.

So by pigeonhole principle

$$\left\lfloor \frac{n-1}{4} \right\rfloor + 1 = 3$$

$$\left\lfloor \frac{n-1}{4} \right\rfloor = 2$$

$$\lfloor n-1 \rfloor = 8$$

$$n \geq 9$$

So the minimum number of cards to be dealt from an arbitrarily shuffled deck of 52 cards to guarantee that three cards are from some same suit is 9.

3.5 (b)

The digits are given to be distinct i.e. no repetition. 4 digit even numbers cannot start with 0 and must end with 0, 2, 4, 6 or 8.

Since there is a condition for 0 in starting as well as ending we will count the even numbers ending with 0 seperately.

So the total number of 4 digit even number = 4 digit even numbers ending with zero + 4 digit even numbers ending with 2, 4, 6, 8.

4 digit even numbers ending with 0

$$9 \times 8 \times 7 \times 1 = 504 \text{ ways}$$

= | 0̸ | | | 0 |

4 digit even numbers ending with 2, 4, 6, 8

$$8 \times 8 \times 7 \times 4 = 1792 \text{ ways}$$

= | 0̸ | | | ← 2,4,6,8 |

So the total number of 4 digit even numbers

$$= 504 + 1792 = 2296$$

3.6 (b)

$$T(2^k) = 3T(2^{k-1}) + 1$$

Let, $T(2^k) = x_n$

$\Rightarrow$ $x_n = 3x_{n-1} + 1$

$\Rightarrow$ $x_n - 3x_{n-1} = 1$

So for **Homogenous solution**

$$x_n - 3x_{n-1} = 0$$

$$n - 3 = 0$$

$$n = 3$$

Homogenous solution is

$$x_n = C_1(3)^n$$

$$T(2^k) = C_1(3)^k$$

For **Particular solution**

Let d be the particular solution

$$d - 3d = 1$$

$$2d = -1$$

$$d = -\frac{1}{2}$$

Therefore, the complete solution is

$$T(2^k) = C_1(3)^k - \frac{1}{2}$$

Given, $T(1) = 1$

$$1 = C_1(3)^0 - \frac{1}{2} = C_1 - \frac{1}{2}$$

$$C_1 = \frac{3}{2}$$

So the complete solution is

$$T(2^k) = \frac{3}{2}(3)^k - \frac{1}{2} = \frac{3^{k+1} - 1}{2}$$

3.7 (d)

The minimum number of colours required to colour the vertices of a cycle with n nodes = 2, when n is even = 3, when n is odd

Therefore $n - 2\left\lfloor \frac{n}{2} \right\rfloor + 2$ gives 2 when n is even, and 3 when n is odd.

3.8 (c)

This corresponds to an ordered partition of 8 elements into two groups, the first with 5 elements and second with 3 elements. The number of ways of doing this is

$$P(8; 5, 3) = \frac{8!}{5!\,3!} = 56$$

3.9 (b)

For each of the n couples invited to the party one of three thing is possible:

1. Both husband and wife attend the party.
2. Wife only attends the party.
3. Neither husband nor wife attends the party. Since there are n such couples, total number of possibilities = 3^n.

3.10 (d)

The problem reduces to finding how many distinct ordered colour pairs (C_1, C_2) are possible with k colors.

Since the first color C_1 can be any one of the k colours and the second color C_2 also can be any one of the k colors (both prints of a letter can be colored with same color), the total no. of such order color pairs is equal to $k \times k = k^2$.

Since each pair of letters must be colored with different color pairs, at least 26 color pairs are required to do this.

Therefore the requirement is $k^2 \geq 26$.

The minimum value of k that satisfies this equation is k = 6.

3.11 (a)

We want every one of the 5 balls to be in the wrong box. This is nothing but the number of derangements of a set of 5 elements = D_5.

i.e. we need to compute D_5

$$D_n = \sum_{r=2}^{n} (-1)^r \frac{n!}{r!}$$

$$D_5 = \sum_{r=2}^{5} (-1)^r \frac{5!}{r!} = \frac{5!}{2!} - \frac{5!}{3!} + \frac{5!}{4!} - \frac{5!}{5!}$$

$$= 60 - 20 + 5 - 1 = 44 \text{ ways}$$

3.12 (d)

The number of numbers from 1 to n, which are relatively prime to n i.e., gcd $(m, n) = 1$, is given by the Euler Totient function $\phi(n)$. If n is broken down into its prime factors as $n = p_1^{n_1} \cdot p_2^{n_2} \ldots$. where p_1, p_2 etc. are distinct prime numbers, then

$$\phi(n) = \phi(p_1^{n_1})\,\phi(p_2^{n_2})\ldots$$

Here, $n = p^2 q$

So, $\phi(n) = \phi(p^2) \times \phi(q)$ ...(i)

Now, using the property $\phi(p^k) = p^k - p^{k-1}$,

$$\phi(p^2) = p^2 - p^1 = p^2 - p \text{ and}$$

$$\phi(q) = q^1 - q^0 = q - 1$$

Substituting these in eq. (i), we get

$$\phi(n) = (p^2 - p)(q - 1) = p(p - 1)(q - 1)$$

3.13 (c)

The number of combinations of pairs (a mod 3, b mod 5) is 3 × 5 = 15 (since a mod 3 can be 0, 1, or 2) and b mod 5 can be 0, 1, 2, 3 or 4)

∴ If 16 different ordered pairs are chosen at least 2 of them must have (a mod 3, b mod 5) as same (basic pigeon hole principle).

Let such two pairs be (a, b) and (c, d) then

a mod 3 ≡ c mod 3 ⇒ a ≡ c mod 3,

and b mod 5 ≡ d mod 5 ⇒ b ≡ d mod 5

3.14 (b)

$$\frac{1}{(1-x)^n} = \sum_{r=0}^{\infty} {}^{(n-1+r)}C_r\, x^r$$

Put, n = 2,

$$\frac{1}{(1-x)^2} = \sum_{r=0}^{\infty} {}^{(2-1+r)}C_r\, x^r = \sum_{r=0}^{\infty} {}^{(r+1)}C_r\, x^r$$

$$= \sum_{r=0}^{\infty} {}^{(r+1)}C_1\, x^r = \sum_{r=0}^{\infty} (r+1)\, x^r$$

$$= \sum_{i=0}^{\infty} (i+1)\, x^i$$

(Since r is a dummy variable, r can be replaced by i)

$$= \sum_{r=0}^{\infty} g(i)\, x^i$$

$$\therefore \quad g(i) = i + 1$$

3.15 (a)

The probability that exactly n elements are chosen = The probability of getting n heads out of 2n tosses

$$= {}^{2n}C_n \left(\frac{1}{2}\right)^n \left(\frac{1}{2}\right)^{2n-n} \quad \text{(Binomial formula)}$$

$$= {}^{2n}C_n \left(\frac{1}{2}\right)^n \left(\frac{1}{2}\right)^n$$

$$= \frac{{}^{2n}C_n}{2^{2n}} = \frac{{}^{2n}C_n}{(2^2)^n} = \frac{{}^{2n}C_n}{4^n}$$

3.16 (a)

Consider the following diagram:

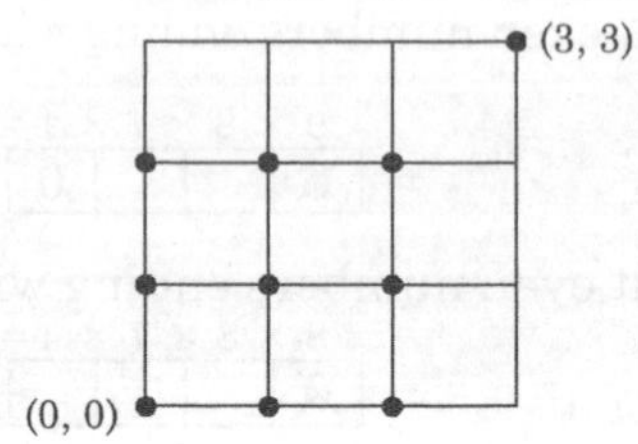

The robot can move only right or up as defined in problem.

Let us denote right move by 'R' and up move by 'U'. Now to reach (3, 3) from (0, 0), the robot has to make exactly 3 'R' moves and 3 'U' moves in any order.

Similarly to reach (10, 10) from (0, 0), the robot has to make 10 'R' moves and 10 'U' moves in any order. The number of ways this can be done is same as number of per mutations of a word consisting of 10 'R's and 10 'U's.

Applying formula of permutation with limited repetitions we get the answer as $\dfrac{20!}{10!\,10!} = {}^{20}C_{10}$.

3.17 (d)

The robot can reach (4, 4) from (0, 0) in ${}^8C_4$ ways as argued in previous problem. Now after reaching (4, 4), robot is not allowed to go to (5, 4). Let us count how many paths are there from (0, 0) to (10, 10) if robot goes from (4, 4) to (5, 4) and then we can subtract this from total number of ways to get the answer.

Now there are $8C_4$ ways for robot to reach (4, 4) from (0, 0) and then robot takes the 'U' move from (4, 4) to (5, 4). Now from (5, 4) to (10, 10) the robot has to make 5 'U' moves and 6 'R' moves in any order which can be done in $\dfrac{11!}{5!\,6!}$ ways $= {}^{11}C_5$ ways.

∴ The number of ways robot can move from (0, 0) (10, 10) via (4, 4) – (5, 4) move is

$${}^8C_4 \times {}^{11}C_5 = \binom{8}{4} \times \binom{11}{5}$$

$\therefore$ Number of ways robot can move from (0, 0) to (10, 10) without using (4, 4) to (5, 4) move is

$$\binom{20}{10}-\binom{8}{4}\times\binom{11}{5} \text{ ways.}$$

which is option (d).

3.18 (a)

$P = \Sigma i = 1 + 3 + 5 + 7 + ... + (2k - 1)$

$1 \le i \le 2k$, i is odd

$Q = \Sigma i = 2 + 4 + 6+...2k$, $1 \le i \le 2k$, i is even.

P is in A.P with a = 1, d = 2 and n = k

Q is in A.P with a = 2, d = 2 and n = k

$$P = \frac{n}{2}(2a + (n-1)d)$$

$$= \frac{k}{2}(2 \times 1 + (k-1) \times 2)$$

$$= \frac{k}{2}(2 + 2k - 2) = k^2$$

$$Q = \frac{n}{2}(2a + (n-1)d)$$

$$= \frac{k}{2}(2 \times 2 + (k-1) \times 2)$$

$$= \frac{k}{2}(4 + 2k - 2) = \frac{k}{2}(2k + 2) = k^2 + k$$

Clearly, $P = Q - k$

3.19 (d)

The ✓ represents those strings with no consecutive 0's.

Let, $n = 1$

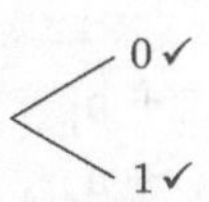

$x_1 = 2$

Let, $n = 2$

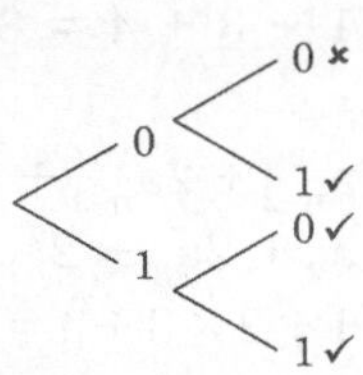

$x_2 = 3$

Let, $n = 3$

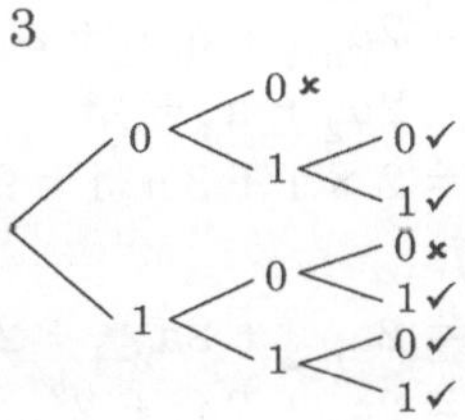

$x_3 = 5$

Now, substituting $n = 3$ in all of the answers only choice (d) $x_n = x_{n-1} + x_{n-2}$ satisfies the numbers obtained from the tree counting.

3.20 (d)

$$x_1 = 2$$
$$x_2 = 3$$
$$\Rightarrow x_3 = x_1 + x_2 = 2 + 3 = 5$$
$$\Rightarrow x_4 = x_2 + x_3 = 3 + 5 = 8$$
$$\Rightarrow x_5 = x_3 + x_4 = 5 + 8 = 13$$

So option (d) is correct.

3.21 (c)

In 300!, distinct numbers divisible by

$$11 = \left\lfloor \frac{300}{11} \right\rfloor = 27$$

Since these 27 numbers contains, same number which can be further divisible by 11 i.e.

$$\left\lfloor \frac{27}{11} \right\rfloor = 2 \text{ (121 and 242)}$$

So, total number of 11's = 27 + 2 = 29

Hence exponent of 11 is 29.

3.22 (a)

Number of ways to distribute 'b' blue balls in n distinct boxes:

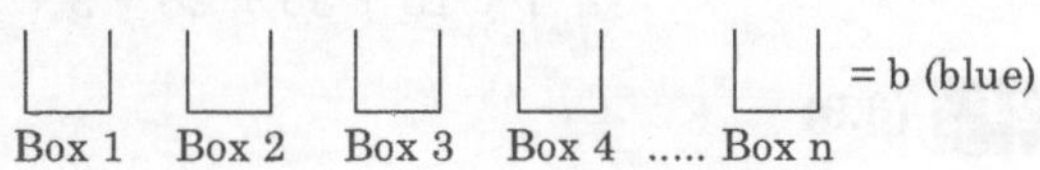

Since each box can contain 0 to n blue balls, so number of ways

$$= {}^{(n+b-1)}C_{(n-1)} = \frac{(n+b-1)!}{(n-1)! \times b!}$$

Number of ways to distribute 'r' red balls in n distinct boxes:

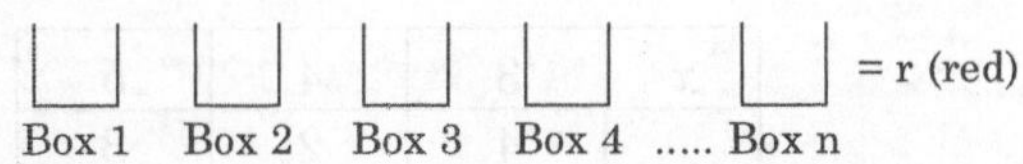

Since each box can contain 0 to n red balls. So number of ways

$$= {}^{(n+r-1)}C_{(n-1)} = \frac{(n+r-1)!}{(n-1)! \times r!}$$

Since both are independent of each other

So total ways = Number of ways to distribute 'b' blue balls × Number of ways to distribute 'r' red balls

$$= \frac{(n+b-1)!\,(n+r-1)!}{(n-1)! \times b!\,(n-1)! \times r!}$$

3.23 (89)

In a 10-pennant let there we x_1 ones and x_2 twos. So we need to find all the solutions of

$$x_1 + 2x_2 = 10$$

Put, $x_1 = 0 \Rightarrow x_2 = \frac{10}{2} = 5$

So, (0, 5) is a solution i.e. a 10-pennant could have 0 ones and 5 twos. The number of ordered permutations of 0 ones and 5 twos $= \frac{5!}{5!} = 1$

Now x_1 cannot be 1 since in that case $x_2 = \frac{9}{2} = 4.5$ (is not an integer).

Put, $x_1 = 2 \Rightarrow x_2 = \frac{8}{2} = 4$

So, (2, 4) is a solution i.e. a 10-pennant could have 2 ones and 4 twos. The number of ordered permutations of 2 ones and 4 twos $= \frac{6!}{2!\,4!} = 15$

Similarly (4, 3), (6, 2), (8, 1) and (10,0) are the other four solutions and the number of pennants for each is respectively

$$\frac{7!}{4!\,3!} = 35,\ \frac{8!}{6!\,2!} = 28,\ \frac{9!}{8!\,1!} = 9,\ \frac{10!}{10!\,0!} = 1$$

So the total number of 10-pennants

$= 1 + 15 + 35 + 28 + 9 + 1 = 89$

3.24 (3.9)

"The quick brown fox jumps over the lazy dog"
(3) (5) (5) (3) (5) (4) (3) (4) (3)

Now let x be the number of letters in the word that is randomly picked.

Now, we make a probability distribution table for *x*

x	3	4	5
$p(x)$	$\frac{4}{9}$	$\frac{2}{9}$	$\frac{3}{9}$

From this table we can easily find the expected value of x.

$$E(x) = \sum x\, p(x)$$

$$= 3\times\frac{4}{9} + 4\times\frac{2}{9} + 5\times\frac{3}{9}$$

$$= \frac{35}{9} = 3.88 = 3.9$$

(after rounding to one decimal accuracy).

3.25 (8)

Factorizing 2014 in primes by successively dividing by primes, we get

$2014 = 2^1 \times 19^1 \times 53^1$

Now we use the **formula**:

If $n = p_1^{n_1} \cdot p_2^{n_2} \ldots p_r^{n_r}$ is the prime factorization of n, then the number of distint factors of n is given by $(n_1+1) \times (n_2+1) \ldots (n_r+1)$.

Now since $2014 = 2^1 \times 19^1 \times 53^1$, the number of distint factors of 2014 = (1 + 1) (1 + 1) (1 + 1)

$= 2 \times 2 \times 2 = 8$

3.26 (0.99)

$$\sum_{x=1}^{99} \frac{1}{x(x+1)} = \sum_{x=1}^{99}\left[\frac{1}{x} - \frac{1}{x+1}\right]$$

$$= \left[\frac{1}{1} - \frac{1}{2}\right] + \left[\frac{1}{2} - \frac{1}{3}\right] + \ldots$$

$$+\left[\frac{1}{97} - \frac{1}{98}\right] + \left[\frac{1}{98} - \frac{1}{99}\right] + \left[\frac{1}{99} - \frac{1}{100}\right]$$

$= 1 - \frac{1}{100}$ (all the terms in above series except the first and last terms cancels out)

$$= \frac{99}{100} = 0.99$$

3.27 (a)

$a_1 = 0$ [no strings of length 1 contain two consecutive 1's]

$a_2 = 1$ [∴ strings = 11]

$a_3 = 3$ [∴ strings are : 011, 110, 111]

$a_4 = 8$ [∴ strings are : 0011, 0110, 0111, 1011, 1100, 1101, 1110, 1111]

Option (a):

$a_n = a_{n-2} + a_{n-1} + 2^{n-2}$

$\Rightarrow a_4 = a_{4-2} + a_{4-1} + 2^{4-2}$

$= a_2 + a_3 + 2^2$

$= 1 + 3 + 4 = 8$ which is True.

Option (b):

$a_n = a_{n-2} + 2a_{n-1} + 2^{n-2}$

$\Rightarrow a_4 = a_2 + 2a_3 + 2^2$

$= 1 + 2 \times 3 + 4 = 11$ which is False.

Option (c):

$a_n = 2a_{n-2} + a_{n-1} + 2^{n-2}$

$\Rightarrow a_4 = 2a_2 + a_3 + 2^2$

$= 2 \times 1 + 3 + 4 = 9$ which is False.

Option (d):

$a_n = 2a_{n-2} + 2a_{n-1} + 2^{n-2}$

$\Rightarrow a_4 = 2a_2 + 2a_3 + 2^2$

$= 2\times 1 + 2\times 3 + 4 = 12$ which is False.

∴ Option (a): $a_n = a_{n-2} + a_{n-1} + 2^{n-2}$ is correct.

3.28 (36)

Dividing 2100 successively by prime numbers, we get that, the prime factors of 2100 are $2 \times 2 \times 3 \times 5 \times 5 \times 7$

$\Rightarrow 2^2 \times 3^1 \times 5^2 \times 7^1$

If $N = p^{k_1} \times q^{k_2} \times r^{k_3} ... \times y^{k_n}$ then the number of factors of N are $(k_1+1)\ (k_2+1)\ ...\ (k_n+1)$

$\therefore$ For 2100 we have $(2 + 1)\ (1 + 1)\ (2 + 1)\ (1 + 1) = 36$ factors.

3.29 (15)

To satisfy the non-decreasing order condition allow 1, 2, 3 after 1, allow only 2, 3 after 2 and allow only 3 after 3. The following tree diagram gives all the allowed numbers satisfying the given condition.

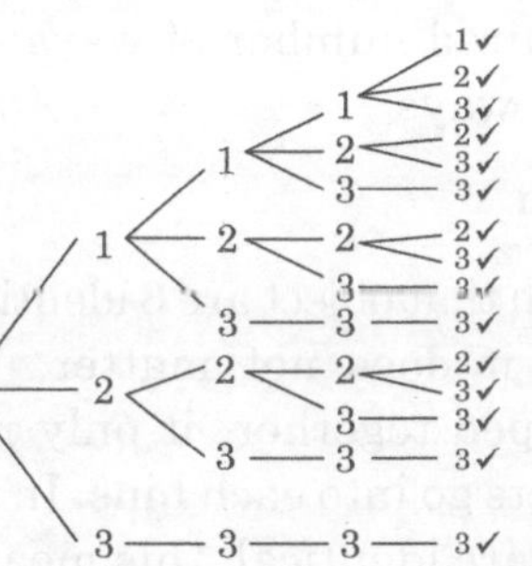

All the tick marked numbers satisfy the non-decreasing order condition.

$\Rightarrow$ Number of 4-digit numbers = Number of tick marks = 15.

3.30 (b)

Let a_n be the number of n-bit strings that do not contain two consecutive 1's. we wish to develop a recurrence relation for a_n.

Consider 1 bit strings 0, 1

So, $a_1 = 2$

Consider 2 bit strings 00, 01, 10, 11

Out of minimum only 00, 01, 10 do not contain two consecutive 1's.

So, $a_2 = 3$

Consider 3 bit strings

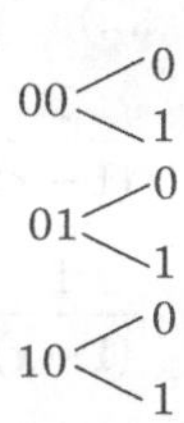

Out of minimum six strings only 000, 001, 010, 100 and 101 five strings satisfy do not contain two consecutive 1's.

So, $a_3 = 5$. Three numbers a_1, a_2, a_3 satisfy clearly only (b) $a_n = a_{n-1} + a_{n-2}$ is correct.

3.31 (10)

We wish to find coefficient of x^{12} in $(x^3 + x^4 + x^5 + ...)^3$

$$= (x^3(1 + x^1 + x^2 + ...))^3$$

$$= x^9(1 + x + x^2 ......)^3$$

$$= \frac{x^9}{(1-x)^3} = x^9 \sum_{r=0}^{\infty} {}^{3-1+r}C_r X^r$$

$$= x^9 \sum_{r=0}^{\infty} {}^{r+2}C_r X^r$$

Now to make x^{12} we need to put $r = 3$

So coefficient of x^{12} is ${}^{3+2}C_3 = {}^5C_3 = {}^5C_2 = 10$

3.32 (198)

Given, $a_n = 6n^2 + 2n + a_{n-1}$ and $a_1 = 8$

We wish to find a_{99}

Now,

$$a_2 = 6 \times 2^2 + 2 \times 2 + a_1$$

$$a_3 = 6 \times 3^2 + 2 \times 3 + a_2$$

$$= 6 \times 3^2 + 2 \times 3 + 6 \times 2^2 + 2 \times 2 + a_1 ...$$

$$a_{99} = 6 \times 99^2 + 2 \times 99 + 6 \times 98^2 + 2 \times 98 ... ... + 6 \times 2^2 + 2 \times 2 + a_1$$

Since, $a_1 = 8$

$$a_{99} = 6 \times 99^2 + 2 \times 99 + 6 \times 98^2 + 2 \times 98 ... ... + 6 \times 2^2 + 2 \times 2 + 8$$

$$= 6 \times 99^2 + 2 \times 99 + 6 \times 98^2 + 2 \times 98 ... ... 6 \times 2^2 + 2 \times 2 + 6 \times 1^2 + 2 \times 1$$

$$= 6(1^2 + 2^2 + 3^2 ... 99^2) + 2.(1 + 2 + 3 ... 99)$$

$$= 6 \cdot \frac{(99(99+1)(2 \times 99+1))}{6} + 2\left(\frac{99(99+1)}{2}\right)$$

$$= 99 \times 100 \times 199 + 99 \times 100$$

$$= 100 \times 99\ (199 + 1) = 100 \times 99 \times 200$$

$$= 2 \times 99 \times 10^4 = 198 \times 10^4$$

So if $a_{99} = K \times 10^4$ then $K = 198$

3.33 (15)

Given that generating function:

$$\{a_n\}_{n=0}^{\infty} \text{ is } \frac{1+z}{(1-z)^3}$$

$$A(z) = \sum a_r z^r = \frac{1+z}{(1-z)^3}$$

We can replace z by x, also

$$A(x) = \sum a_r x^r = \frac{1+x}{(1-x)^3} = \frac{1}{(1-x)^3} + \frac{x}{(1-x)^3}$$

$$= \sum_{r=0}^{\infty} {}^{3-1+r}C_r x^r + x \sum_{r=0}^{\infty} {}^{3-1+r}C_r x^r$$

$$= \sum_{r=0}^{\infty} {}^{r+2}C_2 x^r + \sum_{r=0}^{\infty} {}^{r+2}C_2 x^{r+1}$$

Now we read to find a_0 and a_3 which are nothing but the coefficient of x^0 and x^3 respectively.

a_0 = Coefficient $x^0 = {}^{(0+2)}C_2 = {}^2C_2 = 1$

a_3 = Coefficient $x^3 = {}^{(3+2)}C_2 + {}^{(2+2)}C_2$

$= {}^5C_2 + {}^4C_2 = 16$

So, $a_3 - a_0 = 16 - 1 = 15$

3.34 (d)

Given, $a_n = 2n + 3$

Since generating function for 1 is $\frac{1}{1-x}$ and n is $\frac{x}{(1-x)^2}$, the generating function for a_n is

$$A(x) = \frac{2x}{(1-x)^2} + \frac{3}{1-x}$$

$$= \frac{2x + 3(1-x)}{(1-x)^2} = \frac{3-x}{(1-x)^2}$$

3.35 (a)

$A = \{(x, X), x \in X \text{ and } X \subseteq U\}$

The number of k element subsets of a set U with n elements $= \binom{n}{k} = {}^nC_k$.

The number of possible ordered pairs (x, X) where $x \in X$ is $k \cdot {}^nC_k$ for a given value of k from 1 to n.

So total number of ordered pairs in A

$$= |A| = \sum_{k=1}^{n} k \cdot {}^nC_k = \sum_{k=1}^{n} k \cdot \binom{n}{k}$$

So II is correct. (Note that $k = 0$ is excluded since empty set has no elements and cannot form an order pair such as (x, X)).

But since by the combinational identity

$$\sum_{k=1}^{n} k \cdot \binom{n}{k} = n.2^{n-1}$$

So I is also correct. So both I and II are correct.

3.36 (12)

Since both L's are indistinguishable.

First L's can be arranged in 3 positions 2, 3, or 5 in ${}^3C_2 = 3$ ways as follows:

_ L _ L _

or _ L _ _ L

or _ _ _ L L

Now the letters I, A, C can be deranged in $2 \times 2!$ ways. Example in _ L _ L _. C cannot occupy 5th position, so only 2 ways.

Remaining I and A can be arranged in remaining 2 position in 2! ways = 2 ways.

So answer is $3 \times 2 \times 2! = 12$

3.37 (65)

Let computers be A, B and C

Toughest job assigned to fastest computer (Say, A) is 1 way.

Easiest job assigned to shortest computer (Say, B) is 1 way.

Remaining 4 jobs to be assigned to 3 computers so that the computer C gets at least one job, since A and B already assigned a job.

Number of ways 4 jobs assigned to 3 computers $= 3^4$.

Number of ways 4 jobs assigned to 3 computers, so that computer C does not get any job $= 2^4$.

Required number of ways $= 3^4 - 2^4 = 81 - 16 = 65$ ways

3.38 Sol.

Given that object are 6 identical balls this means that it does not matter. Which objects are grouped together, it only matters how many objects go into each bins. In addition to this the bins are identical, this means that ordering of bins does not matter. This problem can be modeled as partition of '6' into exactly '3' parts as follows.

(6, 0, 0) (5, 1, 0) (4, 2, 0) (4, 1, 1) (3, 3, 0) (3, 2, 1) (2, 2, 2)

3.39 (a)

$a_n = n + 1$ if 'n' is odd

$= 1$ otherwise

$$\frac{(n+k)(n-k+1)}{2} = a_0 + a_1 x + a_2 x^2 + a_3 x^3 + a_4 x^4 + a_5 x^5 + a_6 x^6 + a_7 x^7 + \ldots$$

$= 1 + 2x + x^2 + 4x^3 + x^4 + 6x^5 + x^6 + 8x^7 + \ldots)$

$= (1 + x + x^2 + x^3 + x^4 + \ldots) + x(1 + 3x^2 + 5x^4 + 7x^6 + \ldots)$

$$= \frac{1}{(1-x)} + x\frac{d}{dx}(x + x^3 + x^5 + x^7 + \ldots)$$

$$= \frac{1}{(1-x)} + x\frac{d}{dx}\left[\frac{x}{(1-x^2)}\right]$$

$$= \frac{1}{(1-x)} + x\left[\frac{(1-x^2) - x(-2x)}{(1-x^2)^2}\right]$$

$$= \frac{1}{(1-x)} + \frac{x(1+x^2)}{(1-x^2)^2}$$

3.40 (a, b, c)

Given functions:

$f(2n) = 2f(n) - 1$ // $2n$ is always even

$f(2n+1) = 2f(n) + 1$ // $2n+1$ is always odd

(b)
$$\begin{aligned} f(2^n) &= f(2.2^{n-1}) \\ &= 2f(2.2^{n-2}) - 1 \\ &= 2[2f(2^{n-2}) - 1] - 1 \\ &= 2^2 f(2^{n-2}) - 2 - 1 \\ &\vdots \\ &= 2^k f(2^{n-k}) - [2^k - 1] \\ &[\because 2^{n-k} = 1 \Rightarrow 2^k = 2^n] \\ &= 2^n - [2^n - 1] = 1 \end{aligned}$$

True

(c) $f(5.2^n) = 2^{n+1} + 1$

(a) $f(2^n - 1)$
$$\begin{aligned} f(2(2^{n-1} - 1) + 1) &= 2f(2^{n-1} - 1) + 1 \\ &= 2[2f(2^{n-2} - 1) + 1] + 1 \\ &= 2^2 f(2^{n-2} - 1) + 2 + 1 \\ &\vdots \\ &= 2^k f(2^{n-k} - 1) + 2^{k-1} + ... + 2 + 1 \\ &[\because 2^{n-k} - 1 = 1 \Rightarrow k = n - 1] \\ &= 2^{n-1} + 2^{n-1} - 1 = 2^{n-1} \end{aligned}$$

(d)
$$\begin{aligned} f(2^n + 1) &= f(2.2^{n-1} + 1) \\ &= 2f(2^{n-1}) + 1 \\ &= 2f(2.2^{n-2}) + 1 \\ &= 2[2f(2^{n-2}) + 1] + 1 \\ &= 2^2 f(2^{n-2}) + 2 + 1 \\ &\vdots \\ &= 2^k f(2^{n+k}) + (2^k - 1) \\ &[\because 2^n = 2^k] \\ &= 2^n + 2^n - 1 = 2^{n+1} - 1 \end{aligned}$$

3.41 (a)

Given that $L_n = L_{n-1} + L_{n-2}$ for $n \geq 3$

By replacing 'n' by $(n + 2)$ we have

$$L_{n+2} - L_{n+1} - L_n = 0 \quad ...(*)$$

Characteristics equation for equation (*) is

$$t^2 - t - 1 = 0$$

Characteristics roots are

$$t = \frac{1 \pm \sqrt{5}}{2}$$

$\therefore$ Complementary functions $= C_1 t_1^n + C_2 t_2^n$

$$\therefore \quad L_n = C_1\left(\frac{1+\sqrt{5}}{2}\right)^n + C_2\left(\frac{1-\sqrt{5}}{2}\right)^n$$

Put $n = 1, 1 = C_1\left(\frac{1+\sqrt{5}}{2}\right) + C_2\left(\frac{1-\sqrt{5}}{2}\right)$ ...(i)

Put $n = 2, 3 = C_1\left(\frac{1+\sqrt{5}}{2}\right)^2 + C_2\left(\frac{1-\sqrt{5}}{2}\right)^2$ ...(ii)

By solving equation (i) and (ii) we get, $C_1 = 1$, $C_2 = 1$

$$\therefore \quad L_n = \left(\frac{1+\sqrt{5}}{2}\right)^n + \left(\frac{1-\sqrt{5}}{2}\right)^n$$

Another approach (Substitution method)

Put $n = 1$, $n = 2$ and eliminate option (b), (c), (d) then correct answer is option (a).

For n =1, we have $L_1 = 1$

For n =2, we have $L_2 = 3$ in option (a)

3.42 (d)

$U = \{1, 2, 3, 4, ..., n\}$ where $n > 1000$

$A \subseteq U, B \subseteq U$ and $|A| = |B| = k$ and $A \cap B = \phi$

Case (i): If all elements of A appear before the elements of 'B' then number of permutations

$$= {}^nC_{2k} * (n - 2k)!\ k!\ k!$$
$$= {}^nC_{2k} (n - 2k)!\ (k!)^2$$

Case (ii): If all elements of B appear before the elements of 'A' then number of permutations

$$= {}^nC_{2k} (n - 2k)!\ (k!)^2$$

$\therefore$ Total number of permutations = Case (i) + Case (ii)

$$= 2 * {}^nC_{2k} (n - 2k)!\ (k!)^2$$

■■■■

4 Graph Theory

4.1 Prove that in a finite graph, the number of vertices of odd degree is always even. **[1995 : 2 M]**

4.2 Let G be an undirected connected graph with distinct edge weights. Let e_{max} be the edge with maximum weight and e_{min} be the edge with minimum weight. Which of the following statements is false?

(a) Every minimum spanning tree of G must contain e_{min}
(b) If e_{max} is in a minimum spanning tree, then its removal must disconnect G
(c) No minimum spanning tree contains e_{max}
(d) G has a unique minimum spanning tree

[2000 : 2 M]

4.3 How many undirected graphs (not necessary connected) can be constructed out of a given set $V = \{v_1, v_2, ... v_n\}$ of n vertices?

(a) $\frac{n(n-1)}{2}$ (b) 2^n
(c) n! (d) $2^{\frac{n(n-1)}{2}}$

[2001 : 2 M]

4.4 Maximum number of edges in a n-node undirected graph without self loops is

(a) n^2 (b) $\frac{n(n-1)}{2}$
(c) $n-1$ (d) $\frac{(n+1)(n)}{2}$

[2002 : 1 M]

4.5 Let G be an arbitrary graph with n nodes and k components. If a vertex is removed from G, the number of components in the resultant graph must necessarily lie between.

(a) k and n
(b) $k-1$ and $k+1$
(c) $k-1$ and $n-1$
(d) $k+1$ and $n-k$

[2003 : 1 M]

4.6 How many perfect matchings are there in a complete graph of 6 vertices?

(a) 15 (b) 24
(c) 30 (d) 60 **[2003 : 2 M]**

4.7 A Graph G = (V, E) satisfies $|E| \le 3|V| - 6$. The min-degree of G is defined as $\min_{v \in V}$ {degree (v)}. Therefore, min-degree of G cannot be

(a) 3 (b) 4
(c) 5 (d) 6

[2003 : 2 M]

4.8 The minimum number of colours required to colour the following graph, such that no two adjacent vertices are assigned the same colour, is

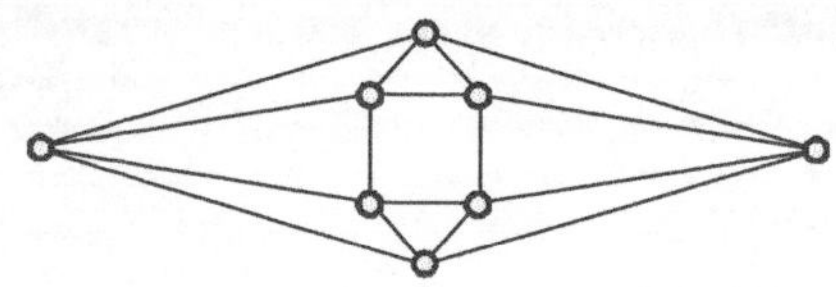

(a) 2 (b) 3
(c) 4 (d) 5

[2004 : 2 M]

4.9 How many graphs on n labeled vertices exist which have at least $\frac{(n^2-3n)}{2}$ edges?

(a) ${}^{(n^2-n)/2}C_{(n^2-3n)/2}$ (b) $\sum_{k=0}^{(n^2-3n)/2} {}^{(n^2-n)/2}C_k$

(c) ${}^{(n^2-n)/2}C_n$ (d) $\sum_{k=0}^{n} {}^{(n^2-n)/2}C_k$

[2004 : 2 M]

4.10 What is the maximum number of edges in an acyclic undirected graph with n vertices?

(a) $n-1$ (b) n
(c) $n+1$ (d) $2n-1$

[2004 : 1 M]

4.11 What is the number of vertices in an undirected connected graph with 27 edges, 6 vertices of degree 2, 3 vertices of degree 4 and remaining of degree 3?

(a) 10 (b) 11
(c) 18 (d) 19 **[2004 : 2 M]**

4.12 Let G be a simple connected planar graph with 13 vertices and 19 edges. Then, the number of faces in the planar embedding of the graph is

(a) 0 (b) 8
(c) 9 (d) 13 **[2005 : 1 M]**

4.13 Which one of the following graphs is NOT plannar?

(a) G_1: 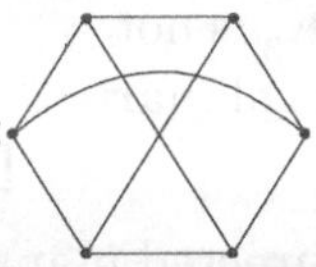

(b) G_2: 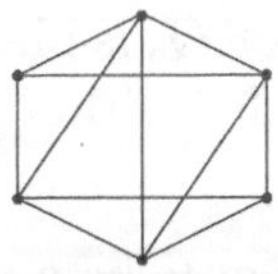

(c) G_3: 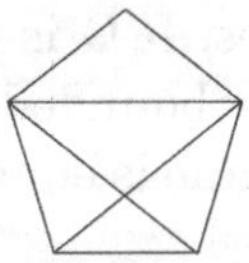

(d) G_4: 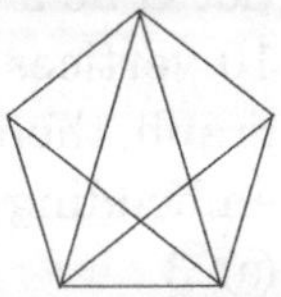

[2005 : 2 M]

4.14 If all the edge weights of an undirected graph are positive, then any subset of edges that connects all the vertices and has minimum total weight is a

(a) Hamiltonian cycle (b) Grid
(c) Hypercube (d) Tree

[2006 : 1 M]

4.15 Consider the undirected graph G defined as follows. The vertices of G are bit strings of length n. We have an edge between vertex u and vertex v if and only if u and v differ in exactly one bit position (in other words, v can be obtained from u by flipping a single bit). The ratio of the chromatic number of G to the diameter of G is

(a) $\frac{1}{(2^{n-1})}$ (b) $\frac{1}{n}$
(c) $\frac{2}{n}$ (d) $\frac{3}{n}$ **[2006 : 2 M]**

Common Data for Q.4.16 & Q.4.18:

The 2^n vertices of graph G correspond to all subsets of a set of size n, for $n \geq 6$. Two vertices of G are adjacent if and only if the corresponding sets intersect in exactly two elements.

4.16 The number of vertices of degree zero in G is

(a) 1 (b) n
(c) n + 1 (d) 2^n **[2006 : 2 M]**

4.17 The maximum degree of a vertex in G is

(a) $\binom{n/2}{2} 2^{n/2}$ (b) 2^{n-2}
(c) $2^{n-3} \times 3$ (d) 2^{n-1} **[2006 : 2 M]**

4.18 The number of connected components in G is

(a) n (b) n + 2
(c) $2^{n/2}$ (d) $\frac{2^n}{n}$ **[2006 : 2 M]**

4.19 Let G be the non-planar graph with the minimum possible number of edges. Then G has

(a) 9 edges and 5 vertices
(b) 9 edges and 6 vertices
(c) 10 edges and 5 vertices
(d) 10 edges and 6 vertices **[2007 : 1 M]**

4.20 Consider the DAG with V = {1, 2, 3, 4, 5, 6}, shown below:

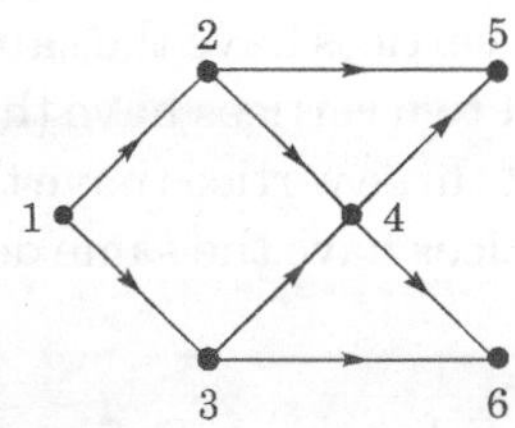

Which of the following is NOT a topological ordering?

(a) 1 2 3 4 5 6 (b) 1 3 2 4 6 5
(c) 1 3 2 4 5 6 (d) 3 2 4 1 6 5

[2007 : 1 M]

4.21 Which of the following graphs has an Eulerian circuit?

(a) Any k-regular graph where k is an even number
(b) A complete graph on 90 vertices
(c) The complement of a cycle on 25 vertices
(d) None of the above **[2007 : 2 M]**

4.22 What is the size of the smallest MIS (Maximal Independent Set) of a chain of nine nodes?

(a) 5 (b) 4
(c) 3 (d) 2

[2008 : 1 Mark]

4.23 What is the chromatic number of the following graph?

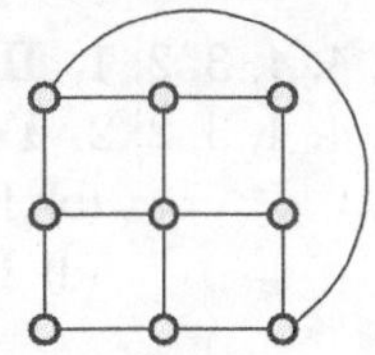

(a) 2 (b) 3
(c) 4 (d) 5 **[2008 : 1 M]**

4.24 G is a simple, connected, undirected graph. Some vertices of G are of odd degree. Add a node v to G and make it adjacent to each odd degree vertex of G. The resultant graph is sure to be

(a) Regular
(b) Complete
(c) Hamiltonian
(d) Euler **[2008 : 2 M]**

4.25 What is the chromatic number of an n-vertex simple connected graph which does not contain any odd length cycle? Assume $n \geq 2$.
(a) 2 (b) 3
(c) $n - 1$ (d) n **[2009 : 1 M]**

4.26 Which one of the following is TRUE for any simple connected undirected graph with more than 2 vertices?
(a) No two vertices have the same degree
(b) At least two vertices have the same degree
(c) At least three vertices have the same degree
(d) All vertices have the same degree
[2009 : 1 M]

4.27 Let, $G = (V, E)$ be a graph. Define $\xi(G) = \sum_d i_d \times d$, where i_d is the number of vertices of degree d in G. If S and T are two different trees with $\xi(S) = \xi(T)$, then
(a) $|S| = 2|T|$ (b) $|S| = |T| - 1$
(c) $|S| = |T|$ (d) $|S| = |T| + 1$
[2010 : 1 M]

4.28 In a binary tree with n nodes, every node has an odd number of descendants. Every node is considered to be its own descendant. What is the number of nodes in the tree that have exactly one child?
(a) 0 (b) 1
(c) $\frac{(n-1)}{2}$ (d) $n - 1$ **[2010 : 1 M]**

4.29 The degree sequence of a simple graph is the sequence of the degrees of the nodes in the graph in decreasing order. Which of the following sequences can not be the degree sequence of any graph?
I. 7, 6, 5, 4, 4, 3, 2, 1 **II.** 6, 6, 6, 6, 3, 3, 2, 2
III. 7, 6, 6, 4, 4, 3, 2, 2 **IV.** 8, 7, 7, 6, 4, 2, 1, 1
(a) I and II (b) III and IV
(c) IV only (d) II and IV
[2010 : 2 M]

4.30 K_4 and Q_3 are graphs with the following structures:

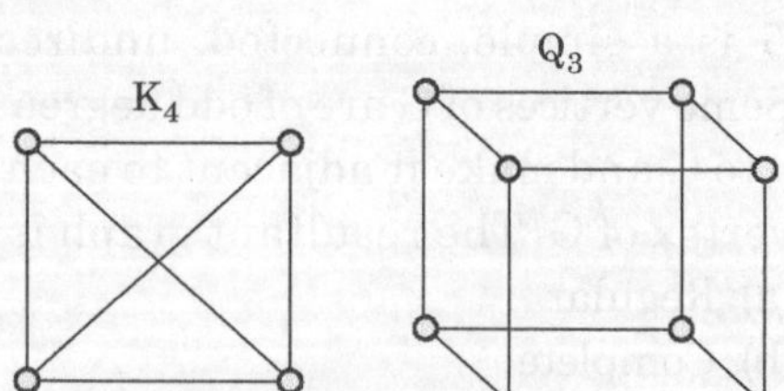

Which one of the following statements is TRUE in relation to these graphs?
(a) K_4 is planar while Q_3 is not
(b) Both K_4 and Q_3 are planar
(c) Q_3 is planar while K_3 is not
(d) Neither K_4 nor Q_3 is planar
[2010 : 2 M]

4.31 Let G be a simple undirected planar graph on 10 vertices with 15 edges. If G is a connected graph, then the number of bounded faces in any embedding of G on the plane is equal to
(a) 3 (b) 4
(c) 5 (d) 6 **[2012 : 1 M]**

4.32 Which of the following graphs is isomorphic to

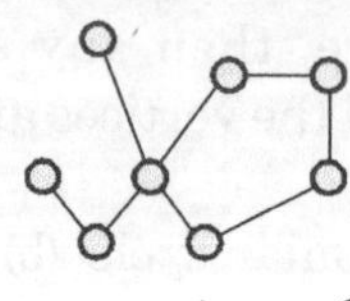

(a) 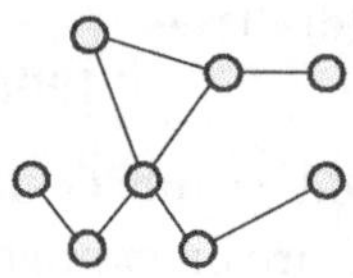 (b) 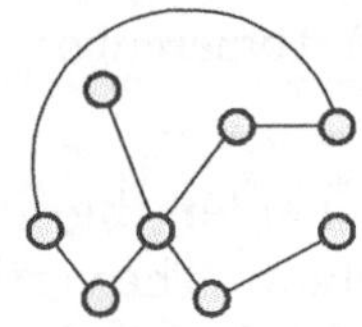

(c) 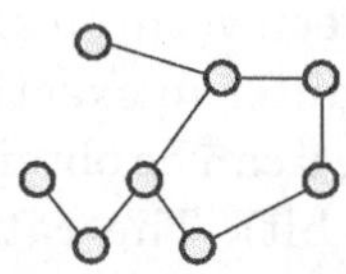 (d) 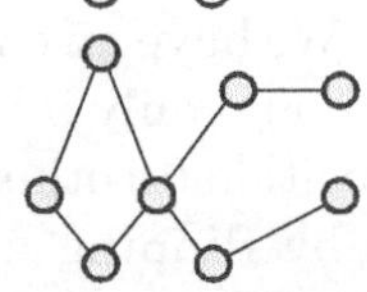

[2012 : 2 M]

4.33 Let G be a complete undirected graph on 6 vertices. If vertices of G are labeled, then the number of distinct cycles of length 4 in G is equal to
(a) 15 (b) 30
(c) 45 (d) 360 **[2012 : 2 M]**

4.34 Consider an undirected random graph of eight vertices. The probability that there is an edge between a pair of vertices is $\frac{1}{2}$. What is the expected numbers of unordered cycles of length three?
(a) $\frac{1}{8}$ (b) 1
(c) 7 (d) 8 **[2013 : 1 M]**

4.35 Which of the following statements is/are TRUE for undirected graph?
P : Number of odd degree vertices is even.
Q : Sum of degrees of all vertices is even.
(a) P only
(b) Q only
(c) Both P and Q
(d) Neither P nor Q
[2013 : 1 M]

4.36 Let $G = (V, E)$ be a directed graph where V is the set of vertices and E the set of edges. Then which one of the following graphs has the same strongly connected components as G?
(a) $G_1 = (V, E_1)$ where $E_1 = \{(u,v) \mid (u,v) \notin E\}$
(b) $G_2 = (V, E_2)$ where $E_2 = \{(u,v) \mid (v,u) \in E\}$
(c) $G_3 = (V, E_3)$ where $E_3 = \{(u,v) \mid$ there is a path of length ≤ 2 from u to v in $E\}$
(d) $G_4 = (V_4, E)$ where V_4 is the set of vertices in G which are not isolated

[2014 (Set-1) : 1 M]

4.37 Consider an undirected graph G where self-loops are not allowed. The vertex set of G is $\{(i, j) : 1 \leq i \leq 12, 1 \leq j \leq 12\}$. There is an edge between (a, b) and (c, d) if $|a - c| \leq 1$ and $|b - d| \leq 1$. The number of edges in this graph is ______.

[2014 (Set-1) : 2 M]

4.38 An ordered n-tuple $(d_1, d_2, ..., d_n)$ with $d_1 \geq d_2 \geq ... \geq d_n$ is called *graphic* if there exists a simple undirected graph with n vertices having degrees $d_1, d_2, ..., d_n$ respectively. Which of the following 6-tuples is NOT graphic?
(a) (1, 1, 1, 1, 1, 1) (b) (2, 2, 2, 2, 2, 2)
(c) (3, 3, 3, 1, 0, 0) (c) (3, 2, 1, 1, 1, 0)

[2014 (Set-1) : 2 M]

4.39 The maximum number of edges in a bipartite graph on 12 vertices is ______.

[2014 (Set-2) : 1 M]

4.40 A cycle on n vertices is isomorphic to its complement. The value of n is ______.

[2014 (Set-2) : 2 M]

4.41 The number of distinct minimum spanning trees for the weighted graph below is ______.

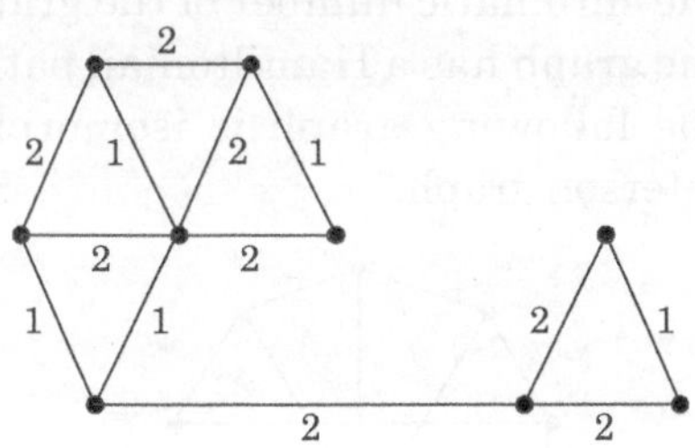

[2014 (Set-2) : 2 M]

4.42 If G is a forest with n-vertices and k connected components, how many edges does G have?
(a) $\left\lfloor \frac{n}{k} \right\rfloor$ (b) $\left\lceil \frac{n}{k} \right\rceil$
(c) $n - k$ (d) $n - k + 1$

[2014 (Set-3) : 2 M]

4.43 Let δ denote the minimum degree of a vertex in a graph. For all planar graphs on n vertices with $\delta \geq 3$ which one of the following is **TRUE**?
(a) In any planar embedding, the number of faces is at least $\frac{n}{2} + 2$
(b) In any planar embedding, the number of faces is less than $\frac{n}{2} + 2$
(c) There is a planar embedding in which the number of faces is less than $\frac{n}{2} + 2$
(d) There is a planar embedding in which the number of faces is at most $\frac{n}{\delta + 1}$

[2014 (Set-3) : 2 M]

4.44 Let G be a connected planar graph with 10 vertices. If the number of edges on each face is three, then the number of edges in G is ______.

[2015 (Set-1) : 2 M]

4.45 A graph is self-complementary if it is isomorphic to its complement. For all self-complementary graphs on n vertices, n is
(a) A multiple of 4
(b) Even
(c) Odd
(d) Congruent to 0 mod 4, or 1 mod 4

[2015 (Set-2) : 2 M]

4.46 In a connected graph, a bridge is an edge whose removal disconnects a graph. Which one of the following statements is True?
(a) A tree has no bridges
(b) A bridge cannot be part of a simple cycle
(c) Every edge of a clique with size ≥ 3 is a bridge (A clique is any complete subgraph of a graph)
(d) A graph with bridges cannot have a cycle

[2015 (Set-2) : 2 M]

4.47 Consider a binary tree T that has 200 leaf nodes. Then, the number of nodes in T that have exactly two children are ______. **[2015 (Set-3) : 1 M]**

4.48 Let G be connected undirected graph of 100 vertices and 300 edges. The weight of a minimum spanning tree of G is 500. When the weight of each edge of G is increased by five, the weight of a minimum spanning tree becomes ______.

[2015 (Set-3) : 2 M]

4.49 The minimum number of colours that is sufficient to vertex-colour any planar graph is ______.

[2016 (Set-2) : 1 M]

4.50 Let T be a tree with 10 vertices. The sum of the degrees of all the vertices in T is ________.
[2017 (Set-1) : 1 M]

4.51 G is an undirected graph with n vertices and 25 edges such that each vertex of G has degree at least 3. Then the maximum possible value of n is ________. **[2017 (Set-2) : 1 M]**

4.52 The chromatic number of the following graph is ________.

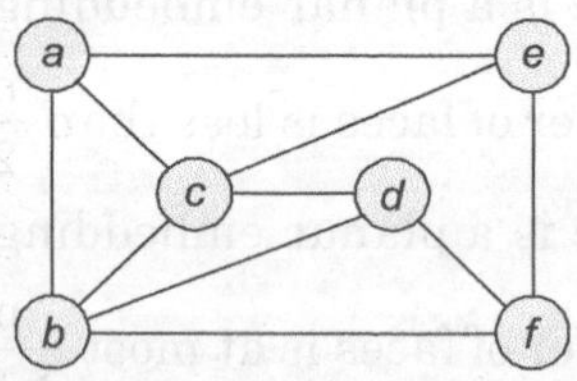

[2018 : 1 M]

4.53 Let G be a graph with 100! vertices, with each vertex labelled by a distinct permutation of the numbers 1, 2, ... 100. There is an edge between vertices u and v if and only if the label of u can be obtained by swapping two adjacent numbers in the label of v. Let y denote the degree of a vertex in G and z denote the number of connected components in G.
Then, $y + 10z$ = ________. **[2018 : 2 M]**

4.54 Let G be an undirected complete graph, on n vertices, where $n > 2$. Then, the number of different Hamiltonian cycles in G is equal to
(a) $n\,!$ (b) $(n-1)!$
(c) 1 (d) $\frac{(n-1)!}{2}$
[2019 : 1 M]

4.55 Graph G is obtained by adding vertex s to $K_{3,4}$ and making s adjacent to every vertex of $K_{3,4}$. The minimum number of colours required to edge-colour G is ______. **[2020 : 2 M]**

4.56 In an undirected connected planar graph G, there are eight vertices and five faces. The number of edges in G is ________.
[2021 (Set-1) : 1 M]

4.57 Let $G = (V, E)$ be an undirected unweighted connected graph. The diameter of G is defined as $\text{diam}(G) = \max_{u,v \in V}$ {the length of shortest path between u and v}
Let M be the adjacency matrix of G.
Define graph G_2 on the same set of vertices with adjacency matrix N, where

$$N_{ij} = \begin{cases} 1 & \text{if } M_{ij} > 0 \text{ or } P_{ij} > 0, \text{ where } P = M^2 \\ 0 & \text{Otherwise} \end{cases}$$

Which one of the following statements is true?
(a) $\lceil \text{diam}(G)/2 \rceil < \text{diam}(G_2) < \text{diam}(G)$
(b) $\text{diam}(G_2) \le \lceil \text{diam}(G)/2 \rceil$
(c) $\text{diam}(G) < \text{diam}(G_2) \le 2\,\text{diam}(G)$
(d) $\text{diam}(G_2) = \text{diam}(G)$
[2021 (Set-1) : 2 M]

4.58 Consider a simple undirected graph of 10 vertices. If the graph is disconnected, then the maximum number of edges it can have is ________.
[2022 : 1 M]

4.59 Consider a simple undirected unweighted graph with at least three vertices. If A is the adjacency matrix of the graph, then the number of 3-cycles in the graph is given by the trace of
(a) A^3
(b) A^3 divided by 2
(c) A^3 divided by 3
(d) A^3 divided by 6
[2022 : 2 M]

4.60 The following simple undirected graph is referred to as the Peterson graph.

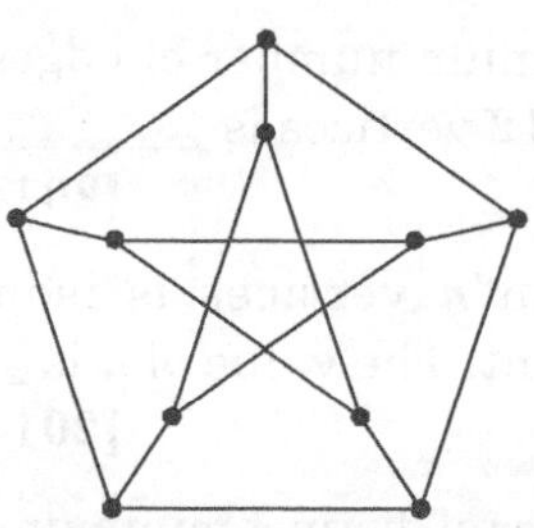

Which of the following statements is/are TRUE?
(a) The chromatic number of the graph is 3.
(b) The graph has a Hamiltonian path.
(c) The following graph is isomorphic to the Peterson graph.

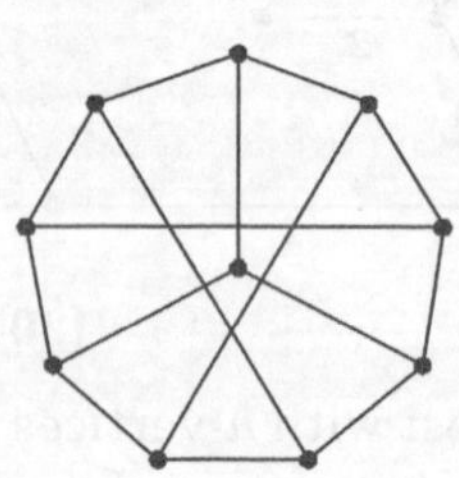

(d) The size of the largest independent set of the given graph is 3. (A subset of vertices of a graph form an independent set if no two vertices of the subset are adjacent.)
[2022 : 2 M]

4.61 Which of the properties hold for the adjacency matrix A of a simple undirected unweighted graph having n vertices?
(a) The diagonal entries of A^2 are the degrees of the vertices of the graph.
(b) If the graph is connected, then none of the entries of $A^{n-1} + I_n$ can be zero.
(c) If the sum of all the elements of A is at most $2(n-1)$, then the graph must be acyclic.
(d) If there is at least a 1 in each of A's rows and columns, then the graph must be connected.
[2022 : 2 M]

4.62 Let G be a simple, finite, undirected graph with vertex set $\{v_1, ..., v_n\}$. Let $\Delta(G)$ denote the maximum degree of G and let $N = \{1, 2, ...\}$ denote the set of all possible colors. Color the vertices of G using the following greedy strategy:
for $i = 1, ..., n$
color(v_i) $\leftarrow$ min$\{j \in N$: no neighbour of v_i is colored $j\}$
Which of the following statements is/are TRUE?
(a) This procedure results in a proper vertex coloring of G.
(b) The number of colors used is at most $\Delta(G) + 1$.
(c) The number of colors used is at most $\Delta(G)$.
(d) The number of colors used is equal to the chromatic number of G. **[2023 : 2 M]**

4.63 The chromatic number of a graph is the minimum number of colours used in a proper colouring of the graph. Let G be any graph with n vertices and chromatic number k. Which of the following statements is/are always TRUE?
(a) G contains a complete subgraph with k vertices
(b) G contains an independent set of size at least $\frac{n}{k}$.
(c) G contains a vertex of degree at least k.
(d) G contains at least $\frac{k(k-1)}{2}$ edges
[2024 (Set-1) : 2 M]

4.64 Let A be the adjacency matrix of a simple undirected graph G. Suppose A is its own inverse. Which one of the following statements is always TRUE?
(a) G is a cycle
(b) G is a perfect matching
(c) G is a complete graph
(d) There is no such graph G
[2024 (Set-2) : 1 M]

4.65 The chromatic number of a graph is the minimum number of colours used in a proper colouring of the graph. The chromatic number of the following graph is ______.

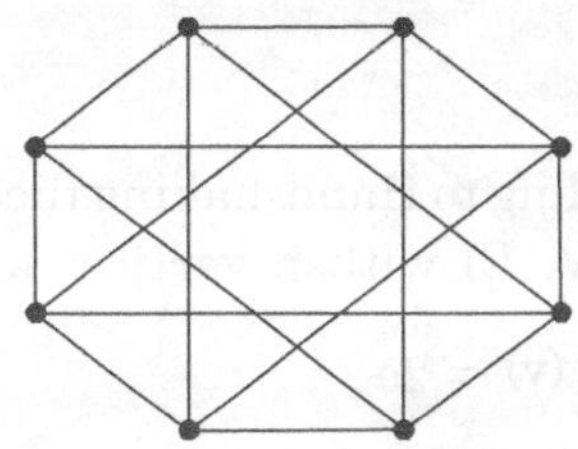

[2024 (Set-2) : 2 M]

■■■■

Answers Graph Theory

4.1	Sol.	4.2	(c)	4.3	(d)	4.4	(b)	4.5	(c)	4.6	(a)	4.7	(d)	4.8	(c)	4.9	(d)
4.10	(a)	4.11	(d)	4.12	(b)	4.13	(a)	4.14	(d)	4.15	(c)	4.16	(c)	4.17	(c)	4.18	(b)
4.19	(b)	4.20	(d)	4.21	(c)	4.22	(b)	4.23	(b)	4.24	(d)	4.25	(a)	4.26	(b)	4.27	(c)
4.28	(a)	4.29	(d)	4.30	(b)	4.31	(d)	4.32	(b)	4.33	(c)	4.34	(c)	4.35	(c)	4.36	(b)
4.37	(506)	4.38	(c)	4.39	(36)	4.40	(5)	4.41	(6)	4.42	(c)	4.43	(a)	4.44	(24)	4.45	(d)
4.46	(b)	4.47	(199)	4.48	(995)	4.49	(4)	4.50	(18)	4.51	(16)	4.52	(3)	4.53	(109)	4.54	(c, d)
4.55	(7)	4.56	(11)	4.57	(b)	4.58	(36)	4.59	(d)	4.60	(a,b,c)	4.61	(a)	4.62	(a, b)	4.63	(b, d)
4.64	(b)	4.65	(2)														

Explanations Graph Theory

4.1 Sol.

According to Handshaking theorem, for a graph G = (V, E) with n vertices and e edges then

$$\sum_{v \in V} \deg(v) = 2e$$

Let V_e and V_0 respectively the set of vertices of even degree and the set of vertices of odd degree in an undirected graph G = (V, E) then

$$\sum_{v \in V} \deg(v) = \sum_{v \in V_e} \deg(v) + \sum_{v \in V_0} \deg(v) = 2e$$

Since deg (v) is even for $v \in V_e$, the first sum in the right hand side of the equality is even. The total sum must be 2e, which is even, so the second sum must be even too. But all of its terms are odd, so there must be an even number of them.

4.2 (c)

(a) All the edge weights are distinct, so every minimum spanning tree of G must contain $e_{min.}$ (kruskal's algorithm will pick e_{min} first)

(b) If e_{max} is in a minimum spanning tree, then surely its removal must disconnect G (i.e. e_{max} must be a cut edge).

(c) A minimum spanning tree can contain e_{max} if its removal disconnect G (i.e. it is a cut edge). So this option is false.
For example, in the graph given below, the edge *cd* is a maximum weight edge but still is present in the minimum spanning tree.

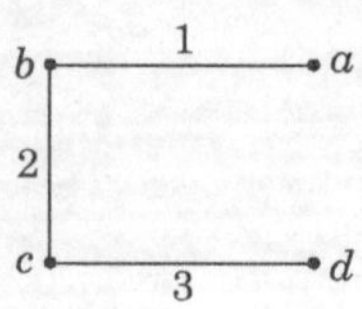

(d) Since all edge weights are distinct G has a unique minimum spanning tree.

4.3 (d)

In a graph G with n vertices, maximum number of edges possible $= \dfrac{n(n-1)}{2}$.

There are two ways for a edge, (the edge may appear in graph or may absent in graph). So there are two options for each edge.

Total number of graphs with n vertices $= 2^{\frac{n(n-1)}{2}}$.

4.4 (b)

The graph containing maximum number of edges in a *n*-node undirected graph without self loops is complete graph.

The number of edges in complete graph with *n*-node, k_n is $\dfrac{n(n-1)}{2}$.

4.5 (c)

Maximum components will result after removal of a node if graph G is a star graph as shown below.

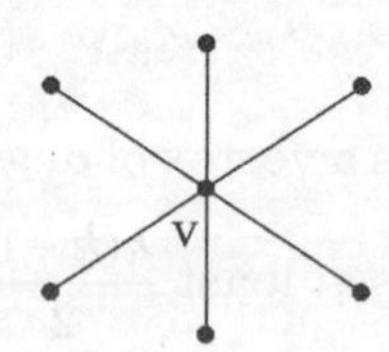

or a null graph of *n* vertices as shown below:

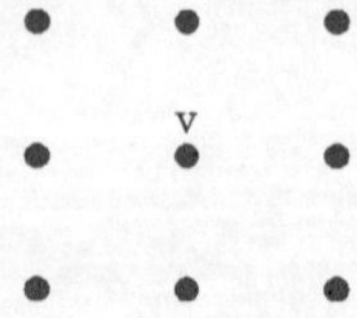

In either case, if node v is removed, the number of components will be $n-1$, where n is the total number of nodes in the star graph.

$\therefore$ $n-1$ is the maximum number of components possible. Minimum components will result if the node being removed is a lone vertex in which case, the number of components will be $k-1$.

$\therefore$ The number of components must necessarily lie between $k-1$ and $n-1$.

4.6 (a)

The number of perfect matchings in a complete graph of n vertices, where n is even, reduces to the problem of finding unordered partitions of the vertex set of the type p(2n; 2, 2, 2, ... n times)

$$= \frac{(2n)!}{(2!)^n\, n!}$$

For n = 3, 2n = 6, i.e. complete graph K_6, we have

$$\text{Number of perfect matchings} = \frac{6!}{(2!)^3\, 3!}$$

$$= \frac{6\times5\times4\times3\times2\times1}{2\times2\times2\times6} = 15$$

4.7 (d)

Given $\quad |E| \le 3|V| - 6 \quad$ ...(i)

Let δ be the minimum degree of the graph.

Now δ cannot exceed the average degree of the graph.

so, $\quad \delta \le \frac{2|E|}{|V|} \quad$ ...(ii)

Substitute eq. (i) in eq. (ii) and get

$$\delta \le \frac{2}{|V|}(3|V|-6)$$

$$\Rightarrow \quad \delta \le 6 - \frac{12}{|V|}$$

Clearly the minimum degree has to be less than 6 always and hence cannot be equal to 6.

4.8 (c)

An assignment of the colors 1, 2, 3 and 4 to the vertices of the graph is shown below such that the graph is properly colored.

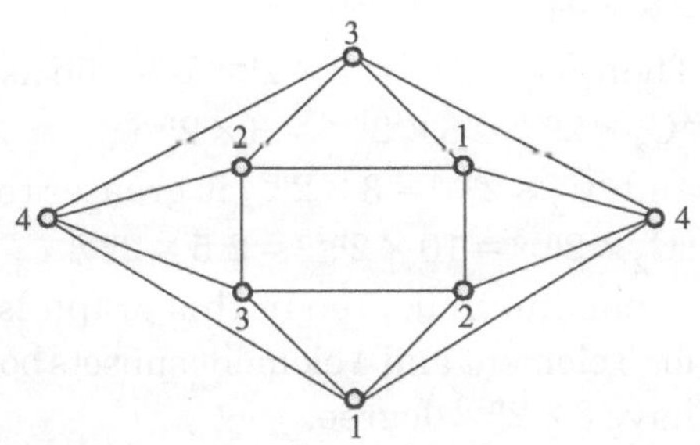

So 4 colours are required.

(**Note:** The graph is a planar graph. The four colour theorem says that the chromatic number of a planar graph is at most = 4).

4.9 (d)

Take number of edges available in n labelled vertices is

$$nC_2 = \frac{n(n-1)}{2} = \frac{n^2-n}{2} \text{ edges}$$

Now from this we need to choose $\frac{n^2-3n}{2}$ edges or more upto a maximum of $\frac{n^2-n}{2}$ edges. Each such choice of edges represents a distinct graph on n labelled vertices.

Total number of such graphs

$$= {}^{\frac{n^2-n}{2}}C_{\frac{n^2-3n}{2}} + {}^{\frac{n^2-n}{2}}C_{\frac{n^2-3n}{2}+1} \ldots + {}^{\frac{n^2-n}{2}}C_{\frac{n^2-n}{2}}$$

Now since ${}^nC_r = {}^nC_{n-r}$

$${}^{\frac{n^2-n}{2}}C_{\frac{n^2-3n}{2}} = {}^{\frac{n^2-n}{2}}C_{\left[\frac{n^2-n}{2}\right]-\left[\frac{n^2-3n}{2}\right]}$$

$$= {}^{\frac{n^2-n}{2}}C_n$$

Similarly, $\quad {}^{\frac{n^2-n}{2}}C_{\left[\frac{n^2-3n}{2}+1\right]} = {}^{\frac{n^2-n}{2}}C_{n-1}$

and so on until,

$${}^{\frac{n^2-n}{2}}C_{\frac{n^2-n}{2}} = {}^{\frac{n^2-n}{2}}C_0$$

So the required summation reduced to

$${}^{\frac{n^2-n}{2}}C_n + {}^{\frac{n^2-n}{2}}C_{n-1} + \ldots + {}^{\frac{n^2-n}{2}}C_0$$

$$= {}^{\frac{n^2-n}{2}}C_0 + {}^{\frac{n^2-n}{2}}C_1 + \ldots + {}^{\frac{n^2-n}{2}}C_n$$

$$= \sum_{k=0}^{n} {}^{\frac{n^2-n}{2}}C_k$$

which is choice (d).

4.10 (a)

Acyclic graph are graphs which does not contain cycle.

For maximum number of edges each vertex connect to other vertex only if it does not from a cycle i.e. from a tree with n vertices, which has maximum n – 1 edges. Ex. with 4 vertices, 3 edges maximum.

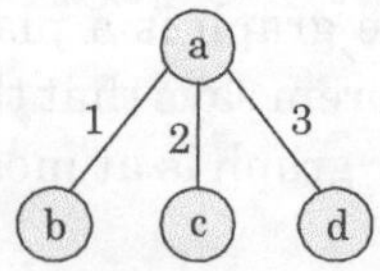

4.11 (d)

Sum of degree of all vertices = 2e (using Handshaking lemma)

So, $6 \times 2 + 3 \times 4 + x \times 3 = 2 \times 27$

$$12 + 12 + 3x = 54$$

$$3x = 54 - 24$$

$$x = \frac{30}{3}$$

$$x = 10$$

So, total number of vertices

$\Rightarrow \quad x + 6 + 3 = 10 + 6 + 3 = 19$

4.12 (b)

Given V = 13, E = 19

Let R be the number of regions.

$$R = E - V + 2$$

$\Rightarrow \quad R = 19 - 13 + 2 = 8$

4.13 (a)

G_1 is same as $K_{3,3}$ which is known to be non planar. G_2, G_3 and G_4 can be redrawn as follows so that they are planar.

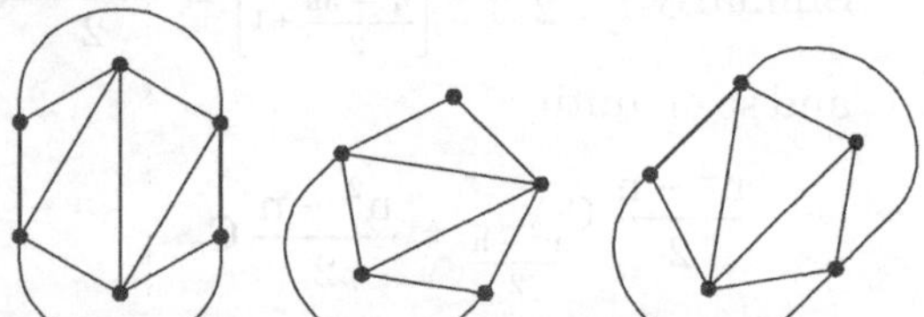

G_2 G_3 G_4

4.14 (d)

If subset of edges that connects all the vertices and it has minimum total weight then it never forms a cycle. So it is called as tree.

4.15 (c)

The hamming distance relation on bit strings has a digraph which will be always an n-cube where n is the number of bits.

Chromatic number of n-cube = 2 (since n-cube is always bipartite). Also the diameter of n-cube = n. So the ratio of chromatic number to diameter of the n-cube $= \frac{2}{n}$.

4.16 (c)

Let S contains n elements then S have 2^n subsets. Graph G contains 2^n vertices.

Let $S = \{v_1, v_2, \ldots, v_n\}$. Two vertices of G are adjacent if and only if the corresponding sets intersect in exactly two elements.

So $|\{V_i\} \cap \{V_j\}| = 2$

For this to happen, the subset must have at least 2 elements.

There are n sets which contains a single elements for V_1 to V_n who doesn't intersect another set such that it contains two elements. Therefore the degree of all these n vertices is zero. G also contains a vertex ϕ whose degree is zero. So the number of vertices whose degree is zero is n + 1.

4.17 (c)

Let the set be S = {1, 2, 3, 4, ..., n}

Consider a subset containing 2 elements of the form {1, 2}. Now {1, 2} will be adjacent to any subset with which it has exactly 2 elements in common. These sets can be formed by adding zero or more elements from remaining n – 2 elements, to the set {1, 2}. Since each of these elements may be either added or not added, number of ways of making such sets containing 1 and 2 is 2^{n-2}.

$\therefore$ Vertices with 2 elements will have 2^{n-2} degrees.

Now consider subsets of 3 elements say {1, 2, 3}. Since we want exactly 2 elements in common, we choose these in ${}^3C_2$ ways and then we can add or not add remaining n–3 elements. This can be done in 2^{n-3} ways.

$\therefore$ Total number of subsets with at least 2 elements common with {1, 2, 3} is given by ${}^3C_2 \times 2^{n-3}$.

Similarly, we can argue that the number of degrees of 4 element subsets is ${}^4C_2 \times 2^{n-4}$ and for 5 element subsets is ${}^5C_2 \times 2^{n-5}$ and so on.

Out of these $2^{n-2} = 2 \cdot 2^{n-3}$ is less than ${}^3C_2 \times 2^{n-3} = 3 \times 2^{n-3}$.

Then ${}^3C_2 \times 2^{n-3} = 3 \times 2^{n-3}$ is same as

${}^4C_2 \times 2^{n-4} = 6 \times 2^{n-4} = 3 \times 2^{n-3}$

and ${}^4C_2 \times 2^{n-4} = 3 \times 2^{n-3}$ is greater than

${}^5C_2 \times 2^{n-5} = 10 \times 2^{n-5} = 2.5 \times 2^{n-3}$

$\therefore$ maximum degree in this graph is occurring for 3 element and 4 element subsets both of which have $3 \times 2^{n-3}$ degree.

4.18 (b)

The number of connected component of G is determined by the degree and edges of vertices there are $n + 1$ vertices whose degree is zero, so they can form $n + 1$ connected component. The remaining vertices of the graph G are all connected as a single component. So total number of connected component is $n + 2$.

4.19 (b)

K_5 and $K_{3,3}$ are the smallest non planar graphs. K_5 has 5 vertices and ${}^5C_2 = 10$ edges and $K_{3,3}$ has 6 vertices and $3 \times 3 = 9$ edges. So, the non planar graph with minimum number of edges is $K_{3,3}$ with 9 edges and 6 vertices.

Note: K_5 is the non planar graph with minimum number of vertices.

4.20 (d)

In topological sorting the partial ordering of the DAG, must be preserved i.e. if $a \le b$ in the DAG, then in the topological order, b must come after a, not before. Consider the ordering 3 2 4 1 6 5. $1 \le 4$ in the given DAG but 4 is coming before 1 in 3 2 4 1 6 5 order which means that 3 2 4 1 6 5 is not a topological order of the given DAG.

4.21 (c)

Whenever in a graph all vertices have even degrees, it will surely have an Euler circuit.

(a) Since in a k-regular graph, every vertex has exactly k degrees and if k is even, every vertex in the graph has even degrees. k-regular graph need not be connected, hence k-regular may not contain Euler circuit.

(b) Complete graph on 90 vertices not contains an Euler circuit, because every vertex degree is odd (89)

(c) C_{25} has 24 edges and each vertex has exactly 2 degrees. So every vertex in the complement of C_{25} will have $24 - 2 = 22$ degrees which is an even number.

Since every vertex in the complement of C_{25} has even degrees. Also, the complements of all C_n with $n \ge 5$ is connected. So, it is an Euler graph.

4.22 (b)

Maximum independent sets: $\{a, c, e, g, i\}$ or $\{b, d, f, h\}$ or $\{b, e, h\}$

Smallest maximum independent set $\{b, e, h\}$

Size = 3. $\therefore$ option (c)

If we add any other node to the set then it will not be MIS.

4.23 (b)

Since there is a 5-cycle (odd cycle) in this graph, this graph is not bipartite and hence its chromatic number is atleast 3.

A proper colouring with 3 colours, is shown below:

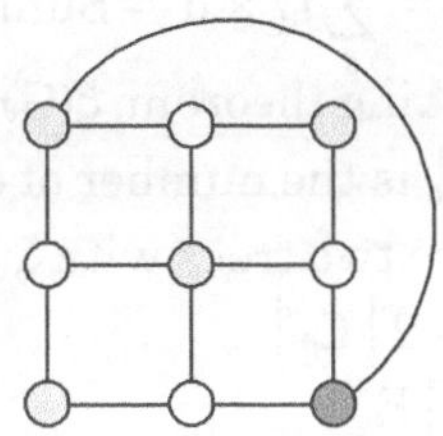

Hence minimum number of colors needed to color given graph is equal to 3.

4.24 (d)

In an undirected simple, connected graph number of vertices must be even of odd degree (using Handshaking lemma).

Adding a vertex v, adjacent to all odd degree vertices in graph, so degree of all odd degree vertices now become even and degree of vertex v is also even (since number of odd degree vertex are even).

Now all vertices in the graph are of even degree and graph is connected, so it must be Eular graph.

4.25 (a)

If a n-vertex simple connected graph contains no cycles of odd length, then its chromatic number is two, since the vertices can be alternately colored with the first color, then the second color, then the first color and then the second color and so on.

Alternatively, since a simple connected graph with no cycles of odd length must be bipartite, and since the chromatic number of a non-null bipartite graph is always 2 (in a bipartite graph each partition requires one color (there are no edges within a partition of a bipartite graph) and there are only two partitions).

Note: Since it is given that the graph is connected, it cannot be a null graph (ϕ_n)

4.26 (b)

In a simple connected undirected graph (with more than two vertices), at least 2 vertices must have same degree, since if this is not true, then all vertices would have different degrees. A graph with all vertices having different degrees is not possible to construct (can be proved as a corollary to the Havell-Hakimi theorem). Notice that it is possible to construct graphs satisfying choices a, c and d.

4.27 (c)

Given, $\xi(G) = \sum_{d} i_d \times d$ = Sum of degrees

By handshaking theorem, $\xi(G) = 2|E_G|$

where, $|E_G|$ is the number of edges in G.

If S and T are two trees with $\xi(S) = \xi(T)$.

$\Rightarrow \quad 2|E_S| = 2|E_T|$

$\Rightarrow \quad |E_S| = |E_T|$

In a tree, $|E_S| = |S| - 1$ and $|E_T| = |T| - 1$

Where $|S|$ is number of vertices of tree S and $|T|$ is number of vertices of tree T.

$\therefore \quad |S| - 1 = |T| - 1$

$\Rightarrow |S| = |T|$

4.28 (a)

A tree with 1 node is not possible, since it is given that every node has exactly 1 child.

Now consider a tree with 2 nodes (a is the root)

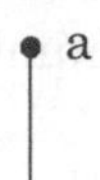

Now a has exactly one child. Number of descendents of a = 2. But this contradicts the given fact that every node has an odd number of descendents.

Now consider a tree with 3 nodes. Since, every-node has exactly one child, it must be of the form,

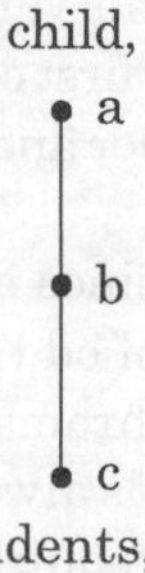

Here a has 3 descendents, b has 2 descendents and c has one. Again we have contradiction in that b does not have odd number of descendents. Similarly can show that for tree with 4, 5, 6 ... nodes, it is not possible to have all nodes with odd number of descendents. So correct answer is the tree has 0 nodes, i.e., choice (a).

4.29 (d)

Havell-Hakimi algorithm can be used to check whether a given degree sequence is a graph or not.

The algorithm is

1. remove top node of the sequence.
2. subtract "1" from as many nodes in remaining sequence as the degree of top node that was removed.
3. rearrange this sequence in non increasing order.
4. check if resulting sequence is a graph.
5. proceed again to step 1.

If the given sequence is not a graph we will see a violation in step 4, such as presence of negative degrees in the sequence. Otherwise the algorithm will bottom out with a degree sequence consisting of only even number of 1's and any number of 0's.

Now applying the algorithm to the degree sequences I, II, III and IV, one by one:

I. 7, 6, 5, 4, 4, 3, 2, 1

6, 5, 4, 4, 3, 2, 1	(Step 1)
5, 4, 3, 3, 2, 1, 0	(Step 2)
5, 4, 3, 3, 2, 1, 0	(Step 3)
Sequence is a graph	(Step 4)
4, 3, 3, 2, 1, 0	(Step 1)
3, 2, 2, 1, 0, 0	(Step 2)
3, 2, 2, 1, 0, 0	(Step 3)
Sequence is a graph	(Step 4)
2, 2, 1, 0, 0	(Step 1)
1, 1, 0, 0, 0	(Step 2)
1, 1, 0, 0, 0	(Step 3)
Sequence is a graph	(Step 4)

Now, the algorithm ends, since the sequence has only 0's and even number of 1's.

The final sequence corresponds to following valid graph

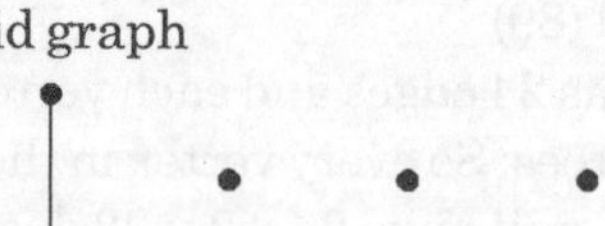

Similarly for sequence II.

II. 6, 6, 6, 6, 3, 3, 2, 2

6, 6, 6, 3, 3, 2, 2	(Step 1)
5, 5, 5, 2, 2, 1, 2	(Step 2)
5, 5, 5, 2, 2, 2, 1	(Step 3)
Sequence is a graph	(Step 4)
5, 5, 2, 2, 2, 1	(Step 1)
4, 4, 1, 1, 1, 1	(Step 2)
4, 4, 1, 1, 1, 1	(Step 3)
Sequence is a graph	(Step 4)

4, 1, 1, 1, 1	(Step 1)
3, 0, 0, 0, 1	(Step 2)
3, 1, 0, 0, 0	(Step 3)
Sequence is a graph	(Step 4)
1, 0, 0, 0	(Step 1)
0, −1, −1, 0	(Step 2)
0, 0, −1, −1	(Step 3)

The sequence is not a graph (Step 4), since negative degrees not possible in a valid graph.
So, algorithm ends.
II is cannot be the degree sequence of any graph.
Similarly we can show that III is degree sequence of some graph and IV is not a degree sequence of any graph.
So, option (d) is correct.

4.30 (b)

The planar embedding of K_4 and Q_3 is shown below:

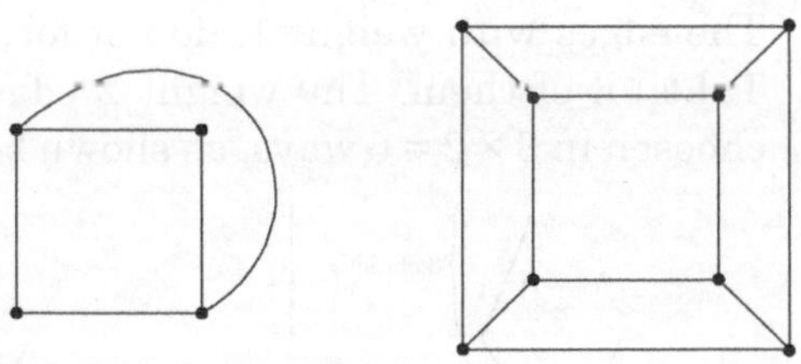

So both graphs are planar.

4.31 (d)

$$n = 10, \quad e = 15$$

In a simple connected planar graph, Euler's formula gives the total number of regions as

$$e - n + 2 = 15 - 10 + 2 = 7$$

Out of this, one region is unbounded and the other 6 are bounded.
So correct answer is 6, which is option (d).

4.32 (b)

Check invariants are one by one.
Step 1: All 4 choice have same number of vertices and edges as given graph.
Step 2: So we find degree sequence of given graph which is (1, 1, 2, 2, 2, 2, 2, 4).
Degree sequence of graph in option (a) is (1, 1, 1, 2, 2, 2, 3, 4).
Degree sequence of graph in option (b) is (1, 1, 2, 2, 2, 2, 2, 4)
Degree sequence of graph in option (c) is (1, 1, 2, 2, 2, 2, 3, 3)
Degree sequence of graph in option (d) is (1, 1, 2, 2, 2, 2, 2, 4).
So only options (a) and (c) are not isomorphic to given graph, since degree sequence of these graphs is not same as given graph.
Step 3: Now to decide between options (b) and (d), which one is isomorphic to given graph, we check the number of cycles.
In given Graph there is one cycle of length 5 but in Graph (d), there is no cycle of length 5.
Graph (b) has one cycle of length 5.
So only Graph (b) can be isomorphic to given Graph.

4.33 (c)

From 6 vertices we can select 4 distinct vertices in ${}^6C_4 = 15$ ways.
Now, with 4 vertices, we can form only 3 distinct cycles (see below) so, total number of distinct cycles of length 4 = 15 × 3 = 45.
Note: Number of cyclic permutations of n objects = $(n-1)!$ and for $n = 4$, we get 3! = 6 ways.
But number of distinct cycles in a graph is exactly half the number of cyclic permutations as there is no left/right ordering in a graph.

4.34 (c)

We need to find unordered cycle of length 3 so we choose any 3 vertices from 8 vertices. This can be done in ${}^8C_3$ ways.
To make cycle we need to choose edge between the selected vertices probability of choosing any edge is $\frac{1}{2}$.

So for three edges $= \frac{1}{2} \times \frac{1}{2} \times \frac{1}{2} = \frac{1}{8}$

Expected number of cycle = np

Here, $n = {}^8C_3$ and $p = \frac{1}{8}$

Expected number of cycles $= {}^8C_3 \times \frac{1}{8} = 7$

4.35 (c)

Q is Hand-shaking theorem and hence true.
P is a corollory to the Hand-shaking theorem and hence also true.

4.36 (b)

$G = (V, E)$ is directed graph
$G_2 = (V, E_2)$ where $E_2 = \{(u, v) \mid (v, u) \in E\}$
G and G_2 has same strongly connected components only the difference in G_2 is that all edges of G have been reversed the direction.

4.37 (506)

The given condition translates into the graph shown below where every vertex is connected only with its neighbours.

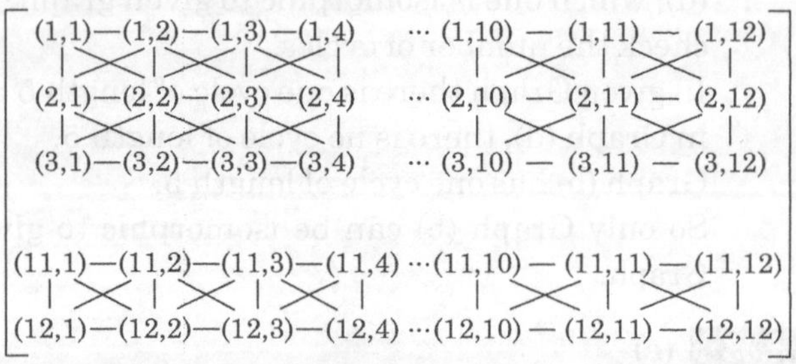

From above diagram:

(i) The four corner vertices have each 3 degrees which gives $4 \times 3 = 12$ degrees.

(ii) The 40 side vertices have 5 degrees each contributing a total of $40 \times 5 = 200$ degrees.

(iii) The 100 interior vertices each have 8 degrees contributing a total of $100 \times 8 = 800$ degrees.

So total degree of the graph

$12 + 200 + 800 = 1012$ degrees

Now the number of edges in any undirected graph

$$= \frac{\text{Total degrees}}{2}$$

Therefore the number of edges in this graph

$$= \frac{1012}{2} = 506$$

4.38 (c)

Use the **Havell-Hakimi** algorithm.

The sequence of steps in the algorithm for each graph is shown below:

(a) (1,1,1,1,1,1) is simple graph

(b) (2,2,2,2,2,2) ⇒ (1,1,2,2,2)
⇒ (2,2,2,1,1)
⇒ (1,1,1,1) is simple graph

(c) (3,3,3,1,0,0) ⇒ (2,2,0,0,0)
⇒ (1,–1,0,0)
cannot be a graph since negative degree not possible.

(d) (3,2,1,1,1,0) ⇒ (1,0,0,1,0)
⇒ (1,1,0,0,0) is simple graph

4.39 (36)

Bipartite graph {*u*, *v*} where *u* and *v* are partition on vertices:

(a) {1, 11} ⇒ $1 \times 11 = 11$ edges maximum

(b) {2, 10} ⇒ $2 \times 10 = 20$ edges maximum

(c) {3, 9} ⇒ $3 \times 9 = 27$ edges maximum

(d) {4, 8} ⇒ $4 \times 8 = 32$ edges maximum

(e) {5, 7} ⇒ $5 \times 7 = 35$ edges maximum

(f) {6, 6} ⇒ $6 \times 6 = 36$ edges maximum

∴ Maximum number of edges in a bipartite graph on 12 vertices = 36 edges.

4.40 (5)

If a graph is isomorphic to its own complement, then it is a self complementary graph. In a self complementary graph

$$e = \frac{n(n-1)}{4}$$

but in the cycle C_n, $e = n$.

Som if a cycle is self complementary, then

$$n = \frac{n(n-1)}{4}$$

The solution of the above equation is $n = 5$.

So, C_5 is the only cycle graph which is self complementary.

4.41 (6)

The edges with weight 1, do not form a cycle. Take all of them. The weight 2 edges can be choosen in $3 \times 2 = 6$ ways, as shown below.

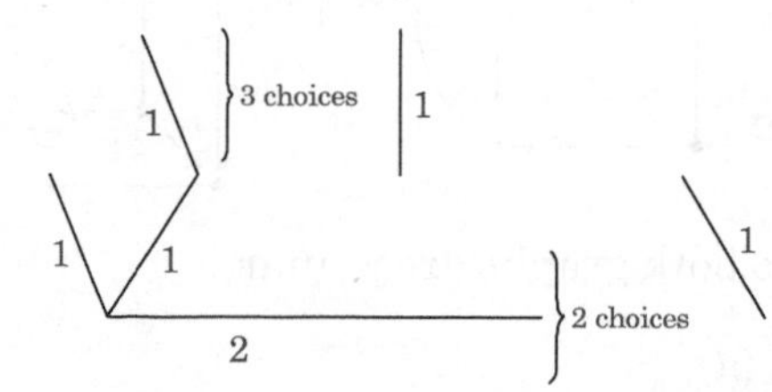

Total there are six possible minimum spanning trees. [3 × 2 = 6]

4.42 (c)

The number of edges in spanning forest of a graph G with *n* vertices and *k* components = Rank (G) $= n - k$.

4.43 (a)

$$\sum \deg = 2e$$

Given $\delta \geq 3$ also $\delta \leq \frac{2e}{n}$

$$\Rightarrow e \geq \frac{n\delta}{2}$$

$$\Rightarrow e \geq \frac{3n}{2}$$

Euler's formula:

$$r = e - n + 2 \geq \frac{3n}{2} - n + 2 \geq \frac{n}{2} + 2$$

So the number of faces is atleast $\frac{n}{2} + 2$

4.44 (24)

Number of vertices (n) = 10

$d(r_i) = 3$

Number of edges (e) = ?

$$r = e - v + 2$$

(euler's formula for connected planar graphs)

$$r = e - 10 + 2 = e - 8 \quad ...(1)$$

Since every region is bounded by exactly 3 edges and since every edge is exactly double counted, we have the equation

$$e = \frac{3r}{2} \Rightarrow r = \frac{2e}{3}$$

Substituting this in equation (1) we get,

$$\frac{2e}{3} = e - 8$$

$$2e = 3e - 24 \Rightarrow e = 24$$

4.45 (d)

An n-vertex self-complementary graph has exactly half number of edges of the complete graph i.e. $\frac{n(n-1)}{4}$ edges.

Since n(n – 1) must be divisible by 4, n must be congruent to 0 mod 4 or 1 mod 4.

4.46 (b)

An edge is called bridge iff it's removal will disconnect the graph into components.

In a cycle there are two paths between every pair of vertices and so removal of an edge from a cycle does not disconnect the cycle. So a bridge cannot be part of a simple cycle.

4.47 (199)

n_0 : Number of leaf nodes 0 children.

n_1 : Number of nodes with 1 child.

n_2 : Number of nodes with 2 children.

We have following two equations for binary tree

$n = n_0 + n_1 + n_2$ (where n is the total number of nodes in the tree) ...(1)

Accounting for total nodes as sum of those nodes which are children and the root which is not a child of any node,

We get, $n = n_1 \times 1 + n_2 \times 2 + 1$

$= n_1 + 2n_2 + 1 \quad ....(2)$

Combining equations (1) and (2) we get

$$n_0 + n_1 + n_2 = n_1 + 2n_2 + 1$$

$$\Rightarrow \quad n_2 = n_0 - 1$$

Therefore the number of nodes of degree two in any binary tree = number of leaf nodes –1.

$$n_2 = 200 - 1 = 199$$

4.48 (995)

G has |V| = 100 vertices and

|E| = 300 edges

the weight of MST of G = 500

In MST:

Number of vertices = 100

Number of edges = 99

If each edge of G is increased by 5 then MST weight also increased. For each edge of old MST increased with 5.

Total 99 edges.

So 99 × 5 = 445 is increased

Total weight of new MST = 500 + 445 = 995

4.49 (4)

Four color theorem says that every planar graph can color with 4 colors i.e. four colors are sufficient to properly color any planar graph.

4.50 (18)

Here in this tree, n = 10

Σdegree − 2|E|

where |E| represents number of edges.

In a tree, |E| = n – 1

= 10 – 1 = 9

So, Σdegree = 2 × 9 = 18

4.51 (16)

Number of vertices:

$$n \le ?$$

Number of vertices:

$$|E| = 25$$

Now since each vertex has at least 3 degree

and $2|E| = \Sigma\text{degree}$

i.e., $2|E| \ge 3n$

$$n \le 2|E|/3$$

$$\Rightarrow \quad n \le \frac{2 \times 25}{3} \le 16.66$$

So n is at most 16.

4.52 (3)

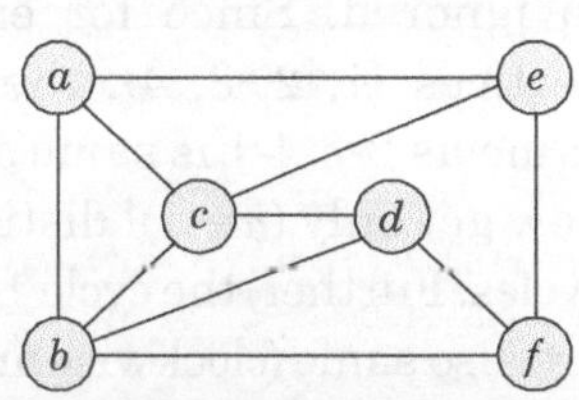

Since the largest complete sub graph is K_3, chromatic number is atleast 3.

We can try for a chromatic number of 3 by using 3 colors, as follows:

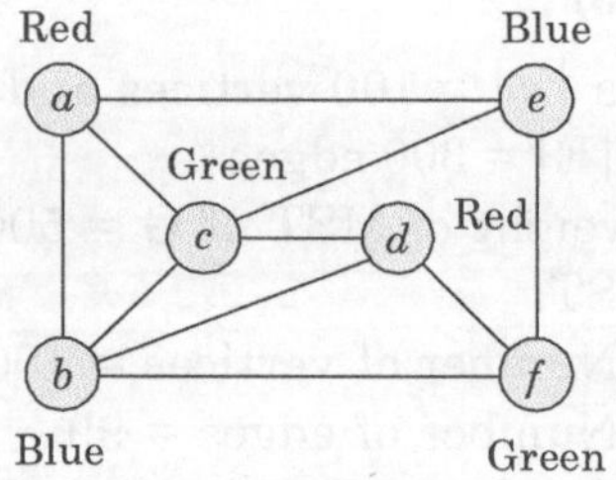

Since we have successfully, properly coloured all vertices with only 3 colors, the chromatic number of this graph is 3.

4.53 (109)

The graph has 100! vertices in which each vertex labelled by one of the 100! permutation.
Let us find degree of each vertex.
Let us take a vertex whose labelling is say 1,2,3,4, ..., 100.
Now it will be connected to all vertices where exactly 2 of the adjacent numbers all swapped. The two swapped numbers could be (1, 2), (2, 3), (3, 4) ..... etc. upto (99, 100) which makes for 99 edges for each such vertex.
So the graph is a regular graph with each vertex connected to 99 other vertices.

So $\quad y = 99$

The number of connected components = $z = 1$ since we can go from any vertex to any other vertex by only swapping 2 adjacent number at a time, many times i.e. there is a path from any vertex to any other vertex.
Graph is connected.

So $\quad z = 1$

So $\quad y + 10z = 99 + 10 \times 1 = 109$

4.54 (c, d)

- **When vertices are labelled.**
In a complete graph we can traverse the n vertices in any order and return to the starting vertex and form a Hamiltonian cycle. The number of such cycles will be $n!$ However, since circular rotations will have to ignored. Since for example K_4 with vertices {1, 2, 3, 4}, the cycle 1-2-3-4 is same as 2-3-4-1 is same as 3-4-1-2 etc. we now get only $(n-1)!$ distinct Hamiltonian cycles. Further, the cycle 1-2-3-4 and 1-4-3-2 are also same (clockwise and anticlockwise). So ignoring this orientation also we finally get $\frac{(n-1)!}{2}$ distinct Hamiltonian cycles which is option (d).

- **When vertices are unlabelled.**
All the Hamiltonian cycles are formed which is isomorphic to each other. So only 1 Hamiltonian cycle will be the answer in this case.
So both option (c) and (d) are correct.

4.55 (7)

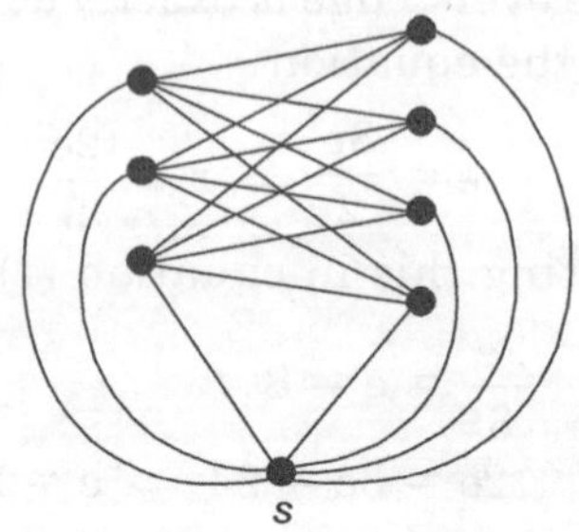

7 edges are touching at S each of which needs to be colors by different color.
So 7 colors minimum are required. All other vertices have less than 7 edges touching.
So 7 colors only are required for edge coloring.

4.56 (11)

In a connected planar graph

$$r = e - n + 2$$

Here, $\quad n = 8, \quad r = 5$

$\therefore \quad 5 = e - 8 + 2$

$$e = 11$$

4.57 (b)

$$N_{ij} = \begin{cases} 1 & \text{if } M_{ij} > 0 \text{ or } P_{ij} > 0, \text{ where } P = M^2 \\ 0 \end{cases}$$

Means G_2 will have not only all the edges of in G, but also will have edges connecting vertices in G which have a path of length 2, since M^2 will have all edges between u and v if there is a path of length 2 between shown in G.

Option (a) $\lceil \text{diam}(G)/2 \rceil < \text{diam}(G_2) < \text{diam}(G)$

Let G be the graph shown below with dia$(G) = 2$

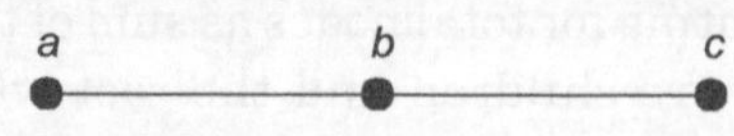

Now G_2 will be

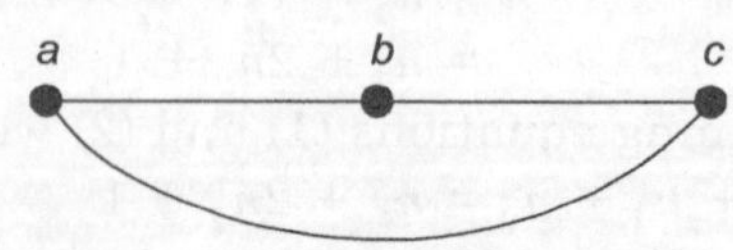

Since (a, c) has a path of length 2 in G, G_2 will have an edge connecting a and c.
Now diameter of $G_2 = 1$

This violates option (a) condition that $\text{dia}(G_2) > \text{dia}(G)/2$.
So option (a) false.

Option (b) $\text{diam}(G_2) \le \lceil \text{diam}(G)/2 \rceil$

Consider graph below with dia(G) = 3

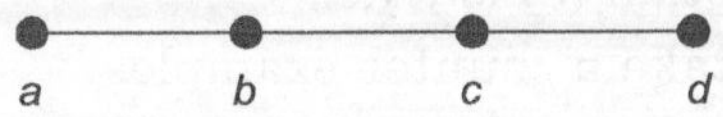

Now G_2 will connect all paths of length 2 with edges and will be

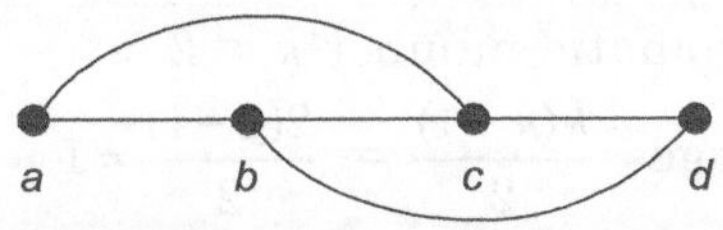

Now $\text{dia}(G_2) = 2$
But $\text{dia}(G) = 3$

$$\text{dia}(G_2) \le \lceil \text{dia}(G)/2 \rceil$$

$$2 \le \lceil 1.5 \rceil \le 2$$

Satisfied.
Option (b) is correct.

Option (c) $\text{diam}(G) < \text{diam}(G_2) \le 2\ \text{diam}(G)$

Taking previous option example where dia(G) = 3 and $\text{dia}(G_2) = 2$

$$3 \le 2 \le 6$$

Is false.
So option (c) is false.

Option (d) $\text{diam}(G_2) = \text{diam}(G)$

Taking previous example where dia(G) = 3 and $\text{dia}(G_2) = 2$

$$2 = 3 \text{ is false}$$

So option (d) is false.
So correct option is option (b).

4.58 (36)

If 'G' is a group with 'n' vertices and having 'k' connected components then maximum number of edges in 'G'

$$= \frac{(n-k)(n-k+1)}{2}$$

In the given problem,

$$n = 10,\ k = 2$$

$$= \frac{(10-2)(10-2+1)}{2} = 36$$

4.59 (d)

Represent graph in adjacency matrix format. If adjacency matrix multiply by itself 3 times matrix.
Adjacency matrix × Adjacency matrix × Adjacency matrix

A^3_{ii} [diagonal element] represents cycle of length 3 with beginning and ending with vertex i.

Trace of $A^3 = A^3_{11} + A^3_{22} + A^3_{33} + \ldots A^3_{nn}$ {sum of diagonals}

Since cycle has 3 vertices it counted for every vertex we need to divide by 3 for trace of A^3 and because and undirected graph $A - B - C - A$ same as $A - C - B - A$ cycle. To eliminate repeated possibility divide by 2.

$$\text{Number of cycles of 3 vertices} = \frac{A^3}{6}$$

4.60 (a, b, c)

(a) Chromatic number of Peterson graph is 3.
(b) Peterson graph has Hamiltonian path.
(c) The given graph is isomorphic to Peterson graph.
(d) The size of the largest independent set of Peterson graph is 4.

4.61 (a)

(a) Assume adjacency matrix representation of undirected graph is "A".

$$A^2 = A \times A$$

Diagonal elements of A^2 are degree of vertices or $(A^2)_{ii}$ represents degree of node i.
So (a) is true.

(b) Take following connected graph with $n = 3$,

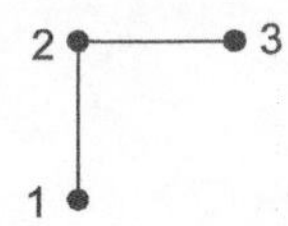

$$\text{Adjacency matrix, } (A) = \begin{bmatrix} 0 & 1 & 0 \\ 1 & 0 & 1 \\ 0 & 1 & 0 \end{bmatrix}$$

$$A^{n-1} = A^{3-1} = A^2 = \begin{bmatrix} 0 & 1 & 0 \\ 1 & 0 & 1 \\ 0 & 1 & 0 \end{bmatrix} \begin{bmatrix} 0 & 1 & 0 \\ 1 & 0 & 1 \\ 0 & 1 & 0 \end{bmatrix} = \begin{bmatrix} 1 & 0 & 1 \\ 0 & 2 & 0 \\ 1 & 0 & 1 \end{bmatrix}$$

$$A^{n-1} + I_n = A^2 + I_3 = \begin{bmatrix} 1 & 0 & 1 \\ 0 & 2 & 0 \\ 1 & 0 & 1 \end{bmatrix} + \begin{bmatrix} 1 & 0 & 0 \\ 0 & 1 & 0 \\ 0 & 0 & 1 \end{bmatrix} = \begin{bmatrix} 2 & 0 & 1 \\ 0 & 3 & 0 \\ 1 & 0 & 2 \end{bmatrix}$$

We can see that $A^{n-1} + I_n$ have some entries as zero. So (b) false.

(c) Consider following acyclic graph with $n = 5$

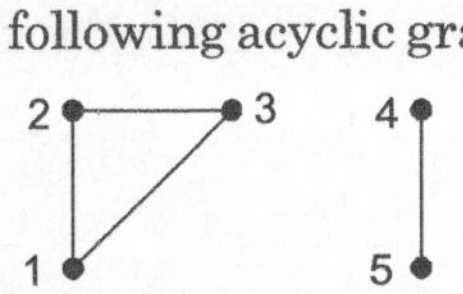

Here, $A = \begin{bmatrix} 0 & 1 & 1 & 0 & 0 \\ 1 & 0 & 1 & 0 & 0 \\ 1 & 1 & 0 & 0 & 1 \\ 0 & 0 & 0 & 1 & 0 \end{bmatrix}$

Sum of all elements in $A = 8$
Here $n = 5$, So $2(n-1) = 2(5-1) = 8$
Here above graph satisfying 'if' condition of graph, but above graph is not acyclic. So, (c) is false.

(d) Consider above graph in option (c), here, in A all rows and columns have atleast a 1, but it is disconnected. So, (d) is false.

4.62 (a, b)

Let, $v = \{v_1, v_2, ..., v_n\}$
$\Delta(G)$ = Maximum degree of G
$N = \{11, 2, 3, ...\}$ set of all possible colors

Greedy strategy:

color(v_i) ← min{$j \in N$: no neighbour of v_i is colored j}

By using the above strategy, no two adjacent vertex have same color so it is proper vertex colouring hence option (a) is true.

By using Leonard Brooke's theorem, the chromatic number of G is almost $\Delta + 1$ hence option (b) is true.

4.63 (b, d)

∵ We know that, the number of vertices in the largest independent set of a graph G.

$$\beta(G) \geq \frac{n}{k}$$

∴ Option (b) is true.

If we take a counter example.

Let,

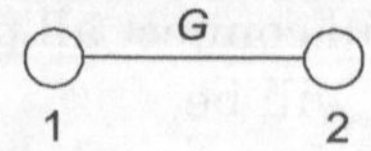

its chromatic number $k = 2$

and edges $\dfrac{k(k-1)}{2} = \dfrac{2(2-1)}{2} = 1$

∴ Graph 'G' contains atleast $\dfrac{k(k-1)}{2}$ edges whose $k = 2$.

∴ Option (d) is also true.

And no vertex has degree 2 in this graph. So, option (c) is false.

4.65 (2)

$$x(G) = 2$$

■■■■

5 Probability

5.1 A bag contains 10 white balls and 15 black balls. Two balls are drawn in succession. The probability that one of them is black and the other is white is

(a) $\frac{2}{3}$ (b) $\frac{4}{5}$

(c) $\frac{1}{2}$ (d) $\frac{1}{3}$ **[1995 : 2 M]**

5.2 Two dice are thrown simultaneously. The probability that at least one of them will have 6 facing up is

(a) $\frac{1}{36}$ (b) $\frac{1}{3}$

(c) $\frac{25}{36}$ (d) $\frac{11}{36}$ **[1996 : 1 M]**

5.3 The probability that top and bottom cards of a randomly shuffled deck are both aces is

(a) $\frac{4}{52} \times \frac{4}{52}$ (b) $\frac{4}{52} \times \frac{3}{52}$

(c) $\frac{4}{52} \times \frac{3}{51}$ (d) $\frac{4}{52} \times \frac{4}{51}$

[1996 : 2 M]

5.4 The probability that it will rain today is 0.5. The probability that it will rain tomorrow is 0.6. The probability that it will rain either today or tomorrow is 0.7. What is the probability that it will rain today and tomorrow?

(a) 0.3 (b) 0.25

(c) 0.35 (d) 0.4 **[1997 : 1 M]**

5.5 A die is rolled three times. The probability that exactly one odd number turns up among the three outcomes is

(a) $\frac{1}{6}$ (b) $\frac{3}{8}$

(c) $\frac{1}{8}$ (d) $\frac{1}{2}$ **[1998 : 1 M]**

5.6 Suppose that the expectation of a random variable X is 5. Which of the following statements is true?

(a) There is a sample point at which X has the value 5.

(b) There is a sample point at which X has value greater than 5.

(c) There is a sample point at which X has a value greater than or equal to 5.

(d) None of the above **[1999 : 1 M]**

5.7 Consider two events E_1 and E_2 such that probability of E_1, Pr $[E_1] = \frac{1}{2}$, probability of E_2, Pr $[E_2] = \frac{1}{3}$, and probability of E_1 and E_2, Pr$[E_1$ and $E_2] = \frac{1}{5}$. Which of the following statements is/are true?

(a) Pr$[E_1$ or $E_2]$ is $\frac{2}{3}$

(b) Events E_1 and E_2 are independent

(c) Events E_1 and E_2 are not independent

(d) $\Pr\left[\frac{E_1}{E_2}\right] = \frac{4}{5}$ **[1999 : 2 M]**

5.8 E1 and E2 are events in a probability space satisfying the following constraints:

- Pr(E1) = Pr(E2)
- Pr(E1 ∪ E2) = 1
- E1 and E2 are independent

The value of Pr(E1), the probability of the event E1, is

(a) 0 (b) $\frac{1}{4}$

(c) $\frac{1}{2}$ (d) 1 **[2000 : 2 M]**

5.9 Seven (distinct) car accidents occurred in a week. What is the probability that they all occurred on the same day?

(a) $\frac{1}{7^7}$ (b) $\frac{1}{7^6}$

(c) $\frac{1}{2^7}$ (d) $\frac{7}{2^7}$ **[2001 : 2 M]**

5.10 Four fair coins are tossed simultaneously. The probability that at least one head and one tail turn up is

(a) $\frac{1}{16}$ (b) $\frac{1}{8}$

(c) $\frac{7}{8}$ (d) $\frac{15}{16}$ **[2002 : 2 M]**

5.11 Let P(E) denote the probability of the event E. Given P(A) = 1, P(B) = $\frac{1}{2}$, and A and B are independent, the values of P(A|B) and P(B|A) respectively are

(a) $\frac{1}{4}, \frac{1}{2}$
(b) $\frac{1}{2}, \frac{1}{4}$
(c) $\frac{1}{2}, 1$
(d) $1, \frac{1}{2}$

[2003 : 1 M]

5.12 A program consists of two modules executed sequentially. Let $f_1(t)$ and $f_2(t)$ respectively denote the probability density functions of time taken to execute the two modules. The probability density function of the overall time taken to execute the program is given by

(a) $f_1(t)+f_2(t)$
(b) $\int_0^t f_1(x)f_2(x)dx$
(c) $\int_0^t f_1(x)f_2(t-x)dx$
(d) $\max\{f_1(t), f_2(t)\}$

[2003 : 2 M]

5.13 If a fair coin is tossed four times. What is the probability that two heads and two tails will result?

(a) $\frac{3}{8}$
(b) $\frac{1}{2}$
(c) $\frac{5}{8}$
(d) $\frac{3}{4}$

[2004 : 1 M]

5.14 In a population of N families, 50% of the families have three children, 30% of the families have two children and the remaining families have one child. What is the probability that a randomly picked child belongs to a family with two children?

(a) $\frac{3}{23}$
(b) $\frac{6}{23}$
(c) $\frac{3}{10}$
(d) $\frac{3}{5}$

[2004 : 1 M]

5.15 Let X and Y be two exponentially distributed and independent random variables with mean α and β, respectively. If Z = min (X,Y), then the mean of Z is given by

(a) $\left(\frac{1}{(\alpha+\beta)}\right)$
(b) min (α, β)
(c) $\left(\frac{\alpha\beta}{(\alpha+\beta)}\right)$
(d) α + β

[2004 : 2 M]

5.16 An examination paper has 150 multiple-choice questions of one mark each, with each question having four choices. Each incorrect answer fetches – 0.25 mark. Suppose 1000 students choose all their answers randomly with uniform probablity. The sum total of the expected marks obtained by all these students is

(a) 0
(b) 2550
(c) 7525
(d) 9375

[2004 : 2 M]

5.17 Two n bit binary strings, S_1 and S_2 are chosen randomly with uniform probablity. The probability that the Hamming distance between these strings (the number of bit positions where the two strings differ) is equal to d is

(a) $\frac{^nC_d}{2^n}$
(b) $\frac{^nC_d}{2^d}$
(c) $\frac{d}{2^n}$
(d) $\frac{1}{2^d}$

[2004 : 2 M]

5.18 A point is randomly selected with uniform probability in the X-Y. Plane within the rectangle with corners at (0, 0), (1, 0), (1, 2) and (0, 2). If p is the length of the position vector of the point, the expected value of p^2 is

(a) $\frac{2}{3}$
(b) 1
(c) $\frac{4}{3}$
(d) $\frac{5}{3}$

[2004 : 2 M]

5.19 Let f(x) be the continuous probability density function of a random variable X. The probability that a < X ≤ b, is

(a) f (b – a)
(b) f(b) – f(a)
(c) $\int_a^b f(x)dx$
(d) $\int_a^b xf(x)dx$

[2005 : 1 M]

5.20 A bag contains 10 blue marbles, 20 green marbles and 30 red marbles. A marble is drawn from the bag, its colour recorded and it is put back in the bag. This process is repeated 3 times. The probability that no two of the marbles drawn have the same colour is

(a) $\frac{1}{36}$
(b) $\frac{1}{6}$
(c) $\frac{1}{4}$
(d) $\frac{1}{3}$

[2005 : 1 M]

5.21 An unbiased coin is tossed repeatedly until the outcome of two successive tosses is the same. Assuming that the trials are independent, the expected number of tosses is
(a) 3 (b) 4
(c) 5 (d) 6 **[2005 : 2 M]**

5.22 In a certain town, the probability that it will rain in the afternoon is known to be 0.6. Moreover, meteorological data indicates that if the temperature at noon is less than or equal to 25°C, the probability that it will rain in the afternoon is 0.4. The temperature at noon is equally likely to be above 25°C, or at/below 25°C.
What is the probability that it will rain in the afternoon on a day when the temperature at noon is above 25°C?
(a) 0.4 (b) 0.6
(c) 0.8 (d) 0.9 **[2006 : 1 M]**

5.23 When a coin is tossed, the probability of getting a Head is p, $0 < p < 1$. Let N be the random variable denoting the number of tosses till the first Head appears, including the toss where the Head appears. Assuming that successive tosses are independent, the expected value of N is
(a) $\frac{1}{p}$ (b) $\frac{1}{(1-p)}$
(c) $\frac{1}{p^2}$ (d) $\frac{1}{(1-p^2)}$
[2006 : 2 M]

5.24 Suppose there are two coins. The first coin gives heads with probability $\frac{5}{8}$ when tossed, while the second coin gives heads with probability $\frac{1}{4}$. One of the two coins is picked up at random with equal probability and tossed. What is the probability of obtaining heads?
(a) $\frac{7}{8}$ (b) $\frac{1}{2}$
(c) $\frac{7}{16}$ (d) $\frac{5}{32}$ **[2007 : 1 M]**

5.25 Suppose we uniformly and randomly select a permutation from the 20! permutations of 1, 2, 3, ..., 20. What is the probability that 2 appears at an earlier position that any other even number in the selected permutation?
(a) $\frac{1}{2}$ (b) $\frac{1}{10}$
(c) $\frac{9!}{20!}$ (d) None of these
[2007 : 2 M]

5.26 In a multi-user operating system on an average, 20 requests are made to use a particular resource per hour. The arrival of requests follows a Poisson distribution. The probability that either one, three or five requests are made in 45 minutes is given by :
(a) $6.9 \times 10^6 \times e^{-20}$ (b) $1.02 \times 10^6 \times e^{-20}$
(c) $6.9 \times 10^3 \times e^{-20}$ (d) $1.02 \times 10^3 \times e^{-20}$
[2007 : 2 M]

5.27 A sample space has two events A and B such that probabilities
$P(A \cap B) = 1/2$, $P(\bar{A}) = 1/3$, $P(\bar{B}) = 1/3$.
What is $P(A \cup B)$?
(a) $\frac{11}{12}$ (b) $\frac{10}{12}$
(c) $\frac{9}{12}$ (d) $\frac{8}{12}$ **[2008 : 1 M]**

5.28 What is the probability that in a randomly chosen group of r people at least three people have the same birthday?
(a) $1-\frac{365.364...(365-r+1)}{365^r}$
(b) $1-\frac{365.364...(365-r+1)}{365^r}+{}^rC_2 \cdot 365 \cdot \frac{364.363...(364-(r-2)+1)}{364^{r-2}}$
(c) $1-\frac{365.364...(365-r+1)}{365^r}-{}^rC_2 \cdot 365 \cdot \frac{364.363...(364-(r-2)+1)}{364^{r-2}}$
(d) $\frac{365.364...(365-r+1)}{365^r}$
[2008 : 2 M]

5.29 Aishwarya studies either computer science or mathematics everyday. If she studies computer science on a day, then the probability that she studies mathematics the next day is 0.6. If she studies mathematics on a day, then the probability that she studies computer science the next day is 0.4. Given that Aishwarya studies computer science on Monday, what is the probablity that she studies computer science on Wednesday?

(a) 0.24 (b) 0.36
(c) 0.4 (d) 0.6 **[2008 : 2 M]**

5.30 Let X be a random variable following normal distribution with mean +1 and variance 4. Let Y be another normal variable with mean –1 and variance unknown. If $P(X \le -1) = P(Y \ge 2)$ the standard deviation of Y is

(a) 3 (b) 2
(c) $\sqrt{2}$ (d) 1 **[2008 : 2 M]**

5.31 An unbalanced dice (with 6 faces, numbered from 1 to 6) is thrown. The probability that the face value is odd is 90% of the probability that the face value is even. The probability of getting any even numbered face is the same.
If the probability that the face is even given that it is greater than 3 is 0.75, which one of the following options is closest to the probability that the face value exceeds 3?

(a) 0.453 (b) 0.468
(c) 0.485 (d) 0.492 **[2009 : 2 M]**

5.32 Consider a company that assembles computers. The probability of a faulty assembly of any computer is p. The company therefore subjects each computer to a testing process. This testing process gives the correct result for any computer with a probability of q. What is the probability of a computer being declared faulty?

(a) $pq + (1 - p)(1 - q)$ (b) $(1 - q)p$
(c) $(1 - p)q$ (d) pq **[2010 : 2 M]**

5.33 What is the probability that a divisor of 10^{99} is a multiple of 10^{96}?

(a) $\frac{1}{625}$ (b) $\frac{4}{625}$
(c) $\frac{12}{625}$ (d) $\frac{16}{625}$ **[2010 : 2 M]**

5.34 If two fair coins are flipped and at least one of the outcomes is known to be a head, what is the probability that both outcomes are heads?

(a) $\frac{1}{3}$ (b) $\frac{1}{4}$
(c) $\frac{1}{2}$ (d) $\frac{2}{3}$ **[2011 : 2 M]**

5.35 If the difference between the expectation of the square of a random variable $(E[x^2])$ and the square of the expectation of the random variable $(E[x])^2$ is denoted by R, then

(a) $R = 0$ (b) $R < 0$
(c) $R \ge 0$ (d) $R > 0$ **[2010 : 2 M]**

5.36 Consider the finite sequence of random values $X = [x_1, x_2, \ldots, x_n]$. Let μ_x be the mean and σ_x be the standard deviation of X. Let another finite sequence Y of equal length be derived from this as $y_i = a * x_i + b$, where a and b are positive constant. Let μ_y be the mean and σ_y be the standard deviation of this sequence. Which one of the following statements INCORRECT?

(a) Index position of mode of X in X is the same as the index position of mode of Y in Y.
(b) Index position of median of X in X is the same as the index position of median of Y in Y
(c) $\mu_y = a\mu_x + b$
(d) $\sigma_y = a\sigma_x + b$ **[2011 : 2 M]**

5.37 A deck of 5 cards (each carrying a distinct number from 1 to 5) is shuffled thoroughly. Two cards are then removed one at a time from the deck. What is the probability that the two cards are selected with the number on the first card being one higher than the number on the second card?

(a) $\frac{1}{5}$ (b) $\frac{4}{25}$
(c) $\frac{1}{4}$ (d) $\frac{2}{5}$ **[2011 : 2 M]**

5.38 Consider a random variable X that takes values +1 and –1 with probability 0.5 each. The values of the cumulative distribution function F(x) at x = – 1 and +1 are

(a) 0 and 0.5 (b) 0 and 1
(c) 0.5 and 1 (d) 0.25 and 0.75 **[2012 : 1 M]**

5.39 Suppose a fair six-sided die is rolled once. If the value on the die is 1, 2 or 3 then die is rolled a second time. What is the probability that the sum total of values that turn up is at least 6?

(a) $\frac{10}{21}$ (b) $\frac{5}{12}$
(c) $\frac{2}{3}$ (d) $\frac{1}{6}$ **[2012 : 2 M]**

5.40 Suppose p is the number of cars per minute passing through a certain road junction between 5 PM, and p has Poisson distribution with mean 3. What is the probability of observing fewer than 3 cars during any given minute in this interval?

(a) $\frac{8}{(2e^3)}$ (b) $\frac{9}{(2e^3)}$
(c) $\frac{17}{(2e^3)}$ (d) $\frac{26}{(2e^3)}$ **[2013 : 1 M]**

5.41 Suppose you break a stick of unit length at a point chosen uniformly at random. Then the expected length of the shorter stick is________. **[2014 (Set-1) : 1 M]**

5.42 Four fair six-sided dice are rolled. The probability that the sum of the results being 22 is $\frac{x}{1296}$. The value of x is ______. **[2014 (Set-1) : 2 M]**

5.43 The security system at an IT office is composed of 10 computers of which exactly four are working. To check whether the system is functional, the officials inspect four of the computers picked at random (without replacement). The system is deemed functional if at least three of the four computers inspected are working. Let the probability that the system is deemed functional be denoted by p.
Then $100p$ = ________. **[2014 (Set-2) : 1 M]**

5.44 The probability that a given positive integer lying between 1 and 100 (both inclusive) is NOT divisible by 2, 3 or 5 is ______. **[2014 (Set-2) : 2 M]**

5.45 Let S be a sample space and two mutually exclusive events A and B be such that $A \cup B = S$. If $P(\cdot)$ denotes the probability of the event, the maximum value of $P(A)\, P(B)$ is ______. **[2014 (Set-3) : 2 M]**

5.46 Suppose X_i for i = 1, 2, 3 are independent and identically distributed random variables whose probability mass functions are

$$Pr[X_i = 0] = Pr[X_i = 1] = \frac{1}{2} \text{ for } i = 1, 2, 3$$

Define another random variable $Y = X_1X_2 \oplus X_3$, where $\oplus$ denotes XOR. Then $Pr[Y = 0 \mid X_3 = 0]$ = ______. **[2015 (Set-3) : 2 M]**

5.47 A probability density function on the interval $[a, 1]$ is given by $1/x^2$ and outside this interval the value of the function is zero. The value of a is ______. **[2016 (Set-1) : 1 M]**

5.48 Consider the following experiment:
Step 1. Flip a fair coin twice.
Step 2. If the outcomes are (TAILS, HEADS) then output Y and stop.
Step 3. If the outcomes are either (HEADS, HEADS) or (HEADS, TAILS), then output N and stop.
Step 4. If the outcomes are (TAILS, TAILS), then go to Step 1.
The probability that the output of the experiment is Y is _____ (up to two decimal places). **[2016 (Set-1) : 2 M]**

5.49 Suppose that a shop has an equal number of LED bulbs of two different types. The probability of an LED bulb lasting more than 100 hours given that it is of Type 1 is 0.7, and given that it is of Type 2 is 0.4. The probability that an LED bulb chosen uniformly at random lasts more than 100 hours is______. **[2016 (Set-2) : 1 M]**

5.50 Let X be a Gaussian random variable with mean 0 and variance σ^2. Let Y = max $(X, 0)$ where max (a, b) is the maximum of a and b. The median of Y is ______. **[2017 (Set-1) : 1 M]**

5.51 P and Q are considering to apply for a job. The probability that P applies for the job is $\frac{1}{4}$, the probability that P applies for the job given that Q applies for the job is $\frac{1}{2}$, and the probability that Q applies for the job given that P applies for the job is $\frac{1}{3}$. Then the probability that P does not apply for the job given that Q does not apply for the job is

(a) $\frac{4}{5}$ (b) $\frac{5}{6}$
(c) $\frac{7}{8}$ (d) $\frac{11}{12}$

[2017 (Set-2) : 2 M]

5.52 For any discrete random variable X, with probability mass function $P(X = j) = p_j, p_j \geq 0, j \in \{0,..., N\}$, and $\sum_{j=0}^{N} p_j = 1$, define the polynomial function $g_X(z) = \sum_{j=0}^{N} p_j z^j$. For a certain discrete random variable Y, there exists a scalar $\beta \in [0, 1]$ such that $g_Y(z) = (1 - \beta + \beta z)^N$. The expectation of Y is
(a) $N\beta\,(1 - \beta)$
(b) $N\beta$
(c) $N\,(1 - \beta)$
(d) Not expressible in terms of N and β alone
[2017 (Set-2) : 2 M]

5.53 If a random variable X has a Poisson distribution with mean 5, then the expectation $E[(X + 2)^2]$ equals ________. **[2017 (Set-2) : 2 M]**

5.54 Two people, P and Q, decide to independently roll two identical dice, each with 6 faces, numbered 1 to 6. The person with the lower number wins. In case of a tie, they roll the dice repeatedly until there is no tie. Define a trial as a throw of the dice by P and Q. Assume that all 6 numbers on each dice are equi-probable and that all trials are independent. The probability (rounded to 3 decimal places) that one of them wins on the third trial is ______.

[2018 : 1 M]

5.55 Consider Guwahati (G) and Delhi (D) whose temperatures can be classified as high (H), medium (M) and low (L). Let $P(H_G)$ denote the probability that Guwahati has high temperature. Similarly, $P(M_G)$ and $P(L_G)$ denotes the probability of Guwahati having medium and low temperatures respectively. Similarly, we use $P(H_D)$, $P(M_D)$ and $P(L_D)$ for Delhi.

The following table gives the conditional probabilities for Delhi's temperature given Guwahati's temperature.

	H_D	M_D	L_D
H_G	0.40	0.48	0.12
M_G	0.10	0.65	0.25
L_G	0.01	0.50	0.49

Consider the first row in the table above. The first entry denotes that if Guwahati has high temperature (H_G) then the probability of Delhi also having a high temperature (H_D) is 0.40; i.e., $P(H_D | H_G) = 0.40$. Similarly, the next two entries are $P(M_D | H_G) = 0.48$ and $P(L_D | H_G) = 0.12$. Similarly for the other rows.

If it is known that $P(H_G) = 0.2$, $P(M_G) = 0.5$ and $P(L_G) = 0.3$, then the probability (correct to two decimal places) that Guwahati has high temperature given that Delhi has high temperature is ______.

[2018 : 2 M]

5.56 Two numbers are chosen independently and uniformly at random from the set {1, 2, ..., 13}. The probability (rounded off to 3 decimal places) that their 4-bit (unsigned) binary representations have the same most significant bit is ______.

[2019 : 1 M]

5.57 Suppose Y is distributed uniformly in the open interval (1,6). The probability that the polynomial $3x^2 + 6xY + 3Y + 6$ has only real roots is (rounded off to 1 decimal place) ______. **[2019 : 2 M]**

5.58 For $n > 2$, let $a \in \{0,1\}^n$ be a non-zero vector. Suppose that x is chosen uniformly at random from $\{0, 1\}^n$. Then, the probability that $\sum_{i=1}^{n} a_i x_i$ is an odd number is ______. **[2020 : 2 M]**

5.59 Consider the two statements:

S_1 : There exist random variables X and Y such that $(E[X - E(X)) (Y - E(Y))]^2) > Var[X] Var[Y])$

S_2 : For all random variables X and Y $Cov[X, Y] = E[|X - E[X]| |Y - E[Y]|]$

Which one of the following choices is correct?

(a) Both S_1 and S_2 are false.
(b) Both S_1 and S_2 are true.
(c) S_1 is true, but S_2 is false.
(d) S_1 is false, but S_2 is true.

[2021 (Set-1) : 2 M]

5.60 The lifetime of a component of a certain type is a random variable whose probability density function is exponentially distributed with parameter 2. For a randomly picked component of this type, the probability that its lifetime exceeds the expected lifetime (rounded to 2 decimal places) is ______.

[2021 (Set-1) : 1 M]

5.61 A sender (S) transmits a signal, which can be one of the two kinds: H and L with probabilities 0.1 and 0.9 respectively, to a receiver (R).

In the graph below, the weight of edge (u, v) is the probability of receiving v when u is transmitted, where $u, v \in \{H, L\}$. For example, the probability that the received signal is L given the transmitted signal was H, is 0.7.

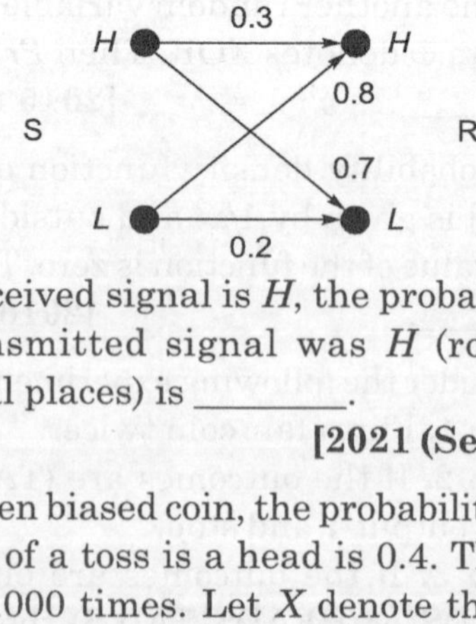

If the received signal is H, the probability that the transmitted signal was H (rounded to 2 decimal places) is ______.

[2021 (Set-1) : 2 M]

5.62 For a given biased coin, the probability that the outcome of a toss is a head is 0.4. This coin is tossed 1,000 times. Let X denote the random variable whose value is the number of times that head appeared in these 1,000 tosses. The standard deviation of X (rounded to 2 decimal places) is ______.

[2021 (Set-2) : 1 M]

5.63 In an examination, a student can choose the order in which two questions (QuesA and QuesB) must be attempted.

- If the first question is answered wrong, the student gets zero marks.
- If the first questions is answered correctly and the second question is not answered correctly, the student gets the marks only for the first question.
- If both the questions are answered correctly, the student gets the sum of the marks of the two questions.

The following table shows the probability of correctly answering a question and the marks of the question respectively.

Question	Probability of answering correctly	Marks
QuesA	0.8	10
QuesB	0.5	20

Assuming that the student always wants to maximize her expected marks in the examination, in which order should she attempt the questions and what is the expected marks for that order (assume that the questions are independent)?

(a) First QuesA and then QuesB. Expected marks 14.
(b) First QuesB and then QuesA. Expected marks 22.
(c) First QuesB and then QuesA. Expected marks 14.
(d) First QuesA and then QuesB. Expected marks 16. **[2021 (Set-2) : 2 M]**

5.64 A bag has r red balls and b black balls. All balls are identical except for their colours. In a trial, a ball is randomly drawn from the bag, its colour is noted and the ball is placed back into the bag along with another ball of the same colour. Note that the number of balls in the bag will increase by one, after the trial. A sequence of four such trials is conducted. Which one of the following choices gives the probability of drawing a red ball in the fourth trial?

(a) $\frac{r}{r+b}$ (b) $\frac{r}{r+b+3}$

(c) $\left(\frac{r}{r+b}\right)\left(\frac{r+1}{r+b+1}\right)\left(\frac{r+2}{r+b+2}\right)\left(\frac{r+3}{r+b+3}\right)$

(d) $\frac{r+3}{r+b+3}$ **[2021 (Set-2) : 2 M]**

5.65 Consider a random experiment where two fair coins are tossed. Let A be the event that denotes HEAD on both the throws, B be the event that denotes HEAD on the first throw, and C be the event that denotes HEAD on the second throw. Which of the following statements is/are TRUE?

(a) A and B are independent.
(b) A and C are independent.
(c) B and C are independent.
(d) $\text{Prob}(B \mid C) = \text{Prob}(B)$ **[2023 : 2 M]**

5.66 Let A and B be two events in a probability space with $P(A) = 0.3$, $P(B) = 0.5$, and $P(A \cap B) = 0.1$. Which of the following statements is/are TRUE?

(a) $P(A \cap B^c) = 0.2$, where B^c is the complement of the event B
(b) The two events A and B are independent
(c) $P(A \cup B) = 0.7$
(d) $P(A^c \cap B^c) = 0.4$, where A^c and B^c are the complements of the events A and B, respectively

[2024 (Set-1) : 1 M]

5.67 Consider a permutation sampled uniformly at random from the set of all permutations of {1, 2, 3, ... , n} for some $n \geq 4$. Let X be the event that 1 occurs before 2 in the permutation, and Y the event that 3 occurs before 4. Which one of the following statements is TRUE?

(a) Event X is more likely than event Y
(b) The events X and Y are independent
(c) Either event X or Y must occur
(d) The events X and Y are mutually exclusive

[2024 (Set-1) : 1 M]

5.68 A bag contains 10 red balls and 15 blue balls. Two balls are drawn randomly without replacement. Given that the first ball drawn is red, the probability (rounded off to 3 decimal places) that both balls drawn are red is ________.

[2024 (Set-1) : 2 M]

5.69 When six unbiased dice are rolled simultaneously, the probability of getting all distinct numbers (i.e. 1, 2, 3, 4, 5 and 6) is

(a) $\frac{5}{324}$ (b) $\frac{11}{324}$

(c) $\frac{1}{324}$ (d) $\frac{7}{324}$

[2024 (Set-2) : 1 M]

■■■■

Answers Probability

5.1 (c)	**5.2** (d)	**5.3** (c)	**5.4** (d)	**5.5** (b)	**5.6** (c)	**5.7** (c)	**5.8** (d)	**5.9** (b)
5.10 (c)	**5.11** (d)	**5.12** (c)	**5.13** (a)	**5.14** (b)	**5.15** (c)	**5.16** (d)	**5.17** (a)	**5.18** (d)
5.19 (c)	**5.20** (b)	**5.21** (a)	**5.22** (c)	**5.23** (a)	**5.24** (c)	**5.25** (d)	**5.26** (b)	**5.27** (b)
5.28 (c)	**5.29** (c)	**5.30** (a)	**5.31** (b)	**5.32** (a)	**5.33** (a)	**5.34** (a)	**5.35** (c)	**5.36** (d)
5.37 (a)	**5.38** (c)	**5.39** (b)	**5.40** (c)	**5.41** (0.25)	**5.42** (10)	**5.43** (11.90)	**5.44** (0.26)	**5.45** (0.25)
5.46 (0.75)	**5.47** (0.5)	**5.48** (0.33)	**5.49** (0.55)	**5.50** (0)	**5.51** (a)	**5.52** (b)	**5.53** (54)	**5.54** (0.023)
5.55 (0.62)	**5.56** (0.503)	**5.57** (0.8)	**5.58** (0.5)	**5.59** (a)	**5.60** (0.367)	**5.61** (0.04)	**5.62** (15.49)	**5.63** (d)
5.64 (a)	**5.65** (c, d)	**5.66** (a, c)	**5.67** (b)	**5.68** (0.375)	**5.69** (a)			

Explanations Probability

5.1 (c)

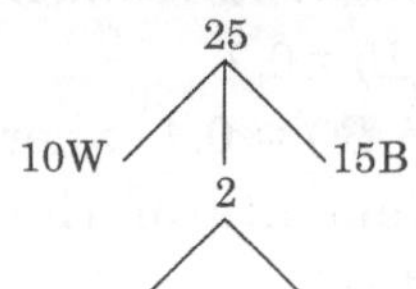

Bag contains 10 white balls and 15 black balls.
Required probability

$$= \frac{{}^{10}C_1 \times {}^{15}C_1}{{}^{25}C_2} = \frac{10 \times 15}{\frac{25 \times 24}{2}} = \frac{1}{2}$$

5.2 (d)

P(atleast one of dice will have 6 facing
$= 1 - $P(none of dice have 6 facing up)

$$= 1 - \left[\frac{5}{6} \times \frac{5}{6}\right] = 1 - \frac{25}{36} = \frac{11}{36}$$

5.3 (c)

The probability that the bottom card of a randomly shuffled deck is ace $= \frac{4}{52}$ because there are 4 aces out of total 52 cards.
From the remaining 3 aces out of 51 cards the probability that the top card is also an ace $= \frac{3}{51}$.

So required probability $= \frac{4}{52} \times \frac{3}{51}$.

5.4 (d)

P(Rain today) = 0.5
P(Rain tomorrow) = 0.6
P(Rain today $\cup$ Rain tomorrow) = 0.7
P(Rain today $\cap$ Rain tomorrow) = ?
P(rain today $\cap$ Rain tomorrow)
= P(Rain today) + P(rain tomorrow) – P(Rain today $\cap$ Rain tomorrow)
So; 0.7 = 0.5 + 0.6 – P(Rain today $\cap$ Rain tomorrow)
P(Rain today $\cap$ Rain tomorrow)
$= 0.5 + 0.6 - 0.7 = 0.4$

So, the probability that it will rain today and tomorrow is 0.4.

5.5 (b)

Probability of getting an odd number in rolling of a die $= \frac{3}{6} = \frac{1}{2}$.

Now using binomial distribution
P(Exactly one odd number among three outcomes)

$$= {}^3C_1 \left(\frac{1}{2}\right)^1 \left(\frac{1}{2}\right)^2 = 3 \times \left(\frac{1}{2}\right)^3 = \frac{3}{8}$$

5.6 (c)

If all the points have $X < 5$, then expectation of a random variable X is surely less than 5. So according to this there should be atleast a sample point at which $X \geq 5$.

5.7 (c)

$$P(E_1) = \frac{1}{2},\ P(E_2) = \frac{1}{3}$$

and $P(E_1 \cap E_2) = \frac{1}{5}$

(a) P(E_1 or E_2)

$$= P(E_1) + P(E_2) - P(E_1 \cap E_2)$$

$$= \frac{1}{2} + \frac{1}{3} - \frac{1}{5} = \frac{19}{30}$$

However given in option (a) is $\frac{2}{3}$.
So option (a) is not true.

(b) For independent events

$P(E_1 \cap E_2) = P(E_1)\,P(E_2)$

Here, $P(E_1 \cap E_2) = \frac{2}{5}$

$P(E_1)\,P(E_2) = \frac{1}{2} \times \frac{1}{3} = \frac{1}{6}$

So; $(P(E_1 \cap E_2) \neq P(E_1)\,P(E_2))$

So event E_1 and E_2 are not independent.

Option (b) is not true.

(c) Since E_1 and E_2 are not independent

So option (c) is true.

(d) $P(E_1/E_2) = \frac{P(E_1 \cap E_2)}{P(E_2)} = \frac{3}{5}$

So option (d) $P(E_1/E_2) = \frac{4}{5}$ is false.

5.8 (d)

Constraints are

(i) $P(E_1) = P(E_2) = x$

(ii) $P(E_1 \cup E_2) = 1$

(iii) E_1 and E_2 are independent so

$P(E_1 \cap E_2) = P(E_1)\,P(E_2)$

$= x \times x = x^2$

Now,

$$P(E_1 \cup E_2) = P(E_1) + P(E_2) - P(E_1 \cap E_2)$$

$$1 = x + x - x^2 = 2x - x^2$$

$$x^2 - 2x + 1 = 0$$

$$(x-1)^2 = 0$$

$$x = 1$$

So; $P(E_1) = P(E_2) = x = 1$

5.9 (b)

Sample space $= 7^7$

All accidents on the same day = 7 ways (all on Monday, all on Tuesday...)

So, required probability $= \frac{7}{7^7} = \frac{1}{7^6}$.

5.10 (c)

p(atleast one head and one tail)

= 1 – p(no head or no tail)

= 1 – [p(no head) + p(no tail) – p(no head and no tail)]

= 1– [p (all tails) + p(all heads) – 0]

$$= 1 - \left[\frac{1}{2^4} + \frac{1}{2^4} - 0\right]$$

$$= 1 - \frac{2}{16} = 1 - \frac{1}{8} = \frac{7}{8}$$

or

Alternate Method:

We need atleast one head (≥ 1 H) and atleast one tail (≥ 1 T). First we satisfy ≥ 1H as follows.

1 H, 3 T

2 H, 2 T

3 H, 1 T

and 4 H, 0 T

now to satisfy the second condition of ≥ 1 T, we have to remove 4 H, 0 T.

So, the favourable cases are only 1 H, 2 H and 3 H.

The probability of this by binomial distribution formula is

$$= \frac{{}^4C_1}{2^4} + \frac{{}^4C_2}{2^4} + \frac{{}^4C_3}{2^4} = \frac{7}{8}$$

5.11 (d)

Given, $P(A) = 1$

$P(B) = \frac{1}{2}$

Since both events are independent

$P(A\,|\,B) = P(A) = 1$

$P(B\,|\,A) = P(B) = \frac{1}{2}$

5.12 (c)

Let the time taken for first and second modules be represented by x and y and total time = t.

∴ t = x + y is a random variable.

Now the joint density function

$$g(t) = \int_0^t f(x, y)\,dx$$

$$= \int_0^t f(x, t-x)\,dx$$

$$= \int_0^t f_1(x)\, f_2(t-x)\,dx$$

which is also called as convolution of f_1 and f_2, abbreviated as $f_1 * f_2$.

Correct answer is therefore, choice (c).

5.13 (a)

Given, $P(H) = \frac{1}{2}$, $P(T) = \frac{1}{2}$

Apply Bernoulli's formula for binomial distribution,

$$P(X=2) = {}^4C_2\left(\frac{1}{2}\right)^2\left(1-\frac{1}{2}\right)^{4-2}$$

$$= {}^4C_2\left(\frac{1}{2}\right)^2\left(\frac{1}{2}\right)^2 = \frac{{}^4C_2}{2^4} = \frac{6}{16} = \frac{3}{8}$$

5.14 (b)

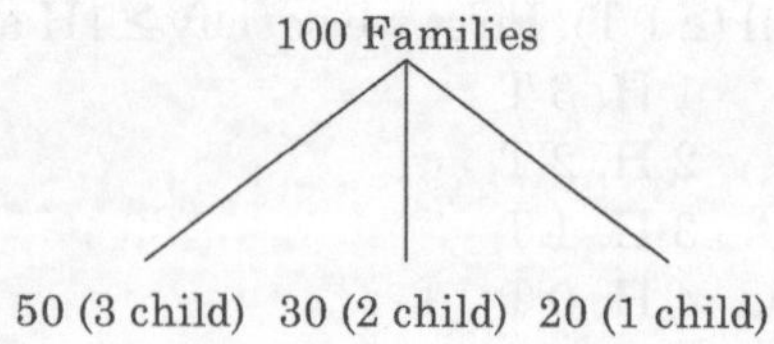

Total number of children

$= 50 \times 3 + 30 \times 2 + 20 \times 1 = 230$

Favorable cases = The number of children who come from 2 children families = $30 \times 2 = 60$

So the probability that a randomly picked child belongs to a 2 children families $= \frac{60}{230} = \frac{6}{23}$

5.15 (c)

$$f(x) = \lambda . e^{-\lambda x}, \; x \geq 0$$

X and Y are two independent exponentially distributed random variables. Let λ_1 and λ_2 parameters of X and Y respectively.

$P(X \geq x) = e^{-\lambda_1 x}, \quad x > 0$

$P(Y \geq x) = e^{-\lambda_2 x}, \quad x > 0$

Given $Z = \min(X, Y)$

$P(Z \geq x) = P(X \geq x, \; Y \geq x)$

$= P(X \geq x) \; P(Y \geq x)$

$= e^{-\lambda_1 x} \cdot e^{-\lambda_2 x} = e^{-(\lambda_1 + \lambda_2)x}$

Since mean of exponential distribution = 1/Parameter

So, $\alpha = \frac{1}{\lambda_1} \Rightarrow \lambda_1 = \frac{1}{\alpha}$

$\beta = \frac{1}{\lambda_2} \Rightarrow \lambda_2 = \frac{1}{\beta}$

$\therefore$ Z is random variable with parameter

$$\text{Mean of } Z = \frac{1}{\lambda_1 + \lambda_2} = \frac{1}{\frac{1}{\alpha} + \frac{1}{\beta}} = \frac{\alpha\beta}{\alpha + \beta}$$

5.16 (d)

Let the marks obtained per question be a random variable X.

Its probability distribution table is given below:

X	1	–0.25
p(X)	1/4	3/4

Expected marks per question

$= E(x) = \Sigma X \, P(X)$

$= 1 \times \frac{1}{4} + (-0.25) \times \frac{3}{4}$

$= \frac{1}{4} - \frac{3}{16} = \frac{1}{16}$ marks

Total marks expected for 150 questions

$= \frac{1}{16} \times 150 = \frac{75}{8}$ marks per student

Total expected marks of 1000 students

$= \frac{75}{8} \times 1000 = 9375$ marks

So correct answer is (d).

5.17 (a)

If hamming distance between two n bit strings is d, we are asking that d out of n trials to be success (success here means that the bits are different). So this is a binomial distribution with n trials and d successes and probability of success

$$p = \frac{2}{4} = \frac{1}{2}$$

(Since out of the 4 possibilities {(0, 0), (0, 1), (1, 0), (1, 1)} only two of them (0, 1) and (1, 0) are success)

So $\quad p(X = d) = {}^nC_d \left(\frac{1}{2}\right)^d \left(\frac{1}{2}\right)^{n-d} = \frac{{}^nC_d}{2^n}$

Correct choice is therefore (a).

5.18 (d)

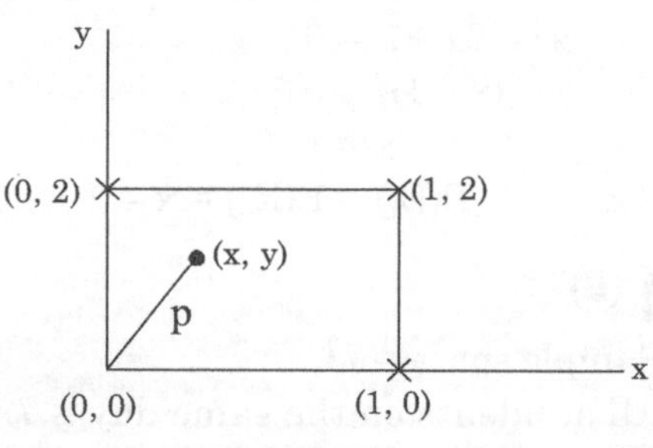

Length of position vector of point $= p = \sqrt{x^2 + y^2}$

$p^2 = x^2 + y^2$

$E(p^2) = E(x^2 + y^2) = E(x^2) + E(y^2)$

Now x and y are uniformly distributes $0 \leq x \leq 1$ and $0 \leq y \leq 2$

Probability density function of

$$x = \frac{1}{1-0} = 1$$

Probability density function of

$$y = \frac{1}{2-0} = \frac{1}{2}$$

$$E(x^2) = \int_0^1 x^2 \, p(x) \, dx = \int_0^1 x^2 \cdot 1 \cdot dx$$

$$= \left[\frac{x^3}{3}\right]_0^1 = \frac{1}{3}$$

$$E(y^2) = \int_0^2 y^2\, p(y)\, dy = \int_0^2 y^2 \cdot 1/2 \cdot dy$$

$$= \left[\frac{y^3}{6}\right]_0^2 = \frac{8}{6} = \frac{4}{3}$$

$$\therefore \quad E(p^2) = E(x^2) + E(y^2) = \frac{1}{3} + \frac{4}{3} = \frac{5}{3}$$

5.19 (c)

If f(x) is the continuous probability density function of a random variable X then,

$$p(a < x \le b) = p(a \le x \le b)$$

$$= \int_a^b f(x)dx$$

5.20 (b)

The given condition corresponds to sampling with replacement and with order.
No 2 marbles have the same color i.e. Drawn 3 different marble.
So total number of ways for picking 3 different marbles = 3! = 6.
Probability of getting blue, green, red in order

$$= \frac{10}{60} \times \frac{20}{60} \times \frac{30}{60} \times 6$$

[Since 6 ways to get the marbles]

$$= \frac{1}{6}$$

5.21 (a)

$$E(X) = S(X_i \times P_i)$$

Where X_i = number of tosses until we get successive HEAD or TAIL.
P_i = Probability that we get in X_i tosses.
Total possibilities by tossing a coin two times i.e. HT, HH, TH, TT. Out of which favourable cases are HH, TT.

$$P(X = 2) = \frac{2}{4}$$

Similarly for $X = 3$ only THH and HTT are favourable out of total 8 outcomes.

$$\text{So,} \quad P(X = 3) = \frac{2}{8}$$

Similarly we can see that in every case we will have only 2 favourable cases and 2^n sample space.

So for n^{th} toss probability = $P(X = n) = \frac{2}{2^n}$

So the probability distribution table for X (the number of Tosses) is given below:

X	2	3	4	5	...
$P(X)$	2/4	2/8	2/16	2/32	

$$E(X) = 2 \times \frac{2}{4} + 3 \times \frac{2}{8} + 4 \times \frac{2}{16} + \ldots$$

It is combination of AP and GP form. So multiplying E(X) by $\frac{1}{2}$ and subtracting from E(X)

$$E(X) = 2 \times \frac{1}{2} + 3 \times \frac{1}{4} + 4 \times \frac{1}{8} + \ldots$$

$$\frac{1}{2} \times E(X) = 2 \times \frac{1}{4} + 3 \times \frac{1}{8} + 4 \times \frac{1}{16} + \ldots$$

After solving both equation we get,

$$\frac{1}{2} \times E(X) = 1 + \frac{1}{4} + \frac{1}{8} + \frac{1}{16} + \ldots$$

$$\frac{1}{2} \times E(X) = 1 + \frac{\frac{1}{4}}{\left(1 - \frac{1}{2}\right)}$$

$$\Rightarrow \quad E(X) = \frac{6}{4} \times 2 = 3$$

5.22 (c)

Let RA : Rain in the afternoon
T > 25 : Temperature more than 25°C
Let the desired probability = $P(RA \mid T \le 25) = x$
The tree diagram for this problem is given below:

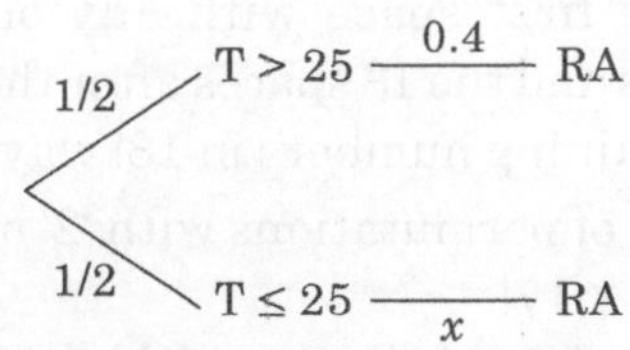

Given P(RA) = 0.6
by rule of total probability P(RA)

$$= \frac{1}{2} \times 0.4 + \frac{1}{2} \times x = 0.6$$

$$\Rightarrow \quad x = 0.8$$

5.23 (a)

The number of attempts to first success follows geometric distribution. It is well known that the expected value in geometric distribution

$$E(X) = \frac{1}{p}$$

where p is the probability of success in any one attempt.

Alternate Method:

Let X be the number of attempts to first success.
Let p be the probability of success in any one attempt.
Now the probability distribution table of X is given below:

X	1	2	3	4	
$P(X)$	p	$(1-p)p$	$(1-p)^2 p$	$(1-p)^3 p$	...

$$E(X) = \sum xp(x)$$
$$= 1\times p + 2\times(p-1)p + 3\times(p-1)^2 p + \ldots$$

This is arithmetico-geometric series which can be solved as $E(X) = \dfrac{1}{p}$

5.24 (c)

The tree diagram for the problem is given below:

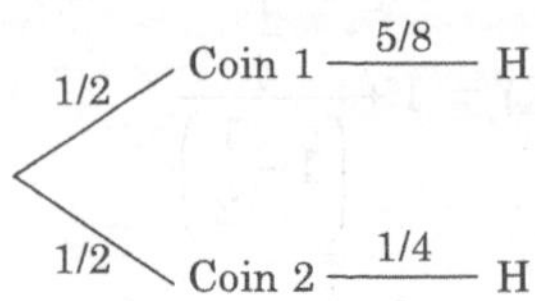

$$P(H) = \frac{1}{2}\times\frac{5}{8}+\frac{1}{2}\times\frac{1}{4} = \frac{7}{16}$$

5.25 (d)

Number of permutations with '2' in the first position = 19!
Number of permutations with '2' in the second position = 10 × 18!
(Fill the first space with any of the 10 odd numbers and the 18 spaces after the 2 with 18 of the remaining numbers in 18! ways)
Number of permutations with '2' in 3rd position = 10 × 9 × 17!
(Fill the first 2 places with 2 of the 10 odd numbers and then the remaining 17 places with remaining 17 numbers) and so on until '2' is in 11th place. After that it is not possible to satisfy the given condition, since there are only 10 odd numbers available to fill before the '2'.
So the desired number of permutations which satisfies the given condition is
19! + 10 × 18! + 10 × 9 × 17! + 10 × 9 × 8 × 16! + ... + 10! × 9!
Now the probability of this happening is given by

$$\frac{19!+10\times18!+10\times9\times17!\ldots+10!\times9!}{20!}$$

Which is clearly not choices (a), (b) or (c)
∴ Answer is (d) none of these.

5.26 (b)

The requests follow Poisson distribution with α = 20 requests/hr
The observation period = ΔT = 45 minutes

$$= \frac{45}{60} = \frac{3}{4}\text{ hr}$$

So the parameter for the poisson distribution = λ = αΔT

$$= 20\times\frac{3}{4} = 15$$

The required probability
P(X = 1) + P(X = 3) + P(X = 5)

$$= \frac{e^{-15}15^1}{1!}+\frac{e^{-15}15^3}{3!}+\frac{e^{-15}15^5}{5!}$$
$$= 1.02\times10^6\times e^{-20}$$

5.27 (b)

We know that:

$$P(A\cup B) = P(A)+P(B)-P(A\cap B)$$
$$\overline{P(A)} = 1-P(A) \text{ and } \overline{P(B)} = 1-P(B)$$

So,
$$P(A) = (1-\overline{P(A)}) \text{ and } P(B) = (1-\overline{P(B)})$$
$$P(A\cup B) = (1-\overline{P(A)})+(1-\overline{P(B)})-P(A\cap B)$$
$$= \left(1-\frac{1}{3}\right)+\left(1-\frac{1}{3}\right)-\left(\frac{1}{2}\right)$$
$$= \left(\frac{2}{3}\right)+\left(\frac{2}{3}\right)-\left(\frac{1}{2}\right)$$
$$= \left(\frac{4}{3}\right)-\left(\frac{1}{2}\right)=\left(\frac{5}{6}\right)=\left(\frac{10}{12}\right)$$

5.28 (c)

P(at least three people have the same birthday) = 1 – P (all have different b'days) – P(exactly two people have same b'day)
Now, P(all have different b'days)

$$= \frac{365.364\ldots(365-r+1)}{365^r}$$

P(exactly two people have same b'day)

$$= {}^rC_2\cdot365\cdot\frac{364.363\ldots(364-(r-2)+1)}{364^{r-2}}$$

[${}^rC_2$ ways to choose who those two people with same b'day are, 365 ways to choose what the b'day is]
Now, P (at least three people have the same birthday)

$$= 1 - \frac{365.364...(365 - r + 1)}{365^r}$$

$$- {}^rC_2 \cdot 365 \cdot \frac{364.363...(364 - (r-2) + 1)}{364^{r-2}}$$

Which is option (c).

5.29 (c)

Let C denote computes science study and M denotes maths study. The tree diagram for the problem can be represented as shown below:

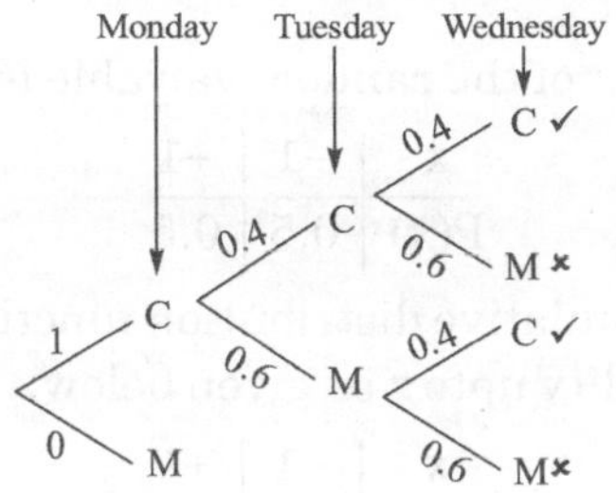

Now by rule of total probability we total up the desired branches (✓) and get the answer as shown below:

p(C on monday and C on wednesday) = p(C on monday, C on tuesday and C on wednesday) + p(C on monday, M on tuesday and C on wednesday)

$$= 1 \times 0.4 \times 0.4 + 1 \times 0.6 \times 0.4$$

$$= 0.16 + 0.24 = 0.40$$

5.30 (a)

Given $\mu_X = 1, \sigma_x{}^2 = 4 \Rightarrow \sigma_x = 2$ and $\mu_Y = -1$. σ_Y is unknown

Given, $p(X \le -1) = p(Y \ge 2)$

Converting into standard normal variates,

$$p\left(z \le \frac{-1 - \mu_x}{\sigma_x}\right) = p\left(z \ge \frac{2 - \mu_y}{\sigma_y}\right)$$

$$p\left(z \le \frac{-1-1}{2}\right) = p\left(z \ge \frac{2-(-1)}{\sigma_y}\right)$$

$$p(z \le -1) = p\left(z \ge \frac{3}{\sigma_y}\right) \quad ... \text{(i)}$$

Now since us know that in standard normal distribution,

$$p(z \le -1) = p(z \ge 1) \quad ... \text{(ii)}$$

Comparing (i) and (ii) we can say that

$$\frac{3}{\sigma_y} - 1 \Rightarrow \sigma_y = 3$$

5.31 (b)

It is given that,

$$p(\text{odd}) = 0.9\,p(\text{even})$$

Now since, $\Sigma p(x) = 1$

$\therefore$ p(odd) + p (even) = 1

$\Rightarrow$ 0.9 p (even) + p (even) = 1

$$\Rightarrow \quad p(\text{even}) = \frac{1}{1.9} = 0.5263$$

Now, it is given that p (any even face) is same

i.e. p(2) = p(4) = p(6)

Now since,

$$p(\text{even}) = p(2) \text{ or } p(4) \text{ or } p(6)$$

$$= p(2) + p(4) + p(6)$$

$$\therefore \quad p(2) = p(4) = p(6)$$

$$= \frac{1}{3}\,p(\text{even}) = \frac{1}{3}(0.5263)$$

$$= 0.1754$$

It is given that

$$p(\text{even} \mid \text{face} > 3) = 0.75$$

$$\Rightarrow \frac{p(\text{even} \cap \text{face} > 3)}{p(\text{face} > 3)} = 0.75$$

$$\Rightarrow \frac{p(\text{face} = 4, 6)}{p(\text{face} > 3)} = 0.75$$

$$\Rightarrow \quad p(\text{face} > 3) = \frac{p(\text{face} = 4, 6)}{0.75} = \frac{p(4) + p(6)}{0.75}$$

$$= \frac{0.1754 + 0.1754}{0.75}$$

$$= 0.4677 \simeq 0.468$$

5.32 (a)

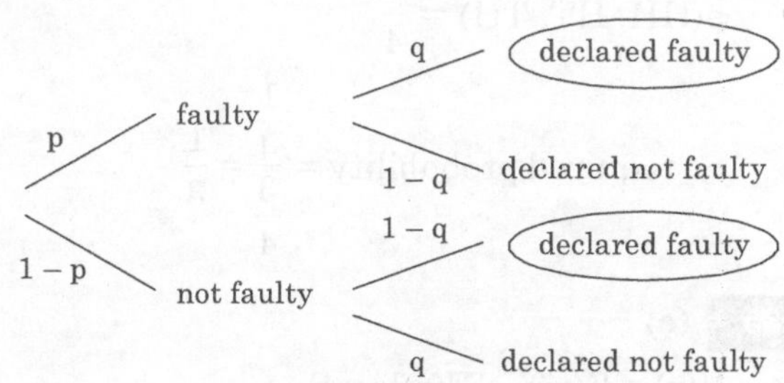

The tree diagram of probabilities is shown above.

From above tree, by rule of total probability,

p(declared faulty) = pq + (1 – p) (1 – q)

5.33 (a)

$$p(\text{multiple of } 10^{96} \mid \text{divisor of } 10^{99})$$

$$= \frac{p(\text{multiple of } 10^{96} \text{ \& divisor of } 10^{99})}{p(\text{divisor of } 10^{99})}$$

$$= \frac{n(\text{multiple of } 10^{96} \text{ \& divisor of } 10^{99})}{n(\text{divisor of } 10^{99})}$$

Since, $10 = 2 . 5$

$$10^{99} = 2^{99} . 5^{99}$$

Any divisor of 10^{99} is of the form $2^a \cdot 5^b$ where $0 \le a \le 99$ and $0 \le b \le 99$.
The number of such divisors is given by (99+1) × (99 + 1) = 100 × 100.
So, no. of divisors of 10^{99} = 100 × 100.
Any number which is a multiple of 10^{96} as well as divisor of 10^{99} is of the form $2^a \cdot 5^b$ where $96 \le a \le 99$ and $96 \le b \le 99$. The number of such combinations of 4 values of a and 4 values of b is 4 × 4 combinations, each of which will be a multiple of 10^{96} as well as a divisor of 10^{99}.
∴ p(multiple of 10^{96} | divisor of 10^{99})

$$= \frac{4 \times 4}{100 \times 100} = \frac{1}{625}$$

5.34 (a)

Let A be the event of head in one coin.
B be the event of head in second coin.
The required probability is

$$p(A \cap B \mid A \cup B) = \frac{p\{(A \cap B) \cap (A \cup B)\}}{p(A \cup B)}$$

$$= \frac{p(A \cap B)}{p(A \cup B)}$$

$p(A \cap B)$ = p(both coin heads)

$$= p(H, H) = \frac{1}{4}$$

$p(A \cup B)$ = p(at least one head)

$$p(HH, HT, TH) = \frac{3}{4}$$

$$\text{So required probability} = \frac{\frac{1}{4}}{\frac{3}{4}} = \frac{1}{3}$$

5.35 (c)

$V(x) = E(x^2) - [E(x)]^2 = R$
where V(x) is the variance of x,
Since variance is σ_x^2 and hence never negative, $R \ge 0$.

5.36 (d)

Standard deviation is affected by scale but not by shift of origin.
So, $y_i = ax_i + b$
⇒ $\sigma_y = a\sigma_x$
If a could be negative then $\sigma_y = |a| \sigma_x$ is more correct since standard deviation cannot be negative)
Clearly, $\sigma_y = a\sigma_x + b$ is false
So (d) is incorrect.

5.37 (a)

The five cards are {1, 2, 3, 4, 5}
Sample space = 5 × 4 ordered pairs.
[Since there is a I[st] card and IInd card we have to take ordered pairs]

$$p(\text{I}^{st}\text{ card} = \text{II}^{nd}\text{ card} + 1) = P\{(2, 1), (3, 2), (4, 3), (5, 4)\}$$

$$= \frac{4}{5 \times 4} = \frac{1}{5}$$

5.38 (c)

The p.d.t of the random variable is

x	−1	+1
P(x)	0.5	0.5

The cumulative distribution function F(x) is the probability upto x as given below:

x	−1	+1
F(x)	0.5	1.0

So correct option is (c).

5.39 (b)

If first throw is 1, 2 or 3 then sample space is only 18 possible ordered pairs. Out of this only (1, 5), (1, 6), (2, 4), (2, 5), (2, 6), (3, 3), (3, 4), (3, 5) and (3, 6) i.e. 9 out of 18 ordered pairs gives a Sum ≥ 6.
If first throw is 4, 5 or 6 then second throw is not made and therefore the only way Sum ≥ 6 is if the throw was 6. Which is one out of 3 possible.
So the tree diagram becomes as follows:

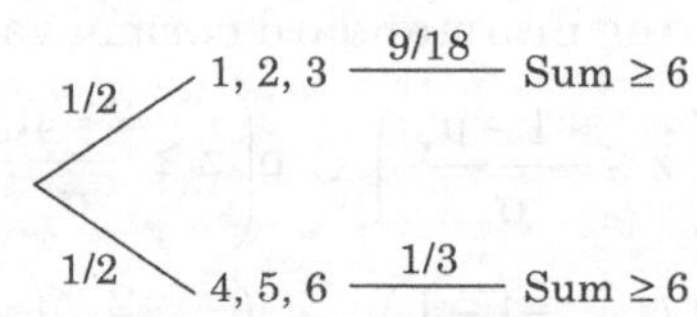

From above diagram,

$$p(\text{sum} \ge 6) = \frac{1}{2} \times \frac{9}{18} + \frac{1}{2} \times \frac{1}{3} = \frac{15}{36} = \frac{5}{12}$$

5.40 (c)

Poisson formula for (P = x) given as $\frac{e^{-\lambda} \lambda^x}{x!}$

α = 3 cars/minute, ΔT = 1 minute
So, $\lambda = \alpha \Delta T = 3 \times 1 = 3$
Probability of observing fewer than 3 cars

$$= P(X < 3) = P(X \le 2)$$

$$= P(X = 0) + P(X = 1) + P(X = 2)$$

$$= \frac{e^{-3} 3^0}{0!} + \frac{e^{-3} 3^1}{1!} + \frac{e^{-3} 3^2}{2!} = \frac{17}{2e^3}$$

Option (c) is correct.

5.41 (0.25)

Suppose you break a stick of unit length at a point chosen uniformly at random, then let the length of the shorter stick = x

x has uniform distribution in the interval $\left[0, \frac{1}{2}\right]$

i.e. $\alpha = 0$ and $\beta = \frac{1}{2}$

In uniform distribution, $E(x) = \frac{(\beta+\alpha)}{2}$

So the expected length of the shorter stick = E(x)

$$= \frac{\left(\frac{1}{2}+0\right)}{2} = \frac{1}{4} = 0.25$$

5.42 (10)

Sample space = 6^4 = 1296

6, 6, 6, 4 $\Rightarrow$ 4 ways $\left(\frac{4!}{3!} = 4\right)$

6, 6, 5, 5 $\Rightarrow$ 6 ways $\left(\frac{4!}{2!2!} = 6\right)$

Probability of sum to be 22 = $\frac{6+4}{1296} = \frac{x}{1296}$

$\Rightarrow \quad x = 10$

5.43 (11.90)

The tree diagram for the problem is shown below:

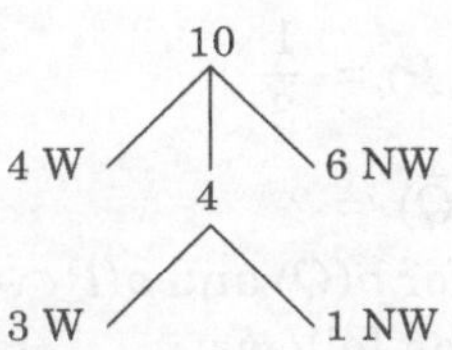

Required probability

$$= \frac{{}^4C_3 \cdot {}^6C_1}{{}^{10}C_4} + \frac{{}^4C_4 \cdot {}^6C_0}{{}^{10}C_4}$$

$$= \frac{24}{210} + \frac{1}{210} = \frac{25}{210}$$

$p = 0.1190$

$\Rightarrow \quad 100p = 11.90$

5.44 (0.26)

$1 \le x \le 100$

P(x is not divisible by 2, 3 or 5) = 1 – P(x is divisible by 2, 3 or 5)

$$= 1 - \left[\left\lfloor\frac{100}{2}\right\rfloor + \left\lfloor\frac{100}{3}\right\rfloor + \left\lfloor\frac{100}{5}\right\rfloor - \left\lfloor\frac{100}{6}\right\rfloor - \left\lfloor\frac{100}{15}\right\rfloor - \left\lfloor\frac{100}{10}\right\rfloor + \left\lfloor\frac{100}{30}\right\rfloor\right]$$

$$1 - \frac{74}{100} = 0.26$$

5.45 (0.25)

It is given that A and B are mutually exclusive also it is given that $A \cup B = S$ which means that A and B are collectively exhaustive.

Now if two events A and B are both mutually exclusive and collectively exhaustive, then P(A) + P(B) = 1 $\Rightarrow$ P(B) = 1 – P(A)

Now we wish to maximize P(A) P(B)

= P(A) (1 – P(A))

Let, P(A) = x

Now P(A) (1 – P(A)) = $x(1 - x) = x - x^2$

Say $y = x - x^2$

$$\frac{dy}{dx} = 1 - 2x = 0 \Rightarrow x = \frac{1}{2}$$

$$= \frac{d^2y}{dx^2} = -2 < 0; \left(\frac{d^2y}{dx^2}\right)_{x=\frac{1}{2}} = -2 < 0$$

y has maximum at $x = \frac{1}{2}$

$$y_{max} = \frac{1}{2} - \left(\frac{1}{2}\right)^2 = 0.25$$

5.46 (0.75)

The p.d.t. for X_1, X_2 and X_3 as given in the problem is shown below:

X_1	0	1
$p(X_1)$	$\frac{1}{2}$	$\frac{1}{2}$

X_2	0	1
$p(X_2)$	$\frac{1}{2}$	$\frac{1}{2}$

X_3	0	1
$p(X_3)$	$\frac{1}{2}$	$\frac{1}{2}$

Given, $Y = X_1X_2 \oplus X_3$.

The required probability = $p(Y = 0 \mid X_3 = 0)$

$$= \frac{p(Y = 0 \cap X_3 = 0)}{p(X_3 = 0)} \quad ...(1)$$

Now, $p(X_3 = 0) = \frac{1}{2}$ (from p.d.t. of X_3) ...(2)

$p(Y = 0 \cap X_3 = 0)$

$= p(X_1X_2 \oplus X_3 = 0 \cap X_3 = 0)$

$= p((X_1X_2 \oplus 0) = 0 \cap X_3 = 0)$

$= p(X_1X_2 = 0 \cap X_3 = 0)$

$= p(X_1 = 0, X_2 = 0, X_3 = 0) + p(X_1 = 0, X_2 = 1, X_3 = 0) + p(X_1 = 1, X_2 = 0, X_3 = 0)$

$$= \frac{1}{2}\times\frac{1}{2}\times\frac{1}{2}+\frac{1}{2}\times\frac{1}{2}\times\frac{1}{2}+\frac{1}{2}\times\frac{1}{2}\times\frac{1}{2}=\frac{3}{8}$$

So, $p(Y = 0 \cap X_3 = 0) = \frac{3}{8}$ ...(3)

Now substituting (2) and (3) in (1), we get

The required probability = $p(Y = 0 | X_3 = 0)$

$$= \frac{\frac{3}{8}}{\frac{1}{2}} = \frac{3}{4} = 0.75$$

5.47 (0.8)

Given, $f(x) = \frac{1}{x^2}$ $a \le x \le 1$

$= 0$ elsewhere

So, $\int_a^1 f(x) = 1$

$\Rightarrow \int_a^1 \frac{1}{x^2} = 1$

$\Rightarrow \left[\frac{-1}{x}\right]_a^1 = 1,\quad -\left[\frac{1}{1}-\frac{1}{a}\right] = 1$

$\Rightarrow \frac{1}{a} = 2$

$\Rightarrow a = \frac{1}{2} = 0.5$

5.48 (0.33)

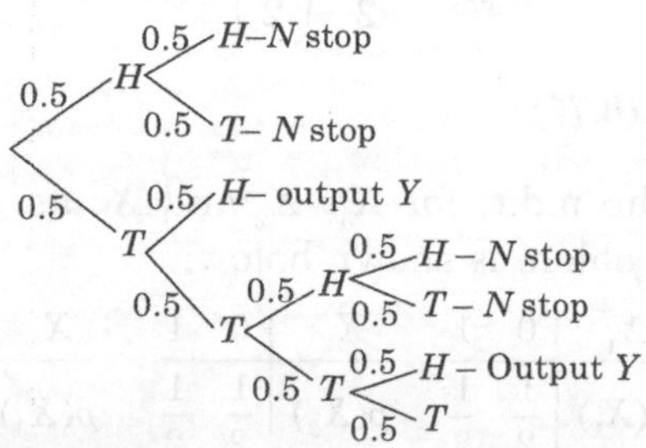

The tree diagram for the problem is given above.

The desired output is Y.

Now by rule of total probability

$p(\text{output} = Y)$

$= 0.5 \times 0.5 + 0.5 \times 0.5 \times 0.5 \times 0.5 + ...$

Infinite geometric series with

$a = 0.5 \times 0.5$ and $r = 0.5 \times 0.5$

So $p(\text{output} = Y)$

$$= \frac{0.5\times0.5}{1-0.5\times0.5} = \frac{0.25}{0.75}$$

$\frac{1}{3} = 0.33$ (upto 2 decimal places)

5.49 (0.55)

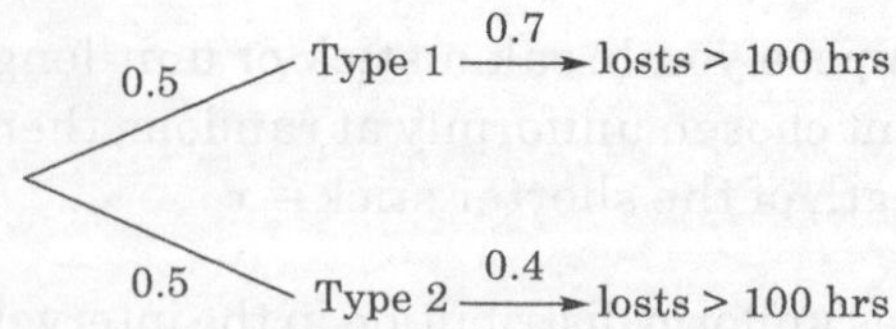

P(losts > 100 hr) = 0.5 × 0.7 + 0.5 × 0.4

= 0.35 + 0.2 = 0.55

5.50 (0)

Gaussian random variable is same as normal random variable.

So, the distribution of X is N(0, σ^2)

For X, Median = Mean = Mode = 0

For Y = Max (X, 0)

Now half the data are below 0 and half the data above 0 for X.

If we apply Y = Max(X, 0) now, all the negative values will becomes 0 and all the positive values will remains positive.

So, Median being positional average will not be affected and will remain at 0, since now half will be at 0 and half will be positive.

So, Median (Y) = 0.

5.51 (a)

Given that,

$p(P) = \frac{1}{4}$ ...(1)

$p(P|Q) = \frac{1}{2}$ ...(2)

$p(Q \mid P) = \frac{1}{3}$ ...(3)

$p(\bar{P} | \bar{Q}) = ?$

First solve for $p(Q)$ and $p(P \cap Q)$ from equation (2) and (3) as follows:

From equation (2)

$p(P \mid Q) = \frac{p(P \cap Q)}{p(Q)} = \frac{1}{2}$ ...(4)

From equation (3)

$$p(Q \mid P) = \frac{p(P \cap Q)}{p(P)} = \frac{1}{3}$$

$$\Rightarrow p(P \cap Q) = \frac{1}{3}\times p(P) = \frac{1}{3}\times\frac{1}{4} = \frac{1}{12}$$

Now substitute in equation (4) and get

$$p(Q) = \frac{p(P \cap Q)}{\frac{1}{2}} = \frac{\frac{1}{12}}{\frac{1}{2}} = \frac{2}{12} = \frac{1}{6}$$

So now we have

$$p(P) = \frac{1}{4},\quad p(Q) = \frac{1}{6}$$

and $p(P \cap Q) = \dfrac{1}{12}$

we need to find

$$p(\bar{P} \mid \bar{Q}) = \frac{p(\bar{P} \cap \bar{Q})}{p(\bar{Q})}$$

$$= \frac{1-(P \cup Q)}{1-p(Q)} = 1 - \frac{[p(P)+p(Q)-p(P \cap Q)]}{1-p(Q)}$$

$$= \frac{1-\left[\frac{1}{4}+\frac{1}{6}-\frac{1}{12}\right]}{1-\frac{1}{6}} = \frac{\frac{2}{3}}{\frac{5}{6}} = \frac{4}{5}$$

So, $p(\bar{P} \mid \bar{Q}) = \dfrac{4}{5}$

5.52 (b)

$$g_Y(z) = ((1-\beta)+\beta z)^N$$

If $g_Y(z)$ is expanded, we would get a bionomial distribution with $n = N$ and $p = \beta$

So the $E[Y] = np = N\beta$

5.53 (54)

Given, Poisson distribution $\lambda = 5$

We know that in Poisson distribution

$$E(X) = V(X) = \lambda$$

So here $E(X) = V(X) = 5$

Now, we need $E[(X + 2)^2]$

$$= E(X^2 + 4X + 4)$$

$$= E(X^2) + 4E(X) + 4$$

To find $E(X^2)$ we write, $V(X) = E(X^2) - (E(X))^2$

$$5 = E(X^2) - 5^2$$

So, $E(X^2) = 5^2 + 5 = 30$

Required value $= 30 + 4 \times 5 + 4 = 54$

5.54 (0.023)

P(one of them wins in 3rd trial)

= P(Ist trial is Tie) × P(IInd trial is Tie) × P(one of them wins 3rd trial)

P(Tie in any trial) = P(P = 1 and Q = 1) + P(P = 2 and Q = 2) + ... + P(P = 6 and Q = 6)

$$= \frac{1}{36}+\frac{1}{36}+\frac{1}{36}+\frac{1}{36}+\frac{1}{36}+\frac{1}{36} = \frac{6}{36} = \frac{1}{6}$$

$$\text{P(one of them wins)} = 1 - \text{P(Tie)} = 1 - \frac{1}{6} = \frac{5}{6}$$

$$\text{So required probability} = \frac{1}{6} \times \frac{1}{6} \times \frac{5}{6} = \frac{5}{216} = 0.023$$

(rounded to 3 decimal places)

5.55 (0.62)

The condition probability table given is

	H_D	M_D	L_D
H_G	0.40	0.48	0.12
M_G	0.10	0.65	0.25
L_G	0.01	0.50	0.49

$P(H_G) = 0.2,\ P(M_G) = 0.5,\ P(L_G) = 0.3$

Drawing the tree diagram for HD we get,

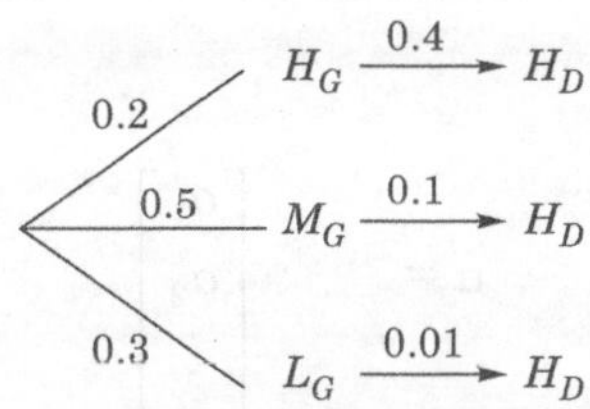

$$P(H_G \mid H_D) = \frac{P(H_G \cap H_D)}{P(H_D)}$$

From diagram, $P(H_G \cap H_D) = 0.2 \times 0.4$

$P(H_D) = 0.2 \times 0.4 + 0.5 \times 0.1 + 0.3 \times 0.01 = 0.133$

Required probability,

$$P(H_G \mid H_D) = \frac{0.2 \times 0.4}{0.133} = 0.60$$ (rounding upto 2 decimal place)

5.56 (0.503)

	MSB			LSB	Decimal
[1 - 7] having same MSB bit '0' Number of elements $MSB_{'0'} = 7$	0	0	0	1	1
	0	0	1	0	2
	0	0	1	1	3
	0	1	0	0	4
	0	1	0	1	5
	0	1	1	0	6
	0	1	1	1	7
[8 - 13] having same MSB bit '1' Number of elements $MSB_{'1'} = 6$	1	0	0	0	8
	1	0	0	1	9
	1	0	1	0	10
	1	0	1	1	11
	1	1	0	0	12
	1	1	0	1	13

P(2 elements out of 13 elements chosen randomly and independently) = (Number of $MSB_{'0'}$ × Number of $MSB_{'0'}$ + Number of $MSB_{'1'}$ × Number of $MSB_{'1'}$) / (Total × Total)

$$= \frac{7\times7+6\times6}{13\times13} = \frac{49+36}{169} = \frac{85}{169} = 0.503$$

5.57 (0.8)

Polynomial $3x^2 + 6xY + 3Y + 6$ has only real roots

$$b^2 - 4ax \geq 0$$

$$(6Y)^2 - 4(3)(3Y + 6) \geq 0$$

$$Y^2 - Y + 2 \geq 0$$

$Y \in (-\infty, -1] \cap [2, \infty)$

$\Rightarrow Y \in [2, 6)$

Since y is uniformly distributed in (1, 6)

Probability distributed function

$$f(Y) = \frac{1}{5} \quad 1 < y < 6$$

$$p(2 \le y < 6) = \int_2^6 f(Y)dy = \frac{1}{5}[Y]_2^6 = \frac{4}{5} = 0.8$$

5.58 (0.5)

$$a = \begin{bmatrix} a_1 \\ a_2 \\ \vdots \\ a_n \end{bmatrix}, \; x = \begin{bmatrix} x_1 \\ x_2 \\ \vdots \\ x_n \end{bmatrix}$$

Fixed non-zero

$a_i = 0$ or 1, $x_i = 0$ or 1

$\sum_{i=1}^{n} a_i x_i$ value lies between 0 to n.

Total number cases $= 2^n$

$a_i x_i$	0	1	2	....	n
fav	${}^nC_0$	${}^nC_1$	${}^nC_2$	....	${}^nC_n$

$$P\left[\sum_{i=1}^{n} a_i x_i \text{ is odd}\right] = \frac{{}^nC_1 + {}^nC_3 + \ldots}{2^n}$$

$$= \frac{2^{n-1}}{2^n} = \frac{1}{2} = 0.5$$

5.59 (a)

S_2 : $\text{Cov}(x, y) = E\{|x - \bar{x}||y - \bar{y}|\}$ is false

Case-I: If $x > \bar{x}$ and $y > \bar{y}$ then above is true.

Case-II: If $x < \bar{x}$ and $y < \bar{y}$ then above is true.

Case-III: If $x > \bar{x}$ but $y < \bar{y}$ then above is false.

Case-IV: If $x < \bar{x}$ but $y > \bar{y}$ then above is false.

$\because$ Given expression is not always true. So we can conclude that it is false.

S_1 : It is obviously false.

$\because$ True statement is

$[E\{(x - \bar{x})(y - \bar{y})\}]^2 < \text{var}(x) \cdot \text{Var}(y)$

So both S_1 and S_2 are false.

5.60 (0.367)

Let, t = {lifetime of component} and $\mu = 2$

Expected lifetime $= \frac{1}{\mu}$

and $f(t) = \begin{cases} \mu e^{-\mu t}, & t > 0 \\ 0, & \text{otherwise} \end{cases}$

$$P\left(t > \frac{1}{\mu}\right) = \int_{\frac{1}{\mu}}^{\infty} f(t)dt = \int_{\frac{1}{\mu}}^{\infty} \mu e^{-\mu t} dt$$

$$= -\left[e^{-\mu t}\right]_{\frac{1}{2}}^{\infty} = -\left[0 - e^{-1}\right]$$

$$= \frac{1}{e} = \frac{1}{2.713} = 0.367$$

5.61 (0.04)

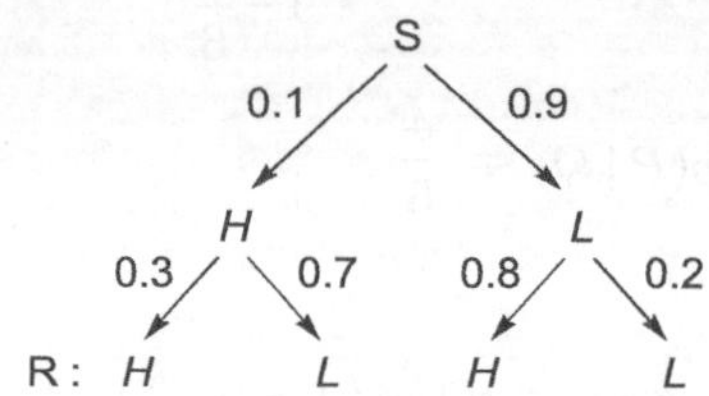

$$P\left(\frac{H_s}{H_R}\right) = \frac{P(H_s \cap H_R)}{P(H_R)}$$

$$= \frac{0.1 \times 0.3}{0.1 \times 0.3 + 0.9 \times 0.8}$$

$$= \frac{1}{25} = 0.04$$

5.62 (15.49)

$n = 1000$, $p = 0.4$, $q = 0.6$

It is a binomially distributed random variable.

So, $\text{S.D.} = \sqrt{npq}$

$= \sqrt{1000 \times 0.4 \times 0.6} = \sqrt{240} = 15.49$

5.63 (d)

$X \rightarrow$ Random variable which represents total marks record.

$P(x) \rightarrow$ Probability of getting those marks.

X ⟶	0	10	20	30
P(x) ⟶	0.2 × 0.5	0.8 × 0.5	0.5 × 0.2	0.8 × 0.5
	11	11	11	11
	0.1	0.4	0.1	0.4
	ΣP(x) = 1			

Now, if QuestionA is attempted first and it is correct.

Case-I:

$E(x) = \Sigma(x) \cdot P(x)$

$= 0.4 \times 10 + 0.4 \times 30$

$= 4 + 12 = 16$

Case-II:

If QuestionB is attempted first and is correct.

$$E(x) = \Sigma(x) \cdot P(x)$$
$$= 0.1\ (20) + 0.4\ (30)$$
$$= 2 + 12 = 14$$

So case-I is giving maximum expected marks. Hence option (d) is correct.

5.64 (a)

There are 10 favourable ways to calculate the probability of red ball in 4th trial.

(RFR)R ≈ R or (BRR)R ≈ 1 way or (RRR)R ≈ 3 ways or (BBR)R ≈ 3 ways

$$P(RRRR) = \frac{r}{r+b} \times \frac{r+1}{r+1+b} \times \frac{r+2}{r+2+b} \times \frac{r+3}{r+3+b} \quad ...(1)$$

$$P(BBBR) = \frac{b}{r+b} \times \frac{b+1}{r+b+1} \times \frac{b+2}{r+b+2} \times \frac{r}{r+b+3} \quad ...(2)$$

$$P(RRBR) = \frac{3!}{2!} \times \frac{r}{r+b} \times \frac{r+1}{r+b+1} \times \frac{b}{r+b+2} \times \frac{r+2}{r+b+3} \ ...(3)$$

$$P(BBRR) = \frac{3!}{2!} \times \frac{b}{r+b} \times \frac{b+1}{r+b+1} \times \frac{r}{r+b+2} \times \frac{r+1}{r+b+3} \ ...(4)$$

Required probability = (1) + (2) + (3) + (4)

$$= \frac{r(r+1)(r+2)(r+3) + b(b+1)(b+2)r + 3r(r+1)b(r+2) + 3b(b+1)r(r+1)}{(r+b)(r+b+1)(r+b+2)(r+b+3)}$$

On solving it we get,

$$= \frac{r(r+1+b)}{(r+b)(r+b+1)} = \frac{r}{r+b}$$

5.65 (c, d)

When two coins tossed we have

Sample space S = {HH, HT, TH, TT}

$P(A)$ = Probability of getting head in both toss

$$= \frac{1}{4}$$

$P(B)$ = Probability getting head on first toss

$$= \frac{2}{4} = \frac{1}{2}$$

$P(C)$ = Probability of getting head on second toss $= \frac{2}{4} = \frac{1}{2}$

$$P(A \cap B) = \frac{1}{4} \neq P(A) \cdot P(B)$$

∴ A and B are not independent.

$$P(B \cap C) = \frac{1}{4} = P(B) \cdot P(C)$$

∴ B and C are not independent.

$$P(A \cap C) = \frac{1}{4} \neq P(A) \cdot P(C)$$

∴ A, C are not independent.

$P(B|C) = P(B)$ is true

$$P(B|C) = \frac{P(B \cap C)}{P(C)} = \frac{\left(\frac{1}{4}\right)}{\frac{1}{2}} = \frac{1}{2} = P(B)$$

5.67 (b)

$$n(s) = n! = 4! = 24$$

X = 1 occurs before '2'

1				Can be done in 3! ways
3	1			Can be done in 2! ways
4	1			Can be done in 2! ways
3	4	1	2	2 ways
4	3	1	2	

$$n(X) = 6 + 2 + 2 + 2 = 12 \text{ ways}$$

Y = 3 occurs before 4

3				Can be filled in 3! ways
1	3			Can be filled in 2! ways
2	3			Can be filled in 2! ways
		3	4	Can be filled in 2! ways

∴ $n(Y) = 6 + 2 + 2 + 2 = 12$ ways

$(X \cap Y)$ = Both X and Y occurs

1	3			→ 2! ways
1	2	3	4	→ 1 way
3	1	4	2	→ 1 way
3	1	2	4	→ 1 way
3	4	1	2	→ 1 way

$n(X \cap Y) = 6$ ways

$$P(X \cap Y) = \frac{6}{24},\ P(X) = \frac{12}{24},\ P(Y) = \frac{12}{24}$$

∵ $P(X)\,P(Y) = \frac{12}{24} \times \frac{12}{24} = \frac{6}{24} = P(X \cap Y)$

∴ X and Y are independent.

5.68 **(0.375)**

10R
15B

It is given that first ball drawn is red, then remaining balls are

9R
15B

$P(\text{both ball is red}) \Rightarrow \dfrac{9}{24}$

or

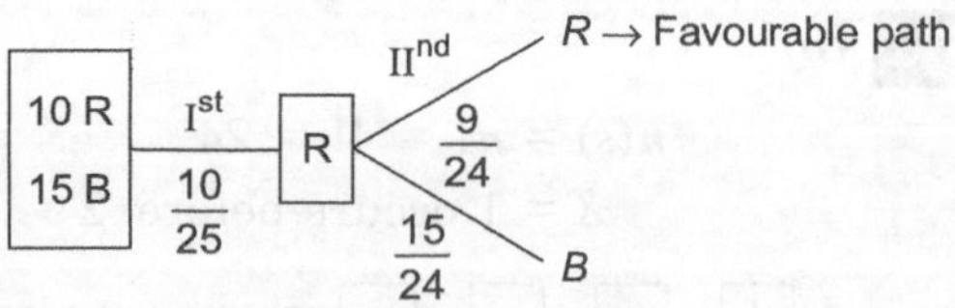

$$\text{Prop. (both red ball)} = \frac{\text{Favourable}}{\text{Total}}$$

$$= \frac{\frac{10}{25}\times\frac{9}{24}}{\frac{10}{25}\times\frac{9}{24}+\frac{10}{25}\times\frac{15}{24}} = \frac{10\times 9}{10\times 9+10\times 15} = \frac{9}{24}$$

5.69 **(a)**

Total outcomes = 6 × 6 × 6 × 6 × 6 × 6 = 6^6

Favourable outcomes (R.N.A.) = 6 × 5 × 4 × 3 × 2 × 1 = 6!

$$\text{So, required probability} = \frac{\text{Favourable}}{\text{Total}}$$

$$= \frac{6!}{6^6} = \frac{5}{324}$$

■■■■

6 Linear Algebra

6.1 The rank of the following (n + 1) × (n + 1) matrix, where a is a real number is

$$\begin{bmatrix} 1 & a & a^2 & \cdot & \cdot & \cdot & a^n \\ 1 & a & a^2 & \cdot & \cdot & \cdot & a^n \\ \cdot & \cdot & \cdot & & & & \cdot \\ \cdot & \cdot & \cdot & & & & \cdot \\ \cdot & \cdot & \cdot & & & & \cdot \\ 1 & a & a^2 & \cdot & \cdot & \cdot & a^n \end{bmatrix}$$

(a) 1
(b) 2
(c) n
(d) Depends on the value of a **[1995 : 1 M]**

6.2 Let $AX = B$ be a system of linear equations where A is an $m \times n$ matrix and b is a $m \times 1$ column vector and X is a $n \times 1$ column vector of unknown. Which of the following is false?
(a) The system has a solution if and only if, both A and the augmented matrix $[A, B]$ have the same rank.
(b) If m < n and B is the zero vector, then the system has infinitely many solutions.
(c) If m = n and B is non-zero vector, then the system has a unique solution.
(d) The system will have only a trivial solution when m = n, B is the zero vector and rank (A) = n.

[1996 : 1 M]

6.3 The matrices $\begin{bmatrix} \cos\theta & -\sin\theta \\ \sin\theta & \cos\theta \end{bmatrix}$ and $\begin{bmatrix} a & 0 \\ 0 & b \end{bmatrix}$ commute under multiplication
(a) if $a = b$ or $\theta = n\pi$, n is an integer
(b) always
(c) never
(d) if a cos θ ≠ b sin θ **[1996 : 2 M]**

6.4 The determinant of the matrix

$$\begin{bmatrix} 6 & -8 & 1 & 1 \\ 0 & 2 & 4 & 6 \\ 0 & 0 & 4 & 8 \\ 0 & 0 & 0 & -1 \end{bmatrix} \text{ is}$$

(a) 11 (b) – 48
(c) 0 (d) – 24 **[1997 : 1 M]**

6.5 Let $a = (a_{ij})$ be an n-rowed square matrix and I_{12} be the matrix obtained by interchanging the first and second rows of the n-rowed Identity matrix. Then AI_{12} is such that its first
(a) row the same as its second row
(b) row is the same as the second row of A
(c) column is the same as the second column A
(d) row is all zero

[1997 : 2 M]

6.6 Consider the following set of equations:

$$x + 2y = 5$$
$$4x + 8y = 12$$
$$3x + 6y + 3z = 15$$

This set
(a) has a unique solution
(b) has no solution
(c) has finite number of solutions
(d) has infinite number of solutions

[1998 : 1 M]

6.7 The rank of the matrix given below is

$$\begin{bmatrix} 1 & 4 & 8 & 7 \\ 0 & 0 & 3 & 0 \\ 4 & 2 & 3 & 1 \\ 3 & 12 & 24 & 21 \end{bmatrix}$$

(a) 3 (b) 1
(c) 2 (d) 4 **[1998 : 2 M]**

6.8 Consider the following determinant:

$$\Delta = \begin{vmatrix} 1 & a & bc \\ 1 & b & ca \\ 1 & c & ab \end{vmatrix}$$

Which of the following is a factor of Δ?
(a) a + b (b) a – b
(c) a + b + c (d) abc

[1998 : 2 M]

6.9 An n × n array v is defined as follows:
v[i, j] = i – j for all i, j, 1 ≤ i ≤ n, 1 ≤ j ≤ n
The sum of the elements of the array v is
(a) 0 (b) n – 1
(c) $n^2 - 3n + 2$ (d) $n^2 \dfrac{(n+1)}{2}$

[2000 : 1 M]

6.10 The determinant of the matrix $\begin{bmatrix} 2 & 0 & 0 & 0 \\ 8 & 1 & 7 & 2 \\ 2 & 0 & 2 & 0 \\ 9 & 0 & 6 & 1 \end{bmatrix}$ is

(a) 4 (b) 0
(c) 15 (d) 20 **[2001 : 1 M]**

6.11 Consider the following statements:
S_1 : The sum of two singular n × n matrices may be non-singular
S_2 : The sum of two n × n non-singular matrices may be singular
Which of the following statements is correct?
(a) S_1 and S_2 are both true
(b) S_1 is true, S_2 is false
(c) S_1 is flase, S_2 is true
(d) S_1 and S_2 are both false

[2002 : 1 M]

6.12 The rank of the matrix $\begin{bmatrix} 1 & 1 \\ 0 & 0 \end{bmatrix}$ is

(a) 4 (b) 2
(c) 1 (d) 0 **[2002 : 1 M]**

6.13 Obtain the eigen values of the matrix

$$A = \begin{bmatrix} 1 & 2 & 34 & 49 \\ 0 & 2 & 43 & 94 \\ 0 & 0 & -2 & 104 \\ 0 & 0 & 0 & -1 \end{bmatrix}$$ **[2002 : 2 M]**

6.14 Consider the following system of linear equations

$$\begin{bmatrix} 2 & 1 & -4 \\ 4 & 3 & -12 \\ 1 & 2 & -8 \end{bmatrix} \begin{bmatrix} x \\ y \\ z \end{bmatrix} = \begin{bmatrix} \alpha \\ 5 \\ 7 \end{bmatrix}$$

Notice that the second and the third columns of the coefficient matrix are linearly dependent. For how many values of α, does this system of equations have infinitely many solutions?
(a) 0 (b) 1
(c) 2 (d) Infinitely many
[2003 : 2 M]

6.15 Let A, B, C, D be n × n matrices, each with non-zero determinant, If ABCD = I, then B^{-1} is
(a) $D^{-1} C^{-1} A^{-1}$
(b) CDA
(c) ADC
(d) does not necessarily exist
[2004 : 1 M]

6.16 What values of *x*, *y* and *z* satisfy the following system of linear equations?

$$\begin{bmatrix} 1 & 2 & 3 \\ 1 & 3 & 4 \\ 2 & 2 & 3 \end{bmatrix} \begin{bmatrix} x \\ y \\ z \end{bmatrix} = \begin{bmatrix} 6 \\ 8 \\ 12 \end{bmatrix}$$

(a) $x = 6, y = 3, z = 2$
(b) $x = 12, y = 3, z = -4$
(c) $x = 6, y = 6, z = -4$
(d) $x = 12, y = -3, z = 0$ **[2004 : 1 M]**

6.17 Let A be an $n \times n$ matrix of the following form.

$$A = \begin{bmatrix} 3 & 1 & 0 & 0 & 0 & \dots & 0 & 0 & 0 \\ 1 & 3 & 1 & 0 & 0 & \dots & 0 & 0 & 0 \\ 0 & 1 & 3 & 1 & 0 & \dots & 0 & 0 & 0 \\ 0 & 0 & 1 & 3 & 1 & \dots & 0 & 0 & 0 \\ \dots \\ \dots \\ 0 & 0 & 0 & 0 & 0 & \dots & 1 & 3 & 1 \\ 0 & 0 & 0 & 0 & 0 & \dots & 0 & 1 & 3 \end{bmatrix}_{n \times n}$$

What is the value of the determinant of *A*?

(a) $\left(\frac{5+\sqrt{3}}{2}\right)^{n-1}\left(\frac{5\sqrt{3}+7}{2\sqrt{3}}\right)+\left(\frac{5-\sqrt{3}}{2}\right)^{n-1}\left(\frac{5\sqrt{3}-7}{2\sqrt{3}}\right)$

(b) $\left(\frac{7+\sqrt{5}}{2}\right)^{n-1}\left(\frac{7\sqrt{5}+3}{2\sqrt{5}}\right)+\left(\frac{7-\sqrt{5}}{2}\right)^{n-1}\left(\frac{7\sqrt{5}-3}{2\sqrt{5}}\right)$

(c) $\left(\frac{3+\sqrt{7}}{2}\right)^{n-1}\left(\frac{3\sqrt{7}+5}{2\sqrt{7}}\right)+\left(\frac{3-\sqrt{7}}{2}\right)^{n-1}\left(\frac{3\sqrt{7}-5}{2\sqrt{7}}\right)$

(d) $\left(\frac{3+\sqrt{5}}{2}\right)^{n-1}\left(\frac{3\sqrt{5}+7}{2\sqrt{5}}\right)+\left(\frac{3-\sqrt{5}}{2}\right)^{n-1}\left(\frac{3\sqrt{5}-7}{2\sqrt{5}}\right)$

[2004 : 2 M]

6.18 If matrix $X = \begin{bmatrix} a & 1 \\ -a^2+a-1 & 1-a \end{bmatrix}$ and $X^2 - X + I = O$ (*I* is the identity matrix and *O* is the zero matrix), then the inverse of *X* is

(a) $\begin{bmatrix} 1-a & -1 \\ a^2 & a \end{bmatrix}$

(b) $\begin{bmatrix} 1-a & -1 \\ a^2-a+1 & a \end{bmatrix}$

(c) $\begin{bmatrix} -a & 1 \\ -a^2+a-1 & 1-a \end{bmatrix}$

(d) $\begin{bmatrix} a^2-a+1 & a \\ 1 & 1-a \end{bmatrix}$

[2004 : 2 M]

6.19 How many solutions does the following system of linear equations have?

$$-x + 5y = -1$$
$$x - y = 2$$
$$x + 3y = 3$$

(a) Infinitely many
(b) Two distinct solutions
(c) Unique
(d) None of these

[2004 : 2 M]

6.20 The determinant of the matrix given below is

$$\begin{bmatrix} 0 & 1 & 0 & 2 \\ -1 & 1 & 1 & 3 \\ 0 & 0 & 0 & 1 \\ 1 & -2 & 0 & 1 \end{bmatrix}$$

(a) –1 (b) 0
(c) 1 (d) 2

[2005 : 1 M]

6.21 Consider the following system of equations in three real variables x_1, x_2 and x_3

$$2x_1 - x_2 + 3x_3 = 1$$
$$3x_1 - 2x_2 + 5x_3 = 2$$
$$-x_1 - 4x_2 + x_3 = 3$$

This system of equations has
(a) no solution
(b) a unique solution
(c) more than one but a finite number of solutions
(d) an infinite number of solutions

[2005 : 2 M]

6.22 What are the eigenvalues of the following 2 × 2 matrix?

$$\begin{bmatrix} 2 & -1 \\ -4 & 5 \end{bmatrix}$$

(a) –1 and 1 (b) 1 and 6
(c) 2 and 5 (d) 4 and –1

[2005 : 2 M]

6.23 F is an n × n real matrix. b is an n × 1 real vector. Suppose there are two n × 1 vectors, u and v such that u ≠ v, and Fu = b, Fv = b. Which one of the following statements is false?
(a) Determinant of F is zero
(b) There are an infinite number of solutions to Fx = b
(c) There is an x ≠ 0 such that Fx = 0
(d) F must have two identical rows

[2006 : 2 M]

6.24 What are the eigenvalues of the matrix P given below:

$$P = \begin{pmatrix} a & 1 & 0 \\ 1 & a & 1 \\ 0 & 1 & a \end{pmatrix}$$

(a) $a, a-\sqrt{2}, a+\sqrt{2}$ (b) a, a, a
(c) 0, a, 2a (d) –a, 2a, 2a

[2006 : 2 M]

6.25 Let A be the matrix $\begin{bmatrix} 3 & 1 \\ 1 & 2 \end{bmatrix}$. What is the maximum value of $x^T Ax$ where the maximum is taken over all x that are the unit eigenvectors of A?

(a) 3 (b) $\frac{(5+\sqrt{5})}{2}$
(c) 3 (d) $\frac{(5-\sqrt{5})}{2}$

[2007 : 1 M]

6.26 Consider the set of (column) vectors defined by $X = \{x \in R^3 \mid x_1 + x_2 + x_3 = 0,$ where $X^T = [x_1, x_2, x_3]^T\}$. Which of the following is TRUE?
(a) $\{[1, -1, 0]^T, [1, 0, -1]^T\}$ is a basis for the subspace X.
(b) $\{[1, -1, 0]^T, [1, 0, -1]^T\}$ is a linearly independent set, but it does not span X and therefore is not a basis of X
(c) X is not a subspace for R^3
(d) None of the above **[2007 : 2 M]**

6.27 The following system of equations

$$x_1 + x_2 + 2x_3 = 1$$
$$x_1 + 2x_3 + 3x_3 = 2$$
$$x_1 + 4x_2 + ax_3 = 4$$

has a unique solution. The only possible value(s) for a is/are
(a) 0
(b) either 0 or 1
(c) one of 0,1 or –1
(d) any real number other than 5

[2008 : 1 M]

6.28 How many of the following matrices have an eigenvalue 1?

$$\begin{bmatrix} 1 & 0 \\ 0 & 0 \end{bmatrix}, \begin{bmatrix} 0 & 1 \\ 0 & 0 \end{bmatrix}, \begin{bmatrix} 1 & -1 \\ 1 & 1 \end{bmatrix} \text{ and } \begin{bmatrix} -1 & 0 \\ 1 & -1 \end{bmatrix}$$

(a) one (b) two
(c) three (d) four **[2008 : 2 M]**

6.29 If M is a square matrix with a zero determinant, which of the following assertion(s) is (are) correct?

S1: Each row of M can be represented as a linear combination of the other rows

S2: Each column of M can be represented as a linear combination of the other columns

S3: MX = O has a nontrivial solution

S4: M has an inverse

(a) S3 and S2 (b) S1 and S4

(c) S1 and S3 (d) S1, S2 and S3

[2008 : 2 M

6.30 Consider the following matrix:

$$A = \begin{bmatrix} 2 & 3 \\ x & y \end{bmatrix}$$

If the eigenvalues of A are 4 and 8, then

(a) x = 4, y = 10 (b) x = 5, y = 8

(c) x = –3, y = 9 (d) x = –4, y = 10

[2010 : 2 M]

6.31 Consider the matrix as given below:

$$\begin{bmatrix} 1 & 2 & 3 \\ 0 & 4 & 7 \\ 0 & 0 & 3 \end{bmatrix}$$

Which one of the following options provides the CORRECT values of the eigenvalues of the matrix?

(a) 1, 4, 3 (b) 3, 7, 3

(c) 7, 3, 2 (d) 1, 2, 3 **[2011 : 2 M]**

6.32 Let A be the 2 × 2 matrix with elements $a_{11} = a_{12} = a_{21} = +1$ and $a_{22} = -1$. Then the eigenvalues of the matrix A^{19} are

(a) 1024 and – 1024

(b) $1024\sqrt{2}$ and $-1024\sqrt{2}$

(c) $4\sqrt{2}$ and $-4\sqrt{2}$

(d) $512\sqrt{2}$ and $-512\sqrt{2}$ **[2012 : 1 M]**

6.33 Which one of the following does NOT equal

$$\begin{vmatrix} 1 & x & x^2 \\ 1 & y & y^2 \\ 1 & z & z^2 \end{vmatrix} ?$$

(a) $\begin{vmatrix} 1 & x(x+1) & x+1 \\ 1 & y(y+1) & y+1 \\ 1 & z(z+1) & z+1 \end{vmatrix}$ (b) $\begin{vmatrix} 1 & x+1 & x^2+1 \\ 1 & y+1 & y^2+1 \\ 1 & z+1 & z^2+1 \end{vmatrix}$

(c) $\begin{vmatrix} 0 & x-y & x^2-y^2 \\ 0 & y-z & y^2-z^2 \\ 1 & z & z^2 \end{vmatrix}$ (d) $\begin{vmatrix} 2 & x+y & x^2+y^2 \\ 2 & y+z & y^2+z^2 \\ 1 & z & z^2 \end{vmatrix}$

[2013 : 1 M]

6.34 Consider the following system of equations:

$$3x + 2y = 1$$
$$4x + 7z = 1$$
$$x + y + z = 3$$
$$x - 2y + 7z = 0$$

The number of solutions for this system is ____.

[2014 (Set-1) : 1 M]

6.35 The value of the dot product of the eigenvectors corresponding to any pair of different eigen values of a 4-*by*-4 symmetric positive definite matrix is ________.

[2014 (Set-1) : 1 M]

6.36 If the matrix A is such that

$$A = \begin{bmatrix} 2 \\ -4 \\ 7 \end{bmatrix} [1 \ 9 \ 5]$$

then the determinant of A is equal to ____.

[2014 (Set-2) : 1 M]

6.37 The product of the non-zero eigenvalues of the matrix

$$\begin{bmatrix} 1 & 0 & 0 & 0 & 1 \\ 0 & 1 & 1 & 1 & 0 \\ 0 & 1 & 1 & 1 & 0 \\ 0 & 1 & 1 & 1 & 0 \\ 1 & 0 & 0 & 0 & 1 \end{bmatrix}$$

is ________. **[2014 (Set-2) : 2 M]**

6.38 Which one of the following statements is TRUE about every $n \times n$ matrix with only real eigenvalues?

(a) If the trace of the matrix is positive and the determinant of the matrix is negative, at least one of its eigenvalues is negative.

(b) If the trace of the matrix is positive, all its eigenvalues are positive.

(c) If the determinant of the matrix is positive, all its eigenvalues are positive.

(d) If the product of the trace and determinant of the matrix is positive, all its eigenvalues are positive. **[2014 (Set-3) : 1 M]**

6.39 If V_1 and V_2 are 4-dimensional subspaces of a 6-dimensional vector space V, then the smallest possible dimension of $V_1 \cap V_2$ is ________.

[2014 (Set-3) : 1 M]

6.40 The minimum number of arithmetic operations required to evaluate the polynomial $P(X) = X^5 + 4X^3 + 6X + 5$ for a given value of X, using only one temporary variable is ________.

[2014 (Set-3) : 1 M]

6.41 Consider the following 2 × 2 matrix A where two elements are unknown and are marked by 'a' and 'b'. The eigenvalues of this matrix are –1 and 7. What are the values of 'a' and 'b'?

$$A = \begin{pmatrix} 1 & 4 \\ b & a \end{pmatrix}$$

(a) a = 6, b = 4
(b) a = 4, b = 6
(c) a = 3, b = 5
(d) a = 5, b = 3

[2015 (Set-1) : 2 M]

6.42 The larger of the two eigenvalues of the matrix $\begin{bmatrix} 4 & 5 \\ 2 & 1 \end{bmatrix}$ is ______.

[2015 (Set-2) : 1 M]

6.43 Perform the following operations on the matrix $\begin{bmatrix} 3 & 4 & 45 \\ 7 & 9 & 105 \\ 13 & 2 & 195 \end{bmatrix}$.

1. Add the third row to the second row.
2. Subtract the third column from the first column.

The determinant of the resultant matrix is______.

[2015 (Set-2) : 2 M]

6.44 In the given matrix $\begin{bmatrix} 1 & -1 & 2 \\ 0 & 1 & 0 \\ 1 & 2 & 1 \end{bmatrix}$, one of the eigenvalues is 1. The eigenvectors corresponding to the eigenvalue 1 are

(a) $\{\alpha(4,2,1) \mid \alpha \neq 0, \alpha \in \mathbb{R}\}$
(b) $\{\alpha(-4,2,1) \mid \alpha \neq 0, \alpha \in \mathbb{R}\}$
(c) $\{\alpha(\sqrt{2},0,1) \mid \alpha \neq 0, \alpha \in \mathbb{R}\}$
(d) $\{\alpha(-\sqrt{2},0,1) \mid \alpha \neq 0, \alpha \in \mathbb{R}\}$

[2015 (Set-3) : 1 M]

6.45 In the LU decomposition of the matrix $\begin{bmatrix} 2 & 2 \\ 4 & 9 \end{bmatrix}$, if the diagonal elements of U are both 1, then the lower diagonal entry l_{22} of L is ______.

[2015 (Set-1) : 1 M]

6.46 If the following system has non-trivial solution,

$$px + qy + rz = 0$$
$$qx + ry + pz = 0$$
$$rx + py + qz = 0$$

then which one of the following options is TRUE?

(a) $p - q + r = 0$ or $p = q = -r$
(b) $p + q - r = 0$ or $p = -q = r$
(c) $p + q + r = 0$ or $p = q = r$
(d) $p - q + r = 0$ or $p = -q = -r$

[2015 (Set-3) : 2 M]

6.47 Two eigenvalues of a 3 × 3 real matrix P are $(2+\sqrt{-1})$ and 3. The determinant of P is ______.

[2016 (Set-1) : 1 M]

6.48 Consider the systems, each consisting of m linear equations in n variables.

I. If $m < n$, then all such systems have a solution.
II. If $m > n$, then none of these sytems has a solution.
III. If $m = n$, then there exists a system which has a solution

Which one of the following is **CORRECT**?

(a) I, II and III are true
(b) Only II and III are true
(c) Only III is true
(d) None of them is true

[2016 (Set-2) : 1 M]

6.49 Suppose that the eigenvalues of matrix A are 1, 2, 4. The determinant of $(A^{-1})^T$ is ______.

[2016 (Set-2) : 1 M]

6.50 Let $c_1, ..., c_n$ be scalars, not all zero, such that $\sum_{i=1}^{n} c_i a_i = 0$ where a_i are column vectors in R^n.

Consider the set of linear equations

$$Ax = b$$

where $A = [a_1, ..., a_n]$ and $b = \sum_{i=1}^{n} a_i$. The set of equations has

(a) a unique solution at $x = J_n$ where J_n denotes a n-dimensional vector of all 1
(b) no solution
(c) infinitely many solutions
(d) finitely many solutions

[2017 (Set-1) : 1 M]

6.51 Let A be $n \times n$ real valued square symmetric matrix of rank 2 with $\sum_{i=1}^{n}\sum_{j=1}^{n} A_{ij}^2 = 50$. Consider the following statements.

I. One eigenvalue must be in [–5, 5]
II. The eigenvalue with the largest magnitude must be strictly greater than 5.

Which of the above statements about eigenvalues of A is/are necessarily CORRECT?

(a) Both I and II (b) I only

(c) II only (d) Neither I nor II

[2017 (Set-1) : 2 M]

6.52 Let $P = \begin{bmatrix} 1 & 1 & -1 \\ 2 & -3 & 4 \\ 3 & -2 & 3 \end{bmatrix}$ and $Q = \begin{bmatrix} -1 & -2 & -1 \\ 6 & 12 & 6 \\ 5 & 10 & 5 \end{bmatrix}$ be two matrices. Then the rank of $P + Q$ is ______.

[2017 (Set-2) : 1 M]

6.53 If the characteristic polynomial of a 3 × 3 matrix M over $\mathbb{R}$ (the set of real numbers) is $\lambda^3 - 4\lambda^2 + a\lambda + 30$, $a \in \mathbb{R}$ and one eigenvalue of M is 2, then the largest among the absolute values of the eigenvalues of M is ______.

[2017 (Set-2) : 1 M]

6.54 Consider a matrix $A = uv^T$ where $u = \begin{pmatrix} 1 \\ 2 \end{pmatrix}$, $v = \begin{pmatrix} 1 \\ 1 \end{pmatrix}$. Note that v^T denotes the transpose of v. The largest eigenvalue of A is ______.

[2018 : 1 M]

6.55 Consider a matrix P whose only eigenvectors are the multiples of $\begin{bmatrix} 1 \\ 4 \end{bmatrix}$.

Consider the following statements:

I. P does not have an inverse.

II. P has a repeated eigenvalue.

III. P cannot be diagonalized.

Which one of the following options is correct?

(a) Only I and III are necessarily true

(b) Only II is necessarily true

(c) Only I and II are necessarily true

(d) Only II and III are necessarily true

[2018 : 2 M]

6.56 Let X be a square matrix. Consider the following two statements on X.

I. X is invertible.

II. Determinant of X is non-zero.

Which one of the following is TRUE?

(a) I implies II; II does not imply I.

(b) II implies I; I does not imply II.

(c) I and II are equivalent statements.

(d) I does not imply II; II does not imply I.

[2019 : 1 M]

6.57 Consider the following matrix:

$$R = \begin{bmatrix} 1 & 2 & 4 & 8 \\ 1 & 3 & 9 & 27 \\ 1 & 4 & 16 & 64 \\ 1 & 5 & 25 & 125 \end{bmatrix}$$

The absolute value of the product of Eigen values of R is ______.

[2019 : 2 M]

6.58 Let A and B be two $n \times n$ matrices over real numbers. Let rank(M) and det(M) denote the rank and determinant of a matrix M, respectively. Consider the following statements:

I. rank(AB) = rank(A) rank(B)

II. det(AB) = det(A) det(B)

III. rank(A + B) ≤ rank(A) + rank(B)

IV. det(A + B) ≤ det(A) + det(B)

Which of the above statements are TRUE?

(a) III and IV only

(b) II and III only

(c) I and IV only

(d) I and II only

[2020 : 2 M]

6.59 Consider the following matrix:

$$\begin{pmatrix} 0 & 1 & 1 & 1 \\ 1 & 0 & 1 & 1 \\ 1 & 1 & 0 & 1 \\ 1 & 1 & 1 & 0 \end{pmatrix}$$

The largest eigenvalue of the above matrix is ______.

[2021 (Set-1) : 2 M]

6.60 Suppose that P is a 4 × 5 matrix such that every solution of the equation $Px = 0$ is a scalar multiple of $[2\ 5\ 4\ 3\ 1]^T$. The rank of P is ______.

[2021 (Set-2) : 1 M]

6.61 Consider the following two statements with respect to the matrices $A_{m \times n}$, $B_{n \times m}$, $C_{n \times n}$ and $D_{n \times n}$.

Statement 1: tr(AB) = tr(BA)

Statement 2: tr(CD) = tr(DC)

where tr( ) represents the trace of a matrix. Which one of the following holds?

(a) Statement 1 is correct and Statement 2 is wrong.

(b) Statement 1 is wrong and Statement 2 is correct.

(c) Both Statement 1 and Statement 2 are correct.

(d) Both Statement 1 and Statement 2 are wrong.

[2022 : 1 M]

6.62 Consider solving the following system of simultaneous equations using LU decomposition.

$$x_1 + x_2 - 2x_3 = 4$$
$$x_1 + 3x_2 - x_3 = 7$$
$$2x_1 + x_2 - 5x_3 = 7$$

where L and U are denoted as

$$L = \begin{pmatrix} L_{11} & 0 & 0 \\ L_{21} & L_{22} & 0 \\ L_{31} & L_{32} & L_{33} \end{pmatrix}, \quad U = \begin{pmatrix} U_{11} & U_{12} & U_{13} \\ 0 & U_{22} & U_{23} \\ 0 & 0 & U_{33} \end{pmatrix}$$

Which one of the following is the correct combination of values for L_{32}, U_{32}, and x_1?

(a) $L_{32} = 2,\ U_{33} = -\dfrac{1}{2},\ x_1 = -1$

(b) $L_{32} = 2,\ U_{32} = 2,\ x_1 = -1$

(c) $L_{32} = -\dfrac{1}{2},\ U_{33} = 2,\ x_1 = 0$

(d) $L_{32} = -\dfrac{1}{2},\ U_{33} = -\dfrac{1}{2},\ x_1 = 0$

[2022 : 2 M]

6.63 Which of the following is/are the eigen vector(s) for the matrix given below?

$$\begin{pmatrix} -9 & -6 & -2 & -4 \\ -8 & -6 & -3 & -1 \\ 20 & 15 & 8 & 5 \\ 32 & 21 & 7 & 12 \end{pmatrix}$$

(a) $\begin{pmatrix} -1 \\ 1 \\ 0 \\ 1 \end{pmatrix}$ (b) $\begin{pmatrix} 1 \\ 0 \\ -1 \\ 0 \end{pmatrix}$

(c) $\begin{pmatrix} -1 \\ 0 \\ 2 \\ 2 \end{pmatrix}$ (d) $\begin{pmatrix} 0 \\ 1 \\ -3 \\ 0 \end{pmatrix}$

[2022 : 2 M]

6.64 Let and $A = \begin{bmatrix} 1 & 2 & 3 & 4 \\ 4 & 1 & 2 & 3 \\ 3 & 4 & 1 & 2 \\ 2 & 3 & 4 & 1 \end{bmatrix}$ and $B = \begin{bmatrix} 3 & 4 & 1 & 2 \\ 4 & 1 & 2 & 3 \\ 1 & 2 & 3 & 4 \\ 2 & 3 & 4 & 1 \end{bmatrix}$

Let det(A) and det(B) denote the determinants of the matrices A and B, respectively.

Which one of the options given below is TRUE?

(a) det(A) = det(B) (b) det(B) = –det(A)

(c) det(A) = 0 (d) det(AB) = det(A) + det(B)

[2023 : 1 M]

6.65 Let A be the adjacency matrix of the graph with vertices {1, 2, 3, 4, 5}.

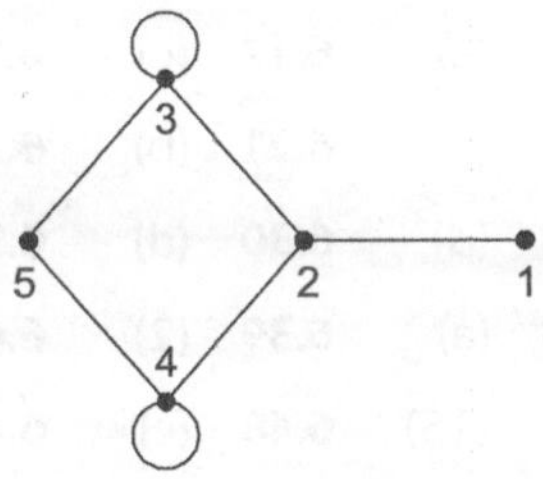

Let $\lambda_1, \lambda_2, \lambda_3, \lambda_4$ and λ_5 be the five eigenvalues of A. Note that these eigenvalues need not be distinct.

The value of $\lambda_1 + \lambda_2 + \lambda_3 + \lambda_4 + \lambda_5 =$ ______.

[2023 : 1 M]

6.66 The product of all eigen values of the matrix $\begin{bmatrix} 1 & 2 & 3 \\ 4 & 5 & 6 \\ 7 & 8 & 9 \end{bmatrix}$ is

(a) 0 (b) 1

(c) –1 (d) 2

[2024 (Set-1) : 1 M]

6.67 Let A be any $n \times m$ matrix, where $m > n$. Which of the following statements is/are TRUE about the system of linear equations $Ax = 0$?

(a) There exist at least $m - n$ linearly independent solutions to this system

(b) There exists a solution in which at least n variables are non-zero

(c) There exists a non-zero solution in which at least $m - n$ variables are 0

(d) There exist $m - n$ linearly independent vectors such that every solution is a linear combination of these vectors

[2024 (Set-1) : 2 M]

6.68 Let A be an $n \times n$ matrix over the set of all real numbers $\mathbb{R}$. Let B be a matrix obtained from A by swapping two rows. Which of the following statements is/are TRUE?

(a) If the trace of A is zero, then the trace of B is also zero.

(b) If A is invertible, then B is also invertible.

(c) If A is symmetric, then B is also symmetric.

(d) The determinant of B is the negative of the determinant of A.

[2024 (Set-2) : 2 M]

■■■■

Answers Linear Algebra

6.1	(a)	**6.2**	(c)	**6.3**	(a)	**6.4**	(b)	**6.5**	(c)	**6.6**	(b)	**6.7**	(a)	**6.8**	(b)	**6.9**	(a)
6.10	(a)	**6.11**	(a)	**6.12**	(c)	**6.13**	Sol.	**6.14**	(b)	**6.15**	(b)	**6.16**	(c)	**6.17**	(d)	**6.18**	(b)
6.19	(c)	**6.20**	(c)	**6.21**	(b)	**6.22**	(b)	**6.23**	(d)	**6.24**	(a)	**6.25**	(b)	**6.26**	(b)	**6.27**	(d)
6.28	(a)	**6.29**	(d)	**6.30**	(d)	**6.31**	(a)	**6.32**	(d)	**6.33**	(a)	**6.34**	(1)	**6.35**	(0)	**6.36**	(0)
6.37	(6)	**6.38**	(a)	**6.39**	(2)	**6.40**	(7)	**6.41**	(d)	**6.42**	(6)	**6.43**	(0)	**6.44**	(b)	**6.45**	(5)
6.46	(c)	**6.47**	(15)	**6.48**	(c)	**6.49**	(0.125)	**6.50**	(c)	**6.51**	(b)	**6.52**	(2)	**6.53**	(5)	**6.54**	(3)
6.55	(d)	**6.56**	(c)	**6.57**	(12)	**6.58**	(b)	**6.59**	(3)	**6.60**	(4)	**6.61**	(c)	**6.62**	(d)	**6.63**	(a, c, d)
6.64	(b)	**6.65**	(2)	**6.66**	(a)	**6.67**	(a)	**6.68**	(b, d)								

Explanations Linear Algebra

6.1 (a)

All the rows of the given matrix is same. So the matrix has only one independent row.

Rank of the matrix = No. of independent rows of the matrix. ∴ Rank of given matrix = 1

6.2 (c)

Following are the possibilities for a system of linear equations:

(i) If matrix A and augmented matrix [AB] have same rank, then the system has solution otherwise there is no solution.

(ii) If matrix A and augmented matrix [AB] have same rank which is equal to the no. of variables, then the system has unique solution and if B is zero vector then the system have only a trivial solution.

(iii) If matrix A and matrix [AB] have same rank which is less than the number of variables, then the system has infinite solution.

Therefore, option (c) is false because if m = n and B is non-zero vector, then it is not necessary that system has a unique solution, because m is the number of equations (quantity) and not the number of linearly independent equations (quality).

6.3 (a)

$$A = \begin{bmatrix} \cos\theta & -\sin\theta \\ \sin\theta & \cos\theta \end{bmatrix}$$

and
$$B = \begin{bmatrix} a & 0 \\ 0 & b \end{bmatrix}$$

$$AB = \begin{bmatrix} \cos\theta & -\sin\theta \\ \sin\theta & \cos\theta \end{bmatrix}\begin{bmatrix} a & 0 \\ 0 & b \end{bmatrix}$$

$$= \begin{bmatrix} a\cos\theta & -b\sin\theta \\ a\sin\theta & b\cos\theta \end{bmatrix}$$

$$BA = \begin{bmatrix} a & 0 \\ 0 & b \end{bmatrix}\begin{bmatrix} \cos\theta & -\sin\theta \\ \sin\theta & \cos\theta \end{bmatrix}$$

$$= \begin{bmatrix} a\cos\theta & -a\sin\theta \\ b\sin\theta & b\cos\theta \end{bmatrix}$$

AB = BA iff $-b\sin\theta = -a\sin\theta$ and $a\sin\theta = b\sin\theta$

Both are same equation which is $(a-b)\sin\theta = 0$ whose solution is $a = b$ or $\theta = \pm n\pi$.

6.4 (b)

The given matrix is

$$\begin{bmatrix} 6 & -8 & 1 & 1 \\ 0 & 2 & 4 & 6 \\ 0 & 0 & 4 & 8 \\ 0 & 0 & 0 & -1 \end{bmatrix}$$

Above matrix is upper triangular matrix and for upper triangular matrix the determinant is product of principle diagonal elements.

Determinant of matrix $= 6 \times 2 \times 4 \times -1 = -48$

6.5 (c)

Let, $A = \begin{bmatrix} a & b \\ c & d \end{bmatrix}$

I_{12} is the matrix obtained by inter-changing the first and second row of the Identity Matrix I.

So $I_{12} = \begin{bmatrix} 0 & 1 \\ 1 & 0 \end{bmatrix}$

$$AI_{12} = \begin{bmatrix} a & b \\ c & d \end{bmatrix}\begin{bmatrix} 0 & 1 \\ 1 & 0 \end{bmatrix} = \begin{bmatrix} b & a \\ d & c \end{bmatrix}$$

AI_{12} is the matrix having first column same as the second column of A.

6.6 (b)

Set of equations is

$$x + 2y = 5$$
$$4x + 8y = 12$$
$$3x + 6y + 3z = 15$$

Above set of equations can be written as

$$\begin{bmatrix} 1 & 2 & 0 \\ 4 & 8 & 0 \\ 3 & 6 & 3 \end{bmatrix}\begin{bmatrix} x \\ y \\ x \end{bmatrix} = \begin{bmatrix} 5 \\ 12 \\ 15 \end{bmatrix}$$

Augmented matrix [AB] is given as

$$\left[\begin{array}{ccc|c} 1 & 2 & 0 & 5 \\ 4 & 8 & 0 & 12 \\ 3 & 6 & 3 & 15 \end{array}\right]$$

Performing gauss-Elimination on the above matrix

$$\left[\begin{array}{ccc|c} 1 & 2 & 0 & 5 \\ 4 & 8 & 0 & 12 \\ 3 & 6 & 3 & 15 \end{array}\right] \xrightarrow[R_3 \to R_3 - 3R_1]{R_2 \to R_2 - 4R_1} \left[\begin{array}{ccc|c} 1 & 2 & 0 & 5 \\ 0 & 0 & 0 & -8 \\ 0 & 0 & 3 & 0 \end{array}\right] \xrightarrow{R_2 \leftrightarrow R_3} \left[\begin{array}{ccc|c} 1 & 2 & 0 & 5 \\ 0 & 0 & 3 & 0 \\ 0 & 0 & 0 & -8 \end{array}\right]$$

The rank of matrix A is 2 and rank of matrix [AB] is 3. Therefore, the system is inconsistent and has no solutions.

6.7 (a)

The given matrix is

$$\begin{bmatrix} 1 & 4 & 8 & 7 \\ 0 & 0 & 3 & 0 \\ 4 & 2 & 3 & 1 \\ 3 & 12 & 24 & 21 \end{bmatrix}$$

Since, $R_4 = 3R_1$, Rank $\neq 4$

Now try for a rank of 3

$$\begin{vmatrix} 1 & 4 & 8 \\ 0 & 0 & 3 \\ 4 & 2 & 3 \end{vmatrix} = -3 \times \begin{vmatrix} 1 & 4 \\ 4 & 2 \end{vmatrix} = -3 \times -14 = 52 \neq 0$$

$\therefore$ Rank of given matrix = 3

6.8 (b)

The determinant of a matrix can't be affected by elementary row operations

So; $\Delta = \begin{vmatrix} 1 & a & bc \\ 1 & b & ca \\ 1 & c & ab \end{vmatrix}$

$R_2 \to R_2 - R_1$
$R_3 \to R_3 - R_1$

$$\Delta = \begin{vmatrix} 1 & a & bc \\ 0 & (b-a) & (ca-bc) \\ 0 & (c-a) & (ab-bc) \end{vmatrix}$$

$$= \begin{vmatrix} 1 & a & bc \\ 0 & (b-a) & -c(b-a) \\ 0 & (c-a) & -b(c-a) \end{vmatrix}$$

$$= \begin{vmatrix} (b-a) & -c(b-a) \\ (c-a) & -b(c-a) \end{vmatrix}$$

$$= 1(b-a)(c-a)\begin{vmatrix} 1 & -c \\ 1 & -b \end{vmatrix}$$

$$= (b-a)(c-a)(-b+c)$$
$$= (b-a)(c-a)(c-b)$$
$$= (a-b)(b-c)(c-a)$$

So $(a-b)$ is a factor of Δ.

6.9 (a)

The matrix V can be defined as

$$\begin{bmatrix} 0 & -1 & -2 & \cdots\cdots & (n-1) \\ 1 & 0 & -1 & \cdots\cdots & (n-2) \\ 2 & 1 & 0 & & \vdots \\ \vdots & \vdots & \vdots & & \vdots \\ (n-1) & (n-2) & \vdots & \cdots\cdots & 0 \end{bmatrix}$$

So above is antisymmetric matrix and the sum of the elements of any antisymmetric matrix is 0.

OR

Alternate Method:

$$\sum_{j=1}^{n}\sum_{i=1}^{n} i - j = \sum_{j=1}^{n} \frac{n(n+1)}{2} - jn$$

$$= \frac{n(n+1)}{2} \times n - \frac{n^2(n+1)}{2} = 0$$

6.10 (a)

The matrix is $\begin{bmatrix} 2 & 0 & 0 & 0 \\ 8 & 1 & 7 & 2 \\ 2 & 0 & 2 & 0 \\ 9 & 0 & 6 & 1 \end{bmatrix}$

Finding the determinant by expanding the first row of the matrix

$$\text{Determinant} = 2\begin{vmatrix} 1 & 7 & 2 \\ 0 & 2 & 0 \\ 0 & 6 & 1 \end{vmatrix} = 2 \times 1 \begin{vmatrix} 2 & 0 \\ 6 & 1 \end{vmatrix}$$

$$= 2 \times 1 \times (2) = 4$$

Therefore, the determinant of given matrix is 4.

6.11 (a)

S_1 is true

Consider two singular matrices

$$A = \begin{bmatrix} 1 & 0 \\ 0 & 0 \end{bmatrix} \text{ and } B = \begin{bmatrix} 0 & 0 \\ 0 & 1 \end{bmatrix}$$

Sum of A and B is given as

However (A + B) is a non-singular matrix $\begin{bmatrix} 1 & 0 \\ 0 & 1 \end{bmatrix}$.

So, S_1 is true.

Now, consider two non-singular matrices

$$C = \begin{bmatrix} 1 & 0 \\ 0 & 1 \end{bmatrix} \text{ and } D = \begin{bmatrix} -1 & 0 \\ 0 & -1 \end{bmatrix}$$

$$C + D = \begin{bmatrix} 0 & 0 \\ 0 & 0 \end{bmatrix}$$

However (C + D) is a singular matrix. So, S_2 is also true.

Therefore, both S_1 and S_2 are true.

6.12 (c)

The given matrix is $\begin{bmatrix} 1 & 1 \\ 0 & 0 \end{bmatrix}$

Above matrix has only 1 independent row, so the given matrix has rank 1.

6.13 Sol.

Matrix is

$$\begin{bmatrix} 1 & 2 & 34 & 49 \\ 0 & 2 & 43 & 94 \\ 0 & 0 & -2 & 104 \\ 0 & 0 & 0 & -1 \end{bmatrix}$$

The given matrix is upper triangular matrix and the eigen values of the matrix are diagonal elements for upper triangular matrix.

So the eigen values of the matrix A are 1, 2, –2, –1.

6.14 (b)

The augmented matrix for the given system is

$$\left[\begin{array}{ccc|c} 2 & 1 & -4 & \alpha \\ 4 & 3 & -12 & 5 \\ 1 & 2 & -8 & 7 \end{array}\right]$$

Performing Gauss-Elimination on the above matrix

$$\left[\begin{array}{ccc|c} 2 & 1 & -4 & \alpha \\ 4 & 3 & -12 & 5 \\ 1 & 2 & -8 & 7 \end{array}\right] \xrightarrow[R_3 - 1/2 R_1]{R_2 - 2R_1} \left[\begin{array}{ccc|c} 2 & 1 & -4 & \alpha \\ 0 & 1 & -4 & 5-2\alpha \\ 0 & 3/2 & -6 & 7-\alpha/2 \end{array}\right]$$

$$\xrightarrow{R_3 - 3/2 R_2} \left[\begin{array}{ccc|c} 2 & 1 & -4 & \alpha \\ 0 & 1 & -4 & 5-2\alpha \\ 0 & 0 & 0 & \frac{5\alpha-1}{2} \end{array}\right]$$

Now for infinite solution it is necessary that at least one row must be completely zero.

$$\therefore \quad \frac{5\alpha - 1}{2} = 0$$

$$\alpha = \frac{1}{5} \text{ is the solution}$$

$\therefore$ There is only one value of α for which infinite solution exists.

6.15 (b)

A, B, C, D is n × n matrix.

Given ABCD = I

$\Rightarrow ABCDD^{-1}C^{-1} = D^{-1}C^{-1}$

$\Rightarrow \quad AB = D^{-1}C^{-1}$

$\Rightarrow \quad A^{-1}AB = A^{-1}D^{-1}C^{-1}$

$\Rightarrow \quad B = A^{-1}D^{-1}C^{-1}$

$B^{-1} = (A^{-1}D^{-1}C^{-1})^{-1}$

$= (C^{-1})^{-1} \cdot (D^{-1})^{-1} \cdot (A^{-1})^{-1} = CDA$

6.16 (c)

$$\begin{bmatrix} 1 & 2 & 3 \\ 1 & 3 & 4 \\ 2 & 2 & 3 \end{bmatrix} \begin{bmatrix} x \\ y \\ z \end{bmatrix} = \begin{bmatrix} 6 \\ 8 \\ 12 \end{bmatrix}$$

$1 \times x + 2 * y + 3 * z = 6$

$1 \times x + 3 * y + 4 * z = 8$

$2 \times x + 2 * y + 3 * z = 12$

Put $x = 6$, $y = 6$, $z = -4$, the above three equations are satisfied.

6.17 (d)

$$A = \begin{bmatrix} 3 & 1 & 0 & 0 & 0 & \dots & 0 & 0 & 0 \\ 1 & 3 & 1 & 0 & 0 & \dots & 0 & 0 & 0 \\ 0 & 1 & 3 & 1 & 0 & \dots & 0 & 0 & 0 \\ 0 & 0 & 1 & 3 & 1 & \dots & 0 & 0 & 0 \\ \dots \\ \dots \\ 0 & 0 & 0 & 0 & 0 & \dots & 1 & 3 & 1 \\ 0 & 0 & 0 & 0 & 0 & \dots & 0 & 1 & 3 \end{bmatrix}_{n \times n}$$

A can be expanded using first row as $|A_n| = 3^*|A_{n-1}|+1^*|A_{n-2}|$.

We find that the remaining sub-matrix is same as the given matrix with a LOWER order.

say $|A_n|$ is denoted as T_n.

Then recurrence relation is

$$T_n = 3T_{n-1} + T_{n-2}$$

$$\Rightarrow T_n - 3T_{n-1} - T_{n-2} = 0$$

This is a Homogenous system whose solution is

$$T_n = C_1\left(\frac{3+\sqrt{5}}{2}\right)^n + C_2\left(\frac{3-\sqrt{5}}{2}\right)^n \quad ...(i)$$

Now putting the initial conditions

$$T_1 = |3| = 3 \text{ and } T_2 = \begin{vmatrix} 3 & 1 \\ 1 & 3 \end{vmatrix} = 8$$

We get,

$$\left(\frac{3+\sqrt{5}}{2}\right)^{n-1}\left(\frac{3\sqrt{5}+7}{2\sqrt{5}}\right) + \left(\frac{3-\sqrt{5}}{2}\right)^{n-1}\left(\frac{3\sqrt{5}-7}{2\sqrt{5}}\right)$$

Alternately, since we know that for $n = 1$ and $n = 2$

$$A_1 = |3| = 3 \text{ and } A_2 = \begin{vmatrix} 3 & 1 \\ 1 & 3 \end{vmatrix} = 8$$

can put n = 1 and n = 2 in those answers which are in the required format i.e. equation (*i*) and see that only choice (d) goes to 3 and 8 respectively.

6.18 (b)

Given, $X = \begin{bmatrix} a & 1 \\ -a^2+a-1 & 1-a \end{bmatrix}$

and $X^2 - X + I = O$ ...(*i*)

we need to find X^{-1}.

Solving (*i*) for I we get,

$$I = X - X^2$$

Multiplying by X^{-1} on both sides

$$X^{-1} = I - X$$

$$= \begin{bmatrix} 1 & 0 \\ 0 & 1 \end{bmatrix} - \begin{bmatrix} a & 1 \\ -a^2+a-1 & 1-a \end{bmatrix}$$

$$= \begin{bmatrix} 1-a & -1 \\ a^2-a+1 & a \end{bmatrix}$$

Which is option (b).

6.19 (c)

$$-x + 5y = -1$$
$$x - y = 2$$
$$x + 3y = 3$$

The augmented matrix is

$$\left[\begin{array}{cc|c} -1 & 5 & -1 \\ 1 & -1 & 2 \\ 1 & 3 & 3 \end{array}\right]$$

Using gauss-elimination on above matrix we get,

$$\left[\begin{array}{cc|c} -1 & 5 & -1 \\ 1 & -1 & 2 \\ 1 & 3 & 3 \end{array}\right] \xrightarrow[R_3+R_1]{R_2+R_1} \left[\begin{array}{cc|c} -1 & 5 & -1 \\ 0 & 4 & 1 \\ 0 & 8 & 2 \end{array}\right]$$

$$\xrightarrow{R_3-2R_2} \left[\begin{array}{cc|c} -1 & 5 & -1 \\ 0 & 4 & 1 \\ 0 & 0 & 0 \end{array}\right]$$

Rank [A | B] = 2 (number of non zero rows in [A | B])

Rank [A] = 2 (number of non zero rows in [A])

Rank [A | B] = Rank [A] = 2 = number of variables

∴ Unique solution exists. Correct choice is (c).

6.20 (c)

$$A = \begin{vmatrix} 0 & 1 & 0 & 2 \\ -1 & 1 & 1 & 3 \\ 0 & 0 & 0 & 1 \\ 1 & -2 & 0 & 1 \end{vmatrix}$$

Expanding A by third row (which contains lot of zeros)

$$A = -1 \times \begin{vmatrix} 0 & 1 & 0 \\ -1 & 1 & 1 \\ 1 & -2 & 0 \end{vmatrix} = -1 \times 1 \times \begin{vmatrix} -1 & 1 \\ 1 & 0 \end{vmatrix}$$

$$= -1 \times 1 \times -1 = 1$$

6.21 (b)

The augmented matrix for the given system is

$$\left[\begin{array}{ccc|c} 2 & -1 & 3 & 1 \\ 3 & -2 & 5 & 2 \\ -1 & -4 & 1 & 3 \end{array}\right]$$

Using gauss-elimination method on above matrix we get,

$$\left[\begin{array}{ccc|c} 2 & -1 & 3 & 1 \\ 3 & -2 & 5 & 2 \\ -1 & -4 & 1 & 3 \end{array}\right] \xrightarrow[R_3+\frac{1}{2}R_1]{R_2-\frac{3}{2}R_1}$$

$$\left[\begin{array}{ccc|c} 2 & -1 & 3 & 1 \\ 0 & -1/2 & 1/2 & 1/2 \\ 0 & -9/2 & 5/2 & 7/2 \end{array}\right]$$

$$\xrightarrow{R_3-9R_2} \left[\begin{array}{ccc|c} 2 & -1 & 3 & 1 \\ 0 & -1/2 & 1/2 & 1/2 \\ 0 & 0 & -2 & -1 \end{array}\right]$$

$$\text{Rank}([A \mid B]) = 3$$
$$\text{Rank}([A]) = 3$$

Since Rank ([A | B]) = Rank ([A]) = number of variables, the system has unique solution.

6.22 (b)

$$A = \begin{bmatrix} 2 & -1 \\ -4 & 5 \end{bmatrix}$$

The characteristic equation of this matrix is given by

$$|A - \lambda I| = 0$$
$$\begin{vmatrix} 2-\lambda & -1 \\ -4 & 5-\lambda \end{vmatrix} = 0$$
$$(2-\lambda)(5-\lambda) - 4 = 0$$
$$\lambda^2 - 7\lambda + 6 = 0$$
$$\lambda = 1, 6$$

∴ The eigen values of A are 1 and 6.

6.23 (d)

Given that Fu = b and Fv = b

If F is non singular, then it has a unique inverse.

Now, $u = F^{-1}b$ and $v = F^{-1}b$

Since F^{-1} is unique u = v but it is given that u ≠ v.

This is a contradiction. So F must be singular.

This means that

(a) Determinate of F is zero is true. Also

(b) There are infinite number of solution to Fx = b is true since |F| = 0.

(c) There is an X ≠ 0 such that F X = 0 is also true, since X has infinite number of solutions, including the X = 0 solution.

(d) F must have 2 identical rows is false, since a determinant may become zero, even if two identical columns are present. It is not necessary that 2 identical rows must be present for |F| to become zero.

6.24 (a)

$$P = \begin{bmatrix} a & 1 & 0 \\ 1 & a & 1 \\ 0 & 1 & a \end{bmatrix}$$

Since P is a square matrix

Sum (eigen values) = Trace (P)

= a + a + a = 3a

Product of eigen values

$= |P| = a(a^2 - 1) - a = a^3 - 2a$

Only choice (a) has sum of eigen values = 3a and product of eigen values = $a^3 - 2a$.

6.25 (b)

Theorem: The maximum value of $x^T Ax$ where the maximum is taken over all x that are the unit eigen-vectors of A is the maximum eigen value of A.

Find the eigen value of A.

$$\begin{vmatrix} 3-\lambda & 1 \\ 1 & 2-\lambda \end{vmatrix} = 0$$
$$\lambda^2 - 5\lambda + 5 = 0$$

The eigen values of A = $\left\{\frac{5+\sqrt{5}}{2}, \frac{5-\sqrt{5}}{2}\right\}$

The required answer is the maximum eigen value of A = $\frac{5+\sqrt{5}}{2}$.

6.26 (b)

$$X = \{x \in R^3 \mid x_1 + x_2 + x_3 = 0\}$$
$$X^T = [x_1, x_2, x_3]^T \text{ then}$$

$\{[1, -1, 0]^T, [1, 0, -1]^T\}$ is a linearly independent set because one cannot be obtained from another by scalar multiplication. However (1, –1, 0) and (1, 0, –1) do not span X, since all such combinations (x_1, x_2, x_3) such that $x_1 + x_2 + x_3 = 0$ cannot be expressed as linear combinations of (1, –1, 0) and (1, 0, –1) [we need (0, 1, –1) also to span X].

6.27 (d)

The augmented matrix for above system is

$$\left[\begin{array}{ccc|c} 1 & 1 & 2 & 1 \\ 1 & 2 & 3 & 2 \\ 1 & 4 & a & 4 \end{array}\right] \xrightarrow[R_3 - R_1]{R_2 - R_1} \left[\begin{array}{ccc|c} 1 & 1 & 2 & 1 \\ 0 & 1 & 1 & 1 \\ 0 & 3 & a-2 & 3 \end{array}\right]$$

$$\xrightarrow{R_3 - 3R_2} \left[\begin{array}{ccc|c} 1 & 1 & 2 & 1 \\ 0 & 1 & 1 & 1 \\ 0 & 0 & a-5 & 0 \end{array}\right]$$

Now as long as a – 5 ≠ 0, rank (A) = rank (A | B) = 3

∴ a can take any real value except 5.

Closest correct answer is (d).

6.28 (a)

Eigen values of $\begin{bmatrix} 1 & 0 \\ 0 & 0 \end{bmatrix}$

$$\begin{vmatrix} 1-\lambda & 0 \\ 0 & 0-\lambda \end{vmatrix} = 0$$
$$(1-\lambda) * -\lambda = 0$$

$$\lambda = 0 \text{ or } \lambda = 1$$

Eigen values of $\begin{bmatrix} 0 & 1 \\ 0 & 0 \end{bmatrix}$

$$\begin{bmatrix} -\lambda & 1 \\ 0 & -\lambda \end{bmatrix} = 0$$

$$\lambda^2 = 0, \ \lambda = 0, 0$$

Eigen values of $\begin{bmatrix} 1 & -1 \\ 1 & 1 \end{bmatrix}$

$$\begin{bmatrix} 1-\lambda & -1 \\ 1 & 1-\lambda \end{bmatrix} = 0$$

$$(1-\lambda)^2 + 1 = 0$$
$$(1-\lambda)^2 = -1$$
$$1-\lambda = \text{i or} -\text{i}$$
$$\lambda = 1-\text{i or } 1+\text{i}$$

Eigen values of $\begin{bmatrix} -1 & 0 \\ 1 & -1 \end{bmatrix}$

$$\begin{bmatrix} -1-\lambda & 0 \\ 1 & -1-\lambda \end{bmatrix}$$

$$(-1-\lambda)(-1-\lambda) = 0$$
$$(1+\lambda)^2 = 0$$
$$\lambda = -1, -1$$

Only one matrix has an eigen value of 1 which is $\begin{bmatrix} 1 & 0 \\ 0 & 0 \end{bmatrix}$.

6.29 (d)

S1 and S2:

Since M has zero determinant, its rank is not full i.e. if M is of size 3 × 3, then its rank is not 3. So there is a linear combination of rows which evaluates to 0 i.e. $k_1R_1 + k_2R_2 + ... + k_nR_n = 0$ and there is a linear combination of columns which evaluates to 0 i.e.

$$k_1C_1 + k_2C_2 + ... + k_nC_n = 0$$

Now any row R_i can be written as linear combination of other rows as:

$$R_i = \frac{k_1}{k_i}R_1 - \frac{k_2}{k_i}R_2 - ...$$

$$-\frac{k_{i-1}}{k_i}R_{i-1} - \frac{k_{i+1}}{k_i}R_{i+1} - ... - \frac{k_n}{k_i}R_n$$

Similar is the case for columns.

So S1 and S2 are true.

S3: MX = O is a homogenous equation and such an equation when |M| = 0 has non-trivial solution.

S4: Since M is singular, M has no inverse. So S4 is false.

So S1, S2 and S3 only are true.

6.30 (d)

Sum of eigen values = Trace (A) = 2 + y

Product of eigen values = |A| = $2y - 3x$

$$\therefore \quad 4 + 8 = 2 + y \quad ...\text{(i)}$$
$$4 \times 8 = 2y - 3x \quad ...\text{(ii)}$$
$$\therefore \quad 2 + y = 12 \quad ...\text{(i)}$$
$$2y - 3x = 32 \quad ...\text{(ii)}$$

∴ Solving (i) and (ii) we get $x = -4$ and $y = 10$.

6.31 (a)

Since the given matrix is upper triangular, its eigen values are the diagonal elements themselves, which are 1, 4 and 3.

6.32 (d)

$$A = \begin{bmatrix} 1 & 1 \\ 1 & -1 \end{bmatrix}$$

Eigen (A) are the roots of the characteristic polynomial given below:

$$\begin{vmatrix} 1-\lambda & 1 \\ 1 & -1-\lambda \end{vmatrix} = 0$$

$$(1-\lambda)(-1-\lambda) - 1 = 0$$
$$-(1-\lambda)(1+\lambda) - 1 = 0$$
$$\lambda^2 - 2 = 0$$
$$\lambda = \pm\sqrt{2}$$

Eigen values of A are $\sqrt{2}$ and $-\sqrt{2}$ respectively.

So eigen values of $A^{19} = (\sqrt{2})^{19}$ and $(-\sqrt{2})^{19}$

$$= 2^{19/2} \text{ and } -2^{19/2} = 2^9.2^{1/2} \text{ and } -2^9.2^{1/2}$$
$$= 512\sqrt{2} \text{ and } -512\sqrt{2}$$

6.33 (a)

The given matrix can be transformed into the matrix given in options (b)(c) and (d) by elementry operations of the type of $R_i \pm kR_j$ or $C_i \pm kC_j$ only as shown below:

Option (b):

$$\begin{vmatrix} 1 & x & x^2 \\ 1 & y & y^2 \\ 1 & z & z^2 \end{vmatrix} \xrightarrow[C_3+C_1]{C_2+C_1} \begin{vmatrix} 1 & x+1 & x^2+1 \\ 1 & y+1 & y^2+1 \\ 1 & z+1 & z^2+1 \end{vmatrix}$$

Option (c):

$$\begin{vmatrix} 1 & x & x^2 \\ 1 & y & y^2 \\ 1 & z & z^2 \end{vmatrix} \xrightarrow[R_2-R_3]{R_1-R_2} \begin{vmatrix} 0 & x-y & x^2-y^2 \\ 0 & y-z & y^2-z^2 \\ 1 & z & z^2 \end{vmatrix}$$

Option (d):

$$\begin{vmatrix} 1 & x & x^2 \\ 1 & y & y^2 \\ 1 & z & z^2 \end{vmatrix} \xrightarrow[R_2+R_3]{R_1+R_2} \begin{vmatrix} 2 & x+y & x^2+y^2 \\ 2 & y+z & y^2+z^2 \\ 1 & z & z^2 \end{vmatrix}$$

Option (a): We can show the given matrix can not be converted into option (a) without doing a column exchange which will change the sign of the determinant as can be seen below:

$$\begin{vmatrix} 1 & x & x^2 \\ 1 & y & y^2 \\ 1 & z & z^2 \end{vmatrix} \xrightarrow[C_3+C_2]{C_2+C_1} \begin{vmatrix} 1 & x+1 & x(x+1) \\ 1 & y+1 & y(y+1) \\ 1 & z+1 & z(z+1) \end{vmatrix}$$

$$= -\begin{vmatrix} 1 & x(x+1) & x+1 \\ 1 & y(y+1) & y+1 \\ 1 & z(z+1) & z+1 \end{vmatrix}$$

6.34 (1)

Performing gauss elimination on the augmented matrix shown below:

$$\left[\begin{array}{ccc|c} 3 & 2 & 0 & 1 \\ 4 & 0 & 7 & 1 \\ 1 & 1 & 1 & 3 \\ 1 & -2 & 7 & 0 \end{array}\right]$$

We can reduce it to

$$\left[\begin{array}{ccc|c} 1 & 1 & 1 & 3 \\ 0 & -1 & -3 & -8 \\ 0 & 0 & 15 & 21 \\ 0 & 0 & 15 & 21 \end{array}\right] \xrightarrow{R_4-R_3} \left[\begin{array}{ccc|c} 1 & 1 & 1 & 3 \\ 0 & -1 & -3 & -8 \\ 0 & 0 & 15 & 21 \\ 0 & 0 & 0 & 0 \end{array}\right]$$

Rank (A) = 3

Rank (A/B) = 3

Rank (A) = Rank (A/B) = Number of variables = 3.

So unique solution.

So, number of solutions = 1.

6.35 (0)

The value of the dot product of the eigenvectors corresponding to any pair of different eigen values of any symmetric positive definite matrix is 0.

6.36 (0)

$$A = \begin{bmatrix} 2 \\ -4 \\ 7 \end{bmatrix} [1 \;\; 9 \;\; 5]$$

$$A = [X]_{3\times 1} [Y]_{1\times 3}$$

So size of A is [3 × 3]

$$A = \begin{bmatrix} 2 & 18 & 10 \\ -4 & -36 & -20 \\ 7 & 63 & 35 \end{bmatrix}$$

Since $R_2 = -2R_1$, rows are linearly dependent, $|A| = 0$.

6.37 (6)

$$A = \begin{bmatrix} 1 & 0 & 0 & 0 & 1 \\ 0 & 1 & 1 & 1 & 0 \\ 0 & 1 & 1 & 1 & 0 \\ 0 & 1 & 1 & 1 & 0 \\ 1 & 0 & 0 & 0 & 1 \end{bmatrix}, \text{ let } X = \begin{bmatrix} x_1 \\ x_2 \\ x_3 \\ x_4 \\ x_5 \end{bmatrix}$$

$Ak = Xk$

$\Rightarrow x_1 + x_5 = kx_1 = kx_5$

$\Rightarrow x_2 + x_3 + x_4 = kx_2 = kx_3 = kx_4$

(i) $k \neq 0$

Say, $x_1 = x_5 = a$

$x_2 = x_3 = x_4 = b$

$\Rightarrow x_1 + x_5 = kx_1 \Rightarrow 2a = ka$

$\Rightarrow k = 2$

$\Rightarrow x_2 + x_3 + x_4 = kx_2 \Rightarrow 3b = kb$

$\Rightarrow k = 3$

(ii) $k = 0$

$\Rightarrow$ Eigen value $k = 0$

$\therefore$ There are 3 distinct eigen values: 0, 2, 3

Product of non-zero eigen values: 2 × 3 = 6

6.38 (a)

Sum (eigen values) = Trace (A)

Product of eigen values = $|A|$

(a) If the trace of the matrix is positive and the determinant of the matrix is negative, at least one of its eigenvalues is negative.

This is true, since if this was is not true then all eigen values would be positive and determinant also therefore positive.

(b) If the trace of the matrix is positive, all its eigenvalues are positive.

This is false, since the sum of positive and negative eigen values could also be positive.

(c) If the determinant of the matrix is positive, all its eigenvalues are positive.

This is false, since product of 2 negative numbers can be positive.

(d) If the product of the trace and determinant of the matrix is positive, all its eigenvalues are positive.

Trace and determinant both could be negative to give a positive product and determinant is negative then all eigen values need not be positive.

6.39 (2)

$$V = \{a, b, c, d, e, f\}$$

$\Rightarrow$ $V_1 = \{a, b, c, d\}$ and $V_2 = \{e, f, \square, \square\}$

Smallest possible dimension of $V_1 \cap V_2 = 2$

6.40 (7)

$$\begin{aligned} P(X) &= X^5 + 4X^3 + 6X + 5 \\ &= X(X^4 + 4X^2 + 6) + 5 \\ &= X(X(X^3 + 4X) + 6) + 5 \\ &= X(X(X(X^2 + 4)) + 6) + 5 \end{aligned}$$

Only one temporary variable can be used.

temp = X*X
temp = temp + 4;
temp = X*temp;
temp = X*temp;
temp = temp + 6;
temp = X*temp;
temp = temp + 5

$\therefore$ 7 arithmetic operations needed.

6.41 (d)

Trace = Sum of eigen values

$$1 + a = 6$$

$\Rightarrow$ $a = 5$

Determinant = Product of eigen values

$$(a - 4b) = -7$$
$$5 - 4b = -7$$
$$-4b = -12$$

$\Rightarrow$ $b = 3$

$\therefore$ $a = 5, b = 3$

6.42 (6)

$$|A - \lambda I| = 0$$

$$\begin{vmatrix} 4-\lambda & 5 \\ 2 & 1-\lambda \end{vmatrix} = (4 - \lambda)(1 - \lambda) - 10 = 0$$

$$\lambda^2 - 5\lambda - 6 = 0$$

$$(\lambda - 6)(\lambda + 1) = 0$$

$\Rightarrow$ $\lambda = 6, -1$

$\therefore$ The Larger eigen value is '6'.

6.43 (0)

Since operations 1 and 2 are elementary operations of the type of $R_i \pm kR_j$ and $C_i \pm kC_j$ respectively, the determinant will be unchanged from the original determinant.

So the required determinant

$$= \begin{vmatrix} 3 & 4 & 45 \\ 7 & 9 & 105 \\ 13 & 2 & 195 \end{vmatrix}$$

$$\begin{vmatrix} 3 & 4 & 45 \\ 7 & 9 & 105 \\ 13 & 2 & 195 \end{vmatrix} \xrightarrow{C_3 - 15C_1} \begin{vmatrix} 3 & 4 & 0 \\ 7 & 9 & 0 \\ 13 & 2 & 0 \end{vmatrix} = 0$$

So the required determinant = 0.

6.44 (b)

Let, $A = \begin{bmatrix} 1 & -1 & 2 \\ 0 & 1 & 0 \\ 1 & 2 & 1 \end{bmatrix}$

Given eigen value $\lambda = 1$.

Lex X be the vector. Then $(A - \lambda I)X = 0$

$$\begin{bmatrix} 1-\lambda & -1 & 2 \\ 0 & 1-\lambda & 0 \\ 1 & 2 & 1-\lambda \end{bmatrix} X = 0$$

Put, $\lambda = 1$

$$\begin{bmatrix} 0 & -1 & 2 \\ 0 & 0 & 0 \\ 1 & 2 & 0 \end{bmatrix}\begin{bmatrix} x_1 \\ x_2 \\ x_3 \end{bmatrix} = 0 \Rightarrow \begin{bmatrix} -x_2 + 2x_3 \\ 0 \\ x_1 + 2x_2 \end{bmatrix} = 0$$

Putting $x_1 = k$

We get $x_2 = -k/2$ and $x_3 = -k/4$.

So the eigen vector $= k\begin{bmatrix} 1 \\ -1/2 \\ -1/4 \end{bmatrix}$

The ratios are $x_1/x_2 = \dfrac{-1}{-1/2} = -2$

and $x_2/x_3 = \dfrac{-1/2}{-1/4} = 2$

Only option (b) (–4, 2, 1) has the same ratios and therefore is a correct eigen vector.

or

Alternate Method:

$$AX = \lambda X$$

Check one-by-one each eigen vector until the equation is satisfied. For example,

Choice (a): $\begin{bmatrix} 1 & -1 & 2 \\ 0 & 1 & 0 \\ 1 & 2 & 1 \end{bmatrix}\begin{bmatrix} 4 \\ 2 \\ 1 \end{bmatrix} = 1 \times \begin{bmatrix} 4 \\ 2 \\ 1 \end{bmatrix}$ **(false)**

Choice (b): $\begin{bmatrix} 1 & -1 & 2 \\ 0 & 1 & 0 \\ 1 & 2 & 1 \end{bmatrix}\begin{bmatrix} -4 \\ 2 \\ 1 \end{bmatrix} = 1 \times \begin{bmatrix} -4 \\ 2 \\ 1 \end{bmatrix}$ **(true)**

and so on...

Only choice (b) satisfies and so it is the correct answer.

6.45 (5)

$$\begin{bmatrix} L_{11} & 0 \\ L_{21} & L_{22} \end{bmatrix}\begin{bmatrix} 1 & U_{12} \\ 0 & 1 \end{bmatrix} = \begin{bmatrix} 2 & 2 \\ 4 & 9 \end{bmatrix}$$

This is crout's LU decomposition, since diagonal elements of U are 1. So we will setup the equations for the elements of the matrix taken column-wise, as follows

$$L_{11} = 2,\ L_{21} = 4$$
$$L_{11} \times U_{12} = 2$$
$$\Rightarrow \quad U_{12} = 1$$
$$L_{21} \times U_{12} + L_{22} = 9$$
$$\Rightarrow \quad 4 + L_{22} = 9$$
$$\Rightarrow \quad L_{22} = 5$$

6.46 (c)

$$px + qy + rz = 0$$
$$qx + ry + pz = 0$$
$$rx + py + qz = 0$$

Let $A = \begin{bmatrix} p & q & r \\ q & r & p \\ r & p & q \end{bmatrix}$. The system is $A\hat{x} = O$

This is a homogenous system. Such a system has non-trivial solution iff $|A| = 0$.

So, $\begin{vmatrix} p & q & r \\ q & r & p \\ r & p & q \end{vmatrix} = 0$

$$p(qr - p^2) - q(q^2 - pr) + r(pq - r^2) = 0$$
$$p^3 + q^3 + r^3 - 3pqr = 0$$

p = q = r satisfies the above equation.

Also if $p + q + r = 0$ then a can be transformed into one of the row as completely 0's as shown below.

$$\begin{vmatrix} p & q & r \\ q & r & p \\ r & p & q \end{vmatrix} \xrightarrow{R_1+R_2+R_3} \begin{vmatrix} p+q+r & p+q+r & p+q+r \\ q & r & p \\ r & p & q \end{vmatrix}$$

$$= \begin{vmatrix} 0 & 0 & 0 \\ q & r & p \\ r & p & q \end{vmatrix} = 0$$

Therefore the correct option is (c) which is $p + q + r = 0$ or $p = q = r$.

6.47 (15)

Two eigen values are $2 + i$ and 3 of a 3×3 matrix. The third eigen value must be $2 - i$

Now, $\Pi\lambda_i = |A|$

$\Rightarrow \quad |A| = (2 + i)(2 - i) \times 3$

$= (4 - i^2) \times 3 = 5 \times 3 = 15$

6.48 (c)

I. $m < n$ (system may still be inconsistent so incorrect)

II. $m > n$ (rank may still be equal to n of hence solution may exist so incorrect).

III. $m = n$ (some system rank may be equal to n and hence may have solution so correct).

So only III is correct.

6.49 (0.125)

Eigen(A) = 1, 2, 4

$\Rightarrow |A| = 1 \times 2 \times 4 = 8$

Now, $|(A^{-1})^T| = |A^{-1}| = \dfrac{1}{|A|} = \dfrac{1}{8} = 0.125$

6.50 (c)

The equation $Ax = b$ becomes according to what is given

$$[a_1, a_2, \ldots a_n]\begin{bmatrix} x_1 \\ x_2 \\ \vdots \\ x_n \end{bmatrix} = a_1 + a_2 + a_3 + \ldots a_n$$

where a_i are column vectors in R^n but since we have C_i (not all zero) such that,

$$\sum C_j a_j = 0$$

it means the n column vectors are not linearly independent and hence

$$\text{rank}(A) < n$$

So we have infinitely many solutions one of which will be J_n, where J_n denotes a n-dimensional vector of all 1.

6.51 (b)

Consider a random matrix which satisfy property of square symmetrix matrix contain real values

i.e., $A = \begin{bmatrix} 0 & 5 \\ 5 & 0 \end{bmatrix}_{2\times2}$ whose eigen values +5 and −5. Since, $|A| = -25$ and sum of trace $|A| = 0$ i.e., $+5 - 5 = 0$.

Since, rank of matrix is 2, so atleast one eigen value would be zero for every square matrix size n > 2. For $n = 2$

$$\lambda_1^2 + \lambda_2^2 \le \sum_{i=1}^{n}\sum_{i=1}^{n} A_{ij}^2$$

$$\lambda_1^2 + \lambda_2^2 \le 50$$

Both λ_1 and λ_2 are real, since A is real symmetric matrix, which means atleast one eigen value would be in range [−5, 5]. But since in our

example no eigen value greater than 5. So IInd statement is wrong.
So, option (b) is only correct.

6.52 (2)

$$P + Q = \begin{bmatrix} 0 & -1 & -2 \\ 8 & 9 & 10 \\ 8 & 8 & 8 \end{bmatrix}$$

$$|P + Q| = -16 + 16 = 0$$

So, rank $\neq 3$

Take the 2×2 minor $\begin{bmatrix} 0 & -1 \\ 8 & 9 \end{bmatrix} = 8 \neq 0$

So, rank of P + Q is 2.

6.53 (5)

$$f(\lambda) = \lambda^3 - 4\lambda^2 + a\lambda + 30 = 0$$

Now 2 is one of roots of this equation

So, $2^3 - 4 \times 2^2 + a \times 2 + 30 = 0$

$\Rightarrow 8 - 16 + 2a + 30 = 0$

$\Rightarrow a = -11$

So, the equation is $\lambda^3 - 4\lambda^2 - 11\lambda + 30 = 0$

Now, by polynomials division we get

$$\frac{\lambda^3 - 4\lambda^2 - 11\lambda + 30}{\lambda - 2} = \lambda^2 - 2\lambda - 15$$

roots of $\lambda^2 - 2\lambda - 15 = 0$ are

$$\lambda = \frac{2 \pm \sqrt{4 + 60}}{2} = \frac{2 \pm 8}{2} = 5 \text{ and } -3$$

So the eigen values are 2, 5 and –3, the maximum absolute eigen value is 5.

6.54 (3)

$$u = \begin{pmatrix} 1 \\ 2 \end{pmatrix}, v = \begin{pmatrix} 1 \\ 1 \end{pmatrix}$$

$$A = uv^T$$

$$= \begin{bmatrix} 1 \\ 2 \end{bmatrix} [1 \ \ 1] = \begin{bmatrix} 1 & 1 \\ 2 & 2 \end{bmatrix}$$

$$(1 - \lambda)(2 - \lambda) - 2 = 0$$

$$\lambda^2 - 3\lambda = 0$$

$$\lambda(\lambda - 3) = 0$$

$$\lambda = 0$$

or, $\lambda = 3$

The largest eigen value is 3.

6.55 (d)

Only Eigen vector is $\begin{bmatrix} 1 \\ 4 \end{bmatrix}$ multiples means that eigen value is repeated since if eigen values were distinct we will get one more independent eigen vector. So, II P has repeated eigen values is true. I need not be true since $\begin{bmatrix} 2 & 0 \\ 0 & 2 \end{bmatrix}$ has repeated eigen values and yet it is invertible. III is true since if a 2×2 matrix has only one linearly independent eigen vector, surely it cannot be diagonalized.

6.56 (c)

Both I and II are equivalent statements.

6.57 (12)

Product of eigen values is same as the determinant of a matrix.

$$\begin{bmatrix} 1 & 2 & 2^2 & 2^3 \\ 1 & 3 & 3^2 & 3^3 \\ 1 & 4 & 4^2 & 4^3 \\ 1 & 5 & 5^2 & 5^3 \end{bmatrix} = \begin{bmatrix} 1 & 2 & 2^2 & 2^3 \\ 0 & 3-2 & 3^2-2^2 & 3^3-2^3 \\ 0 & 4-2 & 4^2-2^2 & 4^3-2^3 \\ 0 & 5-2 & 5^2-2^2 & 5^3-2^3 \end{bmatrix}$$

$$= (3 - 2)(4 - 2)(5 - 2) \begin{bmatrix} 1 & 5 & 19 \\ 1 & 6 & 28 \\ 1 & 7 & 39 \end{bmatrix}$$

$$= 1 \,.\, 2 \,.\, 3 \begin{bmatrix} 1 & 5 & 19 \\ 0 & 1 & 9 \\ 0 & 2 & 20 \end{bmatrix}$$

$$= 1 \,.\, 2 \,.\, 3 \,.\, 2 = 12$$

So the product of eigen value is 12.

6.58 (b)

Statement II and III are correct statements directly based on properties of matrices.

6.59 (3)

$$A = \begin{bmatrix} 0 & 1 & 1 & 1 \\ 1 & 0 & 1 & 1 \\ 1 & 1 & 0 & 1 \\ 1 & 1 & 1 & 0 \end{bmatrix}$$

$$|A - \lambda I| = 0 \Rightarrow \begin{vmatrix} -\lambda & 1 & 1 & 1 \\ 1 & -\lambda & 1 & 1 \\ 1 & 1 & -\lambda & 1 \\ 1 & 1 & 1 & -\lambda \end{vmatrix} = 0$$

$C_1 \to C_1 + C_2 + C_3 + C_4$

$$\begin{vmatrix} -\lambda + 3 & 1 & 1 & 1 \\ -\lambda + 3 & -\lambda & 1 & 1 \\ -\lambda + 3 & 1 & -\lambda & 1 \\ -\lambda + 3 & 1 & 1 & -\lambda \end{vmatrix} = 0$$

$$\Rightarrow \quad (-\lambda + 3) \begin{vmatrix} 1 & 1 & 1 & 1 \\ 1 & -\lambda & 1 & 1 \\ 1 & 1 & -\lambda & 1 \\ 1 & 1 & 1 & -\lambda \end{vmatrix} = 0$$

$R_2 \to R_2 - R_1, R_3 \to R_3 - R_1, R_4 \to R_4 - R_1$

$$\Rightarrow \quad (-\lambda+3)\begin{vmatrix} 1 & 1 & 1 & 1 \\ 0 & -\lambda-1 & 0 & 0 \\ 0 & 0 & -\lambda-1 & 0 \\ 0 & 0 & 0 & -\lambda-1 \end{vmatrix} = 0$$

or $\quad (-\lambda + 3)(-\lambda - 1)^3 = 0$

$\Rightarrow \quad \lambda = 3, -1, -1, -1,$

So largest eigen value is $\lambda = 3$

6.60 (4)

$P_{4\times 5} \Rightarrow$ Number of unknowns $(n) = 5$ in $PX = 0$ ...(1)

Also it is given that Nullity $PX = 0$ is one, i.e., $N(P) = 1$.

Hence, Nullity = Number of unknowns – Rank

$1 = 5 - \rho(P)$

or $\quad \rho(P) = 5 - 1 = 4$

6.63 (a, c, d)

$$|A - \lambda I| = \begin{bmatrix} -9-\lambda & -6 & -2 & -4 \\ -8 & -6-\lambda & -3 & -1 \\ 20 & 15 & 8-\lambda & 5 \\ 32 & 21 & 7 & 12-\lambda \end{bmatrix}$$

$Ax = \lambda x$

(a) $$\begin{bmatrix} -9 & -6 & -2 & -4 \\ -8 & -6 & -3 & -1 \\ 20 & 15 & 8 & 5 \\ 32 & 21 & 7 & 12 \end{bmatrix}\begin{bmatrix} -1 \\ 1 \\ 0 \\ 1 \end{bmatrix} = \lambda \begin{bmatrix} -1 \\ 1 \\ 0 \\ 1 \end{bmatrix}$$

$$= \begin{bmatrix} -1 \\ 1 \\ 0 \\ 1 \end{bmatrix} = \lambda \begin{bmatrix} -1 \\ 1 \\ 0 \\ 1 \end{bmatrix} \Rightarrow \lambda = 1$$

(b) $$\begin{bmatrix} -9 & -6 & -2 & -4 \\ -8 & -6 & -3 & -1 \\ 20 & 15 & 8 & 5 \\ 32 & 21 & 7 & 12 \end{bmatrix}\begin{bmatrix} 1 \\ 0 \\ -1 \\ 0 \end{bmatrix} = \lambda \begin{bmatrix} 1 \\ 0 \\ -1 \\ 0 \end{bmatrix}$$

$$\begin{bmatrix} -7 \\ -5 \\ 12 \\ 25 \end{bmatrix} = \lambda \begin{bmatrix} 1 \\ 0 \\ -1 \\ 0 \end{bmatrix}$$ No real value of λ

Similarly (c)

$$\begin{bmatrix} -9 & -6 & -2 & -4 \\ -8 & -6 & -3 & -1 \\ 20 & 15 & 8 & 5 \\ 32 & 21 & 7 & 12 \end{bmatrix}\begin{bmatrix} -1 \\ 0 \\ 2 \\ 2 \end{bmatrix} = \begin{bmatrix} -3 \\ 0 \\ 6 \\ 6 \end{bmatrix} \lambda = 3$$

6.64 (b)

$$A = \begin{bmatrix} 1 & 2 & 3 & 4 \\ 4 & 1 & 2 & 3 \\ 3 & 4 & 1 & 2 \\ 2 & 3 & 4 & 1 \end{bmatrix}, B = \begin{bmatrix} 3 & 4 & 1 & 2 \\ 4 & 1 & 2 & 3 \\ 1 & 2 & 3 & 4 \\ 2 & 3 & 4 & 1 \end{bmatrix}$$

By interchanging R_1, R_3 of A, we get

$$A = -\begin{bmatrix} 3 & 4 & 1 & 2 \\ 4 & 1 & 2 & 3 \\ 1 & 2 & 3 & 4 \\ 2 & 3 & 4 & 1 \end{bmatrix}$$

$\therefore \quad \det(B) = -\det(A)$

6.65 (2)

	1	2	3	4	5
1	0				
2		0			
3			1		
4				1	
5					0

There is a self loop between 3 to 3 and 4 to 4. Also we know that, sum of the principal diagonal elements is equals to sum of the eigen values.

$\Rightarrow \quad \lambda_1 + \lambda_2 + \lambda_3 + \lambda_4 + \lambda_5 = 2$

6.66 (a)

$$A = \begin{bmatrix} 1 & 2 & 3 \\ 4 & 5 & 6 \\ 7 & 8 & 9 \end{bmatrix}$$

$$|A| = 1(45 - 48) - 2(36 - 42) + 3(32 - 35)$$
$$= -3 + 12 - 9 = 0$$

Since the determinant of matrix is zero, so the product of eigen value = 0

6.68 (b, d)

Let us take $A = \begin{bmatrix} 1 & 2 \\ 2 & -3 \end{bmatrix}$ and $B = \begin{bmatrix} 2 & -3 \\ 1 & 2 \end{bmatrix}$

By $R_1 \leftrightarrow R_2$

$\because \; |A| = -7, \quad |B| = 7$ so (b) is true

$\because \; |B| = -|A|$ so (d) is true

Trace $(A) = -2$ and Trace $(B) = 4$ so (a) is not true

A is symmetric but B is not symmetric so (c) is not true.

■■■■

7 Calculus

7.1 If at every point of a certain curve, the slope of the tangent equals $\frac{-2x}{y}$ the curve is

(a) A straight line (b) A parabola
(c) A circle (d) An ellipse

[1995 : 1 M]

7.2 The formula used to compute an approximation for the second derivative of a function f at a point x_0 is

(a) $\frac{f(x_0+h)+f(x_0-h)}{2}$

(b) $\frac{f(x_0+h)-f(x_0-h)}{2h}$

(c) $\frac{f(x_0+h)+2f(x_0)+f(x_0-h)}{h^2}$

(d) $\frac{f(x_0+h)-2f(x_0)+f(x_0-h)}{h^2}$ **[1996 : 1 M]**

7.3 What is the maximum value of the function $f(x) = 2x^2 - 2x + 6$ in the interval [0, 2]?

(a) 6 (b) 10
(c) 12 (d) 5.5 **[1997 : 2 M]**

7.4 Consider the function $y = |x|$ in the interval [–1, 1]. In this interval, the function is

(a) continuous and differentiable
(b) continuous but not differentiable
(c) differentiable but not continuous
(d) neither continuous nor differentiable

[1998 : 1 M]

7.5 What is the value of $\int_0^{2\pi}(x-\pi)^3(\cos x)dx$

(a) –1 (b) 0
(c) 1 (d) π **[2005 : 2 M]**

7.6 The following definite integral evaluates to

$$\int_{-\infty}^{0} e^{-\left(\frac{x^2}{2}\right)}dx$$

(a) $\frac{1}{2}$ (b) $\pi\sqrt{10}$

(c) $\sqrt{\frac{\pi}{2}}$ (d) π **[2006 : 2 M]**

7.7 If $f(x)$ is defined as follows, what is the minimum value of $f(x)$ for $x \in (0, 2]$?

$$f(x) = \begin{cases} \frac{25}{8x} & \text{when } x \le \frac{3}{2} \\ x + \frac{1}{x} & \text{otherwise} \end{cases}$$

(a) 2 (b) $2\left(\frac{1}{12}\right)$

(c) $2\left(\frac{1}{6}\right)$ (d) $2\left(\frac{1}{2}\right)$ **[2008 : 2 M]**

7.8 Consider the following two statements about the function $f(x) = |x|$

P: f(x) is continuous for all real values of x
Q: f(x) is differentiable for all real values of x

Which of the following is TRUE?

(a) P is true and Q is false
(b) P is false and Q is true
(c) Both P and Q are true
(d) Both P and Q are false

[2007 : 1 M]

7.9 $\lim\limits_{x\to\infty}\frac{x - \sin x}{x + \cos x}$ equals

(a) 1 (b) –1
(c) ∞ (d) – ∞

[2008 : 1 M]

7.10 A point on a curve is said to be an extremum if it is a local minimum or a local maximum. The number of distinct extrema for the curve $3x^4 - 16x^3 + 4x^2 + 37$ is

(a) 0 (b) 1
(c) 2 (d) 3 **[2008 : 2 M]**

7.11 $\int_0^{\pi/4}(1-\tan x)/(1+\tan x)\,dx$ evaluates to

(a) 0 (b) 1
(c) ln 2 (d) $\frac{1}{2}\ln 2$ **[2009 : 2 M]**

7.12 What is the value of $\lim\limits_{n\to\infty}\left(1-\frac{1}{n}\right)^{2n}$?

(a) 0 (b) e^{-2}
(c) $e^{-1/2}$ (d) 1 **[2010 : 1 M]**

7.13 Given $i = \sqrt{-1}$, what will be the evaluation of the definite integral $\int_0^{\pi/2} \frac{\cos x + i \sin x}{\cos x - i \sin x} dx$?

(a) 0 (b) 2
(c) $-i$ (d) i **[2011 : 2 M]**

7.14 Consider the function $f(x) = \sin(x)$ in the interval $x \in \left[\frac{\pi}{4}, \frac{7\pi}{4}\right]$. The number and location(s) of the local minima of this function are

(a) One, at $\frac{\pi}{2}$ $x \in \left[\frac{\pi}{4}, \frac{7\pi}{4}\right]$

(b) One, at $\frac{3\pi}{2}$

(c) Two, at $\frac{\pi}{2}$ and $\frac{3\pi}{2}$

(d) Two, at $\frac{3\pi}{4}$ and $\frac{3\pi}{2}$ **[2012 : 1 M]**

7.15 Which one of the following functions is continuous at x = 3?

(a) $f(x) = \begin{cases} 2, & \text{if } x = 3 \\ x-1, & \text{if } x > 3 \\ \frac{x+3}{3}, & \text{if } x < 3 \end{cases}$

(b) $f(x) = \begin{cases} 4, & \text{if } x = 3 \\ 8-x & \text{if } x \neq 3 \end{cases}$

(c) $f(x) = \begin{cases} x+3, & \text{if } x \leq 3 \\ x-4 & \text{if } x > 3 \end{cases}$

(d) $f(x) = \frac{1}{x^3 - 27}$, if $x \neq 3$ **[2013 : 1 M]**

7.16 Let the function

$$f(q) = \begin{vmatrix} \sin\theta & \cos\theta & \tan\theta \\ \sin(\pi/6) & \cos(\pi/6) & \tan(\pi/6) \\ \sin(\pi/3) & \cos(\pi/3) & \tan(\pi/3) \end{vmatrix}$$

where $\theta \in \left[\frac{\pi}{6}, \frac{\pi}{3}\right]$ and $f'(\theta)$ denote the derivative of f with respect to θ. Which of the following statements is/are TRUE?

I. There exists $\theta \in \left(\frac{\pi}{6}, \frac{\pi}{3}\right)$ such that $f'(\theta) = 0$.

II. There exists $\theta \in \left(\frac{\pi}{6}, \frac{\pi}{3}\right)$ such that $f'(\theta) \neq 0$.

(a) I only (b) II only
(c) Both I and II (d) Neither I nor II
[2014 (Set-1) : 1 M]

7.17 The function $f(x) = x \sin x$ satisfies the following equation: $f''(x) + f(x) + t \cos x = 0$. The value of t is ________. **[2014 (Set-1) : 2 M]**

7.18 A function $f(x)$ is continuous in the interval [0, 2]. It is known that $f(0) = f(2) = -1$ and $f(1) = 1$. Which one of the following statements must be true?

(a) There exists a y in the interval (0, 1) such that $f(y) = f(y+1)$
(b) For every y in the interval (0, 1), $f(y) = f(2-y)$
(c) The maximum value of the function in the interval (0, 2) is 1
(d) There exists a y in the interval (0, 1) such that $f(y) = -f(2-y)$
[2014 (Set-1) : 2 M]

7.19 A non-zero polynomial $f(x)$ of degree 3 has roots at $x = 1$, $x = 2$ and $x = 3$. Which one of the following must be TRUE?

(a) $f(0)f(4) < 0$ (b) $f(0)f(4) > 0$
(c) $f(0) + f(4) > 0$ (d) $f(0) + f(4) < 0$
[2014 (Set-2) : 1 M]

7.20 If $\int_0^{2\pi} |x \sin x| dx = k\pi$, then the value of k is equal to ________. **[2014 (Set-3) : 1 M]**

7.21 The value of the integral given below is

$$\int_0^{\pi} x^2 \cos x \, dx$$

(a) -2π (b) π
(c) $-\pi$ (d) 2π
[2014 (Set-3) : 2 M]

7.22 $\lim_{x \to \infty} x^{1/x}$ is

(a) ∞ (b) 0
(c) 1 (d) Not defined
[2015 (Set-1) : 1 M]

7.23 $\int_{1/\pi}^{2/\pi} \frac{\cos(1/x)}{x^2} dx$ = ________.

[2015 (Set-1) : 2 M]

7.24 Let $f(x) = x^{-(1/3)}$ and A denote the area of the region bounded by $f(x)$ and the X-axis, when x varies from -1 to 1. Which of the following statements is/are True?

1. f is continuous in [–1, 1].
2. f is not bounded in [–1, 1]
3. A is nonzero and finite.

(a) 2 only (b) 3 only
(c) 2 and 3 only (d) 1, 2 and 3
[2015 (Set-2) : 2 M]

7.25 The value of $\lim_{x\to\infty}(1+x^2)^{e^{-x}}$ is

(a) 0 (b) $\frac{1}{2}$

(c) 1 (d) ∞

[2015 (Set-3) : 1 M]

7.26 If for non-zero x, $af(x)+bf\left(\frac{1}{x}\right)=\frac{1}{x}-25$ where $a \neq b$ then $\int_1^2 f(x)\,dx$ is

(a) $\frac{1}{a^2-b^2}\left[a(\ln 2-25)+\frac{47b}{2}\right]$

(b) $\frac{1}{a^2-b^2}\left[a(2\ln 2-25)-\frac{47b}{2}\right]$

(c) $\frac{1}{a^2-b^2}\left[a(2\ln 2-25)+\frac{47b}{2}\right]$

(d) $\frac{1}{a^2-b^2}\left[a(\ln 2-25)-\frac{47b}{2}\right]$

[2015 (Set-3) : 2 M]

7.27 $\lim_{x\to 4}\frac{\sin(x-4)}{x-4}=$ _____.

[2016 (Set-1) : 1 M]

7.28 Let $f(x)$ be a polynomial and $g(x)=f'(x)$ be its derivative. If the degree of $(f(x)+f(-x))$ is 10, then the degree of $(g(x)-g(-x))$ is _______.

[2016 (Set-2) : 1 M]

7.29 The value of $\lim_{x\to 1}\frac{x^7-2x^5+1}{x^3-3x^2+2}$

(a) is 0 (b) is –1

(c) is 1 (d) Does not exist

[2017 (Set-1) : 2 M]

7.30 Consider a quadratic equation $x^2-13x+36=0$ with coefficients in a base 'b'. The solutions of this equation in the same base 'b' are $x=5$ and $x=6$. Then $b=$ _______.

[2017 (Set-2) : 1 M]

7.31 If $f(x)=R\sin\left(\frac{\pi x}{2}\right)+S$, $f'\left(\frac{1}{2}\right)=\sqrt{2}$ and $\int_0^1 f(x)dx=\frac{2R}{\pi}$, then the constants R and S are, respectively

(a) $\frac{2}{\pi}$ and $\frac{16}{\pi}$ (b) $\frac{2}{\pi}$ and 0

(c) $\frac{4}{\pi}$ and 0 (d) $\frac{4}{\pi}$ and $\frac{16}{\pi}$

[2017 (Set-2) : 1 M]

7.32 The value of $\int_0^{\pi/4} x\cos(x^2)\,dx$ correct to three decimal places (assuming that $\pi = 3.14$) is ________.

[2018 : 1 M]

7.33 Compute $\lim_{x\to 3}\frac{x^4-81}{2x^2-5x-3}$

(a) $\frac{108}{7}$ (b) $\frac{53}{12}$

(c) 1 (d) Limit does not exists

[2019 : 1 M]

7.34 Consider the functions:

I. e^{-x} **II.** $x^2-\sin x$

III. $\sqrt{x^3+1}$

Which of the above functions is/are increasing everywhere in [0,1]?

(a) I and III only (b) II and III only

(c) III only (d) II only **[2020 : 1 M]**

7.35 Consider the following expression:

$$\lim_{x\to -3}\frac{\sqrt{2x+22}-4}{x+3}$$

The value of the above expression (rounded to 2 decimal places) is _______.

[2021 (Set-1) : 1 M]

7.36 Suppose that $f: R\to R$ is a continuous function on the interval [–3, 3] and a differentiable function in the interval (–3, 3) such that for every x in the interval. $f'(x)\le 2$. If $f(-3)=7$, then $f(3)$ is at most _________. **[2021 (Set-2) : 1 M]**

7.37 The value of the following limit is ___________.

$$\lim_{x\to 0+}\frac{\sqrt{x}}{1-e^{2\sqrt{x}}}$$

[2022 : 1 M]

7.38 Let $f(x)=x^3+15x^2-33x-36$ be a real-valued function.

Which of the following statements is/are TRUE?

(a) $f(x)$ does not have a local maximum.

(b) $f(x)$ has a local maximum.

(c) $f(x)$ does not have a local minimum.

(d) $f(x)$ has a local minimum. **[2023 : 1 M]**

7.39 The value of the definite integral $\int_{-3}^{3}\int_{-2}^{2}\int_{-1}^{1}(4x^2y-z^3)\,dz\,dy\,dx$ is ______. (Rounded off to the nearest integer)

[2023 : 1 M]

7.40 Let $f : \mathbb{R} \rightarrow \mathbb{R}$ be a function such that $f(x) = \max\{x, x^3\}$, $x \in \mathbb{R}$, where $\mathbb{R}$ is the set of all real numbers. The set of all points where $f(x)$ is NOT differentiable is

(a) $\{0, 1\}$ (b) $\{-1, 1, 2\}$
(c) $\{-2, -1, 1\}$ (d) $\{-1, 0, 1\}$

[2024 (Set-1) : 1 M]

7.41 Let $f(x)$ be a continuous function from $\mathbb{R}$ to $\mathbb{R}$ such that

$$f(x) = 1 - f(2 - x)$$

Which one of the following options is the CORRECT value of $\int_0^2 f(x)\,dx$?

(a) 0 (b) -1
(c) 1 (d) 2

[2024 (Set-2) : 1 M]

7.42 Let x and y be random variables, not necessarily independent, that take real values in the interval $[0,1]$. Let $z = xy$ and let the mean values of x, y, z be $\overline{x}, \overline{y}, \overline{z}$ respectively. Which one of the following statements is TRUE?

(a) $\overline{z} = \overline{x}\,\overline{y}$ (b) $\overline{z} \geq \overline{x}\,\overline{y}$
(c) $\overline{z} \leq \overline{x}\,\overline{y}$ (d) $\overline{z} \leq \overline{x}$

[2024 (Set-2) : 2 M]

■■■■

Answers Calculus

7.1 (d)	7.2 (d)	7.3 (b)	7.4 (b)	7.5 (b)	7.6 (c)	7.7 (b)	7.8 (a)	7.9 (a)
7.10 (b)	7.11 (d)	7.12 (b)	7.13 (d)	7.14 (b)	7.15 (a)	7.16 (c)	7.17 (–2)	7.18 (a)
7.19 (a)	7.20 (4)	7.21 (a)	7.22 (c)	7.23 (–1)	7.24 (c)	7.25 (c)	7.26 (a)	7.27 (1)
7.28 (9)	7.29 (c)	7.30 (6)	7.31 (c)	7.32 (0.289)	7.33 (a)	7.34 (c)	7.35 (0.25)	7.36 (19)
7.37 (–0.5)	7.38 (b, d)	7.39 (0)	7.40 (d)	7.41 (c)	7.42 (d)			

Explanations Calculus

7.1 (d)

Given that at every point of a certain curve, the slope of tangent is $-\frac{2x}{y}$

i.e.,
$$\frac{dy}{dx} = -\frac{2x}{y}$$
$$ydy = -2xdx$$

Integrating both side
$$\frac{y^2}{2} = -2\times\frac{x^2}{2} + c = -x^2 + c$$
$$\frac{y^2}{2} + x^2 = c$$

It is in the form of $\frac{x^2}{a^2} + \frac{y^2}{b^2} = c$

So the curve is an ellipse with $a = 1$ and $b = \sqrt{2}$.

7.2 (d)

The formula used to compute an approximation of first derivative of a function f at a point x_0 is
$$f'(x_0) = \frac{f(x_0+h)-f(x_0)}{h}$$

So an approximation for the second derivative of a function f at a point x_0 is
$$f''(x_0) = \frac{f'(x_0)-f'(x_0-h)}{h}$$
$$= \frac{\left[\left(\frac{f(x_0+h)-f(x_0)}{h}\right)-\left(\frac{f(x_0)-f(x_0-h)}{h}\right)\right]}{h}$$
$$= \frac{f(x_0+h)-2f(x_0)+f(x_0-h)}{h^2}$$

7.3 (b)

We need absolute maximum of
$$f(x) = 2x^2 - 2x + 6 \text{ in the interval } [0, 2]$$

First find local maximum if any by putting $f'(x) = 0$
$$f'(x) = 4x - 2 = 0$$
$$4x = 2$$
$$x = \frac{1}{2}$$
$$f''(x) = 4 > 0$$

So, $x = \frac{1}{2}$

is a point of local minimum. So there is no point of local maximum.

Now tabulate the values of f at end point of interval and at local maximum if any (in this case no point of local maximum).

x	$f(x)$
0	6
2	10

Clearly the absolute maxima is at $x = 2$ and absolute maximum value is 10.

7.4 (b)

The function $y = |x|$ in the interval [–1, 1] is

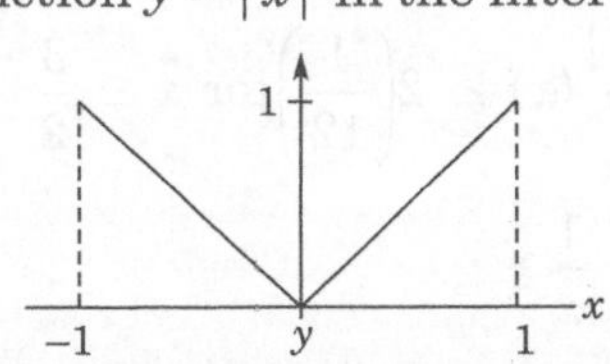

$|x|$ is continuous and differentiable every where except at $x = 0$, where it is continuous but not differentiable.

Since [–1, 1] contains 0, in this interval it is continuous but not differentiable.

7.5 (b)

In the integral $\int_0^{2\pi} (x-\pi)^3 \cos x \, dx$
$$f(x) = (x-\pi)^3 \cos x$$
$$f(2\pi - x) = (2\pi - x - \pi)^3 \cos(2\pi - x)$$
$$= (\pi - x)^3 \cos x = -(x-\pi)^3 \cos x = -f(x)$$

So, $\int_0^{2\pi} (x-\pi)^3 \cos x \, dx = 0$

Since, $\int_0^{2\pi} f(x)dx = 0$ whenever $f(2\pi - x) = -f(x)$

7.6 (c)

$$I = \int_{-\infty}^{0} e^{(-x^2/2)} dx$$

Comparing with area under the standared normal curve from $-\infty$ to 0.

We get

$$\frac{1}{\sqrt{2\pi}} \int_{-\infty}^{0} e^{-\frac{x^2}{2}} dx = 0.5$$

So, the required integral

$$\int_{-\infty}^{0} e^{-\frac{x^2}{2}} dx = 0.5 \times \sqrt{2\pi} = \sqrt{\frac{\pi}{2}}$$

which is option (c).

7.7 (b)

For the function $\frac{25}{8x}$ the minimum value will come when x is maximum since it is a decreasing function.

The maximum value of x is $\frac{3}{2}$.

At $x = \frac{3}{2}$ the function has the value

$$f(x) = \frac{25}{\left(8 \times \frac{3}{2}\right)} = 2.0833 = 2 + \left(\frac{1}{12}\right)$$

So, $\infty \geq f(x) \geq 2\left(\frac{1}{12}\right)$ for $x \leq \frac{3}{2}$

$$x + \frac{1}{x} \geq 2$$

But since for this function $x \geq \frac{3}{2}$, putting $\frac{3}{2}$ in this function we get the minimum of this function which is

$$\frac{3}{2} + \frac{1}{\left(\frac{3}{2}\right)} = \frac{13}{6} = 2.166$$

Now comparing the minimum value 2.0833 of the first function with minimum value 2.166 of the second function, we get the overall minimum of this function to be $2.0833 = 2\left(\frac{1}{12}\right)$ which is option (b).

7.8 (a)

$|x|$ is continuous and differentiable every where except at x = 0, where it is continuous but not differentiable.

Since 0 is a real number, for all real values it is continuous but not differentiable.

7.9 (a)

$$\lim_{x \to \infty} \frac{x - \sin x}{x + \cos x} = \lim_{x \to \infty} \frac{1 - \frac{\sin x}{x}}{1 + \frac{\cos x}{x}}$$

$$= \frac{\lim_{x \to \infty}\left(1 - \frac{\sin x}{x}\right)}{\lim_{x \to \infty}\left(1 + \frac{\cos x}{x}\right)}$$

$$= \frac{1 - \lim_{x \to \infty} \frac{\sin x}{x}}{1 + \lim_{x \to \infty} \frac{\cos x}{x}} = \frac{1-0}{1+0} = 1$$

7.10 (b)

$$y = 3x^4 - 16x^3 + 24x^2 + 37$$

$$\frac{dy}{dx} = 12x^3 - 48x^2 + 48x = 0$$

$$x(12x^2 - 48x + 48) = 0$$

$$x = 0$$

or $\quad 12x^2 - 48x + 48 = 0$

$$x^2 - 4x + 4 = 0$$

$$(x-2)^2 = 0$$

$$\frac{d^2y}{dx^2} = 36x^2 - 96x + 48$$

Now at $\quad x = 0$

$$\frac{d^2y}{dx^2} = 48 > 0$$

$\therefore$ f(x) has a minimum at x = 0

$$\left[\frac{d^2y}{dx^2}\right]_{x=2} = 0 \text{ and } \left[\frac{d^3y}{dx^2}\right]_{x=2} = 48 \neq 0$$

So, x = 2 is a saddle point (point of inflection)

$\therefore$ f(x) has no extremum at x = 2. So f(x) has only one point of extremum (at x = 0).

7.11 (d)

Since

$$\int_0^a f(x)\,dx = \int_0^a f(a-x)\,dx$$

$$\therefore \quad I = \int_0^{\pi/4} \frac{1 - \tan x}{1 + \tan x} dx$$

$$= \int_0^{\pi/4} \frac{1-\tan\left(\frac{\pi}{4}-x\right)dx}{1+\tan\left(\frac{\pi}{4}-x\right)}$$

Since $\tan(A-B) = \dfrac{\tan A - \tan B}{1+\tan A \tan B}$

$$\therefore \quad I = \int_0^{\pi/4} \frac{1-\left[\dfrac{\tan\frac{\pi}{4}-\tan x}{1+\tan\frac{\pi}{4}\tan x}\right]}{1+\left[\dfrac{\tan\frac{\pi}{4}-\tan x}{1+\tan\frac{\pi}{4}\tan x}\right]}dx$$

$$= \int_0^{\pi/4} \frac{1-\left[\dfrac{1-\tan x}{1+\tan x}\right]}{1+\left[\dfrac{1-\tan x}{1+\tan x}\right]}dx$$

$$= \int_0^{\pi/4} \frac{(1+\tan x)-(1-\tan x)}{(1+\tan x)+(1-\tan x)}dx$$

$$= \int_0^{\pi/4} \frac{2\tan x}{2}dx = \int_0^{\pi/4} \tan x\, dx$$

$$= \left[\log(\sec x)\right]_0^{\pi/4}$$

$$= ln\left(\sec\frac{\pi}{4}\right) - ln(\sec 0)$$

$$= ln(\sqrt{2}) - ln(1)$$

$$= ln(2^{1/2}) - 0 = \frac{1}{2} ln 2$$

7.12 (b)

$$\lim_{n\to\infty}\left(1-\frac{1}{n}\right)^{2n} = \lim_{n\to\infty}\left[\left(1-\frac{1}{n}\right)^n\right]^2$$

$$= \left[\lim_{n\to\infty}\left(1-\frac{1}{n}\right)^n\right]^2$$

Now use the standard limit:

$$\lim_{x\to\infty}\left(1+\frac{a}{x}\right)^x = e^a$$

Required limit = $(e^{-1})^2 = e^{-2}$

7.13 (d)

According to **Demoivre's** theorem

$\cos x + i \sin x = e^{ix}$

$$\text{So, } \int_0^{\pi/2} \frac{\cos x + i\sin x}{\cos x - i \sin x}dx = \int_0^{\pi/2} \frac{e^{ix}}{e^{-ix}}dx$$

$$= \int_0^{\pi/2} e^{2ix}\,dx = \left[\frac{e^{2ix}}{2i}\right]_0^{\pi/2} = \frac{1}{2i}\left[e^{i\pi} - e^{o}\right]$$

$$= \frac{1}{2i}[-1-1] (\text{since } e^{i\pi} = \cos\pi + i\sin\pi = -\pi)$$

$$= \frac{-2}{2i} = \frac{-1}{i} = i$$

7.14 (b)

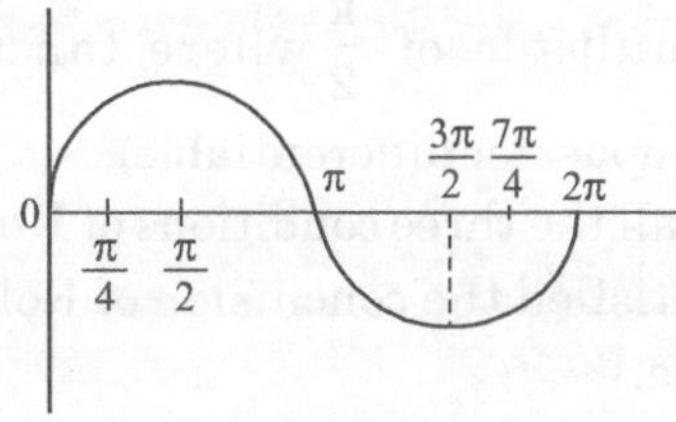

From the plot of sin x given above, we can easily see that in the range $\left[\frac{\pi}{4}, \frac{7\pi}{4}\right]$, there is only one local minima, at $\frac{3\pi}{2}$.

7.15 (a)

$$\begin{cases} 2, & \text{if } x = 3 \\ x-1, & \text{if } x > 3 \\ \dfrac{x+3}{3}, & \text{if } x < 3 \end{cases}$$

$$\lim_{x\to3^-} f(x) = \lim_{x\to3^-} \frac{x+3}{3} = \frac{3+3}{3} = 2$$

$$\lim_{x\to3^+} f(x) = \lim_{x\to3^+} x-1 = 3-1 = 2$$

Also, $f(3) = 2$

So, $\lim_{x\to3^-} f(x) = \lim_{x\to3^+} f(x) = f(3)$

So it is continuous at x = 3

Option (a) is correct.

7.16 (c)

$$f(\theta) = \begin{vmatrix} \sin\theta & \cos\theta & \tan\theta \\ \sin(\pi/6) & \cos(\pi/6) & \tan(\pi/6) \\ \sin(\pi/3) & \cos(\pi/3) & \tan(\pi/3) \end{vmatrix}$$

$$f\left(\frac{\pi}{6}\right) = 0$$

Since if we put $\theta = \frac{\pi}{6}$ in above determinant it will evaluate to zero, since I and II row will become same.

$$f\left(\frac{\pi}{3}\right) = 0$$

Since if we put $\theta = \frac{\pi}{3}$ in above determinant it will evaluate to zero, since I and III row will become same.

So $f\left(\frac{\pi}{6}\right) = f\left(\frac{\pi}{3}\right)$. Also in the interval $\left[\frac{\pi}{6}, \frac{\pi}{3}\right]$ the function f(θ) is continuous and differentiable (**note** that the given interval doesn't contain any odd multiple of $\frac{\pi}{2}$ where tan θ is neither continuous nor differentiable).

Since all the three conditions of **Roll's theorem** are satisfied the conclusion of Rolls theorem is true i.e.

I. $\exists\theta \in \left(\frac{\pi}{6}, \frac{\pi}{3}\right)$ such that $f'(\theta) = 0$ is true

Now the statement

II. $\exists\theta \in \left(\frac{\pi}{6}, \frac{\pi}{3}\right)$ such that $f'(\theta) \neq 0$

is also true, since the only way it can be false is if $f'(\theta) = 0$ for all values of θ, which is possible only if $f(\theta)$ is a constant which is untrue.

Therefore, both (I) and (II) are correct.

7.17 (–2)

$$f(x) = x \sin x$$
$$f'(x) = x \cos x + \sin x$$
$$f''(x) = (-x \sin x + \cos x) + \cos x$$
$$f''(x) + f(x) + t \cos x = 0$$
$$\Rightarrow -x \sin x + \cos x + \cos x + x \sin x + t \cos x = 0$$
$$\Rightarrow (2 + t) \cos x = 0$$
$$\Rightarrow t + 2 = 0$$
$$\Rightarrow t = -2$$

7.18 (a)

Consider option (a)

(a) There exists a y in the interval (0, 1) such that $f(y) = f(y + 1)$.

This can be rewritten as

There exists a y in the interval (0, 1) such that

$$g(y) = f(y) - f(y+1) = 0$$

Now intermediate value theorem says that the above statement is true, if $g(0)\, g(1) < 0$.

Now we check $g(0)\, g(1)$

$$g(0) = f(0) - f(1) = -1 - 1 = -2$$
$$g(1) = f(1) - f(2) = 1 - (-1) = 2$$
$$g(0)\, g(1) = -2 \times 2 = -4 < 0$$

So by intermediate value theorem, option (a) is true.

7.19 (a)

$x = 1$, $x = 2$ and $x = 3$ are roots.

$$f(x) = \alpha(x-1)(x-2)(x-3)$$
$$\Rightarrow f(0) = \alpha(0-1)(0-2)(0-3) = -6\alpha$$
$$\Rightarrow f(4) = \alpha(4-1)(4-2)(4-3) = 6\alpha$$
$$f(0) \times f(4) = -36\alpha^2 < 0$$
$$\therefore f(0) \times f(4) < 0$$

Alternate method:

Since polynomials are continuous and since there is a root (actually 3 roots) in the interval (0, 4), by intermediate value theorem, surely $f(0) \times f(4) < 0$.

7.20 (4)

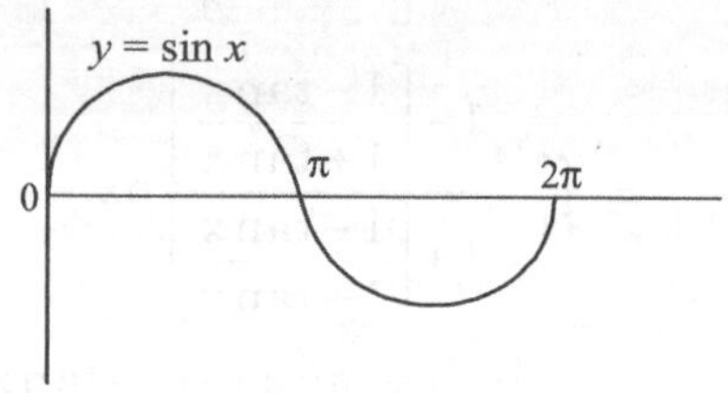

$$\Rightarrow \int_0^{2\pi} |x \sin x|\, dx = k\pi$$
$$\Rightarrow \int_0^{\pi} |x \sin x|\, dx + \int_{\pi}^{2\pi} |x \sin x|\, dx = k\pi$$
$$\Rightarrow \int_0^{\pi} x \sin x\, dx + \int_{\pi}^{2\pi} -(x \sin x)\, dx = k\pi$$
$$\Rightarrow (-x\cos x + \sin x)\Big|_0^{\pi} - (-x\cos x + \sin x)\Big|_{\pi}^{2\pi} = k\pi$$
$$\Rightarrow 4\pi = k\pi$$
$$\Rightarrow k = 4$$

7.21 (a)

Integrating by parts once we get,

$$\int_0^{\pi} x^2 \cos x\, dx = [x^2(\sin x)]_0^{\pi} - 2\int_0^{\pi} x \sin x\, dx$$
$$= 0 - 2\int_0^{\pi} x \sin x\, dx = -2\int_0^{\pi} x \sin x\, dx = -2\pi$$

[Since, integrating by parts again we get

$$\int_0^{\pi} x \sin x\, dx = (-x\cos x + \sin x)\Big|_0^{\pi} = \pi]$$

7.22 (c)

$$y = \lim_{x\to\infty} x^{1/x}$$
$$\log y = \lim_{x\to\infty} \log x^{1/x}$$
$$\log y = \lim_{x\to\infty} \frac{\log x}{x}$$

∞/∞ form, use L' Hospital's rule

$$\log y = \lim_{x\to\infty} \frac{\frac{1}{x}}{1}$$

$$\log y = 0 \Rightarrow y = e^0 = 1$$

7.23 (–1)

$$\int_{1/\pi}^{2/\pi} \frac{\cos\left(\frac{1}{x}\right)}{x^2} dx = (-1)\int_{1/\pi}^{2/\pi} \left(\frac{-1}{x^2}\right)\cdot \cos\left(\frac{1}{x}\right)$$

Let, $\frac{1}{x} = t \Rightarrow dt = -\frac{1}{x^2}$

Substituting in the integral and change limits we get,

$$(-1)\int_{\pi}^{\pi/2} \cos t\, dt = \int_{\pi/2}^{\pi} \cos t\, dt$$

$$= [\sin t]_{\pi 2}^{\pi} = (0 - 1) = -1$$

7.24 (c)

$$f(x) = \frac{1}{\sqrt[3]{x}}$$

Statement 1: f is continuous in [–1, 1]. Let us check this statement.

We need to check continuity at x = 0

$$\text{Let limit} = \lim_{x\to 0^-} \frac{1}{\sqrt[3]{x}} = \lim_{h\to 0} \frac{1}{\sqrt[3]{0-h}}$$

$$= \lim_{h\to 0} \frac{-1}{\sqrt[3]{h}} = -\infty$$

$$\text{Right limit} = \lim_{x\to 0^+} \frac{1}{\sqrt[3]{x}} = \lim_{h\to 0} \frac{1}{\sqrt[3]{0+h}}$$

$$= \lim_{h\to 0} \frac{1}{\sqrt[3]{h}} = +\infty$$

Left limit ≠ Right limit

∴ Statement 1 is false.

Statement 2: f is not bounded is [–1, 1]. Since at x = 0 it goes to –∞ and +∞ the function is not bounded.

∴ Statement 2 is true.

Statement 3: A is non zero and finite.

$$A = \left|\int_{-1}^{0} x^{-1/3} dx\right| + \left|\int_{0}^{1} x^{-1/3} dx\right|$$

$$= \left|\frac{3}{2}\left[x^{2/3}\right]_{-1}^{0}\right| + \left|\frac{3}{2}\left[x^{2/3}\right]_{0}^{1}\right| = \left|\frac{3}{2}\right| + \left|\frac{3}{2}\right| = 3$$

So A is non zero and finite.

∴ Statement 3 is true.

7.25 (c)

Let, $y = \lim_{x\to\infty} (1+x^2)^{e^{-x}}$

$$\log y = \lim_{x\to\infty} \log(1+x^2)^{e^{-x}}$$

$$= \lim_{x\to\infty} \left[\frac{\log(1+x^2)}{e^x}\right]$$

∞/∞ form apply L' Hospital's rule

$$\Rightarrow \log y = \lim_{x\to\infty} \frac{\frac{1}{1+x^2}(2x)}{e^x}$$

$$\Rightarrow \log y = \lim_{x\to\infty} \frac{2x}{(1+x^2)e^x}$$

Again we are getting ∞/∞ form apply L' Hospital's rule

$$\log y = \lim_{x\to\infty} \frac{2}{(1+x^2)e^x + e^x\cdot 2x} = \frac{2}{\infty} = 0$$

$$\Rightarrow y = 1$$

7.26 (a)

$$a\, f(x) + b\, f\left(\frac{1}{x}\right) = \frac{1}{x} - 25 \quad ...(1)$$

Put $x = \frac{1}{x}$ in equation (1)

$$a\, f\left(\frac{1}{x}\right) + b\, f(x) = x - 25 \quad ...(2)$$

Solving equations (1) and (2) simultaneously we get $f(x)$ as follows:

Equation (1) × a – equation (2) × b

$$(1) \times a : \Rightarrow a^2 f(x) + ba\, f\left(\frac{1}{x}\right) = \frac{a}{x} - 25a$$

$$(2) \times b : \Rightarrow ab\, f\left(\frac{1}{x}\right) + b^2 f(x) = bx - 25b$$

$$a^2 f(x) - b^2 f(x) = \frac{a}{x} - 25a - bx + 25b$$

$$\Rightarrow (a^2 - b^2)\,.\, f(x) = \frac{a}{x} - bx + 25(b-a)$$

$$\Rightarrow f(x) = \frac{1}{a^2-b^2}\left[\frac{a}{x} - bx + 25(b-a)\right]$$

$$\Rightarrow \int_1^2 f(x).dx$$

$$= \frac{1}{a^2-b^2}\left[a.\int_1^2 \frac{1}{x}.dx - b\int_1^2 x.dx + 25(b-a)\int_1^2 1.dx\right]$$

$$= \frac{1}{a^2-b^2}\left[a\ln 2 - \frac{3}{2}b + 25(b-a)\right]$$

$$= \frac{1}{a^2-b^2}\left[a\ln 2 - 25a + \frac{47b}{2}\right]$$

$$= \frac{1}{a^2-b^2}\left[a(\ln 2 - 25) + \frac{47b}{2}\right]$$

7.27 (1)

$$\lim_{x\to 4}\frac{\sin(x-4)}{x-4}$$

Let $x - 4 = t$ not as $x \to 4$

So the requires limit is $\lim_{t\to 0}\frac{\sin(t)}{t} = 1$

7.28 (9)

If $f(x) + f(-x)$ is degree 10

$f(x) = a_{10}x^{10} + a_9x^9 \ldots\ldots + a_1x + a_0$

$f(-x) = a_{10}x^{10} - a_9x^9 \ldots - a_1x + a_0$

$f(x) + f(-x) = a_{10}x^{10} + a_8x^8 + \ldots + a_0$

Now $g(x) = f'(x) = 10a_{10}x^9 + 9a_9x^8 + \ldots + a_1$

$g(-x) = f'(-x) = -10a_{10}x^9 + 9a_9x^8 + \ldots + a_1$

$g(x) - g(-x) = 20a_{10}x^9 + \ldots$

Clearly degree of $(g(x) - g(-x))$ is 9.

7.29 (c)

$$\lim_{x\to 1}\frac{x^7-2x^5+1}{x^3-3x^2+2} = \frac{1-2+1}{1-3+2} = \frac{0}{0} \text{ form}$$

So, use L' Hospitals rule

$$= \lim_{x\to 1}\frac{7x^6-10x^4}{3x^2-6x} = \frac{7-10}{3-6} = \frac{-3}{-3} = 1$$

Which is option (c) is correct.

7.30 (6)

$x^2 - 13x + 36 = 0$

In base b,

$13 = 1 \times b^1 + 3 \times b^0 = b + 3$

In base b,

$36 = 3 \times b^1 + 6 \times b^0 = 3b + 6$

So the equation becomes

$x^2 - (b+3)x + (3b+6) = 0$

Now since it is given that $x = 5$ is a solution, so

$5^2 - (b+3)5 + (3b+6) = 0$

$\Rightarrow -2b + 16 = 0$

$\Rightarrow \quad b = 8$

Same can be obtained by putting $x = 6$ also.

7.31 (c)

Given, $f(x) = R\sin\left(\frac{\pi x}{2}\right) + S,$ ...(1)

$f'\left(\frac{1}{2}\right) = \sqrt{2}$ ...(2)

$\int_0^1 f(x)dx = \frac{2R}{\pi}$ ...(3)

Now we need to find R and S.

$$f'(x) = R\cos\left(\frac{\pi x}{2}\right)\frac{\pi}{2}$$

$$f'\left(\frac{1}{2}\right) = R\cos\left(\frac{\pi}{4}\right)\times\frac{\pi}{2} = \sqrt{2}$$

$$\Rightarrow \frac{R}{\sqrt{2}}\times\frac{\pi}{2} = \sqrt{2}$$

$$\Rightarrow R = \frac{4}{\pi}$$

Now, $\int f(x)dx = \int R\sin\frac{\pi x}{2} + S$

Putting $R = \frac{4}{\pi}$

We get $\int f(x)dx = \int \frac{4}{\pi}\sin\left(\frac{\pi x}{2}\right)dx + \int S\,dx$

$$= \frac{4}{\pi}\times - \frac{\cos\left(\frac{\pi x}{2}\right)}{\frac{\pi}{2}} + Sx = \frac{-8}{\pi^2}\cos\left(\frac{\pi x}{2}\right) + Sx$$

Putting limit 0 and 1

$$\int_0^1 f(x)dx = \frac{-8}{\pi^2}\left(\cos\frac{\pi}{2} - \cos(0)\right) + S(1-0) = \frac{2R}{\pi}$$

$$\Rightarrow \frac{-8}{\pi^2}(0-1) + S = \frac{2R}{\pi}$$

Put $R = \frac{4}{\pi}$ and solve for S

$\Rightarrow \quad S = 0$

So, $R = \frac{4}{\pi}$ and $S = 0$ is answer.

7.32 (0.289)

$$\int_0^{\pi/4} x\cos(x^2)\,dx$$

Let, $t = x^2, \quad dt = 2x\,dx$

$$\Rightarrow xdx = \frac{dt}{2}$$

when $x = 0$, $t = 0$ and when $x = \frac{\pi}{4}, t = \left(\frac{\pi}{4}\right)^2$

So required integral reduce to

$$\int_0^{(\pi/4)^2} \cos t\, dt = [\sin t]_0^{(\pi/4)^2}$$

$$= \sin\left(\frac{\pi}{4}\right)^2 - \sin(0) = \sin\left(\frac{\pi}{4}\right)^2 = 0.28898$$

$\simeq$ 0.289 (rounded to 3 decimal places)

7.33 (a)

$$\lim_{x\to 3} \frac{x^4 - 81}{2x^2 - 5x - 3} = \frac{0}{0} \text{ form.}$$

So apply L'H rule

$$\lim_{x\to 3} \frac{4x^3}{4x - 5} = \frac{108}{7}$$

7.34 (c)

$$f(x) = \sqrt{x^3 + 1}$$

$$f'(x) = \frac{1}{\sqrt{x^3+1}} \cdot 3x^2 > 0 \text{ in } [0, 1]$$

$\therefore$ Hence it is increasing function.

7.35 (0.25)

$$\lim_{x\to -3}\left(\frac{\sqrt{2x+22} - 4}{x+3}\right)\left(\frac{0}{0} \text{ form}\right),$$

So apply L Hospital's rule

$$= \lim_{x\to -3}\left[\frac{\frac{2}{2\sqrt{2x+22}}}{1}\right] = \lim_{x\to -3} \frac{1}{\sqrt{2x+22}}$$

$$= \frac{1}{\sqrt{16}} = \frac{1}{4} = 0.25$$

7.36 (19)

$f'(x) \le 2$, $f(-3) = 7$

$$f'(x) = \frac{f(3) - 7}{3 - (-3)} \quad \text{[Using Lagrange]}$$

$$f'(x) = \frac{f(3) - 7}{6}$$

$$6f'(x) = f(3) - 7$$

$$f(3) = 6f'(x) + 7$$

Given max value of $f'(x)$ is 2

So, maximum value of $f(3) = 6 \times 2 + 7 = 19$

7.37 (–0.5)

Apply L Hospital rule,

$$\lim_{x\to 0^+} \frac{\frac{1}{2\sqrt{x}}}{-e^{2\sqrt{x}} \times \frac{2}{2\sqrt{x}}} = \lim_{x\to 0^+} \frac{-1}{2e^{2\sqrt{x}}} = \frac{-1}{2} = -0.5$$

7.38 (b, d)

$$f(x) = x^3 + 15x^2 - 33x - 36$$

$$f'(x) = 3x^2 + 30x - 33$$

$$f''(x) = 6x + 30$$

Let, $f'(x) = 0$

$\Rightarrow$ $3x^2 + 30x - 33 = 0$

$\Rightarrow$ $x^2 + 10x - 11 = 0$

$\Rightarrow$ $(x + 11)(x - 1) = 0$

$\Rightarrow$ $x = 1$, $x = -11$

At $x = 1$, $f''(x) = 36 > 0$

At $x = 11$, $f''(x) = -36 < 0$

$\therefore$ $f(x)$ has local maximum

$f(x)$ has local minimum

7.39 (0)

In the same way that a double integral is used to find the volume under a 3*d*-curve and a normal single variable integral is used to find an area under a specific region below a 2*d*-curve, The volume under a 4*d*-graph can also be determined using triple integrals.

Note: A graph of an odd function is symmetric about the origin. So, if the function is odd and the upper and lower limits are opposite values, the integral equals zero because the net area cancels out each other.

$$\int_{-3}^{3}\int_{-2}^{2}\int_{-1}^{1} (4x^2y - z^3)\,dz\,dy\,dx$$

Integrating w.r.t. z

$$\int_{-3}^{3}\int_{-2}^{2} \left[4x^2yz\right]_{-1}^{1} dy\,dx$$

Note: $\int_{-1}^{1} z^3 dz = 0$ [odd func]

$$\int_{-3}^{3}\int_{-2}^{2} 8x^2 y\,dy\,dz$$

Note: $\int_{-2}^{2} 8x^2 y\,dy = 0$ [odd func]

$$\int_{-3}^{3} 0 \cdot dx = 0$$

7.40 (d)

Putting $x^3 = x$

$$x(x^2 - 1) = 0$$

$$x = 0, -1, 1$$

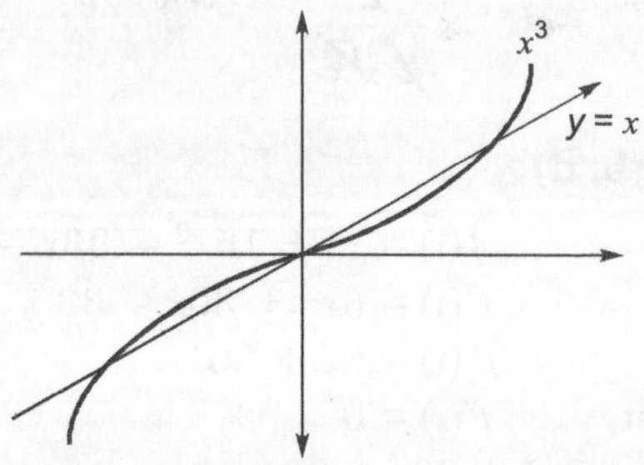

$f(x) = \max(x, x^3)$

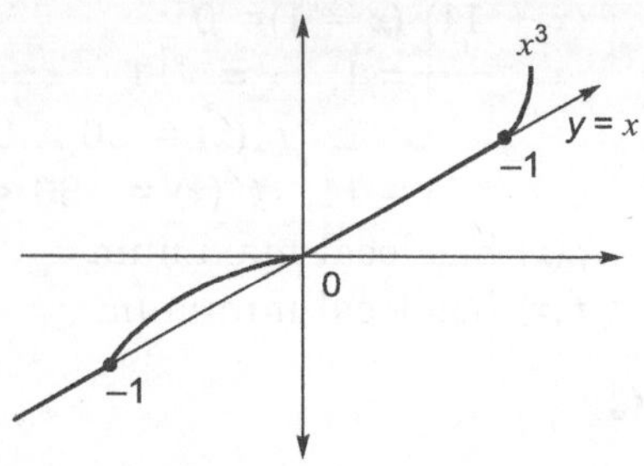

So, we have three sharp points at $x = -1, 0, 1$
So, these are non differentiable prints

Method II:

$$f(x) = \max\{x, x^3\} = \begin{cases} x & ; \quad x < -1 \\ x^3 & ; \quad -1 < x < 0 \\ x & ; \quad 0 < x < 1 \\ x^3 & ; \quad x > 1 \end{cases}$$

$$f'(x) = \begin{cases} 1 & ; \quad x < -1 \\ 3x^2 & ; \quad -1 < x < 0 \\ 1 & ; \quad 0 < x < 1 \\ 3x^2 & ; \quad x > 1 \end{cases}$$

$\because$ At $x = -1$, 0 and 1, LHD $\neq$ RHD so all three points are non differentiable points.

7.41 (c)

$$f(x) = 1 - f(2 - x)$$

or $$f(2 - x) = 1 - f(x)$$

Now, $$I = \int_0^2 f(x)dx = \int_0^2 f(2-x)dx$$

$$= \int_0^2 (1 - f(x))dx$$

$$\int_0^2 f(x)dx = \int_0^2 (1)dx - \int_0^2 f(x)dx$$

$$2I = (x)_0^2 = 2 - 0$$

or $$I = 1$$

■■■■

Theory of Computation

UNIT II

CONTENTS

Theory of Computation

Syllabus

Regular expressions and finite automata. Context-free grammars and push-down automata. Regular and contex-free languages, pumping lemma. Turing machines and undecidability.

Analysis of Previous GATE Papers

Exam Year	1 Mark Ques.	2 Marks Ques.	Total Marks
1995	1	2	**5**
1996	3	2	**7**
1997	1	3	**7**
1998	5	3	**11**
1999	3	1	**5**
2000	2	1	**4**
2001	3	2	**7**
2002	1	6	**18**
2003	3	6	**15**
2004	1	4	**9**
2005	–	7	**14**
2006	2	5	**12**
2007	2	5	**12**
2008	3	5	**13**
2009	4	3	**10**
2010	1	3	**7**
2011	3	3	**9**
2012	4	1	**6**
2013	2	3	**8**

Exam Year	1 Mark Ques.	2 Marks Ques.	Total Marks
2014 Set-1	2	2	**6**
2014 Set-2	2	2	**6**
2014 Set-3	2	2	**6**
2015 Set-1	1	2	**5**
2015 Set-2	1	3	**7**
2015 Set-3	1	1	**3**
2016 Set-1	3	3	**9**
2016 Set-2	3	4	**11**
2017 Set-1	2	4	**10**
2017 Set-2	3	3	**9**
2018	2	3	**8**
2019	2	3	**8**
2020	3	3	**9**
2021 Set-1	2	3	**8**
2021 Set-2	3	4	**11**
2022	2	3	**8**
2023	3	3	**9**
2024 Set-1	–	3	**6**
2024 Set-2	1	3	**7**

1 Finite Automata : Regular Languages

1.1 Which of the following definitions below generates the same language as L
Where $L = \{x^n y^n \mid n \geq 1\}$
(i) $E \rightarrow xEy \mid xy$
(ii) $xy \mid (x^+xyy^+)$
(iii) x^+y^+
(a) (i) only (b) (i) and (ii)
(c) (ii) and (iii) (d) (ii) only
[1995 : 2 M]

1.2 A finite state machine with the follows state table has a single input X and a single output Z.

present state	next state, z	next state, z
	x=1	x=0
A	D,0	B,0
B	B,1	C,1
C	B,0	D,1
D	B,1	C,0

It the initial state is unknown, then the shortest input sequence to reach the final state C is
(a) 01 (b) 10
(c) 101 (d) 110 **[1995 : 2 M]**

1.3 Which two of the following four regular expressions are equivalent?
(i) $(00)^*(\varepsilon + 0)$ (ii) $(00)^*$
(iii) 0^* (iv) $0(00)^*$
(a) (i) and (ii) (b) (ii) and (iii)
(c) (i) and (iii) (d) (iii) and (iv)
[1996 : 1 M]

1.4 Let $L \subseteq \Sigma^*$ where $\Sigma = \{a, b\}$, which of the following is true?
(a) $L = \{x \mid x$ has an equal number of a's and b's$\}$ is regular
(b) $L = \{a^n b^n \mid n \geq 1\}$ is regular
(c) $L = \{x \mid x$ has more a's than b's$\}$ is regular
(d) $L = \{a^m b^n \mid m \geq, n \geq 1\}$ is regular
[1996 : 1 M]

1.5 Given $\Sigma = \{a, b\}$, which one of the following sets is not countable?
(a) Set of all strings over Σ
(b) Set of all languages over Σ
(c) Set of all regular languages over Σ
(d) Set of all languages over Σ accepted by Turing Machines **[1997 : 1 M]**

1.6 Which one of the following regular expressions over {0,1} denotes the set of all strings not containing 100 as a substring?
(a) $0^*(1 + 0)^*$ (b) 0^*1010^*
(c) $0^*1^*01^*$ (d) $0^*(10 + 1)^*$
[1997 : 2 M]

1.7 If the regular set A is represented by $A = (01 + 1)^*$ and the regular set 'B' is represented by $B = ((01)^*1^*)^*$, which of the following is true?
(a) $A \subset B$
(b) $B \subset A$
(c) A and B are incomparable
(d) $A = B$ **[1998 : 1 M]**

1.8 Which of the following set can be recognized by a Deterministic Finite state Automaton?
(a) The numbers 1, 2, 4, 8, ... 2^n, ... written in binary
(b) The numbers 1, 2, 4, ... 2^n, ... written in unary
(c) The set of binary string in which the number of zeros is the same as the number of ones.
(d) The set {1, 101, 11011, 1110111, ...}
[1998 : 1 M]

1.9 The string 1101 does not belong to the set represented by
(a) $110^*(0 + 1)$
(b) $1(0 + 1)^*101$
(c) $(10)^*(01)^*(00 + 11)^*$
(d) $(00 + (11)^*0)^*$
[1998 : 1 M]

1.10 How many substrings of different lengths (non-zero) can be formed from a character string of length n?
(a) n (b) n^2
(c) 2^n (d) $\frac{n(n+1)}{2}$
[1998 : 1 M]

1.11 Let L be the set of all binary strings whose last two symbols are the same. The number of states in the minimum state deterministic finite state automaton accepting L is
(a) 2 (b) 5
(c) 8 (d) 3 **[1998 : 2 M]**

1.12 Consider the regular expression (0 + 1) (0 + 1) ... n times. The minimum state finite automation that recognizes the language represented by this regular expression contains
(a) n states
(b) n + 1 states
(c) n + 2 states
(d) None of these **[1999 : 1 M]**

1.13 Let S and T be language over $\Sigma = \{a, b\}$ represented by the regular expressions $(a + b^*)^*$ and $(a + b)^*$, respectively. Which of the following is true?
(a) $S \subset T$ (b) $T \subset S$
(c) $S = T$ (d) $S \cap T = \phi$
[2000 : 1 M]

1.14 Let L denotes the language generated by the grammar $S \rightarrow 0S0 \mid 00$. Which of the following is true?
(a) $L = 0^+$
(b) L is regular but not 0^+
(c) L is context free but not regular
(d) L is not context free **[2000 : 1 M]**

1.15 What can be said about a regular language L over {a} whose minimal finite state automation has two states?
(a) Can be $\{a^n \mid n \text{ is odd}\}$
(b) Can be $\{a^n \mid n \text{ is even}\}$
(c) Can be $\{a^n \mid \geq 0\}$
(d) Either L can be $\{a^n \mid n \text{ is odd}\}$ or L can be $\{a^n \mid n \text{ is even}\}$ **[2000 : 2 M]**

1.16 Consider the following statements:
S_1 : $\{0^{2n} \mid n \geq 1\}$ is a regular language
S_2 : $\{0^m 1^n 0^{m+n} \mid m \geq 1 \text{ and } n \geq 1\}$ is a regular language
Which of the following is true about S_1 and S_2?
(a) Only S_1 is correct
(b) Only S_2 is correct
(c) Both S_1 and S_2 are correct
(d) None of S_1 and S_2 is correct **[2001 : 1 M]**

1.17 Given an arbitrary non-deterministic finite automaton (NFA) with N states, the maximum number of states in an equivalent minimized DFA is at least
(a) N^2 (b) 2^N
(c) 2N (d) N! **[2001 : 1 M]**

1.18 Consider a DFA over Σ = {a, b} accepting all strings which have number of a's divisible by 6 and number of b's divisible by 8. What is the minimum number of states that the DFA will have?
(a) 8 (b) 14
(c) 15 (d) 48 **[2001 : 2 M]**

1.19 Consider the following languages:
$L_1 = \{ww \mid w \in \{a,b\}^*\}$
$L_2 = \{ww^R \mid w \in \{a, b\}^*, w^R \text{ is the reverse of } w\}$
$L_3 = \{0^{2i} \mid i \text{ is an integer}\}$
$L_4 = \{0^{i^2} \mid i \text{ is an integer}\}$
Which of the languages are regular?
(a) Only L_1 and L_2
(b) Only L_2, L_3 and L_4
(c) Only L_3 and L_4
(d) Only L_3
[2001 : 2 M]

1.20 The finite state machine described by the following state diagram with A as starting state, where an arc label is x/y and x stands for 1-bit input and y stands for 2-bit output

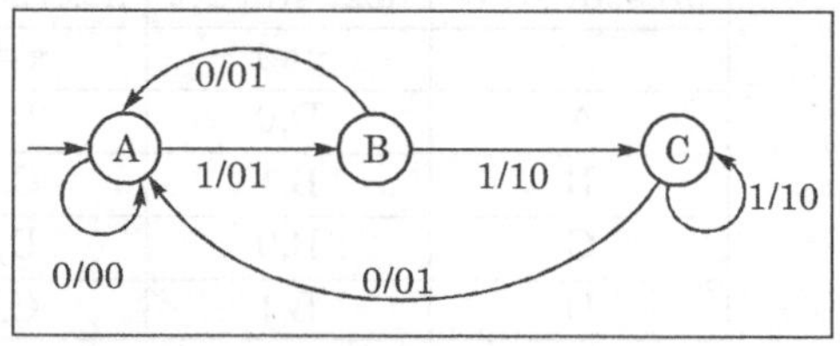

(a) Outputs the sum of the present and the previous bits of the input
(b) Outputs 01 whenever the input sequence contains 11
(c) Output 00 whenever the input sequence contains 10
(d) None of the above
[2002 : 2 M]

1.21 The smallest finite automation which accepts the language $L = \{x \mid \text{length of } x \text{ is divisible by } 3\}$ has
(a) 2 states (b) 3 states
(c) 4 states (d) 5 states
[2002 : 2 M]

1.22 Consider the following deterministic finite state automaton M.

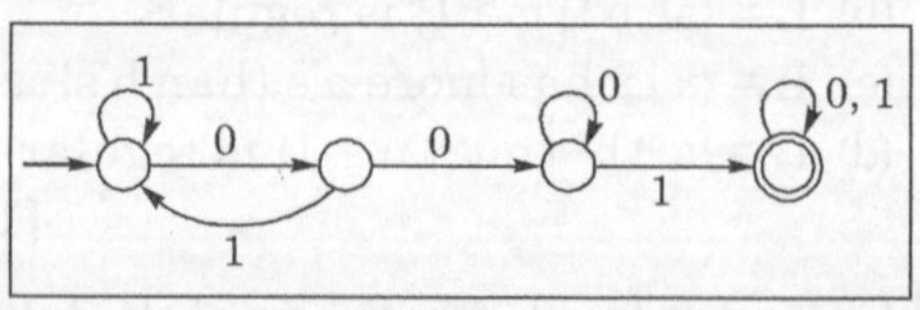

Let S denote the set of seven bit binary strings in which the first, the fourth, and the last bits are 1. The number of strings in S that are accepted by M is
(a) 1 (b) 5
(c) 7 (d) 8 **[2003 : 2 M]**

1.23 Consider the NFA M shown below:

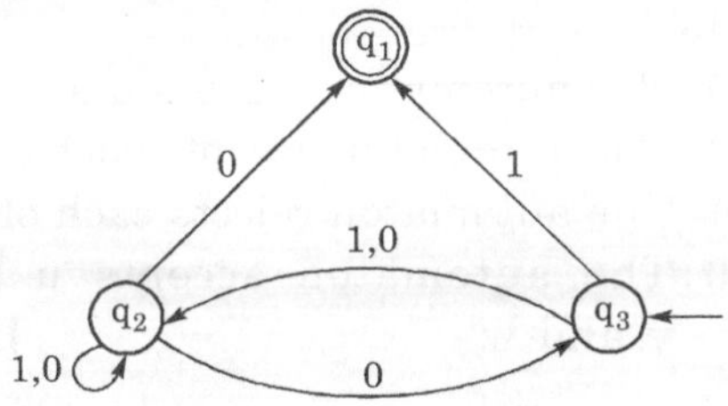

Let the language accepted by M be L. Let L_1 be the language accepted by the NFA M_1, obtained by changing the accepting state of M to a non-accepting state and by changing the non-accepting state of M to accepting states. Which of the following statements is true?

(a) $L_1 = \{0, 1\}^* - L$ (b) $L_1 = \{0, 1\}^*$
(c) $L_1 \subseteq L$ (d) $L_1 = L$ **[2003 : 2 M]**

1.24 Which one of the following regular expressions is NOT equivalent to the regular expression $(a + b + c)^*$?

(a) $(a^* + b^* + c^*)^*$ (b) $(a^*b^*c^*)^*$
(c) $((ab)^* + c^*)^*$ (d) $(a^*b^* + c^*)^*$

[2004 : 1 M]

1.25 Let $M = (K, \Sigma, \delta, s, F)$ be a finite state automaton, where
$K = \{A, B\}$, $\Sigma = \{a, b\}$, $s = A$, $F = \{B\}$,
$\delta(A, a) = A$, $\delta(A, b) = B$, $\delta(B, a) = B$ and $\delta(B, b) = A$
A grammar to generate the language accepted by M can be specified as $G = (V, \Sigma, R, S)$, Where $V = K \cup \Sigma$, and $S = A$
Which one of the following set of rules will make $L(G) = L(M)$?

(a) $\{A \to aB, A \to bA, B \to bA, B \to aA, B \to \varepsilon\}$
(b) $\{A \to aA, A \to bB, B \to aB, B \to bA, B \to \varepsilon\}$
(c) $\{A \to bB, A \to aB, B \to aA, B \to bA, B \to \varepsilon\}$
(d) $\{A \to aA, A \to bA, B \to aB, B \to bA, A \to \varepsilon\}$

[2004 : 2 M]

1.26 The following finite state machine accepts all those binary strings in which the number of 1's and 0's are respectively

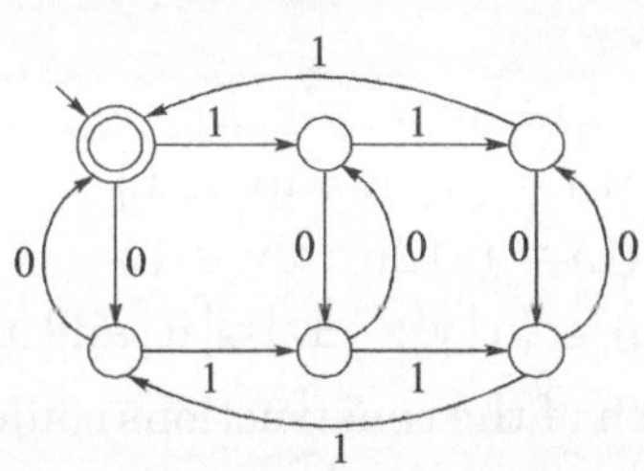

(a) divisible by 3 & 2 (b) odd and even
(c) even and odd (d) divisible by 2 & 3

[2004 : 2 M]

1.27 Which of the following statements is TRUE about the regular expression 01^*0?

(a) It represents a finite set of finite strings.
(b) It represents an infinite set of finite strings.
(c) It represents a finite set of infinite strings.
(d) It represents an infinite set of infinite strings.

[2005 : 1 M]

1.28 The language $\{0^n 1^n 2^n \mid 1 \le n \le 10^6\}$ is

(a) Regular
(b) Context-free but not regular
(c) Context-free but its complement is not context-free
(d) Not context-free **[2005 : 1 M]**

1.29 Consider the non-deterministic finite automaton (NFA) shown in the figure.

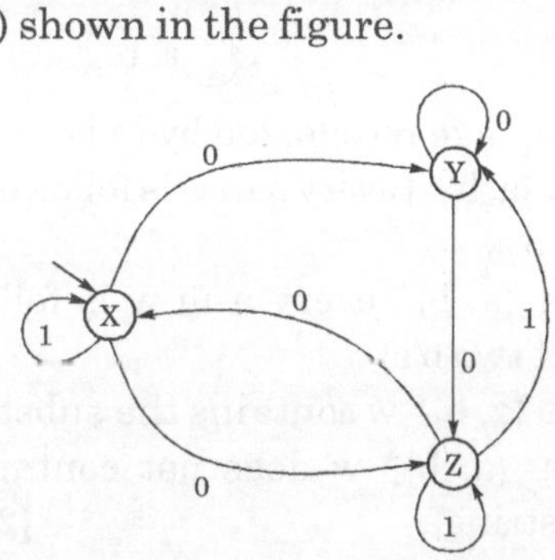

State X is the starting state of the automaton. Let the language accepted by the NFA with Y as the only accepting state be L1. Similarly, let the language accepted by the NFA with Z as the only accepting state be L2. Which of the following statements about L1 and L2 is TRUE?

(a) L1 = L2
(b) $L1 \subset L2$
(c) $L2 \subset L1$
(d) None of these

[2005 : 2 M]

1.30 Consider the regular grammar:

$S \to Xa \mid Ya$
$X \to Za$
$Z \to Sa \mid \varepsilon$
$Y \to Wa$
$W \to Sa$

where S is the starting symbol, the set of terminals is {a} and the set of non-terminals is {S, W, X, Y, Z}. We wish to construct a deterministic finite automaton (DFA) to recognize the same language. What is the minimum number of states required for the DFA?

(a) 2 (b) 3
(c) 4 (d) 5

[2005 : 2 M]

1.31 A language L satisfies the Pumping Lemma for regular languages, and also the Pumping Lemma for context-free languages. Which of the following statements about L is TRUE?
(a) L is necessarily a regular language.
(b) L is necessarily a context-free language, but not necessarily a regular language.
(c) L is necessarily a non-regular language.
(d) None of the above **[2005 : 2 M]**

1.32 Consider the machine M:

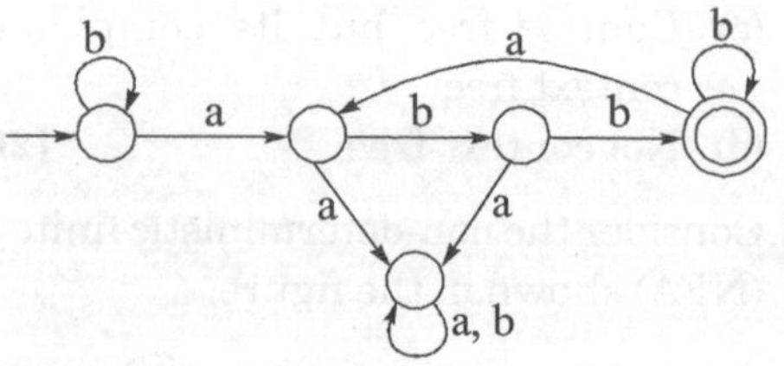

The language recognized by M is
(a) {w ∈ {a, b}* | every a in w is followed by exactly two b's}
(b) {w ∈ {a, b}* every a in w is followed by at least two b's}
(c) {w ∈ {a, b}* w contains the substring 'abb'
(d) {w ∈ {a, b}* w does not contain 'aa' as a substring} **[2005 : 2 M]**

1.33 The following diagram represents a finite state machine which takes as input a binary number from the least significant bit

Which one of the following is TRUE?
(a) It computes 1's complement of the input number
(b) It computes 2's complement of the input number
(c) It increments the input number
(d) It decrements the input number
[2005 : 2 M]

1.34 In the automaton below, s is the start state and t is the only final state.

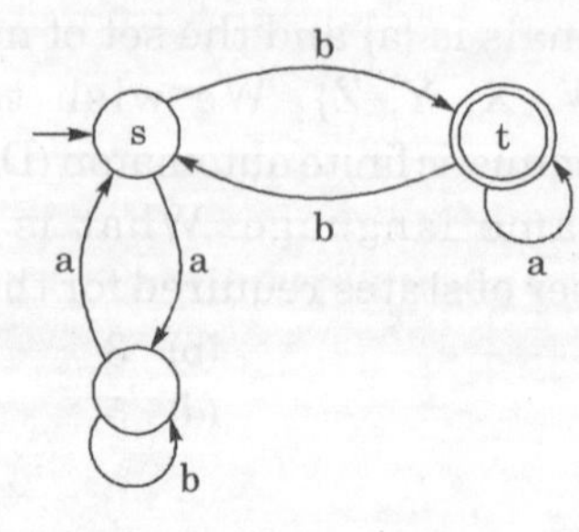

Consider the strings u = abbaba, v = bab, and w = aabb. Which of the following statements is true?
(a) The automaton accepts u and v but not w
(b) The automaton accepts each of u, v, and w
(c) The automaton rejects each of u, v, and w
(d) The automaton accepts u but rejects v and w **[2006 : 1 M]**

1.35 Which regular expression best describes the language accepted by the non-deterministic automaton below?

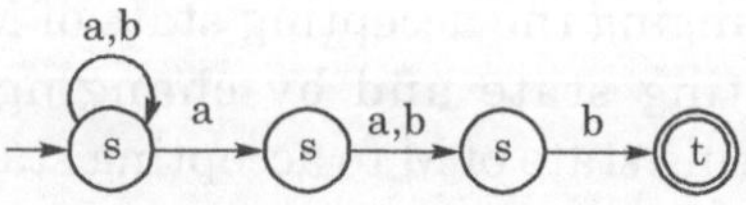

(a) (a + b) * a(a + b)b
(b) (abb)*
(c) (a + b)* a(a + b) * b(a + b)*
(d) (a + b)* **[2006 : 1 M]**

1.36 Consider the regular grammar below:
S → bS | aA | ε
A → aS | bA
The Myhill-Nerode equivalence classes for the language generated by the grammar are
(a) {w ∈ (a + b)* | $\#_a(w)$ is even} and {w ∈ (a + b)* | $\#_a(w)$ is odd}
(b) {w ∈ (a + b)* | $\#_b(w)$ is even} and {w ∈ (a + b)* | $\#_b(w)$ is odd}
(c) {w ∈ (a + b)* | $\#_a(w) = \#_b(w)$} and {w ∈ (a + b)* | $\#_a(w) \neq \#_b(w)$}
(d) {ε}, {w a | w ∈ (a + b)*} and {wb | w ∈ (a + b)*}
[2006 : 2 M]

1.37 Which of the following statements about regular languages is NOT true?
(a) Every language has a regular superset
(b) Every language has a regular subset
(c) Every subset of a regular language is regular
(d) Every subset of a finite language is regular
[2006 : 2 M]

Directions for Q.1.38 to Q.1.39:
Let L be a regular language. Consider the constructions on L below:
I. repeat (L) = {ww | w ∈ L}
II. prefix (L) = {u | ∃v : uv ∈ L}
III. suffix (L) = (v | ∃u : uv ∈ L}
IV. half (L) = {u | v : |v| = |u| and uv ∈ L}

1.38 Which of the constructions could lead to a non-regular language?
(a) Both I and IV
(b) Only I
(c) Only IV
(d) Both II and III
[2006 : 2 M]

1.39 Which choice of L is best suited to support your answer above?
(a) $(a + b)^*$ (b) $\{\varepsilon, a, ab, bab\}$
(c) $(ab)^*$ (d) $\{a^nb^n \mid n \geq 0\}$
[2006 : 2 M]

1.40 If s is a string over $(0 + 1)^*$, then let $n_0(s)$ denote the number of 0's in s and $n_1(s)$ the number of 1's in s. Which one of the following languages is not regular?
(a) $L = \{s \in (0+1)^* \mid n_0(s)$ is a 3-digit prime$\}$
(b) $L = \{s \in (0+1)^* \mid$ for every prefix s' of s, $|n_0(s') - n_1(s')| \leq 2\}$
(c) $L = \{s \in (0+1)^* \mid |n_0(s) - n_1(s)| \leq 4\}$
(d) $L = \{s \in (0+1)^* \mid n_0(s) \bmod 7 = n_1(s) \bmod 5 = 0\}$
[2006 : 2 M]

1.41 Consider the regular language $L = (111 + 11111)^*$. The minimum number of states in any DFA accepting this languages is
(a) 3 (b) 5
(c) 8 (d) 9 **[2006 : 2 M]**

1.42 Which of the following is TRUE?
(a) Every subset of a regular set is regular
(b) Every finite subset of a non-regular set is regular
(c) The union of two non-regular sets is not regular
(d) Infinite union of finite sets is regular
[2007 : 1 M]

1.43 A minimum state deterministic finite automation accepting the language $L = \{w \mid w \in (0, 1)^*$, number of 0s & 1s in w are divisible by 3 and 5, respectively$\}$ has
(a) 15 states (b) 11 states
(c) 10 states (d) 9 states
[2007 : 2 M]

1.44 Which of the following languages is regular?
(a) $\{ww^R \mid w \in \{0, 1\}^+\}$
(b) $\{ww^Rx \mid x, w \in (0, 1)^+\}$
(c) $\{wxw^R \mid x, w \in (0, 1)^+\}$
(d) $\{xww^R \mid x, w \in (0, 1)^+\}$
[2007 : 2 M]

Common Data for Q.1.45 & Q.1.46:

Consider the following Finite State Automation

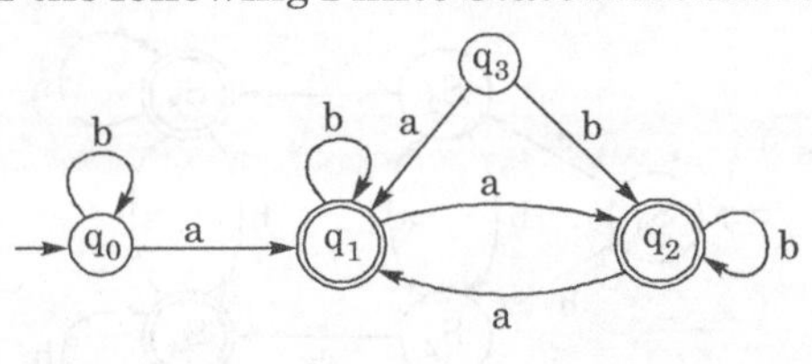

1.45 The language accepted by this automaton is given by the regular expression
(a) $b^*ab^*ab^*ab^*$ (b) $(a + b)^*$
(c) $b^*a(a + b)^*$ (d) $b^*ab^*ab^*$
[2007 : 2 M]

1.46 The minimum state automaton equivalent to the above FSA has the following number of states
(a) 1 (b) 2
(c) 3 (d) 4 **[2007 : 2 M]**

1.47 The two grammars given below generate a language over the alphabet {x, y, z}

G1: $S \to x \mid z \mid xS \mid zS \mid yB$
$B \to y \mid z \mid yB \mid zB$

G2: $S \to y \mid z \mid yS \mid zS \mid xB$
$B \to y \mid yS$

Which one of the following choices describes the properties satisfied by the strings in these languages?
(a) G1 : No y appears before any x
G2 : Every x is followed by at least one y
(b) G1 : No y appears before any x
G2 : No x appears before any y
(c) G1 : No y appears after any x
G2 : Every x is followed by at least one y
(d) G1 : No y appears after any x
G2 : Every y is followed by at least one x
[2007 : 2 M]

1.48 Consider the following DFA in which s_0 is the start state and s_1, s_3 are the final states.

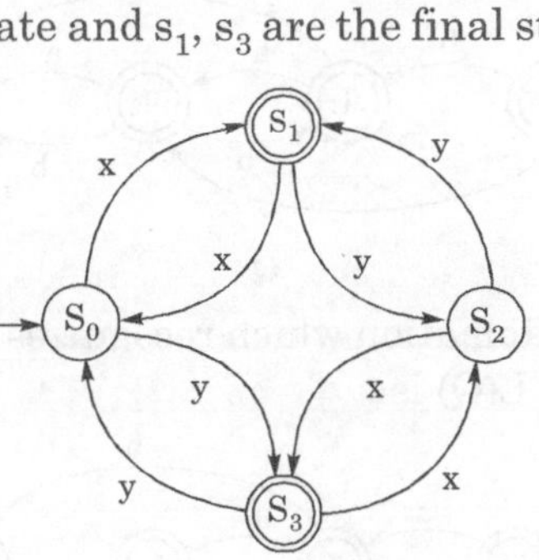

What language does this DFA recognize?
(a) All strings of x and y
(b) All strings of x and y which have either even number of x and even number of y or odd number or x and odd number of y
(c) All strings of x and y which have equal number of x and y
(d) All strings of x and y with either even number of x and odd number of y or odd number of x and even number of y
[2007 : 2 M]

1.49 Consider the grammar given below:

$S \to xB \mid yA$
$A \to x \mid xS \mid yAA$
$B \to y \mid yS \mid yBB$

Consider the following strings.

(i) xxyyx (ii) xxyyxy
(iii) xyxy (iv) yxxy
(v) yxx (vi) xyx

Which of the above strings are generated by the grammar?

(a) (i), (ii), and (iii)
(b) (ii), (v), and (vi)
(c) (iii) and (iv)
(d) (i), (iii), and (iv)

[2007 : 2 M]

1.50 Consider the following finite automata P and Q over the alphabet {a, b, c}. The start states are indicated by a double arrow and final states are indicated by a double circle. Let the languages recognized by them be denoted by L(P) and L(Q) respectively.

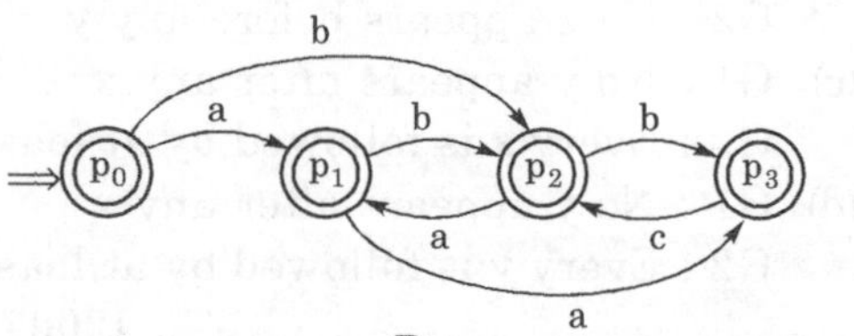

P

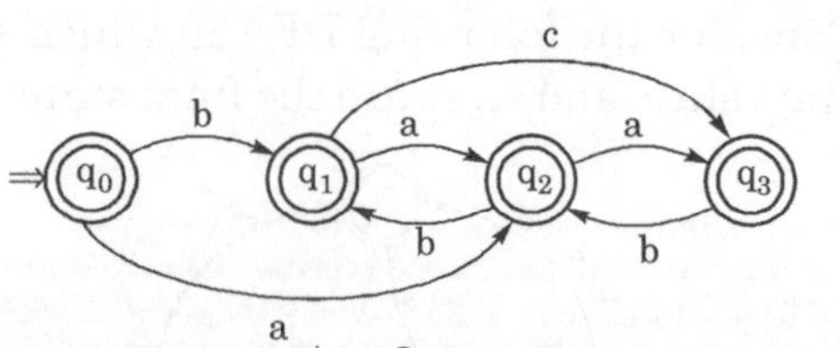

Q

The automation which recognizes the language L(P) ∩ L(Q) is :

(a) 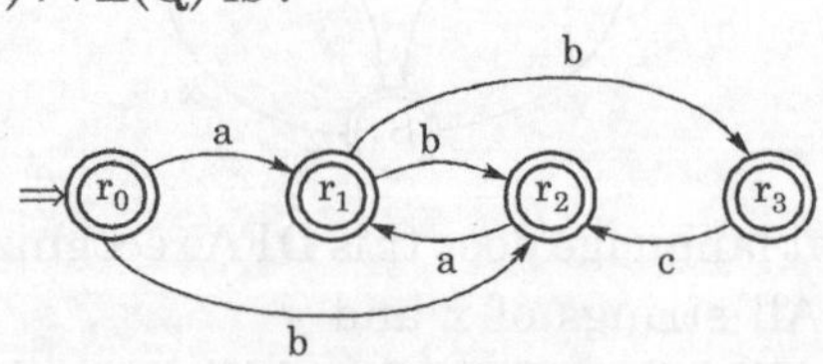

(b) 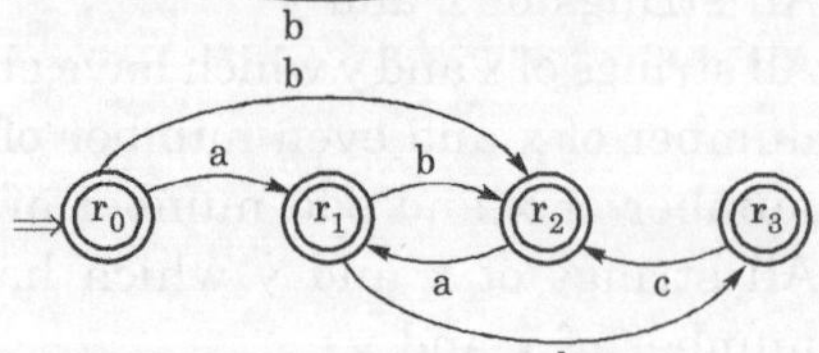

(c) 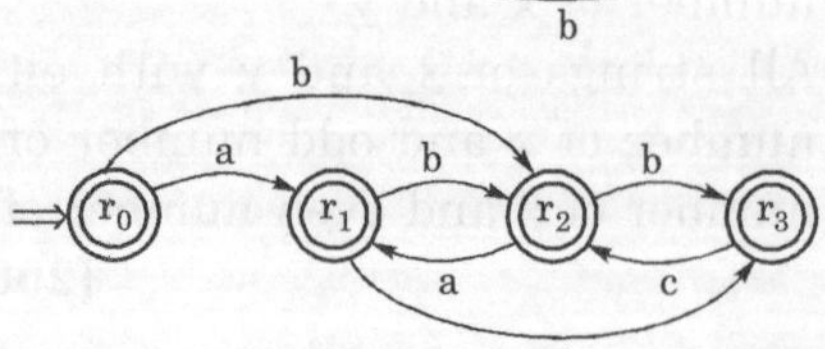

(d) 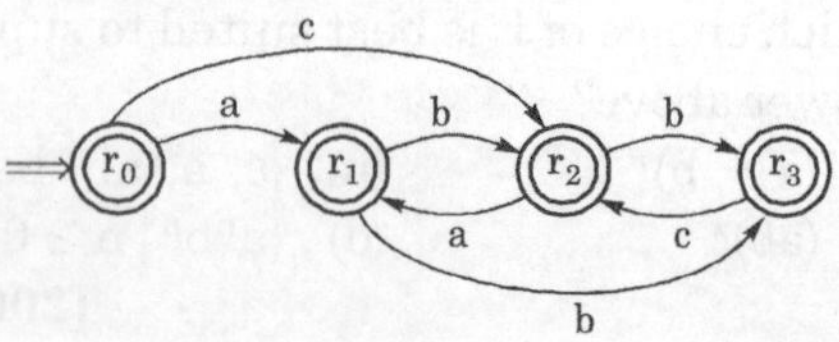

(e) None of the above **[2007 : 2 M]**

Common Data for Q.1.51 to Q.1.53:

Consider the regular expression:

$$R = (a + b)^* (aa + bb) (a + b)^*$$

1.51 Which of the following non-deterministic finite automata recognizes the language defined by the regular expression R? Edges labeled λ denote transitions on the empty string.

(a) 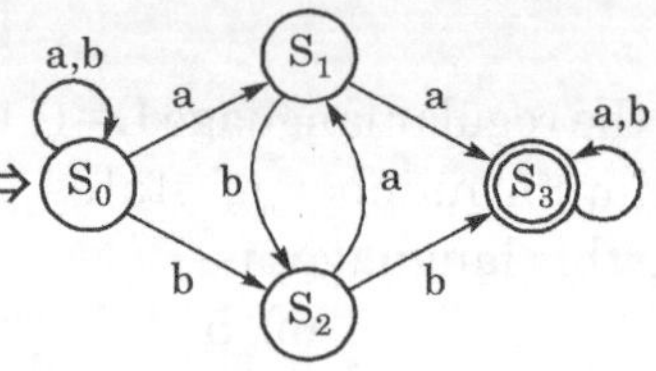

(b) 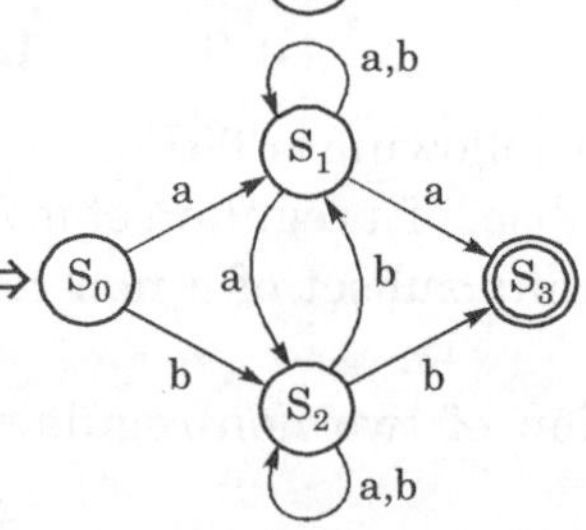

(c) 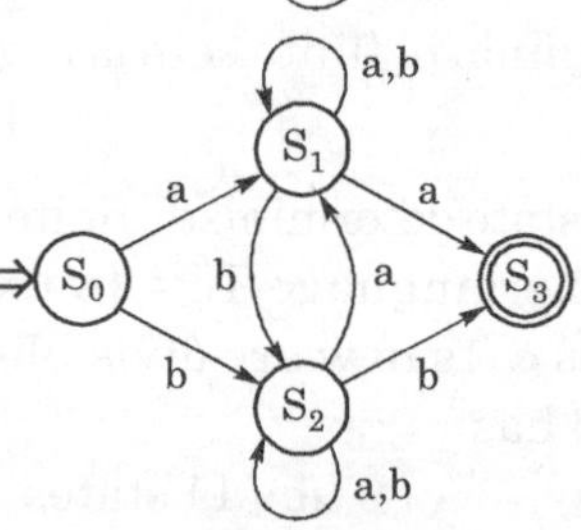

(d) 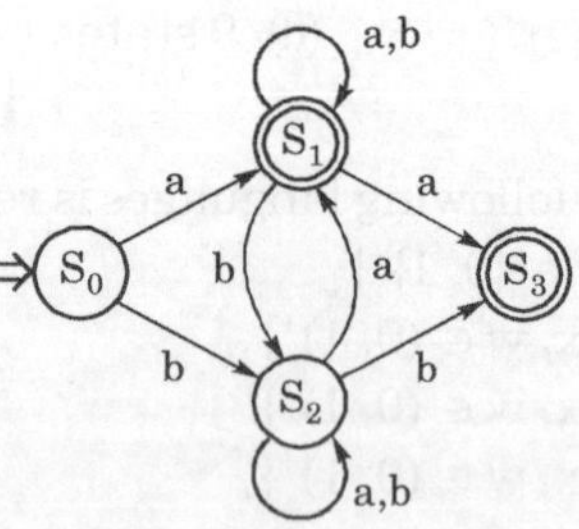

[2007 : 2 M]

1.52 Which deterministic finite automaton accepts the language represented by the regular expression R?

(a) 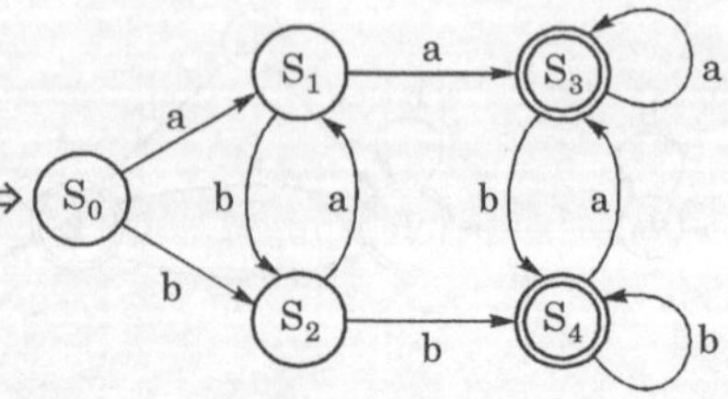

(b) 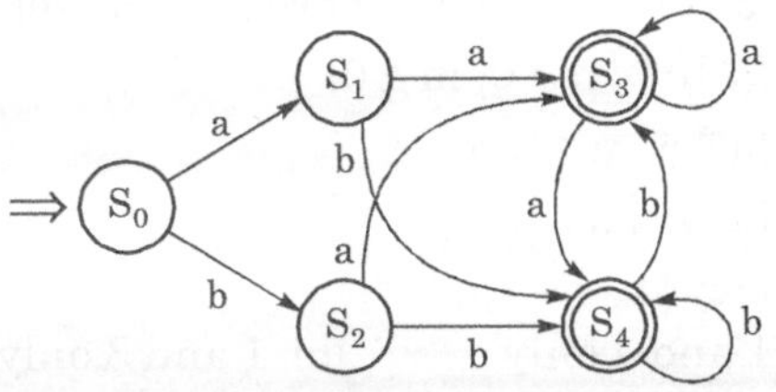

(c) 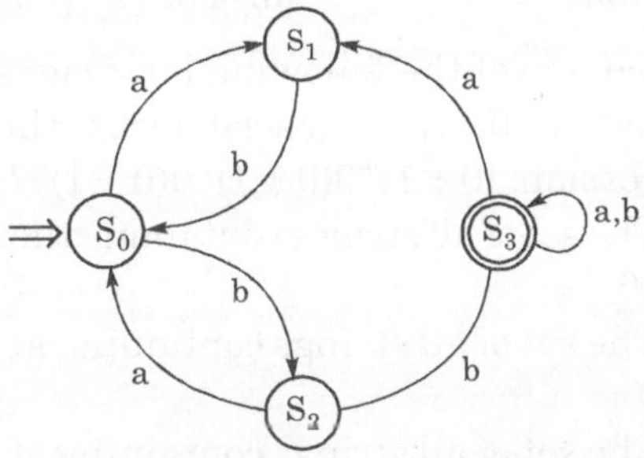

(d) 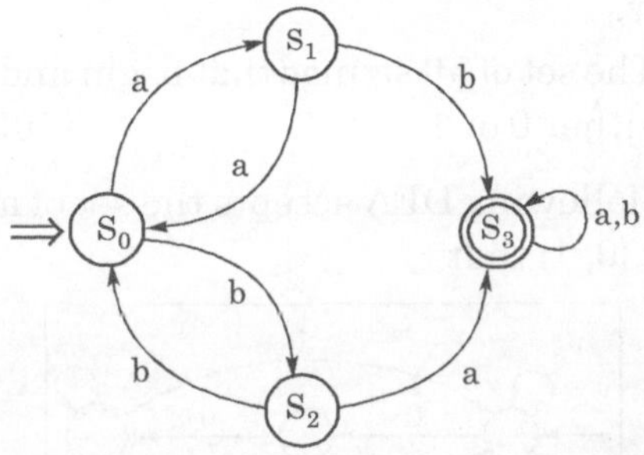

[2007 : 2 M]

1.53 Which one of the regular expressions given below defines the same language as defined by the regular expression R?

(a) $(a(ba)^* + b(ab)^*)(a + b)^+$
(b) $(a(ba)^* + b(ab)^*)^*(a + b)^*$
(c) $(a(ba)^* (a + bb) + b(ab)^*(b + aa))(a + b)^*$
(d) $(a(ba)^*(a + bb) + b(ab)^*(b + aa))(a + b)^+$

[2007 : 2 M]

1.54 Which of the following regular expressions describes the language over {0, 1} consisting of strings that contain exactly two 1's?

(a) $(0 + 1) ^* 11(0 + 1) ^*$
(b) $0 ^* 110 ^*$
(c) $0 ^* 10 ^* 10 ^*$
(d) $(0 + 1) ^* 1(0 + 1) ^* 1 (0 + 1) ^*$

[2008 : 1 M]

1.55 Let N be an NFA with n states and let M be the minimized DFA with m states recogniz-ing the same language. Which of the following in NECESSARILY true?

(a) $m \le 2^n$
(b) $n \le m$
(c) M has one accept state
(d) $m = 2^n$

[2008 : 1 M]

1.56 If the final states and non-final states in the DFA below are interchanged, then which of the following languages over the alphabet {a, b} will be accepted by the new DFA?

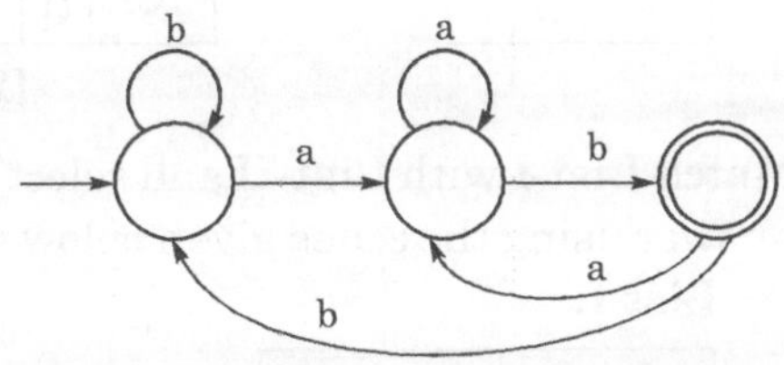

(a) Set of all strings that do not end with ab
(b) Set of all strings that begin with either an a or ab
(c) Set of all strings that do not contain the substring ab,
(d) The set described by the regular expression $b^*aa^*(ba)^*b^*$

[2008 : 2 M]

1.57 Which of the following languages is (are) non-regular?

$L_1 = \{0^m1^n \mid 0 \le m \le n \le 10000\}$
$L_2 = \{w \mid w$ reads the same forward and backward$\}$
$L_3 = \{w \in \{0, 1\}^* \mid w$ contains an even number of 0's and an even number of 1's$\}$

(a) L_2 and L_3 only (b) L_1 and L_2 only
(c) L_3 only (d) L_2 only

[2008 : 2 M]

1.58 Consider the following two finite automata. M_1 accepts L_1 and M_2 accepts L_2. Which one of the following is TRUE?

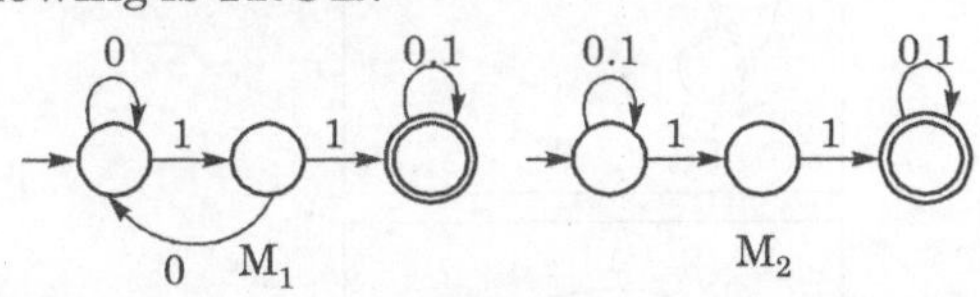

(a) $L_1 = L_2$ (b) $L_1 \cap L_2 = \phi$
(c) $L_1 \cap \overline{L_2} = \phi$ (d) $L_1 \cup L_2 \ne L_1$

[2008 : 2 M]

1.59 Given below are two finite state automata (→ indicates the start and F indicates a final state)

Y:

	a	b
→1	1	2
2(F)	2	1

Z:

	a	b
→1	2	2
2(F)	1	1

Which of the following represents the product automaton Z × Y?

(a)

	a	b
→P	S	R
Q	R	S
R(F)	Q	P
S	P	Q

(b)

	a	b
→P	S	Q
Q	R	S
R(F)	Q	P
S	P	Q

(c)

	a	b
→P	Q	S
Q	R	S
R(F)	Q	P
S	Q	P

(d)

	a	b
→P	S	Q
Q	S	R
R(F)	Q	P
S	Q	P

[2008 : 2 M]

1.60 Match **List-I** with **List-II** and select the correct answer using the codes given below the lists:

List-I:

A. 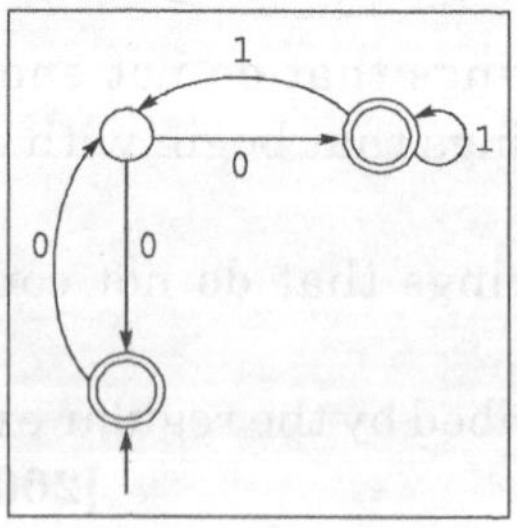

B. 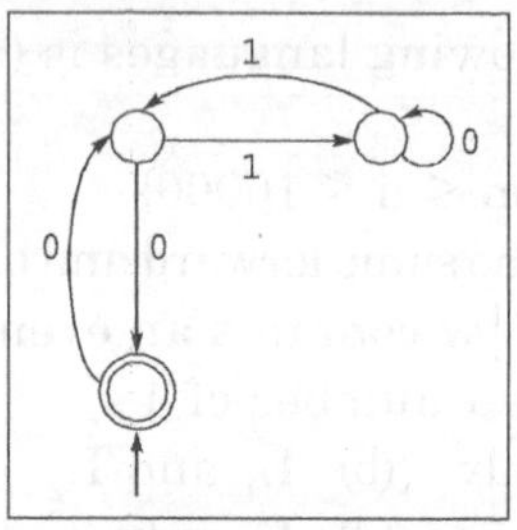

C. 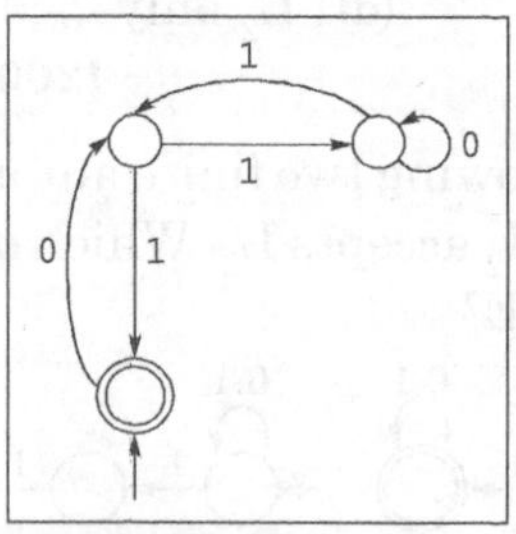

D. 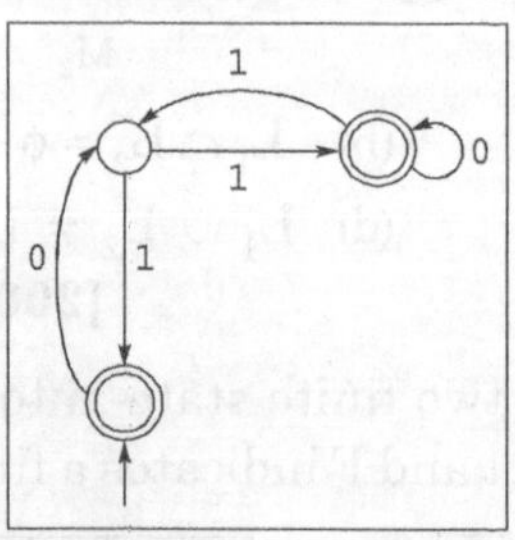

List-II

1. $\varepsilon + 0(01*1+00)*01*$
2. $\varepsilon + 0(10*1+00)*0$
3. $\varepsilon + 0(10*1+10)*1$
4. $\varepsilon + 0(10*1+10)*10*$

Codes:

	A	B	C	D
(a)	2	1	3	4
(b)	1	3	2	4
(c)	1	2	3	4
(d)	3	2	1	4

[2008 : 2 M]

1.61 Which of the following are regular sets?
1. $\{a^n b^{2m} \mid n \geq 0, m \geq 0\}$
2. $\{a^n b^m \mid n = 2m\}$
3. $\{a^n b^m \mid n \neq m\}$
4. $\{xcy \mid x, y \in \{a, b\}^*\}$
(a) 1 and 4 only
(b) 1 and 3 only
(c) 1 only
(d) 4 only **[2008 : 2 M]**

1.62 Which one of the following languages over the alphabet {0, 1} is described by the regular expression: $(0+1)*0(0+1)*0(0+1)*$?
(a) The set of all strings containing the substring 00
(b) The set of all strings containing at most two 0's
(c) The set of all strings containing at least two 0's
(d) The set of all strings that begin and end with either 0 or 1 **[2009 : 1 M]**

1.63 The following DFA accepts the set of all strings over {0, 1} that

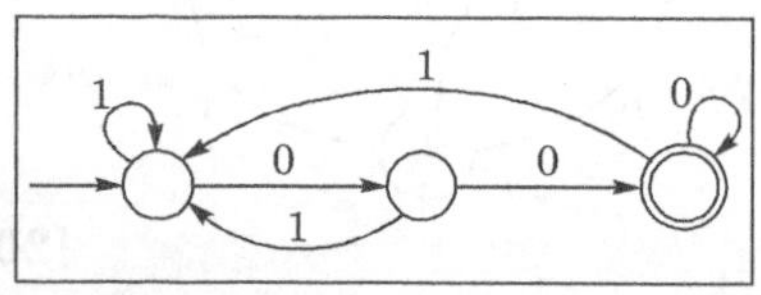

(a) begin either with 0 or 1
(b) end with 0
(c) end with 00
(d) contain the substring 00 **[2009 : 2 M]**

1.64 Let $L = \{\omega \in (0+1)^* \mid \omega$ has even number of 1s$\}$, i.e., L is the set of all bit strings with even number of 1s. Which one of the regular expressions below represents L?
(a) $(0*10*1)*$
(b) $0*(10*10*)*$
(c) $0*(10*1)*0*$
(d) $0*1(10*1)*10*$ **[2010 : 2 M]**

1.65 Let ω be any string of length n in $\{0, 1\}^*$. Let L be the set of all substrings of ω. What is the minimum number of states in a non-deterministic finite automaton that accepts L?
(a) $n - 1$
(b) n
(c) $n + 1$
(d) 2^{n+1} **[2010 : 2 M]**

1.66 The lexical analysis for a modern computer language such as Java needs the power of which one of the following machine models in a necessary and sufficient sense?
(a) Finite state automata
(b) Deterministic pushdown automata
(c) Non-deterministic pushdown automata
(d) Turing machine **[2011 : 1 M]**

1.67 Let P be a regular language and Q be a context-free language such that $Q \subseteq P$. (For example, let P be the language represented by the regular expression p^*q^* and Q be $\{p^n q^n \mid n \in N\}$. Then which of the following is ALWAYS regular?

(a) $P \cap Q$ (b) $P - Q$
(c) $\Sigma^* - P$ (d) $\Sigma^* - Q$

[2011 : 1 M]

1.68 Definition of a language L with alphabet {a} is given as follows:

$L = \{a^{nk} \mid k > 0$, and n is a positive integer constant$\}$.

What is the minimum number of states needed in a dfa to recognize L?

(a) $k + 1$ (b) $n + 1$
(c) 2^{n+1} (d) 2^{k+1} **[2011 : 2 M]**

1.69 A deterministic finite automaton (DFA) D with alphabet $\Sigma = \{a, b\}$ is given below:

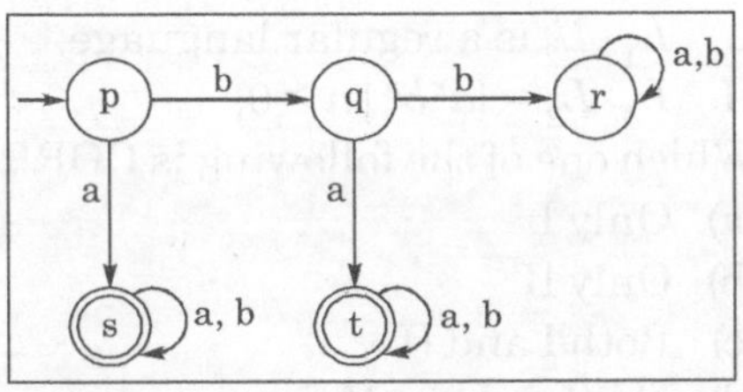

Which of the following finite state machines is a valid minimal DFA which accepts the same language as D?

(a)
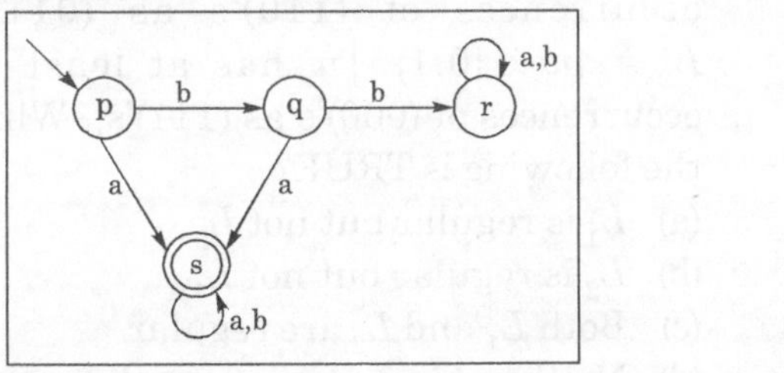

(b)
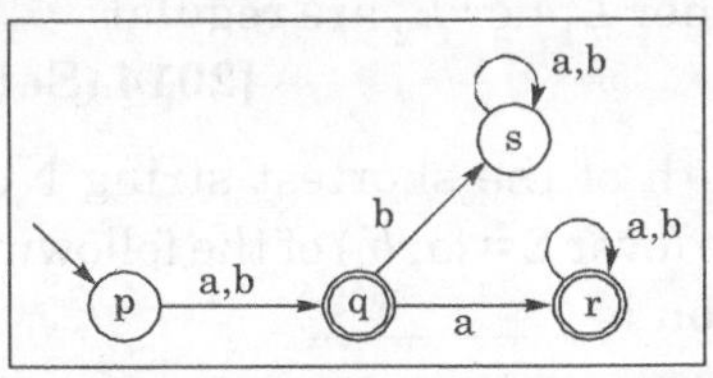

(c)
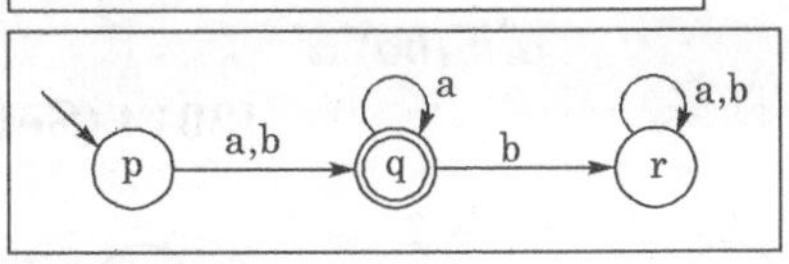

(d)
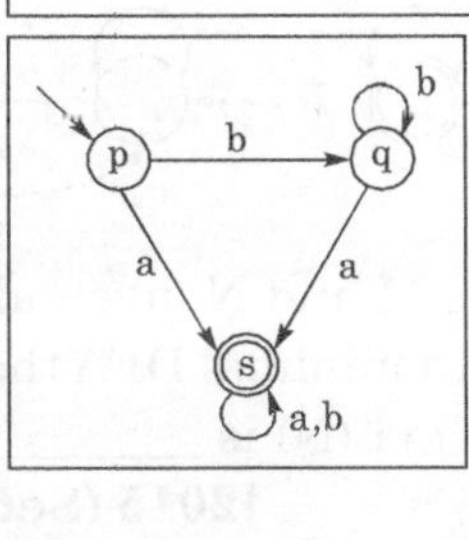

[2011 : 2 M]

1.70 What is the complement of the language accepted by the NFA shown below? Assume $\Sigma = \{a\}$ and ε is the empty string.

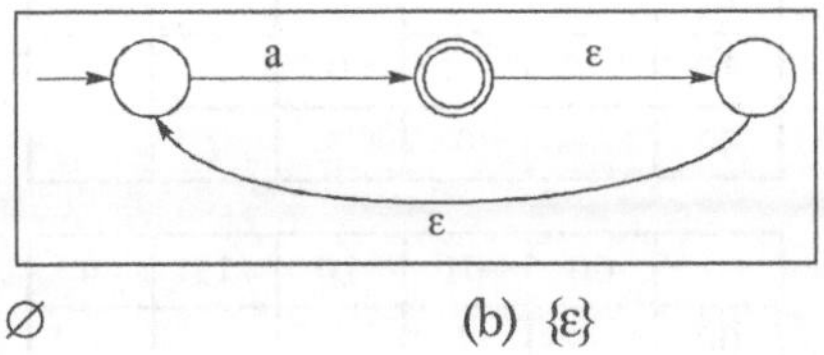

(a) $\varnothing$ (b) $\{\varepsilon\}$
(c) a^* (d) $\{a, \varepsilon\}$ **[2012 : 1 M]**

1.71 Given the language L = {ab, aa, baa}, which of the following strings are in L^*?

1. abaabaaabaa
2. aaaabaaaa
3. baaaaabaaaab
4. baaaabaa

(a) 1. 2 and 3 (b) 2, 3 and 4
(c) 1, 2 and 4 (d) 1, 3 and 4

[2012 : 1 M]

1.72 Consider the set of strings on {0, 1} in which, every substring of 3 symbols has at most two zeros. For example, 001110 and 011001 are in the language, but 100010 is not. All strings of length less than 3 are also in the language. A partially complete DFA that accepts this language is shown below.

The missing arcs in the DFA are

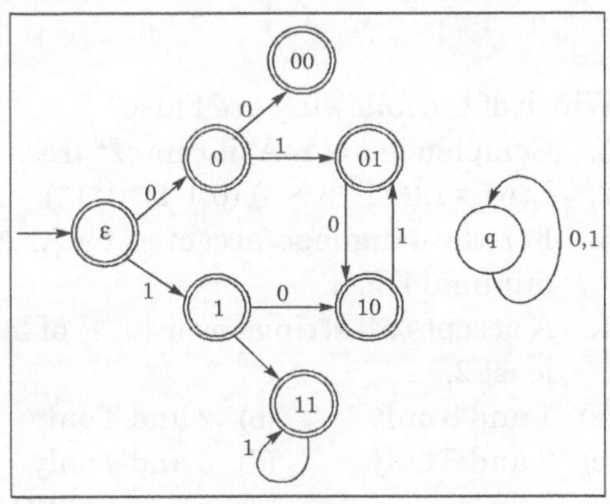

(a)

	00	01	10	11	q
00	1	0			
01				1	
10	0				
11			0		

(b)

	00	01	10	11	q
00		0			1
01		1			
10				0	
11		0			

(c)

	00	01	10	11	q
00		1			0
01		1			
10			0		
11		0			

(d)

	00	01	10	11	q
00		1			0
01				1	
10	0				
11			0		

[2012 : 2 M]

1.73 Consider the languages $L_1 = \phi$ and $L_2 = \{a\}$. Which one of the following represents $L_1L_2^* \cup L_1^*$?

(a) $\{\varepsilon\}$ (b) ϕ
(c) a^* (d) $\{\varepsilon, a\}$

[2013 : 1 M]

1.74 Consider the DFA A is given below:

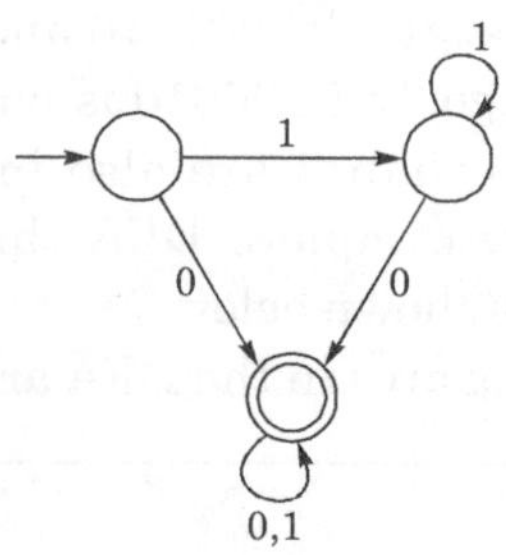

Which of the following are False?

1. Complement of L(A) is context-free.
2. L(A) = L((11 *0 + 0) (0 + 1)*0*1*)
3. For the language accepted by A, A is the minimal DFA.
4. A accepts all strings over {0, 1} of length at least 2.

(a) 1 and 3 only (b) 2 and 4 only
(c) 2 and 3 only (d) 3 and 4 only

[2013 : 2 M]

1.75 Which one of the following is TRUE?

(a) The language $L = \{a^n b^n \mid n \geq 0\}$ is regular.
(b) The language $L = \{a^n \mid n \text{ is prime}\}$ is regular.
(c) The language $L = \{w \mid w \text{ has } 3k + 1\ b\text{'s for some } k \in \mathbb{N} \text{ with } \Sigma = \{a, b\}\}$ is regular.
(d) The language $L = \{ww \mid w \in \Sigma^* \text{ with } \Sigma = \{0, 1\}\}$ is regular. **[2014 (Set-1) : 1 M]**

1.76 Consider the finite automaton in the following figure.

What is the set of reachable states for the input string 0011?

(a) $\{q_0, q_1, q_2\}$ (b) $\{q_0, q_1\}$
(c) $\{q_0, q_1, q_2, q_3\}$ (d) $\{q_3\}$

[2014 (Set-1) : 1 M]

1.77 Which of the regular expressions given below represent the following DFA?

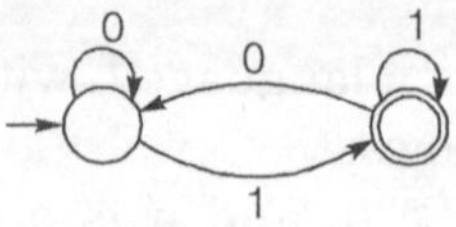

I. 0*1 (1 + 00* 1)*
II. 0*1*1 +11* 0* 1
III. (0 + 1)* 1

(a) I and II only (b) I and III only
(c) II and III only (d) I, II, and III

[2014 (Set-1) : 2 M]

1.78 If $L_1 = \{a^n \mid n \geq 0\}$ and $L_2 = \{b^n \mid n \geq 0\}$, consider

I. $L_1 \cdot L_2$ is a regular language
II. $L_1 \cdot L_2 = \{a^n b^n \mid n \geq 0\}$

Which one of the following is CORRECT?

(a) Only I
(b) Only II
(c) Both I and II
(d) Neither I nor II

[2014 (Set-2) : 1 M]

1.79 Let $L_1 = \{w \in \{0,1\}^* \mid w$ has at least as many occurrences of (110)'s as (011)'s$\}$. Let $L_2 = \{w \in \{0,1\}^* \mid w$ has at least as many occurrences of (000)'s as (111)'s$\}$. Which one of the following is TRUE?

(a) L_1 is regular but not L_2
(b) L_2 is regular but not L_1
(c) Both L_1 and L_2 are regular
(d) Neither L_1 nor L_2 are regular

[2014 (Set-2) : 2 M]

1.80 The length of the shortest string NOT in the language (over $\Sigma = \{a, b\}$) of the following regular expression is ______.

$$a^*b^*(ba)^*a^*$$

[2014 (Set-3) : 1 M]

1.81

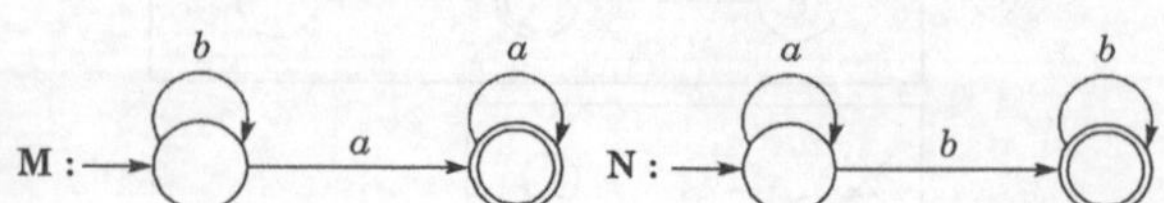

Consider the DFAs M and N given above. The number of states in a minimal DFA that accepts the language L(M) ∩ L(N) is ______.

[2015 (Set-1) : 2 M]

1.82 Consider alphabet $\Sigma = \{0, 1\}$, the null/empty string λ and the sets of strings X_0, X_1 and X_2 generated by the corresponding non-terminals of a regular grammar. X_0, X_1 and X_2 are related as follows:

$$X_0 = 1X_1$$
$$X_1 = 0X_1 + 1X_2$$
$$X_2 = 0X_1 + \{\lambda\}$$

Which one of the following choices precisely represents the strings in X_0?
(a) $10\,(0^* + (10)^*)1$
(b) $10\,(0^* + (10)^*)^*1$
(c) $1\,(0 + 10)^*1$
(d) $10\,(0 + 10)^*1 + 110\,(0 + 10)^*1$

[2015 (Set-2) : 2 M]

1.83 Which of the following languages is/are regular?
L_1: $\{wxw^R \mid w, x \in \{a, b\}^* \text{ and } |w|, |x| > 0\}$ w^R is the reverse of string w
L_2: $\{a^nb^m \mid m \neq n \text{ and } m, n \geq 0\}$
L_3: $\{a^pb^qc^r \mid p, q, r \geq 0\}$
(a) L_1 and L_3 only (b) L_2 only
(c) L_2 and L_3 only (d) L_3 only

[2015 (Set-2) : 2 M]

1.84 The number of states in the minimal deterministic finite automaton corresponding to the regular expression $(0 + 1)^*\,(10)$ is ______.

[2015 (Set-2) : 2 M]

1.85 Let L be the language represented by the regular expression $\Sigma^*0011\Sigma^*$ where $\Sigma = \{0, 1\}$. What is the minimum number of states in a DFA that recognizes $\overline{L}$ (complement of L)?
(a) 4 (b) 5
(c) 6 (d) 8

[2015 (Set-3) : 1 M]

1.86 Which of the following languages is generated by the given grammar?

$$S \rightarrow aS \mid bS \mid \varepsilon$$

(a) $\{a^nb^m \mid n, m \geq 0\}$
(b) $\{w \in \{a, b\}^* \mid w$ has equal number of a's and b's$\}$
(c) $\{a^n \mid n \geq 0\} \cup \{b^n \mid n \geq 0\} \cup \{a^nb^n \mid n \geq 0\}$
(d) $\{a, b\}^*$

[2016 (Set-1) : 1 M]

1.87 Which of the following decision problems are undecidable?
I. Given NFAs N_1 & N_2, is $L(N_1) \cap L(N_2) = \phi$?
II. Given a CFG $G = (N, \Sigma, P, S)$ and a string $x \in \Sigma^*$, does $x \in L(G)$?
III. Given CFGs G_1 and G_2, is $L(G_1) = L(G_2)$?
IV. Given a TM M, is $L(M) = \phi$?
(a) I and IV only
(b) II and III only
(c) III and IV only
(d) II and IV only

[2016 (Set-1) : 1 M]

1.88 Which one of the following regular expressions represents the language: the set of all binary strings having two consecutive 0s and two consecutive 1s?
(a) $(0 + 1)^*0011(0 + 1)^* + (0 + 1)^*1100(0 + 1)^*$
(b) $(0 + 1)^*(00(0 + 1)^*11 + 11(0 + 1)^*00)(0 + 1)^*$
(c) $(0 + 1)^*00(0 + 1)^* + (0 + 1)^*11(0 + 1)^*$
(d) $00(0 + 1)^*11 + 11(0 + 1)^*00$

[2016 (Set-1) : 1 M]

1.89 The number of states in the minimum sized DFA that accepts the language defined by the regular expression $(0 + 1)^*\,(0 + 1)\,(0 + 1)^*$ is ______.

[2016 (Set-2) : 1 M]

1.90 Consider the following two statements:
I. If all states of an NFA are accepting states then the langauge accepted by the NFA is Σ^*.
II. There exists a regular language A such that for all languages B, $A \cap B$ is regular.
Which one of the following is **CORRECT**?
(a) Only I is true
(b) Only II is ture
(c) Both I and II are true
(d) Both I and II are false

[2016 (Set-2) : 2 M]

1.91 Consider the language L given by the regular expression $(a + b)^*\,b(a + b)$ over the alphabet $\{a, b\}$. The smallest number of states needed in a deterministic finite-state automaton (DFA) accepting L is ______.

[2017 (Set-1) : 1 M]

1.92 If G is a grammar with productions:

$$S \rightarrow SaS \mid aSb \mid bSa \mid SS \mid \in$$

Where S is the start variable, then which one of the following strings is not generated by G?
(a) *abab*
(b) *aaab*
(c) *abbaa*
(d) *babba*

[2017 (Set-1) : 2 M]

1.93 The minimum possible number of states of a deterministic finite automaton that accepts the regular language $L = \{w_1aw_2 \mid w_1, w_2 \in \{a, b\}^*, |w_1| = 2, |w_2| \geq 3\}$ is ______.

[2017 (Set-2) : 1 M]

1.94 Consider the context-free grammars over the alphabet $\{a, b, c\}$ given below. S and T are non-terminals.

$G_1 : S \rightarrow aSb \mid T,\ T \rightarrow cT \mid \epsilon$

$G_2 : S \rightarrow bSa \mid T,\ T \rightarrow cT \mid \epsilon$

The language $L(G_1) \cap L(G_2)$ is

(a) Finite
(b) Not finite but regular
(c) Context-Free but not regular
(d) Recursive but not context-free

[2017 (Set-1) : 1 M]

1.95 Let δ denote the transition function and $\hat{\delta}$ denote the extended transition function of the ε-NFA whose transition table is given below:

δ	ϵ	a	b
$\rightarrow q_0$	$\{q_2\}$	$\{q_1\}$	$\{q_0\}$
q_1	$\{q_2\}$	$\{q_2\}$	$\{q_3\}$
q_2	$\{q_0\}$	ϕ	ϕ
q_3	ϕ	ϕ	$\{q_2\}$

The $\hat{\delta}(q_2, aba)$ is

(a) ϕ
(b) $\{q_0, q_1, q_3\}$
(c) $\{q_0, q_1, q_2\}$
(d) $\{q_0, q_2, q_3\}$

[2017 (Set-2) : 2 M]

1.96 Let N be an NFA with n states. Let k be the number of states of a minimal DFA which is equivalent to N. Which one of the following is necessarily true?

(a) $k \geq 2^n$
(b) $k \geq n$
(c) $k \leq n^2$
(d) $k \leq 2^n$

[2018 : 1 M]

1.97 Given a language L, define L^i as follows:

$$L^0 = \{\epsilon\}$$
$$L^i = L^{i-1} \cdot L \text{ for all } i > 0$$

The order of a language L is defined as the smallest k such that $L^k = L^{k+1}$.

Consider the language L_1 (over alphabet 0) accepted by the following automaton.

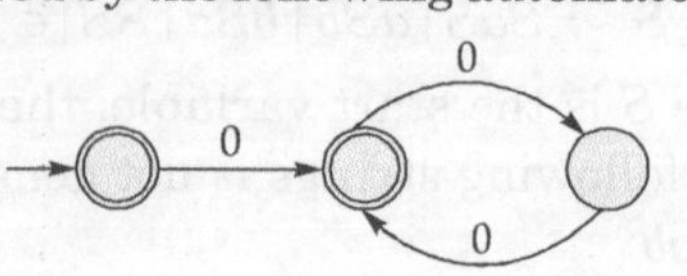

The order of L_1 is ________. **[2018 : 2 M]**

1.98 For $\Sigma = \{a, b\}$, let us consider the regular language $L = \{x \mid x = a^{2+3k} \text{ or } x = b^{10+12k},\ k \geq 0\}$. Which one of the following can be a pumping length (the constant guaranteed by the pumping lemma) for L?

(a) 9
(b) 24
(c) 3
(d) 5

[2019 : 1 M]

1.99 If L is a regular language over $\Sigma = \{a, b\}$, which one of the following languages is NOT regular?

(a) $L \cdot L^R\ \{xy \mid x \in L, y^R \in L\}$
(b) Suffix $(L) = \{y \in \Sigma^* \mid \exists x \in \Sigma^* \text{ such that } xy \in L\}$
(c) Prefix $(L) = \{x \in \Sigma^* \mid \exists y \in \Sigma^* \text{ such that } xy \in L\}$
(d) $\{ww^R \mid w \in L\}$

[2019 : 1 M]

1.100 Let Σ be the set of all bijections from $\{1, ..., 5\}$ to $\{1, ..., 5\}$, where id denotes the identity function, i.e. $\text{id}(j) = j,\ \forall j$. Let $\circ$ denote composition on functions.

For a string $x = x_1 x_2 \dots x_n \in \Sigma^n$, $n \geq 0$, let $\pi(x) = x_1 \circ x_2 \circ \dots \circ x_n$.

Consider the language $L = \{x \in \Sigma^* \mid \pi(x) = \text{id}\}$. The minimum number of states in any DFA accepting L is ________. **[2019 : 2 M]**

1.101 Consider the following statements:

I. If $L_1 \cup L_2$ is regular, then both L_1 and L_2 must be regular.
II. The class of regular languages is closed under infinite union.

Which of the above statements is/are TRUE?

(a) Neither I nor II
(b) II only
(c) I only
(d) Both I and II

[2020 : 1 M]

1.102 Which one of the following regular expressions represents the set of all binary strings with an odd number of 1's?

(a) (0*10*10*)*0*1
(b) 10* (0*10*10*)*
(c) ((0 + 1)* 1(0 + 1)*1)*10*
(d) (0*10*10*)*10*

[2020 : 1 M]

1.103 Consider the following language:

$L = \{x \in \{a, b\}^* \mid$ number of a's in x is divisible by 2 but not divisible by 3$\}$

The minimum number of states in a DFA that accepts L is ________.

[2020 : 2 M]

1.104 Consider the following language:

$L = \{w \in \{0,1\}^* \mid w \text{ ends with the substring } 011\}$

Which one of the following deterministic finite automata accepts L?

(a) 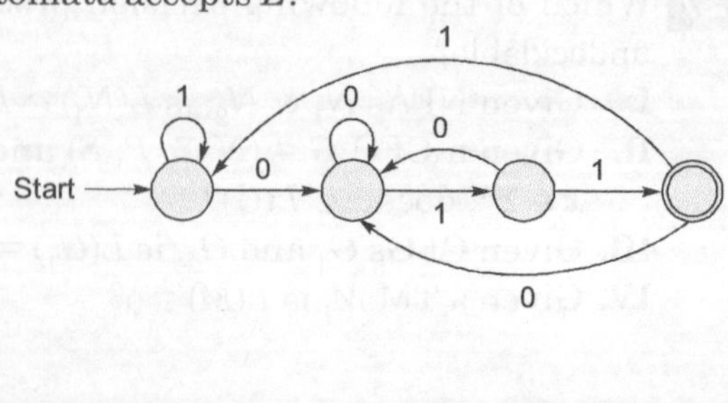

(b) 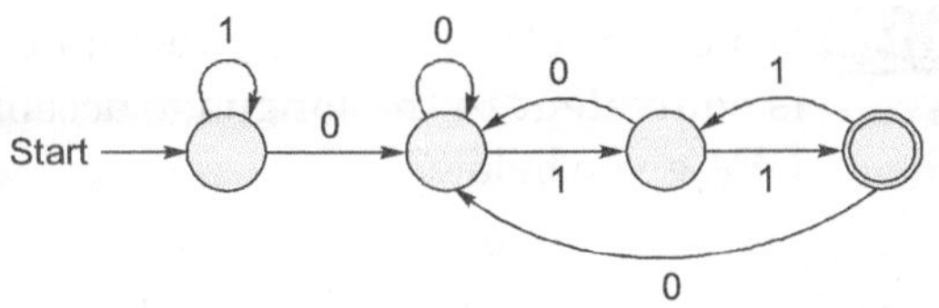

(c) 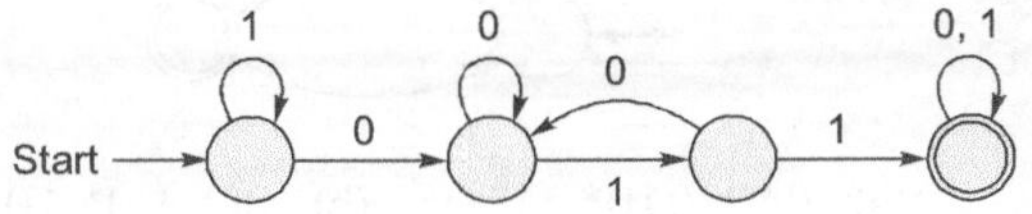

(d) 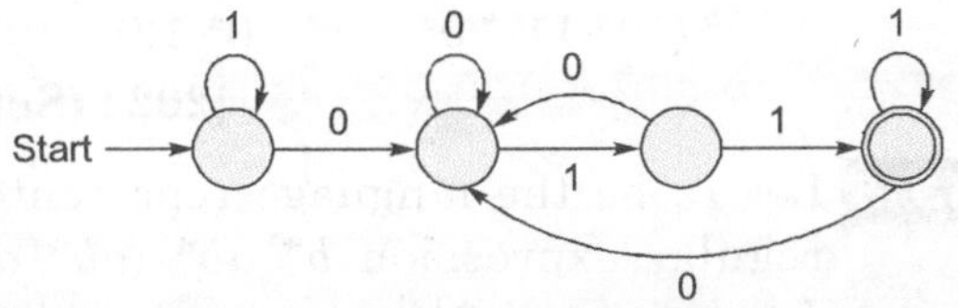

[2021 (Set-1) : 2 M]

1.105 Let $L \subseteq \{0, 1\}^*$ be an arbitrary regular language accepted by a minimal DFA with k states. Which one of the following languages must necessarily be accepted by a minimal DFA with k states?

(a) $\{0, 1\}^* - L$ (b) $L \cup \{01\}$
(c) $L.L$ (d) $L - \{01\}$

[2021 (Set-2) : 1 M]

1.106 Consider the following deterministic finite automation (DFA):

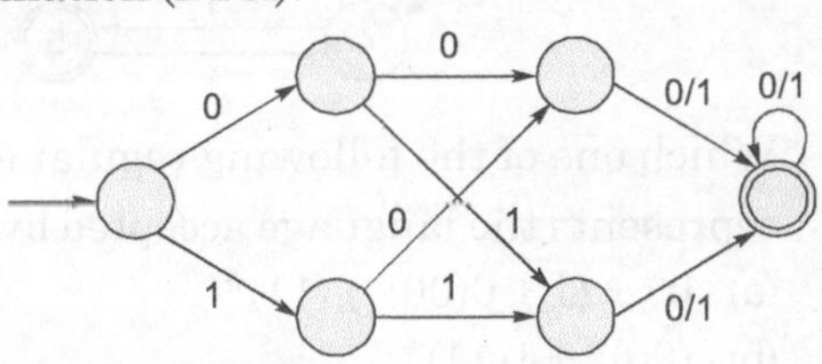

The number of strings of length 8 accepted by the above automation is ________.

[2021 (Set-2) : 1 M]

1.107 Which of the following regular expressions represent(s) the set of all binary numbers that are divisible by three? Assume that the string e is divisible by three.

(a) (0*(1(01*0)*1)*)*
(b) (0 +1(01*0)* 1)*
(c) (0 + 11 + 10(1 + 00)*01)*
(d) (0 + 11 + 11(1 + 00)*00)*

[2021 (Set-2) : 2 M]

1.108 Which one of the following regular expressions correctly represents the language of the finite automaton given below?

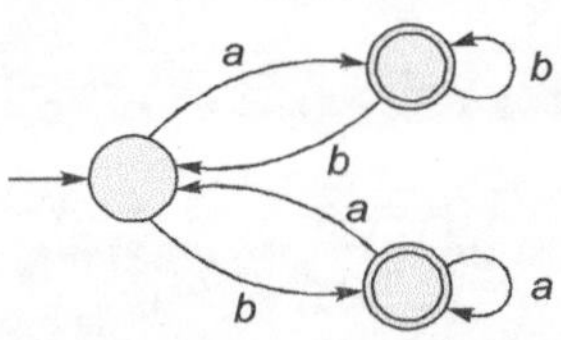

(a) $ab^*bab^* + ba^*aba^*$
(b) $(ab^*b)^*ab^* + (ba^*a)^*ba^*$
(c) $(ab^*b + ba^*a)^*(a^* + b^*)$
(d) $(ba^*a + ab^*b)^*(ab^* + ba^*)$ **[2022 : 1 M]**

1.109 Consider the Deterministic Finite-state Automaton (DFA) A shown below. The DFA runs on the alphabet {0, 1}, and has the set of states {s, p, q, r}, with s being the start state and p being the only final state.

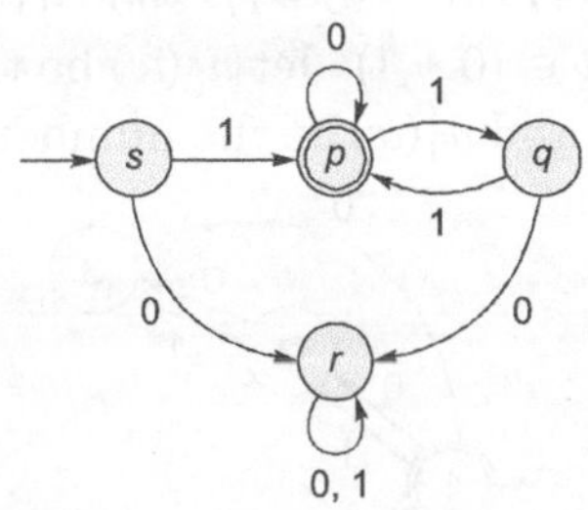

Which one of the following regular expressions correctly describes the language accepted by A?

(a) 1(0*11)* (b) 0(0 + 1)*
(c) 1(0 + 11)* (d) 1(110*)*

[2023 : 1 M]

1.110 Consider the following definition of a lexical token id for an identifier in a programming language, using extended regular expressions:

letter → [A – Za – a]
digit → [0 – 9]
id → letter (letter | digit)*

Which one of the following Non-deterministic Finite-state Automata with ∈-transitions accepts the set of valid identifiers? (A double-circle denotes a final state)

(a) 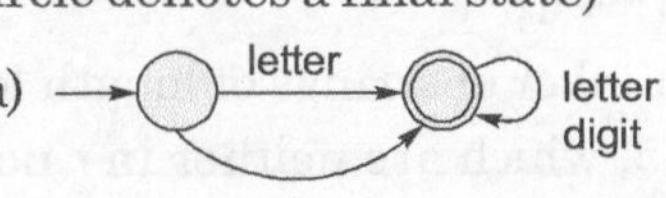

(b) 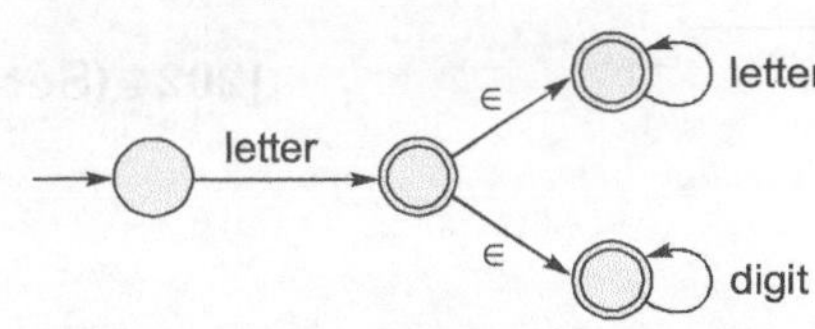

(c) 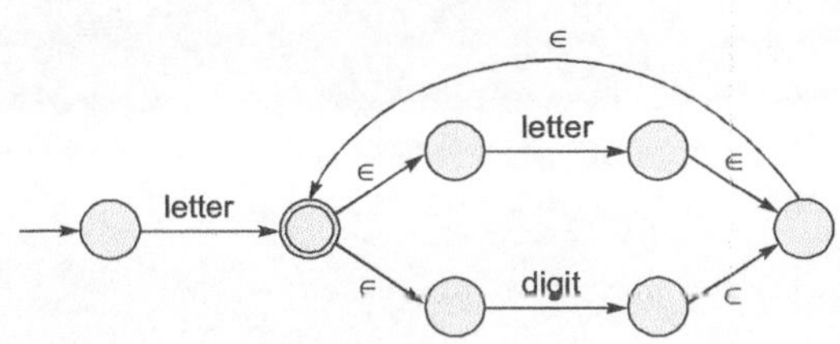

(d) 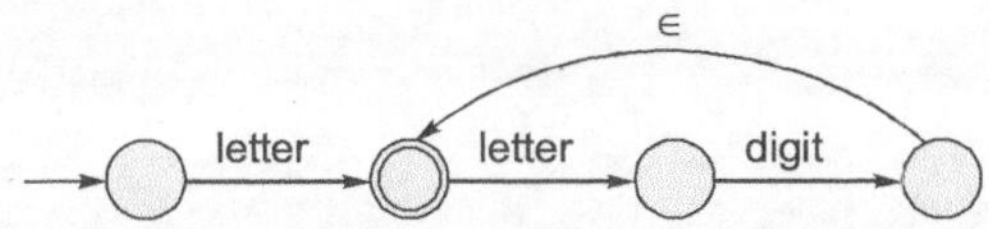

[2023 : 1 M]

1.111 Consider the language L over the alphabet {0, 1}, given below:

$L = \{w \in \{0, 1\}^* \mid w$ does not contain three or more consecutive 1's$\}$.

The minimum number of states in a Deterministic Finite-State Automaton (DFA) for L is ________. **[2023 : 2 M]**

1.112 Consider the 5-state DFA M accepting the language $L(M) \subset (0 + 1)^*$ shown below. For any string $w \in (0 + 1)^*$ let $n_0(w)$ be the number of 0's in w and $n_1(w)$ be the number of 1's in w.

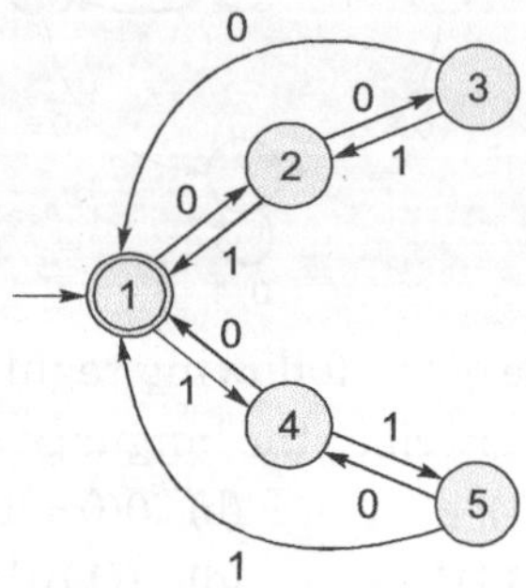

Which of the following statements is/are FALSE?

(a) States 2 and 5 are distinguishable in M
(b) States 2 and 4 are distinguishable in M
(c) States 3 and 4 are distinguishable in M
(d) Any string w with $n_0(w) = n_1(w)$ is in $L(M)$

[2024 (Set-1) : 2 M]

1.113 Consider the following two regular expressions over the alphabet {0, 1}:

$r = 0^* + 1^*$

$s = 01^* + 10^*$

The total number of strings of length less than or equal to 5, which are neither in r nor is s, is ________.

[2024 (Set-1) : 2 M]

1.114 Which one of the following regular expressions is equivalent to the language accepted by the DFA given below?

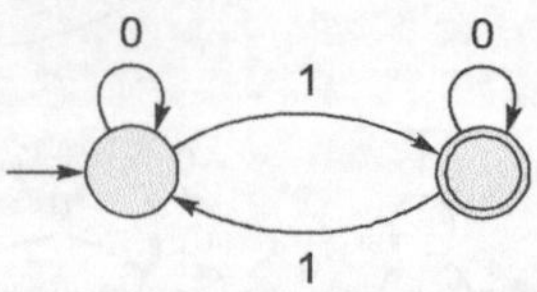

(a) $0^*\,1(010^*\,1)^*\,0^*$ (b) $0(1 + 0^*\,10^*\,1)^*\,0^*$
(c) $0^*\,(10^*\,11)^*\,0^*$ (d) $0^*\,1(0 + 10^*\,1)^*$

[2024 (Set-2) : 1 M]

1.115 Let L_1 be the language represented by the regular expression $b^*\,ab^*\,(ab^*\,ab^*)^*$ and $L_2 = \{w \in (a + b)^* \mid |w| \le 4\}$, where $|w|$ denotes the length of string w. The number of strings in L_2 which are also in L_1 is ________.

[2024 (Set-2) : 2 M]

1.116 Let M be the 5-state NFA with $\in$-transitions shown in the diagram below.

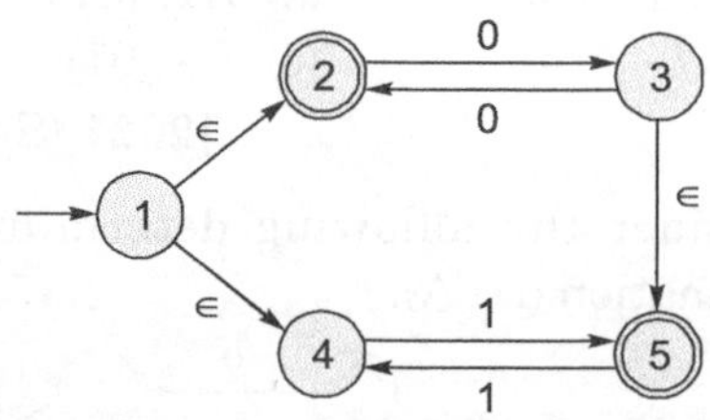

Which one of the following regular expressions represents the language accepted by M?

(a) $0^* + (1 + 0(00)^*)\,(11)^*$
(b) $(00)^* + 1(11)^*$
(c) $(00)^* + (1 + (00)^*)\,(11)^*$
(d) $0^+ + 1(11)^* + 0(11)^*$

[2024 (Set-2) : 2 M]

■■■■

Answers Finite Automata : Regular Languages

1.1	(a)	**1.2**	(b)	**1.3**	(c)	**1.4**	(d)	**1.5**	(b)	**1.6**	(d)	**1.7**	(d)	**1.8**	(a)	**1.9**	(c, d)
1.10	(d)	**1.11**	(b)	**1.12**	(b)	**1.13**	(c)	**1.14**	(b)	**1.15**	(d)	**1.16**	(a)	**1.17**	(b)	**1.18**	(d)
1.19	(d)	**1.20**	(a)	**1.21**	(b)	**1.22**	(c)	**1.23**	(b)	**1.24**	(c)	**1.25**	(b)	**1.26**	(a)	**1.27**	(b)
1.28	(a)	**1.29**	(a)	**1.30**	(b)	**1.31**	(d)	**1.32**	(b)	**1.33**	(b)	**1.34**	(d)	**1.35**	(a)	**1.36**	(a)
1.37	(c)	**1.38**	(b)	**1.39**	(a)	**1.40**	(c)	**1.41**	(d)	**1.42**	(b)	**1.43**	(a)	**1.44**	(c)	**1.45**	(c)
1.46	(b)	**1.47**	(a)	**1.48**	(d)	**1.49**	(c)	**1.50**	(b)	**1.51**	(a)	**1.52**	(a)	**1.53**	(c)	**1.54**	(c)
1.55	(a)	**1.56**	(a)	**1.57**	(d)	**1.58**	(a)	**1.59**	(a)	**1.60**	(c)	**1.61**	(a)	**1.62**	(c)	**1.63**	(c)
1.64	(b)	**1.65**	(c)	**1.66**	(a)	**1.67**	(c)	**1.68**	(b)	**1.69**	(a)	**1.70**	(b)	**1.71**	(c)	**1.72**	(d)
1.73	(a)	**1.74**	(d)	**1.75**	(c)	**1.76**	(a)	**1.77**	(b)	**1.78**	(a)	**1.79**	(a)	**1.80**	(3)	**1.81**	(1)
1.82	(c)	**1.83**	(a)	**1.84**	(3)	**1.85**	(b)	**1.86**	(d)	**1.87**	(c)	**1.88**	(b)	**1.89**	(2)	**1.90**	(b)
1.91	(4)	**1.92**	(d)	**1.93**	(8)	**1.94**	(b)	**1.95**	(c)	**1.96**	(d)	**1.97**	(2)	**1.98**	(b)	**1.99**	(d)
1.100	(2)	**1.101**	(a)	**1.102**	(*)	**1.103**	(6)	**1.104**	(a)	**1.105**	(a)	**1.106**	(256)	**1.107**	(a, b, c)	**1.108**	(d)
1.109	(c)	**1.110**	(c)	**1.111**	(4)	**1.112**	(a, c)	**1.113**	(44)	**1.114**	(d)	**1.115**	(15)	**1.116**	(a)		

Explanations Finite Automata : Regular Languages

1.1 (a)

$L = \{x^n y^n \mid n \geq 1\}$

The language L produces all strings having equal number of x and y and y follows the x.

(i) $E \rightarrow x\,Ey \mid xy$: The language produced by above grammar contain all strings having equal no. of x & y and y follows x.

(ii) $xy \mid x^+xy\,y^+$: The above can generate string xxxxyy which doesn't contain equal no. of x and y

(iii) x^+y^+ : The above regular expression generates all strings in which y follows x and contain x and y greater than one. Above RE can generate string xxxy which doesn't contain equal no of x & y.

So only (i) generates the same language as L.

1.2 (b)

The state diagram which represents the given state table is

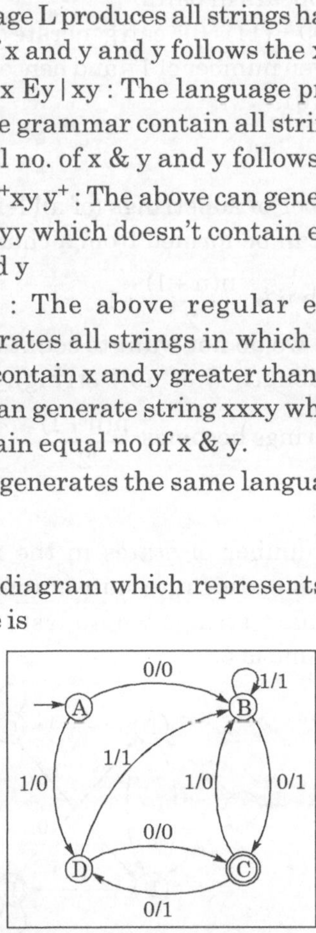

From state A to reach C, the shortest sequence is

$$A \xrightarrow{0} B \xrightarrow{0} C$$
$$A \xrightarrow{1} D \xrightarrow{0} C$$

From B to reach C

$$B \xrightarrow{0} C$$
$$B \xrightarrow{1} B \xrightarrow{0} C$$

From C to reach C

$$C \xrightarrow{0} D \xrightarrow{0} C$$
$$C \xrightarrow{1} B \xrightarrow{0} C$$

From D to reach C

$$D \xrightarrow{0} C$$
$$D \xrightarrow{1} B \xrightarrow{0} C$$

So 10 is the shortest input sequence to reach the final state C, whatever be the initial state.

1.3 (c)

(i) $(00)^*(\varepsilon + 0)$
$\equiv (00)^*\,\varepsilon + (00)^*\,0 = 0^*$
Even number of 0's and Odd number of 0's i.e., any number of 0's

(ii) $(00)^* \equiv$ Even number of 0's

(iii) $0^* \equiv$ any number of 0's

(iv) $0(00)^* \equiv$ Odd number of 0's

So (i) and (iii) are same.

1.4 (d)

(a) L = {x|x has an equal number of a's and b's} is context free language (since there is comparison between number of a's and number of b's).

(b) L = $\{a^nb^n|n \geq 1\}$ is context free language (since there is comparison between number of a's and number of b's).

(c) L = {x|x has more a's than b's} is context free language (since there is comparison between number of a's and number of b's).

(d) L = $\{a^mb^n|m \geq 1, n \geq 1\}$ is regular (since m and n are independent and hence there is no comparison). The regular expression is aa*bb*.

1.5 (b)

(a) The set of all strings over Σ is Σ* which is countably infinite.

(b) Set of all languages over Σ is 2^{Σ^*}. According to Cantor's theorem if S be an countably infinite set, then its power set 2^S is uncountable. So 2^{Σ^*} is uncountable because Σ* is countably infinite.

(c) Set of all regular languages over Σ is countably infinite.

(d) Set of all languages over Σ accepted by Turing machine is the set of all RE languages which is countably infinite.

1.6 (d)

(a) r.e. (regular expression) = 0*(1 + 0)* can generate string 100, which contains substring 100

(b) r.e. (regular expression) = 0* 1010* can generate string 10100, which contain 100 as a substring. Also, this regular expression cannot generate ε which is in the given language.

(c) r.e. (regular expression) = 0*1*01* generates strings which doesn't contain 100 as substring. However, ε is the smallest string which doesn't contain 100 as substring but above RE can't generate ε.

(d) r.e. (regular expression) = 0*(10 + 1)* generates all strings which doesn't contain 100 as substring.

1.7 (d)

(a*b*)* ≡ (a + b)*

So, B = ((01)* 1*)*

≡ (01 + 1)* ≡ A

So, A = B

1.8 (a)

The numbers 1, 2, 4, 8,...2^n ... are represented by 1, 10, 100, 1000,....

The pattern in the regular expression is 1 followed by 0's.

The regular expression for above is 10*

The DFA for above language is

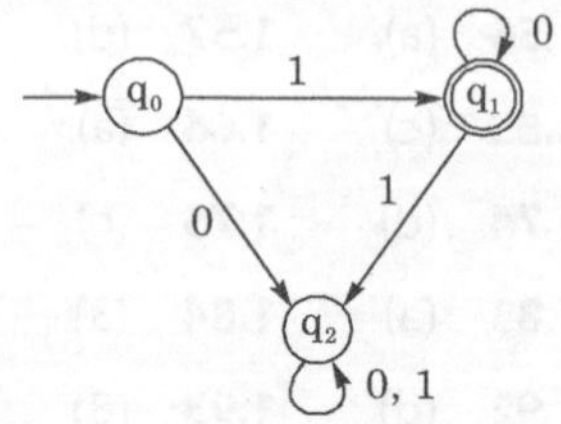

So the numbers 1, 2, 4, 8, ... 2^n ... written in binary can be recognized by a deterministic finite state automaton.

1.9 (c, d)

(c) The string 1101 doesn't belong to set represented by (10)* (01)* (00 + 11)* because once 11 appears in string then 1 and 0 only appears in pairs.

(d) (00 + (11)* 0)* can generate only strings with even number of 1's and hence cannot generate 1101.

1.10 (d)

Number of substrings (of all lengths inclusive) that can be formed from a character strings of length n is $\frac{n(n+1)}{2}+1$.

Since we do not want to count the substring of zero length (i.e. null string), the number of substrings becomes $\frac{n(n+1)}{2}$.

1.11 (b)

The number of states in the minimum state deterministic finite state automation accepting all binary strings whose last two symbols are the same is 5.

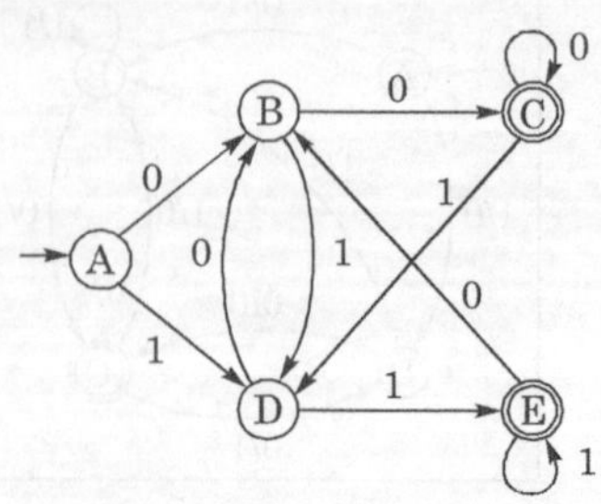

1.12 (b)

The minimum state finite automation that recognizes the language represented by regular expression (0 + 1) (0 + 1) ... n times is n + 1.
This language contains strings with exactly length n.
(n + 1) states are required to count length upto n. No trap state is required since we are making minimal FA, not minimal DFA.
For example, for n = 2 the design is shown below.

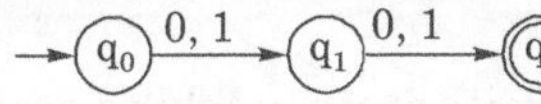

1.13 (c)

According to rules of regular expressions

$$(r_1 + r_2)^* \equiv (r_1 + r_2^*)^*$$

Therefore $(a + b^*)^* \equiv (a + b)^*$

So $S = T$

1.14 (b)

The language generated by the grammar

$S \rightarrow 0S0 \mid 00$ is

$L = \{0^2, 0^4, 0^6, 0^8, ...\}$

$= \{0^{2n+2} \mid n \geq 0\}$

$\Rightarrow \quad = \{0^{2n} \mid n \geq 1\} = 00(00)^*$

So above language is regular but not 0^+.

1.15 (d)

$L = \{a^n \mid n \text{ is odd}\}$

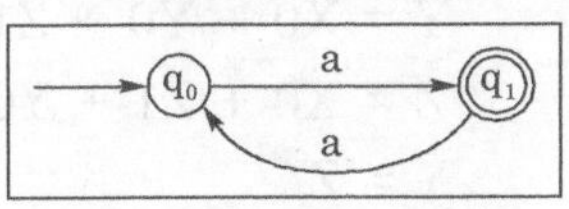

$L = \{a^n \mid n \text{ is even}\}$

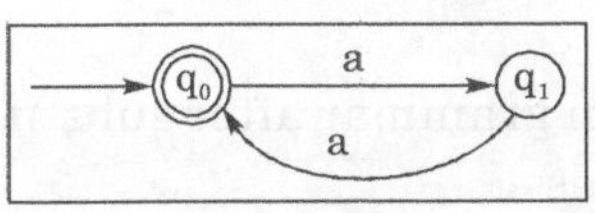

$L = \{a^n \mid n \geq 0\}$

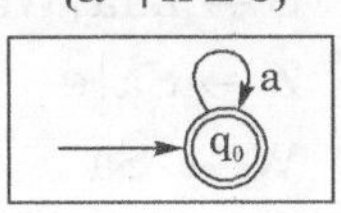

So for two states minimal finite state automation, L can be $\{a^n \mid n \text{ is odd}\}$ or L can be $\{a^n \mid n \text{ is even}\}$

1.16 (a)

$$L_1 = \{0^{2n} \mid n \geq 1\}$$

L_1 produces language having even number of 0's which is regular language.
Regular expression for S_1 is $00(00)^*$
$L_2 = \{0^m 1^n 0^{m+n} \mid m \geq 1 \text{ and } n \geq 1\}$ is context free language not regular. (m occurs in two places, so there is comparison of count).
So, S_1 is correct but S_2 is not correct.

1.17 (b)

For an arbitrary NFA with N states, the maximum number of states in an equivalent minimized DFA is 2^N.

1.18 (d)

A DFA over $\Sigma = \{a, b\}$ accepting all strings which have no. of a's divisible by 6 and number of b's divisible by 8 is a grid machine (product automata) having $6 \times 8 = 48$ states.

1.19 (d)

$$L_1 = \{ww \mid w \in \{a, b\}^*\}$$

is context sensitive language (CSL) (since there is infinite string matching in straight order).

$$L_2 = \{ww^R \mid w \in (a, b)^*, w^R$$

is the reverse of w}
is context free language (since there is infinite string matching in reverse order).

$$L_3 = \{0^{2i} \mid i \text{ is an integer}\} = (00)^*$$

is regular language which contains all strings having even number of 0's.

$$L_4 = \{0^{i^2} \mid i \text{ is an integer}\}$$

is context sensitive language (CSL) (since the power is infinite and non linear).

1.20 (a)

The state diagram represents the FSM which outputs the sum of the present and previous bits of the input.
State A represents previous bit is a 0 and B and C represents previous bit is a 1.

1.21 (b)

The minimal finite automation with 3 states which accepts the language $L = \{x \mid \text{length of } x \text{ is divisible by } 3\}$ is as follows:

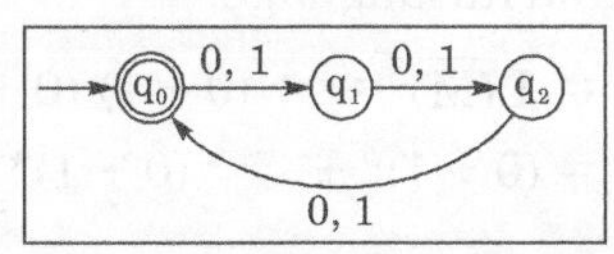

1.22 (c)

The given bit pattern can be represented as:
1 _ _ 1 _ _ 1
The four blanks can be filled in $2^4 = 16$ ways. Therefore there are 16 such strings in this pattern. Not all of these are accepted by the machine. The strings and its acceptance is given below:

							accepted
1	0	0	1	0	0	1	✓
1	0	0	1	0	1	1	✓
1	0	0	1	1	0	1	✓
1	0	0	1	1	1	1	✓
1	0	1	1	0	0	1	✓
1	1	0	1	0	0	1	✓
1	1	1	1	0	0	1	✓

Only these seven strings given above are accepted. The other strings (9 of them) in this pattern are rejected, since they dont reached the final state.

∴ Correct answer is (c).

1.23 (b)

The given machine M is

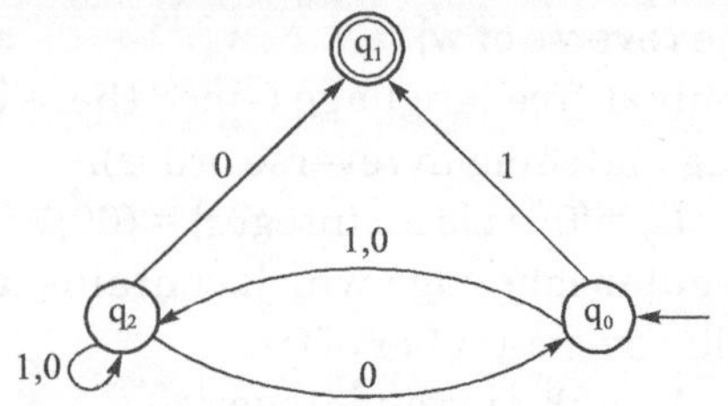

Now the complementary machine $\overline{M}$ is

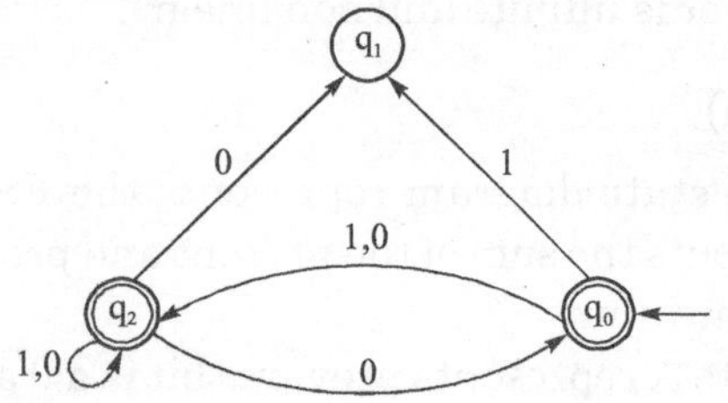

In the case of dfa, $L(\overline{M}) = \overline{L(M)}$ but in the case of nfa this is not true. In fact $L(\overline{M})$ and L(M) have no connection.

∴ To find $L_1 = L(\overline{M})$ we have to look at $\overline{M}$ and directly find its language.

$$L_1 = L(\overline{M}) = \varepsilon + (0 + 1)(0 + 1)^* + ...$$
$$= (0 + 1)^* + ... = (0 + 1)^*$$

1.24 (c)

In choices (a), (b) and (d), inside the parenthesis we can generate "a", "b" and "c" seperately and hence all three are same as (a + b + c)*.

In choice (c) the strings "a" and "b" cannot be generated seperately since "ab" is always together.

So, choice (c) is not same as (a + b + c)*.

1.25 (b)

$$\delta(A, a) = A \equiv A \to aA$$
$$\delta(A, b) = B \equiv A \to bB$$
$$\delta(B, a) = B \equiv B \to aB$$
$$\delta(B, b) = A \equiv B \to bA$$

Since B is final state, so we need to put B→ ε. So the correct grammar is choice (b) which is {A → aA, A → bB, B → aB, B → bA, B → ε}.

1.26 (a)

The given finite state machine accepts any string w ∈ (0, 1)* in which the number of 1s is multiple of 3 and the number of 0s is multiple of 2.

1.27 (b)

Given regular expression is infinite set (because of *) of finite strings. A regular expression cannot generate any infinite string (since string is always finite in length by definition).

1.28 (a)

Given language is finite. Hence it is regular language.

1.29 (a)

Writing Y and Z interms of incoming arrows (Arden's method), we get

$$Y = X0 + Y0 + Z1$$
$$Z = X0 + Z1 + Y0$$

Clearly, $Y = Z$

1.30 (b)

The given grammar after substitution of X and Y becomes

$$S \to Zaa \mid Waa$$
$$Z \to Sa \mid \in$$
$$W \to Sa$$

Which after substituting Z and W is equivalent to S → Saaa | aa | Saaa.

Which is equivalent to S → Saaa | aa.

So, L(G) = (aaa)*aa

So the language generated by the grammar is the set of strings with *a*'s such that number of *a* mod 3 is 2.So the number of states required should be 3 to maintain the count of number of *a*'s mod 3.

1.31 **(d)**

If L is regular ⇒ L satisfies the pumping lemma for regular languages.
If L is CFL ⇒ L satisfies the pumping lemma for CFLs.
By satisfying pumping lemma, we can never say that a language is regular or CFL. It can only be used to prove that a certain language is not regular or not CFL in case the language violates the corresponding pummping lemma.
So, both regular and non-regular languages can satisfy pumping lemma for regular language.
Similarly, both CFLs and non-CFLs can satisfy pumping lemma for CFLs.
So satisfying pumping lemma doesn't prove anything about the type of langauge.

1.32 **(b)**

(a) is false since M is accepting "abbb".
(b) is true.
(c) is false since "abba" contains "abb" as a substring, but is being rejected by the machine.
(d) is false, since λ does not contain "aa" as a substring, but λ is being rejected by M.

1.33 **(b)**

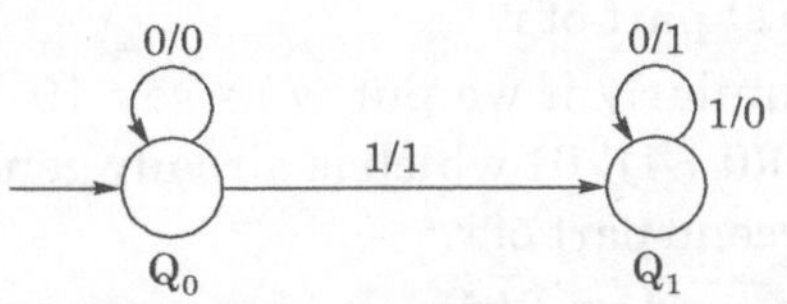

The given machine, executes the algorithm for 2's complement when input is given from LSB.

1.34 **(d)**

u = abbaba: Accepted by automata.
v = bab: Not accepted by automata.
w = aabb: Not accepted by automata.

1.35 **(a)**

Given NFA:

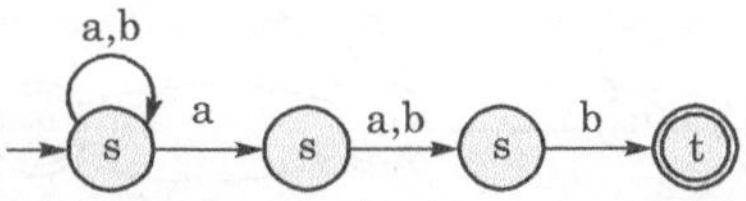

Regular expression equivalent to given NFA is $(a + b)^*a\ (a + b)b$.

1.36 **(a)**

The given right-linear grammar can be converted to the following DFA.

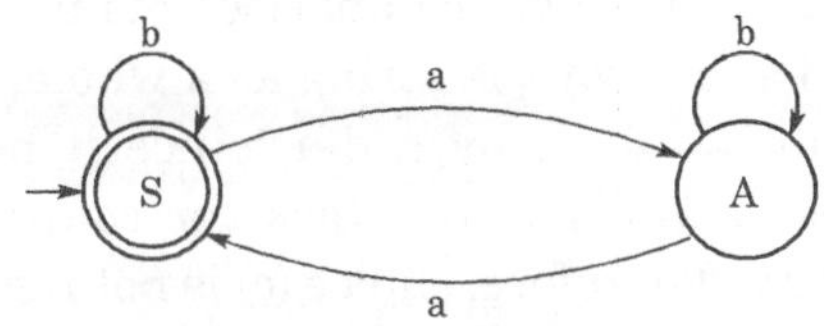

The machine accepts all strings over the alphabet $\{a, b\}$ which have an even number of a's. It is a minimal DFA.
So Myhill-Nerode equivalnce classes for the language is nothing but the set of strings reaching S and A respectively.

$$S = \{w \in (a + b)^* \mid \#_a(w) \text{ is even}\}$$
$$A = \{w \in (a + b)^* \mid \#_a(w) \text{ is odd}\}$$

1.37 **(c)**

- Σ^* is the regular superset of every language.
- ϕ is a regular subset of every language.
- Every subset of a regular language is regular is false since $a^nb^n \subseteq \Sigma^*$, but a^nb^n is not Regular but DCFL.
- Since every subset of a finite language is finite and every finite set is regular. Hence, every subset of a finite language is regular.

1.38 **(b)**

Prefix (L), suffix (L) and Half (L) are regular languages.
Repeat (L) is not a regular language but a CSL.

1.39 **(a)**

Option (a): If $L = (a + b)^*$, then repeat $(L) = \{ww \mid w \in (a + b)^*\}$ is clearly not regular. So option (a) is best suited to show that repeat (L) need not be regular.
Option (b): If $L = \{\varepsilon, a, ab, bab\}$, then repeat $(L) = \{ww \mid w \in L\}$ becomes finite and hence regular. So option (b) is not suited to show that repeat (L) need not be regular.
Option (c): If $L = (ab)^*$, then repeat $(L) = \{ww \mid w \in (ab)^*\} = (ab)^*$ which is regular. So option (c) is not suited to show that repeat (L) need not be regular.
Option (d): $L = \{a^nb^n \mid n \geq 0\}$, is not suited since it is not regular.

1.40 **(c)**

Choice (a) is regular since it is finite.
Choice (b) is regular since although comparison is made between 0's and 1's, it is for all prefixes and this can be done by DFA.
Note: $|n_0(s') - n_1(s')| \leq 2$ is same as $n_0(s') - n_1(s') \geq 2$ or $n_1(s') - n_0(s') \geq 2$.

Choice (c) involves comparison of number of 0's and 1's, but for the string as a whole, and this cannot be done by a dfa, since it has finite memory and has no stack for counting upto infinity. Therefore, choice (c) is not regular.

Choice (d) is regular since $n_0(s) \bmod 7 = n_1(s) \bmod 5 = 0$ means number of 0's is divisible by 7 and number of 1's is divisible by 5 and this can be accepted by a dfa with $7 \times 5 = 35$ states.

A minimal DFA that will accept the language of choice (b) is shown below:

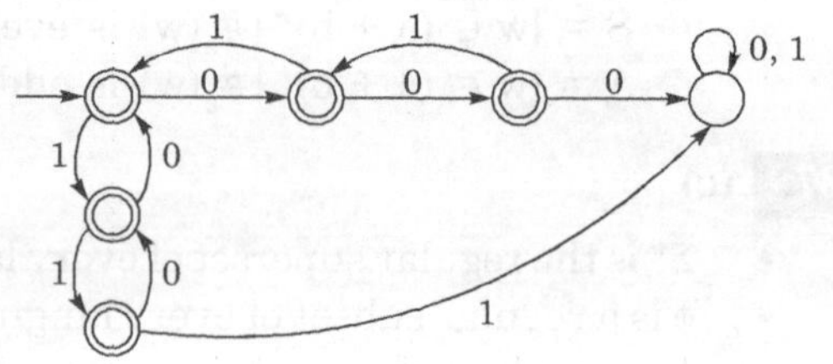

1.41 (d)

$L = (111 + 11111)^*$

Here $\Sigma = \{1\}$

Note that $\lambda \in L$, $1 \notin L$, $11 \notin L$, $111 \in L$, $1111 \notin L$, $11111 \in L$, $111111 \in L$, $1111111 \notin L$, $11111111 \in L$.

Notice also that L includes all w such that $n_1(w) \geq 8$, since all words with more than eight 1's can be generated by some combination of 111 and 11111.

Therefore, the required dfa is shown below:

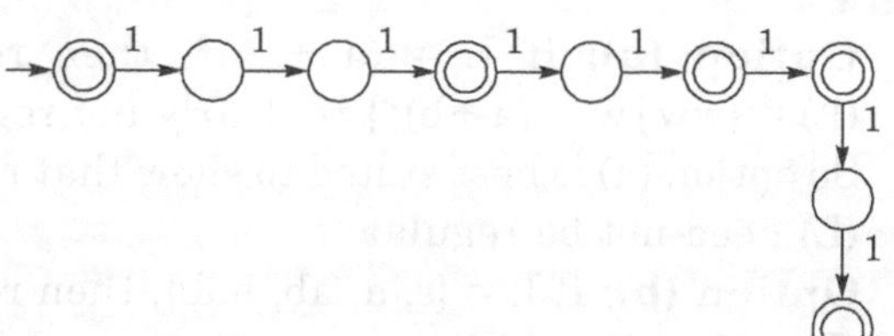

This dfa has 9 states.

1.42 (b)

(a) "Every subset of a regular set is regular" is false, since $L_1 = \Sigma^*$ and $L_2 = \{a^n b^n, n \geq 0\}$. Here, $L_2 \subseteq L$, but L_2 is not regular.

(b) "Every finite subset of a non-regular set is regular", is true, since all finite sets are regular.

(c) "The union of two non-regular sets is not regular" is false, since if you take $L_1 = \{a^n b^n; n \geq 0\}$ and L_1^c, neither of these is regular but $L_1 \cup L_1^c = \Sigma^*$, is regular.

(d) "Infinite union of finite sets is regular" is false, since regular sets are not closed under infinite union.

1.43 (a)

L = {w | s∈{0, 1}* number if 0's and 1's in w are divisible by 3 and 5 respectively}

The minimum state deterministic finite automation accepting the language L has $3 \times 5 = 15$ states.

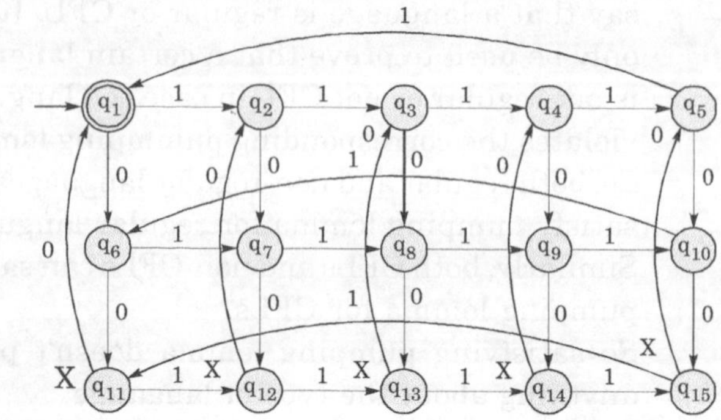

1.44 (c)

$L = \{w x w^R \mid x, w, \in (0, 1)^+\}$ is regular. Put w as 0 and 1 which are its minimal string values and get a minimal expression

$$r = [0(0 + 1)^+ 0] + [1(0 + 1)^+ 1]$$

Now putting w as any other string will not create any new string not generated by r.

Example if we put w as say 01, then $wxw^R = 01(0 + 1)^+ 10$ which is already generated by the first part of r.

Similarly if we put w as say 10, then $wxw^R = 10(0 + 1)^+ 01$ which is already generated by the second part of r.

So, L = L(r)

So the regular expression for L is

$$r = [0(0 + 1)^+ 0] + [1(0 + 1)^+ 1]$$

That is, L reduces to the language of words which starts and ends with the same symbol. Since we are able to write a regular expression for L, therefore L is regular.

1.45 (c)

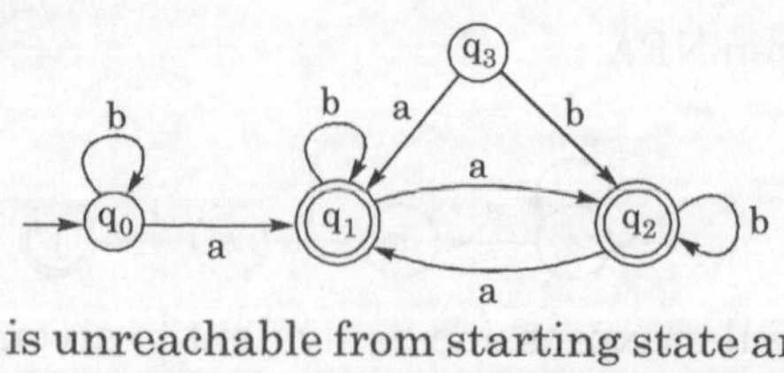

q_3 is unreachable from starting state and hence can be deleted to give the following diagram.

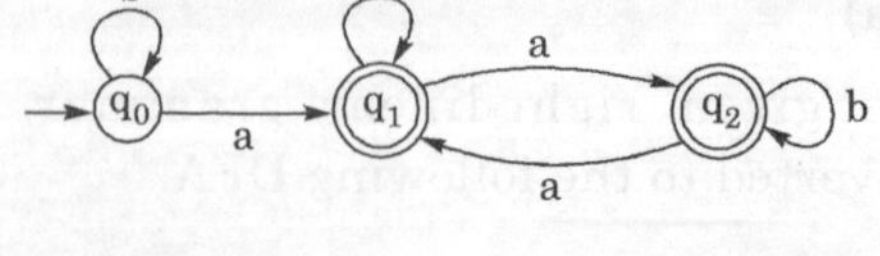

q_0 is the initial state. From q_0 to q_1 the regular expression is b* a. After that, every combination of a and b is accepted by q_1 or q_2. So q_1 and q_2 together can be collapsed into a permanent accept state and now the diagram becomes.

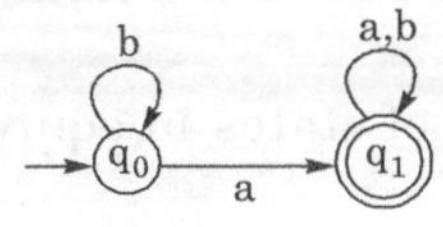

$\therefore \quad r = b^*a(a+b)^*$

1.46 (b)

The minimum state FSA is given below for the regular expression b*a(a + b)*

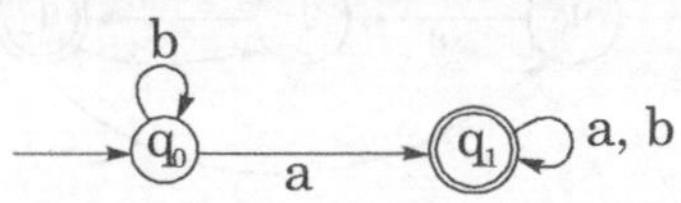

So FSA contains minimum two states q_0 and q_1.

1.47 (a)

G1: $S \rightarrow x \mid z \mid xS \mid zS \mid yB$

$B \rightarrow y \mid z \mid yB \mid zB$

$B \rightarrow (y + z)^+$

Substitute in S to get

$S \rightarrow x \mid z \mid xS \mid zS \mid y(y + z)^+$

Now solution of S is

$S \rightarrow (x + z)^*S$

$S \rightarrow (x + z)^*(x + z + y\ (y + z)^+)$

So, $L(G1) = S = (x+z)^+ + (x + z)^*\ y(y + z)^+$

G1 generates every string in which "no y appears before any x".

G2: $S \rightarrow y \mid z \mid yS \mid zS \mid xB$

$B \rightarrow y \mid yS$

Substitute B in S to get

$S \rightarrow y \mid z \mid yS \mid zS \mid xy \mid xyS$

Now solution of S is

$S \rightarrow (y + z + xy)^*S$

$S \rightarrow (y + z + xy)^*\ (y + z + xy)$

So, $L(G2) = S = (y + z + xy)^+$

G2 generates every string in which "every x followed by atleast one y".

1.48 (d)

Given DFA can be redesigned as S0 as q00, S1 as q10, S2 as q11, S3 as q01.

Each state is q_{ab} [$a = n_a$ mod2, $b = n_b$ mod2].

q00 as $n_a$mod2 = 0, $n_b$mod2 = 0 [number of x is even, number of y is even].

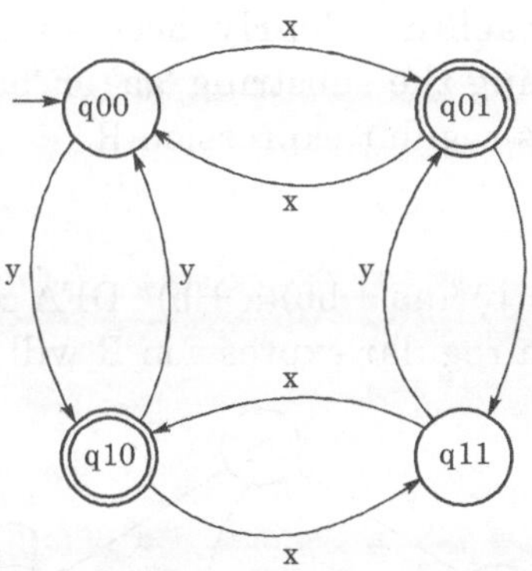

q01 is final state mean where number of x is even and number of y is odd.

q10 is final state mean where number of x is odd and number of y is even.

1.49 (c)

This grammar cannot generate any string starting with *xx* or ending with *xx* so (i), (ii) and (v) cannot be generated by the grammar. *xyx* also cannot be generated by the grammar. The derivation for (iii) *xyxy* and (iv) *yxxy* is shown below.

$S \rightarrow xB \rightarrow xyS \rightarrow xyxB \rightarrow xyxy$

$S \rightarrow yA \rightarrow yxS \rightarrow yxxS \rightarrow yxxy$

So only (iii) and (iv) can we derived from this grammar.

1.50 (b)

$L(P) \cap L(Q)$ must contain all strings common to P and Q. Note that 'aa' is common to both P and Q and hence must be accepted by any FA accepting $L(P) \cap L(Q)$. However, all of the given machine reject 'aa'. Therefore option (e) i.e. none of these is the right answer.

1.51 (a)

The language of the given regular expression R is 'containing the substring aa or bb'. Option (a) is the correct machine for this language.

1.52 (a)

In option (a) S3 and S4 together act as a permanent accept and can therefore be collapsed into a single permanent accept state as shown below.

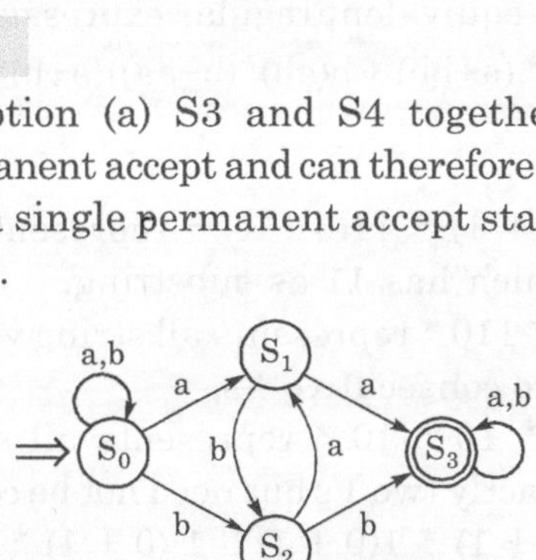

This machine clearly accepts all strings containing the substring 'aa' or 'bb', which is same as regular expression R.

1.53 (c)

$R = (a + b)^*(aa + bb)(a + b)^*$ DFA corresponds to given regular expression R will be:

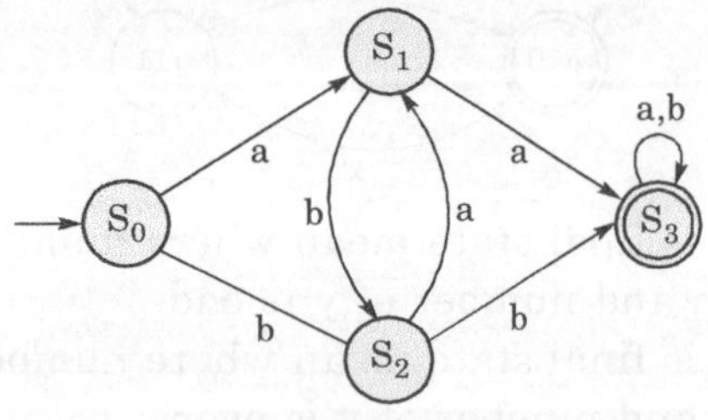

Which is equivalent to:

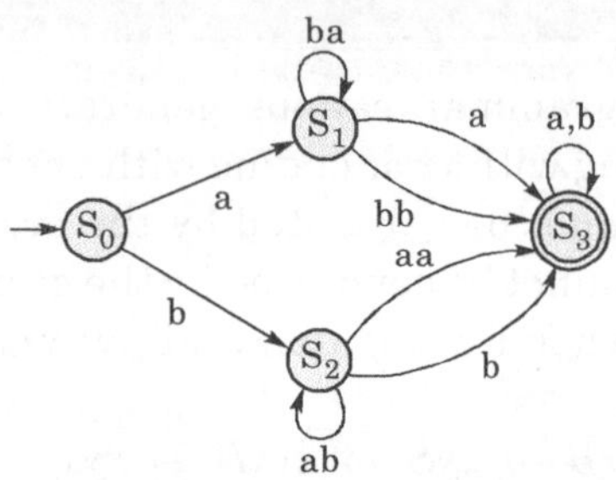

Which is equivalent to:

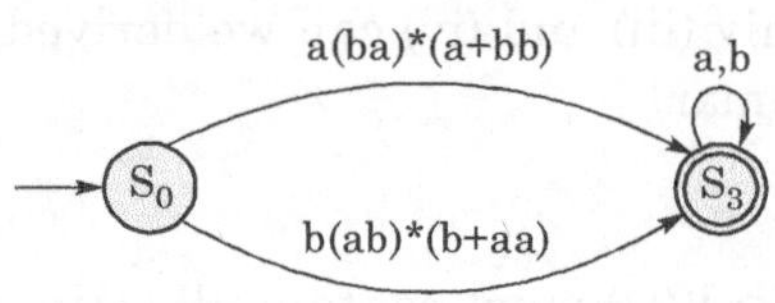

Which is equivalent to:

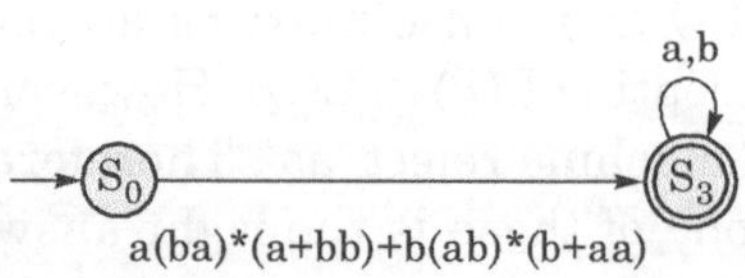

which is equivalent to:

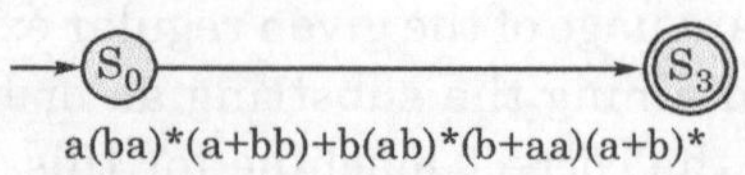

Hence equivalent regular expression is $(a(ba)^* (a+bb) + b(ab)^*(b+aa))(a+b)^*$.

1.54 (c)

- $(0 + 1)^* 11(0 + 1)^*$ represent all string which has 11 as substring.
- $0^* 110^*$ represents all string with exactly two consecutive 1's.
- $0^* 10^* 10^*$ represents all string with exactly two 1's but need not be consecutive.
- $(0 + 1)^* 1(0 + 1)^* 1 (0 + 1)^*$ represents all string with atleast two 1's.

1.55 (a)

A state in a DFA will be a subset of the set of states of the equivalent NFA.

The maximum number of states in the equivalent DFA of an NFA with n number of states will be 2^n.

So, number of states in equivalent DFA $\leq 2^n$.

1.56 (a)

To find complement of a DFA we need to interchange final and non-final states of DFA.

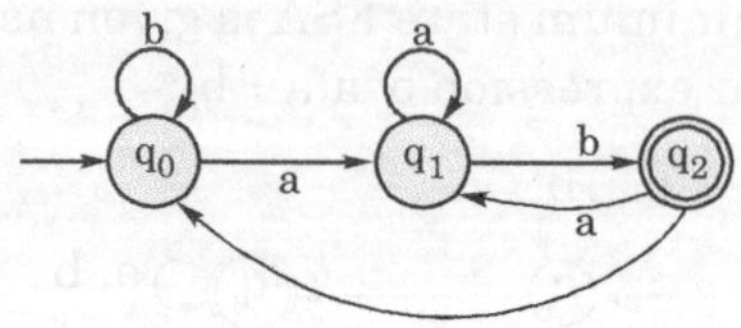

Above DFA generates all strings end with ab.

Complement of DFA accepts all strings do not end with ab.

Complement of given DFA is:

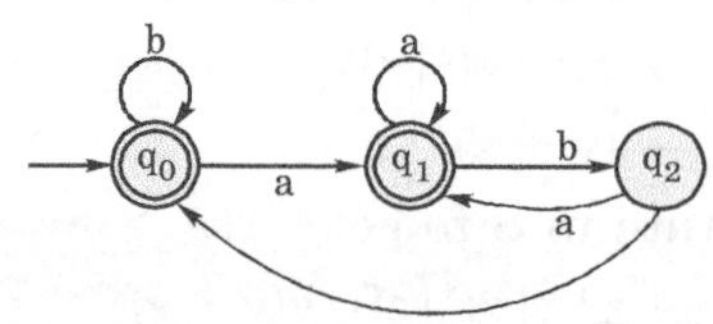

It accepts all strings that do not end with ab.

1.57 (d)

- Since m, n are restricted to 10000 which is finite hence L_1 is regular.
- $L_2 = \{w \mid w = w^R\}$ which is the Palindrome language which is known to be non regular.
- L_3 is regular (since we can design mod machine for it).

1.58 (a)

- M_1 is a DFA accepting all string containing two consecutive 1's.
- M_2 is a NFA accepting the same language i.e. containing two consecutive 1's.

So, $L_1 = L_2$.

1.59 (a)

The transition tables for Y and Z is as follows:

Y.

	a	b
→1	1	2
2(F)	2	1

Z.

	a	b
→1	2	2
2(F)	1	1

Z contains 2 states and Y contains 2 states so the product automation $Z \times Y$ contains $2 \times 2 = 4$ states.

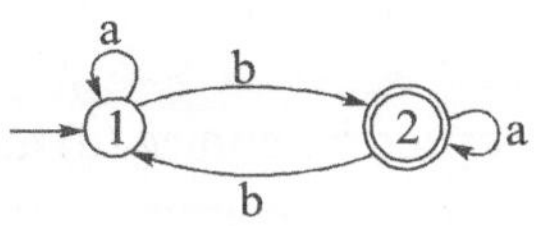

Y

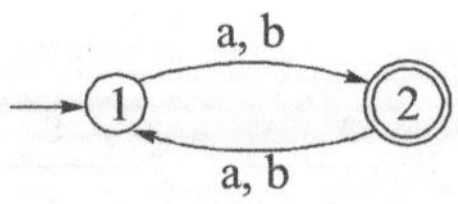

Z

The table for Z × Y is shown below:

	a	b
→(1, 1)	(2, 1)	(2, 2)
(1, 2)	(2, 2)	(2, 1)
(2, 1)	(1, 1)	(1, 2)
(F)(2, 2)	(1, 2)	(1, 1)

State (1, 1) has to be P since (1, 1) is starting state and P is also starting state in answer given.

State (2, 2) has to be R since (2, 2) is the final state and R is also final state in answer given.

Now the remaining 2 states (1, 2) & (2, 1) can be assigned the letters Q & S or S & Q resulting in 2 possible tables.

Table 1.

	a	b
P	S	R
Q(1, 2)	R	S
S(2, 1)	P	Q
R	Q	P

Table 2.

	a	b
P	S	R
S(1, 2)	R	Q
Q(2, 1)	P	S
R	Q	P

Rewriting these two table in same order (P, Q, R, S) as given in answers we get

Table 1.

	a	b
→P	S	R
Q	R	S
R(F)	Q	P
S	P	Q

Table 2.

	a	b
→P	S	R
Q	P	S
R(F)	Q	P
S	R	Q

Now choice (a) is exactly same as the table (1) shown above and hence choice (a) is acceptable as a correct answer.

1.60 (c)

A : $\in + 0(01^*1+00)^*\ (01^*+0)$ which is same as expression 1.

B : Same as expression 2.

C : Same as expression 3.

D : $\in + 0(10^*1+10)^*\ (10^*+1)$ which is same as expression 4.

A-1, B-2, C-3, D-4

Note: In each NFA the initial state of NFA is also the final state of NFA so it contains ε in each regular expression.

1.61 (a)

1. $\{a^n b^{2m} \mid n \geq 0, m \geq 0\}$ is regular, since we can write L as a regular expression $a^*\ (bb)^*$.
2. $\{a^n b^m \mid n = 2m\}$ is a DCFL, but not regular since here, we need to count the a's and compare with b's.
3. $\{a^n b^m \mid n \neq m\}$ is same as $\{a^n b^m \mid n < m\} \cup \{a^n b^m \mid n > m\}$ each of which is a CFL and the union is also a CFL (Since CFLs are closed under union). However this language is not regular since we have to count a's and b's and compare them which cannot be done by a finite state machine.
4. $\{xcy \mid x, y \in (a, b)^*\}$ is regular since we can write a regular expression $(a + b)^*c\ (a + b)^*$ for it.

1.62 (c)

$(0 + 1)^*\ 0\ (0 + 1)^*\ 0\ (0 + 1)^*$ is the regular expression for atleast two zero's.

1.63 (c)

The given DFA is the standard construction for language given by choice (c) "ending with 00".

1.64 (b)

Choice (a) $(0^*\ 1\ 0^*\ 1)^*$

will always generate strings ending with 1. But we want an expression for bit strings with even no of 1's, which includes strings like "110" which ends with 0.

So, choice (a) is not correct.

Choice (c) $0^*\ (1\ 0^*\ 1)^*\ 0^*$

"1010101" $\notin 0^*(1\ 0^*\ 1)^*\ 0^*$

but "1010101" has even number of 1's. So choice (c) is incorrect.

Choice (d) $\lambda \notin 0^*\ 1(1\ 0^*\ 1)^*\ 10^*$

but "λ" is a bit string with even no of 1's (zero 1's). So choice (d) is incorrect.

Choice (b) can generate all bit strings with even no of 1's. So choice (b) is correct.

1.65 (c)

Let w = "0110". To design a minimal NFA to accept all the substrings of this string, first accept the string itself as follows.

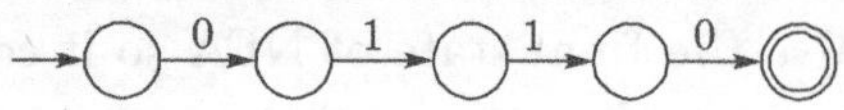

Now, to accept all the other substrings of "0110", make every state as starting state by using null moves and also make every state as final state. The final NFA is shown below.

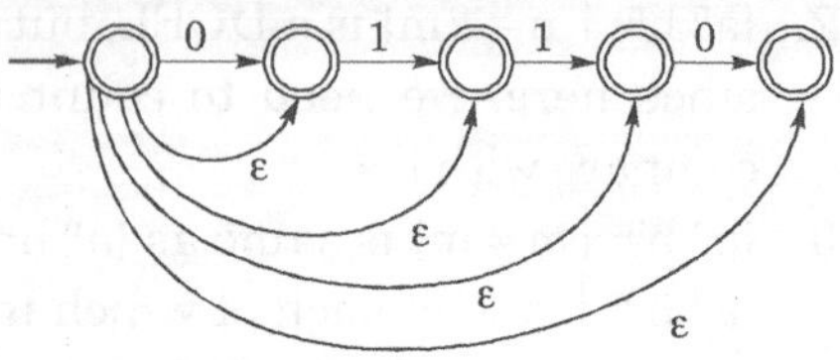

Since a string of length 4 requires 5 states, a string of length n will require n + 1 states.

1.66 (a)

Lexical analysis only requires the power of FA.

1.67 (c)

$$\Sigma^* - P = P^c$$

If P is regular, then P^c is also regular, since regular language is closed under complementation.

1.68 (b)

Let, n = 3 (say)

Then $L = \{a^{3k} \mid k > 0\} = \{a^{3k} \mid k \geq 1\}$

The dfa for this will be

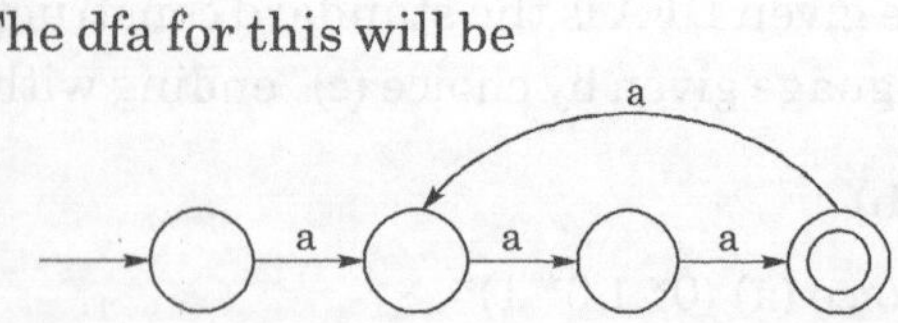

which has 4 states.

Similarly for $L = \{a^{nk} \mid k > 0\}$, we will need n + 1 states.

1.69 (a)

The language accepted by given dfa is

$L = a(a + b)^* + ba(a + b)^*$

Choice (a) is accepting same language and is minimal.

"aba" is rejected by choice (b) and it belongs to L. So choice (b) is incorrect.

Choice (c) is an NFA and hence not the right answer.

"bba" is accepted by choice (d) and is rejected by L. So choice (d) is incorrect.

1.70 (b)

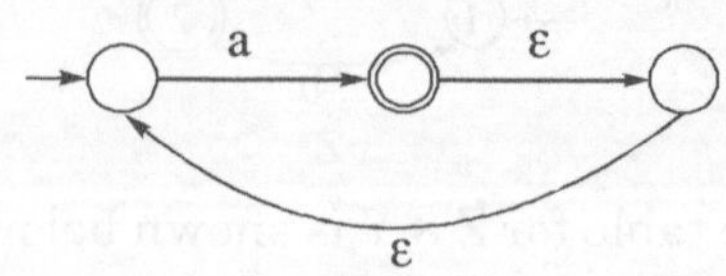

Language for NFA $L = a^+$

Now $\bar{L} = \Sigma^* - L$

$\Rightarrow \bar{L} = a^* - a^+$

$\Rightarrow \bar{L} = \{\varepsilon\}$

1.71 (c)

L = {ab, aa, baa}

The breakdown of the strings 1, 2, 4 in terms of ab, aa and baa is shown below:

1. ab aa baa ab aa 2. aa aab aa aa

3. baa aa ab aa

String no (3) has no breakdown in terms of strings in L and hence string (3) does not belong to L*.

Only 1, 2 and 4 belongs to L*.

1.72 (d)

Notice that the state names are given based on ending bits of the string, which has been processed. The arc from 00 labeled "0" should go to trap state q (since at most 2 zeros are allowed in any substring).

Based on this fact, option (a) and (b) are incorrect.

Between option (c) and (d), if you look at arc labeled "1" from state 01, this arc should go to state 11 since the string at this point is ending with 11. So option (c) is wrong and option (d) is correct.

The dfa corresponding to correct option (d) is shown below with missing arrows shown in dotted lines.

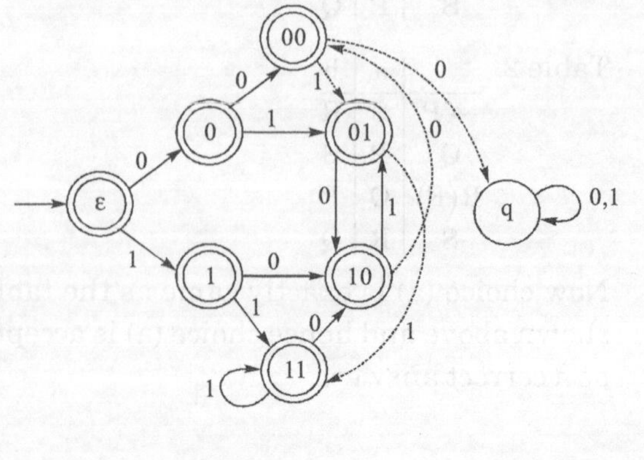

1.73 (a)

$L_1 = \phi$
$L_2 = \{a\}$

So, $L_1L_2^* \cup L_1^* = \phi \cdot a^* \cup \phi^* = \phi \cup \varepsilon = \varepsilon$

1.74 (d)

1. A is a NFA. So L(A) is regular. Complement of L(A) is also regular, since regular languages are closed under complement. So, L(A) is context-free is true.
2. L(A) = L((11 *0 + 0) (0 + 1)*0*1*) : true because the r.e. for the language accepted by above automata is (11*0 + 0) (0 + 1)*.
3. For the language accepted by A, A is the minimal DFA : false
   Writing the r.e. for the language accepted by the given DFA and simplifying, we get,
   $L(A) = 11^*0\,(0+1)^* + 0(0+1)^*$
   $= (11^* + \varepsilon)\,0(0+1)^* = 1^*0(0+1)^*$
   The minimal DFA for the above language is
   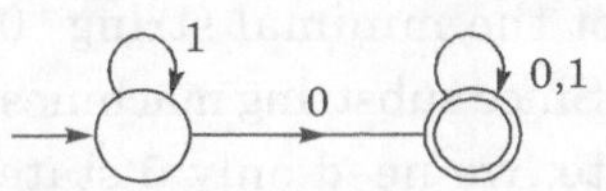

4. Machine A accepts all the string over {0, 1} of length at least 2 : False
   Above DFA accepts string "0" whose length is less than 2.

1.75 (c)

$L = \{a^n b^n \mid n \geq 0\}$ is not regular
$L = \{a^n \mid n \text{ is prime}\}$ is not regular
$L = \{ww \mid w \in \Sigma^*, \Sigma = \{0, 1\}$ is not regular

$L = \{w \mid w$ has $3k+1$ b's for some $k \in N$, $\Sigma = \{a, b\}\}$
$= \{w \mid$ Number of b's in w are 1 modulo 3$\}$ is regular since it can be accepted by a mod 3 machine as shown below.

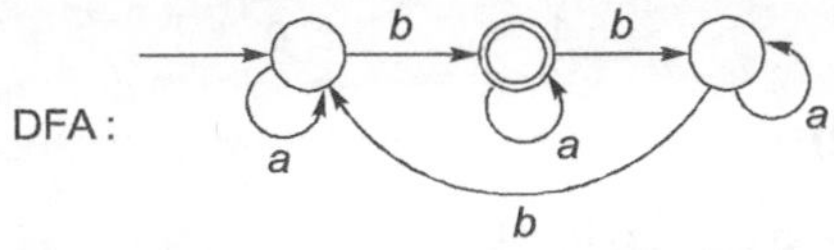

1.76 (a)

Given FA is

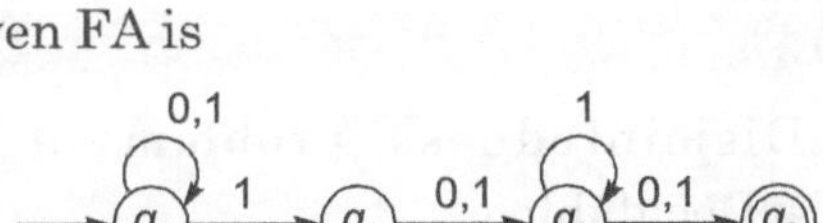

For input string 0011

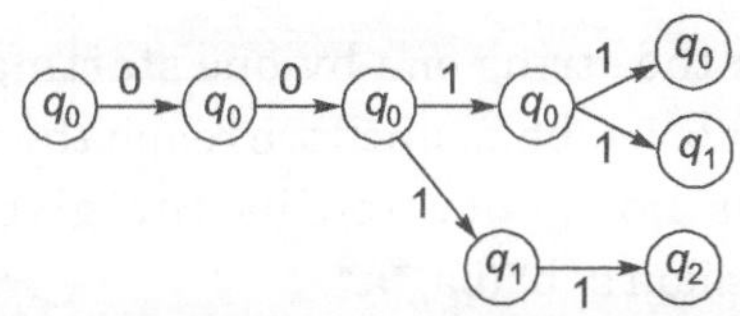

$\therefore$ $\{q_0, q_1, q_2\}$ are reachable states.

1.77 (b)

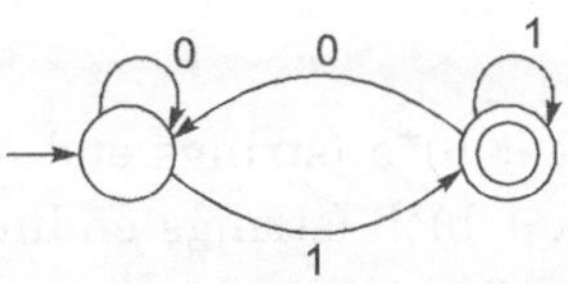

I. 0*1 (1 + 00*1)* is a correct r.e. for the given machine.
II. 0*1*1+11*0*1 is not correct. 0101 is accepted by DFA, but can not be generated by this regular expression
III. (0 + 1)* 1 is correct since the language accepted by the given DFA is all strings ending with 1.
$\therefore$ I and III are equivalent to given DFA.

1.78 (a)

$L_1 = \{a^n \mid n \geq 0\}$
$L_2 = \{b^n \mid n \geq 0\}$

I. $L_1 \cdot L_2 = \{a^m b^n \mid m, n \geq 0\} = a^*b^*$ is regular.
II. $L_1 \cdot L_2 \neq \{a^n b^n \mid n \geq 0\}$
$\Rightarrow$ II is not correct.
$\therefore$ Only I is correct.

1.79 (a)

A machine which accepts L_1 needs only finite memory since there is no need to keep the number of 110's in memory. The reason is because whenever two consecutive 110's come in a row then a 011 always come in between as shown below.

1 1 0 1 1 0

So the difference between number of 110 and number of 011 can only be 1, 0, –1.
So only finite memory needed and therefore L_1 is regular.
L_2 requires infinite memory since any number of 000's can come in a row without any 111 in between. So the difference between the number of 000's and 111's can go upto ∞ or $-\infty$. So this language requires inifinite memory. But FA has only finite memory.
So L_2 is not regular.

1.80 (3)

Check the string one-by-one starting from ε, a, b, aa, ab, ba, bb,... until we reach the first string that is not generated by the given regular expression a*b*(ba)*a*.
In this case smallest string NOT generated by the given regular expression is 'bab', whose length is 3.

1.81 (1)

L(M) = (a + b)*a (strings ending with a)
L(N) = (a + b)*b (strings ending with b)
L(M) ∩ L(N) = ϕ (nothing in common)
∴ Number of states = 1

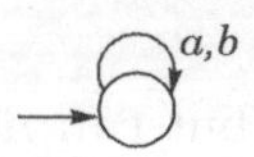

minimal DFA which accepts empty language.

1.82 (c)

Converting the given transitions to a state diagram we get,

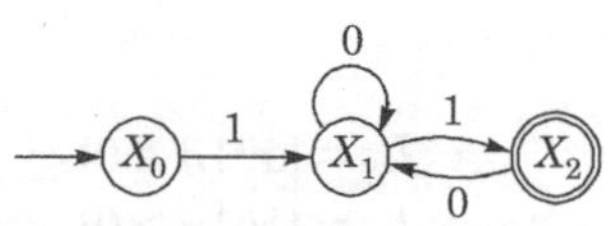

From this diagram we can write,

$$X_0 = 1(0+10)^*1$$

1.83 (a)

$L_1 = \{wxw^R \mid w, x \in \{a, b\}^* \text{ and } |w|, |x| > 0\}$

w cannot be put as "∈" since |w| > 0. So we put w as its smallest string which is 'a' and 'b' and get the regular expression:

$$r = a(a + b)^+a + b(a + b)^+ b.$$

Now putting w as any other string like say "ab" will not add any new string to the expression r, since any such string so generated will be either already generated by either $a(a+b)^+a$ or $b(a+b)^+b$.
So the given language = $a(a+b)^+a + b(a+b)^+b$ which is clearly regular.
∴ L_1 is regular.
$L_2 = \{a^nb^m \mid m \neq n\}$. This language has infinite comparisons between number of a's and b's. L_2 is not regular language.
L_3: $\{a^pb^qc^r \mid p, q, r \geq 0\} = a^*b^*c^*$ is a regular language.

1.84 (3)

The given regular expression (0+1)*(10) corresponds to binary string ending with "10". To accept the minimal string "10", we need 3 states. No trap state is required since this is a machine which accepts "ending with" type strings.
So, we need only 3 states. The design of the minimal DFA is shown below.

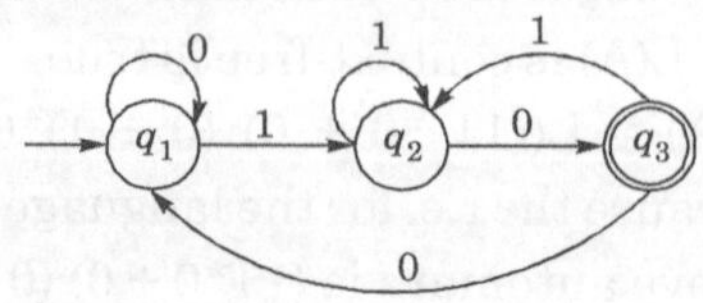

∴ Minimum number of states required for DFA = 3.

1.85 (b)

$\Sigma^*0011\ \Sigma^* = (0 + 1)^*\ 0011(0 + 1)^*$
Every string of the language L contains the substring 0011.
To accept the minimal string "0011" we need 5 states. Since substring machines don't required trap state, we need only 5 states.
Since the complementary DFA will accept complementary language, and since complementary DFA has same number of states as the original DFA, therefore $\bar{L}$ also has only 5 states as shown below.
Minimized DFA for

$L =$ 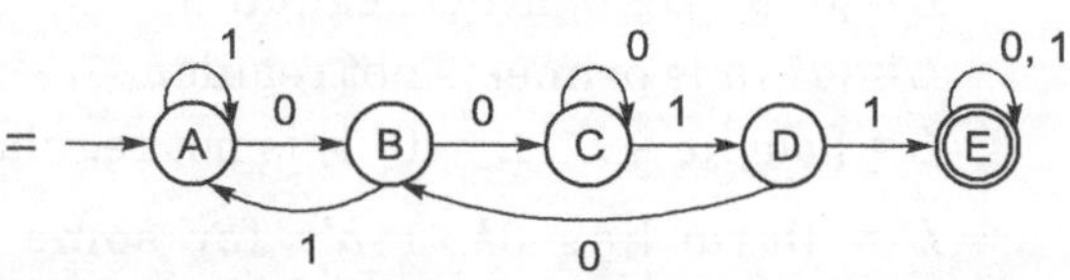

Minimized DFA for

$\bar{L} =$ 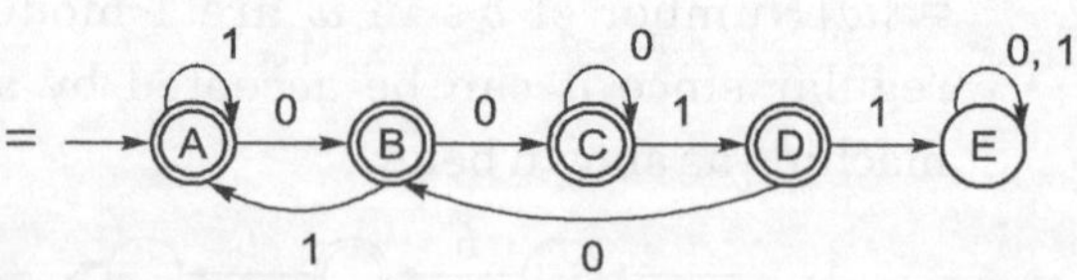

1.86 (d)

$G : S \to aS \mid bS \mid \varepsilon$
$L(G) = \{a, b\}^*$

1.87 (c)

I. Disjointedness problem of regular = Decidable
II. Membership of CFL's = Decidable

III. Equivalence of CFL's = Undecidable
IV. Emptiness of RE language's = Undecidable
So, III and IV only is correct answer.

1.88 (b)

We wish to find regular expression "for all binary strings containing two consecutive 0's and two consecutive 1's".

Now, choice (a) cannot generate "001011"
Choice (b) is correct
Choice (c) can generate "00" which does not belong to given language.
Choice (d) always ends with 11 or 00 and hence cannot generate "001101".

1.89 (2)

L = (0 + 1)*(0 + 1)(0 + 1)* = $(0 + 1)^+$

Min DFA : 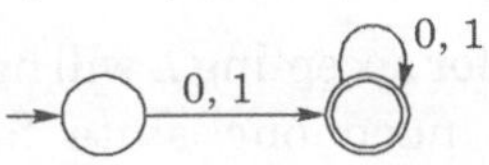

2 states.

1.90 (b)

I. Incorrect, since even if all states of NFA are accepting, dead configuration may be there and L need not be Σ^*.

II. Correct, since ϕ is regular and $\phi \cap B = \phi$ which is regular for all B.

1.91 (4)

The r.e. is (a + b)* b (a + b)
This is all strings whose second last bit is 'b'.
Minimal NFA is

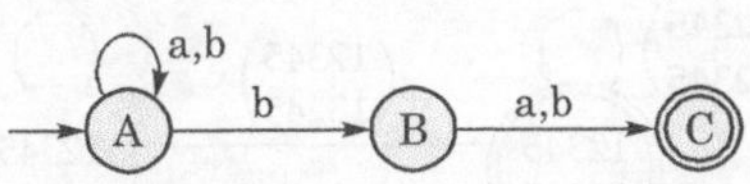

Converting it into minimal DFA by subset construction algorithm, we get

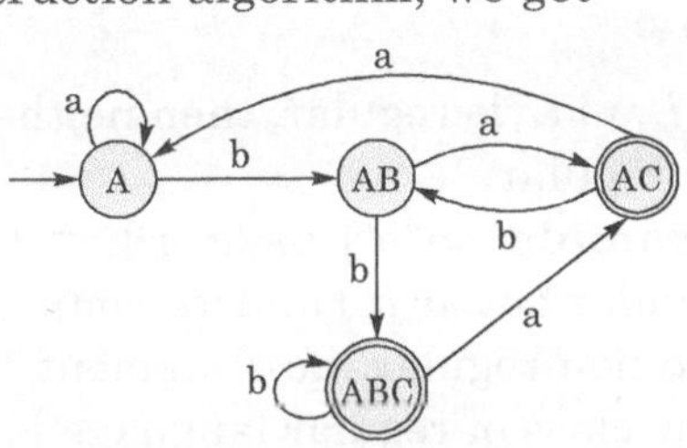

which is a minimal DFA with 4 states.

1.92 (d)

The given grammar is S → SaS | aSb | bSa | SS | ∈
Now, S → aSb | bSa | SS | ∈ generates all strings with equal number of 'a' and 'b'.
Now, S → SaS can only generate strings where 'a' is more than 'b', since on left and right of 'a' in SaS, S will have only strings with $n_a = n_b$ or $n_a > n_b$.
So, option (d) "babba" which is a string with $n_b > n_a$ is not possible to generate by the given grammar.

1.93 (8)

Min DFA for $L = \{w_1 a w_2 \mid w_1, w_2 \in \{a, b\}^*, |w_1| = 2, |w_2| \geq 3\}$
The regular expression for L is $(a + b)(a + b)a(a + b)(a + b)(a + b)(a + b)^*$
The minimal DFA is

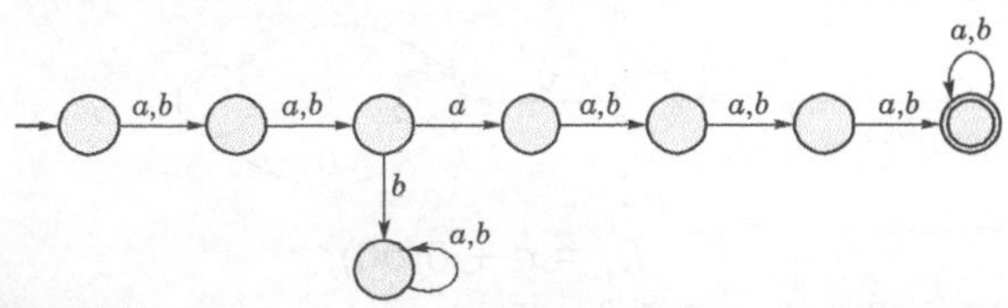

1.94 (b)

G_1: $S \to aSb \mid T,\ T \to cT \mid \in$
G_2: $S \to bSa \mid T,\ T \to cT \mid \in$

$$L(G_1) = \{a^n c^m b^n \mid m, n \geq 0\}$$
$$L(G_2) = \{b^n c^m a^n \mid m, n \geq 0\}$$
$$L(G_1) \cap L(G_2) = \{a^n c^m b^n\} \cap \{b^n c^m a^n\}$$
$$= \{c^m \mid m \geq 0\} = c^*$$

Since the only common strings will be those strings with only 'c', since in the first language all the other strings start with 'a' and in the second language all the other strings start with 'b'.
Clearly the intersection is not finite but regular.

1.95 (c)

Converting the table to a state diagram, we get,

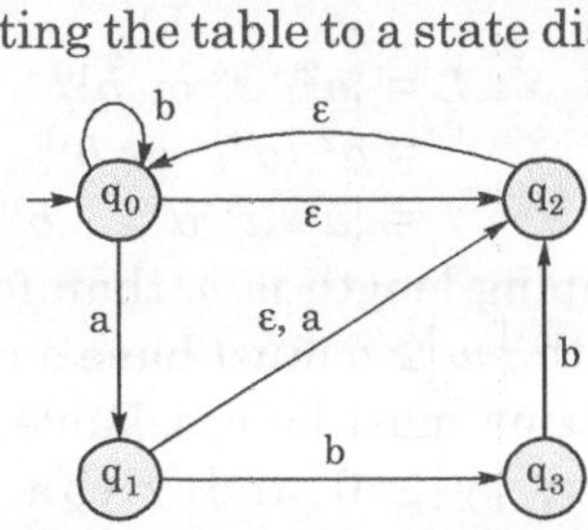

$\hat{\delta}(q_2, aba)$ = All states reachable from q_2 by "aba"
If aba is broken as $\varepsilon.a.\varepsilon.\varepsilon.b.a$. Then from q_2 we can reach q_1 and from there by null transition we can reach state q_2 as well as q_0.

$$\hat{\delta}(q_2, aba) = \{q_0, q_1, q_2\}$$

1.96 (d)

n is number of states of given NFA (may not be minimal) k is number of states of equivalent min DFA. First we convert NFA to DFA using subset construction algorithm and we get an equivalent DFA which will have atmost 2^n states. Then we can convert this DFA to a minimal DFA and get a minimal DFA with k states where $k \le 2^n$.

1.97 (2)

We need to find smallest value of k which satisfies

$$L_1{}^k = L_1{}^{k+1}$$

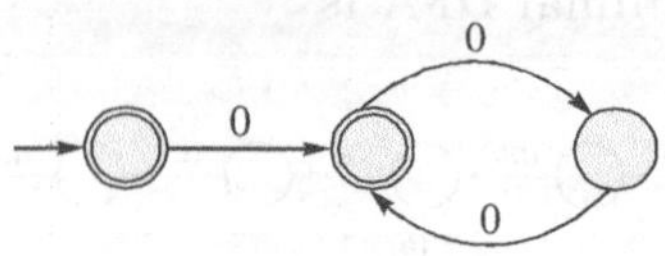

$$L_1 = \in + 0(00)^*$$

Try $k = 0 : L_1{}^0 = L_1{}^1$

$\Rightarrow \quad \in = L_1$ which is false.

So order is not 0.

Try $k = 1 : L_1{}^1 = L_1{}^2$

$\Rightarrow \quad L_1{}^2 = L_1$

Now, $\quad L_1{}^2 = (\in + 0\,(00)^*)\,(\in + 0\,(00)^*)$

$\quad = \in + 0\,(00)^* + 00\,(00)^* = 0^*$

Clearly $\quad L_1{}^2 \ne L_1$

So order is not 1.

Try $k = 2 : L_1{}^2 = L_1{}^3$

Now, $\quad L_1{}^3 = L_1{}^2 \,.\, L_1$

$\quad = 0^*(\in + 0\,(00)^*) = 0^*$

Clearly $\quad L_1{}^3 = L_1{}^2 = 0^*$

(So order of L_1 is 2)

1.98 (b)

$$L = \{a^{2+3k} \text{ or } b^{10+12k}\} \text{ for } k \ge 0$$
$$= a^2\,(a^3)^* \text{ or } b^{10}\,(b^{12})^*$$
$$= \{a^2, a^5, a^8, \ldots, b^{10}, b^{22}, b^{34} \ldots\}$$

The pumping length is p, than for any string $w \in L$ with $|w| \ge p$ must have a repetition i.e. such a string must be breakable into $w = xyz$ such that $|y| \ge 0$ and y can be pumped indefinitely, which is same as saying $xyz \in L \Rightarrow xy^*z \in L$.

The minimum pumping length in this language is clearly 11, since b^{10} is a string which has no repetition number, so upto 10 no number can serve as a pumping length. Minimum pumping length is 11. Any number at or above minimum pumping length can serve as a pumping length. The only number at or above 11, in the choice given is 24. So correct answer is option (b).

1.99 (d)

- If L is regular, $L \cdot L^R$ is also regular by closure property.
- Suffix (L) and Prefix (L) are also regular by closure property.

However option (d) $\{ww^R \mid w \in L\}$ need not be regular since if L is an infinite regular language, then $\{ww^R \mid w \in L\}$ will not only be infinite, but also non-regular. Since it involves string matching and we can increase in length indefinitely and then finite automata FA will run out of memory.

So answer is option (d).

1.100 (2)

The DFA for accepting L will have $5! = 120$ states, since we need one state for every possible permutation function on 5 elements. The starting state will be "id" state, named as $\begin{pmatrix}12345\\12345\end{pmatrix}$ and from there $n!$ arrows will go the $n!$ states each named with a distinct permutation of the set $\{1, 2, 3, 4, 5\}$. Since composition of permutation function is closed every arrow has to go to some permutation and hence some state.

Since the language only has those strings where $\pi(x) =$ id only the starting state ("id" state) will be the final state. Sample machine with only 2 states is shown below:

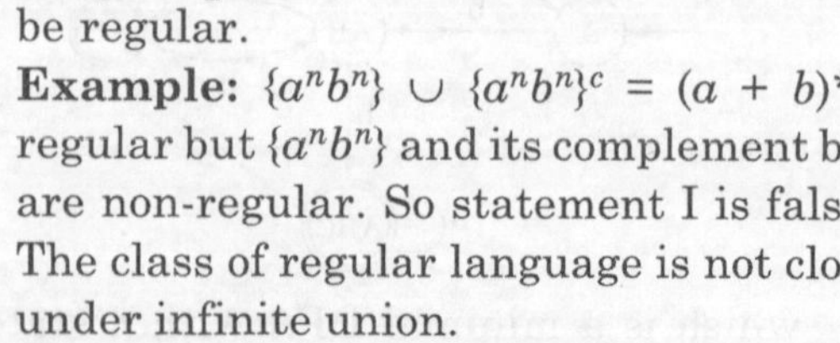

1.101 (a)

- If $L_1 \cup L_2$ is regular, then neither needs to be regular.

  Example: $\{a^nb^n\} \cup \{a^nb^n\}^c = (a + b)^*$ is regular but $\{a^nb^n\}$ and its complement both are non-regular. So statement I is false.
- The class of regular language is not closed under infinite union.

  Proof: It is was closed under infinite union then

  $a^nb^n = \{\in\} \cup \{ab\} \cup \{aabb\} \cup$ .... will be infinite union of finite languages (which are regular) and hence will become regular. But we know that $\{a^n b^n \mid n \geq 0\}$ is non-regular. So II is false

So option (a), neither I nor II is the correct answer.

1.102 (*)

- Regular expression in option (a) is incorrect because it will force the strings to end with 1 and a string of odd number of 1's need not to end with 1.
- Regular expression in option (b) will force it to start with 1 and hence it is incorrect.
- Regular expression in option (c) can create odd number of 1's as well as even number of 1's and hence it is incorrect.
- Regular expression in option (d) is incorrect as it does not generate strings '01' or 1 or more 0 followed by 1 which is having an odd number of 1's.

  Note: Option (d) would be correct only when if the expression were **(0*10*10*)*10* + (0*10*)** means **(0*10*)** missing from the option. Hence option (d) is also incorrect.

  Note: In the official answer key marks were given to all.

1.103 (6)

The minimum number of states in a DFA that accepts:

$L = \{x \in \{a, b\}^* \mid$ number of a's is divisible by 2 but not divisible by 3$\}$ is 6 states as shown below:

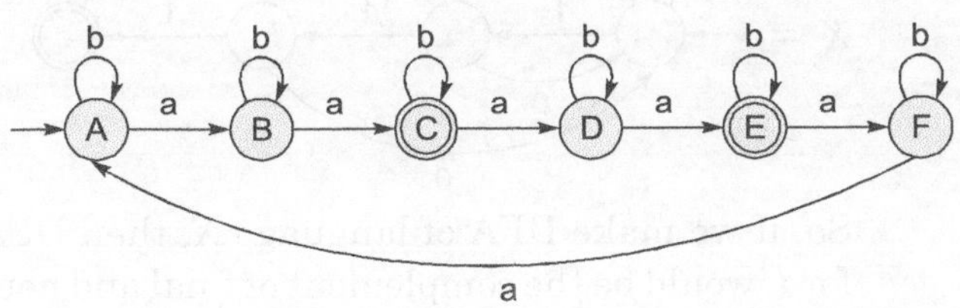

(or) alternatively it can be designed by taking a product automata of

$L_1 = \{x \in \{a, b\}^* \mid$ number of a's divisible by 2$\}$ and $L_2 = \{x \in \{a, b\}^* \mid$ number of a's not divisible by 3$\}$ as shown below:

Minimum DFA for L_1:

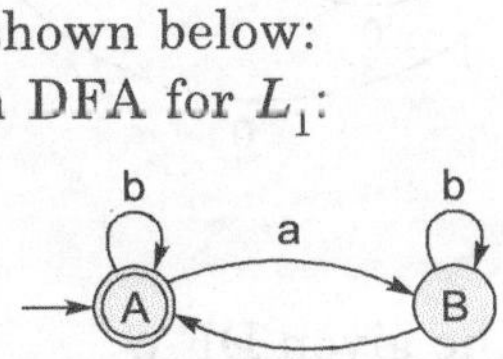

Minimum DFA for L_2:

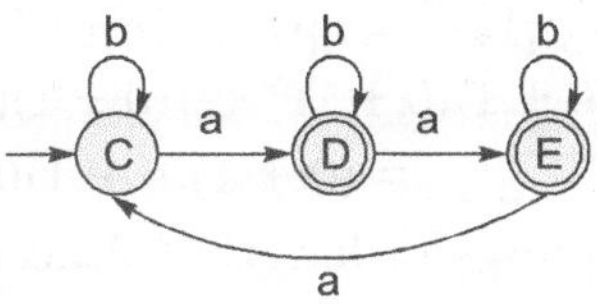

Product automata $L_1 \cap L_2$ having six states, shown below:

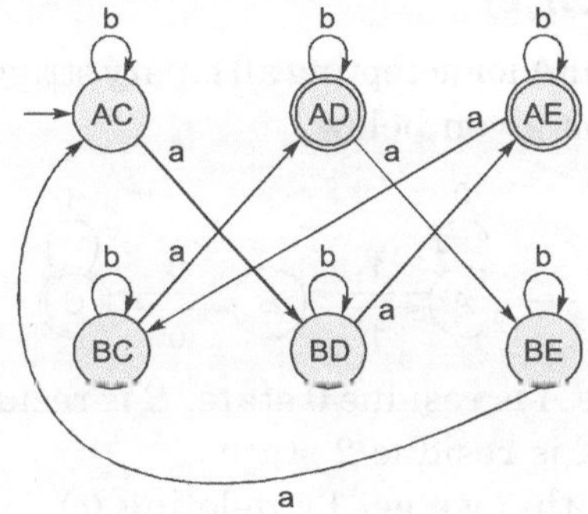

Which is same as our directly designed machine with 6 states.

1.104 (a)

Since 4 states are required in minimal DFA for this language "Ending with 011", and since all given DFA's have 4 states, the answer must be same as minimal DFA.

The minimal DFA for this language is given below:

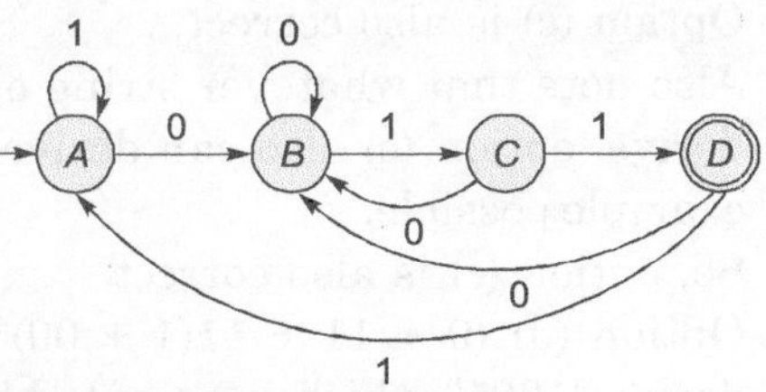

Which is same as option (a).

1.105 (a)

If L is accepted by a min DFA with k states, by exchanging final and non-final states, we can make a minimal DFA with k states which accepts $\{0, 1\}^* - L = \bar{L}$.

So, option (a) is true.

1.106 (256)

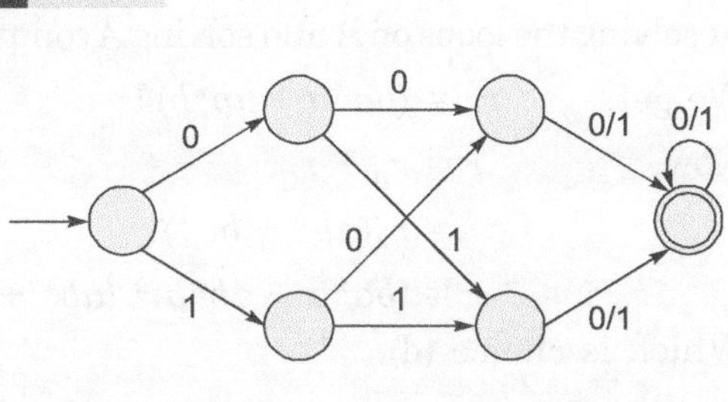

The regular expression for $L(M)$ is
$0(0+1)(0+1)(0+1)^* + 1(0+1)(0+1)(0+1)^*$
$= (0+1)(0+1)(0+1)(0+1)^*$

So, all strings of length ≥ 3 accepted.

Therefore number of strings of length 8 is $2^8 = 256$.

1.107 (a, b, c)

The DFA for accepting all binary strings divisible by 3 is given below:

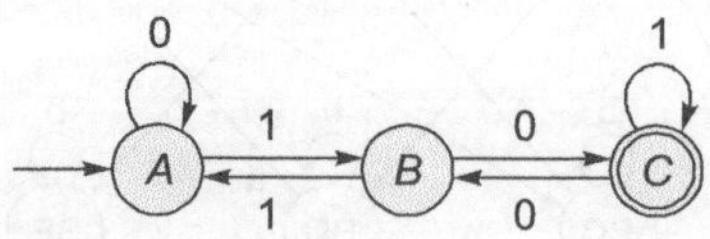

where A is residue 0 state, B is residue 1 state and C is residue 2 state.

From this we get by deleting (c)

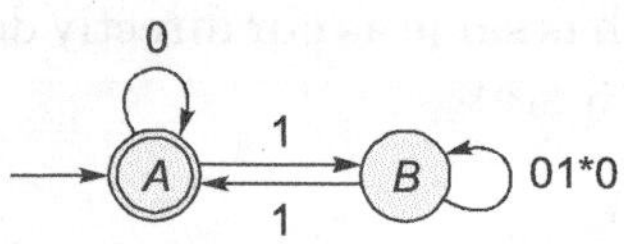

We get option (b) is correct $(0 + 1(01^*0)^*1)^*$

Now option (a) = option (b)

because $(r^*s^*)^* = (r + s)^*$

So, option (a) is also correct.

Option (c) can be obtained by resolving the loop between B and C on "C" instead of on "B".

Option (c) is also correct.

Also note that whatever string option (b) can derive, option (c) also can derive. No counter example possible.

So, option (c) is also correct.

Option (d) $(0 + 11 + 11(1 + 00)^*00)^*$ cannot derive "1001" which is accepted by machine.

So, option (d) is incorrect.

So, options (a), (b) and (c) are correct.

1.108 (d)

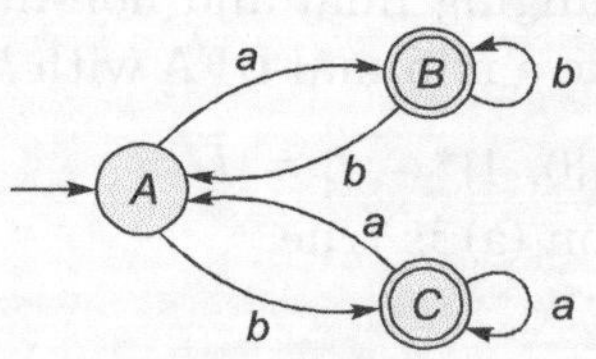

Resolving the loops on A and solving A completely,

We get, $r_A = (ba^*a + ab^*b)^*$

Now, $r = r_B + r_C$

$= r_A(ab^* + ba^*)$

$= (ba^*a + ab^*b)^*(ab^* + ba^*)$

Which is choice (d).

1.109 (c)

s is initial state and p is final state

From the given transition diagram,

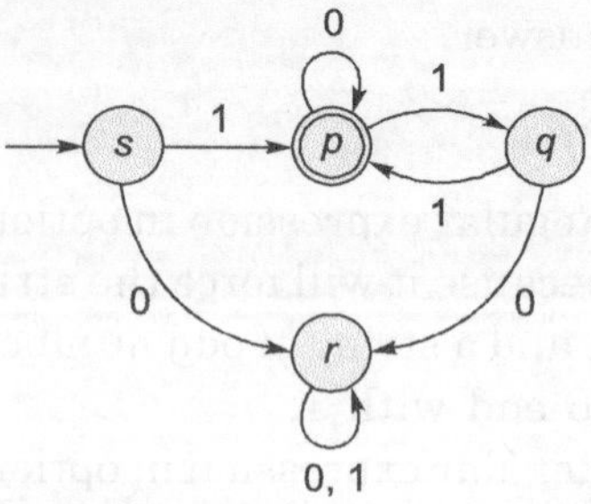

$s = \epsilon$

$p = s1 + p0 + q^1$

$q = p1$

Substituting s and q in p

⇒ $p = \epsilon 1 + p0 + p11$

⇒ $p = 1 + p(0 + 11)$

⇒ $p = 1(0+11)^*$ (Using Arden's theorem)

1.110 (c)

- Option (a) does accept identifiers starting with digit too. Hence, false
- Option (b) (letter)* + letter (digit)* this is wrong.
- Option (c) correct.
- Option (d) (letter (digit)*) this is false.

1.111 (4)

Language L is the complement of the language $X = \{w \in (0, 1)^* \mid w$ does contains three or more consecutive 1's$\}$.

$X =$ 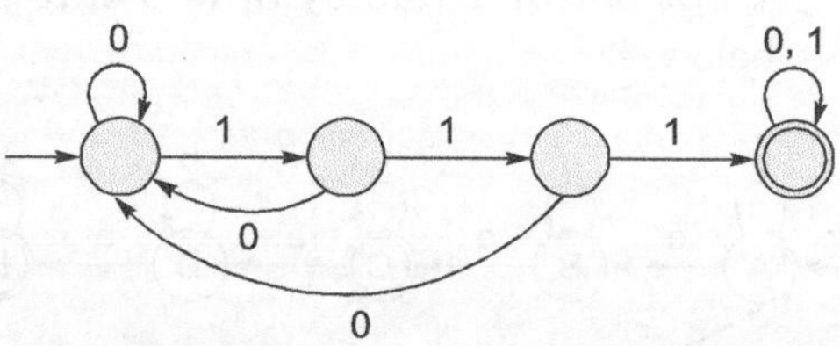

So, if we make DFA of language X, then DFA for L would be the complement of final and non-final states.

$L =$ 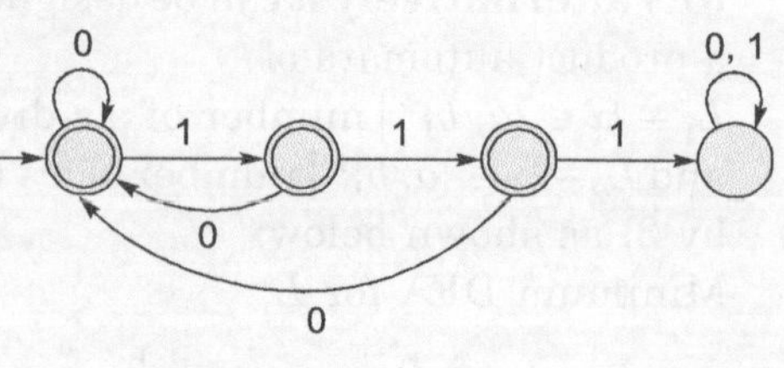

1.112 (a, c)

Minimize the given DFA:

Step 1: Separate the final and non final states.

$S_1 = \{1\}, \{2, 3, 4, 5\}$

Step 2: States 2, 5 are 0-closure and also 1-closure.

States 3, 4 are 0-closure and also 1-closure.

$\therefore \quad S_2 = \{1\}, \{2, 5\}, \{3, 4\}$

Step 3:

$S_3 = \{1\}, \{2, 5\}, \{3, 4\}$

$\therefore$ The states 2 & 5 and 3 & 4 are not distinguishable but the states 2 and 4 distinguishable. We can't construct a FA for $n_0(w) = n_1(w)$ but the given machine accepts all the strings having equal number of 0's and 1's.

$\therefore$ (a), (c) are false (b), (d) are true.

1.113 (44)

The number of strings of length less than or equal to 5 over {0, 1} is

$2^0 + 2^1 + 2^2 + 2^3 + 2^4 + 25 = 63$

Given, $R = 0^* + 1^*$

The strings in R of length less than or equal to 5 are 11,

i.e.$\in$, 0, 00, 000, 0000, 00000

and 1, 11, 111, 1111, 11111

Given, $S = 01^* + 10^*$

The strings in S of length less than or equal to 5 are 10,

i.e., 0, 01, 011, 0111, 01111

and 1, 10, 100, 1000, 10000

The total number of strings in R and S of length less than or equal to 5 is

$= 11 + 10 - 2 = 19$

(The string 0 and 1 are common in R and S)

$\therefore$ The number of strings over {0, 1} of length less than or equal to 5 neither in R nor in S is = 63 − 19 = 44

1.114 (d)

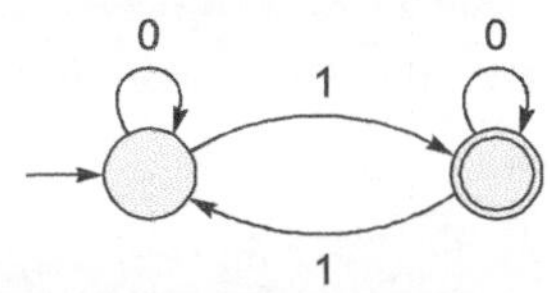

Reach the final state first then if we run the loop at the final state we get then regular expression $0^*\ 1(0 + 10^*\ 1)^*$.

1.115 (15)

All the strings $|w| \leq 4$ in L_1 are also in L_2

$L_1 = b^*\ ab^*\ (ab^*\ ab^*)^*$

L_1 is $b^*\ ab^*$ or $b^*\ ab^*\ ab^*\ ab^*$

The strings that can be formed from $b^*\ ab^*$ with $|w| \leq 4$ are a, ba, b^2 a, b^3 a, ab, ab^2, ab^3, bab, bbab, babb i.e. 10 strings.

The strings that can be format from $b^*\ ab^*\ ab^*\ ab^*$ with $|w| \leq 4$ are aaa, baaa, abaa, aaba, aaab i.e. 5 strings.

So, total strings that are common in L_1 and L_2 are 15.

1.116 (a)

2 and 5 are final states.

The given machine accepts all zeros i.e. 0^*.

We can reach the final state 5 by either $0(00)^*\ (11)^*$ or $1(11)^*$

$\therefore$ The result is

$= 0^* + 0(00)^*\ (11)^* + 1(11)^*$

$= 0^* + (0\ (00)^* + 1)\ (11)^*$

■■■■

2 Push Down Automata: CFL & DCFL

2.1 Consider the grammar with the following productions.

$S \to a\,\alpha\,b \mid b\,\alpha\,c \mid ab$
$S \to \alpha\,S \mid b$
$S \to \alpha\,bb \mid ab$
$S\alpha \to bdb \mid bd$

the above grammar is
(a) context free grammar
(b) regular grammar
(c) context sensitive grammar
(d) LR(k) **[1995 : 1 M]**

2.2 If L_1 and L_2 are context free languages and R a regular set, one of the languages below is not necessarily a context free language. Which one?
(a) L_1L_2
(b) $L_1 \cap L_2$
(c) $L_1 \cap R$
(d) $L_1 \cup L_2$ **[1996 : 2 M]**

2.3 Let G be a context free grammar with productions given below.

$S \to ABAC$
$A \to aA \mid \varepsilon$
$B \to bB \mid \varepsilon$
$C \to d$

Transform the grammar G to an equivalent form G′ that has no ε productions and no unit productions. **[1996 : 2 M]**

2.4 Which of the following language over {a,b,c} is accepted by a deterministic push down automata?
(a) $\{wcw^R \mid w \in \{a, b\}^*\}$
(b) $\{ww^R \mid w \in \{a, b, c\}^*\}$
(c) $\{a^n b^n c^n \mid n \geq 0\}$
(d) {w | w is a palindrome over {a,b,c}} **[1997 : 2 M]**

2.5 Which of the following statement is false?
(a) Every finite subset of a non-regular set is regular
(b) Every subset of a regular set is regular
(c) Every finite subset of a regular set is regular
(d) The intersection of two regular sets is regular **[1998 : 2 M]**

2.6 Context free languages are closed under
(a) union, intersection
(b) union, kleene closure
(c) intersection, complement
(d) complement, kleene closure **[1999 : 1 M]**

2.7 Let L_D be the set of all language accepted by a PDA by final state and L_E the set of all languages accepted by empty stack. Which of the following is true?
(a) $L_D = L_E$
(b) $L_D \supset L_E$
(c) $L_D \subset L_E$
(d) None of the above **[1999 : 1 M]**

2.8 If L_1 is a context free language and L_2 is a regular language which of the following is are false?
(a) $L_1 - L_2$ is not context free
(b) $L_1 \cap L_2$ is context free
(c) $\sim L_1$ is context free
(d) ~ L2 is regular **[1999 : 2 M]**

2.9 Which of the following statement is true?
(a) if a language is context free it can always be accepted by a deterministic push-down automation
(b) the union of two context free language is context free
(c) the intersection of two context free languages is context free
(d) the complement of a context free languages is context free **[2001 : 1 M]**

2.10 The language accepted by a Pushdown Automation in which the stack is limited to 10 items is best described as
(a) context free
(b) regular
(c) deterministic context free
(d) recursive **[2002 : 1 M]**

2.11 Let G = ({S}, {a, b}, R, S) be a context free grammar where the rule set R is

$S \to a\,S\,b \mid S\,S \mid \varepsilon$

Which of the following statements is true?
(a) G is not ambiguous
(b) There exist x, y, ∈ L(G) such that $xy \notin L(G)$
(c) There is a deterministic pushdown automaton that accepts L(G)
(d) We can find a deterministic finite state automaton that accepts L(G) **[2003 : 2 M]**

2.12 Which one of the following statements is FALSE?
(a) There exist context free languages such that all the context free grammars generating them are ambiguous

(b) An unambiguous context-free grammar always has a unique parse tree for each string of the language generated by it
(c) Both deterministic and non-deterministic pushdown automata always accept the same set of languages
(d) A finite set of strings from some alphabet is always a regular language

[2004 : 1 M]

2.13 Let M = (K, Σ, Γ, Δ, s, F) be a pushdown automaton, where
K = (s, f), F = {f}, Σ = {a, b}, Γ = {a} and
Δ = {((s, a, ε), (s, a)),((s, b, ε),(s, a)),((s, a, ε), (f, ε)), ((f, a, a), (f, ε)), ((f, b, a), (f, ε))}
Which one of the following strings is not a member of L(M)?
(a) aaa (b) aabab
(c) baaba (d) bab

[2004 : 2 M]

2.14 The language $\{a^m b^n c^{m+n} \mid m, n \geq 1\}$ is
(a) regular
(b) context-free but not regular
(c) context sensitive but not context free
(d) type-0 but not context sensitive

[2004 : 2 M]

2.15 Consider the following grammar C:
$S \rightarrow bS \mid aA \mid b$
$A \rightarrow bA \mid aB$
$B \rightarrow bB \mid aS \mid a$
Let $N_a(w)$ and $N_b(w)$ denote the number of a's and b's in a string w respectively. The language $L(G) \subseteq \{a, b\}^+$ generated by G is
(a) $\{w \mid N_a(w) > 3N_b(w)\}$
(b) $\{w \mid N_b(w) > 3N_a(w)\}$
(c) $\{w \mid N_a(w) = 3k, k \in \{0, 1, 2, ...\}\}$
(d) $\{w \mid N_b(w) = 3k, k \in \{0, 1, 2, ...\}\}$

[2004 : 2 M]

2.16 Let L be a regular language and M be a context-free language, both over the alphabet Σ. Let L^c and M^c denote the complements of L and M respectively.
Which of the following statements about the language $L^c \cup M^c$ is TRUE?
(a) It is necessarily regular but not neccssarily context-free.
(b) It is necessarily context-free.
(c) It is necessarily non-regular.
(d) None of the above

[2005 : 1 M]

2.17 Let P be a non-deterministic push-down automaton (NPDA) with exactly one state, q, and exactly one symbol, Z, in its stack alphabet. State q is both the starting as well as the accepting state of the PDA. The stack is initialized with one Z before the start of the operation of the PDA. Let the input alphabet of the PDA be Σ.
Let L(P) be the language accepted by the PDA by reading a string and reaching its accepting state. Let N(P) be the language accepted by the PDA by reading a string and emptying its stack.
Which of the following statements is TRUE?
(a) L(P) is necessarily Σ* but N(P) is not necessarily Σ*.
(b) N(P) is necessarily Σ* but L(P) is not necessarily Σ*.
(c) Both L(P) and N(P) is necessarily Σ*.
(d) Neither L(P) nor N(P) are necessarily Σ*.

[2005 : 2 M]

2.18 Let N_f and N_p denote the classes of languages accepted by non-deterministic finite automata and non-deterministic push-down automata, respectively. Let D_f and D_p denote the classes of languages accepted by deterministic finite automata and deterministic push-down automata, respectively. Which one of the following is TRUE?
(a) $D_f \subset N_f$ and $D_p \subset N_p$
(b) $D_f \subset N_f$ and $D_p = N_p$
(c) $D_f = N_f$ and $D_p = N_p$
(d) $D_f = N_f$ and $D_p \subset N_p$

[2005 : 2 M]

2.19 Consider the languages:
$L_1 = \{a^n b^n c^m \mid n, m > 0\}$ and $L_2 = \{a^n b^m c^m \mid n, m > 0\}$
Which one of the following statements is FALSE?
(a) $L_1 \cap L_2$ is a context-free language
(b) $L_1 \cup L_2$ is a context-free language
(c) L_1 and L_2 are context-free language
(d) $L_1 \cap L_2$ is a context sensitive language

[2005 : 2 M]

2.20 Consider the languages:
$L_1 = \{ww^R \mid w \in \{0, 1\}^*\}$
$L_2 = \{w \# w^R \mid w \in \{0, 1\}^*\}$, where # is a special symbol
$L_3 = \{ww \mid w \in \{0, 1\}^*\}$
Which one of the following is TRUE?
(a) L_1 is a deterministic CFL
(b) L_2 is a deterministic CFL
(c) L_3 is a CFL, but not a deterministic CFL
(d) L_3 is a deterministic CFL

[2005 : 2 M]

2.21 Let $L_1 = \{0^{n+m} 1^n 0^m \mid n, m \ge 0\}$,
$L_2 = \{0^{n+m} 1^{n+m} 0^m \mid n, m \ge 0\}$, and
$L_3 = \{0^{n+m} 1^{n+m} 0^{n+m} \mid n, m \ge 0\}$.
Which of these languages are NOT context free?
(a) L_1 only (b) L_3 only
(c) L_1 and L_2 (d) L_2 and L_3
[2006 : 1 M]

2.22 In the context-free grammar below, S is the start symbol, a and b are terminals, and ε denotes the empty string
$S \to aSa \mid bSb \mid a \mid b \mid \varepsilon$
Which of the following strings is NOT generated by the grammar?
(a) aaaa (b) baba
(c) abba (d) babaaabab
[2006 : 1 M]

2.23 Which of the following languages is accepted by a non-deterministic pushdown automaton (PDA) but NOT by a deterministic PDA?
(a) $\{a^n b^n c^n \mid n \ge 0\}$
(b) $\{a^l b^m c^n \mid l \ne m \text{ or } m \ne n\}$
(c) $\{a^n b^n \mid n \ge 0\}$
(d) $\{a^m b^n \mid m, n \ge 0\}$ **[2006 : 2 M]**

2.24 Let L be a context-free language and M a regular language. Then the language $L \cap M$ is
(a) always regular
(b) never regular
(c) always a deterministic context-free language
(d) always a context-free language
[2006 : 2 M]

2.25 Consider the pushdown automaton (PDA) below which runs over the input alphabet (a, b, c). It has the stack alphabet $\{Z_0, X\}$ where Z_0 is the bottom-of-stack marker. The set of states of the PDA is {s, t, u, f} where s is the start state and f is the final state. The PDA accepts by final state. The transitions of the PDA given below are depicted in a standard manner. For example, the transition $(s, b, X) \to (t, XZ_0)$ means that if the PDA is in state s and the symbol on the top of the stack is X, then it can read b from the input and move to state t after popping the top of stack and pushing the symbols Z_0 and X (in that order) on the stack.
$(s, a, Z_0) \to (s, XXZ_0)$
$(s, \varepsilon, Z_0) \to (f, \varepsilon)$
$(s, a, X) \to (s, XXX)$
$(s, b, X) \to (t, \varepsilon)$
$(t, b, X) \to (t, \varepsilon)$
$(t, c, X) \to (u, \varepsilon)$
$(u, c, X) \to (u, \varepsilon)$
$(u, \varepsilon, Z_0) \to (f, \varepsilon)$
The language accepted by the PDA is
(a) $\{a^l b^m c^n \mid l = m = n\}$
(b) $\{a^l b^m c^n \mid l = m\}$
(c) $\{a^l b^m c^n \mid 2l = m + n\}$
(d) $\{a^l b^m c^n \mid m = n\}$ **[2006 : 2 M]**

2.26 In the context-free grammar below, S is the start symbol, a and b are terminals, and ε denotes the empty string.
$S \to aSAb \mid \varepsilon$
$A \to bA \mid \varepsilon$
The grammar generates the language
(a) $((a + b)^* b)^*$ (b) $\{a^m b^n \mid m \le n\}$
(c) $\{a^m b^n \mid m = n\}$ (d) $a^* b^*$
[2006 : 2 M]

2.27 Consider the following statements about the context-free grammar,
$G = \{S \to SS, S \to ab, S \to ba, S \to \varepsilon\}$
1. G is ambiguous.
2. G produces all strings with equal number of a's and b's.
3. G can be accepted by a deterministic PDA.

Which combination below expresses all the true statements about G?
(a) 1 only (b) 1 and 3 only
(c) 2 and 3 only (d) 1, 2 and 3
[2006 : 2 M]

2.28 Consider an ambiguous grammar G and its disambiguated version D. Let the language recognized by the two grammars be denoted by L(G) and L(D) respectively. Which one of the following is true?
(a) $L(D) \subset L(G)$ (b) $L(D) \supset L(G)$
(c) $L(D) = L(G)$ (d) L(D) is empty
[2007 : 1 M]

2.29 Consider the following grammars. Names representing terminals have been specified in capital letters.
G1: stmnt → WHILE (expr) stmnt
stmnt → OTHER
expr → ID
G2: stmnt → WHILE (expr) stmnt
stmnt → OTHER
expr → expr + expr
expr → expr * expr
expr → ID
Which one of the following statements is true?
(a) G1 is context-free but not regular and G2 is regular

(b) G2 is context-free but not regular and G1 is regular
(c) Both G1 and G2 are regular
(d) Both G1 and G2 are context-free but neither of them is regular

[2007 : 2 M]

2.30 The language $L = \{0^i 21^i \mid i \geq 0\}$ over the alphabet {0, 1, 2} is
(a) not recursive
(b) is recursive and is a deterministic CFL
(c) is a regular language
(d) is not a deterministic CFL but a CFL

[2007 : 2 M]

2.31 Which of the following statements is false?
(a) Every NFA can be converted to an equivalent DFA
(b) Every non-deterministic Turing machine can be converted to an equivalent deterministic Turing machine
(c) Every regular language is also a context-free language
(d) Every subset of a recursively enumerable set is recursive

[2008 : 2 M]

2.32 Which of the following statements are true?
1. Every left-recursive grammar can be converted to a right-recursive grammar and vice-versa.
2. All ε-productions can be removed from any context-free grammar by suitable transformations.
3. The language generated by a context-free grammar all of whose productions are of the form $X \to w$ or $X \to wY$ (where, w is a string of terminals and Y is a non-terminal), is always regular.
4. The derivation trees of strings generated by a context-free grammar in Chomsky Normal Form are always binary trees.

(a) 1, 2, 3 and 4 (b) 2, 3 and 4 only
(c) 1, 3 and 4 only (d) 1, 2 and 4 only

[2008 : 2 M]

2.33 Match **List-I** with **List-II** and select the correct answer using the codes given below the lists:

List-I

A. Checking that identifiers are declared before their use
B. Number of formal parameters in the declaration of a function agrees with the number of actual parameters in a use of that function
C. Arithmetic expressions with matched pairs of parentheses
D. Palindromes

List-II

1. $L = \{a^n b^m c^n d^m \mid n \geq 1, m \geq 1\}$
2. $X \to XbX \mid XcX \mid dXf \mid g$
3. $L = \{wcw \mid w \in (a \mid b)^*\}$
4. $X \to bXb \mid cXc \mid \varepsilon$

Codes:

	A	B	C	D
(a)	1	3	2	4
(b)	3	1	4	2
(c)	3	1	2	4
(d)	1	3	4	2

[2008 : 2 M]

2.34 Consider the following languages:
$L_1 = \{a^i b^j c^k \mid i = j, k \geq 1\}$
$L_2 = \{a^i b^j \mid j = 2i, i \geq 0\}$
Which of the following is true?
(a) L_1 is not a CFL but L_2 is
(b) $L_1 \cap L_2 = \phi$ and L_1 is non-regular
(c) $L_1 \cup L_2$ is not a CFL but L_2 is
(d) There is a 4 state PDA that accepts L_1, but there is no DPDA that accepts L_2

[2008 : 2 M]

2.35 Consider a CFG with the following productions.
$S \to AA \mid B$
$A \to 0A \mid A0 \mid 1$
$B \to 0B00 \mid 1$
S is the start symbol, A and B are non-terminals and 0 and 1 are the terminals. The language generated by this grammar is
(a) $\{0^n 10^{2n} \mid n \geq 1\}$
(b) $\{0^i 10^j 10^k \mid i, j, k \geq 0\} \cup \{0^n 10^{2n} \mid n \geq 0\}$
(c) $\{0^i 10^j \mid i, j \geq 0\} \cup \{0^n 10^{2n} \mid n \geq 1\}$
(d) The set of all strings over {0, 1} containing at least two 0's

[2008 : 2 M]

Directions for Question 2.36 to 2.37:

A CFG G is given with the following productions where S is the start symbol, A is a non-terminal and a and b are terminals.
$S \to aS \mid A$
$A \to aAb \mid bAa \mid \varepsilon$

2.36 Which of the following strings is generated by the grammar above?
(a) aabbaba (b) aabaaba
(c) abababb (d) aabbaab

[2008 : 2 M]

2.37 For the correct answer in above question, how many steps are required to derive the string and how many parse trees are there?

(a) 6 and 1 (b) 6 and 2
(c) 7 and 2 (d) 4 and 2

[2008 : 2 M]

2.38 $S \rightarrow aSa \mid bSb \mid a \mid b$

The language generated by the above grammar over the alphabet {a, b} is the set of

(a) all palindromes
(b) all odd length palindromes
(c) strings that begin and end with the same symbol
(d) all even length palindromes **[2009 : 1 M]**

2.39 Which one of the following is FALSE?

(a) There is a unique minimal DFA for every regular language
(b) Every NFA can be converted to an equivalent PDA
(c) Complement of every context-free language is recursive
(d) Every nondeterministic PDA can be converted to an equivalent deterministic PDA

[2009 : 1 M]

2.40 Match all items in **Group-1** with correct options from those given in **Group-2**:

Group 1
P. Regular expression
Q. Pushdown automata
R. Dataflow analysis
S. Register allocation

Group 2
1. Syntax analysis
2. Code generation
3. Lexical analysis
4. Code Optimization

(a) P-4, Q-1, R-2, S-3 (b) P-3, Q-1, R-4, S-2
(c) P-3, Q-4, R-1, S-2 (d) P-2, Q-1, R-4, S-3

[2009 : 1 M]

2.41 Given the following state table of an FSM with two states A and B, one input and one output:

Present State A	Present State B	Input	Next State A	Next State B	Output
0	0	0	0	0	1
0	1	0	1	0	0
1	0	0	0	1	0
1	1	0	1	0	0
0	0	1	0	1	0
0	1	1	0	0	1
1	0	1	0	1	1
1	1	1	0	0	1

If the initial state is A = 0, B = 0, what is the minimum length of an input string which will take the machine to the state A = 0, B = 1 with Output =1?

(a) 3 (b) 4
(c) 5 (d) 6 **[2009 : 2 M]**

2.42 Let $L = L_1 \cap L_2$, where L_1 and L_2 are languages as defined below:

$L_1 = \{a^m b^m c a^n b^n \mid m, n \geq 0\}$
$L_2 = \{a^i b^j c^k \mid i, j, k \geq 0\}$

Then L is

(a) not recursive
(b) regular
(c) context-free but not regular
(d) recursively enumerable but not context-free

[2009 : 2 M]

2.43 Consider the language:

$L1 = \{0^i 1^j \mid i \neq j\}$, $L2 = \{0^i 1^j \mid i = j\}$,
$L3 = (0^i 1^j \mid i = 2j + 1\}$, $L4 = \{0^i 1^j \mid i \neq 2j\}$

Which one of the following statements is true?

(a) Only L2 is context free
(b) Only L2 and L3 are context free
(c) Only L1 and L2 are context free
(d) All are context free

[2010 : 2 M]

2.44 Consider the languages L1, L2 and L3 are given below:

$L1 = \{0^p 1^q \mid p, q \in N\}$,
$L2 = \{0^p 1^q \mid p, q \in N \text{ and } p = q\}$ and
$L3 = \{0^p 1^q 0^r \mid p, q, r \in N \text{ and } p = q = r\}$

Which of the following statements is NOT TRUE?

(a) Push Down Automata (PDA) can be used to recognize L1 and L2
(b) L1 is a regular language
(c) All the three languages are context free
(d) Turing machines can be used to recognize all the languages

[2011 : 2 M]

2.45 Consider the following languages:

$L_1 = \{0^p 1^q 0^r \mid p, q, r \geq 0\}$
$L_2 = \{0^p 1^q 0^r \mid p, q, r \geq 0, p \neq r\}$

Which one of the following statements is FALSE?

(a) L_2 is context-free
(b) $L_1 \cap L_2$ is context-free
(c) Complement of L_2 is recursive
(d) Complement of L_1 is context-free but not regular

[2013 : 2 M]

2.46 Consider the following languages over the alphabet $\Sigma = \{0, 1, c\}$:

$L_1 = \{0^n1^n \mid n \geq 0\}$

$L_2 = \{wcw^r \mid w \in \{0, 1\}^*\}$

$L_3 = \{ww^r \mid w \in \{0, 1\}^*\}$

Here, w^r is the reverse of the string w. Which of these languages are deterministic Context-free languages?

(a) None of the languages
(b) Only L_1
(c) Only L_1 and L_2
(d) All the three languages

[2014 (Set-3) : 2 M]

2.47 Consider the NPDA $\langle Q = \{q_0, q_1, q_2\}, \Sigma = \{0, 1\}, \Gamma = \{0, 1, Z\}, \delta, q_0, Z, F = \{q_2\}\rangle$, where (as per usual convention) Q is the set of states, Σ is the input alphabet, Γ is stack alphabet, δ is the state transition function, q_0 is the initial state, Z is the initial stack symbol, and F is the set of accepting states, The state transition is as follows:

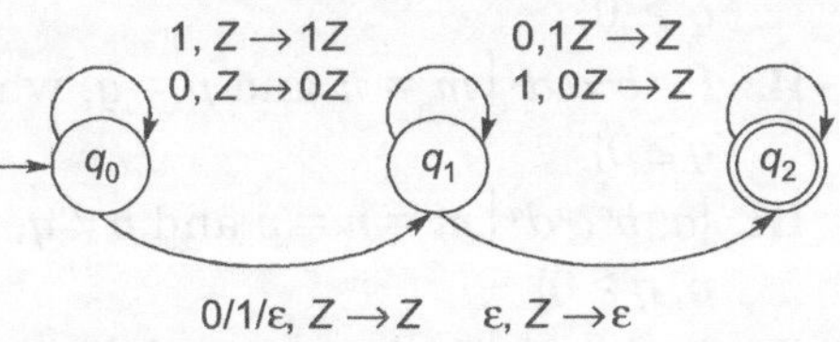

Which one of the following sequences must follow the string 101100 so that the overall string is accepted by the automaton?

(a) 10110
(b) 10010
(c) 01010
(d) 01001

[2015 (Set-1) : 2 M]

2.48 Which of the following languages are context-free?

$L_1 = \{a^mb^na^nb^m \mid m, n \geq 1\}$

$L_2 = \{a^mb^na^mb^n \mid m, n \geq 1\}$

$L_3 = \{a^mb^n \mid m = 2n + 1\}$

(a) L_1 and L_2 only
(b) L_1 and L_3 only
(c) L_2 and L_3 only
(d) L_3 only

[2015 (Set-3) : 1 M]

2.49 Consider the following context-free grammars:

$G_1: S \to aS \mid B, B \to b \mid bB$

$G_2: S \to aA \mid bB, A \to aA \mid B \mid \varepsilon, B \to bB \mid \varepsilon$

Which one of the following pairs of languages is generated by G_1 and G_2, respectively?

(a) $\{a^mb^n \mid m > 0 \text{ or } n > 0\}$ and $\{a^mb^n \mid m > 0 \text{ and } n > 0\}$
(b) $\{a^mb^n \mid m > 0 \text{ and } n > 0\}$ and $\{a^mb^n \mid m > 0 \text{ or } n \geq 0\}$
(c) $\{a^mb^n \mid m \geq 0 \text{ or } n > 0\}$ and $\{a^mb^n \mid m > 0 \text{ and } n > 0\}$
(d) $\{a^mb^n \mid m \geq 0 \text{ and } n > 0\}$ and $\{a^mb^n \mid m > 0 \text{ or } n > 0\}$

[2016 (Set-1) : 2 M]

2.50 Consider the transition diagram of a PDA given below with input alphabet $\Sigma = \{a, b\}$ and stack alphabet $\Gamma = \{X, Z\}$. Z is the initial stack symbol. Let L denote the language accepted by the PDA.

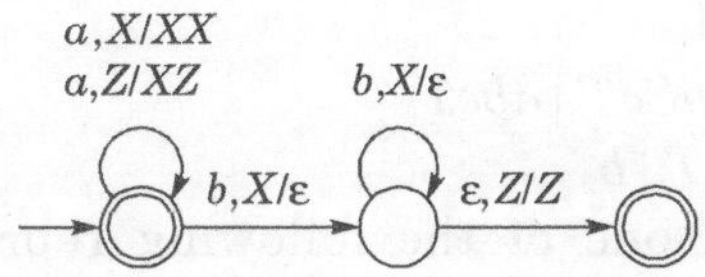

Which one of the following is TRUE?

(a) $L = \{a^nb^n \mid n \geq 0\}$ and is not accepted by any finite automata
(b) $L = \{a^n \mid n \geq 0\} \cup \{a^nb^n \mid n \geq 0\}$ and is not accepted by any deterministic PDA
(c) L is not accepted by any Turing machine that halts on every input
(d) $L = \{a^n \mid n \geq 0\} \cup \{a^nb^n \mid n \geq 0\}$ and is deterministic context-free

[2016 (Set-1) : 2 M]

2.51 Consider the following langauges:

$L_1 = \{a^nb^mc^{n+m} : m, n \geq 1\}$

$L_2 = \{a^nb^nc^{2n} : n \geq 1\}$

Which one of the following is **TRUE**?

(a) Both L_1 and L_2 are context-free.
(b) L_1 is context-free while L_2 is not context-free
(c) L_2 is context-free while L_1 is not context-free
(d) Neither L_1 nor L_2 is context-free

[2016 (Set-2) : 2 M]

2.52 Language L_1 is defined by the grammar:

$S_1 \to aS_1b \mid \varepsilon$

Language L_2 is defined by the grammar:

$S_2 \to abS_2 \mid \varepsilon$

Consider the following statements:

P : L_1 is regular

Q : L_2 is regular

Which one of the following is TRUE?

(a) Both P and Q are true
(b) P is true and Q is false
(c) P is false and Q is true
(d) Both P and Q are false

[2016 (Set-2) : 1 M]

2.53 Which one of the following grammars is free from left recursion?

(a) $S \rightarrow AB$
$A \rightarrow Aa \mid b$
$B \rightarrow c$

(b) $S \rightarrow Ab \mid Bb \mid c$
$A \rightarrow Bd \mid \varepsilon$
$B \rightarrow e$

(c) $S \rightarrow Aa \mid B$
$A \rightarrow Bb \mid Sc \mid \varepsilon$
$B \rightarrow d$

(d) $S \rightarrow Aa \mid Bb \mid c$
$A \rightarrow Bd \mid \varepsilon$
$B \rightarrow Ae \mid \varepsilon$

[2016 (Set-2) : 2 M]

2.54 Consider the following context-free grammar over the alphabet $\Sigma = \{a, b, c\}$ with S as the start symbol:

$S \rightarrow abScT \mid abcT$
$T \rightarrow bT \mid b$

Which one of the following represents the language generated by the above grammar?

(a) $\{(ab)^n (cb)^n \mid n \geq 1\}$
(b) $\{(ab)^n cb^{m_1} cb^{m_2} ... cb^{m_n} \mid n, m_1, m_2,..., m_n \geq 1\}$
(c) $\{(ab)^n (cb^m)^n \mid m, n \geq 1\}$
(d) $\{(ab)^n (cb^n)^m \mid m, n \geq 1\}$

[2017 (Set-1) : 1 M]

2.55 Consider the following languages over the alphabet $\Sigma = \{a, b, c\}$.

Let $L_1 = \{a^n b^n c^m \mid m, n \geq 0\}$ and
$L_2 = \{a^m b^n c^n \mid m, n \geq 0\}$.

Which of the following are context-free languages?

I. $L_1 \cup L_2$
II. $L_1 \cap L_2$

(a) I only (b) II only
(c) I and II (d) Neither I nor II

[2017 (Set-1) : 2 M]

2.56 Identify the language generated by the following grammar, where S is the start variable.

$S \rightarrow XY$
$X \rightarrow aX \mid a$
$Y \rightarrow aYb \mid \in$

(a) $\{a^m b^n \mid m \geq n, n > 0\}$
(b) $\{a^m b^n \mid m \geq n, n \geq 0\}$
(c) $\{a^m b^n \mid m > n, n \geq 0\}$
(d) $\{a^m b^n \mid m > n, n > 0\}$ **[2017 (Set-2) : 1 M]**

2.57 Let L_1, L_2 be any two context-free languages and R be any regular language. Then which of the following is/are CORRECT?

I. $L_1 \cup L_2$ is context-free.
II. $\bar{L}_1$ is context-free.
III. $L_1 - R$ is context-free.
IV. $L_1 \cap L_2$ is context-free.

(a) I, II and IV only
(b) I and III only
(c) II and IV only
(d) I only

[2017 (Set-2) : 1 M]

2.58 Consider the following languages:

$L_1 = \{a^p \mid p \text{ is a prime number}\}$
$L_2 = \{a^n b^m c^{2m} \mid n \geq 0, m \geq 0\}$
$L_3 = \{a^n b^n c^{2n} \mid n \geq 0\}$
$L_4 = \{a^n b^n \mid n \geq 1\}$

Which of the following are CORRECT?

I. L_1 is context-free but not regular.
II. L_2 is not context-free.
III. L_3 is not context-free but recursive.
IV. L_4 is deterministic context-free.

(a) I, II and IV only (b) II and III only
(c) I and IV only (d) III and IV only

[2017 (Set-2) : 2 M]

2.59 Consider the following languages:

I. $\{a^m b^n c^p d^q \mid m + p = n + q, \text{ where } m, n, p, q \geq 0\}$
II. $\{a^m b^n c^p d^q \mid m = n \text{ and } p = q, \text{ where } m, n, p, q \geq 0\}$
III. $\{a^m b^n c^p d^q \mid m = n = p \text{ and } p \neq q, \text{ where } m, n, p, q \geq 0\}$
IV. $\{a^m b^n c^p d^q \mid mn = p + q, \text{ where } m, n, p, q \geq 0\}$

Which of the language above are context-free?

(a) I and IV only
(b) I and II only
(c) II and III only
(d) II and IV only

[2018 : 2 M]

2.60 Which one of the following languages over $\Sigma = \{a, b\}$ is NOT context-free?

(a) $\{a^n b^i \mid i \in \{n, 3n, 5n\}, n \geq 0\}$
(b) $\{wa^n w^R b^n \mid w \in \{a, b\}^*, n \geq 0\}$
(c) $\{ww^R \mid w \in \{a, b\}^*\}$
(d) $\{wa^n b^n w^R \mid w \in \{a, b\}^*, n \geq 0\}$ **[2019 : 2 M]**

2.61 Consider the language
$L = \{a^n \mid n > 0\} \cup \{a^n b^n \mid n \geq 0\}$ and the following statements.

I. L is deterministic context-free.
II. L is context-free but not deterministic context-free.
III. L is not LL(k) for any k.

Which of the above statements is/are TRUE?

(a) I and III only (b) III only
(c) I only (d) II only **[2020 : 1 M]**

2.62 Consider the following languages:

$L_1 = \{w\,xyx \mid w, x, y \in (0+1)^+\}$

$L_2 = \{xy \mid x, y \in (a+b)^*, \ |x| = |y|, x \neq y\}$

Which one of the following is TRUE?

(a) L_1 is regular and L_2 is context-free.

(b) L_1 is context-free hut L_2 is not context-free.

(c) Neither L_1 nor L_2 is context-free.

(d) L_1 is context-free but not regular and L_1 is context-free.

[2020 : 2 M]

2.63 In a pushdown automaton $P = (Q, \Sigma, \Gamma, \delta, q_0, F)$, a transition of the form,

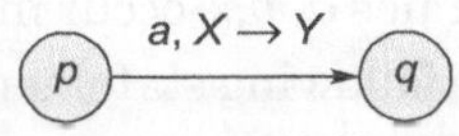

where $p, q, \in Q$, $a \in \Sigma \cup \{\varepsilon\}$, and $X, Y \in \Gamma \cup \{\varepsilon\}$, represents $(q, Y) \in \delta(p, a, X)$

Consider the following pushdown automaton over the input alphabet $\Sigma\{a, b\}$ and stack alphabet $\Gamma = \{\#, A\}$.

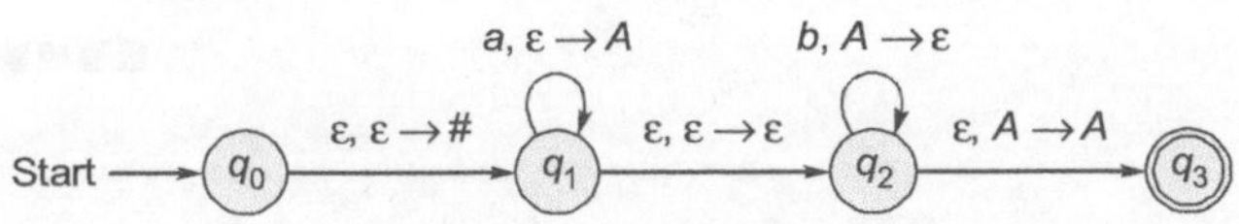

The number of strings of length 100 accepted by the above pushdown automaton is ______.

[2021 (Set-1) : 2 M]

2.64 Suppose that L_1 is a regular language and L_2 is a context-free language. Which one of the following languages is NOT necessarily context-free?

(a) $L_1 \cdot L_2$ (b) $L_1 \cup L_2$

(c) $L_1 - L_2$ (d) $L_1 \cap L_2$

[2021 (Set-1) : 1 M]

2.65 For a string w, we define w^R to be the reverse of w. For example, if $w = 01101$ then $w^R = 10110$. Which of the following languages is/are context-free?

(a) $\{w\,x\,w^R x R \mid w, x \in \{0, 1\}^*\}$

(b) $\{w\,x\,w^R \mid w, x \in \{0, 1\}^*\}$

(c) $\{ww^R x x^R \mid w, x \in \{0, 1\}^*\}$

(d) $\{w\,x\,x^R w^R \mid w, x \in \{0, 1\}^*\}$

[2021 (Set-2) : 2 M]

2.66 Let L_1 be a regular language and L_2 be a context-free language. Which of the following languages is/are context-free?

(a) $L_1 \cap \overline{L_2}$ (b) $\overline{\overline{L_1} \cup \overline{L_2}}$

(c) $L_1 \cup (L_2 \cup \overline{L_2})$ (d) $(L_1 \cap L_2) \cup (\overline{L_1} \cap L_2)$

[2021 (Set-2) : 1 M]

2.67 Consider the following languages:

$L_1 = \{a^n w a^n \mid w \in \{a, b\}^*\}$

$L_2 = \{w x w^R \mid w, x \in \{a, b\}^*, \ |w|, |x| > 0\}$

Note that w^R is the reversal of the string w. Which of the following is/are TRUE?

(a) L_1 and L_2 are regular.

(b) L_1 and L_2 are context-free.

(c) L_1 is regular and L_2 is context-free.

(d) L_1 and L_2 are context-free but not regular.

[2022 : 2 M]

2.68 Consider the following languages:

$L_1 = \{ww \mid w \in \{a, b\}^*\}$

$L_2 = \{a^n b^n c^m \mid m, n \geq 0\}$

$L_3 = \{a^m b^n c^n \mid m, n \geq 0\}$

Which of the following statements is/are FALSE?

(a) L_1 is not context-free but L_2 and L_2 are deterministic context-free.

(b) Neither L_1 nor L_2 is context-free.

(c) L_2, L_3 and $L_2 \cap L_3$ all are context-free.

(d) Neither L_1 nor its complement is context-free.

[2022 : 2 M]

2.69 Which of the following statements is/are CORRECT?

(a) The intersection of two regular languages is regular.

(b) The intersection of two context-free languages is context-free.

(c) The intersection of two recursive languages is recursive.

(d) The intersection of two recursively enumerable languages is recursively enumerable.

[2023 : 1 M]

2.70 Consider the context-free grammar G below:

$S \to aSb \mid X$

$X \to aX \mid Xb \mid a \mid b$

where S and X are non-terminals, and a and b are terminal symbols. The starting non-terminal is S.

Which one of the following statements is CORRECT?

(a) The language generated by G is $(a + b)^*$

(b) The language generated by G is $a^*(a + b)b^*$

(c) The language generated by G is $a^*b^*(a + b)$

(d) The language generated by G is not a regular language

[2023 : 2 M]

2.71 Consider the pushdown automaton (PDA) P below, which runs on the input alphabet $\{a, b\}$, has stack alphabet $\{\perp, A\}$, and has three states $\{s, p, q\}$, with s being the start state. A transition from state u to state v, labelled $c / X / \gamma$, where c is an input symbol or $\in$, X is a stack symbol, and γ is a string of stack symbols, represents the fact that in state u, the PDA can read c from the input, with X on the top of its stack, pop X from the stack, push in the string γ on the stack, and go to state v. In the initial configuration, the stack has only the symbol $\perp$ in it. The PDA accepts by empty stack.

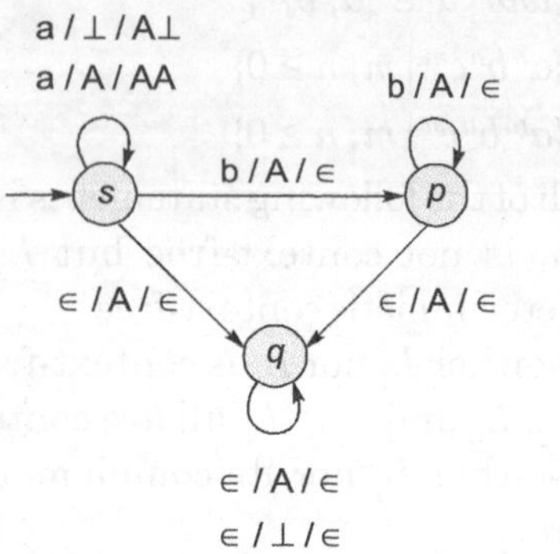

Which one of the following options correctly describes the language accepted by P?

(a) $\{a^m b^n \mid 1 \le m \text{ and } n < m\}$
(b) $\{a^m b^n \mid 0 \le n \le m\}$
(c) $\{a^m b^n \mid 0 \le m \text{ and } 0 \le n\}$
(d) $\{a^m \mid 0 \le m\} \cup \{b^n \mid 0 \le n\}$ **[2023 : 2 M]**

2.72 Consider a context-free grammar G with the following 3 rules.

$S \rightarrow aS$

$S \rightarrow aSbS$

$S \rightarrow c$

Let $w \in L(G)$. Let $n_a(w)$, $n_b(w)$, $n_c(w)$ denote the number of times a, b, c occur in w, respectively. Which of the following statements is/are TRUE?

(a) $n_a(w) > n_b(w)$
(b) $n_c(w) = n_b(w) + 1$
(c) $n_a(w) > n_c(w) - 2$
(d) $n_c(w) = n_b(w) * 2$

[2024 (Set-2) : 2 M]

■■■■

Answers Push Down Automata: CFL & DCFL

2.1	(c)	**2.2**	(b)	**2.3**	Sol.	**2.4**	(a)	**2.5**	(b)	**2.6**	(b)	**2.7**	(a)	**2.8**	(a, c)	**2.9**	(b)
2.10	(b)	**2.11**	(c)	**2.12**	(c)	**2.13**	(b)	**2.14**	(b)	**2.15**	(c)	**2.16**	(d)	**2.17**	(d)	**2.18**	(d)
2.19	(a)	**2.20**	(b)	**2.21**	(d)	**2.22**	(b)	**2.23**	(b)	**2.24**	(d)	**2.25**	(c)	**2.26**	(b)	**2.27**	(b)
2.28	(c)	**2.29**	(d)	**2.30**	(b)	**2.31**	(d)	**2.32**	(c)	**2.33**	(c)	**2.34**	(b)	**2.35**	(b)	**2.36**	(d)
2.37	(a)	**2.38**	(b)	**2.39**	(d)	**2.40**	(b)	**2.41**	(a)	**2.42**	(c)	**2.43**	(d)	**2.44**	(c)	**2.45**	(d)
2.46	(c)	**2.47**	(b)	**2.48**	(b)	**2.49**	(d)	**2.50**	(d)	**2.51**	(b)	**2.52**	(c)	**2.53**	(b)	**2.54**	(b)
2.55	(a)	**2.56**	(c)	**2.57**	(b)	**2.58**	(d)	**2.59**	(b)	**2.60**	(b)	**2.61**	(a)	**2.62**	(a)	**2.63**	(50)
2.64	(c)	**2.65**	(b, c, d)	**2.66**	(b, c, d)	**2.67**	(a, b, c)	**2.68**	(b, c, d)	**2.69**	(a, c, d)	**2.70**	(b)				
2.71	(a)	**2.72**	(b, c)														

Explanations Push Down Automata: CFL & DCFL

2.1 (c)

$S \to a\alpha b \mid b \to c \mid ab$
$S \to \alpha S \mid b$
$S \to abb \mid ab$
$S\alpha \to bdb \mid bd$

Since the LHS of the production $S\alpha \to bdb \mid bd$ which is $S\alpha$ does not belong to variable set V, the grammar is not a CFG (type 2). However since in all the productions length of LHS ≤ length of RHS, the grammar is type 1 (CSG).

2.2 (b)

(a) If L_1 and L_2 are context free language then L_1L_2 is also context free language because context free languages are closed under concatenation.

(b) Since context free languages are not closed under intersection, $L_1 \cap L_2$ is not necessarily a context free language.

(c) If L_1 is context free language and R is regular set then $L_1 \cap R$ is surely context free language because context fee languages are closed under regular intersection.

(d) If L_1 and L_2 are context free languages then $L_1 \cup L_2$ is surely context free language because context free languages are closed under union.

2.3 Sol.

Nullable variables are $V_N = \{A, B\}$

Putting A and B to null in every possible combination, we get

G′: $S \to ABAC \mid ABC \mid BAC \mid AAC \mid AC \mid C \mid BC$
$A \to aA \mid a$
$B \to bB \mid b$
$C \to d$

The above equivalent grammar G' contain unit production
$S \to C$

Mix $S \to C$ with $C \to d$ and get $S \to d$.

So replacing above unit production as
$S \to d$

So resultant grammar G′ without ε-production and unit production is

G′: $S \to ABAC \mid ABC \mid BAC \mid AAC \mid AC \mid BC \mid d$
$A \to aA \mid a$
$B \to bB \mid b$
$C \to d$

2.4 (a)

The language

$$L = \{wcw^R \mid w \in (a, b)^*\}$$

is accepted by DPDA (since there is a clear clue "c" available in the string that tells the machine exactly where the pop starts) and so it is DCFL. For above language until "c" appears everything is pushed into the stack and when c appears it is skipped and from then every symbol in w^R is matched with top of stack and popped. At the end when input string is empty and stack contain initial stack symbol, the string is accepted.

The DPDA for above language is

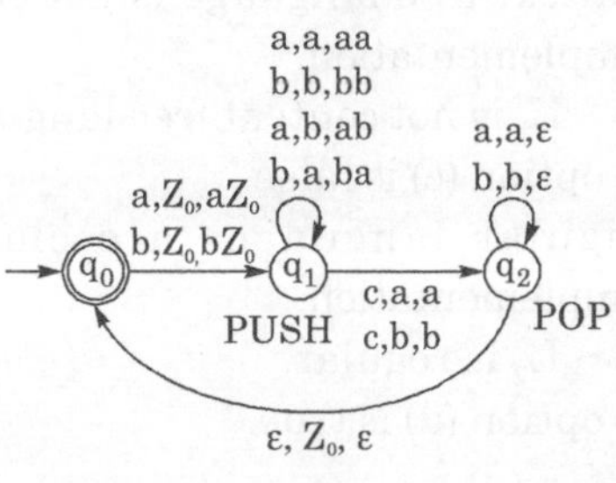

2.5 (b)

(a) and (c) Since for every finite set we can create DFA by brute force. So every finite subset of a non-regular or regular is regular.

(b) Let $\Sigma = (a, b)$. Σ^* is regular set

$L_1 = \{a^n b^n \mid n \geq 0\}$

$L_2 = \{a^n b^m \mid nm \geq 0\}$

Both L_1 & L_2 are subset of Σ^*

$L_1 \subset \Sigma^*$ & $L_2 \subset \Sigma^*$

However L_1 is CFL and L_2 is regular. So by above illustration every subset of a regular set need not to be regular.

(d) Since regular languages are closed under intersection. So the intersection of two regular sets is regular.

2.6 (b)

Since context free language are closed under union and kleene closure and not closed under intersection and complement.

2.7 (a)

There are two approaches by which PDA accepts its input

(a) Acceptance by empty stack.

(b) Accepting by entering final state.

Since there exist algorithm to convert PDA accepting by final state to PDA accepting by empty stack and vice-versa, the acceptance or recognition power of PDA of both approaches are same.

Therefore $L_D = L_E$

2.8 (a, c)

(a) Since CFLs are closed under regular difference, so $L_1 - L_2$ = CFL − regular = CFL. So option (a) is false.

(b) Since context free language is closed under regular intersection so $L_1 \cap L_2$ = CFL $\cap$ REG = CFL.
So option (b) is true.

(c) Context free language is not closed under complementation.
So $\sim L_1$ is not context free language.
So option (c) is false.

(d) Regular languages are closed under complementation.
So $\sim L_2$ is regular.
So option (d) is true.

2.9 (b)

(a) False, CFL's can be divided into two categories

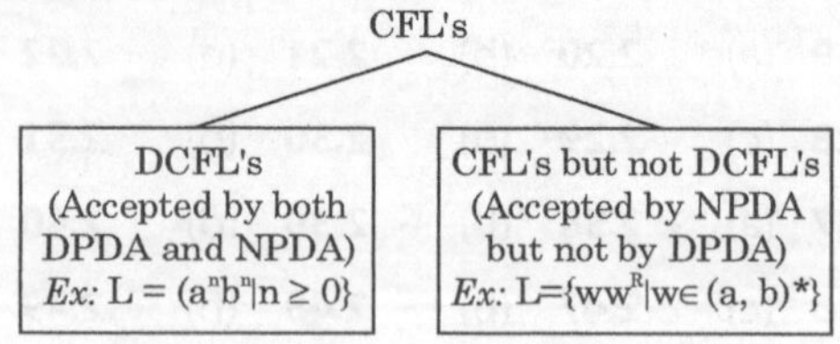

So, if a language is context free it is not always true that it is accepted by a deterministic pushdown automation.

(b) True, since CFL's are closed under union, the union of two CFL is context free

(c) & (d) False, since CFL's are not closed under intersection and complement.

2.10 (b)

A PDA with a stack limited to containing 10 items, has finite memory and hence is equivalent to a DFA. So the language accepted by a PDA in which the stack is limited to 10 items is best described as regular.

2.11 (c)

The grammar is $S \rightarrow aSb \mid SS \mid \varepsilon$

(a) G is not ambiguous is false, since ε which belongs to L(G), has infinite number of derivation trees, which makes G ambiguous. Some derivation trees are given below:

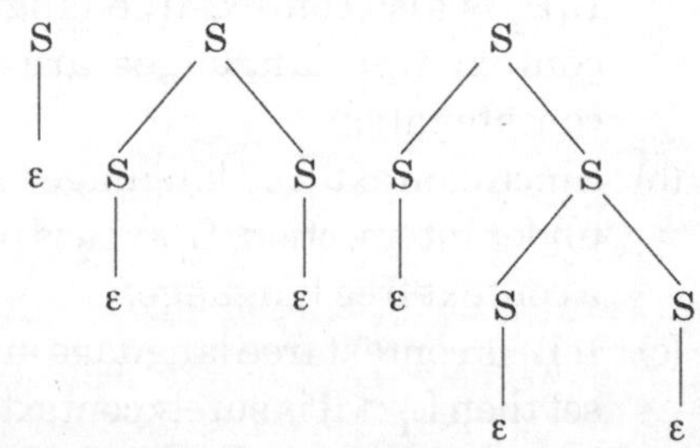

(b) there exists $x, y \in L(G)$ such that $xy \notin L(G)$ is false, since $S \rightarrow SS$ can be used to derive xy, whenever $x \in L(G)$ and $y \in L(G)$.

(c) is true since this language is L(G) = $\{w \mid n_a(w) = n_b(w)$ and $n_a(v) \geq n_b(v)$ where v is any prefix of w$\}$ which is equal to $(a^n b^n)^*$.
This language happens to be a deterministic context free language.
$\therefore$ There exists a DPDA that accepts it.

(d) is false, as the given language is not regular (comparison between number of 'a's and number of 'b's is there).
$\therefore$ No DFA exists to accept it.

2.12 (c)

- Ambiguity starts from CFL, there exist some CFL which are inherently ambiguous. Hence true.
- If a grammar generate unique parse tree for all string present in the language then that grammar will be an unambiguous grammar. Hence true.
- The set of all languages accepted by all NPDAs is corresponds to the set of CFLs, which is a proper superset of the set of all languages accepted by all DPDAs, which is corresponds to the set of DCFLs. Hence false.
- If a language represents Finite set of string then that language will always be regular language but regular language need not be finite language. Hence true.

2.13 (b)

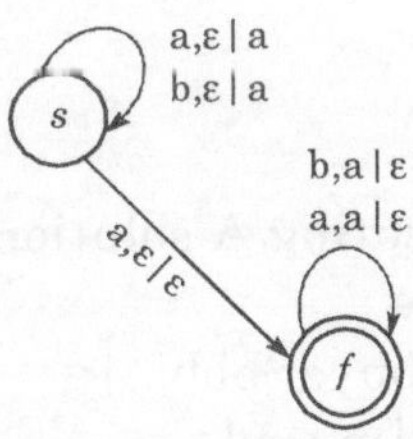

This machine accepts the language

$L = \{w_1 a w_2 \mid |w_1| = |w_2|\}$

The language allows only strings whose centre bit is "a" and the strings on either side have equal length i.e., strings of odd length with centre bit "a".

The string "aabab" is odd length but centre bit is not "a".

So, this string is not a member of L(M).

2.14 (b)

$L = \{a^m b^n c^{m+n} \mid m, n \geq 1\}$

A DPDA can accept this language. When a's and b's are in input, these are pushed into the stack and when c's appear in input the a's and b's are popped out. After all c's are finished, if stack is empty, this means $n_c(w) = n_a(w) + n_b(w)$ and the string is accepted. Else it is rejected. Now since a DPDA exists, the language is DCFL and hence, context-free. Clearly the language is not regular, since we must count and compare c's with a's and b's which cannot be done by any FA.

2.15 (c)

Draw the fa of the given grammar as shown below:

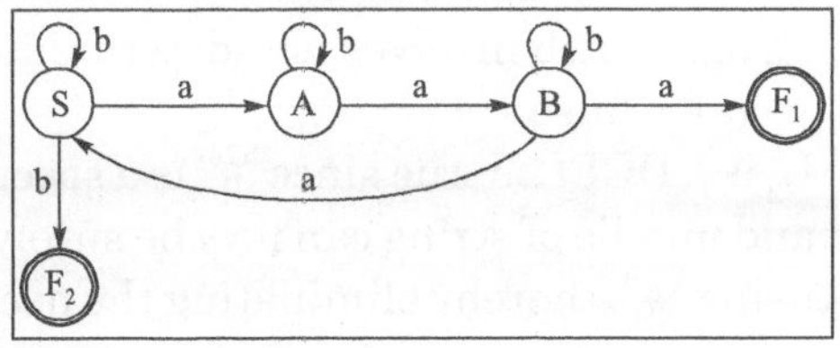

Clearly, the machine is accepting $N_a(w) = 3k$, where $k \in \{0, 1, 2, 3, ...\}$

2.16 (d)

$L^C \cup M^C = (Reg)^C \cup (CFL)^C = Reg \cup (CSL)^C$
$= Reg \cup CSL = CSL$

A CSL may or may not be regular. So, options (a) and (c) are false.

A CSL need not be a CFL. So, option (b) is false. So, answer is (d) none of these.

2.17 (d)

In NPDA may have a dead configuration i.e. transition for every alphabet not present from this state. So it is not necessary that both L(P) and N(P) are represents Σ^*.

2.18 (d)

Nf : Non-deterministic finite automata.
Np: Non-deterministic push-down automata.
Df : Deterministic finite automata.
Dp : Deterministic push down automata.

According to "Subset Construction" theorem every language accepted by Non-deterministic-finite automata (Nf) is also accepted by some Deterministic-Finite automata (Df) so Df = Nf. Deterministic push-down automata (Dp) recognizes the DCFLs which is a proper subset of the language of context-free languages and the non-deterministic push down automata recognizes the context-free languages.

So, Dp ⊂ Np.

2.19 (a)

L_1 and L_2 are context free languages and therefore $L_1 \cap L_2$ may or may not be context free, since CFL's are not closed under intersection. Now let us look at $L_1 \cap L_2$.

$L_1 \cap L_2 = \{a^n b^n c^n \mid n > 0\}$

Which is clearly not context free but is context sensitive.

∴ (a) is false.

2.20 (b)

L_1 is CFL but not a DCFL, since accepting ww^R necessarily involves finding the middle of the string, which involves non-determinism.

∴ (a) is false.

L_2 is a DCFL is true since "#" is a special symbol and middle of string can now be surely found by using "#", thereby eliminating the need for non-deterministic guessing. So (b) is true.

L_3 is a CSL and not a CFL. So (c) is false.

L_3 is not a DCFL either. So (d) is false.

2.21 (d)

$L_1 = \{0^{n+m} 1^n 0^m \mid n, m \geq 0\}$ is context free (first n + m 0's are pushed into the stack, then for each of the 1s and 0s following, we will pop the 0's in the stack one by one until at the end of the word if the stack is empty then the word is accepted).

$L_2 = \{0^{n+m} 1^{n+m} 0^m \mid n, m \geq 0\}$ is not context free, since two comparisons have to be made here to determine if $w \in L_2$.

1. 0's and 1's are equal.
2. Since $m < (n + m)$, we have to ensure that, 0's which come after $0^{n+m} 1^{n+m}$, are less in number, compared to $n + m$.

$L_3 = \{0^{n+m} 1^{n+m} 0^{n+m} \mid n_1\ m \geq 0\}$ is clearly context sensitive since have also 2 comparisons have to be made. Infact this language is same as $L_3 = \{0^x 1^x 0^x \mid x \geq 0\}$, which is context sensitive.

2.22 (b)

L(G) = Set of all palindrome strings.

"baba" is not a palindrome string.

2.23 (b)

To be accepted by NPDA but not DPDA, the language must be a CFL but not DCFL.

(a) is wrong, it is not context free.

(b) $L = \{a^l b^m c^n \mid l \neq m \text{ or } m \neq n\}$

This language has double comparison on m with 'or'. So the language is CFL but not DCFL.

(c) In this language only one comparison is there and push and pop position are clear. This is a DCFL.

(d) a*b* is regular. Every regular language is a DCFL. So this language is also a DCFL.

2.24 (d)

L is CFL and M is Regular. Intersection of CFL and Regular is always a CFL, since all languages are closed under regular intersection.

2.25 (c)

Equivalent PDA of is given transition below:

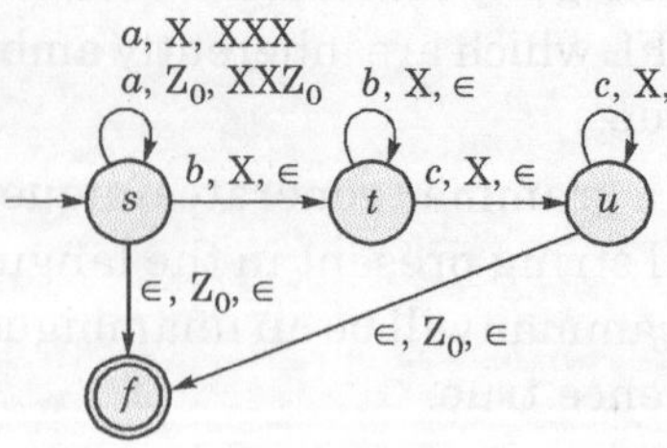

At state *s* for every *a*, we push two X in stack. And when first *b* is come then for every *b* we pop out one X and reaches to state *t*. And after all *b* are over if we get *c* then for every *c* we pop out one *X* and reaches to state *u*.

If all *X* are popped out then reached to final state *f*, means that, sum of number of *b's* and number of *c's* = twice of number of *a's*

i.e. $L = \{a^l b^m c^n \mid 2l = m + n\}$

2.26 (b)

$A \rightarrow bA \mid \in$

$L(A) = b^*$

$S \rightarrow aSAb \mid \in$

Now, substituting A solution into this gives

$S \rightarrow aSb^*b \mid \in$

$S \rightarrow aSb \mid aSbb \mid aSbbb\ldots \mid \in$

$L(S) = \{a^m b^n \mid m \leq n\}$

2.27 (b)

$G = \{S \rightarrow SS, S \rightarrow ab, S \rightarrow ba, S \rightarrow \varepsilon\}$

1. "G is ambiguous", is true since "ε" has infinite number of derivations some of which are shown below:

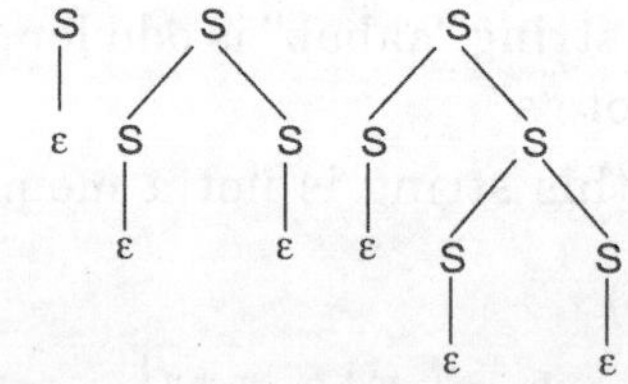

2. "G produces all strings with equal number of a's and b's", is false since L(G) = (ab + ba)* and this language cannot produce all strings with equal number of a's and b's. For example aabb has equal number of a's and b's but aabb ∉ L(G).
3. "G can be accepted by a DPDA" is true, since L(G) = (ab + ba)* and this is a regular language, since we have written a regular expression for it. Since every regular language is also a DCFL, a DPDA exists, which accepts this language.

2.28 (c)

Grammar may change but language remain the same in the grammar and its disambiguated version.

$$L(D) = L(G)$$

2.29 (d)

Using capital letters for variables and small letters for terminals, the given grammar G1 becomes

$S \to w(E)S$
$S \to o$
$E \to i$

Since LHS of every production is a single variable, G1 is context free grammar.

G2 becomes

$S \to w(E)S$
$S \to o$
$E \to E+E$
$E \to E*E$
$E \to i$

Since LHS of every production is a single variable, G2 is also a context free grammar.

RHS of some of the productions in G1 and G2 has two non-terminals, hence it is not regular grammar.

2.30 (b)

$\{0^i 21^i \mid i \geq 0\}$ is a DCFL since, a DPDA can accept this language. 0's are pushed into the stack and then when 2 appears in input, state is changed and immediately after that for every 1, a zero is popped out of the stack. In the end, if stack has start stack symbol only, then the string is accepted. Else it is rejected. Since every DCFL is recursive, we can say that the language is recursive, and is a DCFL.

2.31 (d)

(a) True
(b) True
(c) True
(d) Every subset of a recursively enumerable set is recursive, is false, since a set is a subset of itself and there are RE languages, which are not REC.

2.32 (c)

1. True
2. All ε-productions can be removed from only context free grammars that produce λ-free CFL's.
If $\lambda \in L(G)$, then all ε-productions cannot be successfully removed. So 2 is false.
3. True
4. True, since in Chomsky normal form, every production is of the form of $A \to BC$ or $A \to a$.

An example of a binary tree generated by CNF derivation is shown below:

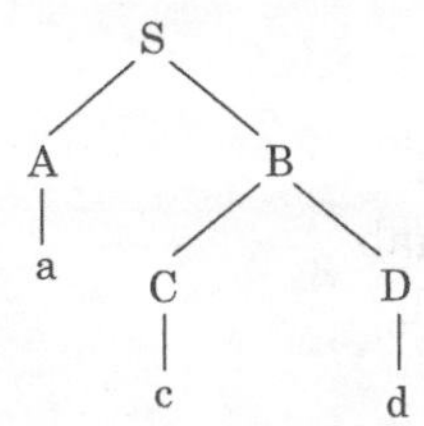

2.33 (c)

In $\{w c w \mid w \in (a, b)^*\}$, leftmost w is the identifier checking and rightmost w is the use of w.

In $a^n b^m c^n d^m$ the actual parameters are a^n and b^m and the formal parameters are c^n and d^n such that the number of arguments of a and b are equal to c and d respectively.

The grammar $\{X \to XbX \mid XcX \mid dXf \mid g\}$ generates the arithmetic expressions with matched pair of parentheses (d is left parenthesis and f is right parenthesis).

The grammar $\{X \to bXb \mid cXc \mid \varepsilon\}$ generates the palindromic strings.

2.34 (b)

L1 is CFL [push a's onto stack, and pop a with each b]

L2 is CFL [push b's onto stack and pop b with each c]. So, option (a) is false.

In L1, every string has atleast one "c".

In L2, every string has no "c".

So, $L1 \cap L2 = \phi$. So option (b) is true.

Union of two CFLs is a CFL and so option (c) is false.

L2 is DCFL and hence acceptable by a DPDA. So, option (d) is false.

2.35 (b)

$S \to B$
$B \to 0B00 \mid 1$

So, S generates $\{0^n 1 0^{2n} \mid n \geq 1\}$

$S \to AA$
$A \to 0A \mid A0 \mid 1$
$A \to \{0^m 1 0^n \mid m, n \geq 0\}$

So, S generates $\{0^m 1 0^n 0^p 1 0^q \mid m,n,p,q \geq 0\}$.

$= \{0^m 1 0^x 1 0^q \mid m, x, q \geq 0\}$

$= \{0^i 1 0^j 1 0^k \mid i, j, k \geq 0\}$

$L(S) = \{0^i 1 0^j 1 0^k \mid i, j, k \geq 0\} \cup \{0^n 1 0^{2n} \mid n \geq 1\}$
Which is same as option (b).

2.36 (d)

Given grammar will generate string aabbaab whose derivation is given below:

$S \rightarrow aS$
$S \rightarrow aA$
$S \rightarrow aaAb$
$S \rightarrow aabAab$
$S \rightarrow aabbAaab$
$S \rightarrow aabbaab$

2.37 (a)

Given grammar will generate string aabbaab whose derivation is given below:

$S \rightarrow aS$ 1
$S \rightarrow aA$ 2
$S \rightarrow aaAb$ 3
$S \rightarrow aabAab$ 4
$S \rightarrow aabbAaab$ 5
$S \rightarrow aabbaab$ 6

Thus 6 steps are needed and only one parse tree shown below:

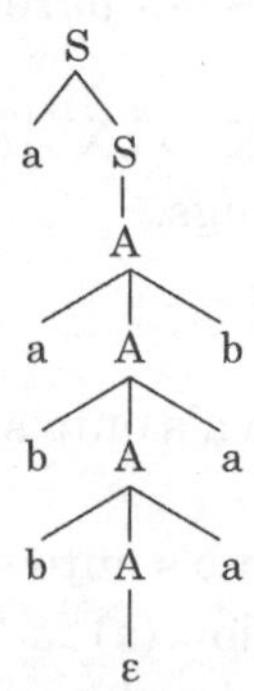

2.38 (b)

$S \rightarrow aSa \mid bSb \mid a \mid b$

is the grammar corresponding to all odd length palindromes.
Notice that choice (c) is not correct since "abaa" is a string that begins and ends with the same symbol but cannot be generated by above grammar.
Since all strings generates by this grammar have odd length, choices (a) and (d) are also incorrect.

2.39 (d)

Since it is known that the set of DCFL's are a proper sub class of the set of CFL's, NPDA's which corresponds in general to class of CFL's cannot in general be converted to equivalent DPDA's, which correspond to the class of DCFL's.

2.40 (b)

P-3, Q-1, R-4, S-2

2.41 (a)

The transition diagram for the given FSM is given below.

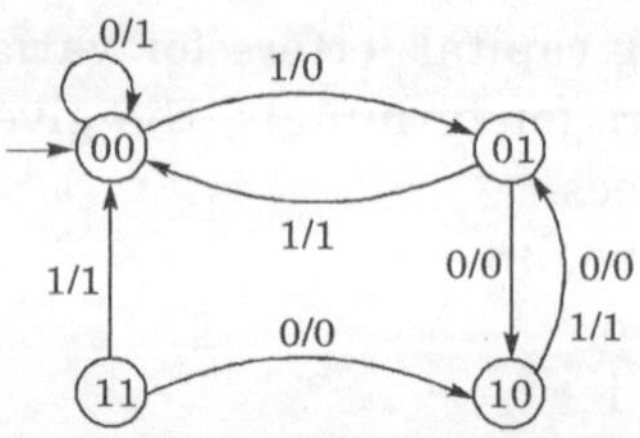

As can be seen from above diagram the minimum length of the input string that will take machine from 00 to 01 with output 1, is three.
In fact the minimum length input string is "101". Which will give an output of 1, while reaching state 01.

2.42 (c)

$$L_1 = \{a^m b^m c a^n b^n \mid m, n \geq 0\}$$
$$L_2 = \{a^i b^j c^k \mid i, j, k \geq 0\}$$
$$L = L_1 \cap L_2$$

L_1 is CFL. L_2 is regular. First use closure property to get a estimate.

$$L = L_1 \cap L_2 = \text{CFL} \cap \text{Reg} = \text{CFL}$$

However, since one of the option (b) is regular is stronger than CFL answer obtained by closure property, we need to find the actual intersection. Any string belonging to both must have equal number of a's & b's followed by a single c followed by no a's or b's; which is the only string allowed by both L_1 and L_2.
i.e. $L = L_1 \cap L_2 = \{a^m b^m c\}$
Now this is clearly context free, but not regular.

2.43 (d)

$$L_1 = \{0^i 1^j ; i \neq j\}$$
$$L_2 = \{0^i 1^j ; i = j\}$$
$$L_3 = \{0^i 1^j ; i = 2j + 1\}$$
$$L_4 = \{0^i 1^j ; i \neq 2j\}$$

All of L_1, L_2, L_3 and L_4 are context free, since, each of these languages has a single linear comparison between i and j and any linear comparison between i and j can be performed, in a PDA.

2.44 (c)

Since L_3 involves two comparison p = q and q = r, it cannot be context free.
So statement (c) is not true.

2.45 (d)

$L_1 = \{0^p 1^q 0^r \mid p, q, r \geq 0\}$: regular language
$L_2 = \{0^p 1^q 0^r \mid p, q, r \geq 0, p \neq r\}$: CFL

(a) L_2 is CFL : True

(b) $L_1 \cap L_2$ is context free : True, because CFL is closed on regular intersection.

(c) Complement of L_2 is recursive:

$\overline{L_2} = \overline{CFL} = \overline{CSL} = CSL$

Now, every CSL is REC. So, this statement is true.

(d) Complement of L_1 is context free but not regular: false, because regular language are closed under complement and so the complement of L_1 is regular.

2.46 (c)

$L_1 = \{0^n 1^n \mid n \geq 0\}$ is DCFL
$L_2 = \{wcw^r \mid w \in \{0, 1\}^*\}$ is DCFL
$L_3 = \{ww^r \mid w \in \{0, 1\}^*\}$ is CFL but not DCFL

2.47 (b)

The language accepted is
L(M) = $\{wx\overline{w}^R \mid x \in \{0, 1, \varepsilon\}$, where $\overline{w}$ is the 1's complement of $w\}$.

Only the string "101100 10010[" is in the language and hence will be accepted.

The following diagram shows how this string will be accepted by the PDA.

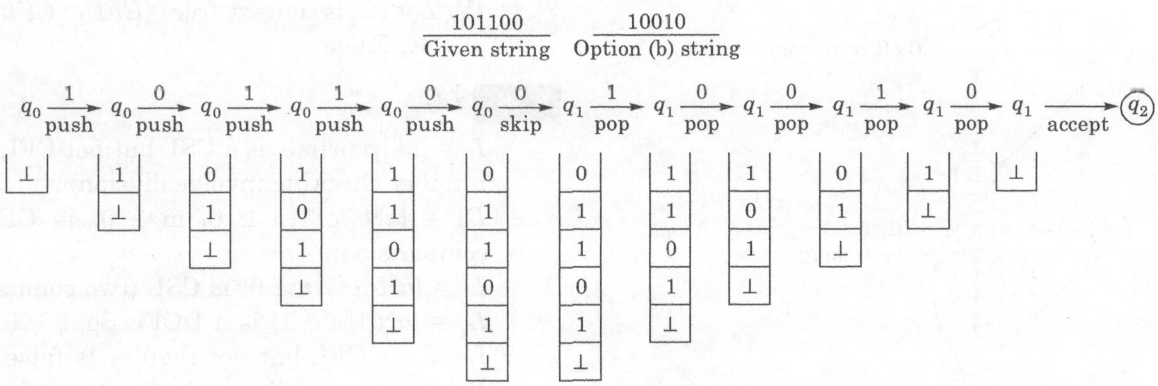

2.48 (b)

$L_1 = \{a^m b^n a^n b^m \mid m, n \geq 1\}$ is CFL (push "a" and "b" and then first "a" pops "b" and then "b" pops "a").

$L_2 = \{a^m b^n a^m b^n \mid m, n \geq 1\}$ is non-CFL (Alternate comparison not possible in a PDA)

$L_3 = \{a^m b^n \mid m = 2n + 1\}$ is CFL (Single comparison is possible in a PDA)

2.49 (d)

G_1: $S \to aS \mid B$
$B \to b \mid bB$

G_2: $S \to aA \mid bB$
$A \to aA \mid B \mid \varepsilon$
$B \to bB \mid \varepsilon$

G: $B \to b \mid bB \Rightarrow B \to b^+$

Now substitute in $S \to aS \mid B$

We get $S \to aS \mid b^+ \Rightarrow S \to a^* b^+$

So, $L(G_1) = \{a^m b^n \mid m \geq 0$ and $n > 0\}$

$G_2 = B \to bB \mid \varepsilon \Rightarrow B \to b^*$

Substitute in $A \to aA \mid B \mid \varepsilon \Rightarrow A \to aA \mid b^* \mid \varepsilon$
$\Rightarrow A \to aA \mid b^*$
$\Rightarrow A \to a^* b^*$

Now substitue A and B in $S \to aA \mid bB$
$\Rightarrow S \to aa^* b^* \mid bb^*$

$S \to aa^* b^* + bb^*$

So $L(G_2) = \{a^m b^n \mid m > 0$ or $n > 0\}$

So correct answer is choice (d).

2.50 (d)

The first state accepts only a's.
The third state accepts $\{a^n b^n \mid n \geq 0\}$
So, $L = \{a^n \mid n \geq 0\} \cup \{a^n b^n \mid n \geq 0\}$
It is a DCFL.

2.51 (b)

L_1 has only one comparisons and hence is CFL, but L_2 has 2 comparisons and hence not a CFL, but a CSL. So answer (b) is correct.

2.52 (c)

$G_1 : S_1 \to aS_1b \mid \varepsilon$

$G_2 : S_2 \to abS_2 \mid \varepsilon$

$$L_1 = L(G_1) = \{a^nb^n \mid n \geq 0\}$$

$$L_2 = L(G_2) = \{(ab)^n \mid n \geq 0\} = (ab)^*$$

Clearly L_1 is regular is false of L_2 is regular is True. So correct answer is (c).

2.53 (b)

Draw left recussion variable dependency graph

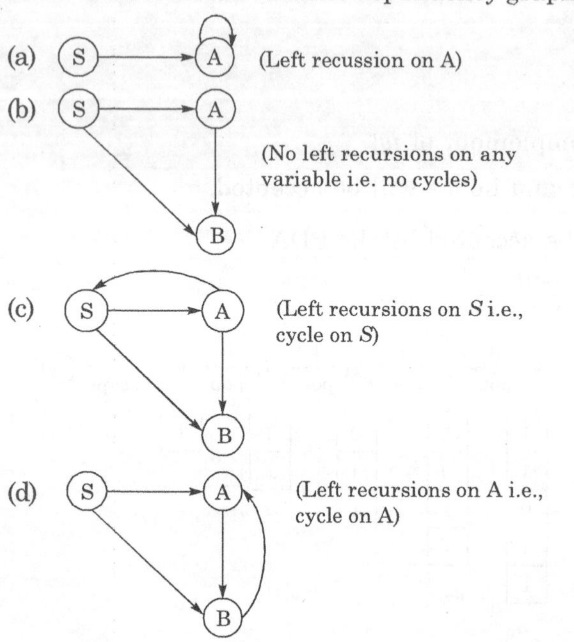

So only (b) is free from left recursion on any variable.

2.54 (b)

$S \to abScT \mid abcT$

$T \to bT \mid b$

Solving $T \to bb^*$ will gives b^+.

Substitute in S to get

$S \to abScb^+ \mid abcbb^+$

So, solution of S would be $\{(ab)^n\,(cb^+)^n \mid n \geq 1\}$ which is nothing but $\{(ab)^n\, cb^{m_1} cb^{m_2} \ldots cb^{m_n} \mid n, m_1, m_2 \ldots m_n \geq 1\}$

2.55 (a)

L_1 is a CFL

L_2 is also a CFL.

Union of 2 CFL is always a CFL, but intersaction may or may not be.

So, I is clearly true.

Let us intersect and check II

$\{a^nb^nc^m\} \cap \{a^nb^mc^m\} = \{a^nb^nc^n\}$

which is clearly not a CFL but a CSL.

So option (a) which is I only, is the answer.

2.56 (c)

$S \to XY$

$X \to aX \mid a \Rightarrow X \to \{a^m \mid m \geq 1\}$

$Y \to aYb \mid \in \Rightarrow Y \to \{a^nb^n \mid n \geq 0\}$

$S \to XY \Rightarrow S \to \{a^mb^n \mid m > n,\ n \geq 0\}$

$m > n$ because at least 1 will be attached on left of a^nb^n.

2.57 (b)

I. $L_1 \cup L_2$ is context-free $= CFL \cup CFL = CFL$. So, True

II. $\bar{L}_1$ is context-free $= \overline{CFL} = CSL$ but not CFL. So, false

III. $L_1 - R$ is context-free $= CFL \cap \overline{\text{Regular}} = CFL$. So, True

IV. $L_1 \cap L_2$ is context-free $= CFL \cap CFL = CSL$. So, False

2.58 (d)

$L_1 = \{a^p \mid p \text{ prime}\}$ is a CSL but not CFL (prime number checking involve divsion)

$L_2 = \{a^nb^mc^{2m} \mid n \geq 0,\ m \geq 0\}$ is CFL (one comparision)

$L_3 = \{a^nb^nc^{2n} \mid n \geq 0\}$ is CSL (two comparision)

$L_4 = \{a^nb^n \mid n \geq 1\}$ is a DCFL So,

I. L_1 is CFL but not regular is false.

II. L_2 is not CFL is false.

III. L_3 is not CFL but recursive is true since every CSL is recursive.

IV. L_4 is DCFL is true.

So, only III and IV are true and correct.

2.59 (b)

I. $\{a^mb^nc^pd^q \mid m + p = n + q\}$ is clearly CFL since, we can rearrange the equation as $m - n + p - q = 0$ which can be done by push, pop, push and pop and check if stack is empty at end.

II. $\{a^mb^nc^pd^q \mid m = n \text{ and } p = q\}$ is clearly CFL since, one comparison at a time can be done by pda.

III. $\{a^mb^nc^pd^q \mid m = n = p \text{ and } p \neq q\}$ is not CFL since $m = n = p$ is a double comparison which cannot be done by PDA.

IV. $\{a^mb^nc^pd^q \mid mn = p + q\}$ is not a CFL, since mn involves multiplying number of a's and number b's which cannot be done by a PDA.

So, only I and II are CFL's.

2.60 (b)

(a) $\{a^n b^i \mid i \in \{n, 3n, 5n\}, n \geq 0\}$
$= a^n b^n \cup a^n b^{3n} \cup a^n b^{5n}$ is CFL since each of the three parts is a CFL and closure under union guarantees that result also is a CFL.

(b) $\{w a^n w^R b^n \mid w \in \{a, b\}^*, n \geq 0\}$ is a not CFL because after pushing w, we need to push a's into stack which will stop the w from being matched with w^R. If we don't push a's after w, than later we cannot match with b^n. So this language is not acceptable by a PDA and hence not a CFL.

(c) $\{ww^R \mid w \in \{a, b\}^*\}$ is a CFL

(d) $\{w a^n b^n w^R \mid w \in \{a, b\}^*, n \geq 0\}$ is a CFL, since we can first push w, then a's, b's pop with a's and w^R pops with the w. So PDA can accept the language.

2.61 (a)

- Statement I: L in DCFL is true, since $\{a^n\} \cup \{a^n b^n\}$ = regular $\cup$ DCFL = DCFL, by closure property. So II is false and I is true.
- Statement III is true because we cannot write LL(k) grammar, for any value of k, since no matter how many a's are shown to compiler it will be impossible to distinguish between whether the string presented is in the form of $\{a^n\}$ or $\{a^n b^n\}$ and hence the production cannot be chosen uniquely for any value of k.

  So option (a) is correct, I and III only are true.

2.62 (a)

$L_1 = \{wxyz \mid w, x, y \in (0 + 1)^+\}$ is regular because by putting x as "0" and then "1" we can create a subset $w0y0 + w1y1$

$r = (0 + 1)^+ 0 (0 + 1)^+ 0 + (0 + 1)^+ 1 (0 + 1)^+ 1$

Which can be shown to cover all other strings obtained by putting x as "00", "01", "10" and "11" etc.

Hence we can write regular expression r to represent given language "L_1" and hence it is regular.

L_2 has strings which can be broken into two parts x and y with not only equal length but also $x \neq y$. The $x \neq y$ involves, string matching which FA cannot do and hence L_2 is not regular.

It is however, context free because even length strings always have $|x| = |y|$. So only one strings matching required and hence CFL.

2.63 (50)

The machine is pushing a "A" for every 'a' in input and popping a "A" for every 'b' in input which follows a's. No a's are allowed after the b's and also at end of input, the stack must have at least one "A" for string to be accepted, which means number of a's > number of b's.

So language accepted is

$$L = \{a^m b^n \mid m > n \geq 0\}$$

Now strings of length 100 which satisfy above condition are $\{a^{100}, a^{99}b, a^{98}b^2, ..., a^{51}b^{48}\}$.

Number of such strings is therefore 50.

2.64 (c)

$L_1 \rightarrow$ Reg, $L_2 \rightarrow$ Context-free

(a) $L_1 \cdot L_2$ = Reg $\cdot$ CFL = CFL $\cdot$ CFL = CFL

(b) $L_1 \cup L_2$ = Reg $\cup$ CFL = CFL

(c) $L_1 - L_2 = L_1 \cap \bar{L}_2$
$= \text{Reg} \cap \overline{\text{CFL}} = \text{Reg} \cap \overline{\text{CSL}}$
$= \text{Reg} \cap \text{CSL} = \text{CSL}$

(d) $L_1 \cap L_2$ = Reg $\cap$ CFL = CFL

So (c) is not necessarily CFL.

2.65 (b, c, d)

Option (a) : $\{wxw^Rx^R \mid w, x \in \{0, 1\}^*\}$

By putting w as "$\in$" we will get $\{xx^R \mid x \in \{0, 1\}^*\}$ which still has string matching. So, this will not be regular. Similarly, by putting x as $\in$ it will be $\{ww^R \mid w \in \{0, 1\}^*\}$ which still has string matching and will not become regular. So, we need to do string matching but alternate order string matching is not possible in PDA. So, it is a CSL. Option (a) is a CSL but not CFL.

Option (b) : $L = \{wxw^R \mid w, x \in \{0, 1\}^*\}$

By putting w as "$\in$" we get $\{x \mid x \in \{0, 1\}^*\} = (0 + 1)^*$

Since a subset of L is $(0 + 1)^*$, L itself must be $(0 + 1)^*$ which is regular and hence CFL. Option (b) is a CFL.

Option (c) : $\{ww^Rxx^R \mid w, x \in \{0, 1\}^*\}$

Here by putting x or w as $\in$, we cannot remove string matching. So, it is not regular. But it is CFL since in a NPDA we can push w, pop for w^R match it and then push x and pop for x^R and match it again and so this language is a CFL.

Option (d) : $\{wxx^Rw^R \mid w, x \in \{0, 1\}^*\}$

Here, also by putting w or x as $\in$, we cannot make it regular. NPDA can do this, push both w and x and then x^R pop and w^R pop and match. By push, push, pop, pop this can be accepted by NPDA. So, option (d) is CFL.

2.66 **(b, c, d)**

$L_1 \rightarrow$ Regular, $L_2 \rightarrow$ CFL

1. $L_1 \cap L_2$ = Reg $\cap$ CFL = Reg $\cap$ $\overline{\text{CSL}}$
 = Reg $\cap$ CSL
 = CSL (need not be CFL)
2. $\overline{\overline{L_1} \cup \overline{L_2}} = L_1 \cap L_2$ = Reg $\cap$ CFL = CFL
3. $L_1 \cup (L_2 \cup \overline{L_2}) = L_1 \cup \Sigma^* = \Sigma^*$
 = Regular and hence CFL
4. $(L_1 \cap L_2) \cup (\overline{L_1} \cap L_2) = (L_1 \cup \overline{L_1}) \cap L_2$
 $= \Sigma^* \cap L_2$
 $= L_2$ = CFL

2.67 **(a, b, c)**

$$L_1 = \{a^n wa^n \mid w \in \{a, b\}^*\}$$
$$L_2 = \{wxw^R \mid w, x \in \{a, b\}^*, \ |w|, |x| > 0\}$$

- L_1 is regular because by putting $n = 0$, we create a subset $\{w \mid w \in \{a, b\}^*\}$ which contains all possible strings. So if subset of L_1 is $(a + b)^*$, then $L_1 = (a + b)^*$.
- L_2 is regular because by putting w as "a" and "b" we get a regular expression $a(a + b)^+a + b(a + b)^+b$, which covers all other string which can be obtained by putting w as "aa", "ab", "ba", "bb", etc.

So, $L_2 = a(a + b)^+a + b(a + b)^+b$, which is regular.
So L_1 and L_2 both are regular.
Now every regular is also context-free.
So, option (a), (b), (c) are all true and option (d) is false.

2.68 **(b, c, d)**

$$L_1 = \{ww \mid w \in \{a, b\}^*\}$$
$$L_2 = \{a^nb^nc^m \mid m, n \geq 0\}$$
$$L_3 = \{a^mb^nc^n \mid m, n \geq 0\}$$

- L_1 is not context free because it has string matching in straight order, which PDA cannot do.
- L_2 and L_3 are clearly DCFL's, since they have only one comparison and DPDA can accept both.

(a) is therefore true.
(b) is false since L_2 is DCFL and every DCFL is a CFL.
(c) is false because $L_2 \cap L_3 = \{a^nb^nc^n \mid n \geq 0\}$ is not a CFL.
(d) is false because complement of "ww" has CFG and is CFL.

2.69 **(a, c, d)**

(a) Regular languages are closed under intersection.
(b) $L_1 = \{a^n b^n c^m \mid n, m \geq 1\}$ and $L_2 = \{a^n b^m c^m \mid n, m \geq 1\}$
 $L_1 \cap L_2 = \{a^n b^n c^n \mid n \geq 1\}$
 which is not context free.
(c) Recursive closed under intersection.
(d) REL are closed under intersection.

2.70 **(b)**

$S \rightarrow aSb \mid X$
$X \rightarrow aX \mid Xb \mid a \mid b$
$S \rightarrow a\underline{S}b$
$\overset{*}{\Rightarrow} a^n \underline{S} b^n \mid n \geq 0$
$\Rightarrow a^n X b^n$
$\Rightarrow a^n a X b^n$
$\overset{*}{\Rightarrow} a^n a^* \underline{X} b^n$
$\Rightarrow a^n a^* \underline{X} bb^n$
$\overset{*}{\Rightarrow} a^n a^* \underline{X} b^* b^n$
$\Rightarrow a^n a^* (a + b) b^* b^n$
which is equal to $a^*(a + b)b^*$.

2.71 **(a)**

From the given diagram it is clear that the starting symbol in the input string must be 'a' and on every 'a' it pushes 'A' into the stack. If the input string becomes empty it goes to the state q and it empties the stack. This means the PDA accepts the language

$$L = \{a^n \mid n \geq 1\}$$

If after a's in the input string if b's comes then for every 'b' it will delete 'A' from the stack. When the string becomes empty iff the topmost symbol in the stack in 'A' then the PDA empties the stack. This means the PDA accepts the language

$$L = \{a^m b^n \mid n < m\}$$

$\therefore$ The language accepted by the PDA is

$$L = \{a^m \mid m \geq 1\} \cup \{a^m b^n \mid n < m\}$$

or

$$L = \{a^m b^n \mid 1 \leq m \text{ and } n < m\}$$

2.72 **(b, c)**

S $\rightarrow$ aS, S $\rightarrow$ aSbS, S $\rightarrow$ c
S $\rightarrow$ aS
 $\rightarrow$ ac
S $\rightarrow$ a$\underline{\text{S}}$bS
 $\rightarrow$ aa$\underline{\text{S}}$b$\underline{\text{S}}$b$\underline{\text{S}}$
 $\rightarrow$ aacbcbc

By verification process

$$n_c(w) = n_b(w) + 1$$

3 Turing Machine: RE, REC and Undecidability

3.1 Which of the following statements is false?
(a) The halting problem for Turing machine is undecidable
(b) Determining whether a context free grammar is ambiguous is undecidable
(c) Given two arbitrary context free grammars G_1 and G_2, it is undecidable whether $L(G_1) = L(G_2)$
(d) Given two regular grammars G_1 and G_2, it is undecidable whether $L(G_1) = L(G_2)$
[1996 : 1 M]

3.2 Which one of the following is not decidable?
(a) Given a Turing machine M, a strings s and an integer k, M accepts s within k steps
(b) Equivalence of two Turing machines
(c) Language accepted by a given finite state machine is non empty
(d) Language accepted by a context free grammar is non empty **[1997 : 2 M]**

3.3 Regarding the power of recognition of languages, which of the following statements is false?
(a) The non-deterministic finite-state automata are equivalent to deterministic finite-state automata
(b) Non-deterministic push-down automata are equivalent to deterministic push-down automata
(c) Non-deterministic Turning machines are equivalent to deterministic Turning machines
(d) Multi-tape Turning machines are equivalent to Single-tape Turning machines.
[1998 : 1 M]

3.4 Which of the following is true?
(a) The complement of a recursive language is recursive
(b) The complement of a recursively enumerable language is recursively enumerable
(c) The complement of a recursively enumerable language is either recursive or recursively enumerable
(d) The complement of a context-free language is context-free **[2002 : 2 M]**

3.5 The C language is
(a) A context free language
(b) A context sensitive language
(c) A regular language
(d) Parsable fully only by a Turing machine
[2002 : 2 M]

3.6 Consider the following decision problems:
P_1: Does a given finite state machine accept a given string
P_2: Does a given context free grammar generate an infinite number of strings.
Which of the following statement is true?
(a) P_1 and P_2 are decidable
(b) Neither P_1 nor P_2 are decidable
(c) Only P_1 is decidable
(d) Only P_2 is decidable **[2002 : 2 M]**

3.7 Consider the following Problem X.
"Given a Turing machine M over the input alphabet Σ, any state q of M and a word Σ^*, does the computation of M on w visit the state q"
Which of the following statements about X is correct?
(a) X is decidable
(b) X is undecidable but partially decidable
(c) X is undecidable and not even partially decidable
(d) X is not a decision problem **[2002 : 2 M]**

3.8 Nobody knows yet if P = NP. Consider the language L defined as follows

$$L = \begin{cases} (0+1)^* \text{ if } P = NP \\ \phi \text{ otherwise} \end{cases}$$

Which of the following statements is true?
(a) L is recursive
(b) L is recursively enumerable but not recursive
(c) L is not recursively enumerable
(d) Whether L is recursive or not will be known after we find out if P = NP **[2003 : 1 M]**

3.9 If the strings of a language L can be effectively enumerated in lexicographic (i.e., alphabetic) order, which of the following statements is true?
(a) L is necessarily finite
(b) L is regular but not necessarily finite
(c) L is context free but not necessarily regular
(d) L is recursive but not necessarily context free
[2003 : 1 M]

3.10 A single tape Turing Machine M has two states q_0 and q_1, of which q_0 is the starting state. The tape alphabet of M is {0, 1, B} and its input alphabet is {0, 1}. The symbol B is the blank symbol used to indicate end of an input string. The transition function of M is described in the following table:

	0	1	B
q_0	q_1, 1, R	q_1, 1, R	Halt
q_1	q_1, 1, R	q_0, 1, L	q_0, B, L

The table is interpreted as illustrated below. The entry (q_1, 1, R) in row q_0 and column 1 signifies that if M is in state q_0 and reads 1 on the current tape square, then it writes 1 on the same tape square, moves its tape head one position to the right and transitions to state q_1. Which of the following statements is true about M?

(a) M does not halt on any string in $(0 + 1)^+$
(b) M does not halt on any string in $(00 + 1)^*$
(c) M halts on all string ending in a 0
(d) M halts on all string ending in a 1

[2003 : 2 M]

3.11 Consider three decision problems P_1, P_2 and P_3. It is known that P_1 is decidable and P_2 is undecidable. Which one of the following is TRUE?

(a) P_3 is decidable if P_1 is reducible to P_3
(b) P_3 is undecidable if P_3 is reducible to P_2
(c) P_3 is undecidable if P_2 is reducible to P_3
(d) P_3 is decidable if P_3 is reducible to P_2's complement

[2005 : 2 M]

3.12 Let L_1 be a recursive language, and let L_2 be a recursively enumerable but not a recursive language. Which one of the following is TRUE?

(a) $\overline{L_1}$ is recursive and $\overline{L_2}$ is recursively enumerable.
(b) $\overline{L_1}$ is recursive and $\overline{L_2}$ is not recursively enumerable.
(c) $\overline{L_1}$ and $\overline{L_2}$ are recursively enumerable.
(d) $\overline{L_1}$ is recursively enumerable and $\overline{L_2}$ is recursive.

[2005 : 2 M]

3.13 For $s \in (0 + 1)^*$ let d(s) denote the decimal value of s (e.g. d (101) = 5).
Let $L = \{s \in (0 + 1)^* \mid d(s) \bmod 5 = 2 \text{ and } d(s) \bmod 7 \neq 4\}$
Which one of the following statements is true?

(a) L is recursively enumerable, but not recursive
(b) L is recursive, but not context-free
(c) L is context-free, but not regular
(d) L is regular

[2006 : 2 M]

3.14 Let L_1 be regular language, L_2 be a deterministic context-free language and L_3 a recursively enumerable, but not recursive, language. Which one of the following statements is false?

(a) $L_1 \cap L_2$ is a deterministic CFL
(b) $L_3 \cap L_1$ is recursive
(c) $L_1 \cup L_2$ is context free
(d) $L_1 \cap L_2 \cap L_3$ is recursively enumerable

[2006 : 2 M]

3.15 Which of the following problems is undecidable?

(a) Membership problem for CFGs
(b) Ambiguity problem for CFGs
(c) Finiteness problem for FSAs
(d) Equivalence problem for FSAs

[2007 : 1 M]

3.16 Which of the following is true for the language $\{a^p \mid p \text{ is a prime}\}$?

(a) It is not accepted by a Turning Machine
(b) It is regular but not context-free
(c) It is context-free but not regular
(d) It is neither regular nor context-free, but accepted by a Turing machine

[2008 : 1 M]

3.17 If L and $\overline{L}$ are recursively enumerable then L is

(a) regular (b) context-free
(c) context-sensitive (d) recursive

[2008 : 1 M]

3.18 Which of the following are decidable?

1. Whether the intersection of two regular languages is infinite
2. Whether a given context-free language is regular
3. Whether two push-down automata accept the same language
4. Whether a given grammar is context-free

(a) 1 and 2 (b) 1 and 4
(c) 2 and 3 (d) 2 and 4

[2008 : 1 M]

3.19 Let L_1 be a recursive language. Let L_2 and L_3 be language that are recursively enumerable but not recursive. Which of the following statements is not necessarily true?

(a) L_2-L_1 is recursively enumerable
(b) L_1-L_3 is recursively enumerable
(c) $L_2 \cap L_3$ is recursively enumerable
(d) $L_2 \cup L_3$ is recursively enumerable

[2010 : 1 M]

3.20 Which of the following pairs have **DIFFERENT** expressive power?
(a) Deterministic finite automata (DFA) and Non-deterministic finite automata (NFA)
(b) Deterministic push down automata (DPDA) and Non-deterministic push down automata (NPDA)
(c) Deterministic single-tape Turing machine and Non-deterministic single-tape Turing machine
(d) Single-tape Turing machine and multi-tape Turing machine **[2011 : 1 M]**

3.21 Which of the following problems are decidable?
1. Does a given program ever produce an output?
2. If L is a context-free language, then, is $\bar{L}$ also context-free?
3. If L is a regular language, then, is $\bar{L}$ also regular
4. If L is a recursive language, then, is $\bar{L}$ also recursive?

(a) 1, 2, 3, 4 (b) 1, 2
(c) 2, 3, 4 (d) 3, 4 **[2012 : 1 M]**

3.22 Which of the following statements is/are FALSE?
1. For every non-deterministic Turing machine, there exists an equivalent deterministic Turing machine.
2. Turing recognizable languages are closed under union and complementation.
3. Turing decidable languages are closed under intersection and complementation.
4. Turing recognizable languages are closed under union and intersection.

(a) 1 and 4 only
(b) 1 and 3 only
(c) 2 only
(d) 3 only **[2013 : 1 M]**

3.23 Which of the following is/are undecidable?
1. G is CFG. Is $L(G) = \phi$?
2. G is a CFG. Is $L(G) = \Sigma^*$?
3. M is a Turing machine. Is L(M) regular?
4. A is a DFA and N is an NFA. Is L(A) = L(N)?

(a) 3 only (b) 3 and 4 only
(c) 1, 2 and 3 only (d) 2 and 3 only
[2013 : 2 M]

3.24 Let L be a language and $\bar{L}$ be its complement. Which one of the following is NOT a viable possibility?
(a) Neither L nor $\bar{L}$ is recursively enumerable (r.e.).
(b) One of L and $\bar{L}$ is r.e. but not recursive; the other is not r.e.
(c) Both L and $\bar{L}$ are r.e. but not recursive.
(d) Both L and $\bar{L}$ are recursive.
[2014 (Set-1) : 2 M]

3.25 Let <M> be the encoding of a Turing machine as a string over $\Sigma = \{0,1\}$. Let $L = \{\langle M \rangle | M$ is a Turing machine that accepts a string of length 2014}. Then, L is
(a) decidable and recursively enumerable
(b) undecidable but recursively enumerable
(c) undecidable and not recursively enumerable
(d) decidable but not recursively enumerable
[2014 (Set-2) : 2 M]

3.26 Let $A \leq_m B$ denotes that language A is mapping reducible (also known as many-to-one reducible) to language B. Which one of the following is FALSE?
(a) If $A \leq_m B$ and B is recursive then A is recursive.
(b) If $A \leq_m B$ and A is undecidable then B is undecidable.
(c) If $A \leq_m B$ and B is recursively enumerable then A is recursively enumerable.
(d) If $A \leq_m B$ and B is not recursively enumerable then A is not recursively enumerable.
[2014 (Set-2) : 1 M]

3.27 Let Σ be a finite non-empty alphabet and let 2^{Σ^*} be the power set of Σ^*. Which one of the following is TRUE?
(a) Both 2^{Σ^*} and Σ^* are countable
(b) 2^{Σ^*} is countable and Σ^* is uncountable
(c) 2^{Σ^*} is uncountable and Σ^* is countable
(d) Both 2^{Σ^*} and Σ^* are uncountable
[2014 (Set-3) : 1 M]

3.28 Which one of the following problems is undecidable?
(a) Deciding if a given context-free grammar is ambiguous.
(b) Deciding if a given string is generated by a given context-free grammar.
(c) Deciding if the language generated by a given context-free grammar is empty.
(d) Deciding if the language generated by a given context-free grammar is finite.
[2014 (Set-3) : 2 M]

3.29 For any two languages L_1 and L_2 such that L_1 is context free and L_2 is recursively enumerable but not recursive, which of the following is/are necessarily true?

1. $\bar{L}_1$ (complement of L_1) is recursive
2. $\bar{L}_2$ (complement of L_2) is recursive
3. $\bar{L}_1$ is context-free
4. $\bar{L}_1 \cup L_2$ is recursively enumerable

(a) 1 only (b) 3 only
(c) 3 and 4 only (d) 1 and 4 only

[2015 (Set-1) : 1 M]

3.30 Consider the following statements:

1. The complement of every Turning decidable language is Turning decidable.
2. There exists some language which is in NP but is not Turing decidable.
3. If L is a language in NP, L is Turing decidable.

Which of the above statements is/are True?

(a) Only 2 (b) Only 3
(c) Only 1 and 2 (d) Only 1 and 3

[2015 (Set-2) : 1 M]

3.31 Let X be a recursive language and Y be a recursively enumerable but not recursive language. Let W and Z be two languages such that $\bar{Y}$ reduces to W, and Z reduces to $\bar{X}$ (reduction means the standard many-one reduction). Which one of the following statements is TRUE?

(a) W can be recursively enumerable and Z is recursive.
(b) W can be recursive and Z is recursively enumerable.
(c) W is not recursively enumerable and Z is recursive.
(d) W is not recursively enumerable and Z is not recursive.

[2016 (Set-1) : 2 M]

3.32 Consider the following types of languages L_1: Regular, L_2: Context-free, L_3: Recursive, L_4: Recursively enumerable.

Which of the following is/are **TRUE**?

I. $\overline{L_3} \cup L_4$ is recursively enumerable.
II. $\overline{L_2} \cup L_3$ is recursive.
III. $L_1^* \cap L_2$ is context-free.
IV. $L_1 \cup \overline{L_2}$ is context-free.

(a) I only (b) I and III only
(c) I and IV only (d) I, II and III only

[2016 (Set-2) : 1 M]

3.33 Consider the following languages:

- $L_1 = \{\langle M\rangle \mid M$ takes at least 2016 steps on some input$\}$
- $L_2 = \{\langle M\rangle \mid M$ takes at least 2016 steps on all inputs$\}$ and
- $L_3 = \{\langle M\rangle \mid M$ accepts $\varepsilon\}$

where for each turning machine M, $\langle M\rangle$ denotes a specific encoding of M. Which one of the following is **TRUE**?

(a) L_1 is recursive and L_2, L_3 are not recursive
(b) L_2 is recursive and L_1, L_3 are not recursive
(c) L_1, L_2 are recursive and L_3 is not recursive
(d) L_1, L_2, L_3 are recursive

[2016 (Set-2) : 2 M]

3.34 Let $L(R)$ be the language represented by regular expression R. Let $L(G)$ be the language generated by a context free grammar G. Let $L(M)$ be the language accepted by a Turing machine M.

Which of the following decision problems are undecidable?

I. Given a regular expression R and a string w, is $w \in L(R)$?
II. Given a context-free grammar G, is $L(G) = \phi$?
III. Given a context-free grammar G, is $L(G) = \Sigma^*$ for some alphabet Σ?
IV. Given a Turing machine M and a string w, is $w \in L(M)$?

(a) I and IV only
(b) II and III only
(c) II, III and IV only
(d) III and IV only

[2017 (Set-2) : 2 M]

3.35 Let A and B be finite alphabets and let # be a symbol outside both A and B. Let f be a total function from A^* to B^*. We say f is computable if there exists a Turing machine M which given an input x in A^*, always halts with $f(x)$ on its tape. Let L_f denote the language $\{x \# f(x) \mid x \in A^*\}$. Which of the following statements is true:

(a) f is computable if and only if L_f is recursive.
(b) f is computable if and only if L_f is recursively enumerable.
(c) If f is computable then L_f is recursive, but not conversely.
(d) If f is computable then L_f is recursively enumerable, but not conversely.

[2017 (Set-1) : 2 M]

3.36 The set of all recursively enumerable languages is
(a) closed under complementation
(b) closed under intersection
(c) a subset of the set of all recursive languages
(d) an uncountable set **[2018 : 1 M]**

3.37 Consider the following problems L(G) denotes the language generated by a grammar G. L(M) denotes the language accepted by a machine M.
I. For an unrestricted grammar G and a string w, whether $w \in L(G)$.
II. Given a Turing Machine M, whether L(M) is regular.
III. Given two grammars G_1 and G_2, whether $L(G_1) = L(G_2)$.
IV. Given an NFA N, whether there is a deterministic PDA P such that N and P accept the same language.

Which one of the following statements is correct?
(a) Only I and II are undecidable
(b) Only III is undecidable
(c) Only II and IV are undecidable
(d) Only I, II and III are undecidable
[2018 : 2 M]

3.38 Consider the following sets:
S_1: Set of all recursively enumerable languages over the alphabet {0, 1}.
S_2: Set of all syntactically valid C programs.
S_3: Set of all languages over the alphabet {0, 1}.
S_4: Set of all non-regular languages over the alphabet {0, 1}.

Which of the above sets are uncountable?
(a) S_1 and S_2 (b) S_3 and S_4
(c) S_1 and S_4 (d) S_2 and S_3
[2019 : 2 M]

3.39 Which of the following languages are undecidable? Note that ⟨M⟩ indicates encoding of the Turing machine M.
$L_1 = \{\langle M\rangle \mid L(M) = \phi]$
$L_2 = \{\langle M, w, q\rangle \mid$ M on input w reaches state q in exactly 100 steps}
$L_3 = \{\langle M\rangle \mid$ L(M) is not recursive)
$L_4 = \{\langle M\rangle \mid$ L(M) contains at least 21 members)
(a) L_2, L_3 and L_4 only
(b) L_1, L_3 and L_4 only
(c) L_1 and L_3 only
(d) L_2 and L_3 only **[2020 : 2 M]**

3.40 For a Turing machine M, $\langle M\rangle$ denotes an encoding of M. Consider the following two languages:
$L_1 = \{\langle M\rangle \mid M$ takes more than 2021 steps on all inputs}
$L_2 = \{\langle M\rangle \mid M$ takes more than 2021 steps on some input}
Which one of the following options is correct?
(a) Both L_1 and L_2 are decidable.
(b) L_1 is undecidable and L_2 is decidable.
(c) Both L_1 and L_2 are undecidable.
(d) L_1 is decidable and L_2 is undecidable.
[2021 (Set-1) : 2 M]

3.41 Let $\langle M\rangle$ denote an encoding of an automaton M. Suppose that $\Sigma\{0, 1\}$. Which of the following languages is/are NOT recursive?
(a) $L = \{\langle M\rangle \mid M$ is a DFA such that $L(M) = \Sigma^*\}$
(b) $L = \{\langle M\rangle \mid M$ is a PDA such that $L(M) = \Sigma^*\}$
(c) $L = \{\langle M\rangle \mid M$ is a PDA such that $L(M) = \phi\}$
(d) $L = \{\langle M\rangle \mid M$ is a DFA such that $L(M) = \phi\}$
[2021 (Set-1) : 1 M]

3.42 Consider the following two statements about regular languages:
S_1 : Every infinite regular language contains an undecidable language as a subset.
S_2 : Every finite language is regular.
Which one of the following choices is correct?
(a) Both S_1 and S_2 are true
(b) Only S_2 is true
(c) Only S_1 is true
(d) Neither S_1 nor S_2 is true
[2021 (Set-2) : 2 M]

3.43 Suppose we want to design a synchronous circuit that processes a string of 0's and 1's. Given a string, it produces another string by replacing the first 1 in any subsequence of consecutive 1's by a 0. Consider the following example.
Input sequence : 00100011000011100
Output sequence : 00000001000001100
A Mealy Machine is a state machine where both the next state and the output are functions of the present state and the current input.
The above mentioned circuit can be designed as a two-state Mealy machine. The states in the Mealy machine can be represented using Boolean values 0 and 1. We denote the current state, the next state, the next incoming bit, and the output bit of the Mealy machine by the variables s, t, b and y respectively.
Assume the initial state of the Mealy machine is 0.
What are the Boolean expressions corresponding to t and y in terms of s and b?

(a) $t = b$
$y = s\bar{b}$

(b) $t = s + b$
$y = sb$

(c) $t = s + b$
$y = s\bar{b}$

(d) $t = b$
$y = sb$

[2021 (Set-2) : 2 M]

3.44 Which of the following statements is/are TRUE?

(a) Every subset of a recursively enumerable language is recursive.

(b) If a language L and its complement $\bar{L}$ are both recursively enumerable, then L must be recursive.

(c) Complement of a context-free language must be recursive.

(d) If L_1 and L_2 are regular, then $L_1 \cap L_2$ must be deterministic context-free.

[2022 : 1 M]

3.45 Which of the following is/are undecidable?

(a) Given two Turing machines M_1 and M_2, decide if $L(M_1) = L(M_2)$.

(b) Given a Turing machine M, decide if $L(M)$ is regular.

(c) Given a Turing machine M, decide if M accepts all strings.

(d) Given a Turing machine M, decide if M takes more than 1073 steps on every string.

[2022 : 2 M]

3.46 Let $G = (V, \Sigma, S, P)$ be a context-free grammar in Chomsky Normal Form with $\Sigma = \{a, b, c\}$ and V containing 10 variable symbols including the start symbol S. The string $w = a^{30}b^{30}c^{30}$ is derivable from S. The number of steps (application of rules) in the derivation $S \rightarrow^* w$ is ________.

[2024 (Set-1) : 2 M]

■■■■

Answers Turing Machine: RE, REC and Undecidability

3.1 (d)	**3.2** (b)	**3.3** (b)	**3.4** (a)	**3.5** (b)	**3.6** (a)	**3.7** (b)	**3.8** (a)	**3.9** (d)
3.10 (a)	**3.11** (c)	**3.12** (b)	**3.13** (d)	**3.14** (b)	**3.15** (b)	**3.16** (d)	**3.17** (d)	**3.18** (b)
3.19 (b)	**3.20** (b)	**3.21** (d)	**3.22** (c)	**3.23** (d)	**3.24** (c)	**3.25** (b)	**3.26** (d)	**3.27** (c)
3.28 (a)	**3.29** (d)	**3.30** (d)	**3.31** (c)	**3.32** (d)	**3.33** (c)	**3.34** (d)	**3.35** (a)	**3.36** (b)
3.37 (d)	**3.38** (b)	**3.39** (b)	**3.40** (a)	**3.41** (b)	**3.42** (a)	**3.43** (d)	**3.44** (b, c, d)	
3.45 (a, b, c)	**3.46** (179)							

Explanations Turing Machine: RE, REC and Undecidability

3.1 (d)

(a) True, the halting problem which states that: "Given a turing machine M and an input w, does M eventually halt when run with w or will run forever" is undecidable problem.

(b) True, an ambiguous grammar is a grammar for which there exists a string that can have more than one parse tree. Since there is no algorithm for finding whether context free grammar is ambiguous or not that why, this problem is undecidable.

(c) True, since equivalence problem for CFLs is undecidable.

(d) False, because there exist an algorithm for determining whether $L(G_1) = L(G_2)$ for any two regular grammars G_1 and G_2. So, equivalence of two regular grammars is a decidable problem.

3.2 (b)

(a) Given a Turing machine M, a strings S and an integer k, M accepts S within k steps is decidable problem, since the turing machine M can be simulated upto a finite number of steps in finite time on a universal turing machine (UTM).

(b) Equivalence of two turing machines is undecidable problem.

(c) Emptiness of regular language is decidable problem. So the non emptiness problem of regular language is also decidable, since the complement of a decidable (REC) problem is always decidable (REC).

(d) Emptiness of CFL's is decidable problem.

3.3 (b)

(a) Since there exist an algorithm to convert any NFA into equivalent DFA. So the recognition power of NFA and DFA is same.

(b) Since there exist no algorithm for converting every NPDA to DPDA and there exist some context free languages which are accepted by NPDA but not by DPA. So the recognition power of NPDA and DPDA is not same.
Example: $L = \{ww^R \mid w \in (a, b)^*\}$ is accepted by NPDA but not by DPDA.

(c) Nondeterministic turing machines are equivalent with deterministic Turing machines. So the recognition power of NTM is equal to the DTM.

(d) Mutlitape turing machine is a turing machine with several tapes. The multitape turing machine is equivalent to the standard turing machine which is a single tape Turing machine. So the recognition power of multitape and single tape turing machine is same.

3.4 (a)

The complement of recursive language is recursive because recursive languages are closed under complementation. RE and CFL are not closed under complementation.
So, options (b), (c) and (d) are false.

3.5 (b)

The C language is context sensitive language. All the feature of C language comes under CFL except following two which makes the C language as CSL:
(a) Declaration of variable before use
(b) The matching of actual and formal parameters.

3.6 (a)

P_1 is membership of FSM i.e., regular language.
P_2 is finite/infiniteness of context free grammar.
Membership of FSM and finiteness of CFG, both are decidable problems.
So P_1 and P_2 both are decidable.

3.7 (b)

Problem X is state entry problem. Halting problem of turing machine can be reduced to state entry problem. Since halting problem of turing machine is undecidable, the state-entry problem must also be undecidable. The language corresponding to state-entry problem is RE but not REC. So X is undecidable but partially decidable.

3.8 (a)

If P = NP then L = (0+1)* which is regular.
If P ≠ NP then L = ϕ which is also regular.
Either way L is regular and hence recursive.

3.9 (d)

Theorem: A language is recursive if and only if it can be effectively enumerated in lexicographic order.

Based on this theorem clearly the answer is (d).

3.10 (a)

Transition function of M:

	0	1	B
q_0	$q_1, 1, R$	$q_1, 1, R$	Halt
q_1	$q_1, 1, R$	$q_0, 1, L$	q_0, B, L

The graphic representation of M is as follows:

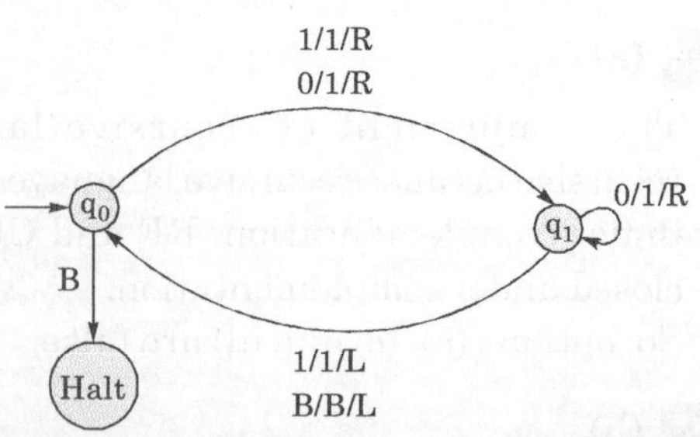

This machine hangs on every string in $(0 + 1)^+$. Example is shown below for the strings "00" and "01".

B00B ⟶ B10B ⟶ B11B ⟶ B11B ⟶ B11B ⟶ Hangs
q_0 q_1 q_1 q_0 q_1

B01B ⟶ B11B ⟶ B11B ⟶ B11B ⟶ Hangs
q_0 q_1 q_0 q_1

3.11 (c)

Note that in general if $L_1 \leq L_2$,
L_2 decidable ⇒ L_1 decidable and
L_1 undecidable ⇒ L_2 is undecidable.
Given that P_1 is decidable and P_2 is undecidable.

Consider (a) P_3 is decidable if P_1 is reducible to P_3 i.e. $P_1 \leq_p P_3$ and if P_1 is undecidable, then P_3 would be undecidable.
But it is given that P_1 is decidable, therefore we cannot use this theorem.
(a) is false.
Consider (b) P_3 is undecidable if P_3 is reducible to P_2
i.e. $P_3 \leq_p P_2$
Now if P_3 is undecidable then P_2 is undecidable but nothing is given regarding P_3.
Also if P_2 is decidable, then P_3 would be decidable, but it is given that P_2 is undecidable, so we can't use this.
(b) is false.
Consider (c) P_3 is undecidable if P_2 is reducible to P_3
i.e. $P_2 \leq_p P_3$
Now if P_2 is undecidable, this would mean P_3 is undecidable. Since it is given that P_2 is undecidable, therefore, surely P_3 is undecidable is correct. Choice (c) is correct.

3.12 (b)

Given that L_1 is REC and L_2 is RE but not REC.
Consider choice (a) $\overline{L_1}$ is REC and $\overline{L_2}$ is RE.
Since L_1 is REC, clearly $\overline{L_1}$ is also REC.
Since L_2 is RE but not REC, its complement has to be not RE.
∴ But (a) claims that $\overline{L_2}$ is RE.
Therefore (a) is false.
Consider (b) $\overline{L_1}$ is REC and $\overline{L_2}$ is not RE.
If L_1 is REC, then $\overline{L_1}$ is REC.
Also, if L_2 is RE but not REC, then $\overline{L_2}$ is not RE.
∴ (b) is true.
Clearly (c) and (d) false since $\overline{L_2}$ is not RE.

3.13 (d)

There is a DFA corresponding to D(S) mod 5 = 2 as well as D(S) mod 7 = 4. Therefore, both of them are regular languages. Let they be L_1 and L_2. The given language is $L_1 \cap L_2^c$. Since regular languages are closed under intersection and complementation, the given language is also regular.

3.14 (b)

Given L_1 be regular, L_2 is DCFL and L_3 is RE but not REC.

1. "$L_1 \cap L_2$ is a DCFL" is true since DCFL's are closed under regular intersection.

2. "$L_3 \cap L_1$ is recursive" is false since L_3 is RE but not REC and therefore $L_3 \cap L_1$ is surely RE according to closure of RE under regular intersection, but we cannot be sure that $L_3 \cap L_1$ is REC.
3. "$L_1 \cup L_2$ is CFL" is true, since L_2 is DCFL and hence a CFL and L_1 is regular. Therefore, by closure of CFL's on regular union, we can say that $L_1 \cup L_2$ is surely CFL.
4. "$L_1 \cap L_2 \cap L_3$ is RE" is true since all three are surely RE and RE languages are closed under intersection.

3.15 (b)

(a) Membership problem for CFG's is decidable (CYK algorithm exists).
(b) Ambiguity problem of CFG's is undecidable.
(c) Finiteness problem of FSA's is decidable, since there exists an algorithm to check if a given regular language L is finite or infinite.
(d) Equivalence problem for FSA's is decidable, since there exists an algorithm to check where $L(M_1) = L(M_2)$ or not.

3.16 (d)

Let $L = \{a^p \mid p \text{ is a prime}\}$. There is no DFA which recognizes L, since to check whether a number is prime or not, requires division to be performed, which neither FA nor PDA can do. It is a CSL and can be accepted by a turing machine.

3.17 (d)

There is a theorem which states that whenever L and $\overline{L}$ are both RE, then both will be REC.

3.18 (b)

1. "Intersection of two regular languages is infinite" is decidable, since we can construct a product dfa of the 2 given dfas and then, using the algorithm to check finiteness or infiniteness on this dfa, we can solve the problem.
2. "Whether a given CFL is regular" is undecidable.
3. "Whether two PDA's accept the same language" is undecidable, since equivalence of CFL's is undecidable.
4. "Whether a given grammar is context-free" is decidable since we can easily check using a TM, whether the LHS of every production has a single variable and, then it is a CFG, else it is not a CFG.

$\therefore$ 1 and 4 are decidable.

3.19 (b)

Given, L_1 is REC and L_2 and L_3 are RE but not REC.

Choice (a): $L_2 - L_1 = L_2 \cap L_1^C$
$= (\text{RE but not REC}) \cap (\text{REC})^C$
$= (\text{RE but not REC}) \cap (\text{REC})$
$= \text{RE} \cap \text{REC} = \text{RE} \cap \text{RE} = \text{RE}$

Choice (a) is true.

Choice (b): $L_1 - L_3 = L_1 \cap L_3^C$
$= \text{REC} \cap (\text{RE but not REC})^C$

Now the complement of a RE but not REC language has to be "not RE".

$\therefore$ $L_1 - L_3 = \text{REC} \cap \text{not RE} = \text{not RE}$

$\therefore$ Choice (b) is false.

Choice (c):
$L_2 \cap L_3 = (\text{RE but not REC}) \cap (\text{RE but not REC})$
$= \text{RE} \cap \text{RE} = \text{RE}$

So, choice (c) is true.

Choice (d):
$L_2 \cup L_3 = (\text{RE but not REC}) \cup (\text{RE but not REC})$
$= \text{RE} \cup \text{RE} = \text{RE}$

$\therefore$ Choice (d) is true.

3.20 (b)

DFA ≡ NFA
DPDA ≠ NPDA
DTM ≡ NTM
Single tape TM ≡ Multi-tape TM

3.21 (d)

Statement (1) is **undecidable** (since it is the halting problem of TM).
Complement of a context free language, may or may not be context free. So statement number (2) is not only a non trivial problem but is also **undecidable**. If L is regular, $\overline{L}$ is surely regular.
So statement number (3) is (trivially) **decidable**.
Also if L is recursive, $\overline{L}$ is also surely recursive.
So statement number (4) is also (trivially) **decidable**.
So, correct option is (d) i.e. only 3 and 4 are **decidable**.

3.22 (c)

1. Non-deterministic Tuning machine can be simulated into an equivalent deterministic Turing machine. So for every non deterministic Turing machine there exits an equivalent deterministic Turing machine. So true.
2. Turing recognizable languages (RE languages)

are not closed under complementation so false.

3. Turing decidable languages (REC languages) are closed under both inter-section and complementation. So, true.
4. Turing recognizable languages (RE languages) are closed under union and intersection. So true.

3.23 (d)

1. G is CFG, Is L(G) = ϕ?
   This is the emptiness problem of CFLs which is **decidable**.
2. G is CFG, Is L(G) = Σ^*
   This is the Kleene closedness problem of CFLs which is **undecidable**.
3. M is turning machine, Is L(M) regular.
   This is the regularity problem of REs which is **undecidable**.
4. A is a DFA and N is NFA, Is L(A) = L(N)?.
   This is the equivalence problem of regular languages which is **decidable**.

So, only 2 and 3 are **undecidable**.

3.24 (c)

Option (c): Both L and $\bar{L}$ are r.e. but not recursive, is not a viable possibility, because if both L and $\bar{L}$ were r.e. then both would become recursive due to the L, $\bar{L}$ theorem.
Option (a), (b) and (d) are allowed by the L, $\bar{L}$ theorem.

3.25 (b)

Whether a TM accepts a finite length string or not is a non trivial question and hence by Rice's theorem the problem is undecidable.
But the problem however falls in recursively enumerable since the condition given is a positive condition.

3.26 (d)

If A $\le_m$ B and B is RE $\rightarrow$ A is RE is a one way theorem, which is true.
Option (d): B is not RE $\rightarrow$ A is not RE is the inverse of this theorem and hence false.

3.27 (c)

Cantor's Theorem: If a set S is countably infinite then 2^S is uncountably infinite.
Now, Σ^* is countably infinite. So by Cantor's theorem, 2^{Σ^*} is uncountably infinite.

3.28 (a)

Ambiguity problem of context free grammar is undecidable.

3.29 (d)

L_1 is CFL

$\Rightarrow \bar{L}_1 = \overline{CFL} = \overline{CSL} = CSL$

L_2 is RE but not REC $\Rightarrow$ $\bar{L}_2$ is not RE

Statement 1: $\bar{L}_1$ is recursive is **true** because, compliment of CFL is always a CSL and hence recursive.

Statement 2: $\bar{L}_2$ is recursive is **false** because, $\bar{L}_2$ is not RE.

Statement 3: $\bar{L}_1$ is CFL is **false** ($\bar{L}_1$ is need not be CFL).

Statement 4: $\bar{L}_1 \cup L_2$ is RE is **true** because,

$$\overline{CFL} \cup RE = \overline{CSL} \cup RE = CSL \cup RE$$
$$= RE \cup RE = RE$$

So only statements 1 and 4 are **true**.

3.30 (d)

Statement 1: True, because decidable languages (REC) are closed under complement operation.
Statement 2: False, because P $\subseteq$ NP $\subseteq$ REC. So every P and every NP language is decidable (REC).

Statement 3: True, every language in NP is decidable.

3.31 (c)

$X \rightarrow$ REC

$Y \rightarrow$ RE but not REC

$\bar{Y} \le W$ and $Z \le \bar{X}$

Now we know that RE, REC both go in reverse direction on reduction.

Now, if Y is RE but not REC, then $\bar{Y}$ is not RE.

Now $\bar{Y} \le W$

If W is RE $\Rightarrow$ $\bar{Y}$ is RE

Contrapositive is $\bar{Y}$ is not RE $\Rightarrow$ W is not RE

Now $Z \le \bar{X}$

If X is REC, $\bar{X}$ is also REC

So $\bar{X}$ REC $\Rightarrow$ Z is REC

So W is not RE and Z is REC.

3.32 (d)

$L_1 \to$ Regular, $L_2 \to$ CFL
$L_3 \to$ REC, $L_4 \to$ RE

I. $\bar{L}_3 \cup L_4$ is RE

$$\bar{L}_3 \cup L_4 = \overline{\text{REC}} \cup \text{RE} = \text{REC} \cup \text{RE}$$
$$= \text{RE} \cup \text{RE} = \text{RE}$$

So I is true

II. $\bar{L}_2 \cup L_3$ is recusive

$$\bar{L}_2 \cup L_3 = \overline{\text{CFL}} \cup \text{REC} = \overline{\text{CSL}} \cup \text{REC}$$
$$= \text{CSL} \cup \text{REC} = \text{REC} \cup \text{REC} = \text{REC}$$

So II is True.

III. $L_1^* \cap L_2$ is CFL

$$L_1^* \cap L_2 = (\text{REG})^* \cap \text{CFL}$$
$$= \text{REG} \cap \text{CFL} = \text{CFL}$$

So III is True.

IV. $L_1 \cup \bar{L}_2$ is CFL

$$L_1 \cup \bar{L}_2 = \text{REG} \cup \overline{\text{CFL}} = \text{REG} \cup \overline{\text{CSL}}$$
$$= \text{REG} \cup \text{CSL} = \text{CSL}$$

Since, A CSL need not be a CFL, so IV is False.
So only I, II and III are true.

3.33 (c)

$$L_3 = \{M \mid \varepsilon \in L(M)\}$$

Since the question is a non-trivial question on RE language, it is undecidable by Rice's theorem. So L_3 is not recursive.

L_1, L_2 are decidable (recursive) since 2016 is finite and that many steps can be simulated on a UTM in finite time.

3.34 (d)

I. Membership of regular language (Decidable)
II. Emptiness of CFL (Decidable)
III. L = Σ* problem of CFL (Undecidable)
IV. Membership of RE language (Undecidable)

So, only III and IV are undecidable.

3.35 (a)

Since the way computable is defined based on halting TM, it means computable is same as Recursive.

So, clearly f is computable iff L_f is recursive.

3.36 (b)

The set of RE languages is closed under intersection, not closed under complementation, is not a subset of set of REC language and is a countable set.

3.37 (d)

I. Membership problem for RE → undecidable
II. Regularity problem for RE → undecidable
III. Equivalence problem for RE → undecidable
IV. Since DPDA P exists for every NFA N and equivalent to it, this problem is trivially decidable.

3.38 (b)

S_1: The set L_{RE} is known to be countably infinite since it corresponds with set of Turing machines.

S_2: Since syntactically valid C programs surely run on Turing machines, this set is also a subset of set of Turing machines, which is countable.

S_3: Set of all languages = 2^{Σ^*} which is known to be uncountable. Σ* countably infinite $\Rightarrow 2^{\Sigma^*}$ is uncountable.

S_4: Set of all non-regular languages includes set $L_{\text{NOT RE}}$ which is uncountable infinite and hence is uncountable.

So, S_3 and S_4 are uncountable.

3.39 (b)

Rice's theorem applies to L_1, L_3 and L_4 since then have condition based on L(M), which are non-trivial.

L_1: L(M)= ϕ is non-trivial condition because some Turing machines accept ϕ and some don't.

L_3: L(M)= non-recursive, is non-trivial because some Turing machines accept non-recursive and some accept recursive.

L_4: $|\text{L(M)}| \geq 21$ is non-trivial since some Turing machines accept more than 21 strings, some accept less.

L_2: We cannot apply Rice's theorem since condition is not on L(M) but on M reaching a state q in exactly 100 steps which can be checked on a UTM. So L_2 is decidable.

So, option (b), L_1, L_3 and L_4 are undecidable is correct.

3.40 (a)

$L_1 = \{\langle M \rangle \mid M \text{ takes more than 2021 steps on all inputs}\}$

$L_2 = \{\langle M \rangle \mid M \text{ takes more than 2021 steps on some input}\}$

A Turing machine reads at most 2021 bits of input while making 2021 steps. So the halting behaviour is completely determined by the first 2021 bits of input. Now the number of strings with 2021 bits is finite and so generate all of them and in finite amount of time we can check if the given TM, M halts on any of these strings.

For L_1, the algorithm will be as follows,

Does not halts on all of these strings → Yes

Halts on at least one of these strings → No

For L_2, the algorithm will be as follows,

does not halt on at least one of these strings → Yes

Halts on all of these strings → No

So, both L_1 and L_2 are decidable.

3.41 (b)

Option (a): Completeness problem for regular → Decidable → Recursive

Option (b): Completeness problem for CFL's → Undecidable → Not recursive

Option (c): Emptiness problem for CFL's → Decidable → Recursive

Option (d): Emptiness problem for Regular → Decidable → Recursive

So only option (b) is not recursive.

3.42 (a)

S_1: Every infinite regular language contains an undecidable language as a subset.

S_2: Every finite language is regular.

Clearly, S_2 is true, since for finite language, we can design FA by brute force, with a finite number of states.

Since, any language can be subset of an infinite language (No infinite language is closed under subset operation).

So, an infinite regular language can have any type of language as a subset including undecidable (non-REC) languages.

So, S_1 is also true. So, both S_1 and S_2 are true.

So, option (a) is correct.

3.43 (d)

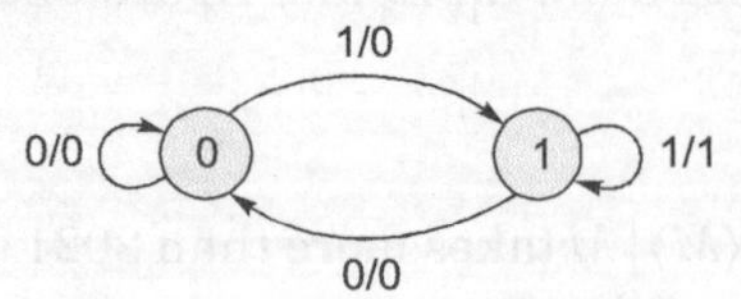

PS	Next State t, O/P y,	
s	$b = 0$	$b = 1$
0	0,0	1,0
1	0,0	1,1

$$t = \bar{s}b + sb = b,\ y = sb$$

3.44 (b, c, d)

(a) No language is closed under subset operation. So subset of an RE language may or may not be REC. So option (a) is false.

(b) If L and $\overline{L}$ are both RE $\Rightarrow$ L is REC is a theorem and is true.

(c) $(\text{CFL})^C = (\text{CSL})^C = \text{CSL}$

Every CSL is recursive.

So complement of CFL is recursive is true.

(d) $L_1 \rightarrow$ regular, $L_2 \rightarrow$ regular

$\Rightarrow \quad L_1 \cap L_2 = \text{Regular} \cap \text{Regular} = \text{Regular}$

Now every regular language is a DCFL.

So option (d) is true.

3.45 (a, b, c)

A, B, C choices are all non-trivial properties of RE language (language of TM's) and hence by Rice's theorem are all UNDECIDABLE.

Choice (d) is DECIDABLE. Why?

A Turing Machine sees only at most the first 1073 symbols of input in its first 1073 steps. Hence whether it stops on first 1073 steps depends only on the first 1073 symbols of input. Since the number of strings of length 1073 is finite, it gives a way to decide this.

Run the input machine M on all inputs of length 1073 and check whether any of them stops within 1073 steps. If so, reject, otherwise accept.

3.46 (179)

If the grammar is in CNF, then for any string of length 'n' it takes $2n - 1$ steps to derive that string.

Given string $\quad w = a^{30}\,b^{30}\,c^{30}$

$\Rightarrow \quad |w| = 90$

$\therefore$ It takes $2 \times 90 - 1$ i.e. 179 steps.

■■■■

Digital Logic

UNIT

III

CONTENTS

Digital Logic

Syllabus

Boolean algebra, Logic functions, Minimization, Design and synthesis of combinational and sequential circuits; Number representation and computer arithmetic (fixed and floating point).

Analysis of Previous GATE Papers

Exam Year	1 Mark Ques.	2 Marks Ques.	5 Marks Ques.	Total Marks
1995	1	–	–	**1**
1996	3	4	1	**16**
1997	1	3	–	**7**
1998	–	–	–	**–**
1999	2	2	–	**6**
2000	1	4	–	**9**
2001	1	3	–	**7**
2002	5	5	–	**15**
2003	1	4	–	**9**
2004	4	5	–	**14**
2005	4	4	–	**12**
2006	1	5	–	**11**
2007	3	5	–	**13**
2008	4	1	–	**9**
2009	2	–	–	**2**
2010	3	2	–	**7**
2011	2	3	–	**8**
2012	2	–	–	**2**
2013	3	–	–	**3**

Exam Year	1 Mark Ques.	2 Marks Ques.	Total Marks
2014 Set-1	2	1	**4**
2014 Set-2	3	1	**5**
2014 Set-3	2	2	**6**
2015 Set-1	1	2	**5**
2015 Set-2	1	2	**5**
2015 Set-3	–	3	**6**
2016 Set-1	3	2	**7**
2016 Set-2	3	–	**3**
2017 Set-1	2	1	**4**
2017 Set-2	1	3	**7**
2018	2	2	**6**
2019	3	2	**7**
2020	–	2	**4**
2021 Set-1	2	2	**6**
2021 Set-2	3	2	**7**
2022	1	2	**5**
2023	2	3	**8**
2024 Set-1	2	2	**6**
2024 Set-2	2	2	**6**

1 Logic Functions and Minimization

1.1 What is the equivalent Boolean expression in product-of-sums form for the Karnaugh map given in figure

CD \ AB	00	01	11	10
00		1	1	
01	1			1
11	1			1
10		1	1	

(a) $B\overline{D}+\overline{B}D$
(b) $(B+\overline{C}+D)(\overline{B}+C+\overline{D})$
(c) $(B+D)(\overline{B}+\overline{D})$
(d) $(B+\overline{D})(\overline{B}+D)$ **[1996 : 2 M]**

1.2 Let $f(x, y, z)=\overline{x}+\overline{y}x+xz$ be a switching function. Which one of the following is valid?
(a) $\overline{y}x$ is a prime implicant of f
(b) xz is a minterm of f
(c) xz is an implicant of f
(d) y is a prime implicant of f **[1997 : 2 M]**

1.3 Let * be defined as $x*y=\overline{x}+y$. Let $z=x*y$ value of $z*x$ is
(a) $\overline{x}+y$ (b) x
(c) 0 (d) 1 **[1997 : 1 M]**

1.4 Let $f(x, y, z)=\overline{x}+\overline{y}z+xz$ be a switching function. Which one of the following is valid?
(a) $\overline{y}x$ is a prime implicant of f
(b) xz is a minterm of f
(c) xz is an implicant of f
(d) y is a prime implicant of f
[1997 : 2 M]

1.5 Consider the logic circuit shown in figure below. The function f_1, f_2 and f(in canonical sum of products form in decimal notation) are

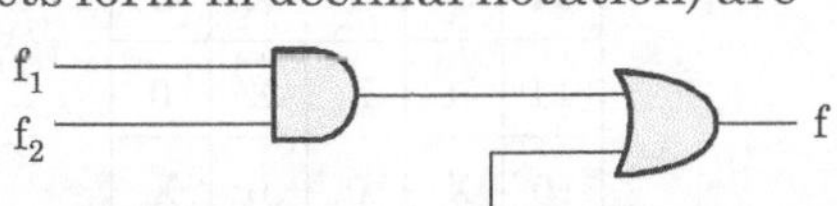

$f_1(w, x, y, z)=\Sigma(8, 9, 10)$
$f_2(w, x, y, z)=\Sigma(7, 8, 12, 13, 14, 15)$
$f(w, x, y, z)=\Sigma(8, 9)$

The function f_3 is
(a) $\Sigma 9, 10$ (b) $\Sigma 9$
(c) $\Sigma 1, 8, 9$ (d) $\Sigma 8, 10, 18$
[1997 : 2 M]

1.6 Which of the following operations is commutative but not associative?
(a) AND (b) OR
(c) NAND (d) EXOR **[1998 : 1 M]**

1.7 What happens when a bit string is XORed with itself n-times as shown:
$[B\oplus(B\oplus(B\oplus(B \ldots n \text{ times}))]$
(a) Complements when n is even
(b) Complements when n is odd
(c) Divides by 2^n always
(d) Remains unchanged when n is even
[1998 : 1 M]

1.8 The function represented by the map given below is:

A \ BC	00	01	10	11
0	1	0	0	1
1	1	0	0	1

(a) A.B (b) AB + BC + CA
(c) $\overline{B\oplus C}$ (d) A.BC **[1998 : 1 M]**

1.9 Which of the following expressions is not equivalent to $\overline{x}$?
(a) x NAND x (b) x NOR x
(c) x NAND 1 (d) x NOR 1
[1999 : 1 M]

1.10 Which of the following functions implements the Karnaugh map shown below?

AB \ CD	00	01	11	10
00	0	0	1	0
01	X	X	1	X
11	0	1	1	0
10	0	1	1	0

(a) $\overline{A}B+CD$
(b) $D(C+A)$
(c) $AD+\overline{A}B$
(d) $(C+D)(\overline{C}+D)(A+B)$ **[1999 : 1 M]**

1.11 The simultaneous equations on the Boolean variables x, y, z and w,

$$x + y + z = 1$$
$$xy = 0$$
$$xz + w = 1$$
$$xy + \bar{z}\bar{w} = 0$$

have the following solution for x, y, z and w, respectively:

(a) 0100 (b) 1101
(c) 1011 (d) 1000 **[2000 : 2 M]**

1.12 Which function does NOT implement the Karnaugh map given below?

xy \ wz	00	01	11	10
00	0	×	0	0
01	0	×	1	1
11	1	1	1	1
10	0	×	0	0

(a) $(w + x)y$
(b) $xy + yw$
(c) $(w + x)(w + y)(x + y)$
(d) None of the above **[2000 : 2 M]**

1.13 Given the following Karnaugh map, which one of the following represents the minimal Sum-of-Products of the map?

yz \ wx	00	01	11	10
00	0	X	0	X
01	X	1	X	1
11	0	X	1	0
10	0	1	X	0

(a) $xy + y'z$
(b) $wx'y' + xy + xz$
(c) $w'x + y'z + xy$
(d) $xz + y$ **[2001 : 1 M]**

1.14 Minimum SOP for f(w, x, y, z) shown in Karnaugh-map below is

yz \ wx	00	01	11	10
00	0	1	1	0
01	x	0	0	1
11	x	0	0	1
10	0	1	1	x

(a) $xz + y'z$ (b) $xz' + zx'$
(c) $x'y + zx'$ (d) None of these **[2002 : 1 M]**

1.15 Consider the following logic circuit whose inputs are functions f_1, f_2, f_3 and output is f.

Given that

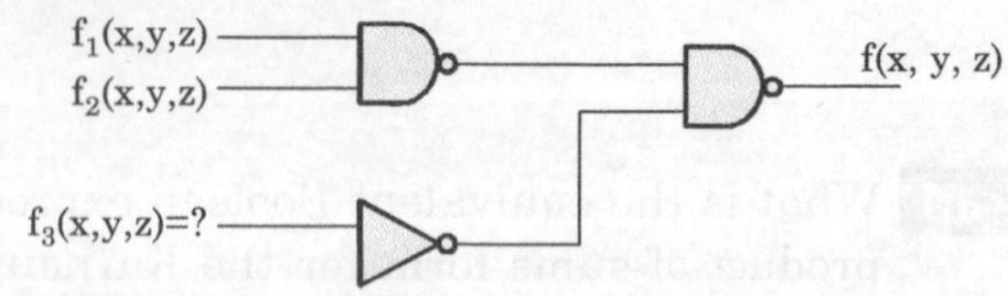

$f_1(x, y, z) = \Sigma(0, 1, 3, 5)$,
$f_2(x, y, z) = \Sigma(6, 5)$, and
$f(x, y, z) = \Sigma(1, 4, 5)$,
f_3 is

(a) $\Sigma(1,4, 5)$ (b) $\Sigma(6, 7)$
(c) $\Sigma(0, 1, 3, 5)$ (d) None of these **[2002 : 2 M]**

1.16 $f(A, B) = A' + B$. Simplified expression for function $f(f(x + y, y), z)$ is

(a) $x' + z$ (b) xyz
(c) $xy' + z$ (d) None of these **[2002 : 2 M]**

1.17 Transform the following logic circuit (without expressing its switching function) into an equivalent logic circuit that employs only 6 NAND gates each with 2-inputs.

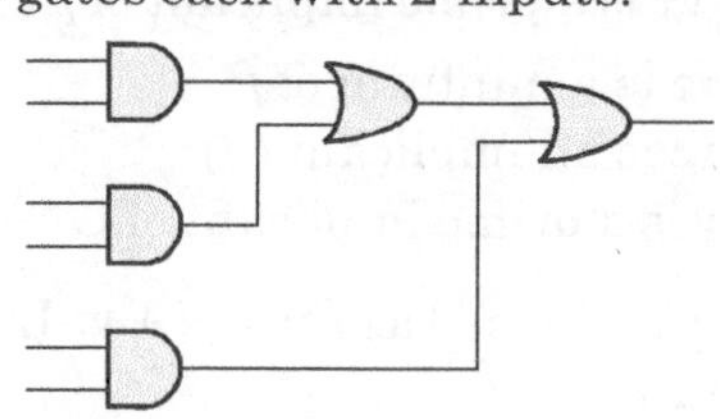

[2002 : 2 M]

1.18 The literal count of a boolean expression is the sum of the number of times each literal appears in the expression. For example, the literal count of (xy + xz) is 4. What are the minimum possible literal counts of the product-of-sum and sum-of-product representations respectively of the function given by the following Karnaugh map? Here, X denotes "don't care"

xy \ zw	00	01	11	10
00	X	1	0	1
01	0	1	X	0
11	1	X	X	0
10	X	0	0	X

(a) (11, 9) (b) (9, 13)
(c) (9, 10) (d) (11, 11) **[2003 : 2 M]**

1.19 Consider the following circuit composed of XOR gates and non-inverting buffers.

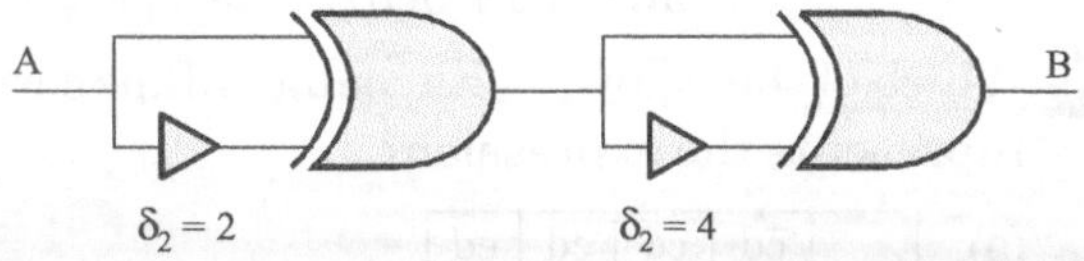

The non-inverting buffers have delays $\delta_1 = 2$ ns and $\delta_2 = 4$ ns as shown in the figure. Both XOR gates and all wires have zero delay. Assume that all gate inputs, outputs and wires are stable at logic level 0 at time 0. If the following waveform is applied at input A, how many transition (s) (change of logic levels) occur (s) at B during the interval from 0 to 10 ns?

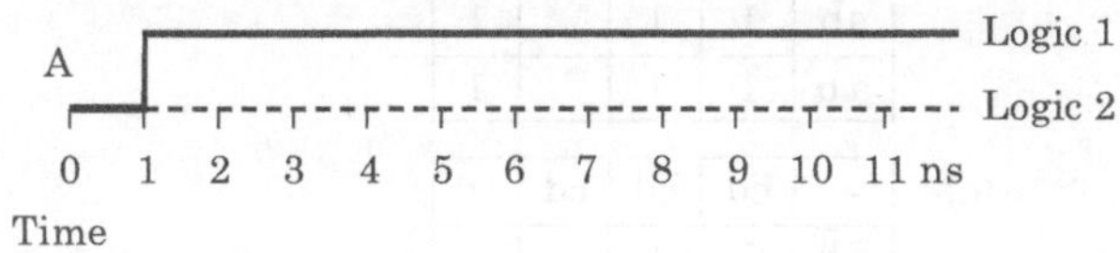

(a) 1 (b) 2
(c) 3 (d) 4 **[2003 : 2 M]**

1.20 The Boolean function x′ y′ + xy + x′ y is equivalent to
(a) x′ + y′ (b) x + y
(c) x + y′ (d) x′ + y **[2004 : 1 M]**

1.21 What is the minimum number of NAND gates required to implement a 2-input EXCLUSIVE-OR function without using any other logic gate?
(a) 3 (b) 4
(c) 5 (d) 6 **[2004 : 1 M]**

1.22 The function AB′C + A′BC + ABC′+A′B′C + AB′C′ is equivalent to
(a) AC′ + AB + A′C
(b) AB′ + AC′ + A′C
(c) A′B + AC′ + AB′
(d) A′B + AC + AB′
[2004 : 2 M]

1.23 Which are the essential prime implicants of the following Boolean function?

f(a, b, c) = a′c + ac′ + b′c

(a) a′c and ac′ (b) a′c and b′c
(c) a′c only (d) ac′ and bc′
[2004 : 2 M]

1.24 Which of the following expressions is equivalent to $(A \oplus B) \oplus C$?
(a) $(A + B + C)(\bar{A} + \bar{B} + \bar{C})$
(b) $(A + B + C)(\bar{A} + \bar{B} + C)$
(c) $ABC + \bar{A}(B \oplus C) + \bar{B}(A \oplus C)$
(d) None of these **[2005 : 1 M]**

1.25 A two-way switch has three terminals a, b and c. In ON position (logic value 1), *a* is connected to b, and in OFF position, *a* is connected to c. Two of these two-way switches S1 and S2 are connected to a bulb as shown below.

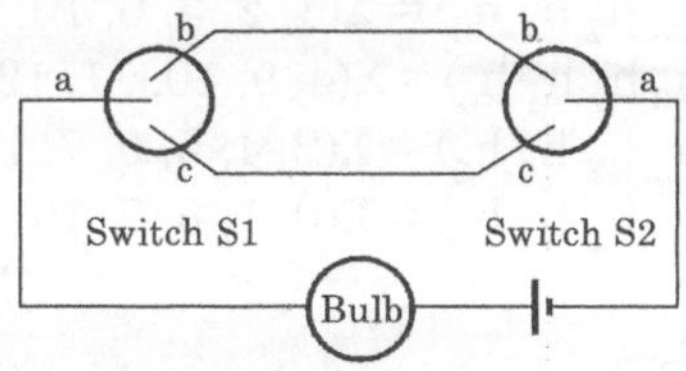

Which of the following expressions, if true, will always result in the lighting of the bulb?
(a) $S1 \cdot \overline{S2}$ (b) S1 + S2
(c) $\overline{S1 \oplus S2}$ (d) $S1 \oplus S2$
[2005 : 1 M]

1.26 Consider the following circuit:

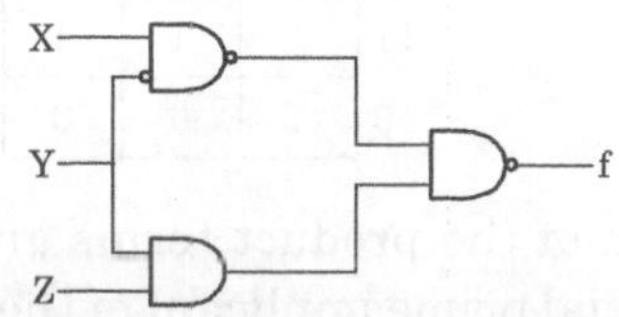

Which one of the following is TRUE?
(a) f is independent of X
(b) f is independent of Y
(c) f is independent of Z
(d) None of X, Y, Z is redundant **[2005 : 1 M]**

1.27 The switching expression corresponding to f(A, B, C, D) = Σ(1, 4, 5, 9, 11, 12) is
(a) BC′D′ + A′C′D + AB′D
(b) ABC′ + ACD + B′C′D
(c) ACD′ + A′BC′ + AC′D′
(d) A′BD + ACD′ + BCD′
[2005 : 1 M]

1.28 Consider a boolean function f(w, x, y, z). Suppose that exactly one of its inputs is allowed to change at a time. If the function happens to be true for two input vectors $i_1 = < w_1, x_1, y_1, x_1 >$ and $i_2 = < w_2, x_2, y_2, z_2 >$, we would like the function to remain true as the input changes from i_1 to i_2 (i_1 and i_2 differ in exactly one bit position), without becoming false momentarily. Let f(w, x, y, z) = Σ(5, 7, 11, 12, 13, 15). Which of the following cube covers of f will ensure that the required property is satisfied?
(a) $\bar{w}xz$, $wx\bar{y}$, $x\bar{y}z$, xyz, wyz
(b) wxy, $\bar{w}xz$, wyz
(c) $wx\overline{yz}$, xz, $w\bar{x}yz$
(d) wyz, wxz, $\bar{w}xz$, $x\bar{y}z$, xyz
[2006 : 2 M]

1.29 Consider numbers represented in 4-bit gray code. Let $h_3h_2h_1h_0$ be the gray code representation of a number n and let $g_3g_2g_1g_0$ be the gray code of (n + 1) (modulo 16) value of the number. Which one of the following functions is correct?

(a) $g_0(h_3\ h_2\ h_1\ h_0) = \Sigma(1, 2, 3, 6, 10, 13, 14, 15)$
(b) $g_1(h_3\ h_2\ h_1\ h_0) = \Sigma(4, 9, 10, 11, 12, 13, 14, 15)$
(c) $g_2(h_3\ h_2\ h_1\ h_0) = \Sigma(2, 4, 5, 6, 7, 12, 13, 15)$
(d) $g_3(h_3\ h_2\ h_1\ h_0) = \Sigma(0, 1, 6, 7, 10, 11, 12, 13)$

[2006 : 2 M]

1.30 The boolean function for a combinational circuit with four inputs is represented by the following Karnaugh map.

RS \ PQ	00	01	11	10
00	1	0	0	1
01	0	0	1	1
11	1	1	1	0
10	1	0	0	1

Which of the product terms given below is an essential prime implicant of the function?

(a) QRS (b) PQS
(c) PQ′S′ (d) Q′S′ **[2006 : 2 M]**

1.31 What is the maximum number of different Boolean functions involving in 'n' Boolean variables?

(a) n^2 (b) 2^n
(c) 2^{2^n} (d) 2^{n^2} **[2007 : 1 M]**

1.32 Consider the following Boolean function of four variables f(w, x, y, z) = Σ(1, 3, 4, 6, 9, 11, 12, 14)
The function is
(a) independent of one variable
(b) independent of two variables
(c) independent of three variables
(d) dependent on all the variables

[2007 : 1 M]

1.33 The following expression was to be realized using 2-input AND and OR gates. However, during the fabrication all 2-input AND gates were mistakenly substituted by 2-input NAND gates.

$$(a.b).c + (\bar{a}.c).d + (b.c).d + a.\ d$$

What is the function finally realized?
(a) 1
(b) $\bar{a} + \bar{b} + \bar{c} + \bar{d}$
(c) $\bar{a} + b + \bar{c} + \bar{d}$
(d) $\bar{a} + \bar{b} + c + \bar{d}$

[2007 : 2 M]

Linked Answer Q.1.34 & Q.1.35:
Consider the following expression:

$$a\bar{d} + \bar{a}\,\bar{c} + b\bar{c}d$$

1.34 Which of the following Karnaugh Maps correctly represents the expression?

(a)

	$\bar{c}\bar{d}$	$\bar{c}d$	cd	$c\bar{d}$
$\bar{a}\bar{b}$	1	1		
$\bar{a}b$	1	1		
ab	1	1		1
$a\bar{b}$	1			1

(b)

	$\bar{c}\bar{d}$	$\bar{c}d$	cd	$c\bar{d}$
$\bar{a}\bar{b}$	1	1		
$\bar{a}b$	1			
ab	1	1		1
$a\bar{b}$	1	1		1

(c)

	$\bar{c}\bar{d}$	$\bar{c}d$	cd	$c\bar{d}$
$\bar{a}\bar{b}$	1	1		
$\bar{a}b$	1	1		1
ab	1	1		1
$a\bar{b}$	1			1

(d)

	$\bar{c}\bar{d}$	$\bar{c}d$	cd	$c\bar{d}$
$\bar{a}\bar{b}$	1	1		
$\bar{a}b$	1	1		
ab	1	1		1
$a\bar{b}$	1		1	1

[2007 : 2 M]

1.35 Which of the following expressions does not correspond to the Karnaugh Map obtained in above question?

(a) $\bar{c}\bar{d} + a\bar{d} + ab\bar{c} + \bar{a}\,\bar{c}\,d$
(b) $\bar{a}\,\bar{c} + \bar{c}\bar{d} + a\bar{d} + ab\bar{c}d$
(c) $\bar{a}\,\bar{c} + a\bar{d} + ab\bar{c} + \bar{c}d$
(d) $\bar{b}\bar{c}\bar{d} + ac\bar{d} + \bar{a}\,\bar{c} + ab\bar{c}$ **[2007 : 2 M]**

1.36 Let f(w, x, y, z) = Σ(0, 4, 5, 7, 8, 9, 13, 15). Which of the following expressions are NOT equivalent to f?
(a) x′y′z′ + w′xy′ + wy′z + xz
(b) w′y′z′ + wx′y′ + xz
(c) w′y′z′ + wx′y′ + xyz + xy′z
(d) x′y′z′ + wx′y′ + w′y **[2007 : 2 M]**

1.37 Define the connective * for the Boolean variable X and Y as : X* Y = XY + X′Y′

Let Z = X * Y. Consider the following expression P, Q and R.

P : X = Y * Z
Q : Y = X * Z
R : X * Y * Z = 1

Which of the following is TRUE?
(a) Only P and Q are valid
(b) Only Q and R are valid
(c) Only P and R are valid
(d) All P, Q, R are valid **[2007 : 2 M]**

1.38 The line T in the following figure is permanently connected to the ground.

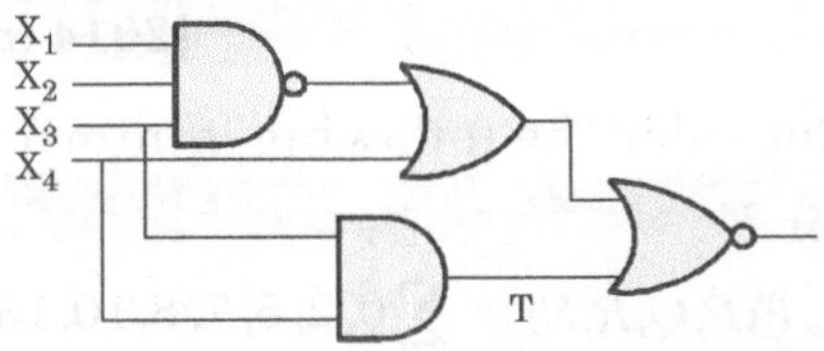

Which of the following inputs (X_1 X_2 X_3 X_4) will detect the fault?
(a) 0000 (b) 0111
(c) 1111 (d) None of these
[2007 : 2 M]

1.39 In the Karnaugh map shown below, X denoted a don't care term. What is the minimal form of the function represented by the Karnaugh map

cd \ ab	00	01	11	10
00	1	1		1
01	X			
11	X			
10	1	1		X

(a) $\bar{b}.\bar{d} + \bar{a}.\bar{d}$ (b) $\bar{a}.\bar{b} + \bar{b}.\bar{d} + \bar{a}.b.\bar{d}$
(c) $\bar{b}.\bar{d} + \bar{a}.b.\bar{d}$ (d) $\bar{a}.\bar{b} + \bar{b}.\bar{d} + \bar{a}.\bar{d}$.
[2008 : 1 M]

1.40 For the given circuit f_1, f_3 and f are given below:

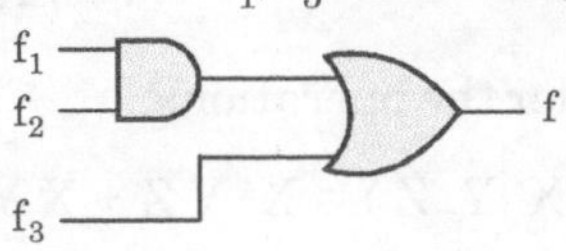

$f_1 = \Sigma m(4, 5, 6, 7, 8)$
$f_3 = \Sigma m(1, 6, 15)$
$f = \Sigma m(1, 6, 8, 15)$
Then f_2 is
(a) $\Sigma m(4, 6)$ (b) $\Sigma m(4, 8)$
(c) $\Sigma m(6, 8)$ (d) $\Sigma m(4, 6, 8)$
[2008 : 1 M]

1.41 A set of Boolean connectives is functionally complete if all Boolean functions can be synthesized using those. Which of the following sets of connectives is NOT functionally complete?
(a) EX-NOR (b) Implication, negation
(c) OR, negation (d) NAND **[2008 : 1 M]**

1.42 If P, Q, R are Boolean variables, then $(P + \bar{Q})(P.\bar{Q} + P.R)(\bar{P}.\bar{R} + \bar{Q})$ simplifies to
(a) $P.\bar{Q}$ (b) $P.\bar{R}$
(c) $P.\bar{Q} + R$ (d) $P.\bar{R} + Q$
[2008 : 2 M]

1.43 Consider the following Boolean function of four variables:
$f(A,B,C,D) = \Sigma(2, 3, 6, 7, 8, 9, 10, 11, 12, 13)$
The function is
(a) independent of one variable
(b) independent of two variables
(c) independent of three variable
(d) dependent on all the variables
[2008 : 1 M]

1.44 What is the minimum number of gates required to implement the Boolean function (AB + C) if we have to use only 2-input NOR gates?
(a) 2 (b) 3
(c) 4 (d) 5 **[2009 : 1 M]**

1.45 The minterm expansion of
$f(P, Q, R) = PQ + Q\bar{R} + P\bar{R}$ is
(a) $m_2 + m_4 + m_6 + m_7$
(b) $m_0 + m_1 + m_3 + m_5$
(c) $m_0 + m_1 + m_6 + m_1$
(d) $m_2 + m_3 + m_4 + m_5$ **[2010 : 1 M]**

1.46 The simplified SOP (Sum of Product) form of the Boolean expression
$(P + \bar{Q} + \bar{R}) \cdot (P + \bar{Q} + R) \cdot (P + Q + \bar{R})$ is
(a) $(P + \bar{Q} + \bar{R})$ (b) $(P + \bar{Q} \cdot \bar{R})$
(c) $(\bar{P}.Q + R)$ (d) $(P \cdot Q + R)$
[2010 : 2 M]

1.47 What is the boolean expression for the output f of the combinational logic circuit of NOR gates given below?

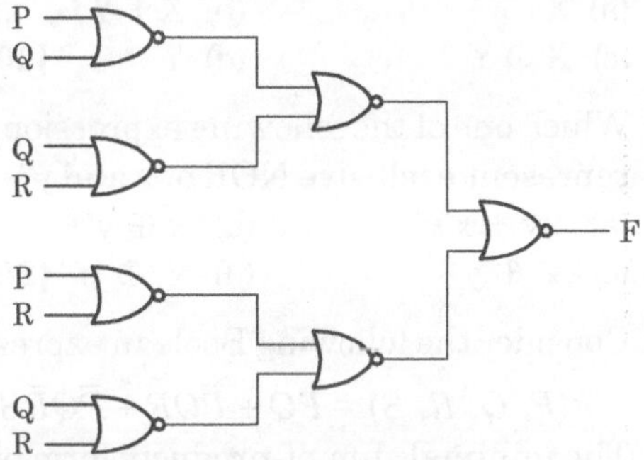

(a) $\overline{Q + R}$ (b) $\overline{P + Q}$
(c) $\overline{P + R}$ (d) $\overline{P + Q + R}$
[2010 : 2 M]

1.48 Which one of the following circuits is NOT equivalent to a 2-input XNOR (exclusive NOR) gate?

(a) 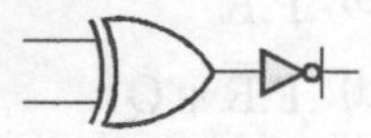

(b) 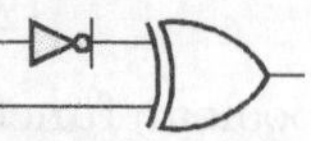

(c) 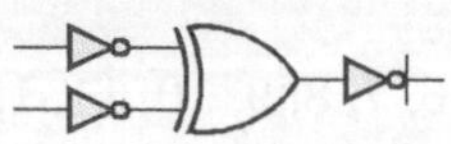

(d) 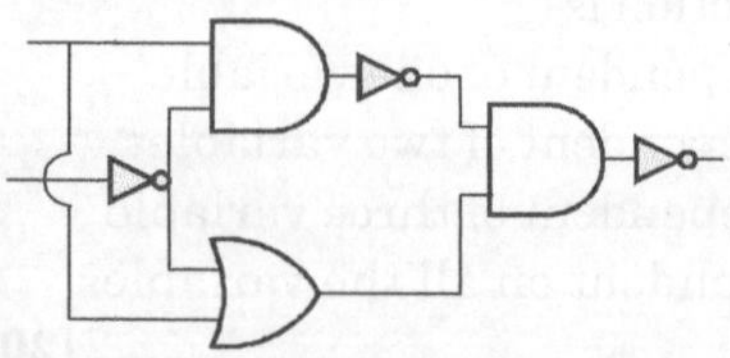

[2011 : 1 M]

1.49 What is the minimal form of the Karnaugh map shown below? Assume that X denotes a don't care term.

cd \ ab	00	01	11	10
00	1	X	X	1
01	X			1
11				
10	1			X

(a) $\bar{b}\,\bar{d}$ (b) $\bar{b}\,\bar{d}+\bar{b}\,\bar{c}$

(c) $\bar{b}\,\bar{d}+a\,\bar{b}\,\bar{c}\,d$ (d) $\bar{b}\,\bar{d}+\bar{b}\,\bar{c}+\bar{c}\,\bar{d}$

[2012 : 2 M]

1.50 The truth table:

X	Y	f(X, Y)
0	0	0
0	1	0
1	0	1
1	1	1

Represents the Boolean function

(a) X (b) X + Y

(c) X ⊕ Y (d) Y **[2012 : 1 M]**

1.51 Which one of the following expression does NOT represent exclusive NOR of x and y?

(a) $xy + x'y'$ (b) $x \oplus y'$

(c) $x' \oplus y$ (d) $x' \oplus y'$ **[2013 : 1 M]**

1.52 Consider the following Boolean expression for *F*:

$$F(P, Q, R, S) = PQ + \bar{P}QR + \bar{P}Q\bar{R}S$$

The minimal sum-of-products form of *F* is

(a) $PQ + QR + QS$ (b) $P + Q + R + S$

(c) $\bar{P} + \bar{Q} + \bar{R} + \bar{S}$ (d) $\bar{P}R + \bar{P}\bar{R}S + P$

[2014 (Set-1) : 1 M]

1.53 The dual of a Boolean function $F(x_1, x_2, \ldots, x_n, +, \cdot, ')$, written as F^D, is the same expression as that of F with + and · swapped. F is said to be self-dual if F = F^D. The number of self-dual functions with n Boolean variables is

(a) 2^n (b) 2^{n-1}

(c) 2^{2^n} (d) $2^{2^{n-1}}$

[2014 (Set-2) : 1 M]

1.54 Consider the following minterm expression for *F*:

$$F(P,Q,R,S) = \sum 0, 2, 5, 7, 8, 10, 13, 15$$

The minterms 2, 7, 8 and 13 are 'do not care' terms. The minimal sum-of-products form for *F* is

(a) $Q\bar{S} + \bar{Q}S$

(b) $\bar{Q}\bar{S} + QS$

(c) $\bar{Q}\bar{R}\bar{S} + \bar{Q}R\bar{S} + Q\bar{R}S + QRS$

(d) $\bar{P}\bar{Q}\bar{S} + \bar{P}QS + PQS + P\bar{Q}\bar{S}$

[2014 (Set-3) : 1 M]

1.55 Let ⊕ denote the Exclusive OR (XOR) operation. Let '1' and '0' denote the binary constants. Consider the following Boolean expression for *F* over two variables *P* and *Q*:

$$F(P, Q) = ((1 \oplus P) \oplus (P \oplus Q)) \oplus ((P \oplus Q) \oplus (Q \oplus 0))$$

The equivalent expression for *F* is

(a) $P + Q$ (b) $\overline{P + Q}$

(c) $P \oplus Q$ (d) $\overline{P \oplus Q}$

[2014 (Set-3) : 2 M]

1.56 Consider the operations:

$$f(X, Y, Z) = X'YZ + XY' + Y'Z' \text{ and}$$
$$g(X, Y, Z) = X'YZ + X'YZ' + XY$$

Which one of the following is correct?

(a) Both {f} and {g} are functionally complete

(b) Only {f} is functionally complete

(c) Only {g} is functionally complete

(d) Neither {f} nor {g} is functionally complete

[2015 (Set-1) : 2 M]

1.57 The number of min-terms after minimizing the following Boolean expression is ______.

$$[D' + AB' + A'C + AC'D + A'C'D]'$$

[2015 (Set-2) : 2 M]

1.58 The total number of prime implicants of the function $f(w,x,y,z) = \Sigma(0,2,4,5,6,10)$ is ______.

[2015 (Set-3) : 2 M]

1.59 Given the function $F = P' + QR$, where F is a function in three Boolean variables P, Q and R and $P' = !P$, consider the following statements.

S1: $F = \Sigma\,(4, 5, 6)$

S2: $F = \Sigma\,(0, 1, 2, 3, 7)$

S3: $F = \Pi\,(4, 5, 6)$

S4: $F = \Pi\,(0, 1, 2, 3, 7)$

Which of the following is true?

(a) S1-False, S2-True, S3-True, S4-False

(b) S1-True, S2-False, S3-False, S4-True

(c) S1-False, S2-False, S3-True, S4-True

(d) S1-True, S2-True, S3-False, S4-False

[2015 (Set-3) : 2 M]

1.60 Consider the Boolean operator # with the following properties: $x\#0 = x,\ x\#1 = \bar{x},\ x\#x = 0$ and $x\#\bar{x} = 1$. Then $x\#y$ is equivalent to

(a) $x\bar{y} + \bar{x}y$ (b) $x\bar{y} + \bar{x}\,\bar{y}$

(c) $\bar{x}y + xy$ (d) $xy + \bar{x}\,\bar{y}$

[2016 (Set-1) : 1 M]

1.61 Let $x_1 \oplus x_2 \oplus x_3 \oplus x_4 = 0$ where x_1, x_2, x_3, x_4 are Boolean variables, and $\oplus$ is the XOR operator. Which one of the following must always be TRUE?

(a) $x_1 x_2 x_3 x_4 = 0$ (b) $x_1 x_3 + x_2 = 0$

(c) $\bar{x}_1 \oplus \bar{x}_3 = \bar{x}_2 \oplus \bar{x}_4$ (d) $x_1 + x_2 + x_3 + x_4 = 0$

[2016 (Set-2) : 1 M]

1.62 Consider the Karnaugh map given below, where X represents "don't care" and blank represents 0.

dc \ ba	00	01	11	10
00		X	X	
01	1			X
11	1			1
10		X	X	

Assume for all inputs (a, b, c, d), the respective complements $(\bar{a}, \bar{b}, \bar{c}, \bar{d})$ are also available. The above logic is implemented using 2-input NOR gates only. The minimum number of gates required is ________. **[2017 (Set-1) : 1 M]**

1.63 Given $f(w, x, y, z) = \Sigma_m\{0, 1, 2, 3, 7, 8, 10) + \Sigma_d(5, 6, 11, 15)$, where d represents the *don't-care* condition in Karnaugh maps. Which of the following is a minimum product-of-sums (POS) form of $f(w, x, y, z)$?

(a) $f = (\bar{w}+\bar{z})(\bar{x}+z)$ (b) $f = (\bar{w}+z)(x+z)$

(c) $f = (w+z)(\bar{x}+z)$ (d) $f = (w+\bar{z})(\bar{x}+z)$

[2017 (Set-2) : 2 M]

1.64 If w, x, y, z are Boolean variables, then which one of the following is INCORRECT?

(a) $wx + w(x + y) + x(x + y) = x + wy$

(b) $\overline{w\bar{x}(y+\bar{z}) + \bar{w}x} = \bar{w} + x + \bar{y}z$

(c) $(w\bar{x}(y + x\bar{z}) + \bar{w}\bar{x})y = x\bar{y}$

(d) $(w + y)\,(wxy + wyz) = wxy + wyz$

[2017 (Set-2) : 2 M]

1.65 Let $\oplus$ and $\odot$ denote the Exclusive OR and Exclusive NOR operations, respectively. Which one of the following is NOT CORRECT?

(a) $\overline{P \oplus Q} = P \odot Q$

(b) $\bar{P} \oplus Q = P \odot Q$

(c) $\bar{P} \oplus \bar{Q} = P \oplus Q$

(d) $(P \oplus \bar{P}) \oplus Q = (P \odot \bar{P}) \odot \bar{Q}$ **[2018 : 1 M]**

1.66 Consider the minterm list form of a Boolean function F given below:

$$F(P, Q, R, S) = \Sigma m(0, 2, 5, 7, 9, 11) + d(3, 8, 10, 12, 14)$$

Here, m denotes a minterm and d denotes a don't care term. The number of essential prime implicants of the function F is ________.

[2018 : 2 M]

1.67 Which one of the following is NOT a valid identity?

(a) $x \oplus y = (xy + x'y')'$

(b) $x \oplus y = x + y$, if $xy = 0$

(c) $(x \oplus y) \oplus z = x \oplus (y \oplus z)$

(d) $(x + y) \oplus z = x \oplus (y + z)$ **[2019 : 1 M]**

1.68 What is the minimum number of 2-input NOR gates required to implement 4-variable function expressed in sum-of-minterms from as $f = \Sigma(0, 2, 5, 7, 8, 10, 13, 15)$? Assume that all the inputs and their complements are available ________.

[2019 : 2 M]

1.69 Consider three 4-variable functions f_1, f_2 and f_3, which are expressed in sum-of-minterms as

$f_1 = \Sigma(0, 2, 5, 8, 14)$

$f_2 = \Sigma(2, 3, 6, 8, 14, 15)$

$f_3 = \Sigma(2, 7, 11, 14)$

For the following circuit with one AND gate and one XOR gate, the output function f can be expressed as:

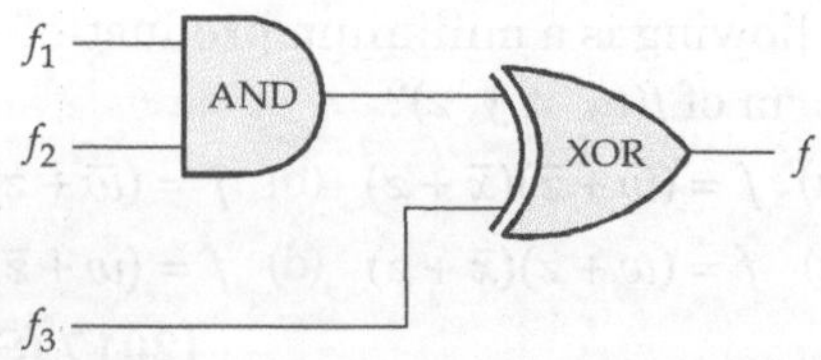

(a) $\Sigma(7, 8, 11)$
(b) $\Sigma(2, 14)$
(c) $\Sigma(0, 2, 3, 5, 6, 7, 8, 11, 14, 15)$
(d) $\Sigma(2, 7, 8, 11, 14)$ **[2019 : 2 M]**

1.70 Consider the Boolean function $z(a, b, c)$:

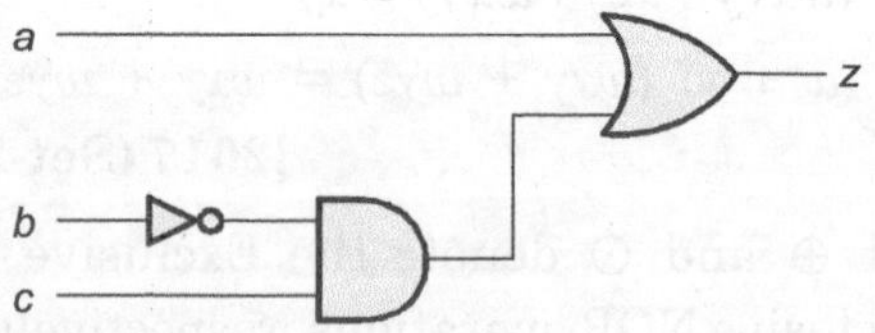

Which one of the following minterm lists represents the circuit given above?

(a) $z = \sum(2, 4, 5, 6, 7)$
(b) $z = \sum(1, 4, 5, 6, 7)$
(c) $z = \sum(0, 1, 3, 7)$
(d) $z = \sum(2, 3, 5)$ **[2020 : 2 M]**

1.71 Consider the following Boolean expression:

$$F = (X + Y + Z)(\overline{X} + Y)(\overline{Y} + Z)$$

Which of the following Boolean expressions is/are equivalent to $\overline{F}$?

(a) $X\overline{Y} + Y\overline{Z} + \overline{X}\overline{Y}\overline{Z}$
(b) $(\overline{X} + \overline{Y} + \overline{Z})(X + \overline{Y})(Y + \overline{Z})$
(c) $(X + \overline{Z})(\overline{Y} + \overline{Z})$
(d) $X\overline{Y} + \overline{Z}$ **[2021 (Set-1) : 2 M]**

1.72 Consider a Boolean function $f(w, x, y, z)$ such that

$$f(w, 0, 0, z) = 1$$
$$f(1, x, 1, z) = x + z$$
$$f(w, 1, y, z) = wz + y$$

The number of literals in the minimal sum-of-products expression of f is **[2021 (Set-2) : 2 M]**

1.73 Consider the circuit shown below where the gates may have propagation delays. Assume that all signal transitions occur instantaneously and that wires have no delays. Which of the following statements about the circuit is/are CORRECT?

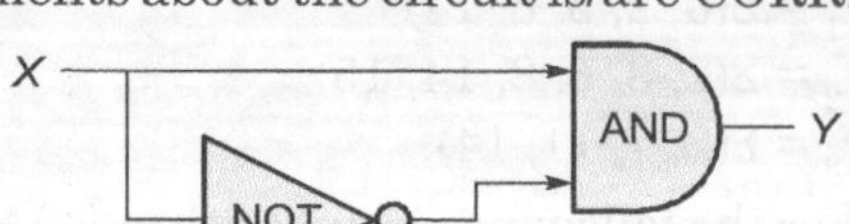

(a) With no propagation delays, the output Y is always logic Zero
(b) With no propagation delays, the output Y is always logic One
(c) With propagation delays, the output Y can have a transient logic Zero after X transitions from logic One to logic Zero
(d) With propagation delays, the output Y can have a transient logic One after X transitions from logic Zero to logic One **[2024 (Set-1) : 1 M]**

1.74 Consider a Boolean expression given by $F(X, Y, Z) = \Sigma(3, 5, 6, 7)$

Which of the following statement is/are CORRECT?

(a) $F(X, Y, Z) = \Pi(0, 1, 2, 4)$
(b) $F(X, Y, Z)$ is independent of input Y
(c) $F(X, Y, Z) = XY + YZ + XZ$
(d) $F(X, Y, Z)$ is independent of input X **[2024 (Set-1) : 2 M]**

1.75 For a Boolean variable x, which of the following statements is/are FALSE?

(a) $x \cdot x = 0$ (b) $x + \overline{x} = 1$
(c) $x + 1 = x$ (d) $x \cdot 1 = x$ **[2024 (Set-2) : 1 M]**

1.76 Consider 4-variable function $f1, f2, f3, f4$ expressed in sum-of-minterms form as given below:

$f1 = \Sigma(0, 2, 3, 5, 7, 8, 11, 13)$
$f2 = \Sigma(1, 3, 5, 7, 11, 13, 15)$
$f3 = \Sigma(0, 1, 4, 11)$
$f4 = \Sigma(0, 2, 6, 13)$

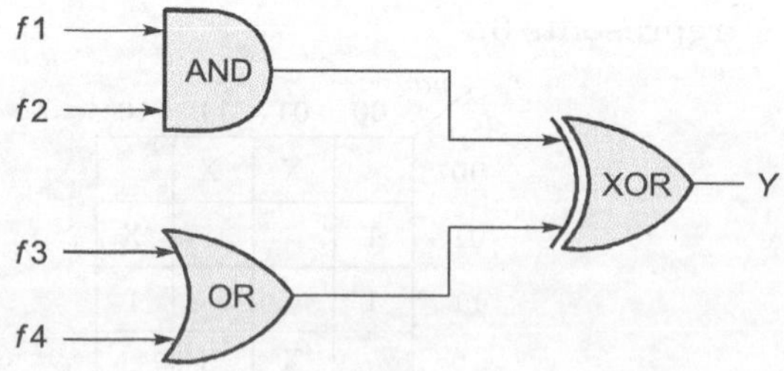

With respect to the circuit given above, which of the following options is/are CORRECT?

(a) $Y = \Sigma(0, 1, 2, 11, 13)$
(b) $Y = \Sigma(0, 1, 2, 3, 4, 5, 6, 7)$
(c) $Y = \Pi(8, 9, 10, 11, 12, 13, 14, 15)$
(d) $Y = \Pi(3, 4, 5, 6, 7, 8, 9, 10, 12, 14, 15)$ **[2024 (Set-2) : 2 M]**

Answers Logic Functions and Minimization

1.1 (c)	**1.2** (c)	**1.3** (b)	**1.4** (c)	**1.5** (b)	**1.6** (c)	**1.7** (d)	**1.8** (c)	**1.9** (d)
1.10 (b)	**1.11** (c)	**1.12** (c)	**1.13** (a)	**1.14** (b)	**1.15** (a)	**1.16** (c)	**1.17** Sol.	**1.18** (*)
1.19 (d)	**1.20** (d)	**1.21** (b)	**1.22** (b)	**1.23** (a)	**1.24** (c)	**1.25** (c)	**1.26** (a)	**1.27** (a)
1.28 (*)	**1.29** (c)	**1.30** (d)	**1.31** (c)	**1.32** (b)	**1.33** (c)	**1.34** (a)	**1.35** (c)	**1.36** (d)
1.37 (d)	**1.38** (d)	**1.39** (a)	**1.40** (c)	**1.41** (a)	**1.42** (a)	**1.43** (a)	**1.44** (b)	**1.45** (a)
1.46 (b)	**1.47** (a)	**1.48** (d)	**1.49** (b)	**1.50** (a)	**1.51** (d)	**1.52** (a)	**1.53** (d)	**1.54** (b)
1.55 (d)	**1.56** (b)	**1.57** (1)	**1.58** (3)	**1.59** (a)	**1.60** (a)	**1.61** (c)	**1.62** (1)	**1.63** (a)
1.64 (c)	**1.65** (d)	**1.66** (3)	**1.67** (d)	**1.68** (3)	**1.69** (a)	**1.70** (b)	**1.71** (a, c, d)	
1.72 (6)	**1.73** (a, d)	**1.74** (a, c)	**1.75** (a, c)	**1.76** (b, c)				

Explanations Logic Functions and Minimization

1.1 (c)

The Karnaugh map for SOP form is given as

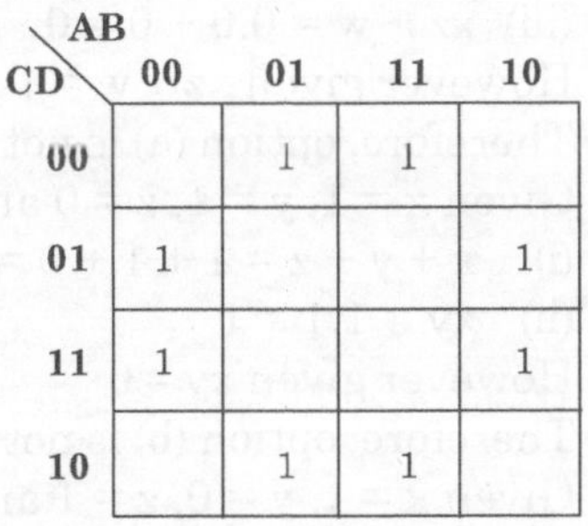

The Karnaugh map for POS form can be given by

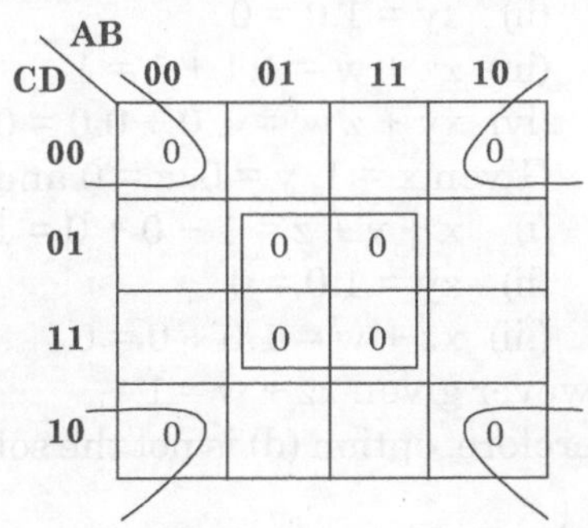

POS form of f is $f(A, B, C, D) = (B + D)(\bar{B} + \bar{D})$

1.2 (c)

$$f(x, y, z) = \overline{\bar{x} + x\bar{y}} + xz$$

$$= \overline{\bar{x} + \bar{y} + xz} = \bar{x} + \bar{y} + z$$

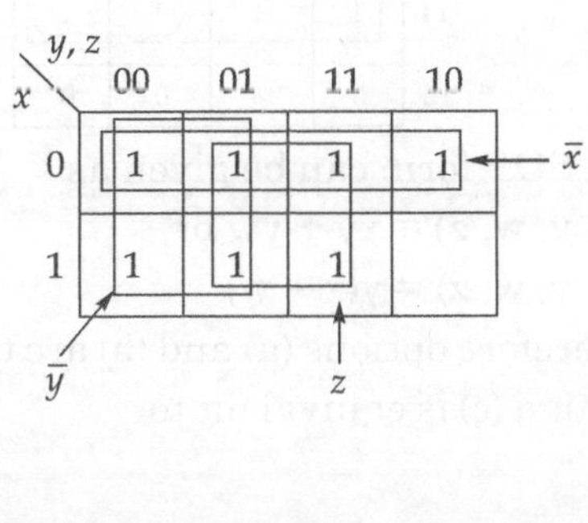

$$\text{PI} = \{\bar{x}, \bar{y}, z\}$$

xz is an implicant of f.

1.3 (b)

$$z = x * y$$

Then value of $z * x$

$$z * x = (x * y) * x = (\bar{x} + y) * x$$

$$= \overline{(\bar{x} + y)} + x = x.\bar{y} + x$$

By absorption law $a + ab = a$

$$x.\bar{y} + x = x$$

Therefore

$$z * x = x$$

1.4 (c)

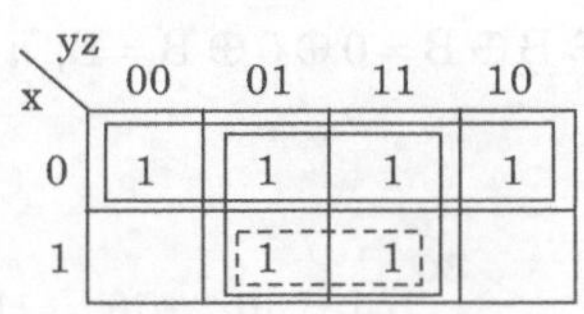

There are two prime implicants, x′ and z for the function f(x, y, z). Both are essential prime implicants.

xz is an implicant of f.

1.5 (b)

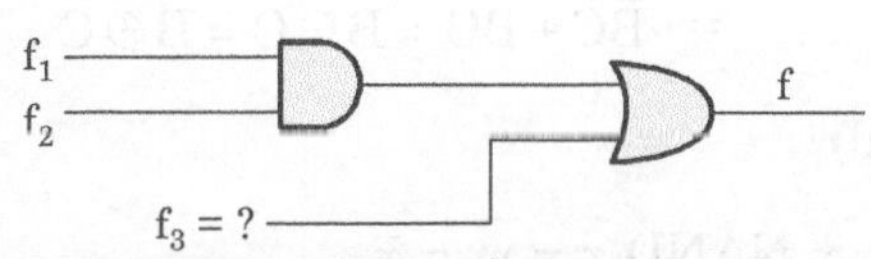

From the logic circuit, function f is

$$f = f_1 f_2 + f_3$$

$$f_1 f_2 = \Sigma(8, 9, 10)\Sigma(7, 8, 12, 13, 14, 15)$$

The common minterm in f_1 and f_2 is 8

Therefore, $f_1 f_2 = \Sigma(8)$

For $f = \Sigma(8, 9)$, f_3 can be any one of the following

(i) $f_3 = \Sigma(8, 9)$,

(ii) $f_3 = \Sigma 9$

Therefore option (b) is correct answer.

1.6 (c)

(a) AND
   (i) $x.y = y.x \Rightarrow$ Commutative
   (ii) $x.(y.z) = (x.y).z \Rightarrow$ Associative

(b) OR
   (i) $x + y = y + x \Rightarrow$ Commutative
   (ii) $x + (y + z) = (x + y) + z \Rightarrow$ Associative

(c) NAND
   (i) x NAND y = y NAND x

   $\overline{xy} = \overline{yx} \Rightarrow$ Associative

   (ii) $(x$ NAND $y)$ NAND $z \neq x$ NAND $(y$ NAND $z)$

   $\overline{xy}$ NAND $z \neq x$ NAND $(\overline{yz})$

   $\overline{\overline{xy}z} \neq \overline{x\overline{yz}}$

   $(xy + \overline{z}) \neq (\overline{x} + yz) \Rightarrow$ Not associative

(d) EXOR
   (i) $x \oplus y = y \oplus x \Rightarrow$ Commutative
   (ii) $(x \oplus y) \oplus z = x \oplus (y \oplus z) \Rightarrow$ Associative

1.7 (d)

$B \oplus B = 0;\ n = 1$

$\underline{B \oplus B} \oplus \underline{B \oplus B} = 0 \oplus 0 = 0;\ n = 3$ — n is odd

$\underline{B \oplus B} \oplus B = 0 \oplus B = B;\ n = 2$

$\underline{B \oplus B} \oplus \underline{B \oplus B} \oplus B = 0 \oplus 0 \oplus B = B;\ n = 4$ — n is even

1.8 (c)

A \ BC	00	01	10	11
0	1			1
1	1			1

$\overline{B}\overline{C}$ (column 00), BC (column 11)

$$= \overline{B}\overline{C} + BC = B \odot C = \overline{B \oplus C}$$

1.9 (d)

(a) x NAND $x = \overline{xx} = \overline{x}$

(b) x NOR $x = \overline{x + x} = \overline{x}$

(c) x NAND $1 = \overline{x.1} = \overline{x}$

(d) x NOR $1 = \overline{x + 1} = \overline{1} = 0$

1.10 (b)

Karnaugh map of the function is given as follows:

AB \ CD	00	01	11	10
00	0	0	1	0
01	×	×	1	×
11	0	1	1	0
10	0	1	1	0

Function from Karnaugh map is

$$f(A, B, C, D) = CD + AD = D(C + A)$$

1.11 (c)

Trying each option

(a) Given x = 0, y = 1, z = 0 and w = 0
   (i) x + y + z = 0 + 1 + 0 = 1
   (ii) xy = 0.1 = 0
   (iii) xz + w = 0.0 + 0 = 0
   However given xz + w = 1
   Therefore, option (a) is not the solution.

(b) Given x = 1, y = 1, z = 0 and w = 1
   (i) x + y + z = 1 + 1 + 0 = 1
   (ii) xy = 1.1 = 1
   However given xy = 0
   Therefore, option (b) is not the solution.

(c) Given x = 1, y = 0, z = 1 and w = 1
   (i) x + y + z = 1 + 0 + 1 = 1
   (ii) xy = 1.0 = 0
   (iii) xz + w = 1.1 + 1 = 1
   (iv) xy + z'w' = 1.0 + 0.0 = 0

(d) Given x = 1, y = 0, z = 0 and w = 0
   (i) x + y + z = 1 + 0 + 0 = 1
   (ii) xy = 1.0 = 0
   (iii) xz + w = 1.0 + 0 = 0

However given xz + w = 1

Therefore, option (d) is not the solution.

1.12 (c)

xy \ wz	00	01	11	10
00	0	×	0	0
01	0	×	1	1
11	1	1	1	1
10	0	×	0	0

f in SOP form can be given as

f(x, y, w, z) = xy + yw or,

f(x, y, w, z) = y(x + w)

Therefore options (a) and (b) are true,

Option (c) is equivalent to

(w + x) (w + y) (x + y) = (w + xy) (x + y)

= wx + wy + xy

= w(x + y) + xy

is not matching to the given Karnaugh map.

1.13 (a)

yz \ wx	00	01	11	10
00	0	x	0	x
01	x	1	x	1
11	0	x	1	0
10	0	1	x	0

Minimal SOP of the given k-map is

= xy + y'z

1.14 (b)

yz \ wx	00	01	11	10
00	0	1	1	0
01	x	0	0	1
11	x	0	0	1
10	0	1	1	x

Minimum SOP for f(w, x, y, z)

= xz' + zx'

1.15 (a)

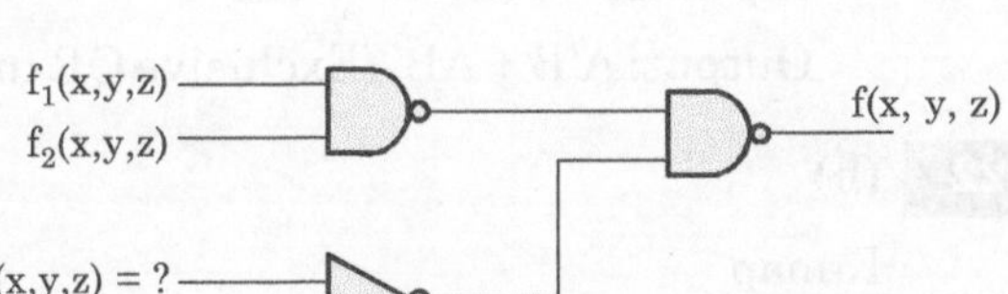

$$f = \overline{\overline{f_1 \cdot f_2} \cdot \overline{f_3}} = f_1 \cdot f_2 + f_3 \quad ...(1)$$

Given $f_1(x, y, z) = \Sigma m$ (0,1,3,5) and

$f_2(x, y, z) = \Sigma m(6, 5)$

$f_1 \cdot f_2 = \Sigma m$ (5) ...(2)

Common minterms of function f_1 and f_2 are minterms in $f_1 \cdot f_2$.

Hence f_3 can be $f_3(x, y, z) = \Sigma m$ (1, 4, 5)

or $f_3 = (x, y, z) = \Sigma m$ (1, 4)

From the given options answer is option (a).

1.16 (c)

= f(f(x + y, y), z)

= f((x + y)' + y, z)

= f(x'y' + y, z)

= f(x' + y, z)

= (x' + y)' + z = xy' + z

1.17 Sol.

Given circuit is

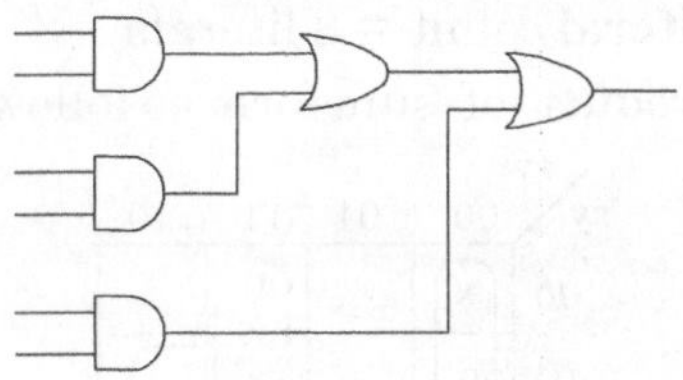

Step-1: Converting AND into NAND

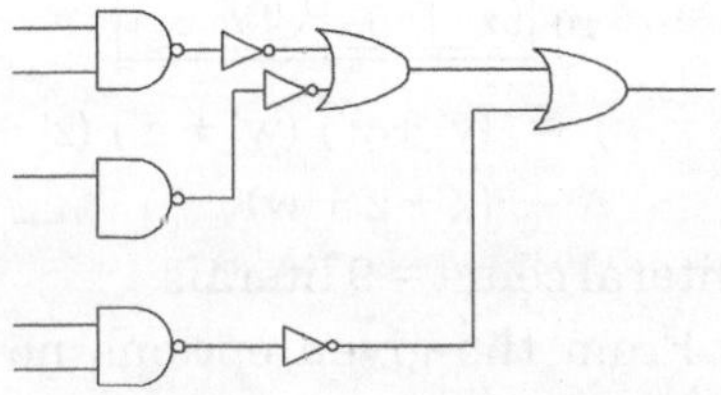

Step-2: Converting OR into bubbled OR

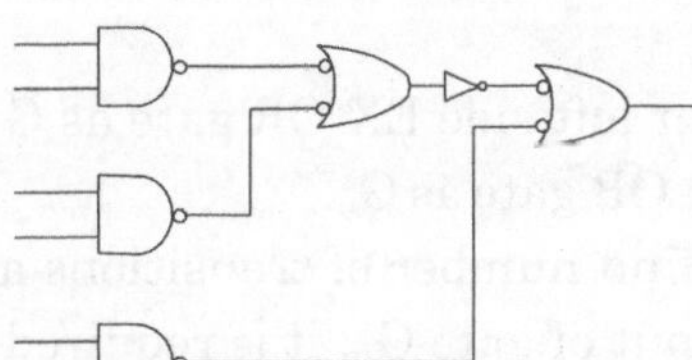

Since bubbled OR is equivalent to NAND

So given circuit can be given by

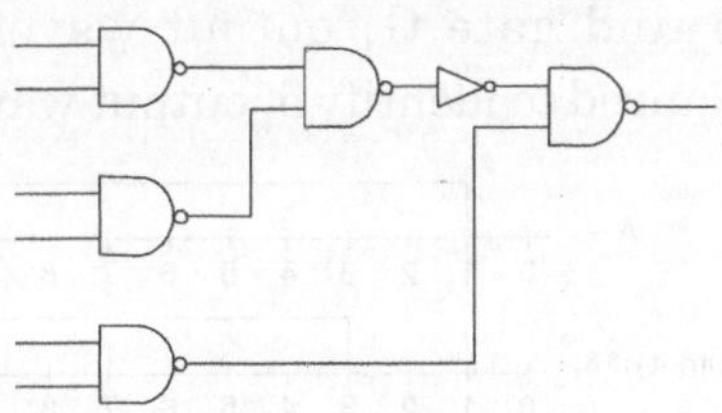

Step-3: Converting NOT into equivalent NAND gate

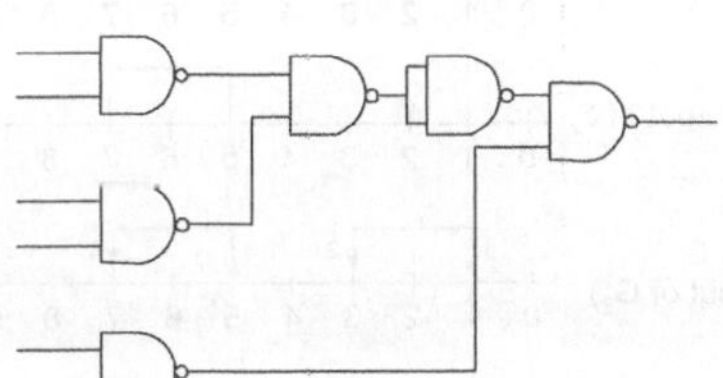

1.18 (*)

Find the sum-of-product form as follows:

xy \ zw	00	01	11	10
00	x	1		1
01		1	x	
11	1	x	x	
10	x			x

= w′y′ + z′wx′ + xyz′ ...(i)

Total literal count = 8 literals

Find product -of- sum form as follows

xy \ zw	00	01	11	10
00	x		0	
01	0		x	0
11		x	x	0
10	x	0	0	x

f(z, w, x, y) = (y′ + z′) (w′ + z′) (z′ + y) (x + z + w)

Total literal count = 9 literals

Note: From the given options no answer is matching.

1.19 (d)

Consider left side EX-OR gate as G_1 and right side EX-OR gate as G_2.

1. To find number of transitions at B i.e. the output of gate G_2, it is required to identify the inputs of gate G_2.
2. To identify gate G_2 inputs it is required to find gate G_1 output waveform.
3. To find gate G_1 output waveform, it is required to identify δ_1 output waveform.

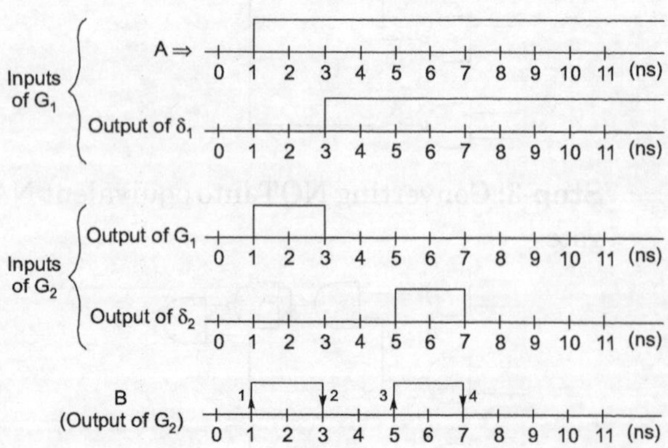

Total numbers of transitions at B during interval from 0 to 10 ns are '4'.

Hence option (d).

1.20 (d)

x′y′ + xy + x′y = x′y′ + x′y + xy

= x′(y′ + y) + xy [a + a′ = 1]

= x′ + xy [a + a′b = a + b]

= x′ + y

1.21 (b)

Minimum four 2 input NAND gates required to realise 2 input EX-OR function.

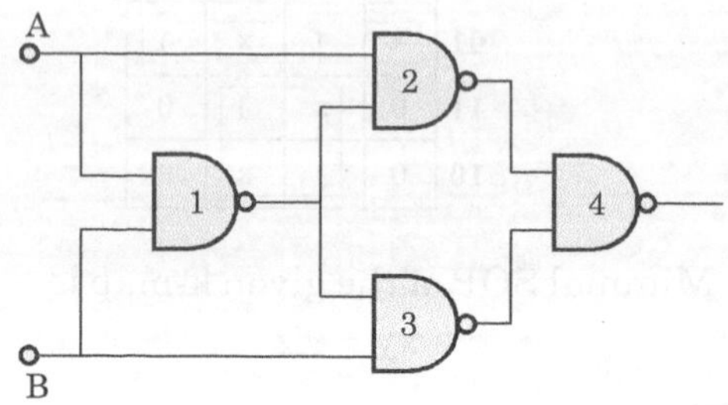

1st NAND gate:

Input: A, B

Output: (A B)′

2nd NAND gate:

Input: (AB)′, A

Output: A′ + AB

3rd NAND gate:

Input: (AB)′, B

Output: B′ + AB

4th NAND gate:

Input: A′ +AB, B′ + AB

Output: A′B + AB′ (Exclusive OR function)

1.22 (b)

K-map

C \ AB	A′B′	A′B	AB	AB′
C′	0	0	1	1
C	1	1	0	1

Equivalent expression of above K-map is A′C + AC′ + AB′.

1.23 (a)

$$f(a, b, c) = \bar{a}c + a\bar{c} + \bar{b}c$$

$$= \Sigma m\ (1, 3, 4, 5, 6)$$

The K-map of the function is

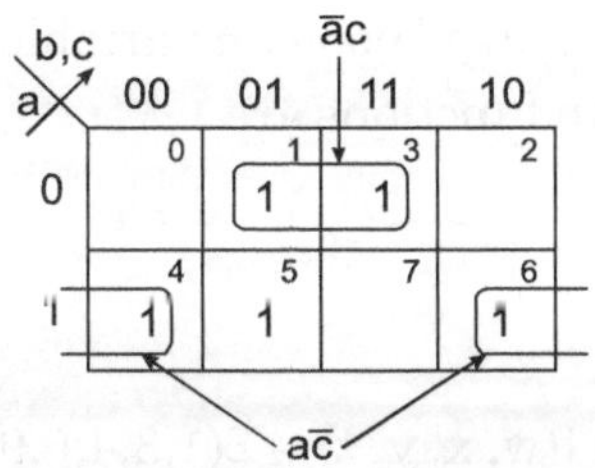

To cover the minterm m_3 the prime implicant $\bar{a}c$ is an essential prime implicant. To cover the minterm m_6 the prime implicant $a\bar{c}$ is an essential prime implicant. Hence answer is option (a).

1.24 (c)

$(A \oplus B) \oplus C$

$\left[A\bar{B} + B\bar{A}\right] \oplus C$

$\left[A\bar{B} + B\bar{A}\right]\bar{C} + \left[\overline{A\bar{B} + \bar{A}B}\right]C$

$A\bar{B}\bar{C} + \bar{A}B\bar{C} + \left[(\bar{A}+B)\cdot(A+\bar{B})\right]C$

$A\bar{B}\bar{C} + \bar{A}B\bar{C} + \bar{A}\bar{B}C + ABC$

$= ABC + \bar{A}[B \oplus C] + \bar{B}(A \oplus C)$

1.25 (c)

Bulb is On when both switch s1 and s2 are in same state, either off or on.

S1	S2	Bulb
0	0	On
0	1	Off
1	0	Off
1	1	On

Above truth table derives Ex-NOR operation.

1.26 (a)

The given circuit is

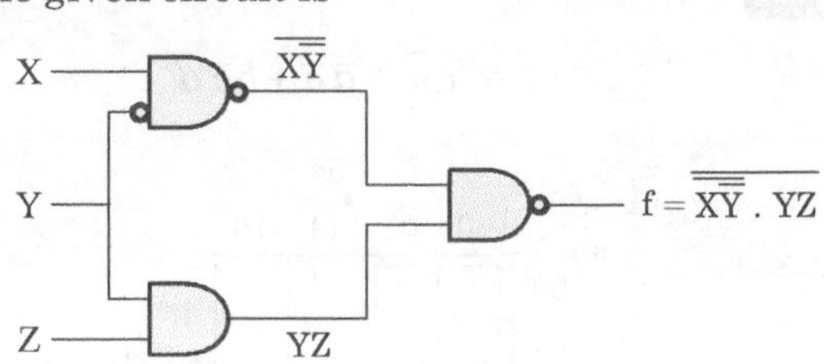

$$f = \overline{\overline{X\overline{Y}} + YZ}$$
$$= \overline{(\overline{X}+Y)\cdot YZ}$$
$$= \overline{\overline{X}YZ + YZ}$$
$$= \overline{YZ}$$

f is independent of X.

1.27 (a)

$f(A, B, C, D) = \Sigma(1, 4, 5, 9, 11, 12)$

The K-map for f is

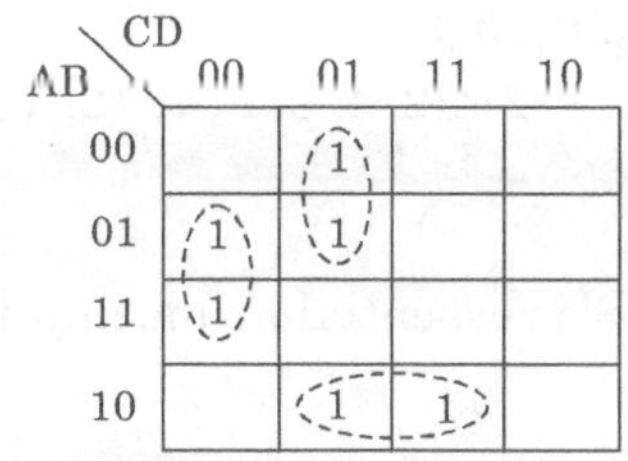

$B\bar{C}\bar{D} + \bar{A}\bar{C}D + A\bar{B}D$

1.28 (*)

Static 1 hazard free solution is required.

$f(w, x, y, z) = \Sigma(5, 7, 11, 12, 13, 15)$

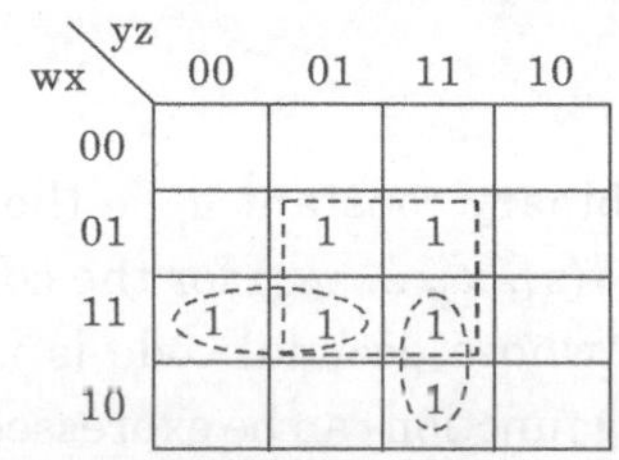

$f = \{xz, wx\bar{y}, wyz\}$

Static 1 hazard free set.

It is not given in the options.

Note: From the the given options suitable answer not available.

1.29 (c)

For a binary number $b_3\,b_2\,b_1\,b_0$ the corresponding gray code $h_3\,h_2\,h_1\,h_0$ is given is Table-1.

	0	1	2	3	4	5	6	7	8	9	10	11	12	13	14	15
b_3	0	0	0	0	0	0	0	0	1	1	1	1	1	1	1	1
b_2	0	0	0	0	1	1	1	1	0	0	0	0	1	1	1	1
b_1	0	0	1	1	0	0	1	1	0	0	1	1	0	0	1	1
b_0	0	1	0	1	0	1	0	1	0	1	0	1	0	1	0	1
h_3	0	0	0	0	0	0	0	0	1	1	1	1	1	1	1	1
h_2	0	0	0	0	1	1	1	1	1	1	1	1	0	0	0	0
h_1	0	0	1	1	1	1	0	0	0	0	1	1	1	1	0	0
h_0	0	1	1	0	0	1	1	0	0	1	1	0	0	1	1	0

Table-1

For a binary number 'n' the gray code is $h_3\,h_2\,h_1\,h_0$, the required output is $g_3\,g_2\,g_1\,g_0$ (it is corresponding gray code of binary number n+1), is given in Table-2.

	0	1	3	2	6	7	5	4	12	13	15	14	10	11	9	8
h_3	0	0	0	0	0	0	0	0	1	1	1	1	1	1	1	1
h_2	0	0	0	0	1	1	1	1	1	1	1	1	0	0	0	0
h_1	0	0	1	1	1	1	0	0	0	0	1	1	1	1	0	0
h_0	0	1	1	0	0	1	1	0	0	1	1	0	0	1	1	0
g_3	0	0	0	0	0	0	0	1	1	1	1	1	1	1	1	0
g_2	0	0	0	1	1	1	1	1	1	1	1	0	0	0	0	0
g_1	0	1	1	1	1	0	0	0	0	1	1	1	1	0	0	0
g_0	1	1	0	0	1	1	0	0	1	1	0	0	1	1	0	0

Table-2

From the table-2, read the equations for g_3, g_2, g_1 and g_0.
The answer g_2 will match.

$g_2(h_3, h_2, h_1, h_0)$
$= \Sigma m(2, 4, 5, 6, 7, 12, 13, 15)$

1.30 (d)

Here Q'S' is essential prime implicant.

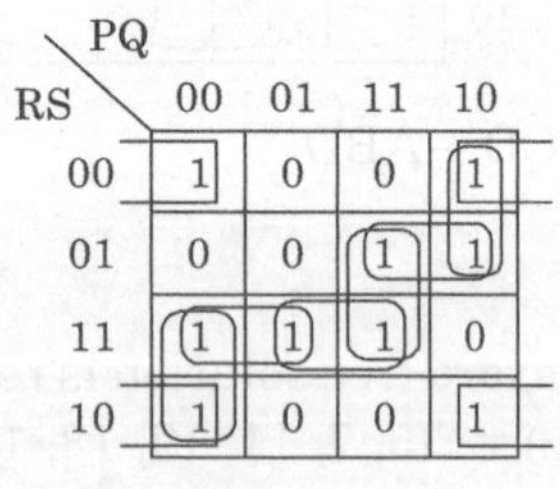

1.31 (c)

Let the binary constant a_i be the value of the function $f(x_1, x_2, ..., x_n)$ for the combination of variable whose decimal code is i. Then every switching function can be expressed in the form.
$f(x_1,x_2,...,x_n) = a_0x'_1...x'_n + a_1x'_1x'_2...x_n + a_rx_1x_2...x_n$.
A factor a_2 is either 0 or 1 if the corresponding minterm is contained in canonical from of the function. Then there are $r = 2^n$ coefficients each of which can have values either o or 1 hence there are 2^{2^n} possible assignments.

OR

Let us consider n = 2, A and B are two boolean variables. The possible combinations are four, they are

	0	1	2	3
A	0	0	1	1
B	0	1	0	1

1. Taking only one combination as a minterms the number of boolean functions are ${}^4C_1$.
2. Taking any two combinations as minterms the number of boolean functions are ${}^4C_2$.
3. Taking any three combinations are ${}^4C_3$.
4. Taking all four combinations as minterms, the number of boolean functions are ${}^4C_4$.
5. Taking no combinations as minterms, the number of boolean functions are ${}^4C_0$.

Total number of functions on '2' boolean variables are

$${}^4C_0 + {}^4C_1 + {}^4C_2 + {}^4C_3 + {}^4C_4$$

$$= 1 + 4 + 6 + 4 + 1 = 16 = 2^{2^2}$$

Similar on 'n' boolean variables total number of boolean functions are

$${}^{2^n}C_0 + {}^{2^n}C_1 + {}^{2^n}C_2 + ... + {}^{2^n}C_{2^n} \text{ is equal to } 2^{2^n}.$$

1.32 (b)

Given f(w, x, y, z) = Σ(1,3,4,6,9,11,12,14)
Corresponding K-map for f is

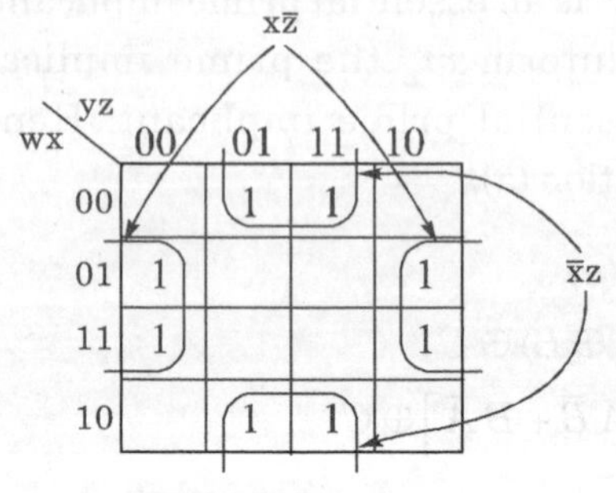

$$f(w, x, y, z) = xz' + x'z$$

So f(w, x, y, z) is independent of two variables w and y.

1.33 (c)

$$(a \cdot b)c + (\bar{a} \cdot c)d + (bc)d + ad$$

Substitude AND by NAND.

$$= \overline{\overline{ab}\,c} + \overline{\overline{\bar{a}c}\,d} + \overline{\overline{bc}\,d} + \overline{ad}$$

$$= ab + \bar{c} + \bar{a}c + \bar{d} + bc + \bar{d} + \bar{a} + \bar{d}$$

$$= b + \bar{c} + \bar{d} + \bar{a}$$

$$= \bar{a} + b + \bar{c} + \bar{d}$$

Note: We used the following properties.

$$\bar{a} + \bar{a}c = \bar{a}$$
$$\bar{a} + ab = \bar{a} + b$$
$$\bar{c} + bc = \bar{c} + b$$

1.34 (a)

$$f = a\bar{d} + \bar{a}\bar{c} + b\bar{c}d$$

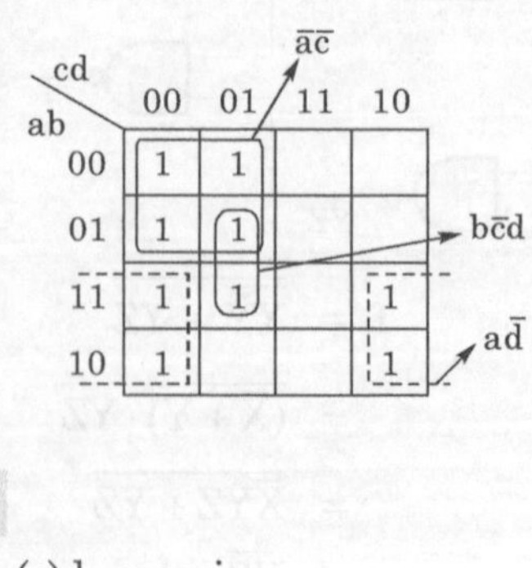

1.35 (c)

Option (c) k-map is

$$f = \bar{a}\bar{c} + a\bar{d} + ab\bar{c} + \bar{c}d$$

	c'd'	c'd	cd	cd'
a'b'	1	1		
a'b	1	1		
ab	1	1		1
ab'	1	1		1

It is not equivalent to given k-map.

1.36 (d)

For minterm m_{15}, the given function output must be 1.

Verify each option while substituting w = 1, x = 1, y = 1, z = 1.

(a) $x'y'z' + w'xy' + wy'z + xz = 0·0·0 + 0·1·0 + 1·0·1 + 1·1 = 1$

(b) $w'y'z' + wx'y' + xz = 0·0·0 + 1·0·0 + 1·1 = 1$

(c) $w'y'z' + wx'y' + xyz + xy'z = 0·0·0 + 1·0·0 + 1·1·1 + 1·0·1 = 1$

(d) $x'y'z' + wx'y' + w'y = 0·0·0 + 1·0·0 + 0·1 = 0$

So option (d) is not equivalent to f.

1.37 (d)

$$X * Y = XY + X'Y'$$
$$\& \quad Z = X * Y = XY + X'Y'$$
$$\bar{Z} = X'Y + XY'$$
$$P : Y * Z = YZ + Y'Z'$$
$$= Y[XY + X'Y'] + Y'[X'Y + XY']$$
$$= XY + XY'$$
$$= X \quad \text{True}$$
$$Q : X * Z = XZ + X'Z'$$
$$= X[XY + X'Y'] + X'[X'Y + XY']$$
$$= XY + X'Y = Y \text{ True}$$
$$R : \underline{X * Y} * Z = Z * Z$$
$$= 1 \quad \text{True}$$

1.38 (d)

$$o/p = \overline{\overline{X_1X_2X_3X_4} + X_4 + X_3X_4}$$
$$= \overline{\overline{X_1X_2X_3X_4} + X_4} \quad \text{...(1) without fault}$$
$$o/p = \overline{\overline{X_1X_2X_3X_4} + X_4 + T}$$

with fault T = 0

$$\therefore o/p = \overline{\overline{X_1X_2X_3X_4} + X_4 + 0}$$
$$= \overline{\overline{X_1X_2X_3X_4} + X_4} \quad \text{...(2) with fault}$$

equation 1 = equation 2

∴ There is no possible set to detect the fault.

1.39 (a)

The K-Map is

cd \ ab	00	01	11	10
00	1	1		1
01	x			
11	x			
10	1	1		x

$a'd' + b'd'$

1.40 (c)

$$f = f_1 \cdot f_2 + f_3$$

f_1 and f_2 are connected with AND gates so when $f_2 = \Sigma m(6, 8)$ then f is true for $\Sigma m(1, 6, 8, 15)$

1.41 (a)

EX-NOR is not functionally complete {Implication, Negation} is functionally complete {OR, Negation} is functionally complete NAND is functionally complete.

1.42 (a)

$$(P + \bar{Q})\,(\underline{P\bar{Q} + PR})(\underline{\bar{P}\bar{R} + \bar{Q}}) = (P + \bar{Q})(P\bar{Q} + P\bar{Q}R)$$
$$= (P + \bar{Q})P\bar{Q} = P\bar{Q}$$

1.43 (a)

$f(A, B, C, D) = \Sigma m(2, 3, 6, 7, 8, 9, 10, 11, 12, 13)$

CD \ AB	00	01	11	10
00			1	1
01			1	1
11	1	1		1
10	1	1		1

$$f = AC' + AB' + A'C.$$

Out of 4 variables A, B and C is present and 1 variable D is not present.

Hence, independent of D variable.

1.44 (b)

AB + C = (A+C)(B+C) using distributive property. The modified expression is in POS from. Hence 2 level NOR-NOR circuit is required using 3 NOR gates.

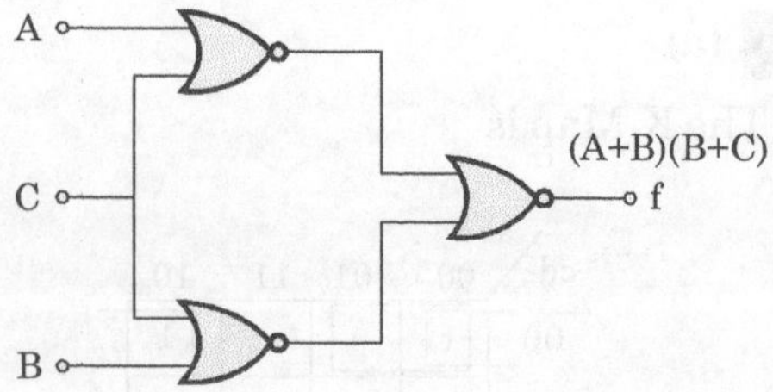

1.45 (a)

$f(P, Q, R) = PQ + Q\overline{R} + P\overline{R}$

Make this as a standard form

$= PQ(R+\overline{R}) + (P+\overline{P})Q\overline{R} + P(Q+\overline{Q})\overline{R}$

$= PQR + PQ\overline{R} + PQ\overline{R} + \overline{P}Q\overline{R} + PQ\overline{R} + P\overline{Q}\overline{R}$

$= PQR + PQ\overline{R} + \overline{P}Q\overline{R} + P\overline{Q}\overline{R}$ $\quad [\because A + A = A]$

$= m_7 + m_6 + m_2 + m_4$

1.46 (b)

$(P + \overline{Q} + \overline{R})\ (P + \overline{Q} + R)\ (P + Q + \overline{R})$

$(P+\overline{Q})(P+Q+\overline{R})$

$P + PQ + P\overline{R} + P\overline{Q} + \overline{Q}\overline{R}$

$(P + \overline{Q}\overline{R})$

1.47 (a)

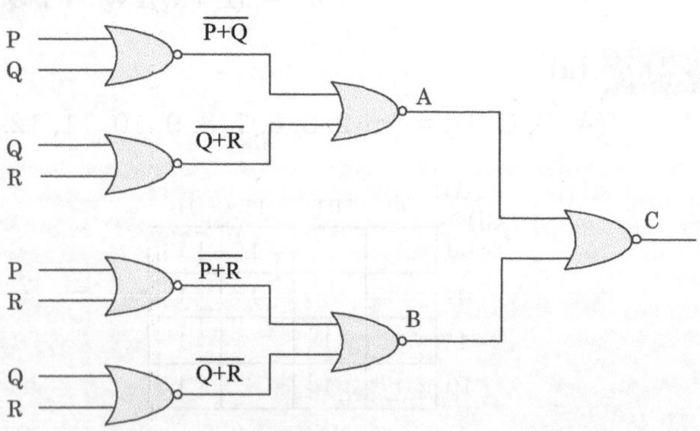

A and B are the outputs from 2 level NOR-NOR circuits.

$A = (P + Q)\ (Q + R)$

$= Q + PR \leftarrow$ using distributivity property

$B = (P + R)\ (Q + R)$

$= R + PQ \leftarrow$ using distributivity property

$C = \overline{A + B}$

$= \overline{Q + PR + R + PQ}$

$= \overline{Q + R}$

1.48 (d)

(a)

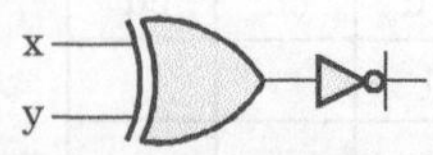

$\overline{x \oplus y} = \overline{\overline{x}y + \overline{y}x} = (x+\overline{y})(y+\overline{x})$

$= xy + x\overline{x} + \overline{y}y + \overline{y}\overline{x} = xy + \overline{x}\overline{y} = x \odot y$

(b)

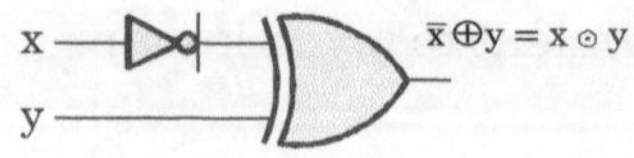

(c)

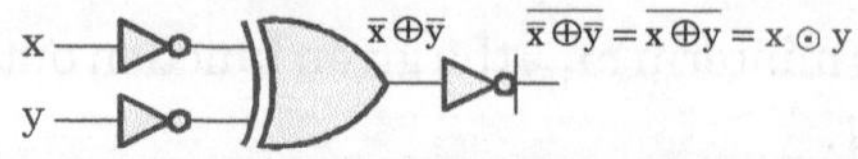

(d)

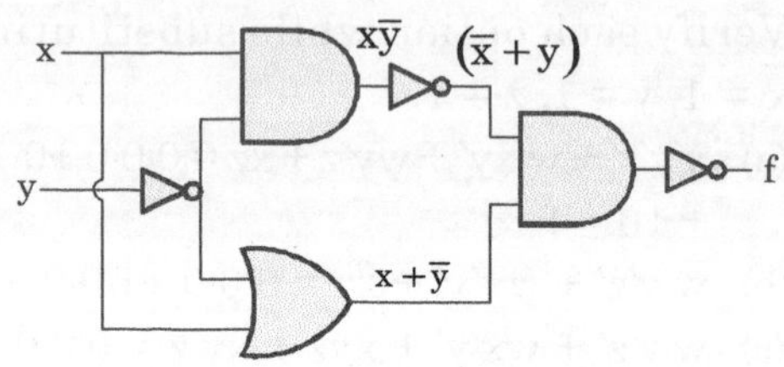

$f = \overline{(\overline{x}+y)(x+\overline{y})} = \overline{(\overline{x}+y)} + \overline{(x+\overline{y})}$

$= x.\overline{y} + \overline{x}y = x \oplus y$

$\neq x \odot y$

$\therefore$ Option (d) is not XNOR.

1.49 (b)

Two quads are possible here. So $\overline{b}\,\overline{d} + \overline{b}\,\overline{c}$ is the right answer.

cd \ ab	00	01	11	10
00	1	×	×	1
01	×			1
11				
10	1			×

1.50 (a)

X	Y	f(X, Y)
0	0	0
0	1	0
1	0	1
1	1	1

Now equation for $f(X, Y) = X\overline{Y} + XY$

$\Rightarrow X(\overline{Y} + Y) \Rightarrow X$

1.51 (d)

Exclusive NOR of x and y

$$= \overline{(x \oplus y)}$$

$$\overline{(x \oplus y)} = (xy' + x'y)'$$
$$= ((x' + y)(x + y'))$$
$$= x'y' + xy \quad \ldots(i)$$

Option (a) : is same as (i)

Options (b) : $x \oplus y' = x'y' + xy$ same as (i)

Option (c) : $x' \oplus y = xy + x'y'$ same as (i)

Option (d) $x \oplus y = x'y + xy'$ not same as (i)

Option (d) is correct.

1.52 (a)

$$F = PQ + \overline{P}QR + \overline{P}Q\overline{R}S$$
$$F = PQ + QR + QS$$

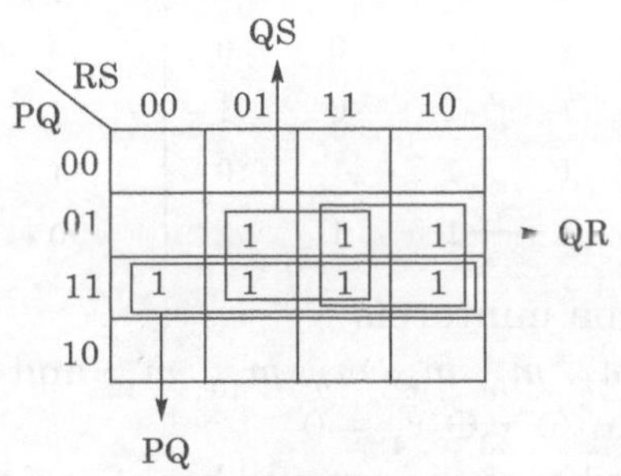

F = PQ + QR + QS

1.53 (d)

n = 3 (3 boolean variables)

	A	B	C
0	0	0	0
1	0	0	1
2	0	1	0
3	0	1	1
4	1	0	0
5	1	0	1
6	1	1	0
7	1	1	1

$\left.\begin{matrix} 0 \rightarrow & \bar{x}\,\bar{y}\,\bar{z} \\ 7 \rightarrow & x\,y\,z \end{matrix}\right\}$ (0, 7) mutual exclusive pair

$\left.\begin{matrix} 1 \rightarrow & \bar{x}\,\bar{y}\,z \\ 6 \rightarrow & x\,y\,\bar{z} \end{matrix}\right\}$ (0, 6) mutual exclusive pair

$\left.\begin{matrix} 2 \rightarrow & \bar{x}\,y\,\bar{z} \\ 5 \rightarrow & x\,\bar{y}\,z \end{matrix}\right\}$ (2, 5) mutual exclusive pair

$\left.\begin{matrix} 3 \rightarrow & \bar{x}\,y\,z \\ 4 \rightarrow & x\,\bar{y}\,\bar{z} \end{matrix}\right\}$ (3, 4) mutual exclusive pair

In a self dual function

Number of minterms = Number of maxterms

(i.e., the function must be neutral function)

(0, 7), (1, 6), (2, 5), (3, 4) from these each mutual exclusional pair, only one combination must present to make the function is to be self dual.

Hence (0, 7) (1, 6) (2, 5) (3, 4)

Posibilities: $2 \times 2 \times 2 \times 2 = 16 = 2^{2^{3-1}}$

similarly with n boolean variables.

$(i, j), (k, l), (\,) \ldots, (\,) \Rightarrow 2^{n-1}$ number of mutual exclusive pairs

Hence $2 \times 2 \times 2 \times \ldots \times 2 = 2^{2^{n-1}}$

Answer = $2^{2^{n-1}}$.

1.54 (b)

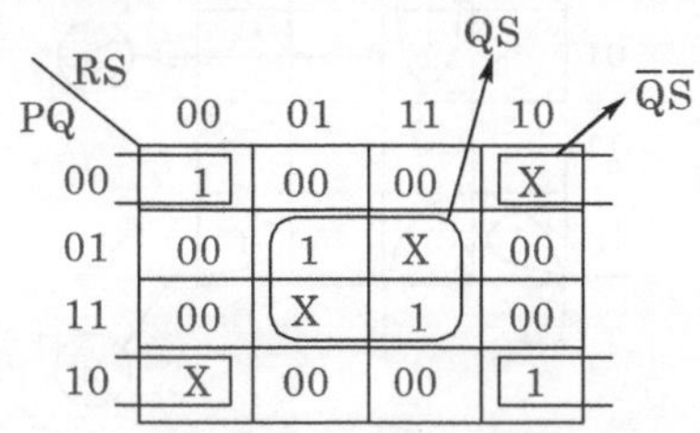

Minimal SOP form = $QS + \overline{Q}\overline{S}$

1.55 (d)

$$F(P, Q) = [(1 \oplus P) \oplus (P \oplus Q)] \oplus [(P \oplus Q) \oplus (Q \oplus 0)]$$
$$= [\overline{P} \oplus P \oplus Q] \oplus [P \oplus Q \oplus Q]$$
$$= [1 \oplus Q] \oplus [P \oplus 0]$$
$$= \overline{Q} \oplus P = \overline{P \oplus Q}$$

1.56 (b)

f(X,Y,Z) = X′ Y Z + X Y′ +Y′ Z′

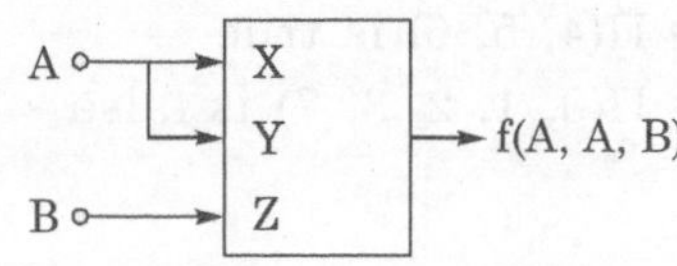

f(A, A, B) = A′AB + AA′ + A′B′ = A′B′ = (A + B)′

it is NOR operation.

Hence it is functionally complete.

Similarly if we try on g(x, y, z), it is not functionally complete.

1.57 (1)

[D′ + AB′ + A′C + AC′D + A′C′D]′

= [D′ + AC′D + AB′ + A′C + A′C′D]′

= [D′ + AC′ + AB′ + A′ [C + C′D]]′

= [D′ + AC′ + AB′ + A′ [C + D]]′

= [D′ + AC′ + AB′ + A′C + A′D]′

($\because$ D′ + A′D = D′ + A′)
= [D′ + A′ + AC′ + AB′ + A′C]′
($\because$ A′ + A′C = A′)
($\because$ A′ + AC′ + AB′ = A′+A(C′+B′)
= A′+C′ + B′)
= [D′ + A′ + C′ + B′]′ = ABCD

1.58 (3)

$f(w, x, y, z) = \Sigma(0, 2, 4, 5, 6, 10)$

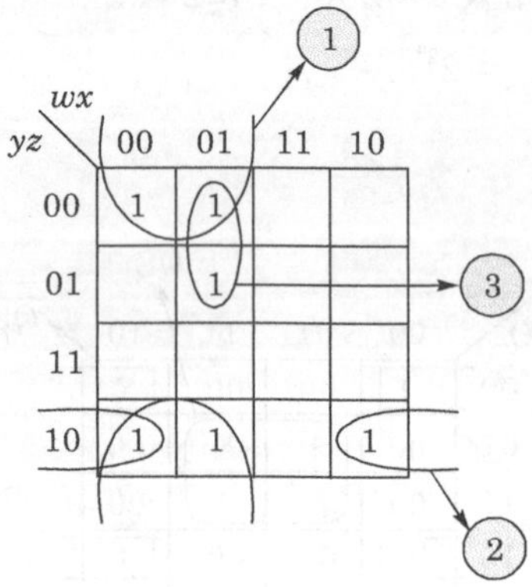

$\therefore$ 3 prime implicants.

1.59 (a)

$$F = P' + QR$$

P \ QR	00	01	11	10
0	1^0	1^1	1^3	1^2
1	0^4	0^5	1^7	0^6

$\Rightarrow \quad F = \Sigma(0, 1, 2, 3, 7) = \pi(4, 5, 6)$

S1: $F = \Sigma(4, 5, 6)$ is false
S2: $F = \Sigma(0, 1, 2, 3, 7)$ is true
S3: $F = \Pi(4, 5, 6)$ is true
S4: $F = \Pi(0, 1, 2, 3, 7)$ is false

1.60 (a)

x	y	$x \# y$
0	0	0
0	1	1
1	0	1
1	1	0

$x \# 0 = x$ ...1

$x \# 1 = \bar{x}$ ...2

$x \# x = 0$ ...3

$x \# \bar{x} = 1$ ...4

$x \# y = x\bar{y} + \bar{x}y$

1.61 (c)

It is given $x_1 \oplus x_2 \oplus x_3 \oplus x_4 = 0$

	x_1	x_2	x_3	x_4	$x_1 \oplus x_2 \oplus x_3 \oplus x_4$
0	0	0	0	0	0 ←
1	0	0	0	1	1
2	0	0	1	0	1
3	0	0	1	1	0 ←
4	0	1	0	0	1
5	0	1	0	1	0 ←
6	0	1	1	0	0 ←
7	0	1	1	1	1
8	1	0	0	0	1
9	1	0	0	1	0 ←
10	1	0	1	0	0 ←
11	1	0	1	1	1
12	1	1	0	0	0 ←
13	1	1	0	1	1
14	1	1	1	0	1
15	1	1	1	1	0 ←

For the minterms
$m_0, m_3, m_5, m_6, m_9, m_{10}, m_{12}$ and m_{15}
$x_1 \oplus x_2 \oplus x_3 \oplus x_4 = 0$
[i.e. whenever even number of variables are 1, $x_1 \oplus x_2 \oplus x_3 \oplus x_4 = 0$]
Option (a) fails whenever

$$x_1 = x_2 = x_3 = x_4 = 1$$

Option (b) fails whenever
x_1, x_2, x_3 and x_4 are 0 1 0 0 [and for few more combinations]
Options (d) fails whenver
x_1, x_2, x_3 and x_4 are 0 0 1 1 [and for few more combinations]
Options (c) satisfies for every combination.
Hence option (c) is correct.

1.62 (1)

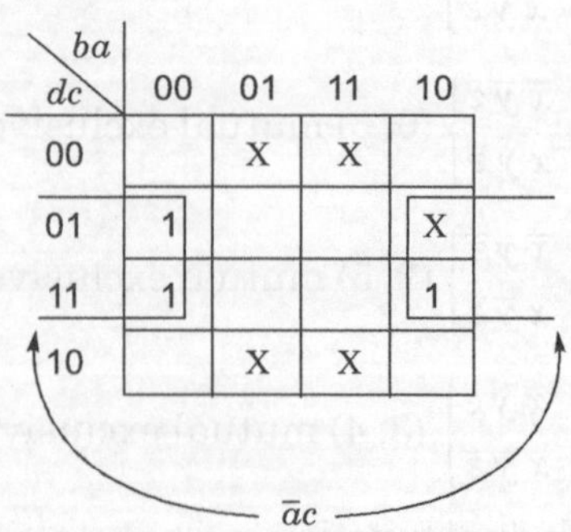

$$f = \bar{a}c = \overline{a + \bar{c}}$$

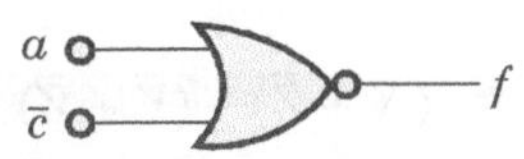

$\therefore$ Number of NOR gates required = 1.

1.63 (a)

$$f(w, x, y, z) = \Sigma_m\{0,1,2,3,7,8,10) + d(5, 6, 11, 15)$$
$$= \Pi_M\{4,9,12,13,14) \,.\, d(5,6,11,15)$$

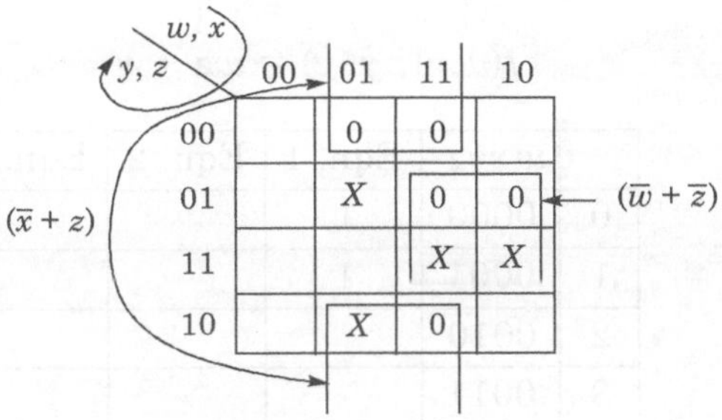

$$f(w, x, y, z) = (\overline{w}+\overline{z})(\overline{x}+z)$$

1.64 (c)

(a) $wx + w(x + y) + x(x + y) = x + wy$

$wx + wx + wy + x + xy$

$x(w + 1 + y) + wy$

$x + wy = x + wy$ True

(b) $\overline{w\overline{x}(y+\overline{z})}+\overline{w}x = \overline{w}+x+\overline{y}z$

$\overline{w}+\overline{w}x+x+\overline{y}z$

$\overline{w}+x+\overline{y}z = \overline{w}+x+\overline{y}z$ True

(c) $(w\overline{x}(y+x\overline{z})+\overline{w}\overline{x})y = x\overline{y}$

$(w\overline{x}y+\overline{w}\overline{x})y$

$w\overline{x}y+\overline{w}\overline{x}y$

$\overline{x}y(w+\overline{w})$, since $(w+\overline{w})$ evaluate to 1

$\overline{x}y \neq x\overline{y}$ False

(d) $(w + y)(wxy + wyz) = wxy + wyz$

$wxy + wyz = wxy + wyz$ True

1.65 (d)

(A) $\overline{P\oplus Q} = P\odot Q$ true

(B) $\overline{P}\oplus Q = P\odot Q$ true

(C) $\overline{P}\oplus\overline{Q} = P\oplus Q$ true

(D) $(P\oplus\overline{P})\oplus Q = (P\odot\overline{P})\odot\overline{Q}$ false

$(P\oplus\overline{P})\oplus Q = 1\oplus Q = \overline{Q}$

$(P\odot\overline{P})\odot\overline{Q} = 0\oplus\overline{Q} = Q$

$\therefore$ $(P\oplus\overline{P})\oplus Q \neq (P\odot\overline{P})\odot\overline{Q}$

1.66 (3)

F(P,Q,R,S) = Σm(0, 2, 5, 7, 9, 11) + d(3, 8, 10, 12, 14)

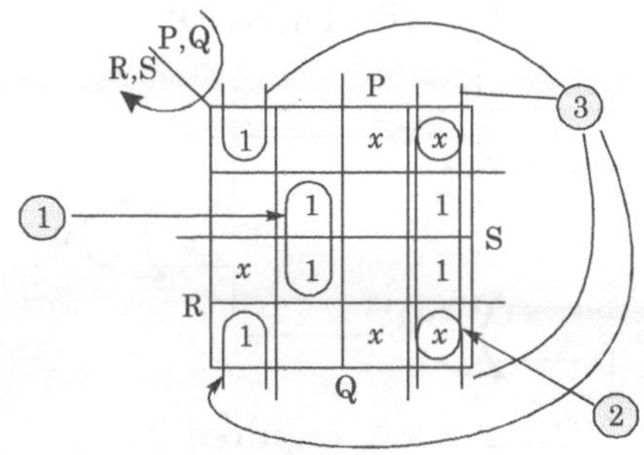

Number of EPI = 3

1.67 (d)

(a) $x \oplus y = (xy+x'y')' = (x\odot y)'$

$= x \oplus y$, it is valid.

(b) $x \oplus y = (x+y)(\overline{x}+\overline{y})$

$= (x+y)\,\overline{xy} = (x+y)\overline{0}$

$= (x + y)$, so it is valid.

(c) $(x \oplus y) \oplus z = x \oplus (y \oplus z)$

Associativity is true on Ex-OR operator so it valid.

(d) $(x + y) \oplus z = \overline{(x+y)}\cdot z+(x+y)\overline{z}$

$= \overline{x}\,\overline{y}z + x\overline{z} + y\overline{z}$

1   4,6   2,6

$= \Sigma m(1, 2, 4, 6)$

$x \oplus (y + z) = \overline{x}(y+z)+x\overline{(y+z)}$

$= \overline{x}y+\overline{x}z+x\overline{y}\,\overline{z}$

2,3   1,3   4

$= \Sigma m(1, 2, 3, 4)$

$(x + y) \oplus z \neq x \oplus (y + z)$

So option (d) is invalid.

1.68 (3)

$$f = \Sigma_m(0, 2, 5, 7, 8, 10, 13, 15)$$
$$f = \Pi_M(1, 3, 4, 6, 9, 11, 12, 14)$$

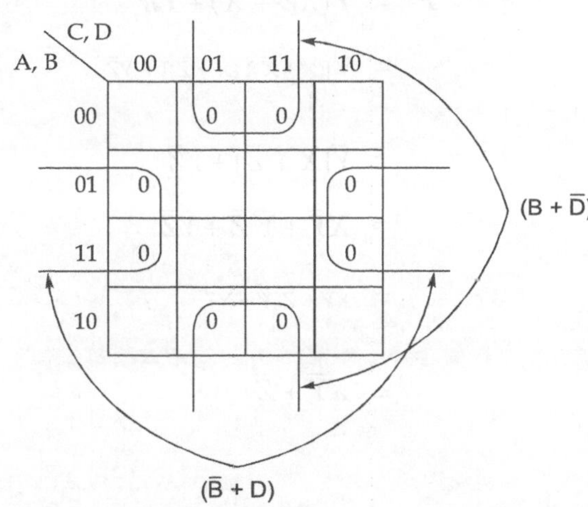

$$f = (\bar{B}+D)(B+\bar{D})$$

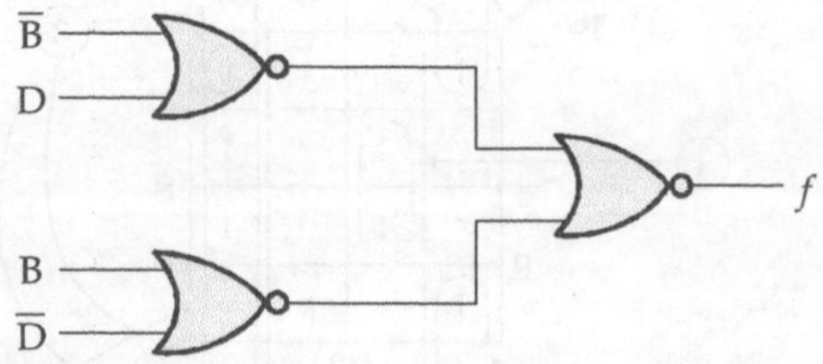

So 3 NOR gates is required.

1.69 (a)

$$f_1 \cdot f_2 = \Sigma(2, 8, 14)$$
$$f = f_3 \oplus (f_1 \cdot f_2) = \Sigma(7, 8, 11)$$

1.70 (b)

$$\therefore \quad z(a, b, c) = a + \bar{b}c$$

K-map of the output z is

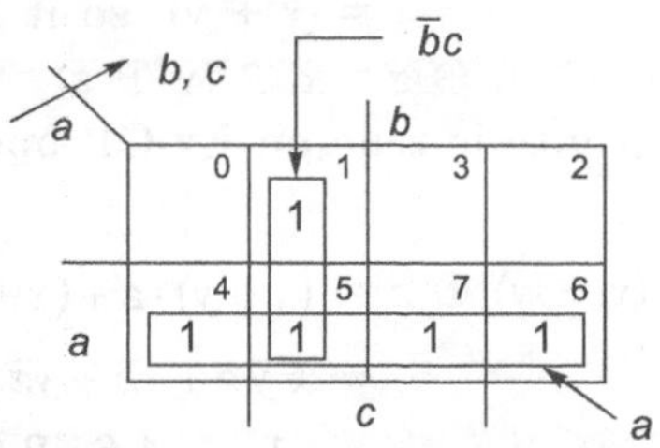

$$\therefore \quad z(a, b, c) = \Sigma m(1, 4, 5, 6, 7)$$

1.71 (a, c, d)

$$F = (X+Y+Z)(\bar{X}+Y)(\bar{Y}+Z)$$
$$\bar{F} = \overline{(X+Y+Z)(\bar{X}+Y)(\bar{Y}+Z)}$$
$$= \overline{(X+Y+Z)} + \overline{(\bar{X}+Y)} + \overline{(\bar{Y}+Z)}$$

Using Demorgan's theorem

$$= \bar{X}\bar{Y}\bar{Z} + X\bar{Y} + Y\bar{Z}$$

Option (a) is matching.

$$\bar{F} = \bar{Y}(\bar{X}\bar{Z} + X) + Y\bar{Z}$$
$$= \bar{Y}[\underbrace{(X+\bar{X})}_{1}](X+\bar{Z}) + Y\bar{Z}$$
$$= \bar{Y}[X+\bar{Z}] + Y\bar{Z}$$
$$= X\bar{Y} + \bar{Y}\bar{Z} + Y\bar{Z}$$
$$= X\bar{Y} + \bar{Z}(\underbrace{\bar{Y}+Y}_{1})$$
$$= X\bar{Y} + \bar{Z}$$

Option (d) is matching.

$$X\bar{Y} + \bar{Z} = (X+\bar{Z}) + (\bar{Y}+\bar{Z})$$

Using distributing property.
So option (c) is matching.
∴ So answer is (a), (c), (d).

1.72 (6)

$$f(w, 0\ 0, z) = 1 \quad \ldots(1)$$
$$f(1, x, 1, z) = x + z \quad \ldots(2)$$
$$f(w, 1, y, z) = wz + y \quad \ldots(3)$$

	wxyz	Eqn. 1	Eqn. 2	Eqn. 3	f
0	0000	1			1
1	0001	1			1
2	0010				×
3	0011				×
4	0100			0	0
5	0101			0	0
6	0110			1	1
7	0111			1	1
8	1000	1			1
9	1001	1			1
10	1010		0		0
11	1011		1		1
12	1100			0	0
13	1101			1	1
14	1110		1	1	1
15	1111		1	1	1

$$f(w, x, y, z) = \Sigma m(0, 1, 6, 7, 8, 9, 11, 13, 14, 15) + d(2, 3)$$

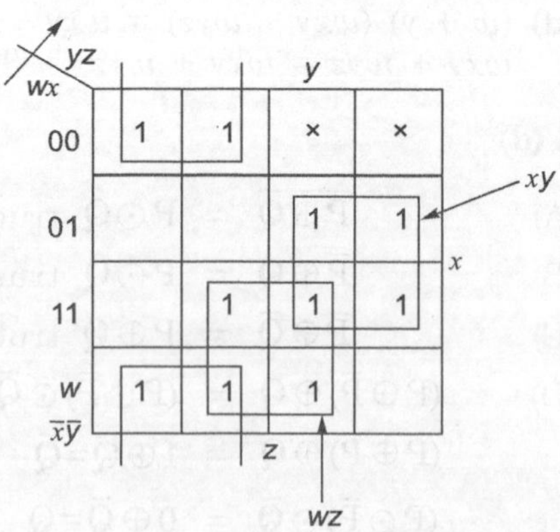

$$\therefore \quad f = \bar{x}\bar{y} + xy + wz$$

The number of literals in the minimal SOP expression is 6.

1.73 **(a, d)**

(a) If no propagation delay for AND and NOT gates, always one of the input of AND gate is '0', so output is always '0'.

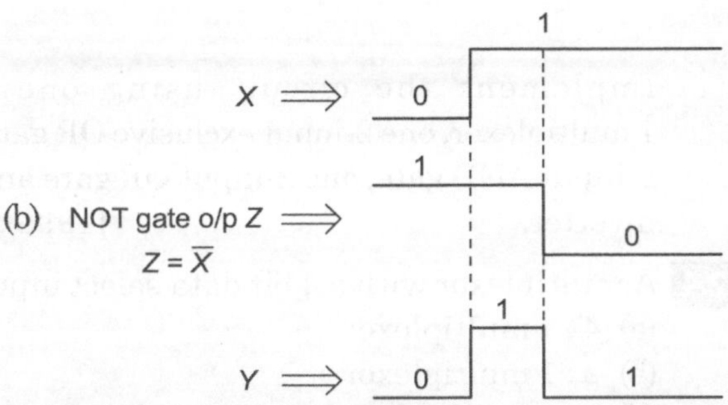

Output Y is logic 1 for small time interval whenever X changes from logic '0' to logic '1', assuming gates are having some delay. Answer is option (a), (d).

1.74 **(a, c)**

Given function $F(X, Y, Z) = \Sigma m(3, 5, 6, 7)$

In terms of max terms function is

$$F(X, Y, Z) = \Pi M(0, 1, 2, 4)$$

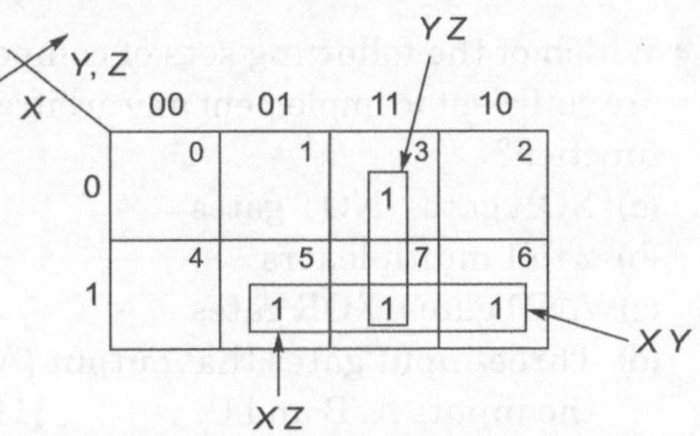

$$F(X, Y, Z) = XY + YZ + ZX$$

1.75 **(a, c)**

(a) $x \cdot x = 0$ False

(b) $x + \bar{x} = 1$ True

(c) $1 + x = x$ False

(d) $x \cdot 1 = x$ True

1.76 **(b, c)**

Output and AND gate $x = f_1 \cdot f_2$

$= \Sigma m(3, 5, 7, 11, 13)$

Output and OR gate $z = f_3 + f_4$

$= \Sigma m(0, 1, 2, 4, 6, 11, 13)$

Output of EX-OR gate $y = x \oplus z$

$= \Sigma m(0, 1, 2, 3, 4, 5, 6, 7)$

$= \Pi M(8, 9, 10, 11, 12, 13, 14, 15)$

■■■■

2 Combinational Circuits

2.1 Consider the circuit in figure which has a four bit binary number $b_3\,b_2\,b_1\,b_0$ as input and a five bit binary number $d_4\,d_3\,d_2\,d_1\,d_0$ as output. The circuit implements:

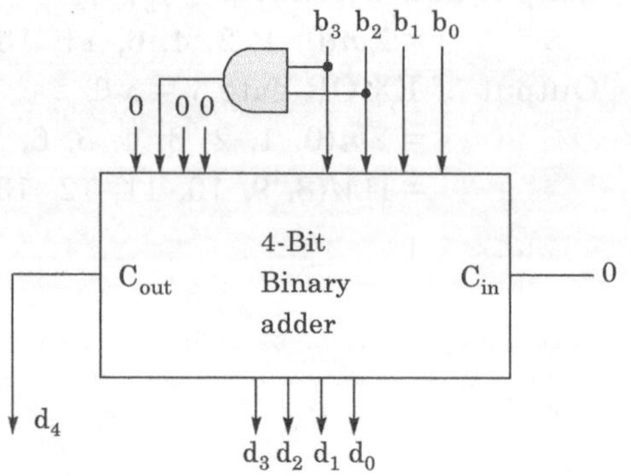

(a) Binary to Hex conversion
(b) Binary to BCD conversion
(c) Binary to grey code conversion
(d) Binary to radix-12 conversion **[1996 : 2 M]**

2.2 Consider the circuit in figure shown. f implements

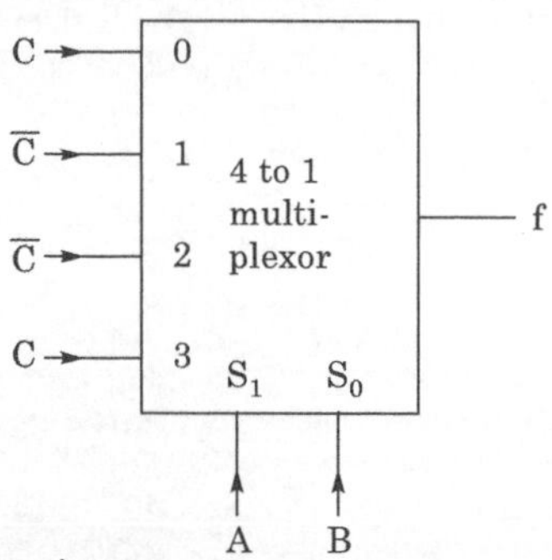

(a) $\bar{A}\bar{B}C + \bar{A}B\bar{C} + ABC$
(b) $A + B + C$
(c) $A \oplus B \oplus C$
(d) $AB + BC + CA$ **[1996 : 2 M]**

2.3 A logic network has two data inputs A and B and two control inputs C_0 and C_1. It implements the function F according to the following table.

C_1	C_0	F
0	0	$\overline{A+B}$
0	1	$A+B$
1	0	$A \oplus B$
1	1	AB

Implement the circuit using one 4 to 1 multiplexer, one 2-input exclusive OR gate, one 2-input AND gate, one 2-input OR gate and one inverter. **[1996 : 2 M]**

2.4 A multiplexor with a 4 bit data select input is a
(a) 4 : 1 multiplexor
(b) 2 : 1 multiplexor
(c) 16 : 1 multiplexor
(d) 8 : 1 multiplexor
[1998 : 1 M]

2.5 The number of full and half-adders required to add 16-bit numbers is:
(a) 8 half-adders, 8 full-adders
(b) 1 half-adders, 15 full-adders
(c) 16 half-adders, 0 full-adders
(d) 4 half-adders, 12 full-adders
[1999 : 1 M]

2.6 Which of the following sets of component(s) is/are sufficient to implement any arbitrary Boolean function?
(a) XOR gates, NOT gates
(b) 2 to 1 multiplexors
(c) AND gates, XOR gates
(d) Three-input gates that output (A.B) + C for the inputs A. B and C. **[1999 : 2 M]**

2.7 Consider the circuit shown below. The output of a 2 : 1 Mux is given by the function (ac' + bc).

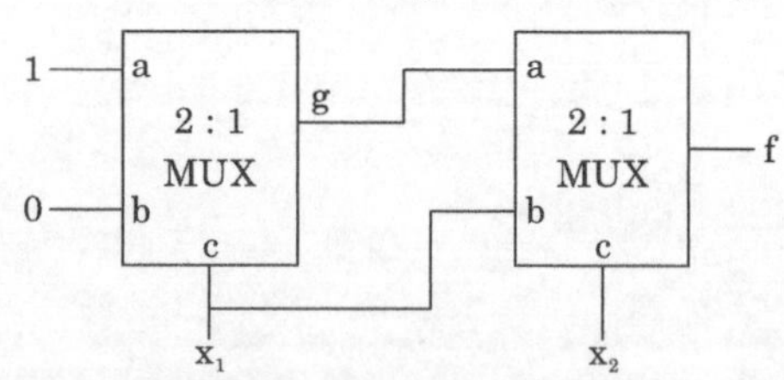

Which of the following is true?
(a) $f = x_1' + x_2$
(b) $f = x_1\,x_2 + x_1x_2'$
(c) $f = x_1\,x_2 + x_1'\,x_2'$
(d) $f = x_1 + x_2$
[2001 : 2 M]

2.8 Consider the following multiplexer where I0, I1, I2, I3 are four data input lines selected by two address line combinations A1A0 = 00, 01, 10, 11 respectively and f is the output of the multiplexer, EN is the Enable input.

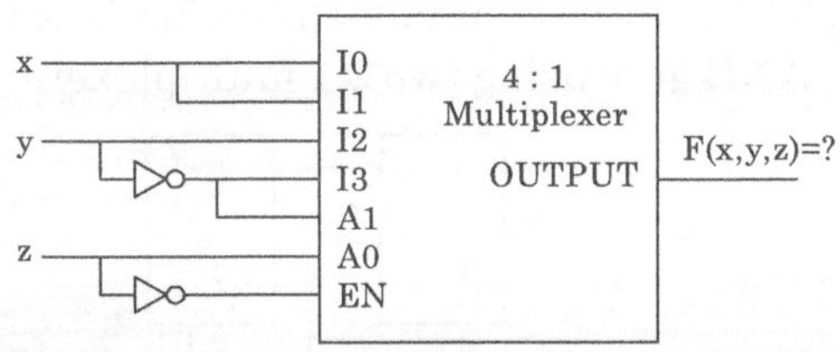

The function f(x, y, z) implemented by the above circuit is

(a) xyz′ (b) xy + z

(c) x + y (d) None of these

[2002 : 2 M]

2.9 Consider the ALU shown below:

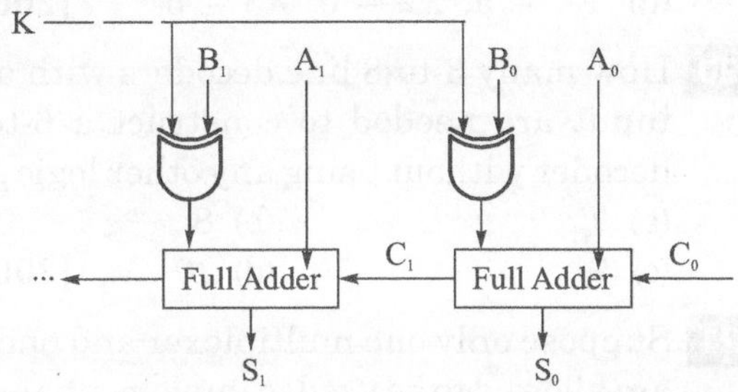

If the operands are in 2's complement representation, which of the following operations can be performed by suitably setting the control lines K and C_0 only (+ and – denote addition and subtraction respectively)?

(a) A + B, and A – B, but not A + 1

(b) A + B, and A + 1, but not A – B

(c) A + B, but not A – B, or A + 1

(d) A + B, and A – B, and A + 1 **[2003 : 2 M]**

2.10 A 4-bit carry lookahead adder, which adds two 4-bit numbers, is designed using AND, OR, NOT, NAND, NOR gates only. Assuming that all the inputs are available in both complemented and uncomplemented forms and the delay of each gate is one time unit, what is the overall propagation delay of the adder? Assume that the carry network has been implemented using two-level AND-OR logic.

(a) 4 time units (b) 6 time units

(c) 10 time units (d) 12 time units

[2004 : 2 M]

2.11 A circuit outputs a digit in the form of 4 bits. 0 is represented by 0000, 1 by 0001, ..., 9 by 1001. A combinational circuit is to be designed which takes these 4 bits as input and outputs 1 if the digit ≥ 5, and 0 otherwise. If only AND, OR and NOT gates may be used, what is the minimum number of gates required?

(a) 2 (b) 3

(c) 4 (d) 5 **[2004 : 2 M]**

2.12 Consider a multiplexer with X and Y as data inputs and Z as control input. If Z = 0 selects input X and Z = 1 selects input Y. What are the connections required to realize the 2-variable Boolean function *f* = T + R without using any additional hardware?

(a) R to X, 1 to Y, T to Z

(b) T to X, R to Y, T to Z

(c) T to X, R to Y, O to Z

(d) R to X, O to Y, T to Z **[2004 : 1 M]**

2.13 The circuit shown below implements a 2-input NOR gate using two 2 : 1 MUX (control signal 1 selects the upper input). What are the values of signals x, y and z?

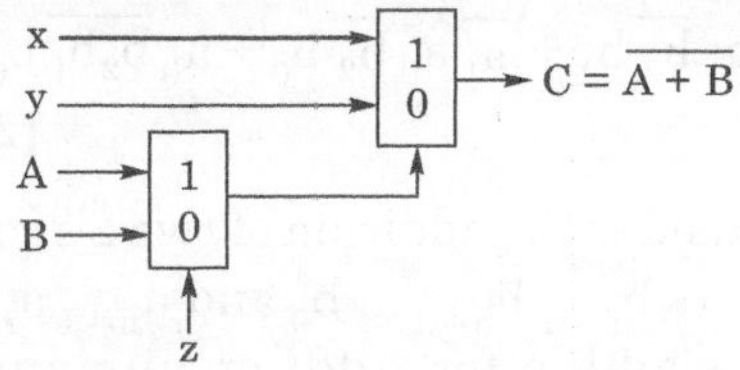

(a) 1, 0, B (b) 1, 0, A

(c) 0, 1, B (d) 0, 1, A

[2005 : 2 M]

2.14 The majority function is a Boolean function f(x, y, z) that takes the value 1 whenever a majority of the variables x, y, z and 1. In the circuit diagram for the majority function shown below, the logic gates for the boxes labeled P and Q are, respectively,

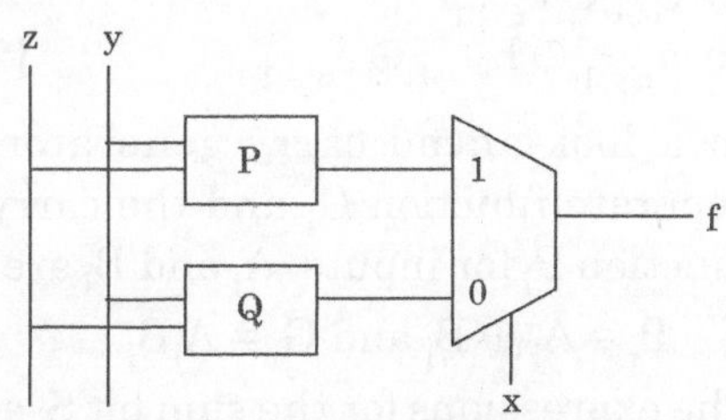

(a) XOR, AND (b) XOR, XOR

(c) OR, OR (d) OR, AND

[2006 : 2 M]

2.15 Consider the circuit below. Which one of the following options correctly represents f(x, y, z)?

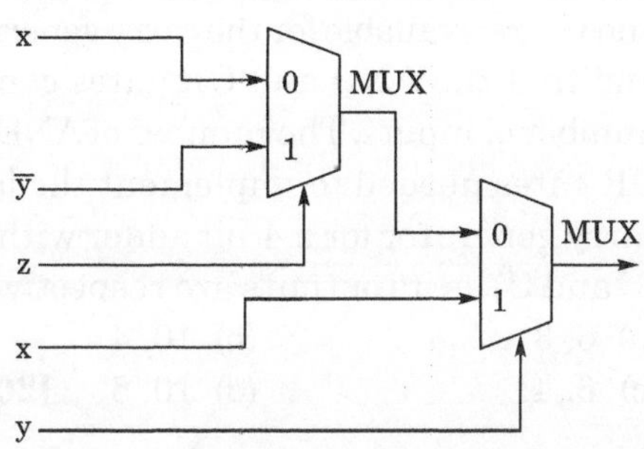

(a) $x\bar{z} + xy + \bar{y}z$ (b) $x\bar{z} + xy + y\bar{z}$

(c) $xz + xy + \overline{yz}$ (d) $xz + x\bar{y} + \bar{y}z$

[2006 : 2 M]

2.16 Given two 3-bit numbers $a_2a_1a_0$ and $b_2b_1b_0$ and c, the carry in, the function that represents the carry generate function when these two numbers are added is

(a) $a_2b_2 + a_2a_1b_1 + a_2a_1a_0b_0 + a_2a_0b_1b_0 + a_1b_2b_1 + a_1a_0b_2b_0 + a_0b_2b_1b_0$

(b) $a_2b_2 + a_2b_1b_0 + a_2a_1b_1b_0 + a_1a_0b_2b_1 + a_1a_0b_2 + a_1a_0b_2b_0 + a_2a_0b_1b_0$

(c) $a_2 + b_2 + (a_2 \oplus b_2)\,[a_1 + b_1 + (a_1 \oplus b_1)\,(a_0 + b_0)]$

(d) $a_2b_2 + \overline{a_2}\,a_1b_1 + \overline{a_2a_1}\,a_0b_0 + \overline{a_2}\,a_0\,\overline{b_1}\,b_0 + a_1\,\overline{b_2}\,b_1 + \overline{a_1}\,a_0\,\overline{b_2}\,b_0 + a_0\,\overline{b_2b_1}\,b_0$

[2006 : 2 M]

2.17 We consider the addition of two 2's complement numbers $b_{n-1}\,b_{n-2} \ldots b_0$ and $a_{n-1}\,a_{n-2} \ldots a_0$. A binary adder for adding unsigned binary numbers is used to add the two numbers. The sum is denoted by $c_{n-1}\,c_{n-2} \ldots c_0$ and the carryout by c_{out}.

Which one of the following options correctly identifies the overflow condition?

(a) $c_{out}\,\overline{(a_{n-1} \oplus b_{n-1})}$

(b) $a_{n-1}\,b_{n-1}\,\overline{c_{n-1}} + \overline{a_{n-1}\,b_{n-1}}\,c_{n-1}$

(c) $c_{out} \oplus c_{n-1}$

(d) $a_{n-1} \oplus b_{n-1} \oplus c_{n-1}$ **[2006 : 2 M]**

2.18 In a look-ahead carry generator, the carry generate function G_i and the carry propagate function P_i for inputs, A_i and B_i are given by

$$P_i = A_i \oplus B_i \text{ and } G_i = A_iB_i$$

The expressions for the sum bit S_i and carry bit C_{i+1} of the look-ahead carry adder are given by $S_i = P_i \oplus C_i$ and $C_{i+1} = G_i + P_iC_i$, where C_0 is the input carry.

Consider a two-level logic implementation of the look-ahead carry generator. Assume that all P_i and G_i are available for the carry generator circuit and that the AND and OR gates can have any number of inputs. The number of AND gates and OR gates needed to implement the look-ahead carry generator for a 4-bit adder with S_3, S_2, S_1, S_0 and C_4 as its outputs are respectively

(a) 6, 3 (b) 10, 4

(c) 6, 4 (d) 10, 5 **[2007 : 2 M]**

2.19 The following circuit implements a two-input AND gate using two 2-1 multiplexers.

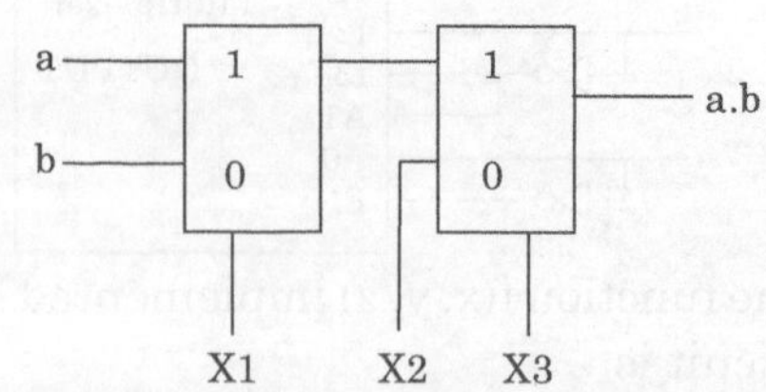

What are the values of X1, X2, X3?

(a) X1 = b, X2 = 0, X3 = a

(b) X1 = b, X2 = 1, X3 = b

(c) X1 = a, X2 = b, X3 = 1

(d) X1 = a, X2 = 0, X3 = b **[2007 : 1 M]**

2.20 How many 3-to-8 line decoders with an enable input are needed to construct a 6-to-64 line decoder without using any other logic gates?

(a) 7 (b) 8

(c) 9 (d) 10 **[2007 : 1 M]**

2.21 Suppose only one multiplexer and one inverter are allowed to be used to implement any Boolean function of n variables. What is the minimum size of the multiplexer needed?

(a) 2^n line to 1 line (b) 2^{n+1} line to 1 line

(c) 2^{n-1} line to 1 line (d) 2^{n-2} line to 1 line

[2007 : 2 M]

2.22 What Boolean function does the circuit below realize?

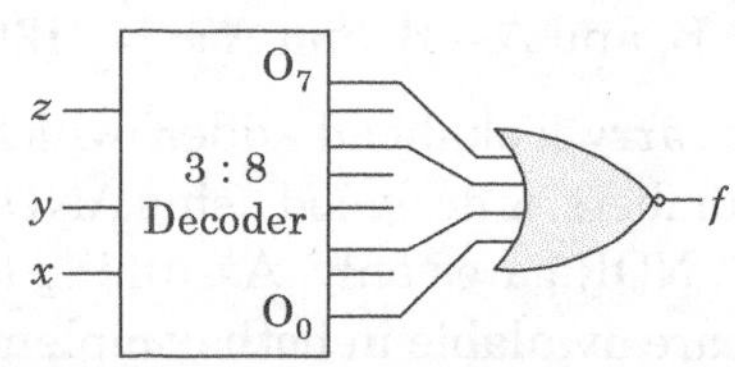

(a) $xz + \bar{x}\,\bar{z}$ (b) $x\bar{z} + \bar{x}z$

(c) $\bar{x}\,\bar{y} + yz$ (d) $xy + \bar{y}\,\bar{z}$

[2008 : 1 M]

2.23 The Boolean expression for the output f of the multiplexer shown below is

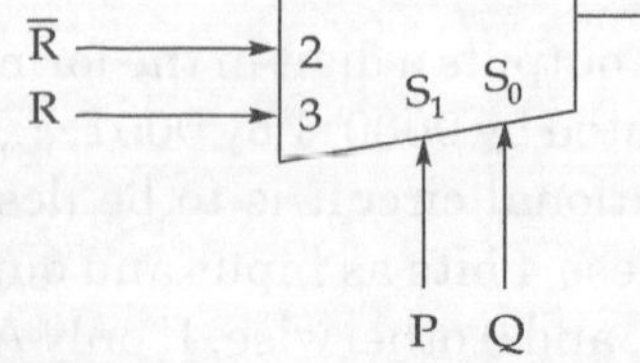

(a) $\overline{P \oplus Q \oplus R}$ (b) $P \oplus Q \oplus R$

(c) $P + Q + R$ (d) $\overline{P + Q + R}$

[2010 : 1 M]

2.24 In the following truth table, V = 1 if and only if the input is valid.

Inputs				Outputs		
D_0	D_1	D_2	D_3	X_0	X_1	V
0	0	0	0	x	x	0
1	0	0	0	0	0	1
x	1	0	0	0	1	1
x	x	1	0	1	0	1
x	x	x	1	1	1	1

What function does the truth table represent?

(a) Priority encoder (b) Decoder
(c) Multiplexer (d) Demultiplexer

[2013 : 1 M]

2.25 Consider the 4 to 1 multiplexer with two select lines S_1 and S_0 given below:

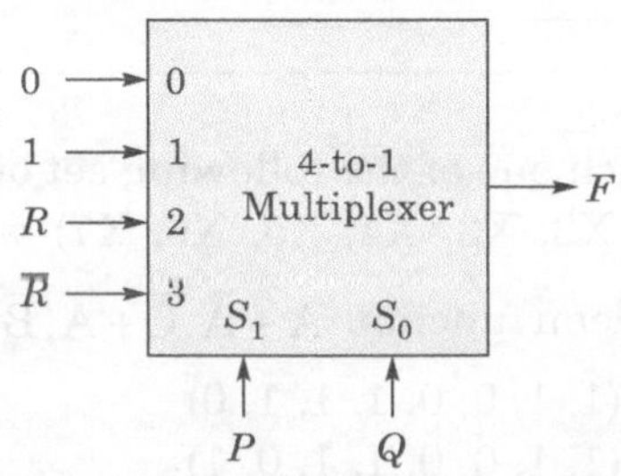

The minimal sum-of-products form of the Boolean expression for the output F of the multiplexer is

(a) $\bar{P}Q + Q\bar{R} + P\bar{Q}R$
(b) $\bar{P}Q + \bar{P}Q\bar{R} + PQ\bar{R} + P\bar{Q}R$
(c) $\bar{P}QR + \bar{P}Q\bar{R} + Q\bar{R} + P\bar{Q}R$
(d) $PQ\bar{R}$ **[2014 (Set-1) : 2 M]**

2.26 Consider the following combinational function block involving four Boolean variables x, y, a, b where x, a, b are inputs and y is the output.

```
f (x, y, a, b)
{
    if (x is 1) y = a;
    else y = b;
}
```

Which one of the following digital logic blocks is the most suitable for implementing this function?

(a) Full adder (b) Priority encoder
(c) Multiplexor (d) Flip-flop

[2014 (Set-3) : 1 M]

2.27 A half adder is implemented with XOR and AND gates. A full adder is implemented with two half adders and one OR gate. The propagation delay of an XOR gate is twice that of an AND/OR gate. The propagation delay of an AND/OR gate is 1.2 microseconds.

A 4-bit ripple-carry binary adder is implemented by using four full adders. The total propagation time of this 4-bit binary adder in microseconds is ________.

[2015 (Set-2) : 2 M]

2.28 Consider the two cascaded 2-to-1 multiplexers as shown in the figure.

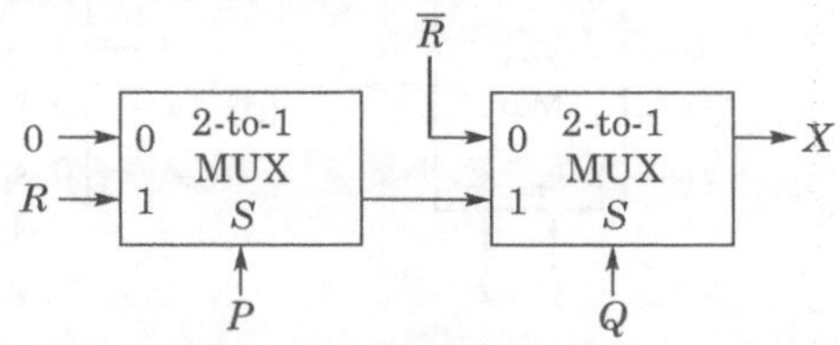

The minimal sum of products form of the output X is

(a) $\bar{P}\bar{Q} + PQR$ (b) $\bar{P}Q + QR$
(c) $PQ + \bar{P}\bar{Q}R$ (d) $\bar{Q}\bar{R} + PQR$

[2016 (Set-1) : 2 M]

2.29 Consider a carry lookahead adder for adding two n-bit integers, built using gates of fan-in atmost two. The time to perform addition using this adder is

(a) $\Theta(1)$ (b) $\Theta(\log(n))$
(c) $\Theta(\sqrt{n})$ (d) $\Theta(n)$

[2016 (Set-1) : 2 M]

2.30 Consider an eight bit ripple-carry adder for computing the sum of A and B, where A and B are integers represented in 2's complement form. If the decimal value of A is one, the decimal value of B that leads to the longest latency for the sum to stabilize is ________.

[2016 (Set-2) : 1 M]

2.31 When two 8-bit numbers $A_7 \ldots A_0$ and $B_7 \ldots B_0$ in 2's complement representation (with A_0 and B_0 as the least significant bits) are added using a **ripple-carry adder**, the sum bits obtained are $S_7 \ldots S_0$ and the carry bits are $C_7 \ldots C_0$. An overflow is said to have occured if

(a) the carry bit C_7 is 1
(b) all the carry bits ($C_7, \ldots, C_0$) are 1
(c) $\left(A_7 \cdot B_7 \cdot \bar{S}_7 + \bar{A}_7 \cdot \bar{B}_7 \cdot S_7\right)$ is 1
(d) $\left(A_0 \cdot B_0 \cdot \bar{S}_0 + \bar{A}_0 \cdot \bar{B}_0 \cdot S_0\right)$ is 1

[2017 (Set-1) : 1 M]

2.32 Which one of the following circuits implements the Boolean function given below?

$f(x, y, z) = m_0 + m_1 + m_3 + m_4 + m_5 + m_6$, where m_i is the ith minterm.

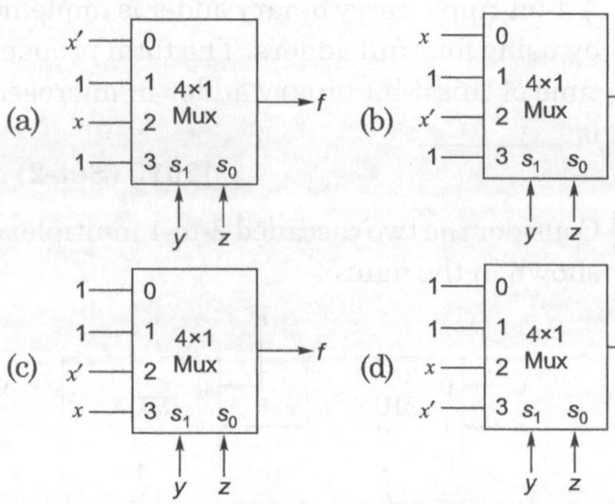

[2021 (Set-2) : 1 M]

2.33 Consider a digital display system (DDS) shown in the figure that displays the contents of register X. A 16-bit code word is used to load a word in X, either from S or from R. S is a 1024-word memory segment and R is a 32-word register file. Based on the value of mode bit M, T selects an input word to load in X. P and Q interface with the corresponding bits in the code word to choose the addressed word. Which one of the following represents the functionality of P, Q, and T?

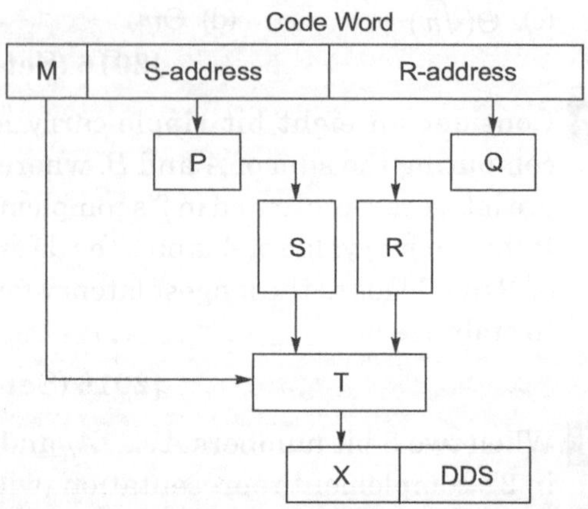

(a) P is 10 : 1 multiplexer; Q is 5 : 1 multiplexer; T is 2 : 1 multiplexer
(b) P is 10 : 2^{10} decoder; Q is 5 : 2^5 decoder; T is 2 : 1 encoder
(c) P is 10 : 2^{10} decoder; Q is 5 : 2^5 decoder; T is 2 : 1 multiplexer
(d) P is 1 : 10 de-multiplexer; Q is 1 : 5 de-multiplexer; T is 2 : 1 multiplexer

[2022 : 2 M]

2.34 A Boolean digital circuit is composed using two 4-input multiplexers (M1 and M2) and one 2-input multiplexer (M3) as shown in the figure. X0 – X7 are the inputs of the multiplexers M1 and M2 and could be connected to either 0 or 1. The select lines of the multiplexers are connected to Boolean variables A, B and C as shown.

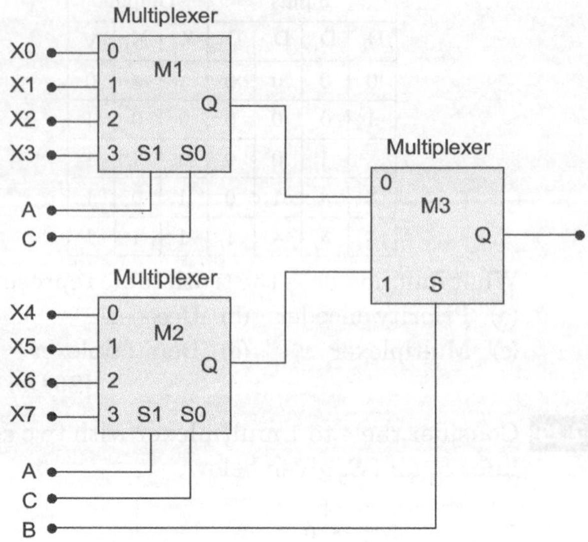

Which one of the following set of values of (X0, X1, X2, X3, X4, X5, X6, X7) will realise the Boolean function $A + \bar{A}.\bar{C} + A.\bar{B}.C$?

(a) (1, 1, 0, 0, 1, 1, 1, 0)
(b) (1, 1, 0, 0, 1, 1, 0, 1)
(c) (1, 1, 0, 1, 1, 1, 0, 0)
(d) (0, 0, 1, 1, 0, 1, 1, 1)

[2023 : 2 M]

2.35 Consider a digital logic circuit consisting of three 2-to-1 multiplexers M1, M2, and M3 as shown below. X1 and X2 are inputs of M1. X3 and X4 are inputs of M2. *A*, *B*, and *C* are select lines of M1, M2, and M3, respectively.

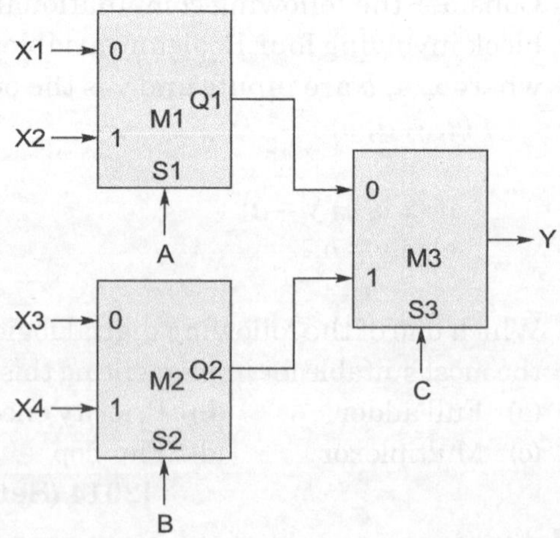

For an instance of inputs X1 = 1, X2 = 1, X3 = 0, and X4 = 0, the number of combinations of *A*, *B*, *C* that give the output Y = 1 is ______.

[2024 (Set-1) : 2 M]

■■■■

Answers Combinational Circuits

2.1	(d)	**2.2**	(c)	**2.3**	Sol.	**2.4**	(c)	**2.5**	(b)	**2.6**	(b, c)	**2.7**	(c)	**2.8**	(a)	**2.9**	(a)
2.10	(a)	**2.11**	(b)	**2.12**	(a)	**2.13**	(d)	**2.14**	(d)	**2.15**	(a)	**2.16**	(a)	**2.17**	(c)	**2.18**	(b)
2.19	(a)	**2.20**	(c)	**2.21**	(c)	**2.22**	(b)	**2.23**	(b)	**2.24**	(a)	**2.25**	(a)	**2.26**	(c)	**2.27**	(12)
2.28	(d)	**2.29**	(b)	**2.30**	(–1)	**2.31**	(c)	**2.32**	(d)	**2.33**	(c)	**2.34**	(c)	**2.35**	(4)		

Explanations Combinational Circuits

2.1 (d)

Whenever both b_3 and b_2 is 1, then 01 00 is added to the number.

$b_3b_2b_1b_0 = (1100)_B = (12)_{10}$

So in all the numbers greater than or equal to 12, 01 00 is added.

Therefore the circuit is binary to radix –12 converter.

2.2 (c)

The circuit for function f(A, B, C) is given as follows:

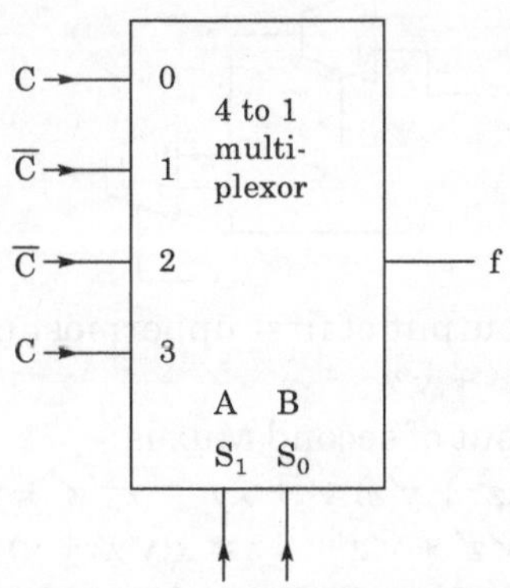

$$f(A, B, C) = \overline{A}\,\overline{B}C + \overline{A}B\overline{C} + A\overline{B}\,\overline{C} + ABC$$
$$= \Sigma m\,(1, 2, 4, 7) = A \oplus B \oplus C$$

2.3 Sol.

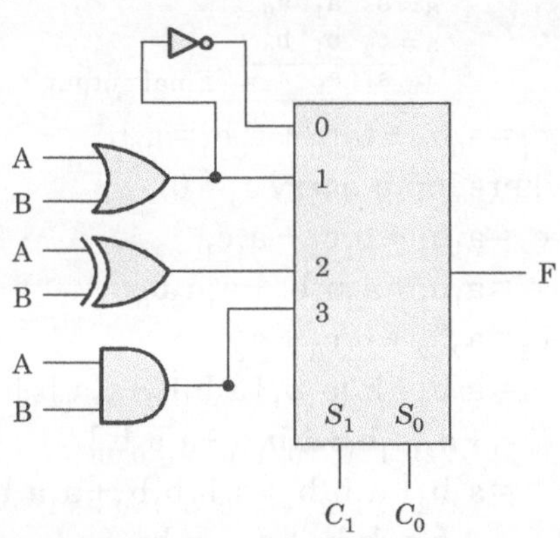

2.4 (c)

Number of select inputs = 4

MUX size $\Rightarrow 2^4 \times 1 = 16 : 1$ MUX

2.5 (b)

At LSB H.A, remaining places F.A are used.

Number of H.A = 1

Number of F.A = 15

2.6 (b, c)

Functionally complete operation set is a set of logic functions from which any arbitrary boolean function can be realized. Examples of functionally complete operation set are:

FC_1 = {OR, AND, NOT}

FC_2 = {NAND}

FC_3 = {NOR}

FC_4 = {XOR, AND} and 2 to 1 multiplexers.

2.7 (c)

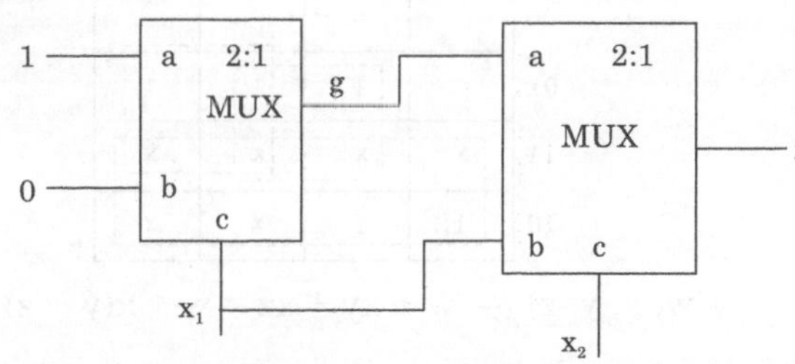

The output g is

$$g = 1\,x_1' + 0\,x_1 = x_1'$$

The output f is

$$f = gx_2' + x_1x_2$$
$$f = x_1'x_2' + x_1x_2$$

2.8 (a)

The output function f(x, y, z) is given as

f(x, y, z)

$$= EN\left(\overline{A}_1\overline{A}_0I_0 + \overline{A}_1A_0I_1 + A_1\overline{A}_0I_2 + A_1A_0I_3\right)$$
$$= z'(yz'x + yzx + y'z'y + y'zy')$$
$$= z'(xyz' + xyz + 0 + y'z)$$
$$= xyz' + xyzz' + 0.z' + z'.zy')$$
$$= xyz' + 0 + 0 + 0 = xyz'$$

Therefore, option (a) is true.

2.9 (a)

If K = 0 and $C_0 = 0$ than it is A + B
If K = 1 and $C_0 = 1$ than it is A – B
So it computes A + B, A – B and not A + 1

2.10 (a)

As per the problem description for generating carry the delay is 2 time units, and 2 time units delay is involved for generating sum output after knowing carry.
Hence the total delay is 4 time units.

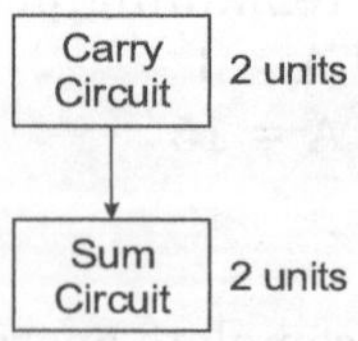

2.11 (b)

Let the 4-input is w, x, y, z then
$f(w, x, y, z) = \Sigma(5, 6, 7, 8, 9)$
The decimal values 10, 11, 12, 13, 14, 15 are don't care condition so
$f(w, x, y, z) = \Sigma d(10, 11, 12, 13, 14, 15)$
K-map:

wx \ yz	00	01	11	10
00				
01		1	1	1
11	x	x	x	x
10	1	1	x	x

$$f(w, x, y, z) = w + xy + xz = w + x(y + z)$$

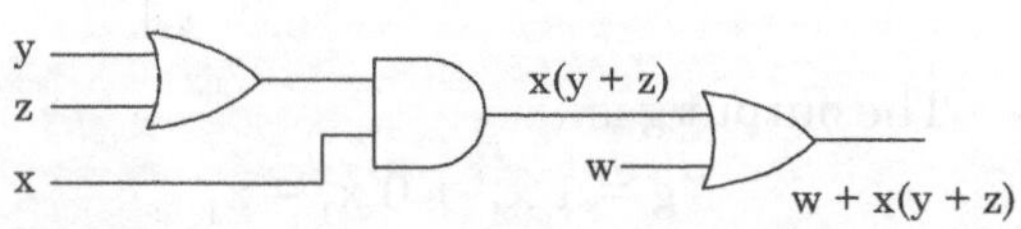

So, 3-Gates are required.

2.12 (a)

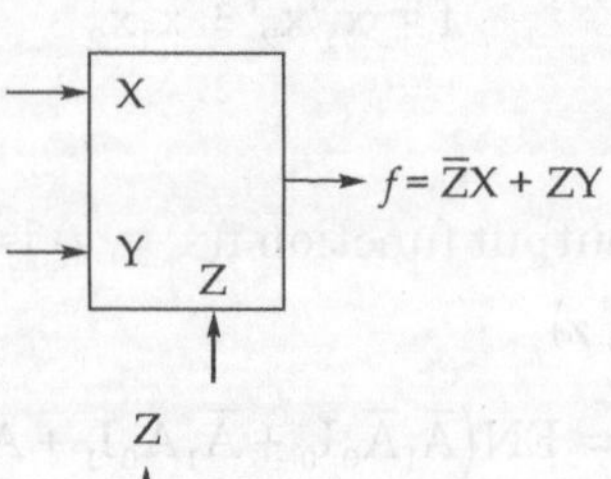

	T	R	f = T + R	
X →	0	0	0	R
T = 0	0	1	1	
Y →	1	0	1	1
T = 1	1	1	1	

∴ X = R, Y = 1 and T = Z

2.13 (d)

The first multiplexer output f = Az + Bz′ (A is selected when z is high).
The second multiplexer output C = xf + yf′
= x(Az + Bz′) + y(Az + Bz′)′
Putting x = 0, y = 1, z = A
⇒ f = (A + A′B)′ = (A + B)′

2.14 (d)

x	y	z	f	
0	0	0	0	yz AND gate
0	0	1	0	
0	1	0	0	
0	1	1	1	
1	0	0	0	(y + z) OR gate
1	0	1	1	
1	1	0	1	
1	1	1	1	

(s ↑ above x)

P is OR gate and Q is AND gate.

2.15 (a)

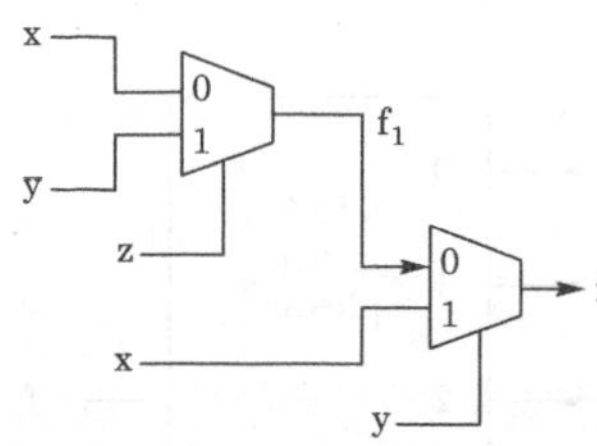

Let the output of first uppermost mux is f_1
$f_1 = xz' + y'z$
The output of second Mux is
$$f = (xz' + y'z)\,y' + xy = xy'z' + y'y'z + xy$$
$$= xy'z' + y'z + xy = x(y'z' + y) + y'z$$
$$= x(z' + y) + y'z = xz' + xy + y'z$$

2.16 (a)

$$\begin{array}{rccc l} & c_2 & c_1 & c_0 & \leftarrow \text{Input carry} \\ a = & a_2 & a_1 & a_0 & \\ a = & b_2 & b_1 & b_0 & \\ \hline c_3 & s_2 & s_1 & s_0 & \leftarrow \text{Final output} \end{array}$$

$$c_1 = a_0b_0 + b_0c_0 + a_0c_0 = a_0b_0 \quad ...(1)$$
(where input carry $c_0 = 0$)
$$c_2 = a_1b_1 + b_1c_1 + a_1c_1$$
$$= a_1b_1 + a_0b_1b_0 + a_1a_0b_0 \quad ...(2)$$
$$c_3 = a_2b_2 + b_2c_2 + a_2c_2$$
$$= a_2b_2 + b_2[a_1b_1 + a_0b_1b_0 + a_1a_0b_0]$$
$$+ a_2[a_1b_1 + a_0b_1b_0 + a_1a_0b_0]$$
$$= a_2b_2 + a_1b_2b_1 + a_0b_2b_1b_0 + a_1a_0b_2b_0 + a_2a_1b_1$$
$$+ a_2a_0b_1b_0 + a_2a_1a_0b_0$$

2.17 (c)

$B = b_{n-1}\,b_{n-2}\,...b_0,\quad A = a_{n-1}a_{n-2}...a_0$
$S = c_{n-1}c_{n-2}...c_0$

The over flow condition V

$$V = b'_{n-1}a'_{n-1}c_{n-1} + b_{n-1}a_{n-1}c'_{n-1}$$

or $\quad V = c_{out} \oplus c_{n-1}$

So the over flow condition is $c_{out} \oplus c_{n-1}$

2.18 (b)

The number of AND gates in carry generator circuit in 'n' bit adder $= \dfrac{n(n+1)}{2}$

If $n = 4 \Rightarrow \dfrac{4(5)}{2} = 10$

The number of OR gates in carry generator circuit in 'n' bit adder $= n$. If $n = 4 \Rightarrow 4$

2.19 (a)

$$f = X3(\overline{X1}\,b + X1\,a) + \overline{X3}X2$$

Put $\quad X1 = b,\ X2 = 0,\ X3 = a$

$$f = a(\bar{b}\cdot b + ba) + \bar{a}\cdot 0 = a\cdot b$$

2.20 (c)

For the construction of a 6 × 64 decoder by using 3 × 8 line decoder

$\dfrac{64}{8} = 8 + 1 \rightarrow$ extra decoder for combining the result $6 \times 64 \xrightarrow{8+1} 3 \times 8$

2.21 (c)

To implement n variable function we require a data selector with $n-1$ select inputs and 2^{n-1} data inputs. So minimum size of multiplexer needed is 2^{n-1} line to 1 line.

2.22 (b)

$$\left[\bar{z}y\bar{x} + \bar{z}\bar{y}\bar{x} + z\bar{y}x + zyx\right]$$

$$\left[\bar{x}\bar{z}(y+\bar{y}) + zx(y+\bar{y})\right]$$

$$\left[\bar{x}\bar{z} + xz\right]$$

$$\left[\bar{x}z + x\bar{z}\right]$$

2.23 (b)

In the given Multiplexer the select inputs S_1 and S_0 are connected to variables P and Q respectively.

S_0(P)	S_1(Q)	f
0	0	R
0	1	$\bar{R}$
1	0	$\bar{R}$
1	1	R

$$f = \bar{P}\bar{Q}R + \bar{P}Q\bar{R} + P\bar{Q}\bar{R} + PQR$$

$$= \Sigma m(1, 2, 4, 7) = P \oplus Q \oplus R$$

2.24 (a)

Truth table given represent the priority encoder i.e. 4 × 2 priority encoder V = 0 indicates output is invalid and V = 1 indicates output is valid.

2.25 (a)

From the given multiplexer:

$$F = \bar{P}Q + P\bar{Q}R + PQ\bar{R} = Q[\bar{P} + P\bar{R}] + P\bar{Q}R$$

$$= Q[(\bar{P} + P)(\bar{P} + \bar{R})] + P\bar{Q}R$$

$$= Q[\bar{P} + \bar{R}] + P\bar{Q}R = \bar{P}Q + Q\bar{R} + P\bar{Q}R$$

$$= \bar{P}Q + Q\bar{R} + P\bar{Q}R$$

2.26 (c)

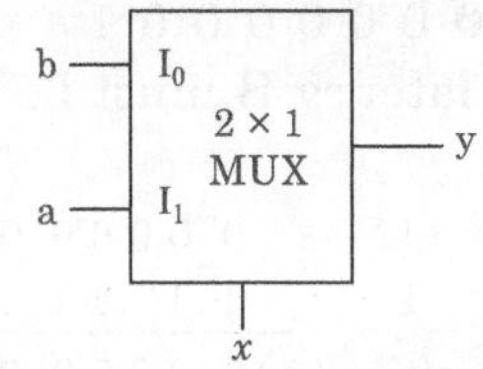

2.27 (12)

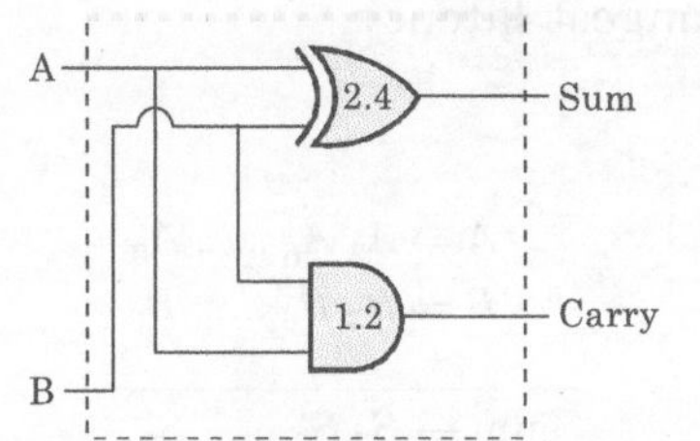

Half Adder (2.4 μs)

F.A. is implemented using 2 H.A. and 1 OR gate

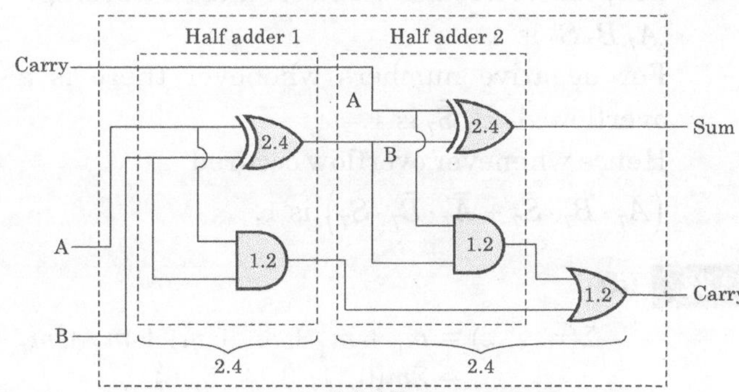

Full Adder

(All half adders at level-1 are computed parallely)

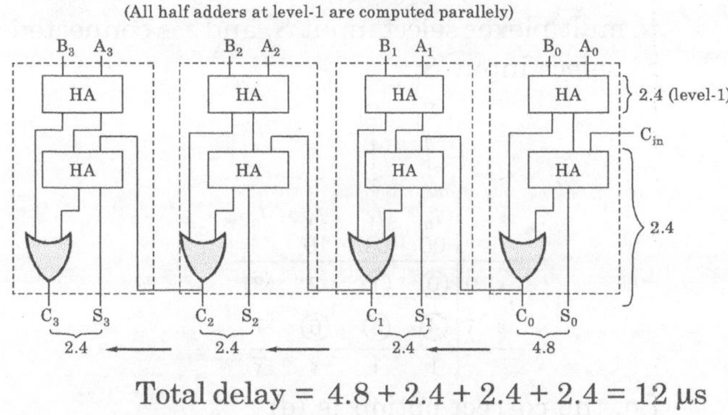

Total delay = 4.8 + 2.4 + 2.4 + 2.4 = 12 μs

2.28 (d)

MUX-1 output $\Rightarrow \bar{P}(0) + P(R) = PR$

MUX-2 output $\Rightarrow X = \bar{Q}(\bar{R}) + Q(PR)$

$= \bar{Q}\bar{R} + PQR$

2.29 (b)

The gates to be used in CLA adder with the fan-in at most '2'. Therefore, time to perform addition using this adder is Θ(log (n)).

2.30 (–1)

A and B are 8 bit numbers in 2′s complement form.

$A = +1 \Rightarrow$ **0** 0 0 0 0 0 0 1

for longest latency B must be equal to –1.

Verification

A = +1	0 0000 001
B = –1	1 1111 111
Carry out → (1)	0 0000 000

carry is propagating from LSB to MSB. Hence it is longest latency.

2.31 (c)

$A \Rightarrow A_7\, A_6 \ldots\ldots A_0$

$B \Rightarrow B_7\, B_6 \ldots\ldots B_0$

+

sum $\rightarrow S_7\, S_6 \ldots\ldots S_0$

carry $\rightarrow C_7\, C_6 \ldots\ldots C_0$

For positive nubmer whenever there is a overflow $\bar{A}_7\, \bar{B}_7\, S_7$ is 1.

For negative numbers whenever there is a overflow $A_7\, B_7\, \bar{S}_7$ is 1.

Hence whenever overflow occured

$\left(A_7 \cdot B_7 \cdot \bar{S}_7 + \bar{A}_7 \cdot \bar{B}_7 \cdot S_7\right)$ is 1.

2.32 (d)

$\Sigma(x, y, z) = m_0 + m_1 + m_3 + m_4 + m_5 + m_6$

$= \Sigma m(0, 1, 3, 4, 5, 6)$

As per the given options, variable y is connected to multiplexer select input S_1 and z is connected to select input S_0.

$S_1 \leftarrow y$, $S_0 \leftarrow z$

	I_0 00	I_1 01	I_2 10	I_3 11
x 0	(0)	(1)	2	(3)
x 1	(4)	(5)	(6)	7
	1	1	x	$\bar{x}$

So, the correct option is (d).

2.33 (c)

S is 1024 word memory segment, it needs 10 address lines.

∴ P must be a decoder, with 10 input lines and 1024 output lines. (10 : 2^{10} decoder)

R is 32 word register file, it needs 5 address lines.

∴ Q must be a decoder, with 5 input lines and 32 output lines. (5 : 2^5 decoder)

Based on mode bit M, T selects an input word to load X.

∴ T must be 2 : 1 multiplexer, M is select input to the multiplexer.

2.34 (c)

$F(A, B, C) = \bar{A} + \bar{A}\bar{C} + A\bar{B}C$

$= \bar{A} + A\bar{B}C$

(0, 1, 2, 3)   5

$= \Sigma m(0, 1, 2, 3, 4)$

B is select input of M3 i.e. 2 × 1 MUX.

If B = 0; 2 × 1 MUX output is M1 output.

If B = 1; 2 × 1 MUX output is M2 output.

A and C are select inputs to both M1 and M2.

	A	B	C	F
0	0	0	0	1 = X_0
1	0	0	1	1 = X_1
2	0	1	0	1 = X_4
3	0	1	1	1 = X_5
4	1	0	0	0 = X_2
5	1	0	1	1 = X_3
6	1	1	0	0 = X_6
7	1	1	1	0 = X_7

$X_0 = 1,\quad X_1 = 1,\quad X_2 = 0,\quad X_3 = 1$

$X_4 = 1,\quad X_5 = 1,\quad X_6 = 0,\quad X_7 = 0$

2.35 (4)

2 : 1 MUX o/p $= \bar{S}_0 I_0 + S_0 I_1$

MUX M_1 o/p $= \bar{A} \cdot 1 + A \cdot 1 = 1$

MUX M_2 o/p $= \bar{B} \cdot 0 + B \cdot 0 = 0$

MUX M_3 o/p $= \bar{C} \cdot 1 + C \cdot 0 = \bar{C}$

∴ $Y = \bar{C}$

Whenever $C = 0$, the output $Y = 1$

A and B variables may be with any values.

A	B	C	Y
0	0	0	1
0	1	0	1
1	0	0	1
1	1	0	1

For '4' combinations the output $Y = 1$.

3 Sequential Circuits

3.1 Consider the synchronous sequential circuit in the figure.

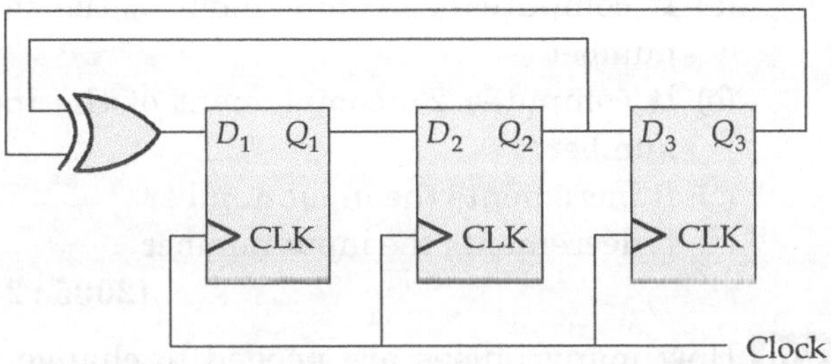

(a) Draw a state diagram which is implemented by the circuit. Use the following names for the states corresponding to the values of flip-flops as given below.

Q_1	Q_2	Q_3	State
0	0	0	S_0
0	0	1	S_1
⋮	⋮	⋮	⋮
1	1	1	S_7

(b) Given that the initial state of the circuit is S_4, identify the set of states which are not reachable.

[1996 : 5 M]

3.2 The following arrangement of master-slave flip flops.

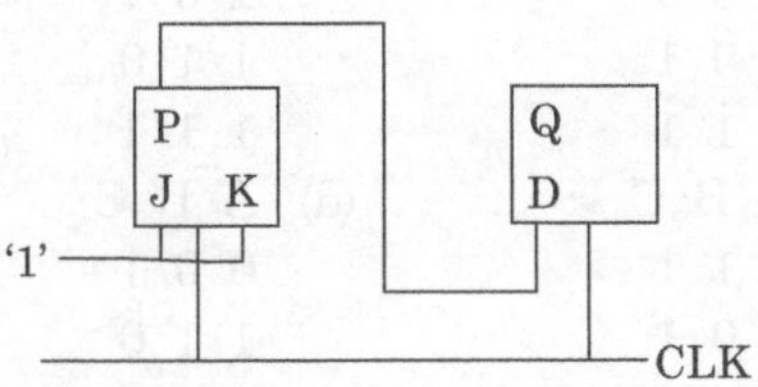

Has the initial state of P, Q as 0, 1 (respectively). After the clock cycles the output state P, Q is (respectively),

(a) 1, 0
(b) 1, 1
(c) 0, 0
(d) 0, 1

[2000 : 2 M]

3.3 Consider the following circuit with initial state $Q_0 = Q_1 = 0$. The D flip-flops are positive edged triggered and have set up time 20 ns and hold time is 0 ns.

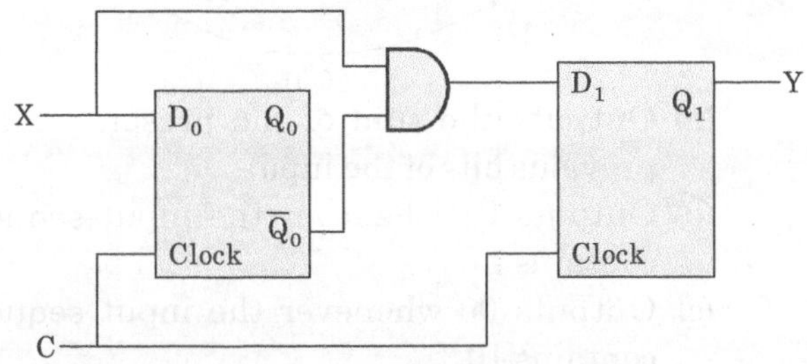

Consider the following timing diagrams of X and C; the clock period of C ≥ 40 nanosecond. Which one is the correct plot of Y?

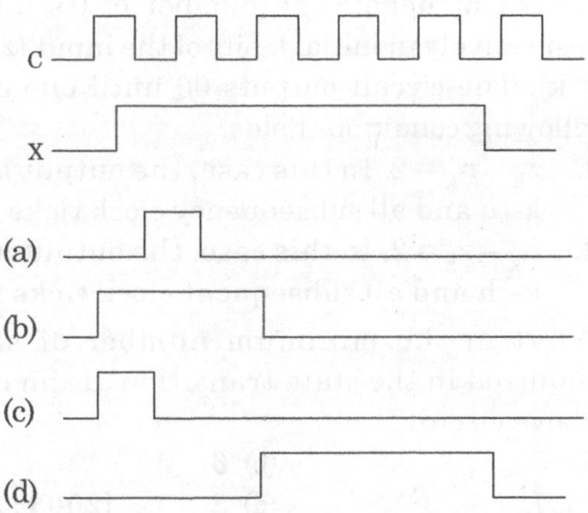

[2001 : 2 M]

3.4 Consider the circuit given below with initial state $Q_0 = 1$, $Q_1 = Q_2 = 0$. The state of the circuit is given by the value $4Q_2 + 2Q_1 + Q_0$

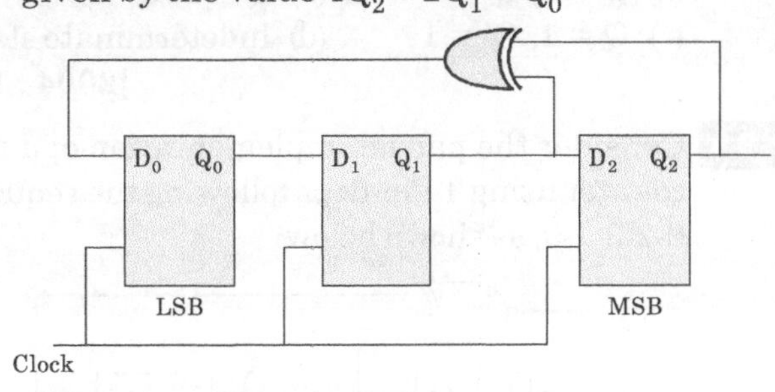

Which one of the following is the correct state sequence of the circuit

(a) 1, 3, 4, 6, 7, 5, 2
(b) 1, 2, 5, 3, 7, 6, 4
(c) 1, 2, 7, 3, 5, 6, 4
(d) 1, 6, 5, 7, 2, 3, 4

[2001 : 2 M]

3.5 The finite state machine described by the following state diagram with A as starting state, where an arc label x/y and x stands for 1-bit input and y stands for 2-bit output

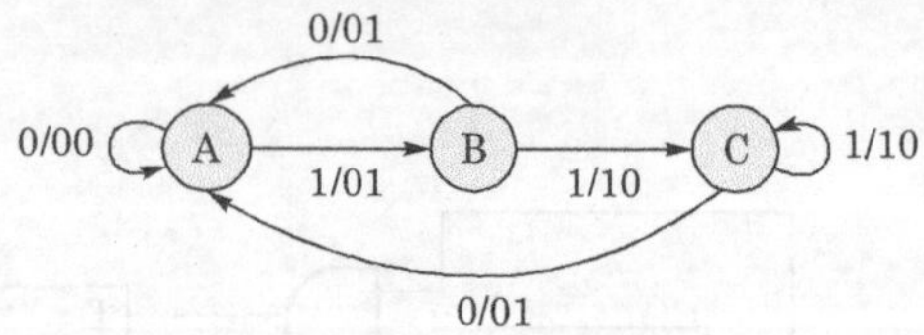

(a) Outputs the sum of the present and the previous bits of the input.
(b) Outputs 01 whenever the input sequence contains 11.
(c) Outputs 00 whenever the input sequence contains 10.
(d) None of the above **[2002 : 2 M]**

3.6 A 1-input, 2-output synchronous sequential circuit behaves as follows:

Let z_k, n_k denote the number of 0's and 1's respectively in initial k bits of the input ($z_k + n_k = k$). The circuit outputs 00 until one of the following conditions holds.

1. $z_k - n_k = 2$. In this case, the output at the k-th and all subsequency clock ticks is 10.
2. $n_k - z_k = 2$. In this case, the output at the k-th and all subsequent clock ticks is 01.

What in the minimum number of states required in the state transition graph of the above circuit?

(a) 5 (b) 6
(c) 7 (d) 8 **[2003 : 2 M]**

3.7 In an SR latch made by cross-coupling two NAND gates, if both S and R inputs are set to 0, then it will result in

(a) $Q = 0, Q' = 1$ (b) $Q = 1, Q' = 0$
(c) $Q = 1, Q' = 1$ (d) Indeterminate states **[2004 : 1 M]**

3.8 Consider the partial implementation of a 2-bit counter using T flip-flops following the sequence 0-2-3-1-0, as shown below:

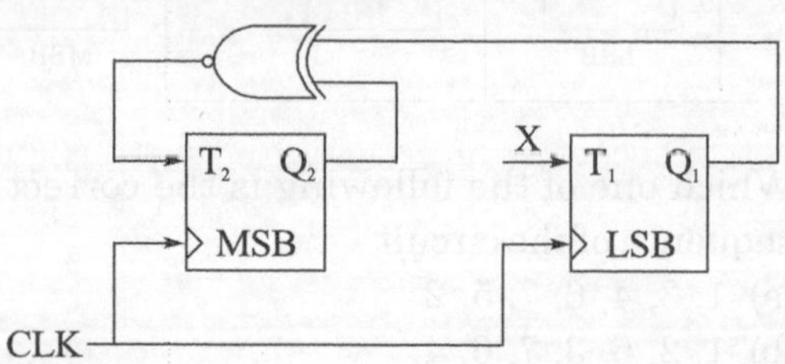

To complete the circuit, the input X should be

(a) Q_2' (b) $Q_2 + Q_1$
(c) $(Q_1 \oplus Q_2)'$ (d) $Q_1 \oplus Q_2$ **[2004 : 2 M]**

3.9 The flowing diagram represents a finite state machine which takes input a binary number from the least significant bit.

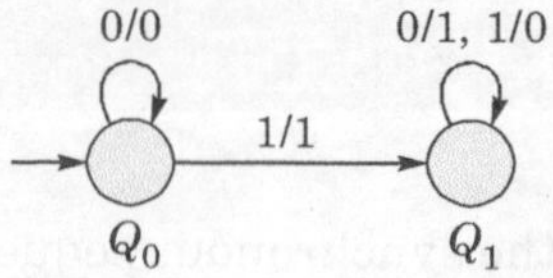

Which one of the following is TRUE?

(a) It computes 1's complement of the input number
(b) It computes 2's complement of the input number
(c) It increments the input number
(d) It decrements the input number **[2005 : 2 M]**

3.10 How many pulses are needed to change the contents of a 8-bit upcounter from 10101100 to 00100111 (rightmost bit is the LSB)?

(a) 134 (b) 133
(c) 124 (d) 123 **[2005 : 1 M]**

3.11 Which of the following input sequences will always generate a 1 at the output z at the end of the third cycle?

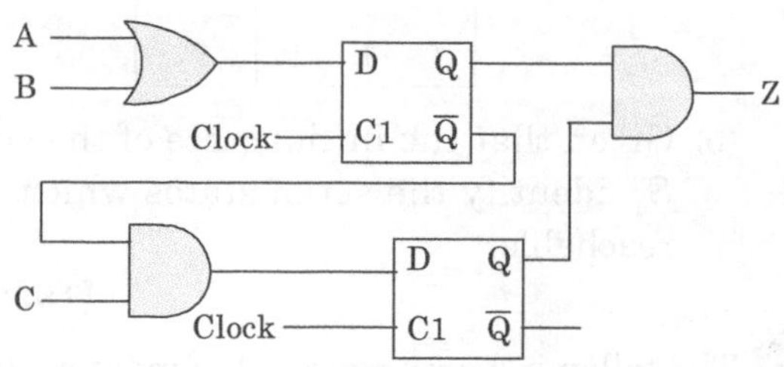

(a) A B C
0 0 0
1 0 1
1 1 1

(b) A B C
1 0 1
1 1 0
1 1 1

(c) A B C
0 1 1
1 0 1
1 1 1

(d) A B C
0 0 1
1 1 0
1 1 1

[2005 : 2 M]

3.12 Consider the following circuit involving a positive edge triggered D FF.

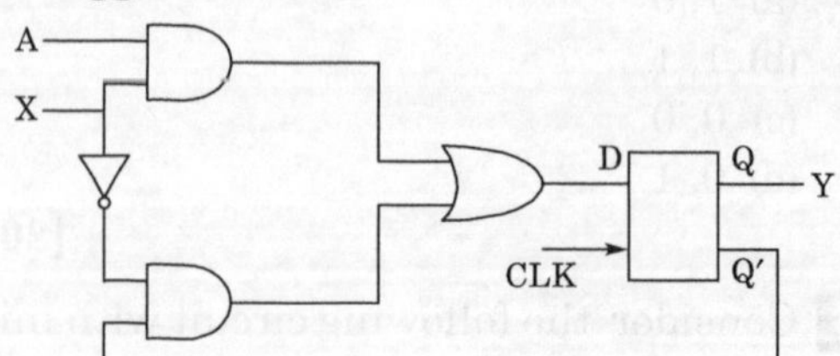

Consider the following timing diagram. Let A_i represent the logic level on the line A in the i-th clock period.

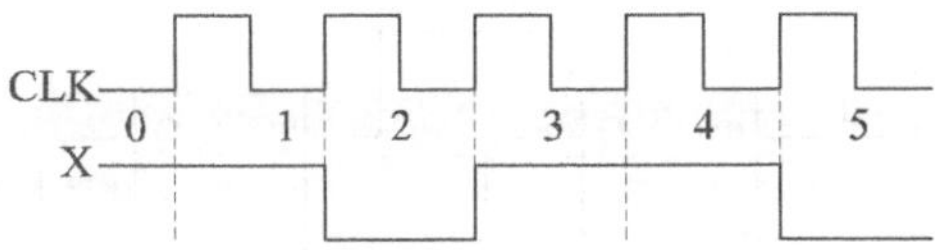

Let A′ represent the complement of A. The correct output sequence on Y over the clock periods 1 through 5 is

(a) $A_0\,A_1\,A_1'\,A_3\,A_4$
(b) $A_0\,A_1\,A_2'\,A_3\,A_4$
(c) $A_1\,A_2\,A_2'\,A_3\,A_4$
(d) $A_1\,A_2'\,A_3\,A_4\,A_5'$

[2005 : 2 M]

3.13 Consider the following circuit:

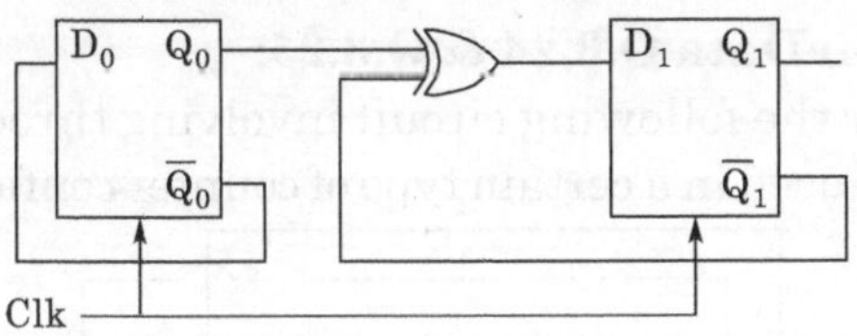

The flip-flops are positive edge triggered D FFs. Each state is designated as a two bit string Q_0Q_1. Lot the initial state be 00. The state transition sequence is

(a) 00 → 11 → 01
(b) 00 → 11
(c) 00 → 10 → 01 → 11
(d) 00 → 11 → 01 → 10

[2005 : 2 M]

3.14 Consider the circuit in the diagram. The ⊕ operator represents Ex-OR. The D flip-flops are initialized to zero (cleared).

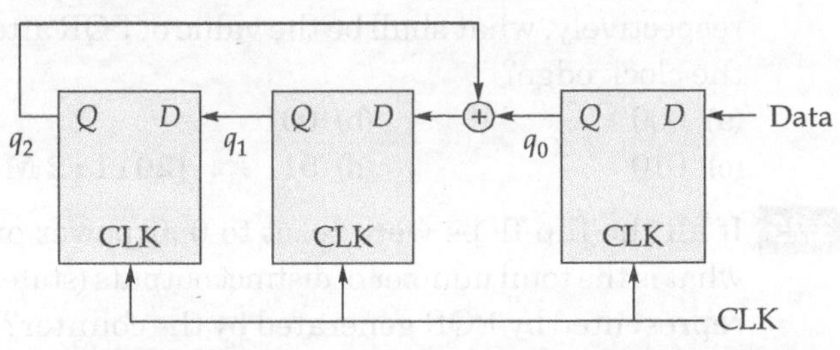

The following data: 100110000 is supplied to the "data" terminal in nine clock cycles. After that the values of $q_2q_1q_0$ are

(a) 000
(b) 001
(c) 010
(d) 101

[2006 : 2 M]

3.15 You are given a free running clock with a duty cycle of 50% and a digital waveform f which changes only at the negative edge of the clock. Which one of the following circuits (using clocked D flip-flops) will delay the phase of f by 180°?

(a)
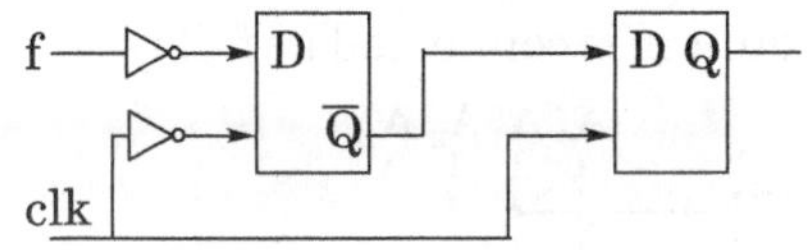

(b)
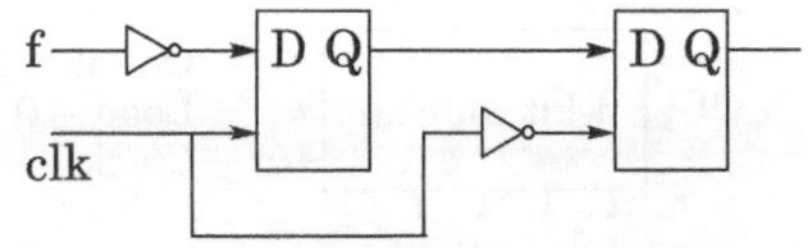

(c)
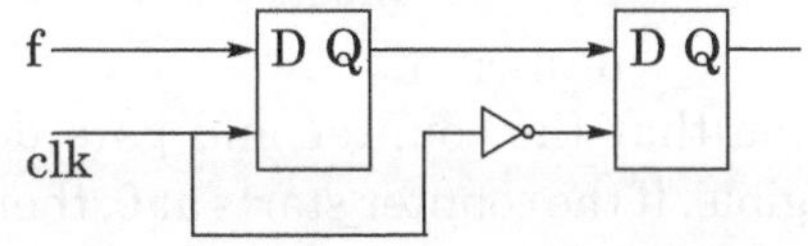

(d)
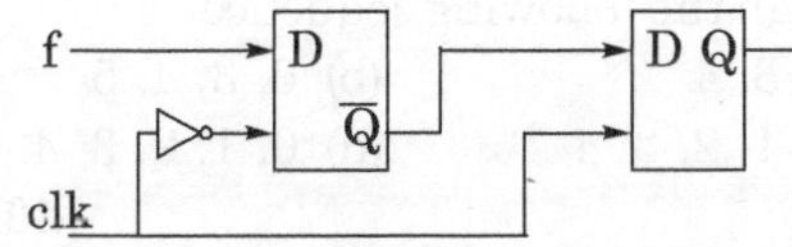

[2006 : 1 M]

3.16 For a state machine with the following state diagram the expression for the next state S^+ in terms of the current state S and the input variables x and y is

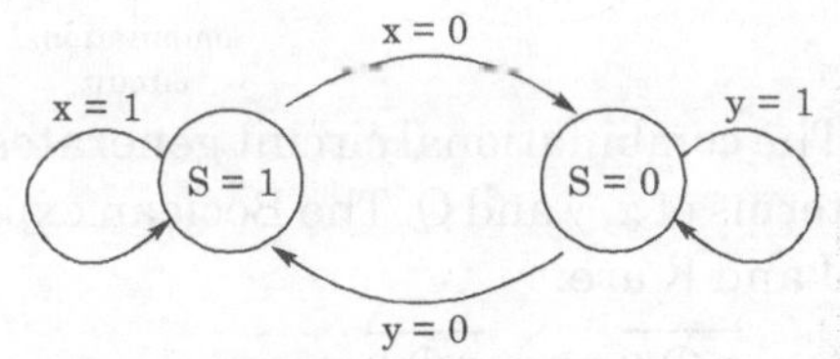

(a) $S^+ = S'\cdot y' + S\cdot x$
(b) $S^+ = S\cdot x\cdot y' + S'\cdot y\cdot x'$
(c) $S^+ = x\cdot y'$
(d) $S^+ = S'\cdot y + S\cdot x'$

[2006 : 2 M]

3.17 Which of the following input sequences for a cross-coupled R-S flip-flop realized with two NAND gates may lead to an oscillation?

(a) 11, 00
(b) 01, 10
(c) 10, 01
(d) 00, 11

[2007 : 1 M]

3.18 What is the final value stored in the linear feedback shift register if the input is 101101?

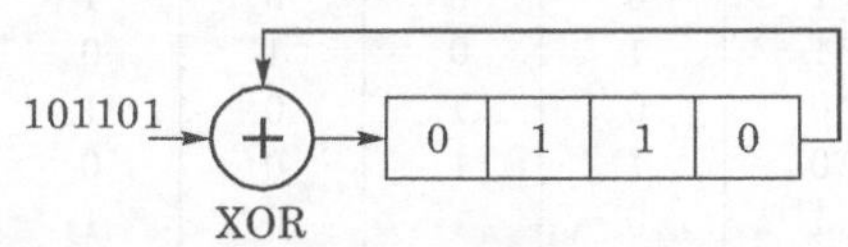

(a) 0110
(b) 1011
(c) 1101
(d) 1111

[2007 : 2 M]

3.19 The control signal functions of 4-bit binary counter are given below (where X is "don't care")

Clear	Clock	Load	Count	Function
1	X	X	X	Clear to 0
0	X	0	0	No change
0	↑	1	X	Load input
0	↑	0	1	Count next

The counter is connected as follows:

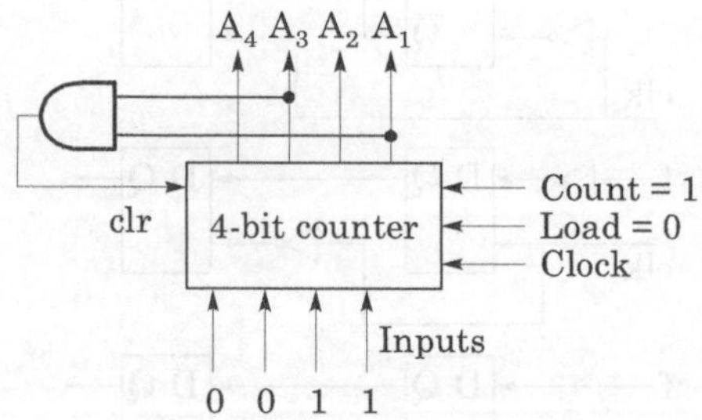

Assume that the counter and gate delays are negligible. If the counter starts at 0, then it cycles through the following sequence

(a) 0, 3, 4
(b) 0, 3, 4, 5
(c) 0, 1, 2, 3, 4
(d) 0, 1, 2, 3, 4, 5

[2007 : 2 M]

3.20 Consider the following state diagram and its realization by a JK flip flop.

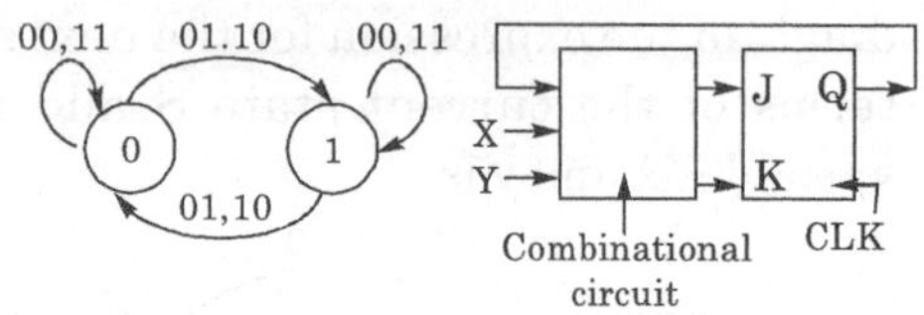

The combinational circuit generates J and K in terms of x, y and Q. The Boolean expressions for J and K are:

(a) $\overline{x \oplus y}$ and $\overline{x \oplus y}$
(b) $\overline{x \oplus y}$ and $x \oplus y$
(c) $x \oplus y$ and $\overline{x \oplus y}$
(d) $x \oplus y$ and $x \oplus y$

[2008 : 2 M]

3.21 Given the following state table of an FSM with two states A and B, one input and one output:

Present State A	Present State B	Input	Next State A	Next State B	Output
0	0	0	0	0	1
0	1	0	1	0	0
1	0	0	0	1	0
1	1	0	1	0	0
0	0	1	0	1	0
0	1	1	0	0	1
1	0	1	0	1	1
1	1	1	0	0	1

If the initial state is A = 0, B = 0, what is the minimum length of an input string which will take the machine to the state A = 0, B = 1 with output = 1?

(a) 3
(b) 4
(c) 5
(d) 6

[2009 : 2 M]

3.22 In the sequential circuit shown below, if the initial value of the output Q_1Q_0 is 00, what are the next four values of Q_1Q_0?

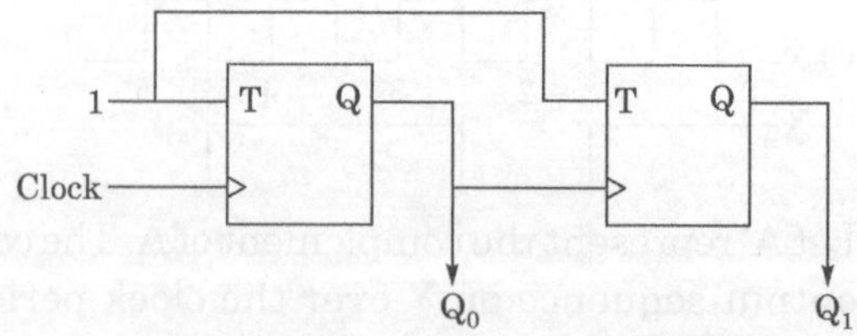

(a) 11,10,01,00
(b) 10,11,01,00
(c) 10,00,01,11
(d) 11,10,00,01

[2010 : 2 M]

3.23 The minimum number of D flip-flops needed to design a mod-258 counter is

(a) 9
(b) 8
(c) 512
(d) 258

[2011 : 1 M]

Common Data Q.3.24 & Q.3.25:

Consider the following circuit involving three D-type flip-flops used in a certain type of counter configuration

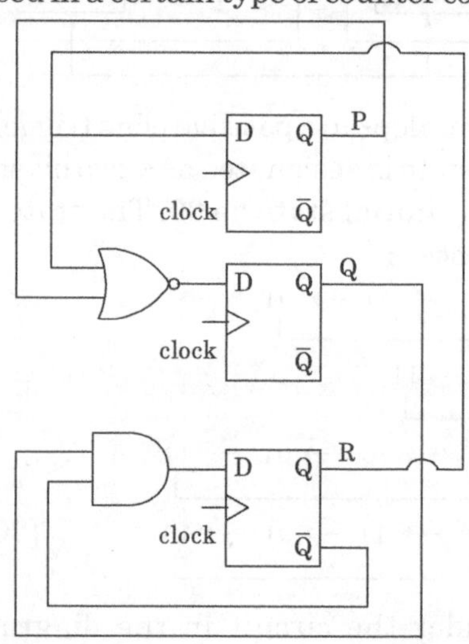

3.24 If at some instance prior to the occurrence of the clock edge, P, Q and R have a value 0, 1 and 0 respectively, what shall be the value of PQR after the clock edge?

(a) 000
(b) 001
(c) 010
(d) 011

[2011 : 2 M

3.25 If all the flip-flops were reset to 0 at power on, what is the total number of distinct outputs (states) represented by PQR generated by the counter?

(a) 3
(b) 4
(c) 5
(d) 6

[2011 : 2 M]

3.26 Let $k = 2^n$. A circuit is built by giving the output of an n-bit binary counter as input to an n-to-2^n bit decoder. This circuit is equivalent to a

(a) k-bit binary up counter.
(b) k-bit binary down counter.
(c) k-bit ring counter.
(d) k-bit Johnson counter.

[2014 (Set-2) : 1 M]

3.27

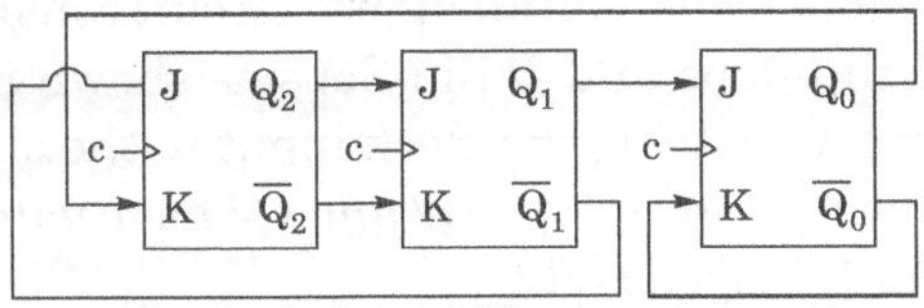

The above synchronous sequential circuit built using JK flip-flops is initialized with $Q_2Q_1Q_0 = 000$. The state sequence for this circuit for the next 3 clock cycles is
(a) 001, 010, 011 (b) 111, 110, 101
(c) 100, 110, 111 (d) 100, 011, 001
[2014 (Set-3) : 2 M]

3.28 Consider a 4 bit Johnson counter with an initial value of 0000. The counting sequence of this counter is
(a) 0, 1, 3, 7, 15, 14, 12, 8, 0
(b) 0, 1, 3, 5, 7, 9, 11, 13, 15, 0
(c) 0, 2, 4, 6, 8, 10, 12, 14, 0
(d) 0, 8, 12, 14, 15, 7, 3, 1, 0
[2015 (Set-1) : 1 M]

3.29 A positive edge-triggered D flip-flop is connected to a positive edge-triggered JK flip-flop as follows. The Q output of the D flip-flop is connected to both the J and K inputs of the JK flip-flop, while the Q output of the JK flip-flop is connected to the input of the D flip-flop. Initially, the output of the D flip-flop is set to logic one and the output of the JK flip-flop is cleared.
Which one of the following is the bit sequence (including the initial state) generated at the Q output of the JK flip-flop when the flip-flops are connected to a free-running common clock? Assume that J = K = 1 is the toggle mode and J = K = 0 is the state-holding mode of the JK flip-flop. Both the flip-flops have non-zero propagation delays.
(a) 0110110... (b) 0100100...
(c) 011101110... (d) 011001100...
[2015 (Set-1) : 2 M]

3.30 The minimum number of JK flip-flops required to construct a synchronous counter with the count sequence (0, 0, 1, 1, 2, 2, 3, 3, 0, 0,...) is____.
[2015 (Set-2) : 1 M]

3.31 We want to design a synchronous counter that counts the sequence 0-1-0-2-0-3 and then repeats. The minimum number of J-K flip-flops required to implement this counter is ______.
[2016 (Set-1) : 1 M]

3.32 Consider a combination of *T* and *D* flip-flops connected as shown below. The output of the *D* flip-flop is connected to the input of the *T* flip-flop and the output of the *T* flip-flop is connected to the input of the *D* flip-flop.

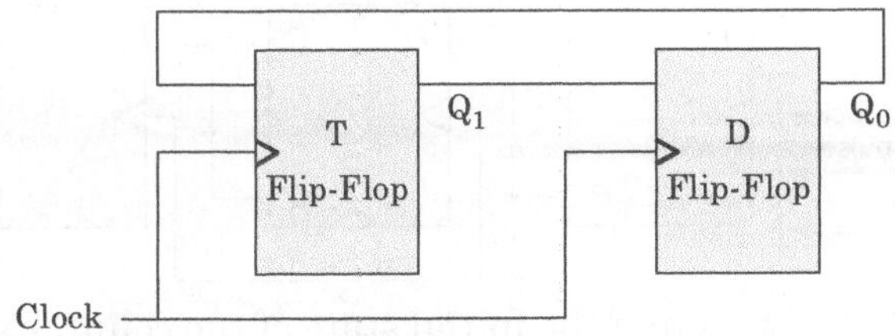

Initially, both Q_0 and Q_1 are set to 1 (before the 1st clock cycle). The outputs
(a) $Q_1\,Q_0$ after the 3rd cycle are 11 and after the 4th cycle are 00 respectively.
(b) $Q_1\,Q_0$ after the 3rd cycle are 11 and after the 4th cycle are 01 respectively.
(c) $Q_1\,Q_0$ after the 3rd cycle are 00 and after the 4th cycle are 11 respectively.
(d) $Q_1\,Q_0$ after the 3rd cycle are 01 and after the 4th cycle are 01 respectively.
[2017 (Set-1) : 2 M]

3.33 The next state table of a 2-bit saturating up-counter is given below:

Q_1	Q_0	Q_1^+	Q_0^+
0	0	0	1
0	1	1	0
1	0	1	1
1	1	1	1

The counter is built as a synchronous sequential circuit using T flip-flops. The expressions for T_1 and T_0 are
(a) $T_1 = Q_1Q_0,\quad T_0 = \bar{Q}_1\bar{Q}_0$
(b) $T_1 = \bar{Q}_1Q_0,\quad T_0 = \bar{Q}_1 + \bar{Q}_0$
(c) $T_1 = Q_1 + Q_0,\quad T_0 = \bar{Q}_1 + \bar{Q}_0$
(d) $T_1 = \bar{Q}_1Q_0,\quad T_0 = Q_1 + Q_0$
[2017 (Set-2) : 2 M]

3.34 Consider the sequential circuit shown in the figure, where both flip-flops used are positive edge-triggered D flip-flops.

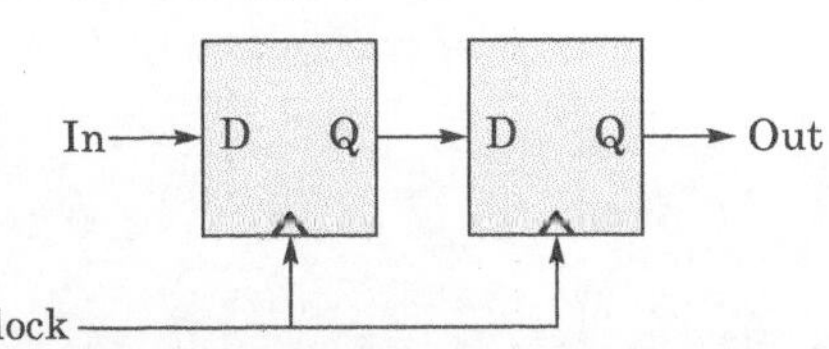

The number of states in the state transition diagram of the circuit that have a transition back to the same state on some value of "in" is________.
[2018 : 1 M]

3.35 Consider a 3-bit counter, designed using T flip-flops, as shown below:

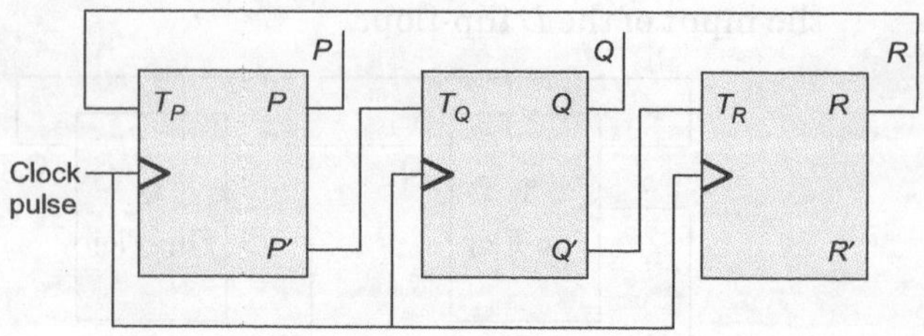

Assuming the initial state of the counter given by *PQR* as 000. What are the next three states?

(a) 011, 101, 111 (b) 001, 010, 000
(c) 001, 010, 111 (d) 011, 101, 000

[2021 (Set-1) : 2 M]

3.36 The output of a 2-input multiplexer is connected back to one of its inputs as shown in the figure.

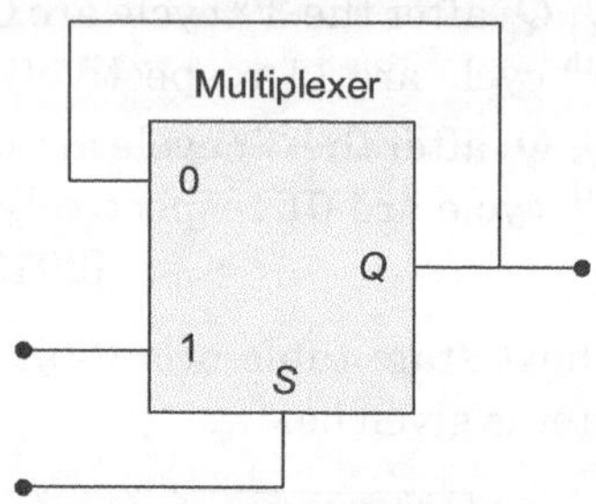

Match the functional equivalence of this circuit to one of the following options.

(a) *D* Flip-flop (b) *D* Latch
(c) Half-adder (d) Demultiplexer

[2023 : 1 M]

3.37 Consider a sequential digital circuit consisting of *T* flip-flops and *D* flip-flops as shown in the figure. CLKIN is the clock input to the circuit. At the beginning, Q1, Q2 and Q3 have values 0, 1 and 1, respectively.

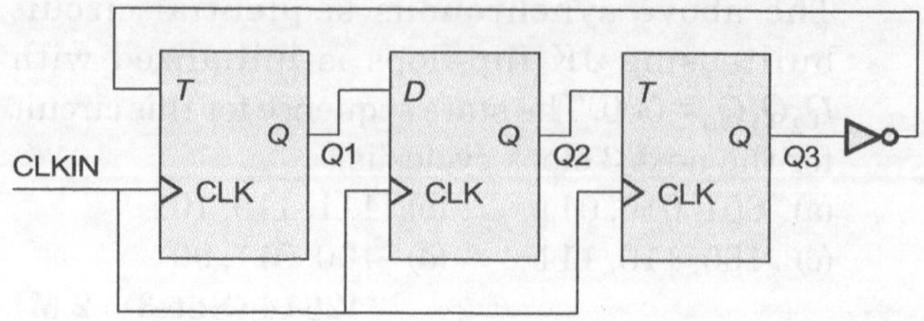

Which one of the given values of (Q1, Q2, Q3) can NEVER be obtained with this digital circuit?

(a) (0, 0, 1) (b) (1, 0, 0)
(c) (1, 0, 1) (d) (1, 1, 1)

[2023 : 2 M]

■■■■

Answers Sequential Circuits

3.1	Sol.	**3.2**	(a)	**3.3**	(a)	**3.4**	(b)	**3.5**	(a)	**3.6**	(a)	**3.7**	(c)	**3.8**	(d)	**3.9**	(b)
3.10	(d)	**3.11**	(*)	**3.12**	(a)	**3.13**	(d)	**3.14**	(c)	**3.15**	(c)	**3.16**	(a)	**3.17**	(a)	**3.18**	(a)
3.19	(c)	**3.20**	(d)	**3.21**	(a)	**3.22**	(a)	**3.23**	(a)	**3.24**	(d)	**3.25**	(b)	**3.26**	(c)	**3.27**	(c)
3.28	(d)	**3.29**	(a)	**3.30**	(3)	**3.31**	(4)	**3.32**	(b)	**3.33**	(b)	**3.34**	(2)	**3.35**	(d)	**3.36**	(b)
3.37	(a)																

Explanations Sequential Circuits

3.1 Sol.

(b)

	CLK	Q_1	Q_2	Q_3	$D_1 = Q_2 \oplus Q_3$	$D_2 = Q_1$	$D_3 = Q_2$
S_4	0	1	0	0	0	1	0
S_2	1	0	1	0	1	0	1
S_5	2	1	0	1	1	1	0
S_6	3	1	1	0	1	1	1
S_7	4	1	1	1	0	1	1
S_3	5	0	1	1	0	0	1
S_1	6	0	0	1	1	0	0
S_4	7	1	0	0			

$$\rightarrow S_4 \rightarrow S_2 \rightarrow S_5 \rightarrow S_6 \rightarrow S_7 \rightarrow S_3 \rightarrow S_1 \text{ ─┐ (back to } S_4)$$

The state which is not reachable is S_0.

3.2 (a)

J	K	D = P	P	Q
			0	1
1	1	0	1	0
1	1	1	0	1

The circuit output PQ alternates between 01 and 10 states.

3.3 (a)

Flip-flops are having setup time 20 ns and hold time is 0 ns.
Flip-flops are positive edged triggered.
The input 'X' and clock 'C' waveform are given.
The output waveform Y i.e. the output of flip-flop 1, Q_1 is to be identified.

Clock	Q_0	Q_1	FF0 $D_0 = X$	FF1 $D_1 = X\overline{Q_0}$
Initial state	0	0	0	0
1	0	0	1	1
2	1	1	1	0
3	1	0	1	0
4	1	0	1	0
5	1	0	0	0
6	0	0		

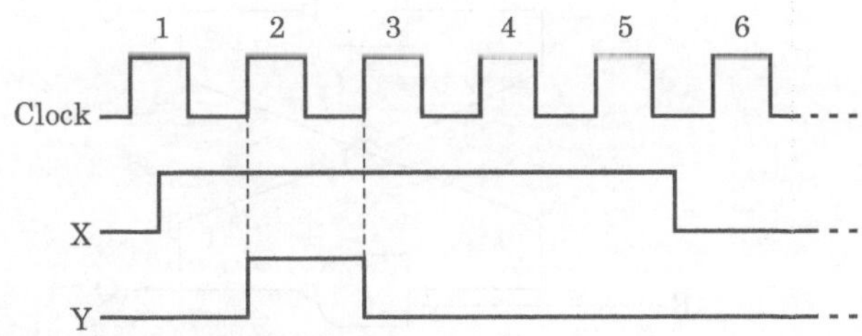

3.4 (b)

Clock	$D_2 = Q_1$	$D_1 = Q_0$	$D_0 = Q_1 \oplus Q_2$	Q_2	Q_1	Q_0	
				0	0	1	(1)
1	0	1	0	0	1	0	(2)
2	1	0	1	1	0	1	(5)
3	0	1	1	0	1	1	(3)
4	1	1	1	1	1	1	(7)
5	1	1	0	1	1	0	(6)
6	1	0	0	1	0	0	(4)
7	0	0	1	0	0	1	(1)

Therefore, the correct state sequence of the circuit is 1, 2, 5, 3, 7, 6, 4.

3.5 (a)

For any input bit stream the output is resulting the sum of the present and the previous bits of the input.

3.6 (a)

1. If $z_k - n_k = 2$. In the case, the output at the kth and subsequency clock ticks is 10.
2. If $n_k - z_k = 2$. In this case, the output at the kth and subsequency clock ticks is 01.

Based on sequential circuit description, the state diagram of the circuit is

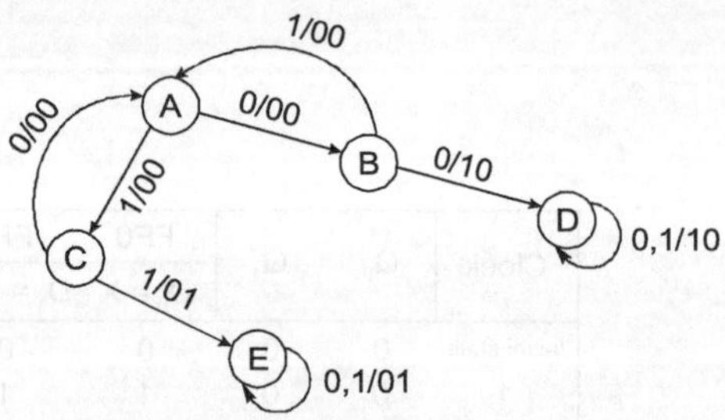

∴ Total numbers of states are 5.

3.7 (c)

SR latch by cross-coupling two NAND gates

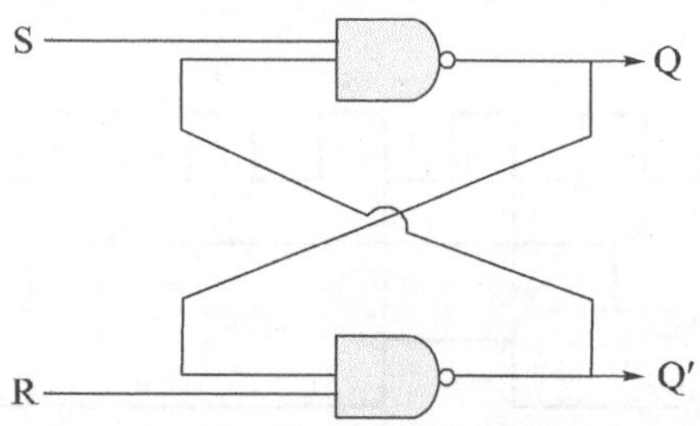

NAND LATCH

If the both S and R inputs are set 0 then the result will be:

Table for the SR Latch using the NAND gates:

S	R	Q
0	0	Invalid state (Q = Q′ = 1)
0	1	1
1	0	0
1	1	Previous state

∴ Q = 1 and Q′ = 1

3.8 (d)

	Clock	Q_2	Q_1	FF1 T_1
0		0	0	0
2	1	1	0	1
3	2	1	1	0
1	3	0	1	1
	4	0	0	

The required switching sequence of the counter is $0 \rightarrow 2 \rightarrow 3 \rightarrow 1 \rightarrow 0$.

$$T1(Q_2, Q_1) = \Sigma m\,(1, 2)$$

$$= \overline{Q}_2 Q_1 + Q_2 \overline{Q}_1 = Q_2 \oplus Q_1$$

3.9 (b)

First bit entering in to the machine is LSB. Suppose input is 010100 then the output is 101100. It is 2's complement of the input string. So, it computes 2's complement.

3.10 (d)

In a 2^8 Counter the range would be from 0-255. Hence to go from 10101100 (172) to 00100111 (39), the counter has to go initially from 172 to 255 and then from 0 to 39.
Hence to go from 172 to 255, 255 − 172 = 83 Clock pulses will be required.
From 255 to 0, 1 clock pulse is required.
Then from 0 to 39, 39 clock pulses will be required.
Hence in total 83 + 1 + 39 = 123 Clock pulses are required.

3.11 (*)

No suitable answer from the given options.

3.12 (a)

Clock and input waveform X are given in problem description.

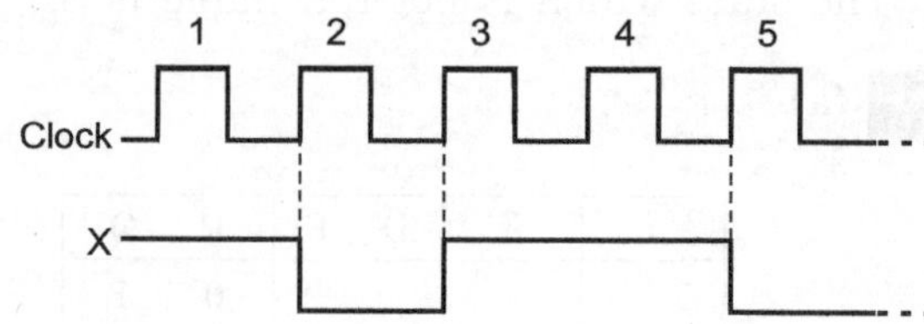

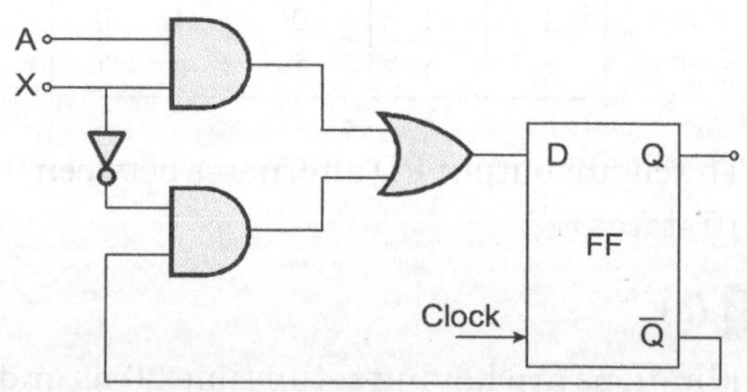

Clock	$Q_{n+1} = Y$	$D = AX + \overline{XQ_n}$
Initial state	0 or 1	A_0
1	A_0	A_1
2	A_1	A_1'
3	A_1'	A_3
4	A_3	A_4
5	A_4	

Before first clock the input is A_0 at 1st clock input is A_1 and so on.
Output Y is $A_0 A'_1 A'_1 A_3 A_4$ over the clock period 1 through 5.

3.13 (d)

Clock	Q_0	Q_1	FF0 $D_0 = \overline{Q}_0$	FF1 $D_1 = \overline{Q}_1 \oplus Q_0$
Initial state	0	0	1	1
1	1	1	0	1
2	0	1	1	0
3	1	0	0	0
4	0	0		

The state transitions sequence is

→ 00 → 11 → 01 → 10 →

3.14 (c)

CLK	Q_2	Q_1	Q_0	$D_2 = Q_1$	$D_1 = Q_2 \oplus Q_0$	D_0 = Data
0	0	0	0	0	0	1
1	0	0	1	0	1	0
2	0	1	0	1	0	0
3	1	0	0	0	1	1
4	0	1	1	1	1	1
5	1	1	1	1	0	0
6	1	0	0	0	1	0
7	0	1	0	1	0	0
8	1	0	0	0	1	0
9	0	1	0			

3.15 (c)

1. In all the options consider left flip-flop as FF0 and right side flip-flop as FF1, FF0 output is Q_0 and FF1 output is Q_1.
2. Digital input 'f' changes at the negative edge of the clock. To get delay the phase of 'f' by 180°, the output i.e., Q_1 must change at the negative edge of the clock.
   Hence option (a) and (d) are not possible answer. [Q_1 is changing at the positive edge of the clock]
3. In option (b) and (c) FF0 will change at the clock rise and FF1 will change at clock falls.
4. The changed output of FF0 after clock rise is the input of FF1 in options (b) and (c).

Let us consider option (b):

Clock	$D_0 = \overline{f}$	Q_0	Q_1	
		0	0	Initial state
1	1	1	1	
2	0	0	0	
3	1	1	1	
4	0	0	0	

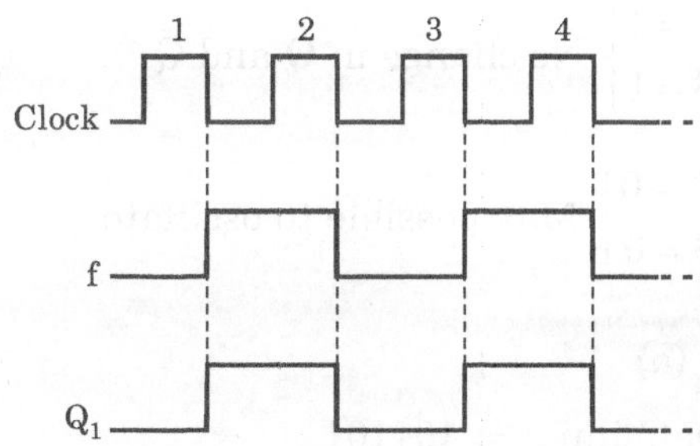

Output Q_1 is in phase with the input f, hence it is not the answer.

Let us consider option (c):

Clock	$D_0 = f$	Q_0	Q_1	
		0	0	Initial state
1	0	0	0	
2	1	1	1	
3	0	0	0	
4	1	1	1	

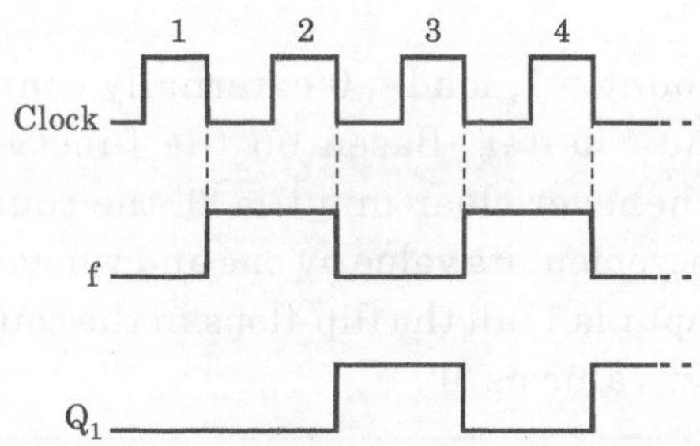

Output Q_1 is out of phase with input f by 180°, hence the output is 180° delayed by input f, hence the answer is option (c).

3.16 (a)

Present State	Inputs		Next state
S	x	y	S^+
0	X	0	1
0	X	1	0
1	0	X	0
1	1	X	1

From the above table: $S^+ = S' \cdot y' + S \cdot x$

3.17 (a)

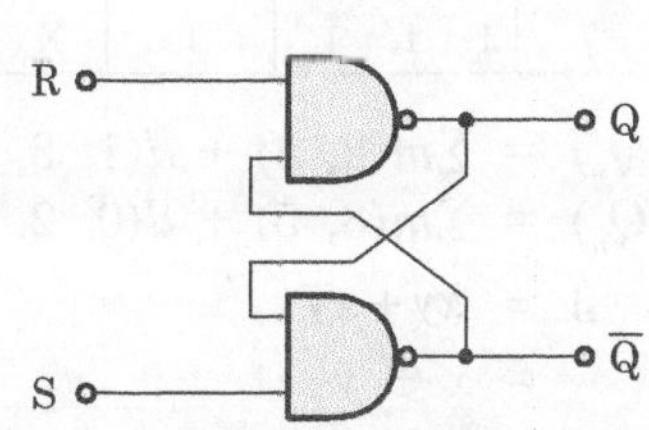

$\left.\begin{matrix} R = 1 \\ S = 1 \end{matrix}\right\}$ No change in Q and $\bar{Q}$

$\left.\begin{matrix} R = 0 \\ S = 0 \end{matrix}\right\}$ May possible to oscillate

3.18 (a)

Data input ⇒ 101101

$$SI = D_i \oplus Q_0$$

CLK	D_i	SI	Q_3	Q_2	Q_1	Q_0
			0	1	1	0
1	1	1 →	1	0	1	1
2	0	1 →	1	1	0	1
3	1	0 →	0	1	1	0
4	1	1 →	1	0	1	1
5	0	1 →	1	1	0	1
6	1	0 →	0	1	1	0 ← Final value

3.19 (c)

Count = 1, load = 0 externally connected to the counter. Based on the function table, whenever clear input is '0' the counter will increment its value by one and whenever clear input is 1, all the flip-flops in the counter will get value as '0'.

Clock	A_4	A_3	A_2	A_1	Clear = $A_3 \times A_1$
Initial state	0	0	0	0	0
1	0	0	0	1	0
2	0	0	1	0	0
3	0	0	1	1	0
4	0	1	0	0	0
5	0	1	0	1	1 ← Transition
	0	0	0	0	

Hence the switching sequence is 0,1,2,3,4.

3.20 (d)

	x	y	Q_n	Q_{n+1}	J	K
0	0	0	0	0	0	X
1	0	0	1	1	X	0
2	0	1	0	1	1	X
3	0	1	1	0	X	1
4	1	0	0	1	1	X
5	1	0	1	0	X	1
6	1	1	0	0	0	X
7	1	1	1	1	X	0

$$J(x, y, Q_n) = \Sigma m(2, 4) + d(1, 3, 5, 7) \quad ...(i)$$
$$K(x, y, Q_n) = \Sigma m(3, 5) + d(0, 2, 4, 5) \quad ...(ii)$$
$$J = \bar{x}y + x\bar{y}$$
$$= x \oplus y \quad ...(i)$$
$$K = \bar{x}y + x\bar{y}$$
$$= x \oplus y \quad ...(ii)$$

3.21 (a)

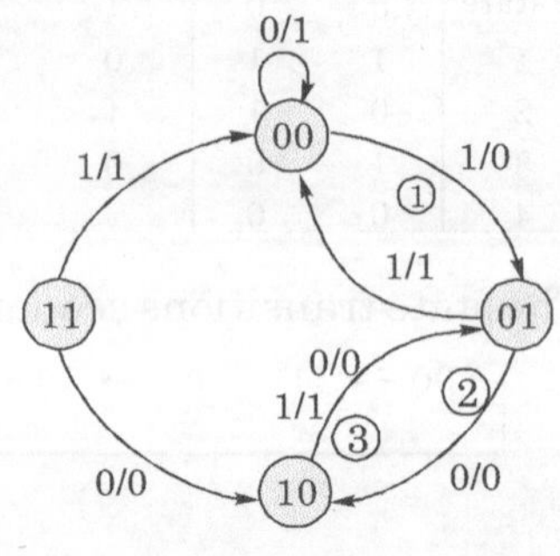

The machine to switch from A = 0, B = 0 state to A = 0, b = 1 with output = 1, the minimum length of the string required is 3.

3.22 (a)

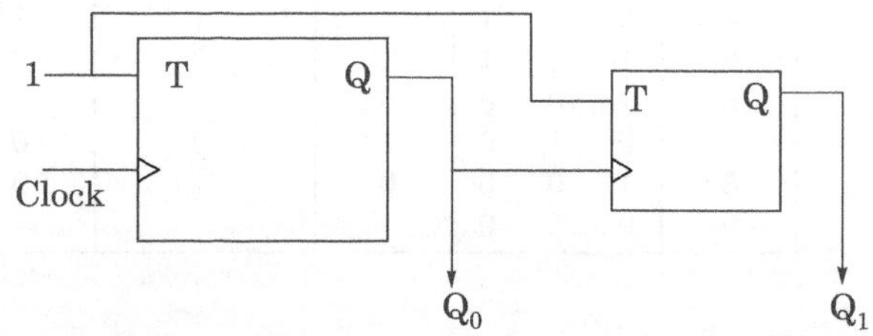

The counter is a ripple counter with positive edged triggered T flip-flops, with T inputs are connected to 1.

Q_0 will change for every clock input.

Q_1 will change whenever Q_0 is changing from '0' to '1'.

Clock	Q_1	Q_0
Initial state	0	0
1	1	1
2	1	0
3	0	1
4	0	0

The sequence is 11, 10, 01, 00

3.23 (a)

$\lceil \log_2 258 \rceil = 9$ D flip-flops

3.24 (d)

	P Q R	$D_p = R$	$D_Q = \overline{P+R}$	$D_R = Q\bar{R}$
Initial state	0 1 0	0	1	1
1	0 1 1			

3.25 (b)

Clk	P Q R	$D_p = R$	$D_Q = \overline{P+R}$	$D_R = Q\overline{R}$
	0 0 0	0	1	0
1	0 1 0	0	1	1
2	0 1 1	1	0	0
3	1 0 0	0	0	0
4	0 0 0			

The counter switching sequence is

→ 000 → 010 → 011 → 100 → (back to 000)

The number of distinct states are 4.

3.26 (c)

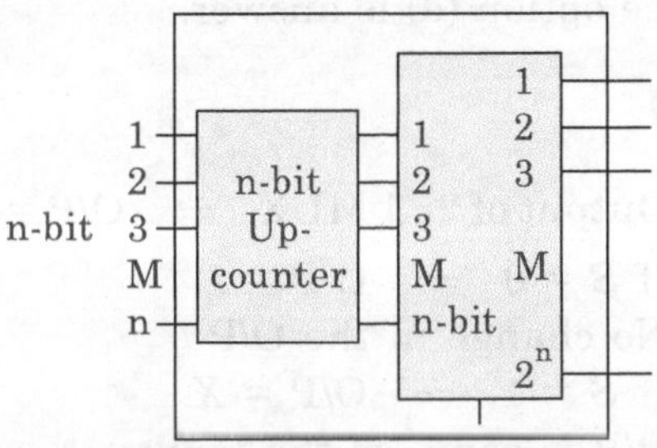

Ring counter of 2^n bits i.e. K-bit ring counter.

3.27 (c)

Clock	Q_2 Q_1 Q_0	FF2 $J_2=\overline{Q}_1$	FF2 $K_2=Q_0$	FF1 $J_1=Q_2$	FF1 $K_1=\overline{Q}_2$	FF0 $J_0=Q_1$	FF0 $K_0=\overline{Q}_0$
Initial state	0 0 0	1	0	0	1	0	1
1	1 0 0	1	0	1	0	0	1
2	1 1 0	0	0	1	0	1	1
3	1 1 1						

∴ The state sequence is 100, 110, 111.

3.28 (d)

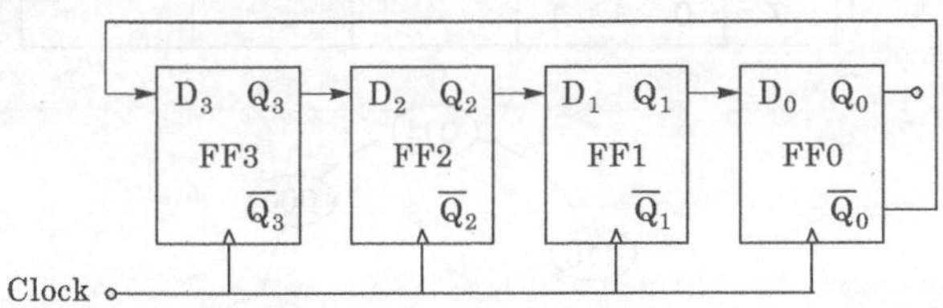

Clock	$D_3 = \overline{Q}_0$	Q_3	Q_2	Q_1	Q_0
		0	0	0	0 ← Initial state
1	1	→1	0	0	0
2	1	→1	1	0	0
3	1	→1	1	1	0
4	1	→1	1	1	1
5	0	→0	1	1	1
6	0	→0	0	1	1
7	0	→0	0	0	1
8	0	→0	0	0	0

Hence the switching sequence is : 0, 8, 12, 14, 15, 7, 3, 1, 0.

3.29 (a)

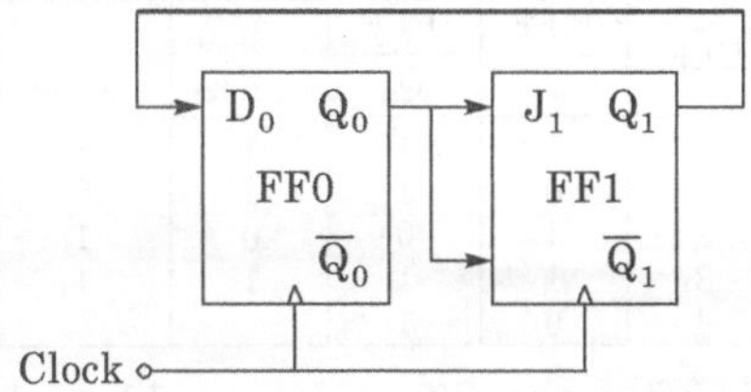

Clock	Q_1	Q_0	FF1 $J_1 = Q_0$	FF1 $K_1 = Q_0$	FF0 $D_0 = Q_1$
Initial state →	0	1	1	1	0
1	1	0	0	0	1
2	1	1	1	1	1
3	0	1	1	1	0
4	1	0			

J-K flip-flop output is 0110110...

3.30 (3)

The required switching sequence of the synchronous counter is

→ 0 → 0 → 1 → 1 → 2 → 2 → 3 → 3 → (back to 0)

1. In the counter switching sequence highest decimal number is '3', to represent state '3' minimum 2 flip-flops are required.
2. In the switching sequence each decimal number is repeating '2' times, to distinguish this repeatation 1 more flip-flop is required.

∴ The minimum number of JK flip-flops required to design the counter = 2 + 1 = 3.

3.31 (4)

The required switching sequence of the synchronous counter is

→ 0 → 1 → 0 → 2 → 0 → 3 → (back to 0)

1. In the counter switching sequence highest decimal number is '3', to represent state '3' minimum 2 flip-flops are required.
2. In the switching sequence decimal '0' is repeating '3' times, to distinguish this repeatation 2 more flip-flop are required.

∴ The minimum number of JK flip-flops required to design the counter = 2 + 2 = 4.

3.32 (b)

CLK	T-FF Q_1	D-FF Q_0	$T_1 = Q_0$	$D_0 = Q_1$
	1	1	1	1
1	0	1	1	0
2	1	0	0	1
3	1	1	1	1
4	0	1		

$Q_1 Q_0$ after the 3rd cycle are 11 and after the 4th cycle are 01 respectively.

3.33 (b)

Q_1	Q_0	Q_1^+	Q_0^+	FF inputs T_1	T_0
0	0	0	1	0	1
0	1	1	0	1	1
1	0	1	1	0	1
1	1	1	1	0	0

$$T_1(Q_1, Q_0) = \bar{Q}_1 Q_0$$

$$T_2(Q_1, Q_0) = \bar{Q}_1 \bar{Q}_0 + \bar{Q}_1 Q_0 + Q_1 \bar{Q}_0$$

$$= \bar{Q}_1 + \bar{Q}_0$$

3.34 (2)

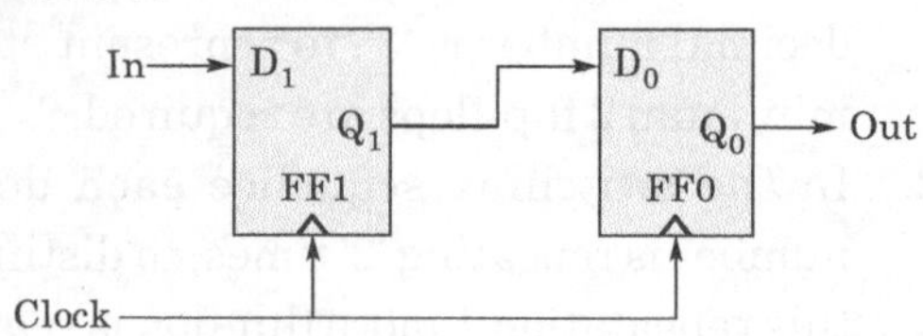

State Table:

P.S. Q_1	Q_0	Input x	FF inputs $D_1 = x$	$D_0 = Q_1$	N.S. Q_1	Q_0	Out = Q_0
0	0	0	0	0	0	0	0
0	0	1	1	0	1	0	0
0	1	0	0	0	0	0	0
0	1	1	1	0	1	0	0
1	0	0	0	1	0	1	1
1	0	1	1	1	1	1	1
1	1	0	0	1	0	1	1
1	1	1	1	1	1	1	1

State Transition Diagram:

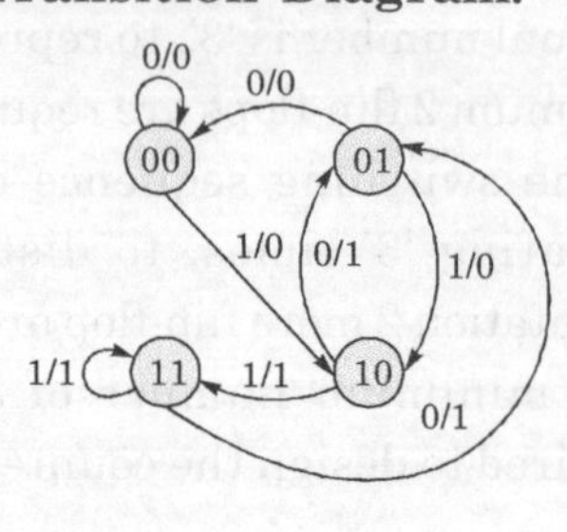

The question is on the number of self loop states.

The number of self loop states are 00 and 11.

Hence answer is 2.

3.35 (d)

Clock	Present state P	Q	R	Flip-flop inputs $T_P = R$	$T_Q = \bar{P}$	$T_R = \bar{Q}$	Next state P^+	Q^+	R^+
1	0	0	0	0	1	1	0	1	1
2	0	1	1	1	1	0	1	0	1
3	1	0	1	1	0	1	0	0	0

∴ Next three states of counter are 011, 101, 000

Hence option (d) is answer.

3.36 (b)

Output of 2 : 1 MUX ⇒ O/P = $\bar{S}I_0 + SI_1$

If $S = 0$ ⇒ O/P = I_0

No change in the O/P

If $S = 1$ ⇒ O/P = X

O/P is same as the input.

∴ Operation is similar to D latch.

∴ Answer (b).

3.37 (a)

CLK	Q_1	Q_2	Q_3	$T_1 = \bar{Q}_3$	$D_2 = Q_1$	$T_3 = Q_2$
	0	1	1	0	0	1
1	0	0	0	1	0	0
2	1	0	0	1	1	0
3	0	1	0	1	0	1
4	1	0	1	0	1	0
5	1	1	1	0	1	1
6	1	1	0	1	1	1
7	0	1	1			

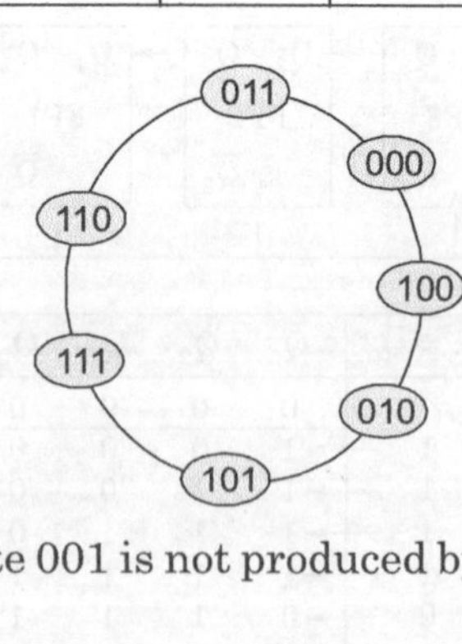

State 001 is not produced by given counter.

4 Number Systems

4.1 The number of 1's in the binary representation of (3 * 4096 + 15 * 256 + 5 * 16 + 3) is
(a) 8 (b) 9
(c) 10 (d) 12 **[1995 : 1 M]**

4.2 A ROM is used to store the table for multiplication of two 8-bit unsigned integers. The size of ROM required is
(a) 256×16 (b) $64K \times 8$
(c) $4K \times 16$ (d) $64K \times 16$
[1996 : 1 M]

4.3 Booth's algorithm for integer multiplication gives worst performance when the multiplier pattern is
(a) 101010......1010 (b) 100000.....001
(c) 111111......1111 (d) 011111.....1110
[1996 : 1 M]

4.4 Consider the following floating point number representation.

31 24	23 0
Exponent	Mantissa

The exponent is in 2's complement representation and mantissa is in the sign magnitude representation. The range of the magnitude of the normalized numbers in this representation is
(a) 0 to 1 (b) 0.5 to 1
(c) 2^{-23} to 0.5 (d) 0.5 to $(1 - 2^{-23})$
[1996 : 1 M]

4.5 Given $\sqrt{(224)_r} = (13)_r$. The value of the radix' r is
(a) 10 (b) 8
(c) 5 (d) 6 **[1997 : 2 M]**

4.6 Booth's coding in 8 bits for the decimal number – 57 is
(a) 0 – 100 + 1000
(b) 0 – 100 + 100 – 1
(c) 0 – 1 + 100 – 10 + 1
(d) 00 – 10 + 100 – 1 **[1999 : 1 M]**

4.7 Zero has two representations in:
(a) Sign magnitude (b) 1's complement
(c) 2's complement (d) None of these
[1999 : 2 M]

4.8 The number 43 in 2's complement representation is
(a) 01010101 (b) 11010101
(c) 00101011 (d) 10101011
[2000 : 1 M]

4.9 Consider the values $A = 2.0 \times 10^{30}$, $B = -2.0 \times 10^{30}$, $C = 1.0$ and the sequence. What will be the values of X and Y if the following sequence of statements are carried out
X : = A + B
Y : = A + C
X : = X + C
Y : = Y + B
(a) X = 1.0, Y = 1.0 (b) X = 0.0, Y = 1.0
(c) X = 1.0, Y = 0.0 (d) X = 0.0, Y = 0.0
[2000 : 2 M]

4.10 The 2's complement representation of $(-539)_{10}$ in hexadecimal is
(a) ABE (b) DBC
(c) DE5 (d) 9E7 **[2001 : 2 M]**

4.11 In 2's complement addition, overflow
(a) is flagged whenever there is carry from sign bit addition
(b) cannot occur when a positive value is added to a negative value
(c) is flagged when the carries from sign bit and previous bit match
(d) None of the above **[2002 : 1 M]**

4.12 The decimal value 0.25
(a) is equivalent to binary 0.1
(b) is equivalent to binary 0.01
(c) is equivalent to binary 0.00111
(d) can not be represented precisely in binary
[2002 : 1 M]

4.13 The 2's complement representation of the decimal value –15 is
(a) 1111 (b) 11111
(c) 111111 (d) 10001 **[2002 : 1 M]**

4.14 Sign extension is the step in
(a) Floating point multiplication
(b) Signed 16 bit integer addition
(d) Arithmetic left shift
(d) Converting a signed integer from one size to another
[2002 : 1 M]

4.15 Assuming all numbers are in 2's complement representation, which of the following numbers is divisible by 11111011?

(a) 11100111 (b) 11100100
(c) 11010111 (d) 11011011

[2003 : 1 M]

4.16 The following is a scheme for floating point number representation using 16 bits.

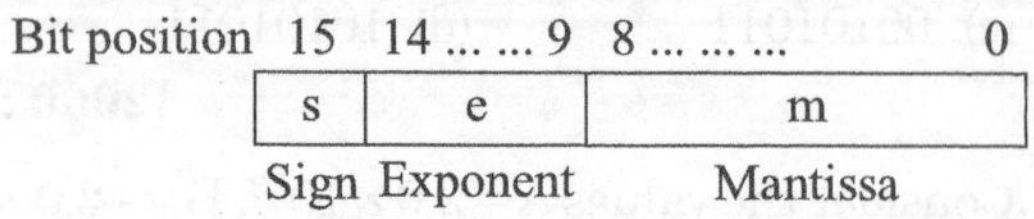

Let s, e, and m be the numbers represented in binary in the sign, exponent, and mantissa fields respectively. Then the floating point number represented is

$$\begin{cases} (-1)^s(1 + m \times 2^{-9})2^{e-31}, \text{ if the exponent} \neq 111111 \\ 0 \text{ otherwise} \end{cases}$$

What is the maximum difference between two successive real numbers representable in this system?

(a) 2^{-40} (b) 2^{-9}
(c) 2^{22} (d) 2^{31} **[2003 : 2 M]**

4.17 If 73_x (in base-x number system) is equal to 54_y, (in base-y number system), the possible values of x and y are

(a) 8, 16 (b) 10, 12
(c) 9, 13 (d) 8, 11 **[2004 : 1 M]**

4.18 What is the result of evaluating the following two expressions using three-digit floating point arithmetic with rounding?

(113. + –111.) + 7.51
113. + (–111. + 7.51)

(a) 9.51 and 10.0 respectively
(b) 10.0 and 9.51 respectively
(c) 9.51 and 9.51 respectively
(d) 10.0 and 10.0 respectively

[2004 : 1 M]

4.19 Let A = 1111 1010 and B = 0000 1010 be two 8-bit 2's complement numbers. Their product in 2's complement is

(a) 1100 0100 (b) 1001 1100
(c) 1010 0101 (d) 1101 0101

[2004 : 2 M]

4.20 The number $(123456)_8$ is equivalent to

(a) $(A72E)_{16}$ and $(22130232)_4$
(b) $(A72E)_{16}$ and $(22131122)_4$
(c) $(A73E)_{16}$ and $(22130232)_4$
(d) $(A62E)_{16}$ and $(22120232)_4$ **[2004 : 2 M]**

4.21 The range of integers that can be represented by an n bit 2's complement number system is

(a) -2^{n-1} to $(2^{n-1} - 1)$
(b) $-(2^{n-1} - 1)$ to $(2^{n-1} - 1)$
(c) -2^{n-1} to 2^{n-1}
(d) $-(2^{n-1} + 1)$ to $(2^{n-1} - 1)$ **[2005 : 1 M]**

4.22 The hexadecimal representation of 657_8 is

(a) 1AF (b) D78
(c) D71 (d) 32F **[2005 : 1 M]**

4.23 Using Booth's Algorithm for multiplication, the multiplier –57 will be recoded as

(a) 0 –1 0 0 1 0 0 –1 (b) 1 1 0 0 0 1 1 1
(c) 0 –1 0 0 1 0 0 0 (d) 0 1 0 0 –1 0 0 1

[2005 : 1 M]

4.24 $(34.4)_8 \times (23.4)_8$ evaluates to

(a) $(1053.6)_8$ (b) $(1053.2)_8$
(c) $(1024.2)_8$ (d) None of these

[2005 : 2 M]

Common Data Q.4.25 & Q.4.26:

Consider the following floating point format:

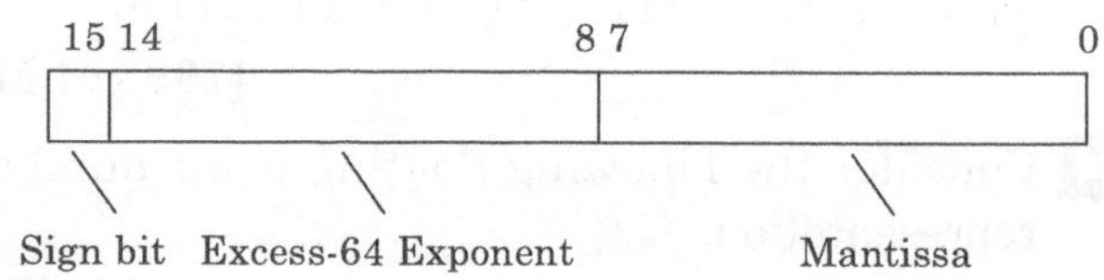

Mantissa is a pure fraction is sign-magnitude form.

4.25 The decimal number 0.239×2^{13} has the following hexadecimal representation without normalization and rounding off

(a) 0D 24 (b) 0D 4D
(c) 4D 0D (d) 4D 3D **[2005 : 2 M]**

4.26 The normalized representation for the above format is specified as follows. The mantissa has an implicit 1 preceding the binary (radix) point. Assume that only 0's are padded in while shifting a field. The normalized representation of the above number (0.239×2^{13}) is

(a) 0A 20 (b) 11 34
(c) 4D D0 (d) 4A E8 **[2005 : 2 M]**

4.27 The addition of 4-bit, two's complement, binary numbers 1101 and 0100 results in

(a) 0001 and an overflow
(b) 1001 and no overflow
(c) 0001 and no overflow
(d) 1001 and an overflow **[2006 : 1 M]**

4.28 When multiplicand Y is multiplied by multiplier $X = x_{n-1}x_{n-2},\ldots, x_0$ using bit-pair recoding in Booth's algorithm, partial products are generated according to the following table.

Row	x_{i+1}	x_i	x_{i-1}	Partial Product
1	0	0	0	0
2	0	0	1	Y
3	0	1	0	Y
4	0	1	1	2Y
5	1	0	0	?
6	1	0	1	–Y
7	1	1	0	–Y
8	1	1	1	?

The partial products for rows 5 and 8 are
(a) 2Y and Y (b) –2Y and 2Y
(c) –2Y and 0 (d) 0 and Y

[2006 : 2 M]

4.29 $(C012.25)_H - (10111001110.101)_B =$
(a) $(135103.412)_O$ (b) $(564411.412)_O$
(c) $(564411.205)_O$ (d) $(135103.205)_O$

[2007 : 2 M]

4.30 The following bit pattern represents a floating point number in IEEE 754 single precision format:
1 10000011 101000000000000000000000
The value of the number in decimal form is
(a) –10 (b) –13
(c) –26 (d) None of these

[2008 : 1 M]

4.31 In the IEEE floating point representation the hexadecimal value 0x00000000 corresponds to
(a) the normalized value 2^{-127}
(b) the normalized value 2^{-126}
(c) the normalized value +0
(d) the special value +0 **[2008 : 1 M]**

4.32 Let a denote number system radix. The only value(s) of r that satisfy the equation $\sqrt{121} = 11$, is/are
(a) decimal 10
(b) decimal 11
(c) decimal 10 and 11
(d) any value > 2

[2008 : 1 M]

4.33 The two numbers given below are multiplied using the Booth's algorithm.
Multiplicand : 0101 1010 1110 1110
Multiplier : 0111 0111 1011 1101
How many additions/Subtractions are required for the multiplication of the above two numbers?
(a) 6 (b) 8
(c) 10 (d) 12

[2008 : 2 M]

4.34 $(1217)_8$ is equivalent to
(a) $(1217)_{16}$ (b) $(028F)_{16}$
(c) $(2297)_{10}$ (d) $(0B17)_{16}$

[2009 : 1 M]

4.35 P is a 16-bit signed integer. The 2's complement representation of P is $(F87B)_{16}$. The 2's complement representation of 8*P is
(a) $(C3D8)_{16}$ (b) $(187B)_{16}$
(c) $(F878)_{16}$ (d) $(987B)_{16}$

[2010 : 1 M]

4.36 The decimal value 0.5 in IEEE single precision floating point representation has
(a) fraction bits of 000 ... 000 and exponent value of 0
(b) fraction bits of 000 ... 000 and exponent value of – 1
(c) fraction bits of 100 ... 000 and exponent value of 0
(d) no exact representation **[2012 : 1 M]**

4.37 The smallest integer that can be represent by and 8-bit number in 2's complement form is
(a) – 256 (b) – 128
(c) – 127 (d) 0 **[2013 : 1 M]**

4.38 The base (or radix) of the number system such that the following equation holds is ________.

$$\frac{312}{20} = 13.1$$

[2014 (Set-1) : 1 M]

4.39 Consider the equation $(123)_5 = (x8)_y$ with x and y as unknown. The number of possible solutions is ________. **[2014 (Set-2) : 1 M]**

4.40 The value of a float type variable is represented using the single-precision 32-bit floating point format of IEEE-754 standard that uses 1 bit for sign, 8 bits for biased exponent and 23 bits for mantissa. A float type variable X is assigned the decimal value of –14.25. The representation of X in hexadecimal notation is
(a) C1640000H
(b) 416C0000H
(c) 41640000H
(d) C16C0000H

[2014 (Set-2) : 2 M]

4.41 Consider the equation $(43)_x = (y3)_8$ where x and y are unknown. The number of possible solutions is ________.

[2015 (Set-3) : 2 M]

4.42 The 16-bit 2's complement representation of an integer is 1111 1111 1111 0101; its decimal representation is ________.

[2016 (Set-1) : 1 M]

4.43 Let X be the number of distinct16-bit integers in 2's complement representation. Let Y be the number of distinct 16-bit integers in sign magnitude representation. Then $X - Y$ is ______.

[2016 (Set-2) : 1 M]

4.44 The n-bit fixed-point representation of an unsigned real number X uses f bits for the fraction part. Let $i = n - f$. The range of decimal values for X in this representation is

(a) 2^{-f} to 2^i
(b) 2^{-f} to $(2^i - 2^{-f})$
(c) 0 to 2^i
(d) 0 to $(2^i - 2^{-f})$

[2017 (Set-1) : 1 M]

4.45 Given the following binary number in 32-bit (single precision) IEEE-754 format:

00111110011011010000000000000000

The decimal value closest to this floating-point number is

(a) 1.45×10^1
(b) 1.45×10^{-1}
(c) 2.27×10^{-1}
(d) 2.27×10^1

[2017 (Set-2) : 1 M]

4.46 The representation of the value of a 16-bit unsigned integer X in hexadecimal number system is BCA9. The representation of the value of X in octal number system is

(a) 571244
(b) 736251
(c) 571247
(d) 136251

[2017 (Set-2) : 1 M]

4.47 Consider the unsigned 8-bit fixed point binary number representation below:

$$b_7\, b_6\, b_5\, b_4\, b_3\, .\, b_2\, b_1\, b_0$$

where the position of the binary point is between b_3 and b_2. Assume b_7 is the most significant bit. Some of the decimal numbers listed below cannot be represented exactly in the above representation:

(i) 31.500
(ii) 0.875
(ii) 12.100
(iv) 3.001

Which one of the following statements is true?

(a) None of (i), (ii), (iii), (iv) can be exactly represented.
(b) Only (ii) cannot be exactly represented.
(c) Only (iii) and (iv) cannot be exactly represented.
(d) Only (i) and (ii) cannot be exactly represented.

[2018 : 2 M]

4.48 In 16-bit 2's complement representation, the decimal number −28 is:

(a) 1111 1111 0001 1100
(b) 1111 1111 1110 0100
(c) 0000 0000 1110 0100
(d) 1000 0000 1110 0100

[2019 : 1 M]

4.49 Consider Z = X − Y, where X, Y and Z are all in sign-magnitude form. X and Y are each represented in n bits. To avoid overflow, the representation of Z would require a minimum of:

(a) $n + 2$ bits
(b) n bits
(c) $n - 1$ bits
(d) $n + 1$ bits

[2019 : 1 M]

4.50 Consider three registers R1, R2, and R3 that store numbers in IEEE-754 single precision floating point format. Assume that R1 and R2 contain the values (in hexadecimal notation) 0x42200000 and 0xC1200000, respectively.

If $R3 = \frac{R1}{R2}$, what is the value stored in R3?

(a) 0xC0800000
(b) 0x40800000
(c) 0x83400000
(d) 0xC8500000

[2020 : 2 M]

4.51 Let the representation of a number in base 3 be 210. What is the hexadecimal representation of the number?

(a) 528
(b) 21
(c) 15
(d) D2

[2021 (Set-1) : 1 M]

4.52 Consider the following representation of a number in IEEE 754 single-precision floating point format with a bias of 127.

S: 1 E: 10000001 F: 11110000000000000000000

Here S, E and F denote the sign, exponent and fraction components of the floating point representation.

The decimal value corresponding to the above representation (rounded to 2 decimal places) is ________.

[2021 (Set-1) : 1 M]

4.53 The format of the single-precision floating-point representation of a real number as per the IEEE 754 standard is as follows:

sign	exponent	mantissa

Which one of the following choices is correct with respect to the smallest normalized positive number represented using the standard?

(a) exponent = 00000000 and maintissa = 00000000000000000000001
(b) exponent = 00000001 and maintissa = 00000000000000000000000
(c) exponent = 00000001 and maintissa = 00000000000000000000001
(d) exponent = 00000000 and maintissa = 00000000000000000000000

[2021 (Set-2) : 1 M]

4.54 If x and y are two decimal digits and $(0.1101)_2 = (0.8xy5)_{10}$, the decimal value of $x + y$ is ________. **[2021 (Set-2) : 1 M]**

4.55 If the numerical value of a 2-byte unsigned integer on a little endian computer is 255 more than that on a big endian computer, which of the following choices represent(s) the unsigned integer on a little endian computer?
(a) 0x6665
(b) 0x0001
(c) 0x0100
(d) 0x4243

[2021 (Set-2) : 2 M]

4.56 Let R1 and R2 be two 4-bit registers that store numbers in 2's complement form. For the operation R1 + R2, which one of the following values of R1 and R2 gives an arithmetic overflow?
(a) R1 = 1011 and R2 = 1110
(b) R1 = 1100 and R2 = 1010
(c) R1 = 0011 and R2 = 0100
(d) R1 = 1001 and R2 = 1111 **[2022 : 1 M]**

4.57 Consider three floating point numbers A, B and C stored in registers R_A, R_B and R_C, respectively as per IEEE-754 single precision floating point format. The 32-bit content stored in these registers (in hexadecimal form) are as follows.

R_A = 0xC1400000	R_B = 0xC42100000	R_C = 0x41400000

Which one of the following is FALSE?
(a) $A + C = 0$
(b) $C = A + B$
(c) $B = 3C$
(d) $(B - C) > 0$

[2022 : 2 M]

4.58 A particular number is written as 132 in radix-4 representation. The same number in radix-5 representation is ________. **[2023 : 1 M]**

4.59 Consider the IEEE-754 single precision floating point numbers
P = 0xC1800000 and Q = 0x3F5C2EF4.
Which one of the following corresponds to the product of these numbers (i.e., $P \times Q$), represented in the IEEE-754 single precision format?
(a) 0x404C2EF4
(b) 0x405C2EF4
(c) 0xC15C2EF4
(d) 0xC14C2EF4

[2023 : 2 M]

4.60 Consider a system that uses 5 bits for representing signed integers in 2's complement format. In this system, two integers A and B are represented as A = 01010 and B = 11010. Which one of the following operations will result in either an arithmetic overflow or an arithmetic underflow?
(a) $B - A$
(b) $A - B$
(c) $A + B$
(d) $2 * B$

[2024 (Set-1) : 1 M]

4.61 The format of a single-precision floating-point number as per the IEEE 754 standard is:

Sign (1 bit)	Exponent (8 bits)	Mantissa (23 bits)

Choose the largest floating-point number among the following options.

(a)

Sign	Exponent	Mantissa
0	0111 1111	1111 1111 1111 1111 1111 111

(b)

Sign	Exponent	Mantissa
0	0111 1111	0000 0000 0000 0000 0000 000

(c)

Sign	Exponent	Mantissa
0	1111 1110	1111 1111 1111 1111 1111 111

(d)

Sign	Exponent	Mantissa
0	1111 1111	1111 1111 1111 1111 1111 111

[2024 (Set-2) : 1 M]

4.62 Which of the following is/are EQUAL to 224 in radix-5 (i.e. base-5) notation?
(a) 121 in radix-7
(b) 100 in radix-8
(c) 64 in radix-10
(d) 50 in radix-16

[2024 (Set-2) : 2 M]

Answers Number Systems

4.1	(c)	**4.2**	(d)	**4.3**	(a)	**4.4**	(d)	**4.5**	(c)	**4.6**	(b)	**4.7**	(a, b)	**4.8**	(c)	**4.9**	(c)
4.10	(c)	**4.11**	(b)	**4.12**	(b)	**4.13**	(d)	**4.14**	(d)	**4.15**	(a)	**4.16**	(c)	**4.17**	(d)	**4.18**	(a)
4.19	(a)	**4.20**	(a)	**4.21**	(a)	**4.22**	(a)	**4.23**	(a)	**4.24**	(a)	**4.25**	(d)	**4.26**	(d)	**4.27**	(c)
4.28	(c)	**4.29**	(a)	**4.30**	(c)	**4.31**	(d)	**4.32**	(d)	**4.33**	(b)	**4.34**	(b)	**4.35**	(a)	**4.36**	(b)
4.37	(b)	**4.38**	(5)	**4.39**	(3)	**4.40**	(a)	**4.41**	(5)	**4.42**	(–11)	**4.43**	(1)	**4.44**	(d)	**4.45**	(c)
4.46	(d)	**4.47**	(c)	**4.48**	(b)	**4.49**	(d)	**4.50**	(a)	**4.51**	(c)	**4.52**	(–7.75)	**4.53**	(b)	**4.54**	(3)
4.55	(a, c)	**4.56**	(b)	**4.57**	(b)	**4.58**	(110)	**4.59**	(c)	**4.60**	(b)	**4.61**	(c)	**4.62**	(a, b, c)		

Explanations Number Systems

4.1 (c)

[3 * 4096 + 15 * 256 + 5 * 16 + 3]

$[3 * 16^3 + 15 * 16^2 + 5 * 16^1 + 3 * 16^0] = [3\text{ F }5\text{ }3]_{16}$

$(0011\ 1111\ 0101\ 0011)_2$

Number of 1's = 10

4.2 (d)

To store the table for multiplication of two 8-bit unsigned integers, size of ROM required is

$$= (2^8 \times 2^8) \times (8 + 8) = 2^{16} \times 16$$
$$= 2^6 \times 2^{10} \times 16 = 64\text{ K} \times 16$$

4.3 (a)

Booth recording of a multiplier are

(a) 1 0 1 0 1 0 . . . . . 1 0 1 0
⇓
–1 +1 –1 +1 –1 +1 . . . . –1 +1 –1 0

(b) 1 0 0 0 0 0 . . . . 0 0 0 1
⇓
–1 0 0 0 0 0 . . . . 0 0 +1 –1

(c) 1 1 1 1 . . . . 1 1 1 1
⇓
0 0 0 0 . . . . 0 0 0 –1

(d) 0 1 1 1 1 1 . . . . 1 1 1 0
⇓
+1 0 0 0 0 0 . . . . 0 0 –1 0

Therefore, option (a) gives worst performance because it contains maximum number of 1's.

4.4 (d)

31 24	23 0
Exponent	Mantissa

The range of the magnitude of the normalized numbers is as follows: 0.5 to $(1 - 2^{-23})$.

4.5 (c)

$$\sqrt{(224)_r} = (13)_r$$

$$\Rightarrow \sqrt{(2r^2 + 2r + 4)_{10}} = (r + 3)_{10}$$

Square both side.

$$\Rightarrow 2r^2 + 2r + 4 = r^2 + 6r + 9$$
$$\Rightarrow r^2 - 4r - 5 = 0$$
$$\Rightarrow r^2 - 5r + r - 5 = 0$$
$$\Rightarrow r(r - 5) + 1(r - 5) = 0$$
$$\Rightarrow (r + 1)(r - 5) = 0$$
$$\Rightarrow r = -1, 5$$

∴ r = 5 and choice (c) is correct.

4.6 (b)

–57 can be represented in 2's complement form in 8-bits as follows:

1 1 0 0 0 1 1 1

Booth's coding for decimal number –57 is

1 1 0 0 0 1 1 1 [0]
⇓
0 –1 0 0 +1 0 0 –1

Alternative Method:

0 – 1 0 0 + 1 0 0 – 1 evaluates to 57 in Booth's coding as follows

$$(0 \times 2^7) - (1 \times 2^6) + (0 \times 2^5) + (0 \times 2^4) + (1 \times 2^3) + (0 \times 2^2) + (0 \times 2^1) - (1 \times 2^0)$$
$$= -64 + 8 - 1$$
$$= -65 + 8 = -57$$

4.7 (a, b)

Zero has two representations in

(a) Sign magnitude as
(i) 0 0 0 0   (ii) 1 0 0 0

(b) 1's complement as
(i) 0 0 0 0   (ii) 1 1 1 1

4.8 (c)

Positive 2's complement numbers are represented as the simple binary.

So, +43 in binary is represented as

0 0 1 0 1 0 1 1

4.9 (c)

$A = 2.0 \times 10^{30}$

$B = -2.0 \times 10^{30}$

$C = 1.0$

(i) $X = A + B$
$= 2.0 \times 10^{30} + (-2.0 \times 10^{30}) = 0.0$

(ii) $Y = A + C$
$= 2.0 \times 10^{30} + (1.0)$
$= 2.0 \times 10^{30} + 0.0000...1 \times 10^{30}$
$= 2.0000...1 \times 10^{30} \approx 2.0 \times 10^{30}$

(iii) $X = X + C = 0.0 + 1.0 = 1.0$

(iv) $Y = Y + B$
$= 2.000...1 \times 10^{30} - 2.0 \times 10^{30} = 0.0$

Therefore, after the sequence value of X and Y are 1.0 and 0.0 respectively.

4.10 (c)

2's complement representation of a negative number is determined as follows:

(a) Compute a binary representation of the magnitude of the number.

(b) Take 2's complement of binary representation.

+ 539 = 0010 0001 1011

–539 = 1101 1110 0101 = $(DE5)_H$

Therefore option (c) is correct.

4.11 (b)

If both numbers are positive or negative then only overflow occurs.

4.12 (b)

$0.25 \times 2 = 0.50; \quad 0.50 \times 2 = 1.00$

Therefore, the decimal value 0.25 is equivalent to binary 0.01.

4.13 (d)

The 2's complement of negative numbers are represented as the binary number that when added to a positive number of the same magnitude equals zero

+15 = 0 1 1 1 1

The 2's complement representation of the decimal value –15 is

–15 in 2's complement = 10001

4.14 (d)

Sign extension is the operation, in computer arithmetic, of increasing the number of bits of a binary number while preserving the number's sign (positive/negative) and value.

4.15 (a)

	2's Complement	Decimal Value
	11111011 (Divider)	5
(a)	11100111	25
(b)	11100100	28
(c)	11010111	41
(d)	11011011	37

So –25 is divisible by –5 so we can say 11100111 is divisible by 11111011.

4.16 (c)

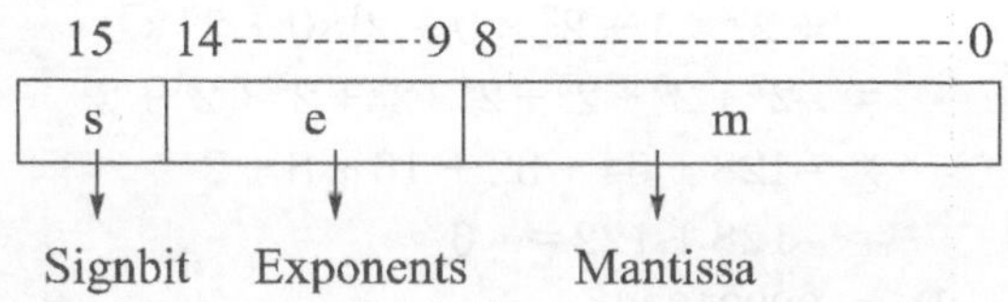

Then the floating point number represented

$$\begin{cases} (-1)^s(1 + m \times 2^{-9})2^{e-31}, & \text{if the exponent } \neq 111111 = 63 \\ 0 \text{ otherwise} \end{cases}$$

Max. exponent = 111110 = 62

$\because \quad e = 62$

Maximum value is

$(-1)^s\,[1 + (m + 1) \times 2^{-9}] \times 2^{e-31}$

Next maximum value is

$(-1)^s\,[1 + m \times 2^{-9}] \times 2^{e-31}$

$\therefore$ Difference of last two successive numbers

$= (-1)^s[1 + (m+1)\times 2^{-9}]\times 2^{62-31} - (-1)^s [1 + m \times 2^{-9}] \times 2^{62-31}$

$= [1 + (m + 1) \times 2^{-9}]\, 2^{31} - [1 + m \times 2^{-9}]2^{31}$

$= 2^{31} \times 2^{-9} = 2^{22}$

4.17 (d)

$(73)_x$ if $x = 8$

$(73)_8 = 7 \times 8^1 + 3 \times 8^o$
$= 56 + 3$
$= 59$

$(54)_y$ if $y = 11$

$(54)_{11} = 5 \times 11^1 + 4 \times 11^o$
$= 55 + 4$
$= 59$

Alternative Method:

$(73)_x = (54)_y$

$\Rightarrow \; 7x + 3 = 5y + 4$

$\Rightarrow \; 7x - 5y = 1$ ...(i)

Put options (a), (b),(c) , and (d) in (i) and check

$7 \times 8 - 5 \times 11 = 1$

$\therefore$ Option (d) is correct.

4.18 (a)

(113. + –111.) + 7.51

113. + (–111. + 7.51)

The exponent value is 10.0.

The Value evaluated by the expression is 9.51 and 10.0.

4.19 (a)

$A = 11111010$

$= -2^7 \times 1 + 2^6 \times 1 + 2^5 \times 1 + 2^4 \times 1 + 2^3 \times 1 + 2^2 \times 0 + 2^1 \times 0 + 2^0 \times 0$

$= -2^7 + 2^6 + 2^5 + 2^4 + 2^3 + 2^4 + 2^3 + 2^1$

$= -128 + 64 + 32 + 16 + 8 + 2$

$= -128 + 122 = -6$

$B = 00001010$

$= 2^3 \times 1 + 2^1 \times 1$

$= 8 + 2 = 10$

$A \cdot B = (-6) \times 10 = -60$

So the value of AB in 2's complement representation

AB = 11000100

4.20 (a)

$(123456)_8 = (001\,010\,011\,100\,101\,110)_2$

$= (00\,1010\,0111\,0010\,1110)_2$

$= (A72E)_{16}$

$= (00\,10\,10\,01\,11\,00\,10\,11\,10)_2$

$= (22130232)_4$

4.21 (a)

The range is -2^{n-1} to $(2^{n-1} - 1)$ including 0.

4.22 (a)

$(657)_8 = (110\,101\,111)_2$

$= (1\,\underline{1010}\,\underline{1111})_2$

$= (1\;A\;F)_{16}$

4.23 (a)

2's complement of –57 is 110001110.

Append 0 at the LSB and start from right end taking pairs of two symbols which are encoded as:

$00 \rightarrow 0$

$01 \rightarrow +1$

$10 \rightarrow -1$

$11 \rightarrow 0$

4.24 (a)

$34.4_8 = 011\,100.1$

$= 16 + 8 + 4 + 0.5$

$= (28.5)_{10}$

$23.4_8 = 010\,011.1$

$= 16 + 2 + 1 + 0.5$

$= (19.5)_{10}$

$28.5 \times 19.5 = (555.75)_{10}$

$= (1053.6)_8$

4.25 (d)

The decimal number is 0.239×2^{13}

We have to find hexadecimal representation without normalization.

Biased exponent = 13 + 64 = 77

Representing 77 in binary

$(77)_{10} = (1001101)_2$

Representing mantissa in binary

$(0.239)_{10} = 0.00111101000101$

Floating point representation is as follows:

Sign	Exponent	Mantissa
0	1001101	00111101

0100 1101 0011 1101

↓ ↓ ↓ ↓

4 D 3 D

$= (4D3D)_H$

4.26 (d)

The decimal number is 0.239×2^{13}

We have to find hexadecimal representation with normalization.

Binary representation of 0.239 is

0.00111101000101

$\therefore$ N is $(0.00111101000101) \times 2^{13}$

After normalization we have

$1.11101000101 \times 2^{10}$

Then, biased exponent = 10 + 64 = 74

Representating biased exponent 74 in binary

$(74)_{10} = (1001010)_2$

Floating point representation of number is as follows:

Sign	Exponent	Mantissa
0	1001010	11101000

0100 1010 1110 1000
↓ ↓ ↓ ↓
4 A E 8

$= (4AE8)_H$

4.27 (c)

$$\begin{array}{rl} 1101 & (-3)_{10} \\ 0100 & (+4)_{10} \\ \hline \boxed{1}0001 & (+1)_{10} \end{array}$$

No overflow

Option (c) is correct.

4.28 (c)

Row	Multiplier bit triplet			Partial product
	X_{i+1}	X_i	X_{i-1}	
1	0	0	0	No operation
2	0	0	1	y add
3	0	1	0	y add
4	0	1	1	2y add
5	1	0	0	–2y (sub)
6	1	0	1	–y (sub)
7	1	1	0	–y (sub)
8	1	1	1	No operation

$$5\ \underbrace{1\ 0}\ 0 \Rightarrow \begin{pmatrix} 2^1 \\ -1 \end{pmatrix}\begin{pmatrix} 2^0 \\ 0 \end{pmatrix} = -2y$$

$$8\ \underbrace{1\ 1}\ 1 \Rightarrow 0\ \ 0 = 0$$

4.29 (a)

$(C012.25)_H - (10111001110.101)_B$

= 1100 0000 0001 0010.0010 0101

– 0000 0101 1100 1110.1010 0000

= 1011 1010 0100 0011.1000 0101

= 1 011 101 001 000 011.100 001 010

$= (135103.412)_o$

Binary subtraction is like decimal subtraction:

0 – 0 = 0, 1 – 1 = 0, 1 – 0 = 1,

0 – 1 = 1 with 1 borrow.

4.30 (c)

Sign bit is 1 implies number is negative.

Exponent bits: 10000011

Exponent is added with 127 bias in IEEE single precision format.

So, Actual exponent = 10000011 – 127 = 131 – 127 = 4

Mantissa bits: 10100000000000000000000

In IEEE format, an implied 1 is before mantissa.

Hence the Number is $-1.101*2^4$

$= -(11010)_2 = -26$

4.31 (d)

An exponent of 0 together with a fraction of 0 with positive sign represents +0 which is a special value.

4.32 (d)

$$\sqrt{121_r} = (11)_r$$

So we will take base r any value that is greater than 2 since "the digit value must be less than radix".

4.33 (b)

Apply booth's algorithm:

0	1	1	1	0	1	1	1	1	0	1	1	1	1	0	1
(1–0)	(1–1)	(1–1)	(0–1)	(1–0)	(1–1)	(1–1)	(1–1)	(0–1)	(1–0)	(1–1)	(1–1)	(1–1)	(1–1)	(0–1)	(0–1)
+1	0	0	–1	+1	0	0	0	–1	+1	0	0	0	–1	+1	–1

4 additions and 4 substractions total 8 operations.

4.34 (b)

$$(1217)_8 = (028F)_{16}$$

$(1\ 2\ 1\ 7)_8$

001 010 001 111

2 8 F

4.35 (a)

2's complement of P is given as $(F87B)_{16}$

2's complement of $8 \times P = 2^3 \times (F87B)_{16}$

Left shift $(F87B)_{16}$ by 3 bits $= (C3D8)_{16}$

$(F87B)_{16} =$ 1111 1000 0111 1011

1100 0011 1101 1000

↓ ↓ ↓ ↓

$(C\ \ 3\ \ D\ \ 8)_{16}$

4.36 (b)

$$0.5 = 0.1000 \times 2^0 = 1.0000 \times 2^{-1}$$

$\therefore$ exponent = – 1 and mantissa = 000000

$\therefore$ (b) is correct.

4.37 (b)

Range of 2's complement no is

$(-2^{n-1}$ to $2^{n-1}-1)$

So minimum no represents by 8 bit is

$$-2^{8-1} = -128$$

maximum no represented by 8 bit is

$$2^{8-1} - 1 = 127$$

4.38 (5)

$$\frac{(3x^2 + 1x^1 + 2x^0)}{(2x^1 + 0x^0)} = \left(\frac{1x^2 + 3x^0 + 1}{x}\right)$$

$$\Rightarrow \; 3x^2 + 1x + 2 = \left(\frac{x+3+1}{x}\right)(2x)$$

$$\Rightarrow \; 3x^2 + x + 2 = 2x^2 + 6x + 2$$

$$\Rightarrow \; x^2 - 5x = 0$$

$$\Rightarrow \; x = 5$$

4.39 (3)

$$1*5^2 + 2*5^1 + 3*5^0 = x*y^1 + 8*y^0$$

$$25 + 10 + 3 = xy + 8$$

$$38 - 8 = xy$$

$$\Rightarrow \; xy = 30$$

Possible combinations: ($y > 8$)

and also the radix y should be greater than x

Possible x-y values

1. $x = 1, y = 30$
2. $x = 2, y = 15$
3. $x = 3, y = 10$

= 3 possibilities

4.40 (a)

$$14.25 = 1110.010 = 1.110010 \times 2^3$$

$$M = 110010$$

$$E = 127 + 3 = 130 = 10000010$$

1	8	23	
S	E	M	= C1640000
1	10000010	11001000...	

4.41 (5)

$$(43)_x = (y3)_8$$

$$\Downarrow \qquad \Downarrow$$

$x > 4$ and $y < 8$

$$\Rightarrow \; 4x + 3 = 8y + 3$$

$$\Rightarrow \; x = 2y$$

(i) $x = 6, y = 3$

(ii) $x = 8, y = 4$

(iii) $x = 10, y = 5$

(iv) $x = 12, y = 6$

(v) $x = 14, y = 7$

$\therefore$ Total 5 possible solutions.

4.42 (–11)

[1]111 1111 1111 0101 ← Given number is in 2's complement form

No. is –ve

0000 0000 0000 1011 ← 2's of the number.

$\therefore$ Decimal equivalent is –11.

4.43 (1)

Number of bits $n = 16$

In sign 2's complement representation the range is

$$= -2^{15} \text{ to } + (2^{15} - 1)$$

$$= -32768 \text{ to } + 32767$$

$$\therefore \; X = 65536$$

In sign magnitude representation the range is

$$= -(2^{15} - 1) \text{ to } + (2^{15} - 1)$$

$$= -32767 \text{ to } + 32767$$

$$\therefore \; Y = 65535$$

Hence

$$X - Y = 1$$

4.44 (d)

Let n = 5 bit

f = 2 bit

So i = 5 – 2 = 3 bit

Min value : 000.00

$\Rightarrow$ 0

Max value : 111.11

$\Rightarrow$ 7.75

$[2^i - 2^{-f}]$

$\Rightarrow$ $[2^3 - 2^{-2}]$

$\Rightarrow$ 8 – 0.25 = 7.75

4.45 (c)

1 bit	8 bit	23 bit
0	01111100	1101101000...
S	BE	M

1. Sign = 0

= +ve

AE = BE − Bias

2 2 2 2 2 2 2 2

BE = 0 1 1 1 1 1 0 0

2. Bias = 0 1 1 1 1 1 1 1

AE = 1 1 1 1 1 1 0 1

Here sign of AE is negative so take two complement of AE.

i.e., 00000010

1

00000011

⇒ −3

3. Mantissa

∴ Normal Mantissa = 1.M = 1.1101101

Data + 1.1101101×2^{-3} {$\pm M \times B^{\pm e}$}

mantissa align to right upto 3 times

↓

+0.0011101101

↓

0.228

↓

2.28×10^{-1}

4.46 (d)

$X = (B \quad C \quad A \quad 9)_{16}$

$= 1011110010101001$

$= (1 \quad 3 \quad 6 \quad 2 \quad 5 \quad 1)_8$

4.47 (c)

Binary code: $(b_7\ b_6\ b_5\ b_4\ b_3\ .\ b_2\ b_1\ b_0)_2$

$(31.5)_{10} = (11111.1)_2$

$(0.875)_{10} = (0.111)_2$

$(12.100)_{10} = (01100.000110\ .....)_2$

↓ Only 3 bits of fraction space available. So can't be stored.

$(3.001)_{10} = (00011.000000\ ......)_2$

It is also not accurate storage.

4.48 (b)

+28 ⇒ 0000 0000 0001 1100

−28 ⇒ 1111 1111 1110 0100 (2's complement form)

4.49 (d)

For example:

$$\text{Let, } \left.\begin{array}{l} X = +6, \quad n = 4 \\ Y = -5, \quad n = 4 \end{array}\right\} \Rightarrow (X - Y) = +11$$

So Z = 11 which required 5 bits which is (n + 1) bits.

4.50 (a)

R1: 0x42200000

R2: 0xC1200000

32 bit format

	S	BE	M
R1 :	0	10000100	010000 ......

32 bit format

	S	BE	M
R2 :	1	10000010	010000 ......

1. Subtract the exponents i.e. divisor exponent from the dividend exponent.

```
        2
   10000100            111111
                   AE : 00000010
   10000010      Bias : 011111111 (+127)
AE : 00000010        BE : 10000001
```

2. Divide the mantissa i.e.

R1 M: 1.010 ...

R2 M: 1.010 ... (Divide)

Result : 1

3. Sign (opposite sign division result sign is −ve)

32 bit format

R3 :	1	10000001	000000 ......

C 0 8 00000

Hence, option (a) is correct answer.

4.51 (c)

$$(210)_3 = 3^2 \times 2 + 3^1 \times 1 + 3^0 \times 0$$
$$= 18 + 3 + 0$$
$$= (21)_{10}$$

$(21)_{10}$ = 16 ⌊21 = $(15)_{16}$

1 → 5

4.52 (−7.75)

Value: $(-1)^S\ (1.M) \times 2^{BE\text{-}Bias}$

$(-1)^1\ (1.11100...) \times 2^{10000001 - 127}$

$-(1.1111) \times 2^{129 - 127}$

-1.1111×2^2

-111.11

$(-7.75)_{10}$

4.53 (b)

$$\left\{\begin{array}{l}\text{All 0's BE} \Rightarrow \text{Used for "0"} \\ \text{All 1's BE} \Rightarrow \text{Used for } (+\infty \text{ and } -\infty)\end{array}\right\}$$

4.54 (3)

$$\begin{aligned}(0.1101)_2 &= 1 \times 2^{-1} + 1 \times 2^{-2} + 0 \\ &\quad \times 2^{-3} + 1 \times 2^{-4} \\ &= 0.5 + 0.25 + 0 + 0.0625 \\ &= (0.8125)_{10} \\ (0.8125)_{10} &= (0.8xy5)_{10} \\ \therefore \quad x &= 1,\ y = 2,\ x + y = 1 + 2 = 3\end{aligned}$$

4.55 (a, c)

Option (a): In little endian 0x6665 on converting it to decimal = 26213

In big endian it will be 6566 on converting it to decimal = 25958

Now little endian – big endian = 26231 – 25958 = 255 which is correct.

Option (b): In little endian 0x0001 which is 1 in decimal in big endian 0x0100 which is greater than little endian in decimal. So this is incorrect option.

Option (c): Little endian = 0x0100 which is 256 in decimal and big endian = 0x0001 which is 1 in decimal an difference will be 255 hence this is also correct option.

Option (d): Little endian = 0x4243 and big endian = 0x4342, big endian value is greater than little endian so this is incorrect.

4.56 (b)

Stored numbers in register R1 and R2 are in 2's complement form.

Register size is 4 bits. The range of numbers in 2's complement form is –8 to + 7.

If R1 + R2, result is out of out of the above range, then it is overflow.

(a) R1 = 1011 = –(0101) = –5
+R2 = 1110 = –(0010) = –2
–7 No over flow

(b) R1 = 1100 = –(0100) = –4
+R2 = 1010 = –(0110) = –6
–10 Over flow

(c) R1 = 0011 = +(0011) = +3
+R2 = 0100 = +(0100) = +4
+7 No over flow

(d) R1 = 1001 = –(0111) = –7
+R2 = 1111 = –(0001) = –1
–8 No over flow

4.57 (b)

IEEE-754 single precision format:

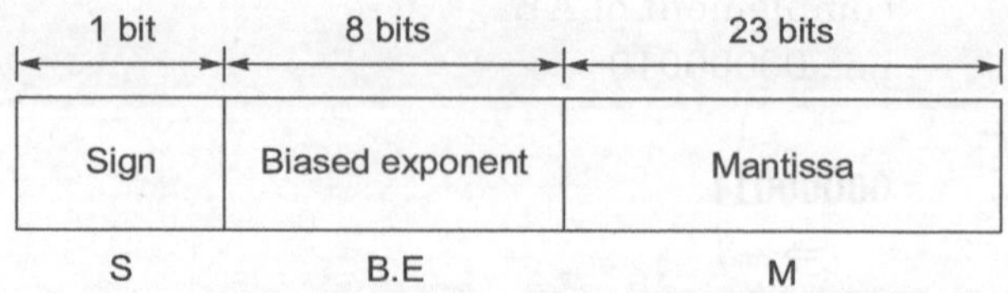

Decimal value (D.V) = $(-1)^S \times 1.M \times 2^{B.E - Bias}$

Bias value in IEEE single precision format is 127.

R_A = 1100 0001 0100 0000 0000 0000 0000 0000 (S = first bit, B.E = next 8 bits, M = remaining bits)

$$\begin{aligned}S &= 1,\ B.E = 130, \\ M &= 1.100\ 0000\ 0000 \ldots\ldots \\ D.V &= (-1)^1 \times 1.1 \times 2^{130-127} \\ &= -1.1 \times 2^3 = -1100 = (-12)_{10} \\ \therefore \quad A &= -12\end{aligned}$$

R_B = 0100 0010 0001 0000 0000 0000 0000 0000 (S, B.E, M)

$$\begin{aligned}S &= 0,\ B.E = 132 \\ M &= 1.001\ 000000 \ldots\ldots \\ D.V &= (-1)^0 \times 1.001 \times 2^{132-127} \\ &= +1.001 \times 2^5 = +100100 = (+36)_{10} \\ \therefore \quad B &= +36\end{aligned}$$

R_C = 0100 0001 0100 0000 0000 0000 0000 0000 (S, B.E, M)

$$\begin{aligned}S &= 0,\ B.E = 130 \\ M &= 1.100\ 0000 \ldots\ldots \\ D.V &= (-1)^0 \times 1.1 \times 2^{130-127} \\ &= +1.1 \times 2^3 = 1100 = (+12)_{10} \\ \therefore \quad C &= +12\end{aligned}$$

(a) $A + C = -12 + 12 = 0$ True

(b) $C = A + B$

$A + B = -12 + 36 = +24 \neq C$ False

(c) $B = 3C$

$= 3 \times +12 = 36 = B$ True

(d) $B - C > 0$

$= 36 - 12 = 24 > 0$ True

4.58 (110)

In radix-4 representation 132 i.e. $(132)_4$

↓ Convert into decimal

$1 * 4^2 + 3 * 4^1 + 2 * 4^0 = (30)_{10}$

↓ radix-5

$$\begin{array}{r|rr} 5 & 30 & \\ \hline 5 & 6 & 0 \\ \hline & 1 & 1 \end{array} = (110)_5$$

4.59 (c)

P = 0xC1800000
Q = 0x3F5C2EF4

P:	1	100 00011	000 0000 0000 ....
	S	BE	M
		[–ve] [131]	[0]

Data: $-(1.0) * 2^{131-127}$ [Bias = 127]
$-(1.0) * 2^{+4}$

Q:	0	01111110	10111000010111011110100
	S	BE	M
		[+ve] [126]	

Data: $+(1.10111000010111011110100) * 2^{126-127}$
$\Rightarrow +(1.10111000010111011110100) * 2^{-1}$

Product: $[P * Q]$
S = (–ve) * (+ve) = –ve
AE = (+4) + (–1) = +3
M = (1.0)* (1.10111000010111011110100)
= 1.10111000010111011110100

Storage:

1	10000010	10111000010111011110100

C 1 5 C 2 E F 4

Answer is $(C15C2EF4)_H$

4.60 (b)

5 bit 2's complement Range
$\{-(2^{5-1})$ to $+(2^{5-1}-1)\}$
$\{-16$ to $+15\}$
A : 01010 ⇒ +10
B : 11010 ⇒ –6
$(B - A) \Rightarrow -6 - (+10) \Rightarrow -16$
$(A - B) \Rightarrow 10 - (-6) \Rightarrow +16$
$(A + B) \Rightarrow 10 + (-6) \Rightarrow +4$
$2B \Rightarrow 2 * -6 \Rightarrow -12$

4.61 (c)

S: 0
BE: 1111 1110
M: All 1's

Except the special exponent, remaining all are data exponents.

4.62 (a, b, c)

$(224)_5 = 2 \times 5^2 + 2 \times 5^1 + 4 \times 5^0$
$= 50 + 10 + 4 = (64)_{10}$

(a) $(121)_7 = 1 \times 7^2 + 2 \times 7^1 + 1 \times 7^0$
$= 49 + 14 + 1 = (64)_{10}$

(d) $(50)_{16} = 5 \times 16^2 + 0 \times 16^0$
$= (80)_{10}$

(b) $(100)_8 = 1 \times 8^2 + 0 \times 8^1 + 0 \times 8^0$
$= (64)_{10}$

Answer is (a, b, c).

Computer Organization and Architecture

UNIT

IV

CONTENTS

Computer Organization and Architecture

Syllabus

Machine instructions and addressing modes, ALU and data-path, CPU control design, Memory interface, I/O interface (Interrupt and DMA mode), Instruction pipelining, Pipeline hazards, Cache and main memory, Secondary storage.

Analysis of Previous GATE Papers

Exam Year	1 Mark Ques.	2 Marks Ques.	Total Marks
1995	1	1	**3**
1996	1	1	**3**
1997	2	–	**2**
1998	–	1	**2**
1999	1	4	**9**
2000	2	1	**4**
2001	5	1	**7**
2002	5	2	**9**
2003	2	3	**8**
2004	3	13	**29**
2005	4	8	**20**
2006	1	7	**15**
2007	2	6	**14**
2008	–	12	**24**
2009	2	14	**10**
2010	1	4	**9**
2011	1	4	**9**
2012	2	2	**6**
2013	1	4	**9**

Exam Year	1 Mark Ques.	2 Marks Ques.	Total Marks
2014 Set-1	1	3	**7**
2014 Set-2	1	3	**7**
2014 Set-3	1	2	**5**
2015 Set-1	1	1	**3**
2015 Set-2	1	2	**5**
2015 Set-3	1	2	**5**
2016 Set-1	1	2	**5**
2016 Set-2	1	5	**11**
2017 Set-1	3	4	**11**
2017 Set-2	1	3	**7**
2018	3	3	**9**
2019	2	1	**4**
2020	5	3	**11**
2021 Set-1	1	2	**5**
2021 Set-2	2	2	**6**
2022	3	2	**7**
2023	2	3	**8**
2024 Set-1	2	3	**8**
2024 Set-2	2	3	**8**

1 CPU Architecture and Addressing Modes

1.1 The most appropriate matching for the following pairs

X. Indirect addressing	**1.** Loops
Y. Immediate addressing	**2.** Pointers
Z. Auto decrement addressing	**3.** Constants

(a) X – 3 Y – 2 Z – 1
(b) X – 1 Y – 3 Z – 2
(c) X – 2 Y – 3 Z – 1
(d) X – 3 Y – 1 Z – 2

[2000 : 1 M]

1.2 Which is the most appropriate match for the items in the first column with the items in the second column

List-I
X. Indirect Addressing
Y. Indexed Addressing
Z. Base Register Addressing

List-II
I. Array implementation
II. Writing re-locatable code
III. Passing array as parameter

(a) (X, III) (Y, I) (Z, II)
(b) (X, II) (Y, III) (Z, I)
(c) (X, III) (Y, II) (Z, I)
(d) (X, I) (Y, III) (Z, II)

[2001 : 2 M]

Common Data for Q.1.3 & Q.1.4:

Consider the following assembly language program for a hypothetical processor. A, B and C are 8 bit registers. The meanings of various instructions are shown as comments.

```
     MOV  B,#0  ;  B ← 0
     MOV  C,#8  ;  C ← 8
Z :  CMP  C,#0  ;  Compare C with 0
     JZ   X     ;  Jump to X if 0 flag is set
     SUB  C,#1  ;  C ← C – 1
     RRC  A,#1  ;  Rotate right A through carry by
                   one bit
     JC   Y     ;  Jump to Y if any carry flag is
                   set
     JMP  Z     ;  Jump to Z
Y :  ADD  B,#1  ;  B ← B + 1
     JMP  Z     ;  Jump to Z
X :
```

1.3 If the initial value of register A is A_0, the value of register B after the program execution will be
(a) the number of 0 bits in A_0
(b) the number of 1 bits in A_0
(c) A_0
(d) 8

[2003 : 2 M]

1.4 Which of the following instructions when inserted at location X will ensure that the value of register A after program execution is the same as its initial value?
(a) RRC A, # 1
(b) NOP; no operation
(c) LRC A, # 1; left rotate A through carry flag by one bit
(d) ADD A, # 1

[2003 : 2 M]

1.5 Which of the following addressing modes are suitable for program relocation at run time?
1. Absolute addressing
2. Based addressing
3. Relative addressing
4. Indirect addressing

(a) 1 and 4
(b) 1 and 2
(c) 2 and 3
(d) 1, 2 and 4

[2004 : 1 M]

1.6 Using a 4-bit 2's complement arithmetic, which of the following additions will result in an overflow?
(i) 1100 + 1100
(ii) 0011 + 0111
(iii) 1111 + 0111

(a) (i) only
(b) (ii) only
(c) (iii) only
(d) (i) and (iii) only

[2004 : 2 M]

1.7 If we use internal data forwarding to speed up the performance of a CPU (R1, R2 and R3 are registers and M[100] is a memory reference), then the sequence of operations

R1 → M[100]
M[100] → R2
M[100] → R3

can be replaced by

(a) R1 → R3
R2 → M[100]

(b) M[100] → R2
R1 → R2
R1 → R3

(c) R1 → M[100]
R2 → R3

(d) R1 → R2
R1 → R3
R1 → M[100] **[2004 : 2 M]**

Common Data for Q.1.8 to Q.1.10:
Consider the following program segment for a hypothetical CPU having three user registers R1, R2 and R3.

Instruction	Operation	Instruction Size (in words)
MOV R1,5000 ;	R1 ← Memory [5000]	2
MOV R2,(R1) ;	R2 ← Memory [(R1)]	1
ADD R2, R3 ;	R2 ← R2 + R3	1
MOV 6000, R2;	Memory [6000] ← R2	2
HALT ;	Machine halts	1

1.8 Consider that the memory is byte addressable with size 32 bits, and the program has been loaded starting from memory location 1000 (decimal). If an interrupt occurs while the CPU has been halted after executing the HALT instruction, the return address (in decimal) saved in the stack will be
(a) 1007 (b) 1020
(c) 1024 (d) 1028 **[2004 : 2 M]**

1.9 Consider that the memory is word addressable with size 32 bits and the program has been loaded starting from memory location 1000 (decimal). If an interrupt occurs during the ADD instruction, what will be the return address pushed on to the stack.
(a) 1007 (b) 1004
(c) 1005 (d) 1016 **[2004 : 2 M]**

1.10 Let the clock cycles required for various operations be as follows:
Register to/from memory transfer:
3 clock cycles
ADD with both operands in register:
1 clock cycle
Instruction fetch and decode:
2 clock cycles per word
The total number of clock cycles required to execute the program is
(a) 29 (b) 24
(c) 23 (d) 20 **[2004 : 2 M]**

1.11 Consider a three word machine instruction
ADD A[R_0], @ B
The first operand (destination) "A [R0]" uses indexed addressing mode with R0 as the index register. The second operand (source) "@B" uses indirect addressing mode. A and B are memory addresses residing at the second and the third words, respectively. The first word of the instruction specifies the opcode, the index register designation and the source and destination addressing modes. During execution of ADD instruction, the two operands are added and stored in the destination (first operand).
The number of memory cycles needed during the execution cycle of the instruction is
(a) 3 (b) 4
(c) 5 (d) 6 **[2005 : 2 M]**

1.12 Match **List-I** with **List-II** and select the correct answer using the codes given below the lists:

List-I	List-II
A. A[I] = B[J];	**1.** Indirect addressing
B. while[*A++];	**2.** Indexed addressing
C. int temp = *x;	**3.** Autoincrement

Codes:

	A	B	C
(a)	3	2	1
(b)	1	3	2
(c)	2	3	1
(d)	1	2	3

[2005 : 2 M]

1.13 A CPU has 24-bit instructions. A program starts at address 300 (in decimal). Which one of the following is a legal program counter (all values in decimal)?
(a) 400 (b) 500
(c) 600 (d) 700 **[2006 : 1 M]**

1.14 Which of the following statements about relative addressing mode is FALSE?
(a) It enables reduced instruction size
(b) It allows indexing of array elements with same instruction
(c) It enables easy relocation of data
(d) It enables faster address calculations than absolute addressing
[2006 : 2 M]

1.15 The memory locations 1000, 1001 and 1020 have data values 18, 1 and 16 respectively before the following program is executed.

MOVI	Rs, 1	Move immediate
LOAD	Rd, 1000(Rs)	Load from memory
ADDI	Rd, 1000	Add immediate
STOREI	0(Rd), 20	Store immediate

Which of the statements below is TRUE after the program is executed?

(a) Memory location 1000 has value 20
(b) Memory location 1020 has value 20
(c) Memory location 1021 has value 20
(d) Memory location 1001 has value 20

[2006 : 2 M]

1.16 Consider a new instruction named branch-on-bit-set (mnemonic bbs). The instruction "bbs reg, pos, label" jumps to label if bit in position pos of register operand reg is one. A register is 32 bits wide and the bits are numbered 0 to 31, bit in position 0 being the least significant. Consider the following emulation of this instruction on a processor that does not have bbs implemented.

temp ← reg and mask
Branch to label if temp is non-zero

The variable temp is a temporary register. For correct emulation, the variable mask must be generated by

(a) mask ← 0x1 << pos
(b) mask ← 0x ffffffff >> pos
(c) mask ← pos
(d) mask ← 0xf **[2006 : 2 M]**

1.17 Data forwarding techniques can be used to speed up the operation in presence of data dependencies. Consider the following replacements of LHS with RHS.

(i) R1 → Loc, Loc → R2 ≡ R1 → R2, R1 → Loc
(ii) R1 → Loc, Loc → R2 ≡ R1 → R2
(iii) R1 → Loc, R2 → Loc ≡ R1 → Loc
(iv) R1 → Loc, R2 → Loc ≡ R2 → Loc

In which of the following options, will the result of executing the RHS be the same as executing the LHS irrespective of the instructions that follow?

(a) (i) and (iii) (b) (i) and (iv)
(c) (ii) and (iii) (d) (ii) and (iv)

[2007 : 2 M]

1.18 Following table indicates the latencies of operations between the instruction producing the result and instruction using the result.

Instruction production the result	Instruction using the result	Latency
ALU Operation	ALU Operation	2
ALU Operation	Store	2
Load	ALU Operation	1
Load	Store	0

Consider the following code segment.

Load R1, Loc1;	Load R1 from memory location Loc1
Load R2, Loc2;	Load R2 from memory location Loc2
Add R1, R2, R1;	Add R1 and R2 and save result in R1
Dec R2;	Decrement R2
Dec R1;	Decrement R1
Mpy R1, R2, R3;	Multiply R1 and R2 and save result in R3
Store R3, Loc3;	Store R3 in memory location Loc3

What is the number of cycles needed to execute the above code segment assuming each instruction takes one cycle to execute?

(a) 7 (b) 10
(c) 13 (d) 14 **[2007 : 2 M]**

Common Data for Q.1.19 to Q.1.21:

Consider the following program segment. Here R1, R2 and R3 are the general purpose registers.

Instruction	Operation	Instruction size (no. of words)
MOV R1, (3000)	R1 ← M[3000]	2
LOOP:		
MOV R2, M[R3]	R2 ← M[R3]	1
ADD R2, R1	R2 ← R1 + R2	1
MOV (R3), R2	M[R3] ← R2	1
INC R3	R3 ← R3 + 1	1
DEC R1	R1 ← R1 – 1	1
BNZ LOOP	Branch on not zero	2
HALT		Stop

Assume that the content of memory location 3000 is 10 and the content of the register R3 is 2000. The content of each of the memory locations from 2000 to 2010 is 100. The program is loaded from the memory location 1000. All the numbers are in decimal.

1.19 Assume that the memory is word addressable. The number of memory references for accessing the data in executing the program completely is

(a) 10 (b) 11
(c) 20 (d) 21 **[2007 : 2 M]**

1.20 Assume that the memory is word addressable. After the execution of this program, the content of memory location 2010 is

(a) 100 (b) 101
(c) 102 (d) 110 **[2007 : 2 M]**

1.21 Assume that the memory is byte addressable and the word size is 32 bits. If an interrupt occurs during the execution of the instruction "INC R3", what return address will be pushed on to the stack?

(a) 1005 (b) 1020
(c) 1024 (d) 1040 **[2007 : 2 M]**

1.22 A processor that has carry, overflow and sign flag bits as part of its program status word (PSW) performs addition of the following two 2's complement numbers 01001101 and 11101001. After the execution of this addition operation, the status of the carry, overflow and sign flags, respectively will be:
(a) 1, 1, 0
(b) 1, 0, 0
(c) 0, 1, 0
(d) 1, 0, 1
[2008 : 1 M]

1.23 Assume that EA = (X)+ is the effective address equal to the contents of location X, with X incremented by one word length after the effective address is calculated; EA = –(X) is the effective address equal to the contents of location X, with X decremented by one word length before the effective address is calculated; EA = (X)– is the effective address equal to the contents of location X, with X decremented by one word length after the effective address is calculated. The format of the instruction is (opcode, source, destination), which means (destination ← source op destination). Using X as a stack pointer, which of the following instructions can pop the top two elements from the stack, perform the addition operation and push the result back to the stack.
(a) ADD (X)–, (X) (b) ADD (X), (X)–
(c) ADD –(X), (X)+ (d) ADD –(X), (X)
[2008 : 2 M]

1.24 Which of the following must be true for the RFE (Return from Exception) instruction on a general purpose processor.
1. It must be a trap instruction
2. It must be a privileged instruction
3. An exception can not be allowed to occur during execution of an RFE instruction.
(a) 1 only
(b) 2 only
(c) 1 and 2 only
(d) 1, 2 and 3 only
[2008 : 2 M]

1.25 The program below uses six temporary variables a, b, c, d, e, f.
a = 1
b = 10
c = 20
d = a + b
e = c + d
f = c + e
b = c + e
e = b + f
d = 5 + e
return d + f
Assuming that all operations take their operands from registers, what is the minimum number of registers needed to execute this program without spilling?
(a) 2 (b) 3
(c) 4 (d) 6 **[2010 : 2 M]**

1.26 Consider a hypothetical processor with a instruction of type LW R1, 20(R2), which during execution reads a 32-bit word from memory and stores it in a 32-bit register R1. The effective address of the memory location is obtained by the addition of a constant 20 and the contents of register R2. Which of the following best reflects the addressing mode implemented by this instruction for the operand in memory?
(a) Immediate Addressing
(b) Register Addressing
(c) Register Indirect Scaled Addressing
(d) Base Indexed Addressing **[2011 : 1 M]**

1.27 Consider evaluating the following expression tree on a machine with load-store architecture in which memory can be accessed only through load and store instructions. The variables a, b, c, d and e are initially stored in memory. The binary operators used in this expression tree can be evaluated by the machine only when the operands are in registers. The instructions produce result only in a register. If no intermediate results can be stored in memory, what is the minimum number of registers needed to evaluate this expression?

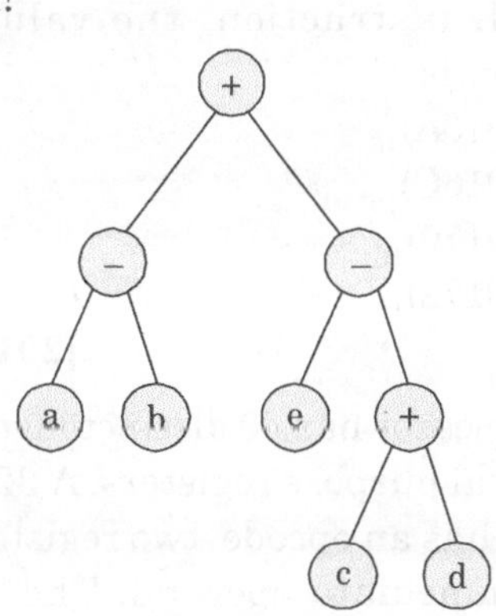

(a) 2 (b) 9
(c) 5 (d) 3 **[2011 : 2 M]**

1.28 A machine has a 32-bit architecture, with 1-word long instructions. It has 64 registers, each of which is 32 bits long. It needs to support 45 instructions, which have an immediate operand in addition to two register operands. Assuming that the immediate operand is an unsigned integer, the maximum value of the immediate operand is ________. **[2014 (Set-1) : 1 M]**

1.29 For computers based on three-address instruction formats, each address field can be used to specify which of the following:

S1: A memory operand
S2: A processor register
S3: An implied accumulator register

(a) Either S1 or S2
(b) Either S2 or S3
(c) Only S2 and S3
(d) All of S1, S2 and S3

[2015 (Set-1) : 1 M]

1.30 Consider a processor with byte-addressable memory. Assume that all registers, including Program Counter (PC) and Program Status Word (PSW), are of size 2 bytes. A stack in the main memory is implemented from memory location $(0100)_{16}$ and it grows upward. The stack pointer (SP) points to the top element of the stack. The current value of SP is $(016E)_{16}$. The CALL instruction is of two words, the first word is the op-code and the second word is the starting address of the subroutine (one word = 2 bytes). The CALL instruction is implemented as follows:

- Store the current value of PC in the stack.
- Store the value of PSW register in the stack.
- Load the starting address of the subroutine in PC.

The content of PC just before the fetch of a CALL instruction is $(5FA0)_{16}$. After execution of the CALL instruction, the value of the stack pointer is

(a) $(016A)_{16}$
(b) $(016C)_{16}$
(c) $(0170)_{16}$
(d) $(0172)_{16}$

[2015 (Set-2) : 2 M]

1.31 A processor has 40 distinct instructions and 24 general purpose registers. A 32-bit instruction word has an opcode, two register operands and an immediate operand. The number of bits available for the immediate operand field is ________. **[2016 (Set-2) : 1 M]**

1.32 Consider a processor with 64 registers and an instruction set of size twelve. Each instruction has five distinct fields, namely, opcode, two source register identifiers, one destination register identifier, and a twelve-bit immediate value. Each instruction must be stored in memory in a byte-aligned fashion. If a program has 100 instructions, the amount of memory (in bytes) consumed by the program text is ________.

[2016 (Set-2) : 2 M]

1.33 Consider the C struct defined below:

```
struct data {
  int marks [100];
  char grade;
  int cnumber;
};
struct data student;
```

The base address of student is available in register Rl. The field student, grade can be accessed efficiently using

(a) Post-increment addressing mode, (R1) +
(b) Pre-decrement addressing mode, – (R1)
(c) Register direct addressing mode, R1
(d) Index addressing mode, X(R1). where X is an offset represented in 2's complement 16-bit representation.

[2017 (Set-1) : 1 M]

1.34 Consider a RISC machine where each instruction is exactly 4 bytes long. Conditional and unconditional branch instructions use PC-relative addressing mode with Offset specified in bytes to the target location of the branch instruction. Further the Offset is always with respect to the address of the next instruction in the program sequence. Consider the following instruction sequence

Instr. No.	Instruction
i	add R2, R3, R4
$i+1$	sub R5, R6, R7
$i+2$	cmp R1, R9, R10
$i+3$	beq R1, Offset

If the target of the branch instruction is i, then the decimal value of the Offset is ________.

[2017 (Set-1) : 2 M]

1.35 The following are some events that occur after a device controller issues an interrupt while process L is under execution.

(P) The processor pushes the process status of L onto the control stack.

(Q) The processor finishes the execution of the current instruction.
(R) The processor executes the interrupt service routine.
(S) The processor pops the process status of L from the control stack.
(T) The processor loads the new PC value based on the interrupt.

Which one of the following is the correct order in which the events above occur?
(a) QPTRS
(b) PTRSQ
(c) TRPQS
(d) QTPRS

[2018 : 1 M]

1.36 Consider the following processor design characteristics:
I. Register-to-register arithmetic operations only.
II. Fixed-length instruction format.
III. Hardwired control unit.

Which of the characteristics above are used in the design of a RISC processor?
(a) I and II only
(b) II and III only
(c) I and III only
(d) I, II and III

[2018 : 1 M]

1.37 A processor has 16 integer registers (R0, R1, ..., R15) and 64 floating point registers (F0, F1, ..., F63). It uses a 2-byte instruction format. There are four categories of instructions: Type-1, Type-2, Type-3 and Type-4. Type-1 category consists of four instructions, each with 3 integer register operands (3Rs). Type-2 category consists of eight instructions, each with 2 floating point register operands (2Fs). Type-3 category consists of fourteen instructions, each with one integer register operand and one floating point register operand (1R+1F). Type-4 category consists of N instructions, each with a floating point register operand (IF).
The maximum value of N is ________.

[2018 : 2 M]

1.38 The chip selects logic for a certain DRAM chip in memory system design is shown below. Assume that memory has 16 address lines denoted by A_{15} to A_0. What is the range of addresses (in hexadecimal) of the memory system that can get enabled by the chip select (CS) signal?

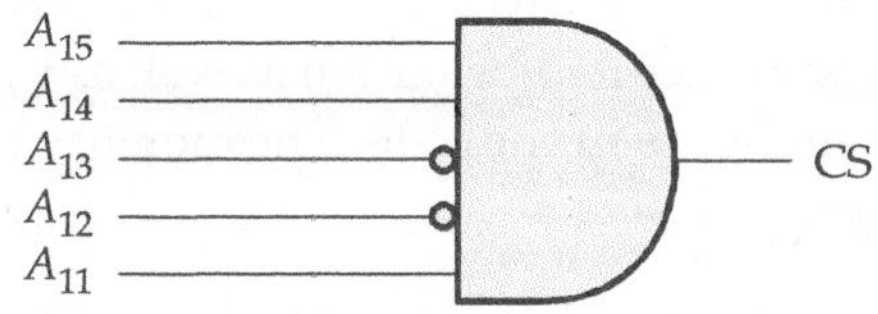

(a) DA00 to DFFF
(b) C800 to CFFF
(c) C800 to C8FF
(d) CA00 to CAFF

[2019 : 1 M]

1.39 If there are m input lines and n output lines for a decoder that is used to uniquely address a byte addressable 1 KB RAM, then the minimum value of $m + n$ is ________.

[2020 : 1 M]

1.40 A processor has 64 registers and uses 16-bit instruction format. It has two types of instructions: I-type and R-type. Each I-type instruction contains an opcode, a register name, and a 4-bit immediate value. Each R-type instruction contains an opcode and two register names. If there are 8 distinct I-type opcodes, then the maximum number of distinct R-type opcodes is ________.

[2020 : 2 M]

1.41 Consider the following instruction sequence where registers R1, R2 and R3 are general purpose and MEMORY[X] denotes the content at the memory location X.

Instruction	Semantics	Instruction Size (bytes)
MOV R1, (5000)	R1 ← MEMORY [5000]	4
MOV R2, (R3)	R2 ← MEMORY [R3]	4
ADD R2, R1	R2 ← R1 + R2	2
MOV (R3), R2	MEMORY [R3] ← R2	4
INC R3	R3 ← R3 + 1	2
DEC R1	R1 ← R1 – 1	2
BNZ 1004	Branch if not zero to the given absolute address	2
HALT	Stop	1

Assume that the content of the memory location 5000 is 10 and the content of the register R3 is 3000. The content of each of the memory locations from 3000 to 3010 is 50. The instruction sequence starts from the memory location 1000. All the numbers are in decimal format. Assume that the memory is byte addressable.
After the execution of the program, the content of memory location 3010 is ________.

[2021 (Set 1) : 2 M]

1.42 Consider the given C-code and its corresponding assembly code, with a few operands U1–U4 being unknown. Some useful information as well as the semantics of each unique assembly instruction is annotated as inline comments in the code. The memory is byte-addressable.

```
//C-code                          ; assembly-code (; indicates comments)
                                      ; r1-r5 are 32-bit integer registers
                                      ; initialize r1=0, r2=10
                                      ; initialize r3, r4 with base address of a, b
int a[10], b[10], i;              L01: jeq r1, r2, end; if(r1==r2) goto end
// int is 32-bit                  L02: lw r5, 0(r4); r5 <- Memory[r4+0]
for (i=0; i<10; i++)                  L03: shl r5, r5, U1; r5 <- r5 << U1
      a[i] = b[i] * 8;                L04: sw r5, 0(r3); Memory[r3+0] <- r5
                                      L05: add r3, r3, U2; r3 <- r3+U2
                                      L06: add r4, r4, U3
                                      L07: add r1, r1, 1
                                      L08: jmp U4; goto U4
                                      L09: end
```

Which one of the following options is a CORRECT replacement for operands in the position (U1, U2, U3, U4) in the above assembly code?

(a) (8, 4, 1, L02)   (b) (3, 4, 4, L01)
(c) (8, 1, 1, L02)   (d) (3, 1, 1, L01)

[2023 : 2 M]

1.43 The baseline execution time of a program on a 2 GHz single core machine is 100 nanoseconds (ns). The code corresponding to 90% of the execution time can be fully parallelized. The overhead for using an additional core is 10 ns when running on a multicore system. Assume that all cores in the multicore system run their share of the parallelized code for an equal amount of time.

The number of cores that minimize the execution time of the program is ________.

[2024 (Set-1) : 2 M]

1.44 A processor uses a 32-bit instruction format and supports byte-addressable memory access. The ISA of the processor has 150 distinct instructions. The instructions are equally divided into two types, namely R-type and I-type, whose formats are shown below.

R-type Instruction Format:

OPCODE	UNUSED	DST Register	SRC Register1	SRC Register2

I-type Instruction Format:

OPCODE	DST Register	SRC Register	# Immediate value/address

In the OPCODE, 1 bit is used to distinguish between I-type and R-type instructions and the remaining bits indicate the operation. The processor has 50 architectural registers, and all register fields in the instructions are of equal size.

Let X be the number of bits used to encode the UNUSED field, Y be the number of bits used to encode the OPCODE field, and Z be the number of bits used to encode the immediate value/address field. The value of $X + 2Y + Z$ is ________.

[2024 (Set-2) : 2 M]

1.45 A processor with 16 general purpose registers uses a 32-bit instruction format. The instruction format consists of an opcode field, an addressing mode field, two register operand fields, and a 16-bit scalar field. If 8 addressing modes are to be supported, the maximum number of unique opcodes possible for every addressing mode is ________.

[2024 (Set-2) : 2 M]

■■■■

Answers CPU Architecture and Addressing Modes

1.1	(c)	**1.2**	(a)	**1.3**	(b)	**1.4**	(a)	**1.5**	(c)	**1.6**	(b)	**1.7**	(d)	**1.8**	(c)	**1.9**	(b)
1.10	(b)	**1.11**	(d)	**1.12**	(c)	**1.13**	(c)	**1.14**	(d)	**1.15**	(d)	**1.16**	(a)	**1.17**	(b)	**1.18**	(d)
1.19	(d)	**1.20**	(a)	**1.21**	(c)	**1.22**	(b)	**1.23**	(a)	**1.24**	(d)	**1.25**	(b)	**1.26**	(d)	**1.27**	(d)
1.28	(16383)	**1.29**	(a)	**1.30**	(d)	**1.31**	(16)	**1.32**	(500)	**1.33**	(d)	**1.34**	(–16)	**1.35**	(a)	**1.36**	(d)
1.37	(32)	**1.38**	(b)	**1.39**	(1034)	**1.40**	(14)	**1.41**	(50)	**1.42**	(b)	**1.43**	(3)	**1.44**	(34)	**1.45**	(32)

Explanations CPU Architecture and Addressing Modes

1.1 (c)

Indirect addressing mode is used for pointers access, immediate addressing mode is used for constant access and auto-decrement addressing mode is used for loops.

1.2 (a)

- Indirect addressing → Passing array as parameter
- Indexed addressing → Array implementation
- Base register addressing → Writing relocatable code

1.3 (b)

$$A \leftarrow A_0$$

The Value of B after execution of the program is number of 1 bits in A_0.

1.4 (a)

Insert the instruction
RRC A, #

1.5 (c)

Based addressing mode & relative addressing mode are suitable for program relocation at run time due to following reason. In relative addressing scheme in which the operand field contains a relative address also called an offset or displacement D. If R_1, R_2...... R_k are CPU registers then A is computed as follows.

$$A := R + D$$

By changing contents of R, the processor can change the absolute address referred to by a block of instructions B. This address modification permits the processor to move (relocate) the entire block B from one region of main memory to another without invalidating the addresses in B. When used in this way R is referred to as a base register and it contents as a base addresses.

1.6 (b)

Over flow in 2's complement arithmetic occur when:

1. There is a carry from MSB but no carry come on MSB.
2. There is a carry on MSB but no carry come outside from MSB.

i.e. $C_{out} \oplus C_{in} = 1$

So,

```
 ① 1 1
 0 0 1 1
 0 1 1 1      No overflow since:
 -------      Cout ⊕ Cin = 0
 1 0 1 0

 ①
 0 0 1 1
 0 1 1 1      Overflow since:
 -------      Cout ⊕ Cin = 1
 1 0 0 1

 ①①①
 1 1 1 1
 0 1 1 1      No overflow since:
 ---------    Cout ⊕ Cin = 0
 ① 0 1 1 0
```

Note: Overflow condition $C_{out} \oplus C_{in}$.

1.7 (d)

Internal data forwarding instead write to memory and then read from memory. We read directly from register itself instead from memory i.e.

R1 → M[100]		R1 → M[100]	
M[100] → R2	⇒	R1 → R2	Order can be changed
M[100] → R3		R1 → R3	

1.8 (c)

Instruction	Instruction Size	Location (Decimal)
MOV R_1,5000	2	1000 to 1007
MOV R_0, (R_1)	1	1008 to 1011
ADD R_0, R_3	1	1012 to 1015
MOV 6000, R_2	2	1016 to 1023
Halt	1	1024 to 1027

If an interrupt occurs the CPU has been halted after executing the HALT instruction the return address 1024 saved in the stack.

1.9 (b)

Instruction	Instruction Size	Location (decimal)
MOV R_1, 5000	2	1000 to 1001
MOV R_2,(R_1)	1	1002
ADD R_2, R_3	1	1003
MOV 6000, R_2	2	1004 – 1005
Halt	1	1006

The return address pushed onto the stack is 1004.

1.10 (b)

Operation	Instruction Size	(IF + ID)	(OF + PD + WB)
R_0, Memory [5000]	2	2 × 2 C	3 C
$R_2 \leftarrow$ Memory [(R_1)]	1	1 × 2 C	3 C
$R_2 \leftarrow (R_2 + R_3)$]	1	1 × 2 C	1 C
Memory [6000] $\leftarrow R_2$	2	2 × 2 C	3 C
Machine Halt	1	1 × 2 C	0 C
		14 C	10 C

So total (14 C + 10 C) = 24 Cycles are required.

1.11 (d)

ADD A[R_0], @ B

A[R_0] uses indexed addressing mode as follow such that R_0 as the index register

Regs[R3] ← Regs[R3] + Mem[Regs[B] + Regs [R0]] ...(i)

And @ B uses indirect as follows

ADD A[R_0], @B

Regs [R_0] ← Regs[R_0] + Mem[Mem[B]] ...(ii)

So the first instruction needed 3 memory cycles and the second needed 2 and one memory cycle is needed for writing the result into the memory.

So total 6 memory cycles are needed to execute ADD A[R_0], @ B.

1.12 (c)

The correct matching is

1. A[I] = B[J] Indexed Addressing Mode
2. While[*A++] Autoincrement
3. int temp = * x ; Indirect Addressing Mode

1.13 (c)

Size of instruction 24 bits

Starting address of the program is 300. The size of instruction is 3 byte long So the address is always the multiple of 3 byte next address is 600 it is also the next instruction of the program.

1.14 (d)

- Absolute addressing contain actual address of location while relative address contain relative address i.e. Actual address = Base address + Relative value require ALU operation. So absolute addressing has fast address calculation than relative.
- Relative addressing (PC relative, based indexed) used for program reallocation of runtime.
- Relative address can be used for indexing of array elements i.e. only relative value changed.
- Relative addressing reduce size of instruction i.e. MOV R_1, O(R_S) instead of MOV R_1, 0x 1234567.

1.15 (d)

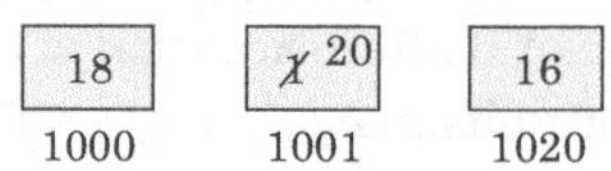

1. MOVI Rs, 1

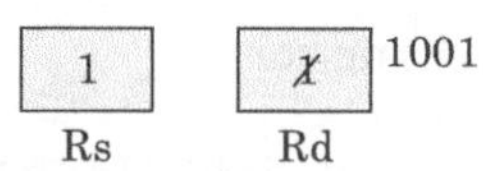

2. LOAD Rd, 1000(Rs)

   i.e. Rd ← [1000 + 1]

   Rd ← [1001]

3. ADDI Rd, 1000

   Rd ← [Rd] +1000

4. STOREI 0(Rd), 20

   [0 + 1001] ← 20

   [1001] ← 20

So, memory location 1001 will contain value 20.

1.16 (a)

bbs reg, pos, label

temp ← reg and mask

Branch to label if temp is non zero. If we shift one position left by pos of ox1 then it evaluate the bbs instruction and mask ← 0x1 << pos

1.17 (b)

For (i):

R1 → Loc ↘ Loc → R2 ≡ R1 → Loc, R1 → R2

Since both Loc and R2 contain R1's data. Hence same result as LHS.

For (ii):

R1 → Loc, Loc → R2 ≢ R1 → R2

Since R2 contains R1's data but no information about Loc content. So, different result from LHS.

For (iii):

R1 → Loc, R2 → Loc ≢ R1 → Loc

Since LHS overwrite the contents of R1's in Loc to R2's contents. So LHS results different from RHS.

For (iv):

R1 → Loc, R2 → Loc ≡ R2 → Loc

Since both side Loc contain R2's contents. So LHS = RHS. So no modification to R1 and R2's contents.

1.18 (d)

Instruction	Latency	No. of cycles
Load R1, Loc1	0	0
Load R2, Loc2	0	0
Add R1, R2, R1	1 + 1 + 2 = 4	4 × 1 cycle = 4
Dec R2	1 + 2 = 3	3 × 1 cycle = 3
Dec R1	1 + 2 = 3	3 × 1 cycle = 3
Mpy R1, R2, R3	1 + 1 + 2 = 4	4 × 1 cycle = 4
Store R3, Loc3	0	0
		Total = 14

1.19 (d)

Given M[3000] = 10

Instruction	Required Memory Reference
$R_1 \leftarrow M[3000]$	1
$R_2 \leftarrow M[R_3]$ or $M[R_3]$	10
$M[R_3] \leftarrow R_2$	10
	Total = 21

1.20 (a)

The register R_3 contain the memory location 2010 whose value is 100 at the end of loop BNZ Loop. Because 2010 cell data is not used in the program.

1.21 (c)

Location	Operation
1000-1007	$R_1 \leftarrow M[3000]$
1008-1011	
1012-1015	$R_2 \leftarrow M[R_3]$
1016-1019	$R_2 \leftarrow R_1 + R_2$
1020-1023	$M[R_3] \leftarrow R_2$
1024-1031	$R_3 \leftarrow R_3 + 1$

So if an interrupt occurs during the execution of the instruction "INC R_0, the return address 1024 will be pushed on to the stack.

1.22 (b)

```
                ①     ①     ①
Number 1:   0 1 0 0 1 1 0 1
Number 2:   1 1 1 0 1 0 0 1
Add       ① 0 0 1 1 0 1 1 0
```

Since, $C_{out} \oplus C_{in} = 0$

Hence no over flow.

Carry present i.e. bits comes out from MSB.

MSB of addition is 0. So positive number.

So, output will be carry = 1, overflow = 0, sign = 0.

1.23 (a)

EA = (X) + works as post increment

EA = – (X) works as pre-decrement

EA = (X) – works as post decrement

To add top 2 elements:

1. Pop (Top) ⇒ Pop (X) ⇒ a

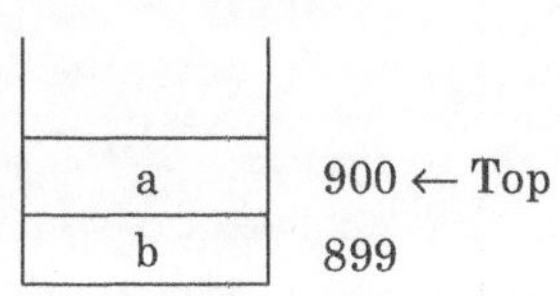

2. Then decrement X i.e. X = X – 1

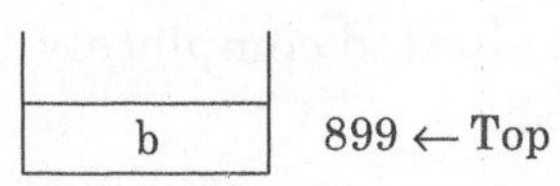

3. Pop (Top) = Pop (X) ⇒ b

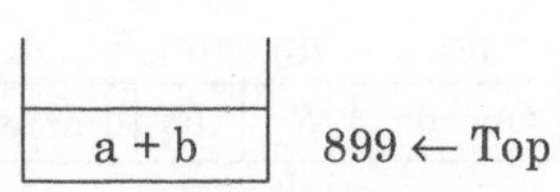

Add a + b and push at address of X.

Which is implemented by ADD (X) –, X.

1.24 (d)

A RFE (Return From Exception) instruction contains a trap instruction, privileged instruction and an exception can't be allowed to occur during the execution of an REF instruction.

1.25 (b)

Register operations

R1 ← a	....	(a = 1)
R2 ← b	....	(b = 10)
R3 ← c	....	(c = 11)
R1 ← R1 + R2	....	(d = a + b)
R1 ← R3 + R1	....	(e = c + d)
R2 ← R3 + R1	....	(f = c + e)
R3 ← R3 + R1	....	(b = c + e)
R1 ← R2 + R3	....	(e = b + f)
R3 ← 5 + R1	....	(d = 5 + e)
return (R2 + R3)	....	return (d + f)

hence 3 registers needed only.

1.26 (d)

LW R1, 20(R2) Instruction shows the source address can be formed by adding the constant with the R2 content .It is the MIPS instruction. R2 points the base address of an array and 20 points the index address of an array. Hence the addressing mode used for the above is Based Indexed Addressing mode. This addressing mode is used to implement the Arrays.

1.27 (d)

Load r1,d
Load r2,c
Add r1,r2
Load r2,e
Sub r2,r1
Load r1,b
Load r3,a
Sub r3,r1
Add r3,r2

Tree is evaluated completely. It takes only 3 registers.

1.28 (16383)

Opcode (6)	R_1 (6)	R_2 (6)	Immediate

32 – (6 + 6 + 6) = 14 bits for immediate field

⇒ $2^{14} - 1 = 16383$ maximum possible value of immediate operand.

1.29 (a)

Address field of 3-address instruction can be either memory operand or processor register. Implied accumulator register can not be address field.

1.30 (d)

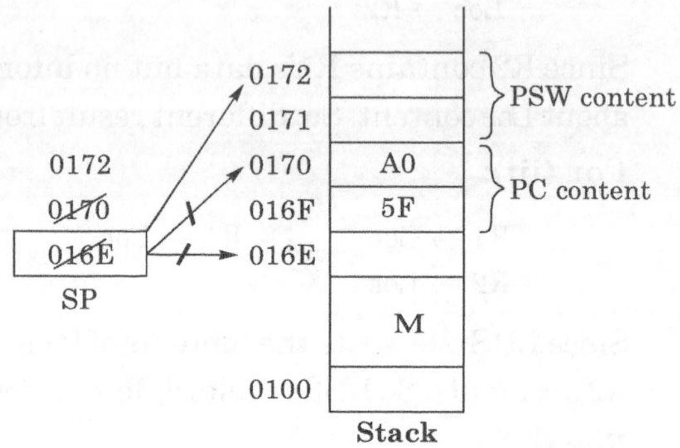

Just before CALL instruction execution, SP contains 016E

While CALL execution:

(i) PC contents are pushed i.e., SP incremented by 2 ⇒ SP = 0170

(ii) PSW contents are pushed i.e., SP incremented by 2 ⇒ SP = 0172

∴ The value of stack pointer is $(0172)_{16}$.

1.31 (16)

32 bit

opcode	Reg	Reg	Immediate
6 bit	5 bit	5 bit	16 bit

The number of bits for immediate operand field is 16 bits.

1.32 (500)

34 bit

opcode	Reg	Reg	Reg	12 bit Immediate
4 bit	6	6	6	12

One instruction size = 34 bit = 5 bytes

100 instruction occupies 500 bytes.

1.33 (d)

Since direct access is possible with only index addressing mode in given options.

So, option (d) is the correct answer.

1.34 (–16)

Program storage, assume the starting address as 1000

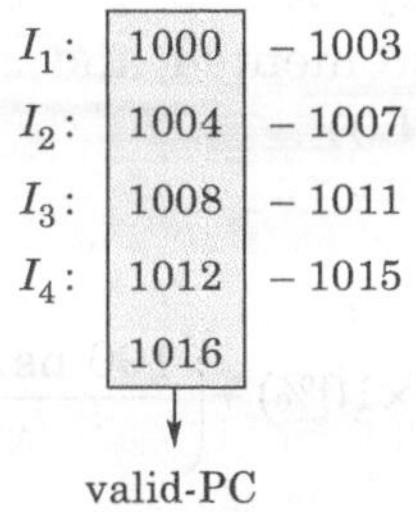

- I_4 is as branch instruction I_1 as a target location.
- PC-relative adressing mode EA = PC + relative value.

$$\text{Relative value} = \text{EA} - \text{PC} = 1000 - 1016 = -16$$

1.35 (a)

Processor required to handle the interrupt:

1. Processor finishes the execution of the current instruction.
2. Processor pushes the process status of L onto the control stack.
3. Processor loads the new PC value based on the interrupt.
4. Processor executes the interrupt service routine.
5. Processor pops the process status of "L" from the control stack.

1.36 (d)

RISC processor characteristics:

1. It supports more registers, so ALU operations are performed only on a register data.
2. It support fixed length instructions.
3. It uses hard-wired control unit.

1.37 (32)

$$\text{Instruction size} = 16 \text{ bit}$$

Type I instruction design:

16 bit

opcode	IR	IR	IR
4 bit	4 bit	4 bit	4 bit

$$\text{Number of operations} = 2^4 = 16$$

↓

$$\text{Free opcodes} = 16 - 4 = 12$$

Type II instruction design:

16 bit

opcode	FR	FR
4 bit	6 bit	6 bit

Free opcodes after type 1 instruction $= 12$

Free opcodes after type 2 instruction $= 12 - 8 = 4$

Type III instruction design:

16 bit

opcode	IR	FR
6 bit	4 bit	4 bit

$$\text{Expand opcode size} = 6 \text{ bit}$$

↓

$$\text{Number of opcodes} = 4 \times 2^2 = 16$$

∴ Number of free opcodes after type 3 instruction $= 16 - 14 = 2$

Type IV instruction design:

16 bit

opcode	FR
10 bit	6 bit

$$\text{Expand opcode size} = 10 \text{ bit}$$

↓

$$\text{Number of opcodes} = 2 \times 2^4 = 2^5 = 32$$

1.38 (b)

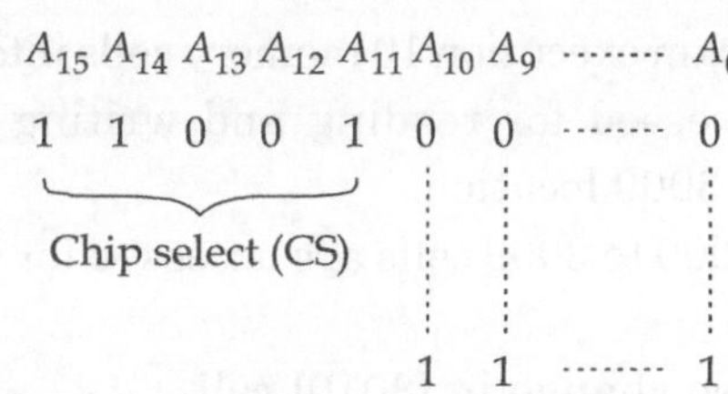

[C800 to CFFF]

1.39 (1034)

We need 2^{10} outputs to map 1 KB RAM.

For this we need 10×2^{10} decoder.

Here $m = 10$

and $n = 2^{10}$

$$m + n = 1034$$

1.40 (14)

Given, 16 bit instruction and 64 registers

I-type:

16 bit

Opcodes	Reg.	AM
6 bit	6 bit	4 bit

: Derived

R-type:

16 bit

Opcodes	Reg.	Reg.
4 bit	6 bit	6 bit

: Primitive

1. Primitive instruction

   R-type:

Opcodes	Reg.	Reg.
4 bit	6 bit	6 bit

2. Number of operation possible = $2^4 = 16$
3. Number of tree opcodes = $(16 - x)$

   Assume x is number of R-type instruction existed.
4. Number of I-type instruction possible

$$= (16 - x) \times 2^2$$
$$8 = 64 - 4x$$
$$4x = 64 - 8 = 56$$
$$x = \frac{56}{4} \Rightarrow 14$$

1.41 (50)

Program execution 10 memory cells information is accessed for reading and writing starting from 3000 location.

So, 3000 to 3009 cells are accessed for read and write.

$\therefore$ No change in [3010] cell.

So, it contain 50 only.

1.42 (b)

Byte addressable memory so after accessing every element address is incremented by "4". After every element processing count register 'r_1' is incremented and compare with 'r_2' so jump to 'L01'.

1.43 (3)

$$= (100 \text{ ns} \times 10\%) + \left(\frac{100 \text{ ns} \times 90\%}{x}\right) + (x-1)10 \text{ ns}$$

$$= 10 \text{ ns} + \frac{90 \text{ ns}}{x} + (x-1)\ 10 \text{ ns}$$

If $x = 1$ then ET = 100 ns

If $x = 2$ then ET = 65 ns

If $x = 3$ then ET = 60 ns

If $x = 4$ then ET = 62.5 ns

If $x = 5$ then ET = 68 ns

So, minimum execution time is possible with $x = 3$

1.44 (34)

32 bit

opcode	unused	Reg	Reg	Reg
8 bit	X	6 bit	6 bit	6 bit

$$X = 6$$

32 bit

opcode	Reg	Reg	Immediate value
8 bit	6 bit	6 bit	Z

$$Z = 12$$
$$Y = \text{opcode (8 bit)}$$
$$X + 2Y + Z = 6 + (2 \times 8) + 12 = 34$$

1.45 (32)

32 bit

opcode	AM	Reg	Reg	16 bit Immediate
?	3 bit	4 bit	4 bit	16 bit

$$\text{opcode} = 5 \text{ bit}$$
$$\text{Number of operations} = 2^5 = 32$$

■■■■

2 Control Unit Design

2.1 RAID configurations of disks are used to provide
(a) fault-tolerance
(b) high speed
(c) high data density
(d) None of these

[1999 : 2 M]

2.2 Arrange the following configuration for CPU in decreasing order of operating speeds: Hard wired control, vertical micro-programming, horizontal micro-programming.
(a) Hard wired control, vertical microprogramming, horizontal micro-programming.
(b) Hard wired control, horizontal micro-programming, vertical micro-programming.
(c) Horizontal micro-programming, vertical micro-programming, Hard wired control.
(d) Vertical micro-programming, horizontal micro-programming, hard wired control.

[1999 : 2 M]

2.3 A processor needs software interrupt to
(a) test the interrupt system of the processor
(b) implement co-routines
(c) obtain system services which need execution of privileged instructions
(d) return from subroutine

[2001 : 1 M]

2.4 Consider the following data path of a simple non-pipelined CPU. The registers A, B, A_1, A_2, MDR, the bus and the ALU are 8-bit wide. SP and MAR are 16-bit registers. The MUX is of size $8 \times (2:1)$ and the DEMUX is of size $8 \times (1:2)$. Each memory operation takes 2 CPU clock cycles and uses MAR (Memory Address Register) and MDR (Memory Data Register). SP can be decremented locally.

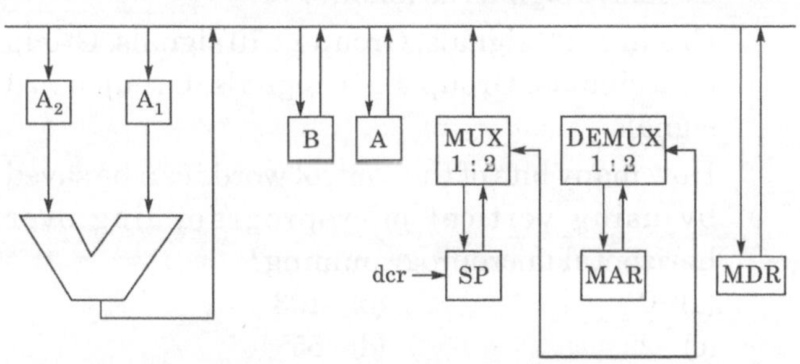

The CPU instruction "push r", where r = A or B, has the specification

M [SP] ← r
SP ← SP – 1

How many CPU clock cycles are needed to execute the "push r" instruction?
(a) 2
(b) 3
(c) 4
(d) 5

[2001 : 2 M]

2.5 A device employing INTR line for device interrupt puts the CALL instruction on the data bus while
(a) $\overline{\text{INTA}}$ is active
(b) HOLD is active
(c) READY is active
(d) None of these

[2002 : 1 M]

2.6 The absolute addressing mode
(a) the operand is inside the instruction
(b) the address of the operand is inside the instruction
(c) the register containing the address of the operand is specified inside the instruction
(d) the location of the operand is implicit

[2002 : 1 M]

2.7 Horizontal microprogramming
(a) does not require use of signal decoders
(b) results in larger sized microinstructions than vertical microprogramming
(c) uses one bit for each control signal
(d) All of the above

[2002 : 2 M]

2.8 Consider an array multiplier for multiplying two n bit numbers. If each gate in the circuit has a unit delay, the total delay of the multiplier is
(a) $\Theta(1)$
(b) $\Theta(\log n)$
(c) $\Theta(n)$
(d) $\Theta(n^2)$

[2003 : 1 M]

2.9 What is the minimum size of ROM required to store the complete truth table of an 8-bit × 8-bit multiplier?
(a) 32 K × 16 bits
(b) 64 K × 16 bits
(c) 16 K × 32 bits
(d) 64 K × 32 bits

[2004 : 1 M]

2.10 A CPU has only three instructions I1, I2 and I3, which use the following signals in time steps T1-T5:

I1: T1 : Ain, Bout, Cin
T2 : PCout, Bin
T3 : Zout, Ain
T4 : PCin, Bout
T5 : End

I2: T1 : Cin, Bout, Din
T2 : Aout, Bin
T3 : Zout, Ain
T4 : Bin, Cout
T5 : End

I3: T1 : Din, Aout
T2 : Ain, Bout
T3 : Zout, Ain
T4 : Dout, Ain
T5 : End

Which of the following logic functions will generate the hardwired control for the signal Ain?
(a) T1.I1 + T2.I3 + T4.I3 + T3
(b) (T1 + T2 + T3).I3 + T1.I1
(c) (T1 + T2).I1 + (T2 + T4).I3 + T3
(d) (T1 + T2).I2 + (T1 + T3).I1 + T3

[2004 : 2 M]

2.11 Normally user programs are prevented from handling I/O directly by I/O instructions in them. For CPUs having explicit I/O instructions, such I/O protection is ensured by having the I/O instructions privileged. In a CPU with memory mapped I/O, there is no explicit I/O instruction. Which one of the following is true for a CPU with memory mapped I/O?
(a) I/O protection is ensured by operating system routine(s)
(b) I/O protection is ensured by a hardware trap
(c) I/O protection is ensured during system configuration
(d) I/O protection is not possible **[2005 : 1 M]**

2.12 Which one of the following is true for a CPU having a single interrupt request line and a single interrupt grant line?
(a) Neither vectored interrupt nor multiple interrupting devices are possible
(b) Vectored interrupts are not possible but multiple interrupting devices are possible
(c) Vectored interrupts and multiple interrupting devices are both possible
(d) Vectored interrupt is possible but multiple interrupting devices are not possible

[2005 : 1 M]

2.13 A hardwired CPU uses 10 control signals S1 to S10 in various time steps T1 to T5 to implement 4 instructions I1 to I4 as shown below:

	T1	T2	T3	T4	T5
I1	S1, S3, S5	S2, S4, S6	S1, S7	S10	S3, S8
I2	S1, S3, S5	S8, S9, S10	S5, S6, S7	S6	S10
I3	S1, S3, S5	S7, S8, S10	S2, S6, S9	S10	S1, S3
I4	S1, S3, S5	S2, S6, S7	S5, S10	S6, S9	S10

Which of the following pairs of expressions represent the circuit for generating control signals S5 and S10 respectively [(IJ + Ik)Tn indicates that the control signal should be generated in time step Tn if the instruction being executed is IJ or lk]?
(a) S5 = T1 + I2.T3 and S10 = (I1 + I3).T4 + (I2 + I4).T5
(b) S5 = T1 + (I2 + I4).T3 and S10 = (I1 + I3).T4 + (I2 + I4).T5
(c) S5 = T1 + (I2 + I4).T3 and S10 = (I2 + I3 + I4).T2 + (I1 + I3).T4 + (I2 + I4).T5
(d) S5 = T1 + (I2+I4).T3 and S10 = (I2 + I3).T2 + I4.T3 + (I1+I3).T4 + (I2+ I4).T5

[2005 : 2 M]

2.14 A line L in a circuit is said to have a stuck-at-0 fault if the line permanently has a logic value 0. Similarly a line L in a circuit is said to have a stuck-at-1 fault if the line permanently has a logic value 1. A circuit is said to have a multiple stuck-at fault if one or more lines have stuck at faults. The total number of distinct multiple stuck-at faults possible in a circuit with N lines is
(a) 3^N (b) $3^N - 1$
(c) $2^N - 1$ (d) 2^N

[2005 : 2 M]

2.15 An instruction set of a processor has 125 signals which can be divided into 5 groups of mutually exclusive signals as follows:

Group 1 : 20 signals, Group 2 : 70 signals, Group 3 : 2 signals, Group 4 : 10 signals, Group 5 : 23 signals.

How many bits of the control words can be saved by using vertical microprogramming over horizontal microprogramming?
(a) 0 (b) 103
(c) 22 (d) 55

[2005 : 2 M]

Common Data for Q.2.16 & Q.2.17:

Consider the following data path of a CPU

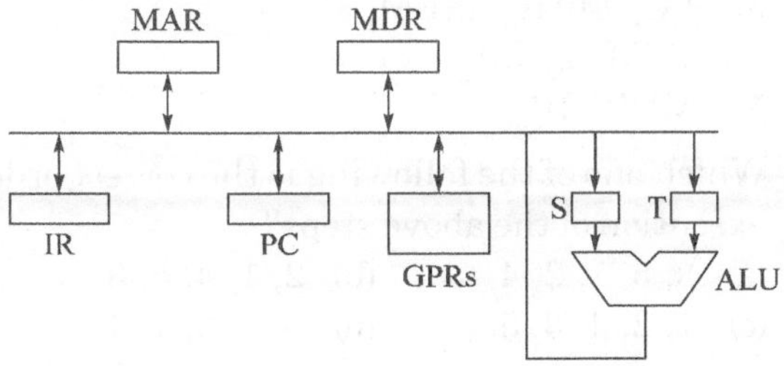

The, ALU, the bus and all the registers in the data path are of identical size. All operations including incrementation of the PC and the GPRs are to be carried out in the ALU. Two clock cycle are needed for memory read operation – the first one for loading address in the MAR and the next one for loading data from the memory but into the MDR.

2.16 The instruction "add R0, R1" has the register transfer interpretation R0 < = R0 + R1. The minimum number of clock cycles needed for execution cycle of this instruction is

(a) 2 (b) 3
(c) 4 (d) 5 **[2005 : 2 M]**

2.17 The instruction "call Rn, sub" is a two word instruction. Assuming that PC is incremented during the fetch cycle of the first word of the instruction, its register transfer interpretation is

Rn ⇐ PC + 1;
PC ⇐ M [PC];

The minimum number of CPU clock cycles needed during the execution cycle of this instruction is

(a) 2 (b) 3
(c) 4 (d) 5 **[2005 : 2 M]**

2.18 The floating point unit of a processor using a design D takes 2t cycles compared to t cycles taken by the fixed point unit. There are two more design suggestions D1 and D2. D1 uses 30% more cycles for fixed point unit but 30% less cycles for floating point unit as compared to design D. D2 uses 40% less cycles for fixed point unit but 10% more cycles for floating point unit as compared to design D. For a given program which has 80% fixed point operations and 20% floating point operations, which of the following ordering reflects the relative performances of three designs? (Di > Dj denotes that Di is faster than Dj)

(a) D1 > D > D2 (b) D2 > D > D1
(c) D > D2 > D1 (d) D > D1 > D2

[2007 : 2 M]

2.19 Consider a CPU where all the instructions require 7 clock cycles to complete execution. There are 140 instructions in the instruction set. It is found that 125 control signals are needed to be generated by the control unit. While designing the horizontal micro-programmed control unit, single address field format is used for branch control logic. What is the minimum size of the control word and control address register?

(a) 125, 7 (b) 125, 10
(c) 135, 9 (d) 135, 10

[2008 : 2 M]

2.20 Which of the following is/are true of the auto increment addressing mode?

1. It is useful in creating self relocating code
2. If it is included in an Instruction Set Architecture, then an additional ALU is required for effective address calculation
3. The amount of increment depends on the size of the data item accessed.

(a) 1 only
(b) 2 only
(c) 3 only
(d) 2 and 3 only

[2008 : 2 M]

2.21 The use of multiple register windows with overlap causes a reduction in the number of memory accesses for

1. function locals and parameters
2. register saves and restores
3. instruction fetches

(a) 1 only (b) 2 only
(c) 3 only (d) 1, 2 and 3

[2008 : 2 M]

2.22 A CPU generally handles an interrupt by executing an interrupt service routine

(a) as soon as an interrupt is raised
(b) by checking the interrupt register at the end of fetch cycle
(c) by checking the interrupt register after finishing the execution of the current instruction
(d) by checking the interrupt register at fixed time intervals **[2009 : 1 M]**

2.23 The amount of ROM needed to implement at 4 bit multiplier is

(a) 64 bits (b) 128 bits
(c) 1 Kbits (d) 2 Kbits

[2012 : 1 M]

2.24 Consider the following sequence of micro operations.

MBR ← PC
MAR ← X
PC ← Y
Memory ← MBR

Which one of the following is a possible operation performed by this sequence

(a) Instruction fetch
(b) Operand fetch
(c) Conditional branch
(d) Initiation of interrupt service

[2013 : 2 M]

2.25 A multiplexer is placed between a group of 32 registers and an accumulator to regulate data movement such that at any given point in time the content of only one register will move to the accumulator, The minimum number of select lines needed for the multiplexer is ______.

[2020 : 1 M]

2.26 Consider the following data path diagram:

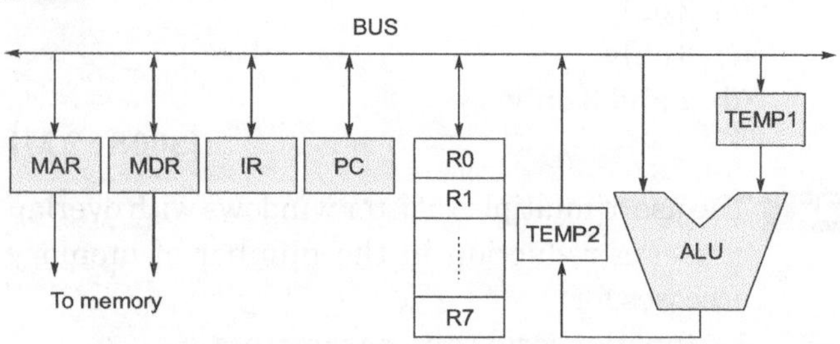

Consider an instruction: R0 ← R1 + R2. The following steps are used to execute it over the given data path. Assume that PC is incremented appropriately. The subscripts *r* and *w* indicate read and write operations, respectively.

1. $R2_r$, $TEMP1_r$, ALU_{add}, $TEMP2_w$
2. $R1_r$, $TEMP1_w$
3. PC_r, MAR_w, MEM_r
4. $TEMP2_r$, $R0_w$
5. MDR_r, IR_w

Which one of the following is the correct order of execution of the above steps?

(a) 3, 5, 1, 2, 4
(b) 2, 1, 4, 5, 3
(c) 1, 2, 4, 3, 5
(d) 3, 5, 2, 1, 4

[2020 : 1 M]

2.27 A keyboard connected to a computer is used at a rate of 1 keystroke per second. The computer system polls the keyboard every 10 ms (milli seconds) to check for a keystroke and consumes 100 μs (micro seconds) for each poll. If it is determined after polling that a key has been pressed, the system consumes an additional 200 μs to process the keystroke. Let T_1 denote the fraction of a second spent in polling and processing a keystroke.

In an alternative implementation, the system uses interrupts instead of polling. An interrupt is raised for every keystroke. It takes a total of 1 ms for servicing an interrupt and processing a keystroke. Let T_2 denote the fraction of a second spent in servicing the interrupt and processing a keystroke.

The ratio $\frac{T_1}{T_2}$ is ______. (Rounded off to one decimal place)

[2023 : 1 M]

■■■■

Answers Control Unit Design

2.1	(a, b)	2.2	(b)	2.3	(c)	2.4	(a)	2.5	(a)	2.6	(b)	2.7	(d)	2.8	(c)	2.9	(b)
2.10	(a)	2.11	(a)	2.12	(b)	2.13	(d)	2.14	(b)	2.15	(b)	2.16	(b)	2.17	(b)	2.18	(b)
2.19	(d)	2.20	(c)	2.21	(c)	2.22	(c)	2.23	(d)	2.24	(d)	2.25	(5)	2.26	(d)	2.27	(10.2)

Explanations Control Unit Design

2.1 (a, b)

RAID (Redundant array of independent disks) is a storage technology that combines multiple disk drive components into a logical unit. RAID configurations of disks are used to provide fault-tolerance and high speed.

2.2 (b)

Configurations for CPU in decreasing order of operating speeds:
Hardwired control > Horizontal micro-programming > Vertical micro-programming

2.3 (c)

A processor needs software interrupt to obtain system services which need execution of privileged instructions.

2.4 (a)

SP is decremented locally.
Therefore memory operation needs 2 clocks.

2.5 (a)

A device employing INTR line for device interrupt puts the CALL instruction on the data bus while $\overline{\text{INTA}}$ is active.

2.6 (b)

In absolute addressing mode, the address of the operand is inside the instruction.

2.7 (d)

Features of horizontal micro-programming are:
(a) doesn't require use of signal decoders
(b) results in larger sized micro-instructions than vertical micro-programming
(c) uses one bit for each control signal.

2.8 (c)

An array (n × n contains) multiplier contains (2n – 1) gate cells unit if each unit (cells) contain a θ(1) then delay then total delay is (2n – 1) θ(1) which is θ(n).

2.9 (b)

Multiplying two 8 bit numbers will give result in maximum 16 bits.
Total number of multiplications possible
$= 2^8 \times 2^8 = 64$ K
Hence space required = 64 K × 16 bits.

2.10 (a)

We need to find hardware control for the signal A_{in}:
In instruction I1 A_{in} present for T1 and T3, so T1 I1 + T3 I1.
In instruction I2 A_{in} present for T3 only, so T3 I2.
In instruction I3 A_{in} present for T2, T3 and T4, so T3 I3 + T4 I3 + T2 I3.
So, logic function in SOP form is:
= T1 I1 + T3 I1 + T3 I2 + T3 I3 + T4 I3 + T2 I3
= T1 I1 + T3 (I1 + I2 + I3) + T4 I3 + T2 I3
= T1 I1 + T3 + T4 I3 + T2 I3 (∵ I1 + I2 + I3 = 1)

2.11 (a)

Memory-mapped I/O requires that memory locations and I/O ports share the same set of addresses, So address bit pattern that is assigned to memory can not also be assigned to an I/O port and vice versa. I/O protection in this approach is ensured by operating systems macros and routines.

2.12 (b)

Consider the following figure:

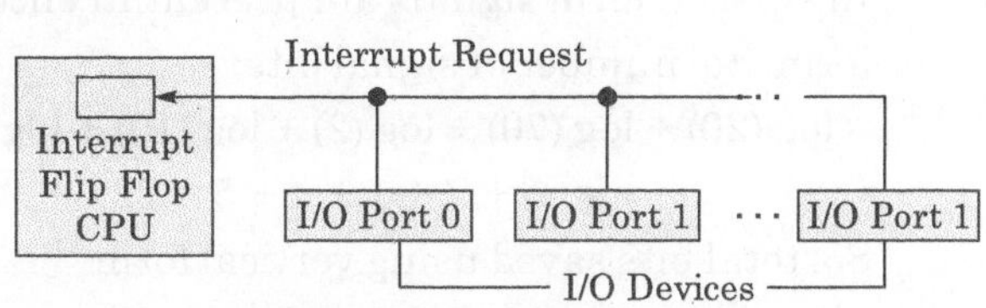

In single line interrupt system contains a single interrupt request line and a interrupt grant line. In this system it may be possible that more than one output devices request interrupt at the same time for example I/O port 0 to I/O port n may request an interrupt at the same time but only one request will be granted according to priority. So, In single interrupt system vectored interrupts are not possible but multiple interrupting devices are possible.

2.13 (d)

First we need to find hardware control signal S5:
In instruction I1, S5 present for T1 only, so T1 I1.
In instruction I2, S5 present for T1, T3, so T1 I2 + T3 I2.
In instruction I3, S5 present for T1 only, so T1 I3.
In instruction I4, S5 present for T1, T3, so T1 I4 + T3 I4.
So, logic function is SOP form is:
= T1 I1 + T1 I2 + T3 I2 + T1 I3 + T1 I4 + T3 I4
= T1 (I1 + I2 + I3 + I4) + T3 I2 + T3 I4
= T1 + T3 I2 + T3 I4 $\quad (\because$ I1 + I2 + I3 + I4 = 1)
= T1 + T3 (I2 + I4)
Similarly for signal S10, SOP form will be:
= T4 I1 + T2 I2 + T5 I2 + T2 I3 + T4 I3 + T3 I4 + T5 I4
= T4 (I1 + I3) + T2 (I2 + I3) + T5 (I2 + I4) + T3 I4

2.14 (b)

Types of possible of distinct multiple stuck:

1. Only stuck at –0
2. Only stuck at –1
3. Both stuck at –0 and stuck at –1.

So 3^N multiple stack possible.

2.15 (b)

In horizontal form signals are present in decoded form. So number of signal bits

= 20 + 70 + 2 + 10 + 23
= 125 bits

In vertical form signals are present in encoded form. So, number of signal bits:
= log (20) + log (70) + log (2) + log (10) + log (23)
= 5 + 7 + 1 + 4 + 5 = 22
So, total bits saved using vertical form
= 125 – 22 = 103 bits

2.16 (b)

RTL $R_0 \Leftarrow R_0 + R_1$
The sequence of microinstruction take place in following cycles.
I cycle: $R_{1\,out}$, S_{in}
II cycle: $R_{2\,out}$, T_{in}
III cycle: S_{out}, T_{out}, ALU_{add}, R_{in}
$\therefore$ 3 cycles will be required to execute.

2.17 (b)

RTL R_n = PC + 1
P_c = M[Pc]
The sequence of microinstrution take place in following cycles.
I cycle: PC_{out}, S_{in}, MAR_{in} (MAR can be loaded with PC_{out})
II cycle: S_{out}, $ALU_{increment}$, $R_{n\,in}$
III cycle: MDR_{out}, PC_{in} (MDR_{out} can be performed once MAR_{in} has been performed)
$\therefore$ 3 cycles will be required to execute.

2.18 (b)

For design D:
Time = 0.8 (Fixed point cycle) + 0.2 (floating point cycle)

= 0.8 (t) + 0.2 (2t)
= 0.8 t + 0.4 t
= 1.2 t

For design D1:
Time = 0.8 (30% more cycle than fixed) + 0.2 (30% less cycle than floating)

= 0.8 (t + 0.3 t) + 0.2 (2t – (0.3) 2 t)
= 0.8 (1.3 t) + 0.2 (1.4 t)
= 1.32 t

For design D2:
Time = 0.8 (40% less cycle than fixed) + 0.2 (10% more cycle than floating)

= 0.8 (t – 0.4 t) + 0.2 (2t + (0.1) × 2 t)
= 0.8 (0.6 t) + 0.2 (2.2 t)
= 0.92 t

So, D2 > D > D1 is correct.

2.19 (d)

Control word format:

← Control word →

Control signal	Address

Number of control address = Number of instruction × Number of clock cycle

$= 140 \times 7 = 980$

Number of bits for (CAR) = log (980)

$= 10$ bits

Since horizontal μ programming used. So number of bits for control signal = 125 bits.
So, control word size in bits = 125 + 10 = 135

2.20 (c)

In auto increment addressing mode the amount of increment depends on the size of the data item accessed. Consider the following example.

$Regs[R_1] \leftarrow Regs[R_1] + Mem\ [Regs[R_2]]$

$Regs[R_2] \leftarrow Reg[R_2] + d$

Where d is the size of the data item.

2.21 (c)

The use of multiple register windows with overlap causes a reduction in the number of memory access for instruction fetches only.

2.22 (c)

Interrupt handled by execution of interrupt service routine. It checks interrupt register after finishing the execution of current instruction.

2.23 (d)

4-bit × 4-bit ⇒ 8-bit

ROM size = $2^{(n+n)} \times (n+n)$

ROM size = $2^4 \times 2^4 \times 8$ bits = 2 Kbits

2.24 (d)

PC holds the value of next instruction to be executed we store the value of PC to MBR and then to memory. We are saving the value of PC in memory and new address value is loaded into PC. This can be done only in interrupt service. Option (d) is correct.

2.25 (5)

Number of registers = $n = 32$

Required multiplexer size is $n : 1$ i.e. 32 : 1

Number of select lines required to the multiplexer = m

$\therefore \quad m = \log_2 n = \log_2 32 = 5$

2.26 (d)

1. Send the address to memory via MAR.
2. Read the opcode into IR from the memory via MBR.
3. Send the first operand to Temp1(ALU).
4. Read the second operand directly from the R2 and process the data in ALU and store the result into TEMP2.
5. Store the result into R0.

2.27 (10.2)

10 ms — 1 poll

1 sec — ? Number of polls generated

i.e. $\dfrac{1 \text{ sec}}{10 \text{ ms}} \Rightarrow \dfrac{1000 \text{ ms}}{10 \text{ ms}} = 100$

1 polling time = 100 μsec

So, total polling time in 1 sec is

$= 100 \times 100\ \mu\text{sec} = 10000\ \mu\text{sec}$

Service time if stroke occurred = 200 μsec

∴ Total time to service the keyboard stroke is

⇒ 10200 μsec

Fraction of time consumed for polling (T_1)

$$= \left[\frac{10200\ \mu\text{sec}}{1000000\ \mu\text{sec}}\right] \times 100 = 1.02$$

Interrupt:

Interrupt time (CPU time) = 1 sec = 1000 ms

Interrupt processing time = 1 ms

∴ Fraction of time consumed for interrupt

$$(T_2) = \left(\frac{1 \text{ ms}}{1000 \text{ ms}}\right) \times 100 = 0.1$$

$$\frac{T_1}{T_2} = \frac{1.02}{0.1} \Rightarrow 10.2$$

■■■■

3 Instruction Pipelining

3.1 Comparing the time T1 taken for a single instruction on a pipelined CPU with time T2 taken on a non-pipelined but identical CPU, we can say that
(a) $T1 \leq T2$
(b) $T1 \geq T2$
(c) $T1 < T2$
(d) T1 plus T2 is the time taken for one instruction fetch cycle **[2000 : 1 M]**

3.2 In serial data transmission, every byte of data is padded with a '0' in the beginning and one or two '1's at the end of byte because
(a) Receiver is to be synchronized for byte reception
(b) Receiver recovers lost '0's and '1's from these padded bits
(c) Padded bits are useful in parity computation
(d) None of the above **[2002 : 1 M]**

3.3 In 2's complement addition, overflow
(a) is flagged whenever there is carry from sign bit addition
(b) cannot occur when a +ve value is added to a –ve value
(c) is flagged when the carries from sign bit and previous bit match
(d) None of the above **[2002 : 1 M]**

3.4 The performance of a pipelined processor suffers if
(a) the pipeline stages have different delays
(b) consecutive instructions are dependent on each other
(c) the pipline stages share hardware resources
(d) All of the above **[2002 : 2 M]**

3.5 For a pipelined CPU with a single ALU, consider the following situations
1. The $j + 1^{st}$ instruction uses the result of the jth instruction as an operand
2. The execution of a conditional jump instruction
3. The jth and $j + 1^{st}$ instructions require the ALU at the same time

Which of the above can cause a hazard?
(a) 1 and 2 only (b) 2 and 3 only
(c) 3 only (d) All the three
[2003 : 1 M]

3.6 Consider a pipeline processor with 4 stages S1 to S4. We want to execute the following loop : for (i = 1; i < = 1000; i++) {I1, I2, I3, I4} where the time taken (in ns) by instructions I1 to I4 for stages S1 to S4 are given below:

	S1	S2	S3	S4
I1:	1	2	1	2
I2:	2	1	2	1
I3:	1	1	2	1
I4:	2	1	2	1

The output of I1 for i = 2 will be available after
(a) 11 ns (b) 12 ns
(c) 13 ns (d) 28 ns
[2004 : 2 M]

3.7 In an enhancement of a design of a CPU, the speed of a floating point unit has been increased by 20% and the speed of a fixed point unit has been increased by 10%. What is the overall speedup achieved if the ratio of the number of floating point operations to the number of fixed point operations is 2 : 3 and the floating point operation used to take twice the time taken by the fixed point operation in the original design?
(a) 1.155 (b) 1.185
(c) 1.255 (d) 1.285 **[2004 : 2 M]**

3.8 A 4-stage pipeline has the stage delays as 150, 120, 160 and 140 nanoseconds respectively. Registers that are used between the stages have a delay of 5 nanoseconds each. Assuming constant clocking rate, the total time taken to process 1000 data items on this pipeline will be
(a) 120.4 microseconds
(b) 160.5 microseconds
(c) 165.5 microseconds
(d) 590.0 microseconds **[2004 : 2 M]**

3.9 A 5 stage pipelined CPU has the following sequence of stages
IF – Instruction fetch from instruction memory,
RD – Instruction decode and register read,
EX – Execute: ALU operation for data and address computation,
MA – Data memory access – for write access the register read at RD stage it used,
WB – Register write back.

Consider the following sequence of instructions:

I_1: L R0, Loc1; R0 ⇐ M[Loc1]

I_2: A R0, R0 ; R0 ⇐ R0 + R0

I_3: A R2, R0 ; R2 ⇐ R2 – R0

Let each stage take one clock cycle.

What is the number of clock cycles taken to complete the above sequence of instructions starting from the fetch of I_1?

(a) 8 (b) 10
(c) 12 (d) 15 **[2005 : 2 M]**

3.10 We have two designs D1 and D2 for a synchronous pipeline processor. D1 has 5 pipeline stages with execution times of 3 nsec, 2 nsec, 4 nsec, 2 nsec and 3 nsec while the design D2 has 8 pipeline stages each with 2 nsec execution time. How much time can be saved using design D2 over design D1 for executing 100 instructions?

(a) 214 nsec (b) 202 nsec
(c) 86 nsec (d) –200 nsec

[2005 : 2 M]

3.11 A CPU has five-stages pipeline and runs at 1 GHz frequency. Instruction fetch happens in the first stage of the pipeline. A conditional branch instruction computes the target address and evaluates the condition in the third stage of the pipeline. The processor stops fetching new instructions following a conditional branch until the branch outcome is known. A program executes 10^9 instructions out of which 20% are conditional branches. If each instruction takes one cycle to complete on average, then total execution time of the program is

(a) 1.0 second (b) 1.2 seconds
(c) 1.4 seconds (d) 1.6 seconds

[2006 : 2 M]

3.12 The data path shown in the figure computes the number of 1s in the 32-bit input word corresponding to an unsigned even integer stored in the shift register. The unsigned counter, initially zero, is incremented if the most significant bit of the shift register is 1.

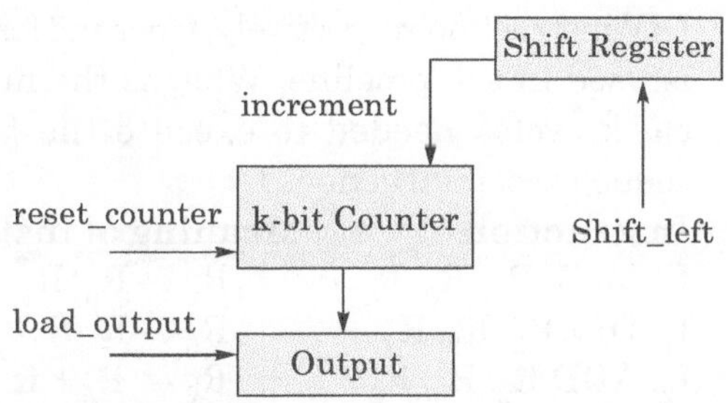

The microprogram for the control is shown in the table below with missing control words for microinstructions $I_1, I_2, ..., I_n$.

Micro-instruction	reset_counter	shift_left	load_output
BEGIN	1	0	0
I_1	?	?	?
:	:		:
I_n	?	?	?
END	0	0	1

The counter width (k), the number of missing microinstructions (n), and the control word for microinstructions $I_1, I_2, ..., I_n$ are, respectively,

(a) 32, 5, 010 (b) 5, 32, 010
(c) 5, 31, 011 (d) 5, 31, 010

[2006 : 2 M]

Common Data for Q.3.13 & 3.14:

A pipelined processor uses a 4-stage instruction pipeline with the following stages: Instruction fetch (IF), Instruction decode (ID), Execute (EX) and Writeback (WB). The arithmetic operations as well as the load and store operations are carried out in the EX stage. The sequence of instructions corresponding to the statement X = (S – R * (P + Q)) / T is given below. The values of variables P, Q, R, S and T are available in the registers R0, R1, R2, R3 and R4 respectively, before the execution of the instruction sequence.

```
ADD    R5, R0, R1  ; R5 → R0 + R1
MUL    R6, R2, R5  ; R6 → R2 * R5
SUB    R5, R3, R6  ; R5 → R3 – R6
DIV    R6, R5, R4  ; R6 → R5/R4
STORE  R6, X       ; X ← R6
```

3.13 The number of Read-After-Write (RAW) dependencies, Write-After-Read (WAR) dependencies, and Write-After-Write (WAW) dependencies in the sequence of instructions are, respectively

(a) 2, 2, 4 (b) 3, 2, 3
(c) 4, 2, 2 (d) 3, 3, 2

[2006 : 2 M]

3.14 The IF, ID and WB stages take 1 clock cycle each. The EX stage takes 1 clock cycle each for the ADD, SUB and STORE operations, and 3 clock cycles each for MUL and DIV operations. Operand forwarding from the EX stage to the ID stage is used. The number of clock cycles required to complete the sequence of instructions is

(a) 10 (b) 12
(c) 14 (d) 16 **[2006 : 2 M]**

3.15 A processor takes 12 cycles to complete an instruction 1. The corresponding pipelined processor uses 6 stages with the execution times of 3, 2, 5, 4, 6 and 2 cycles respectively. What is the asymptotic speedup assuming that a very large number of instructions are to be executed?
(a) 1.83 (b) 2
(c) 3 (d) 6 **[2007 : 1 M]**

3.16 Consider a pipelined processor with the following four stages
IF : Instruction Fetch
ID : Instruction Decode and Operand Fetch
EX : Execute
WB : Write Back
The IF, ID and WB stages take one clock cycle each to complete the operation. The number of clock cycles for the EX stage depends on the instruction. The ADD and SUB instructions need 1 clock cycle and the MUL instruction need 3 clock cycles in the EX stage. Operand forwarding is used in the pipelined processor. What is the number of clock cycles taken to complete the following sequence of instructions?

ADD	R2,	R1,	R0	R2 ← R1 + R0
MUL	R4,	R3,	R2	R4 ← R3 * R2
SUB	R6,	R5,	R4	R6 ← R5 – R4

(a) 7 (b) 8
(c) 10 (d) 14 **[2007 : 2 M]**

3.17 A nonpipelined single cycle processor operating at 100 MHz is converted into a synchronous pipelined processor with five stages requiring 2.5 nsec, 1.5 nsec, 2 nsec, 1.5 nsec and 2.5 nsec, respectivley. The delay of the latches is 0.5 nsec. The speedup of the pipeline processor for a large number of instructions is
(a) 4.5 (b) 4.0
(c) 3.33 (d) 3.0 **[2008 : 2 M]**

3.18 Which of the following are NOT true in a pipelined processor?
1. Bypassing can handle all Raw hazards.
2. Register renaming can eliminate all register carried WAR hazards.
3. Control hazard penalties can be eliminated by dynamic branch prediction.

(a) 1 and 2 only (b) 1 and 3 only
(c) 2 and 3 only (d) 1, 2 and 3
[2008 : 2 M]

Common Data for Q.3.19 & Q.3.20:
Delayed branching can help in the handling of control hazardous.

3.19 For all delayed conditional branch instructions, irrespective of whether the condition evaluates to true or false, A
(a) the instruction following the conditional branch instruction in memory is executed
(b) the first instruction in the fall through path is executed
(c) the first instruction in the taken path is executed
(d) the branch takes longer to execute than any other instruction **[2008 : 2 M]**

3.20 The following code is to run on a pipelined processor with one branch delay slot
I1 : ADDR2 ← R7 + R8
I2 : SUB R4 ← R5 – R6x
I3 : ADD R1 ← R2 + R3
I4 : STORE Memory [R4]←R1
BRANCH to Label if R1 = = 0
Which of the instruction I1, I2, I3 or I4 can legitimately occupy the delay slot without any other program modification?
(a) I1 (b) I2
(c) I3 (d) I4 **[2008 : 2 M]**

3.21 Consider a 4 stage pipeline processor. The number of cycles needed by the four instructions I1, I2, I3 , I4 in stages S1, S2, S3, S4 is shown below:

	S1	S2	S3	S4
I1	2	1	1	1
I2	1	3	2	2
I3	2	1	1	3
I4	1	2	2	2

What is the number of cycles needed to execute the following loop? for (i = 1 to 2) {I1;I2; I3; I4;}
(a) 16 (b) 23
(c) 28 (d) 30 **[2009 : 2 M]**

3.22 A 5-stage pipelined processor has Instruction Fetch (IF), Instruction Decode (ID), Operand Fetch (OF), Perform Operation (PO) and Write Operand (WO) stages. The IF, ID, OF and WO stages take 1 clock cycle each for any instruction. The PO stage takes 1 clock cycle for ADD and SUB instructions 3 clock cycles for MUL instruction, and 6 clock cycles for DIV instruction respectively. Operand forwarding is used in the pipeline. What is the number of clock cycles needed to execute the following sequence of instructions?

Instruction	**Meaning of Instruction**
I_0: MUL R_2, R_0, R_1	$R_2 \leftarrow R_0 * R_1$
I_1: DIV R_5, R_3, R_4	$R_5 \leftarrow R_3 / R_4$
I_2: ADD R_2, R_5, R_2	$R_2 \leftarrow R_5 + R_2$
I_3: SUB R_5, R_2, R_6	$R_5 \leftarrow R_2 - R_6$

(a) 13 (b) 15
(c) 17 (d) 19 **[2010 : 2 M]**

3.23 Consider an instruction pipeline with four stages (S1, S2, S3 and S4) each with combinational circuit only. The pipeline registers are required between each stage and at the end of the last stage. Delays for the stages and for the pipeline registers are as given in the figure

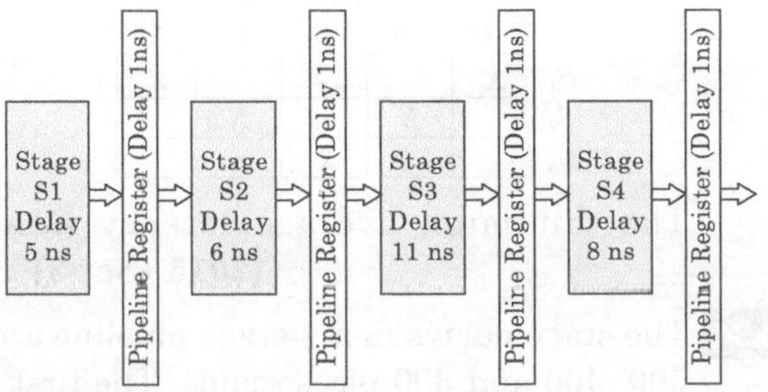

What is the approximate speed up of the pipeline in steady state under ideal conditions when compared to the corresponding non-pipeline implementation?
(a) 4.0 (b) 2.5
(c) 1.1 (d) 3.0 **[2011 : 2 M]**

3.24 Register renaming is done in pipelined processors
(a) as an alternative to register allocation at compile time
(b) for efficient access to function parameters and local variables
(c) to handle certain kinds of hazards
(d) as part of address translation **[2012 : 1 M]**

3.25 Consider an instruction pipeline with five stages without any branch prediction: Fetch Instruction (FI), Decode Instruction (DI), Fetch Operand (FO), Execute Instruction (EI) and Write Operand (WO). The stage delays for FI, DI, FO, EI and WO are 5 ns, 7 ns, 10 ns, 8 ns and 6 ns, respectively. There are intermediate storage buffers after each stage and the delay of each buffer is 1 ns. A program consisting of 12 instruction I_1, I_2, I_3, ... I_{12} is execute in this pipelined processor. Instruction I_4 is the only branch instruction and its branch target is I_9. If the branch is taken during the execution of this program, the time (in ns) needed to complete the program is
(a) 132 (b) 165
(c) 176 (d) 328 **[2013 : 2 M]**

3.26 Consider a 6-stage instruction pipeline, where all stages are perfectly balanced. Assume that there is no cycle-time overhead of pipelining. When an application is executing on this 6-stage pipeline, the speedup achieved with respect to non-pipelined execution if 25% of the instructions incur 2 pipeline stall cycles is ________.
[2014 (Set-1) : 2 Ma]

3.27 Consider two processors P_1 and P_2 executing the same instruction set. Assume that under identical conditions, for the same input, a program running on P_2 takes 25% less time but incurs 20% more CPI (clock cycles per instruction) as compared to the program running on P_1. If the clock frequency of P_1 is 1GHz, then the clock frequency of P_2 (in GHz) is ________. **[2014 (Set-1) : 2 M]**

3.28 Consider the following processors (ns stands for nanoseconds). Assume that the pipeline registers have zero latency.
P1: Four-stage pipeline with stage latencies 1 ns, 2 ns, 2 ns, 1 ns.
P2: Four-stage pipeline with stage latencies 1 ns, 1.5 ns, 1.5 ns, 1.5 ns.
P3: Five-stage pipeline with stage latencies 0.5 ns, 1 ns, 1 ns, 0.6 ns, 1 ns.
P4: Five-stage pipeline with stage latencies 0.5 ns, 0.5 ns, 1 ns, 1 ns, 1.1 ns.
Which processor has the highest peak clock frequency?
(a) P1 (b) P2
(c) P3 (d) P4
[2014 (Set-3) : 1 M]

3.29 An instruction pipeline has five stages, namely, instruction fetch (IF), instruction decode and register fetch (ID/RF), instruction execution (EX), memory access (MEM), and register writeback (WB) with stage latencies 1 ns, 2.2 ns, 2 ns, 1 ns, and 0.75 ns, respectively (ns stands for nanoseconds). To gain in terms of frequency, the designers have decided to split the ID/RF stage into three stages (ID, RF1, RF2) each of latency 2.2/3 ns. Also, the EX stage is split into two stages (EX1, EX2) each of latency 1ns. The new design has a total of eight pipeline stages. A program has 20% branch instructions which execute in the EX stage and produce the next instruction pointer at the end of the EX stage in the old design and at the end of the EX2 stage in the new design. The IF stage stalls after fetching a branch instruction until the next instruction pointer is computed. All instructions other than the branch instruction have an average CPI of one in both the designs. The execution times of this program on the old and the new design are P and Q nanoseconds, respectively. The value of P/Q is ________. **[2014 (Set-3) : 2 M]**

3.30 Consider a non-pipelined processor with a clock rate of 2.5 gigahertz and average cycles per instruction of four. The same processor is upgraded to a pipelined processor with five stages; but due to the internal pipeline delay, the clock speed is reduced to 2 gigahertz. Assume that there are no stalls in the pipeline. The speed up achieved in this pipelined processor is ______.

[2015 (Set-1) : 2 M]

3.31 Consider the sequence of machine instructions given below:

MUL	R5, R0, R1
DIV	R6, R2, R3
ADD	R7, R5, R6
SUB	R8, R7, R4

In the above sequence, R0 to R8 are general purpose registers. In the instructions shown, the first register stores the result of the operation performed on the second and the third registers. This sequence of instructions is to be executed in a pipelined instruction processor with the following 4 stages: (1) Instruction Fetch and Decode (IF), (2) Operand Fetch (OF), (3) Perform Operation (PO) and (4) Write back the Result (WB). The IF, OF and WB stages take 1 clock cycle each for any instruction. The PO stage takes 1 clock cycle for ADD or SUB instruction, 3 clock cycles for MUL instruction and 5 clock cycles for DIV instruction. The pipelined processor uses operand forwarding from the PO stage to the OF stage. The number of clock cycles taken for the execution of the above sequence of instructions is ______.

[2015 (Set-2) : 2 M]

3.32 Consider the following code sequence having five instructions I_1 to I_5. Each of these instructions has the following format.

OP Ri, Rj, Rk

where operation OP is performed on contents of registers Rj and Rk and the result is stored in register Ri.

I_1 : ADD R1, R2, R3 $\quad$ I_2 : MUL R7, R1, R3
I_3 : SUB R4, R1, R5 $\quad$ I_4 : ADD R3, R2, R4
I_5 : MUL R7, R8, R9

Consider the following three statements:

S1: There is an anti-dependence between instructions I_2 and I_5.

S2: There is an anti-dependence between instructions I_2 and I_4.

S3: Within an instruction pipeline an anti-dependence always creates one or more stalls.

Which one of above statements is/are correct?

(a) Only S1 is true
(b) Only S2 is true
(c) Only S1 and S3 are true
(d) Only S2 and S3 are true

[2015 (Set-3) : 2 M]

3.33 Consider the following reservation table for a pipeline having three stages S_1, S_2 and S_3.

Time ⟶

	1	2	3	4	5
S_1	X				X
S_2		X		X	
S_3			X		

The minimum average latency (MAL) is ______.

[2015 (Set-3) : 2 M]

3.34 The stage delays in a 4-stage pipeline are 800, 500, 400 and 300 picoseconds. The first stage (with delay 800 picoseconds) is replaced with a functionally equivalent design involving two stages with respective delays 600 and 350 picoseconds. The throughput increase of the pipeline is ______ percent.

[2016 (Set-1) : 2 M]

3.35 Consider a 3 GHz (gigahertz) processor with a three-stage pipeline and stage latencies τ_1, τ_2, and τ_3 such that $\tau_1 = \frac{3\tau_2}{4} = 2\tau_3$. If the longest pipeline stage is split into two pipeline stages of equal latency, the new frequency is ______ GHz, ignoring delays in the pipeline registers.

[2016 (Set-2) : 2 M]

3.36 Suppose thefunctions F and G can be computed in 5 and 3 nanoseconds by functional units U_F and U_G, respectively. Given two instances of U_F and two instances of U_G, it is required to implement the computation $F(G(X_i))$ for $1 \le i \le 10$. Ignoring all other delays, the minimum time required to complete this computation is ______ nanoseconds.

[2016 (Set-2) : 2 M]

3.37 Instruction execution in a processor is divided into 5 stages. Instruction Fetch (IF), Instruction Decode (ID), Operand Fetch (OF), Execute (EX) and Write Back (WB). These stages take **5, 4, 20, 10 and 3 nanoseconds** (ns) respectively. A pipelined implementation of the processor requires buffering between each pair of consecutive stages with a delay of **2 ns**. Two pipelined implementations of the processor are contemplated:

(i) a naive pipeline implementation (NP) with 5 stages and

(ii) an efficient pipeline (EP) where the OF stage is divided into stages OF1 and OF2 with execution times of **12 ns** and **8 ns** respectively.

The speedup (correct to two decimal places) achieved by EP over NP in executing 20 independent instructions with no hazards is _______.

[2017 (Set-1) : 2 M]

3.38 The instruction pipeline of a RISC processor has the following stages. Instruction Fetch (IF), Instruction Decode (ID), Operand Fetch (OF), Perform Operation (PO) and Writeback (WB). The IF, ID, OF and WB stages take 1 clock cycle each for every instruction. Consider a sequence of 100 instructions. In the PO stage, 40 instructions take 3 clock cycles each, 35 instructions take 2 clock cycles each, and the remaining 25 instructions take 1 clock cycle each. Assume that there are no data hazards and no control hazards.

The number of clock cycles required for completion of execution of the sequence of instructions is _______. **[2018 : 2 M]**

3.39 Consider a non-pipelined processor operating at 2.5 GHz. It takes 5 clock cycles to complete an instruction. You are going to make a 5-stage pipeline out of this processor. Overheads associated with pipelining force you to operate the pipelined processor at 2 GHz. In a given program, assume that 30% are memory instructions. 60% are ALU instructions and the rest are branch instructions. 5% of the memory instructions cause stalls of 50 clock cycles each due to cache misses and 50% of the branch instructions cause stalls of 2 cycles each. Assume that there are no stalls associated with the execution of ALL) instructions. For this program, the speedup achieved by the pipelined processor over the non-pipelined processor (round off to 2 decimal places) is _______. **[2020 : 2 M]**

3.40 A five-stage pipeline has stage delays of 150, 120, 150, 160 and 140 nanoseconds. The registers that are used between the pipeline stages have a delay of 5 nanoseconds each.

The total time to execute 100 independent instructions on this pipeline, assuming there are no pipeline stalls, is _______ nanoseconds.

[2021 (Set 1) : 2 M]

3.41 Consider a pipelined processor with 5 stages. Instruction Fetch (IF), Instruction Decode (ID), Execute (EX), Memory Access (MEM), and Write Back (WB). Each stage of the pipeline, except the EX stage, takes one cycle. Assume that the ID stage merely decodes the instruction and the register read is performed in the EX stage. The EX stage takes one cycle for ADD instruction and two cycles for MUL instruction. Ignore pipeline register latencies. Consider the following sequence of 8 instructions:

ADD, MUL, ADD, MUL,
ADD, MUL, ADD, MUL

Assume that every MUL instruction is data-dependent on the ADD instruction just before it and every ADD instruction (except the first ADD) is data-dependent on the MUL instruction just before it. The Speedup is defined as follows:

$$\text{Speedup} = \frac{\text{Execution time without operand forwarding}}{\text{Execution time with operand forwarding}}$$

The Speedup achieved in executing the given instruction sequence on the pipelined processor (rounded to 2 decimal places) is _______. **[2021 (Set 2) : 2 M]**

3.42 A processor X_1 operating at 2 GHz has a standard 5-stage RISC instruction pipeline having a base CPI (cycles per instruction) of one without any pipeline hazards. For a given program P that has 30% branch instructions, control hazards incur 2 cycles stall for every branch. A new version of the processor X_2 operating at same clock frequency has an additional branch predictor unit (BPU) that completely eliminates stalls for correctly predicted branches. There is neither any savings nor any additional stalls for wrong predictions. There are no structural hazards and data hazards for X_1 and X_2. If the BPU has a prediction accuracy of 80%, the speed up (rounded off to two decimal places) obtained by X_2 over X_1 in executing P is _______. **[2022 : 2 M]**

3.43 Consider a 3-stage pipelined processor having a delay of 10 ns (nanoseconds), 20 ns, and 14 ns, for the first, second, and the third stages, respectively. Assume that there is no other delay and the processor does not suffer from any pipeline hazards. Also assume that one instruction is fetched every cycle.

The total execution time for executing 100 instructions on this processor is _______ ns.

[2023 : 1 Mark]

3.44 Consider a 5-stage pipelined processor with Instruction Fetch (IF), Instruction Decode (ID), Execute (EX), Memory Access (MEM), and Register Writeback (WB) stages. Which of the following statements about forwarding is/are CORRECT?

(a) Forwarding does not require any extra hardware to retrieve the data from the pipeline stages
(b) In a pipelined execution, forwarding means the result from a source stage of an earlier instruction is passed on to the destination stage of a later instruction
(c) In forwarding, data from the output of the MEM stage can be passed on to the input of the EX stage of the next instruction
(d) Forwarding cannot prevent all pipeline stalls

[2024 (Set-1) : 1 M]

3.45 An instruction format has the following structure:
Instruction Number: Opcode destination reg, source reg-1, source reg-2
Consider the following sequence of instructions to be executed in a pipelined processor:

I1 : DIV R3, R1, R2
I2 : SUB R5, R3, R4
I3 : ADD R3, R5, R6
I4 : MUL R7, R3, R8

Which of the following statements is/are TRUE?

(a) There is a WAW dependency on R3 between I3 and I4
(b) There is a WAR dependency on R3 between I1 and I2
(c) There is a RAW dependency on R3 between I1 and I2
(d) There is a RAW dependency on R3 between I2 and I3

[2024 (Set-2) : 1 M]

3.46 A non-pipelined instruction execution unit operating at 2GHz takes an average of 6 cycles to execute an instruction of a program P. The unit is then redesigned to operate on a 5-stage pipeline at 2GHz. Assume that the ideal throughput of the pipelined unit is 1 instruction per cycle. In the execution of program P, 20% instructions incur an average of 2 cycles stall due to data hazards and 20% instructions incur an average of 3 cycles stall due to control hazards. The speedup (rounded off to one decimal place) obtained by the pipelined design over the non-pipelined design is ________.

[2024 (Set-2) : 2 M]

■■■■

Answers Instruction Pipelining

3.1	(b)	**3.2**	(a)	**3.3**	(b)	**3.4**	(d)	**3.5**	(d)	**3.6**	(c)	**3.7**	(a)	**3.8**	(c)	**3.9**	(a)
3.10	(b)	**3.11**	(c)	**3.12**	(d)	**3.13**	(c)	**3.14**	(b)	**3.15**	(b)	**3.16**	(b)	**3.17**	(c)	**3.18**	(d)
3.19	(a)	**3.20**	(d)	**3.21**	(d)	**3.22**	(b)	**3.23**	(b)	**3.24**	(c)	**3.25**	(b)	**3.26**	(4)	**3.27**	(1.6)
3.28	(c)	**3.29**	(1.54)	**3.30**	(3.2)	**3.31**	(13)	**3.32**	(b)	**3.33**	(3)	**3.34**	(33.28)	**3.35**	(4)	**3.36**	(28)
3.37	(1.508)	**3.38**	(219)	**3.39**	(2.16)	**3.40**	(17160)			**3.41**	(1.875)	**3.42**	(1.42)	**3.43**	(2040)	**3.44**	(b, c, d)
3.45	(c)	**3.46**	(3)														

Explanations Instruction Pipelining

3.1 (b)

In pipelined CPU, there will be buffer delay and stage delay. So for 1 instruction nonpipelined CPU takes less time compared to pipelined CPU.

3.2 (a)

Bits in the beginning and at the end of byte is known as start and stop bits respectively. Start and stop bits are used for synchronization purpose.

3.3 (b)

When two signed 2's complement numbers are added, overflow is detected if:

1. both operands are positive and the result is negative, or
2. both operands are negative and the result is positive.

Obviously, adding two numbers of opposite sign can't produce overflow because addition of different signed number is equal to subtraction.

3.4 (d)

Hazards are caused due to dependencies. Different dependencies in pipelined processor are:

1. **Structural dependency:** The pipeline stages have different delays.
2. **Control dependency:** Consecutive instructions are dependent as each other.
3. **Data dependency:** The pipeline stages share hardware resources.

3.5 (d)

(i) The j + 1st instruction uses the result of the jth instruction as an operand then read-after-write (RAW) hazard occurs. It is a part of data dependency.

(ii) The execution of a conditional jump instruction causes a flushing so conditional dependency occurs.

(iii) The jth and j + 1st instructions require the ALU at the same time causes write-after-read (WAR) hazard.

3.6 (c)

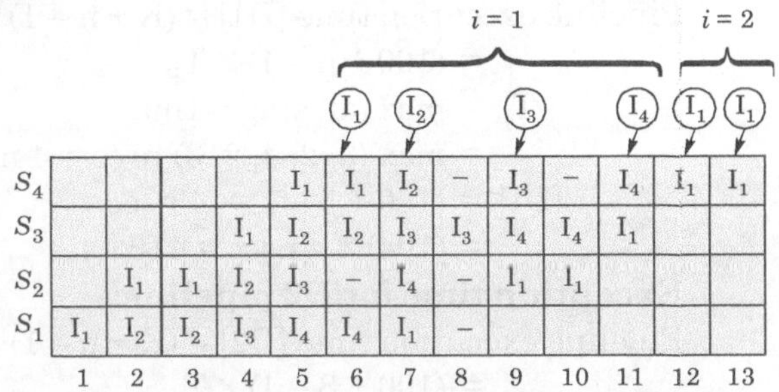

Total time = 13 ns

3.7 (a)

In Old system: 20 floating point operation and 30 fixed point operations.

Floating point takes 2 units and fixed point takes 1 unit.

Total time in old system = 20 × 2 + 30 × 1 = 70 units

In Enhanced system:

$$\text{Floating point takes} = 2 \times \frac{1}{1.2} = 1.\overline{6} \text{ units}$$

$$\text{Fixed point takes} = 1 \times \frac{1}{1.1} = 0.\overline{90} \text{ units}$$

$$\text{Total time} = 20 \times 1.\overline{6} + 30 \times 0.\overline{90} = 60.60 \text{ units}$$

$$\text{Therefore Speedup} = \frac{70}{60.60} = 1.155$$

3.8 (c)

K stage pipeline can process n tasks in T_K time

$$T_K = [K + (n-1)]\,\tau$$

When $\tau = \tau_m + d$ where τ_m is the maximum stage delay so max (150, 120, 160, 140) = 160

$$\tau = 160 + 5 = 165 \text{ ns} = 165 \times 10^{-3}\text{ms}$$

$$K = 4,\ n = 1000$$

$$T = [4 + (1000 - 1)]\tau$$

$$= 1003 \times 165 \times 10^{-3} = 165.5 \text{ ms}$$

3.9 (a)

This is RISC pipeline and in RISC pipeline dataforwarding is used as well as branch penalty is 1.

	CC_1	CC_2	CC_3	CC_4	CC_5	CC_6	CC_7	CC_8
I_1:	IF	ID	Ex	MA	WB			
I_2:		IF	ID	////	EX	MA	WB	
I_3:			IF	////	ID	EX	MA	WB

Number of cycle = 8 cycles.

3.10 (b)

Execution time for D1 pipeline:

Pipeline execution time [D1] = $(K + n - 1) \times T_P$

$$= (100 + 5 - 1) \times T_P$$

$$T_P = \text{max (all stages time)}$$

$$= \text{max (3, 2, 4, 2, 3) nsec} = 4 \text{ nsec}$$

$$[D1] = (100 + 5 - 1) \times 4 \text{ nsec}$$

$$= (104) \times 4 \text{ nsec} = 416 \text{ nsec}$$

Exception time for D2 pipeline:

Pipeline execution time [D2] = $(K + n - 1) \times T_P$

$$= (100 + 8 - 1) \times T_P$$

$$T_P = \text{max (all stages time)}$$

$$= \text{max (2, 2, 2, 2, 2, 2, 2, 2) nsec}$$

$$= 2 \text{ nsec}$$

$$[D2] = (100 + 8 - 1) \times T_P$$

$$= (107) \times 2 \text{ nsec} = 214 \text{ nsec}$$

Time saved by design D2 = [D2] – [D1]

$$= (416 - 214) \text{ nsec} = 202 \text{ nsec}$$

3.11 (c)

$$10^9 \text{ cycles} + \frac{20}{100} \times 2 \times 10^9$$

$$= 10^9 [1 + 0.4] = 1.4 \times 10^9 \text{ cycles}$$

$$= 1.4 \times 1 \text{ sec} = 1.4 \text{ sec}$$

3.12 (d)

- Since the number is even number i.e. LSB bit must be 0, so maximum 31 1's presents, so loop will be runs for 31 times i.e. number of missing instruction are 31.
- The input word is of 32 bits long, so width of counter = log (32) = 5 bits.
- To count number of 1's, reset counter will be reset i.e. 0. Shift left = 1 i.e. decrease number of 1's by 1 in each iteration and load output is 0 for all instructions except "End".

3.13 (c)

I_1 : ADD R_5, R_0, R_1

I_2 : MUL R_6, R_2, R_5

I_3 : SUB R_5, R_3, R_6

I_4 : DIV R_6, R_5, R_4

I_5 : STORE X, R_6

RAW dependency exist between $R_5(I_1) \rightarrow R_5(I_2)$, $R_6(I_2) \rightarrow R_6(I_3)$, $R_5(I_3) \rightarrow R_5(I_4)$ and $R_6(I_4) \rightarrow R_6(I_5)$.

WAR dependency exist between $R_5(I_2) \rightarrow R_5(I_3)$ and $R_6(I_3) \rightarrow R_6(I_4)$.

WAW dependency exist between $R_5(I_1) \rightarrow R_5(I_3)$ and $R_6(I_2) \rightarrow R_6(I_4)$.

3.14 (b)

1	2	3	4	5	6	7	8	9	10	11	12
IF	ID	EX	WB								
	IF	ID	EX	EX	EX	WB					
		IF			ID	EX	WB				
					IF	ID	EX	EX	EX	WB	
						IF			ID	EX	WB

So total 12 cycles are required.

3.15 (b)

Non pipeline processor time (NP) = 12 cycle × Number of instruction = 12 n cycles

Pipeline processor time (P) = $(K + n - 1) \times T_P$

$$= (6 - 1 + n) \times T_P$$

$$T_P = \text{max (all stages times)}$$

$$= \text{max (3, 2, 5, 4, 6, 2) cycles}$$

$$= 6 \text{ cycles}$$

$$(P) = (6 - 1 + n) \times T_P$$

$$= (5 + n) \times 6 \text{ cycles} = 30 + 6n \text{ cycles}$$

$$\text{So, Speed up} = \lim_{n \to \infty} \frac{12n}{30 + 6n}$$

$$= \lim_{n \to \infty} \frac{12}{6} \text{ (Apply L-Hospital rule since}(\infty/\infty))$$

$$= \left(\frac{12}{6}\right) = 2$$

So, option (b) is correct.

3.16 (b)

Pipelined processor has 4 stages IF, ID, EX, WB

Clock Cycles	Instruction
1	ADD
1	SUB
3	MUL

Consider the following diagram:

Clock Cycles	1	2	3	4	5	6	7	8
$R_2 \leftarrow R_1 + R_0$	IF	ID	EX	WB				
$R_4 \leftarrow R_3 * R_2$		IF	ID	EX	EX	EX	WB	
$R_6 \leftarrow R_5 - R_4$			IF	ID			EX	WB

So total required clocks cycle is 8.

3.17 (c)

Time needed for non-pipelined process (NP) = Sum of time for each state

$= (2.5 + 1.5 + 2 + 1.5 + 2.5)$ nsec
$= 10$ nsec

Time needed for pipelined processor (P) = T_P + Buffer delay

T_P = max (all stages time)
= max (2.5, 1.5, 2, 1.5, 2.5) nsec
= 2.5 nsec

P = T_P + Buffer delay
= 2.5 nsec + 0.5 nsec = 3 nsec

$$\text{Speed up} = \frac{\text{Non pipeline time}}{\text{Pipeline time}} = \frac{10 \text{ nsec}}{3 \text{ nsec}} = 3.33$$

3.18 (d)

All statement 1, 2 and 3 are false for a pipelined processor.

3.19 (a)

For all delayed conditional branch instruction the first instruction in the fall through path is executed and evaluates to true or false.

3.20 (d)

No re-arrangement technique given so, take NOP after "I_3." So "I_4" instruction not entered. {With re-arrangment "I_2" instruction lead to delay}.

In pipelining result of 1 instruction is used for next in pipeline. Delay slot will be occupied by the next instruction in the fall through path. The branching instruction R1 = = 0 goto label X.

So in delay slot I_2 will be there.

3.21 (d)

With loop level parallelism 25 cycles required. But no option matching. So use without loop level parallelism.

Number of cycles needed to executed for loop when $i = 1 \Rightarrow 2 + 1 + 3 + 2 + 2 + 3 + 2 = 15$

So total cycle needed to execute the loop
$= 2 \times 15 = 30$

3.22 (b)

	IF	ID	OF	PO	WO
MUL : I_0	1	1	1	3	1
DIV : I_1	1	1	1	6	1
ADD : I_2	1	1	1	1	1
SUB : I_3	1	1	1	1	1

	1	2	3	4	5	6	7	8	9	10	11	12	13	14	15
I_0	IF	ID	OF	PO	PO	PO	WO								
I_1		IF	ID	OF	////	////	PO	PO	PO	PO	PO	PO	WO		
I_2			IF	ID	////	////	OF	////	////	////	////	////	PO	WO	
I_3				IF	///	///	ID	///	///	////	////	////	OF	PO	WO

$\therefore$ Number of cycles required = 15 cycles.

3.23 (b)

Speed up = Nonpipe line execution time/ Pipeline execution time

Non pipeline Execution time
= S1 delay + S2 delay + S3 delay + S4 delay
= 5 + 6 + 11 + 8 = 30 ns

Pipeline Execution time
= Max(stage delay + buffer delay)
= max(6, 7, 12, 9) ns = 12 ns

$$\text{Hence Speedup} = \frac{30}{12} = 2.5$$

3.24 (c)

Register renaming is done in pipelining to handle certain hazards.

Operand forwarding (Bypassing), predictive buffer is also used to handle hazards.

$\therefore$ Option 'c' is true.

3.25 (b)

Number of stages = 5

Stage maximum delay = 10 + 1 = 11ns

I_4 instruction 4th stage output is a target address.

So, number of instruction in pipeline = 11.

So, $ET_{pipe} = (K + n - 1)\, t_p$
$= (5 + 11 - 1) \times 11$ ns = 165 ns

3.26 (4)

$$\text{Speed up(S)} = \frac{\text{Pipe depth}}{(1 + \#\text{stalls/instruction})}$$

Number of stalls / instruction
$= 0.75 \times 0 + 0.25 \times 2 = 0.5$

$$\therefore \quad S = \frac{6}{1 + 0.5} = 4$$

3.27 (1.6)

$T_1 = \text{CPI} \times \text{Cycle time}_1 = \text{CPI} \times 1\ \text{nsec}$

$T_2 = 1.2\ \text{CPI} \times \text{Cycle Time}_2$ ...(1)

$T_2 = 0.75 \times T_1 = 0.75 \times \text{CPI} \times 1\ \text{nsec}$ ...(2)

from (1) and (2)

$1.2\ \text{CPI} \times \text{Cycle time}_2 = 0.75\text{CPI} \times 1\ \text{nsec}$

$$\Rightarrow \text{Cycle time}_2 = \frac{0.75}{1.2} = 0.625\ \text{n sec}$$

$$\text{Frequency of } P_2 = \frac{1}{0.625\ \text{n sec}} = 1.6\ \text{GHz}$$

3.28 (c)

Clock cycle time = Maximum of stage delays.

For P1: clock cycle time = Maximum is 2

For P2: clock cycle time = Maximum is 1.5

For P3: clock cycle time = Maximum is 1

For P4: clock cycle time = Maximum is 1.1

Minimum clock cycle time gives the high clock rate.

Peak clock cycle rate

$$= \text{Maximum}\left(\frac{1}{2}, \frac{1}{1.5}, \frac{1}{1}, \frac{1}{1.1}\right) = 1$$

∴ P3 has peak clock cycle rate.

3.29 (1.54)

$$T_{avg} = (1 + \text{stall freq.} \times \text{stall cycle}) \times T_{clock}$$

$$TP_{avg} = (1 + 0.2 \times 2) \times 2.2\ \text{ns} = 3.08\ \text{ns}$$

$$TQ_{avg} = (1 + 0.2 \times 5) \times 1\ \text{ns} = 2\ \text{ns}$$

$$\frac{P}{Q} = \frac{TP_{avg}}{TQ_{avg}} = \frac{3.08}{2} = 1.54$$

3.30 (3.2)

Non-pipelined processor:

For n instructions execution time

$$= \frac{(n \times 4)}{2.5} = 1.6n \text{ nanoseconds}$$

Pipelined processor:

For n instructions execution time

$$= \frac{n}{2} = 0.5n \text{ nanoseconds}$$

$$\text{Speedup} = \frac{1.6n}{0.5n} = 3.2$$

3.31 (13)

SUB				IF	–	–	–	–	–	–	OF	PO	WB
ADD			IF	–	–	OF	–	–	–	–	PO	WB	
DIV		IF	OF	–	–	PO	PO	PO	PO	PO	WB		
MUL	IF	OF	PO	PO	PO	WB							
	1	2	3	4	5	6	7	8	9	10	11	12	13

∴ It takes 13 clock cycles.

3.32 (b)

S1 is false

S2 is true because there is an anti-dependence between instructions I_2 and I_4.

S3 is false because anti-dependence stalls may be avoided when register renaming is used.

3.33 (3)

- Collision vector (C) = (C_5 C_4 C_3 C_2 C_1)
- Forbiden latency creates collision. So it is non-permissible state i.e., 0.
- If permissible latency is 1 then, will initiate new task in 2nd cycle.

Latency sequence with (1)

	1	2	3	4	5	6	7	8	9	10	11	12	13	14	15
S_1	1	2			1	2	3	4			3	4	5		
S_2		1	2	1	2			3	4	3	4			5	
S_3			1	2					3	4					5

Latency cycle:

$(2 - 1) = 1$

$(7 - 2) = 5$

$(8 - 7) = 1$

$(12 - 7) = 5$

i.e., (1, 5), (1, 5), (1, 5), ...

So, average latency $= \frac{6}{2} = 3$

Latency sequence with (3)

	1	2	3	4	5	6	7	8	9	10	11	12	13	14	15	16
S_1	1			2	1		3	2		4	3		5	4		
S_2		1		1	2		2	3		3	4		4	5		5
S_3			1			2			3			4			5	

Latency cycle:

$(4 - 1) = 3$

$(7 - 4) = 3$

$(10 - 7) = 3$

so, average latency = 3.

Latency sequence with (5)

	1	2	3	4	5	6	7	8	9	10	11	12	13	14	15	16	17	18
S_1	1				1	2	3			2	3	4	5			4	5	
S_2		1		1			2	3	2	3			4	5	4	5		
S_3			1					2	3					4	5			

Latency cycle:

$(6 - 1) = 5$

$(7 - 6) = 1$

$(12 - 7) = 5$

$(13 - 12) = 1$

Latency cyle (5, 1), (5, 1), (5, 1), ...

So, average latency = 3

$\therefore$ MAL value = 3

3.34 (33.28)

P_1 : 4-stage

t_p = Max (stage delay) = 800 ps

1 instruction — 800 ps

? number of instruction — 1 sec

TP_R (throughput) = 1250 instruction/sec

P_2 : 5-stage

t_p = 600 ps

1 instruction — 600 ps

? number of instruction — 1 sec

TP_{P2} (throughput) = 1666 instruction/sec

1250 — 100% (old)

(1666 – 1250) — ? (New)

$$\Rightarrow \frac{416}{1250} = 0.3328 \text{ i.e., } 33.28\%$$

3.35 (4)

Let, $T_3 = x,\ T_1 = 2x$

$$T_2 = \frac{8x}{3}(2.66x)$$

2.66x .... 3 GHz

$$\text{Cycle time} = \frac{1}{3\text{ GHz}} = 0.33 \text{ ns}$$

$\therefore$ $x = 0.125$

New, $T_3 = x,\ T_1 = 2x$

$T_{21} = 1.33x,\ T_{22} = 1.33x$

$\therefore$ Cycle time = $2x$

i.e., Cycle time = $2 \times 0.125 = 0.25$ ns

$$\therefore \text{Clock frequency} = \frac{1}{\text{Cycle time}}$$

$$= \frac{1}{0.25 \text{ ns}} = 4 \text{ GHz}$$

3.36 (28)

Here pipelining concept is used. Given U_F take 5 ns while U_G take 3 ns to complete F and G respectively.

We have to do 10 computations i.e., $1 \le i \le 10$ with 2 instances of U_F and U_G. So,

$$U_F \text{ will take} = \frac{10 \times 5}{2} = \frac{50}{2} = 25 \text{ (since 2 instances are used).}$$

We have to wait 3 ns more to complete U_G to get the output.

So answer is 25 + 3 = 28 ns (rest of the instructions already complete at this time because pipelining is used).

3.37 (1.508)

NP: $k = 5,\ n = 20$

t_p = Max (Stage delay + Buffer delay) = 22 ns

Execution time = $(k + n - 1)t_p$

$= (5 + 20 - 1) \times 22 \text{ ns} = 528 \text{ ns}$

EP: $k = 6,\ n = 20$

t_p = Max (Stage delay + Buffer delay) = 14 ns

Execution time = $(k + n - 1)t_p$

$= (6 + 20 - 1) \times 14 \text{ ns} = 350 \text{ ns}$

$$\textbf{Speedup (S)} = \frac{ET_{NP}}{ET_{EP}} = \frac{528}{350} = 1.508$$

3.38 (219)

Number of instruction = 100

Number of stalls w.r.t. instruction

$= (40 \times 2) + (35 \times 1) + (25 \times 0) = 115$

So, total number of instruction

$= 100 + 115 = 215$

$ET_{pipe} = (K + n - 1)$ cycles

$= (5 + 215 - 1) = 219$ cycles

3.39 (2.16)

Non-pipeline:

Clock frequency = 2.5 GHz

$$\text{Cycle time} = \frac{1}{2.5 \text{ GHz}} = 0.4 \text{ ns}$$

CPI = 5

So, $ET_{non\text{-}pipe}$ = CPI × Cycle time

$= 5 \times 0.4 \text{ ns} = 2 \text{ ns}$

Pipeline:

Clock frequency = 2.5 GHz

$$\text{Cycle time} = \frac{1}{2 \text{ GHz}} = 0.5 \text{ ns}$$

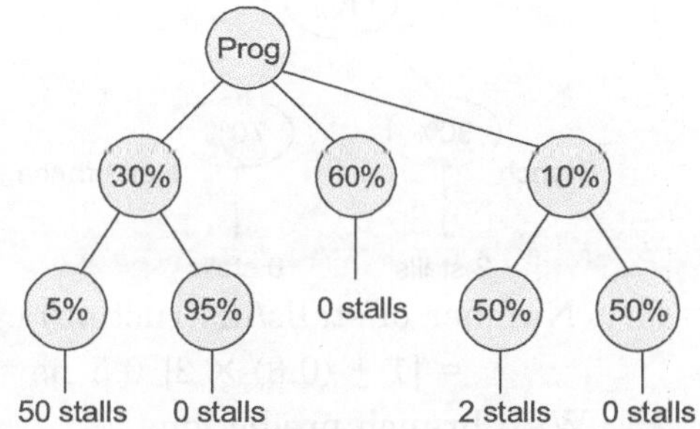

Number of stalls/instruction = 0.85

Average instruction ET_{pipe} = (1 + Number of stalls/instruction) × Cycle time

$= (1 + 0.85) \times 0.5 \text{ ns} = 0.925 \text{ ns}$

$$S = \frac{ET_{non\text{-}pipe}}{ET_{pipe}} = \frac{2}{0.925} = 2.16$$

3.40 (17160)

$K = 5$

$t_p = \max(\text{Stage delay} + \text{Buffer delay}) = 165 \text{ ns}$

$n = 100$ (finite)

↓

$ET_{pipe} = (K + n - 1)t_p$

$= (5 + 100 - 1)\ 165 \text{ ns} = 17160 \text{ ns}$

3.41 (1.875)

With operand forwarding:

8 Instructions + 4 MUL Instruction × 1 Extra Cycle in Ex-stage

⇒ $n = 12$ (finite)

⇒ $K = 5$

$ET_{Pipe} = (K + n - 1)$ Cycles

$= (5 + 12 - 1) = 16$ cycles

Without operand forwarding:

- 8 Instructions + 4 MUL Instruction × 2 Stalls at ID stage for ADD O/P
  + 3 ADD Instruction × 3 Stalls at ID stage for MUL O/P
  + 1 MUL Instruction × 1 Extra Cycle in Ex-Stage (Last Instruction)

⇒ $n = 26$

⇒ $K = 5$

$ET_{Pipe} = (K + n - 1)$ Cycles

$= (5 + 26 - 1)$ Cycles = 30 Cycles

$$\therefore \quad S = \frac{30}{16} = 1.875$$

3.42 (1.42)

Cycle time = 0.5 ns

X_1 : Without branch prediction

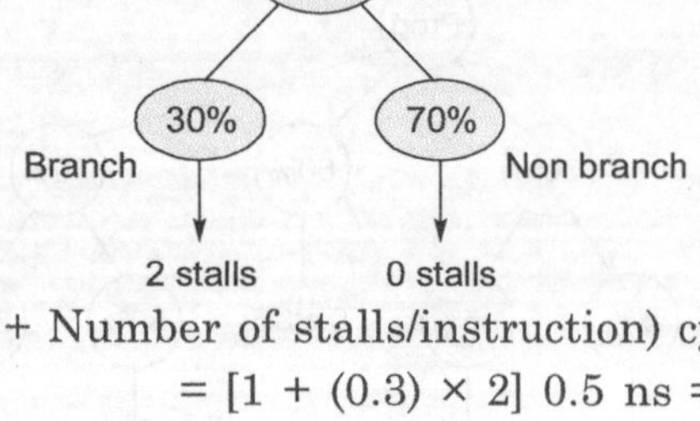

E_T = (1 + Number of stalls/instruction) cycle time

$= [1 + (0.3) \times 2]\ 0.5 \text{ ns} = 0.8 \text{ ns}$

X_2 : With branch prediction

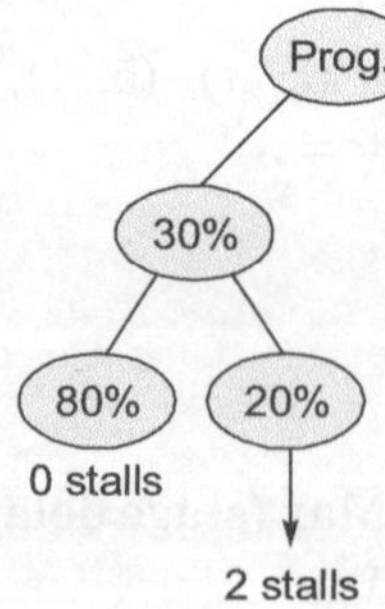

$E_T = [1 + (0.3 \times 0.2 \times 2)]\ 0.5 \text{ ns} = 0.56 \text{ ns}$

$$S = \frac{0.8}{0.56} = 1.42$$

3.43 (2040)

Performance of a pipelined processor: Take into account a pipeline with k segments and a clock cycle time of T_p. Let's say the pipelined processor needs to finish n tasks. The first instruction will now exit the pipeline after 'k' cycles, while the subsequent '$n - 1$' instructions will do so after just '1' cycles, for a total of '$n - 1$' cycles. In a pipelined processor, the time it takes to execute 'n' instructions is as follows:

$ET_{pipe} = (k + n - 1) * T_p$

$ET_{pipe} = (k + n - 1)T_p$

$= (3 + 100 - 1)\ 20 = 2040 \text{ ns}$

Note: $T_p = \max(10, 20, 14) \text{ ns} = 20 \text{ ns}$

3.44 (b, c, d)

- Buffer is used to hold the intermediate result to performed the Data.
- Forwarding can't prevent all the pipeline stalls. This mechanism minimizes only Data stall.
- For the load instructions memory data is forwarded to execute state with stall.

3.45 (c)

RAW existed between I2 and I1 with respect to "R3". Because I2 input register and I1 output register both are same.

3.46 (3)

$ET_{non\ pipe} = \text{CPI} \times \text{Cycle time}$

$= 6 \times 0.5 \text{ ns} = 3 \text{ ns}$

$ET_{pipe} = (1 + \text{Number of stalls/Instruction})t_p$

$= (1 + ((0.2 \times 2) + (0.2 \times 3))\ 0.5 \text{ ns} = 1 \text{ ns}$

$$S = \frac{ET_{non\ pipe}}{ET_{pipe}} = \frac{3 \text{ ns}}{1 \text{ ns}} = 3$$

■■■■

4 Memory Organization

4.1 The principle of locality justifies the use of
(a) Interrupts (b) DMA
(c) Polling (d) Cache Memory
[1995 : 1 M]

4.2 The main memory of a computer has 2 cm blocks while the cache has 2 c blocks. If the cache uses the set associative mapping scheme with 2 blocks per set, then block k of the main memory maps to the set
(a) (k mod m) of the cache
(b) (k mod c) of the cache
(c) (k mod 2 c) of the cache
(d) (k mod 2 cm) of the cache **[1999 : 1 M]**

4.3 Which of the following is/are advantage of virtual memory?
(a) Faster access to memory on an average.
(b) Processes can be given protected address spaces.
(c) Linker can assign addresses independent of where the program will be loaded in physical memory.
(d) Programs larger than the physical memory size can be run. **[1999 : 2 M]**

4.4 A graphics card has on board memory of 1 MB. Which of the following modes can the card not support?
(a) 1600 × 400 resolution with 256 colours on a 17 inch monitor
(b) 1600 × 400 resolution with 16 million colours on a 14 inch monitor
(c) 800 × 400 resolution with 16 million colours on a 17 inch monitor
(d) 800 × 800 resolution with 256 colours on a 14 inch monitor **[2000 : 2 M]**

4.5 Which of the following requires a device driver?
(a) Register (b) Cache
(c) Main memory (d) Disk **[2001 : 1 M]**

4.6 More than one word arc put in onc cachc block to
(a) exploit the temporal locality of reference in a program
(b) exploit the spatial locality of reference in a program
(c) reduce the miss penalty
(d) None of the above **[2001 : 1 M]**

4.7 Which of the following statements is false?
(a) Virtual memory implements the translation of a program's address space into physical memory address space
(b) Virtual memory allows each program to exceed the size of the primary memory
(c) Virtual memory increases the degree of multiprogramming
(d) Virtual memory reduces the context switching overhead **[2001 : 1 M]**

4.8 Which of the following is not a form of memory?
(a) Instruction cache
(b) Instruction Register
(c) Instruction opcode
(d) Translation look aside buffer
[2002 : 1 M]

4.9 Consider a system with 2 level cache. Access times of Level 1 cache, Level 2 cache and main memory are 1 ns, 10 ns, and 500 ns, respectively. The hit rates of Level 1 and Level 2 caches are 0.8 and 0.9, respectively. What is the average access time of the system ignoring the search time within the cache?
(a) 13.0 ns (b) 12.8 ns
(c) 12.6 ns (d) 12.4 ns
[2004 : 1 M]

4.10 Consider a small two-way set-associative cache memory, consisting of four blocks. For choosing the block to be replaced, use the least recently used (LRU) scheme. The number of cache misses for the following sequence of block addresses is 8, 12, 0, 12, 8
(a) 2 (b) 3
(c) 4 (d) 5 **[2004 : 2 M]**

4.11 A dynamic RAM has a memory cycle time of 64 nsec. It has to be refreshed 100 times per msec and each refresh takes 100 nsec. What percentage of the memory cycle time is used for refreshing?
(a) 10 (b) 6.4
(c) 1 (d) 0.64 **[2005 : 1 M]**

4.12 Increasing the RAM of a computer typically improves performance because
(a) Virtual memory increases
(b) Larger RAMs are faster

(c) Fewer page faults occur
(d) Fewer segmentation faults occur

[2005 : 1 M]

4.13 Consider a direct mapped cache of size 32 KB with block size 32 bytes. The CPU generates 32 bit addresses. The number of bits needed for cache indexing and the number of tag bits are respectively

(a) 10, 17 (b) 10, 22
(c) 15, 17 (d) 5, 17 **[2005 : 2 M]**

4.14 Consider a 2-way set associative cache memory with 4 sets and total 8 cache blocks (0-7) and a main memory with 128 blocks (0-127). What memory blocks will be present in the cache after the following sequence of memory block references if LRU policy is used for cache block replacement. Assuming that initially the cache did not have any memory block from the current job?

0 5 3 9 7 0 16 55

(a) 0 3 5 7 16 55 (b) 0 3 5 7 9 16 55
(c) 0 5 7 9 16 55 (d) 3 5 7 9 16 55

[2005 : 2 M]

4.15 A cache line is 64 bytes. The main memory has latency 32ns and bandwidth 1 GB/s. The time required to fetch the entire cache line from the main memory is

(a) 32 ns (b) 64 ns
(c) 96 ns (d) 128 ns

[2006 : 2 M]

4.16 A computer system has a level-1 instruction cache (I-cache), a level-1 data cache (D-cache) and a level-2 cache (L2-cache) with the following specifications:

	Capacity	Mapping method	Block size
I-cache	4K words	Direct mapping	4 Words
D-cache	4K words	2-way set-associative mapping	4 Words
L2-cache	64K words	4-way set-associative mapping	16 Words

Capacity Mapping methodBlock size I-cache 4K words Direct mapping 4 Words D-cache 4K words 2-way set-associative mapping 4 Words L2-cache 64K words 4-way set-associative mapping 16 WordsThe length of the physical address of a word in the main memory is 30 bits. The capacity of the tag memory in the I-cache, D-cache and L2-cache is, respectively,

(a) 1K × 18-bit, 1 K × 19-bit, 4K × 16-bit
(b) 1K × 16-bit, 1 K × 19-bit, 4K × 18-bit
(c) 1K × 16-bit, 512 × 18-bit, 1K × 16-bit
(d) 1K × 18-bit, 512 × 18-bit, 1K × 18-bit

[2006 : 2 M]

4.17 A CPU has a cache with block size 64 bytes. The main memory has k banks, each bank being c bytes wide. Consecutive c-byte chunks are mapped on consecutive banks with warp-around. All the k banks can be accessed in parallel, but two accesses to the same bank must be serialized. A cache block access may involve multiple iterations of parallel bank accesses depending on the amount of data obtained by accessing all the k banks in parallel. Each iteration requires decoding the bank numbers to be accessed in parallel and this takes k/2 ns. The latency of one bank access is 80 ns. If c = 2 and k = 24, then latency of retrieving a cache block starting at address zero from main memory is

(a) 92 ns (b) 104 ns
(c) 172 ns (d) 184 ns

[2006 : 2 M]

Common Data for Q.4.18 & Q.4.19:

Consider two cache organizations : The first one is 32 KB 2-way set associative with 32-byte block size. The second one is of the same size but direct mapped. The size of an address is 32 bits in both cases. A 2-to-1 multiplexer has latency of 0.6 ns while a k-bit comparator has a latency of k/10 ns. The hit latency of the set associative organization is h_1 while that of the direct mapped one is h_2.

4.18 The value of h_1 is

(a) 2.4 ns (b) 2.3 ns
(c) 1.8 ns (d) 1.7 ns

[2006 : 2 M]

4.19 The value of h_2 is

(a) 2.4 ns (b) 2.3 ns
(c) 1.8 ns (d) 1.7 ns

[2006 : 2 M]

Common Data for Q.4.20 & Q.4.21:

A CPU has a 32 KB direct mapped cache with 128-byte block size. Suppose A is a two dimensional array of size 512 × 512 with elements that occupy 8-bytes each. Consider the following two C code segments, P_1 and P_2.

```
P1 : for (i = 0; i < 512; i ++)  {
         for (j = 0; j < 512; j ++) {
             x + = A [i] [j];
         }
     }
P2 : for (i = 0; i < 512; i++)  {
         for (j = 0 ; j < 512; j ++) {
             x + = A [j] [i];
         }
     }
```

P1 and P2 are executed independently with the same initial state, namely, the array A is not in the cache and i, j, x are in registers. Let the number of cache misses experienced by P1 be M_1 and that for P2 be M_2.

4.20 The value of M_1 is
(a) 0 (b) 2048
(c) 16384 (d) 262144
[2006 : 2 M]

4.21 The value of the ratio M_1 / M_2 is
(a) 0 (b) $\frac{1}{16}$
(c) $\frac{1}{8}$ (d) 16 **[2006 : 2 M]**

4.22 Consider a 4-way set associative cache consisting of 128 lines with a line size of 64 words. The CPU generates a 20-bit address of a word in main memory. The number of bits in the TAG, SET and WORD fields are respectively
(a) 9, 6, 5 (b) 7, 7, 6
(c) 7, 5, 8 (d) 9, 5, 6
[2007 : 1 M]

Common Data for Q. 4.23 & Q. 4.24:

Consider a machine with a byte addressable main memory of 2^{16} bytes. Assume that a direct mapped data cache consisting of 32 lines of 64 bytes each is used in the system. A 50 × 50 two-dimensional array of bytes is stored in the main memory starting from memory location 1100 H. Assume that the data cache is initially empty. The complete array is accessed twice. Assume that the contents of the data cache do not change in between the two accesses.

4.23 How many data cache misses will occur in total?
(a) 48 (b) 50
(c) 56 (d) 59 **[2007 : 2 M]**

4.24 Which of the following lines of the data cache will be replaced by new blocks in accessing the array
(a) line 4 to line 11 (b) line 4 to line 12
(c) line 0 to line 7 (d) line 0 to line 8
[2007 : 2 M]

Common Data for Q. 4.25 & 4.26:

Consider a computer with a 4-ways set-associative mapped cache of the following characteristics: a total of 1 MB of main memory, a word size of 1 byte, a block size of 128 words and a cache size of 8 KB.

4.25 The number of bits in the TAG, SET and WORD fields, respectively are:
(a) 7, 6, 7 (b) 8, 5, 7
(c) 8, 6, 6 (d) 9, 4, 7
[2008 : 2 M]

4.26 While accessing the memory location 0C795H by the CPU, the contents of the TAG field of the corresponding cache line is
(a) 000011000 (b) 110001111
(c) 00011000 (d) 110010101
[2008 : 2 M]

4.27 For inclusion to hold between two cache level L1 and L2 in a multilevel cache hierarchy, which of the following are necessary?
1. L1 must be a write-through cache
2. L2 must be write-through cache
3. The associativity of L2 must be greater that of L1
4. The L2 cache must be at least as large as the L1 cache

(a) 4 only (b) 1 and 2 only
(c) 1 and 4 only (d) 1, 2, 3 and 4
[2008 : 2 M]

4.28 In an instruction execution pipeline, the earliest that the data TLB (Translation Look a side Buffer) can be accessed is
(a) before effective address calculation has started
(b) during effective address calculation
(c) after effective address calculation has completed
(d) after data cache lookup has completed
[2008 : 2 M]

Common Data for Q.4.29 to 4.30:

Consider a machine a 2-way set associative data cache of size 64 Kbytes and block size 16 bytes. The cache is managed using 32 bit virtual addressed and the page size is 4 Kbytes. A program to be run on this machine begins as follows:

```
double APR [1024] [1024]
int i, j ;
/ * Initalize array APR to 0.0 */
for ( i = 0; i < 1024; i ++)
    for ( j = 0; j < 1024; j++
        APR [i] [j] = 0.0;
```

The size of double 8 bytes. Array APR is in memory starting at the beginning of virtual page 0xFF000 and stored in row major order. The cache is initially empty and no pre-fetching is done. The only data memory references made by the program are those to array APR.

4.29 The total size of the tags in the cache directory is
(a) 32 kbits (b) 34 kbits
(c) 64 kbits (d) 68 kbits **[2008 : 2 M]**

4.30 Which of the following array elements has the same cache index as APR[0][0]?
(a) APR[0][4] (b) APR[4][0]
(c) APR[0][5] (d) APR[5][0] **[2008 : 2 M]**

4.31 The cache hit ratio for this initialization loop is
(a) 0% (b) 25%
(c) 50% (d) 75% **[2008 : 2 M]**

4.32 How many 32K × 1 RAM chips are needed to provide a memory capacity of 256 K-bytes?
(a) 8 (b) 32
(c) 64 (d) 128 **[2009 : 1 M]**

4.33 Consider a 4-way set associative cache (initially empty) with total 16 cache blocks. The main memory consists of 256 blocks and the request for memory blocks is in the following order:
0, 255, 1, 4, 3, 8, 133, 159, 216, 129, 63, 8, 48, 32, 73, 92, 155
Which one of the following memory block will NOT be in cache if LRU replacement policy is used?
(a) 3 (b) 8
(c) 129 (d) 216 **[2009 : 2 M]**

4.34 A main memory unit with a capacity of 4 megabytes is built using 1M × 1-bit DRAM chips. Each DRAM chip has 1K rows of cells with 1K cells in each row. The time taken for a single refresh operation is 100 nanoseconds. The time required to perform one refresh operation on all the cells in the memory unit is
(a) 100 nanoseconds
(b) $100 * 2^{10}$ nanoseconds
(c) $100 * 2^{20}$ nanoseconds
(d) $3200 * 2^{20}$ nanoseconds **[2010 : 1 M]**

Common Data for Q. 4.35 & Q. 4.36:

A computer system has an L1 and L2 cache, an L2 cache, and a main memory unit connected as shown below. The block size in L1 cache is 4 words. The block size in L2 cache is 16 words. The memory access times are 2 nanoseconds, 20 nanoseconds and 200 nanoseconds for L1 cache, L2 cache and main memory unit respectively.

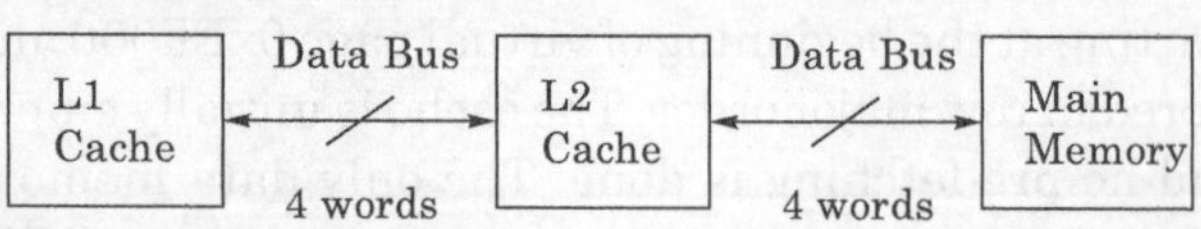

4.35 When there is a miss in L1 cache and a hit in L2 cache, a block is transferred from L2 cache to L1 cache. What is the time taken for this transfer?
(a) 2 nanoseconds (b) 20 nanoseconds
(c) 22 nanoseconds (d) 88 nanoseconds
[2010 : 2 M]

4.36 When there is a miss in both L1 cache and L2 cache, first a block is transferred from main memory to L2 cache, and then a block is transferred from L2 cache to L1 cache. What is the total time taken for these transfers?
(a) 222 nanoseconds (b) 888 nanoseconds
(c) 902 nanoseconds (d) 968 nanoseconds
[2010 : 2 M]

4.37 An 8KB direct-mapped write back cache is organized as multiple blocks, each of size 32 bytes. The processor generates 32 bit addresses. The cache controller maintains the tag information for each cache block comprising of the following.
1 Valid bit
1 Modified bit
As many bits as the minimum needed to identify the memory block mapped in the cache.
What is the total size of memory needed at the cache controller to store meta-data (tags) for the cache?
(a) 4864 bits (b) 6144 bits
(c) 6656 bits (d) 5376 bits
[2011 : 2 M]

Linked Answer Q. 4.38 & Q. 4.39:

A computer has a 256 Kbyte, 4-way set associative, write back data cache with block size of 32 Bytes. The processor sends 32 bit addresses to the cache controller. Each cache tag directory entry contains, in addition to address tag, 2 valid bits, 1 modified bit and 1 replacement bit.

4.38 The number of bits in the tag field of an address is
(a) 11 (b) 14
(c) 16 (d) 27 **[2012 : 2 M]**

4.39 The size of the cache tag directory is
(a) 160 Kbits (b) 136 Kbits
(c) 40 Kbits (d) 32 Kbits
[2012 : 2 M]

4.40 In a k-way set associative cache, the cache is divided into v sets, each of which consists of k lines. The lines of a set are placed in sequence on after another. The lines in set s are sequenced before the lines in set (s + 1). The main memory blocks are numbered 0 onwards. The main memory block numbered 'j' must be mapped to any one of the cache lines from

(a) (j mod v) * k to (j mod v) * k + (k – 1)
(b) (j mod v) * (j mod v) + (k – 1)
(c) (j mod k) to (j mod k) + (v – 1)
(d) (j mod k) * v (j mod k) * v + (v – 1)

[2013 : 1 M]

4.41 A RAM chip has capacity of 1024 words of 8 bits each (1K × 8). The number of 2 × 4 decoders with enable line needed to construct a 16 K × 16 RAM from 1K × 8 RAM is

(a) 4 (b) 5
(c) 6 (d) 7 **[2013 : 2 M]**

4.42 An access sequence of cache block address of length N and contains n unique block addresses. The number of unique block addresses between two consecutive accesses to the same block address is bounded above by k. What is the miss ratio if the access sequence is passed through a cache of associativity $A \geq k$ exercising least-recently used replacement policy?

(a) $\frac{n}{N}$ (b) $\frac{1}{N}$
(c) $\frac{1}{A}$ (d) $\frac{k}{n}$

[2014 (Set-1) : 2 M]

4.43 A 4-way set-associative cache memory unit with a capacity of 16 KB is built using a block size of 8 words. The word length is 32 bits. The size of the physical address space is 4 GB. The number of bits for the TAG field is ______.

[2014 (Set-2) : 1 M]

4.44 In designing a computer's cache system, the cache block (or cache line) size is an important parameter. Which one of the following statements is correct in this context?
(a) A smaller block size implies better spatial locality
(b) A smaller block size implies a smaller cache tag and hence lower cache tag overhead
(c) A smaller block size implies a larger cache tag and hence lower cache hit time
(d) A smaller block size incurs a lower cache miss penalty **[2014 (Set-2) : 2 M]**

4.45 If the associativity of a processor cache is doubled while keeping the capacity and block size unchanged, which one of the following is guaranteed to be NOT affected?
(a) Width of tag comparator
(b) Width of set index decoder
(c) Width of way selection multiplexor
(d) Width of processor to main memory data bus

[2014 (Set-2) : 2 M]

4.46 Consider a main memory system that consists of 8 memory modules attached to the system bus, which is one word wide. When a write request is made, the bus is occupied for 100 nanoseconds (ns) by the data, address, and control signals. During the same 100 ns, and for 500 ns thereafter, the addressed memory module executes one cycle accepting and storing the data. The (internal) operation of different memory modules may overlap in time, but only one request can be on the bus at any time.
The maximum number of stores (of one word each) that can be initiated in 1 millisecond is ______.

[2014 (Set-2) : 2 M]

4.47 The memory access time is 1 nanosecond for a read operation with a hit in cache, 5 nanoseconds for a read operation with a miss in cache, 2 nanoseconds for a write operation with a hit in cache and 10 nanoseconds for a write operation with a miss in cache. Execution of a sequence of instructions involves 100 instruction fetch operations, 60 memory operand read operations and 40 memory operand write operations. The cache hit-ratio is 0.9. The average memory access time (in nanoseconds) in executing the sequence of instructions is_____. **[2014 (Set-3) : 2 M]**

4.48 Assume that for a certain processor, a read request takes 50 nanoseconds on a cache miss and 5 nanoseconds on a cache hit. Suppose while running a program, it was observed that 80% of the processor's read requests result in a cache hit. The average read access time in nanoseconds is ______. **[2015 (Set-2) : 1 M]**

4.49 Consider a machine with a byte addressable main memory of 2^{20} bytes, block size of 16 bytes and a direct mapped cache having 2^{12} cache lines. Let the addresses of two consecutive bytes in main memory be $(E201F)_{16}$ and $(E2020)_{16}$. What are the tag and cache line address (in hex) for main memory address $(E201F)_{16}$?

(a) E, 201 (b) F, 201
(c) E, E20 (d) 2, 01F

[2015 (Set-3) : 1 M]

4.50 A processor can support a maximum memory of 4 GB, where the memory is word-addressable (a word consists of two bytes). The size of the address bus of the processor is at least ______ bits.

[2016 (Set-1) : 1 M]

4.51 The width of the physical address on a machine is 40 bits. The width of the tag field in a 512 KB 8-way set associative cache is _______ bits.

[2016 (Set-2) : 2 M]

4.52 A file system uses an in-memory cache to cache disk blocks. The miss rate of the cache is shown in the figure. The latency to read a block from the cache is 1 ms and to read a block from the disk is 10 ms. Assume that the cost of checking whether a block exists in the cache is negligible. Available cache sizes are in multiples of 10 MB.

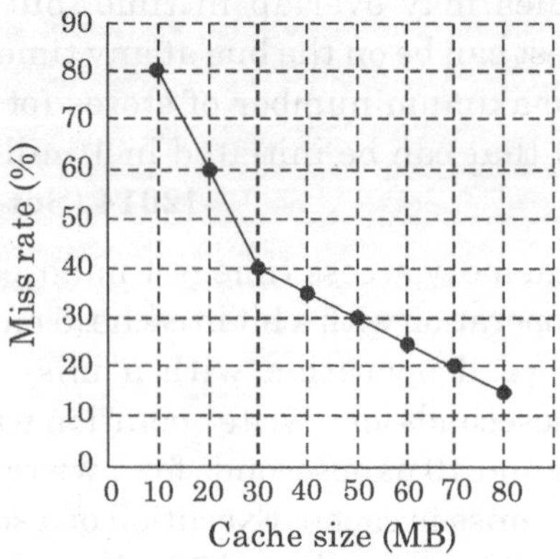

The smallest cache size required to ensure an average read latency of less than 6 ms is _______ MB.

[2016 (Set-2) : 2 M]

4.53 Consider a 2-way set associative cache with 256 blocks and uses LRU replacement. Initially the cache is empty. Conflict misses are those misses which occur due to contention of multiple blocks for the same cache set. Compulsory misses occur due to first time access to the block. The following sequence of accesses to memory blocks (0, 128, 256, 128, 0, 128, 256, 128, 1, 129, 257, 129, 1, 129, 257, 129) is repeated 10 times. The number of conflict misses experienced by the cache is _______.

[2017 (Set-1) : 2 M]

4.54 A cache memory unit with capacity of N words and block size of B words is to be designed. If it is designed as a direct mapped cache, the length of the TAG field is 10 bits. If the cache unit is now designed as a 16-way set-associative cache, the length of the TAG field is _______ bits.

[2017 (Set-1) : 2 M]

4.55 Consider a two-level cache hierarchy with L_1 and L_2 caches. An application incurs 1.4 memory accesses per instruction on average. For this application, the miss rate of L_1 cache is 0.1; the L_2 cache experiences, on average, 7 misses per 1000 instructions. The miss rate of L_2 expressed correct to two decimal places is _______.

[2017 (Set-1) : 1 M]

4.56 Consider a machine with a byte addressable main memory of 2^{32} bytes divided into blocks of size 32 bytes. Assume that a direct mapped cache having 512 cache lines is used with this machine. The size of the tag field in bits is _______.

[2017 (Set-2) : 2 M]

4.57 The read access times and the hit ratios for different caches in a memory hierarchy are as given below:

Code	Read access time (in nanoseconds)	Hit ratio
I-cache	2	0.8
D-cache	2	0.9
L2-cache	8	0.9

The read access time of main memory is 90 nanoseconds. Assume that the caches use the referred word-first read policy and the write back policy. Assume that all the caches are direct mapped caches. Assume that the dirty bit is always 0 for all the blocks in the caches. In execution of a program, 60% of memory reads are for instruction fetch and 40% are for memory operand fetch. The average read access time in nanoseconds (upto 2 decimal places) is _______.

[2017 (Set-2) : 2 M]

4.58 In a two-level cache system, the access times of L_1 and L_2 caches are 1 and 8 clock cycles, respectively. The miss penalty from the L_2 cache to main memory is 18 clock cycles. The miss rate of L_1 cache is twice that of L_2. The average memory access time (AMAT) of this cache system is 2 cycles. The miss rates of L_1 and L_2 respectively are:

(a) 0.111 and 0.056 (b) 0.056 and 0.111

(c) 0.0892 and 0.1784 (d) 0.1784 and 0.0892

[2017 (Set-2) : 2 M]

4.59 A 32-bit wide main memory unit with a capacity of 1 GB is built using 256 M × 4-bit DRAM chips. The number of rows of memory cells in the DRAM chip is 2^{14}. The time taken to perform one refresh operation is 50 nanoseconds. The refresh period is 2 milliseconds. The percentage (rounded to the closest integer) of the time available for performing the memory read/write operations in the main memory unit is _______.

[2018 : 1 M]

4.60 The size of the physical address space of a processor is 2^P bytes. The word length is 2^W bytes. The capacity of cache memory is 2^N bytes, The size of each cache block is 2^M words. For a K-way set-associative cache memory, the length (in number of bits) of the tag field is

(a) $P - N - \log_2 K$
(b) $P - N + \log_2 K$
(c) $P - N - M - W - \log_2 K$
(d) $P - N - M - W + \log_2 K$ **[2018 : 2 M]**

4.61 A certain processor uses a fully associative cache of size 16 kB. The cache block size is 16 bytes. Assume that the main memory is byte addressable and uses a 32-bit address. How many bits are required for the Tag and the Index fields respectively in the addresses generated by the processor?

(a) 28 bits and 0 bits (b) 24 bits and 0 bits
(c) 28 bits and 4 bits (d) 24 bits and 4 bits

[2019 : 1 M]

4.62 A certain processor deploys a single-level cache. The cache block size is 8 words and the word size is 4 bytes. The memory system uses a 60 MHz clock. To service a cache miss, the memory controller first takes 1 cycle to accept the starting address of the block, it then takes 3 cycles to fetch all the eight words of the block, and finally transmits the words of the requested block at the rate of 1 word per cycle. The maximum bandwidth for the memory system when the program running on the processor issues a series of read operations is ________ $\times 10^6$ bytes/sec.

[2019 : 2 M]

4.63 A direct mapped cache memory of 1 MB has a block size of 256 bytes. The cache has an access time of 3 ns and a hit rate of 94%. During a cache miss, it takes 20 ns to bring the first word of a block from the main memory, while each subsequent word takes 5 ns. The word size is 64 bits. The average memory access time in ns (round off to 1 decimal place) is ________.

[2020 : 1 M]

4.64 A computer system with a word length of 32 bits has a 16 MB byte-addressable main memory and a 64 KB, 4-way set associative cache memory with a block size of 256 bytes. Consider the following tour physical addresses represented in hexadecimal notation.

A1 = 0x42C8A4, A2 = 0x546888,
A3 = 0x6A289C, A4 = 0x5E4880

Which one of the following is TRUE?

(a) A1 and A3 are mapped to the same cache set.
(b) A2 and A3 are mapped to the same cache set.
(c) A3 and A4 are mapped to the same cache set.
(d) A1 and A4 are mapped to different cache sets.

[2020 : 2 M]

4.65 Consider a computer system with a byte-addressable primary memory of size 2^{32} bytes. Assume the computer system has a direct-mapped cache of size 32 KB (1 KB = 2^{10} bytes), and each cache block is of size 64 bytes.

The size of the tag field is ________ bits.

[2021 (Set 1) : 1 M]

4.66 Assume a two-level inclusive cache hierarchy. L1 and L2, where L2 is the larger of the two. Consider the following statements.

S_1: Read misses in a write through L1 cache do not result in writebacks of dirty lines to the L2.

S_2: Write allocate policy must be used in conjunction with write through caches and no-write allocate policy is "used with writeback caches.

Which of the following statements is correct?

(a) S_1 is true and S_2 is true.
(b) S_1 is true and S_2 is false.
(c) S_1 is false and S_2 is true.
(d) S_1 is false and S_2 is false.

[2021 (Set 2) : 2 M]

4.67 Consider a set-associative cache of size 2 KB (1 KB = 2^{10} bytes) with cache block size of 64 bytes. Assume that the cache is byte-addressable and a 32-bit address is used for accessing the cache. If the width of the tag field is 22 bits, the associativity of the cache is ________.

[2021 (Set 2) : 1 M]

4.68 Let WB and WT be two set associative cache organizations that use LRU algorithm for cache block replacement. WB is a write back cache and WT is a write through cache. Which of the following statements is/are FALSE?

(a) Each cache block in WB and WT has a dirty bit.
(b) Every write hit in WB leads to a data transfer from cache to main memory.
(c) Eviction of a block from WT will not lead to data transfer from cache to main memory.
(d) A read miss in WB will never lead to eviction of a dirty block from WB. **[2022 : 1 M]**

4.69 A cache memory that has a hit rate of 0.8 has an access latency 10 ns and miss penalty 100 ns. An optimization is done on the cache to reduce the miss rate. However, the optimization results in an increase of cache access latency to 15 ns, whereas the miss penalty is not affected. The minimum hit rate (rounded off to two decimal places) needed after the optimization such that it should not increase the average memory access time is ________. **[2022 : 1 M]**

4.70 Consider a system with 2 KB direct mapped data cache with a block size of 64 bytes. The system has a physical address space of 64 KB and a word length of 16 bits. During the execution of a program, four data words P, Q, R, and S are accessed in that order 10 times (i.e., PQRSPQRS ...). Hence, there are 40 accesses to data cache altogether. Assume that the data cache is initially empty and no other data words are accessed by the program. The addresses of the first bytes of P, Q, R, and S are 0xA248, 0xC28A, 0xCA8A, and 0xA262, respectively. For the execution of the above program, which of the following statements is/are TRUE with respect to the data cache?

(a) Every access to S is a hit.
(b) Once P is brought to the cache it is never evicted.
(c) At the end of the execution only R and S reside in the cache.
(d) Every access to R evicts Q from the cache.

[2022 : 2 M]

4.71 A 4 kilobyte (KB) byte-addressable memory is realized using four 1 KB memory blocks. Two input address lines (IA4 and IA3) are connected to the chip select (CS) port of these memory blocks through a decoder as shown in the figure. The remaining ten input address lines from IA11 – IA0 are connected to the address port of these blocks. The chip select (CS) is active high.

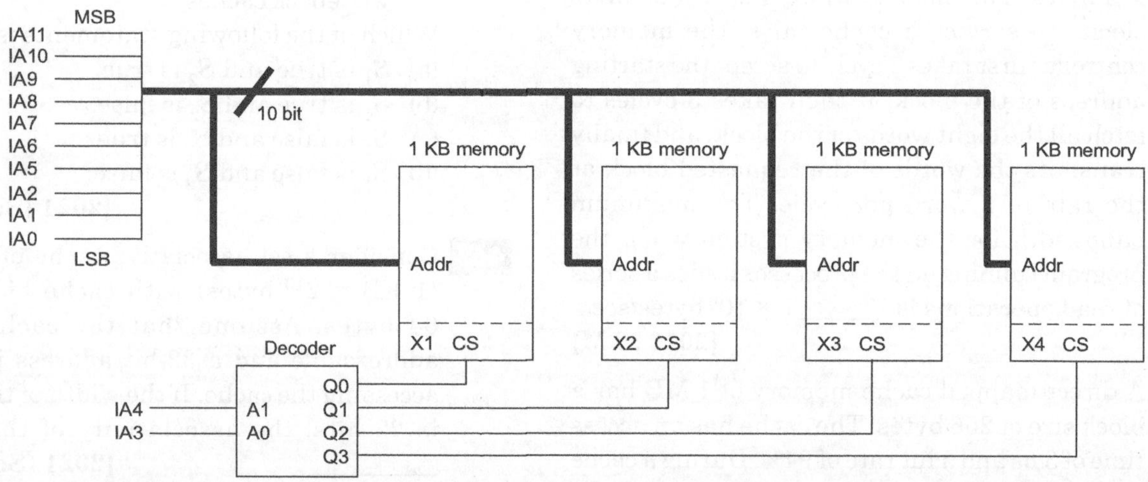

The input memory addresses (IA11 – IA0), in decimal, for the starting locations (Addr=0) of each block (indicated as X1, X2, X3, X4 in the figure) are among the options given below. Which one of the following options is CORRECT?

(a) (0, 1, 2, 3)
(b) (0, 1024, 2048, 3072)
(c) (0, 8, 16, 24)
(d) (0, 0, 0, 0)

[2023 : 2 M]

4.72 An 8-way set associative cache of size 64 KB (1 KB = 1024 bytes) is used in a system with 32-bit address. The address is sub-divided into TAG, INDEX, and BLOCK OFFSET.
The number of bits in the TAG is ________.

[2023 : 2 M]

4.73 Consider two set-associative cache memory architectures: **WBC**, which uses the write back policy, and **WTC**, which uses the write through policy. Both of them use the LRU (Least Recently Used) block replacement policy. The cache memory is connected to the main memory. Which of the following statements is/are TRUE?

(a) A write hit in **WBC** can modify the value of the dirty bit of a cache block

(b) A read miss in **WTC** never triggers a write back operation of a cache block to main memory

(c) A write miss in **WTC** always writes the victim cache block to main memory before loading the missed block to the cache

(d) A read miss in **WBC** never evicts a dirty block

[2024 (Set-1) : 2 M]

4.74 A given program has 25% load/store instructions. Suppose the ideal CPI (cycles per instruction) without any memory stalls is 2. The program exhibits 2% miss rate on instruction cache and 8% miss rate on data cache. The miss penalty is 100 cycles. The speedup (rounded off to two decimal places) achieved with a perfect cache (i.e., with NO data or instruction cache misses) is ________.

[2024 (Set-1) : 2 M]

■■■■

Answers Memory Organization

4.1 (d)	**4.2** (b)	**4.3** (d)	**4.4** (b)	**4.5** (d)	**4.6** (b)	**4.7** (d)	**4.8** (c)	**4.9** (a)
4.10 (c)	**4.11** (c)	**4.12** (c)	**4.13** (a)	**4.14** (c)	**4.15** (c)	**4.16** (a)	**4.17** (d)	**4.18** (a)
4.19 (d)	**4.20** (c)	**4.21** (b)	**4.22** (d)	**4.23** (c)	**4.24** (a)	**4.25** (d)	**4.26** (a)	**4.27** (c)
4.28 (b)	**4.29** (d)	**4.30** (b)	**4.31** (c)	**4.32** (c)	**4.33** (d)	**4.34** (b)	**4.35** (d)	**4.36** (d)
4.37 (d)	**4.38** (c)	**4.39** (a)	**4.40** (a)	**4.41** (b)	**4.42** (a)	**4.43** (20)	**4.44** (d)	**4.45** (d)
4.46 (10000)	**4.47** (1.68)	**4.48** (14)	**4.49** (a)	**4.50** (31)	**4.51** (24)	**4.52** (30)	**4.53** (76)	**4.54** (14)
4.55 (0.05)	**4.56** (18)	**4.57** (4.72)	**4.58** (a)	**4.59** (59)	**4.60** (b)	**4.61** (a)	**4.62** (160)	**4.63** (13.5)
4.64 (b)	**4.65** (17)	**4.66** (b)	**4.67** (2)	**4.68** (a, b, d)	**4.69** (0.85)	**4.70** (a, b, d)	**4.71** (c)	**4.72** (19)
4.73 (a, b)	**4.74** (2.56)							

Explanations Memory Organization

4.1 (d)

The principle of locality justifies the use of cache memory.

4.2 (b)

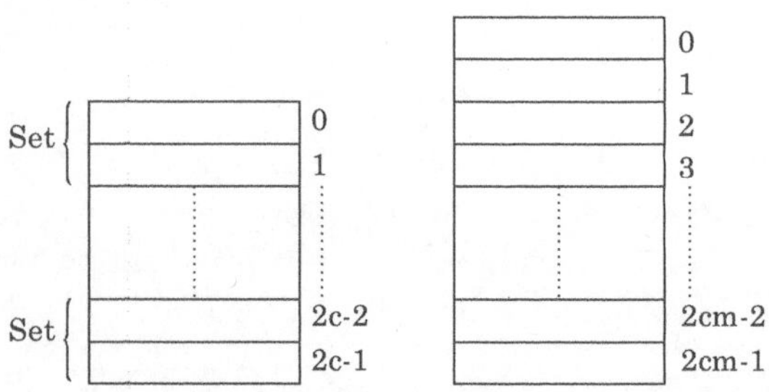

Set Associative Cache   Main Memory

# of set in set associative cache

$$= \frac{\text{\# of blocks in cache}}{\text{\# of blocks in one set}}$$

$$= \frac{2c}{2} = c$$

# of sets in cache = c

Therefore, the block k of the main memory maps to the set (k mod c) of the cache.

4.3 (d)

Virtual memory is a concept which allows programs larger than the physical memory size to run.

4.4 (b)

1 MB of memory supported by graphics card

(a) [1600*400*$\log_2$ 256 bits] < 1 MB

(b) [1600*400*$\log_2 10^6$ bits] > 1 MB

(c) [800*400*$\log_2 10^6$ bits] < 1 MB

(d) [800*800*$\log_2$ 256 bits] < 1 MB

$\therefore$ Option (b) can not supported by card.

4.5 (d)

Disk is the IO device attached externally to the processor. Therefore, disk requires a device driver.

4.6 (b)

There are two types of locality of reference:

1. **Temporal:** A recently executed instruction is likely to be executed again very soon.
2. **Spatial:** Instructions in close proximity to a recently executed instruction are also likely to be executed soon.

The spatial aspect suggests that instead of fetching just one item from the main memory to the cache, it is useful to fetch several items that reside at adjacent address as well.

Therefore more than one work are put in one cache block to exploit the spatial locality of reference in the program.

4.7 (d)

Techniques that automatically move program and data blocks into the physical main memory when they are required for execution are called virtual memory techniques. Programs, and hence the processor, reference an instruction and data space that is independent of the available physical main memory space. Virtual memory increases the degree of multi-programming. However virtual memory increases the context switching overhead.

4.8 (c)

An opcode is the portion of a machine language instruction that specifies the operation to be performed.

4.9 (a)

Access time (Hierarchial)

$\Rightarrow$ $H_1T_1 + (1 - H_1) H_2(T_2 + T_1) + (1 - H_1)(1 - H_2) H_3(T_3 + T_2 + T_1)$

$= 0.8 + 1.98 + 10.22 = 13$ ns.

4.10 (c)

Sequence 8, 12, 0, 12, 8

Apply the LRU as follows:

Page Frame

Frame					
3					
2					
1		12	12	12	12
0	8	8	0	0	8
	Miss	Miss	Miss	No Miss	Miss

or

Set 0	~~8~~	~~0~~	8
	12		
Set 1			

Total number of miss = 4

4.11 (c)

Memory cycle time = 64 nsec

Each refresh time = 100 nsec

In 1 msec = 100 times refreshed

$$\text{So in 64 msec} = \frac{64 \text{ nsec} \times 100 \text{ times}}{1 \text{ msec}}$$

$$= 64 \times 10^{-9} \times 10^2 \times 10^3$$

$$= 64 \times 10^{-4}$$

So, time to refresh in memory cycle

$$= 100 \text{ nsec} \times 64 \times 10^{-4}$$

$$= 0.64 \text{ nsec}$$

So, percentage of time spent for refreshing

$$= \frac{0.64 \text{ nsec}}{64 \text{ nsec}} \times 100 = 1$$

4.12 (c)

Increasing the RAM means increase the primary memory which reduce the swapping so fewer page faults occur.

4.13 (a)

Size of cache = 32 KB

$= 32 \times 2^{10}$ byte $= 2^5 \times 2^{10}$ byte

$= 2^{15}$ byte = 15 bits

Size of tag = 32 – 15 = 17 bits

Cache index = LO + WO = 15 bit

4.14 (c)

Number of bits requires for 128 main memory block

$= \log_2 (128)$

= 7 bits

Main memory format:

Block number	Offset
7 bits	

Since main memory and cache memory offset will be same.

Number of bits for set $= \log_2 (4) = 2$

So, cache memory format will be:

Tag	Set	Offset
← 5 bits →	2 bits	

(Tag + Set = ← 7 bits →)

So, each set can maps 32 blocks and at one time can hold only 2 different block.

	Block 1st	Block 2nd
Set 0	0	16
Set 1	5	9
Set 2		
Set 3	~~3~~ 55	7

For block 0 = 0000000 = set 0

For block 5 = 0000101 = set 1

For block 3 = 0000011 = set 3

For block 9 = 0001001 = set 1

For block 7 = 0000111 = set 3

For block 0 = 0000000 = set 0

For block 16 = 0010000 = set 0

For block 55 = 0110111 = set 3

So, for block 55, block 3 will be replaced since 3 is come before block 7.

So, memory block 0, 5, 7, 9, 16 and 55 will be present in cache.

4.15 (c)

A cache line size = 64 Byte

$$\text{So cache access time} = \frac{64 \text{ B}}{1 \text{ GBps}} = 64 \text{ nsec}$$

Total time to fetch the entire cache = Main m/m latency + Transfer time

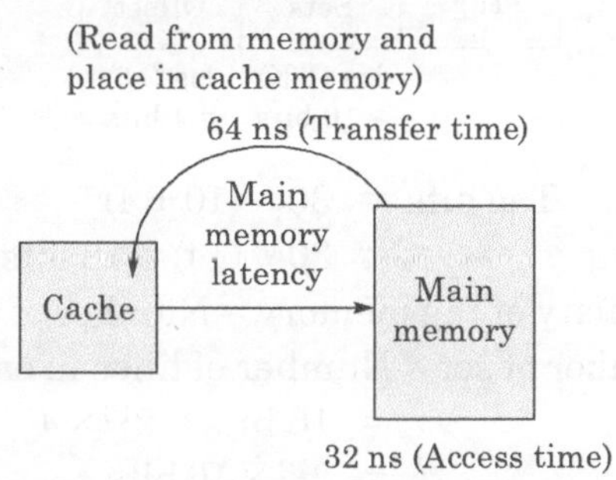

= 32 nsec + 64 nsec = 96 nsec

4.16 (a)

For I-cache:

$$\text{Number of lines} = \frac{4\,K}{4} = 1\,K$$

Tag	Line	Offset
	log (1 K) = 10 bits	log (4) = 2 bits

(30 bits)

$$\text{Tag bits} = 30 - (10 + 2) = 30 - (12) = 18 \text{ bits}$$

Capacity of tag memory = Tag bits × line size
= 18 bits × 1 K

For D-cache:

$$\text{Number of lines} = \frac{4\,K}{4} = 1\,K$$

$$\text{Number of sets} = \frac{1\,K}{2} = 2^9$$

Tag	Sets	Offset
	log (2^9) = 9 bits	log (4) = 2 bits

(30 bits)

$$\text{Tag bits} = 30 - (9 + 2) = 30 - (11) = 19 \text{ bits}$$

Capacity of tag memory = Number of tag bits × Number of set × Number of lines in each set
= 19 bits × 2^9 × 2
= 1 K × 19 bits

For L2-cache:

$$\text{Number of lines} = \frac{64\,K}{16} = 4\,K$$

$$\text{Number of sets} = \frac{4\,K}{2^2} = 1\,K$$

Tag	Sets	Offset
	log (1 K) = 10 bits	log (16) = 4 bits

(30 bits)

$$\text{Tag bits} = 30 - (10 + 4) = 30 - (14) = 16 \text{ bits}$$

Capacity of tag memory = Number of tag bits × Number of set × Number of lines in each set
= 16 bits × 2^{10} × 4
= 2^{12} × 16 bits
= 4 K × 16 bits

4.17 (d)

Cache block size = 64 bytes
Main memory has K banks or k = 24
Each bank is 2 byte long because c = 2
Total time for one parallel access

$$T = \frac{K}{2} + \text{Latency}$$

$$= \frac{24}{2} + 80 = 12 + 80 = 92 \text{ ns}$$

Total latency time = CT
= 2 × 92 = 184 ns

4.18 (a)

Tag	Set	Block
18	9	5

So, $h_1 = \frac{18}{10} + 0.6$ ns
= 1.8 + 0.6 ns = 2.4 ns

4.19 (d)

Tag	Set	Block
17	10	5

So, $h_2 = \frac{17}{10}$ ns = 1.7 ns

4.20 (c)

$$\begin{bmatrix} a[0][0], a[0][1], \ldots\ldots, a[0][511] \\ a[1][0], a[1][1], \ldots\ldots, a[1][511] \\ \vdots \ \ldots\ldots\ldots \\ a[511][0], a[511][1], \ldots\ldots, a[511][511] \end{bmatrix}$$

One element size = 8 B
Block size = 128 B

$\therefore$ One block holds = $\frac{128}{8}$ = 16 element

1 row contains = 512 elements
$\therefore$ 32 block are required to carry row
Cache memory size = 32 kB
Block size = 128 B

$$\#\text{lines} = \frac{32\,\text{kB}}{128\,\text{B}} = \frac{2^{15}}{2^7} = 2^8 = 256$$

P1: In this array is accessed in row-major order like
a[0][0], a[0][1], a[0][2], ...

Accessing a[0][0] = Miss
Accessing a[0][1] = hit
Accessing a[0][2] = hit
⋮
Accessing a[0][15] = hit
Accessing a[0][16] = Miss
Accessing a[0][17] = hit
⋮
Accessing a[0][31] = hit
Accessing a[0][32] = Miss
⋮

∴ One row accessing = 32 miss operations

Total miss $(M_1) = 32 \times 512$
$= 16384$

4.21 (b)

P_2: In this array is accessed in column-major order like.

a[0][0], a[1][0], a[2][0], ...

Accessing a[0][0] = Miss
Accessing a[1][0] = Miss
Accessing a[2][0] = Miss
⋮
Accessing a[511][0] = Miss
Accessing a[0][1] = Miss
⋮
Accessing a[511][1] = Miss
⋮

One column accessing = 512 miss operation

Total miss $(M_2) = 512 * 512$

$$\therefore \quad \frac{M_1}{M_2} = \frac{16384}{512 \times 512} = \frac{1}{16}$$

4.22 (d)

4-way set associative

lines = 128

Block size = 64 words

← 20 bits →

TAG	SET OFFSET	WORD OFFSET
⇓	⇓	⇓
20 – (5 + 6) = 9 bits	$\log_2 32$ = 5 bits	$\log_2 (64)$ = 6 bits

$$\text{Number of set} = \frac{\#\text{lines}}{4} = \frac{128}{4} = 32$$

4.23 (c)

$$\begin{bmatrix} a[0][0], a[0][1], \ldots\ldots\ldots, a[0][49] \\ a[1][0], a[1][1], \ldots\ldots\ldots, a[1][49] \\ \vdots \\ a[49][0], a[49][1], \ldots\ldots\ldots, a[49][49] \end{bmatrix}$$

One element size = 1 B

Total array elements = 50 × 50 = 2500

So total array size = 2500 × 1 B = 2500 B

Block size = 64 B

∴ # block required to store the array

$$= \frac{2500\text{ B}}{64\text{ B}} = 39.04 \approx 40$$

lines in cache = 32, Block size = 64 B

Physical address = 1100 H

← 16 bits →

TAG	LINE OFFSET	WORD OFFSET
5 bit	5 bit	6 bit

$$(1100)_H = (0001000100000000)_B$$

TAG	LO	WO
00010	00100	000000

$$\therefore \quad LO = (00100)_B = 4$$

∴ Array is stored in the main memory starting from memory location of line 4.

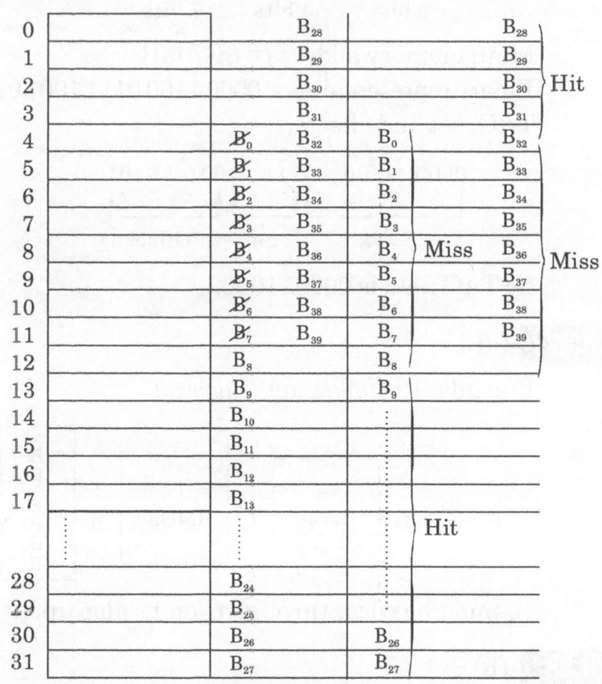

1st access 40 Miss   2nd access 16 Miss (8 + 8)

∴ Total data cache misses will occur in total = 40 + 16 = 56

4.24 (a)

The lines of the data cache that is replaced by new blocks in accessing the array = line 4 to line 11.

4.25 (d)

Cache memory size = 8 KB

Block size = 128 words × 1 byte = 128 byte

$$\text{Number of block in cache} = \frac{2^{13}\text{ B}}{2^7\text{ B}} = 2^6$$

$$\text{Number of sets} = \frac{\text{Number of blocks}}{\text{Number of ways}} = \frac{2^6}{2^2} = 2^4$$

← 20 →

Tag	Set	Offset
	log (2^4) = 4 bits	log (128) = 7 bits

Tag bits = 20 – (4 + 7) = 20 – 11 = 9 bits

So, TAG, set and word fields are 9, 4 and 7 respectively.

4.26 (a)

Format of memory:

Tag	Set	Offset
9 bits	4 bits	7 bits

Main memory address = 0C795H

Binary representation = 0000 1100 0111 1001 0101

TAG, set and offset:

0000 1100 0 | 111 1 | 001 0101

Tag | Set | Offset

So TAG field is 000011000.

4.27 (c)

Consider the following diagram:

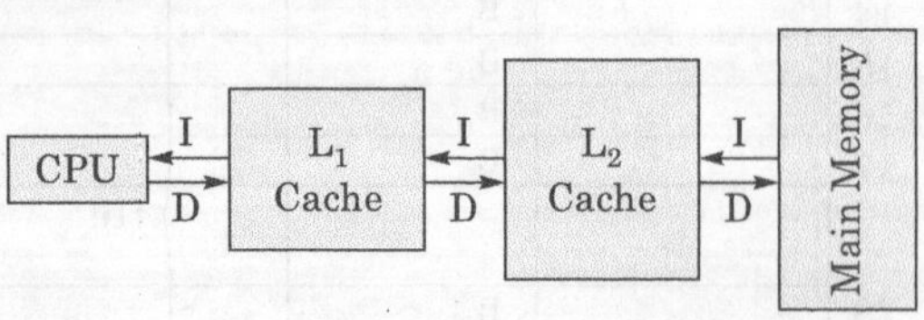

L_1 must be write through then L_2 also updated.

4.28 (b)

In instruction execution pipeline, the earliest that the data TLB can be accessed during effective address calculation has started.

4.29 (d)

2 way set associative

Cache memory size = 64 kB

Block size = 16 B

$$\#\text{lines} = \frac{64\text{ kB}}{16\text{ B}}$$

$$= 4\text{ K} = 2^2 \times 2^{10} = 2^{12}$$

$$\#\text{sets} = \frac{2^{12}}{2} = 2^{11}$$

32-bit virtual address

← 32 bits →

TAG	SO	WO
⇓	⇓	⇓
32 – (4 + 11) = 17 bits	$\log_2(2^{11})$ = 11 bits	$\log_2(16)$ = 4 bits

Tag memory size = Number of sets * No. of lines in a set * Number of tag bits

$= 2^{11} \times 2 \times 17$ bits

$= 68 \times 2^{10}$ bits = 68 K bits

4.30 (b)

Array is APR [1024][1024]

$$\begin{bmatrix} a[0][0],\ a[0][1],\ \ldots\ldots,\ a[0][1023] \\ a[1][0],\ a[1][1],\ \ldots\ldots,\ a[1][1023] \\ \vdots \\ a[1023][0],\ a[1023][1],\ \ldots a[1023][1023] \end{bmatrix}$$

1 element size = 8 B

Block size = 16 B

∴ One block holds 2 elements

One row size = 1024 element

∴ Number of blocks required to store one row = 512 blocks

Sets:

0	APR [0], APR [0][1]	APR [4][0], APR [4][1]
1	APR [0][2], APR [0][3]	APR [4][2], APR [4][3]
2	APR [0][4], APR [0][5]	APR [4][4], APR [4][5]
⋮	1st Row	5th Row
511		
512	APR [0][1022], APR [0][1023]	APR [4][4], APR [4][5]
	APR [1][0], APR [1][1]	
⋮	2nd Row	6th Row
1023	a, APR[1][1023]	
1024	3rd Row	7th Row
1535		
1536		
⋮	4th Row	8th Row
2047		

APR[0][0] has the same cache index as APR [4][0]

4.31 (c)

Row major access

a[0] [0] = Miss
a[0] [1] = hit
a[0] [2] = Miss
a[0] [3] = hit

⋮

Every miss is followed by a hit

$$\therefore \quad \text{Hit ratio} = \frac{1024}{1024+1024}$$

$$= \frac{1024}{2048} = \frac{1}{2} = 50\%$$

4.32 (c)

$$\text{Basic RAM} = 32\text{ K} \times 1$$

$$\text{Design RAM} = 256\text{ K} \times 8$$

$$\text{No. of RAM} = \frac{256\text{ K} \times 8}{32\text{ K} \times 1}$$

$$= \frac{2^{18} \times 2^3}{2^{15} \times 2^0} = 64$$

It require 8 parallel line and in each parallel line 8 serial RAM chip are required.

4.33 (d)

There are total 4 sets in the cache and each set contains 4 blocks.

Set	Block
Set 0	~~0~~ 48
	~~4~~ 32
	8
	~~216~~ 92
Set 1	1
	133
	129
	73
Set 2	
Set 3	~~255~~ 155
	3
	159
	63

0 mod 4 = 0 set 0
255 mod 4 = 3 set 3
1 mod 4 = 1 set 1
4 mod 4 = 0 set 0
3 mod 4 = 3 set 3
8 mod 4 = 0 set 0
133 mod 4 = 1 set 1
159 mod 4 = 3 set 3
216 mod 4 = 0 set 0
129 mod 4 = 1 set 1
63 mod 4 = 3 set 3
8 mod 4 = 0 set 0 (already in cache)
48 mod 4 = 0 set 0 48 will replace block 0 using LRU
32 mod 4 = 0 set 0 32 will replace block 4
73 mod 4 = 1 set 1
92 mod 4 = 0 set 0 92 will replace block 216
155 mod 4 = 3 set 3 155 will replace 255

4.34 (b)

Given MM capacity = 4 MB = 4 M × 8 bit

Given DRAM chip = 1 M × 1 bit

Total number of DRAM chips required

$$= \frac{4\text{ M} \times 8\text{ bit}}{1\text{ M} \times 1\text{ bit}}$$

$$= \underset{\text{Verticle}}{\underset{\downarrow}{4}} \times \underset{\text{Horizontal}}{\underset{\downarrow}{8}} = 32\text{ DRAM chips}$$

Now, as given number of rows / DRAM chip = 1 K

Time taken for single refresh = Time to refresh 1 row = 100 ns

So, time taken to refresh 1 chip = 1 row refresh time × Number of rows/chip

$$= 100\text{ ns} \times 2^{10} = 2^{10} \times 100\text{ ns}$$

Now, as all the chips refresh parallerly.

So, 1 chip refresh time = 32 chips refresh time

$$= 2^{10} \times 100\text{ ns}$$

4.35 (d)

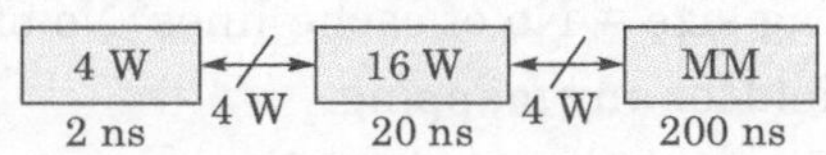

Memory access time to access L1 = 2 nanoseconds

Memory access time to access L2 = 20 nanoseconds

When there is a miss in L1 cache and hit in L2 cache so we have to transfer block from L2 to L1.

i.e,

L2 – L1 (4W)
L2 – L1 (4W)
L2 – L1 (4W)
L2 – L1 (4W)

4T2 = 80 ns

No option matching so take the serial accessing i.e., 4T2 + 4T1 = 88 ns.

4.36 (d)

When there is a miss in L1 and L2 cache then block transfer for mm to L2 cache, and then block is transferred from L2 cache to L1 cache- So total time for these transfer

Total Access time = Block transfer time from main memory (MM) to L2 cache + block transfer time from L2 to L1.

MM – L2 → L2 – L1
MM – L2 → L2 – L1
MM – L2 → L2 – L1
MM – L2 → L2 – L1

4 MM + 8T2 + 4T1
= 968 ns

4.37 (d)

Cache memory size = 8KB
Block size = 32 bytes
Mapping technique is Direct.
Physical address size is 32 bits
cache lines= $2^{13} / 2^5 = 2^8 = 256$ lines
Memory address interpretation in direct mapping is:

Modified bit	Valid bit	tag	Line offset	Word offset

Word offset requires = 5 bits
Line offset requires = 8 bits
Tag requires = 32bits – (8 + 5)bits = 19 bits
The metadata present in cache memory is = data memory + tag size
Tag size = No of cache lines*No of bits in tag field for any mapping.
Hence the total size of memory needed at the cache controller to store metadata tags for the write back cache is:
= Number of cache lines*(tag bits + valid bit + modified bit)
= 256*(19 + 1 + 1) = 256 * 21
= 5376 bits

4.38 (c)

$$\text{Number of cache lines} = \frac{2^8 \times 2^{10}}{2^5} = 2^{13}$$

$$\text{Number of sets} = \frac{2^{13}}{2^2} = 2^{11}$$

32

16	11	5
Tags	So	Word offset

Number of tag bits are 16 bits
∴ Option 'c' is correct.

4.39 (a)

Tag directory size = Number of sets × no. of tag s in each sets × (no. of tag bits in each tag +valid bits + modified bit + replacement bit)
$= 2^{11} \times 4 \times (16 + 2 + 1 + 1) = 160$ KBits
∴ Option (a) is correct.

4.40 (a)

Main memory block 'j' mapped to any of cache line. Cache line range is:
(j mod v)*k → (j mod v)*k + (k – 1)

4.41 (b)

We need 16 K × 16 RAM from 1 K × 8 RAM
So, number of chips required

$$= \frac{16\text{K} \times 16}{1\text{K} \times 8} = 16 \times 2$$

So we need 16 output lines

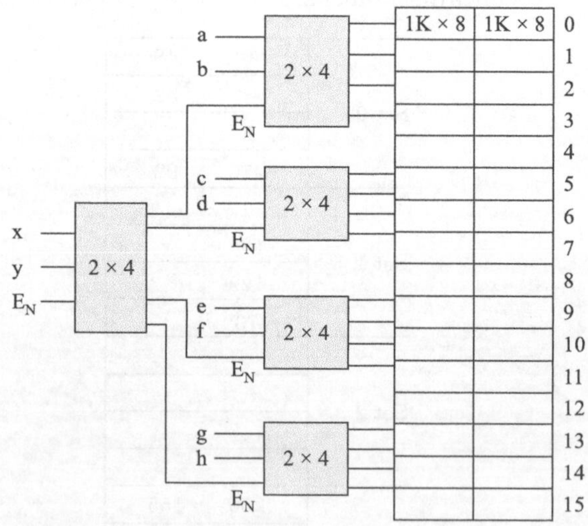

∴ We need 5 decoders.

4.42 (a)

Miss ratio $= \frac{n}{N}$, if the access sequence is passed through a cache of associativity $A \geq k$ exercising least-recently used replacement policy.

4.43 (20)

TAG	SET	WORD
20	7	5

← 32 bit →

WORD offset = 5 bits

[∵ word length = 32 bits]

SET offset = 7 bits

$$\because \quad \#\text{blocks} = \frac{16\text{kB}}{8\,\text{Words}} = 512 \text{ blocks}$$

$$\#\text{sets} = \frac{512}{4} = 128$$

$$\therefore \quad \#\text{TAG bits} = 32 - (7 + 5) = 20 \text{ bits}$$

4.44 (d)

A smaller cache block implies a larger cache tag and hence higher cache hit time.
So it incurs lower cache miss penalty.

4.45 (d)

By doubling the associativity of the cache, width of the processor to main memory databus is guaranteed to be not affected.

4.46 (10000)

To initiate one word data to the bus takes 100 ns.
So, one words store → 100 ns
? Number of words store → 1 msec
∴ Number of words initiated for store

$$= \frac{1 \text{ msec}}{100 \text{ ns}} = 10000$$

4.47 (1.68)

Average read time
$= 0.9 \times 1 \text{ ns} + 0.1 \times 5 \text{ ns}$
$= 0.9 + 0.5 = 1.4 \text{ ns}$

In the execution sequence number of read operations = 160
So total time required for read operation
$= 160 \times 1.4 \text{ ns} = 224 \text{ ns}$

Average write time $= 0.9 \times 2 \text{ ns} + 0.1 \times 10 \text{ ns}$
$= 1.8 + 1 = 2.8 \text{ ns}$

In the execution sequence number write operation = 40.
So, the total time required for write operation
$= 40 \times 2.8 \text{ ns} = 112 \text{ ns}$

Total time for instruction execution time for both read and write = 224 + 112 = 336 ns.

200 times, access takes ______ 336 ns.
1 time, access takes ______ ?

Average memory access time

$$= \frac{336}{200} = 1.68 \text{ ns}$$

4.48 (14)

$T_{readmiss} = 50$ nsec, $T_{readhit} = 5$ nsec
$h = 0.8$
$T_{avgread} = h \times T_{readhit} + (1 - h) \times \text{Miss penality}$
$= 0.8 \times 5 + 0.2 \times 50$
$= 4 + 10 = 14$ nsec

4.49 (a)

Block size = 16 bytes = 2^4 bytes
Offset = 4 bits
Index = 12 bits

[∵ number of lines = 2^{12}]
Main memory size = 2^{20} bytes
⇒ Physical address = 20 bits

TAG (4)	INDEX (12)	OFFSET (4)

20 bits

E	2	0	1	F
1110	0010	0000	0001	1111

∴ E is tag and 201 is line address (index).

4.50 (31)

Main memory size = 4 GB
Word size = 2 byte
1 word — 2 byte
? words — 4 GB
∴ Total 2 G words are present.
So, address size 2 G = 31 bits.

4.51 (24)

← 40 bit →

Tag	Set offset	word offset
21		

cache index ($\log_2 512$ K = 19 bits)
Block size not given so, cache index is considered

or

Hypothetically, line size considered as cell size.
∴ Number of lines = 512 K

So, number of sets = $\frac{512\text{K}}{8}$

$$\Rightarrow \quad \frac{2^{19}}{2^{3}} = 2^{16}$$

40 bit

Tag	Set offset	Word offset
24	16	Block size not given. So, no word offset.

So, TAG size is 24 bits.

4.52 (30)

1. Miss rate = 0.8, 10 MB
   $T_c = 1$ ms
   $T_d = 10$ ms
   $T_{avg} = HT_c + (1 - H)(T_d + T_c) = 9$ ns
2. Miss rate = 0.6, 20 MB
   $T_{avg} = 7$ ms
3. Miss rate = 0.4, 30 MB
   $T_{avg} = 5$ ms
4. Miss rate = 0.35, 40 MB
   $T_{avg} = 4.5$ ms
5. Miss rate = 0.3, 50 MB
   $T_{avg} = 4$ ms
6. Miss rate = 0.25, 60 MB
   $T_{avg} = 3.5$ ms
7. Miss rate = 0.2, 70 MB
   $T_{avg} = 3$ ms
8. Miss rate = 0.15, 80 MB
   $T_{avg} = 2.5$ ms

So, 30 MB will the answer.

4.53 (76)

Number of lines = 256

$$\text{Number of sets (S)} = \frac{256}{2} = 128$$

Cache memory

0	~~0~~ ~~256~~ ~~0~~ 256	128
1	~~1~~ ~~257~~ ~~1~~ 257	129
2		
⋮		
127		

{K mode S = *i*}
LRU

Compulsary miss: Very 1st access to anyblock is miss then it is called compulsary miss.
Conflict miss: After every block come atleast one time in memory, if miss occure due to conflict i.e., block need to replace called conflict miss.

So for 1st access of block number 0, 256, 128 and 1, 257 and 129 are compulsary miss after that any miss is conflict miss.
After 1st access = 4 (conflict miss)
After 2nd access = 8 (conflict miss)
So total conflict misses $= 4 + 8 \times 9$
$= 4 + 72 = 76$

4.54 (14)

Cache memory size = N words
Block size = B words, Tag = 10 bits
Direct cache memory
Address binding

tag	LO	WO
10 bit		

When, cache is configured as 16-way set associative then set offset value decreases. So tag size incrases.

tag	SO	WO
14 bit		

Eg., Cache memory size = 8B
Block size = 2B, MM size = 16B
(a) Direct Cache memory

tag	LO	WO
1 bit	2 bit	1 bit

$$\#\text{lines} = \frac{8}{2} = 4$$

(b) 2-way set associate Cache memory

$$\#\text{sets} = \frac{N}{\text{P-way}} = \frac{4}{2} = 2$$

tag	SO	WO
2 bit	1 bit	1 bit

4.55 (0.05)

1.4 memory accesses —— 1 instruction
? # memory accesses —— 1000 instructions

⇒ Number of memory access in 1000 instructions = 1400
Number of misses in L_2 cache

$$= \frac{7 \text{ miss}}{1400 \text{ memory access}} = 0.005$$

Miss rate L_2

$$= \frac{\#\text{ misses in } L_2}{\#\text{ misses in } L_1} = \frac{0.005}{0.1} = 0.05$$

4.56 **(18)**

MM size = 2^{32} B, Block size = 32 B

Direct CM

Number of lines = 512

Address format is (32 bit)

Tag	LO	WO
18 bit	$\log_2 512 = 9$ bit	$\log_2 32 = 5$ bit

4.57 **(4.72)**

Instruction read cycle time:

$\Rightarrow H_1T_1 + (1 - H_1)\, H_2(T_2 + T_1) + (1 - H_1)(1 - H_2)\, H_3\,(T_m + T_2 + T_D)$

$\Rightarrow (0.8 \times 2) + (1 - 0.8)\; 0.9\; (8 + 2) + (1 - 0.8)(1 - 0.9)\;(90 + 8 + 2)$

$\Rightarrow 5.4$ ns

$T_{avg\ read\ (inst)}$ = (Frequency of instruction × Read cycle time)

= 60% × 5.4 ns = 3.24 ns

Data read cycle time:

$\Rightarrow H_1T_D + (1 - H_1)\, H_2\,(T_2 + T_D) + (1 - H_1)(1 - H_2)\, H_3\,(T_m + T_2 + T_D)$

$\Rightarrow (0.9 \times 2) + (1 - 0.9)\,(0.9)\,(8 + 2) + (1 - 0.9)(1 - 0.9)\,(90 + 8 + 2)$

$\Rightarrow 3.7$ ns

$T_{avg\ read\ (data)}$ = (Frequency of data × Read cycle time) = 40% × 3.7 ns = 1.48 ns

Total time = (3.24 + 1.48) = 4.72 ns

4.58 **(a)**

Hit time L_1 = 1 cycle

Hit time L_2 = 8 cycles

Miss penality L_2 = 18 cycles

T_{avg} = 2 ns

Miss rate $L_1 = x = 2x$

Formula:

- Tavg = Hit time L_1 + (Miss rate L_1 × Miss penality L_1)
- Miss penality L_1 = Hit time L_2 + (Miss rate L_2 × Miss penality L_2)
- Substitute the above data and verifying with respect to the given options.
- In this context after substitute the option (a). Data, Tavg becomes 2ns.

4.59 **(59)**

Number of chips required to designed 1 GB

$$\text{memory} = \frac{1\text{ GB}}{256\text{M}\times 4} = \frac{2^{33}}{2^{30}} = 2^3 = 8 \text{ Chips}$$

So, we can arrange them in 1 row to make the word size as 32 bits.

Rows : C1 C2 C3.........C8 | C : Chip | 256M × 4

Refresh circuit is designed for memory unit to refresh the entire chip data.

∴ To minimize this overhead, Design the chips into a banks i.e. 4 rows of chips grouped into 1 bank. So, refresh occurs in parallel.

$$\text{i.e. Refresh overhead} = \frac{\text{Time required for refresh}}{\text{Refresh Interval}}$$

$$= \frac{2^{14}\times 50\text{ ns}}{2\text{ ms}} = \frac{16384\times 50\text{ ns}}{2\text{ ms}}$$

= 0.4096 = 40.96%

% time available for read/write operations in the main memory unit is

⇒ 1 – 0.4096

⇒ 0.5904 = 59.04% = 59%

4.60 **(b)**

MM space = 2^P bytes

Physical Address (PA) size = P bits

CM size = 2^N bytes

Block size = 2^M words

= 2^M words * 2^W bytes/word

= $2^{(M+W)}$ bytes

$$\text{Number of lines} = \frac{\text{CM size}}{\text{Block size}} \Rightarrow \frac{2^N}{2^{M+W}}$$

$\Rightarrow 2^{N-M-W}$

$$\text{Number of sets} = \frac{\text{Number of lines in cache}}{\text{P-way}}$$

$$= \frac{2^{N-M-W}}{K}$$

Address format:

PA ← P bits →

tag	so	wo

so: $\log_2 \dfrac{2^{N-M-W}}{K}$ ↓ $\dfrac{\log_2 2^{N-M-W}}{\log_2 K}$ ↓ $\log_2 2^{N-M-W} - \log_2 K$

wo: $\log_2 2^{M+W}$ ↓ (M + W) bits

$\Rightarrow$ (N – M – W – $\log_2$ K)

$\therefore$ Tag size

$\Rightarrow$ P – (N – M – W – $\log_2$ K + M + W)

$\Rightarrow$ P – N + $\log_2$ K

4.61 (a)

Cache memory size = 16 kB

Block size = 16 B

Main memory address = 32 bit

$$\text{Number of lines (N)} = \frac{16\text{ K}}{16} \Rightarrow \frac{2^{14}}{2^4} = 2^{10}$$

Fully associative cache memory (N-way)

$$\text{So, number of sets (S)} = \frac{N}{\text{P-way}} \Rightarrow \frac{2^{10}}{2^{10}} = 1$$

$\therefore$ Address format:

TAG	WO
28 bit	$\log_2 16 = 4$ bit

(32 bit)

So, TAG = 28 bit

Index = 0 bit (No address)

4.62 (160)

Total time to transfer a cache block = 1 + 3 + 8 = 12 cycles

8 W ——— 12 cycles

8 × 4 bytes ——— 12 cycles

? B ——— 1 sec

$$= \frac{32\text{B}}{12\times\left(\frac{1}{60}+10^{-6}\right)\text{sec}} = 160 \times 10^6 \text{ bytes/sec}$$

4.63 (13.5)

Word size = 64 bit = 8B

Block size = 256B

$$\therefore \frac{\text{Number of words}}{\text{Block}} = \frac{256}{8} = 2^5(32)$$

$$T_{avg} = (0.94 \times 3) + (1 - 0.94)\,[3 + 20 + (31 \times 5)]$$

$$= 13.5 \text{ ns}$$

4.64 (b)

Word length = 32 bit, MM size = 16 MB

Physical address = 24 bit, CM size = 64 KB

4-way set associative

Block size = 256 B

$$\text{Number of lines (N)} = \frac{64\text{ K}}{256} \Rightarrow \frac{2^{16}}{2^8} = 2^8$$

$$\text{Number of sets (S)} = \frac{N}{\text{P-way}} \Rightarrow \frac{2^8}{4} = 2^6$$

Address format:

TAG	No. of set	Word offset
10	6	8
10 bit	$\log_2 2^6$	$\log_2 2^8$

(24 bit)

A1 : $(42C8A4)_H$

4211	001000	A4
	Set offset	

A2 : $(546888)_H$

5401	101000	88
	Set offset	

A3 : $(6A289C)_H$

6A00	101000	9C
	Set offset	

A2 and A3 are in same set

A4 : $(5E4880)_H$

5E01	001000	80
	Set offset	

A1 and A4 are in same set

Hence, option (b) is correct choice.

4.65 (17)

MM size = 2^{32} B

Addressable size = $\log_2 2^{32}$ = 32 bit

Direct mapped

CM size = 32 KB, Block size = 64 B

$$\therefore \text{Number of lines} = \frac{32\text{ K}}{64} \Rightarrow \frac{2^{15}}{2^6} = 2^9$$

Address format:

Tag	LO	WO
17 bit	$\log_2 2^9$ (9 bit)	$\log_2 64 = 6$ bit

(32 bit)

4.67 (2)

Set Associative Map

CM Site = 2 KB, Block Size = 64 B

$$\text{Number of lines} = \frac{2^{11}}{2^6} \Rightarrow 2^5(32)$$

MM Adder = 32 bit, Tag field size = 22 bits

Set associative CM adder format

32-bit

tag	S_o	w_o
22-bit	4-bit	$\log_2 64 = 6$ bit

Set offset (S_o) = 4 bit

$\therefore$ Number of sets $(S) = 2^4(16)$

$$\text{Number of Sets } (S) = \frac{N}{P\text{-way}}$$

$$16 = \frac{32}{P\text{-way}},\ P\text{-way} = \frac{32}{16} = 2$$

4.69 (0.85)

Memory w/o optimization

$H_C = 0.8$

$T_C = 10$ ns

Miss penalty = 100 ns

$T_{avg} = T_C + (1 - H_C)$ miss penalty

$= (10) + (1 - 0.8)\ 100$

$= 30$ ns

Memory with optimization

$T_C = 15$ ns

Miss penalty = 100 ns

$H_C = ?$

$T_{avg} = 28$ ns

$T_{avg} = T_C + (1 - H_C)$ miss penalty

$30 = (15) + (1 - H_C)\ 100$ ns

$H_C = 0.85$

4.70 (a, b, d)

- Direct cm

CM size = 2 KB

MM size = 64 KB

Block size = 64 B

$$\text{Number of lines} = \frac{2\text{K}}{64} \Rightarrow \frac{2^{11}}{2^6} = 2^5$$

16 bit

tag	LO	WO
5 bit	5 bit	6 bit

P : (A248)H 1010 0010 0100 1000; line 9

Q : (C284)H 1100 0010 1000 0100; line 10

R : (CA8A)H 1100 1010 1000 1010; line 10

S : (A262)H 1010 0010 0110 0010; line 9

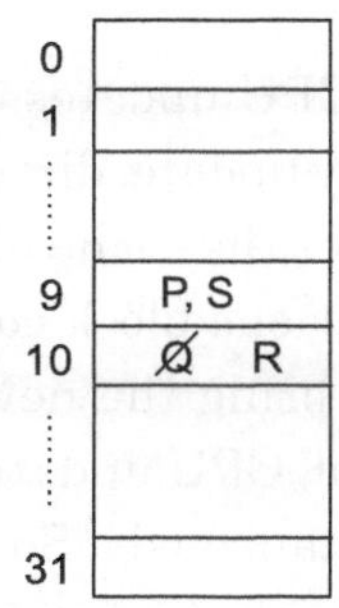

1st	2nd
P : M	P : Hit
Q : M	Q : M
R : M	R : M
S : Hit	S : Hit

4.71 (c)

$A_4\ A_3$

$(0)_{10}$: $[0000\ 0000\ 0000]_2$ → Chip select (CS) → X_1 is enabled

$(8)_{10}$: $[0000\ 0000\ 1000]_2$ → X_2 is enabled

$(16)_{10}$: $[0000\ 0001\ 0000]_2$ → CS → X_3 is enabled

$(24)_{10}$: $[0000\ 0001\ 1000]_2$ → CS → X_4 is enabled

So (0, 8, 16, 24) address enables the starting address of every bank respectively (block).

4.72 (19)

Cache size = 64 KB

Word offset is not given, so take it is 1 B.

$$\text{Number of cache lines} = \frac{2^{16}\text{ B}}{1\text{ B}} = 2^{16}$$

$$\text{Number of sets} = \frac{2^{16}}{\text{P-way}} = \frac{2^{16}}{8} = 2^{13}$$

Given,

Tag	SO	WO
19	13	0

32 bits

4.73 (a, b)

In write back protocol, CPU updates only the cache blocks. So, for every update, dirty bit will be affected. Before replace the cache block due to a miss operation the cache block copy back to main memory before bring the new Block. In write through protocol, CPU updates cache and main memory simultaneously. So, no need to copy back the cache blocks into a main memory before brine the new blocks due to a miss operations.

4.74 (2.56)

Normal cache:

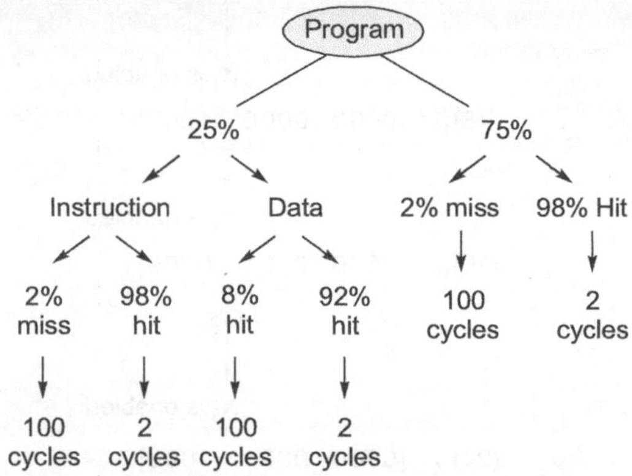

$$ET = (0.25 \times\ 0.02 \times 100) + (0.2 \times 0.98 \times 2) + (0.25 \times 0.08 \times 100) + (0.25 \times 0.92 \times 2) + (0.75 \times 0.02 \times 100) + (0.75 \times 0.98 \times 2)$$

$$ET = 0.5 + 0.49 + 2 + 0.46 + 1.5 + 1.47 = 6.42$$

Perfect cache:

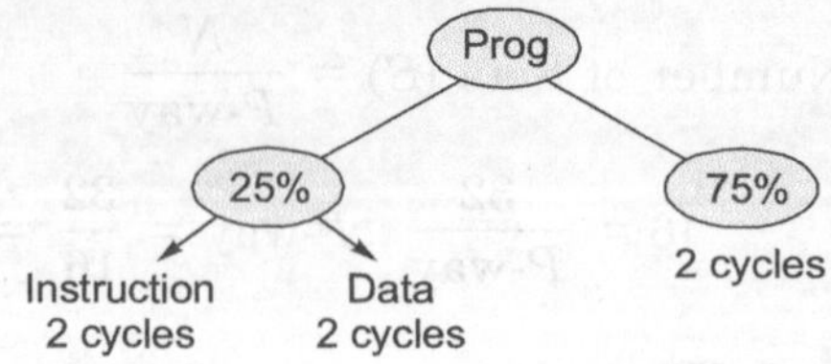

$$ET = 0.25 \times 2 + 0.25 \times 2 + 0.75 \times 2$$
$$= 0.5 + 0.5 + 1.5 = 2.5$$

$$S = \frac{ET_{\text{Normal cache}}}{ET_{\text{Perfect cache}}} = \frac{6.42}{2.5} = 2.56$$

5 IO Organization

5.1 If the disk is rotating at 3600 rpm, determine the effective data transfer rate which is defined as the number of bytes transferred per second between disk and memory. (Given size of track = 512 bytes) **[1995 : 2 M]**

5.2 For the daisy chain scheme of connecting I/O devices, which of the following statements is true?
(a) It gives non-uniform priority to various devices
(b) It gives uniform priority to all devices
(c) It is only useful for connecting slow devices to a processor device
(d) It requires a separate interrupt pin on the processor for each device **[1996 : 1 M]**

5.3 A micro program control unit is required to generate a total of 25 control signals. Assume that during any microinstruction at most two control signals are active. Minimum number of bits required in the control word to generate the required control signal
(a) 2 (b) 25
(c) 10 (d) 12 **[1996 : 2 M]**

5.4 The correct matching for the following pairs is

A. DMA I/O **1.** High speed RAM
B. Cache **2.** Disk
C. Interrupt I/O **3.** Printer
D. Condition code Register **4.** ALU

Codes:
(a) A – 4, B – 3, C – 1, D – 2
(b) A – 2, B – 1, C – 3, D – 4
(c) A – 4, B – 3, C – 2, D – 1
(d) A – 2, B – 3, C – 4, D – 1 **[1997 : 1 M]**

5.5 I/O redirection
(a) implies changing the name of a file
(b) can be employed to use an existing file as input file for a program
(c) implies connection 2 programs through a pipe
(d) None of the above **[1997 : 1 M]**

5.6 Formatting of floppy disk refers to
(a) arranging the data on the disk in contiguous fashion
(b) writing the directory
(c) erasing the system area
(d) writing identification information on all tracks and sectors **[1998 : 2 M]**

5.7 The main differences(s) between a CISC and A RISC processor is/are that a RISC processor typically
(a) has fewer instructions
(b) has fewer addressing modes
(c) has more registers
(d) is easier to implement using hard-wired control logic **[1999 : 2 M]**

5.8 The microinstructions stored in the control memory of a processor have a width of 26 bits. Each microinstruction is divided into three fields: a micro-operation field of 13 bits, a next address field (X), and a MUX select field (Y). There are 8 status bits in the inputs of the MUX.

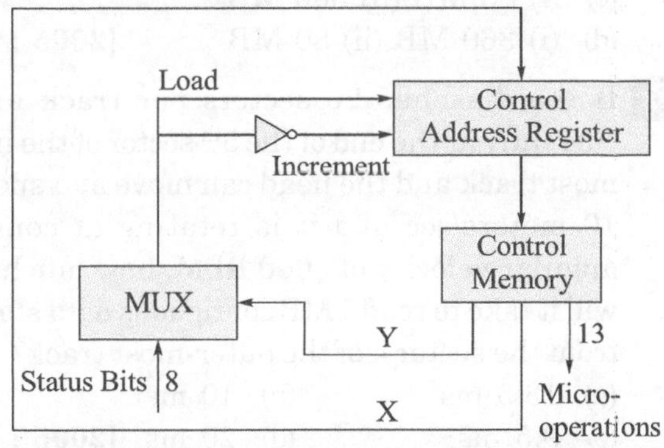

How many bits are there in the X and Y fields, and what is the size of the control memory in number of words?
(a) 10, 3, 1024 (b) 8, 5, 256
(c) 5, 8 , 2048 (d) 10, 3, 512
[2004 : 2 M]

5.9 A hard disk with a transfer rate of 10 M bytes/second is constantly transferring data to memory using DMA. The processor runs at 600 MHz, and takes 300 and 900 clock cycles to initiate and complete DMA transfer respectively. If the size of the transfer is 20 Kbytes, what is the percentage of processor time consumed for the transfer operation?
(a) 5.0% (b) 1.0%
(c) 0.5% (d) 0.1% **[2004 : 2 M]**

5.10 The storage area of a disk has innermost diameter of 10 cm and outermost diameter of 20 cm. The maximum storage density of the disk is 1400bits/cm. The disk rotates at a speed of 4200 RPM. The main memory of a computer has 64-bit word length and 1μs cycle time. If cycle stealing is used for data transfer from the disk, the percentage of memory cycles stolen for transferring one word is
(a) 0.5% (b) 1%
(c) 5% (d) 10% **[2004 : 2 M]**

Linked Answer Questions 5.11 and 5.12:

A disk has 8 equidistant tracks. The diameters of the innermost and outermost tracks are 1 cm and 8 cm respectively. The innermost track has a storage capacity of 10 MB.

5.11 What is the total amount of data that can be stored on the disk if it is used with a drive that rotates it with (i) Constant Linear Velocity (ii) Constant Angular Velocity?
(a) (i) 80 MB (ii) 2040 MB
(b) (i) 2040 MB (ii) 80 MB
(c) (i) 80 MB (ii) 360 MB
(d) (i) 360 MB (ii) 80 MB **[2005 : 2 M]**

5.12 If the disk has 20 sectors per track and is currently at the end of the 5th sector of the inner-most track and the head can move at a speed of 10 meters/sec and it is rotating at constant angular velocity of 6000 RPM, how much time will it take to read 1 MB contiguous data starting from the sector 4 of the outer-most track?
(a) 13.5 ms (b) 10 ms
(c) 9.5 ms (d) 20 ms **[2005 : 2 M]**

5.13 What is the swap space in the disk used for?
(a) Saving temporary html pages
(b) Saving process data
(c) Storing the superblock
(d) Storing device drivers
[2005 : 1 M]

5.14 A device with data transfer rate 10 KB/sec is connected to a CPU. Data is transferred byte-wise. Let the interrupt overhead be 4 μsec. The byte transfer time between the device interface register and CPU or memory is negligible. What is the minimum performance gain of operating the device under interrupt mode over operating it under program controlled mode?
(a) 15 (b) 25
(c) 35 (d) 45 **[2005 : 2 M]**

5.15 Consider a disk drive with the following specifications.
16 surfaces, 512 tracks/surface, 512 sectors/track, 1 KB/sector, rotation speed 3000 rpm. The disk is operated in cycle stealing mode whereby whenever one byte word is ready it is sent to memory; similarly, for writing, the disk interface reads a 4 byte word from the memory in each DMA cycle. Memory cycle time is 40 nsec. The maximum percentage of time that the CPU gets blocked during DMA operation is
(a) 10 (b) 25
(c) 40 (d) 50 **[2005 : 2 M]**

5.16 Which of the following DMA transfer modes and interrupt handling mechanisms will enable the highest I/O band-width?
(a) Transparent DMA and Polling interrupts
(b) Cycle-stealing and Vectored interrupts
(c) Block transfer and Vectored interrupts
(d) Block transfer and Polling interrupts
[2006 : 1 M]

5.17 Which of the following systems is a most likely candidate example of a pipe and filter architecture?
(a) Expert system
(b) DB repository
(c) Aircraft flight controller
(d) Signal processing
[2007 : 1 M]

5.18 Consider a disk pack with 16 surfaces, 128 tracks per surface and 256 sectors per track. 512 bytes of data are stored in a bit serial manner in a sector. The capacity of the disk pack and the number of bits required to specify a particular sector in the disk are respectively
(a) 256 Mbyte, 19 bits (b) 256 Mbyte, 28 bits
(c) 512 Mbyte, 20 bits (d) 64 Gbyte, 28 bits
[2007 : 1 M]

5.19 Consider a Direct Mapped Cache with 8 cache blocks (numbered 0-7). If the memory block requests are in the following order 3, 5, 2, 8, 0, 63, 9,16, 20, 17, 25, 18, 30, 24, 2, 63, 5, 82,17, 24. Which of the following memory blocks will not be in the cache at the end of the sequence?
(a) 3 (b) 18
(c) 20 (d) 30
[2007 : 2 M]

5.20 A hard disk system has the following parameters:
Number of tracks = 500
Number of sectors/track = 100
Number of bytes /sector = 500

Time taken by the head to move from one track to adjacent track = 1 ms Rotation speed = 600 rpm. What is the average time taken for transferring 250 bytes from the disk?

(a) 300.5 ms (b) 255.5 ms
(c) 255 ms (d) 300 ms

[2007 : 2 M]

5.21 For a magnetic disk with concentric circular tracks, the latency is not linearly proportional to the seek distance due to

(a) non-uniform distribution of requests
(b) arm starting and stopping inertia
(c) higher capacity of tracks on the periphery of the platter
(d) use of unfair arm scheduling policies.

[2008 : 2 M]

Common Data for Q. 5.22 and Q.5.23:

A hard disk has 63 sectors per track, 10 platters each with 2 recording surfaces and 1000 cylinders. The address of a sector is given as a triple $\langle c, h, s \rangle$, where c is the cylinder number, h is the surface number and s is the sector number. Thus, the 0th sector is addressed as $\langle 0, 0, 0 \rangle$, the 1st sector as $\langle 0, 0, 1 \rangle$, and so on.

5.22 The address $\langle 400, 16, 29 \rangle$ corresponds to sector number:

(a) 505035 (b) 505036
(c) 505037 (d) 505038

[2009 : 2 M]

5.23 The address of 1039th sector is

(a) $\langle 0, 15, 31 \rangle$ (b) $\langle 0, 16, 30 \rangle$
(c) $\langle 0, 16, 31 \rangle$ (d) $\langle 0, 17, 31 \rangle$

[2009 : 2 M]

5.24 An application loads 100 libraries at start-up. Loading each library requires exactly one disk access. The seek time of the disk to a random location is given as 10 ms. Rotational speed of disk is 6000 rpm. If all 100 libraries are loaded from random locations on the disk, how long does it take to load all libraries? (The time to transfer data from the disk block once the head has been positioned at the start of the block may be neglected.)

(a) 0.50 s (b) 1.50 s
(c) 1.25 s (d) 1.00 s

[2011 : 2 M]

5.25 On a non-pipelined sequential processor, a program segment, which is a part of the interrupt service routine, is given to transfer 500 bytes from an I/O device to memory.

Initialize the address register
Initialize the count to 500
LOOP: Load a byte from device
Store in memory at address given by address register
Increment the address register
Decrement the court
If count! = 0 go to LOOP

Assume that each statement in this program is equivalent to a machine instruction which takes one clock cycle to execute if it is a non-load/store instruction. The load-store instructions take two clock cycles to execute.

The designer of the system also has an alternate approach of using the DMA controller to implement the same transfer. The DMcontroller requires 20 clock cycles for initialization and other overhead. Each DMA transfer cycle takes two clock cycles to transfer one byte of data from interrupt driven program based input-output?

(a) 3.4 (b) 4.4
(c) 5.1 (d) 6.7 **[2011 : 2 M]**

5.26 Consider a hard disk with 16 recording surfaces (0-15) having 16384 cylinders (0-16383) and each cylinder contain 64 sectors (0-63). Data storage capacity in each sector is 512 bytes. Data are organized cylinder-wise and addressing format is <cylinder no., surface no., sector no>. A file of size 42797 KB is stored in the disk and the starting disk location of the file is < 1200, 9, 40>. What is the cylinder number of the last sector of the file, if it is stored in a contiguous manner?

(a) 1281 (b) 1282
(c) 1283 (d) 1284 **[2013 : 2 M]**

5.27 Consider a disk pack with a seek time of 4 milliseconds and rotational speed of 10000 rotations per minute (RPM). It has 600 sectors per track and each sector can store 512 bytes of data. Consider a file stored in the disk. The file contains 2000 sectors. Assume that every sector access necessitates a seek, and the average rotational latency for accessing each sector is half of the time for one complete rotation. The total time (in milliseconds) needed to read the entire file is _______.

[2015 (Set-1) : 2 M]

5.28 Consider a typical disk that rotates at 15000 rotations per minute (RPM) and has a transfer rate of 50×10^6 bytes/sec. If the average seek time of the disk is twice the average rotational delay and the controller's transfer time is 10 times the disk transfer time, the average time (in milliseconds) to read or write a 512 byte sector of the disk is ______.

[2015 (Set-2) : 2 M]

5.29 The size of the data count register of a DMA controller is 16 bits. The processor needs to transfer a file of 29,154 kilobytes from disk to main memory. The memory is byte addressable. The minimum number of times the DMA controller needs to get the control of the system bus from the processor to transfer the file from the disk to main memory is ______.

[2016 (Set-1) : 2 M]

5.30 Consider the following statements:

I. Daisy chaining is used to assign priorities in attending interrupts.

II. When a device raises a vectored interrupt, the CPU does polling to identify the source of interrupt.

III. In polling, the CPU periodically checks the status bits to know if any device needs its attention.

IV. During DMA, both the CPU and DMA controller can be bus masters at the same time.

Which of the above statements is/are TRUE?

(a) I and III only (b) I and II only
(c) III only (d) I and IV only

[2020 : 1 M]

5.31 Consider a computer system with DMA support. The DMA module is transferring one 8-bit character in one CPU cycle from a device to memory through cycle stealing at regular intervals. Consider a 2 MHz processor. If 0.5% processor cycles are used for DMA, the data transfer rate of the device is ____ bits per second.

[2021 (Set 2) : 1 M]

5.32 Which one of the following facilitates transfer of bulk data from hard disk to main memory with the highest throughput?

(a) DMA based I/O transfer
(b) Interrupt driven I/O transfer
(c) Polling based I/O transfer
(d) Programmed I/O transfer

[2022 : 1 M]

5.33 Which one of the following statements is FALSE?

(a) In the cycle stealing mode of DMA, one word of data is transferred between an I/O device and main memory in a stolen cycle
(b) The CPU can start executing an interrupt service routine faster with vectored interrupts than with non-vectored interrupts
(c) Programmed I/O mechanism has a better CPU utilization than the interrupt driven I/O mechanism
(d) For bulk data transfer, the burst mode of DMA has a higher throughput than the cycle stealing mode

[2024 (Set-1) : 1 M]

5.34 Consider a computer with a 4 MHz processor. Its DMA controller can transfer 8 bytes in 1 cycle from a device to main memory through cycle stealing at regular intervals. Which one of the following is the data transfer rate (in bits per second) of the DMA controller if 1% of the processor cycles are used for DMA?

(a) 3,200 (b) 25,60,000
(c) 32,000 (d) 2,56,000

[2024 (Set-2) : 1 M]

■■■■

Answers IO Organization

5.1	(30)	**5.2**	(a)	**5.3**	(c)	**5.4**	(b)	**5.5**	(c)	**5.6**	(d)	**5.7**	Sol.	**5.8**	(a)	**5.9**	(d)
5.10	(c)	**5.11**	(d)	**5.12**	(a)	**5.13**	(b)	**5.14**	(b)	**5.15**	(b)	**5.16**	(c)	**5.17**	(d)	**5.18**	(a)
5.19	(b)	**5.20**	(d)	**5.21**	(c)	**5.22**	(c)	**5.23**	(c)	**5.24**	(b)	**5.25**	(a)	**5.26**	(d)	**5.27**	(14020)
5.28	(6.1)	**5.29**	(456)	**5.30**	(a)	**5.31**	(80000)			**5.32**	(a)	**5.33**	(c)	**5.34**	(b)		

Explanations IO Organization

5.1 (30)

3600 Rotation — 60 sec

$$1 \text{ Rotation} — \frac{60}{3600} = \frac{1}{60} \text{ sec}$$

$$\text{Transfer rate} = 60 \times (\text{size of track})/\text{sec.}$$
$$= 60 \times 512 \text{ B/sec} = 30 \text{ Kbps}$$

5.2 (a)

The daisy-chinning method of establishing priority consists of a serial connection of all devices that request an interrupt. The device with the highest priority is placed in the first position, followed by lower-priority devices up to the device with the lowest priority, which is placed last in the chain. The farther the device is from the first position, the lower is its priority. Therefore daisy-chain gives non-uniform priority to various devices.

5.3 (c)

5 bits	5 bits

5 bits to identify first and 5 bits to identify second including the case when one of them not present. So total bits required = 10

5.4 (b)

DMA I/O — Disk
Cache — High speed RAM
Interrupt I/O — Printer
Condition code register – ALU

5.5 (c)

I/O redirection implies connection two programs through a pipe.

5.6 (d)

Formatting of floppy disk refers to writing identification information on all tracks and sectors.

5.7 Sol.

(a), (b), (c), (d)

The major characteristics of a RISC processor are:

1. Relatively few instructions.
2. Relatively few addressing modes.
3. More registers.
4. Hardwired rather than micro-programmed control.

5.8 (a)

Total Size of micro instruction = 26 bits
Size of micro-operation = 13 bits
Mux input = 8 bit
So select line field size Y $\Rightarrow 8 = 2^3 \Rightarrow 3$ bits
Next address field size = 13 – 3 = 10 bits
Size of control memory = $2^{10} = 1024$

5.9 (d)

$$\text{Data transfer rate} = 10\text{M bytes/sec}$$
$$= 10 \times 2^{10} \text{ K byte/sec}$$
$$\text{Size of transfer} = 20 \text{ Kbytes}$$
$$10 \times 2^{10} \times x\% = 20$$
$$10 \times 2^{10} \times x/100 = 20$$
$$x = \frac{20 \times 100}{10 \times 2^{10}}$$
$$x = \frac{200}{1024} = 0.19\%$$
$$\text{or} \quad x = 0.10\%$$

5.10 (c)

$$\text{The capacity of track} = \pi \times d \times \text{Density}$$
$$= 3.14 \times 10 \times 1400 \text{ bits} = 14000\pi \text{ bits}$$
$$4200 \text{ rotation} = 60 \text{ sec}$$
$$\text{So, 1 rotation time} = \frac{60}{4200} = \frac{1}{70} \text{ sec}$$

DMA controller transfer 64 bits = 1 μsec
In 1 sec = 64×10^6 bits.

In 1 sec the disk can read $= \dfrac{14000\pi}{\dfrac{1}{70}}$

$= 3.08 \times 10^6$ bits

Total memory cycle stolen $= \dfrac{3.08\times10^6}{64\times10^6}\% = 5\%$

5.11 (d)

In Constant Linear Velocity, the density of bits is increases from outer most cylinder to inner most cylinder. So, Constant Linear Velocity = 10 + 20 + 30 + 40 +...80 = 360 MB
In Constant Angular Velocity, the disk rotates at a constant angular speed, with the bit density decreasing on outer cylinders. So, Constant Linear Velocity = Number of track × Capacity of each track

= 8 × 10 MB = 80 MB

5.12 (a)

Diameter of inner track = 1 cm

So radius of inner track $= \dfrac{1}{2} = 0.5$ cm

Diameter of outer track = 8 cm

So radius of outer track $= \dfrac{8}{2} = 4$ cm

So the header has to seek (4 – 0.5) = 3.5 cm.
Seek time for header = Seek distance × Speed of header

= 3.5 cm × 10 m/s = 3.5 ms

600 rotation = 60 sec

So, 1 rotation time $= \dfrac{60}{600} = 10$ ms

i.e. 20 sector reading 10 ms.
Angular velocity is constant and header is now end of 5th sector. To keep the head at the beginning of 4th sector it need to rotate 18 sectors.
20 sector reading = 10 ms
18 sector required = 9 ms
In 1 rotation, we can transfer the whole track.
In 10 MB = 10 ms

For 1 MB $= \dfrac{10\times1}{10} = 1$ ms

Total time required = Time to reach inner most track to outer most track + Time required to reach at desired sector of outer most track + Data transfer time

= 3.5 ms + 9 ms + 1 ms = 13.5 ms

5.13 (b)

The swap space is basically used for saving process data. Consider CPU contains two process P_1 and P_2 if P_1 is running on the CPU then P_2 is swap out and all data (PC for process, content etc) are saved in the disk, when CPU executes process P_2 then all data items of P_1 saved on a disk.

5.14 (b)

$ET_{prog\text{-}IO}$: 10 kB ⇒ 1 sec

1 B ⇒ ? sec

$ET_{prog\text{-}IO} = 100\ \mu sec$

$ET_{INT\text{-}IO} = 4\ \mu sec$

$S = \dfrac{100}{4} = 25$

5.15 (b)

Revolution Per minute = 3000 RPM

or $\dfrac{3000}{60} = 50$RPS

In one track rotation it can read = 512 KB
In 50 RPS it can read = 512 × 50

For 1 byte read $= \dfrac{1}{(512\times50)} = 39.06$ ns

⇒ For 4 bytes it takes 156 ns
Percentage of time that the CPU gets blocked during DMA operation $= \dfrac{40}{156}\times100 \simeq 25\%$.

5.16 (c)

In block transfer mode, DMA transfer the entire block of data then gives control of bus to CPU.
In vectored Interrupts, I/O device send vector address of Interrupt Service routine along with interrupts which guides CPU to execute for a specific I/O device.
Hence high bandwidth will be required.

5.17 (d)

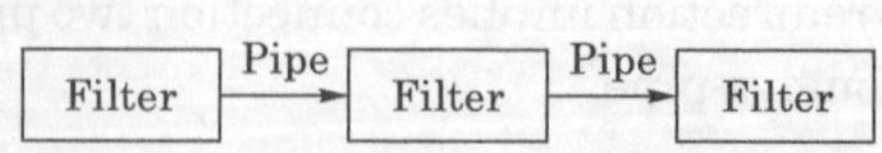

Pipe is communication channel carring a signal messages, and filters are signal processing componants such as amplifiers noise filters etc.

5.18 (a)

No. of surfaces = 16
No. of tracks/sector = 128
No. of sector/track = 256

Total size of disk = $16 \times 128 \times 256 \times 512$ bytes
$= 2^4 \times 2^7 \times 2^8 \times 2^9$ bytes
$= 2^8 \times 2^{20}$ bytes $= 2^8$ Mega bytes
$=$ 256 MB

Total Number of Sector in the disk
$= 16 \times 128 \times 256$ bytes
$= 2^4 \times 2^7 \times 2^8$ bytes $= 2^{19}$ bytes

So 19 bits are needed.

5.19 (b)

In direct mapped cache main memory block location in cache memory will be decide using below formula:
Cache location = (main memory block) % (number of cache blocks).

	Cache memory
0	~~8~~ ~~0~~ ~~16~~ 24
1	~~9~~ ~~17~~ ~~25~~ 17
2	~~2~~ ~~18~~ ~~2~~ 82
3	3
4	20
5	5
6	30
7	63

Cache memory

3%8 = 3	16%8 = 0	2%8 = 2
5%8 = 5	20%8 = 4	63%8 = 7
2%8 = 2	17%8 = 1	5%8 = 5
8%8 = 0	25%8 = 1	82%8 = 2
0%8 = 0	18%8 = 2	17%8 = 1
63%8 = 7	30%8 = 6	24%8 = 0
9%8 = 1	24%8 = 0	

So memory block 18 will not be present in the cache memory.

5.20 (d)

600 rotation = 60 sec

So, 1 rotation time $= \frac{60}{600} = 0.1$ sec

Rotational time $= \frac{\text{Rotation latency}}{2}$

$= \frac{0.1}{2} = 50$ ms

Capacity of track = Number of sector /track × Number of bytes/sector
$= 100 \times 500 = 50{,}000$ bytes

In 1 rotation, we can transfer the whole track.
In 50,000 bytes = 0.1 sec

For 250 bytes $= \frac{0.1 \times 250}{50000} = 0.5$ ms

Average seek time $= \frac{0+1+2+3...+499}{500}$

$= \frac{499 \times 250}{500} = 249.5$ ms

Average time to transfer = Average seek time + Average rotational delay + Average data transfer time for transferring 250 bytes
$= 249.5 + 50 + 0.5 = 300$ ms

5.21 (c)

In magnetic disk once head is in position, the desired cell may be in wrong part due to a high capacity of tracks. Some time is required for this cell to reach the read-write head so that data transfer can begin. The average time for this movement to take place is seek latency which is not linearly proportional to the seek distance because a higher capacity of tracks on the periphery of the platter.

5.22 (c)

The address <400, 16, 29> corresponds to sector no.
$400 \times 2 \times 10 \times 63 + 16 \times 63 + 29 = 505037$

5.23 (c)

The address <0, 16, 31> corresponds to sector no.
$16 \times 63 + 31 = 1039$

5.24 (b)

Rotational speed = 6000 rpm

Average latency $= \left(\frac{60\,\text{sec}}{6000}\right)/2 = 5$msec

Avg access time = Avg latency + Avg seek time
= 5 ms + 10 ms = 15 msec

Time to load 1 library is 15 msec
Time to load 100 libraries
= 100 × 15 msec =1.5 sec

5.25 (a)

Speedup = Execution time of the program in Interrupt driven IO/Execution time of the program in DMA.

Execution time of Interrupt Driven IO:
The program consists of 7 Instructions. The first 2 Instructions are in out of LOOP. So these Instructions takes 2 clock cycles.

Next 5 Instructions are in the LOOP. This LOOP will be executed upto 500 times. Each time 1 byte will be transferred to memory. We need to transfer 500 bytes from Disk to main memory. In the 5 Instructions 1 LOAD and 1 STORE Instruction present.
According to problem statement these instructions takes 2 clock cycles each. Hence These instructions takes 2000 clock cycles.
The remaining 3 Instruction in the LOOP will takes 1500 clock cycles.
The total time required for the above program in Interrupt driven IO is : 3502 clock cycles.
Execution time of DMA transfer:
According to the problem statement DMA initialization takes 20 clock cycles.
To transfer 1 Byte takes 2 clock cycles.
We need to transfer 500 bytes so the takes required for the complete transfer is :
2 * 500 = 1000 clock cycles.
Hence the total time required in DMA is: 1020 clock cycles
SPEED UP = INTERRUPT DRIVEN IO EXECUTION TIME / DMA EXECUTION TIME

i.e., speedup $= \frac{3502}{1020} = 3.4$

5.26 (d)

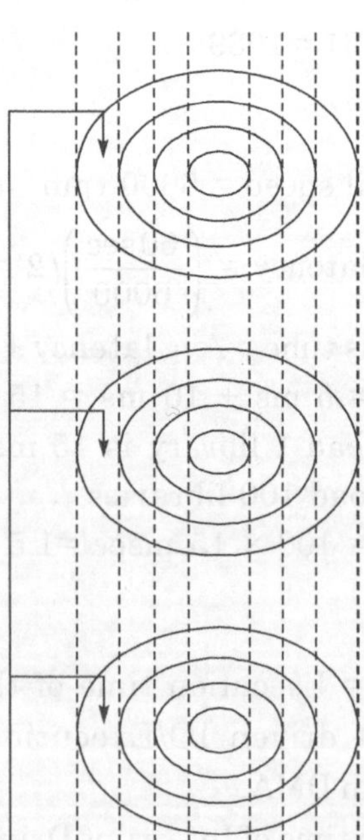

When head is moving then it is reading data from all 16 surface simultaneously.

Currently head is on 9th surface, sector no. 40.
How much data it read from 9th surface

$= 24 \times 512 = 12288$ bytes.

How much data it read from surface 10th to 15th

$= 5 \times 512 \times 64 = 163840$ bytes.

How much data it read from cylinder 1201 to 1283
$82 \times 16 \times 64 \times 512 = 42991616$ bytes.
Total data read from 1200th to 1283th cylinder
$12288 + 163840 + 42991616$

$= 13167744$ bytes $= 42156$ KB

But we need to real 42797 KB. So we need to go on 1284th cylinder.

5.27 (14020)

Seek time = 4 ms

Rotational latency $= \frac{60}{10000}$ sec $= 6$ ms

Average rotational latency = R/2 = 3 ms

For each sector, we require Seek time + Rotational Latency + Transfer time.

In one rotation, 512 B × 600 data is read in 6 ms

One sector of 512 B can read in $\frac{6\text{ms}}{600}$

= 0.01 ms (transfer time for 1 sector)

Total time required for 1 sector

= 4 ms + 3 ms + 0.01 = 7.01

For 2000 sectors 7.01 × 2000 = 14020 ms

5.28 (6.1)

1500 rotations → 1 minute

1 rotation → ?

$$\text{Rotational latency} = \frac{1}{15000} \times 1 \text{ minute}$$

$$= \frac{60\,\text{sec}}{15000} = 4 \text{ msec}$$

(i) Average rotational latency

$$= \frac{1}{2} \times 4 \text{ msec} = 2 \text{ msec}$$

(ii) Average seek time = 2 × Average rotational delay

= 2 × 2 = 4 msec

Transfer rate = 500×10^6 bytes/sec

Sector size = 512 byte

512 byte → ?

(iii) Transfer time = $\frac{512}{50\times10^6} \times 1$ sec

$= 0.01024$ msec

(iv) Controller's transfer time = 10 × Transfer time

$= 10 \times 0.01024 = 0.1024$ msec

∴ Average time = Average rotational delay + Average seek time + Controller's time + Transfer time

$= 2 + 4 + 0.01024 + 0.1024 = 6.11264$

$\cong 6.1$ msec

5.29 (456)

Data count register = 16 bits.

So, count value = 2^{16} = 64 K bytes

One time control, transfer — 64 K bytes

Number of controls to transfer — 29154 K bytes

So, number of time bus control required

$$= \left\lceil \frac{29154}{64} \right\rceil = 456$$

5.30 (a)

I and III are the only true statements.

So option (a) is correct answer.

5.31 (80000)

$y = 0.5$ μsec (Transfer Time)

x = Preparation time

$$\% \text{ time CPU blocked} = \left(\frac{y}{x+y}\right)\times100$$

$$0.5 = \left(\frac{0.5\ \mu\text{sec}}{x+0.5\ \mu\text{sec}}\right)\times100$$

$0.005x + 0.0025 = 0.5$

$0.005x = 0.4975$

$x = 99.5$ μsec

(99.5 μsec + 0.5 μsec) Total time ———— 8 bit

1 sec ———————— ?

$$= \frac{8\text{ bit}}{100\ \mu\text{sec}} = 80000 \text{ bits/sec}$$

5.32 (a)

In DMA, bulk amount of data will be transferred from the secondary memory to main memory without involvement of a CPU.

5.33 (c)

In programmed I/O mode processor utilization is very poor because, I/O operations are programmed in the CPU.

5.34 (b)

$$\% \text{ CPU time blocked} = \left(\frac{Y}{X}\right)100$$

$$1 = \left(\frac{Y}{X}\right)100$$

$$\frac{Y}{X} = 0.01$$

4 MHz CLK given

So, Cycle time $= \frac{1}{4\text{ M}}$ sec ⇒ 0.25 μsec

$$\frac{0.25\ \mu\text{sec}}{X} = 0.01$$

$$X = 25\ \mu\text{sec}$$

25 μsec ⟶ 8 byte (64 bits)

1 sec ⟶ ?

$$\Rightarrow \frac{64\text{ bits}}{25\ \mu\text{sec}}$$

$$\Rightarrow \frac{64}{25}\times10^6 \text{ bits/sec} = 2560000$$

Programming and Data Structures

UNIT

V

CONTENTS

Programming and Data Structures

Syllabus

Programming in C; Functions, Recursion, Parameter passing, Scope, Binding; Abstract data types, Arrays, Stacks, Queues, Linked Lists, Trees, Binary search trees, Binary heaps.

Analysis of Previous GATE Papers

Exam Year	1 Mark Ques.	2 Marks Ques.	Total Marks
1995	1	2	**3**
1996	4	4	**12**
1997	2	2	**6**
1998	1	7	**15**
1999	–	2	**4**
2000	8	4	**16**
2001	–	4	**8**
2002	3	2	**7**
2003	5	7	**19**
2004	9	13	**35**
2005	5	6	**17**
2006	1	7	**15**
2007	2	7	**16**
2008	3	–	**17**
2009	1	4	**9**
2010	1	5	**11**
2011	1	3	**7**
2012	1	4	**9**
2013	–	3	**6**

Exam Year	1 Mark Ques.	2 Marks Ques.	Total Marks
2014 Set-1	3	3	**9**
2014 Set-2	3	3	**9**
2014 Set-3	3	3	**9**
2015 Set-1	3	3	**6**
2015 Set-2	3	2	**4**
2015 Set-3	4	6	**16**
2016 Set-1	3	5	**13**
2016 Set-2	3	6	**15**
2017 Set-1	4	5	**14**
2017 Set-2	3	5	**13**
2018	4	5	**14**
2019	2	5	**12**
2020	3	2	**7**
2021 Set-1	3	4	**11**
2021 Set-2	3	2	**7**
2022	4	3	**10**
2023	3	4	**11**
2024 Set-1	3	3	**9**
2024 Set-2	2	4	**10**

1 Programming

1.1 The correct matching for the following pairs is

List-I	List-II
A. Activation record	**1.** Linking loader
B. Location counter	**2.** Garbage collection
C. Reference counts	**3.** Subroutine call
D. Address relocation	**4.** Assembler

Codes:
(a) A - 3, B - 4, C - 1, D - 2
(b) A - 4, B - 3, C - 1, D - 2
(c) A - 4, B - 3, C - 2, D - 1
(d) A - 3, B - 4, C - 2, D - 1 **[1996 : 2 M]**

1.2 Heap allocation is required for languages
(a) that support recursion
(b) that support dynamic data structures
(c) that use dynamic scope rules
(d) None of the above **[1997 : 1 M]**

1.3 Faster access to non-local variables is achieved using an array of pointers to activation records called a
(a) stack (b) heap
(c) display (d) activation tree
[1998 : 2 M]

1.4 Consider the following program in a language that has dynamic scoping:

```
var x: real;
procedure show:
        begin print(x);end;
procedure small;
        var x: real;
        begin x: = 0.125; show; end;
        begin
                x: = 0.25;
                show; small
        end.
```

Then the output of the program is:
(a) 0.125 0.125
(b) 0.25 0.25
(c) 0.25 0.125
(d) 0.125 0.25
[1999 : 2 M]

1.5 A certain processor supports only the immediate and the direct addressing modes. Which of the following programming language features cannot be implemented on this processor?
(a) Pointers
(b) Arrays
(c) Records
(d) Recursive procedures with local variable
[1999 : 2 M]

1.6 Consider the following C function definition:

```
int Trial (int a, int b, int c)
{
    if ((a > = b) && (c < b) return b;
        else if (a > = b) return Trial (a,c,b);
        else return Trial (b,a,c);
}
```

The function Trial:
(a) Finds the maximum of a, b and c
(b) Finds the minimum of a, b and c
(c) Finds the middle number of a, b and c
(d) None of the above
[1999 : 2 M]

1.7 The most appropriate matching for the following pairs:

List-I
X. Indirect addressing
Y. Immediate addressing
Z. Auto decrement addressing

List-II
1. Loops
2. Pointers
3. Constants

(a) X – 3 Y – 2 Z – 1
(b) X – 1 Y – 3 Z – 2
(c) X – 2 Y – 3 Z – 1
(d) X – 3 Y – 1 Z – 2
[2000 : 1 M]

1.8 The following C declarations:

```
struct node{
    int i:
  float j;
};
struct node *s[10];
```

define s to be
(a) An array, each element of which is a pointer to a structure of type node

(b) A structure of 2 fields, each field being a pointer to an array of 10 elements
(c) A structure of 3 fields: an integer, a float, and an array of 10 elements
(d) An array, each element of which is a structure of type node

[2000 : 1 M]

1.9 The most appropriate matching for the following pairs

List-I

X. m=malloc(5); m = NULL;

Y. free(n); n->value=5;

Z. char *p; *p='a';

List II

1. using dangling pointers

2. using uninitialized pointers

3. lost memory

Codes:

(a) X – 1 Y – 3 Z – 2
(b) X – 2 Y – 1 Z – 3
(c) X – 3 Y – 2 Z – 1
(d) X – 3 Y – 1 Z – 2

[2000 : 1 M]

1.10 Aliasing in the context of programming languages refers to
(a) multiple variables having the same memory location
(b) multiple variables having the same value
(c) multiple variables having the same identifier
(d) multiple uses of the same variable

[2000 : 1 M]

1.11 Consider the following C declaration:

```
struct {
    short s[5];
    union {
    float y;
    long z;
    } u;
} t;
```

Assume that objects of the type short, float and long occupy 2 bytes, 4 bytes and 8 bytes, respectively. The memory requirement for variable t, ignoring alignment considerations, is

(a) 22 bytes (b) 14 bytes
(c) 18 bytes (d) 10 bytes

[2000 : 1 M]

1.12 The value of j at the end of the execution of the following C program is _______.

```
int incr (int i) {
    static int count = 0;
    count = count + i;
    return (count);
}
main () {
    int i, j;
    for (i = 0; i <=4; i++)
    j = incr(i);
}
```

(a) 10 (b) 4
(c) 6 (d) 7

[2000 : 2 M]

1.13 What is printed by the print statements in the program P1 assuming call by reference parameter passing?

```
Program P1()
{
    x = 10;
    y = 3;
    func1(y, x, x);
    print x;
    print y;
}
func1(x, y, z)
{
    y = y + 4;
    z = x + y + z;
}
```

(a) 10, 3 (b) 31, 3
(c) 27, 7 (d) None of these

[2001 : 2 M]

1.14 Consider the following three C functions:

```
[P1]    int *g (void)
        {
            int x = 10;
            return (&x);
        }
[P2]    int *g (void)
        {
            int *px;
            *px = 10;
            return px;
        }
[P3]    int *g(void)
        {
            int *px;
                px = (int *)malloc (size of (int));
            *px = 10;
            return px;
        }
```

Which of the above three functions are likely to cause problems with pointers?

(a) Only P3 (b) Only P1 and P3
(c) Only P1 and P2 (d) P1, P2 and P3

[2001 : 2 M]

1.15 Consider the following program:

```
Program P2
    var n:int;
    procedure W(var x:int)
    begin
        x = x + 1;
        print x;
    end
    procedure D
    begin
        var n:int;
        n = 3;
        W(n);
    end
    begin // begin P2
        n = 10;
        D;
    end
```

If the language has dynamic scoping and parameters are passed by reference, what will be printed by the program?

(a) 10 (b) 11
(c) 3 (d) None of these

[2001 : 2 M]

1.16 The results returned by function under value-result and reference parameter passing conventions

(a) Do not differ
(b) Differ in the presence of loops
(c) Differ in all cases
(d) May differ in the presence of exception

[2002 : 1 M]

1.17 In the C language

(a) at most one activation record exists between the current activation record and the activation record for the main
(b) the number of activation records between the current activation record and the activation record for the main depends on the actual function calling sequence
(c) The visibility of global variables depends on the actual function calling sequence
(d) Recursion requires the activation record for the recursive function to be saved on a different stack before the recursive function can be called.

[2002 : 1 M]

1.18 Consider the following declaration of a two-dimensional array in C

char a[100][100];

Assuming that the main memory is byte-addressable and that the array is stored starting from memory address 0, the address of a [40] [50] is

(a) 4040 (b) 4050
(c) 5040 (d) 5050 [2002 : 2 M]

1.19 Assume the following C variable declaration

int*A[10], B[10][10];

Of the following expressions

1. A[2]
2. A[2] [3]
3. B[1]
4. B[2] [3]

Which will not give compile-time errors if used as left hand sides of assignment statements in a C program?

(a) 1, 2 and 4 only (b) 2, 3 and 4 only
(c) 2 and 4 only (d) 4 only [2003 : 1 M]

1.20 In the following C program fragment j, k, n and TwoLog_n are integer variables, and A is an array of integers. The variable n is intialized to an integer ≥ 3, and TwoLog_n is initialized to the value of $2 * \lfloor \log_2(n) \rfloor$

```
for (k = 3; k < = n; k++)
    A [k] = 0;
for (k = 2; k < = TwoLog_n; k++)
    for (j = k + 1; j < = n; j++)
    A [j] = A [j] || (j%k);
for (j = 3; j < = n; j++)
    if (!A[j]) print f ("%d", j);
```

The set of numbers printed by this program fragment is

(a) $\{m \mid m \le n, (\exists i)\ [m = i!]\}$
(b) $\{m \mid m \le n, (\exists i)\ [m = i^2]\}$
(c) $\{m \mid m \le n, m \text{ is prime}\}$
(d) { } [2003 : 2 M]

1.21 Consider the C program shown below:

```
# include <stdio.h>
# define print (x) printf ("%d", x)
int x;
void Q (int z) {
        z + = x; print (z);

}
void p (int *y) {
        int x = *y + 2;
        Q (x); *y = x – 1;
        print (x);
```

```
}
main (void) {
        x = 5;
        p (&x);
        print (x);
}
```

The output of this program is

(a) 12 7 6 (b) 22 12 11
(c) 14 6 6 (d) 7 6 6 **[2003 : 2 M]**

1.22 The goal of structured programming is to
(a) have well indented programs
(b) be able to infer the flow of control from the compiled code
(c) be able to infer the flow of control from the program text
(d) avoid the use of GOTO statements

[2004 : 1 M]

1.23 Consider the following C function:

```
void swap (int a, int b)
    { int temp ;
    temp = a ;
    a = b ;
    b = temp ;
    }
```

In order to exchange the values of two variables x and y.
(a) call swap (x, y)
(b) call swap (&x, &y)
(c) swap (x, y) cannot be used as it does not return any value
(d) swap (x, y) cannot be used as the parameters are passed by value **[2004 : 1 M]**

1.24 Let x be an integer which can take a value of 0 or 1. The statement

```
  if (x = = 0) x = 1;
  else x = 0;
```

is equivalent to which one of the following?

(a) x = 1 + x; (b) x = 1 – x;
(c) x = x – 1; (d) x = 1% x;

[2004 : 1 M]

1.25 Consider the following C program which is supposed to compute the transpose of a given 4 × 4 matrix M. Note that, there is an X in the program which indicates some missing statements. Choose the correct option to replace X in the program.

```
#include <stdio.h>
#define ROW 4
#define COL 4
int M[ROW][COL] = {1, 2, 3, 4, 5, 6, 7, 8, 9, 10,
11, 12, 13, 14, 15, 16};
main( )
{
    int i, j, t;
    for (i = 0; i < 4; ++i)
    {
        X
    }
    for (i = 0; i < 4; ++i)
        for (j = 0; j < 4; ++j)
            printf ("%d", M[i][j]);
}
```

(a)
```
for(j = 0; j < 4; ++j)
{
    t = M[i][j];
    M[i][j] = M[j][i];
    M[j][i] = t;
}
```

(b)
```
for(j = 0; j < 4; ++j)
{
    M[i][j] = t;
    t = M[j][i];
    M[j][i] = M[i][j];
}
```

(c)
```
for(j = i; j < 4; ++j)
{
    t = M[j][i];
    M[i][j] = M[j][i];
    M[j][i] = t;
}
```

(d)
```
for(j = i; j < 4; ++j)
{
    M[i][j] = t;
    t = M[j][i];
    M[j][i] = M[i][j];
}
```

[2004 : 2 M]

1.26 What is the output of the following program?

```
#include <stdio.h>
int funcf (int x);
int funcg (int y);
main()
{
    int x = 5, y = 10, count;
    for(count = 1; count < = 2; ++count)
    {
        y + = funcf(x) + funcg(x);
        printf ("%d", y);
    }
}
funcf(int x)
{
```

```
    int y;
    y = funcg(x);
    return (y);
}
funcg(int x)
{
    static int y = 10;
    y + = 1;
    return (y + x);
}
```

(a) 43 80
(b) 42 74
(c) 33 37
(d) 32 32 [2004 : 2 M]

1.27 Consider the following C program:

```
#include <stdio.h>
typedef struct
{
    char *a;
    char *b;
} t;
void f1(t s);
void f2(t *p);
main( )
{
    static t s = {"A", "B"};
    printf("%s %s\n", s.a, s.b);
    f1(s);
    printf("%s %s\n", s.a, s.b);
    f2(&s);
}
void f1(t s)
{
    s.a = "U";
    s.b = "V"
    printf(%s %s\n", s.a, s.b);
    return;
}
    void f2(t *p)
{
    p → a = "V";
    p → b = "W"
    printf("%s %s\n", p → a, p → b);
    return;
}
```

What is the output generated by the program?

(a) A B
U V
V W
V W

(b) A B
U V
A B
V W

(c) A B
U V
U V
V W

(d) A B
U V
V W
U V [2004 : 2 M

1.28 Consider the following C function:

```
int f (int n)
{       static int i = 1 ;
        if (n > = 5) return n ;
        n = n + i ;
        i ++;
        return f (n) ;
}
```

The value returned by f(1) is

(a) 5
(b) 6
(c) 7
(d) 8 [2004 : 2 M]

1.29 Consider the following C program:

```
main()
{
    int x, y, m, n;
    scanf ("%d %d", &x  &y);
    / * Assume x > 0 and y > 0* /
    m = x; n = y;
    while(m! = n)
    {
      if (m > n)
         m = m – n;
      else
         n = n – m;
    }
    printf ("% d", n);
}
```

The program computes

(a) x ÷ y, using repeated subtraction
(b) x mod y using repeated subtraction
(c) the greatest common divisor of x and y
(d) the least common multiple of x and y

[2004 : 2 M]

1.30 Consider the following program fragment for reversing the digits in a given integer to obtain a new integer. Let $n = d_1 d_2 \ldots d_m$.

```
int n, rev ;
rev = 0 ;
while (n > 0) {
        rev = rev * 10 + n % 10 ;
        n = n / 10 ;
}
```

The loop invariant condition at the end of the ith iteration is

(a) $n = d_1 d_2 \ldots d_{m-i}$ and $rev = d_m d_{m-1} \ldots d_{m-i+1}$
(b) $n = d_{m-i+1} \ldots d_{m-1} d_m$ or $rev = d_{m-i} \ldots d_2 d_1$
(c) $n \neq rev$
(d) $n = d_1 d_2 \ldots d_m$ or $rev = d_m \ldots d_2 d_1$

[2004 : 2 M]

1.31 Consider the following C program segment:

```
char p [20] ;
char * s = "string" ;
int length = strlen (s) ;
for (i = 0 ; i < length; i++)
    p[i] = s[length – i];
printf ("%s", p);
```

The output of the program is

(a) gnirts
(b) string
(c) gnirt
(d) no output is printed

[2004 : 2 M]

1.32 What does the following C-statement declare?

```
int (*f) (int *);
```

(a) A function that takes an integer pointer as argument and returns an integer
(b) A function that takes an integer pointer as argument and returns an integer pointer
(c) A pointer to a function that takes an integer pointer as argument and returns an integer
(d) A function that takes an integer pointer as argument returns a function pointer

[2005 : 1 M]

1.33 An Abstract Data Type (ADT) is

(a) same as an abstract class
(b) a data type that cannot be instantiated
(c) a data type for which only the operations defined on it can be used, but none else
(d) all of the above **[2005 : 1 M]**

1.34 Consider the following C-program:

```
double foo (double); /* Line 1*/
  int main () {
  double da, db;
  // input da
  db = foo (da);
  }
double foo (double a){
  return a;
  }
```

The above code complied without any error or warning. If Line 1 is deleted, the above code will show

(a) no compile warning or error
(b) some complier-warning not leading to unitended results
(c) Some complier-warning due to type-mismatch eventually leading to unitended results
(d) Complier errors **[2005 : 2 M]**

1.35 Consider the following C-program:

```
void foo (int n, int sum ) {
      int k = 0, j = 0;
      if (n = = 0) return;
      k = n % 10;
      j = n / 10;
      sum = sum + k;
      foo ( j, sum);
      printf ("% d", k);
      }
int main() {
      int a = 2048, sum = 0;
      foo (a, sum);
      printf ("%d\n", sum);
      }
```

What does the above program print?

(a) 8, 4, 0, 2, 14 (b) 8, 4, 0, 2, 0
(c) 2, 0, 4, 8, 14 (d) 2, 0, 4, 8, 0

[2005 : 2 M]

Linked Answer Questions 1.36 & 1.37:

A sink in a directed graph is a vertex i such that there is an edge from every vertex j ≠ i to i and there is no edge from i to any other vertex. A directed graph G with n vertices is represented by its adjacency matrix A, where A[i][j] = 1 if there is an edge directed from vertex i to j and 0 otherwise. The following algorithm determines whether there is a sink in the graph G.

```
i = 0;
do
{
    j = i + 1;
    while ((j < n) && E1) j++;
    if (j < n) E2;
} while (j < n);
flag = 1;
for (j = 0; j < n; j++)
if ((j! = i) && E3) flag = 0;
if (flag) printf("Sink exists");
else printf ("Sink does not exist");
```

1.36 Choose the correct expressions for E_1 and E_2

(a) E_1 : A[i][j] and E_2 : i = j;
(b) E_1 : !A[i][j] and E_2 : i = j + 1;
(c) E_1: !A[i][j] and E_2 : i = j;
(d) E_1 : A[i][j] and E_2 : i = j + 1;

[2005 : 2 M]

1.37 Choose the correct expression for E_3

(a) (A[i][j] && !A[j][i])
(b) (!A[i][j] && !A[j][i])
(c) (!A[i][j] || A[j][i])
(d) (A[i][j] || !A[j][i]) **[2005 : 2 M]**

1.38 The following C function takes two ASCII strings and determines whether one is an anagram of the other. An anagram of a string s is a string obtained by permuting the letters in s.

```
int anagram (char *a, char *b)
{
    int count [128], j;
    for (j = 0; j < 128; j++)
        count[j] = 0;
    j = 0;
    while (a[j] && b[j])
    {
        A ;
        B ;
    }
    for (j = 0; j < 128; j++)
        if (count [j]) return 0;
    return 1;
}
```

Choose the correct alternative for statements A and B.

(a) A: count [a[j]]++ and
B: count[b[j]]– –

(b) A: count [a[j]]++ and
B: count[b[j]]++

(c) A: count [a[j++]]++ and
B: count[b[j]]– –

(d) A: count [a[j]]++ and
B: count[b[j++]]– – **[2005 : 2 M]**

1.39 What is the output printed by the following program?

```
# include <stdio.h>
int f(int n, int k)
{
    if (n = = 0) return 0;
    else if (n% 2) return f(n/2, 2*k)+k;
    else return f(n/2, 2*k) – k;
}
int main( )
{
    printf("%d", f(20, 1));
    return 0;
}
```

(a) 5 (b) 8
(c) 9 (d) 20 **[2005 : 2 M]**

1.40 Which one of the choices given below would be printed when the following program is executed?

```
#include <stdio.h>
struct test
{
    int i;
    char *c;
} st[ ] = {5, "become", 4, "better", 6, "jungle", 8,
"ancestor", 7, "brother"};
main( )
{
    struct test *p = st;
    p+ = 1;
        ++p → c;
    printf("%s", p++ → c);
    printf("%c", *++p → c);
    printf("%d", p[0].i);
    printf("%s \n", p → c);
```

(a) jungle, n, 8, ncestor
(b) etter, u, 6, ungle
(c) cetter, k, 6, jungle
(d) etter, u, 8, ncestor **[2006 : 2 M]**

1.41 Which one of the choices given below would be printed when the following program is executed?

```
#include <stdio.h>
void swap (int *x, int *y)
{
    static int *temp;
    temp = x;
    x = y;
    y = temp;
}
void printab( )
{
    static int i, a = –3, b = –6;
    i = 0;
    while (i < = 4)
    {
        if ((i++)%2 = = 1) continue;
        a = a + i;
        b = b + i;
    }
    swap (&a, &b);
    printf("a = %d, b = %d\n", a, b);
}
main( )
{
    printab( );
    printab( );
}
```

(a) a = 0, b = 3
a = 0, b = 3

(b) a = 3, b = 0
a = 12, b = 9

(c) a = 3, b = 6
a = 3, b = 6

(d) a = 6, b = 3
a = 15, b = 12 **[2006 : 2 M]**

1.42 Which one of the choices given below would be printed when the following program is executed?

```
#include <stdio.h>
int a1[] = {6, 7, 8, 18, 34, 67};
int a2[]= {23, 56, 28, 29};
int a3[] = {-12, 27, -31};
int *x[] = {a1, a2, a3};
void print(int *a[])
{
    printf("%d,", a[0][2]);
    printf("%d,", *a[2]);
    printf("%d,", *++a[0]);
    printf("%d,", *(++a)[0]);
    printf("%d\n", a[-1][+1]);
}
main()
{
    print(x);
}
```

(a) 8, −12, 7, 23, 8
(b) 8, 8, 7, 23, 7
(c) −12, −12, 27, −31, 23
(d) −12, −12, 27, −31, 56

[2006 : 2 M]

1.43 The following function computes the value of $\binom{m}{n}$ correctly for all legal values m and n ($m \geq 1$, $n \geq 0$ and $m > n$)

```
int func(int m, int n)
{
    if (E) return 1;
    else return(func(m-1, n) + func(m - 1, n - 1));
}
```

In the above function, which of the following is the correct expression for E?
(a) (n = = 0) | | (m = = 1)
(b) (n = = 0) && (m = = 1)
(c) (n = = 0) | | (m = = n)
(d) (n = = 0) && (m = = n)

[2006 : 2 M]

1.44 Consider the following C-function in which a [n] and b [m] are two sorted integer arrays and c [n + m] be another integer array.

```
void xyz (int a[ ], int b[ ], int c[ ]) {
    int i, j, k;
    i = j = k = 0;
    while ((i < n) && (j < m))
        if (a[i] < b[j]) c[k++] = a[i ++];
        else c[k ++] = b[j ++];
}
```

Which of the following condition (s) hold (s) after the termination of the while loop?
(i) $j < m$, $k = n + j - 1$, & a [n −1] < b [j] if i = n
(ii) $i < n$, $k = m + i - 1$, & b [m − 1] ≤ a [i] if j = m
(a) only (i)
(b) only (ii)
(c) either (i) or (ii) but not both
(d) neither (i) nor (ii)

[2006 : 2 M]

1.45 Consider these two functions and two statements S1 and S2 about them.

```
int work1(int* a, int i, int j)
{
    int x=a[i+2];
    a[j]=x+1;
    return a[i+2]-3;
}
```

```
int work2(int* a, int i, int j)
{
    int t1=i+2;
    int t2=a[t1];
    a[j] = t2+1;
    return t2-3;
}
```

S1 : The transformation from work1 to work2 is valid, i.e., for any program state and input arguments, work2 will compute the same output and have the same effect on program state as work1

S2: All the transformations applied to work1 to get work2 will always improve the performance (i.e. reduce CPU time) of work2 compared to work1

(a) S1 is false and S2 is false
(b) S1 is false and S2 is true
(c) S1 is true and S2 is false
(d) S1 is true and S2 is true

[2006 : 2 M]

1.46 Consider this C code to swap two integers and these five statements : the code

```
void swap (int*px, int* py) {
    *px = *px - *py;
    *py = *px + *py;
    *px =*py -*px;
}
```

S1: will generate a compilation error

S2: may generate a segmentation fault at runtime depending on the arguments passed

S3: Correctly implements the swap procedure for all input pointers referring to integers stored in memory locations accessible to the process

S4: implements the swap procedure correctly for some but not all valid input pointers

S5: may add or subtract integers and pointers

(a) S1
(b) S2 and S3
(c) S2 and S4
(d) S2 and S5

[2006 : 2 M]

1.47 Consider the following C function:

```
int f(int n)
{       static int r = 0;
        if(n <= 0) return 1;
        if(n > 3)
        {
           r = n;
           return f(n – 2) + 2;
        }
        return f(n –1) + r;
}
```

What is the value of f(5)?

(a) 5 (b) 7
(c) 9 (d) 18 **[2007 : 2 M]**

1.48 The function f is defined as follows:

```
int f(int n)
{
    if (n < = 1)
    return 1;
    else if (n % 2 = = 0) return f(n/2);
    else return f(3n – 1);
```

Assuming that arbitrarily large integers can be passed as a parameter to the function, consider the following statements.

(i) The function f terminates for finitely many different values of $n \geq 1$.
(ii) The function f terminates for infinitely many different values of $n \geq 1$.
(iii) The function f does not terminate for finitely many different values of $n \geq 1$.
(iv) The function f does not terminate for infinitely many different values of $n \geq 1$.

Which one of the following options is true of the above?

(a) (i) and (iii)
(b) (i) and (iv)
(c) (ii) and (iii)
(d) (ii) and (iv)

[2007 : 2 M]

1.49 Consider the C program given below:

```
# include <stdio.h>
int main( )
{
    int sum = 0, maxsum = 0, i, n = 6;
    int a( ) = {2, –2, –1, 3, 4, 2};
    for(i = 0; i < n; i++)
    {
        if(i = = 0 | | a[i] < 0 | | a[i] < a[i – 1])
        {
            if(sum > maxsum)
                maxsum = sum;
            sum = (a[i] > 0) ? a[i] : 0;
        }
        else sum + = a[i];
        if(sum > maxsum) maxsum = sum;
        printf ("%d\n", maxsum);
}
```

What is the value printed out when this program is executed?

(a) 9 (b) 8
(c) 7 (d) 6 **[2007 : 2 M]**

1.50 Consider the following C program:

```
# include <stdio.h>
# define EOF –1
void push (int); /* push the argument on the stack */
int pop  (void); /* pop the top of the stack */
void flagError ();
int main()
{
    int c, m, n, r;
    while ((c = getchar ()) ! = EOF)
    {   if (isdigit (c)) push (c);
        else if ((c = = '+') | | (c = = '*'))
        {
            m = pop ();
            n = pop ();
            r = (c = = '+') ? n + m : n*m;
            push (r);
        }
        else if (c ! = ' ')
        flagError ();
    }
    printf("% c", pop ());
}
```

What is the output of the program for the following input?

5 2 * 3 3 2 + * +

(a) 15 (b) 25
(c) 30 (d) 150 **[2007 : 2 M]**

1.51 Consider the program below in a hypothetical language which allows global variables and a choice of call by reference or call by value methods of parameter passing.

```
int i ;
program main( )
{   int j = 60;
    i = 50;
    call f(i, j);
    print i, j;
}
procedure f(x, y)
{
    i = 100;
    x = 10;
    y = y + i ;
}
```

Which one of the following options represents the correct output of the program for the two parameter passing mechanisms?

(a) Call by value : i = 70, j = 10; Call by reference : i = 60, j = 70
(b) Call by value : i = 50, j = 60; Call by reference : i = 50, j = 70
(c) Call by value : i = 10, j = 70; Call by reference : i = 100, j = 60
(d) Call by value : i = 100, j = 60; Call by reference : i = 10, j = 70

[2007 : 2 M]

1.52 Consider the program below in a hypothetical programming language which allows global variables and a choice of static or dynamic scoping.

```
int i;
program main( )
{
    i = 10;
    call f( );
}
procedure f( )
{
    int i = 20;
    call g( );
}
procedure g( )
{
    print i;
}
```

Let x be the value printed under static scoping and y be the value printed under dynamic scoping. Then, x and y are

(a) x = 10, y = 10 (b) x = 20, y = 10
(c) x = 10, y = 20 (d) x = 20, y = 20

[2007 : 2 M]

1.53 Which combination of the integer variables x, y and z makes the variable a get the value 4 in the following expression?

a = (x > y) ? ((x > z) ? x : z) : ((y>z) ? y : z)

(a) x = 3, y = 4, z = 2
(b) x = 6, y = 5, z = 3
(c) x = 6, y = 3, z = 5
(d) x = 5, y = 4, z = 5 **[2008 : 1 M]**

1.54 What is the output printed by the following C code?

```
# include <stdio.h>
int main( )
{
    char a[6] = "world";
    int i, j;
    for (i = 0, j = 5; i < j; a[i++] = a [j – –]);
    printf ("%s\n", a);
}
```

(a) dlrow (b) Null string
(c) dlrld (d) worow

[2008 : 2 M]

1.55 Consider the C program below. What does it print?

```
# include <stdio.h>
# define swapl (a,b) tmp = a; a = b; b =tmp;
void swap2 ( int a, int b)
{
    int tmp;
    tmp = a; a = b; b = tmp;
}
void swap3 (int*a, int*b)
{
    int tmp;
    tmp = *a; *a = *b; *b = tmp;
}
int main( )
{
    int num1 = 5, num2 = 4, tmp;
    if(num1 > num2)
    { swap1(num1, num2); }
    if (num1 < num2)
    { swap2(num1 + 1, num2); }
    if(num1 > = num2)
    { swap3 (&num1, &num2); }
    printf ("%d, %d", num1, num2);
}
```

(a) 5, 5 (b) 5, 4
(c) 4, 5 (d) 4, 4 **[2008 : 2 M]**

1.56 Consider the C program given below. What does it print?

```
# include <stdio.h>
int main( )
{
    int i, j;
    int a[8] = {1, 2, 3, 4, 5, 6, 7, 8};
    for(i = 0; i < 3; i++)
    {
        a[i] = a[i]+1;
        i++;
    }
    i – –;
    for (j = 7; j > 4, j– –)
    {
        int i = j/2;
        a[i] = a[i] –1;
    }
    printf (“%d, %d”, i, a [i]);
}
```

(a) 2, 3
(b) 2, 4
(c) 3, 2
(d) 3, 3

[2008 : 2 M]

1.57 A C program is given below:

```
# include <stdio.h>
int main( )
{
    int i, j;
    char a[2][3] = {{‘a’, ‘b’, ‘c’}, {‘d’, ‘e’, ‘f’}};
    char b[3][2];
    char *p = *b;
    for (i = 0; i < 2; i++)
    {
        for (j = 0; j < 3; j++)
        {
            *(p + 2*j + i) = a [i] [j];
        }
    }
}
```

What should be the contents of the array b at the end of the program?

(a)
```
a b
c d
e f
```
(b)
```
a d
b e
c f
```
(c)
```
a c
e b
d f
```
(d)
```
a e
d c
b f
```

[2008 : 2 M]

Directions for Question 1.58 to 1.59:

Consider the code fragment written in C below:

```
void f(int n)
{
    if(n ≤ 1 )
    {
        printf(“%d”, n);
    }
    else
    {
        f(n/2);
        printf(“%d”, n%2);
    }
}
```

1.58 What does f(173) print?

(a) 010110101 (b) 010101101
(c) 10110101 (d) 10101101

[2008 : 2 M]

1.59 Which of the following implementations will produce the same output for f(173) as the one from above Question?

P1:	P2:
void f(int n)	void f(int n)
{	{
if(n/2)	if (n < = 1)
{	{
f(n/2);	printf(“%d”, n); }
}	else
printf(“%d”, n%2);	{ printf(“%d”, n%2);
}	f(n/2); }
	}

(a) Both P1 and P2 (b) P2 only
(c) P1 only (d) Neither P1 nor P2

[2008 : 2 M]

1.60 What is printed by the following C program?

```
int f(int x, int *py, int ** ppz)
{ int y, z;
  **ppz + = 1; z = **ppz;
  *py + = 2; y = *py;
  x + = 3;
  return x + y + z;
}
void main( )
{ int c, *b, **a;
  c = 4; b = &c; a = & b;
  printf (“% d”, f (c, b, a));
}
```

(a) 18 (b) 19
(c) 21 (d) 22 **[2008 : 2 M]**

1.61 Choose the correct option to fill ?1 and ?2 so that the program below prints an input string in reverse order. Assume that the input string is terminated by a newline character.

```
void reverse (void) {
    int c;
    if (?1) reverse ();
    ?2
}
main () {
    printf ("Enter Text") ; printf ("\n");
    reverse (); printf ("\ n");
}
```

(a) ? 1 is (getchar ( ) ! = '\n')
   ? 2 is getchar (c);
(b) ? 1 is (c = getchar ( )) ! = '\n')
   ? 2 is getchar (c)
(c) ? 1 is (c ! = '\n')
   ? 2 is putchar (c);
(d) ? 1 is ((c = getchar ( )) ! = '\n')
   ? 2 is putchar (c); **[2008 : 2 M]**

1.62 Consider the program below:

```
# include <stdio.h>
int fun (int n, int *f_p) {
    int t, f;
    if (n < = 1) {
     *f_p = 1 ;
     return 1 ;
    }
    t = fun(n–1, f_p) ;
    f = t + * f_p ;
    *f_p = t ;
    return f ;
}
int main( )
{
    int x = 15 ;
    printf ("% d\n", fun (5, &x)) ;
    return 0;
}
```

The value printed is

(a) 6 (b) 8
(c) 14 (d) 15 **[2009 : 1 M]**

1.63 What does the following program print?

```
#include < stdio.h>
void f(int *p, int *q)
{
    p = q;
    *p = 2;
}
int i = 0, j = 1;
int main( )
{
    f(&i, &j);
    printf ("%d %d\n", i, j);
    return 0;
}
```

(a) 2 2 (b) 2 1
(c) 0 1 (d) 0 2 **[2010 : 1 M]**

1.64 What is the value printed by the following C program?

```
#include < stdio.h>
int f(int *a, int n)
{
    if (n <= 0) return 0;
    else if (*a % 2 == 0)
        return *a + f(a + 1,  n – 1);
    else return *a – f(a + 1, n – 1);
}
int main( )
{
    int a[ ] = {12, 7, 13, 4, 11, 6};
    printf("%d", f(a, 6));
    return 0;
}
```

(a) –9 (b) 5
(c) 15 (d) 19 **[2010 : 2 M]**

1.65 The following program is to be tested for statement coverage:

```
begin
    if (a = = b) {S1; exit;}
    else if (c = = d) {S2;}
    else {S3; exit;}
    S4;
end
```

The test cases T1, T2, T3 and T4 given below are expressed in terms of the properties satisfied by the values of variables a, b, c and d. The exact values are not given.

T1 : a, b, c and d are all equal
T2 : a, b, c and d are all distinct
T3 : a = b and c != d
T4 : a !=b and c = d

Which of the test suites given below ensures coverage of statements S1, S2, S3 and S4?

(a) T1, T2, T3 (b) T2, T4
(c) T3, T4 (d) T1, T2, T4
[2010 : 2 M]

1.66 What does the following fragment of C program print?

```
char c[ ] = "GATE2011";
char *p = c;
printf("%s", p + p[3] – p[1]);
```

(a) GATE 2011 (b) E2011
(c) 2011 (d) 011

[2011 : 1 M]

Common Data Question 1.67 and 1.68:

Consider the following recursive C function that takes two arguments.

```
unsigned int foo(unsigned int n, unsigned int r)
{
  if(n>0) return ((n%r)+foo(n/r,r));
  else return 0;
}
```

1.67 What is the return value of the function foo when it is called as foo(345, 10)?

(a) 345 (b) 12
(c) 5 (d) 3

[2011 : 2 M]

1.68 What is the return value of the function foo when it is called as foo(513, 2)?

(a) 9 (b) 8
(c) 5 (d) 2

[2011 : 2 M]

1.69 What will be the output of the following C program segment?

```
char inChar = 'A';
switch(inChar)
{
  case 'A' : printf ("Choice A\ n") ;
  case 'B' :
  case 'C' : printf ("Choice B") ;
  case 'D' :
  case 'E' :
  default  : printf (" NO Choice") ; }
```

(a) No Choice
(b) Choice A
(c) Choice A
Choice B No Choice
(d) Program gives no output as it is erroneous

[2012 : 1 M]

Common Data for Questions 1.70 and 1.71:

Consider the following C code segment.

```
int a, b, c = 0;
void prtFun (void);
main()
{static int a = 1;/* Line 1*/
prtFun();
a+=1;
prtFun();
printf("\n %d %d ",a,b);}
void prtFun(void)
{static int a = 2;/*Line 2*/
int b = 1;
a += ++b;
printf("\n %d %d ",a,b);}
```

1.70 What output will be generated by the given code segment?

(a) 3 1
4 1
4 2

(b) 4 2
6 1
6 1

(c) 4 2
6 2
2 0

(d) 3 1
5 2
5 2

[2012 : 2 M]

1.71 What output will be generated by the given code segment if:
Line 1 is replaced by auto int a=1;
Line 2 is replaced by register int a = 2;

(a) 3 1
4 1
4 2

(b) 4 2
6 1
6 1

(c) 4 2
6 2
2 0

(d) 4 2
4 2
2 0

[2012 : 2 M]

1.72 What is the return value of f(p, p), if the value of p is initialized to 5 before the call? Note that the first parameter is passed by reference, whereas the second parameter is passed by value.

```
int f (int &x, int c)
{  c = c – 1;
   if (c = = 0) return 1;
   x = x + 1;
   return f(x, c) * x; }
```

(a) 3024 (b) 6561
(c) 55440 (d) 161051

[2013 : 2 M]

1.73 Consider the following program in C language:

```
#include <stdio.h>
main()
{
  int i;
  int *pi = &i;
  scanf("%d",pi);
  printf("%d\n", i+5);
}
```

Which one of the following statements is **TRUE**?
(a) Compilation fails.
(b) Execution results in a run-time error.
(c) On execution, the value printed is 5 more than the address of variable *i*.
(d) On execution, the value printed is 5 more than the integer value entered.

[2014 (Set-1) : 1 M]

1.74 Consider the following pseudo code. What is the total number of multiplications to be performed?

```
D = 2
for i = 1 to n do
    for j = i to n do
        for k = j + 1 to n do
            D = D * 3
```

(a) Half of the product of the 3 consecutive integers.
(b) One-third of the product of the 3 consecutive integers.
(c) One-sixth of the product of the 3 consecutive integers.
(d) None of the above

[2014 (Set-1) : 2 M]

1.75 Consider the function func shown below:

```
int func(int num)
{   int count = 0;
    while (num)
    {   count++; num>>= 1;
    }
    return (count);
}
```

The value returned by function (435) is ________.

[2014 (Set-2) : 1 M]

1.76 Suppose n and p are unsigned int variables in a C program. We wish to set p to ${}^nC_3$. If n is large, which one of the following statements is most likely to set p correctly?
(a) $p = n*(n-1)*(n-2)/6;$
(b) $p = n*(n-1)/2*(n-2)/3;$
(c) $p = n*(n-1)/3*(n-2)/2;$
(d) $p = n*(n-1)*(n-2)/6.0;$

[2014 (Set-2) : 1 M]

1.77 Consider the following function:

```
double f(double x) {
    if(abs(x*x – 3) < 0.01) return x;
    else return f(x/2 + 1.5/x);
}
```

Give a value *q* (to 2 decimals) such that *f*(*q*) will return *q* : ________. **[2014 (Set-2) : 2 M]**

1.78 Consider the C function given below:

```
int f(int j) {
    static int i = 50;
    int k;
    if (i == j) {
        printf("something");
        k = f(i);
        return 0;
    }
    else return 0;
}
```

Which one of the following is TRUE?
(a) The function returns **0** for all values of ***j***.
(b) The function prints the string something for all values of ***j***.
(c) The function returns **0** when ***j* = 50**.
(d) The function will exhaust the runtime stack or run into an infinite loop when ***j* = 50**.

[2014 (Set-2) : 2 M]

1.79 The output of the following C program is ______.

```
void f1 (int a, int b) {
    int c;
    c=a; a=b; b=c;
}
void f2 (int *a, int *b) {
    int c;
    c=*a; *a=*b; *b=c;
}
int main() {
    int a=4, b=5, c=6;
    f1 (a, b);
    f2 (&b, &c);
    printf ("%d", c – a – b);
}
```

[2015 (Set-1) : 1 M]

1.80 Consider the following pseudo code, where x and y are positive integers.

```
begin
    q := 0;
    r := x;
    while r ≥ y do
      begin r := r – y ; q := q + 1 ; end
end
```

The post condition that needs to be satisfied after the program terminates is
(a) $\{r = qx + y \wedge r < y\}$
(b) $\{x = qy + r \wedge r < y\}$
(c) $\{y = qx + r \wedge 0 < r < y\}$
(d) $\{q + 1 < r - y \wedge y > 0\}$

[2015 (Set-1) : 2 M]

1.81 Consider the following C function:

```
int fun (int n) {
    int x=1, k;
    if (n==1) return x;
    for (k=1; k<n; ++k)
        x=x + fun (k) * fun (n – k);
    return x;
}
```

The return value of fun (5) is ________.

[2015 (Set-2) : 1 M]

1.82 Consider the following function written in the C programming language.

```
void foo (char *a) {
    if (*a && *a ! = ' ') {
        foo (a+1);
        putchar (*a);
    }
}
```

The output of the above function on input "ABCD EFGH" is

(a) ABCD EFGH (b) ABCD
(c) HGFE DCBA (d) DCBA

[2015 (Set-2) : 1 M]

1.83 A Young tableau is a 2D array of integers increasing from left to right and from top to bottom. Any unfilled entries are marked with ∞, and hence there cannot be any entry to the right of, or below a ∞. The following Young tableau consists of unique entries.

1	2	5	14
3	4	6	23
10	12	18	25
31	∞	∞	∞

When an element is removed from a Young tableau, other elements should be moved into its place so that the resulting table is still a Young tableau (unfilled entries may be filled in with a ∞). The minimum number of entries (other than 1) to be shifted, to remove 1 from the given Young tableau is ________.

[2015 (Set-2) : 2 M]

1.84 Consider the following C program segment:

```
# include <stdio.h>
int main( ) {
    char s1[7] = "1234", *p;
    p = s1 + 2;
    *p = '0' ;
    printf ("%s", s1);
}
```

What will be printed by the program?

(a) 12 (b) 120400
(c) 1204 (d) 1034

[2015 (Set-3) : 1 M]

1.85 Consider the following C program:

```
# include < stdio.h>
int main( ) {
    static int a[ ] = {10, 20, 30, 40, 50};
    static int *p[ ] = {a, a+3, a+4, a+1, a+2};
    int **ptr = p; ptr++;
    printf ("%d%d", prt-p, **ptr);
}
```

The output of the program is ______.

[2015 (Set-3) : 2 M]

1.86 Consider the following recursive C function:

```
void get (int n) {
    if (n < 1) return;
    get (n – 1);
    get (n – 3);
    printf ("%d", n);
}
```

If get(6) function is being called in main() then how many times will the get( ) function be invoked before returning to the main( )?

(a) 15 (b) 25
(c) 35 (d) 45

[2015 (Set-3) : 2 M]

1.87 Consider the following C program:

```
# include <stdio.h>
int main( ) {
    int i, j, k = 0;
    j = 2 * 3 / 4 + 2.0 / 5 + 8 / 5;
    k – = – –j;
    for(i = 0; i < 5; i++) {
        switch(i + k)
        {
            case 1:
            case 2: printf("\n%d", i + k);
            case 3: printf("\n%d", i + k);
            default: printf("\n%d", i + k);
        }
    }
    return 0;
}
```

The number of times printf statement is executed is ________. **[2015 (Set-3) : 2 M]**

1.88 Consider the following C program:

```
# include <stdio.h>
int f1(void);
int f2(void);
int f3(void);
```

```
int x = 10;
int main( )
{
    int x = 1;
    x +=f1( ) + f2( ) + f3( ) + f2( );
    pirntf("%d", x);
    return 0;
}
int f1( ) { int x = 25; x++; return x; }
int f2( ) { static int x = 50; x++; return x; }
int f3( ) { x *= 10; return x; }
```

The output of the program is ________.

[2015 (Set-3) : 2 M]

1.89 Consider the following C program:

```
void f (int, short);
void main()
{
   int i = 100;
   short s = 12;
   short *p = &s;
   ______;    // call to f ()
}
```

Which one of the following expressions, when placed in the blank above, will NOT result in a type checking error?

(a) $f(s, *s)$ (b) $i = f(i, s)$
(c) $f(i, *s)$ (d) $f(i, *p)$

[2016 (Set-1) : 1 M]

1.90 Consider the following C program:

```
#include<stdio.h>
void mystery (int *ptra, int *ptrb)
{   int *temp;
    temp = ptrb;
    ptrb = ptra;
    ptra = temp;
}
int main()
{   int a = 2016, b = 0, c = 4, d = 42;
    mystery(&a, &b);
    if (a < c)
        mystery(&c, &a);
    mystery(&a, &d);
    printf("%d\n", a);
}
```

The output of the program is _____.

[2016 (Set-1) : 1 M]

1.91 The following function computes the maximum value contained in an integer array $p[\ ]$ of size $n (n >= 1)$.

```
int max(int *p, int n)
{   int a = 0, b = n – 1;
    while (________)
    {
        if (p[a] <= p[b])
        {
            a = a + 1;
        }
        else
        {
            b = b – 1;
        }
    }
    return p[a];
}
```

The missing loop condition is

(a) $a\ != n$ (b) $b\ != 0$
(c) $b > (a + 1)$ (d) $b\ != a$

[2016 (Set-1) : 2 M]

1.92 What will be the output of the following C program?

```
void count(int n)
{  static int d = 1;
   printf("%d", n);
   printf("%d", d);
   d++;
   if (n > 1) count (n – 1);
   printf("%d", d);
}
void main()
{
   count (3);
}
```

(a) 3 1 2 2 1 3 4 4 4 (b) 3 1 2 1 1 1 2 2 2
(c) 3 1 2 2 1 3 4 (d) 3 1 2 1 1 1 2

[2016 (Set-1) : 2 M]

1.93 What will be the output of the following pseudo-code when parameters are passed by reference and dynamic scoping is assumed?

```
a = 3;
void n(x)
{
    x = x * a; print(x);
}
void m(y)
{   a = 1; a = y – a; n(a);
    print(a);
}
void main()
{
    m(a);
}
```

(a) 6, 2 (b) 6, 6
(c) 4, 2 (d) 4, 4

[2016 (Set-1) : 2 M]

1.94 The value printed by the following program is ______.

```
void f(int* p, int m)
{    m = m + 5;
     *p = *p + m;
     return;
}
void main ()
{    int i = 5, j = 10;
     f(&i, j);
     printf ("%d", i + j) ;
}
```

[2016 (Set-2) : 1 M]

1.95 The following function computes X^Y for positive integers X and Y.

```
int exp (int X, int Y)
{    int res = 1, a = X, b = Y;
     while (b ! = 0)
     {   if (b% 2 == 0)
         {  a = a*a; b = b/2;  }
            else {res = res*a; b = b – 1;
         }
     }
     return res;
}
```

Which one of the following conditions is TRUE befor every iteration of the loop?

(a) $X^Y = a^b$
(b) $(\text{res} * a)^Y = (\text{res} * X)^b$
(c) $X^Y = \text{res} * a^b$
(d) $X^Y = (\text{res} * a)^b$

[2016 (Set-2) : 2 M]

1.96 Consider the following program:

```
int f(int *p, int n)
{    if (n <= 1) return 0;
     else return max (f(p+1, n–1), p[0] –p[1]);
}
int main ()
{    int a[ ] = {3, 5, 2, 6, 4};
     print f("%d", f(a, 5));
}
```

Note: max (x, y) returns the maximum of x and y.

The value printed by this program is ______.

[2016 (Set-2) : 2 M]

1.97 The output of executing the following C program is ______.

```
#include <stdio.h>
int total (int v)
{    static int count = 0;
     while (v)
     {    count+ = v & 1;
          v >> = 1;
     }
     return count;
}
void main ()
{    static int x = 0;
     int i = 5;
     for (; i > 0; i ––) {
          x = x + total (i);
     }
     printf ("%d\n", x);
}
```

[2017 (Set-1) : 2 M]

1.98 Consider the C functions foo and bar given below:

```
int foo (int val)
{
     int x = 0;
     while (val > 0)
     {
          x = x + foo (val ––);
     }
     return val;
}
int bar (int val)
{
     int x = 0;
     {
          while (val > 0)
          x = x + bar (val – 1);
     }
return  val;
}
```

Invocations of foo (3) and bar (3) will result in :

(a) Return of 6 and 6 respectively.
(b) Infinite loop and abnormal termination respectively.
(c) Abnormal termination and infinite loop respectively.
(d) Both terminating abnormally.

[2017 (Set-1) : 2 M]

1.99 Consider the following C program:

```
#include <stdio.h>
#include <string.h>
void printlength (char *s, char *t)
{
    unsigned int c = 0;
    int len = ((strlen(s) – strlen(t)) > c) ? strlen(s):
    strlen(t);
    printf("%d\n", len);
}
void main( )
{
    char *x = "abc";
    char *y = "defgh";
    printlength (x, y);
```

Recall that strlen is defined in string.h as returning a value of type size_t, which is an unsigned int. The output of the program is ______.

[2017 (Set-1) : 2 M]

1.100 Consider the following two functions:

```
void fun1 (int n)
{   if (n = = 0) return;
    printf ("%d", n);
    fun2 (n – 2);
    printf ("%d", n);
}
void fun2 (int n)
{   if (n = = 0) return;
    printf ("%d", n);
    fun1(++n);
    printf ("%d", n);
}
```

The output printed when fun1 (5) is called is
(a) 53423122233445
(b) 53423120112233
(c) 53423122132435
(d) 53423120213243

[2017 (Set-1) : 2 M]

1.101 Match the following:

List-I

P. static char var;
Q. m = malloc(10); m = NULL;
R. char *ptr[10];
S. register int var1;

List-II

(i) Sequence of memory locations to store addresses.
(ii) A variable located in data section of memory.
(iii) Request to allocate a CPU register to store data.
(iv) A lost memory which cannot be freed.

(a) P → (ii), Q → (iv), R → (i), S → (iii)
(b) P → (ii), Q → (i), R → (iv), S → (iii)
(c) P → (ii), Q → (iv), R → (iii), S → (i)
(d) P → (iii), Q → (iv), R → (i), S → (ii)

[2017 (Set-2) : 1 M]

1.102 Consider the following function implemented in C:

```
void printxy (int x, int y)
{
    int *ptr;
    x = 0;
    ptr = &x;
    y = *ptr;
    *ptr = 1;
    printf ("%d, %d", x, y);
}
```

The output of invoking printxy (1, 1) is
(a) 0, 0 (b) 0, 1
(c) 1, 0 (d) 1, 1

[2017 (Set-2) : 1 M]

1.103 Consider the following C program:

```
# include <stdio.h>
# include <string.h>
int main( )
{   char* c = "GATECSIT2017";
    char* p = c;
    printf("%d", (int) strlen (c+2[p] – 6[p]–1));
    return 0;
}
```

The output of the program is ______.

[2017 (Set-2) : 2 M]

1.104 Consider the following C program:

```
# include <stdio.h>
int main( )
{
    int m = 10;
    int n, n1;
    n = ++m;
    n1 = m++;
    n ––;
    –– n1;
    n – = n1;
    printf("%d", n);
    return 0;
}
```

The output of the program is ______.

[2017 (Set-2) : 2 M]

1.105 Consider the following snippet of a C program. Assume that swap (&x, &y) exchanges the contents of x and y.

```
int main( )
{
    int array[ ] = {3, 5, 1, 4, 6, 2};
    int done = 0;
    int i;
    while (done == 0)
    {
        done = 1;
        for (i = 0; i <= 4; i++)
        {
            if (array[i] < array[i + 1])
            {
                swap(&array[i], &array[i + 1]);
                done = 0;
            }
        }
    }
    for (i = 5; i >= 1; i --)
    {
        if (array[i] > array[i – 1])
        {
            swap(&array[i], &array[i – 1]);
            done = 0;
        }
    }
    printf("%d", array[3]);
}
```

The output of the program is ________.

[2017 (Set-2) : 2 M]

1.106 Consider the following C code:

```
#include <stdio.h>
int *assignval (int *x, int val)
{   *x = val;
    return x;
}
void main( )
{   int *x = malloc (size of (int));
    if (NULL == x) return;
    x = assignval(x, 0);
    if(x)
    {  x = (int*) malloc (size of (int));
       if (NULL == x) return;
       x = assignval (x, 10);
    }
    printf ("%d"\n", *x);
    free (x);
}
```

The code suffers from which one of the following problems:

(a) compiler error as the return of malloc is not typecast appropriately
(b) compiler error because the comparison should be made as x == NULL and not as shown
(c) compiles successfully but execution may result in dangling pointer
(d) compiles successfully but execution may result in memory leak

[2017 (Set-1) : 1 M]

1.107 Consider the C program fragment below which is meant to divide x by y using repeated subtractions. The variables x, y, q and r are all unsigned int.

```
while
(r >= y)
{  r = r – y;
   q = q + 1;
}
```

Which of the following conditions on the variables x, y, q and r before the execution of the fragment will ensure that the loop terminates in a state satisfying the condition $x == (y \times q + r)$?

(a) $(q == r)$ && $(r == 0)$
(b) $(x > 0)$ && $(r == x)$ && $(y > 0)$
(c) $(q == 0)$ && $(r == x)$ && $(y > 0)$
(d) $(q == 0)$ && $(y > 0)$

[2017 (Set-2) : 2 M]

1.108 Consider the following C program:

```
#include <stdio.h>
struct ournode {
    char x, y, z;
};
int main( ) {
    struct ournode p = {'1', '0', 'a' + 2};
    struct ournode *q = &p;
    printf("%c, %c", *((char*) q + 1),
    *((char*) q + 2));
    return 0;
}
```

The output of this program is:

(a) 0, c
(b) 0, a + 2
(c) '0', 'a + 2'
(d) '0', 'c'

[2018 : 1 M]

1.109 Consider the following C program:

```
#include <stdio.h>
int counter = 0;
int calc (int a, int b) {
    int c;
    counter ++;
    if (b = = 3) return (a × a × a);
    else {
        c = calc (a, b/3);
        return (c × c × c);
    }
}
int main() {
    calc(4, 81);
    printf("%d", counter);
}
```

The output of this program is ________.

[2018 : 1 M]

1.110 Consider the following program written in pseudo-code. Assume that x and y are integers:

```
Count (x, y) {
    if (y ! = 1) {
        if (x ! = 1) {
            print ("*");
            Count (x/2, y);
        }
        else {
            y = y – 1;
            Count (1024, y);
        }
    }
}
```

The number of times that the print statement is executed by the call Count (1024, 1024) is ________.

[2018 : 1 M]

1.111 Consider the following C program:

```
#include <stdio.h>
void fun1 (char *s1, char *s2) {
    char *tmp;
    tmp = s1;
    s1 = s2;
    s2 = tmp;
}
void fun2 (char **s1, char **s2) {
    char *tmp;
    tmp = *s1;
    *s1 = *s2;
    *s2 = tmp;
}
int main() {
    char *str1 = "Hi", *str2 = "Bye";
    fun1 (str1, str2); printf("%s %s", str1, str2);
    fun2 (&str1, &str2); printf("%s %s", str1, str2);
    return 0;
}
```

The output of the program above is

(a) Hi Bye Bye Hi
(b) Hi Bye Hi Bye
(c) Bye Hi Hi Bye
(d) Bye Hi Bye Hi

[2018 : 2 M]

1.112 Consider the following C code. Assume that unsigned long int type length is 64 bits.

```
unsigned long int fun (unsigned long int n) {
    unsigned long int i, j = 0, sum = 0;
    for (i = n; i > 1; i = i/2) j++;
    for (; j > 1; j = j/2) sum++;
    return (sum);
}
```

The value returned when we call fun with the input 2^{40} is

(a) 4 (b) 5
(c) 6 (d) 40

[2018 : 2 M]

1.113 Consider the following C program:

```
#include <stdio.h>
int main() {
    int arr[ ] = {1, 2, 3, 4, 5, 6, 7, 8, 9, 0, 1,
    2, 5}, *ip = arr + 4;
    printf ("%d\n", ip[1])'
    return 0;
}
```

The number that will be displayed on execution of the program is ________.

[2019 : 1 M]

1.114 Consider the following C program:

```
#include <stdio.h>
int jumble(int x, int y) {
    x = 2 * x + y;
    return x;
}
int main() {
    int x = 2, y = 5;
    y = jumble(y, x);
    x = jumble(y, x);
    printf("%d\n", x);
    return 0;
}
```

The value printed by the program is ________.

[2019 : 1 M]

1.115 Consider the following C function:

```
void convert (int n) {
    if (n < 0)
        printf("%d", n);
    else {
        convert (n/2);
        printf("%d", n % 2);
    }
}
```

Which one of the following will happen when the function convert is called with any positive integer n as argument?

(a) It will print the binary representation of n in the reverse order and terminate.
(b) It will print the binary representation of n but will not terminate.
(c) It will not print anything and will not terminate.
(d) It will print the binary representation of n and terminate.

[2019 : 2 M]

1.116 Consider the following C programs:

```
#include <stdio.h>
int main( ) {
    float sum = 0.0, j = 1.0, i = 2.0;
    while (i/j > 0.0625) {
        j = j + j;
        sum = sum + i/j;
        printf("%f\n", sum);
    }
    return 0;
}
```

The number of times the variable sum will be printed, when the above program is executed, is ________.

[2019 : 2 M]

1.117 Consider the following C program:

```
#include<stdio.h>
int r( )
{
    Static int num = 7;
    return num --;
}
int main( )
{
    for ( r( ); r( ); r( ) )
        printf("%d",r( ));
    return 0;
}
```

Which one of the following values will be displayed on execution of the programs?

(a) 41
(b) 630
(c) 63
(d) 52

[2019 : 2 M]

1.118 Consider the following C program:

```
#include <stdio.h>
int main()
{
    int a[ ] = {2, 4, 6, 8, 10};
    int i, sum = 0, *b = a + 4;
    for (i = 0; i < 5; i++)
        sum = sum + (*b - i) - *(b - i);
    printf ("%d\n", sum);
    return 0;
}
```

The output of the above C program is ________.

[2019 : 2 M]

1.119 Consider the following C functions:

```
int tob(int b, int* arr) {
    int i;
    for(i = 0, b > 0; i++) {
        if(b%2)  arr[i] = 1;
        else     arr[i] = 0;
        b = b/2;
    }
    return(i);
}
```

```
int pp(int a, int b) {
    int arr[20];
    int i, tot = 1, ex, len;
    ex = a;
    len = tob(b,arr);
    for(i = 0; i < len; i++) {
        if(arr[i] == 1)
        tot = tot * ex;
        ex = ex * ex;
    }
    return(tot);
}
```

The value returned by pp(3, 4) is ________.

[2020 : 2 M]

1.120 Consider the following C functions:

```
int fun1(int n) {                  int fun2(int n) {
    static int i = 0;                  static int i = 0;
    if (n > 0) {                       if (n > 0) {
        ++i;                               i = i + fun1(n);
        fun1(n – 1);                       fun2(n – 1);
    }                                  }
    return(i);                         return(i);
}                                  }
```

The return value of fun2 (5) is ________. **[2020 : 2 M]**

1.121 Consider the following ANSI C function:

```
int SimpleFunction (int Y[ ], int n, int x)
{
    int total = Y[0], loopIndex;
    for (loopIndex = 1; loopIndex <= n – 1; loopIndex++)
    total = x × total + Y[loopIndex];
    return total;
}
```

Let Z be an array of 10 elements with Z[i] = 1, for all i such that $0 \le i \le 9$. The value returned by SimpleFunction (Z, 10, 2) is ________. **[2021 (Set-1) : 2 M]**

1.122 Consider the following ANSI C program:

```
#include <stdio.h>
int main( )
{
    int i, j, count;
    count = 0;
    i = 0;
    for (j = –3; j <= 3; j++)
    {
        if ((j >= 0) && (i++))
            count = count + j;
    }
    count = count + i;
    printf("%d", count);
    return 0;
}
```

Which one of the following options is correct?

(a) The program will compile successfully and output 8 when executed.
(b) The program will compile successfully and output 13 when executed.
(c) The program will not compile successfully.
(d) The program will compile successfully and output 10 when executed **[2021 (Set-1) : 2 M]**

1.123 Consider the following ANSI C function:

```
int SomeFunction(int x, int y)
{
    if ((x == 1) || (y == 1)) return 1;
    if (x == y) return x;
    if (x > y) return SomeFunction(x – y, y);
    if (y > x) return SomeFunction(x, y – x);
}
```

The value returned by SomeFunction(15, 255) is ________. **[2021 (Set-2) : 1 M]**

1.124 Consider the following ANSI C program:

```
#include <stdio.h>
int foo(int x, int y, int q)
{
    if ((x <= 0) || (y <= 0))
        return q;
    if (x <= 0)
        return foo(x, y – q, q)
    if (y <= 0)
        return foo(x – q, y, q);
    return foo(x, y – q, q) + foo(x – q, y, q);
}
int main()
{
    int r = foo(15, 15, 10);
    printf("%d", r);
    return 0;
}
```

The output of the program upon execution is ________. **[2021 (Set-2) : 2 M]**

1.125 Consider the following ANSI C program:

```
#include <stdio.h>
int main() {
    int arr[4] [5] ;
    int i, j;
    for (i = 0; i < 4; i++) {
        for (j = 0; j < 5; j++) {
            arr[i] [j] = 10 * i + j;
        }
    }
    printf("%d", * (arr[i] + 9));
    return 0;
}
```

What is the output of the above program?

(a) 24 (b) 20
(c) 14 (d) 30

[2021 (Set-2) : 1 M]

1.126 What is printed by the following ANSI C program?

```
#include<stdio.h>
int main(int argc, char *argv[ ])
{
    int x = 1, z[2] = {10, 11};
    int *p = NULL;
    p = &x;
    *p = 10;
    p = &z[1];
    *(&z[0] + 1) += 3;
    printf("%d, %d, %d\n", x, z[0], z[1]);
    return 0;
}
```

(a) 1, 10, 11 (b) 1, 10, 14
(c) 10, 14, 11 (d) 10, 10, 14

[2022 : 1 M]

1.127 What is printed by the following ANSI C program?

```
#include<stdio.h>
int main(int argc, char *argv[ ])
{
        int a[3][3][3] = {{1, 2, 3, 4, 5, 6, 7, 8, 9},
        {10, 11, 12, 13, 14, 15, 16, 17, 18},
        {19, 20, 21, 22, 23, 24, 25, 26, 27}};
    int i = 0, j = 0, k = 0;
    for(i = 0; i < 3; i++) {
        for(k = 0; k < 3; k++)
            printf("%d"", a[i][j][k]);
        printf("\n");
    }
    return 0;
}
```

(a)
1 2 3
10 11 12
19 20 21

(b)
1 4 7
10 13 16
19 22 25

(c)
1 2 3
4 5 6
7 8 9

(d)
1 2 3
13 14 15
25 26 27

[2022 : 2 M]

1.128 What is printed by the following ANSI C program?

```
#include <stdio.h>
int main(int argc, char *argv[ ]) {
    char a = 'P';
    char b = 'x';
    char c = (a & b) + '*';
    char d = (a | b) – '–';
    char e = (a ^ b) + '+';
    printf("%c %c %c\n", c, d, e);
    return 0;
}
```

ASCII encoding for relevant characters is given below:

A	B	C	...	Z
65	66	67	...	90

a	b	c	...	z
97	98	99	...	122

*	+	–
42	43	46

(a) z K S (b) 122 75 83
(c) * – + (d) P x +

[2022 : 2 M

1.129 The integer value printed by the ANSI-C program given below is ________.

```
#include<stdio.h>
int funcp() {
        static int x = 1;
        x++;
        return x;
}
int main(){
        int x, y;
        x = funcp();
        y = funcp() + x;
        printf("%d\n", (x + y));
    return 0;
}
```

[2023 : 1 M]

1.130 Consider the following program:

```
int main()
{
    f1();
    f1(2);
    f3();
    return (0);
}

int f1()
{
        return (1);
}

int f2(int X)
{
    f3();
    if (X==1)
        return f1();
    else
        return (X * f2(X-1));
}

int f3()
{
    return(5);
}
```

Which one of the following options represents the activation tree corresponding to the main function?

(a) 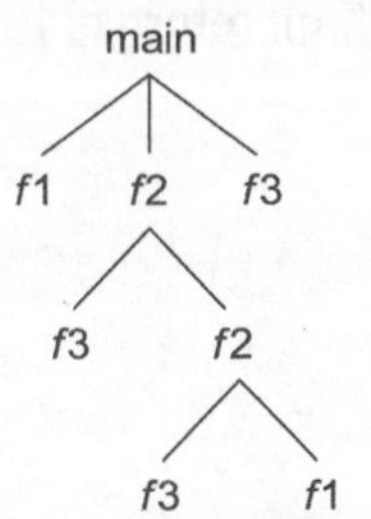

(b) 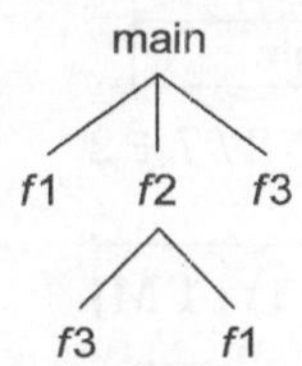

(c) 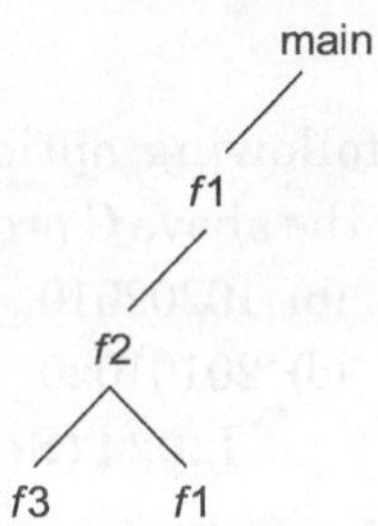

(d) 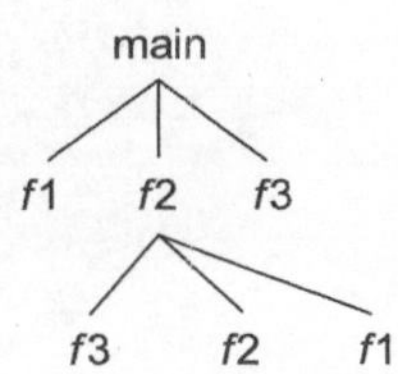

[2023 : 2 M]

1.131 Consider the following C program:

```
# include <stdio.h>
int main( ) {
    int a = 6;
    int b = 0;
    while (a < 10) {
        a = a / 12 + 1;
        a += b; }
    printf("%d", a);
    return 0; }
```

Which one of the following statements is CORRECT?

(a) The program prints 6 as output
(b) The program prints 10 as output
(c) The program prints 9 as output
(d) The program gets stuck in an infinite loop

[2024 (Set-1) : 1 M]

1.132 Consider the following program:

```
# include <stdio.h>          void fX( ) {
void fX( );                     char a;
int main( ) {                   if ((a=getchar ( )) != '\n' )
   fX( );                          fX( );
   return 0; }                  if (a != '\n' )
                                   putchar (a); }
```

Assume that the input to the program from the command line is 1234 followed by a newline character. Which one of the following statements is CORRECT?

(a) The program will terminate with 1234 as output
(b) The program will not terminate
(c) The program will terminate with no output
(d) The program will terminate with 4321 as output **[2024 (Set-1) : 1 M]**

1.133 Consider the operator precedence and associativity rules for the integer arithmetic operators given in the table below.

Operator	Precedence	Associativity
+	Highest	Left
–	High	Right
*	Medium	Right
/	Low	Right

The value of the expression 3 + 1 + 5 * 2 / 7 + 2 – 4 – 7 – 6 / 2 as per the above rules is ________.

[2024 (Set-1) : 1 M]

1.134 Consider the following C function definition.

```
int f (int x, int y) {
    for (int i = 0; i < y; i++) {
        x = x + x + y;
    }
    return x;
}
```

Which of the following statements is/are TRUE about the above function?

(a) If the inputs are $x = 20$, $y = 10$, then the return value is greater than 2^{20}
(b) If the inputs are $x = 20$, $y = 20$, then the return value is greater than 2^{20}
(c) If the inputs are $x = 20$, $y = 10$, then the return value is less than 2^{10}
(d) If the inputs are $x = 10$, $y = 20$, then the return value is greater than 2^{20}

[2024 (Set-1) : 2 M]

1.135 Consider the following C function definition:

```
int fX(char *a) {
    char *b = a;
    while (*b)
        b++;
    return b – a; }
```

Which of the following statements is/are TRUE?

(a) Assuming a character array c is declared as char c[ ] = "abcd" in main( ), the function call fX(c) will always return a value.
(b) The function call fX("abcd") will always return a value.
(c) The code of the function will not compile.
(d) Assuming a character pointer c is declared as char *c = "abcd" in main( ), the function call fX(c) will always return a value.

[2024 (Set-2) : 1 M]

1.136 Consider the following C program. Assume parameters to a function are evaluated from right to left.

```
# include <stdio.h>
int g(int p) { printf ("%d", p); return p; }
int h(int q) { printf ("%d", q); return q; }
void f(int x, int y) {
    g(x);
    h(y);
}
int main( ) {
    f(g(10), h(20));
}
```

Which one of the following options is the CORRECT output of the above C program?

(a) 20102010 (b) 10202010
(c) 10201020 (d) 20101020

[2024 (Set-2) : 1 M]

1.137 What is the output of the following C program?

```
#include <stdio.h>
int main( ) {
    double a[2]={20.0,25.0},*p, *q;
    p = a;
    q = p + 1;
    printf ("%d, %d",(int)(q - p),(int)(*q - *p));
    return 0; }
```

(a) 4, 8 (b) 1, 5
(c) 1, 8 (d) 8, 5

[2024 (Set-2) : 2 M]

1.138 Consider an array X that contains n positive integers. A subarray of X is defined to be a sequence of array locations with consecutive indices.

The C code snippet given below has been written to compute the length of the longest subarray of X that contains at most two distinct integers. The code has two missing expressions labelled (P) and (Q).

```
int first = 0, second = 0, len1 = 0, len2 = 0, maxlen = 0;
for (int i = 0; i < n; i++) {
    if (X[i] == first) {
        len2++; len1++;
    } else if (X[i] == second) {
        len2++;
        len1 = (P);
        second = first;
    } else {
        len2 = (Q);
        len1 = 1; second = first;
    }
    if (len2 > maxlen) {
        maxlen = len2;
    }
    first = x[i];
}
```

Which one of the following options gives the CORRECT missing expressions?

(**Hint:** At the end of the ith iteration, the value of len1 is the length of the longest subarray ending with X[i] that contains all equal values, and len2 is the length of the longest subarray ending with X[i] that contains at most two distinct values.)

(a) (P) len2 + 1 (Q) len1 + 1
(b) (P) 1 (Q) len1 + 1
(c) (P) len1 + 1 (Q) len2 + 1
(d) (P)1 (Q) len2 + 1

[2024 (Set-2) : 2 M]

Answers Programming

1.1 (d)	**1.2** (b)	**1.3** (d)	**1.4** (c)	**1.5** (a)	**1.6** (d)	**1.7** (c)	**1.8** (a)	**1.9** (d)
1.10 (a)	**1.11** (c)	**1.12** (a)	**1.13** (b)	**1.14** (d)	**1.15** (d)	**1.16** (b)	**1.17** (b)	**1.18** (b)
1.19 (a)	**1.20** (d)	**1.21** (a)	**1.22** (c)	**1.23** (d)	**1.24** (b)	**1.25** (c)	**1.26** (a)	**1.27** (b)
1.28 (c)	**1.29** (c)	**1.30** (a)	**1.31** (d)	**1.32** (c)	**1.33** (c)	**1.34** (d)	**1.35** (d)	**1.36** (c)
1.37 (d)	**1.38** (d)	**1.39** (c)	**1.40** (b)	**1.41** (d)	**1.42** (a)	**1.43** (c)	**1.44** (d)	**1.45** (b)
1.46 (c)	**1.47** (d)	**1.48** (d)	**1.49** (c)	**1.50** (b)	**1.51** (d)	**1.52** (c)	**1.53** (a)	**1.54** (b)
1.55 (c)	**1.56** (c)	**1.57** (b)	**1.58** (d)	**1.59** (c)	**1.60** (b)	**1.61** (d)	**1.62** (b)	**1.63** (d)
1.64 (c)	**1.65** (d)	**1.66** (c)	**1.67** (b)	**1.68** (d)	**1.69** (c)	**1.70** (c)	**1.71** (d)	**1.72** (b)
1.73 (d)	**1.74** (c)	**1.75** (9)	**1.76** (b)	**1.77** (1.73)	**1.78** (d)	**1.79** (–5)	**1.80** (b)	**1.81** (51)
1.82 (d)	**1.83** (5)	**1.84** (c)	**1.85** (140)	**1.86** (b)	**1.87** (10)	**1.88** (230)	**1.89** (d)	**1.90** (2016)
1.91 (d)	**1.92** (a)	**1.93** (d)	**1.94** (30)	**1.95** (c)	**1.96** (3)	**1.97** (23)	**1.98** (c)	**1.99** (3)
1.100 (a)	**1.101** (a)	**1.102** (c)	**1.103** (2)	**1.104** (0)	**1.105** (3)	**1.106** (d)	**1.107** (c)	**1.108** (a)
1.109 (4)	**1.110** (10230)		**1.111** (a)	**1.112** (b)	**1.113** (6)	**1.114** (26)	**1.115** (c)	**1.116** (5)
1.117 (d)	**1.118** (10)	**1.119** (81)	**1.120** (55)	**1.121** (1023)	**1.122** (d)	**1.123** (15)	**1.124** (24)	**1.125** (a)
1.126 (d)	**1.127** (a)	**1.128** (a)	**1.129** (7)	**1.130** (a)	**1.131** (d)	**1.132** (d)	**1.133** (6)	**1.334** (b, d)
1.135 (a, b, d)		**1.136** (d)	**1.137** (b)	**1.138** (b)				

Explanations Programming

1.1 (d)

Activation record → Subroutine call

Location counter → Assembler

Reference counts → Garbage collection

Address Relocation → Linking loader

1.2 (b)

Heap allocation is required to support dynamic data structure where memory can be allocated and deallocated at run time.

1.3 (d)

Activation tree helps to access the non-local variables in faster using an array of pointers to activation records.

1.4 (c)

Starting value of 'x' stores in global variable. Var x = 0.25; real, after that put value according to question.

1.5 (a)

Pointers can not be implemented if the processor supports only immediate and direct addressing modes. Pointer requires indirect addressing modes.

1.6 (d)

Take some values and verify the options as 10, 6, 7.

1.7 (c)

Indirect addressing → Pointer;

MOV A, [100]

Immediate addressing → Constants;

MOV A, 10

Auto decrement → Loops;

for (i = 0; i < 10; i– –);

1.8 (a)

It defines an array each element of which is a pointer to a structure of type node.

1.9 (d)

m = malloc (S);
m = NULL;
} → Lost memory: (memory is lost for m)

---

free (n);
n → values = 5;
} → Dangling pointer: (No memory for n)

```
char * p;  } → Uninitialized pointer:
*p = 'a';  }
```

(p is not initilized yet. So trying for *p will fail)

1.10 (a)

Definition of aliasing.

1.11 (c)

Here structure creates the minimum of array and union, but union only creates the minimum for only 'long z' which is max.
So total minimum required = 18.

1.12 (a)

Count is static variable so every time when value is changed it becomes the new value of count.
So, first count value is 0 then 1, then 3, then 6 at last 10.

1.13 (b)

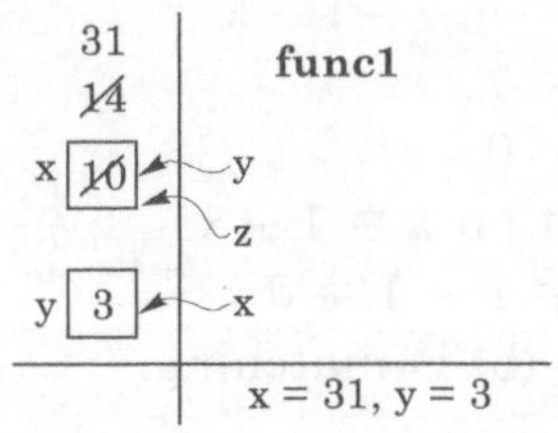

1.14 (d)

- P[1] will give segmentation fault, since when function is over, space allocated to local variable will be deallocated and accessing that location will give segmentation fault because address of local variable is return.
- P[2] also give segmentation fault, since px is returned which contain address of memory 10, when function is over accessing that memory will give segmentation fault because may be memory 10 will be allocated to some other pointer variable.
- P[3] may give segmentation error, when system run out of memory and dynamic memory will not allocated to pointer px.

1.15 (d)

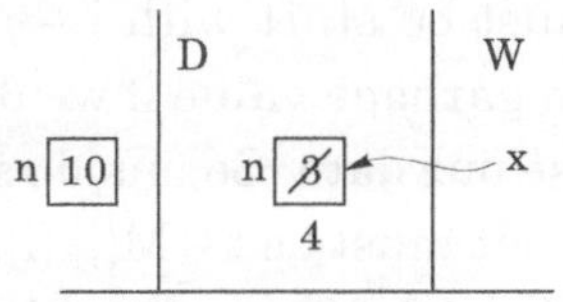

1.16 (b)

Call by value result and call by reference may differ in the presence of aliasing (sending the same parameter more than once).

1.17 (b)

In C language number of activation records between the current activation and the activation record for the main depends on the actual function calling sequence.

Example 1:

```
main(){     → f(){     → g(){     → h(){
f();          g();       h();
}             }          }          }
```

Calling sequence:

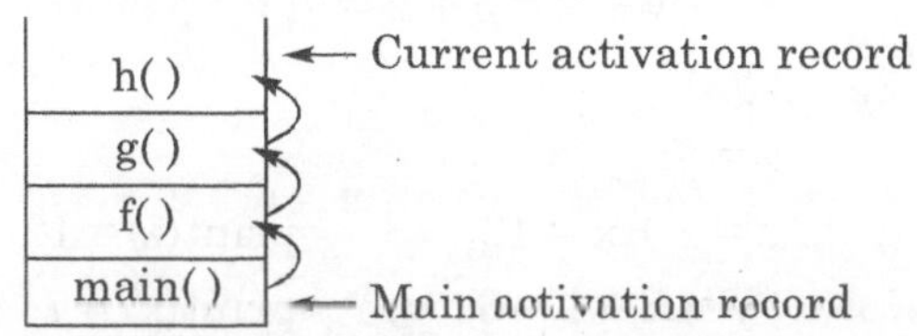

Example 2:

```
main(){     → g(){     → h(){     → f(){
g();          h();       f();
}             }          }          }
```

Calling sequence:

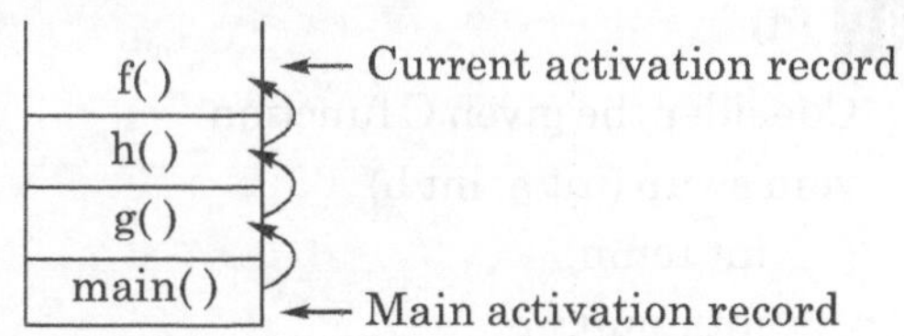

1.18 (b)

Address of a [40][50] = Base address + 40 × 100 × element size + 50 * element size
= 0 + 4000 × 1 + 50 × 1
= 4050
Therefore, option (b) is correct.

1.19 (a)

Here int *A[10] is an array of 10 pointers.
Which can be A[2][3] as its left hand side.
It can also be A[2] as its left hand side and also B[2][3] can obviously be used as left hand side as it is given in declaration of B[10][10] only, but B[1] can not be used because it is declared as 2-dimensional array.

1.20 (d)

1. First for loop fill array with '0' from index 3 to n.
2. Second for loop location where index j is not multiple of k i.e. non multiple index of k and update index of non multiple index of 4 as 1.

Since we are updating one array A as 1 because of we are doing step 2 for every value of 2 to $2 \times \log_2 n$. So if (! A[j] will return false for each index of array 4 because we set each index in step 2. So it will print nothing.

1.21 (a)

x = 5;
p (& x);
x = *y + 2 = 5 + 2 = 7;
Q(x);
z = 7;
z = z + x = 12 ⇒ print(z) = 12
*y = x – 1 = 6 ⇒ print(x) = 7
print(x) = 6

1.22 (c)

The goal of structured programming is able to infer the flow of control from the program text.

1.23 (d)

Consider the given C function

```
void swap (int a, int b)
{   int temp;
    int temp;
    temp = a;
    a = b;
    b = temp;
}
```

The main program for given code is

```
# include <stdio.h>
void main( )
{   int x, y;
    x = 1;
    y = 2;
    swap(x, y);
    printf("%d\n", a);
    printf("%d\n", y);
}
```

If we call swap (x, y) then there is no interchange in the value of x and y because the parameters are passed by value. There is interchange in formal parameters a and b but no interchange in actual parameter x and y because the scope of a and b lies with in the function but not in the main program so the value of x and y remain unchanged. If we will call it by reference then the value of x and y will be interchanged as below code.

```
void swap (int * a, int * b)
{   int temp ;
    temp = *a;
    *a = *b;
    *b = temp;
}
```

So option (d) is the current.

1.24 (b)

Put x = 0
then x = 1
so use option x =1 – x
Put x = 1
then x = 0
so use option x = 1 – x
= 1 – 1 = 0

So option (b) is matching.

1.25 (c)

Code for Transpose of matrix:

```
for(i = 0; i < 4; ++i)
    {
    for(j = i; j < 4; ++j)
        {
            t = M[j][i];
            M[i][j] = M[j][i];
            M[j][i] = t;
        }
    }
```

Key part here is both index i.e. i and j should not start with same index. Otherwise transpose will be done 2 time which results original matrix. So here first for loop start with i = 0. So second for loop must be start with j = i. Since t initially contain garbage value if we do M[i] [j] = t then we loose our data. So, inside second for loop 1st statement must be t = M[j] [i].

Note: Take 2 × 2 matrix and run the code on it. Will save time in exam.

1.26 (a)

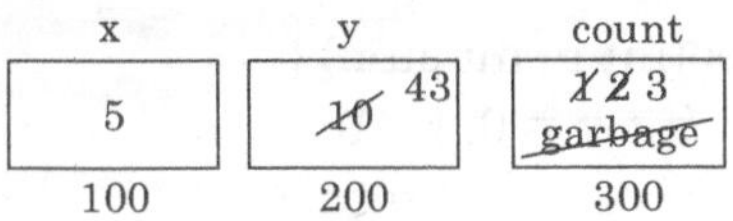

1. y = 10 + funcf(5) + funcg(5)

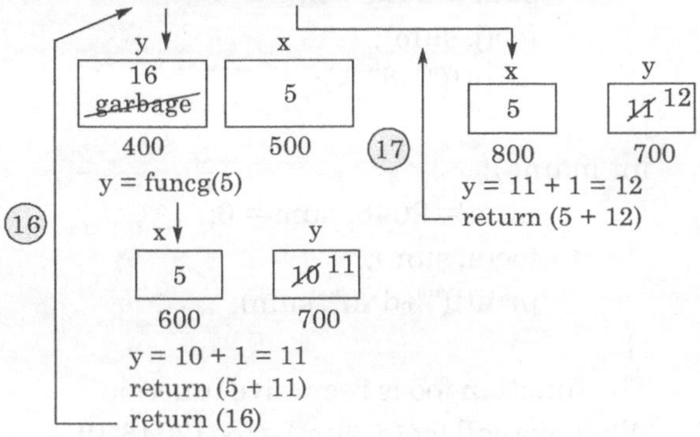

y = 10 + 16 + 17 = 43

2. y = 43 + funcf(5) + funcg(5)

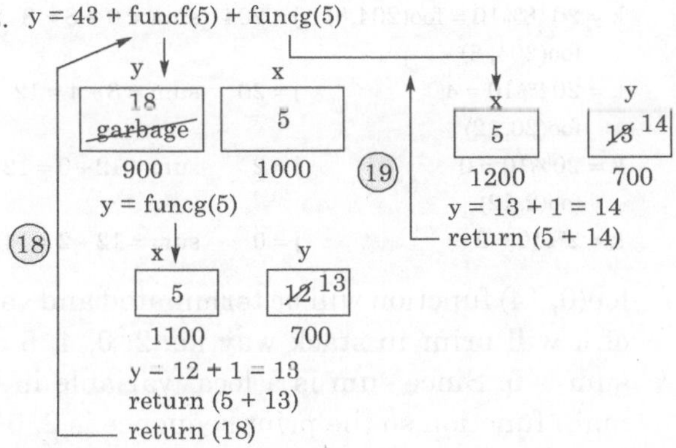

y = 43 + 18 + 19 = 80

1.27 (b)

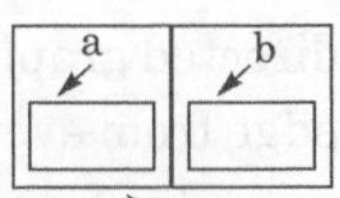

1. main()
   Struct t s = {"A", "B"}; so structure s will be created of same structure as t and value assign A to S → a and B is assign to S → b. printf (s.a, s.b) will print A.B.

2. f1(s) → f1(t s)

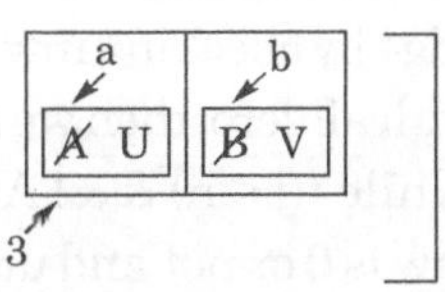

s.a = 'U'
s.b = 'V'
Printf (s.a, s.b) will print U and V.

3. printf (s.a, s.b) will print A.B.

4. f2(&s) → f2(t * p)

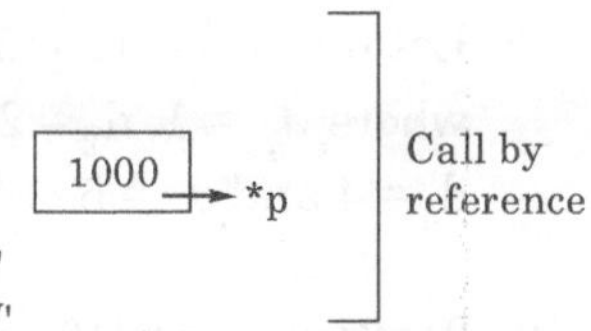

P → a = 'V'
P → b = 'W'
Printf (P → a, P → b) will print V and W.

1.28 (c)

```
int f(int n)
{    static int i = 1;
     if (n >= 5) return n;
     n = n + i;
     i ++;
     return f(n)
}
```

	Iteration	n	i
	0	1	1
Initilize	1	1 + 1 = 2	2
	2	2 + 2 = 4	3
	3	4 + 3 = 7	4

So when we call f(1), the value i = 4 and n = 7, if condition is $7 \geq 5$ so returns the value 7.

1.29 (c)

The given C program is non-recursive. So we will get the value of n after some iteration. The main thing about the program is that.

1. If m = n then n is the GCD
2. If m > n then m = m – n
3. If m < n then n = n – m

For example let x = 40 and y = 32

m = x = 40 and n = y = 32
m > n so apply the 2nd condition
m = m – n
m = 40 – 32 = 8
m = 8 and n = 8

apply the first condition
GCD (40, 32) = 8

1.30 (a)

Loop invariant is the part of code motion. Loop invariant for while loop is a condition, if we assign this condition before the while loop then there is no effect in the code and produces same output. Consider the given code.

```
int n, rev;
rev = 0;
while (n > 0){
    rev = rev * 10 + n%10;
    n = n/10;
}
```

Let input n = 12345
where $d_1 = 1, d_2 = 2, d_3 = 3,$
$d_4 = 4$ and $d_5 = 5$

Iteration	n = n/10	rev = rev * 10 + n%10
	$d_1d_2d_3d_4d_5$	
0	1 2 3 4 5	rev = 0
	$d_1d_2d_3d_4$	d_5
1	1 2 3 4	rev = 0 * 10 + 5 = 5
	$d_1d_2d_3$	d_5d_4
2	1 2 3	rev = 5 * 10 + 5 = 5 4
	d_1d_2	$d_5d_4d_3$
3	1 2	rev = 54 * 10 + 2 = 5 4 3
	d_1	$d_5d_4d_3d_2$
4	1	rev = 543 * 10 + 2 = 5 4 3 2
		$d_5d_4d_3d_2d_1$
5	0	rev = 5432 * 10 + 1 = 5 4 3 2 1

Lent $n = d_1d_2 \ldots d_m$ in the ith iteration
$rev = d_m.d_m - i + 1$
For example if m = 5 and i = 3 the above example produces $n = d_1d_{5-3} = 12$ and
$rev = d_5d_4d_{5-3+1} = 543$

1.31 (d)

Because of first character is '\0' nothing will be printed.

1.32 (c)

int (*f) (int*);

return type int, (*f) is a pointer to a function the argument is (int*) an integer pointer So, int (*f) (int*) means a pointer to a function that takes an integer pointer as an argument and returns an integer.

1.33 (c)

The abstract data type (ADT) refers to a programmer defined data type together with a set of operations that can be performed on that data so the choice (c) is correct.

1.34 (d)

Without declaration we can't call any function. If you call it will assume by default return data type as integer. But here return data type is double. So it will give compile time error type mis-match.

1.35 (d)

```
void foo(int n, int sum) {
    int k = 0, j = 0;
    if (n = = 0) return;
    k = n%10; j = n/10;
    sum = sum + k;
    foo(j, sum);
    printf("%d", k);
}
int main(){
    int a = 2048, sum = 0;
    foo(a, sum);
    printf("%d\n", sum);
}
```

The function foo is recursive function
When we call foo(a, sum) = foo (2048, 0)

k	j	sum
k = 2048%10 = foo(204,8)	j = 204	sum = 0 + 8 = 8
foo(204, 8)		
k = 204%10 = 4	j = 20	sum = 8 + 4 = 12
foo(20, 12)		
k = 20%10 = 0	j = 2	sum = 12 + 0 = 12
foo(2, 12)		
k = 2%10 = 2	j = 0	sum = 12 + 2 = 14

foo(0, 14) function will be terminated and value of k will print in stack way i.e. 2, 0, 4, 8 and sum = 0. Since sum is a local variable in the main function so the print sequence is 2, 0 , 4, 8, 0.

1.36 (c)

A sink in a directed graph is a vertex i such that there is an edge from every other vertex $j \neq i$ to i and there is no edge from i to any other vertex.

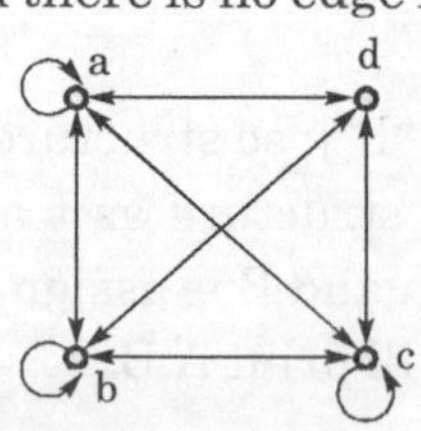

$$\begin{array}{c} \\ a \\ b \\ c \\ d \end{array} \begin{array}{c} \begin{array}{cccc} a & b & c & d \end{array} \\ \begin{bmatrix} 1 & 1 & 1 & 1 \\ 1 & 1 & 1 & 1 \\ 1 & 1 & 1 & 1 \\ 0 & 0 & 0 & 0 \end{bmatrix} \end{array}$$

In above image 'd' is sink i.e. there is no outgoing edge by checking row of all vertex if we find any row with all zero, then we find a potential vertex for sink. While ((j < n) &&! A[i] [j]) will check every entry row is 0 or not and until it find 1 in row, increment j. If it find 1 in row skip a row by which it does not has any incoming edge by doing i = j.

1.37 (d)

In this program initially flag is set to 1 i.e. assume sink is present in the graph. And by using For loop if flag is reset to 0 then sink is not present otherwise sink is present.

For loop will set flag = 0 if the coloum entries are all 1 for sink (vertex) and the entries for row is all 0.

If we look at above diagram for vertex 'd', i = 3 and j = 0.

So while ((j! = i) && (A[3] [0] || !A[0] [3])

True && (0 || !1))

True && false = false. so flag not set to 0.

Then while ((j! = j) && (A[3] [1] || !A[1] [3])

True && false = false, so flag not set to 0.

Try for other vertex we will set flag = 0 i.e. not exist sink.

Like this for vertex d flag will not set to set to 0 i.e. sink exist.

1.38 (d)

Definition of anagram: a string s is a string obtained by permuting the letter in s.

1st for loop just create an array count and initialize to 0 for each element i.e. for 128 element.

While loop run till string reaches to end of strings. In while loop, for every element a[i] count is increased which is decreased by array b[i] when same element arrive in array b[i] and increment loop with j++. If both strings are anagram of each other the count value will be zero for each elements.

So, count[a[j]] ++; and count[b[j++]] – – ;

Example:

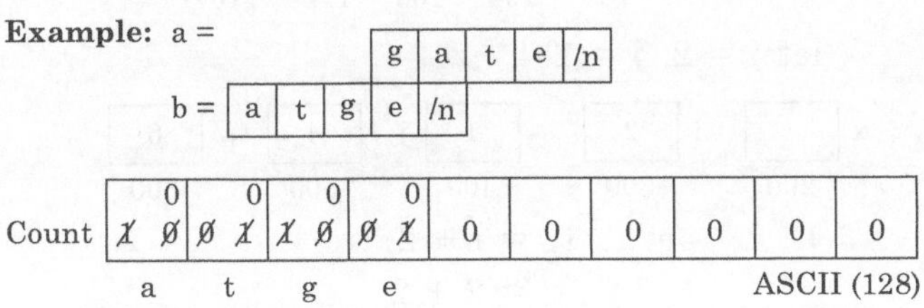

Inside while loop

(1) count[g]++;
count[a]– –;
j++;

(2) count[a]++;
count[t]– –;
j++;

(3) count[t]++;
count[g]– –;
j++;

(4) count[e]++;
count[e]– –;
j++;

While loop fail for (5).

Since atlast count will contain all zero. So both are anagram of each other.

1.39 (c)

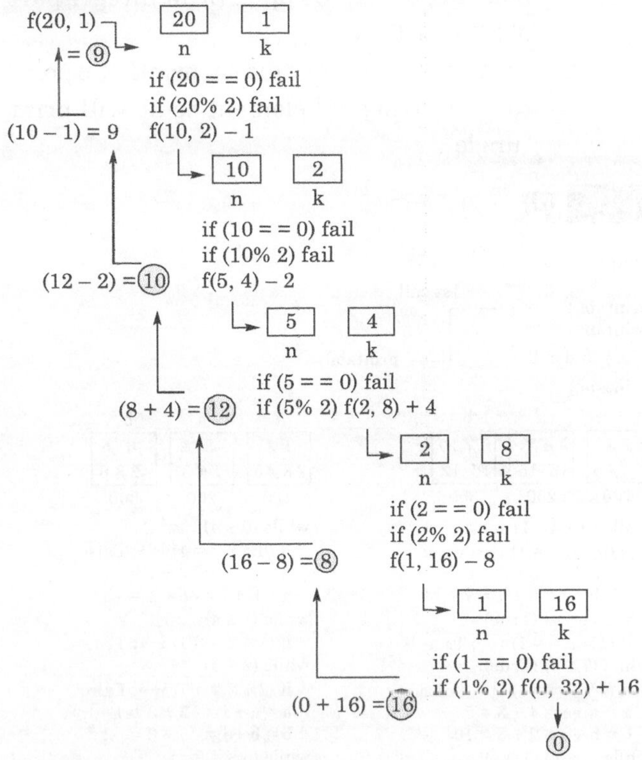

So, 9 will be printed.

1.40 (b)

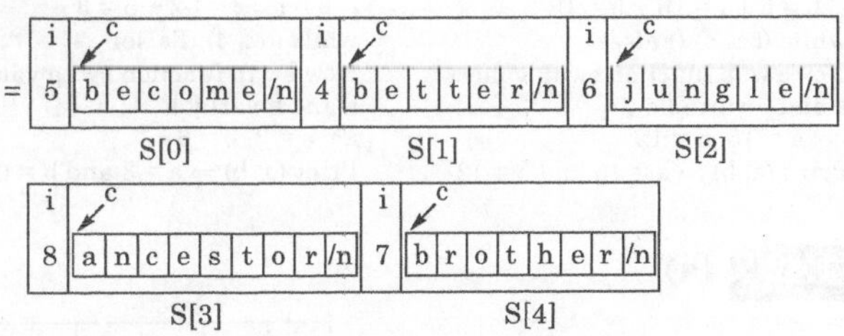

test *p = str; here p is pointing to str[ ] now.

p+ = 1; p = p + 1

Now p is pointing to S[1].

++p → c evaluate as ++(p → c) i.e. now p is pointing to | b | e | t | t | e | r | /n | (p at e).

So,

1. Printf ("%s", p++ → c); prints "etter".
   p++; p points to S[2]
2. Printf ("%c", * ++ p → c); evaluated as *(++ (p → c)) i.e. p is now points to | j | u | n | g | l | e | /n | (p at j) now p is increment to 1 character i.e | j | u | n | g | l | e | /n | (p at u). Which is printed i.e. 'u'.

3. Printf ("%d", p[0].i); since after printf p is pointing to S[2]. So p[0].i print integer part of S[2] i.e. 6.
4. Printf ("%s", p → c); since in printf '2' p point to char string 2nd element so at will print "ungle".

1.41 (d)

```
main()
{
printab()  ──────── 1st call ────────→ printab()
printab()
   ↓ 2nd call
printab()
```

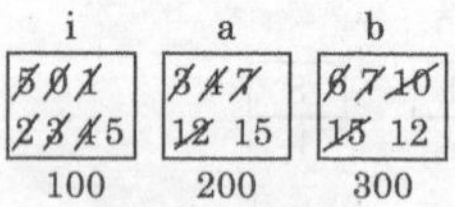

```
while (0 ≤ 4) (True)
   if (0% 2 = = 1) i++; False
   a = a + i = 3 + 1 = 4
   a = b + i = 6 + 1 = 7
while (1 ≤ 4) (True)
   if (1% 2 = = 1) i++; True
while (2 ≤ 4) (True)
   if (2% 2 = = 1) i++; False
   a = a + i = 4 + 3 = 7
   b = b + i = 7 + 3 = 10
while (3 ≤ 4) (True)
   if (3% 2 = = 1) i++; True
while (4 ≤ 4) (True)
   if (4% 2 = = 1) i++; False
   a = a + i = 7 + 5 = 12
   b = b + i = 10 + 5 = 15
while (5 ≤ 4) (False)
```

Now swap function swap value of a and b variable
i.e. a = 15, b = 12
Print (a, b) ⇒ a = 15 and b = 12

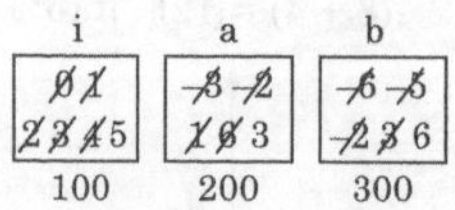

```
while (0 ≤ 4)
   if (0% 2 = = 1) i++; False
   a = a + i = –3 + 1 = –2
   a = b + i = –6 + 1 = –5
while (1 ≤ 4)
   if (1% 2 = = 1) i++; True
while (2 ≤ 4)
   if (2% 2 = = 1) i++; False
   a = a + i = –2 + 3 = 1
   b = b + i = –5 + 3 = –2
while (3 ≤ 4)
   if (3% 2 = = 1) i++; True
while (4 ≤ 4)
   if (4% 2 = = 1) i++; False
   a = a + i = 1 + 5 = 6
   b = b + i = –2 + 5 = 3
while (5 ≤ 4) (False)
```

Now swap function swap value of a and b variable
i.e. a = 3, b = 6
Print (a, b) ⇒ a = 3 and b = 6

1.42 (a)

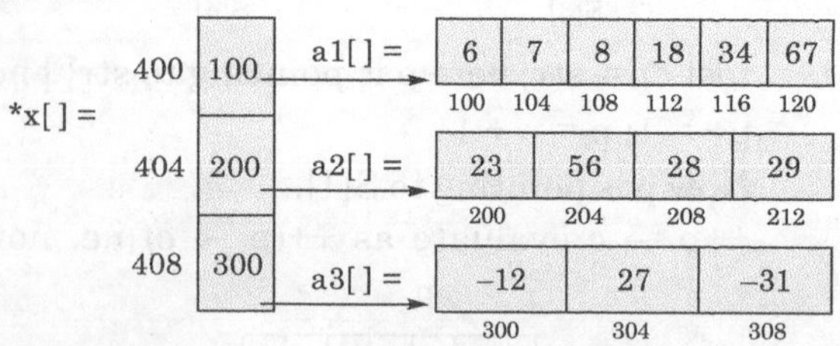

*a[ ] = &x;
*a[ ] contain base address of array x;

1. Printf (a[0] [2]) = *(*(a + 0) + 2) = *(100 + 2 × 4) = *(108) = 8
2. Printf (*a[2]) = *(*(a + 2) + 0) = *(408 + 0) = *(408) = –12
3. Printf (*++ a[0]) = *(++(a + 0)) = *(++(100)) = *(104) = 7
4. Printf (*(++ a)[0]) = *(*(a + 1) + 0) = *(*(404) + 0) = *(200 + 0) = *(200) = 23
5. Printf (a[–1] [+1]) = *(*(a – 1) + 1) = *(*(404 – 4) +1) = *(104+1) = *(104 + 1 × 4) = *(108) = 8

Note: In step 3 'a1' point to 'a1' = 104 and in step 4 'a' point to 404 as starting address used in step 5.

1.43 (c)

When $n = 0,\ {}^mC_n = {}^mC_0 = 1$
$m = n,\ {}^mC_n = {}^mC_m = 1$

1.44 (d)

If array a and b contain identical elements then neither (i) nor (ii) holds.

1.45 (b)

Let a[ ] be an array:

Working of work 1:

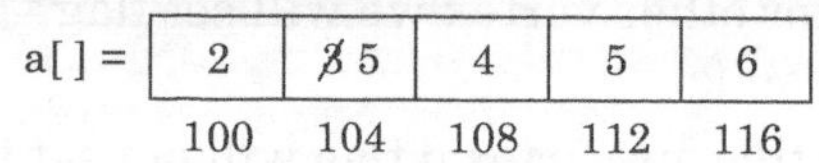

let i = 2, j = 1

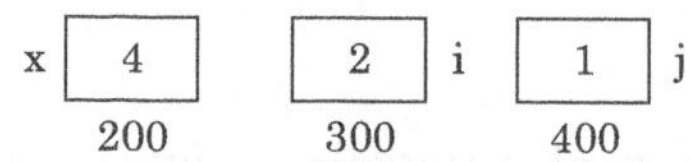

1. x = a[100 + 2 × 4] = a[108]
2. a[1] = 4 + 1 = 5
3. return a[i + 2] – 3
⇒ a[100 + 2 × 4] – 3
⇒ a[108] – 3
⇒ 4 – 3 = 1

Working of work 2:

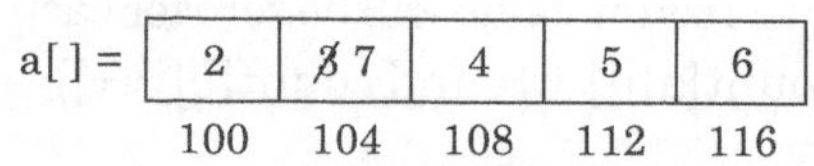

let i = 2, j = 1

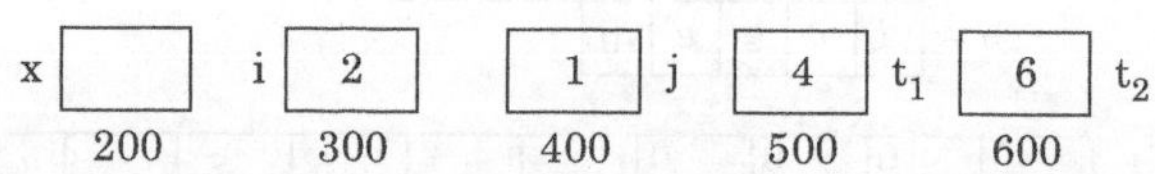

1. $t_1 = i + 2 = 2 + 2$
2. $t_2 = a[4] = a[116] = 6$
3. $a[1] = t_2 + 1 = 6 + 1 = 7$
4. return $t_2 - 3$
⇒ 6 – 3 = 3

Both work 1 and work 2 will produce different output some time and work 2 is compute i + 2 one time, hence take less time to compute where as work 1 will compute 2 times.

1.46 (c)

- Given code will not give any compilation error.
- When an address passed to pointer variable px and py will be deallocated by some other function, then access those memory location will generate segmentation fault.

Eg:
```
fun () {
        int b = 10;
        int * p = (int *) malloc (size of (int));
              *p = &b;
        int * a = (int *) malloc (size of (int));
              *a = p;
        free (p);
        call (int &a, &a);
}
```
will generate segmentation fault because memory allocated to p will be deallocated which is pointed by pointer variable a, which is passed as parameter.

- Swap procedure work correctly when px and py will have different memory location but when both px and py will contain same memory address then it will not work correctly or lost 1 variable value.

1.47 (d)

To compute f(5), initially r = 5.

f(3) + 2 = 18   [(n > 3) : f(n – 2) + 2]
↑
f(2) + 5 = 16   [f(n – 2) + r]
↑
f(1) + 5 = 11   [f(n – 2) + r]
↑
f(0) + 5 = 6   [f(n – 2) + r]
↑
1   [n ≤ 0]

1.48 (d)

- By analysing code we see when n = 2^k i.e. power of 2 then f(n/2) will definitely terminate since n% 2 always equal to 0. Hence after $\log_2 n$ operation n = i and function will terminate.
- When we input number like 5 and 7 it will enter into infinite loop i.e.

for n = 5; f(5) → f(14) → f(7) → f(20) → f(10) (→ back to f(5))

for n = 7; f(7) → f(20) → f(10) → f(5) → f(14) (→ back to f(7))

So, for any power of 2^k function terminates which is infinite so option (b) is true and for infinite values of n for which function will not terminate i.e. number 5×2^k, 14×2^k, 7×2^k, 20×2^k and 10×2^k for any value of k.

So option (d) is true.

1.49 (c)

Syntax error in above code segment. Right code given below:
```
# include <stdio.h>
int main( )
{
    int sum = 0, maxsum = 0, i, n = 6;
    int a[ ] = {2, –2, –1, 3, 4, 2};
    for (i = 0; i < n; i++) {
        if (i == 0 | | a[i] < 0 | | a[i] < a[i – 1])
        {
            if (sum > maxsum) maxsum = sum;
            sum = (a[i] > 0) ? a[i] : 0;
        }
        else
        sum += a[i];
    }
    if (sum > maxsum) maxsum = sum;
    printf (“%d\n”, maxsum);
}
```
Simply execute and trace value of sum and maxsum for i = 0 to 5

i →	initially	0	1	2	3	4	5
sum	0	0	2	0	3	7	2
maxsum	0	0	2	2	2	2	7

so maxsum = 7

1.50 (b)

push 5
push 2
push 5*2 = 10. (pops 5 and 2)
push 3
push 3
push 2

push 3 + 2 = 5 (pops 2 and 3)

push 5 * 3 = 15 (pops (5 and 3)

push 15 + 10 = 25 (pops (15 and 10)

1.51 (d)

Using call by value: In call by value actual value are passed to formal parameter and operation applied on formal parameter so no change made to actual parameter.

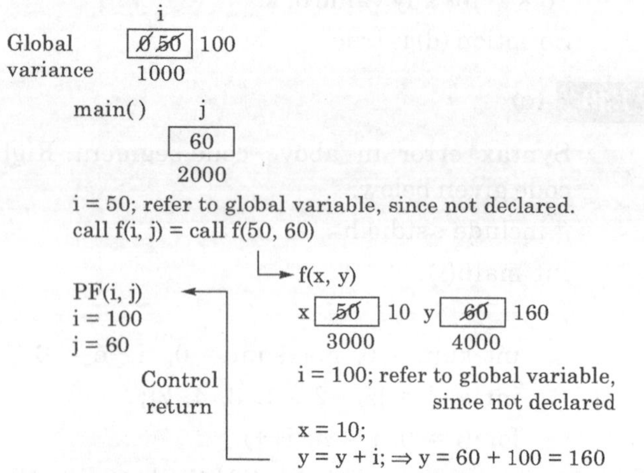

Using call by reference: In call by reference address of actual parameter are passed to call function and any operation performed on it reflect to actual parameter.

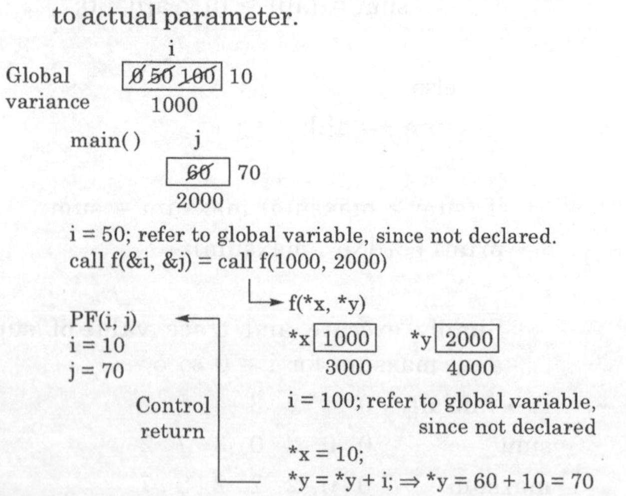

1.52 (c)

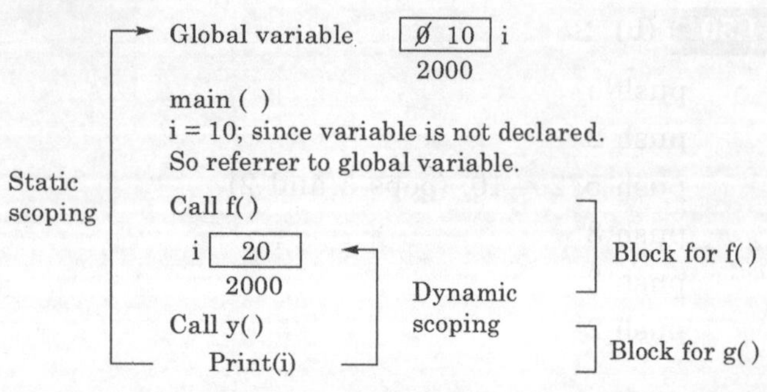

In dynamic scoping undeclared variable referred to declared variable in block from which call is made i.e. here call f( ). But in static scooping undeclared variable always take value from global variable.

So static scooping x = 10

Dynamic scooping x = 20

Note: Global variable by default initialize to 0.

1.53 (a)

a = (x ≤ y) ? ((x ≤ z) ? x : z) : ((y > z) ? y : z)

In C "? : " is a ternary operator the syntax is (exp1 ? exp2 : exp3). It means if exp1 is true print exp2 else exp3.

Let x = 3, y = 4, z = 2.

⇒ a = (3 > 4) ? ((3 > 2) ? 3 : 2) : ((4 > 2) ? 4 : 2)

a = (3 > 4) ? 3 : 4 = 4

1.54 (b)

char a[6] =

W	o	r	l	d	\0
a[0]	1	2	3	4	5

When Loop will execute first time,

a[0] = a[5]

a[0] = '\0'

printf("%s", a), the string starting at address a prints the string starting with '/0' and it indicates the end of string, so it will print null string.

1.55 (c)

Num 1 = 5 num 1 = 4

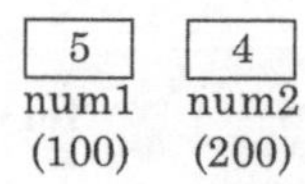

1st condition fail

2nd condition fail

3rd condition execute because (num1> = num2)

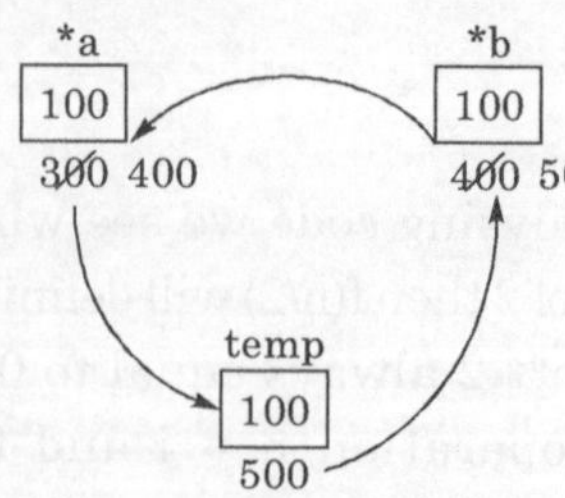

So value of a = 4

value of b = 5

i.e., swapping done via call by reference.

1.56 (c)

Inside main function:

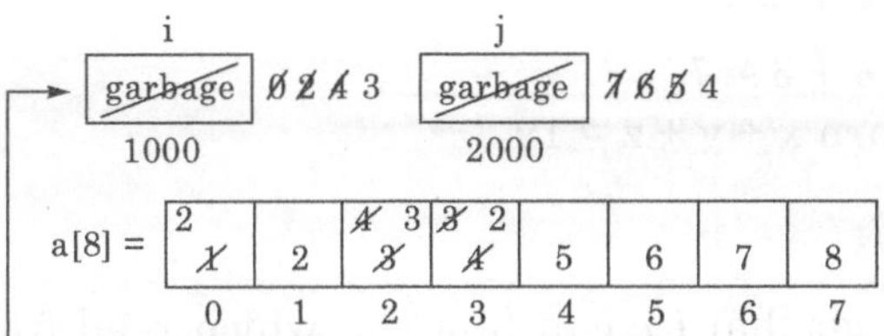

B_ock scope of i and j

for: i < 3 we need to perform a[i] = a[i] + 1; i+ +;

i = 0; a[0] = a[0] + 1; i+ +;

i = 2; a[2] = a[2] + 1; i+ +;

i – –; | 3 3 2 | i
3000

for: j > 4 we need to perform i = j/2; a[i] = a[i] – 1;

j = 7; i = j/2 = 3; a[3] = a[3] – 1;

j = 6; i = j/2 = 3; a[3] = a[3] – 1;

j = 5; i = j/2 = 2; a[2] = a[2] – 1;

Block scope of inner 'i'

printf ("%d%d", i, a[i])

i = 3

a[3] = 2

Note: When any variable not declared inside any block then it can referred to outer block for same variable.

1.57 (b)

char [2] [3] = {{'a', 'b', 'c', } {'d', 'e', 'f' }}

a →

	0	1	2
0	1000 a	1001 b	1002 c
1	1003 d	1004 e	1005 f

char*p = *b make new char pointer p which point to array b.

p →

	0	1
0	3000 garbage	3001 garbage
1	3002 garbage	3003 garbage
2	3004 garbage	3005 garbage

*(p + 2 * j + i) = a[i] [j]; will assign element of array a to array b, by scanning array 'd' row wise and sorting character in array 'b' column wise.

When i = 0 : i < 2 (true)

j = 0; j < 3 (condition true)
*(3000 + 2*0 + 0) = a[0][0];
*(3000) = a;
j++ ;
j = 1; j < 3 (condition true)
*(3000 + 2*1 + 0) = a[0][1];
*(3002) = b;
j++;
j = 2; j < 3 (condition true)
*(3000 + 2*2 + 0) = a[0][2];
*(3004) = c;
j++;
j = 3; j < 3 (condition false)
i++;

When i = 1 : i < 2 (true)

j = 0; j < 3 (condition true)
*(3000 + 2*0 + 1) = a[1][0];
*(3001) = d;
j++;
j = 1; j < 3 (condition true)
*(3000 + 2*1 + 1) = a[1][1];
*(3003) = e;
j++;
j = 2; j < 3 (condition true)
*(3000 + 2*2 + 1) = a[1][2];
*(3005) = f;
j++;
j = 3; j < 3 (condition false)
i++;

Final array b[3] [2] will be:

b →

	0	1
0	3000 a	3001 d
1	3002 b	3003 e
2	3004 c	3005 f

1.58 (d)

10101101 i.e Option (d)

since recusive funtion calls will be

1st function call $\frac{173}{2}$ = 86(int part only)

2nd function call $\frac{86}{2}$ = 43

3rd function call $\frac{43}{2}$ = 21(int part only)

4th function call $\frac{21}{2}$ = 10(int part only),

5th function call $\frac{10}{2}$ = 5,

6th function call $\frac{5}{2}$ = 2 (integer part only),

7th function call $\frac{2}{2}$ = 1,

In 7th function call condition if(n <=1) will become true so n will be printed (i.e 1 will be printed) now while returning every function will execute its remaining part of code i.e. printf ("%d", n%2);

So 6th function will print 2 mod 2 = 0,
5th function will print 5 mod 2 = 1,
4th function will print 10 mod 2 = 0,
3rd function will print 21 mod 2 = 1,
2nd function will print 43 mod 2 = 1,
1st function will print 86 mod 2 = 0,
Function call f(173) will print
173 mod 2 = 1

1.59 (c)

Execution of P1:

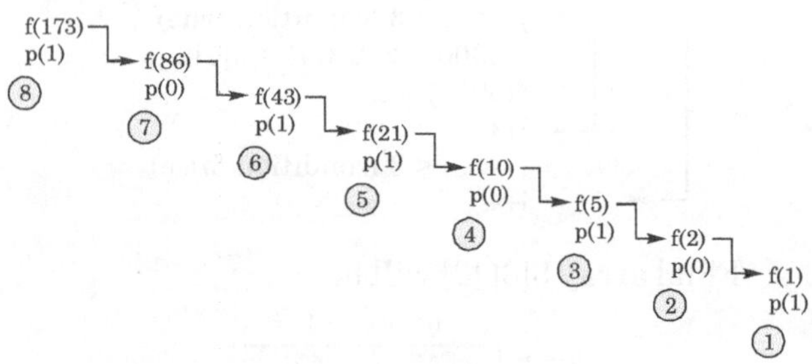

So output will be 10101101.

Execution of P2:

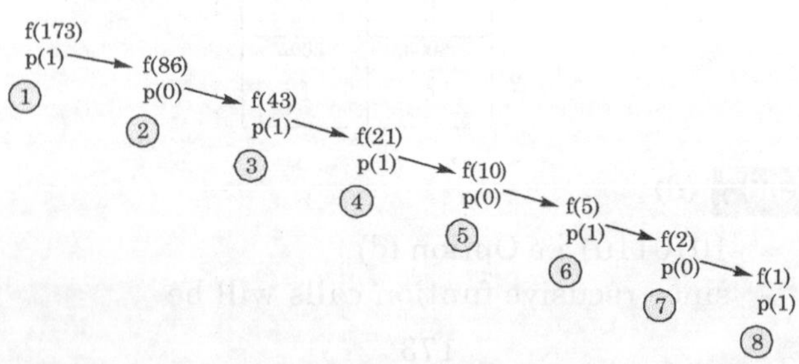

So output will be 10110101.

1.60 (b)

f(c, b, a) is called by the main( ) function the graphical execution of the program is given below.

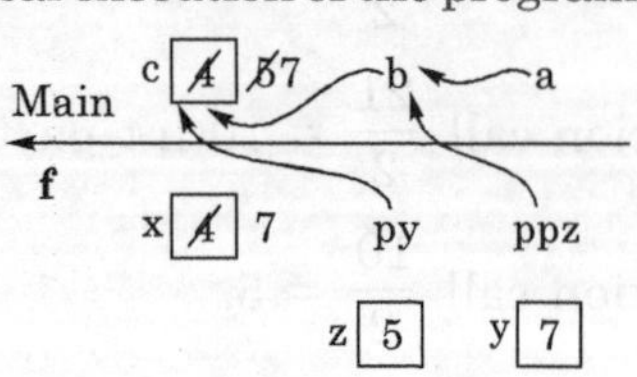

⇒ x + y + z = 7 + 7 + 5 = 19

int y, z;
**ppz = **ppz + 1;
z = **ppz = 5;
*py = *py + 2;
y = *py = 7
x = 4 + 3 = 7
return x + y + z = 19

1.61 (d)

- getchar ( ) is a function which read from standard input and return integer value.
- putchar ( ) is a function which used to write a character on standard output/screen.

Here if condition need to check input is finished or not by doing ((c = getchar ( ))! = '\n') i.e. when character = '\n', '\n' represent end of input.
After reverse ( ) function we need to print the output by using putchar (c).

1.62 (b)

x = 15
fun (5, &x) let &x = 100
↓
fun (5, 100)
↓ t = 5, f = 5 + 3, *f_p = 5
return 8
fun (4, 100)
↓ t = 3, f = 3 + 2, *f_p = 3
return 5
fun (3, 100)
↓ t = 2, f = 2 + 1, *f_p = 2
return 3
fun (2, 100)
↓ t = 1, f = 1 + –1, *f_p = 1
return 2
fun (1, 100)
*f_p = 1
return 1

So fun(5, & x) will **return** 8 and it will be printed.

1.63 (d)

In the given program it begin from main function, i and j are globally initialized by 0 and 1. So when we call function f(& i, & j) the address of i and j are passed.
when p = q and *p = 2 means
*q = 2, so value of *q is passed.
and value of *q return to j.
Value of i and j are 0 and 2 respected.
So printf("% d %d", i, j) gives output (0, 2)

1.64 (c)

a[ ] =	12	7	13	4	11	6

So f(a, 6) → first this call

Thus is the number stored in the array and we can access these by using

*a – f(a, 6)
↓ 12

So12 + f(a + 1, n – 1) {even}
↓
7 – f(a + 1, n – 1) {odd}
↓
13 – f(a + 1, n – 1) {odd}
↓
4 + f(a + 1, n – 1) {even}
↓
11 – f(a + 1, n – 1) {odd}
↓
6 + f(a + 1, 0) {even}
↓
0

= 12 + [7 – [13 – [4 + [11 –(6 + 0)]]]]
= 12 + [7 – [13 – [4 + [11 – 6]]]
= 12 + [7 – [13 – [4 + 5]]] = 12 + (7 – (13 – 9))
= 12 + (7 – 4) = 12 + 3 = 15

1.65 (d)

In given program we take the test cases and apply

From T_1 if all values are equal means a = b = c = d, a = b condition satisfied. S_1 and S_4 executed.

From T_2 when all a, b, c, d distinct:
S_1, S_2 not execute, S_3 executes.

From T_3 when a = b, S_1 execute but c = d.
S_2 will not execute but S_3 and S_4 executes.
Here no need of T_3 because we get all result from above two.

From T_4 if a! = b and c = d, S_1 not executes and S_2 and S_4 executes.

All of S_1, S_2, S_3, S_4 will execute and covered by T_1, T_2 and T_4.

1.66 (c)

C ↘

Address	2000	01	02	03	04	05	06	07	08
Value	G	A	T	E	2	0	1	1	10
Index	0	1	2	3	4	5	6	7	8

P

2000

P + P[3] – P[1] = 2000 + 69 – 65 = 2004

⇒ 2011 string will be printed.
(From 2004 Address)

1.67 (b)

Consider the following recursive C function foo (n, r)

```
{
    if (n > 0) return (n% r) + foo (n/r, r);
    else return 0;
}
```

Value return by the function foo (345, 10):

foo (345, 10)
⇓
5 + foo (34, 10)
⇓
4 + foo (3, 10)
⇓
3 + foo (0, 10)
⇓
0

The return value is 5 + 4 + 3 = 12

1.68 (d)

Call	Return (↑)
foo (513, 2)	
⇓ 1 + foo (256, 2)	= 1 + 1 = 2
⇓ 0 + foo (128, 2)	= 0 + 1 = 1
⇓ 0 + foo (64, 2)	= 0 + 1 = 1
⇓ 0 + foo (32, 2)	= 0 + 1 = 1
⇓ 0 + foo (16, 2)	= 0 + 1 = 1
⇓ 0 + foo (8, 2)	= 0 + 1 = 1
⇓ 0 + foo (4, 2)	= 0 + 1 = 1
⇓ 0 + foo (2, 2)	= 0 + 1 = 1
⇓ 0 + foo (1, 2)	= 0 + 1 = 1
⇓ 1 + foo (0, 2)	= 1 + 0 = 1
⇓ 0	

foo(513, 2) = 2

1.69 (c)

There is no break statement present in the code. Hence the switch case will jump to case : A and execute every statement from there.

∴ Choice A
Choice B No choice.

1.70 (c)

a b c (Global)

| 0 | 0 | 0 |

Static int a (Main)

| ~~1~~ 2 |

Static ⇒ (ptrFun)

a | ~~2~~ | ~~4~~ 6 b | + | 2 b | + | 2

a = a + ++b

O/P 4 2
6 2 ∴ Option 'C' is correct.
2 0

1.71 (d)

a b c

| 0 | 0 | 0 |

a (Main)

| + | 2

a | ~~2~~ | 4 b | + | 2 a | ~~2~~ | 4 b | ~~1~~ | 2

a = a+ ++b

O/P 4 2
4 2
2 0

1.72 (b)

Value of x is passed by reference so every time changes are reflected to original value

f(5, 5)
↓
x * f(6 * 4) ⇒ 9^4
↓
x * f(7 * 3) ⇒ 9^3
↓
x * f(8, 2) ⇒ 9^2
↓
x * f(9, 1) ⇒ 9
↓
1

when c = 1, x = 9
So x * x * x * x = $x^4 = 9^4 = 6561$
Option (b) is correct.

1.73 (d)

pi = &i; pi → | | i

scanf("%d", pi); pi → | 10 | i

printf("%d\n", i + 5); pi → | 15 | i

The value entered is 10. It prints 15.

1.74 (c)

$$\sum_{i=0}^{n}\sum_{j=0}^{n}\sum_{k=j+1}^{n} 1 = \frac{1}{6} n \times (n+1)(n+2)$$

1.75 (9)

In fun( ), num >>= 1 will be evaluated as num = num >> 1 i.e. left shift 1 bit of bit representation of num or num divide by 2.
Binary representation of 435 = 110110011 here while loop will terminate when all bit became 0 i.e. left shift 9 time, so bit now becomes 000000000. Hence count = 9.

1.76 (b)

While calculating $^nC_3 = \frac{n(n-1)(n-2)}{6}$ i.e. multiplying n × (n – 1) × (n – 2) then divide by 6. Since n value can be large such that n × (n – 1) × (n – 2) will generate value which cannot be representable by bits assigned to unsigned int.

So, idea is to calculate $^nC_3$ as $\left(\frac{n(n-1)}{2}\right)\frac{(n-2)}{3}$.

Since n(n – 1) always gives an even number greater than 0, so always divisible by 2 but if n(n – 1) divide by 3 first then it may be possible that answer comes in floating point, since we use integer variable so part after decimal point need to be discarded i.e. will not get exact answer as $^nC_3$. So most appropriate choice is

$$= \left(\frac{n(n-1)}{2}\right) \times \frac{(n-2)}{3}$$

1.77 (1.73)

$|x^2 - 3| < 0.01$

if $x = \sqrt{3}$, the given program returns x.

Therefore, $x = \sqrt{3} = 1.732$

For $q = 1.73 \Rightarrow f(q)$ returns q.

1.78 (d)

$f(50)$ goes into infinite loop by calling $f(i)$ recursively for $i = j = 50$.

1.79 (–5)

Here both f_1 and f_2 swaps the two arguments. f_1 is call by value but f_2 is called by reference hence changes made by only f_2 are reflected in main function.

After f_2 is called, the values of a, b and c are 4, 6 and 5 respectively.

Hence the output is 5 – 4 – 6 = –5

1.80 (b)

q = 0
r = x
while (r ≥ y) do
  begin
    r = r – y
    q = q + 1
  end

Post condition: [after program termination]

r < y and r = r – qy

$\Rightarrow$ r = x – qy

$\Rightarrow$ x = r + qy

$\therefore$ Post condition is: $(r < y) \wedge (x = qy + r)$

1.81 (51)

Initially compute fun(1), fun(2), fun(3) and fun(4)

(i) fun(1) = 1

(ii) fun(2)

$k = 1: x = x + \text{fun}(1) \times \text{fun}(1) = 1 + 1 \times 1 = 2$

$\therefore$ fun(2) = 2

(iii) fun(3):

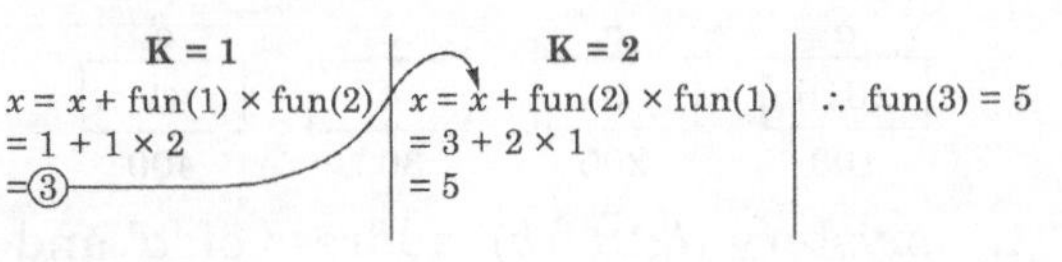

(iv) fun(4):

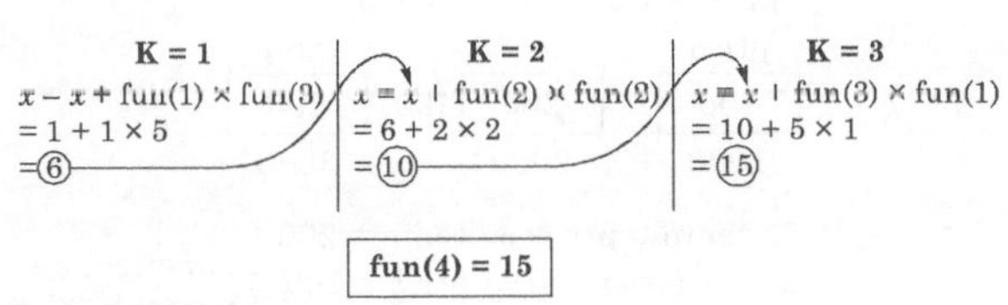

Compute now fun(5)

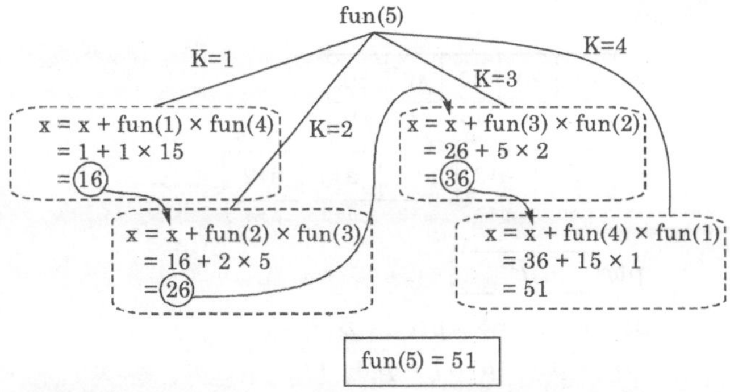

1.82 (d)

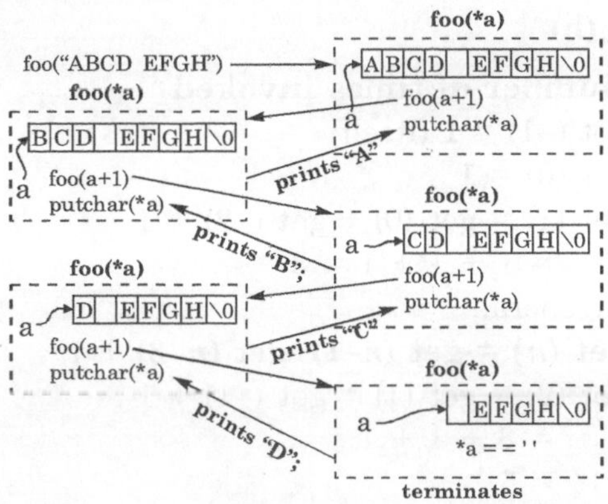

$\therefore$ Output is DCBA

1.83 (5)

The minimum number of movements are 5. The following sequences will give minimum movements, after 1 moving.

Seq 1: Move 2 left, move 5 left, move 6 up, move 18 up, move 25 left.

Seq 2: Move 2 left, move 4 up, move 6 left, move 18 up, move 25 left.

2 ←	4 ↑	5	14
3	6 ←	18 ↑	23
10	12	25 ←	∞
31	∞	∞	∞

1.84 (c)

1	2	3	4	\0		
1000	1001	1002	1003	1004	1005	1006

s_1 [1000]

p = [1002]

$*p$ = '0'; $\Rightarrow$

1	2	0	4	\0		
1000	1001	1002	1003	1004	1005	1006

Therefore the output is 1204

1.85 (140)

a	10	20	30	40	50
	a+0	a+1	a+2	a+3	a+4

p	a	a+3	a+4	a+1	a+2
	p+0	p+1	p+2	p+3	p+4

ptr [*P*]

ptr ++ ; ⇒ *ptr* is *p* + 1.

ptr – *p* = (*p*+1) – *p* = 1

***ptr* = *(*(*p* + 1)) = *(a + 3) = **40**

∴ Output is 1, 40 which is **140**.

1.86 (b)

Number of times invoked

get (–1) = **1** (itself)

get (0) = **1**

get (1) = get (0) + get (–2) + 1
= 1 + 1 + 1 = **3**

Ingeneral,

get (*n*) = get (*n*–1)+get (*n*–3) + 1

get (2) = get (1) + get (–1) + 1
= 3 + 1 + 1
= **5**

get (3) = get (2) + get (0) + 1
= 5 + 1 + 1
= **7**

get (4) = get (3) + get (2) + 1
= 7 + 3 + 1
= **11**

get (5) = get (4) + get (2) + 1
= 11 + 5 + 1
= **17**

get (6) = get (5) + get (3) + 1
= 17 + 7 + 1 = **25**

∴ Number of times get function invoked on get (6) = **25**

1.87 (10)

i = 0, *k* = –1 only default executes –*i* + *k* = –1 ⇒ executes 1 time.

i = 1, *k* = –1 only default executes –*i* + *k* = 0 ⇒ executes 1 time.

i = 2, *k* = –1 – *i* + *k* = 1 ⇒ executes 3 times (cases 1, 2, 3 and default)

i = 3, *k* = –1 –*i* + *k* = 2 ⇒ executes 3 times (cases 2,3 and default)

i = 4, *k* = –1 –*i* + *k* = 3 ⇒ executes 2 times (cases 3 and default)

∴ Total 10 times printf is executed.

1.88 (230)

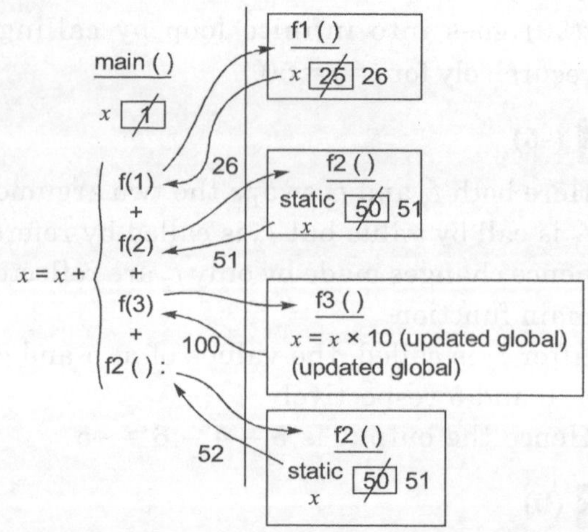

x = 1 + 26 + 51 + 100 + 52 = 230

1.89 (d)

f(*s*, **s*) since there is no pointer variable define with **s*. So this function will give type checking error.

i = *f*(*i*, s) since function prototype is void in void *f* (int, short) i.e., *f* is accepting argument int type and short type and its return type should be integer because we store the output of function in variable *i* which is integer type. So there must be type casting into integer type but type casting is not present. So this function will give type checking error.

f(*i*, **s*) since there is no pointer variable define with **s*. So this function will give type checking error.

f(*i*, **p*) in this function to argument are pass one is int type and another is short type defined in the main function. So this function will not give type checking error.

1.90 (2016)

a	*b*	*c*	*d*
2016	0	4	42
100	200	300	400

1. mystery (&*a*, &*b*); address of '*a*' and '*b*' is passed.

*ptr *a* = &*a*; *ptr *b* = &*b*;

ptr *a*	ptr *b*	temp
100	~~200~~ 100	200
1000	2000	3000

1. temp = ptr *b*; ⇒ temp = 200
2. ptr *b* = ptr *a*; ⇒ ptr *b* = 100
3. ptr *a* = temp; ⇒ ptr *a* = 200

} No effect on variable '*a*' and '*b*'

2. if (2016 < 4) false
   So, mystery (&a, &b);

   *ptr a = &a; *ptr b = &d;

ptr a	ptr b	temp
100	~~200~~ 100	400
1000	2000	3000

1. temp = ptr b; ⇒ temp = 400
2. ptr b = ptr a; ⇒ ptr b = 100
3. ptr a = temp; ⇒ ptr a = 400

} No effect on variable 'a' & 'd'

Print$f$("a"); ⇒ 2016
So, output of above program is 2016.

1.91 (d)

Option (a) is fail for descending array. Since n value is not decreament in program.
Option (b) is fail for ascending array. Since value of 'a' is increment every time and in while condition we check for value of 'b'. So it is incorrect.
Option (c) is fail for input 5 6 7 1 2 3. Since pointer a is pointing to 6 and 'b' will pointing to 7 i.e., 'a' and 'b' are adjacent to each other. Then while condition $b > (a + 1)$ is fail and it return 6 is maximum value which is incorrect.
Option (d) at the end of the program, pointer 'a' and 'b' pointed to same element then only while condition is false and return the maximum value of array.

1.92 (a)

D ~~1~~ ~~2~~ ~~3~~ 4

count (3)
n [3]
1. Print$f$(n) = 3
2. Print$f$(d) = 1
   d++;
   if (3 > 1) true
   count (2)
   n [2]
   3. Print$f$(n) = 2
   4. Print$f$(d) = 2
      d++;
      if (2 > 1) true
      count (1)
      n 1
      5. Print$f$(n) = 1
      6. Print$f$(d) = 3
         d++;
         if (1 > 1) false
         count (1)
      7. Print$f$(d) = 4
   8. Print$f$(d) = 4
9. Print$f$(d) = 4

So output will be printed in serial order given in solution i.e., 312213444.

1.93 (d)

a [~~3~~ 1] globaly initialize.

1. m(3)

   y [3]

   1. a = ~~1~~ ~~2~~ 4
   2. a = 3 – 1 = 2
      n(a);

      n(2)

      x [2]

      $x = x * a$;
      = 2 ∗ 2 (here a is taken from global variable)
      = 4

   Print$f$(4); = 4
3. Print$f$(a) = 4 since dynamic scoping is used. So, take value of inner variable 'a'. So answer will be 4, 4.

1.94 (30)

i [~~5~~ 20] (1000) j [10] (2000)

f($\$i, j$);

*P [1000] m [~~10~~ 15]

1. $m = m + 5$; ∵ $m = 15$
2. $*P = *P + m$; ∵ $*P = 20$
3. Print$f$($i + j$); ∵ 20 + 10 = 30

So value printed by above code is 30.

1.95 (c)

Take random value of X and Y i.e., X = 5 and Y = 3.
Initially X = 5, Y = 3, res = 1, a =X and b =Y.

Option (a): $X^Y = a^b$
$X^Y = a^b \equiv 5^3 = 5^3 \equiv 125 = 125$

After iteration 1
res = 5; a = 5; b = 2; X = 5; Y = 3
$X^Y = a^b \equiv 5^3 \neq 5^2 \equiv 125 \neq 25$
So, case fail. Option (a) cannot be answer.

Option (b): $(res * a)^Y = (res * X)^b$
$(1 \times 5)^3 = (1 \times 5)^3 \equiv 125 = 125$

After iteration 1
res = 5; a = 5; b = 2; X = 5; Y = 3
$(res * a)^Y = (res * X)^b \equiv (5 \times 5)^3 = (5 \times 5)^2$
15625 ≠ 625 So, case fail. Option (b) cannot be answer.

Option (d): $X^Y = (res * a)^b$
$5^3 = (1 \times 5)^3 \equiv 125 = 125$

After iteration 1

res = 5; a = 5; b = 2; X = 5; Y = 3

$X^Y = (res * a)^b \equiv 5^3 = (5 \times 5)^2$

$125 \neq 625$ So, case fail.

Option (d) cannot be answer.

Option (c): $X^Y = res * a^b$

$5^3 = 1 \times 5^3 \equiv 125 = 125$

After iteration 1

res = 5; a = 5; b = 2; X = 5; Y = 3

$X^Y = res * a^b \equiv 5^3 = 5 \times 5^2 \equiv 125 = 125$

After iteration 2

res = 25; a = 5; b = 1; X = 5; Y = 3

$X^Y = res * a^b \equiv 5^3 = 25 \times 5^1 \equiv 125 = 125$

So, all cases are passes.

So option (c) will the answer.

1.96 (3)

a	3	5	2	6	4

1. *f*(*a*, 5) is a function contain 2 parameter one contain starting address of array and second parameter tell number of element in the array.
2. Every time 'n' value compare with 1 when it is less than equal to 1 return 0 and stop the program otherwise continue with recursive function call.

1. *f*(*a*, 5)
   *P = *a*; *P* pointed to same address pointed by '*a*'.
   $n = 5$; *n* value greater than 1.
   So, max (*f*(*P* + 1, 5 – 1), 3 – 5); or
   max (*f*(*P* + 1, 4) – 2);
2. *f*(*P* + 1, 4)
   *$P = P + 1$; P is pointed to next element of array i.e., 5.
   $n = 4$; *n* value greater than 1.
   So, max (max(*f*(*P* + 1, 4 – 1), 5 – 2), –2) or max (max(*f*(*P* + 1) 3), 3), –2)
3. *f*(*P* + 1, 3)
   *$P = P + 1$; P is pointed to next element of array i.e., 2.
   $n = 3$; *n* value greater than 1.
   So, max(max(max(*f*(*P* + 1, 3 – 1), 2 – 6), 3) –2) or max(max(max(*f*(*P*+1, 2), – 4), 3) –2);
4. *f*(*P* + 1, 2);
   *$P = P+1$; *P* is pointed to next element of array i.e., 6.
   $n = 2$; *n* value greater than 1.
   So, max(max(max(max(*f*(*P*+ 1, 2 – 1), 2),–4), 3), –2) or
   max(max(max(max(*f*(*P*+1, 1), 2), –4), 3),–2)
5. *f*(*P* + 1, 1);
   *$P = P+1$; *P* is pointed to next element of array i.e., 4.
   $n = 1$; *n* value equal to 1 so, return 0.
   So max(max(max(max(0, 2), –4), 3), –2)
   max(max(max(2, –4), 3), –2)
   max(max(2, 3), –2)
   max(3, –2) = 3.

So the value printed by given code is 3.

1.97 (23)

Here total () function will return number of bits are set i.e., 1 binary representation of value of variable 'v'.

The output of above program is:

	5	4	3	2	1	0
	(101)	(100)	(011)	(010)	(001)	(000)
	↓	↓	↓	↓	↓	↓
Static count:	2	3	5	6	7	
	+	+	+	+	+	
Static *x*:	0 / 2 ↗	2 / 5 ↗	5 / 10 ↗	10 / 16 ↗	16 / 23 ↗	23

1.98 (c)

In foo (int var) function, variable 'val' is created for every function call and foo (val ––) means first call is made after that 'val' values is decremented by 1 i.e.,

```
foo (3)
int val = 3̸2;
int x = 0;
x = x + foo (3 ––);
        ↓
        foo (3)
        int val = 3;
        int x = 0;
        x = x + foo (3 ––);
```

which makes stack overflow, so abnormal termination.

In bar (int val) function, variable 'val' is created for every function call and bar (val – 1) means new call is made with 1 less than previous value of 'val' i.e.,

```
bar (3)
int val = 3̸2;
int x = 0;
x = x + bar (2); {infinite loop}
        ↓
        bar (2)
        int val = 3;
        int x = 0;
    x = x + bar (1); {infinite loop}
            ↓
            bar (1)
            int val = 3;
            int x = 0;
            x = x + bar (0); terminate
```

If we see while (var > 0) is make infinite loop since 'val' value is not decremented in bar( ) function for value of variable val > 0, it created infinite loop.

1.99 (3)

Strlen (s) = 3 and Strlen (t) = 5
Value of c is 0
[Strlen (s) – Strlen (t) > 0]: Strlen (s) : Strlen (t);
[Strlen (s) – Strlen (t)] = –2
Since, we are comparing with a unsigned integer.
So, [Strlen (s) – Strlen (t)] = [–2] = 2.
2 > 0 : Strlen (s) : Strlen (t); since 2 > 0 will evaluate True so it will give output as Strlen (s) = 3.
Hence '3' will be printed.

1.100 (a)

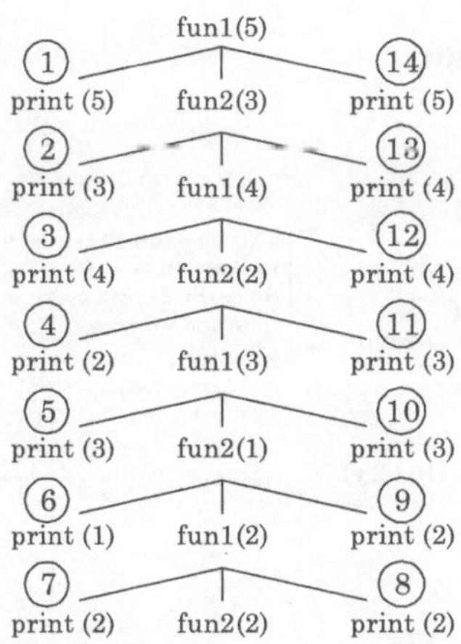

Hence the output is : 53423122233445

1.101 (a)

- **static char var;** : Initialization of a variable located in data section of memory.
- ***m* = malloc(10); *m* = NULL;** : A lost memory which can't be freed because free (*m*) is missed in code.
- **char *ptr[10];** : Sequence of memory locations to store addresses.
- **register int var1;** : Request to allocate a CPU register to store data.

1.102 (c)

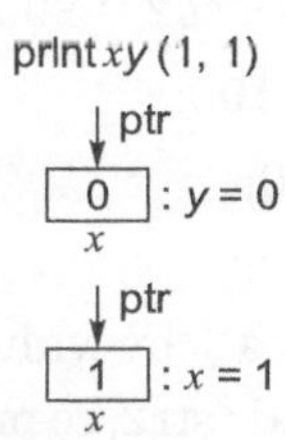

Hence the output will be (1, 0).

1.103 (2)

(C + 2[p] – 6[p] – 1)

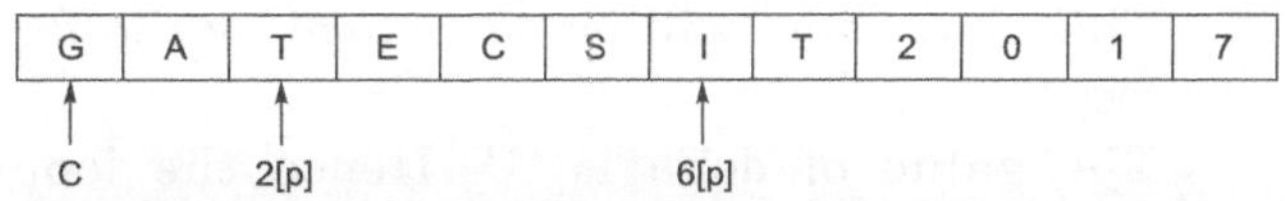

$$\begin{aligned} & C + \text{'T'} - \text{'I'} - 1 \\ &= C + 11 - 1 \\ &= C + 10 \end{aligned}$$

Hence printf will print '2'.

1.104 (0)

```
1. int m = 10;     // m = 10
2. int n, n1;
3. n = ++m;        // n = 11
4. n1 = m++;       // n1 = 11, m = 12
5. n – –;          // n = 10
6. – – n1;         // n1 = 10
7. n – = n1;       // n = 0
8. printf("%d", n);
```

The output will be 0.

1.105 (3)

3	5	1	4	6	2
0	1	2	3	4	5

First for loop:

(*i* = 0)	5	3	1	4	6	2
(*i* = 1)	5	3	1	4	6	2
(*i* = 2)	5	3	4	1	6	2
(*i* = 3)	5	3	4	6	1	2
(*i* = 4)	5	3	4	6	2	1

Second for loop:

(*i* = 5)	5	3	4	6	2	1
(*i* = 4)	5	3	4	6	2	1
(*i* = 3)	5	3	6	4	2	1
(*i* = 2)	5	6	3	4	2	1
(*i* = 1)	6	5	3	4	2	1

Now since done is '0', hence the for loops will execute again.

First for loop:

(*i* = 0)	6	5	3	4	2	1
(*i* = 1)	6	5	3	4	2	1
(*i* = 2)	6	5	4	3	2	1
(*i* = 3)	6	5	4	3	2	1
(*i* = 4)	6	5	4	3	2	1

Second for loop:

(*i* = 5)	6	5	4	3	2	1
(*i* = 4)	6	5	4	3	2	1
(*i* = 3)	6	5	4	3	2	1
(*i* = 2)	6	5	4	3	2	1
(*i* = 1)	6	5	4	3	2	1

Value of done is still '0', hence the for loop will execute again.
First for loop:
This time there will be no change by the for loop.
The value of done is '1'. Hence the loop terminates as

6	5	4	3	2	1
0	1	2	3	4	5

The output of the program will be '3'.

1.106 (d)

In **main( )** function, **malloc( )** function return a void type address. So no need of type casting i.e., **(int*)** malloc (size of (int));, inplace of type casting malloc (size of (int)); will not give compilation error. **Assignval( )** function assign 0 in memory location pointed by 'x' and return address of memory location pointed by 'x'. Since next **malloc( )** function can take memory location assign by previous **malloc( )** function because there is no **free(m)** is given to freed previously assign memory location.
So this can give memory leak problem if variable 'x' non-zero value.

1.107 (c)

Here $\quad x == (y \times q + r)$
$\quad\quad q$ = Quotient and r = Remainder
In given code indirectly we are dividing 'r' by 'y' in form of substraction and 'q' count the number of time number gets divide successfully before reaching $r \geq y$.
So, initial condition will be $q = 0$ i.e., $r = x$ and final condition $(q == 0)$ && $(r == x)$ && $(y > 0)$.

1.108 (a)

	100	101	102
Struct p =	1	0	c

Since we know 'a' + 2 = skip 'a', 'b', i.e. 'c'
Now $\quad$ *q = &p
i.e. now q point to struct p.
So, *((char*) q + 1)) will typecast structure into char and skip one char and points to '0' i.e. addresses '101' and *(101) = 0.
And *((char*) q + 2) will typecast structure into char and skip two char and points to 'c' i.e. addresses '102' and *(102) = c.
Final output by given code is 0, c.

1.109 (4)

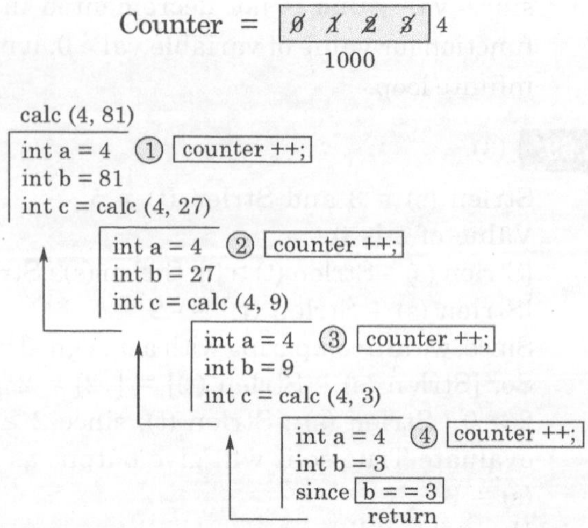

So, $\quad$ Counter = 4
Hence 4 will be printed.

1.110 (10230)

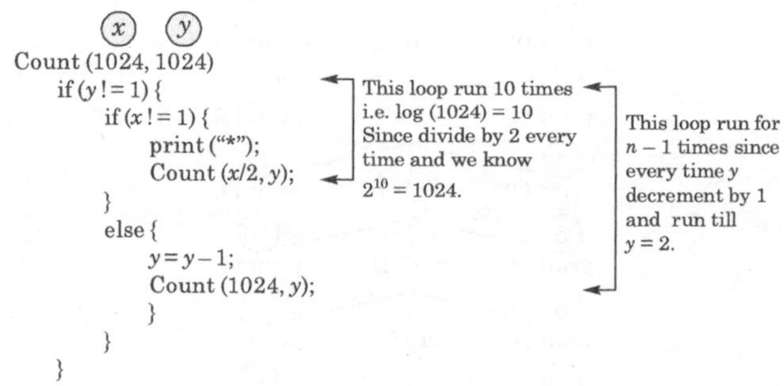

So, inner if loop will run 10 times and y run for 1023 times, which make $x = 1024$ for every y.
So count (1024, 1024) = $10 \times 1023 = 10230$.

1.111 (a)

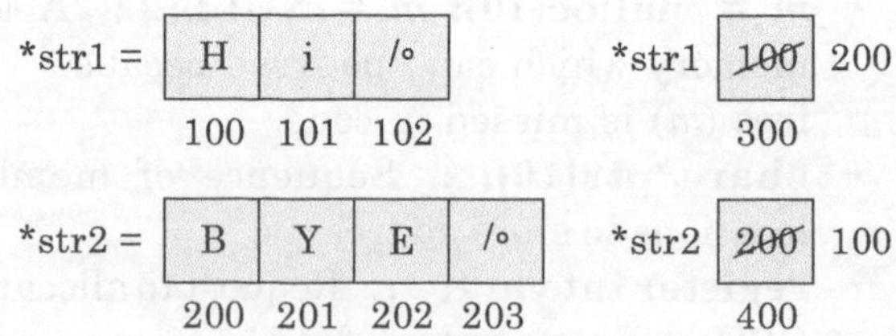

1. fun1(100, 200) **"Call By Value"**
   *s1 = 100; *s2 = 200;
   *temp = 100;
   *s1 = 200;
   *s2 = 100;

Since, fun1 is call by value. So no change to original *str1 and *str2, so printf will print Hi Bye

2. fun2(&str1, &str2) **"Call By Refference"**
   **s1 = 300; **s2 = 400;
   *temp = 300;
   **s1 = 400;
   **s2 = 300;

Since, fun2 is call by refference. So it will swap the value of str1 and str2. So, str1 points to "Bye" and str2 points to "Hi". So printf will point Bye Hi.

Final output: Hi Bye Bye Hi

1.112 (b)

Ø40	j	Ø5	sum
100		200	

First for loop deal with value of 'j':

1. for ($i = n$; $i > 1$, $i = i/2$)
   j++;
   When $n = 2^{40}$ will make $j = \log(2^{40}) = 40$.

Second for loop deal with valuc of 'sum':

1	j = 40	Since j > 1	So sum++ = 1	Then j = $\frac{40}{2} = 20$
2	j = 20	Since j > 1	So sum++ = 2	Then j = $\frac{20}{2} = 10$
3	j = 10	Since j > 1	So sum++ = 3	Then j = $\frac{10}{2} = 5$
4	j = 5	Since j > 1	So sum++ = 4	Then j = $\frac{5}{2} = 2$
5	j = 2	Since j > 1	So sum++ = 5	Then j = $\frac{2}{2} = 1$

Now for $j = 1$ loop will terminate and return sum = 5.

1.113 (6)

Let size of integer is 4 byte.

arr[ ] = {1, 2, 3, 4, 5, 6, 7, 8, 9, 0, 1, 2, 5}

Address	100	104	108	112	116	120	124	128	132	136	140	148	152
	1	2	3	4	5	6	7	8	9	0	1	2	5
					↑ ip								

ip = arr + 4

ip[1] = *(ip + 1)

= *(starting address of ip + 1 * size of integer)

= *(116 + 1 × 4)

= *(120) = 6

6 will be displayed on execution.

1.114 (26)

int x = 2, y = 5

Function jumble is call by value

$$y = \text{jumble}(y, x)$$

In function jumble

$$x = 2 \times x + y$$
$$x = 2 \times 5 + 2$$
$$x = 12$$

It will return 12 and y will hold the return value.

Now value of $x = 2, y = 12$

$$x = \text{jumble}(y, x)$$
$$= \text{jumble}(12, 2)$$

Function return 26 and x hold the return value x will be printed so 26 is printed by the program.

1.116 (5)

Initially $i = 2.0, j = 1.0$

while $\left(\frac{i}{j} > 0.0625\right)$

1. $\frac{i}{j} = \frac{2}{1} > 0.0625$
   $j = j + j = 1 + 1 = 2$
   First time sum will be printed.
2. $\frac{i}{j} = \frac{2}{2} > 0.0625$
   $j = j + j = 2 + 2 = 4$
   Second time sum will be printed.
3. $\frac{i}{j} = \frac{2}{4} > 0.0625$
   $j = j + j = 4 + 4 = 8$
   Third time sum will be printed.
4. $\frac{i}{j} = \frac{2}{8} > 0.0625$
   $j = j + j = 8 + 8 = 16$
   Fourth time sum will be printed.
5. $\frac{i}{j} = \frac{2}{16} > 0.0625$
   $j = j + j = 16 + 16 = 32$
   Fifth time sum will be printed.
6. $\frac{i}{j} = \frac{2}{32} = 0.0625$
   So while condition become false and break.

Total 5 time sum will be printed.

1.117 (d)

for (statement 1; statement 2; statement 3)
{
  // code block to be executed
}

- **Statement 1** is executed (one time) before the execution of the code block for initialization.
- **Statement 2** defines the condition and executed (every time) before executing the code block.
- **Statement 3** is executed (every time) after the code block has been executed.

Static initialization num

num = ~~7~~ ~~6~~ ~~5~~ ~~4~~ ~~3~~ ~~2~~ ~~1~~ ~~0~~ –1

1st iteration of for loop

- **Statement 1:** r( ) ⇒ return 7 and post decrement num to 6
- **Statement 2:** r( ) ⇒ return 6 ⇒ (condition true) and post decrement num to 5

  printf("%d", r( )) ⇒ return 5 and post decrement num to 4 and **print 5**

2nd iteration

- **Statement 3:** r( ) ⇒ return 4 and post decrement num to 3
- **Statement 2:** r( ) ⇒ return 3 ⇒ (condition true) and post decrement num to 2

  printf("%d", r( )) ⇒ return 2 and post decrement num to 1 and **print 2**

3rd iteration

- **Statement 3:** r( ) ⇒ return 1 and post decrement num to 0
- **Statement 2:** r( ) ⇒ return 0 ⇒ (Condition false) ⇒ Exit loop and post decrement num to –1

Output on the screen : 52

1.118 (10)

a[ ] = {2, 4, 6, 8, 10}
sum = 0, *b = a + 4

	100	104	108	112	116
a	2	4	6	8	10
					↑ b

- **1st iteration**

  $i = 0$

  sum = sum + (*b – i) – * (b – i)

  = 0 + (10 – 0) – 10 = 0
- **2nd iteration**

  $i = 1$

  sum = 0 + (10 – 1) – (8) = 1
- **3rd iteration**

  $i = 2$

  sum = 1 + (10 – 2) – 6 = 3
- **4th iteration**

  $i = 3$

  sum = 3 + (10 – 3) – 4 = 6
- **5th iteration**

  $i = 4$

  sum = 6 + (10 – 4) – 2 = 10

Output of the program is 10.

1.119 (81)

pp(3, 4) a = 3, b = 4, tot = 1

len = length (4, array) ⇒ i will return 3 with array set as [001] (array updated 1 only when b%2! = 0)

Now pp will run loop 3 times:

1. Since arr[0] = 0.

   So ex = 3 × 3 = 9
2. Since arr[1] = 0.

   So ex = 9 × 9 = 81
3. Since arr[1] = 1.

   So tot = 1 × 81 = 81

After executing the given C program integer 81 is returned.

1.120 (55)

- Variable i is static in both the functions thus its value will persist throughout the program.
- When fun2 (5) gets called, it will first call fun1 with value of n as 5 and then fun2 will gets called with n = 4.
- Similarly, same steps are taken for all the upcoming values of n.

1.121 (1023)

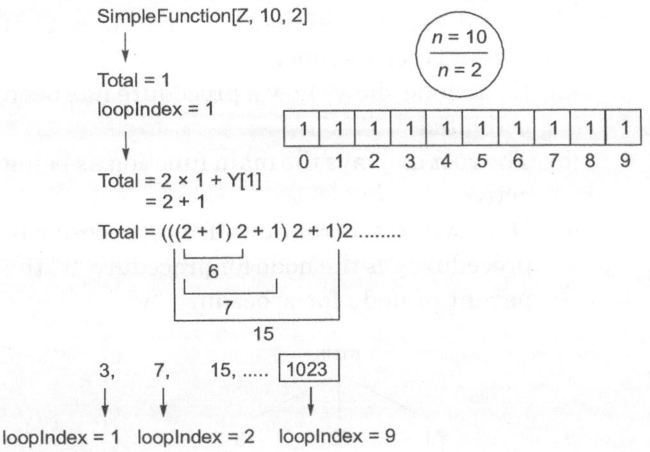

1.122 (d)

$j = -3, j = -2, j = -1$

⇒ Short circuiting in if((j >= 0) && (i++))

$j = 0; i = 0$ used but i is 1

$j = 1; i = 1$ used but i is 2

count $= 0 + 1 = 1$

$j = 2; i = 2$ used but i is 3

count $= 1 + 2$

$j = 3; i = 3$ used but i is 4

count $= 3 + 3 = 6$

count $= 6 + 4 = 10]$

corresponding to count = count + i

Option (d) is correct.

1.123 (15)

This function will keep on subtracting till both x and y becomes equal that is 15.

1.124 (24)

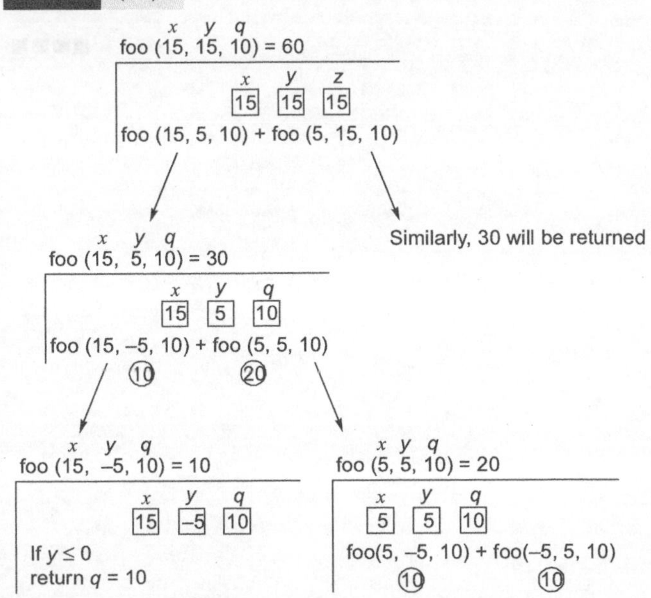

1.125 (a)

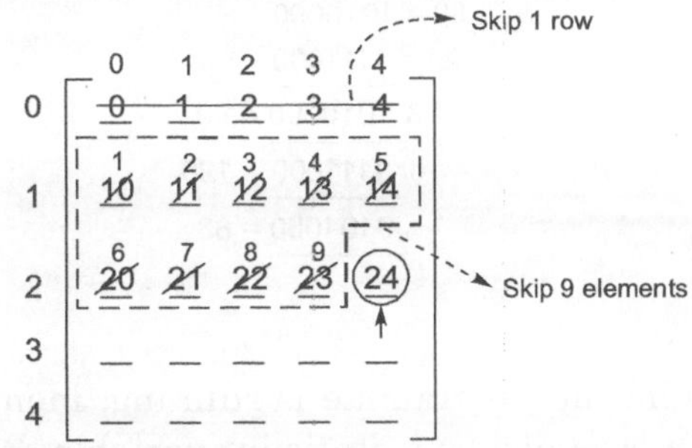

int a[4][5]

[*(a + 1) + 9]

It means skip one row and then skip 9 elements.

So, the resultant value is 24.

1.126 (d)

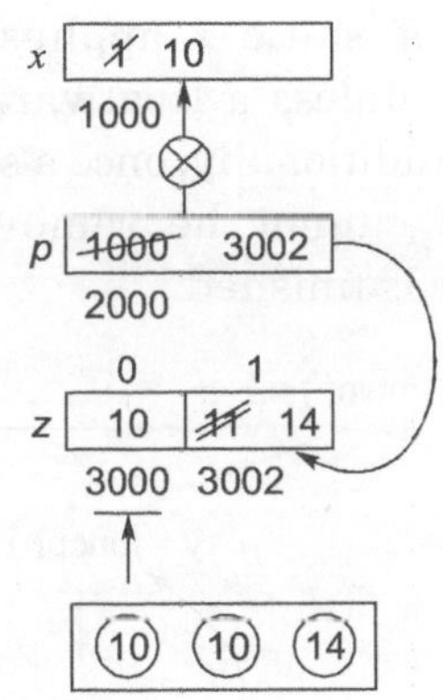

1.127 (a)

int a[3] [3] [3] = {{**1, 2, 3,** 4, 5, 6, 7, 8, 9},

{**10, 11, 12,** 13, 14, 15, 16, 17, 18},

2D 1D 3 {**19, 20, 21,** 22, 23, 24, 25, 26, 27}}

int i = 0, j = 0; k = 0;

for (i = 0; i < 3; i ++)

for (k = 0; k < 3, k ++)

printf(a[i] [j] [k]);

0 0 0	0 0 1	0 0 2	1 0 0	1 0 1
i j k	i j k	i j k	i j k	i j k

and so on.

= 1 2 3

10 11 12

19 20 21

This is the answer.

1.128 (a)

$a \rightarrow 80 = 1010000$
$b \rightarrow 120 = 1111000$
& $1010000 = 80$
or $1111000 = 120$
$\wedge\ 0101000 = 83$

1.129 (7)

While the programme is running, memory is still occupied by a static int variable. When a function call where the variable was declared is finished, a normal or auto variable is destroyed. Memory for static variables is allotted in the data segment rather than the stack segment.

Note: While the scope of the variables **x** and **y** declared in main( ) only applies to the function, the scope of static x applies to the entire programme unless a local variable is defined explicitly. Additionally, once a static variable is defined, it cannot be removed before the programme is finished.

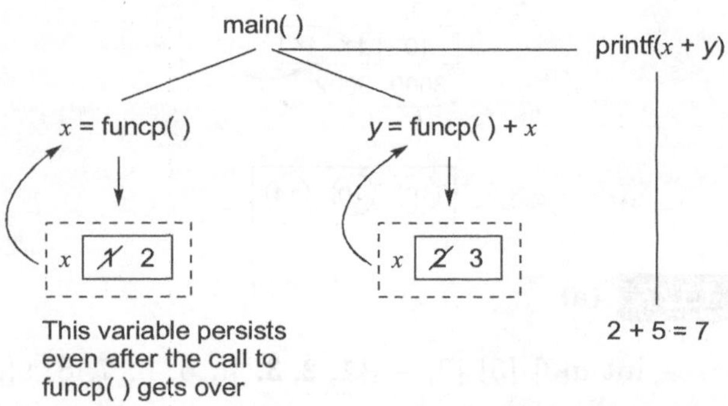

1.130 (a)

The way control enters and exits activations is displayed using an activation tree. Aspects of activation trees include:

(a) Each node shows how a procedure has been activated.
(b) The root displays the main function as being active.
(c) Only when control flows from process x to procedure y is the node for procedure "x" the parent of node for procedure "y".

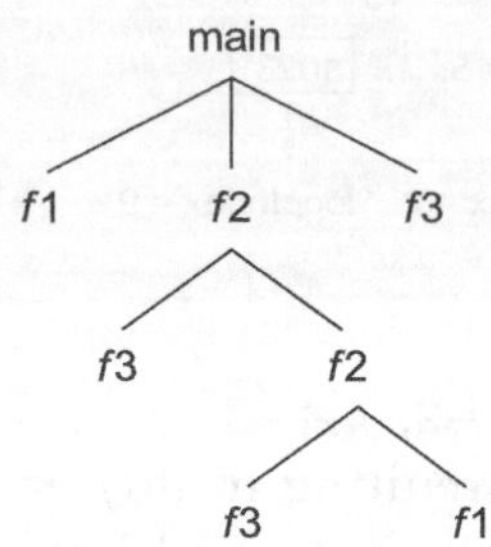

1.133 (6)

Given expression is

$3 + 1 + 5 * 2 / 7 + 2 - 4 - 7 - 6 / 2$

+ is highest and left associative

$((3 + 1) + 5) * 2 / (7 + 2) - 4 - 7 - 6 / 2$
$= 9 * 2 / 9 - 4 - 7 - 6 / 2$

– is high and right associative

$\Rightarrow 9 * 2 / (9 - (4 - (7 - 6))) / 2 = 9 * 2 / 6 / 2$

* is high precedence

$\Rightarrow\ (9 * 2) / 6 / 2 = 18 / 6 / 2$

/ is right associative

$$\Rightarrow\ (18 / (6 / 2)) = \frac{18}{3} = 6$$

2 Arrays

2.1 Let A be a two-dimensional array declared as follows:
A : array [1 ... 10] [1 ... 15] of integer;
Assuming that each integer takes one memory location. The array is stored in row-major order and the first element of the array is stored at location 100, what is the address of the element A[i][j]?
(a) $15i + j + 84$ (b) $15j + i + 84$
(c) $10i + j + 89$ (d) $10j + i + 89$
[1998 : 2 M]

2.2 An $n \times n$ array v is defined as follows:
$v[i, j] = i - j$ for all i, j, $1 \le i \le n$, $1 \le j \le n$
The sum of the elements of the array v is
(a) 0 (b) $n - 1$
(c) $n^2 - 3n + 2$ (d) $\frac{n^2(n+1)}{2}$
[2000 : 1 M]

2.3 Suppose you are given an array s[1...n] and a procedure reverse(s, i, j) which reverses the order of elements in between positions i and j (both inclusive). What does the following sequence do, where $1 \le k \le n$:

```
reverse (s, 1, k);
reverse (s, k + 1, n);
reverse (s, 1, n);
```

(a) Rotates s left by k positions
(b) Leaves s unchanged
(c) Reverses all elements of s
(d) None of the above **[2000 : 1 M]**

2.4 A program P reads in 500 integers in the range (0, 100) representing the scores of 500 students. If prints the frequency of each score above 50, what would be the best way for P to store the frequencies?
(a) An array of 50 numbers
(b) An array of 100 numbers
(c) An array of 500 numbers
(d) A dynamically allocated array of 550 numbers
[2005 : 1 M]

2.5 Let a be an array containing n integers in increasing order. The following algorithm determines whether there are two distinct numbers in the array whose difference is a specified number $S > 0$.

```
i = 0; j = 1;
while (j < n) {
    if (E) j++;
    else if (a[j] – a[i] = = S) break;
    else i++;
}
if (j < n) printf("yes");
else printf ("no");
```

Choose the correct expression for E.
(a) $a[j] - a[i] > S$ (b) $a[j] - a[i] < S$
(c) $a[i] - a[j] < S$ (d) $a[i] - a[j] > S$
[2005 : 2 M]

2.6 Let a and b be two sorted arrays containing n integers each, in non-decreasing order. Let c be a sorted array containing 2n integers obtained by merging the two arrays a and b. Assuming the arrays are indexed starting from 0, consider the following four statements:
I. $a[i] \ge b[i] \Rightarrow c[2i] \ge a[i]$
II. $a[i] \ge b[i] \Rightarrow c[2i] \ge b[i]$
III. $a[i] \ge b[i] \Rightarrow c[2i] \le a[i]$
IV. $a[i] \ge b[i] \Rightarrow c[2i] \le b[i]$
Which of the following is TRUE
(a) Only I and II (b) Only I and IV
(c) Only II and III (d) Only III and IV
[2005 : 2 M]

2.7 Consider the following C function in which size is the number of elements in the array E:

```
int MyX(int *E, unsigned int size) {
    int Y = 0;
    int Z;
    int i, j, k;
    for(i = 0; i < size; i++)
        Y = Y + E[i];
    for(i = 0; i < size; i++)
        for(j = i; j < size; j++)
        {
            Z = 0;
            for(k = i; k <= j; k++)
                Z = Z + E[k];
            if (Z > Y)
                Y = Z;
        }
    return Y;
}
```

The value returned by the function **MyX** is the
(a) maximum possible sum of elements in any sub-array of array **E**.
(b) maximum element in any sub-array of array **E**.
(c) sum of the maximum elements in all possible sub-arrays of array **E**.
(d) the sum of all the elements in the array **E**.

[2014 (Set-1) : 2 M]

2.8 Let A be a square matrix of size $n \times n$. Consider the following pseudocode. What is the expected output?

```
C = 100;
for i=1 to n do
    for j=1 to n do
    {
        Temp = A[i][j] + C;
        A[i][j] = A[j][i];
        A[j][i] = Temp – C;
    }
for i=1 to n do
    for j =1 to n do
        output (A[i][j]);
```

(a) The matrix A itself
(b) Transpose of the matrix A
(c) Adding 100 to the upper diagonal elements and subtracting 100 from lower diagonal elements of A
(d) None of the above

[2014 (Set-3) : 1 M]

2.9 What is the output of the following C code? Assume that the address of x is 2000 (in decimal) and an integer requires four bytes of memory.

```
int main()
{    unsigned int x[4][3] =
  {{1, 2, 3}, {4, 5, 6}, {7, 8, 9}, {10, 11, 12}};
  printf("%u,%u, %u", x+3, *(x+3),*(x+2)+3);
}
```

(a) 2036, 2036, 2036 (b) 2012, 4, 2204
(c) 2036, 10, 10 (d) 2012, 4, 6

[2015 (Set-1) : 2 M]

2.10 Consider the following two C code segments. Y and X are one and two dimensional arrays of size n and $n \times n$ respectively, where $2 \le n \le 10$. Assume that in both code segments, elements of Y are initialized to 0 and each element $X[i][j]$ of array X is initialized to $i + j$. Further assume that when stored in main memory all elements of X are in same main memory page frame.

Code segment 1:

```
// initialize elements of Y to 0
// initialize elements X[i][j] of X to i+j
    for (i = 0; i < n; i++)
    Y[i] + = X[0][i];
```

Code segment 2:

```
// initialize elements of Y to 0
// initialize elements X[i][j] of X to i+j
    for (i = 0; i < n; i++)
    Y[i] + = X[i][0];
```

Which of the following statements is/are correct?
S1: Final contents of array Y will be same in both code segments.
S2: Elements of array X accessed inside the for loop shown in code segment 1 are contiguous in main memory.
S3: Elements of array X accessed inside the for loop shown in code segment 2 are contiguous in main memory.
(a) Only S2 is correct
(b) Only S3 is correct
(c) Only S1 and S2 are correct
(d) Only S1 and S3 are correct

[2015 (Set-3) : 2 M]

2.11 Suppose $c = \langle c[0], \dots, c[k-1] \rangle$ is an array of length k, where all the entries are from the set {0, 1}. For any positive integers a and n, consider the following pseudocode.

DOSOMETHING (c, a, n)
$z \leftarrow 1$
for $i \leftarrow 0$ to $k - 1$
  do $z \leftarrow z^2 \bmod n$
  if $c[i] = 1$
    then $z \leftarrow (z \times a) \bmod n$
return z

If $k = 4$, $c = \langle 1, 0, 1, 1 \rangle$, $a = 2$ and $n = 8$, then the output of DOSOMETHING (c, a, n) is ______.

[2015 (Set-3) : 2 M]

2.12 Consider the following C program:

```
#include <stdio.h>
int main() {
        int a[4][5] = {{1, 2, 3, 4, 5},
                       {6, 7, 8, 9, 10},
                       {11, 12, 13, 14, 15},
                       16, 17, 18, 19, 20});
        printf("%d"\n,*(*(a+**a+2)+3));
        return(0);
}
```

The output of the program is ______.

[2020 : 1 M]

■■■■

Answers Arrays

2.1 (a) **2.2** (a) **2.3** (a) **2.4** (a) **2.5** (b) **2.6** (c) **2.7** (a) **2.8** (a) **2.9** (a)
2.10 (c) **2.11** (0) **2.12** (19)

Explanations Arrays

2.1 (a)

Let r be number of elements in a row.
Address of the element A[i][j]

$= \text{Base address} + (i-1)*r + (j-1)$
$= 100 + (i-1) \times 15 + (j-1)$
$= 100 + 15i - 15 + j - 1$
$= 100 + 15i + j - 16$
$= 84 + 5i + j$

2.2 (a)

Array is

$$\begin{bmatrix} 0 & -1 & -2 \;\dots & 1-n \\ 1 & 0 & -1 \;\dots & \\ 2 & 1 & 0 & \\ 3 & 2 & 1 & \\ \vdots & \vdots & \vdots & \\ -(1-n) & -(2-n) & & 0 \end{bmatrix}$$

Sum = 0 + 1 + 2 + ... + (–1) + (–2) + ... = 0.

2.3 (a)

Effect of the given 3 reversals for any k is equivalent to left rotation of the array of size n by k.

Let, S[1 .... 7]

1	2	3	4	5	6	7

∴ n = 7, k = 2
reverse (S, 1, 2) we get [2, 1, 3, 4, 5, 6, 7]
reverse (S, 3, 7) we get [2, 1, 7, 6, 5, 4, 3]
reverse (S, 1, 7) we get [3, 4, 5, 6, 7, 1, 2]

2.4 (a)

There are 500 students the score range is 0 to 100. Print the frequency of those student whose score above 50. So frequency range contains score from 50 to 100, so an array of 50 numbers is suitable for representing the frequency.

2.5 (b)

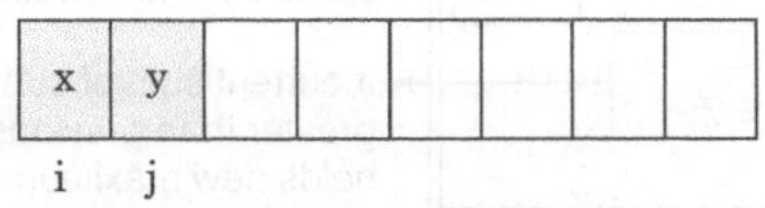

So, if condition will check a[j] – a[i] < S then increment j so that difference between two element increase and when a[j] – a[i] > S then increment i for making difference between two element indexed at i and j respectively is decrease. So, a[j] – a[i] < S is correct condition.

2.6 (c)

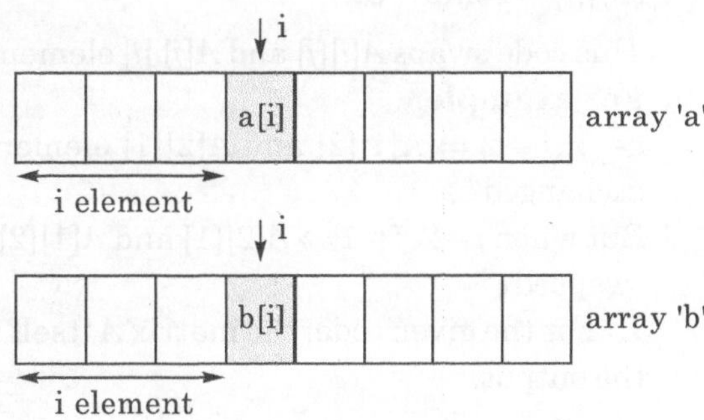

- When a[i] ≥ b[i] then 2i elements smaller than equal to a[i]. Since 'i' elements to array 'b' that are less than equal to b[i] and 'i' element of array 'a' which are less than equal to a[i]. So a[i] will comes after c[2i] i.e. c[2i] ≤ a[i] is true.
- When a[i] ≥ b[i] then b[i] always be less than equal to c[2i] i.e. when 'i' elements of array 'a' and 'i' elements of array 'b' are same then maximum index of b[i] is c[2i]. So, c[2i] ≥ b[i] is true.

2.7 (a)

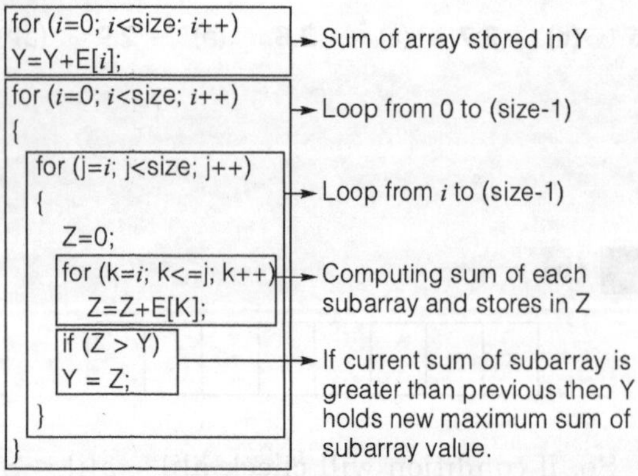

∴ The value (Y) returned by function is the maximum possible sum of elements in any subarray of array E.

2.8 (a)

Temp = $A[i][j] + C$;

$A[i][j] = A[j][i]$;

$A[j][i] = \text{Temp} - C$;

This code swaps $A[i][j]$ and $A[j]\,[i]$ elements

For example

$i = 1, j = 2 \Rightarrow$ A[1][2] and A[2][1] elements are exchanged

But when $i = 2, j = 1 \Rightarrow A$[2][1] and A[1][2] again swapped

∴ For the given code, the matrix A itself will be the output.

2.9 (a)

0th row	2000 1	2004 2	2008 3
1st row	2012 4	2016 5	2020 6
2nd row	2024 7	2028 8	2032 9
3rd row	2036 10	2040 11	2044 12

Arithmatic of two-dimensional array addresses.

'x' is the base address of the 0th row.

'x+1' is the base address of the 1st row.

'x+i' is the base address of the ith row.

Similarly *(x+i) is the ith row. So *(x+i) the address of the 0th element in the 0th row (by putting i = 0)

The outputs printed by printf statements are:

x+3 → 2036 (starting address of 3rd row)

*(x+3) → 2036 (address of 0th element in 3rd row)

*(x+2)+3 → 2036 (gives address in 2nd row but +3 makes the starting address of 2nd row (2024) incremented by 3 × 4 = 12)

∴ Output is 2036, 2036, 2036.

2.10 (c)

Y is one dimensional array

X is two dimensional array

Y	0	0	0	0	...	0
	0	1				$n-1$

$$X = \begin{array}{c} \\ 0 \\ 1 \\ 2 \\ 3 \\ \\ n-1 \end{array}\begin{array}{c}\begin{matrix} 0 & 1 & 2 & 3 & \dots & n-1 \end{matrix}\\ \left[\begin{matrix} 0 & 1 & 2 & 3 & \dots & n-1 \\ 1 & 2 & 3 & 4 & \dots & n \\ 2 & 3 & 4 & 5 & \dots & n+1 \\ 3 & 4 & 5 & 6 & \dots & n+2 \\ \vdots & & & & & \\ n-1 & n & n+1 & n+2 & \dots & n+n-2 \end{matrix}\right]_{n\times n}\end{array}$$

Code segment 1:

```
for (i = 0; i < n; i++)
    Y[i] + = X[0] [i];
```

Code segment 2:

```
for (i = 0; i < n; i++)
    Y[i] + = X[i] [0];
```

S1: Final contents of array Y will be same in both code segments.

Y	0	1	2	3	4	5	6	7		$n-1$
	0	1	2	3	4	5	6	7	. . .	$n-1$

is the output of both segments.

S2: X contents are stored row wise in C-program hence code segment row '0' accessed contiguous in memory.

S3: In code segment 2, content of X are not accessed in contiguous.

2.11 (0)

$K = 4$

0	1	2	3
1	0	1	1

a = 2, n = 8, Z = 1,

<u>$i = 0$</u> Z = 1^2 mod 8 = 1

if (C[0] = = 1) Z = (1 × 2) mod 8 = 2

<u>$i = 1$</u> Z = 2^2 mod 8 = 4

<u>$i = 2$</u> Z = 4^2 mod 8 = 0

if (C[2] = = 1) Z = (0 × 2) mod 8 = 0

<u>$i = 3$</u> Z = 0^2 mod 8 = 0

if (C[3] = = 1) Z = (0 × 2) mod 8 = 0

Returns 0

2.12 (19)

'a' is a two dimensional array as shown below:

0, 0 **1**	0, 1 **2**	0, 2 **3**	0, 3 **4**	0, 4 **5**
1, 0 **6**	1, 1 **7**	1, 2 **8**	1, 3 **9**	1, 4 **10**
2, 0 **11**	2, 1 **12**	2, 2 **13**	2, 3 **14**	2, 4 **15**
3, 0 **16**	3, 1 **17**	3, 2 **18**	3, 3 **19**	3, 4 **20**

a = address of 0th index 1-D array
*a = address of 0th index element of 0th index 1-D array
**a = value of 0th index element of 0th index 1-D array
**a = 1
**a + 2 = 1 + 2 = 3
a + 3 = address of 3rd index 1-D array
*(a + 3) = address of 0th index element of 3rd index 1-D array
*(a + 3) + 3 = address of 3rd index element of 3rd index 1-D array
((a + 3) + 3) = value of 3rd index element of 3rd index 1-D array = 19
Hence it will print 19.

■■■■

3 Stacks and Queues

3.1 The postfix expression for the infix expression A + B * (C + D) / F + D * E is
(a) AB + CD + *F/D+ E*
(b) ABCD+*F/ + DE*++
(c) A *B + CD/F *DE++
(d) A+*BCD/F* DE++

[1995 : 2 M]

3.2 Consider the following statements:
(i) First-in-first out types of computations are efficiently supported by STACKS.
(ii) Implementing LISTS on linked lists is more efficient than implementing LISTS on an array for almost all the basic LIST operations.
(iii) Implementing QUEUES on a circular array is more efficient than implementing QUEUES on a linear array with two indices.
(iv) Last-in-first-out type of computations are efficiency supported by QUEUES.
(a) (ii) & (iii) are true (b) (i) & (ii) are true
(c) (iii) & (iv) are true (d) (ii) & (iv) are true

[1996 : 1 M]

3.3 Which of the following is essential for converting an infix expression to the post fix form efficiently?
(a) An operator stack
(b) An operand stack
(c) An operand stack and an operator stack
(d) A parse tree **[1997 : 1 M]**

3.4 A priority queue Q is used to implement a stack that stores characters. PUSH (C) is implemented INSERT (Q, C, K) where K is an appropriate integer key chosen by the implementation. POP is implemented as DELETEMIN(Q). For a sequence of operations, the keys chosen are in
(a) non-increasing order
(b) non-decreasing order
(c) strictly increasing order
(d) strictly decreasing order

[1997 : 2 M]

3.5 Compute the postfix equivalent of the following expression:

$$3 * \log(x+1) - \frac{a}{2}$$

[1998 : 2 M]

3.6 What value would the following function return for the input x = 95?
Function fun (x:integer): integer;
Begin
If x > 100 then fun = x – 10
Else fun : = fun(fun (x + 11))
End;
(a) 89 (b) 90
(c) 91 (d) 92 **[1998 : 2 M]**

3.7 Let S be a stack of size n ≥ 1. Starting with the empty stack, suppose we push the first n natural numbers in sequence, and then perform n pop operations. Assume that Push and Pop operation take X seconds each, and Y seconds elapse between the end of one such stack operation and the start of the next operation. For m ≥ 1, define the stack-life of m is the time elapsed from the end of push(m) to the start of the pop operation that removes m from S. The average stack-life of an element of this stack is
(a) n (X + Y)
(b) 3Y + 2X
(c) n (X + Y) – X
(d) Y + 2X

[2003 : 2 M]

3.8 A single array A[1...MAXSIZE] is used to implement two stacks. The two stacks grow from opposite ends of the array. Variables top1 and top2 (top1 < top2) point to the location of the topmost element in each of the stacks. If the space is to be used efficiently, the condition for "stack full"is
(a) (top1 = MAXSIZE/2) and (top2 = MAXSIZE/2 + 1)
(b) top1 + top2 = MAXSIZE
(c) (top1 = MAXSIZE/2) or (top2 = MAXSIZE)
(d) top1 = top2 – 1

[2004 : 1 M]

3.9 The best data structure to check whether an arithmetic expression has balanced parentheses is a
(a) queue (b) stack
(c) tree (d) list

[2004 : 1 M]

3.10 A circularly linked list is used to represent a Queue. A single variable p is used to access the Queue. To which node should p point such that both the operations enQueue and deQueue can be performed in constant time?

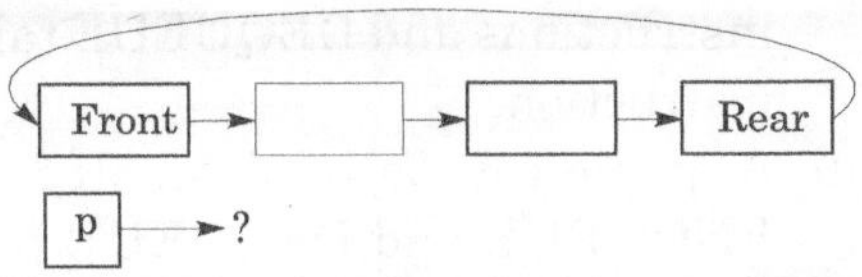

(a) rear node
(b) front node
(c) not possible with a single pointer
(d) node next to front

[2004 : 2 M]

3.11 Assume that the operators +, –, × are left associative and ^ is right associative. The order of precedence (from highest to lowest) is ^, ×, +, –. The postfix expression corresponding to the infix expression a + b × c – d ^ e ^ f is
(a) abc × + def ^ ^ –
(b) abc × + de ^ f ^
(c) ab + c × d – e ^ f ^
(d) – + a × bc ^ ^ def **[2004 : 2 M]**

3.12 A program attempts to generate as many permutations as possible of the string, 'abcd' by pushing the characters a, b, c, d in the same order onto a stack, but it may pop off the top character at any time. Which one of the following strings CANNOT be generated using this program?
(a) abcd (b) dcba
(c) cbad (d) cabd

[2004 : 2 M]

3.13 A function f defined on stacks of integers satisfies the following properties. f(ϕ) = 0 and f(push (S, i)) = max (f(S), 0) + i for all stacks S and integers i. If a stack S contains the integers 2, –3, 2, –1, 2 in order from bottom to top, what is f(S)?
(a) 6 (b) 4
(c) 3 (d) 2

[2005 : 1 M]

3.14 An implementation of a queue Q, using two stacks S1 and S2, is given below

```
void insert(Q, x)
{
    push(S1, x);
}
void delete(Q, x)
{
    if (stack-empty(S2)) then
        if (stack-empty(S1)) then
        {
            print("Q is empty");
            return;
        }
    else while (! stack-empty(S1))
    {
        x = pop(S1);
        push(S2, x);
    }
    x = pop(S2);
}
```

Let n insert and m (≤ n) delete operations be performed in an arbitrary order on an empty queue Q. Let x and y be the number of push and pop operations performed respectively in the processes. Which one of the following is true for all m and n?
(a) n + m ≤ x ≤ 2n and 2m ≤ y ≤ n + m
(b) n + m ≤ x < 2n and 2m ≤ y ≤ 2n
(c) 2m ≤ x < 2n and 2m ≤ y ≤ n + m
(d) 2m ≤ x < 2n and 2m ≤ y ≤ 2n

[2006 : 2 M]

3.15 The following postfix expression with single digit operands in evaluated using a stack

8 2 3 ∧ / 2 3* + 5 1* –

Note that ∧ is the exponentiation operator. The top two elements of the stack after the first* is evaluated are
(a) 6, 1 (b) 5, 7
(c) 3, 2 (d) 1, 5 **[2007 : 2 M]**

3.16 Suppose you are given an implementation of a queue of integers. The operations that can be performed on the queue are:
(i) isEmpty (Q) : returns true if the queue is empty, false otherwise.
(ii) delete (Q) : deletes the element at the front of the queue and returns its value.
(iii) insert (Q, i) : inserts the integer i at the rear of the queue.

Consider the following function:

```
void f (queue Q)
{
    int i;
    if (! is Empty (Q))
    {
        i = delete (Q);
        f(Q);
        insert (Q, i);
    }
}
```

What operation is performed by the above function f?

(a) Leaves the queue Q unchanged
(b) Reverses the order of the elements in the queue Q
(c) Deletes the element at the front of the queue Q and inserts it at the rear keeping the other elements in the same order
(d) Empties the queue Q

[2007 : 2 M]

3.17 Suppose a circular queue of capacity (n – 1) elements is implemented with an array of n elements. Assume that the insertion and deletion operations are carried out using REAR and FRONT as array indeed variables, respectively. Initially, REAR = FRONT = 0. The conditions to detect queue full and queue empty are

(a) full : (REAR + 1) mod n == FRONT
empty : REAR == FRONT
(b) full : (REAR + 1) mod n == FRONT
empty : (FRONT + 1) mod n == REAR
(c) full : REAR == FRONT
empty : (REAR + 1) mod n == FRONT
(d) full : (FRONT + 1) mod n == REAR
empty : REAR == FRONT

[2012 : 2 M]

3.18 Consider the following operation along with Enqueue and Dequeue operations on queues, where k is a global parameter.

```
Multi-Dequeue (Q) {
    m = k;
    while (Q is not empty) and (m > 0) {
        Dequeue (Q);
        m = m – 1;
    }
}
```

What is the worst case time complexity of a sequence of n queue operations on an initially empty queue?

(a) Θ(n) (b) Θ(n + k)
(c) Θ(nk) (d) $\Theta(n^2)$

[2013 : 2 M]

3.19 Suppose a stack implementation supports an instruction REVERSE, which reverses the order of elements on the stack, in addition to the PUSH and POP instructions. Which one of the following statements is TRUE *with respect to this modified stack*?

(a) A queue cannot be implemented using this stack.
(b) A queue can be implemented where ENQUEUE takes a single instruction and DEQUEUE takes a sequence of two instructions.
(c) A queue can be implemented where ENQUEUE takes a sequence of three instructions and DEQUEUE takes a single instruction.
(d) A queue can be implemented where both ENQUEUE and DEQUEUE take a single instruction each.

[2014 (Set-2) : 2 M]

3.20 Consider the C program below:

```
# include <stdio.h>
int *A, stkTop;
int stkFunc (int opcode, int val)
{
    static int size=0, stkTop = 0;
    switch (opcode)
    {
        case –1:  size = val; break;
        case 0:   if (stkTop < size)
                  A[stkTop++] = val; break;
        default:  if (stkTop)
                  return A[– – stkTop];
    }
    return –1;
}
int main( )
{
    int B[20]; A=B; stkTop = –1
    stkFunc (–1, 10);
    stkFunc (0, 5);
    stkFunc (0, 10);
    printf ("%d\n", stkFunc (1, 0)+stkFunc (1, 0));
}
```

The value printed by the above program is ______.

[2015 (Set-2) : 2 M]

3.21 The result evaluating the postfix expression 10 5 + 60 6 / * 8 – is

(a) 284 (b) 213
(c) 142 (d) 71

[2015 (Set-3) : 1 M]

3.22 A queue is implemented using an array such that ENQUEUE and DEQUEUE operations are performed efficiently. Which one of the following statements is CORRECT (*n* refers to the number of items in the queue)?

(a) Both operations can be performed in O(1) time
(b) At most one operation can be performed in O(1) time but the worst case time for the other operation will be Ω(*n*)

(c) The worst case time complexity for both operations will be $\Omega(n)$

(d) Worst case time complexity for both operations will be $\Omega(\log n)$

[2016 (Set-1) : 1 M]

3.23 Let Q denote a queue containing sixteen numbers and S be an empty stack. Head(Q) returns the element at the head of the queue Q without removing it from Q. Similarly Top(S) returns the element at the top of S without removing it from S. Consider the algorithm given below.

while *Q is not Empty* do
 if *S* is *Empty OR Top* (*S*) ≤ Head(*Q*) then
  x : = Dequeue (*Q*);
  Push (*S*, *x*);
 else
  x : = Pop (*S*);
  Enqueue (*Q*, *x*);
 end
end

The maximum possible number of iterations of the **while** loop in the algorithm is ______.

[2016 (Set-1) : 2 M]

3.24 Consider the following New-order stragety for traversing a binary tree:

- Visit the root;
- Visit the right subtree using New-order;
- Visit the left subtree using New-order;

The New-order traversal of the expression tree corresponding to the reverse polish expression 3 4 * 5 – 2 ^ 6 7 * 1 + – is given by

(a) + – 1 6 7 * 2 ^ 5 – 3 4 *
(b) – + 1 * 6 7 ^ 2 – 5 * 3 4
(c) – + 1 * 7 6 ^ 2 – 5 * 4 3
(d) 1 7 6 * + 2 5 4 3 * – ^ –

[2016 (Set-2) : 2 M]

3.25 A circular queue has been implemented using a singly linked list where each node consists of a value and a single pointer pointing to the next node. We maintain exactly two external pointers **FRONT** and **REAR** pointing to the front node and the rear node of the queue, respectively. Which of the following statements is/are CORRECT for such a circular queue, so that insertion and deletion operations can be performed in O(1) time?

I. Next pointer of front node points to the rear node.

II. Next pointer of rear node points to the front node.

(a) I only
(b) II only
(c) Both I and II
(d) Neither I nor II

[2017 (Set-2) : 1 M]

3.26 A queue is implemented using a non-circular singly linked list. The queue has a head pointer and tail pointer, as shown in the figure. Let *n* denote of number of nodes in the queue. Let enqueue be implemented by inserting a new node at the head and dequeue be implemented by deletion of a node from the tail.

Which one of the following is the time complexity of the most time-efficient implementation of enqueue and dequeue, respectively, for this data structure?

(a) $\theta(1), \theta(1)$
(b) $\theta(1), \theta(n)$
(c) $\theta(n), \theta(1)$
(d) $\theta(n), \theta(n)$

[2018 : 1 M]

3.27 Consider the following sequence of operations on an empty stack.

push(54); push(52); pop( ); push(55);

push(62); *s* = pop( ):

Consider the following sequence of operations on an empty queue.

enqueue(21); enqueue(24); dequeue( ); enqueue(28); enqneue(32); *q* = dequeue( ):

The value of *s* + *q* is ________.

[2021 (Set-1) : 1 M]

3.28 Consider the queues Q_1 containing four elements and Q_2 containing none (shown as the Initial State in the figure). The only operations allowed on these two queues are Enqueue(*Q*, element) and Dequeue(*Q*). The minimum number of Enqueue operations on Q_1 required to place the elements of Q_1 in Q_2 in reverse order (shown as the Final State in the figure) without using any additional storage is ________.

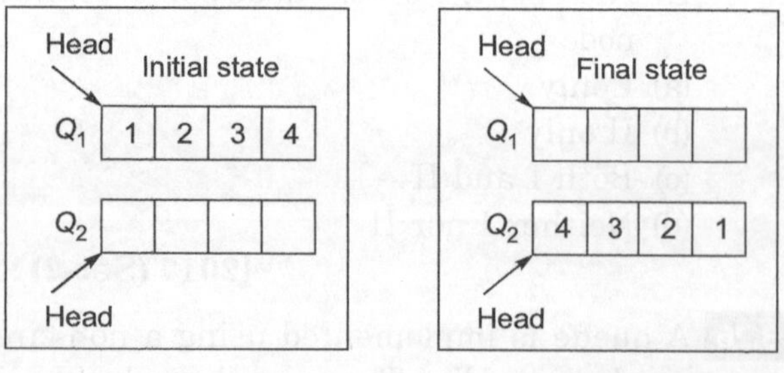

[2022 : 2 M]

3.29 Consider a sequence a of elements $a_0 = 1$, $a_1 = 5$, $a_2 = 7$, $a_3 = 8$, $a_4 = 9$, and $a_5 = 2$. The following operations are performed on a stack S and a queue Q, both of which are initially empty.

I: push the elements of a from a0 to a5 in that order into S.
II: enqueue the elements of a from a0 to a5 in that order into Q.
III: pop an element from S.
IV: dequeue an element from Q.
V: pop an element from S.
VI: dequeue an element from Q.
VII: dequeue an element from Q and push the same element into S.
VIII: Repeat operation VII three times.
IX: pop an element from S.
X: pop an element from S

The top element of S after executing the above operations is ________.

[2023 : 2 M]

3.30 Let S1 and S2 be two stacks. S1 has capacity of 4 elements. S2 has capacity of 2 elements. S1 already has 4 elements: 100, 200, 300, and 400 , whereas S2 is empty, as shown below.

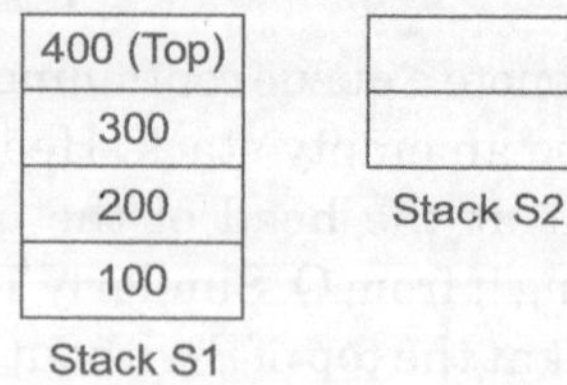

Only the following three operations are available:
PushToS2: Pop the top element from S1 and push it on S2.
PushToS1: Pop the top element from S2 and push it on S1.
GenerateOutput: Pop the top element from S1 and output it to the user.
Note that the pop operation is not allowed on an empty stack and the push operation is not allowed on a full stack.
Which of the following output sequences can be generated by using the above operations?

(a) 400, 200, 100, 300
(b) 200, 300, 400, 100
(c) 100, 200, 400, 300
(d) 300, 200, 400, 100

[2024 (Set-2) : 2 M]

■■■■

Answers Stacks and Queues

3.1	(b)	**3.2**	(a)	**3.3**	(a)	**3.4**	(d)	**3.5**	Sol.	**3.6**	(c)	**3.7**	(b)	**3.8**	(d)	**3.9**	(b)
3.10	(c)	**3.11**	(a)	**3.12**	(d)	**3.13**	(c)	**3.14**	(a)	**3.15**	(a)	**3.16**	(b)	**3.17**	(a)	**3.18**	(a)
3.19	(c)	**3.20**	(15)	**3.21**	(c)	**3.22**	(a)	**3.23**	(256)	**3.24**	(c)	**3.25**	(b)	**3.26**	(b)	**3.27**	(86)
3.28	(0)	**3.29**	(8)	**3.30**	(a, b, d)												

Explanations Stacks and Queues

3.1 (b)

Infix: A+B*(C+D)/F+D*E

Postfix: ABCD + *F/ + DE*++

Therefore, option (b) is correct.

3.2 (a)

Stack is LIFO system so FIFO not supported by stack hence option statement 1 is incorrect and queue is using FIFO so statement 4 is also wrong.

3.3 (a)

Operands never change its position during expression conversion so only operator stack is needed.

3.4 (d)

In stack last element will be deleted first but according to question, it should be work like queue. So last element deleted at last, for that we give lower priority to last element. So all elements should be in strictly decreasing order.

3.5 Sol.

$$3 * \log(x+1) - \frac{a}{2}$$

Step by step conversion to postfix

= 3 * (x + 1) log – a/2

= 3 * x1+ log – a2/

= 3x1+ log * – a2/

= 3x1 + log * a2/–

3.6 (c)

fun(95)
fun (fun (106))
fun (96)
fun (fun (107))
fun (97)
fun (fun (108))
fun (98)
fun (fun (109))
fun (99)
fun (fun (200))
fun (100)
fun (fun (111))
fun (101)
fun = 101 – 10 = 91

So the return value is 91.

3.7 (b)

Let we have n = 3 sized stack:

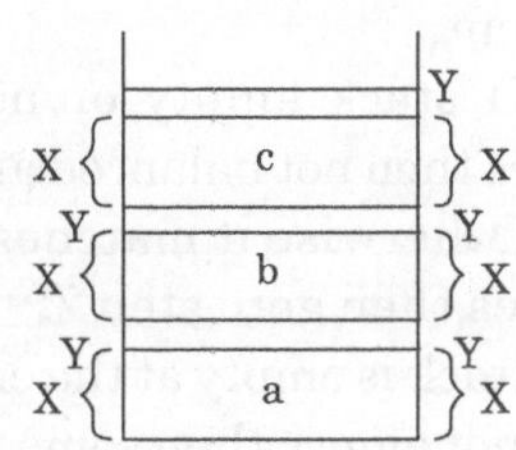

So, lifetime of c : Y

lifetime of b : Y + X + Y + X + Y

= 3Y + 2X

lifetime of a : Y + X + Y + X + Y + X + Y + X + Y

= 5Y + 4X

So, average lifetime of an element

$$= \frac{Y + 3Y + 2X + 5Y + 4X}{3} = \frac{9Y + 6X}{3}$$

= 3Y + 2X = (3Y + 3X) – X

= 3(Y + X) – X

So, in general : n(Y + X) – X

3.8 (d)

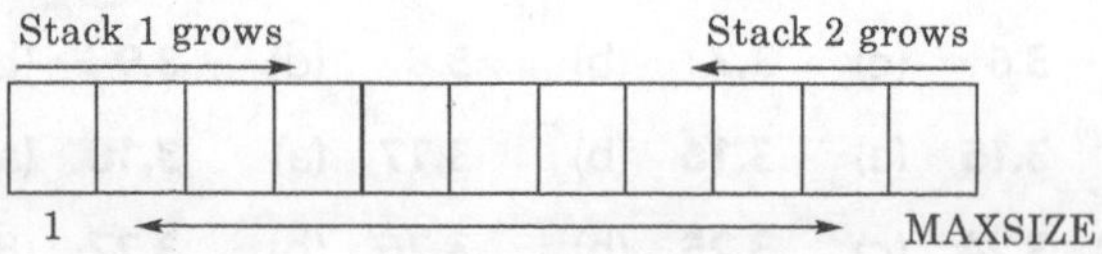

- When stack 1 contain 1 element and stack 2 contain 1 element then top1 = 1 and top2 = MAXSIZE – 1.
  Here top1 + top2 = MAXSIZE. Since both stack are not full. Hence no overflow. So false
- (top1 = MAXSIZE/2) and (top2 = MAXSIZE/2 + 1) will true when both stack has n/2 element but space is not effectively used i.e. when stack 1 has no element and stack 2 has n – 1 elements then it will not detect overflow. So false
- top1 = top2 – 1 is true, since it cover all cases when both stack are with full capacity, stack1 is empty and stack2 is full and vice-versa. Hence true

3.9 (b)

The stack is a data structure in which each push (insert) or pop (delete) operation takes constant θ(1) time. The best data structure to check whether an arithmetic expression has a balanced parentheses.

Step 1: Start with empty stack of currently open parentheses.
Step 2: Process each char of an expression string.
Step 3: If it is open, push it on the stack.
Step 4: If it is closed, check it against the top of stack element.
Step 4.1: If stack empty or not a matching parentheses then not balanced and return false.
Step 4.2: Otherwise it matches. pop the open parentheses then goto step 2.
Step 5: If stack is empty at the end, return true.
Step 6: If not empty then some parentheses is unmatched, return false.

3.10 (c)

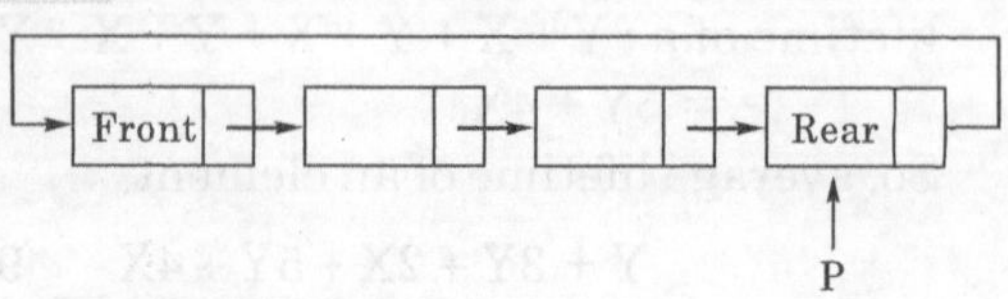

To add a new node:

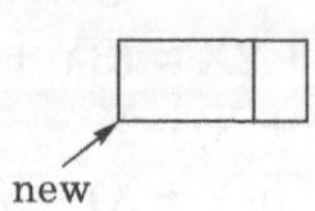

new → next = Rear → next;
Rear → next = new;
Rear = new;

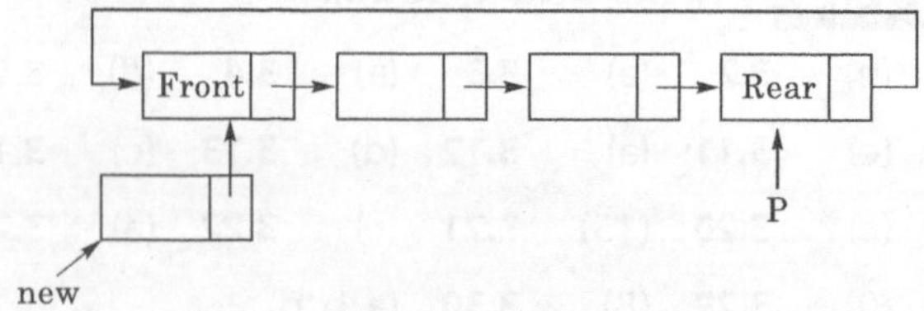

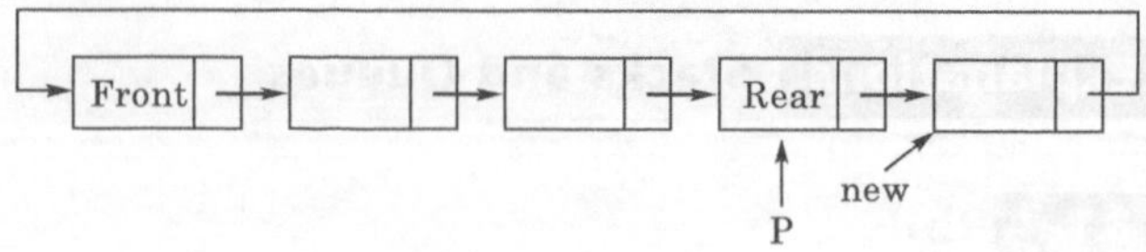

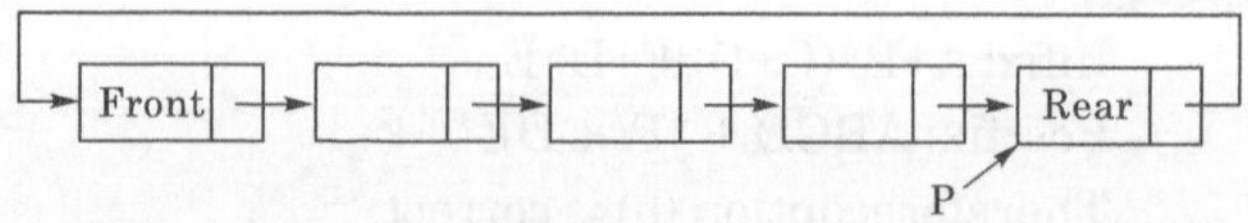

It will take constant time.
Similarly delete node from Front code will be:
Front = Rear → next;
Rear → next = Front → next;
free (Front);
It will also take constant time.

3.11 (a)

According to conditions given:
⇒ [(a + (b × c)) – (d ^ (e ^ f))]
⇒ [(a + (b c ×)) – (d ^ (e f ^))]
⇒ [(abc × +) – (def ^^)]
⇒ abc × + def ^^ –

3.12 (d)

- For every element push an element and pop immediately i.e. push (a) then pop will give permutation abcd.
- For all elements push (a), push (b), push (c) and push (d) then pop, pop, pop, pop will give us dcba.
- Push (a), push (b), push (c) then pop, pop, pop will give is cba, then push (d) and pop will gives us dcba.
- Option (d) is not possible because a cannot pop after c and before 'b'.

3.13 (c)

i: The element to be pushed
S: Stack
Initially f(S) = 0

f(S)	max(f(S),0)	i	Updated_f(S)=max(f(S),0)+i
0	0	2	2
2	2	–3	–1
–1	0	2	2
2	2	–1	1
1	1	2	**3**

3.14 (a)

Number of push operations

= n(insert) + m(delete) = n + m

So, $n + m \leq x$ but there are maximum 2n insert operations so $n + m \leq x \leq 2n$ ...(1)

Number of pop operations = n + m

But there are 2m delete operations which are less than no. of pop operations, hence

$2m \leq n + m$ ...(2)

From (1) and (2): $n + m \leq x \leq 2n$ & $2m \leq n + m$

3.15 (a)

Given postfix expression is WW

8 2 3 ∧ / 23 * + 5 1 * –

Expression	Opl	Op2	Value	top S(RL)
8				8
2				8, 2
3				8, 2, 3
^	2	3	8	8, 8
/	8	8	1	1
2				1, 2
3				1, 2, 3
*	2	3	6	1, 6

So the top two elements of the stack are 6, 1 after the first * is evaluated.

3.16 (b)

Case-1: When queue is empty then no change would happen on queue.

Case-2: 1st delete all elements one by one from queue then last deleted element enter at 1st position, 2nd last deleted element at 2nd position etc.

So finally we get reverse order of given (input queue).

3.17 (a)

In actual implementation of Q of n-elements, we can store n elements in the array and initially F = R = –1. Here, we are implementing Q in (n – 1) slots of n element array and given, initially F = R = 0

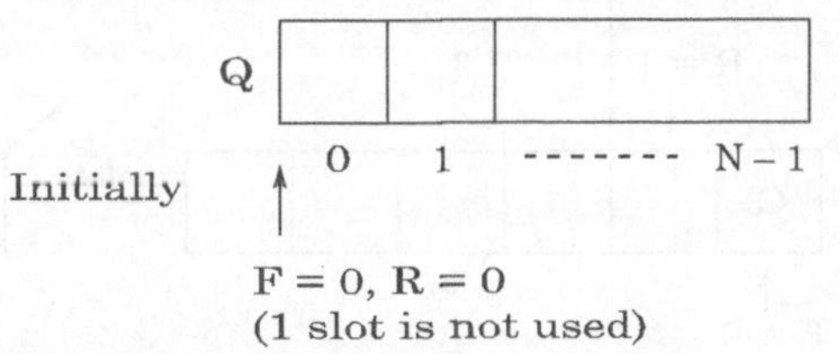

Initially R = 0 means R contains the position of next element to be inserted. (In actual implementation of Q, R contains the position of newly inserted element). (Also in actual implementation of Q, we increment R and then store the element) Here, firstly we will insert the element then R will get incremented, because initially R = 0.

i.e. | Q[R] = a
(R++) % N | ⟹ Enqueue

ex:

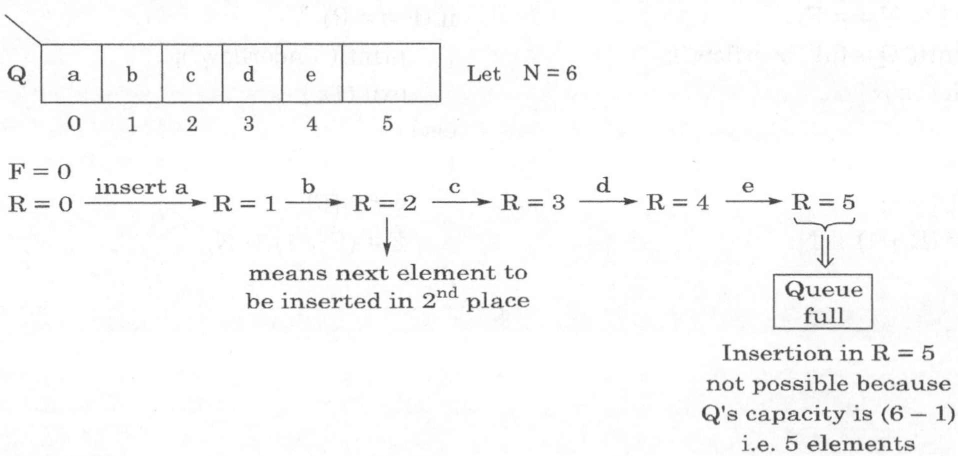

So, queue will be full when:

(R + 1) % 6 = F

{(5 + 1) % 6 = 0}

Now, for deleting:

y = x[F] (delete) / store

(F++)% N (increment)

Return y

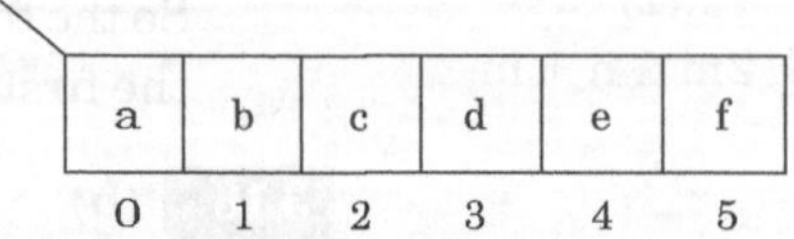

R = 5 —a→ F = 1 +f→ R = 0

{ (R + 1) % 6 ≠ F}

Delete element and increment F:

F = 1 —del→ F = 2 —del→ F = 3 —del→ F = 4 —del→ F = 5 —del→ F = 0

(and R is also 0)

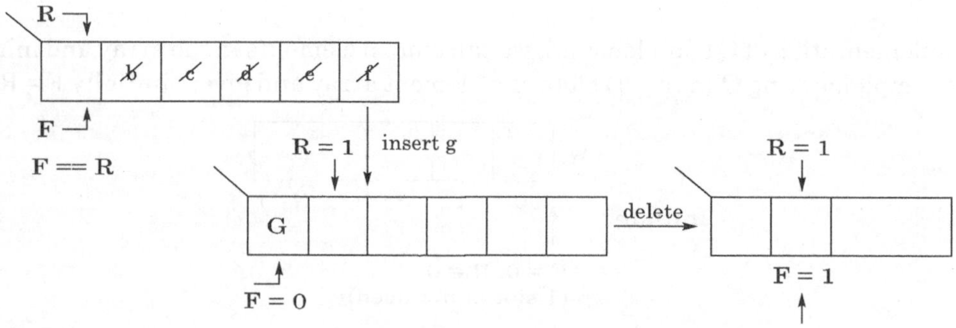

So, at anywhere, if F = = R means Q is empty.

So, in short initially R = = 0, means no element present

So store the element then increment R and after insertion if F = = 0 then delete the element and increment F

1 slot empty as queue capacity is n – 1.

```
Enqueue(x)
{
    if (R + 1 % N = = F)
    {   printf("Q is full, overflow");
        exit (1); }
    else
    {
        Q[R] = x;
        R = (R + 1) % N;
    }
}
```

```
Dequeue(x)
{
    if (F = = R)
    {   printf("underflow");
        exit (1); }
    else
    {
        y = Q[F];
        F = (F + 1) % N;
    }
}
```

3.18 (a)

Since their are 3 queue operation i.e., enqueue, dequeue and Multi-dequeue 'n' enqueue operation will take $\Theta(n)$ time, 'n' dequeue operation will take $\Theta(n)$ time, 'n' Multi-dequeue operation will take $\Theta(n)$ time i.e., Min (n, k) but for worst case when n = k time complexity will be $\Theta(n)$.

3.19 (c)

Consider an queue with elements inserted in same order as element present in queue.

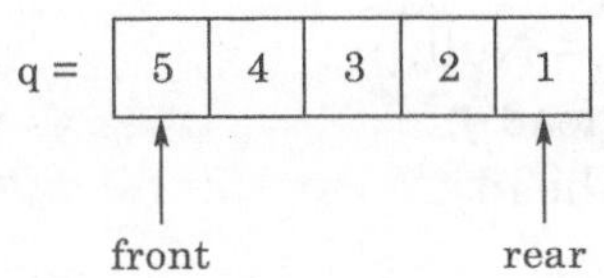

Now to implement stack same as queue:

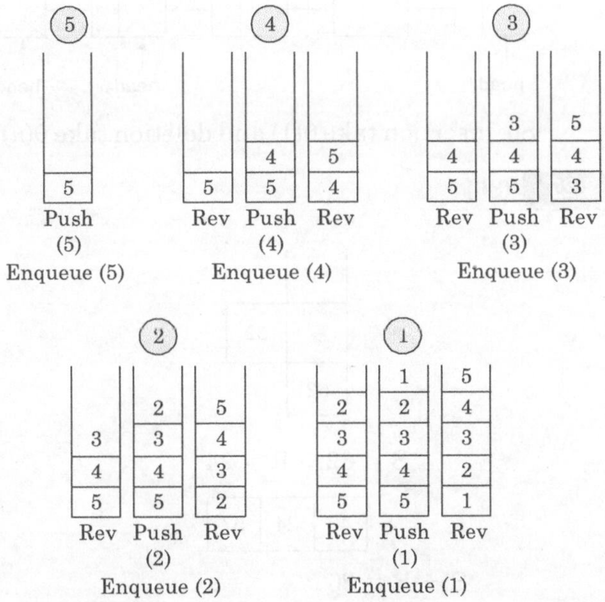

A queue can be implemented where ENQUEUE takes a sequence of three instructions and DEQUEUE takes a single instruction.

3.20 (15)

StkFunc (–1, 10); ⇒ size = 10; stkTop = 0 [case –1 executed]

StkFunc (0, 5); ⇒ A[0] sets to 5 and stkTop = 1 [case 0 executed]

StkFunc (0, 10); ⇒ A[1] sets to 10 and stkTop = 2 [case 0 executed]

StkFunc (1, 0); ⇒ returns **10**, stkTop = 1 [default case executed]

StkFunc (1, 0); ⇒ returns **5**, stkTop = 0 [default case executed]

∴ Program prints 10 + 5 = 15

3.21 (c)

10 5 + 60 6 / ∗ 8 –

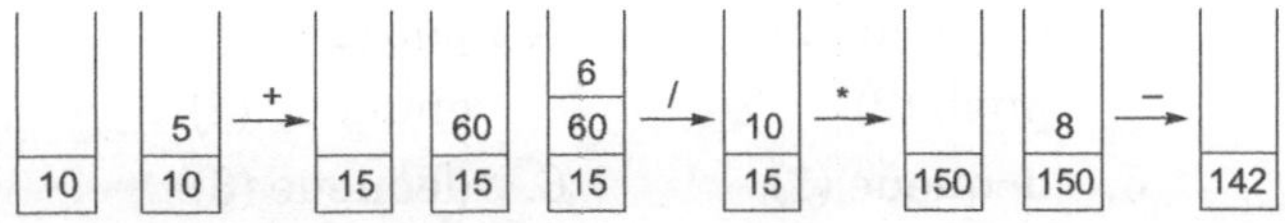

Result is 142.

3.22 (a)

Implementing queue using array:

ENQUEUE Operation:

Check array full or not
if array is full
stop
else enter the element in the end of array;
which will take O(1) time.

DEQUEUE Operation:

Check array empty or not
if array is empty
stop

olso delete the element from front of array and increment the head value (pointer to the starting element of array).

which will take O(1) time.

So for array implementation of queue, ENQUEUE and DEQUEUE operation takes O(1) time.

3.23 (256)

The minimum number of iterations of the while loop in algorithm when use take queue contain element in ascending order i.e., 1, 2, 3, 4,..., 16 is 16.

The maximum number of iterations of while loop in algorithm when we take queue containing elements in descending order i.e., 16, 15, 14, ..., 1.

First 16 will push into stack and then enqueue it in the end of the queue. This process do till we get 1 as head element. When head point to 1 then simple push the 1 in stack. In this manner we have to push all element in stack in assending order, until queue is empty it will take 256 of iterations.

Example:

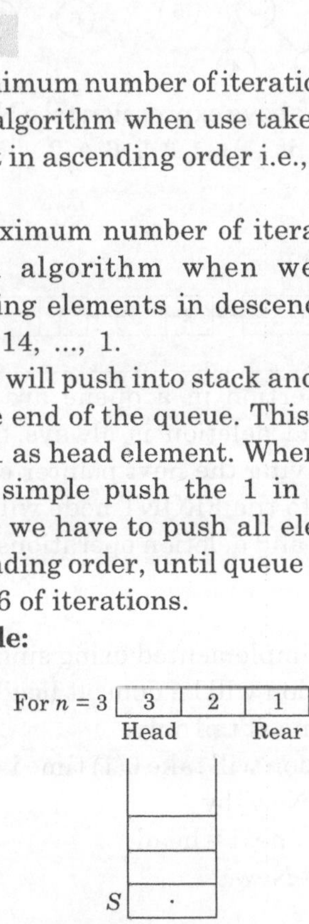

Sequence of operation with while loop execution.

1. dequeue (3) push (3)
2. pop (3) enqueue (3)
3. dequeue (2) push (2)
4. pop (2) enqueue (2)
5. dequeue (1) push (1)
6. dequeue (3) push (3)
7. pop (3) enqueue (3)
8. dequeue (2) push (2)
9. dequeue (3) push (3)

So for n = 3 it takes 3 × 3 = 9 iterations of while loop in algorithm.

So, for n = 16 it will take 16 × 16 = 256 iterations of while loop.

3.24 (c)

The expression is given in reverse polish notation i.e., post order.

Expression is 3 4 * 5 – 2 ^ 6 7 * 1 + –

Expression tree for above post order expression is

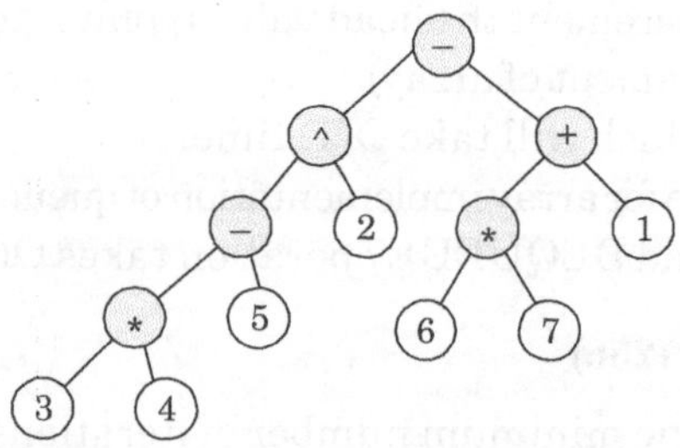

From the above expression tree **NEW ORDER** traversal is – + 1 * 7 6 ^ 2 – 5 * 4 3.

3.25 (b)

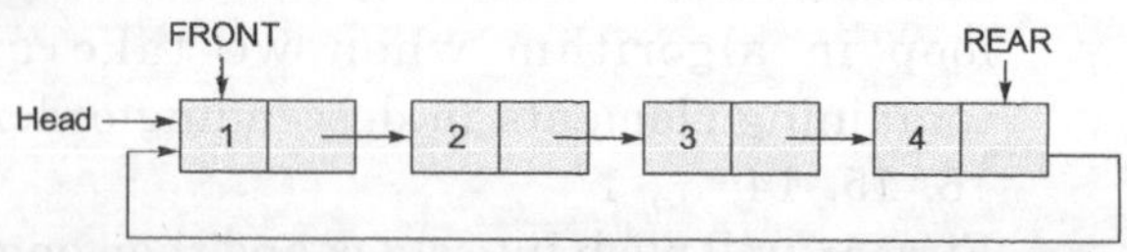

Since insertion in a queue are always from REAR and deletion is always from FRONT. Hence having the next pointer of REAR node pointing to the FRONT node will lead to both insertion and deletion operations in O(1) time.

3.26 (b)

Queue is implemented using singly linked list. So, insertion will be done at head and deletion will be done at tail only.

- Insertion will take θ(1) time i.e. adding new node "New" by
  New → next = head
  head = New

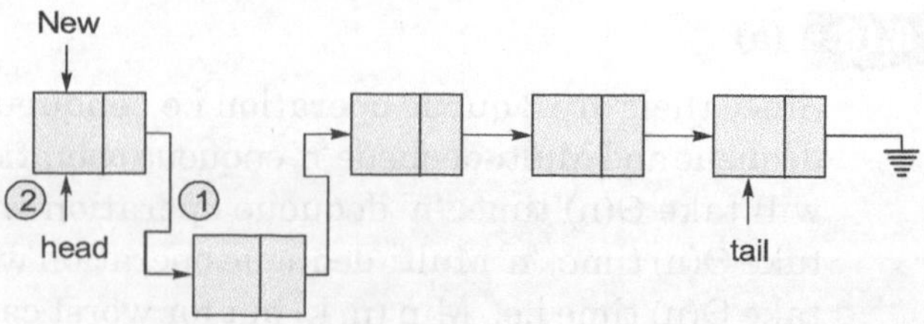

- Deletion will take θ(n) time i.e. we have to read one node before last node, otherwise we loose tail pointer i.e.

while ($head_1$ → next! = Null)
$head_2$ = $head_1$;
$head_1$ = $head_1$ → next;
$head_2$ = Null;
free ($head_1$);
tail = $head_2$;

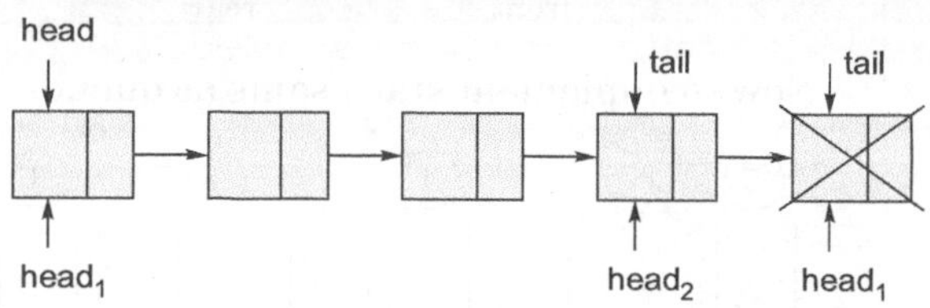

So, insertion take θ(1) and deletion take θ(n).

3.27 (86)

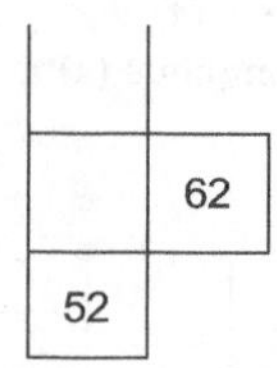

S = 62, R = 24

43	24	57

S + R = 86

3.28 (0)

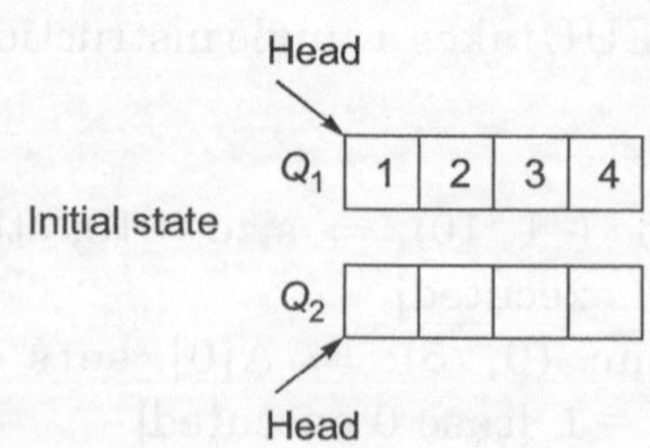

No additional storage.

Enqueues on Q_1 = ?

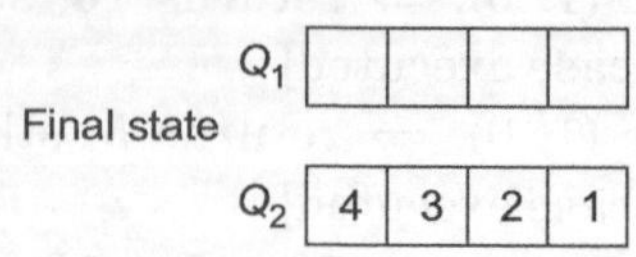

Consider Q_1 and Q_2 as the following

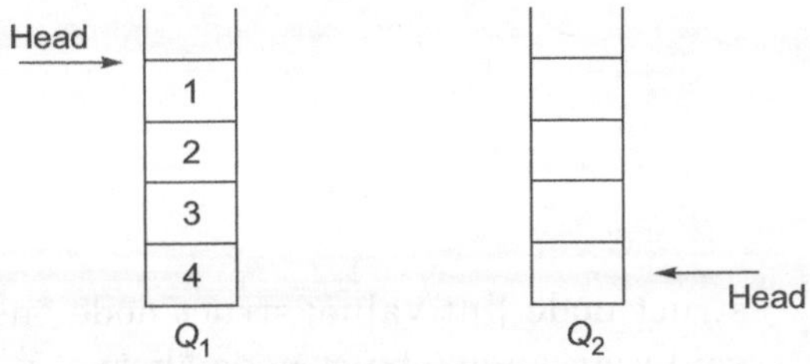

Now, Enqueue (Q_2, \$) 4 times,

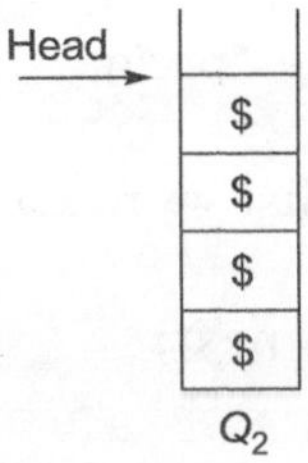

Now, one by one perform

1. x = Dequeue (Q_1)
2. Enqueue (Q_2, x) until Q_1 is empty.

3.29 (8)

I. S is stack and follows LIFO order. Thus, after pushing a_0 to a_5, a_5 will come at the top. Top of the stack points to a_5.

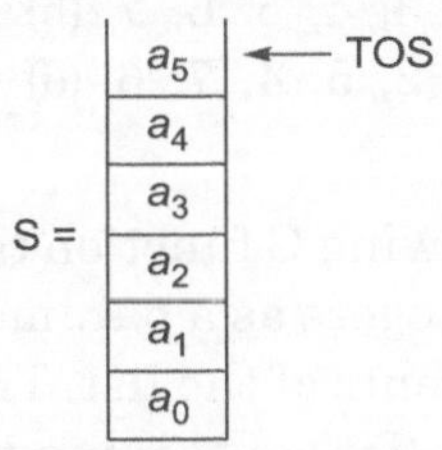

II. Q is queue and follows FIFO order. Thus, after enqueuing a_0 to a_5 order will persist. Front will point to a_0 and rear will point to a_5.

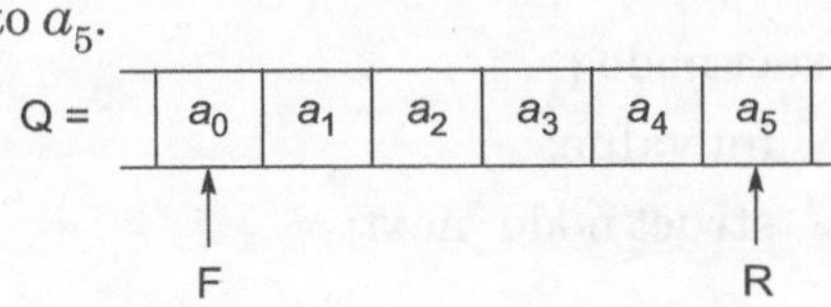

III. Popping an element will remove top of the stack S i.e. a_5 will be removed. Now, TOS is a_4.

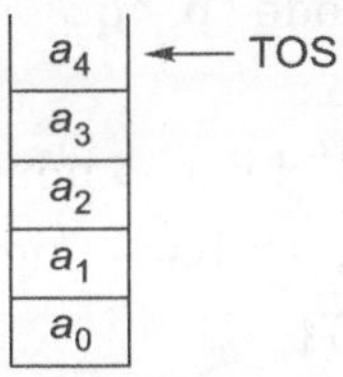

IV. Dequeuing an element will be from Front of the queue Q i.e. a_0 will be removed. Now, Front will point to a_1.

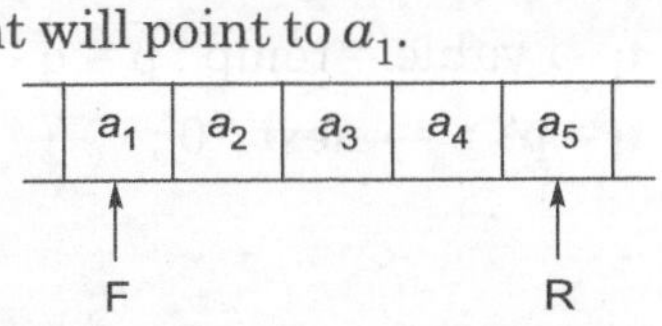

V. Popping an element will remove top of the stack S i.e. a_4 will be removed. Now, TOS is a_3.

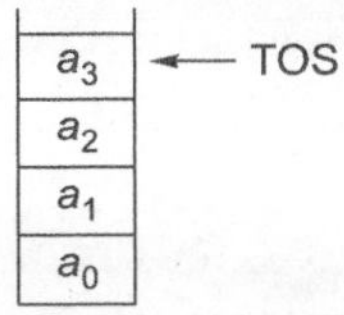

VI. Dequeuing an element will be from front of the queue Q i.e. a_1 will be removed. Now, front will point to a_2.

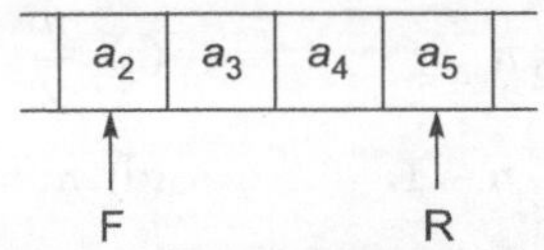

VII. 3 times dequeuing an element will be from front of the queue Q i.e. a_2, a_3, a_4 will be removed and pushed into stack S in the same order. So, TOS will be a_4.

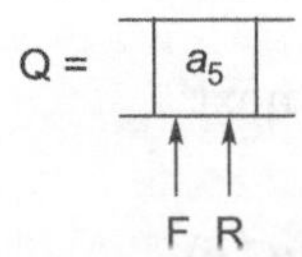

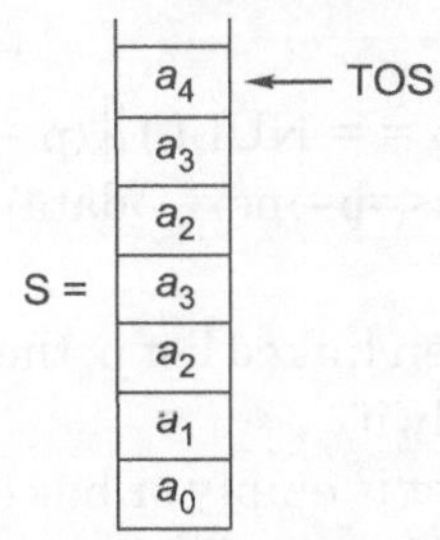

VIII. Popping an element will remove top of the stack S i.e. a_4 will be removed. Now, TOS is a_3.

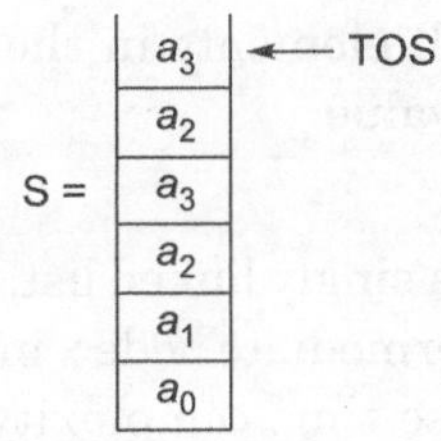

IX. Popping an element will remove top of the stack S i.e. a_3 will be removed. Now, TOS is a_2.

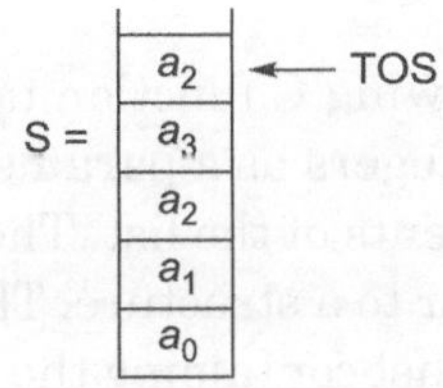

Hence, a_2 i.e. 8 will be at the top of the stack.

4 Linked List

4.1 In the worst case, the number of comparisons needed to search singly linked list of length n for a given element is

(a) $\log_2 n$ (b) $\frac{n}{2}$

(c) $\log_2 n - 1$ (d) n

[2002 : 1 M]

4.2 Consider the function f defined below:

```
struct item
{
  int data;
  struct item* next;
};
int f(struct item *p)
{
return((p = = NULL) || (p → next = =NULL) ||
((p→data<=p→next→data)&& f(p→next)));
}
```

For a given linked list p, the function f returns 1 if and only if

(a) the list is empty or has exactly one element
(b) the elements in the list are sorted in non-decreasing order of data value
(c) the elements in the list are sorted in non-increasing order of data value
(d) not all elements in the list have the same data value

[2003 : 2 M]

4.3 Let P be a singly linked list. Let Q be the pointer to an intermediate node x in the list. What is the worst-case time complexity of the best-known algorithm to delete the node x from the list?

(a) O(n) (b) O($\log^2$ n)

(c) O(log n) (d) O(1)

[2004 : 1 M]

4.4 The following C function takes a singly-linked list of integers as a parameter and rearranges the elements of the list. The list is represented as pointer to a structure. The function is called with the list containing the integers 1, 2, 3, 4, 5, 6, 7 in the given order. What will be the contents of the list after the function completes execution?

```
struct node {int value; struct node *next;);
void rearrange (struct node *list)
{
  struct node *p, *q;
  int temp;
  if (!list | | !list → next) return;
  p = list;
  q = list → next;
  while (q)
  {
    temp = p → value;
    p → value = q → value;
    q → value = temp;
    p = q → next;
    q = p ? p → next : 0;
  }
}
```

(a) 1, 2, 3, 4, 5, 6, 7 (b) 2, 1, 4, 3, 6, 5, 7

(c) 1, 3, 2, 5, 4, 7, 6 (d) 2, 3, 4, 5, 6, 7, 1

[2005 : 2 M]

4.5 The following C function takes a singly-linked list of integers as a parameter and rearranges the elements of the list. The function is called with the list containing the integers 1, 2, 3, 4, 5, 6, 7 in the given order. What will be the contents of the list after the function completes execution?

```
struct node {
  int value;
  struct node *next;
};
void rearrange (struct node *list) {
  struct node *p, *q ;
  int temp ;
  if (! list|| ! list → next ) return;
  p = list ; q = list → next;
  while (q) {
    temp = p → value ;
    p → value = q → value;
    q → value = temp ; p = q → next;
    q = p? p → next : 0 ;
  }
}
```

(a) 1, 2, 3, 4, 5, 6, 7 (b) 2, 1, 4, 3, 6, 5, 7
(c) 1, 3, 2, 5, 4, 7, 6 (d) 2, 3, 4, 5, 6, 7, 1

[2008 : 2 M]

4.6 The following C function takes a singly-linked list as input argument. It modified the list by moving the last element to the front of the list and returns the modified list. Some part of the code is left blank.

```
typed of struct node
{
  int value;
  struct node *next;
} Node;
Node *mode_to_front (Node *head)
{
  Node *p, *q;
  if((head==NULL)||(head->next==NULL))
      return head;
  q = NULL;
  p = head;
  while (p->next != NULL)
  {
     q = p;
     p = p->next;
  }
  ____________
  return head;
}
```

Choose the correct alternative to replace the blank line.

(a) q = NULL; p→next = head; head = p;
(b) q→next = NULL; head = p; p→next = head;
(c) head = p; p->next = q; q→next = NULL;
(d) q→next = NULL; p→next = head; head = p;

[2010 : 2 M]

4.7 Consider the C code fragment given below:

```
typedef struct node
{  int data;
   node* next;
} node;
void join (node* m, node* n)
{
   node* p = n;
   while (p -> next != NULL)
   {
      p = p -> next;
   }
   p -> next = m;
}
```

Assuming that m and n point to valid NULL-terminated linked lists, invocation of join will

(a) append list m to the end of list n for all inputs.
(b) either cause a null pointer dereference or append list m to the end of list n.
(c) cause a null pointer dereference for all inputs.
(d) append list n to the end of list m for all inputs.

[2017 (Set-1) : 1 M]

4.8 Consider the following ANSI C program:

```
#include <stdio.h>
#include <stdlib.h>
struct Node{
        int value;
        struct Node *next;};
int main() {
    struct Node *boxE, *head, *boxN; int index = 0;
    boxE = head = (struct Node *)
    malloc(sizeof(struct Node));
    head → value = index;
    for (index = 1; index <= 3; index++) {
        boxN = (struct Node *) malloc(sizeof(struct Node));
        boxE → next = boxN;
        boxN → value = index;
        boxE = boxN; }
    for (index = 0; index <= 3; index++) {
        printf("Value at index %d is %d\n", index, head → value);
        head = head → next;
        printf("Value at index %d is %d\n", index+1, head → value); } }
```

Which one of the statements below is correct about the program?

(a) Upon execution, the program creates a linked-list of five nodes.
(b) It has a missing return which will be reported as an error by the compiler.
(c) It dereferences an uninitialized pointer that may result in a run-time error.
(d) Upon execution, the program goes into an infinite loop.

[2021 (Set-2) : 2 M]

4.9 Consider the problem of reversing a singly linked list. To take an example, given the linked list below:

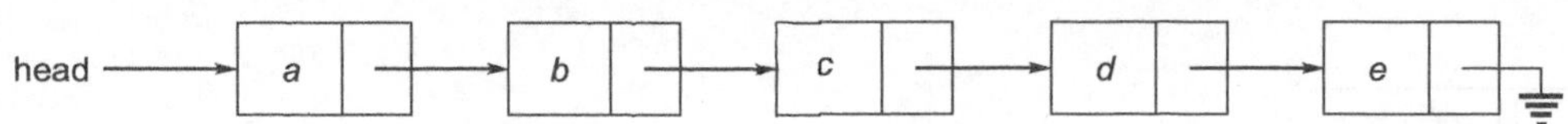

the reversed linked list should look like

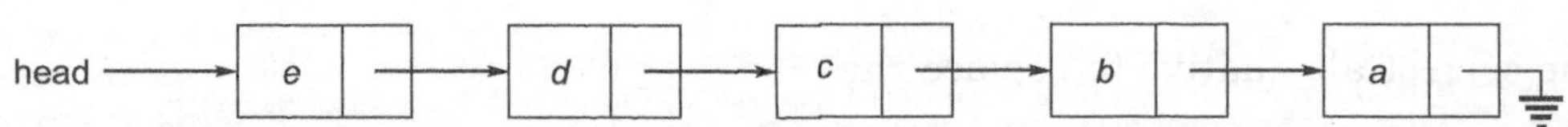

Which one of the following statements is TRUE about the time complexity of algorithms that solve the above problem in O(1) space?

(a) The best algorithm for the problem takes $\theta(n)$ time in the worst case.
(b) The best algorithm for the problem takes $\theta(n \log n)$ time in the worst case.
(c) The best algorithm for the problem takes $\theta(n^2)$ time in the worst case.
(d) It is not possible to reverse a singly linked list in O(1) space.

[2022 : 1 M]

4.10 Let SLLdel be a function that deletes a node in a singly-linked list given a pointer to the node and a pointer to the head of the list. Similarly, let DLLdel be another function that deletes a node in a doubly-linked list given a pointer to the node and a pointer to the head of the list.

Let n denote the number of nodes in each of the linked lists. Which one of the following choices is TRUE about the worst-case time complexity of SLLdel and DLLdel?

(a) SLLdel is O(1) and DLLdel is O(n)
(b) Both SLLdel and DLLdel are O(log(n))
(c) Both SLLdel and DLLdel are O(1)
(d) SLLdel is O(n) and DLLdel is O(1)

[2023 : 1 M]

Answers Linked List

4.1 (d) **4.2** (b) **4.3** (a) **4.4** (b) **4.5** (b) **4.6** (d) **4.7** (b) **4.8** (c) **4.9** (a)
4.10 (d)

Explanations Linked List

4.1 (d)

Linear search number of comparisons: O(n).

4.2 (b)

For a given linked list p, the function f returns 1 if and only if the elements in the list are stored in non-decreasing (increasing) order of the data value.
f returns 1 if p == NULL or p → Next == NULL or p → data <= p → next → data,
satisfies in linked list for every node p.

4.3 (a)

To delete node *x*, we should findout the previous node to the *x*. To find the previous nodes to the *x* will take O(n).
Therefore it takes O(n).

4.4 (b)

2 1 4 3 6 5 7: as p and q are swapping each other where q is p → next all the time.

4.5 (b)

The given C code is nonrecursive which interchange the two consecutive elements of the list. So the input sequence 1, 2, 3, 4, 5, 6, 7 produces the output 2, 1, 4, 3, 6 , 5, 7

4.6 (d)

This is the program by moving last element to front of list and returns modified list.
When while (P → next ! = NULL) execute then p pass to q and now p point to next. When "while" condition is false then we get q → next = NULL; p → next = head; head = p; means next to q made null, and points p → next is head and then make p as the head.

4.7 (b)

Consider two linked lists:

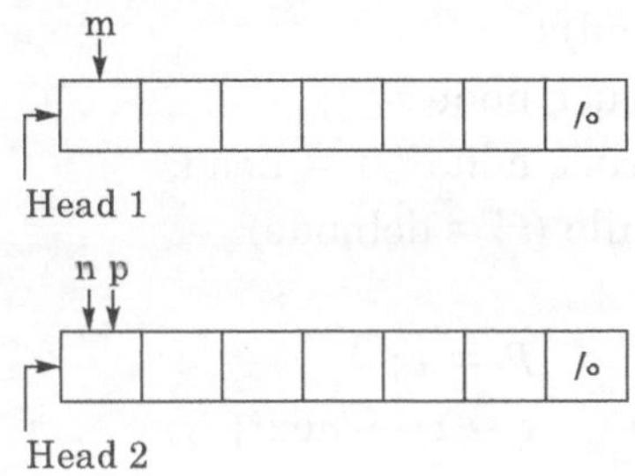

After the code execution, list 'm' will be appended to the end of list 'n'. But in some cases it can cause a null pointer dereference also.

4.8 (c)

As we can see in the loop, *i* runs from 1 to 3. So, four nodes will be created because one node is already created with value 0.

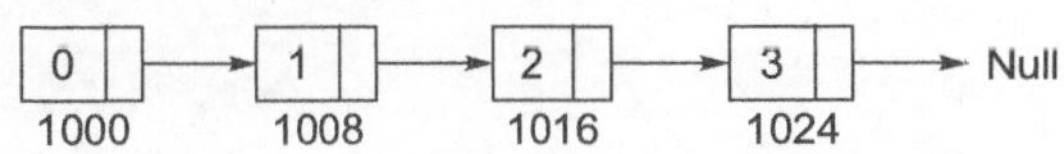

When index = 3, then head value = 3
Head = Head → Next **(Now head will point to NULL)**
Head → Value [which is in print. So it will generate Run time Error].
So, option (c) is correct.

4.9 **(d)**

In order to delete a node in a singly linked list we need to traverse the nodes sequentially, this will take O(*n*).

```
struct node {
    int val;
    struct node * next;
};
SLLdel (struct node * delnode, struct node *
    head) {
    struct node * P;
    struct node * t = head;
    while (t ! = delnode)
      {
        P = t;
        t = t → next;
      }
    P → next = delnode → next;
    free (delnode);
    }
```

Deletion in double linked list takes O(1) time.

```
struct DLLnode {
    int val;
    struct DLLnode * next;
    struct DLLnode * prev;
};
DLLdel (struct DLLnode * delnode, struct
    DLLnode * head) {
    struct DLLnode * t = delnode → prev;
    t → next = delnode → next;
    delnode → next → prev = t;
    free (delnode);
}
```

Note: Corners cases in the above codes are not handled.

■■■■

5 Trees

5.1 A binary tree T has n leaf nodes. The number of nodes of degree 2 in T is

(a) $\log_2 n$ (b) $n - 1$
(c) n (d) 2^n **[1995 : 1 M]**

5.2 In the balanced binary tree in the figure given below, how many nodes will become unbalanced when a node is inserted as a child of the node "g"?

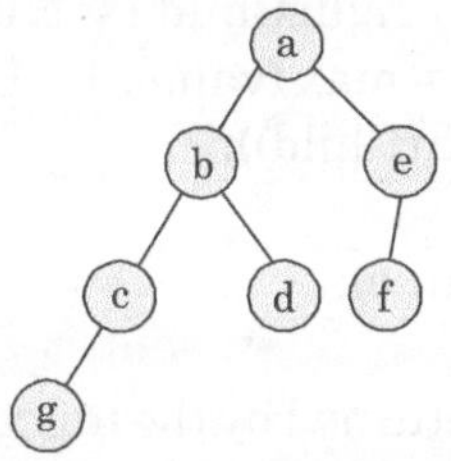

(a) 1 (b) 3
(c) 7 (d) 8 **[1996 : 1 M]**

5.3 Which of the following sequences denotes the post order traversal sequence of the tree of above question?

(a) fegcbdba (b) gcbdafe
(c) gcdbfea (d) fedgcba

[1996 : 1 M]

5.4 A binary search tree is generated by inserting in order of the following integers: 50, 15, 62, 5, 20, 58, 91, 3, 8, 37, 60, 24. The number of nodes in the left subtree and right subtree of the root respectively is

(a) (4, 7) (b) (7, 4)
(c) (8, 3) (d) (3, 8) **[1996 : 2 M]**

5.5 A binary search tree is used to locate the number 43. Which of the following probe sequences are possible and which are not? Explain.

(a) 61 52 14 17 40 43 (b) 2 3 50 40 60 43
(c) 10 65 31 48 37 43 (d) 81 61 52 14 41 43
(e) 17 77 27 66 18 43 **[1996 : 2 M]**

5.6 A binary search tree is generated by inserting in order the following integers: 50, 15, 62, 5, 20, 58, 91, 3, 8, 37, 60, 24. The number of nodes in the left subtree and right subtree of the root respectively is

(a) (4, 7) (b) (7, 4)
(c) (8, 3) (d) (3. 8) **[1996 : 2 M]**

5.7 A binary search tree contains the values 1, 2, 3, 4, 5, 6, 7, 8. The tree is traversed in pre-order and the values are printed out. Which of the following sequences is a valid output?

(a) 53124786 (b) 53126487
(c) 53241678 (d) None of these

[1997 : 2 M]

5.8 Which of the following statements is false?

(a) A tree with n nodes has (n – 1) edges
(b) A labeled rooted binary tree can be uniquely constructed given its postorder and preorder traversal results.
(c) A complete binary tree with n internal nodes has (n + 1) leaves
(d) The maximum number of nodes in a binary tree of height h is $(2^{h+1} - 1)$ **[1998 : 1 M]**

5.9 A complete n-ary tree is one in which every node has 0 or n sons. If x is the number of internal nodes of a complete n-ary tree, the number of leaves in it is given by

(a) x (n – 1) + 1 (b) xn – 1
(c) xn + 1 (d) x(n + 1)

[1998 : 2 M]

5.10 Consider the following nested representation of binary trees: (X Y Z) indicates Y and Z are the left and right subtrees, respectively, of node X. Note that Y and Z may be NULL, or further nested. Which of the following represents a valid binary tree?

(a) (1 2 (4 5 6 7))
(b) (1 ((234) 56)7)
(c) (1(234) (567))
(d) (1(23NULL) (45))

[2000 : 1 M]

5.11 Let LASTPOST, LASTIN and LASTPRE denote the last vertex visited in a postorder, inorder and preorder traversal. Respectively, of a complete binary tree. Which of the following is always true?

(a) LASTIN = LASTPOST
(b) LASTIN = LASTPRE
(c) LASTPRE = LASTPOST
(d) None of the above **[2000 : 2 M]**

5.12 The number of leaf nodes in a rooted tree of n nodes, with each node having 0 or 3 children is

(a) $\frac{n}{2}$ (b) $\frac{(n-1)}{3}$

(c) $\frac{(n-1)}{2}$ (d) $\frac{(2n+1)}{3}$

[2002 : 2 M]

5.13 Suppose the numbers 7, 5, 1, 8, 3, 6, 0, 9, 4, 2 are inserted in that order into an initially empty binary search tree. The binary search tree uses the usual ordering on natural numbers. What is the inorder traversal sequence of the resultant tree?

(a) 7 5 1 0 3 2 4 6 8 9
(b) 0 2 4 3 1 6 5 9 8 7
(c) 0 1 2 3 4 5 6 7 8 9
(d) 9 8 6 4 2 3 0 1 5 7

[2003 : 1 M]

5.14 A data structure is required for storing a set of integers such that each of the following operations can be done in O(log n) time, where n is the number of elements in the set.

1. Deletion of the smallest element.
2. Insertion of an element if it is not already present in the set.

Which of the following data structures can be used for this purpose?

(a) A heap can be used but not a balanced binary search tree
(b) A balanced binary search tree can be used but not a heap
(c) Both balanced binary search tree and heap can be used
(d) Neither balanced binary search tree nor heap can be used

[2003 : 2 M]

5.15 The following numbers are inserted into an empty binary search tree in the given order: 10, 1, 3, 5, 15, 12, 16. What is the height of the binary search tree (the height is the maximum distance of a leaf node from the root)?

(a) 2 (b) 3
(c) 4 (d) 6 **[2004 : 1 M]**

5.16 Level order traversal of a rooted tree can be done by starting from the root and performing

(a) preorder traversal
(b) inorder traversal
(c) depth first search
(d) breadth first search **[2004 : 1 M]**

5.17 Consider the following C program segment:

```
struct CellNode
{
    struct CellNode *leftChild;
    int element;
    struct CellNode *rightChild;
};
int DoSomething (struct CellNode *ptr)
{
    int value = 0;
    if (ptr ! = NULL)
    {
      if (ptr → leftChild ! = NULL)
          value = 1 + DoSomething (ptr → leftChild);
      if (ptr → rightChild ! = NULL)
          value = max (value, 1 + DoSomething (ptr
          → rightChild));
    }
    return (value);
}
```

The value returned by the function DoSomething when a pointer to the root of a non-empty tree is passed as argument is

(a) The number of leaf nodes in the tree
(b) The number of nodes in the tree
(c) The number of internal nodes in the tree
(d) The height of the tree **[2004 : 2 M]**

5.18 Consider the label sequences obtained by the following pairs of traversals on a labeled binary tree. Which of these pairs identify a tree uniquely?

1. Preorder and postorder
2. Inorder and postorder
3. Preorder and inorder
4. Level order and postorder

(a) 1 only (b) 2 and 3
(c) 3 only (d) 4 only **[2004 : 2 M]**

5.19 Which one of the following binary trees has its inorder and preorder traversals as BCAD and ABCD, respectively?

(a) 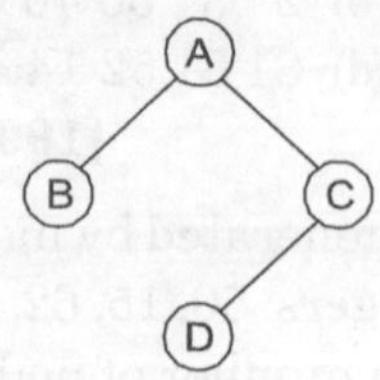

(b) 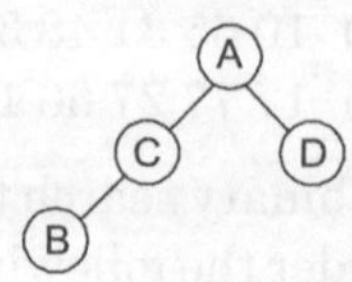

(c) 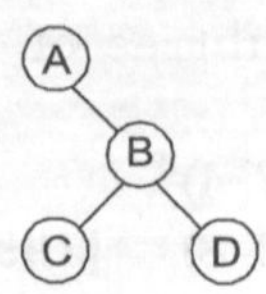

(d) 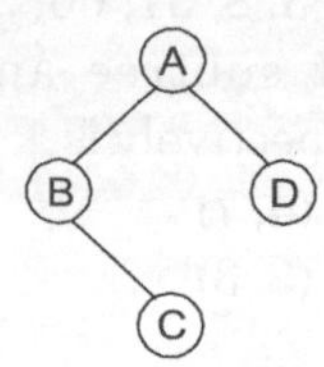

[2004 : 2 M]

5.20 The numbers 1, 2,..., n are inserted in a binary search tree in some order. In the resulting tree, the right subtree of the root contains p nodes. The first number to be inserted in the tree must be

(a) p (b) p + 1
(c) n − p (d) n − p + 1

[2005 : 1 M]

5.21 In a binary tree, for every node the difference between the number of nodes in the left and right subtrees is at most 2. If the height of the tree is h > 0, then the minimum number of nodes in the tree is

(a) 2^{h-1} (2) $2^{h-1} + 1$
(c) 2^h 1 (4) 2^h **[2005 : 2 M]**

5.22 A binary search tree contains the numbers 1, 2, 3, 4, 5, 6, 7, 8. When the tree is traversed in preorder and the values in each node printed out, the sequence of values obtained is 5, 3, 1, 2, 4, 6, 8, 7. If the tree is traversed in postorder, the sequence obtained would be

(a) 8, 7, 6, 5, 4, 3, 2, 1
(b) 1, 2, 3, 4, 8, 7, 6, 5
(c) 2, 1, 4, 3, 6, 7, 8, 5
(d) 2, 1, 4, 3, 7, 8, 6, 5

[2005 : 2 M]

5.23 Postorder traversal of a given binary search tree, T produces the following sequence of keys
10, 9, 23, 22, 27, 25, 15, 50, 95, 60, 40, 29
Which one of the following sequences of keys can be the result of an inorder traversal of the tree T?

(a) 9, 10, 15, 22, 23, 25, 27, 29, 40, 50, 60, 95
(b) 9, 10, 15, 22, 40, 50, 60, 95, 23, 25, 27, 29
(c) 29, 15, 9, 10, 25, 22, 23, 27, 40, 60, 50, 95
(d) 95, 50, 60, 40, 27, 23, 22, 25, 10, 0, 15, 29

[2005 : 2 M]

5.24 In a complete k-ary, every internal node has exactly k children. The number of leaves in such a tree with n internal nodes is

(a) n k (b) (n − 1) k +1
(c) n(k − 1) + 1 (d) n (k −1)

[2005 : 2 M]

5.25 In a binary tree, the number of internal nodes of degree 1 is 5, and the number of internal nodes of degree 2 is 10. The number of leaf nodes in the binary tree is

(a) 10 (b) 11
(c) 12 (d) 15 **[2006 : 1 M]**

5.26 Suppose that we have numbers between 1 and 100 in a binary search tree and want to search for the number 55. Which of the following sequences CANNOT be the sequence of nodes examined?

(a) {10, 75, 64, 43, 60, 57, 55}
(b) {90, 12, 68, 34, 62, 45, 55}
(c) {9, 85, 47, 68, 43, 57, 55}
(d) {79, 14, 72, 56, 16, 53, 55} **[2006 : 2 M]**

Directions for Question 5.27 to 5.29:

An array X of n distinct integers is interpreted as a complete binary tree. The index of the first element of the array is 0.

5.27 Tho indox of thc parcnt of clement X [i], i ≠ 0, is

(a) $\left\lfloor \frac{i}{2} \right\rfloor$ (b) $\left\lceil \frac{(i-1)}{2} \right\rceil$

(c) $\left\lceil \frac{i}{2} \right\rceil$ (d) $\left\lceil \frac{i}{2} \right\rceil - 1$

[2006 : 2 M]

5.28 If only the root node does not satisfy the heap property, the algorithm to convert the complete binary tree into a heap has the best asymptotic time complexity of

(a) O(n) (b) O(log n)
(c) O(n log n) (d) O(n loglog n)

[2006 : 2 M]

5.29 If the root node is at level 0, the level of element X[i], i ≠ 0, is

(a) $\lfloor \log_2 i \rfloor$ (b) $\lceil \log_2(i+1) \rceil$
(c) $\lfloor \log_2(i+1) \rfloor$ (d) $\lceil \log_2 i \rceil$

[2006 : 2 M]

5.30 The height of a binary tree is the maximum number of edges in any root to leaf path. The maximum number of nodes is a binary tree of height h is

(a) 2^h (b) $2^{h-1} - 1$
(c) $2^{h+1} - 1$ (d) 2^{h+1} **[2007 : 1 M]**

5.31 The maximum number of binary trees that can be formed with three unlabeled nodes is

(a) 1 (b) 5
(c) 4 (d) 3 **[2007 : 1 M]**

5.32 The inorder and preorder traversal of a binary tree are *dbeafcg* and *abdecfg* respectively. The postorder traversal of the binary tree is

(a) debfgca (b) edbgfca
(c) edbfgca (d) defgbca

[2007 : 2 M]

5.33 A complete n-ary tree is a tree in which each node has n children or no children. Let I be the number of internal nodes and L be the number of leaves in a complete n-ary tree. If L = 41, and I = 10, what is the value of n

(a) 3 (b) 4
(c) 5 (d) 6 **[2007 : 2 M]**

5.34 Consider the following C program segment where CellNode represents a node in a binary tree

```
struct CellNode {
struct CelloNode * leftchild;
int element;
struct CellNode *rightchild;
};
int GetValue (struct CellNode * ptr) {
    int value = 0;
if (ptr! = NULL) {
    if ((ptr → leftChild = = NULL) &&
        (ptr → rightChild = = NULL))
        Value = 1;
else
    value = value +
        GetValue (ptr→leftChild)
        + GetValue(ptr→ rightChild);
}
    return (value);
}
```

The value returned by Get Value when a pointer to the root of a binary tree is passed as its argument is

(a) the number of nodes
(b) the number of internal nodes in the tree
(c) the number of leaf nodes in the tree
(d) the height of the tree

[2007 : 2 M]

5.35 When searching for the key value 60 in a binary search tree, nodes containing the key values 10, 20, 40, 50, 70 80, 90 are traversed, not necessarily in the order given. How many different orders are possible in which these key values can occur on the search path from the root to the node containing the value 60?

(a) 35 (b) 64
(c) 128 (d) 5040

[2007 : 2 M]

5.36 Which of the following is TRUE?

(a) The cost of searching an AVL tree is Θ(log n) but that of a binary search tree is O(n)
(b) The cost of searching an AVL tree is Θ(log n) but that of a complete binary tree is Θ(n log n)
(c) The cost of searching a binary search tree is O (log n) but that of an AVL tree is Θ(n)
(d) The cost of searching an AVL tree is Θ(n log n) but that of a binary search tree is O(n)

[2008 : 1 M]

Linked Questions 5.37 & 5.38:

A binary tree with n > 1 nodes has n_1, n_2 and n_3 nodes of degree one, two and three respectively. The degree of a node is defined as the number of its neighbours.

5.37 n_3 can be expressed as:

(a) $n_1 + n_2 - 1$ (b) $n_1 - 2$
(c) $\left[\left(\frac{(n_1 + n_2)}{2}\right)\right]$ (d) $n_2 - 1$

[2008 : 2 M]

5.38 Starting with the above tree, while there remains a node v of degree two in the tree, add an edge between the two neighbours of v and then remove v from the tree.
How many edges will remain at the end of the process?

(a) $2 * n_1 - 3$ (b) $n_2 + 2 * n_1 - 2$
(c) $n_3 - n_2$ (d) $n_2 + n_1 - 2$

[2008 : 2 M]

5.39 You are given the postorder traversal, P of a binary search tree on the n elements 1, 2, ..., n. You have to determine the unique binary search tree that has P as its postorder traversal. What is the time complexity of the most efficient algorithm for doing this?

(a) Θ(log n)
(b) Θ(n)
(c) Θ(n log n)
(d) none of the above, as the tree cannot be uniquely determined. **[2008 : 2 M]**

5.40 We have a binary heap on n elements and wish to insert n more elements (not necessarily one after another) into this heap. The total time required for this is

(a) Θ(log n) (b) Θ(n)
(c) Θ(n log n) (d) $\Theta(n^2)$ **[2008 : 2 M]**

5.41 The following three are known to be the preorder, inorder and postorder sequences of a binary tree. But it is not known which is which.

I. MBCAFHPYK
II. KAMCBYPFH
III. MABCKYFPH

Pick the true statement from the following.

(a) I and II are preorder and inorder sequences, respectively
(b) I and III are preorder and postorder sequences, respectively
(c) II is the inorder sequence, but nothing more can be said about the other two sequences
(d) II and III are the preorder and inorder sequences, respectively **[2008 : 2 M]**

5.42 What is the maximum height of any AVL-tree with 7 nodes? Assume that the height of a tree with a single node is 0.
(a) 2 (b) 3
(c) 4 (d) 5 **[2009 : 2 M]**

5.43 The height of a tree is defined as the number of edges on the longest path in the tree. The function shown in the pseudocode below is invoked as the height (root) to compute the height of a binary tree rooted at the tree pointer root.

```
int height(treeptr n)  {
   if(n==NULL)return -1;
   if(n → left==NULL)
      if(n→right==NULL)return0;
      else return [B1] ;//Box1
   else { h1=height(n → left);
      if(n → right==NULL) return(1+h1);
      else {h2=height(n → right);
          return [B2] ;//Box2 }
   }
}
```

The appropriate expressions for the two boxes B1 and B2 are
(a) B1: (1 + height (n → right))
B2: (1 + max(h1, h2))
(b) B1: (height(n → right))
B2: (1 + max(h1, h2))
(c) B1: height(n → right)
B2: max(h1, h2)
(d) B1: (1 + height(n → right))
B2: max(h1, h2))
[2012 : 2 M]

5.44 The preorder traversal sequence of a binary search tree is 30, 20, 10, 15, 25, 23, 39, 35, 42. Which one of the following is the postorder traversal sequence of the same tree?
(a) 10, 20, 15, 23, 25, 35, 42, 39, 30
(b) 15, 10, 25, 23, 20, 42, 35, 39, 30
(c) 15, 20, 10, 23, 25, 42, 35, 39, 30
(d) 15, 10, 23, 25, 20, 35, 42, 39, 30
[2013 : 2 M]

5.45 Consider a rooted n node binary tree represented using pointers. The best upper bound on the time required to determine the number of subtrees having exactly 4 nodes is $O(n^a \log^b n)$. Then the value of $a + 10b$ is ________.
[2014 (Set-1) : 1 M]

5.46 Consider the following rooted tree with the vertex labeled P as the root:

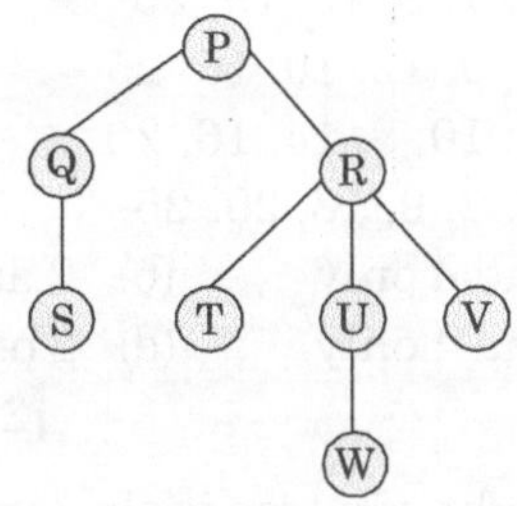

The order in which the nodes are visited during an in-order traversal of the tree is
(a) SQPTRWUV
(b) SQPTUWRV
(c) SQPTWUVR
(d) SQPTRUWV
[2014 (Set-3) : 1 M]

5.47 Consider the pseudocode given below. The function **DoSomething()** takes as argument a pointer to the root of an arbitrary tree represented by the *leftMostChild-rightSibling* representation. Each node of the tree is of type **treeNode**.

```
typedef struct treeNode* treeptr;
struct treeNode
{
    treeptr leftMostChild, rightSibling;
};
int DoSomething (treeptr tree)
{
    int value=0;
    if (tree != NULL)
    {
        if (tree->leftMostChild == NULL)
            value = 1;
        else
            value = DoSomething(tree->
                                leftMostChild);
        value=value+DoSomething(tree->
                                rightSibling);
    }
    return(value);
}
```

When the pointer to the root of a tree is passed as the argument to **DoSomething**, the value returned by the function corresponds to the

(a) number of internal nodes in the tree.
(b) height of the tree.
(c) number of nodes without a right sibling in the tree.
(d) number of leaf nodes in the tree.

[2014 (Set-3) : 2 M]

5.48 Which of the following is/are correct inorder traversal sequence(s) of binary search tree(s)?

1. 3, 5, 7, 8, 15, 19, 25
2. 5, 8, 9, 12, 10, 15, 25
3. 2, 7, 10, 8, 14, 16, 20
4. 4, 6, 7, 9, 18, 20, 25

(a) 1 and 4 only (b) 2 and 3 only
(c) 2 and 4 only (d) 2 only

[2015 (Set-1) : 1 M]

5.49 The height of a tree is the length of the longest root-to-leaf path in it. The maximum and minimum number of nodes in a binary tree of height 5 are

(a) 63 and 6, respectively
(b) 64 and 5, respectively
(c) 32 and 6, respectively
(d) 31 and 5, respectively

[2015 (Set-1) : 1 M]

5.50 A binary tree T has 20 leaves. The number of nodes in T having two children is ________.

[2015 (Set-2) : 1 M]

5.51 While inserting the elements 71, 65, 84, 69, 67, 83 in an empty binary search tree (BST) in the sequence shown, the element in the lowest level is

(a) 65 (b) 67
(c) 69 (d) 83

[2015 (Set-3) : 1 M]

5.52 The number of ways in which the numbers 1, 2, 3, 4, 5, 6, 7 can be inserted in an empty binary search tree, such that the resulting tree has height 6, is ________. **Note:** The height of a tree with a single node is 0.

[2016 (Set-2) : 2 M]

5.53 Let T be a binary search tree with 15 nodes. The minimum and maximum possible heights of T are:

Note: The height of a tree with a single node is 0.

(a) 4 and 15 respectively
(b) 3 and 14 respectively
(c) 4 and 14 respectively
(d) 3 and 15 respectively **[2017 (Set-1) : 1 M]**

5.54 The pre-order traversal of a binary search tree is given by 12, 8, 6, 2, 7, 9, 10, 16, 15, 19, 17, 20. Then the post-order traversal of this tree is:

(a) 2, 6, 7, 8, 9, 10, 12, 15, 16, 17, 19, 20
(b) 2, 7, 6, 10, 9, 8, 15, 17, 20, 19, 16, 12
(c) 7, 2, 6, 8, 9, 10, 20, 17, 19, 15, 16, 12
(d) 7, 6, 2, 10, 9, 8, 15, 16, 17, 20, 19, 12

[2017 (Set-2) : 2 M]

5.55 The postorder traversal of a binary tree is 8, 9, 6, 7, 4, 5, 2, 3, 1. The inorder traversal of the same tree is 8, 6, 9, 4, 7, 2, 5, 1, 3. The height of a tree is the length of the longest path from the root to any leaf. The height of the binary tree above is ________.

[2018 : 1 M]

5.56 Let T be a full binary tree with 8 leaves. (A full binary tree has every level full.) Suppose two leaves a and b of T are chosen uniformly and independently at random. The expected value of the distance between a and b in T (i.e., the number of edges in the unique path between a and b) is (rounded off to 2 decimal places) ________.

[2019 : 2 M]

5.57 The preorder traversal of a binary search tree is 15, 10. 12, 11, 20, 18, 16, 19.
Which one of the following is the postorder traversal of the tree?

(a) 20, 19, 18, 16, 15, 12, 11, 10
(b) 10, 11, 12, 15, 16, 18, 19, 20
(c) 19, 16, 18, 20, 11, 12, 10, 15
(d) 11, 12, 10, 16, 19, 18, 20, 15 **[2020 : 1 M]**

5.58 Consider the following statements:

S_1 : The sequence of procedure calls corresponds to a preorder traversal of the activation tree.

S_2 : The sequence of procedure returns corresponds to a postorder traversal of the activation tree.

Which one of the following options is correct?

(a) S_1 is false and S_2 is false
(b) S_2 is true and S_2 is false
(c) S_1 is true and S_2 is true
(d) S_1 is false and S_2 is true

[2021 (Set-1) : 1 M]

5.59 A binary search tree T contains n distinct elements. What is the time complexity of picking an element in T that is smaller than the maximum element in T?

(a) $\theta(n)$ (b) $\theta(n \log n)$
(c) $\theta(\log n)$ (d) $\theta(1)$

[2021 (Set-1) : 1 M]

5.60 Consider a complete binary tree with 7 nodes. Let A denote the set of first 3 elements obtained by performing Breadth-First Search (BFS) starting from the root. Let B denote the set of first 3 elements obtained by performing Depth-First Search (DFS) starting from the root. The value of $|A - B|$ is ________.

[2021 (Set-2) : 1 M]

5.61 Suppose a binary search tree with 1000 distinct elements is also a complete binary tree. The tree is stored using the array representation of binary heap trees. Assuming that the array indices start with 0, the 3rd largest element of the tree is stored at index ________. **[2022 : 1 M]**

5.62 Consider the C function foo and the binary tree shown:

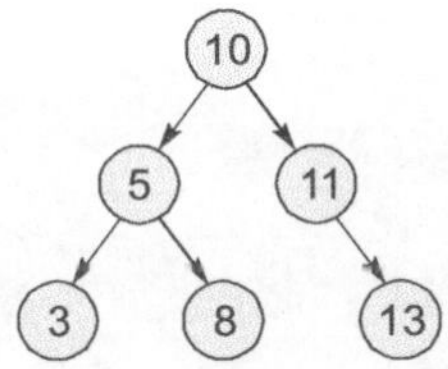


```
typedef struct node {
    int val;
    struct node *left, *right;
} node;
int foo(node *p) {
    int retval;
    if (p == NULL)
        return 0;
    else {
        retval = p->val + foo(p->left) + foo(p->right);
        printf("%d ", retval);
        return retval;
    }
}
```

When foo is called with a pointer to the root node of the given binary tree, what will it print?

(a) 3 8 5 13 11 10 (b) 3 5 8 10 11 13
(c) 3 8 16 13 24 50 (d) 3 16 8 50 24 13

[2023 : 2 M]

5.63 An array [82, 101, 90, 11, 111, 75, 33, 131, 44, 93] is heapified. Which one of the following options represents the first three elements in the heapified array?

(a) 131, 111, 90 (b) 131, 11, 93
(c) 82, 11, 93 (d) 82, 90, 101

[2024 (Set-1) : 2 M]

5.64 Consider a binary min-heap containing 105 distinct elements. Let k be the index (in the underlying array) of the maximum element stored in the heap. The number of possible values of k is

(a) 52 (b) 1
(c) 53 (d) 27

[2024 (Set-1) : 2 M]

5.65 You are given a set V of distinct integers. A binary search tree T is created by inserting all elements of V one by one, starting with an empty tree. The tree T follows the convention that, at each node, all values stored in the left subtree of the node are smaller than the value stored at the node. You are not aware of the sequence in which these values were inserted into T, and you do not have access to T.

Which one of the following statements is TRUE?

(a) Preorder traversal of T can be determined from V.
(b) Inorder traversal of T can be determined from V.
(c) Postorder traversal of T can be determined from V.
(d) Root node of T can be determined from V.

[2024 (Set-2) : 2 M]

■■■■

Answers Trees

5.1	(b)	**5.2**	(b)	**5.3**	(c)	**5.4**	(b)	**5.5**	Sol.	**5.6**	(b)	**5.7**	(d)	**5.8**	(b)	**5.9**	(a)
5.10	(c)	**5.11**	(b)	**5.12**	(d)	**5.13**	(c)	**5.14**	(b)	**5.15**	(b)	**5.16**	(d)	**5.17**	(d)	**5.18**	(b)
5.19	(d)	**5.20**	(c)	**5.21**	(b)	**5.22**	(d)	**5.23**	(a)	**5.24**	(c)	**5.25**	(b)	**5.26**	(c)	**5.27**	(a)
5.28	(b)	**5.29**	(c)	**5.30**	(c)	**5.31**	(b)	**5.32**	(a)	**5.33**	(c)	**5.34**	(c)	**5.35**	(a)	**5.36**	(a)
5.37	(b)	**5.38**	(a)	**5.39**	(c)	**5.40**	(b)	**5.41**	(d)	**5.42**	(b)	**5.43**	(a)	**5.44**	(d)	**5.45**	(1)
5.46	(a)	**5.47**	(d)	**5.48**	(a)	**5.49**	(a)	**5.50**	(19)	**5.51**	(b)	**5.52**	(64)	**5.53**	(b)	**5.54**	(b)
5.55	(4)	**5.56**	(4.25)	**5.57**	(d)	**5.58**	(c)	**5.59**	(d)	**5.60**	(1)	**5.61**	(509)	**5.62**	(c)	**5.63**	(a)
5.64	(c)	**5.65**	(b)														

Explanations Trees

5.1 (b)

Because for 2 degree node, every time '2' leafs are added and number of nodes increases is '1'. So no. of nodes with degree 2 is always one less than number of leafs present in tree.

5.2 (b)

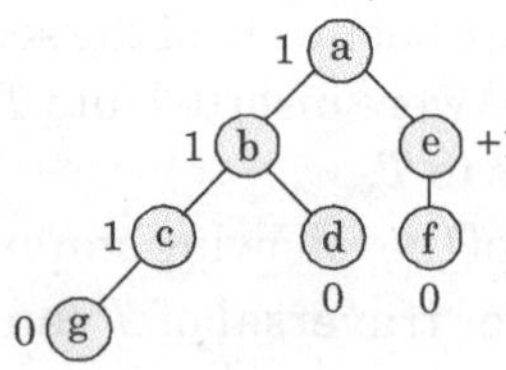

When a node is added as a child of g

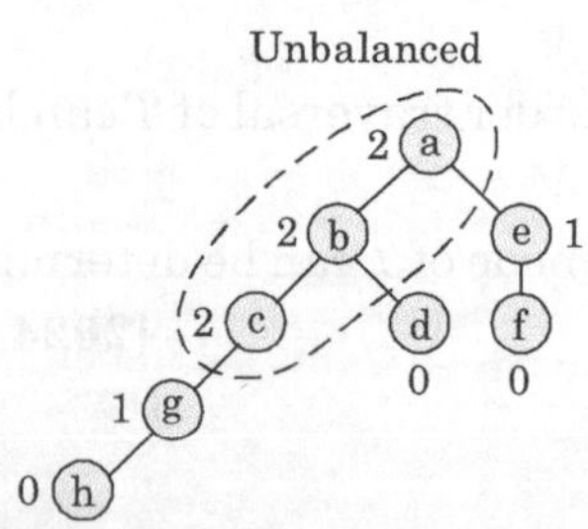

Nodes a, b & c become unbalanced.
Therefore, option (b) is correct.

5.3 (c)

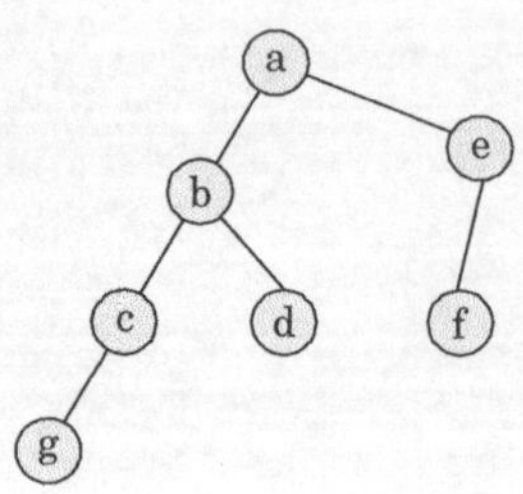

⇒ Postorder: gcdbfea

5.4 (b)

The binary search tree is

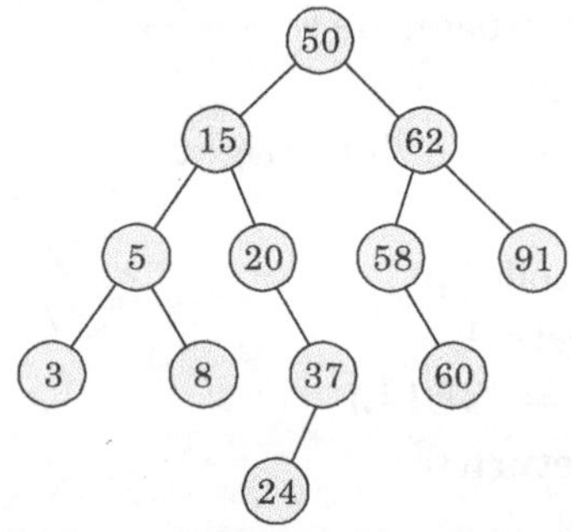

The number of nodes in the left subtree and right subtree of the root respectively is (7, 4)

5.5 Sol.

Possible sequence:

(a) 61 52 14 17 40 43
(c) 10 65 31 48 37 43
(d) 81 61 52 14 41 43

Not possible:

(b) 2 3 50 40 60 43
(e) 17 77 27 66 18 43

5.6 (b)

The binary search tree is as follows:

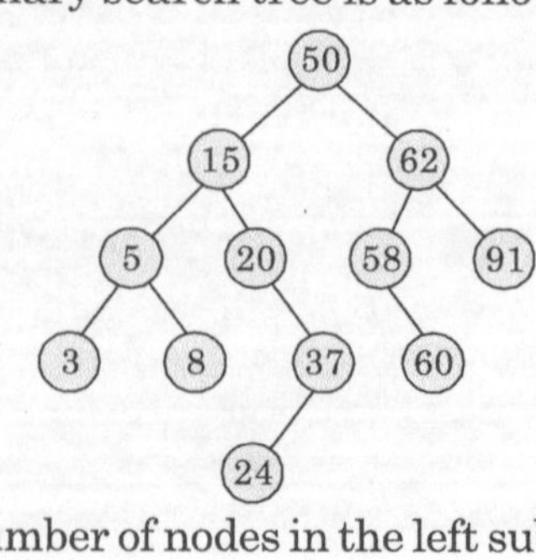

The number of nodes in the left subtree and right subtree of the root respectively is (7, 4). Therefore option (b) is correct.

5.7 (d)

(a) 53124786

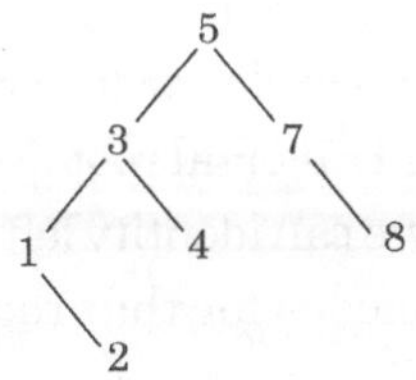

6 can not be child of 8.

(b) 53126487

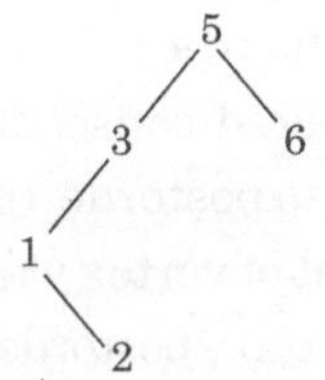

4 can not be child of 6.

(c) 53241678

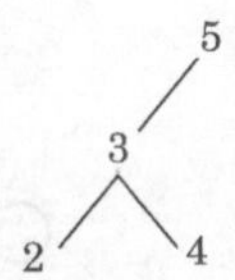

1 can not be child of 4 or 5.

5.8 (b)

A labelled rooted binary tree can not be constructed uniquely when inorder traversal is given along with post-order or pre-order traversal.

5.9 (a)

In n-ary tree the total number of vertices are

$$m = nx + 1$$

Where x: no. of internal nodes and let l be the number of leaf nodes.

$$x + l = nx + 1$$

$$l = (n-1)x + 1$$

So number of leaf nodes in n-ary tree with x internal nodes is $(n - 1) x + 1$.

5.10 (c)

(1 (2 3 4) (5 6 7))

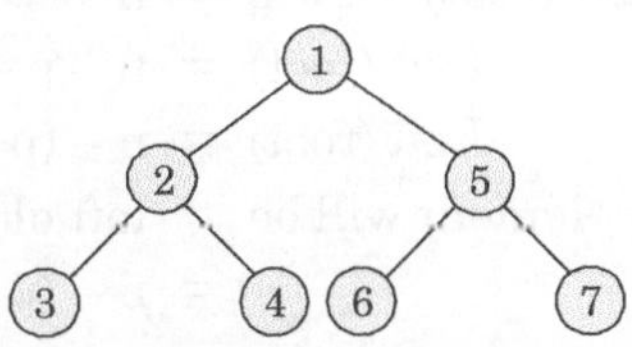

5.11 (b)

Because of complete binary tree option (b) is always correct.

5.12 (d)

In m-ary tree

Total number of nodes = m(internal nodes) + 1

Here, $n = 3i + 1$

Total nodes = internal nodes + leaf nodes

$$n = i + l$$

$$i = n - l$$

$$n = 3(n - l) + 1 \qquad [\because n = 3i + 1 \ \& \ i = n - l]$$

$$n = 3n - 3l + 1$$

$$3l = 2n + 1$$

$$l = \frac{2n+1}{3}$$

Where l is leaf nodes.

5.13 (c)

Input sequence 7, 5, 1, 8, 3, 6, 0, 9, 4, 2

Construct the BST in which the value of left subtree is less then or equal to root and the value of right-subtree is greater than or equal to the root.

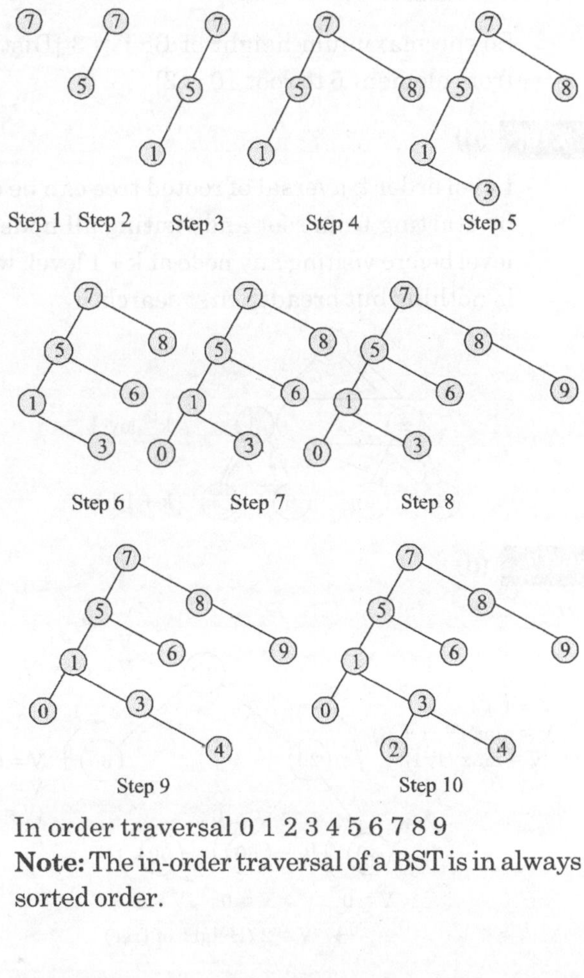

In order traversal 0 1 2 3 4 5 6 7 8 9

Note: The in-order traversal of a BST is in always sorted order.

5.14 (b)

Insertion and deletion in AVL and Min heap will take O(logn). But in this problem we have to check before inserting that element is already there are not, to do this Min heap will take O(n). AVL will take O(logn).

5.15 (b)

The binary search tree is initially empty.

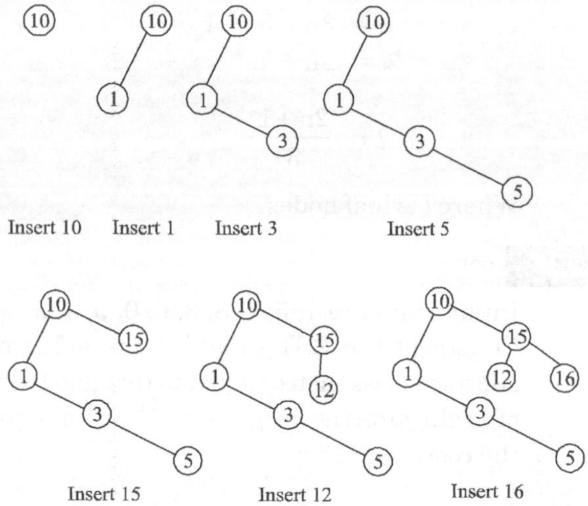

So the maximum height of BST = 3 [Distance from element 5 to root 10 is 3]

5.16 (d)

Level order traversal of rooted tree can be done by starting from root and visiting all node at k level before visiting any node at k + 1 level, which is nothing but breadth first search.

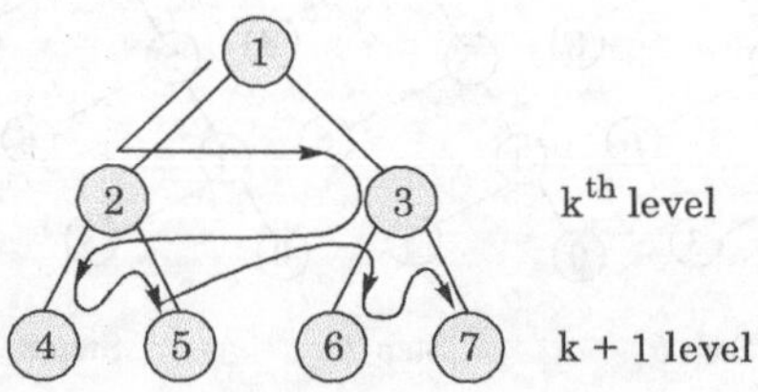

5.17 (d)

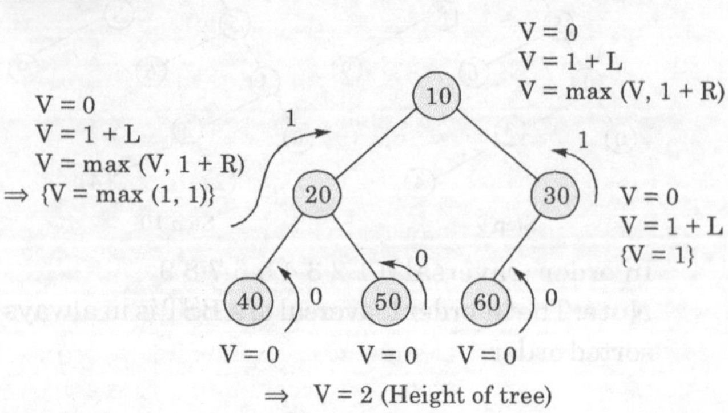

5.18 (b)

In a labeled binary tree preorder traversal produces prefix expression, postorder produces postfix expression and inorder produce infix expression. Preorder traversal root is first element and using inorder we can identity left subtree nodes and right subtree nodes for that root. So we can construct unique Binary tree using preorder and inorder similarly postorder and inorder. The postorder numbers assigned to the nodes have the useful property that the nodes in the subtree with root n are numbered consecutively from (postorder(n) – desc(n)) to postorder(n). To test if a vertex x is a descendent of vertex y then postorder(y) – desc(y) ≤ postorder(x) ≤ postorder(y)

A similar property holds for preorder and doesn't holds for inorder.

5.19 (d)

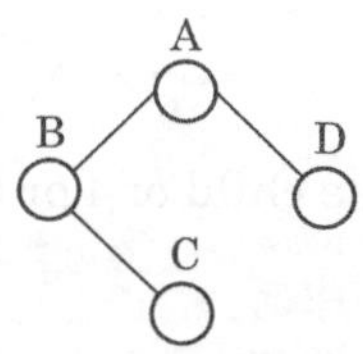

Inorder traversal is BCAD and preorder traversal is ABCD.

5.20 (c)

Number 1, 2, 3, ... n are inserted into BST in some order.

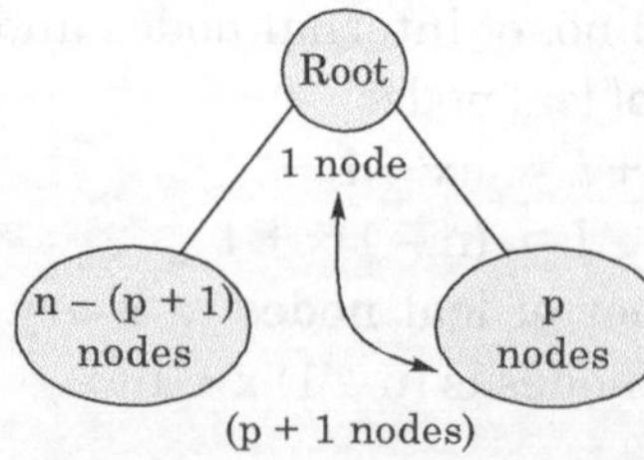

Since we know total nodes = n

So, Left (root) + 1 + p = n

Left (root) = n – p – 1

Left (root) = n – (p + 1)

So, root element will be = left element + 1

= n – (p + 1) + 1

= n – p – 1 + 1

= n – p

5.21 (b)

For h = 1 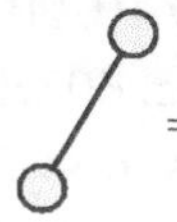 ⇒ Min nodes = 2

h = 2 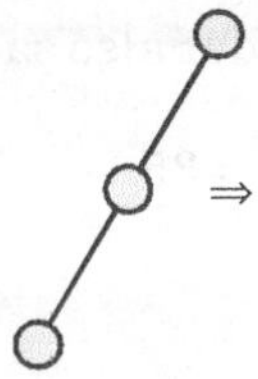 ⇒ Min nodes = 3

So by putting in option we get option (b) is correct.
$2^{h-1} + 1$.

5.22 (d)

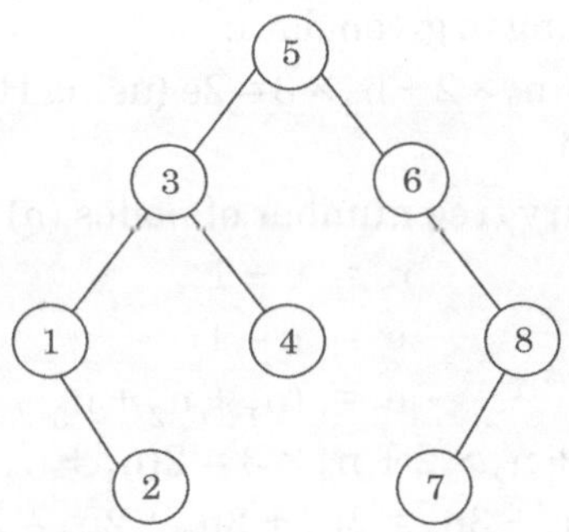

5.23 (a)

The inorder traversal of a binary search tree is always is in sorted order or increasing order of a given sequence so the inorder traversal of the tree T is 9, 10, 15, 22, 23, 25, 27, 29, 40, 50, 60, 55

5.24 (c)

In a complete-k-ary tree. If every internal node has exacity k children then (k – 1) key contains no leaves. If there are n internal nodes, then number of leaves is n(k – 1) + 1.

5.25 (b)

Let number of vertices in binary tree = n
Number of edges = n–1

So $n - 1 = 1 \times 5 + 2 \times 10$

$n - 1 = 5 + 20$

$\Rightarrow \quad n = 26$

Number of leaf nodes = 26 – 5 – 10 = 11

5.26 (c)

According to given options:

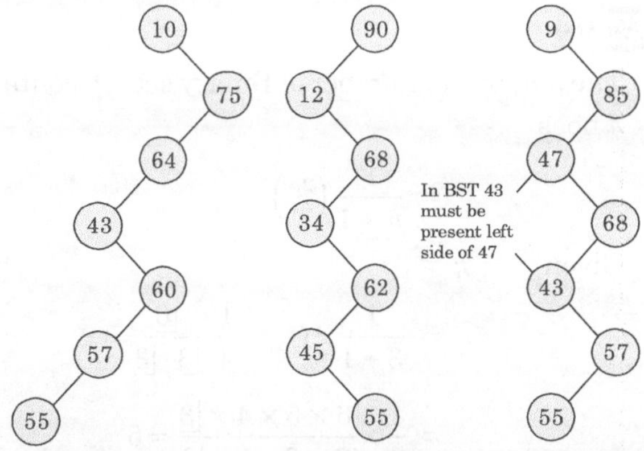

Valid BST, hence we can search 55 from root | Valid BST, hence we can search 55 from root | Invalid BST, hence cannot be valid sequence

5.27 (a)

Left child of ith element will be at 2*i+1 and right child at 2(i+1). The index of the parent of element X [i] is $\lfloor i / 2 \rfloor$.

5.28 (b)

Since only root node does not satisfy the heap property, so applying one time heapify will make whole tree as heap. Heapify will take $O(\log_2 n)$ i.e. height of heap tree.

5.29 (c)

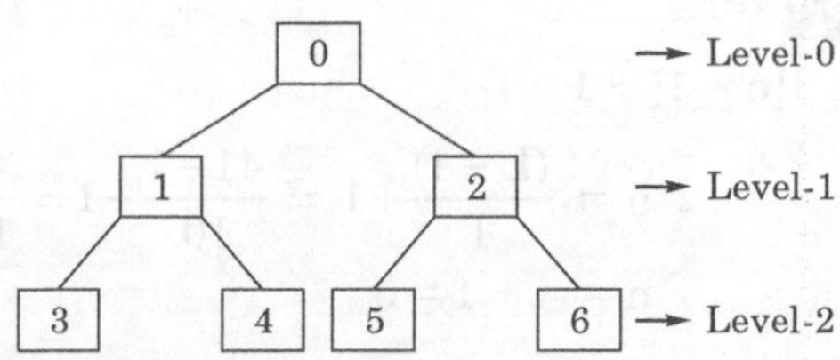

Index of Every node is shown in the box. Floor (log (i + 1)) draw the tree and realise that the last element at each level is the best choice to arrive at a conclusion.

5.30 (c)

The height of a binary tree is the maximum number of edges in any root to leaf path. Apply the induction.

Height No. of Nodes

0 $2^{0+1} - 1 = 2 - 1 = 1$ so it contains root itself

1 $2^{1+1} - 1 = 4 - 1 = 3$ root and one left & right child

h $2^{h+1} - 1$

So maximum number of nodes in a binary tree of height h is $2^{h+1} - 1$.

5.31 (b)

The maximum number of Binary search contain n noes

$$= \frac{1}{n+1}\binom{2n}{n}$$

Given, $n = 3$

$$= \frac{1}{3+1}\,{}^6C_3 = \frac{1}{4}\cdot\frac{\lfloor 6}{\lfloor 3 \cdot \lfloor 3}$$

$$= \frac{1}{4} \times \frac{6 \times 5 \times 4 \times \lfloor 3}{3 \times 2 \times 1 \times \lfloor 3} = 5$$

5.32 (a)

The in order traversal sequence is d b e a f c g and the pre order traversal sequence is a b d e c f g. So the tree is

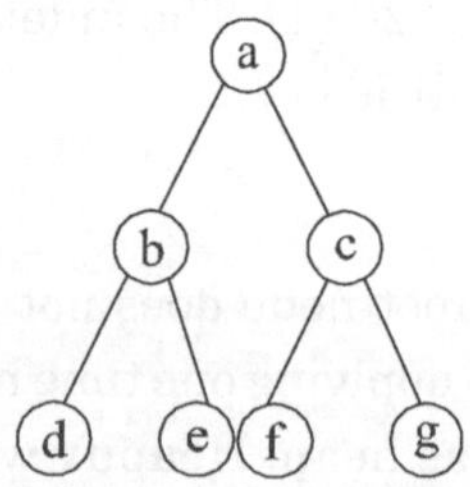

In the post order traversal the sequence is d e b f g c a.

5.33 (c)

$I[n-1] + 1 = L$

$$n = \frac{(L-1)}{I} + 1 = \frac{41-1}{10} + 1 = \frac{40}{10} + 1$$

$$n = 4 + 1 = 5$$

5.34 (c)

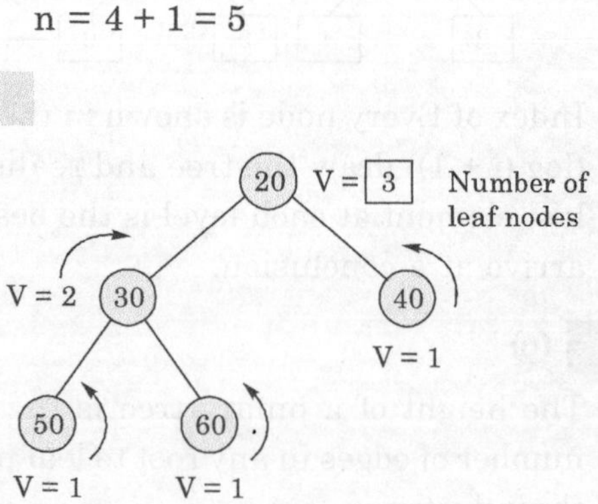

5.35 (a)

In Binary Search Tree to traverse each node, while searching for any key only order of left elements of key and right element of key is matter i.e. 10, 20, 40, 50 [key] 90, 80, 70 because we need to search element 60, so order of elements less than 60 will be only 10, 20, 40, 50 and order of elements greater than 60 will be 90, 80, 70.

So number of ways are = Arrangement of 7 numbers such that 4 element order same and remaining 3 elements order also same

$$= \frac{7!}{3!4!} = 35$$

5.36 (a)

Height of AVL tree always O(log n), so searching of an element will take O(log n) but height of a BST can be O(n) since skew BST also possible. So searching an element will take O(n) time.

5.37 (b)

According to given data:

$n_1 \times 1 + n_2 \times 2 + n_3 \times 3 = 2e$ (using Handshaking lemma)

In binary tree number of nodes (n) = e + 1

$$n = e + 1$$

$$e = n - 1$$

i.e. $e = (n_1 + n_2 + n_3 - 1)$

$$n_1 \times 1 + n_2 \times 2 + n_3 \times 3 = 2(n_1 + n_2 + n_3 - 1)$$

$$n_1 + 2n_2 + 3n_3 = 2n_1 + 2n_2 + 2n_3 - 2$$

$$n_1 = n_3 + 2$$

and $n_3 = n_1 - 2$

5.38 (a)

Since all node of degree 2 are removed after adding an edge between neighbour of nodes degree 2. Ex: Each node showing its degree.

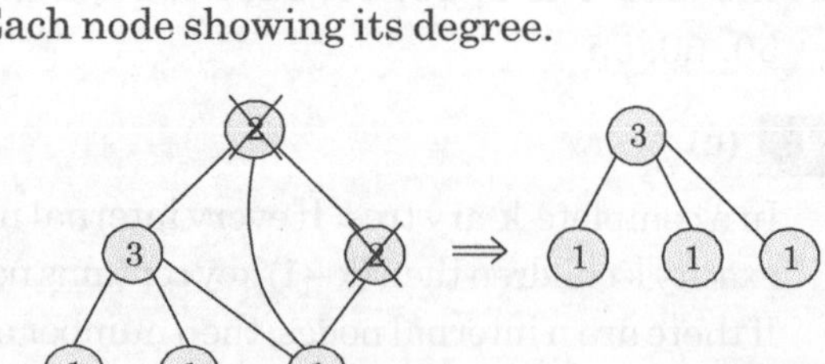

It will not effect the number of nodes of degree 1 and degree 3.

So, according to Handshaking lemma.

$$n_1 \times 1 + n_3 \times 3 = 2e$$

We know, $n_3 = n_1 - 2$

$$n_1 \times 1 + (n_1 - 2) \times 3 = 2e$$

$$n_1 + 3n_1 - 6 = 2e$$

$$4n_1 - 6 = 2e$$

$$e = \frac{4n_1 - 6}{2} = 2n_1 - 3$$

5.39 (c)

We have post order traversal and the tree is binary search tree so in order traversal of binary search tree is assending order.
Using these two we can construct binary search tree with O(nlogn) [we can apply binary search].

5.40 (b)

If we have a binary heap on elements and we wish to insert n more elements then we call a recursive call build heap which is responsible for constructing the heap.
The time spend by build heap is on the order of the sum over all vertices of the heights of the vertices. But at most $\lceil n/2^{i+1} \rceil$ vertices are of height i. So the total time spent by build heap is

$$\sum_{i=1}^{h} \frac{n}{2^i} i = \theta(n).$$

5.41 (d)

II. KAMCBYPFH (Preorder)
III. MABCKYFPH (Inorder)
If II is preorder and III is inorder then it is possible to construct binary tree.

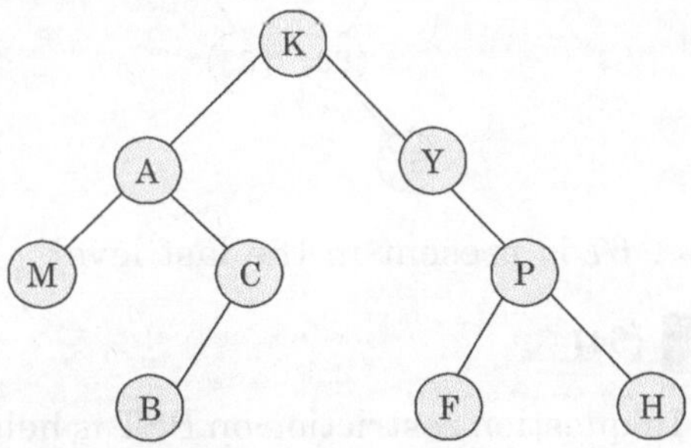

⇒ I. MBCAFHPYK is postorder.

5.42 (b)

Maximum height of any AVL-tree with 7.

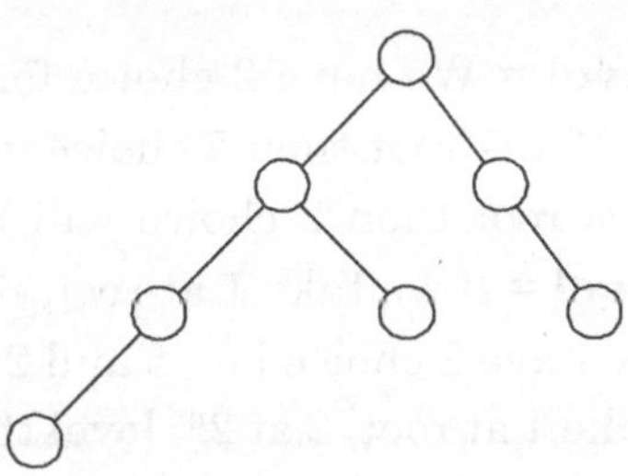

(There may be different way to draw AVL with 7 nodes).

5.43 (a)

e.g.
$h_1 = 1 + \max(0, 0)$, $h_2 = 0$,
return $1 + \max(h_1, h_2)$

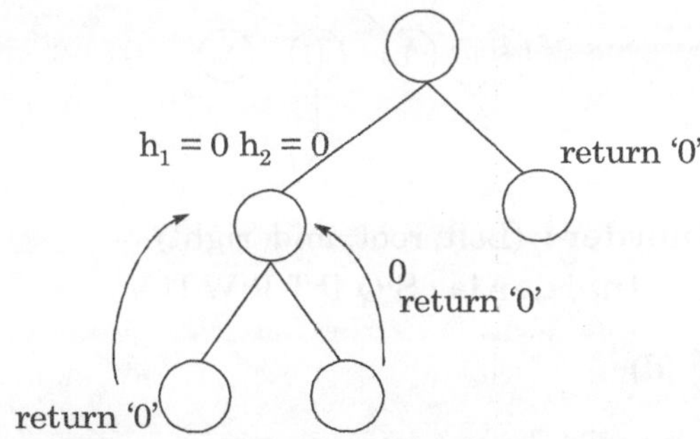

another e.g.

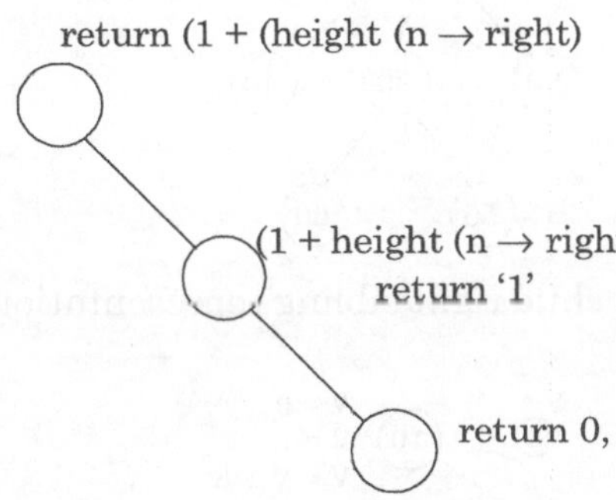

∴ Option (a) is correct by going through option.

5.44 (d)

Binary search tree and preorder is given
30, 20, 10, 15, 25, 23, 39, 35, 42

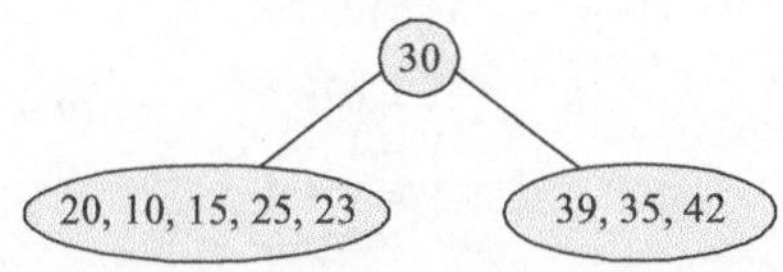

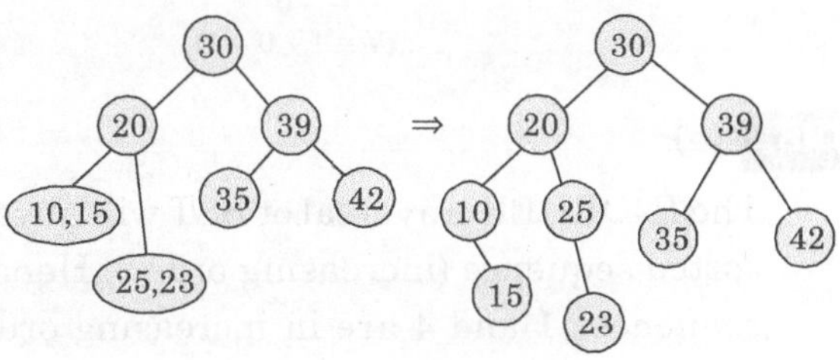

Postorder traversed of resultant tree is
15, 10, 23, 25, 20, 35, 42, 39, 30

5.45 (1)

Upper bound on time required to determine the number of subtrees having exactly 4 nodes = $O(n)$
$= O(n^a \log^b n)$

$\Rightarrow \quad a = 1, b = 0$

$\therefore \quad a + 10b = 1$

5.46 (a)

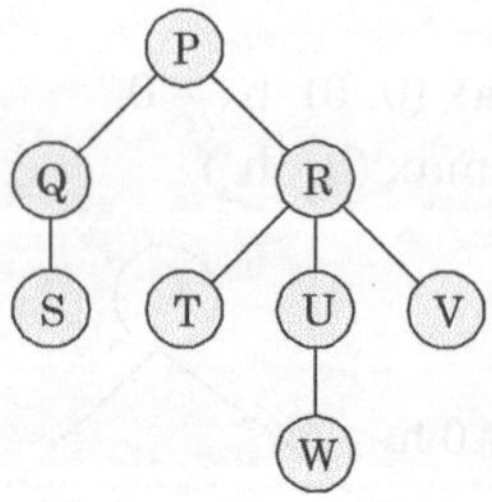

Inorder : (Left, root, mid, right)

$\therefore$ **Inorder is :** S Q P T R W U V

5.47 (d)

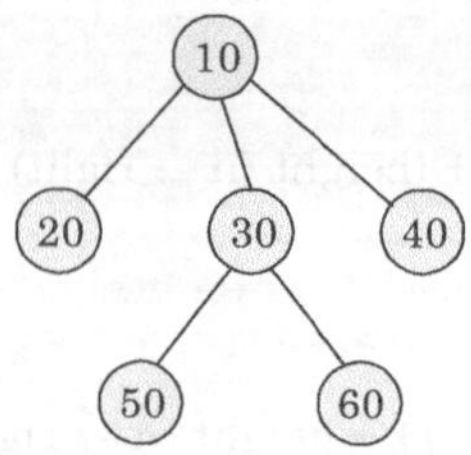

Left most child right sibling representation

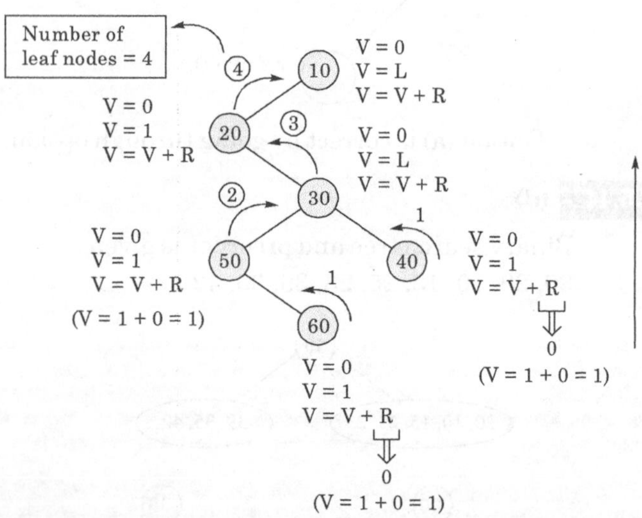

5.48 (a)

The INORDER traversal of BST will always be sorted sequence (increasing order). Hence the sequences 1 and 4 are in increasing order.

5.49 (a)

Height is defined as path length i.e. (height +1) will give number of levels.

In a binary tree of height h = 5:

Maximum number of nodes $= 2^{h+1}-1 = 2^{5+1}-1$
$= 64 - 1 = 63$

Minimum number of nodes = 6 i.e. one node at every level.

5.50 (19)

n_0 : Number of leaf nodes (degree 0)
n_1 : Number of nodes with degree 1
n_2 : Number of nodes with degree 2

We have following two equations for binary tree

$n = n_0 + n_1 + n_2$ (where n is the total number of nodes in the tree) ...(1)

Accounting for total nodes as sum of those nodes which are children and the root which is not a child of any node,

we get

$n = n_1 \times 1 + n_2 \times 2 + 1 = n_1 + 2n_2 + 1$ ...(2)

Combining equations (1) and (2) we get

$n_0 + n_1 + n_2 = n_1 + 2n_2 + 1 \Rightarrow n_2 = n_0 - 1$

Therefore the number of nodes of degree two in any binary tree = number of leaf nodes –1.

$= 20 - 1 = 19$

5.51 (b)

71, 65, 84, 69, 67, 83 are inserted into empty BST. Binary search tree:

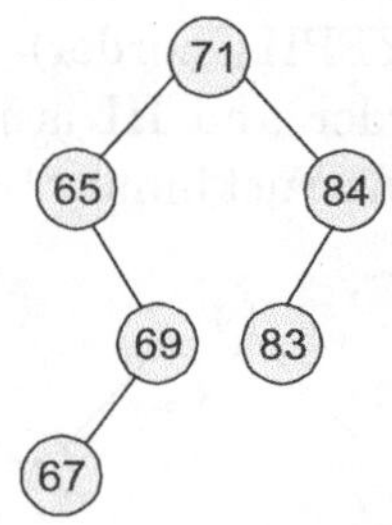

$\therefore$ 67 is present in the last level.

5.52 (64)

In question restriction on BST is height should be ‘6’. So we need 7 levels (given that root at height ‘0’). In creation of BST we have to use all element without repeatation.

At **1 level** = We have 2 choice i.e., either take 1 or 7.

At **2 level** = We have 2 choice for root 1 and 7 each. If 1 is root then 2 choice will be 6 and 2. If 7 is root then 2 choice will be 1 and 6.

At **3 level** = If we take 1 at root, 6 at 2nd level than we have 2 choice i.e., 5 and 2 at 3rd level. If we take 1 at root, 2 at 2nd level than we have 2 choice i.e., 6 and 3 at 3rd level. Similarly if we take 7 as root element.

So till 6th level, we have two choice at every level and for last level we left with only 1 element. So number of BST with height 6

$$= 2 \times 2 \times 2 \times 2 \times 2 \times 2 \times 1 = 2^6 = 64.$$

5.53 (b)

Since there are 15 nodes, hence the minimum height of the tree will be 3 (when tree will be balanced).

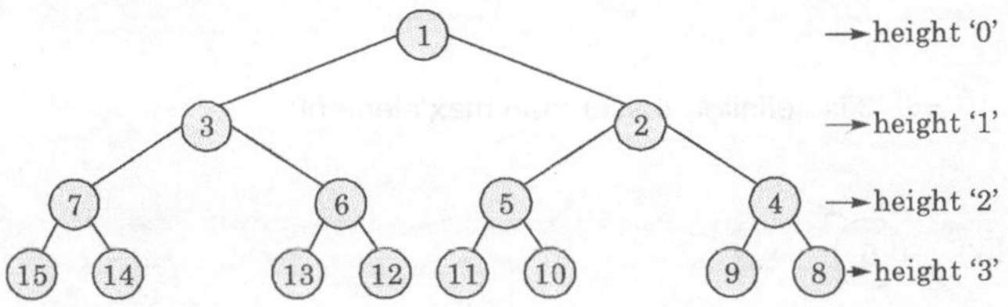

The maximum height will be when the tree is skew tree, which will give rise to height 14.

5.54 (b)

Preorder: 12, 8, 6, 2, 7, 9, 10, 16, 15, 19, 17, 20
Inorder: 2, 6, 7, 8, 9, 10, 12, 15, 16, 17, 19, 20
Tree will be,

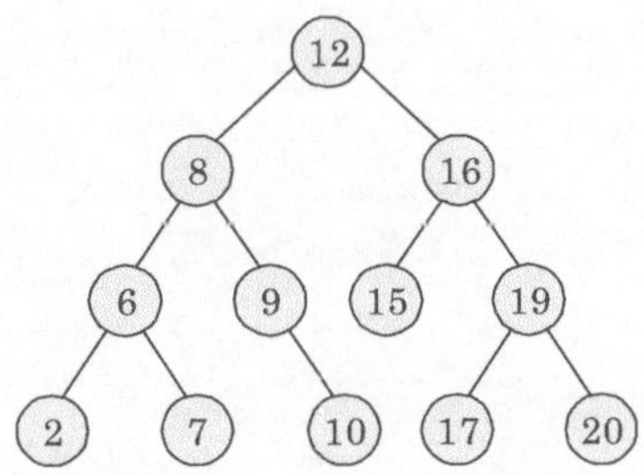

Postorder will be, 2, 7, 6, 10, 9, 8, 15, 17, 20, 19, 16, 12

5.55 (4)

Postorder = 8, 9, 6, 7, 4, 5, 2, 3, **(1)** Root
Inorder = 8, 6, 9, 4, 7, 2, 5, 1, 3

Binary tree will be

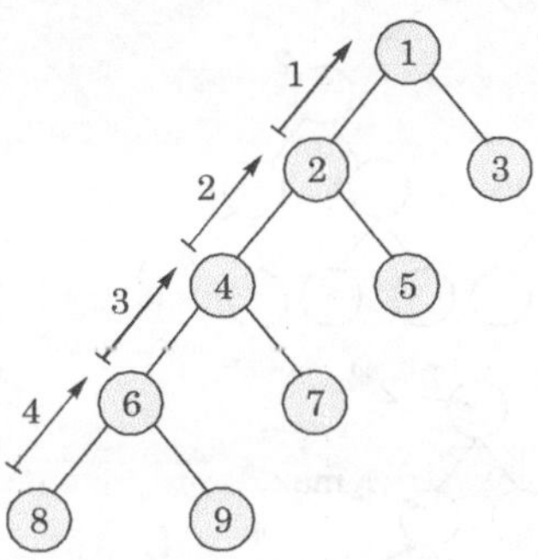

So, height of binary tree will be 4.

5.56 (4.25)

Two nodes can be selected in 8 × 8 = 64 ways.

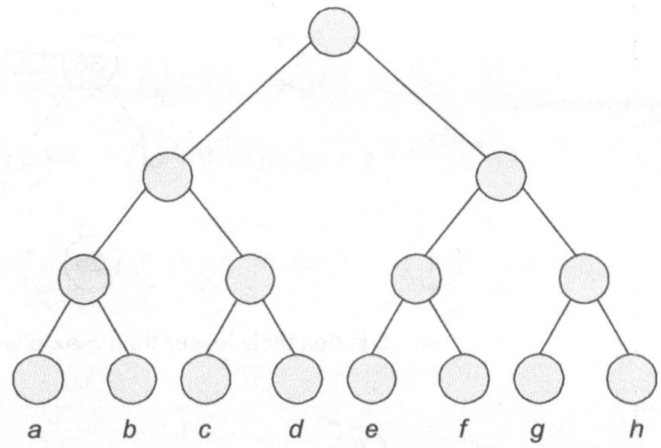

X : length between two nodes selected

X	0	1	2	3	4	5	6
Number of pairs	8	0	8	0	16	0	32

To find expected length between a and b is E[X]

$$E[X] = 2\times\frac{8}{64}+4\times\frac{16}{64}+6\times\frac{32}{64}$$

$$= \frac{16+64+192}{64}$$

$$= \frac{272}{64} = 4.25$$

5.57 (d)

BST inorder: 10, 11, 12, (15), 16, 18, 19, 20
LST: 10, 11, 12 — RST: 16, 18, 19, 20

Preorder: (15), 10, 12, 11, 20, 18, 16, 19
(15) → I^{st} root

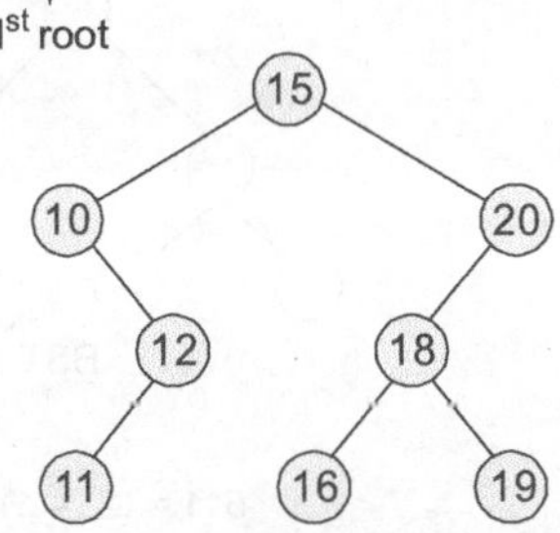

Postorder: 11, 12, 10, 16, 19, 18, 20, 15

5.59 (d)

(i)

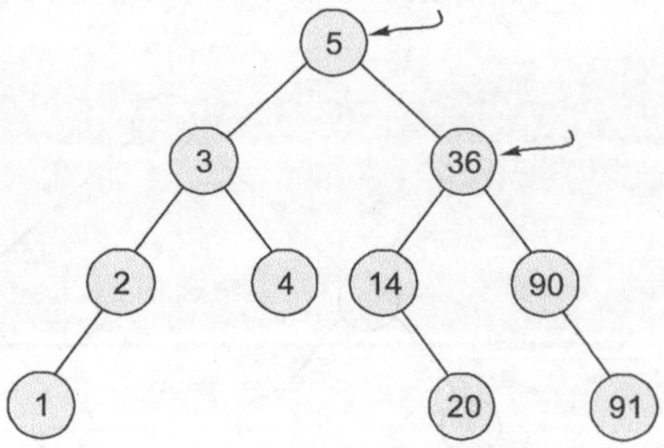

⇒ 5 is definitely lesser than max element

(ii)

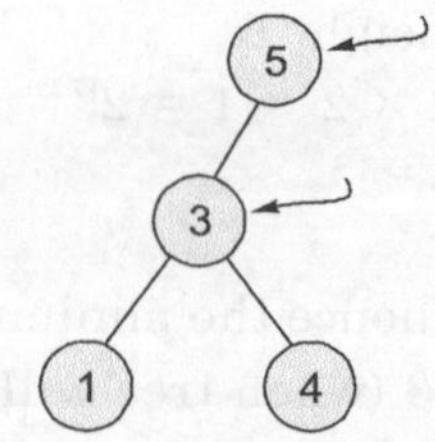

⇒ 3 is definitely lesser than max element

(iii)

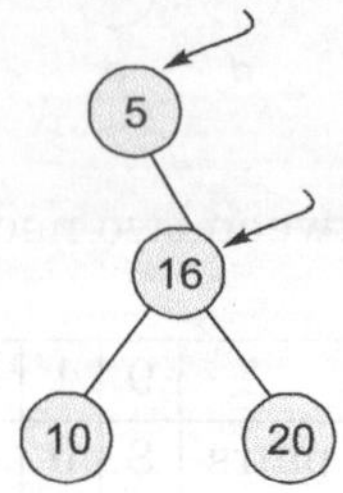

⇒ 5 is definitely lesser than max element

(iv)

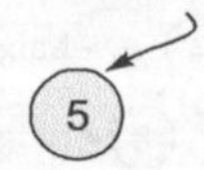

No such element

In any case, we can find one element that is less in just 2 seeks ⇒ $\theta(1)$.

5.60 (1)

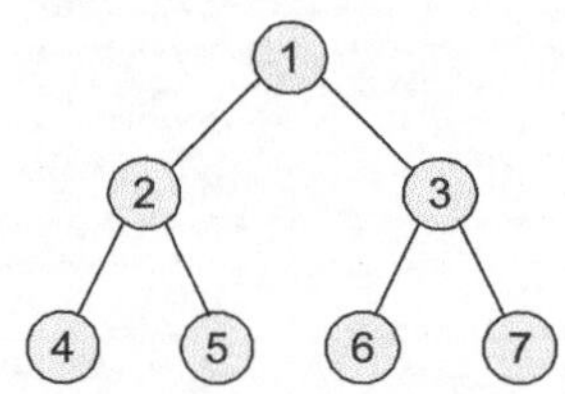

Using BFS, $A = \{1, 2, 3\}$

Using DFS, $B = \{1, 2, 4\}$

$|A - B|$ = Number of elements which are in A but not in B is only element $\{3\}$

So only 1 element present.

5.61 (509)

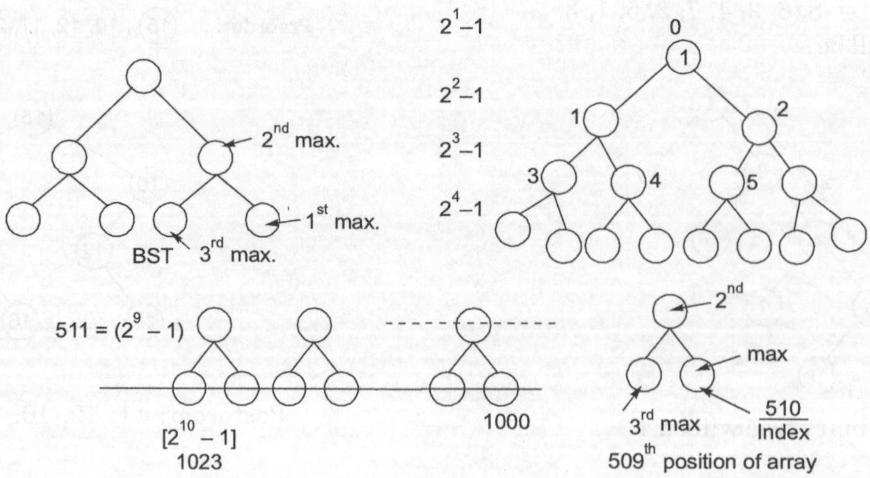

5.62 (c)

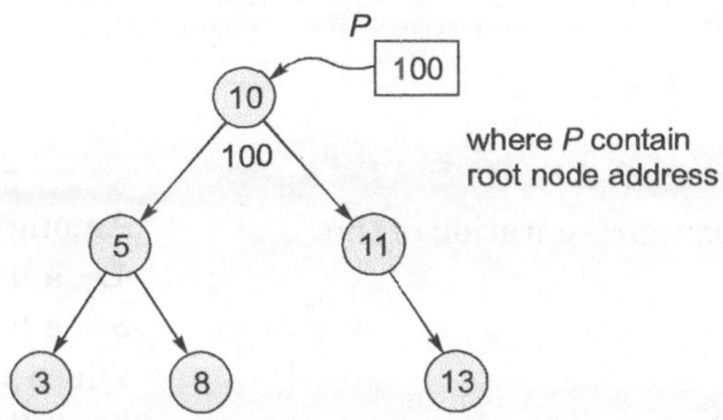

When foo is called with a pointer to the root node of the given binary tree then we reach the node until that node has left pointer and right pointer both pointing to NULL.

Whenever p → Left is NULL and p → Right is NULL then foo (p → left) and foo(p → right) both will return 'O' and variable retval contain the following value.

$$\text{retval} = (p \to \text{val}) + \text{foo}(p \to \text{left}) + \text{foo}(p \to \text{right})$$

for node whose value is 3, retval = 3 + 0 + 0 = 3

for node whose value is 8, retval = 8 + 0 + 0 = 8

for node whose value is 5, retval = 5 + 3 + 8 = 16

for node whose value is 13, retval = 13 + 0 + 0 = 13

for node whose value is 11, retval = 11 + 13 + 0 = 24

for node whose value is 10, retval = 10 + 16 + 24 = 50

5.63 (a)

Heapify: bottom-up construction

(Adjust internal nodes into max heap).

CBT of given 10 elements.

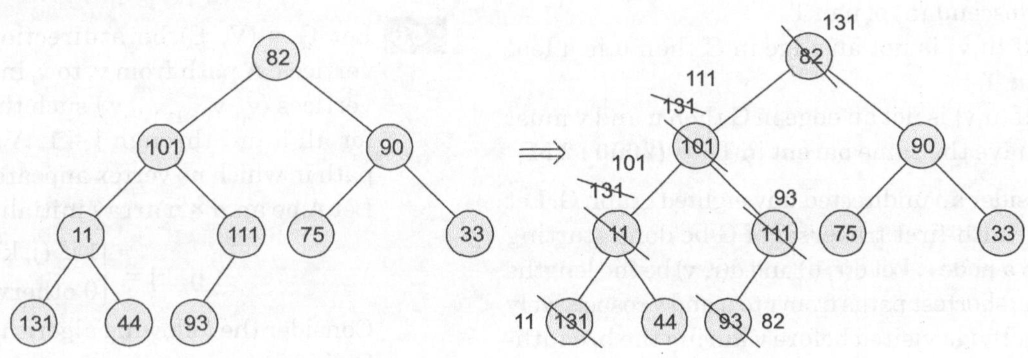

5.64 (c)

In binary min heap of n elements exactly $\left\lfloor \frac{n}{2} \right\rfloor$ internal nodes and $\left\lceil \frac{n}{2} \right\rceil$ leaf nodes.

Max element in min heap of n distinct elements must be one of the leaf node.

$\therefore \left\lceil \frac{105}{2} \right\rceil$ are number of leaf nodes = 53

■■■■

6 Graphs

6.1 Write the adjacency matrix representation of the graph given in figure.

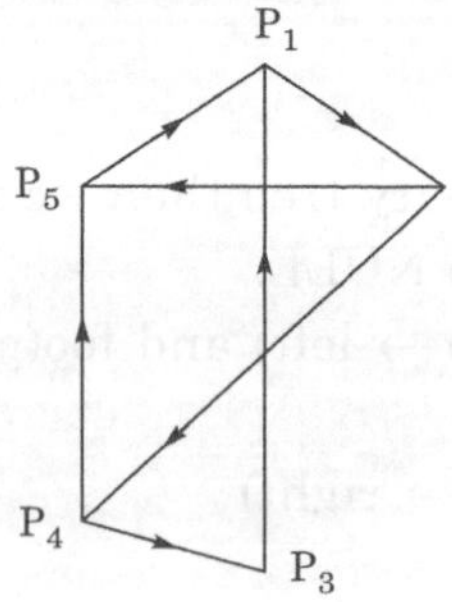

[1998 : 2 M]

6.2 Lt G be an undirected graph. Consider a depth-first traversal of G, and let T be the resulting depth-first search tree. Let u be a vertex in G and let v be the first new (unvisited) vertex visited after visiting u in the traversal. Which of the following statements is always true?

(a) {u,v} must be an edge in G, and u is a descendant of v in T
(b) {u,v} must be an edge in G, and v is a descendant of u in T
(c) If {u,v} is not an edge in G then u is a leaf in T
(d) If {u,v} is not an edge in G then u and v must have the same parent in T **[2000 : 2 M]**

6.3 Consider an undirected unweighted graph G. Let a breadth-first traversal of G be done starting from a node r. Let d(r, u) and d(r, v) be the lengths of the shortest paths from r to u and v respectively in G. If u is visited before v during the breadth-first traversal, which of the following statement is correct?

(a) d(r, u) < d(r, v) (b) d(r, u) > d(r, v)
(c) d(r, u) ≤ d(r, v) (d) None of these

[2001 : 2 M]

6.4 Consider the following graph:

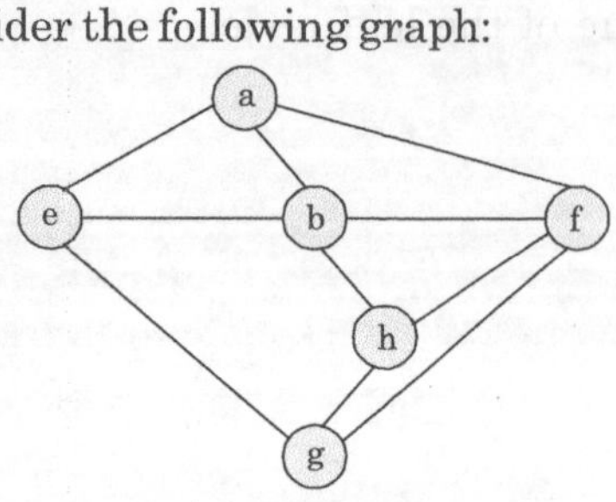

Among the following sequences:

1. a b e g h f 2. a b f e h g
3. a b f h g e 4. a f g h b e

Which are depth first traversals of the above graph?

(a) 1, 2 and 4 only (b) 1 and 4 only
(c) 2, 3 and 4 only (d) 1, 3 and 4 only

[2003 : 1 M]

6.5 Let G = (V, E) be an undirected graph with a subgraph $G_1 = (V_1, E_1)$. Weights are assigned to edges of G as follows:

$$w(e) = \begin{cases} 0 \text{ if } e \in E_1 \\ 1 \text{ otherwise} \end{cases}$$

A single-source shortest path algorithm is executed on the weighted graph (V, E, w) with an arbitrary vertex v_1 of V_1 as the source. Which of the following can always be inferred from the path costs computed?

(a) The number of edges in the shortest paths from v_1 to all vertices of G
(b) G_1 is connected (c) V_1 forms a clique in G
(d) G_1 is a tree **[2003 : 2 M]**

6.6 Let G = (V, E) be a direction graph with n vertices. A path from v_i to v_j in G is sequence of vertices $(v_i, v_{i+1}, ..., v_j)$ such that $(v_k, v_{k+1}) \in E$ for all k in i through j – 1. A simple path is a path in which no vertex appears more than once. Let A be an n × n array initialized as follow:

$$A[j, k] = \begin{cases} 1 \text{ if } (j, k) \in E \\ 0 \text{ otherwise} \end{cases}$$

Consider the following algorithm:

```
for i = 1 to n
for j = 1 to n
for k = 1 to n
A [j, k] = max (A[j, k], (A[j, i] + A[i, k]));
```

Which of the following statements is necessarily true for all j and k after terminal of the above algorithm?

(a) A[j, k] ≤ n
(b) If A [j, j] ≥ n – 1, then G has a Hamiltonian cycle
(c) If there exists a path from j to k, A[j, k] contains the longest path lengths from j to k
(d) If there exists a path from j to k, every simple path from j to k contain most A[j, k] edges

[2003 : 2 M]

6.7 In a depth-first traversal of a graph G with n vertices, k edges are marked as tree edges. The number of connected components in G is
(a) k (b) k + 1
(c) n – k – 1 (d) n – k **[2005 : 1 M]**

6.8 Let G be a directed graph whose vertex set is the set of numbers from 1 to 100. There is an edge from a vertex i to a vertex j iff either j = i + 1 or j = 3i. The minimum number of edges in a path in G from vertex 1 to vertex 100 is
(a) 4 (b) 7
(c) 23 (d) 99 **[2005 : 2 M]**

6.9 Let T be a depth first search tree in a undirected graph G. Vertices u and v are leaves of this tree T. The degrees of both u and v in G are at least 2. Which one of the following statements is true?
(a) There must exist a vertex w adjacent to both u and v in G
(b) There must exist a vertex w whose removal disconnects u and v in G
(c) There must exist a cycle in G containing u and v
(d) There must exist a cycle in G containing u and all its neighbours in G **[2006 : 2 M]**

6.10 Consider the depth-first-search of an undirected graph with 3 vertices P, Q, and R. Let discovery time d(u) represent the time instant when the vertex u is first visited, and finish time f(u) represent the time instant when the vertex u is last visited. Given that
d(P) = 5 units, f(P) = 12 units
d(Q) = 6 units, f(Q) = 10 units
d(R) = 14 unit, f(R) = 18 units
Which one of the following statements is TRUE about the graph?
(a) There is only one connected component
(b) There are two connected components, and P and R are connected
(c) There are two connected components, and Q and R are connected
(d) There are two connected components, and P and Q are connected **[2006 : 2 M]**

6.11 What is the largest integer m such that every simple connected graph with n vertices and n edges contains at least m different spanning trees?
(a) 1 (b) 2
(c) 3 (d) n **[2007 : 2 M]**

6.12 A depth-first search is performed on a directed acyclic graph. Let d[u] denote the time at which vertex u is visited for the first time and f[u] the time at which the dfs call to the vertex u terminates. Which of the following statements is always true for all edges (u, v) in the graph?
(a) $d[u] < d[v]$ (b) $d[u] < f[v]$
(c) $f[u] < f[v]$ (d) $f[u] > f[v]$
[2007 : 2 M]

6.13 The most efficient algorithm for finding the number of connected components in an undirected graph on n vertices and m edges has time complexity.
(a) $\Theta(n)$ (b) $\Theta(m)$
(c) $\Theta(m + n)$ (d) $\Theta(mn)$ **[2008 : 1 M]**

6.14 The Breadth First Search algorithm has been implemented using the queue data structure. One possible order of visiting the nodes of the following graph is

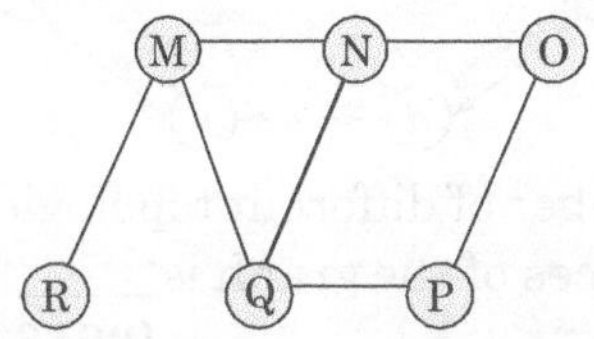

(a) MNOPQR (b) NQMPOR
(c) QMNPRO (d) QMNPOR
[2008 : 1 M]

6.15 Consider the following sequence of nodes for the undirected graph given below:

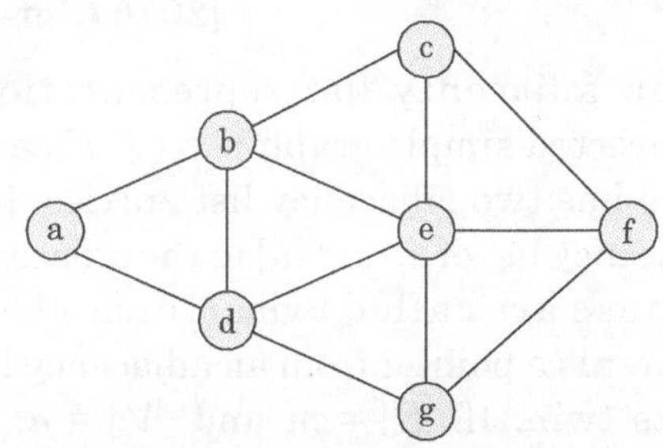

1. a b e f d g c 2. a b e f c g d
3. a d g e b c f 4. a d b c g e f

A Depth First Search (DFS) is started at node *a*. The nodes are listed in the order they are first visited. Which all of the above is (are) possible output(s)?
(a) 1 and 3 only (b) 2 and 3 only
(c) 2, 3 and 4 only (d) 1, 2 and 3 only
[2008 : 2 M]

6.16 Let *G* of be a graph with *n* vertices and *m* edges. What is the tightest upper bound on the running time of Depth First Search on *G*, when *G* is represented as an adjacency matrix ?
(a) $\Theta(n)$ (b) $\Theta(n + m)$
(c) $\Theta(n)^2$ (d) $\Theta(m^2)$
[2014 (Set-1) : 1 M]

6.17 Suppose depth first search is executed on the graph below starting at some unknown vertex. Assume that a recursive call to visit a vertex is made only after first checking that the vertex has not been visited earlier. Then the maximum possible recursion depth (including the initial call) is.

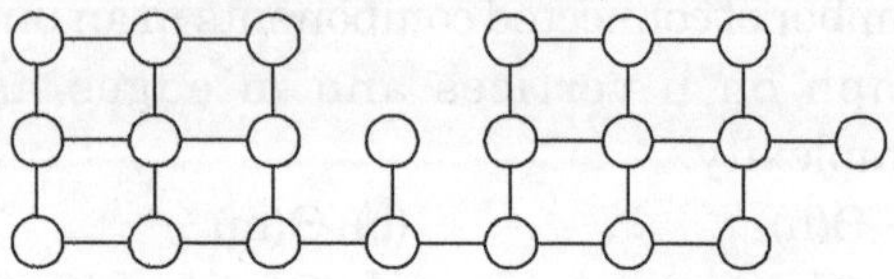

[2014 (Set-3) : 1 M]

6.18 Consider the following directed graph:

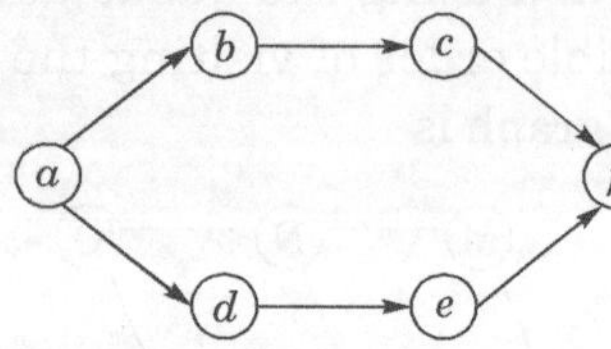

The number of different topological ordering of the vertices of the graph is______.

[2016 (Set-1) : 1 M]

6.19 Breadth First Search (BFS) is started on a binary tree beginning from the root vertex. There is a vertex t at a distance four from the root. If t is the n-th vertex in this BFS traversal, then the maximum possible value of n is______.

[2016 (Set-2) : 1 M]

6.20 In an adjacency list representation of an undirected simple graph $G = (V, E)$, each edge (u, v) has two adjacency list entries: $[v]$ in the adjacency list of u, and $[u]$ in the adjacency listof v. These are called twins of each other. A twin pointer is a pointer from an adjacency list entry to its twin. If $|E| = m$ and $|V| = n$, and the memory size is not a constraint, what is the time complexity of the most efficient algorithm to set the twin pointer in each entry in each adjacency list?

(a) $\Theta(n^2)$ (b) $\Theta(n + m)$
(c) $\Theta(m^2)$ (d) $\Theta(n^4)$

[2016 (Set-2) : 2 M]

6.21 The Breadth First Search (BFS) algorithm has been implemented using the queue data structure. Which one of the following is a possible order of visiting the nodes in the graph below?

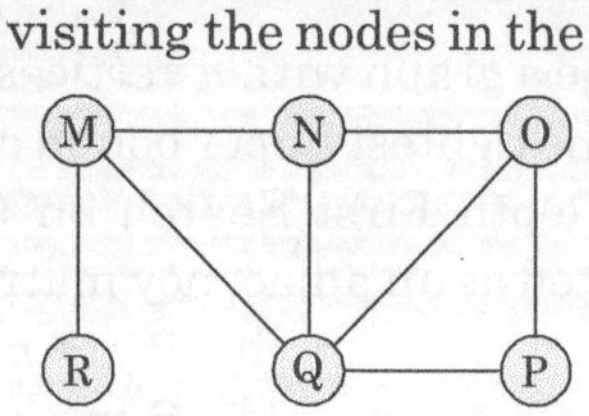

(a) MNOPQR (b) NQMPOR
(c) QMNROP (d) POQNMR

[2017 (Set-2) : 1 M]

6.22 Let G be a simple undirected graph. Let T_D be a depth first search tree of G. Let T_B be a breadth first search tree of G. Consider the following statements:

I. No edge of G is a cross edge with respect to T_D. (A cross edge in G is between two nodes neither of which is an ancestor of the other in T_D).

II. For every edge (u, v) of G, if u is at depth i and v is at depth j in T_B, then $|i - j| = 1$.

Which of the statements above must necessarily be true?

(a) I only (b) II only
(c) Both I and II (d) Neither I nor II

[2018 : 2 M]

6.23 The number of possible min-heaps containing each value from {1, 2, 3, 4, 5, 6, 7} exactly once is ________.

[2018 : 2 M]

6.24 An articulation point in a connected graph is a vertex such that removing the vertex and its incident edges disconnects the graph into two or more connected components.

Let T be a DFS tree obtained by doing DFS in a connected undirected graph G.

Which of the following options is/are correct?

(a) Root of T is an articulation point in G if and only if it has 2 or more children.
(b) A leaf of T can be an articulation point in G.
(c) Root of T can never be an articulation point in G.
(d) If u is an articulation point in G such that x is an ancestor of u in T and y is a descendent of u in T, then all paths from x to y in G must pass through u.

[2021 (Set-1) : 2 M]

6.25 Let $U = \{1, 2, 3\}$. Let 2^U denote the powerset of U. Consider an undirected graph G whose vertex set is 2^U. For any $A, B \in 2^U$, (A, B) is an edge in G if and only if (i) $A \neq B$, and (ii) either $A \subseteq B$ or $B \subseteq A$. For any vertex A in G, the set of all possible orderings in which the vertices of G can be visited in a Breadth First Search (BFS) starting from A is denoted by $B(A)$.

If ϕ denotes the empty set, then the cardinality of $B(\phi)$ is ________.

[2023 : 2 M]

■■■■

Answers Graphs

6.1	Sol.	**6.2**	(c)	**6.3**	(c)	**6.4**	(d)	**6.5**	(b)	**6.6**	(d)	**6.7**	(d)	**6.8**	(b)	**6.9**	(d)
6.10	(d)	**6.11**	(c)	**6.12**	(d)	**6.13**	(c)	**6.14**	(c)	**6.15**	(b)	**6.16**	(c)	**6.17**	(19)	**6.18**	(6)
6.19	(31)	**6.20**	(b)	**6.21**	(d)	**6.22**	(a)	**6.23**	(80)	**6.24**	(a)	**6.25**	(5040)				

Explanations Graphs

6.1 Sol.

Adjacency Matrix

$$\begin{array}{c} \\ P_1 \\ P_2 \\ P_3 \\ P_4 \\ P_5 \end{array}\begin{array}{c} \begin{array}{ccccc} P_1 & P_2 & P_3 & P_4 & P_5 \end{array} \\ \begin{bmatrix} 0 & 1 & 0 & 0 & 0 \\ 0 & 0 & 0 & 1 & 0 \\ 1 & 0 & 0 & 0 & 0 \\ 0 & 0 & 1 & 0 & 1 \\ 1 & 1 & 0 & 0 & 0 \end{bmatrix} \end{array}$$

6.2 (c)

Draw some random graph and try to verify the options.

6.3 (c)

If u is visited before v, $d(r, u) \le d(r, v)$ i.e., r to u is having shortest or equal length compared from r to v.

6.4 (d)

Consider each search

I. Search (a) = b add (b) a,b
   Search (b) = e add (e) a, b, e
   Search (e) = g add (g) a, b, e, g
   Search (g) = h add (h) a, b, e, h
   Search (h) = f add (f) a, b, e, h, f

II. Search (a) = b add (b) a, b
   Search (b) = f add (f) a, b, f
   Search (f) = failure because there is no edge connected to e.

III. Search (a) = b add (b) a, b
   Search (b) = f add (f) a, b, f
   Search (f) = h add (h) a, b, f, h
   Search (h) = g add (g) a, b, f, h, g
   Search (g) = e add (e) a, b, f, h, g, e

IV. Search (a) = f add (f) a, f
   Search (f) = g add (g) a, f, g
   Search (g) = h add (b) a, f, g, h
   Search (h) = b add (b) a, f, g, h, b
   Search (b) = e add (e) a, f, g, h, b, e

6.5 (b)

Let G = (V, E) be an undirected graph $G_1 = (V, E_1)$ such that $E_1 \subseteq E$ and $V_1 \subseteq V$

Consider an example:

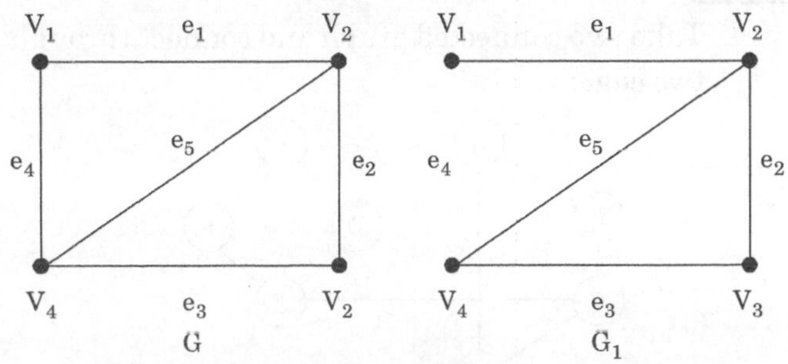

Consider the w (e) =1 if $e \notin E_1$ it means the cost of V_1 to V_4 is only 1 other edges having cost 0. It is notated that G_1 is connected. V_1 does not form of any clique in G a well as G_1 is not a tree. So choice (b) is correct.

6.6 (d)

for i = 1 to n
for j = 1 to n
for k = 1 to n
A[j, k] = max(A[j, k], (A[j, i] + A[i, k]))
A be $n \times n$ array and A[j, k] = 1 if $(j, k) \in E$.
So there exist a path from j to k and path must contain A[j, k] edges.

6.7 (d)

Number of vertices = n
Number of edges = k
Number of connected components = n – k
Ex. 8 vertex with 6 edges

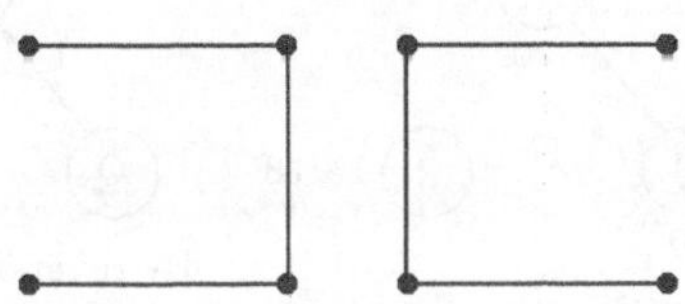

So components = 8 – 6 = 2.

6.8 (b)

In graph G there is a directed edge between i to j when j is either $i + 1$ or $3i$.

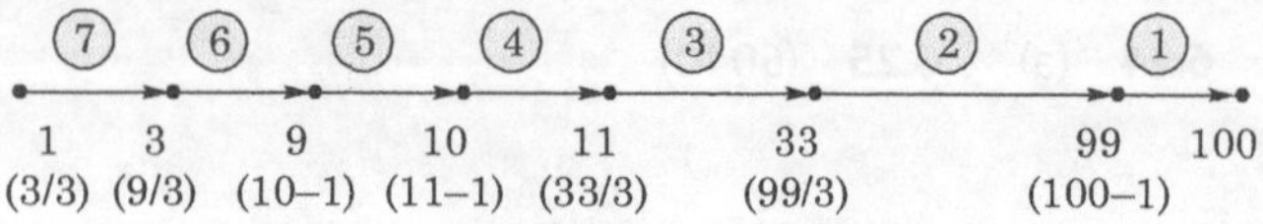

Since minimum value is finding, so we need to make edge which maximum difference in i and j here (99 – 33) = 66 is maximum. So 7 edges are needed.

6.9 (d)

Take two connected graph and connect through two edge:

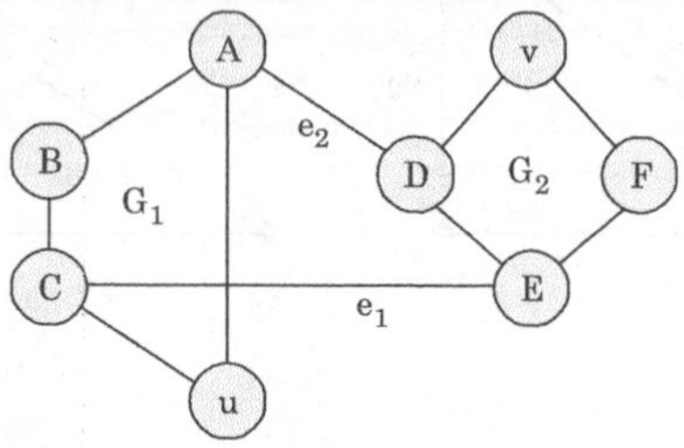

Depth first tree of graph 'G'.

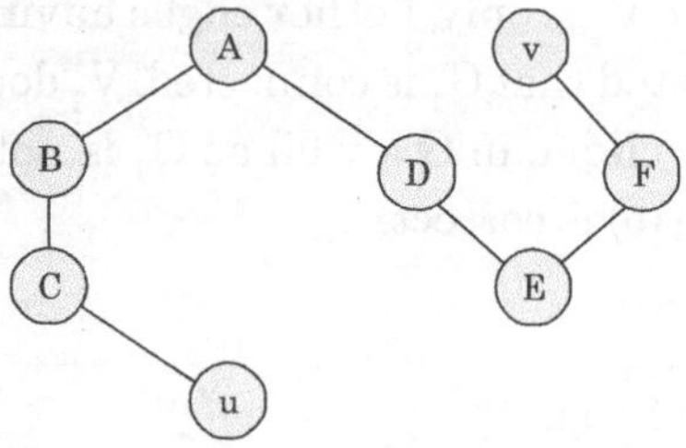

- Option (a) is false, since u and v are not adjacent to single vertex.
- Option (b) is false, since removal of single vertex in G cannot is connect (u) and (v).
- Option (c) is false, since there exist a cycle but not containing both.

6.10 (d)

Consider graph with P, Q and R as vertices and apply dFS algorithm since visiting and leaving time given:

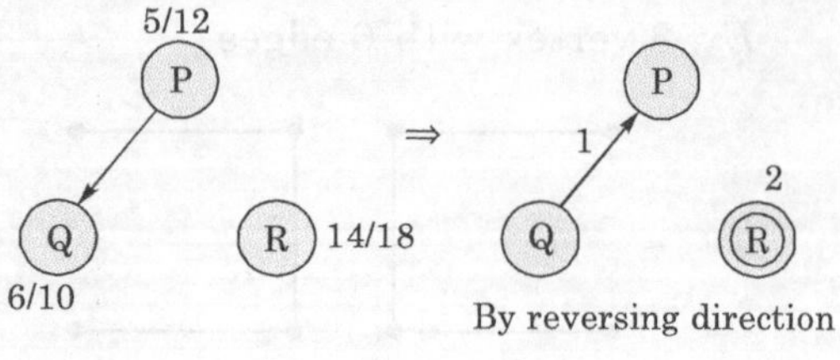

So here 2 strongly connected components are present and P and Q connected.

6.11 (c)

If a graph contain n vertices and n edges and it is simple connected graph then it forms a cycle. Atleast 3 vertices should participate hence the number of spanning trees will be atleast 3.

6.12 (d)

Consider a random graph

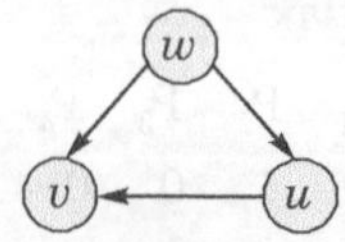

Only option (d) is satisfied since finishing time of vertex 'u' is always greater than finishing time of vertex 'v'.

If we visit vertex 'v' from vertex 'w' then vertex 'u' option (a) is give wrong answer.

6.13 (c)

The most efficient algorithm for finding the number of connected components (articulation point) in an undirected graph on n vertices and m edges using depth-first search takes O(m + n) time. Assume n ≤ m.

6.14 (c)

The BFS using Queue data structure is QMNPRO

6.15 (b)

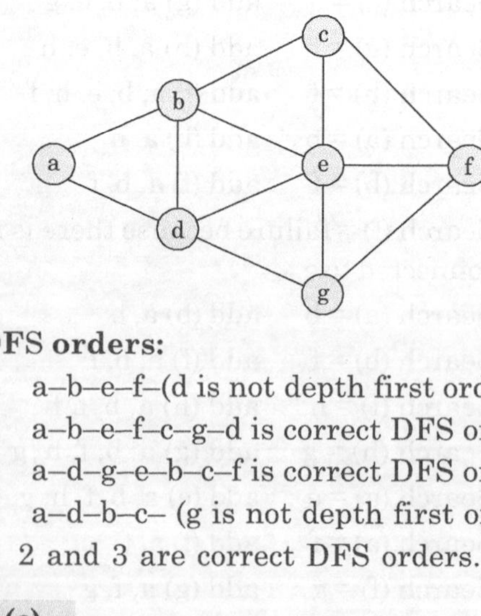

DFS orders:

1. a–b–e–f–(d is not depth first order)
2. a–b–e–f–c–g–d is correct DFS order.
3. a–d–g–e–b–c–f is correct DFS order.
4. a–d–b–c– (g is not depth first order).

∴ 2 and 3 are correct DFS orders.

6.16 (c)

Tightest upper bound on running time of DFS is $\Theta(n)^2$.

6.17 (19)

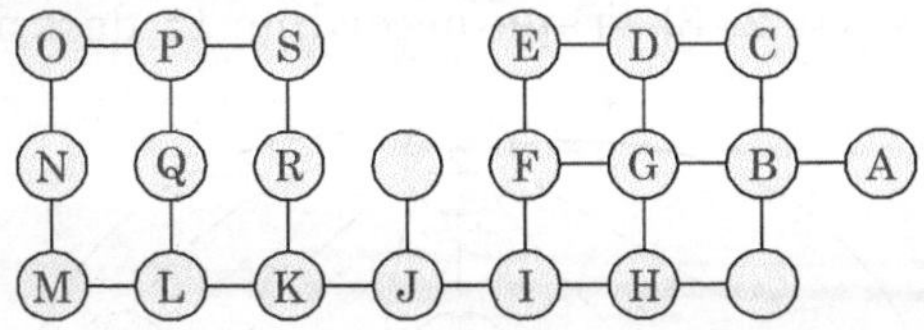

Maximum DFS recursion depth:
$A \to B \to C \to D \to E \to F \to G \to H \to I \to J \to K \to L \to M \to N \to O \to P \to Q \to R \to S = 19$

6.18 (6)

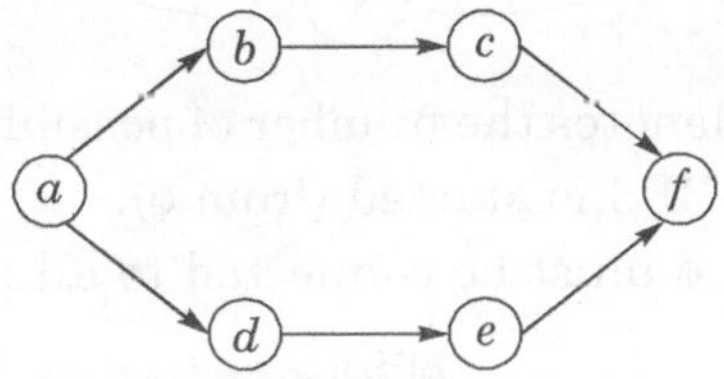

Number of topological orders: 6

a b c d e f
a b d e c f
a d b c e f
a d b c e f
a d e b c f
a d b e c f

6.19 (31)

In worst case binary tree with height 4 (i.e., distance from root node) look like.

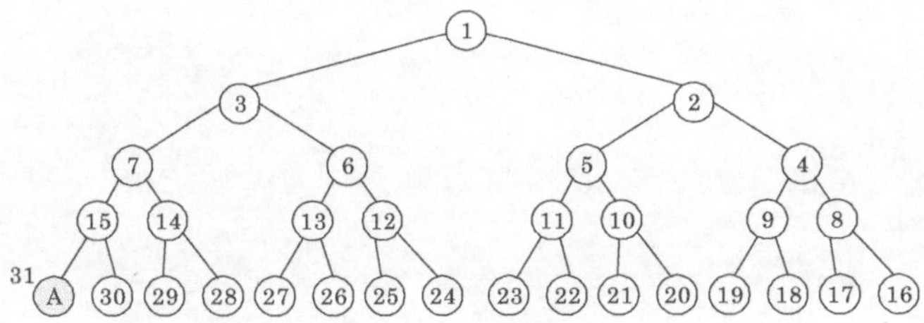

Suppose we have to visit node A.
In worst case of BFS traversal n^{th} node (A) will be the last node at 4th level i.e., we have to visit $2^{4+1} - 1 = 31$ node.
So value of n will be 31.

6.20 (b)

By using BSF (Breadth First Search) traversal we can set the twin pointer in each entry in each adjacency list.
So it will take $\Theta(m + n)$ times (since adjacency list are using).

6.21 (d)

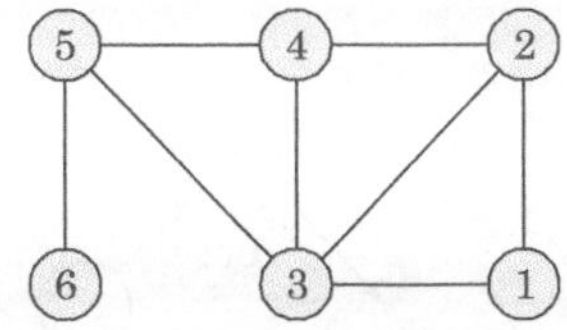

Considering each option:

- After vertices 'M' and 'N', vertex 'O' can't be traversed.
- After visiting vertex 'M', vertex 'O' should be traversed.
- After 'N' either of the vertices 'O' or 'P' should be traversed.

6.22 (a)

- The depth first tree on simple undirected graph never contain cross edge but on directed graph it may or may not contain cross edge.
  So, this statement is true since asked about undirected graph.
- Consider a simple graph:

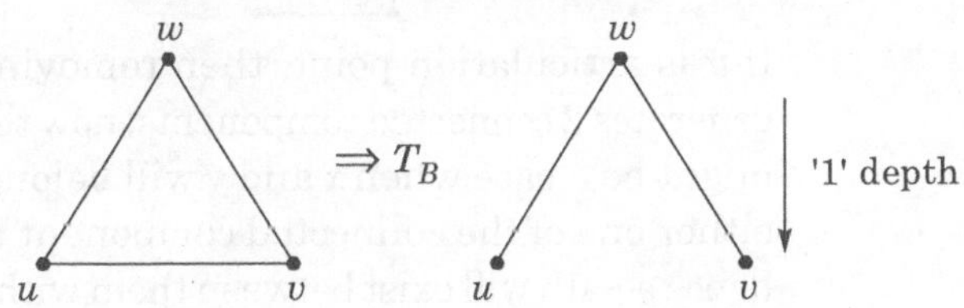

Here (u, v) edge in graph 'G' are on same depth.
Hence it is not always true that $|u_i - u_j| = 1$.
But $|u_i - u_j| \le 1$ i.e. 1 and 0. So, false

Note: To prove statement false one case is sufficient but to prove statement true we need to take care each and every case.

6.23 (80)

Structure of min-heap will be: Since complete Binary Tree

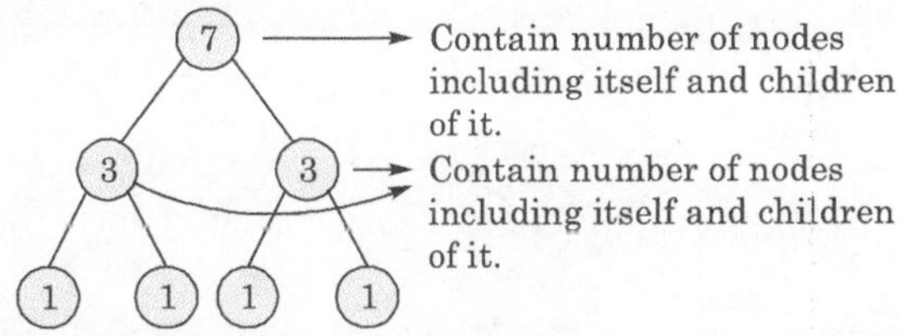

Number of possible min heap

$$= \frac{7!}{7 \times 3 \times 3 \times 1 \times 1 \times 1} = 80$$

6.24 (a)

(a) True

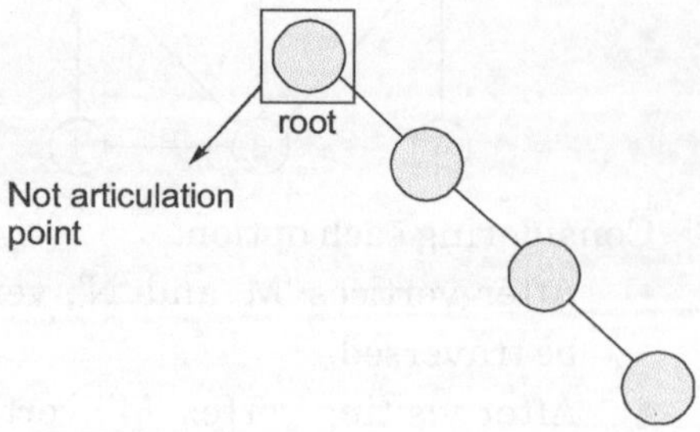

We need at least 2 children so that root is articulation point.

(b) False: This can never happen. Leaf will always have degree = 1.

(c) False: Check option (a) for more information.

(d) False: Below is the reasoning to show how this is false.

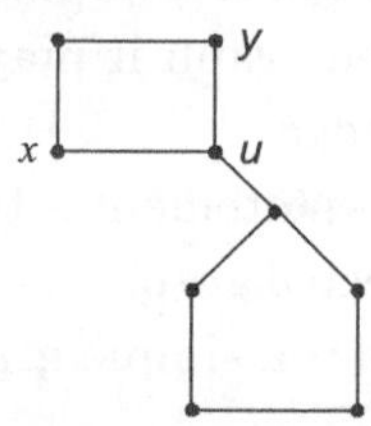

If u is articulation point, then removing u generates 2 connected components, now there might be a case when x and y will belong to either one of the connected component and hence a path will exist between them without passing through u.

Option (a) is correct.

6.25 (5040)

$U = \{1, 2, 3\}$ graph according to description

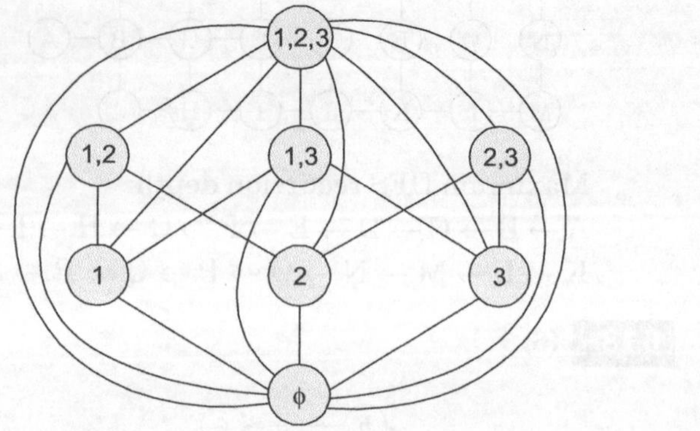

$B(\phi)$ denotes the number of possible orderings when BFS is started (from ϕ).

Since, ϕ must be connected to all the nodes.

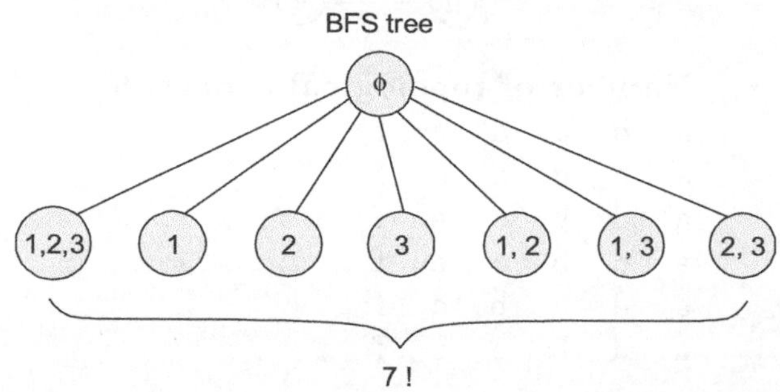

■■■■

7 Hashing

7.1 An advantage of chained hash table (external hashing) over the open addressing scheme is
(a) worst case complexity of search operations is less
(b) space used is less
(c) deletion is easier
(d) None of the above

[1996 : 1 M]

7.2 Given the following input (4322, 1334, 1471, 9679, 1989, 6171, 6173, 4199) and the hash function x mod 10, which of the following statements are true?
1. 9679, 1989, 4199 hash to the same value
2. 1471, 6171 hash to the same value
3. All elements hash to the same value
4. Each element hashes to a different value

(a) 1 only (b) 2 only
(c) 1 and 2 only (d) 3 and 4 only

[2004 : 1 M]

7.3 A hash table contains 10 buckets and uses linear probing to resolve collisions. The key values are integers and the hash function used is key % 10. If the values 43, 165, 62, 123, 142 are inserted in the table, in what location would the key value 142 be inserted?
(a) 2 (b) 3
(c) 4 (d) 6

[2005 : 1 M]

7.4 Which of the following statement(s) is TRUE?
I. A hash function takes a message of arbitrary length and generates a fixed length code.
II. A hash function takes a message of fixed length and generates a code of variable length.
III. A hash function may give the same hash value for distinct messages.

(a) I only (b) II and III only
(c) I and III only (d) II only

[2006 : 1 M]

7.5 Consider a hash function that distributes keys uniformly. The hash table size is 20. After hashing of how many keys will the probability that any new key hashed collides with an existing one exceed 0.5.
(a) 5 (b) 6
(c) 7 (d) 10

[2007 : 2 M]

7.6 Consider the hash table of size seven, with starting index zero, and a hash function (3x + 4) mod 7. Assuming the has table is initially empty, which of the following is the contents of the table when the sequence 1, 3, 8, 10 is inserted into the table using closed hashing? Note that - denotes an empty location in the table
(a) 8, -, -, -, -,-, 10
(b) 1, 8, 10, -, -, -, 3
(c) 1, -, -, -, -, -, 3
(d) 1, 10, 8, -, -, -, 3

[2007 : 2 M]

7.7 Consider a hash table of size 11 that uses open addressing with linear probing. Let h(k) = k mod 11 be the hash function used. A sequence of records with keys 43 36 92 87 11 4 71 13 14 is inserted into an initially empty hash table, the bins of which are indexed from zero to ten. What is the index of the bin into which the last record is inserted?
(a) 3 (b) 4
(c) 6 (d) 7

[2008 : 2 M]

7.8 The keys 12, 18, 13, 2, 3, 23, 5 and 15 are inserted into an initially empty hash table of length 10 using open addressing with hash function h(k) = k mod 10 and linear probing. What is the resultant hash table?

(a)

0	
1	
2	2
3	23
4	
5	15
6	
7	
8	18
9	

(b)

0	
1	
2	12
3	13
4	
5	5
6	
7	
8	18
9	

(c)

0	
1	
2	12
3	13
4	2
5	3
6	23
7	5
8	18
9	15

(d)

0	
1	
2	12,2
3	13,3,23
4	
5	5,15
6	
7	
8	18
9	

[2009 : 2 M]

Linked Answer Questions 7.9 and 7.10:

A has table of length 10 uses open addressing with hash function h(k) = k mod 10, and linear probing. After inserting 6 values into an empty hash table, the table is as shown below.

0	
1	
2	42
3	23
4	34
5	52
6	46
7	33
8	
9	

7.9 Which one of the following choices gives a possible order in which the key values could have been inserted in the table?
(a) 46, 42, 34, 52, 23, 33
(b) 34, 42, 23, 52, 33, 46
(c) 46, 34, 42, 23, 52, 33
(d) 42, 46, 33, 23, 34, 52 **[2010 : 2 M]**

7.10 How many different insertion sequences of the key values using the same hash function and linear probing will result in the hash table shown above?
(a) 10 (b) 20
(c) 30 (d) 40 **[2010 : 2 M]**

7.11 Consider a hash table with 9 slots. The hash function is $h(k) = k$ mod 9. The collisions are resolved by chaining. The following 9 keys are inserted in the order: 5, 28, 19, 15, 20, 33, 12, 17, 10. The maximum, minimum, and average chain lengths in the hash table, respectively, are
(a) 3, 0, and 1 (b) 3, 3, and 3
(c) 4, 0, and 1 (d) 3, 0, and 2
[2014 (Set-1) : 2 M]

7.12 Consider a hash table with 100 slots. Collisions are resolved using chaining. Assuming simple uniform hashing, what is the probability that the first 3 slots are unfilled after the first 3 insertions?
(a) $\frac{(97\times 97\times 97)}{100^3}$ (b) $\frac{(99\times 98\times 97)}{100^3}$
(c) $\frac{(97\times 96\times 95)}{100^3}$ (d) $\frac{(97\times 96\times 95)}{(3!\times 100^3)}$
[2014 (Set-3) : 2 M]

7.13 Which one of the following hash functions on integers will distribute keys most uniformly over 10 buckets numbered 0 to 9 for i ranging from 0 to 2020?
(a) $h(i) = i^2 \bmod 10$
(b) $h(i) = i^3 \bmod 10$
(c) $h(i) = (11 * i^2) \bmod 10$
(d) $h(i) = (12 * i) \bmod 10$
[2015 (Set-2) : 2 M]

7.14 Given a hash table T with 25 slots that stores 2000 elements, the load factor α for T is _____.
[2015 (Set-3) : 1 M]

7.15 Consider a double hashing scheme in which the primary hash function is $h_1(k) = k \bmod 23$ and the secondary hash function is $h_2(k) = 1 + (k \bmod 19)$. Assume that the table size is 23. Then the address returned by probe 1 in the probe sequence (assume that the probe sequence begins at probe 0) for key value $k = 90$ is ________.
[2020 : 1 M]

7.16 Consider a dynamic hashing approach for 4-bit integer keys:
1. There is a main hash table of size 4.
2. The 2 least significant bits of a key is used to index into the main hash table,
3. Initially, the main hash table entries are empty.
4. Thereafter, when more keys are hashed into it, to resolve collisions, the set of all keys corresponding to a main hash table entry is organized as a binary tree that grows on demand.
5. First, the 3rd least significant bit is used to divide the keys into left and right subtrees.
6. To resolve more collisions, each node of the binary tree is further sub-divided into left and right subtrees based on the 4th least significant bit.

7. A split is done only if it is needed, i.e., only when there is a collision.

Consider the following state of the hash table.

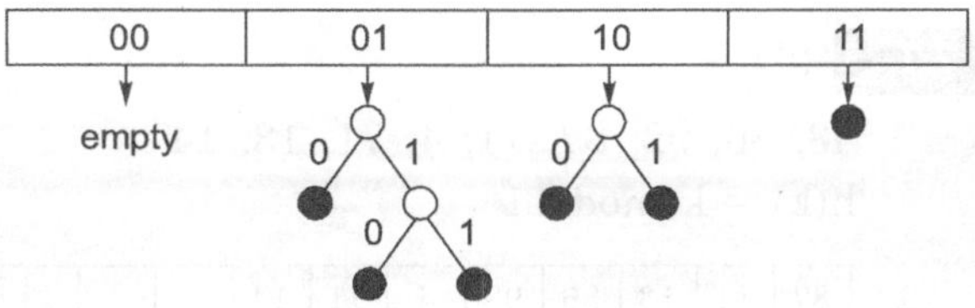

Which of the following sequences of key insertions can cause the above state of the hash table (assume the keys are in decimal notation)?

(a) 9, 5, 13, 6, 10, 14 (b) 9, 5, 10, 6, 7, 1

(c) 10, 9, 6, 7, 5, 13 (d) 5, 9: 4, 13, 10, 7

[2021 (Set-1) : 2 M]

7.17 Suppose we are given n keys, m hash table slots, and two simple uniform hash functions h_1 and h_2. Further suppose our hashing scheme uses h_1 for the odd keys and h_2 for the even keys. What is the expected number of keys in a slot?

(a) $\frac{m}{n}$ (b) $\frac{n}{m}$

(c) $\frac{2n}{m}$ (d) $\frac{n}{2m}$

[2022 : 1 M]

7.18 An algorithm has to store several keys generated by an adversary in a hash table. The adversary is malicious who tries to maximize the number of collisions. Let k be the number of keys, m be the number of slots in the hash table, and $k > m$. Which one of the following is the best hashing strategy to counteract the adversary?

(a) Division method, i.e., use the hash function $h(k) = k \bmod m$.

(b) Multiplication method, i.e., use the hash function $h(k) = \lfloor m(kA - \lfloor kA \rfloor) \rfloor$, where A is a carefully chosen constant.

(c) Universal hashing method.

(d) If k is a prime number, use Division method. Otherwise, use Multiplication method.

[2023 : 1 M]

Answers Hashing

7.1	(c)	**7.2**	(c)	**7.3**	(d)	**7.4**	(c)	**7.5**	(b)	**7.6**	(b)	**7.7**	(d)	**7.8**	(c)	**7.9**	(c)
7.10	(c)	**7.11**	(a)	**7.12**	(a)	**7.13**	(b)	**7.14**	(80)	**7.15**	(13)	**7.16**	(c)	**7.17**	(b)	**7.18**	(c)

Explanations Hashing

7.1 (c)

Deletion is easier, because chained hash table uses linked list.

7.2 (c)

Input (4322, 1334, 1471, 9679, 1989, 6171, 6173, 4199)

Given Hash function,

h(a) = x mod 10

h(a) = h(1)

= {1471, 6171} hash to the same value

h(a) = h(2) = {4322}

h(a) = h(3) = {6173}

h(a) = h(9)

= {9679, 1989, 4199} has to the same value.

h(0), h(5), h(6), h(7), h(8) are empty for given input.

So the statement (i) and (ii) are correct.

7.3 (d)

43, 165, 62 , 123 and 142 are inserted in the table.

43 mapped to location 3, 165 mapped to location 5, 62 mapped to location 2, 123 mapped to location 3 (occupied), so it goes to location 4, 142 mapped to location 2 (occupied), 3, 4, and 5 are also probes, so it goes to location 6.

7.4 (c)

A hash function takes a message of arbitrary length and generates a fixed length code. by taking mod (some value).

A hash function may give the same hash value for distinct messages.

7.5 (b)

Probability of hashed 1st key $= \frac{20}{20} = 1$

Probability of hashed 2nd key with no collision

$= \frac{19}{20}$

Probability of hashed 3rd key with no collision

$= \frac{18}{20}$

Similarly, probability of hashed n^{th} key with no collision $= \frac{20-n+1}{20}$

According to question:

Probability of hashed $(n+1)^{th}$ key with collision > 0.5

Probability of hashed $(n+1)^{th}$ key with collision

P(C) = 1 – P (Till (n^{th}) no collision)

$$P(C) = 1-\left(1\times\frac{19}{20}\times\frac{18}{20}\times\frac{17}{20}\times.....\frac{20-n+1}{20}\right)$$

$$= 1-\left(\frac{19}{20}\times\frac{18}{20}\times\frac{17}{20}.....\frac{20-n+1}{20}\right) > 0.5$$

$$= \left(\frac{19}{20}\times\frac{18}{20}\times\frac{17}{20}.....\frac{20-n+1}{20}\right) < 1-0.5$$

$$= \left(\frac{19}{20}\times\frac{18}{20}\times\frac{17}{20}.....\frac{20-n+1}{20}\right) < 0.5$$

$19 \times 18 \times 17 ..... \times 20 - n + 1 < 0.5 \times (20)^{n-1}$

By put $n = 5$, we get 0.581, for $n = 6$, we get 0.436, which satisfies the equation.

7.6 (b)

Size of hash table = 7

h(x) = (3x + 4) mod 7

h(1) = (3.1 + 4) mod 7 = 7 mod 7
= 0; insert at 0th location.

h(3) = (3.3 + 4) mod 7 = 13 mod 7
= 6; insert at 6th location

h(8) = (3.8 + 4) mod 7 = 28 mod 7
= 0; 0th position is already filled by element 3 so insert 8 at next free location which is 1st position

h(10) = (3.10 + 4) mod 7 = 34 mod 7 = 6 but 6th position is already filled with the element 3 so insert 10 at next free location which is 2nd position

1	8	10	–	–	–	3
0	1	2	3	4	5	6

7.7 (d)

43, 36, 92, 87, 11, 4, 71, 13, 14

h(k) = k mod 11

87	11	13	36	92	4	71	14			43
0	1	2	3	4	5	6	7	8	9	10

∴ At index 7, last record 14 is inserted.

7.8 (c)

12 mod 10 = 2
18 mod 10 = 8
13 mod 10 = 3
2 mod 10 = 2 collision
(2 + 1) mod 10 = 3 again collision (using linear probing)
(3 + 1) mod 10 = 4
3 mod 10 = 3 collision
(3 + 1) mod 10 = 4 collision (using linear probing)
(4 + 1) mod 10 = 5
23 mod 10 = 3 collision
(3 + 1) mod 10 = 4 collision
(4 + 1) mod 10 = 5 again collision
(5 + 1) mod 10 = 6
5 mod 10 = 5 collision
(5 + 1) mod 10 = 6 again collision
(6 + 1) mod 10 = 7
15 mod 10 = 5 collision
(5 + 1) mod 10 = 6 collision
(6 + 1) mod 10 = 7 collision
(7 + 1) mod 10 = 8 collision
(8 + 1) mod 10 = 9

So resulting hosh table.

0	
1	
2	12
3	13
4	2
5	3
6	23
7	5
8	18
9	15

7.9 (c)

In this method we simply checking the options by using linear probing

Option (a): 46, 42, 34, 52, 23, 33

0	
1	
2	42
3	52
4	34
5	23
6	46
7	33
8	
9	

According to linear probing this is hash table

Option (b): 34, 42, 23, 52, 33, 46

Hash table

0	
1	
2	42
3	23
4	34
5	52
6	33
7	46
8	
9	

Option (c): 46, 34, 42, 23, 52, 33

Hash table

0	
1	
2	42
3	23
4	34
5	52
6	46
7	33
8	
9	

Therefore option (c) is correct.

7.10 (c)

Different insertion sequences of key values using same hash function and linear probing.

46, 34, 42, 23, 52, 33

No problem to enter 46, 34, 42, 23

0	
1	
2	42
3	23
4	34
5	
6	46
7	
8	
9	

52 $\longrightarrow$

0	
1	
2	42
3	23
4	34
5	52
6	46
7	
8	
9	

To enter 52 there are 6 possible options. After enter 52 we get hash table. Now to enter 23 there are 5 possible options so total

$= 6 \times 5 = 30$ possible options

7.11 (a)

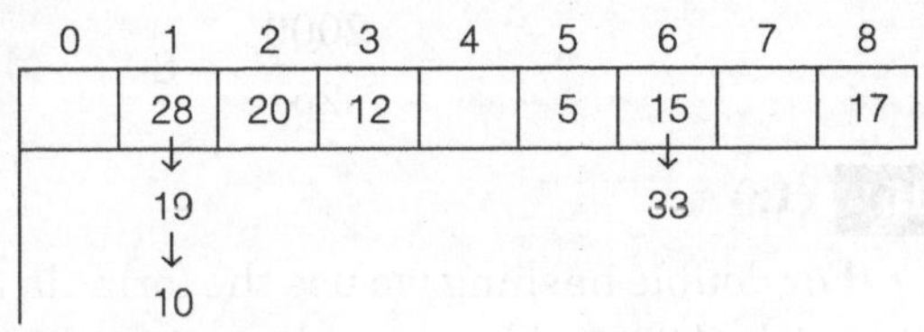

Maximum length = 3

Minimum length = 0

$$\text{Avg length} = \frac{0+3+1+1+0+1+2+0+1}{9} = 1$$

7.12 (a)

We have 100 slots each of which are picked with equal probability by the hash function (since hashing is uniform). So, to avoid first 3 slots, the hash function has to pick from the remaining 97 slots. And repetition is allowed, since chaining is used meaning a list of elements are stored in a slot and not a single element.

$$\text{So, required probability} = \frac{97}{100} \times \frac{97}{100} \times \frac{97}{100}$$

$$= \frac{(97 \times 97 \times 97)}{100^3}$$

7.13 (b)

$h(i) = i^3 \bmod 10$

M	M	M	M	M	M	M	M	M	M
30	31	38	37	34	35	36	33	32	39
20	21	28	27	24	25	26	23	22	29
10	11	18	17	14	15	16	13	12	19
0	1	8	7	4	5	6	3	2	9
0	1	2	3	4	5	6	7	8	9

$h(0) = 0^3 \bmod 10 = \mathbf{0} = h(10) = h(20) = h(30) = ...$
$h(1) = 1^3 \bmod 10 = \mathbf{1} = h(11) = h(21) = h(31) = ...$
$h(2) = 2^3 \bmod 10 = \mathbf{8} = h(12) = h(22) = h(32) = ...$
$h(3) = 3^3 \bmod 10 = \mathbf{7} = h(13) = h(23) = h(33) = ...$
$h(4) = 4^3 \bmod 10 = \mathbf{4} = h(14) = h(24) = h(34) = ...$
$h(5) = 5^3 \bmod 10 = \mathbf{5} = h(15) = h(25) = h(35) = ...$
$h(6) = 6^3 \bmod 10 = \mathbf{6} = h(16) = h(26) = h(36) = ...$
$h(7) = 7^3 \bmod 10 = \mathbf{3} = h(17) = h(27) = h(37) = ...$
$h(8) = 8^3 \bmod 10 = \mathbf{2} = h(18) = h(28) = h(38) = ...$
$h(9) = 9^3 \bmod 10 = \mathbf{9} = h(19) = h(29) = h(39) = ...$

7.14 (80)

$$\text{Load factor } \alpha = \frac{\text{Number of elements}}{\text{Number of slots}}$$

$$= \frac{2000}{25} = 80$$

7.15 (13)

For double hashing we use the formula as $(h_1(k) + i\ h_2(k))\%$ table size where i denotes probe value.

$$h_1(k) = 90\%\ 23 = 21$$
$$h_2(k) = 1 + k\ \%\ 19$$
$$= 1 + 90\ \%\ 19 = 15$$

For probe 1 the value of i is 1 thus, $(21 + 13)\ \%\ 23 = 13$

7.16 (c)

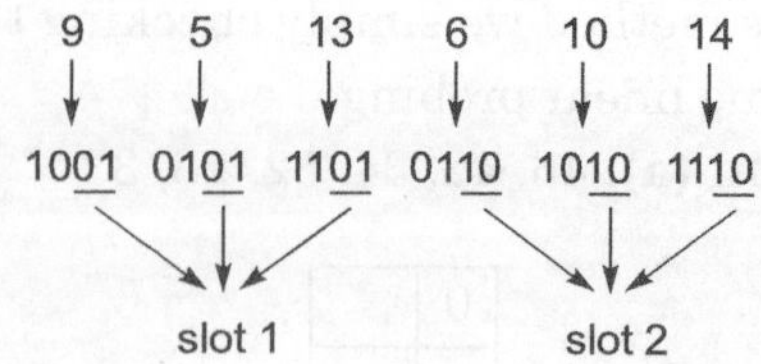

No entry for 11 in this option and 10 → has 3 entries but 2 are required.
So option (a) is wrong.
Similarly, (b), (d) options can also be proved wrong. Option (c) is correct.
Let's check option (c).

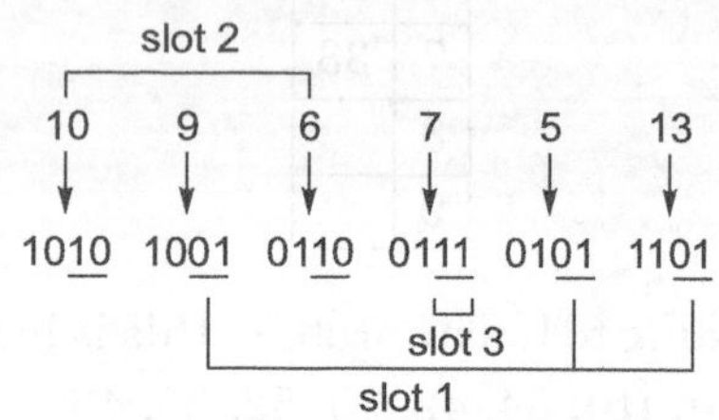

7.18 (c)

The adversary is malicious who tries to maximize the number of collisions by choosing keys that all hash to the same slot. In such case, consider a finite collection H of hash functions that maps universe U of keys into $\{0, 1, 2, 3, ... m - 1\}$. H is such function called universal if for each pair of keys $k, l, \in U$ where $k \neq 1$, the number of hash functions $h \in H$ for which $h(k) = h(l)$ is less than or equal to $\frac{|H|}{m}$.

The otherwords, with a hash function 'h' chosen randomly from H, the probability of collision between two different keys is no more than $\frac{1}{m}$.

The chance of collision when choosing two slots randomly and independently.

■■■■

UNIT

VI

Algorithms

CONTENTS

Algorithms

Syllabus

Analysis, Asymptotic notation, Notions of space and time complexity, Worst and average case analysis; Design: Greedy approach, Dynamic programming, Divide-and-conquer; Tree and graph traversals, Connected components, Spanning trees, Shortest paths; Hashing, Sorting, Searching. Asymptotic analysis (best, worst, average cases) of time and space, upper and lower bounds.

Analysis of Previous GATE Papers

Exam Year	1 Mark Ques.	2 Marks Ques.	Total Marks
1995	3	–	**3**
1996	1	4	**9**
1997	2	1	**4**
1998	2	–	**2**
1999	4	1	**6**
2000	1	1	**3**
2001	3	1	**5**
2002	2	5	**12**
2003	3	5	**13**
2004	1	7	**15**
2005	2	7	**16**
2006	7	5	**17**
2007	2	8	**18**
2008	–	11	**22**
2009	3	5	**13**
2010	1	3	**7**
2011	1	3	**7**
2012	3	3	**9**
2013	4	3	**10**

Exam Year	1 Mark Ques.	2 Marks Ques.	Total Marks
2014 Set-1	2	2	**6**
2014 Set-2	2	2	**6**
2014 Set-3	1	3	**5**
2015 Set-1	3	3	**9**
2015 Set-2	3	3	**9**
2015 Set-3	2	3	**8**
2016 Set-1	2	3	**8**
2016 Set-2	2	2	**6**
2017 Set-1	2	2	**6**
2017 Set-2	3	3	**9**
2018	–	3	**6**
2019	2	3	**8**
2020	3	5	**13**
2021 Set-1	4	2	**8**
2021 Set-2	4	3	**10**
2022	1	2	**5**
2023	2	2	**6**
2024 Set-1	3	3	**9**
2024 Set-2	2	2	**6**

1 Algorithm Analysis and Asymptotic Notations

1.1 Which of the following is false?

(a) $100n \log n = O\left(\frac{n \log n}{100}\right)$

(b) $\sqrt{\log n} = O(\log \log n)$

(c) If $0 < x < y$ then $n^x = O(n^y)$

(d) $2n \neq O(nk)$ **[1996 : 1 M]**

1.2 The concatenation of two lists is to be performed in O(1) time. Which of the following implementations of a list should be used?

(a) Singly linked list
(b) Doubly linked list
(c) Circular doubly linked list
(d) Array implementation of list **[1997 : 1 M]**

1.3 Let $f(n) = n^2 \log n$ and $g(n) = n(\log n)^{10}$ be two positive functions of n. Which of the following statements is correct?

(a) $f(n) = O(g(n))$ and $g(n) \neq O(f(n))$
(b) $g(n) = O(f(n))$ and $f(n) \neq O(g(n))$
(c) $f(n) \neq O(g(n))$ and $g(n) \neq O(f(n))$
(d) $f(n) = O(g(n))$ and $g(n) = O(f(n))$ **[2001 : 1 M]**

1.4 In the worst case, the number of comparisons needed to search singly linked list of length n for a given element is

(a) $\log_2 n$ (b) $\frac{n}{2}$
(c) $\log_2 n - 1$ (d) n **[2002 : 1 M]**

1.5 Consider the following functions:

$$f(n) = 3n^{\sqrt{n}}$$
$$g(n) = 2^{\sqrt{n} \log_2 n}$$
$$h(n) = n!$$

Which of the following is true?

(a) h(n) is O(f(n)) (b) h(n) is O(g(n))
(c) g(n) is not O(f(n)) (d) f(n) is O(g(n))
[2002 : 2 M]

1.6 Consider the following algorithm for searching for a given number x in an unsorted array A[1..... n] having n distinct values:

1. Choose an i uniformly at random from 1...n;
2. If A[i] = x then Stop else Goto 1;

Assuming that x is present on A, what is the expected number of comparisons made by the algorithm before it terminates?

(a) n (b) n – 1
(c) 2n (d) $\frac{n}{2}$ **[2002 : 2 M]**

1.7 The running time of the following algorithm Procedure A(n)

If n < = 2 return (1) else return $\left(A\left(\lceil \sqrt{n} \rceil\right)\right)$;

is best described by

(a) O(n) (b) O(log n)
(c) O(log log n) (d) O(1) **[2002 : 2 M]**

1.8 Consider the following three claims

1. $(n + k)^m = \Theta(n^m)$, where k and m are constants
2. $2^{n+1} = O(2^n)$
3. $2^{2n+1} = O(2^n)$

Which of these claims are correct?

(a) 1 and 2 (b) 1 and 3
(c) 2 and 3 (d) 1, 2 and 3
[2003 : 1 M]

1.9 Consider the following C function:

```
float f (float x, int y)
{
  float p, s; int i;
  for (s = 1, p = 1, i = 1; i < y; i ++)
  {
    p* = x/i;
    s + = p;
  }
  return s;
}
```

For large values of y, the return value of the function f best approximates

(a) x^y (b) e^x
(c) ln (1 + x) (d) x^x **[2003 : 1 M]**

1.10 The cube root of a natural number n is defined as the largest natural number m such that $m^3 \leq n$. The complexity of computing the cube root of n (n is represented in binary notation) is

(a) O(n) but not $O(n^{0.5})$
(b) $O(n^{0.5})$ but not $O((\log n)^k)$ for any constant k > 0.
(c) $O((\log n)^k)$ for some constant k > 0, but not $O((\log \log n)^m)$ for any constant m > 0
(d) $O((\log \log n)^k)$ for some constant k > 0.5, but not $O((\log \log n)^{0.5})$ **[2003 : 2 M]**

1.11 The tightest lower bound on the number of comparisons, in the worst case, for comparison-based sorting is of the order of

(a) n (b) n^2
(c) nlogn (d) $n\log^2 n$ **[2004 : 1 M]**

1.12 What does the following algorithm approximate? (Assume $m > 1$, $\varepsilon > 0$).

```
x = m;
y = 1;
while (x – y > ε)
{
   x = (x+y)/2;
   y = m/x;
}
   print (x);
```

(a) log m (b) m^2
(c) $m^{1/2}$ (d) $m^{1/3}$ **[2004 : 2 M]**

1.13 Let A [1,..., n] be an array storing a bit (1 or 0) at each location, and f(m) is a function whose time complexity is $\Theta(m)$. Consider the following program fragment written in a C like language:

```
counter = 0;
for(i = 1; i < n; i++)
{
   if (A[i] = = 1) counter ++;
   else
   {
     f (counter);
     counter = 0;
   }
}
```

The complexity of this program fragment is

(a) $\Omega(n^2)$ (b) $\Omega(n \log n)$ and $O(n^2)$
(c) $\Theta(n)$ (d) $O(n \log n)$

[2004 : 2 M]

1.14 The time complexity of the following C function is (assume n > 0)

```
int recursive (int n)
{
   if (n = = 1)
     return (1);
   else
     return(recursive(n–1)+ recursive(n–1));
}
```

(a) O(n) (b) O (n log n)
(c) $O(n^2)$ (d) $O(2^n)$ **[2004 : 2 M]**

1.15 The recurrence equation:

$T(1) = 1$
$T(n) = 2T(n-1) + n, n \geq 2$

evaluates to

(a) $2^{n+1} - n - 2$ (b) $2^n - n$
(c) $2^{n+1} - 2n - 2$ (d) $2^n + n$

[2004 : 2 M]

1.16 Let f(n), g(n) and h(n) be functions defined for positive integers such that f(n) = O(g(n), g(n)) $\neq$ O(f(n)), g(n) = O(h(n)), and h(n) = O(g(n)). Which one of the following statements is FALSE?

(a) f(n) + g(n) = O(h(n)) + h(n))
(b) f(n) = O(h(n))
(c) h(n) $\neq$ O(f(n))
(d) f(n)h(n) $\neq$ O(g(n)h(n)) **[2004 : 2 M]**

1.17 The time complexity of computing the transitive closure of a binary relation on a set of n elements is known to be

(a) O(n) (b) O(n log n)
(c) $O(n^{3/2})$ (d) $O(n^3)$ **[2005 : 1 M]**

1.18 Let T(n) be a function defined by the recurrence

$T(n) = 2T\left(\frac{n}{2}\right) + \sqrt{n}$ for $n \geq 2$ and $T(1) = 1$. Which of the following statements is TRUE?

(a) $T(n) = \Theta(\log n)$
(b) $T(n) = \Theta(\sqrt{n})$
(c) $T(n) = \Theta(n)$
(d) $T(n) = \Theta(n \log n)$

[2005 : 2 M]

1.19 Suppose $T(n) = 2T\left(\frac{n}{2}\right) + n$, T(0) = T(1) = 1

Which one of the following is FALSE?

(a) $T(n) = O(n^2)$
(b) $T(n) = \Theta(n \log n)$
(c) $T(n) = \Omega(n^2)$
(d) $T(n) = O(n \log n)$ **[2005 : 2 M]**

Common Data for Q. 1.20 & Q. 1.21

Consider the following C function:

```
double foo(int n)
{
   int i;
   double sum;
   if (n = = 0) return 1.0;
   else
   {
     sum = 0.0;
     for (i = 0; i < n ; i ++)
     sum + = foo(i);
     return sum;
   }
}
```

1.20 The space complexity of the above function is
(a) O(1) (b) O(n)
(c) O(n!) (d) $O(n^n)$ **[2005 : 2 M]**

1.21 The space complexity of the above function is foo( ) and store the values of foo(i), 0 < = i < n, as and when they are computed. With this modification, the time complexity for function foo( ) is significantly reduced. The space complexity of the modified function would be
(a) O(1) (b) O(n)
(c) $O(n^2)$ (d) O(n!) **[2005 : 2 M]**

1.22 Consider the following C-program fragment in which i, j, and n are integer variables.
for (i = n, j = 0; i > 0; i /= 2, j + = i);
Let Val (j) = denote the value stored in the variable j after termination of the for loop. Which one of the following is true?
(a) val (j) = Θ (log n) (b) val (j) = Θ ($\sqrt{n}$)
(c) val (j) = Θ (n) (d) val (j) = Θ (n log n)
[2006 : 1 M]

1.23 Consider the following is true?

$$T(n) = 2T\left(\left\lceil \sqrt{n} \right\rceil\right) + 1,\ T(1) = 1$$

Which one of the following is true?
(a) T (n) = Θ (log log n) (b) T (n) = Θ (log n)
(c) T (n) = $\Theta\left(\sqrt{n}\right)$ (d) T (n) = Θ(n)
[2006 : 2 M]

1.24 Consider the following segment of C code:
```
int j, n;
j = 1;
while (j < = n)
j = j* 2;
```
The number of comparisons made in the execution of the loop for any n > 0 is
(a) $\lceil \log_2 n \rceil + 1$ (b) n
(c) $\lceil \log_2 n \rceil$ (d) $\lfloor \log_2 n \rfloor + 1$
[2007 : 1 M]

1.25 What is the time complexity of the following recursive function:
```
int DoSomething ( int n)
{
  if (n < = 2)
    return 1;
  else
    return DoSomething (floor (sqrt (n)))+n;
}
```
(a) $\Theta(n^2)$ (b) $\Theta(n \log_2 n)$
(c) $\Theta(\log_2 n)$ (d) $\Theta(\log_2 \log_2 n)$
[2007 : 2 M]

1.26 An array of n numbers is given, where n is an even number. The maximum as well as the minimum of these n numbers needs to be determined. Which of the following is true about the number of comparisons needed?
(a) At least 2n – c comparisons, for some constant c, are needed.
(b) At most 1.5n – 2 comparisons are needed.
(c) At least n $\log_2$ n comparisons are needed.
(d) None of the above **[2007 : 2 M]**

1.27 Consider the following C code segment:
```
int IsPrime(n)
{
  int i, n;
  for(i = 2; i < = sqrt(n); i ++)
  {
    if (n % i = = 0)
      printf ("Not Prime\n");
    return 0;
  }
  return 1;
}
```
Let T(n) denote the number of times the for loop is executed by the program on input n. Which of the following is TRUE?
(a) $T(n) = O\left(\sqrt{n}\right)$ and $T(n) = \Omega\left(\sqrt{n}\right)$
(b) $T(n) = O\left(\sqrt{n}\right)$ and $T(n) = \Omega(1)$
(c) $T(n) = O(n)$ and $T(n) = \Omega\left(\sqrt{n}\right)$
(d) None of these
[2007 : 2 M]

1.28 Arrange the following functions in increasing asymptotic order:
A. $n^{1/3}$ **B.** e^n
C. $n^{7/4}$ **D.** $n \log^9 n$
E. 1.0000001^n
(a) A, D, C, E, B (b) D, A, C, E, B
(c) A, C, D, E, B (d) A, C, D, B, E
[2008 : 1 M]

1.29 When $n = 2^{2k}$ for some $k \ge 0$, the recurrence relation $T(n) = \sqrt{2}\, T\left(\frac{n}{2}\right) + \sqrt{n}$, $T(1) = 1$ evaluates to :
(a) $\sqrt{n}$ (log n+1) (b) $\sqrt{n} \log n$
(c) $\sqrt{n} \log \sqrt{n}$ (d) $n \log \sqrt{n}$
[2008 : 2 M]

1.30 Consider the following functions:

$f(n) = 2^n$
$g(n) = n!$
$h(n) = n^{\log n}$

which of the following statements about the asymptotic behaviour of f (n), g(n), and h (n) is true?

(a) $f(n) = O(g(n))$; $g(n) = O(h(n))$
(b) $f(n) = \Omega(g(n))$; $g(n) = O(h(n))$
(c) $g(n) = O(f(n))$; $h(n) = O(f(n))$
(d) $h(n) = O(f(n))$; $g(n) = \Omega(f(n))$ **[2008 : 2 M]**

1.31 The minimum number of comparison required to determine if an integer appears more than $\frac{n}{2}$ times in a sorted array of n integers is

(a) $\Theta(n)$ (b) $\Theta(\log n)$
(c) $\Theta(\log^2 n)$ (d) $\Theta(1)$ **[2008 : 2 M]**

Common Data Questions Q.1.32 and Q.1.33

Consider the following C functions:

```
int f1(int n)
{
  if(n = = 0|| n = = 1) return n;
  else
    return (2* f1( n –1) + 3 * f1(n – 2));
}
int f2( int n)
{
  int i ;
  int X[N] , Y[N] , Z[N] ;
  X[1] = 1; Y[1] = 2; Z[1] = 3;
  for( i = 2; i < = n ; i ++ )
  {
    X[ i ] = Y[ i – 1] + Z[ i – 2] ;
    Y[ i ] = 2 * X[ i ] ;
    Z[ i ] = 3 * X[ i ];
  }
  return X[n] ;
}
```

1.32 The running time of f1(n) and f2(n) are

(a) $\Theta(n)$ and $\Theta(n)$ (b) $\Theta(2^n)$ and $O(n)$
(c) $\Theta(n)$ and $\Theta(2^n)$ (d) $\Theta(2^n)$ and $\Theta(2^n)$

[2008 : 2 M]

1.33 f1(8) and f2(8) return the values

(a) 1661 and 1640 (b) 59 and 59
(c) 1640 and 1640 (d) 1640 and 1661

[2008 : 2 M]

1.34 What is the number of swaps required to sort n elements using selection sort, in the worst case?

(a) $\Theta(n)$ (b) $\Theta(n \log n)$
(c) $\Theta(n^2)$ (d) $\Theta(n^2 \log n)$

[2009 : 1 M]

1.35 The running time of an algorithm is represented by the following recurrence relation:

$$T(n) = \begin{cases} n & n \le 3 \\ T\left(\frac{n}{3}\right) + cn & \text{otherwise} \end{cases}$$

Which one of the following represents the time complexity of the algorithm?

(a) $\Theta(n)$ (b) $\Theta(n \log n)$
(c) $\Theta(n^2)$ (d) $\Theta(n^2 \log n)$

[2009 : 2 M]

1.36 Two alternative packages A and B are available for processing a database having 10^k records. Package A requires $0.0001\ n^2$ time units and package B requires $10n\log_{10}n$ time units to process n records. What is the smallest value of k for which package B will be preferred over A?

(a) 12 (b) 10
(c) 6 (d) 5 **[2010 : 1 M]**

1.37 Let W(n) and A(n) denote respectively, the worst case and average case running time of an algorithm executed on an input of size n. Which of the following is ALWAYS TRUE?

(a) $A(n) = \Omega(W(n))$ (b) $A(n) = \Theta(W(n))$
(c) $A(n) = O(W(n))$ (d) $A(n) = o(W(n))$

[2012 : 1 M]

1.38 The recurrence relation capturing the optimal execution time of the Towers of Hanoi problem with n discs is

(a) $T(n) = 2T(n - 2) + 2$
(b) $T(n) = 2T(n - 1) + n$
(c) $T(n) = 2T\left(\frac{n}{2}\right) + 1$
(d) $T(n) = 2T(n - 1) + 1$ **[2012 : 1 M]**

1.39 Consider the following function:

```
int unknown (int n)
{
  int i, j, k = 0;
  for (i = n/2; i <= n; i++)
    for (j = 2; j <= n; j = j*2)
      k = k + n/2;
  return (k);
}
```

The return value of the function is

(a) $\Theta(n^2)$ (b) $\Theta(n^2 \log n)$
(c) $\Theta(n^3)$ (d) $\Theta(n^3 \log n)$

[2013 : 2 M]

1.40 Which one of the following correctly determines the solution of the recurrence relation with $T(1) = 1$?

$$T(n) = 2T\left(\frac{n}{2}\right) + \log n$$

(a) $\Theta(n)$ (b) $\Theta(n \log n)$
(c) $\Theta(n^2)$ (d) $\Theta(\log n)$

[2014 (Set-2) : 1 M]

1.41 Suppose we have a balanced binary search tree T holding n-numbers. We are given two numbers L and H and wish to sum up all the numbers in T that lie between L and H. Suppose there are m such numbers in T.
If the tightest upper bound on the time to compute the sum is $O(n^a \log^b n + m^c \log^d n)$, the value of $a + 10b + 100c + 1000d$ is ____.

[2014 (Set-3) : 2 M]

1.42 Consider the following C function:

```
int fun1 (int n)
{    int i, j, k, p, q = 0;
     for (i = 1; i<n; ++i)
     {    p = 0;
          for (j=n; j>1; j=j/2)
          ++p;
          for (k=1; k<p; k=k*2)
          ++q;
     } return q;
}
```

Which one of the following most closely approximates the return value of the function fun1?
(a) n^3
(b) $n(\log n)^2$
(c) $n \log n$
(d) $n \log (\log n)$

[2015 (Set-1) : 2 M]

1.43 An algorithm performs $(\log N)^{1/2}$ find operations, N insert operations, $(\log N)^{1/2}$ delete operations, and $(\log N)^{1/2}$ decrease-key operations on a set of data items with keys drawn from a linearly ordered set. For a delete operation, a pointer is provided to the record that must be deleted. For the decrease-key operation, a pointer is provided to the record that has its key decreased. Which one of the following data structures is the most suited for the algorithm to use, if the goal is to achieve the best total asymptotic complexity considering all the operations?
(a) Unsorted array
(b) Min-heap
(c) Sorted array
(d) Sorted doubly linked list

[2015 (Set-1) : 2 M]

1.44 Consider a complete binary tree where the left and the right subtrees of the root are max-heaps. The lower bound for the number of operations to convert the tree to a heap is
(a) $\Omega(\log n)$ (b) $\Omega(n)$
(c) $\Omega(n \log n)$ (d) $\Omega(n^2)$

[2015 (Set-2) : 1 M]

1.45 An unordered list contains n distinct elements. The number of comparisons to find an element in this list that is neither maximum nor minimum is
(a) $\Theta(n \log n)$ (b) $\Theta(n)$
(c) $\Theta(\log n)$ (d) $\Theta(1)$

[2015 (Set-2) : 1 M]

1.46 Consider the equality $\sum_{i=0}^{n} i^3 = X$ and the following choices for X
I. $\Theta(n^4)$ II. $\Theta(n^5)$
III. $O(n^5)$ IV. $\Omega(n^3)$
The equality above remains correct if X is replace by
(a) Only I
(b) Only II
(c) I or III or IV but not II
(d) II or III or IV but not I

[2015 (Set-3) : 1 M]

1.47 Let $f(n) = n$ and $g(n) = n^{(1+\sin n)}$, where n is a positive integer. Which of the following statements is/are correct?
I. $f(n) = O(g(n))$
II. $f(n) = \Omega(g(n))$
(a) Only I
(b) Only II
(c) Both I and II
(d) Neither I nor II

[2015 (Set-3) : 2 M]

1.48 The given diagram shows the flow chart for a recursive function $A(n)$. Assume that all statements, except for the recursive calls, have O(1) time complexity. If the worst case time complexity of this function is $O(n^\alpha)$, then the least possible value (accurate up to two decimal positions) of α is ________.

Flow chart for Recursive Function $A(n)$

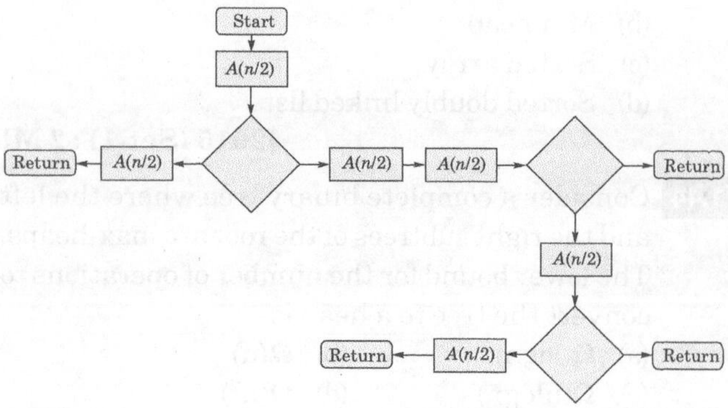

[2016 (Set-2) : 2 M]

1.49 N items are stored in a sorted doubly linked list. For a *delete* operation, apointer is provided to the record to be deleted. For a *decrease-key* operation, a pointer is provided to the record on which the operation is to be performed. An algorithm performs the following operations on the list in this order: $\Theta(N)$ *delete*, $O(\log N)$ insert, $O(\log N)$ *find*, and $\Theta(N)$ *decrease-key*. What is the time complexity of all these operations put together?

(a) $O(\log^2 N)$ (b) $O(N)$
(c) $O(N^2)$ (d) $\Theta(N^2 \log N)$

[2016 (Set-2) : 1 M]

1.50 Match the algorithms with their time complexities:

List-I (Algorithm)
(P) Towers of Hanoi with n disks
(Q) Binary search given n sorted numbers
(R) Heap sort given n numbers at the worst case
(S) Addition of two $n \times n$ matrices

List-II (Time complexity)
(i) $\Theta(n^2)$
(ii) $\Theta(n \log n)$
(iii) $\Theta(2^n)$
(iv) $\Theta(\log n)$

(a) P – (iii), Q – (iv), R – (i), S – (ii)
(b) P – (iv), Q – (iii), R – (i), S – (ii)
(c) P – (iii), Q – (iv), R – (ii), S – (i)
(d) P – (iv), Q – (iii), R – (ii), S – (i)

[2017 (Set-2) : 1 M]

1.51 Consider the recurrence function:

$$T(n) = \begin{cases} 2T(\sqrt{n})+1, & n > 2 \\ 2, & 0 < n \le 2 \end{cases}$$

Then $T(n)$ in terms of Θ notation is

(a) $\Theta(\log\log n)$ (b) $\Theta(\log n)$
(c) $\Theta(\sqrt{n})$ (d) $\Theta(n)$

[2017 (Set-2) : 2 M]

1.52 Consider the following C function:

```
int fun (int n)
{   int i, j;
    for (i = 1; i <= n; i++)
    {  for (j = 1; j < n; j + = i)
       {
           printf("%d %d", i, j);
       }
    }
}
```

Time complexity of fun in terms of Θ notation is

(a) $\Theta(n\sqrt{n})$ (b) $\Theta(n^2)$
(c) $\Theta(n \log n)$ (d) $\Theta(n^2 \log n)$

[2017 (Set-2) : 2 M]

1.53 Consider the following functions from positive integers to real numbers:

$$10, \sqrt{n}, n, \log_2 n, \frac{100}{n}$$

Tire CORRECT arrangement of the above functions in increasing order of asymptotic complexity is:

(a) $\log_2 n, \frac{100}{n}, 10, \sqrt{n}, n$

(b) $\frac{100}{n}, 10, \log_2 n, \sqrt{n}, n$

(c) $10, \frac{100}{n}, \sqrt{n}, \log_2 n, n$

(d) $\frac{100}{n}, \log_2 n, 10, \sqrt{n}, n$

[2017 (Set-1) : 1 M]

1.54 Let A be an array of 31 numbers consisting of a sequence of 0's followed by a sequence of 1's. The problem is to find the smallest index i such that $A[i]$ is 1 by probing the minimum number of locations in A. The worst case number of probes performed by an optimal algorithm is ______.

[2017 (Set-1) : 2 M]

1.55 What is the worst case time complexity of inserting n elements into an empty linked list, if the linked list needs to be maintained in sorted order?

(a) $\Theta(n \log n)$ (b) $\Theta(n)$
(c) $\Theta(1)$ (d) $\Theta(n^2)$ **[2020 : 1 M]**

1.56 What is the worst case time complexity of inserting n^2 elements into an AVL-tree with n elements initially?

(a) $\theta(n^4)$ (b) $\theta(n^2 \log n)$
(c) $\theta(n^3)$ (d) $\theta(n^2)$ **[2020 : 1 M]**

1.57 For parameters a and b, both of which are $\omega(1)$, $T(n) = T(n^{1/a}) + 1$, and $T(b) = 1$. Then $T(n)$ is

(a) $\Theta(\log_2 \log_2 n)$ (b) $\Theta(\log_a \log_b n)$
(c) $\Theta(\log_b \log_a n)$ (d) $\Theta(\log_{ab} n)$

[2020 : 1 M]

1.58 Consider the following recurrence relation:

$$T(n) = \begin{cases} T\left(\frac{n}{2}\right) + T\left(\frac{2n}{5}\right) + 7n & \text{if } n > 0 \\ 1 & \text{if } n = 0 \end{cases}$$

Which one of the following options is correct?

(a) $T(n) = \theta(n \log n)$ (b) $T(n) = \theta(n^{5/2})$
(c) $T(n) = \theta(n)$ (d) $T(n) = \theta((\log n)^{5/2})$

[2021 (Set-1) : 2 M]

1.59 Consider the following three functions:

$$f_1 = 10^n,\ f_2 = n^{\log n},\ f_3 = n^{\sqrt{n}}$$

Which one of the following options arranges the functions in the increasing order of asymptotic growth rate?

(a) f_2, f_3, f_1 (b) f_2, f_1, f_3
(c) f_1, f_2, f_3 (d) f_3, f_2, f_1

[2021 (Set-1) : 1 M]

1.60 For constants $a \geq 1$ and $b > 1$, consider the following recurrence defined on the non-negative integers:

$$T(n) = a \cdot T\left(\frac{n}{b}\right) + f(n)$$

Which one of the following options is correct about the recurrence $T(n)$?

(a) If $f(n)$ is $\Theta(n^{\log_b(a)})$, then $T(n)$ is $\Theta(n^{\log_b(a)})$.
(b) If $f(n)$ is $O(n^{\log_b(a) - e})$ for some $e > 0$, then $T(n)$ is $\Theta(n^{\log_b(a)})$.
(c) If $f(n)$ is $\frac{n}{\log_2(n)}$, then $T(n)$ is $\Theta(\log_2(n))$.
(d) If $f(n)$ is $n \log_2(n)$, then $T(n)$ is $\Theta(n \log_2(n))$.

[2021 (Set-2) : 2 M]

1.61 Which one of the following statements is TRUE for all positive functions $f(n)$?

(a) $f(n^2) = \theta(f(n)^2)$, when $f(n)$ is a polynomial
(b) $f(n^2) = o(f(n)^2)$
(c) $f(n^2) = O(f(n)^2)$, when $f(n)$ is an exponential function
(d) $f(n^2) = \Omega(f(n)^2)$

[2022 : 1 M]

1.62 Let f and g be functions of natural numbers given by $f(n) = n$ and $g(n) = n^2$.

Which of the following statements is/are TRUE?

(a) $f \in O(g)$ (b) $f \in \Omega(g)$
(c) $f \in o(g)$ (d) $f \in \theta(g)$

[2023 : 1 M]

1.63 Consider functions Function 1 and Function 2 expressed in pseudocode as follows:

Function 1
```
while n > 1 do
    for i = 1 to n do
        x = x + 1;
    end for
    n = ⌊n/2⌋;
end while
```

Function 2
```
for i = 1 to 100 * n do
x = x + 1;
end for
```

Let $f_1(n)$ and $f_2(n)$ denote the number of times the statement "$x = x + 1$" is executed in **Function_1** and **Function_2**, respectively.

Which of the following statements is/are TRUE?

(a) $f_1(n) \in \theta(f_2(n))$ (b) $f_1(n) \in o(f_2(n))$
(c) $f_1(n) \in \omega(f_2(n))$ (d) $f_1(n) \in O(n)$

[2023 : 2 M]

1.64 Let L_1, L_2 be two regular languages and L_3 a language which is not regular. Which of the following statements is/are always TRUE?

(a) $L_1 \cup L_3$ is not regular
(b) $\overline{L_3}$ is not regular
(c) $L_1 = L_2$ if and only if $L_1 \cap \overline{L_3} = \phi$
(d) $\overline{L_1} \cup \overline{L_2}$ is regular

[2024 (Set-1) : 1 M]

1.65 Given an integer array of size N, we want to check if the array is sorted (in either ascending or descending order). An algorithm solves this problem by making a single pass through the array and comparing each element of the array only with its adjacent elements. The worst-case time complexity of this algorithm is

(a) $\Omega(N)$ but not $O(N)$
(b) $O(N)$ but not $\Omega(N)$
(c) Both $O(N)$ and $\Omega(N)$
(d) Neither $O(N)$ nor $\Omega(N)$

[2024 (Set-1) : 1 M]

1.66 Consider the following recurrence relation:

$$T(n) = \begin{cases} \sqrt{n}T\left(\sqrt{n}\right)+n & \text{for } n \geq 1 \\ 1 & \text{for } n = 1 \end{cases}$$

Which one of the following options is CORRECT?

(a) $T(n) = \Theta(n \log \log n)$
(b) $T(n) = \Theta(n^2 \log n)$
(c) $T(n) = \Theta(n \log n)$
(d) $T(n) = \Theta(n^2 \log \log n)$

[2024 (Set-1) : 2 M]

1.67 Let A be an array containing integer values. The distance of A is defined as the minimum number of elements in A that must be replaced with another integer so that the resulting array is sorted in non-decreasing order. The distance of the array [2, 5, 3, 1, 4, 2, 6] is ______.

[2024 (Set-2) : 1 M]

1.68 Let $T(n)$ be the recurrence relation defined as follows:

$T(0) = 1$,
$T(1) = 2$ and
$T(n) = 5T(n-1) - 6T(n-2)$ for $n \geq 2$

Which one of the following statement is TRUE?

(a) $T(n) = \Theta(n3^n)$
(b) $T(n) = \Theta(n2^n)$
(c) $T(n) = \Theta(3^n)$
(d) $T(n) = \Theta(2^n)$

[2024 (Set-2) : 1 M]

■■■■

Answers Algorithm Analysis and Asymptotic Notations

1.1	(b, d)	1.2	(c)	1.3	(b)	1.4	(d)	1.5	(d)	1.6	(a)	1.7	(c)	1.8	(a)	1.9	(b)
1.10	(c)	1.11	(c)	1.12	(c)	1.13	(c)	1.14	(d)	1.15	(a)	1.16	(d)	1.17	(d)	1.18	(c)
1.19	(c)	1.20	(b)	1.21	(b)	1.22	(c)	1.23	(b)	1.24	(d)	1.25	(d)	1.26	(b)	1.27	(b)
1.28	(a)	1.29	(a)	1.30	(d)	1.31	(b)	1.32	(b)	1.33	(c)	1.34	(a)	1.35	(a)	1.36	(c)
1.37	(c)	1.38	(d)	1.39	(b)	1.40	(a)	1.41	(110)	1.42	(d)	1.43	(a)	1.44	(a)	1.45	(d)
1.46	(c)	1.47	(d)	1.48	(2.32)	1.49	(c)	1.50	(c)	1.51	(b)	1.52	(c)	1.53	(b)	1.54	(5)
1.55	(d)	1.56	(b)	1.57	(b)	1.58	(c)	1.59	(a)	1.60	(b)	1.61	(a)	1.62	(a, c)	1.63	(a, d)
1.64	(b, d)	1.65	(c)	1.66	(a)	1.67	(3)	1.68	(d)								

Explanations Algorithm Analysis and Asymptotic Notations

1.1 (b, d)

(a) We know that

$f(n) = O(g(n))$ i.e.,

$f(n) \le k.g(n)$

for k, some positive integer and $n > n_0$

$$100\ n \log n \le 10000 \times \frac{n \log n}{100}$$

for k = 10000

$$\therefore 100\ n \log n = O\left(\frac{n \log n}{100}\right)$$

(b) $\sqrt{\log n} \le 1 * \log \log n$

$$\therefore \sqrt{\log n} \ne O(\log \log n)$$

(c) $n^x \le n^y$ as $0 < x < y$

$\therefore n^x = O(n^y)$

(d) $2n \le kn$ for $k \ge 2$

$\therefore 2n = O(nk)$

1.2 (c)

As list concatenation requires traversing at least one list to the end. So singly linked list and doubly linked list requires O(n) time complexity whereas circular doubly linked list required O(1) time.

1.3 (b)

$$f(n) = n^2 \log n$$
$$g(n) = n(\log n)^{10}$$
$$n(\log n)^{10} \le n^2 \log n$$
$$\therefore \quad g(n) = O(f(n))$$

Whereas

$$f(n) \ne O(g(n)) \text{ because}$$
$$n^2 \log n \nleq n(\log n)^{10}$$

1.4 (d)

Worst case of searching singly linked list is when given element doesn't present at all in the singly linked list. Using linear search then require "n" comparisons in worst case.

1.5 (d)

$$f(n) = 3n^{\sqrt{n}},\ g(n) = 2^{\sqrt{n}\log_2 n}$$

$$\Rightarrow g(n) = n^{\sqrt{n}\log_2 2}\left[a^{\log b} = b^{\log a}\right]$$

$$\Rightarrow n^{\sqrt{n}}$$

1.6 (a)

Let expected number of comparisons be E. Value of E is sum of following expression for all the possible cases:

Case-I If A[i] is found in the first attempt,
Number of comparisons = 1
Probability = 1/n.

Case-II If A[i] if found in the second attempt,
Number of comparisons = 2

$$\text{Probability} = \frac{(n-1)}{n} * \frac{1}{n}$$

Case-III If A[i] is found in the third attempt,
Number of comparisons = 3

$$\text{Probability} = \frac{(n-1)}{n} * \frac{(n-1)}{n} * \frac{1}{n}$$

There are actually infinite such cases. So, we have following infinite series for E.

$$E = \frac{1}{n} + \frac{n-1}{n} * \frac{1}{n} * 2 + \frac{(n-1)}{n} * \frac{(n-1)}{n} * \frac{1}{n} * 3 + \dots \text{(i)}$$

After multiplying equation (i) with $\frac{(n-1)}{n}$, we get

$$E\frac{(n-1)}{n} = \frac{n-1}{n}*\frac{1}{n}+\frac{n-1}{n}*\frac{1}{n}*2+\ldots \text{(ii)}$$

Subtracting (ii) from (i), we get

$$\frac{E}{n} = \frac{1}{n}+\frac{n-1}{n}*\frac{1}{n}+\frac{n-1}{n}*\frac{n-1}{n}*\frac{1}{n}+\ldots$$

The expression on right side is a G.P with infinite elements.

So apply the sum formula $\left(\frac{a}{1-r}\right)$

$$\frac{E}{n} = \frac{\left(\frac{1}{n}\right)}{\left(\frac{n-(n-1)}{n}\right)} = 1$$

$$E = n$$

Therefore, option (a) is correct.

1.7 (c)

Recursive relation for procedure A(n) is

$$T(n) = T\left(\sqrt{n}\right) + c1 \text{ if } n > 2$$

Let $n = 2^m$

$\Rightarrow T(n) = T(2^m)$

$\Rightarrow T(2^m) = S(m)$

$\Rightarrow T(n) = S(m)$

$\Rightarrow T\left(\sqrt{n}\right) = T\left(2^{m/2}\right) = S(m/2)$

$$T(n) = T\left(\sqrt{n}\right) + c1 \text{ if } n > 2$$

$\therefore S(m) = S\left(\frac{m}{2}\right) + c1$

$= O(\log m)$

$= O(\log \log n)[\because n=2^m \Rightarrow m = \log_2 n]$

$\Rightarrow T(n) = S(m) = O(\log \log n)$

$\therefore$ Option (c) is correct.

1.8 (a)

Consider each statement separately

I. $f(n) = (n + k)^m$

so, $f(n) = (1 + n)^m$

(Assume k = 1 is constant)

$f(n) = 1 + {}^mC_1 n + {}^mC_2 n^2 + \ldots {}^mC_m n^m$

$f(n) = O(n^m)$

II. $f(n) = 2^{n+1}$

$f(n) = 2^n \cdot 2^1$

$f(n) = 2 \cdot 2^n$

$f(n) = O(2^n)$

III. $f(n) = 2^{2n+1}$

$f(n) = 2^{2n} \cdot 2^1$

$f(n) = 2 \cdot 2^{2n}$

$f(n) = O(2^{2n})$

Therefore I and II are correct.

1.9 (b)

The given function f is not recursive, so consider the following iteration method.

i	p	s
	$p = p*\frac{x}{i}$	$s = s + p$
Initilize 1	1	1
1	$p = x$	$s = 1 + x$
2	$p = x\cdot\frac{x}{2}$	$s = 1 + x + \frac{x^2}{2}$
3	$p = \frac{x^2}{2}\cdot\frac{x}{3}$	$s = 1 + x + \frac{x^2}{2} + \frac{x^3}{6}$
4	$p = \frac{x^3}{6}\cdot\frac{x}{4}$	$s = 1 + x + \frac{x^2}{2} + \frac{x^3}{6} + \frac{x^4}{24}$
5	$p = \frac{x^4}{24}\cdot\frac{x}{5}$	$s = 1 + x + \frac{x^2}{2} + \frac{x^3}{6} + \frac{x^4}{24} + \frac{x^5}{120}$

For large value of y assume $y \to \infty$ then i also tends to infinite it means increment of for loop may tends to infinite. In the given function we choose y as a large integer but not infinite. The return value of the function f is s.

$$s = 1 + x + \frac{x^2}{2} + \frac{x^3}{6} + \frac{x^4}{24} + \frac{x^5}{120} \ldots + \infty$$

$$s = 1 + x + \frac{x^2}{!2} + \frac{x^3}{!3} + \frac{x^4}{!4} + \frac{x^5}{!5} + \ldots + \infty = e^x$$

1.10 (c)

The cube root of a natural number n is defined as the largest natural number m such that $m^3 \leq n$. Complexity of computing the cube root of n is $O((\log n)^k)$ for some constant $k > 0$.

1.11 (c)

Any decision tree that sorts n distinct elements has height at least log $\underline{|n}$. So the tightest lower bound on the number of comparison based sorting is log $\underline{|n}$ but from starling's approximation.

$$\underline{|n} = (n/e)^n$$

Taking log both sides

$$\log \underline{|n} = \log (n/e)^n$$
$$\log \underline{|n} = n\log (n/e)$$
$$\log \underline{|n} = n(\log n - \log e)$$
$$\log \underline{|n} = n(\log n - 1.44)$$
$$\log \underline{|n} = n\log n - 1.44\, n$$

So $\log \underline{|n} = O(n\log n)$

1.12 (c)

Let x = m = 9. The loop will be terminated when x − y = 0 or x − y < 0. Consider the following iteration for x = m = 9, y = 1

x − y > 0 ⇒	x = (x + y)/2,	y = m/x
9 − 1 = 8	x = 5.0,	y = 9/5.0 = 1.8
5.0 − 1.8 = 3.2	x = 3.4,	y = 9/3.4 = 2.6
3.4 − 2.6 = .80	x = 3.0,	y = 9/3.0 = 3.0

x − y = 3.0 − 3.0 = 0, loop terminated

So, m = 9 then x = 3

$$x = (m)^{1/2} = (9)^{1/2} \Rightarrow x = 3$$

So the algorithm computes $x = m^{1/2}$.

1.13 (c)

The given code is

```
1.    Counter = 0;
2. for (i = 1; i < = n; i ++)
3. {  if (A[i] = = 1) counter ++;
4.    else {f(counter); counter = 0;}
5. }
```

The time complexity of the program fragment depends on the frequency (Number of steps) of line 3 and 4. In line 4 the frequency depends on the variable counter and there is no increment in the counter variable which is initialize to 0, so f(0) then counter = 0 it means there is no cell in an array which having a bit 0, so all cells in the array contains 1. Consider the line 3 if (A[i] = = 1) counter ++; the value of i will be increases upto n so the value of counter will be n. Since n is the frequency of the line 3 and the frequency of line 4 is 0. So the time complexity of line 3 is O(n) on average n and f(0) = O(1) is the time complexity of line 4. So the time complexity of the program fragment is maximum of line 3 and 4 which is O(n) on average.

1.14 (d)

The given C function is recursive. The best way to find the time complexity of recursive function is that convert the code (algorithm) into recursion equation and solution of the recursion equation is the time complexity of given algorithm.

```
1. int recursive (int n) {
2. if (n = = 1) return (1);
3. else
4. return (recursive(n–1)+recursive (n– 1));
5. }
```

Let recursive(n) = T(n)

According to line 2 if (n = = 1) return(1) then the recursive equation is

$$T(n) = 1, \quad n = 1$$

According to line 4 the recursion equation is

$$T(n) = T(n-1) + T(n-1),\ n > 1$$

So the complete recursion equation is

$$T(n) = 1, \quad n = 1$$
$$T(n) = T(n-1) + T(n-1),\ n > 1$$

or $T(n) = 2T(n-1),\ n > 1$

$$T(1) = 1 = 2^0$$
$$T(2) = 2T(1) = 2.1 = 2^1$$
$$T(3) = 2T(2) = 2.2 = 2^2$$
$$T(4) = 2T(3) = 2.2^2 = 2^3$$

. . .

. . .

$$T(n) = 2^{n-1}$$

Or, $T(n) = 2^n.\dfrac{1}{2}$

So, $T(n) = O(2^n)$

1.15 (a)

$$T(1) = 1$$
$$T(n) = 2T(n-1) + n \quad n \geq 2$$
$$T(2) = 2T(1) + 2 = 2.1 + 2 = 4$$
$$T(3) = 2T(2) + 3 = 2.4 + 3 = 11$$
$$T(4) = 2T(3) + 4 = 2.11 + 4 = 26$$

⋮

$$T(n-1) = 2T(n-2) + n = 2^n - (n-1) - 2$$

So, $T(n) = 2^{n+1} - n - 2$

1.16 (d)

We can verify as:

$f \leq g$ BUT $g \not\leq f$. Therefore $f < g$

Also $g = h$ as $g = O(h)$ and $h = O(g)$.

Therefore $f < g$ and $g = h$

(a) $f(n) + g(n) = O(h(n)) + h(n))$ is true.

$\Rightarrow f + g = f + h < h + h$

(b) $f(n) = O(h(n))$ is true.

$\Rightarrow f < h$

(c) $h(n) \neq O(f(n))$ is true.

$\Rightarrow h \not< f$ is correct.

(d) $f(n)h(n) \neq O(g(n)h(n))$ is false.

$\Rightarrow f \cdot h < g \cdot h$ implies $fh = O(gh)$

1.17 (d)

The time complexity of computing the transitive closure of a binary relation on a set of n elements is $O(n^3)$ apply the Warshall's algorithm to compute the transitive closure. The algorithm contains three nested for loops each having frequency n so time complexity is $O(n^3)$.

1.18 (c)

$$T(n) = 2T(n/2) + \sqrt{n}$$

By substitution method $T(n) = \Theta(n)$.

$$T(n) = 2 \cdot T(n/2) + \sqrt{n} \quad \ldots(1)$$

$$= 2\left[2 \cdot T(n/4) + \sqrt{n/2}\right] + \sqrt{n}$$

$$= (2)^2 \cdot T(n/2^2) + \sqrt{n}(1+\sqrt{2}) \ldots(2)$$

$$= (2)^3 \cdot T(n/2^3) + \sqrt{n}(1+\sqrt{2}+2) \ldots(3)$$

$$\vdots$$

$$= (2)^k \cdot T(n/2^k) + \sqrt{n}\left(\sum_{i=0}^{k-1}\left(\sqrt{2}\right)^i\right) = \Theta(n)$$

1.19 (c)

$$T(0) = T(1) = 1$$

$$T(n) = 2T(n/2) + n$$

T(n) be computed as follows

$$T(n) = n\sum_{i=0}^{\log n}\left(\frac{2}{2}\right)^i$$

$$T(n) = n\sum_{i=0}^{\log n}(1)^i$$

$$T(n) = \Theta(n\log n)$$

[T(n) can also proved by Master Theorem]

If $T(n) = \Theta(n \log n)$ then it is also $O(n \log n)$ and $O(n^2)$ but it is not $\Omega(n^2)$.

1.20 (b)

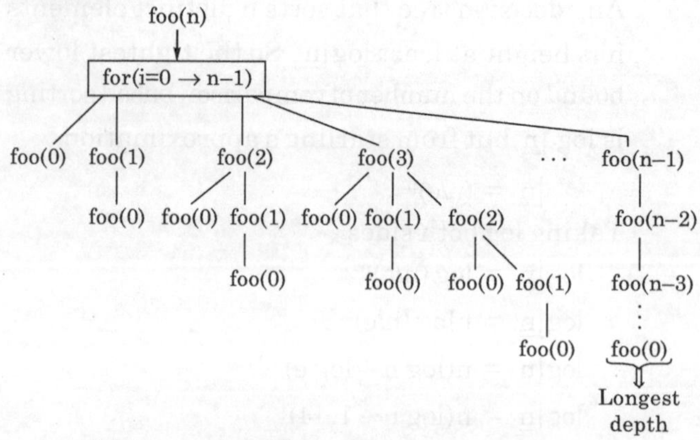

The longest calling sequence:

$$foo(n) \rightarrow foo(n-1) \rightarrow \ldots foo(0)$$

Therefore space complexity = O(n)

[**Note:** Time complexity = $O(n^n)$]

1.21 (b)

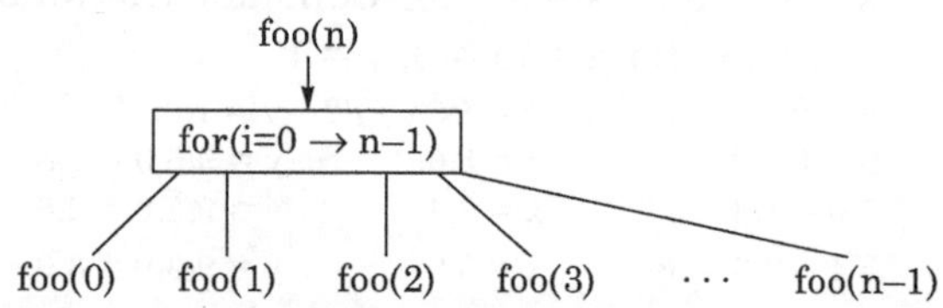

Space required: foo(0), foo(1),...,foo(n–1) = O(n)

Even storing foo(i), 0 <= i < n the space complexity is O(n).

[**Note:** Time complexity = $O(n^2)$]

1.22 (c)

The frequency of the for loop is $i = \frac{i}{2} = \frac{n}{2}$ so after termination of for loop vol(j) = O(n) on average.

$$j = n + \frac{n}{2} + \frac{n}{2^2} + \ldots \frac{n}{2^{\log_2 n}} = \Theta(n)$$

1.23 (b)

Let $n = 2^m$

$$T(2^m) = T\left(2\left(\frac{m}{2}\right)\right) + 1$$

Let $T(2^m) = S(m)$

Then, $S(m) = 2S\left(\frac{m}{2}\right) + 1$

Using Master's theorem

$S(m) = \Theta(m) = \Theta(\log n)$ since $n = 2^m$

Now, $T(n) = T(2^m) = S(m) = \Theta(\log n)$

Therefore, option (b) is correct.

1.24 (d)

Let the increment of j is $2^0, 2^1, ..., 2^i$ for some value of i so according to the question for while loop: $2^i \le n$ **or** $i \le \log_2 n$.
One extra comparison required for the termination of while loop. So total number of comparisons $= i + 1 = \lfloor \log_2 n \rfloor + 1$

1.25 (d)

The given function is recursive so the equivalent recursion equation is

$$T(n) = 1; \quad n \le 2$$

$$T(n) = \left\lfloor \sqrt{n} \right\rfloor + n; \; n > 2$$

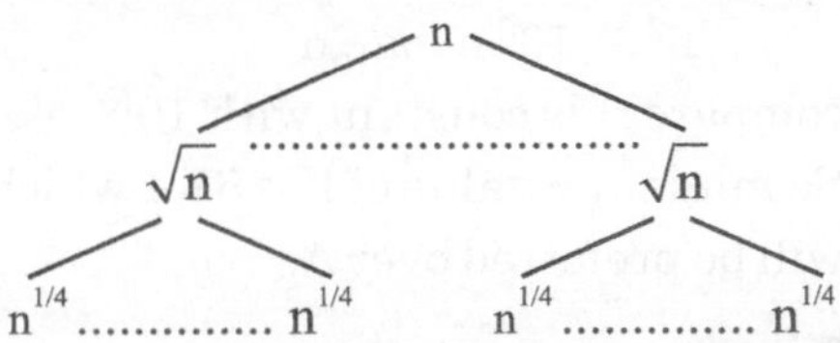

All the level sums are equal to n. The problem size at level k of the recursion tree is $n^{2^{-k}}$ and we stop recursing when this value is a constant. Setting $n^{2^{-k}} = 2$ and solving for k gives us

$$2^{-k}\log_2 n = 1$$

$$\Rightarrow \quad 2^k = \log_2 n$$

$$\Rightarrow \quad k = \log_2\log_2 n$$

So $\quad T(n) = \Theta(\log_2\log_2 n)$

1.26 (b)

When we apply the divide and conquer method such that dived n into two parts of n/2 then recurrence equation is

$$T(n) = 1 \text{ for } n = 2$$

$$T(n) = 2T\left(\frac{n}{2}\right) + 2 \text{ for } n > 2$$

The solution of T(n) is

$$T(n) = \frac{3}{2}n - 2 = 1.5\,n - 2 \text{ comparison}$$

1.27 (b)

T(n) denote the number of times the for loop is executed by the program on input n.

$$T(n) = \sqrt{n} + 1$$

when, n = 1 then best case occurs so T(n) takes constant time i.e., $T(n) = \Omega(1)$.
In worst case,

$$T(n) = \sqrt{n} + 1$$

So, $\quad T(n) = O\left(\sqrt{n}\right)$

1.28 (a)

A. $n^{1/3} = (n)^{0.34}$
B. $e^n = (2.7)^n$
C. $n^{7/4} = (n^{3/4}) \cdot (n^1) = (n)^{1.7} = n(n^{3/4})$
D. $1.0000001^n = (1.0000001)^n$
So, $(n)^{0.34} < n(\log^9 n) < n^{7/4} < (1.0000001)^n < (2.7)^n$
[A] < [D] < [C] < [E] < [B]

1.29 (a)

$$T(n) = \sqrt{2}\cdot T(n/2) + \sqrt{n} \quad ...(1)$$

$$= \sqrt{2}\left[\sqrt{2}\cdot T(n/4) + \sqrt{n/2}\,\right] + \sqrt{n}$$

$$= \left(\sqrt{2}\right)^2 \cdot T\left(n/2^2\right) + 2\sqrt{n} \quad ...(2)$$

$$= \left(\sqrt{2}\right)^3 \cdot T\left(n/2^3\right) + 3\sqrt{n} \quad ...(3)$$

$$\vdots$$

$$= \left(\sqrt{2}\right)^{2k} \cdot T\left(n/2^{2k}\right) + 2k\sqrt{n}$$

$$= \left(\sqrt{2}\right)^{\log n} \cdot 1 + \log n \cdot \sqrt{n} \quad [\because \text{Put } 2k = \log n]$$

$$= \sqrt{n} + \log n \cdot \sqrt{n}$$

$$= \sqrt{n}(\log n + 1)$$

1.30 (d)

$$f(n) = 2^n \Rightarrow f(n) = O(2^n)$$

$$g(n) = n! \Rightarrow g(n) = O(n!)$$

$$h(n) = n^{\log n} \Rightarrow h(n) = O(n^{\log n})$$

[$n < n^2$ means n grows more slowly than n^2]
The asymptotic order of function is as follows:
$1 < \log\log n < \log n < n^{\varepsilon} < n^c < n^{\log n} < c^n < n^n < c^{c^n} < n!$
where $0 < \varepsilon < 1 < c < n$
$\Rightarrow n^{\log n} < c^n < n!$
$\Rightarrow n^{\log n} < 2^n < n!$
Assume c = 2
$h(n) < f(n) < g(n)$

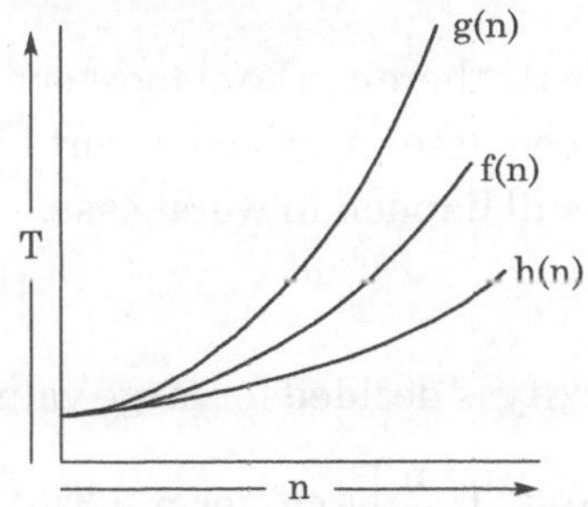

$h(n) = O(f(n)) \Rightarrow h(n) \le k.f(n)$ is true
$g(n) = \Omega(f(n)) \Rightarrow g(n) \ge k.f(n)$ is also true

1.31 (b)

Consider an array of n elements which stored integers in sorted order, A[1], A[2], ..., A[n]. If we want to search an element x with the help of binary search then in expected case comparison is not greater than $\lceil \log(n+1) \rceil$. But if integer appears $\frac{n}{2}$ times in an array and array is already sorted then it take $\Theta(1)$ times for consecutive locations. So total no. of comparisons is not greater than $\Theta(\log n)$ time.

1.32 (b)

The $f_1(n)$ is a recursive function.
The recurrence equation for $f_1(n)$ is

$$T(0) = 0$$
$$T(1) = 1$$
$$T(n) = 2T(n-1) + 3T(n-2)$$

The solution of this linear equation contains a polynomial of 2^n so on average the running time of $f_1(n)$ is $\Theta(2^n)$.

1.33 (c)

$$T(0) = 0$$
$$T(1) = 1$$
$$\mathbf{T(n) = 2T(n-1) + 3T(n-2)}$$
$$T(2) = 2T(1) + 3T(0) = 2.1 + 3.0 = 2$$
$$T(3) = 2T(2) + 3T(1) = 2.2 + 3.1 = 7$$
$$T(4) = 2T(3) + 3T(2) = 2.7 + 3.2 = 20$$
$$T(5) = 2T(4) + 3T(3) = 2.20 + 3.7 = 61$$
$$T(6) = 2T(5) + 3T(4) = 2.61 + 3.20 = 182$$
$$T(7) = 2T(6) + 3T(5) = 2.182 + 3.61 = 547$$
$$T(8) = 2T(7) + 3T(6) = 2.547 + 3.182 = 1640$$

So f(8) returns 1640, $f_2(n)$ is the nonrecursive version of $f_1(n)$ so the output of $f_2(8)$ is 1640.

1.34 (a)

In each pass there can be at most one swap. There are n such passes in selection sort i.e. maximum n swap will happen in worst case.

1.35 (a)

Complexity is decided for large values of n only,

So, $T(n) = T\left(\frac{n}{3}\right) + cn$ for $n > 3$

using master's theorem

here $a = 1, b = 3, \log_b a = \log_3 1 = 0$

$f(n) = cn = \Theta(n^1)$

since $n^{\log_b a} = n^0$ is below $f(n) = \Theta(n^1)$

this belongs to case III of master's theorem, where the solution is $T(n) = \Theta(f(n)) = \Theta(n)$

1.36 (c)

A requires $0.0001\, n^2$ time units. And B require $10\, n \log_{10} n$ time unit to process n records for package B will be preferred over A.

So, $0.0001\, n^2 \geq 10\, n \log_{10} n$

$$\frac{n^2}{10^5} \geq 10\, n \log_{10} n$$

$$n^2 \geq 10^6\, n \log_{10} n$$

compare this constant with 10^K

So minimum value of K = 6 for which package B will be preferred over A.

1.37 (c)

$A(n) \rightarrow$ Average case complexity

$W(n) \rightarrow$ Worst case complexity

We know

Best case $\leq$ Average case complexity $\leq$ Worst case complexity

and always $A(n) \leq k.W(n) \Rightarrow A(n) = O(W(n))$

1.38 (d)

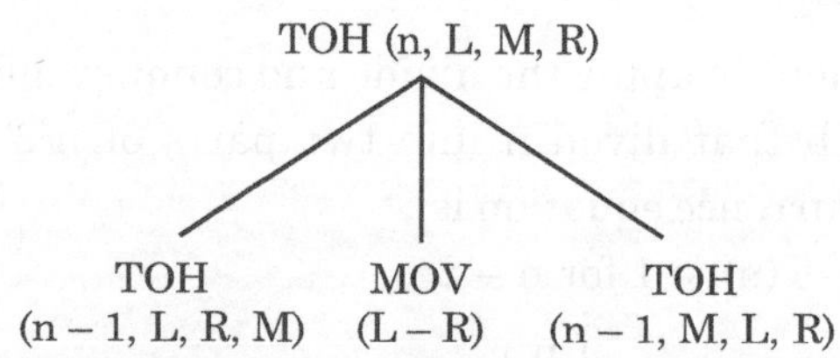

$\Rightarrow T(n) = 2T(n-1) + 1$

1.39 (b)

Outer loop execute for $\frac{n}{2}+1$ iterations. Inner loop executes for $\log_2 n$ iterations. In every iteration of inner loop $\frac{n}{2}$ is added to k.

Return value $= \frac{n}{2} \times$ number of outer loops $\times$ number of inner loops

$$= \frac{n}{2} \times \left(\frac{n}{2} + 1\right)(\log n) = O(n^2 \log n)$$

1.40 (a)

$T(n) = 2.T\left(\frac{n}{2}\right) + \log n$ [by master's theorem case1] $= \Theta(n)$

1.41 (110)

$O(m + \log n) = O(n^0 \log^1 n + m^1 \log^0 n)$

$\Rightarrow a = 0, b = 1, c = 1, d = 0$

$\Rightarrow a + 10b + 100c + 1000d = 10 + 100 = 110$

1.42 (d)

```
int fun1(int n)
{
    int i, j, k, p, q = 20
    for (i = 1; i < n; ++i)
    {  P = 0;
       for (j=n; j<1; j=j/2) ⇒ log n times for each
       i ⇒ Total = n × log n
          ++p;
       for (k=1; k<p; k=k×2) ⇒ log log n times
       for each i ⇒ Total = n log n
          ++q;
    }
return q; ⇒ Return value = O(n log log n)
}
```

Time complexity of above program = O(n log n)

Return value q = O(n log log n)

1.43 (a)

	Find	Insert	Delete	Decrease key	Overall Asymptotic time complement
Unsorted Array	Each find operation in O(N) $O(N\sqrt{\log N})$	O(N)	$O(N\sqrt{\log N})$	$O(\sqrt{\log N})$	$O(N\sqrt{\log N})$
Min heap	$O(N\sqrt{\log N})$	O(N·logN)	$O(N\sqrt{\log N})$	$O(\log N)^{3/2}$	O(N·logN)
Sorted array	$O(\log N)^{3/2}$	$O(N^2)$	$O(N\sqrt{\log N})$	$O(N\sqrt{\log N})$	$O(N^2)$
Sorted doubly linked list	$O(N\sqrt{\log N})$	$O(N^2)$	$O(\sqrt{\log N})$	$O(\sqrt{\log N})$	$O(N^2)$

∴ Unsorted array is best.

1.44 (a)

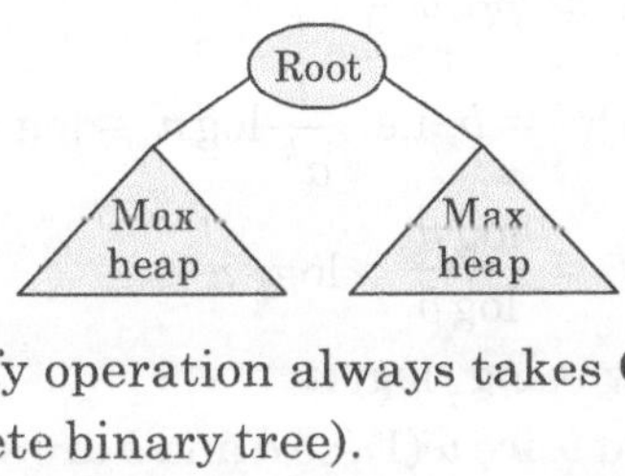

Heapify operation always takes O(logn) (for a complete binary tree).

Since left subtree and right subtree are max heaps it takes Ω(log n) to heapify.

1.45 (d)

Take first 3 elements. The middle of the 3 elements will be the element that is neither minimum nor maximum in the array.

Hence O(1) time to compare 3 elements.

Example: Assume 4, 6, 2 are first three elements. Then 4 is the middle element of 4, 6 and 2. Which is neither minimum nor maximum.

1.46 (c)

$$X = 1 + 2^3 + 3^3 + 4^3 + ... n^3$$

$$X = \left[\frac{n(n+1)}{2}\right]^2 = \frac{n^2(n+1)^2}{4} = \Theta(n^4)$$

$$X = \Theta(n^4) = O(n^5) = \Omega(n^3)$$

1.47 (d)

When ever the value of 'sin n' value is –1, then g(n) = 1. {sin n value ranges from –1 to +1}.

Hence f(n) = O(g(n)) is violated. Hence I is false.

and when ever the value of 'sin n' is +1, then $g(n) = n^2$

Hence f(n) = Ω(g(n)) is violated. Hence II is false.

1.48 (2.32)

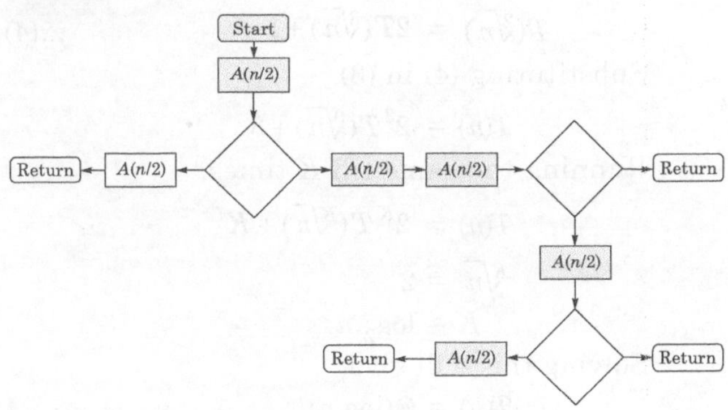

In worst case scenerio, we have $5\,A\left(\frac{n}{2}\right)$ function calls and other statements takes O(1) time. So, the recurrrance relation will be,

$$A(n) = 5\,A\left(\frac{n}{2}\right) + O(1)$$

By using Master theorem, we get $n^{\log_2 5}$ which is equivalent to $n^{2.32}$. So the value of α is 2.32.

1.49 (c)

Data structure used	Find	Insert	Delete	Decrease key	Overall asymptotic time
Sorted doubly linked list	O(N logN)	O(N logN)	O(N)	O(N)	O(N logN)

Sorted doubly linked list take maximum of O(N logN) time, but since not present in option. So, go for nearest value O(N^2).

1.50 (c)

- Towers of Hanoi with n disks = $2T(n-1)+1$ $= \Theta(2^n)$.
- Binary search given n sorted numbers $= T\left(\frac{n}{2}\right) + 1 = \Theta(\log n)$.
- Heap sort given n numbers at the worst case $= 2T\left(\frac{n}{2}\right) + n = \Theta(n \log n)$.
- Addition of two $n \times n$ matrices $= 4T\left(\frac{n}{2}\right) + 1$ $= \Theta(n^2)$.

1.51 (b)

$$T(n) = 2T(\sqrt{n})+1 \quad \text{...(1)}$$
$$T(n) = 2T(\sqrt[2]{n})+1 \quad \text{...(2)}$$

Substituting (2) in (1)

$$T(n) = 2.2T(\sqrt[2]{n})+2$$
$$T(n) = 2^2 T(\sqrt[2]{n})+2 \quad \text{...(3)}$$
$$T(\sqrt[2]{n}) = 2T(\sqrt[3]{n})+1 \quad \text{...(4)}$$

Substituting (4) in (3)

$$T(n) = 2^3 T(\sqrt[3]{n})+3$$

Running the same till K times,

$$T(n) = 2^K T(\sqrt[K]{n})+K$$
$$\sqrt[K]{n} = 2$$
$$K = \log_2 n$$

Solving this will give

$$T(n) = \Theta(\log n)$$

1.52 (c)

First loop will execute 'n' times and the inner loop will execute $\Theta(\log n)$ times.

Hence the compelxity will be $\Theta(n \log n)$.

1.53 (b)

$$\frac{100}{n} < 10 < \log_2 n < \sqrt{n} < n$$

So correct answer is (b).

1.54 (5)

Since 0 are followed by 1, so we have a sorted array of 0 and 1 and we can apply binary search algorithm with some modification, which will take maximum probes $= \lceil \log_2 31 \rceil = 5$.

1.55 (d)

Insert element at the beginning of linked list, take O(1)

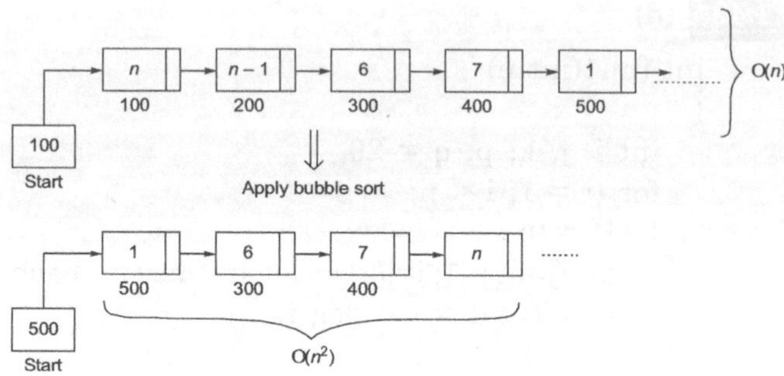

Time complexity = $O(n) + O(n^2) = \Theta(n^2)$

Hence correct answer is option (d).

1.56 (b)

AVL with n element: [height balanced [–1, 0, +1] BST]

log$n$ level due to balanced BST.

(i) Every insertion of element:
log$n$: Find place to insert.
log$n$: If property not satisfied do rotation.

$\therefore$ n^2 element insertion:
For 1 element ≡ 2 log$n$
So, for n^2 element ≡ $\theta(n^2 \log n)$

1.57 (b)

$$T(n) = \begin{cases} T(n^{1/a}) + 1, & \text{when } n \neq b \\ 1, & \text{when } n = b \end{cases}$$

$$T(n) = T(n^{1/a}) + 1$$
$$= T(n^{1/a^2}) + 1 + 1 \; [\because T(n^{1/a}) = T(n^{1/a^2}) + 1]$$
$$= T(n^{1/a^3}) + 1 + 1 + 1 \; [\because T(n^{1/a^2}) = T(n^{1/a^3}) + 1]$$

After k interactions

$$= T(n^{1/a^k}) + k$$

when, $n^{1/a^k} = b$, i.e. $\frac{1}{a^k}\log n = \log b$

$$\Rightarrow \quad a^k = \frac{\log n}{\log b} = \log_b n$$
$$\Rightarrow \quad k = \log_a \log_b n$$

[$\because$ a and b are $w(1)$, so a and b are some function of n and not constant. So a and b can't be replaced with 2]

Now, $T(n) = T(n^{1/a^k}) + k$
$= T(b) + \log_a \log_b n$
$= 1 + \log_a \log_b n = \theta(\log_a \log_b n)$

Hence, option (b) is the answer.

1.58 (c)

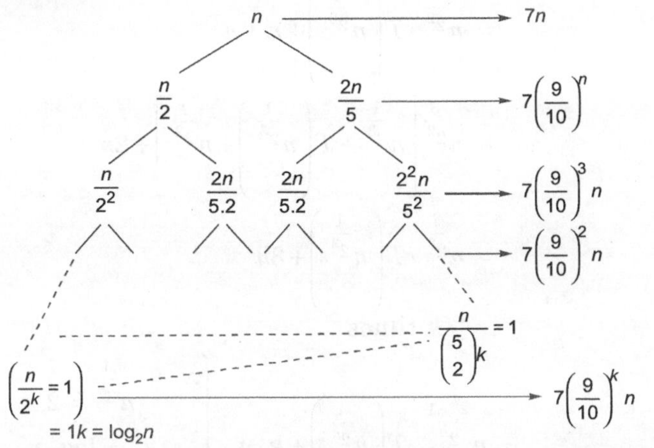

$$T(n) = 7n\left[1+\frac{9}{10}+\left(\frac{9}{10}\right)^2+\left(\frac{9}{10}\right)^3+\ldots+\left(\frac{9}{10}\right)^k\right]$$

$$= \frac{7n\left[1-\left(\frac{9}{10}\right)^{k+1}\right]}{1-\frac{9}{10}} = \frac{7n}{\frac{1}{10}}\left[1-\left(\frac{9}{10}\right)^{\log_2 n+1}\right]$$

Decreasing term

$\therefore \quad T(n) = \theta(n)$

1.59 (a)

$f_2 : n^{\log n} f_3 : n^{\sqrt{n}}$

$\log(n^{\log n}) \log(n^{\sqrt{n}})$

$= (\log n) \cdot (\log n) = \sqrt{n} \cdot \log n$

$\therefore \quad n^{\log n} = O(n^{\sqrt{n}}) \quad (f_2 < f_3)$

$f_1 = 10^n \leftarrow$ exponential function

$\therefore \quad f_1 > f_3 > f_2$

In increasing order of asymptotic growth rate.

1.60 (b)

According to Standard Master Theorem, only option (b) is correct.

1.61 (a)

True (a) $f(n^2) = \theta(f(n)^2)$ $f(n)$ polynomial

$$f(n) = n^3 \Rightarrow \left.\begin{aligned} f(n)^2 &= (n^2)^3 = n^6 \\ f(n)^2 &= (n^3)^2 = n^6 \end{aligned}\right\} \text{asy equal}$$

False (b) $f(n^2) = o(f(n)^2)$

$f(n) = n^3 \Rightarrow f(n)^2 = (n^2)^3 = n^6$

$f(n)^2 = (n^3)^2 = n^6$

False (c) $f(n^2) = \Omega(f(n)^2)$ $f(n)$ exponential

$f(n) = 2^n$

$\Rightarrow f(n)^2 = 2^{(n^2)} \qquad 2^{(n^2)} > 2^{2n}$

$f(n)^2 = (2^n)^2 = 2^{2n}$

$f(n) = n^2 \Rightarrow f(n^2) = f(n^2)$

$f(n) = \log n$

$$\left.\begin{aligned} \Rightarrow f(n)^2 &= \log n^2 = 2\log n \\ f(n)^2 &= (\log n)^2 \end{aligned}\right\} 2\log n < (\log n)^2$$

So, $f(n^2) = \Omega(f(n)^2)$ is false

1.62 (a, c)

$f(n) \in O(g(n))$ iff $f(n)$ asymptotically smaller or equal to $g(n)$

$f(n) \in o(g(n))$ iff $f(n)$ asymptotically smaller than $g(n)$

$f(n) \in \Omega(g(n))$ iff $f(n)$ asymptotically greater or equal to $g(n)$

$f(n) \in \theta(g(n))$ iff $f(n)$ asymptotically equal to $g(n)$

Given, $f(n) = n,\ g(n) = n^2$

$n \in O(n^2)$ ✓

$n \in \Omega(n^2)$ ×

$n \in o(n^2)$ ✓

$n \in \theta(n^2)$ ×

1.63 (a, d)

Function 1

```
while n > 1 do
    for i = 1 to n do
        x = x + 1;
    end for
    n = ⌊n/2⌋;
end while
```

$f_1(n)$: Number of times $x = x + 1$ executes

$$f_1(n) = n + \frac{n}{2} + \frac{n}{2^2} + \ldots + \frac{n}{2^k}$$

where $\frac{n}{2^k} > 1$

$k \approx \log_2 n$

$$f_1(n) = n\left[1 + \frac{1}{2} + \frac{1}{2^2} + \ldots + \frac{1}{2^k}\right] = \theta(n)$$

Function 2

for $i = 1$ to $100 * n$ do
    $x = x + 1;$
end for

$f_2(n)$: Number of times $x = x + 1$ executes

$$f_2(n) = 100 * n = \theta(n)$$

$$\therefore \quad f_1(n) = \theta(f_2(n))$$

$$f_1(n) = O(n) \text{ are correct}$$

1.64 (b, d)

(a) False

Let, $L_1 = \Sigma^*$

$L_1 \cup L_3 = \Sigma^* \cup L_3$

$= \Sigma^*$ which is regular

(b) True: If a language is not regular then its complement is also not regular.

(c) False: Consider the given Venn diagram

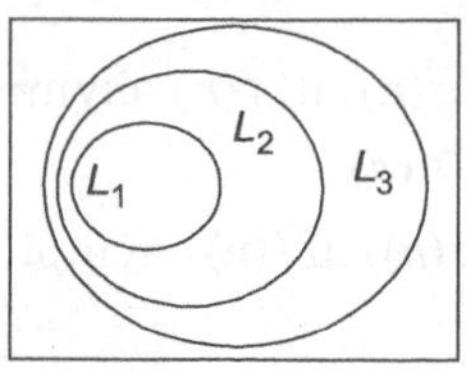

Here, $L_1 \cap \overline{L_3} = \phi$ but L_1 is not equal to L_2.

(d) True:

If L_1 and L_2 are regular

$\Rightarrow$ $\overline{L_1}$ and $\overline{L_2}$ are also regular.

$\Rightarrow$ $\overline{L_1} \cup \overline{L_2}$ are also regular.

1.65 (c)

```
ASc = 0; desc = 0;
for (i = 1; i ≤ n – 1; i++)
{
    if (A[i] > A[i + j])// not Asc order
        Asc = 1;
    if (A[i] < A[i + j])// not desc order
        desc = 1;
}
If (Asc == 0)
    print("Given array ascending order")
else if (desc == 0)
    print("Given array is descending order")
else
    print("Given array unsorted")
```

1.66 (a)

$$T(n) = n^{1/2} \cdot T\left(n^{1/2}\right) + n$$

$$= n^{1/2}\left[n^{\frac{1}{2^2}} T\left(n^{\frac{1}{2^2}}\right) + n^{1/2}\right] + n$$

$$= n^{\frac{3}{2^2}} \cdot T\left(n^{\frac{1}{2^2}}\right) + n + n$$

$$= n^{\frac{3}{2^2}}\left[n^{\frac{1}{2^3}} + T\left(n^{\frac{1}{2^3}}\right) + n^{\frac{1}{2^2}}\right] + 2n$$

$$= n^{\frac{7}{2^3}} \cdot T\left(n^{\frac{1}{2^3}}\right) + 3n$$

$\vdots$ k times

$$= n^{\frac{2^k - 1}{2^k}} \cdot T\left(n^{\frac{1}{2^k}}\right) + k \cdot n \quad \left\{ \begin{aligned} \therefore \quad n^{\frac{1}{2^k}} &= 2 \\ 2^k &= \log_2 n \\ k &= \log_2 \log_2 n \end{aligned} \right.$$

$$= \frac{n}{n^{\frac{1}{2^k}}} \cdot T(2) + n \cdot \log_2 \log_2 n$$

$$T(n) = \frac{n}{2} + n \cdot \log_2 \log_2 n = \theta(n \log_2 \log_2 n)$$

1.67 (3)

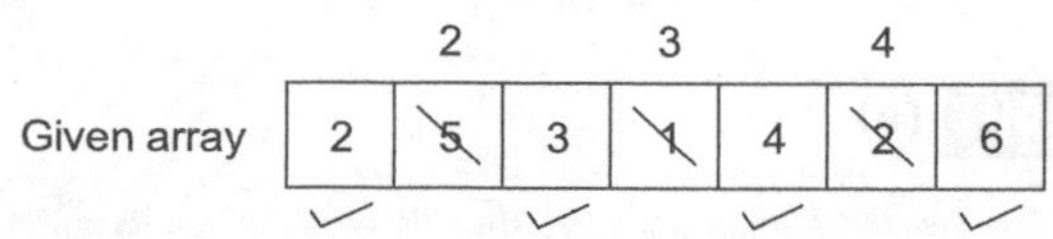

Max sub array which are in sorted are 4 element ex: 2, 3, 4, 6 replace remaining 3 elements by elements either previous or next to it.

1.68 (d)

$$T(n) = 5T(n - 1) - 6T(n - 2)$$

Generating function:

$$t^2 - 5t + 6 = 0$$

Roots = 2, 3

$$T(n) = C_1 2^n + C_3 3^n$$

$$1 = C_1 + C_2 \quad \{T(0) = 1\}$$

$$2 = 2C_1 + 3C_2 \quad \{T(1) = 2\}$$

$$C_1 = 1, C_2 = 0$$

$$\therefore \quad T(n) = 1 \times 2^n + 0 \times 3^n$$

$$T(n) = 2^n$$

2 Divide and Conquer

2.1 Merge sort uses
(a) Divide and conquer strategy
(b) Backtracking approach
(c) Heuristic search
(d) Greedy approach
[1995 : 1 M]

2.2 For merging two sorted lists of sizes m and n into a sorted list of size m+n, we required comparisons of
(a) O(m) (b) O(n)
(c) O(m + n) (d) O(logm + logn)
[1995 : 1 M]

2.3 Which of the following statements is true?
I. As the number of entries in a hash table increases, the number of collisions increases.
II. Recursive programs are efficient
III. The worst case complexity for Quicksort is $O(n^2)$
IV. Binary search using a linear linked list is efficient.
(a) I and II (b) II and III
(c) I and IV (d) I and III
[1995 : 1 M]

2.4 The recurrence relation
$T(1) = 2$
$$T(n) = 3T\left(\frac{n}{4}\right) + n$$
Has the solution T(n) equal to
(a) O(n) (b) O(log n)
(c) $O(n^{3/4})$ (d) None of these
[1996 : 2 M]

2.5 Quicksort is run on two inputs shown below to sort in ascending order
(i) 1, 2, 3 ..... n
(ii) n, n – 1, n – 2, ..., 2, 1
Let C_1 and C_2 be the number of comparisons made for the inputs (i) and (ii) respectively. Then,
(a) C1 < C2
(b) C1 > C2
(c) C1 = C2
(d) we cannot say anything for arbitrary n
[1996 : 2 M]

2.6 Let T(n) be the function defined by T(1) = 1,
$$T(n) = 2T\left(\left\lfloor \frac{n}{2} \right\rfloor\right) + \sqrt{n} \text{ for } n \geq 2.$$ Which of the following statements is true?
(a) $T(n) = O(\sqrt{n})$ (b) T(n) = O(n)
(c) T(n) = O (log n) (d) None of these
[1997 : 2 M]

2.7 A sorting technique is called stable if
(a) it takes O(nlog n) time
(b) it maintains the relative order of occurrence of non-distinct elements
(c) it uses divide and conquer paradigm
(d) it takes O(n) space **[1999 : 1 M]**

2.8 If one uses straight two-way merge sort algorithm to sort the following elements in ascending order:
20, 47, 15, 8, 9, 4, 40, 30, 12, 17
then the order of these elements after second pass of the algorithm is
(a) 8, 9, 15, 20, 47, 4, 12, 17, 30, 40
(b) 8, 15, 20, 47, 4, 9, 30, 40, 12, 17
(c) 15, 20, 47, 4, 8, 9, 12, 30, 40, 17
(d) 4, 8, 9, 15, 20, 47, 12, 17, 30, 40
[1999 : 1 M]

2.9 Let s be a sorted array of n integers. Let t(n) denote the time taken for the most efficient algorithm to determine if there are two elements with sum less than 1000 in s. Which of the following statements is true?
(a) t(n) is O(1)
(b) $n \leq t(n) \leq n \log_2 n$
(c) $n \log_2 n \leq t(n) < \binom{n}{2}$
(d) $t(n) = \binom{n}{2}$ **[2000 : 1 M]**

2.10 Randomized quicksort is an extension of quicksort where the pivot is chosen randomly. What is the worst case complexity of sorting n numbers using randomized quicksort?
(a) O(n) (b) (n log n)
(c) $O(n^2)$ (d) O(n!) **[2001 : 1 M]**

2.11 The usual $\Theta(n^2)$ implementation of Insertion Sort to sort an array uses linear search to identify the position where an element is to be inserted into the already sorted part of the array. If instead, we use binary search to identify the position, the worst case running time will
(a) remain $\Theta(n^2)$ (b) become $\Theta(n (\log n)^2)$
(c) become $\Theta(n \log n)$ (d) become $\Theta(n)$
[2003 : 1 M]

2.12 Let T(n) be the number of different binary search trees on n distinct elements.

Then $T(n) = \sum_{k=1}^{n} T(k-1)\, T(x)$, where x is
(a) $n - k + 1$ (b) $n - k$
(c) $n - k - 1$ (d) $n - k - 2$
[2003 : 1 M]

2.13 Consider a list of recursive algorithms and a list of recurrence relations as shown below. Each recurrence relation corresponds to exactly one algorithm and is used to derive the time complexity of the algorithm.

List-I (Recursive Algorithm)
P. Binary search
Q. Merge sort
R. Quick sort
S. Tower of Hanoi

List-II (Recurrence Relation)
I. $T(n) = T(n - k) + T(k) + cn$
II. $T(n) = 2T(n - 1) + 1$
III. $T(n) = 2T(n/2) + cn$
IV. $T(n) = T(n/2) + 1$

Which of the following is the correct match between the algorithms and their recurrence relations?
Codes:

	P	Q	R	S
(a)	II	III	IV	I
(b)	IV	III	I	II
(c)	III	II	IV	I
(d)	IV	II	I	III

[2004 : 2 M]

2.14 Suppose there are log n sorted lists of $\frac{n}{\log n}$ elements each. The time complexity of producing a sorted list of all these elements is:
(**Hint:** Use a heap data structure)
(a) $O(n \log \log n)$ (b) $\Theta(n \log n)$
(c) $\Omega(n \log n)$ (d) $\Omega(n^{3/2})$
[2005 : 2 M]

2.15 How many distinct binary search trees can be created out of 4 distinct keys?
(a) 5 (b) 14
(c) 24 (d) 35 **[2005 : 2 M]**

2.16 The median of n elements can be found in O(n) time. Which one of the following is correct about the complexity of quick sort, in which remains is selected as pivot?
(a) $\Theta(n)$ (b) $\Theta(n \log n)$
(c) $\Theta(n^2)$ (d) $\Theta(n^3)$ **[2006 : 2 M]**

2.17 Which of the following sorting algorithms has the lowest worst-case complexity?
(a) Merge sort (b) Bubble sort
(c) Quick sort (d) Selection sort
[2007 : 1 M]

2.18 In the following C function, let $n \geq m$.

```
int gcd (n, m)
{    if (n%m = = 0} return m;
     n = n%m;
     return gcd (m, n);
}
```

How many recursive calls are made by this function?
(a) $\Theta(\log_2 n)$ (b) $\Omega(n)$
(c) $\Theta(\log_2 \log_2 n)$ (d) $\Theta(\sqrt{n})$ **[2007 : 2 M]**

2.19 Consider the Quicksort algorithm. Suppose there is a procedure for finding a pivot element which splits the list into sub-lists each of which contains at least one-fifth of the elements. Let T(n) be the number of comparisons required to sort n elements. Then

(a) $T(n) \leq 2T\left(\frac{n}{5}\right) + n$

(b) $T(n) \leq T\left(\frac{n}{5}\right) + T\left(\frac{4n}{5}\right) + n$

(c) $T(n) \leq 2T\left(\frac{4n}{5}\right) + n$

(d) $T(n) \leq 2T\left(\frac{n}{2}\right) + n$ **[2008 : 2 M]**

Linked Answers Q. 2.20 and Q. 2.21
Consider the following C program that attempts to locate an element x in an array Y[ ] using binary search. The program is erroneous.

```
f(int Y[10] , int x) {
    int i, j, k;
    i = 0; j = 9;
    do {
```

```
5.      k = ( i + j) / 2;
6.      if (Y[K] < x)  i = k; else j = k;
7.    } while ((Y[k]  ! = x) && ( i < j));
8.    if (Y[k] = = x )
        printf("x is in the array") ;
9.    else
        printf("x is not in the array");
10. }
```

2.20 On which of the following contents of Y and x does the program fail?
(a) Y is [1 2 3 4 5 6 7 8 9 10] and x < 10
(b) Y is [1 3 5 7 9 11 13 15 17 19] and x < 1
(c) Y is [2 2 2 2 2 2 2 2 2 2] and x > 2
(d) Y is [2 4 6 8 10 12 14 16 18 20] and 2 < x < 20 and x is even

[2008 : 2 M]

2.21 The correction needed in the program to make it work properly is
(a) change line 6 to : if (Y[k] < x) i = k + 1; else j = k –1;
(b) change line 6 to : if (Y[k] < x) i = k –1; else j = k +1;
(c) change line 6 to : if (Y [k] < x) i = k; else j = k;
(d) change line 7 to : while ((Y [k] = = x) && (i < j));

[2008 : 2 M]

Directions for Question 2.22 to 2.24:
A Binary Search Tree (BST) stores values in the range 37 to 573. Consider the following sequence of keys.

I. 81, 537, 102, 439, 285, 376, 305
II. 52, 97, 121, 195, 242, 381, 472
III. 142, 248, 520, 386, 345, 270, 307
IV. 550, 149, 507, 395, 463, 402, 270

2.22 Suppose the BST has been unsuccessfully searched for key 273. Which all of the above sequences list nodes in the order in which we could have encountered them in the search?
(a) II and III only (b) I and III only
(c) III and IV only (d) III only

[2008 : 2 M]

2.23 Which of the following statements is TRUE?
(a) I, II and IV are inorder sequences of three different BSTs
(b) I is a preorder sequence of some BST with 439 as the root
(c) II is an inorder sequence of some BST where 121 is the root and 52 is a leaf
(d) IV is a postorder sequence of some BST with 149 as the root **[2008 : 2 M]**

2.24 How many distinct BSTs can be constructed with 3 distinct keys?
(a) 4 (b) 5
(c) 6 (d) 9

[2008 : 2 M]

2.25 In quick sort, for sorting n elements, the $\left(\frac{n}{4}\right)^{th}$ smallest element is selected as pivot using an O(n) time algorithm. What is the worst case time complexity of the quick sort?
(a) $\Theta(n)$ (b) $\Theta(n \log n)$
(c) $\Theta(n^2)$ (d) $\Theta(n^2 \log n)$

[2009 : 2 M]

2.26 Four matrices M_1, M_2, M_3 and M_4 of dimensions $p \times q$, $q \times r$, $r \times s$ and $s \times t$ respectively can be multiplied in several ways with different number of total scalar multiplications. For example when multiplied as $((M_1 \times M_2) \times (M_3 \times M_4))$, the total number of scalar multiplications is pqr + rst + prt. When multiplied as $(((M_1 \times M_2) \times M_3) \times M_4)$, the total number of scalar multiplications is pqr + prs + pst.
If p = 10, q = 100, r = 20, s = 5 and t = 80, then the minimum number of scalar multiplications needed is
(a) 248000 (b) 44000
(c) 19000 (d) 25000 **[2011 : 2 M]**

2.27 We are given a set of n distinct elements and an unlabeled binary tree with n nodes. In how many ways can we populate the tree with the given set so that it becomes a binary search tree?
(a) 0 (b) 1
(c) n! (d) $\frac{^{2n}C_n}{n+1}$

[2011 : 2 M]

2.28 The worst case running time to search for an element in a balanced binary search tree with $n2^n$ elements is
(a) $\Theta(n \log n)$ (b) $\Theta(n2^n)$
(c) $\Theta(n)$ (d) $\Theta(\log n)$

[2012 : 1 M]

2.29 A list of n strings, each of length n, is sorted into lexicographic order using the merge sort algorithm. The worst case running time of this computation is
(a) $O(n \log n)$ (b) $O(n^2 \log n)$
(c) $O(n^2 + \log n)$ (d) $O(n^2)$

[2012 : 2 M]

2.30 Which one of the following is the tightest upper bound that represents the time complexity of inserting an object into a binary search tree of n nodes?

(a) O(1)
(b) O(log n)
(c) O(n)
(d) O(n log n)

[2013 : 1 M]

2.31 Let P be a quicksort program to sort numbers in ascending order using the first element as the pivot. Let t_1 and t_2 be the number of comparisons made by P for the inputs [1 2 3 4 5] and [4 1 5 3 2] respectively. Which one of the following holds?

(a) $t_1 = 5$
(b) $t_1 < t_2$
(c) $t_1 > t_2$
(d) $t_1 = t_2$

[2014 (Set-1) : 1 M]

2.32 The minimum number of comparisons required to find the minimum and the maximum of 100 numbers is ______.

[2014 (Set-1) : 2 M]

2.33 You have an array of n elements. Suppose you implement quicksort by always choosing the central element of the array as the pivot. Then the tightest upper bound for the worst case performance is

(a) O(n^2)
(b) O(nlogn)
(c) Θ(nlogn)
(d) O(n^3)

[2014 (Set-3) : 1 M]

2.34 Consider the C function given below. Assume that the array list A contains n (> 0) elements, sorted in ascending order.

```
int ProcessArray (int *listA, int x, int n)
{
    int i, j, k;
    i = 0;
    j = n – 1;
    do
    {
        k = (i +j )/2;
        if (x <= listA[k])
            j = k –1;
        if (listA[k] <= x)
            i = k+1;
    } while (i <= j);
    if (listA[k] == x)
        return(k);
    else
        return –1;
}
```

Which one of the following statements about the function **ProcessArray** is **CORRECT**?

(a) It will run into an infinite loop when x is not in **listA**.
(b) It is an implementation of binary search
(c) It will always find the maximum element in **listA**
(d) It will return –1 even when x is present in **listA**.

[2014 (Set-3) : 2 M]

2.35 Which one of the following is the recurrence equation for the worst case time complexity of the Quicksort algorithm for sorting $n(\geq 2)$ numbers? In the recurrence equations given in the options below, c is a constant.

(a) $T(n) = 2T\left(\frac{n}{2}\right) + cn$
(b) $T(n) = T(n-1) + T(1) + cn$
(c) $T(n) = 2T(n-1) + cn$
(d) $T(n) = T\left(\frac{n}{2}\right) + cn$

[2015 (Set-1) :1 M]

2.36 What are the worst-case complexities of insertion and deletion of a key in a binary search tree?

(a) Θ(log$n$) for both insertion and deletion
(b) Θ(n) for both insertion and deletion
(c) Θ(n) for insertion and Θ(log$n$) for deletion
(d) Θ(log$n$) for insertion and Θ(n) for deletion

[2015 (Set-1) : 1 M]

2.37 Suppose you are provided with the following function declaration in the C programming language.

```
int partition (int a[ ], int n);
```

The function treats the first element of a[ ] as a pivot, and rearranges the array so that all elements less than or equal to the pivot is in the left part of the array, and all elements greater than the pivot is in the right part. In addition, it moves the pivot so that the pivot is the last element of the left part. The return value is the number of elements in the left part.

The following partially given function in the C programming language is used to find the kth smallest element in an array a[ ] of size n using the partition function. We assume k ≤ n.

```
int kth_smallest (int a[ ], int n, int k)
{
    int left_end = partition (a, n);
    if (left_end+1==k)
    {
        return a [left_end];
```

```
    }
    if (left_end+1 > k)
    {
        return kth_smallest (______);
    }
    else
    {
        return kth_smallest (______);
    }
}
```

The missing argument lists are respectively

(a) (a, left_end, k) and (a+left_end+1, n–left_end–1, k–left_end–1)
(b) (a, left_end, k) and (a, n–left_end–1, k–left_end–1)
(c) (a+left_end+1, n–left_end–1, k–left_end–1) and(a, left_end, k)
(d) (a, n–left_end–1, k–left_end–1) and (a, left_end, k) **[2015 (Set-2) : 2 M]**

2.38 Assume that a mergesort algorithm in the worst case takes 30 seconds for an input of size 64. Which of the following most closely approximates the maximum input size of a problem that can be solved in 6 minutes?

(a) 256 (b) 512
(c) 1024 (d) 2048

[2015 (Set-3) : 2 M]

2.39 The worst case running times of Insertion sort, Merge sort and Quick sort, respectively, are:

(a) $\Theta(n \log n)$, $\Theta(n \log n)$, and $\Theta(n^2)$
(b) $\Theta(n^2)$, $\Theta(n^2)$, and $\Theta(n \log n)$
(c) $\Theta(n^2)$, $\Theta(n \log n)$, and $\Theta(n \log n)$
(d) $\Theta(n^2)$, $\Theta(n \log n)$, and $\Theta(n^2)$

[2016 (Set-1) : 1 M]

2.40 Assume that the algorithms considered here sort the input sequences in ascending order. If the input is already in ascending order, which of the following are TRUE?

I. Quicksort runs in $\Theta(n^2)$ time
II. Bubblesort runs in $\Theta(n^2)$ time
III. Mergesort runs in $\Theta(n)$ time
IV. Insertion sort runs in $\Theta(n)$ time

(a) I and II only (b) I and III only
(c) II and IV only (d) I and IV only

[2016 (Set-2) : 1 M]

2.41 Consider a sequence of 14 elements: A = [–5, –10, 6, 3, –1, –2, 13, 4, –9, –1, 4, 12, –3, 0]. The subsequence sum $S(i, j) = \sum_{k=1}^{j} A[k]$. Determine the maximum of $S(i, j)$, where $0 \le i \le j < 14$. (Divide and conquer approach may be used.)

[2019 : 1 M]

2.42 An array of 25 distinct elements is to be sorted using quicksort. Assume that the pivot element is chosen uniformly at random. The probability that the pivot element gets placed in the worst possible location in the first round of partitioning (rounded off to 2 decimal places) is ______.

[2019 : 1 M]

2.43 Consider the following statements:

I. The smallest element in a max-heap is always at a leaf node.
II. The second largest element in a max-heap is always a child of the root node.
III. A max-heap can be constructed from a binary search tree in $\Theta(n)$ time.
IV. A binary search tree can be constructed from a max-heap in $\Theta(n)$ time.

Which of the above statements are TRUE?

(a) II, III and IV (b) I, II and III
(c) I, III and IV (d) I, II and IV

[2019 : 2 M]

2.44 There are n unsorted arrays: $A_1, A_2, ..., A_n$. Assume that n is odd. Each of $A_1, A_2, ..., A_n$ contains n distinct elements. There are no common elements between any two arrays. The worst-case time complexity of computing the median of the medians of $A_1, A_2, ..., A_n$ is

(a) $O(n \log n)$ (b) $O(n^2)$
(c) $O(n)$ (d) $\Omega(n^2 \log n)$

[2019 : 2 M]

2.45 In a balanced binary search tree with n elements, what is the worst case time complexity of reporting all elements in range $[a, b]$? Assume that the number of reported elements is k.

(a) $\Theta(\log n)$ (b) $\Theta(n \log k)$
(c) $\Theta(\log n + k)$ (d) $\Theta(k \log n)$

[2020 : 2 M]

2.46 Consider the following array:

23	32	45	69	72	73	89	97

Which algorithm out of the following options uses the least number of comparisons (among the array elements) to sort the above array in ascending order?

(a) Quicksort using the last element as pivot
(b) Selection sort
(c) Mergesort
(d) Insertion sort **[2021 (Set-1) : 1 M]**

2.47 Let P be an array containing n integers. Let t be the lowest upper bound on the number of comparisons of the array elements, required to find the minimum and maximum values in an arbitrary array of n elements. Which one of the following choices is correct?

(a) $t > n$ and $t \le 3\left\lceil \frac{n}{2} \right\rceil$

(b) $t > 3\left\lceil \frac{n}{2} \right\rceil$ and $t \le 2n-2$

(c) $t > 2n-2$

(d) $t > \lceil \log_2(n) \rceil$ and $t \le n$

[2021 (Set-1) : 1 M]

2.48 What is the worst-case number of arithmetic operations performed by recursive binary search on a sorted array of size n?

(a) $\Theta(n)$ (b) $\Theta(n^2)$

(c) $\Theta(\sqrt{n})$ (d) $\Theta(\log_2(n))$

[2021 (Set-2) : 1 M]

2.49 Let G be an undirected connected graph in which every edge has a positive integer weight. Suppose that every spanning tree in G has even weight. Which of the following statements is/are TRUE for every such graph G?

(a) All edges in G have even weight.

(b) In each cycle C in G, all edges in C have even weight.

(c) All edges in G have even weight OR all edges in G have odd weight.

(d) In each cycle C in G, either all edges in C have even weight OR all edges in C have odd weight

[2024 (Set-2) : 2 M]

2.50 The number of distinct minimum-weight spanning trees of the following graph is ________.

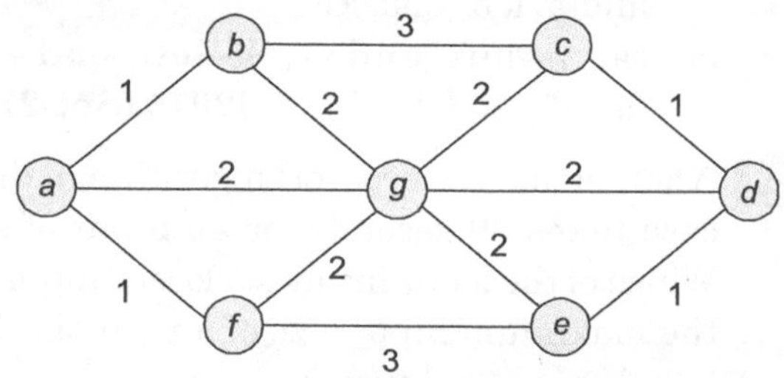

[2024 (Set-2) : 2 M]

■■■■

Answers Divide and Conquer

2.1	(a)	**2.2**	(c)	**2.3**	(d)	**2.4**	(a)	**2.5**	(c)	**2.6**	(b)	**2.7**	(b)	**2.8**	(b)	**2.9**	(a)
2.10	(c)	**2.11**	(a)	**2.12**	(b)	**2.13**	(b)	**2.14**	(a)	**2.15**	(b)	**2.16**	(b)	**2.17**	(a)	**2.18**	(a)
2.19	(b)	**2.20**	(c)	**2.21**	(a)	**2.22**	(d)	**2.23**	(c)	**2.24**	(b)	**2.25**	(b)	**2.26**	(c)	**2.27**	(b)
2.28	(c)	**2.29**	(b)	**2.30**	(c)	**2.31**	(c)	**2.32**	(162)	**2.33**	(a)	**2.34**	(b)	**2.35**	(b)	**2.36**	(b)
2.37	(a)	**2.38**	(b)	**2.39**	(d)	**2.40**	(d)	**2.41**	(29)	**2.42**	(0.08)	**2.43**	(b)	**2.44**	(b)	**2.45**	(c)
2.46	(d)	**2.47**	(a)	**2.48**	(d)	**2.49**	(d)	**2.50**	(9)								

Explanations Divide and Conquer

2.1 (a)

A merge sort is comparison based sorting algorithm and divide-and-conquer algorithm.

2.2 (c)

The number of comparisons required in the worst case is O(m + n).

2.3 (d)

I. In hashing, as the number of entries in a hash table increases, the number of collisions increases.

II. It is not always true that recursive programs are efficient.

III. Quicksort is a comparison sort which is worst case makes $O(n^2)$ comparisons.

IV. Binary search can be used with linked list, however, it is inefficient.

2.4 (a)

Using Master Theorem

We get $n^{\log_4 3} < n$

2.5 (c)

Both of given cases are

Quicksort Worst cases problem, so comparisons are equal.

2.6 (b)

Use Master Theorem $n^{\log_2 2} \geq \sqrt{n}$.

2.7 (b)

A sorting algorithm is called stable if it keeps elements with equal keys in the same relative order in the output as they were in the input. For example in the following input the two 4's are indistinguishable. 1, 4_a, 3, 4_b, 2 and so the output of a stable sorting algorithm must be:

1, 2, 3, 4_a, 4_b

Bubble sort, merge sort, counting sort, insertion sort are stable sorting algorithms.

2.8 (b)

Given: 20, 47, 15, 8, 9, 4, 40, 30, 12, 17

2-way merge sort so group of 2 is taken at once.

2nd pass:

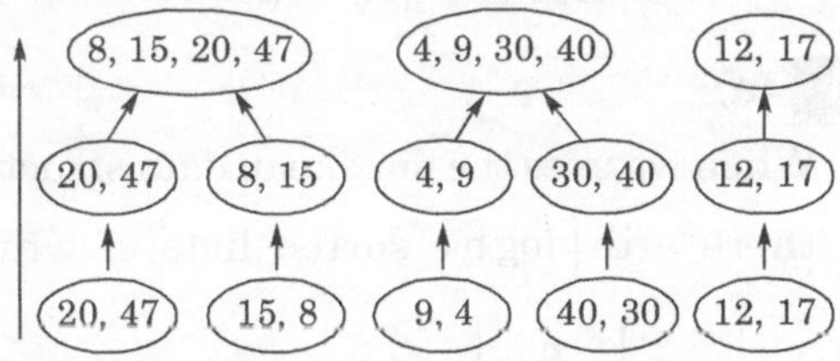

The order of elements after second pass of the algorithm is 8, 15, 20, 47, 4, 9, 30, 40, 12, 17.

2.9 (a)

Take starting 2 elements of array, if sum of these two are less than 1000 then return those elements, otherwise return false.

2.10 (c)

Randomized quick sort worst case time complexity = $O(n^2)$.

2.11 (a)

The usual $O(n^2)$ implementation of insertion sort to sort an array uses binary search to identify the position, the worst case running time will remain $O(n^2)$. The time required to insert an element x into a binary search tree is bounded by a constant times the number of comparisons made between x and elements already in the tree. Thus we can measure time in terms of the number of comparisons made. In worst case implementation of insertion sort takes $O(n^2)$. So in the worst case adding n elements to a binary search tree could require $O(n^2)$ time. In expected case means element are sorted in increasing order adding n elements to a binary search tree could require O(n log n) time.

2.12 (b)

$$\text{Left} + \text{Right} + 1 = n$$

So, $$k - 1 + x + 1 = n$$

$$x = n - k$$

2.13 (b)

Binary search : $T(n) = T\left(\frac{n}{2}\right) + 1$

Merge sort : $T(n) = 2T\left(\frac{n}{2}\right) + cn$

Quick sort : $T(n) = T(n - k) + T(k) + cn$

Tower of Hanoi : $T(n) = 2T(n - 1) + 1$

2.14 (a)

When we uses the max heap data structure then there are $\lceil \log n \rceil$ sorted lists of which each contains $\left\lfloor \frac{n}{\log n} \right\rfloor$ elements. Construction of heap takes O(log logn) time and if we have $\left\lfloor \frac{n}{\log n} \right\rfloor$ elements then worst case time complexity is O(n log log n).

2.15 (b)

Total number of distinct binary search tree with n nodes (key)

$$B_n = \frac{1}{n+1}\binom{2n}{n}$$

$$B_4 = \frac{1}{4+1}\binom{8}{4} = \frac{1}{5}\frac{\underline{|8}}{\underline{|4}\,\underline{|4}}$$

$$B_4 = \frac{1}{5}\frac{8\times7\times6\times5\times\underline{|4}}{4\times3\times2\times1\times\underline{|4}} = 14$$

2.16 (b)

If the median of n elements can be found in O(n) time and we select the median as pivot. Then quicksort sort n elements of an array A[1],..., A[n] as follows. We permute the elements in the array so that for some j all the records with key less than v appear in A[1],..., A[j] and all those with keys v or greater appear in A[j + 1],..., A[n]. We then apply quicksort recursively to A[1],..., A[j] and to A[j + 1],..., A[n].

The recursion equation becomes for some value of i

$$T(n) \le \frac{1}{n}\sum_{i=1}^{n}\left(T(i-1) + T(n-i)\right)$$

The solution of the $T(n) \le \Theta(n \log n)$.

Given to find median O(n) time required.

2.17 (a)

Sorting Algorithm	Worst care complexity (with n inputs)
Merge sort	O(n log n)
Bubble sort	$O(n^2)$
Quick sort	$O(n^2)$
Selection sort	$O(n^2)$

2.18 (a)

Let T(m, n) be the total number of steps.

So T(m, 0) = 0, T(m, n) = T(n, m mod n) on average

$$T_n = \frac{1}{n}\sum_{0\le k\le n} T(k, n)$$

$$T_n \approx 1 + \frac{1}{n}(T_0\ T_1 + \ldots + T_{n-1})$$

$$T_n \approx S_n$$

$$S_n = 1 + \frac{1}{n}(S_0\ S_1 + \ldots + S_{n-1})$$

$$S_n = 1 + \frac{1}{n+1}(S_0\ S_1 + \ldots + S_n)$$

$$= 1 + \frac{1}{n+1}(n(S_{n-1}) + S_n)$$

$$= 1 + \frac{1}{n+1} = S_n + \frac{1}{n+1}$$

So $$T_n \approx \Theta(\log_2 n) + 0(1)$$

$$T \approx \Theta(\log_2 n)$$

2.19 (b)

If we wants to sort n elements with the help of Qucksort algorithm. If pivot elements which split the list into two sub-lists each in which one list contains one-fifth element or $\frac{n}{5}$ and other list contains $\frac{4n}{5}$ and balancing takes n. So

$$T(n) \le T\left(\frac{n}{5}\right) + T\left(\frac{4n}{5}\right) + n$$

[Note : $n - \frac{n}{5} = \frac{5n - n}{5} = \frac{4n}{5}$]

2.20 (c)

When Y is [2 2 2 2 2 2 2 2 2 2] and x > 2, given binary search can't search and the program fails.

2.21 (a)

If we will change line 6 to if (Y[k] < x) i = k + 1, else j = k – 1 then the program works properly for all the choices.

2.22 (d)

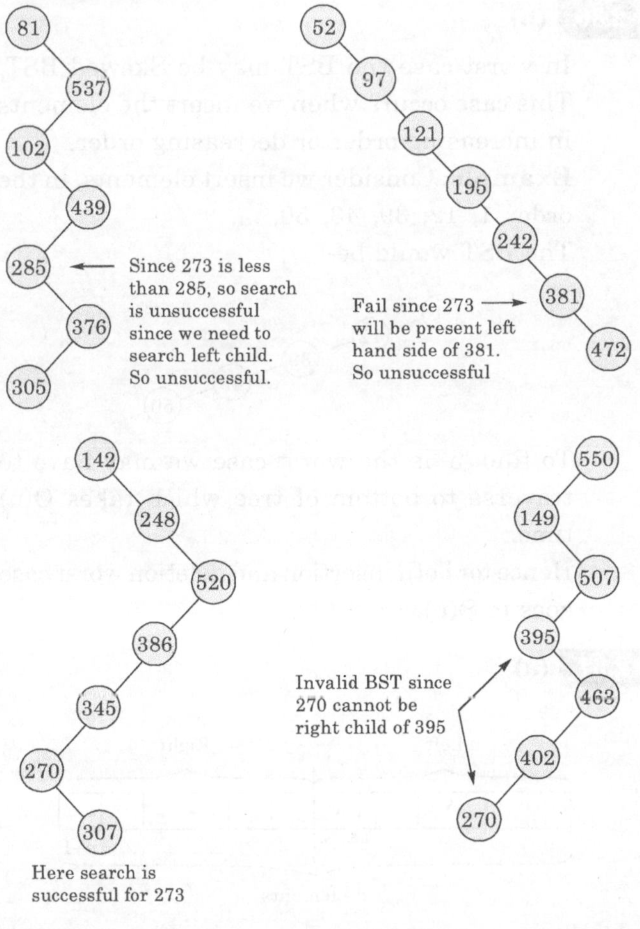

2.23 (c)

- In order sequence of BST is increasing order of given elements or ascending order. So I and IV cannot be in order of BST. So false
- Since in preorder root visited first which is 81 but option (b) says 439. So false
- Since in postorder root visited at the end which is 270 but option (d) says 149. So false.

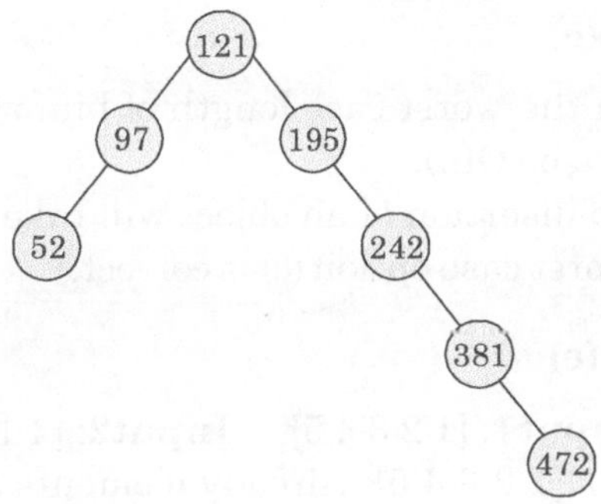

Is BST with root 121 and leaf 52. So option (c) is true.

2.24 (b)

For distinct BST we apply this formula C(2n, n) / n + 1

n = 3 here so C(6, 3) = 20 and 20/4 = 5

2.25 (b)

The relation T(n) = T(n/4) + T(3n/4) + n

The pivot element is selected in such a way that it will divide the array into 1/4th and 3/4th always solving this relation give $\Theta(n \log n)$.

2.26 (c)

We get minimum number of multiplications using ((M1 × (M2 × M3)) × M4).

Total number of multiplication

= 100 * 20 * 5 + 10 * 100 * 5 + 10 * 5 * 80

= 19000

2.27 (b)

n = {1, 2, 3}

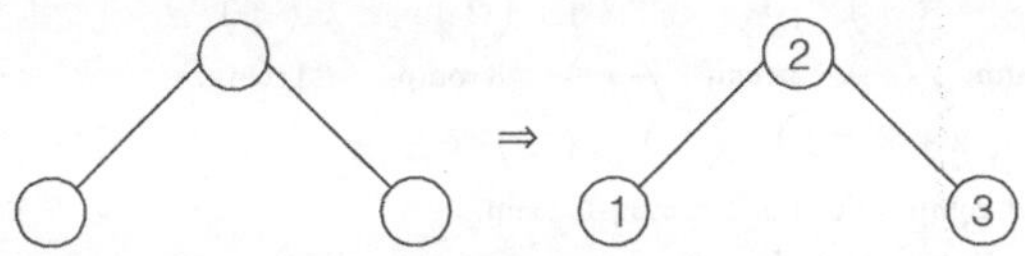

2.28 (c)

We know that in balanced BST there are $\log_2 n$ levels in both worst as well as best case where 'n' is no. of elements

$\therefore \log_2 (n\, 2^n) = \log_2 n + \log_2 2^n$

$\log_2 n + n \log_2 2 = \log_2 n + n = \Theta(n)$

2.29 (b)

The recurrence tree for merge sort will have height n and O(n^2) work will be done at each level of the recurrence tree (Each level involves n comparisons and a comparison takes O(n) time in worst case). So time complexity of this merge sort will be O(n^2 log n).

2.30 (c)

In the worst case length of binary search tree can be O(n).

So insertion of an object will take O(n) time in worst case option (c) is correct.

2.31 (c)

Input1: [1 2 3 4 5], **Input2:** [4 1 5 3 2]

(a) [1 2 3 4 5]: Already elements are in sorted order $\frac{5(5-1)}{2}$ comparisons are required.

$\Rightarrow \quad t_1 = 10$

(b) [4 1 5 3 2]: Elements are in random order. Choosing first element as pivot will give less comparisons compared to t_1.

$\therefore \quad t_2 < t_1$.

2.32 (162)

Given 100 element $\neq 2^k$

$\therefore$ No direct formula.

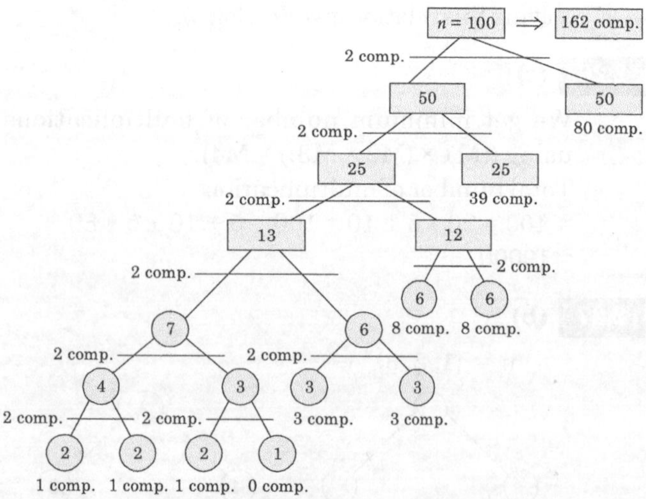

Number of element	Comparison
2	1
3	3
4	4
6	8
7	9
⋮	⋮

2.33 (a)

Choosing the central element (middle) as the pivot always can not make the partition into equal parts.

$\therefore$ The worst case time complexity = O(n^2)

2.34 (b)

Given program is implementation of binary search. x is used as key, K is in middle position of an array/subarray, i and j positions may change based on availability of key with comparison to listA[K] element.

2.35 (b)

In worst case the pivot element position may be either first or last. In that case the elements are always divided in 1 : (n – 1) proportion. The recurrence relation for such a proportional division would be

T(n) = T(1) + T(n–1) + O(n) = O(n^2)

(a) $T(n) = 2T(n/2) + cn$ = O(n log n)

(b) $T(n) = T(n-1) + T(1) + cn$ = O(n^2)

(c) $T(n) = 2T(n-1) + cn$ = O(2^n)

(d) $T(n) = T(n/2) + cn$ = O(n)

2.36 (b)

In worst case the BST may be Skewed BST. This case occurs when we insert the elements in increasing order or decreasing order.

Example: Consider we insert elements, in the order 1, 12, 39, 43, 50, ...

The BST would be

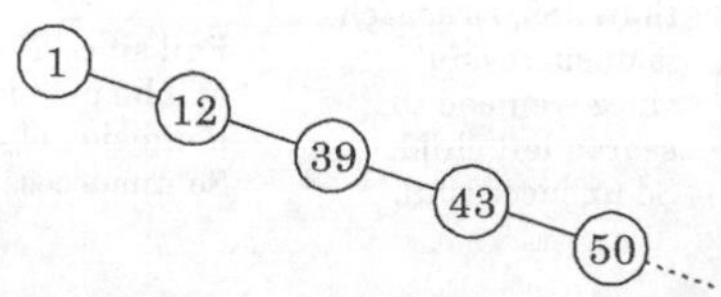

To find 'n' is the worst case we may have to traverse to bottom of tree which takes O(n) time.

Hence for both insertion and deletion worst case goes to Θ(n).

2.37 (a)

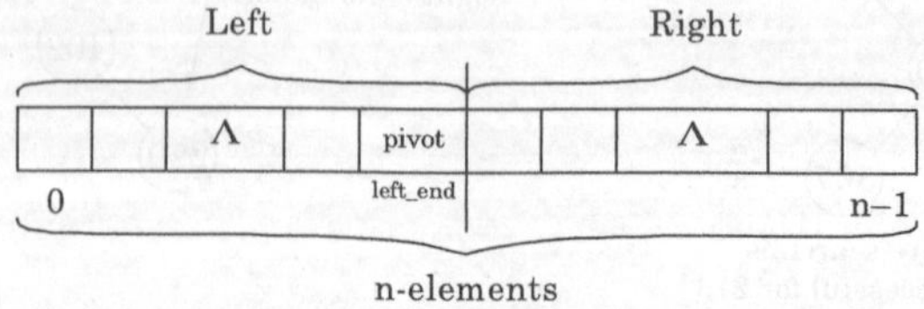

left_end (is pivot position) = partition (a, n);

Note: k^{th} smallest is same as an element at index (k–1) in the sorted array.

if (left_end+1 = = k) // k^{th} smallest found.

return a[left_end]; // return pivot element.

if (left_end+1 > k) // k^{th} smallest is before pivot.

return k^{th}_smallest (a, left_end, k); // search before pivot

else // k^{th} smallest is after pivot (right part)

return kth_smalest $\underbrace{\text{(a+left\_end+1,}}_{\substack{\text{address of first}\\ \text{element after}\\ \text{pivot (right part)}}}$ $\underbrace{\text{n–(left\_end–1), k–(left\_end+1));}}_{\substack{\text{Number of elements}\\ \text{in right part}}}$

2.38 (b)

For $n = 64$

$c_1.n\log n = 30$ sec

$c_1.64.\ \log 64 = 30$ sec

$$c_1 = \frac{5}{64} \quad ...(1)$$

Now, $c_1.n\log n = 6 \times 60$ sec

$$n \log n = \frac{6 \times 60 \text{ sec}}{c_1 \text{ sec}}$$

$$n \log n = \frac{6 \times 60}{5} \times 64$$

$n \log n = 4608$

$n \log n = 512 \times 9 \Rightarrow \quad n = 512$

2.39 (d)

The worst case time complexity of algo given are:

Insertion sort = $\Theta(n^2)$

Merge sort = $\Theta(n \log n)$

Quick sort = $\Theta(n^2)$

2.40 (d)

Quick sort takes $O(n^2)$ time to sort the input (ascending order) into ascending order. by using divide and conquer.

Insertion sort take $O(n)$ time to sort input (ascending order) into ascending order since need one comparision every time and no swaping.

Merge sort take $O(n \log n)$ and bubble sort can take $O(n)$ time.

So, option (d) is most appropriate.

2.41 (29)

Subsequence is a sequence that can be derived from another sequence without changing the order of sequence and by deleting no or some element.

A = [–5, –10, 6, 3, –1, –2, 13, 4, –9, –1, 4, 12, –3, 0]

Subsequence sum $S(i, j) = \sum_{k=i}^{j} A[k]$.

The maximum subsequence can be possible with A is (6, 3, –1, –2, 13, 4, –9, –1, 4, 12)

The sum of subsequence

$$= 6 + 3 - 1 - 2 + 13 + 4 - 9 - 1 + 4 + 12 = 29$$

2.42 Sol.

Worst cases in quick sort when array in divided into $1 : n - 1$ ratio.

Total 25 element in the array and worst case when pivot element is maximum or minimum.

So total 2 cases for worst case

$$\text{Probability} = \frac{\text{Favourable cases}}{\text{Total cases}}$$

$$= \frac{2}{25} = 0.08$$

2.43 (b)

- In max-heap largest element is at the root node (level-0) second largest element at (level-1) which is child of root, third largest can be at level-1 or level-2 and so on.
  So statement I and II are correct.
- A max-heap can be constructed from a binary search tree in $\Theta(n)$ time using build-heap method.
  So statement III is correct.

2.44 (b)

n unsorted arrays is given where n is odd.

Each array contain n elements so finding median will take $O(n)$.

Total n array are given so time take

$$n \times O(n) = O(n^2)$$

Medians of array $A_1, A_2 ....., A_n$ has n element M_i

where $M_i = \sum_{i=1}^{n} M_i$, to find median of M_i time taken is $O(n)$.

Worst case time complexity of computing the median of the medians

$$= O(n^2) + O(n) = O(n^2)$$

2.45 (c)

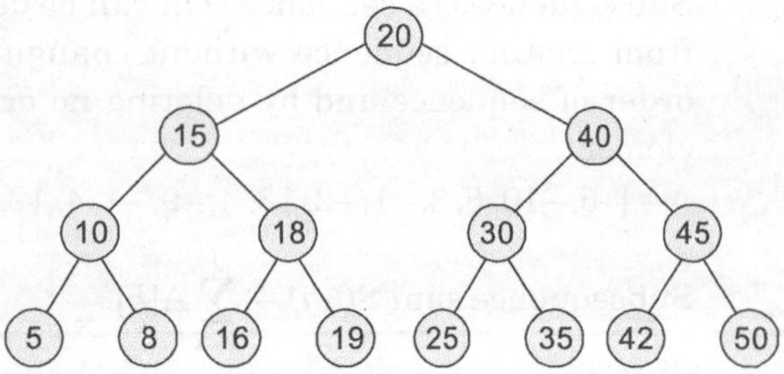

Let $a = 16$, $b = 42$

16: Find the '16' element in the BST = O(log$n$)

42: Find the '42' element in the BST = O(log$n$)

[16, 42]: Inorder sorted element between 16 to 42

{16, 18, 19, 20, 25, 30, 35, 40, 42} ⇒ requires O(k) time for k element

So total time: O(2 log$n$ + k) ≡ O(log$n$ + k)

2.46 (d)

Given array already sorted.

∴ Insertion sort takes least number of comparisons θ(n).

Since, for a number which is to be inserted in the already sorted array, only 1 comparison will be required.

2.47 (a)

Using divide and conquer min-max algo, WC number of comparisons = $\frac{3n}{2} - 2$.

Hence $\frac{3n}{2} - 2$ will lie in the range of n to $3\left\lceil \frac{n}{2} \right\rceil$.

Hence option (a) is the correct answer.

2.48 (d)

Worst case is when the element not present in the sorted array.

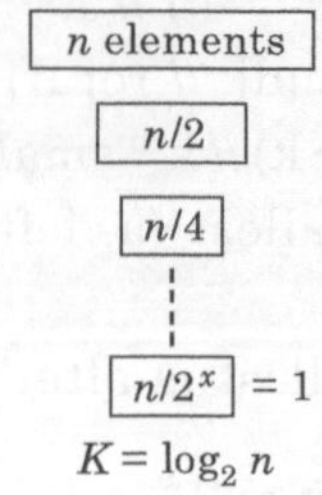

Worst case occurrence relation is

$$T(n) = \begin{cases} T\left(\frac{n}{2}\right) + 1, & n > 1 \\ 1, & n \le 1 \end{cases}$$

∴ $\Theta(\log_2(n))$

2.49 (d)

For each cycle in G.

All edges of cycle even weight (OR).

All edges of cycle odd weight then every spanning tree cost of G is even.

2.50 (9)

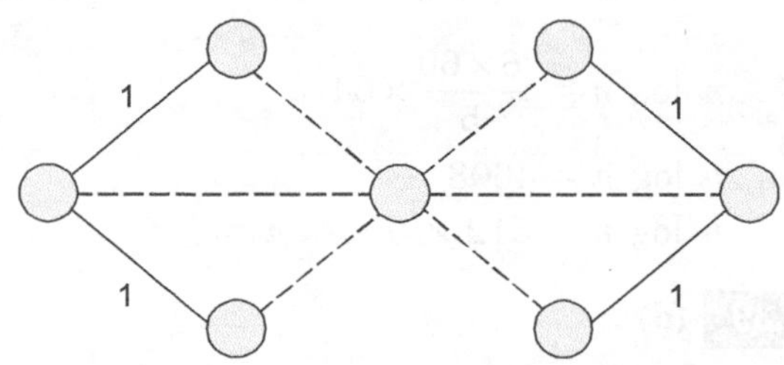

$^6C_2 - 3 - 3 = 9$ MST's

■■■■

3 Greedy Method

3.1 Complexity of Kruskal's algorithm for finding the minimum spanning tree of an undirected graph containing n vertices and m edges if the edges are sorted is ______. **[2002 : 2 M]**

3.2 Let G be an undirected connected graph with distinct edge weights. Let e_{max} be the edge with maximum weight and e_{min} the edge with minimum weight. Which of the following statements is false?

(a) Every minimum spanning tree of G must contain e_{min}

(b) If e_{max} is in a minimum spanning tree, then its removal must disconnect G

(c) No minimum spanning tree contains e_{max}

(d) G has a unique minimum spanning tree

[2002 : 2 M]

3.3 In a heap with n elements with the smallest element at the root, the 7th smallest element can be found in time

(a) $\Theta(n \log n)$ (b) $\Theta(n)$

(c) $\Theta(\log n)$ (d) $\Theta(1)$ **[2003 : 1 M]**

3.4 What is the weight of a minimum spanning tree of the following graph?

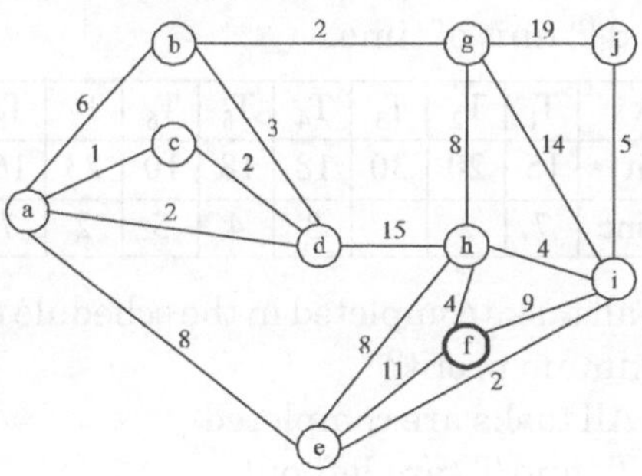

(a) 29 (b) 31

(c) 38 (d) 41 **[2003 : 2 M]**

3.5 The following are the starting and ending times of activities A, B, C, D, E, F, G, and H respectively in chronological order: "a_s b_s c_s a_e d_s c_e e_s f_s b_e d_e g_s e_e f_e h_s g_e h_e" Here, x_s denotes the starting time and x_e denotes the ending time of activity X. W need to schedule the activities in a set of rooms available to us. An activity can be scheduled in a room only if the room is reserved for the activity for its entire duration. What is the minimum number of rooms required?

(a) 3 (b) 4

(c) 5 (d) 6 **[2003 : 2 M]**

3.6 Suppose we run Dijkstra's single source shortest-path algorithm on the following edge-weighted directed graph with vertex P as the source.

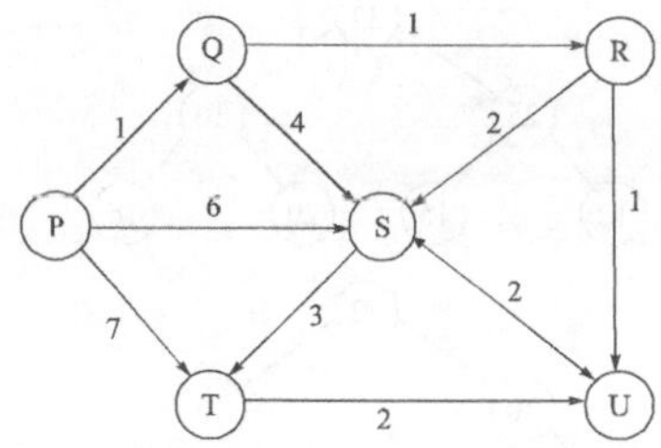

In what order do the nodes get included into the set of vertices for which the shortest path distances are finalized?

(a) P, Q, R, S, T, U (b) P, Q, R, U, S, T

(c) P, Q, R, U, T, S (d) P, Q, T, R, U, S

[2004 : 2 M]

3.7 Consider the undirected graph below:

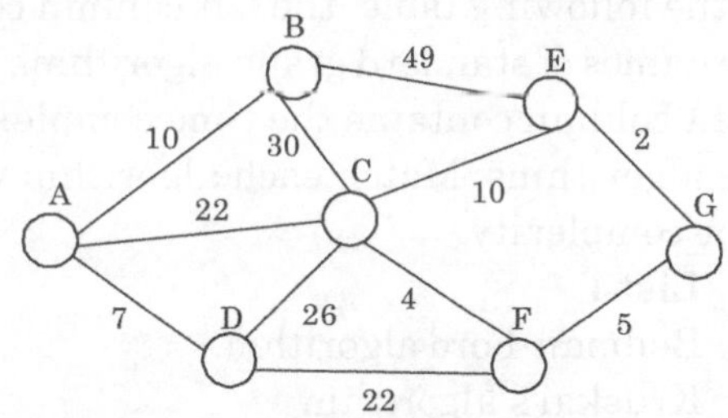

Using Prim's algorithm to construct a minimum spanning tree starting with node A, which one of the following sequences of edges represents a possible order in which the edges would be added to construct the minimum spanning tree?

(a) (E,G), (C,F), (F,G), (A,D), (A,B), (A,C)

(b) (A,D), (A,B), (A,C), (C,F), (G,E), (F,G)

(c) (A,B), (A,D), (D,F), (F,G), (G,E), (F,C)

(d) (A,D), (A,B), (D,F), (F,C), (F,G), (G,E)

[2004 : 2 M]

3.8 An array of integers of size n can be converted into a heap by adjusting the heaps rooted at each internal node of the complete binary tree starting at the node $\lfloor(n-1)/2\rfloor$, and doing this adjustment up to the root node (root node is at index 0) in the order $\lfloor(n-1)/2\rfloor$, $\lfloor(n-3)/2\rfloor$,..., 0. The time required to construct a heap in this manner is

(a) O(log n) (b) O(n)
(c) O(n log log n) (d) O(n log n)
[2004 : 2 M]

3.9 The elements 32, 15, 20, 30, 12, 25, 16 are inserted one by one in the given order into a maxHeap. The resultant maxHeap is

(a)
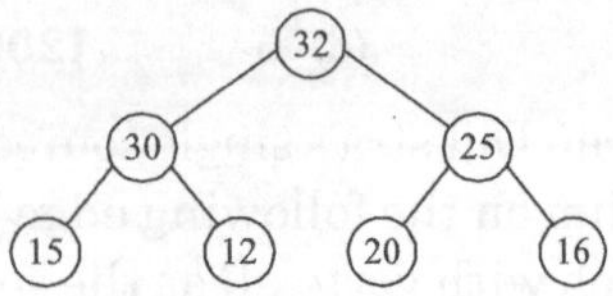

(b)
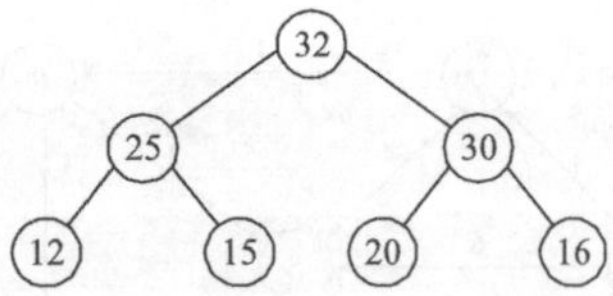

(c)
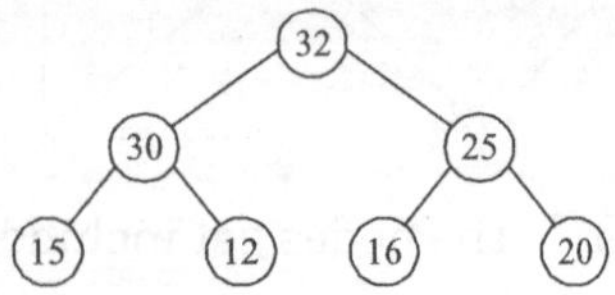

(d)
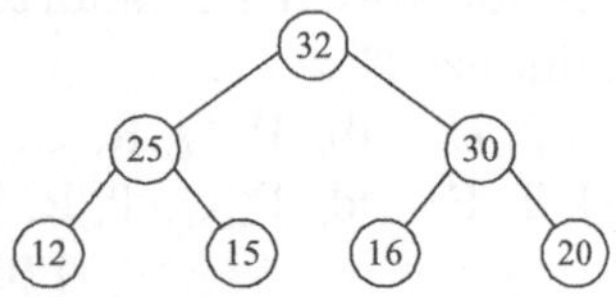

[2004 : 2 M]

3.10 In the following table, the left column contains the names of standard graph algorithms and the right column contains the time complexities of the algorithms. Match each algorithm with its time complexity.

List-I
1. Bellman-Ford algorithm
2. Kruskal's algorithm
3. Floyd-Warshall algorithm
4. Topological sorting

List-II
A. O(mlog n)
B. $O(n^3)$
C. O(nm)
D. O(n + m)

(a) 1-C, 2-A, 3-B, 4-D
(b) 1-B, 2-D, 3-C, 4-A
(c) 1-C, 2-D, 3-A, 4-B
(d) 1-B, 2-A, 3-C, 4-D **[2005 : 1 M]**

3.11 A undirected graph G has n nodes. Its adjacency matrix is given by an n × n square matrix whose
1. diagonal elements are 0's, and
2. non-diagonal elements are 1's.

Which one of the following is TRUE?
(a) Graph G has no minimum spanning tree (MST)
(d) Graph G has a unique MST of cost n-1
(c) Graph G has multiple distinct MST's, each of cost n-1
(d) Graph G has multiple spanning trees of different costs **[2005 : 1 M]**

3.12 Let G be a weighted undirected graph and e be an edge with maximum weight in G. Suppose there is a minimum weight spanning tree in G containing the edge e. Which of the following statements is always TRUE?
(a) There exists a cutset in G having all edges of maximum weight.
(b) There exists a cycle in G having all edges of maximum weight.
(c) Edge e cannot be contained in a cycle.
(d) All edges in G have the same weight.
[2005 : 2 M]

3.13 Let G(V, E) an undirected graph with positive edge weights. Dijkstra's single source-shortest path algorithm can be implemented using the binary heap data structure with time complexity?
(a) $O(|V|^2)$ (b) $O(|E|+|V|\log|V|)$
(c) $O(|V|\log|V|)$ (d) $O((|E|+|V|)\log|V|)$
[2005 : 2 M]

Common Data for Q. 3.14 & Q. 3.15

We are given 9 tasks T_1, T_2 ... T_9. The execution of each task requires one unit of time. We can execute one task at a time. T_i has a profit P_i and a deadline d_i profit P_i is earned if the task is completed before the end of the d_i^{th} unit of time.

Task	T_1	T_2	T_3	T_4	T_5	T_6	T_7	T_8	T_9
Profit	15	20	30	18	18	10	23	16	25
Deadline	7	2	5	3	4	5	2	7	3

3.14 Are all tasks completed in the schedule that gives maximum profit?
(a) All tasks are completed
(b) T_1 and T_6 are left out
(c) T_1 and T_8 are left out
(d) T_4 and T_6 are left out **[2005 : 2 M]**

3.15 What is the maximum profit earned?
(a) 147 (b) 165
(c) 167 (d) 175 **[2005 : 2 M]**

3.16 A priority-Queue is implemented as a Max-Heap. Initially, it has 5 elements. The level-order traversal of the heap is given below:

10, 8, 5, 3, 2

Two new elements '1' and '7' are inserted in the heap in that order. The level-order traversal of the heap after the insertion of the elements is

(a) 10, 8, 7, 5, 3, 2, 1
(b) 10, 8, 7, 2, 3, 1, 5
(c) 10, 8, 7, 1, 2, 3, 5
(d) 10, 8, 7, 3, 2, 1, 5

[2005 : 2 M]

3.17 Consider the polynominal $p(x) = a_0 + a_1x + a_2x^2 + a_3x^3$, where $a_i \neq 0, \forall i$. The minimum number of multiplcations needed to evaluate p on an input x is

(a) 3 (b) 4
(c) 6 (d) 9 [2006 : 1 M]

3.18 Consider a weighted complete graph G on the vertex set $\{v_1, v_2, ..., v_n\}$ such that the weight of the edge (v_i, v_j) is $2|i - j|$. The weight of a minimum spanning tree of G is

(a) n − 1 (b) 2n − 2
(c) $\binom{n}{2}$ (d) n^2 [2006 : 1 M]

3.19 To implement Dijkstra's shortest path algorithm on unweighted graphs so that it runs in linear time, then data structure to be used is

(a) Queue (b) Stack
(c) Heap (d) B-Tree

[2006 : 1 M]

3.20 In a binary max heap containing n numbers, the smallest element can be found in time

(a) θ(n) (b) θ(log n)
(c) θ(log log n) (d) θ(1)

[2006 : 1 M]

3.21 Consider the following graph:

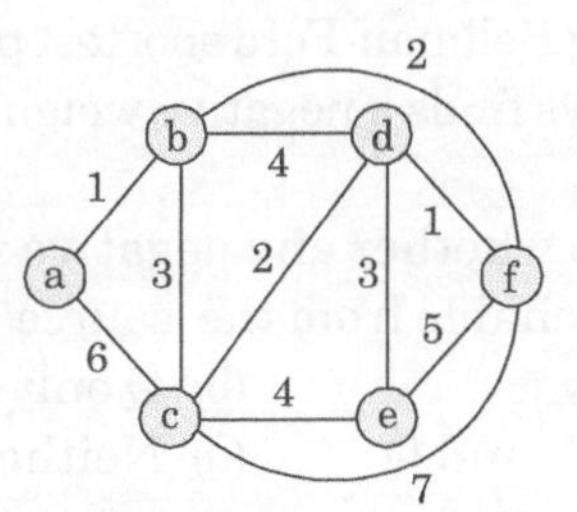

Which one of the following cannot be the sequence of edges added, in that order, to a minimum spanning tree using Kruskal's algorithm?

(a) (a – b), (d – f), (b – f), (d – c), (d – e)
(b) (a – b), (d – f), (d – c), (b –f), (d – e)
(c) (d – f), (a – b), (d – c), (b – f), (d – e)
(d) (d – f), (a – b), (b – f), (d – e), (d – c)

[2006 : 2 M]

3.22 The characters a to h have the set of frequencies based on the first 8 Fibonacci numbers as follows:
a : 1, b : 1, c : 2, d : 3, e : 5, f : 8, g : 13, h : 21

A Huffman code is used to represent the characters. What is the sequence of characters corresponding to the following code?

110111100111010

(a) fdheg (b) ecgdf
(c) dchfg (c) fehdg

[2006 : 2 M]

3.23 Which of the following sequences of array elements forms a heap?

(a) {23, 17, 14, 6, 13, 10, 1, 12, 7, 5}
(b) {23, 17, 14, 6, 13, 10, 1, 5, 7, 12}
(c) {23, 17, 14, 7, 13, 10, 1, 5, 6, 12}
(d) {23, 17, 14, 7, 13, 10, 1, 12, 5, 7}

[2006 : 2 M]

Common Data for Q. 3.24 & Q. 3.25:

A 3-ary max heap is like a binary max heap, but instead of 2 children, nodes have 3 children. A 3-ary heap can be represented by an array as follows: The root is stored in the first location, a [0], nodes in the next level, from left to right, is stored from a [1] to a [3]. The nodes from the second level of the tree from left to right are stored from a [4] location onward. An item x can be inserted into a 3-ary heap containing n items by placing x in the location a [n] and pushing it up the tree to satisfy the heap property.

3.24 Which one of the following is a valid sequence of elements in an array representing 3-ary max heap?

(a) 1, 3, 5, 6, 8, 9 (b) 9, 6, 3, 1, 8, 5
(c) 9, 3, 6, 8, 5, 1 (d) 9, 5, 6, 8, 3, 1

[2006 : 2 M]

3.25 Suppose the elements 7, 2, 10, and 4 are inserted, in that order, into the valid 3-ary max heap found in the above question. Which one of the following is the sequence of items in the array representing the resultant heap?

(a) 10, 7, 9, 8, 3, 1, 5, 2, 6, 4
(b) 10, 9, 8, 7, 6, 5, 4, 3, 2, 1
(c) 10, 9, 4, 5, 7, 6, 8, 2, 1, 3
(d) 10, 8, 6, 9, 7, 2, 3, 4, 1, 5

[2006 : 2 M]

3.26 In an unweighted, undirected connected graph, the shortest path from a node S to every other node is computed most efficiently, in terms of time complexity, by

(a) Dijkstra's algorithm starting from S.
(b) Warshall's algorithm
(c) performing a DFS starting from S
(d) performing a BFS starting from S

[2007 : 2 M]

3.27 Let w be the minimum weight among all edge weights in an undirected connected graph. Let e be a specific edge of weight w. Which of the following is FALSE?

(a) There is a minimum spanning tree containing e.
(b) If e is not in a minimum spanning tree T, then in the cycle formed by adding e to T, all edges have the same weight.
(c) Every minimum spanning tree has an edge of weight w
(d) e is present in every minimum spanning tree

[2007 : 2 M]

Common Data for Q. 3.28 & Q. 3.29

Suppose the letters a, b, c, d, e have probabilities $\frac{1}{2}, \frac{1}{4}, \frac{1}{8}, \frac{1}{16}, \frac{1}{32}$ respectively.

3.28 Which of the following is the Huffman code for the letters a, b, c, d, e?

(a) 0, 10, 110, 1110, 1111
(b) 11, 10, 011, 010, 001
(c) 11, 10, 01, 001, 0001
(d) 110, 100, 010, 000, 001 **[2007 : 2 M]**

3.29 What is the average length of the correct answer to above question?

(a) 3 (b) 2.1875
(c) 2.25 (d) 1.781 **[2007 : 2 M]**

3.30 Consider the process of inserting an element into a Max Heap, where the Max Heap is represented by an array. Suppose we perform a binary search on the path from the new leaf to the root to find the position for the newly inserted element, the number of comparisons performed is

(a) $\theta(\log_2 n)$ (b) $\theta(\log_2 \log_2 n)$
(c) $\theta(n)$ (d) $\theta(n \log_2 n)$

[2007 : 2 M]

3.31 For the undirected, weighted graph given below, which of the following sequences of edges represents a correct execution of Prim's algorithm to construct a Minimum Spanning Tree?

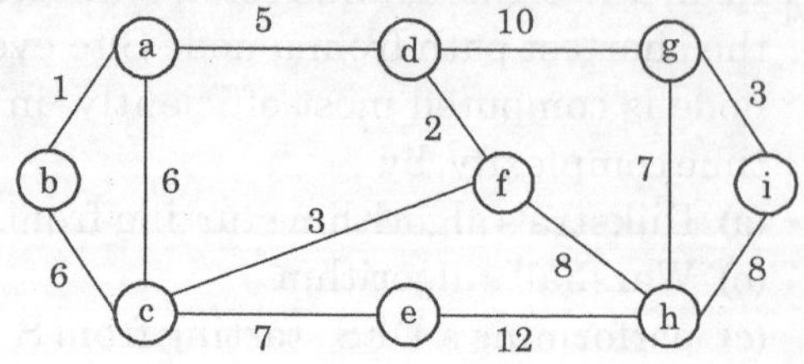

(a) (a, b), (d, f), (f, c), (g, i), (d, a), (g, h), (c, e), (f, h)
(b) (c, e), (c, f), (f, d), (d, a), (a, b), (g, h), (h, f), (g, i)
(c) (d, f), (f, c), (d, a), (a, b), (c, e), (f, h), (g, h), (g, i)
(d) (h, g), (g, i), (h, f), (f, c), (f, d), (d, a), (a, b), (c, e)

[2008 : 2 M]

3.32 G is a graph on n vertices and 2n – 2 edges. The edges of G can be partitioned into two edge-disjoint spanning trees. Which of the following is NOT true for G?

(a) For every subset of k vertices, the induced subgraph has a most 2k – 2 edges.
(b) The minimum cut in G has a least two edges
(c) There are two edges-disjoint paths between every pair of vertices
(d) There are two vertex-disjoint paths between every pair of vertices. **[2008 : 2 M]**

3.33 Dijkstra's single source shortest path algorithm when run from vertex a in the below graph, computes the corrects shortest path distance to

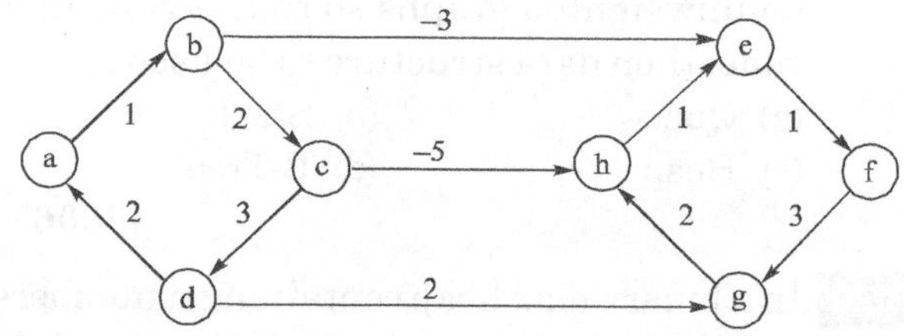

(a) Only vertex a
(b) Only vertices a, e, f, g, h
(c) Only vertices, a, b, c, d
(d) All the vertices **[2008 : 2 M]**

3.34 Which of the following statement(s) is/are correct regarding Bellman-Ford shortest path algorithm?

P. Always finds a negative weighted cycle, if one exists.
Q. Finds whether any negative weighted cycle is reachable from the source

(a) P only (b) Q only
(c) Both P and Q (d) Neither P nor Q

[2009 : 1 M]

3.35 Consider the following graph:

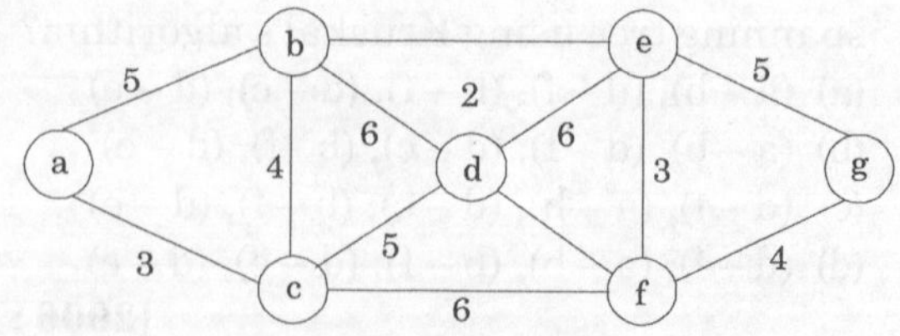

Which one of the following is NOT the sequence of edges added to the minimum spanning tree using Kruskal's algorithm?

(a) (b, e) (e, f) (a, c) (b, c) (f, g) (c, d)
(b) (b, e) (e, f) (a, c) (f, g) (b, c) (c, d)
(c) (b, e) (a, c) (e, f) (b, c) (f, g) (c, d)
(d) (b, e) (e, f) (b, c) (a, c) (f, g) (c, d)

[2009 : 2 M]

Linked Answer Question 3.36 and 3.37:
Consider a binary max-heap implemented using an array.

3.36 Which one of the following array represents a binary max-heap?

(a) {25, 12, 16, 13, 10, 8, 14}
(b) {25, 14, 13, 16, 10, 8, 12}
(c) {25, 14, 16, 13, 10, 8, 12}
(d) {25, 14, 12, 13, 10, 8, 16} **[2009 : 2 M]**

3.37 What is the content of the array after two delete operations on the correct answer to the previous question?

(a) {14, 13, 12, 10, 8} (b) {14, 12, 13, 8, 10}
(c) {14, 13, 8, 12, 10} (d) {14, 13, 12, 8, 10}

[2009 : 2 M]

3.38 The weight of a sequence $a_0, a_1, \dots, a_{n-1}$ of real numbers is defined as $a_0 + \frac{a_1}{2} + \dots + \frac{a_{n-1}}{2^{n-1}}$. A subsequence of a sequence is obtained by deleting some elements from the sequence, keeping the order of the remaining elements the same. Let X denote the maximum possible weight of a subsequence of $a_0, a_1, \dots, a_{n-1}$ and Y the maximum possible weight of a subsequence of $a_1, a_2, \dots, a_{n-1}$. Then X is equal to

(a) $\max(Y, a_0 + Y)$ (b) $\max\left(Y, a_0 + \frac{Y}{2}\right)$
(c) $\max(Y, a_0 + 2Y)$ (d) $a_0 + \frac{Y}{2}$

[2010 : 2 M]

Common Data for Questions 3.39 and 3.40
Consider a complete undirected graph with vertex set {0, 1, 2, 3, 4}. Entry W_{ij} in the matrix W below is the weight of the edge {i, j}.

$$W = \begin{pmatrix} 0 & 1 & 8 & 1 & 4 \\ 1 & 0 & 12 & 4 & 9 \\ 8 & 12 & 0 & 7 & 3 \\ 1 & 4 & 7 & 0 & 2 \\ 4 & 9 & 3 & 2 & 0 \end{pmatrix}$$

3.39 What is the minimum possible weight of a spanning tree T in this graph such that vertex 0 is a leaf node in the tree T?

(a) 7 (b) 8
(c) 9 (d) 10 **[2010 : 2 M]**

3.40 What is the minimum possible weight of a path P from vertex 1 to vertex 2 in this graph such that P contains at most 3 edges?

(a) 7 (b) 8
(c) 9 (d) 10 **[2010 : 2 M]**

3.41 A max-heap is a heap where the value of each parent is greater than or equal to the value of its children. Which of the following is a max-heap?

(a)
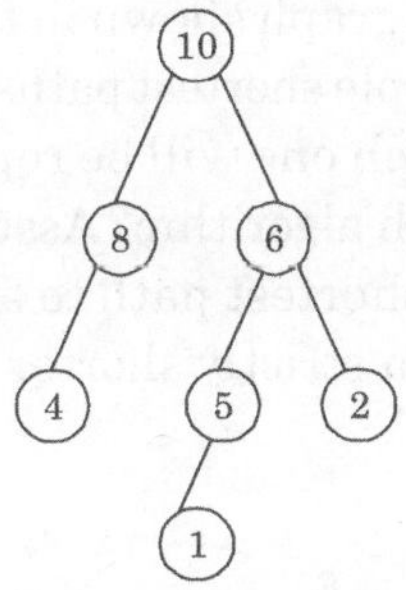

(b)
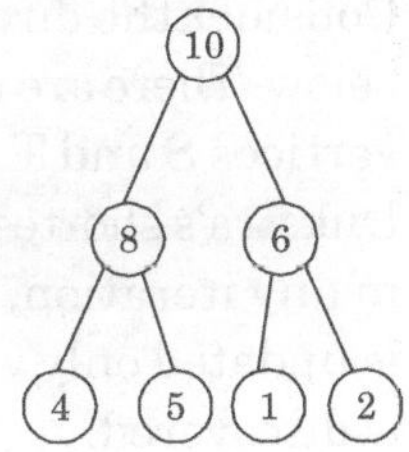

(c)
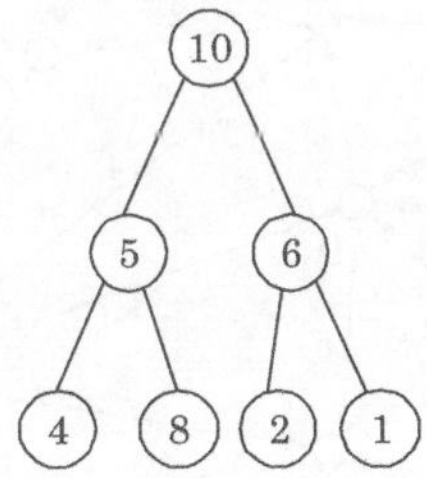

(d)
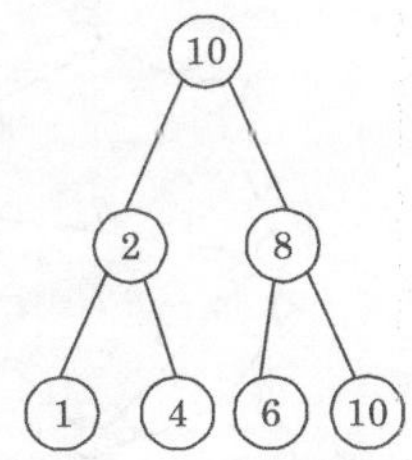

[2011 : 1 M]

Linked Data Question 3.42 and 3.43
An undirected graph G(V, E) contains n(n > 2) nodes named $v_1, v_2, \dots, v_n$. Two nodes v_i, v_j are connected if and only if $0 < |i - j| \le 2$. Each edge (v_i, v_j) is assigned a weight i + j. A sample graph with n = 4 is shown below.

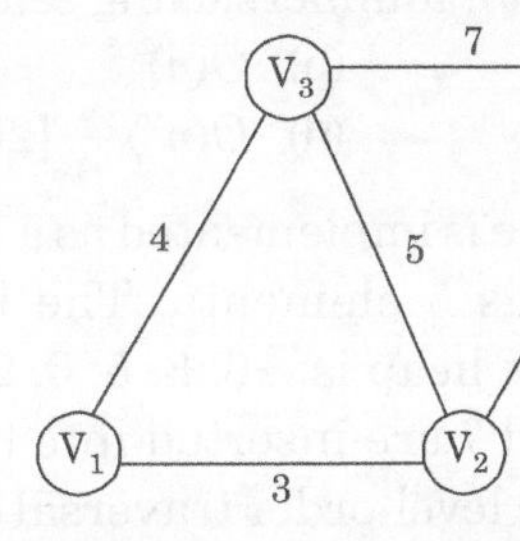

3.42 What will be the cost of the Minimum Spanning Tree (MST) of such a graph with n nodes?

(a) $\frac{1}{12}(11n^2 - 5n)$ (b) $n^2 - n + 1$
(c) $6n - 11$ (d) $2n + 1$ **[2011 : 2 M]**

3.43 The length of the path from v_5 to v_6 in the MST of previous question with n = 10 is

(a) 11 (b) 25
(c) 31 (d) 41 **[2011 : 2 M]**

3.44 Let G be a weighted graph with edge weights greater than one and G′ be the graph constructed by squaring the weights of edge in G. Let T and T′ be the minimum spanning trees of G and G′ respectively, with total weights t and t′. Which of the following statements is TRUE?

(a) T′ = T with total weight $t' = t^2$
(b) T′ = T with total weight $t' < t^2$
(c) T′ ≠ T but total weight $t' = t^2$
(d) None of the above **[2012 : 2 M]**

3.45 Consider the directed graph shown in the figure below. There are multiple shortest paths between vertices S and T. Which one will be reported by Dijkstra's shortest path algorithm? Assume that, in any iteration, the shortest path to a vertex v is updated only when a strictly shorter path to v is discovered.

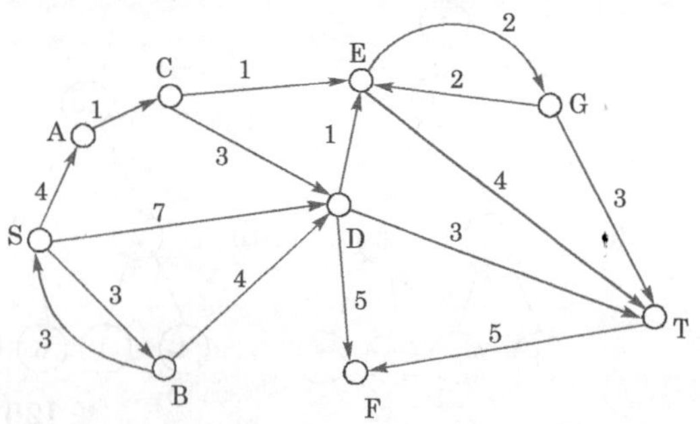

(a) SDT (b) SVDT
(c) SACDT (d) SACET

[2012 : 2 M]

3.46 Which one of the following is the tightest upper bound that represents the number of swaps required to sort *n* numbers using selection sort?

(a) O(log *n*) (b) O(*n*)
(c) O(*n* log *n*) (d) O(n^2) **[2013 : 1 M]**

3.47 A priority queue is implemented as a Max-Heap. Initially, it has 5 elements. The level-order traversal of the heap is: 10, 8, 5, 3, 2. Two new elements 1 and 7 are inserted into the heap in that order. The level-order traversal of the heap after the insertion of the elements is:

(a) 10, 8, 7, 3, 2, 1, 5 (b) 10, 8, 7, 2, 3, 1, 5
(c) 10, 8, 7, 1, 2, 3, 5 (d) 10, 8, 7, 5, 3, 2, 1

[2014 (Set-2) : 1 M]

3.48 Consider the tree arcs of a BFS traversal from a source node **W** in an unweighted, connected, undirected graph. The tree **T** formed by the tree arcs is a data structure for computing

(a) The shortest path between every pair of vertices.
(b) The shortest path from **W** to every vertex in the graph.
(c) The shortest paths from **W** to only those nodes that are leaves of **T**.
(d) The longest path in the graph.

[2014 (Set-2) : 1 M]

3.49 Suppose *P*, *Q*, *R*, *S*, *T* are sorted sequences having lengths 20,24,30,35,50 respectively. They are to be merged into a single sequence by merging together two sequences at a time. The number of comparisons that will be needed in the worst case by the optimal algorithm for doing this is ________.

[2014 (Set-2) : 2 M]

3.50 The graph shown below has 8 edges with distinct integer edge weights. The minimum spanning tree (MST) is of weight 36 and contains the edges: {(A, C), (B, C), (B, E), (E, F), (D, F)}. The edge weights of only those edges which are in the MST are given in the figure shown below. The minimum possible sum of weights of all 8 edges of this graph is ________.

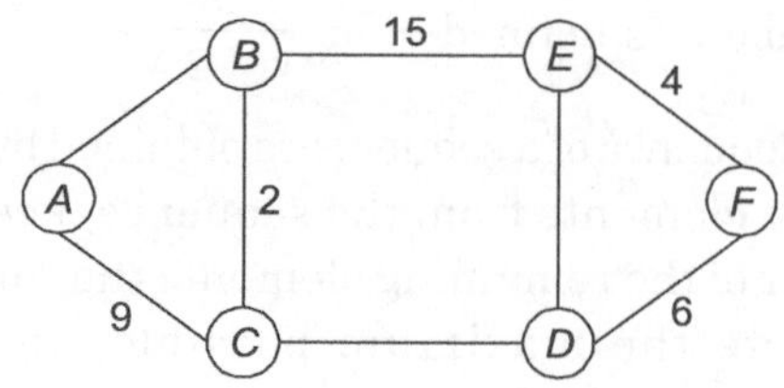

[2015 (Set-1) : 2 M]

3.51 Consider a max heap, represented by the array: 40, 30, 20, 10, 15, 16, 17, 8, 4.

Array Index	1	2	3	4	5	6	7	8	9
Value	40	30	20	10	15	16	17	8	4

Now consider that a value 35 is inserted into this heap. After insertion, the new heap is

(a) 40, 30, 20, 10, 15, 16, 17, 8, 4, 35
(b) 40, 35, 20, 10, 30, 16, 17, 8, 4, 15
(c) 40, 30, 20, 10, 35, 16, 17, 8, 4, 15
(d) 40, 35, 20, 10, 15, 16, 17, 8, 4, 30

[2015 (Set-1) : 2 M]

3.52 Consider the following array of elements ⟨89, 19, 50, 17, 12, 15, 2, 5, 7, 11, 6, 9, 100⟩.
The minimum number of interchanges needed to convert it into a max-heap is

(a) 4 (b) 5
(c) 2 (d) 3

[2015 (Set-3) : 1 M]

3.53 Let G be a weighted connected undirected graph with distinct positive edge weights. If every edge weight is increased by the same value, then which of the following statements is/are TRUE?

P: Minimum spanning tree of G does not change

Q: Shortest path between any pair of vertices does not change

(a) P only (b) Q only

(c) Neither P nor Q (d) Both P and Q

[2016 (Set-1) : 1 M]

3.54 Let G be a complete undirected graph on 4 vertices, having 6 edges with weights being 1, 2, 3, 4, 5, and 6. The maximum possible weight that a minimum weight spanning tree of G can have is ________.

[2016 (Set-1) : 2 M]

3.55 $G = (V, E)$ is an undirected simple graph in which each edge has a distinct weight, and e is a particular edge of G. Which of the following statements about the minimum spanning trees (MSTS) of G is/are TRUE?

I. If e is the lightest edge of *some* cycle in G, then every MST of G *includes* e.

II. If e is the heaviest edge of *some* cycle in G, then every MST of G *excludes* e.

(a) I only (b) II only

(c) Both I and II (d) Neither I nor II

[2016 (Set-1) : 2 M]

3.56 An operator *delete* (i) for a binary heap data structure is to be designed to delete the item in the i-th node. Assume that the heap is implemented in an array and i refers to the i-th index of the array. If the heap tree has depth d (number of edges on the path from the root to the farthest leaf), then what is the time complexity to re-fix the heap efficiently after the removal of the element?

(a) O(1)

(b) $O(d)$ but not O(1)

(c) $O(2^d)$ but not $O(d)$

(d) $O(d2^d)$ but not $O(2d)$

[2016 (Set-1) : 2 M]

3.57 A complete binary min-heap is made by including each integer in [1, 1023] exactly once. The depth of a node in the heap is the length of the path from the root of the heap to that node. Thus, the root is at depth 0. The maximum depth at which integer 9 can appear is ________.

[2016 (Set-2) : 2 M]

3.58 Let $G = (V, E)$ be any connected undirected edge-weighted graph. The weights of the edges in E are positive and distinct. Consider the following statements:

I. Minimum Spanning Tree of G is always unique.

II. Shortest path between any two vertices of G is always unique.

Which of the above statements is/are necessarily true?

(a) I only (b) II only

(c) Both I and II (d) Neither I nor II

[2017 (Set-1) : 2 M]

3.59 A message is made up entirely of characters from the set X = {P, Q, R, S, T}. The table of probabilities for each of the characters is shown below:

Character	Probability
P	0.22
Q	0.34
R	0.17
S	0.19
T	0.08
Total	1.00

If a message of 100 characters over X is encoded using Huffman coding, then the expected length of the encoded message in bits is ________.

[2017 (Set-2) : 2 M]

3.60 Consider the weights and values of items listed below. Note that there is only one unit of each item.

Item Number	Weight in (Kgs)	Value in (Rupees)
1	10	60
2	7	28
3	4	20
4	2	24

The task is to pick a subset of these items such that their total weight is not more than 11 Kgs and their total value is maximized. Moreover, no item may be split. The total value of items picked by an optimal algorithm is denoted by V_{opt}. A greedy algorithm sorts the items by their value-to-weight ratios in descending order and packs them greedily, starting from the first item in the ordered list. The total value of items picked by the greedy algorithm is denoted by V_{greedy}.

The value of $V_{opt} - V_{greedy}$ is ________.

[2018 : 2 M]

3.61 Consider the following undirected graph G:

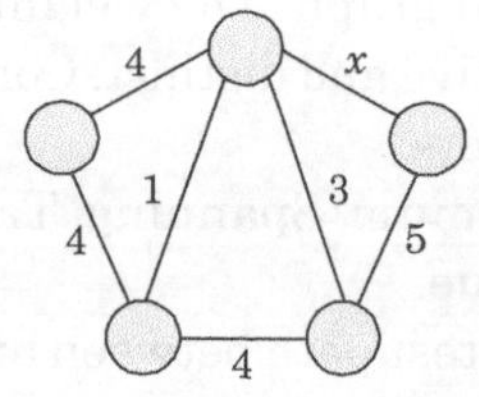

Choose a value of x that will maximize the number of minimum weight spanning trees (MWSTs) of G. The number of MWSTs of G for this value of x is ________. **[2018 : 2 M]**

3.62 Let G be any connected, weighted, undirected graph:

I. G has a unique minimum spanning tree, if no two edges of G have the same weight.

II. G has a unique minimum spanning tree, if for every cut G, there is a unique minimum weight edge crossing the cut.

Which of the above two statements is/are TRUE?

(a) Neither I nor II
(b) I only
(c) II only
(d) Both I and II

[2019 : 2 M]

3.63 Consider the array representation of a binary min-heap containing 1023 elements. The minimum number of comparisons required to find the maximum in the heap ________.

[2020 : 2 M]

3.64 Let $G = (V, E)$ be a weighted undirected graph and let T be a Minimum Spanning Tree (MST) of G maintained using adjacency lists. Suppose a new weighted edge $(u, v) \in V \times V$ is added to G. The worst case time complexity of determining if T is still an MST of the resultant graph is

(a) $\Theta(|E||V|)$
(b) $\Theta(|V|)$
(c) $\Theta(|E| + |V|)$
(d) $\Theta(|E| \log |V|)$

[2020 : 2 M]

3.65 Consider a graph $G = (V, E)$, where $V = \{v_2, v_2, ..., v_{100}\}$, $E = \{(v_i, v_j) \mid 1 \le i < j \le 100\}$ and weight of the edge (v_i, v_j) is $|i - j|$. The weight of minimum spanning tree of G is ________.

[2020 : 2 M]

3.66 Let $G = (V, E)$ be a directed, weighted graph with weight function $w : E \to \mathbb{R}$. For some function $f : V \to \mathbb{R}$, for each edge $(u, v) \in E$, define $w'(u, v)$ as $w(u, v) + f(v)$.

Which one of the options completes the following sentence so that it is TRUE?

"The shortest paths in G under w are shortest paths under w' too, ________."

(a) if and only if $f(u)$ is the distance from s to u in the graph obtained by adding a new vertex s to G and edges of zero weight from s to every vertex of G
(b) if and only if $\forall u \in V, f(u)$ is positive
(c) if and only if $\forall u \in V, f(u)$ is negative
(d) for every $f : V \to \mathbb{R}$

[2020 : 2 M]

3.67 Define R_n to be the maximum amount earned by cutting a rod of length n meters into one or more pieces of integer length and selling them. For $i > 0$, let $p[i]$ denote the selling price of a rod whose length is i meters. Consider the array of prices:

$$p[1] = 1, p[2] = 5, p[3] = 8, p[4] = 9,$$
$$p[5] = 10, p[6] = 17, p[7] = 18$$

Which of the following statements is/are correct about R_7?

(a) $R_7 = 19$
(b) $R_7 = 18$
(c) R_7 is achieved by three different solutions.
(d) R_7 cannot be achieved by a solution consisting of three pieces.

[2021 (Set-1) : 2 M]

3.68 Consider the following undirected graph with edge weights as shown:

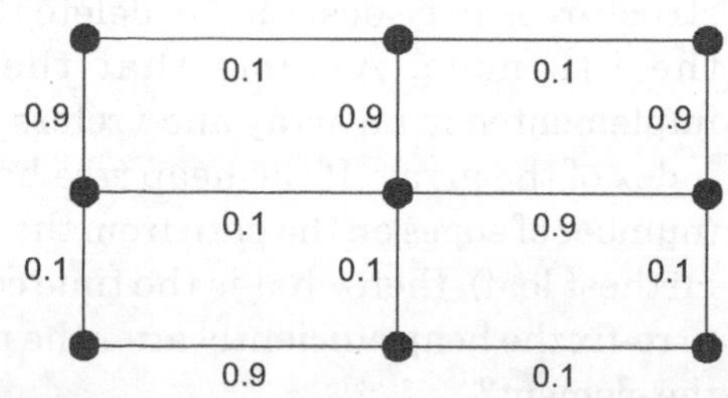

The number of minimum-weight spanning trees of the graph is ________.

[2021 (Set-1) : 1 M]

3.69 In a directed acyclic graph with a source vertex s, the quality-score of a directed path is defined to be the product of the weights of the edges on the path. Further, for a vertex v other than s, the quality-score of v is defined to be the maximum among the quality-scores of all the paths from s to v. The quality-score of s is assumed to be 1.

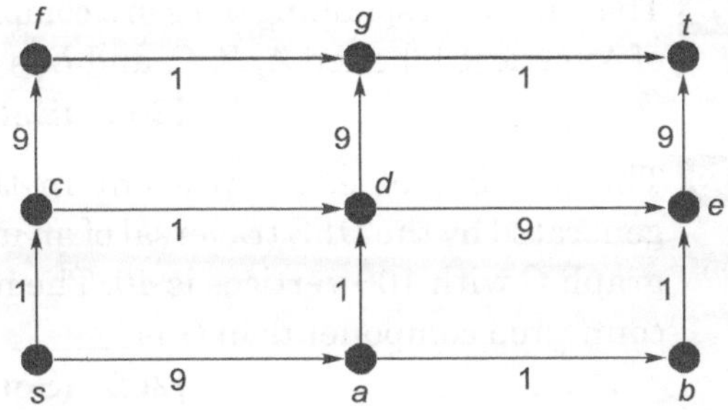

The sum of the quality-scores of all the vertices in the graph shown above is ________.

[2021 (Set-2) : 1 M]

3.70 Let H be a binary min-heap consisting of n elements implemented as an array. What is the worst case time complexity of an optimal algorithm to find the maximum element in H?

(a) $\Theta(\log n)$ (b) $\Theta(1)$
(c) $\Theta(n \log n)$ (d) $\Theta(n)$

[2021 (Set-2) : 1 M]

3.71 Consider the following directed graph:

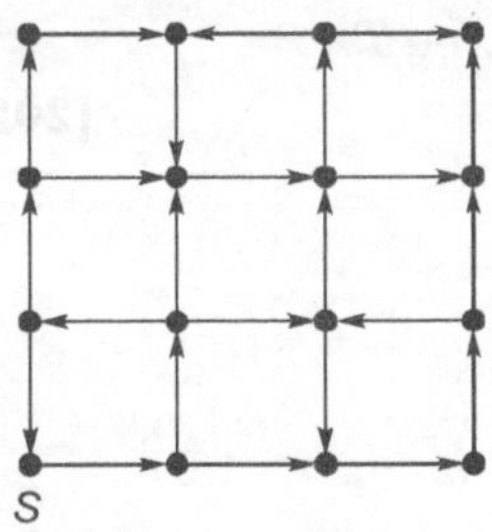

Which of the following is/are correct about the graph?

(a) The graph does not have a topological order.
(b) A depth-first traversal starting at vertex S classifies three directed edges as back edges.
(c) For each pair of vertices u and v, there is a directed path from u to v.
(d) The graph does not have a strongly connected component.

[2021 (Set-2) : 2 M]

3.72 Consider the string abbccddeee. Each letter in the string must be assigned a binary code satisfying the following properties:

1. For any two letters, the code assigned to one letter must not be a prefix of the code assigned to the other letter.
2. For any two letters of the same frequency, the letter which occurs earlier in the dictionary order is assigned a code whose length is at most the length of the code assigned to the other letter.

Among the set of all binary code assignments which satisfy the above two properties, what is the minimum length of the encoded string?

(a) 23 (b) 25
(c) 21 (d) 30

[2021 (Set-2) : 2 M]

3.73 Let G be a connected undirected weighted graph. Consider the following two statements.

S_1: There exists a minimum weight edge in G which is present in every minimum spanning tree of G.

S_2: If every edge in G has distinct weight, then G has a unique minimum spanning tree.

Which one of the following options is correct?

(a) S_1 is false and S_2 is true.
(b) Both S_1 and S_2 are false.
(c) S_1 is true and S_2 is false.
(d) Both S_1 and and S_2 are true.

[2021 (Set-2) : 1 M]

3.74 Consider a simple undirected weighted graph G, all of whose edge weights are distinct. Which of the following statements about the minimum spanning trees of G is/are TRUE?

(a) The edge with the second smallest weight is always part of any minimum spanning tree of G.
(b) One or both of the edges with the third smallest and the fourth smallest weights are part of any minimum spanning tree of G.
(c) Suppose $S \subseteq V$ be such that $S \neq \phi$ and $S \neq V$. Consider the edge with the minimum weight such that one of its vertices is in S and the other in $V \setminus S$. Such an edge will always be part of any minimum spanning tree of G.
(d) G can have multiple minimum spanning trees.

[2022 : 2 M]

3.75 Let $G(V, E)$ be a directed graph, where $V = \{1, 2, 3, 4, 5\}$ is the set of vertices and E is the set of directed edges, as defined by the following adjacency matrix A.

$$A[i][j] = \begin{cases} 1, & 1 \le j \le i \le 5 \\ 0, & \text{otherwise} \end{cases}$$

$A[i][j] = 1$ indicates a directed edge from node i to node j. A directed spanning tree of G, rooted at $r \in V$, is defined as a subgraph T of G such that the undirected version of T is a tree, and T contains a directed path from r to every other vertex in V. The number of such directed spanning trees rooted at vertex 5 is ________.

[2022 : 2 M]

3.76 Which one of the following sequences when stored in an array at locations $A[1]$, ..., $A[10]$ forms a max-heap?

(a) 23, 17, 10, 6, 13, 14, 1, 5, 7, 12
(b) 23, 17, 14, 7, 13, 10, 1, 5, 6, 12
(c) 23, 17, 14, 6, 13, 10, 1, 5, 7, 15
(d) 23, 14, 17, 1, 10, 13, 16, 12, 7, 5

[2023 : 1 M]

3.77 Let A be a priority queue for maintaining a set of elements. Suppose A is implemented using a max-heap data structure. The operation Extract-Max(A) extracts and deletes the maximum element from A. The operation Insert(A, key) inserts a new element key in A. The properties of a max-heap are preserved at the end of each of these operations.

When A contains n elements, which one of the following statements about the worst case running time of these two operations is TRUE?

(a) Both Extract-Max(A) and Insert(A, key) run in O(1).
(b) Both Extract-Max(A) and Insert(A, key) run in O(log(n)).
(c) Extract-Max(A) runs in O(1) whereas Insert(A, key) runs in O(n).
(d) Extract-Max(A) runs in O(1) whereas Insert(A, key) runs in O(log(n)).

[2023 : 2 M]

3.78 The number of spanning trees in a complete graph of 4 vertices labelled A, B, C, and D is ________.

[2024 (Set-1) : 1 M]

3.79 The number of edges present in the forest generated by the DFS traversal of an undirected graph G with 100 vertices is 40. The number of connected components in G is ________.

[2024 (Set-1) : 2 M]

3.80 Let G be a directed graph and T a depth first search (DFS) spanning tree in G that is rooted at a vertex v. Suppose T is also a breadth first search (BFS) tree in G, rooted at v. Which of the following statements is/are TRUE for every such graph G and tree T?

(a) The only edges in G are the edges in T.
(b) There are no cross-edges in G with respect to the tree T.
(c) There are no back-edges in G with respect to the tree T.
(d) There are no forward-edges in G with respect to the tree T.

[2024 (Set-1) : 2 M]

■■■■

Answers Greedy Method

3.1	Sol.	**3.2**	(b)	**3.3**	(c)	**3.4**	(b)	**3.5**	(b)	**3.6**	(b)	**3.7**	(c)	**3.8**	(b)	**3.9**	(a)
3.10	(a)	**3.11**	(c)	**3.12**	(a)	**3.13**	(d)	**3.14**	(d)	**3.15**	(a)	**3.16**	(d)	**3.17**	(a)	**3.18**	(b)
3.19	(c)	**3.20**	(a)	**3.21**	(d)	**3.22**	(a)	**3.23**	(c)	**3.24**	(d)	**3.25**	(a)	**3.26**	(d)	**3.27**	(d)
3.28	(a)	**3.29**	(d)	**3.30**	(b)	**3.31**	(c)	**3.32**	(d)	**3.33**	(d)	**3.34**	(b)	**3.35**	(d)	**3.36**	(c)
3.37	(d)	**3.38**	(d)	**3.39**	(d)	**3.40**	(b)	**3.41**	(b)	**3.42**	(b)	**3.43**	(c)	**3.44**	(d)	**3.45**	(d)
3.46	(b)	**3.47**	(a)	**3.48**	(b)	**3.49**	(358)	**3.50**	(69)	**3.51**	(b)	**3.52**	(d)	**3.53**	(a)	**3.54**	(7)
3.55	(b)	**3.56**	(b)	**3.57**	(8)	**3.58**	(a)	**3.59**	(225)	**3.60**	(16)	**3.61**	(4)	**3.62**	(d)	**3.63**	(511)
3.64	(b)	**3.65**	(99)	**3.66**	(d)	**3.67**	(b, c)	**3.68**	(3)	**3.69**	(929)	**3.70**	(d)	**3.71**	(a, b)	**3.72**	(a)
3.73	(a)	**3.74**	(a, b, c)	**3.75**	(24)	**3.76**	(b)	**3.77**	(b)	**3.78**	(16)	**3.79**	(60)	**3.80**	(d)		

Explanations Greedy Method

3.1 Sol.

O(m + n), edges are sorted so no need to make min-heap. So deletion from min. heap time is saved.

3.2 (b)

All edges are distinct weights.
Removal of e_{max} may not disconnect G, because other edges may cover the vertices incident with e_{max}.

3.3 (c)

In a heap with n elements with the smallest element at the root this data structure is known as min heap or priority queue. We can implement priority queue with the help of 2–3 tree. Finding 7th smallest element means search the position of 7th smallest element. In 2–3 tree we can find the 7th smallest element in O(log *n*) time because member insert or delete instruction can be executed in at most O(log *n*) steps on a 2–3 tree with n leaves.

3.4 (b)

Number of vertices in the graph n = 10 so after adding n – 1 = 9 edges which contains all vertices and doesn't form a circuit in the minimum spanning tree of G.

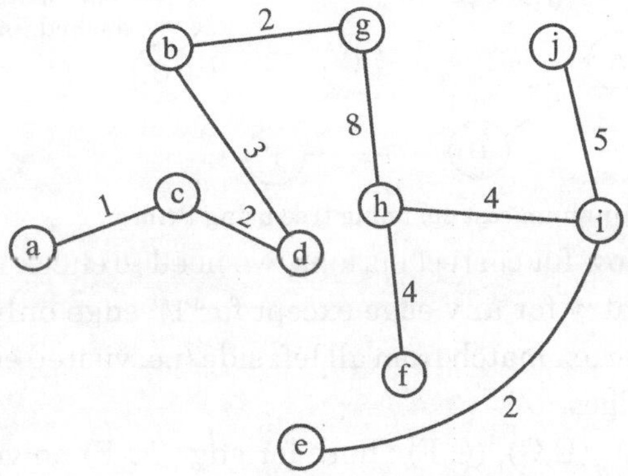

$$\text{Total weight} = \sum_{i=1}^{9} w(i)$$
$$= 1 + 2 + 2 + 2 + 3 + 4 + 4 + 5 + 8 = 31$$

3.5 (b)

The activities are A, B, C, D, E, F, G and H. The starting and ending times are "$a_s\, b_s\, c_s\, a_e\, d_s\, c_e\, e_s\, f_s\, b_e\, d_e\, g_s\, e_s\, f_e\, h_s\, g_e\, h_e$".
The gap between starting and ending time of B is greater than all other activities. Consider the order $b_s c_s a_e d_s c_e e_s f_s b_e$ be the gap between b_s to b_e is 6 in which two activity a_e and c_e ended so minimum number of required rooms is 4.

3.6 (b)

In Dijkstra's single source shortest path the nodes are included in following sequence.

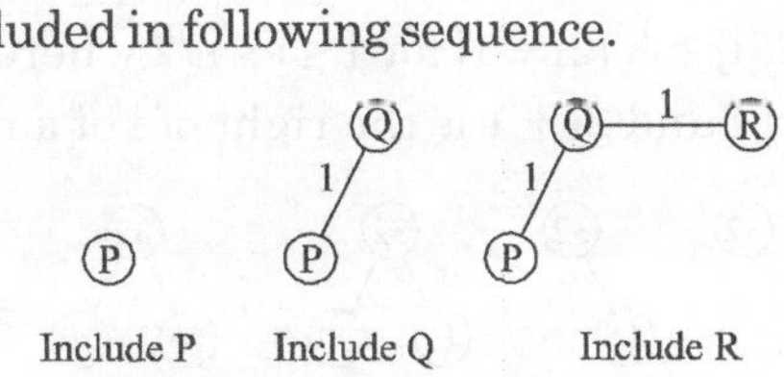

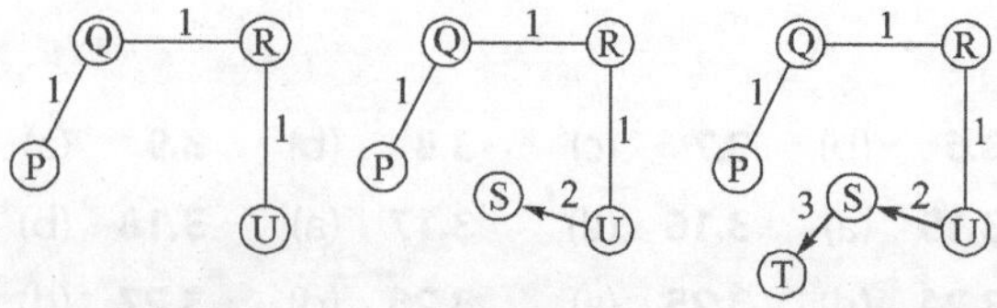

Include U Include R Include T

Total Distance (from source P)

$$= 1 + 1 + 1 + 2 + 3 = 8$$

Node sequence = P, Q, R, U, S, T

3.7 (c)

Prim's algorithm having special property which is, while finding minimum cost spanning tree, graph always be connected.

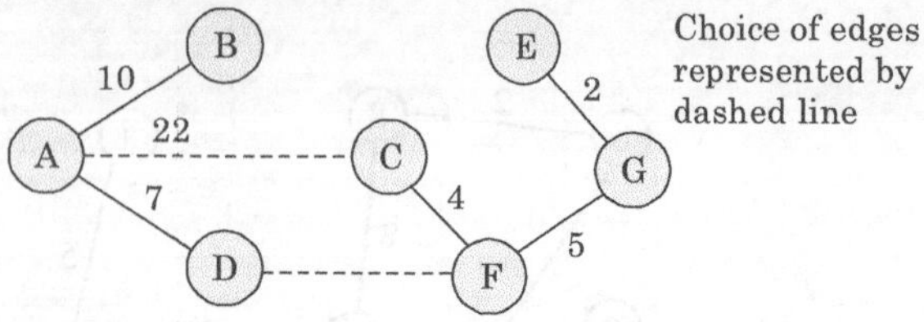

Minimum cost spanning tree using Prim's

Now for correct options, we need to check while entry for any edge except for 1st edge only one vertex match from all left side (i.e. visited edges) edges.

(a) (E,G), (C,F) : here for edge (C,F) no vertex match to (E,G). So not possible using Prim's.

(b) (A,D), (A,B), (A,C), (C,F), (G,E) : here for edge (G,E) no vertex match to any edge present left side of it so not possible using Prim's.

(c) (A,B), (A,D), (D,F), (F,G), (G,E), (F,C) not violated rule i.e. always connected in construction of MST.
So True.

3.8 (b)

Build heap method to make heap on 'n' elements will take O(n) time. Since less than this is not possible.

3.9 (a)

Input sequence 32 ,15, 20, 30, 12, 25, 16

If i is the root then $A[i] \geq A[2i]$ for $1 \leq i \leq n/2$ and $A[i] \geq A[2i + 1]$ for $1 \leq i \leq n/2$ where 2i is the left of i and 2i + 1 is the right of i of a max heap.

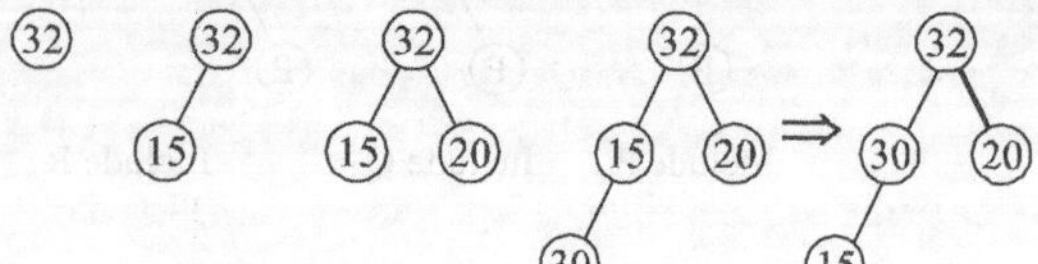

Insert 32 Insert 15 Insert 20 Insert 30 Build Heap

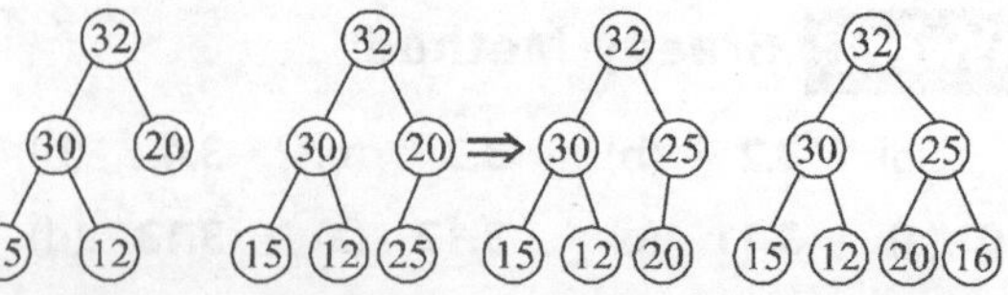

Insert 12 Insert 25 Build Beap Insert 16

3.10 (a)

Bellman-Ford algorithm : O(nm)
Kruskal's algorithm : O(mlog n)
Floyd-Warshall algorithm : $O(n^3)$
Topological sorting : O(n + m)

3.11 (c)

Undirected graph G contains n nodes.

	1	2	3	...	n
1	0	1	1	...	1
2	1	0	1	...	1
3	1	1	0	...	1
⋮					
n	1	1	1	...	0

In undirected graph G, the diagonal elements are 0 means there is no self loop of any vertex. Where each vertex in G is connected through n – 1 vertices in the graph G so G is a complete graph. In a complete graph apply the Cayley's theorem for the total number of minimum spanning trees in a complete graph G with n vertices is n^{n-2} so G has multiple distinct MSTs. The cost of a spanning tree is sum of all edges. If G contains n vertices then we will must stop the algorithm after adding the n – 1 edges so each spanning tree have a cost n – 1.

3.12 (a)

Consider an random graph and analyse:

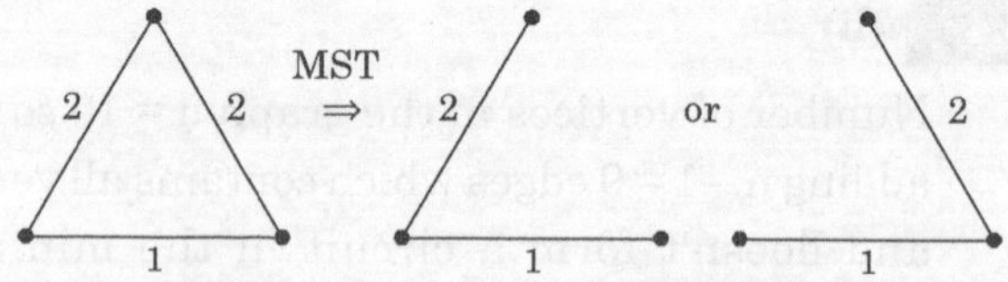

(a) Option (a) true since if we remove (2,2) weighted edges then graph is disconnected and which forms cut set.

(b) There does not exist cycle with all maximum weight. So, cannot be true always.

(c) Since there exist a cycle and maximum edge weight present in it. So, cannot be true always.

(d) Since one edge has weight other than e(2) i.e. 1. So cannot be true always.

Since (b), (c), (d) cannot be true always. So (a) must be true always.

3.13 (d)

Time complexity of Dijkstra's using adjacency list representation of graph and min heap used for greedy selection.

(i) Delete min from priority queue repeated n times $\Rightarrow O(\log V) * V$

(ii) Finding adj of all vertex in graph : $\theta(V + E)$

(iii) $\theta(E)$ times key decr. from priority queue D.S. $\Rightarrow \theta(\log V) * E$

$\therefore$ Overall : $(V \log V) + (V + E) + E(\log V)$
T.C. $= \theta((V + E) \log|V|)$

$\therefore$ Option (d).

3.14 (d)

The given problem is job scheduling problem 9 tasks are $T_1, T_2, ..., T_9$.

J is initially empty then according to deadlines it includes $\{T_1, T_2, T_3, T_5, T_7, T_8, T_9\}$.

So T_4 and T_6 can't be included in J

3.15 (a)

Total profit earn by algorithm
$= 15 + 20 + 30 + 18 + 23 + 16 + 25 = 147$

3.16 (d)

Initially the sequence of Max Heap is

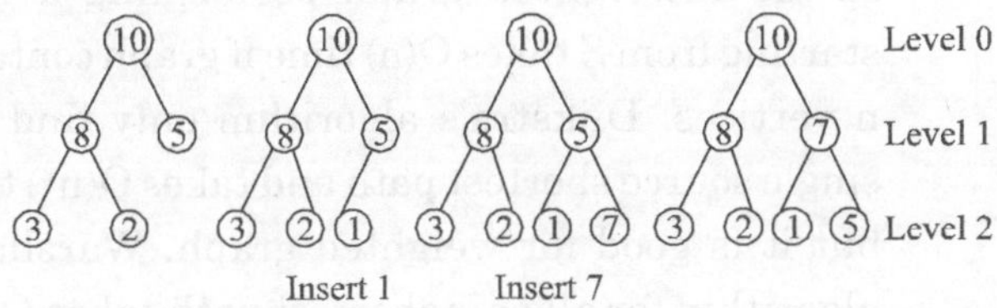

So, the sequence of level order traversal is 10, 8, 7, 3, 2, 1, 5

3.17 (a)

$$p(x) = a_0 + a_1x + a_2x^2 + a_3x^3$$

Apply Horner's rule to evaluate the p(x)

$$p(x) = a_0 + ((a_3x + a_2)x + a_1)x$$
$$t = a_3 * x = t + a_2$$
$$t = t * x = t + a_1$$
$$t = t * x = t + a_0$$

So, 3 minimum number of multiplications needed to evaluate p on an input x.

3.18 (b)

Given vertex set of $G = \{v_1, v_2, ..., v_n\}$

Weight of an edge $= 2|i\text{-}j| : (v_i, v_j) \in G$

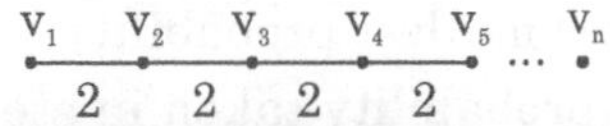

is MST.

In the case of minimum spanning tree of a graph G we will add upto (n – 1) vertices, so the weight of a minimum spanning tree of G

$$= \sum_{i=1}^{n-1} 2\,|V_i - V_{i-1}|$$

$$= 2\sum_{i=1}^{n-1} |V_i - V_{i-1}| = 2\sum_{i=1}^{n-1} |1|$$

$$= 2\,|n - 1| = 2(n - 1) = 2n - 2$$

3.19 (c)

Heap and priority queue are very neat data structures allowing:

- Add an element to heap with an associated priority.
- Remove the element from the heap or priority queue that has the highest priority, and return it.
- Peak at the element with highest priority without removing it.

A simple way to implement a heap or priority queue data type is to keep a list of elements, and search through the list for the highest priority which gives O(n) time to implement Dijkstra's shortest path algorithm on unweighted graph.

3.20 (a)

MAX Heap used to identity max element in O(1) time for to identity min element require O(n), to apply the linear search.

3.21 (d)

In Kruskal's algorithm we will add the edges of minimum weight whether they are adjacent or not. Consider each choice separately

(a) (a – b), (d – f), (b – f), (d – c), (d, e)
1 + 1 + 2 + 2 + 3 = 9

(b) (a – b), (d – f), (d – c), (b – f), (d, e)
1 + 1 + 2 + 2 + 3 = 9

(c) (d – f), (a – b), (d – c), (b – f), (d, e)
1 + 1 + 2 + 2 + 3 = 9

(d) (d – f), (a – b), (b – f), (d – e), (d – c)
1 + 1 + 2 + 3 + 2 = 9

In choice (d), we added (d – e) before (d – c) which violates algorithm, hence option (d) is not correct.

3.22 (a)

According to Huffman coding use Min heap data structure.

Step-1: Take two smallest probability.

Step-2: Add the probability taken in step-1.

Step-3: Put back previous sum in minimum heap.

Step-4: Repeat step-1 until probability becomes 1.

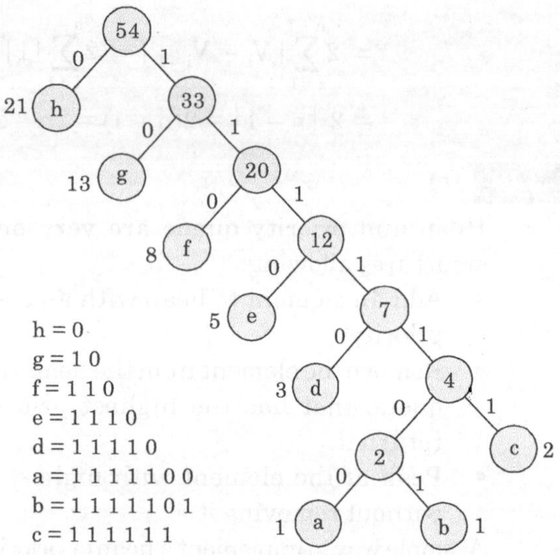

h = 0
g = 1 0
f = 1 1 0
e = 1 1 1 0
d = 1 1 1 1 0
a = 1 1 1 1 1 0 0
b = 1 1 1 1 1 0 1
c = 1 1 1 1 1 1

So, 1 1 0 1 1 1 1 0 0 1 1 1 0 1 0

f d h e g

3.23 (c)

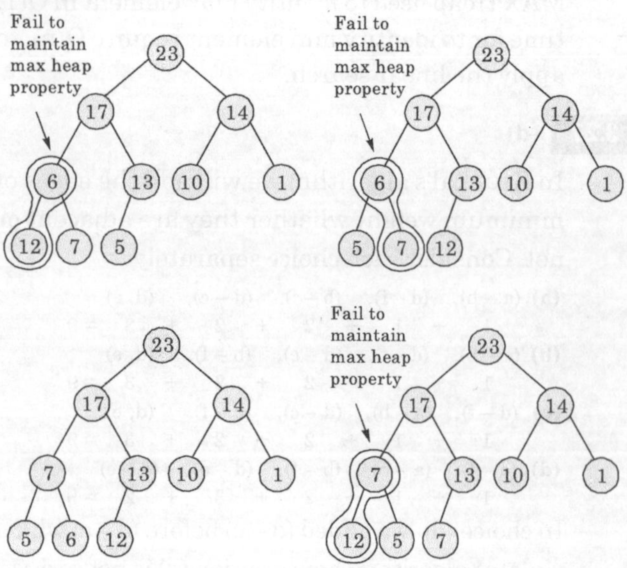

Correct heap tree

Note: In max heap tree every parent node is greater than equal to its left and right child.

3.24 (d)

The 3 ary max heap

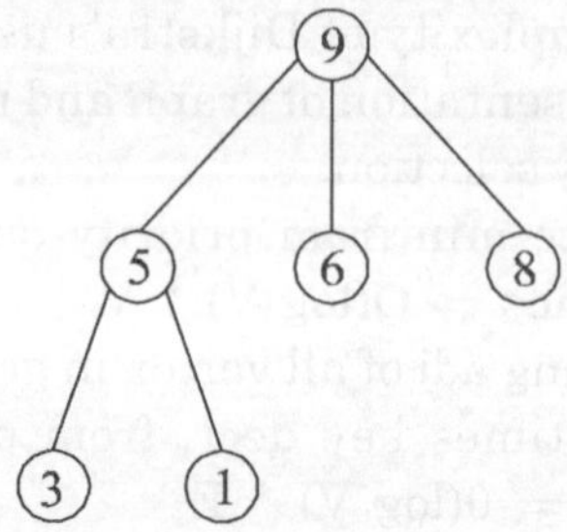

So the valid sequence is 9, 5, 6, 8, 3, 1

3.25 (a)

After inserting 7, 2, 10, 4 the resultant heap become

10, 7, 9, 8, 3, 1, 5, 2, 6, 4

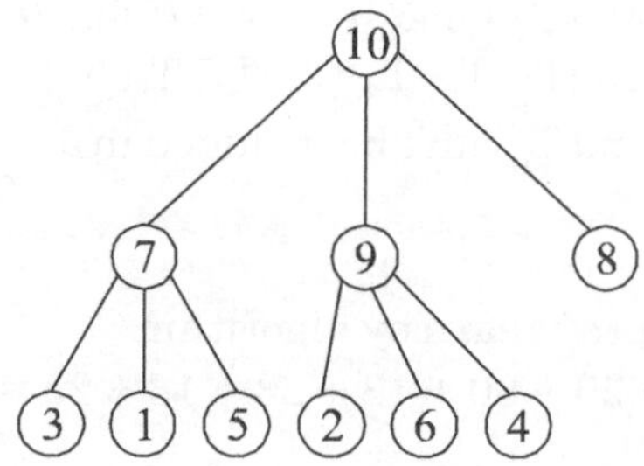

3.26 (d)

In an unweighted graph performing a BFS starting from S takes O(n) time if graph contains n vertices. Dijkstra's algorithm only find the single source shortest path and takes $O(n^2)$ time but it is good for weighted graph. Warshall's algorithm for all pair shortest path takes $O(n^3)$ time but it is also applicable for weighted graph.

3.27 (d)

W be the minimum weight among all edge weights in an undirected connected graph. e is the specific edge of weight w. It may be possible that another edge in the graph having weight w which had been added to minimum spanning tree and when we add e to minimum spanning tree it form a simple circuit. So we can't include e in every minimum spanning tree.

3.28 (a)

The probabilities are $\frac{1}{2}, \frac{1}{4}, \frac{1}{8}, \frac{1}{16}, \frac{1}{32}$

So, the Huffman code according to tree is unique. 0, 10, 110, 1110, 11110, 11111 represents a, b, c, d, e, f respectively.

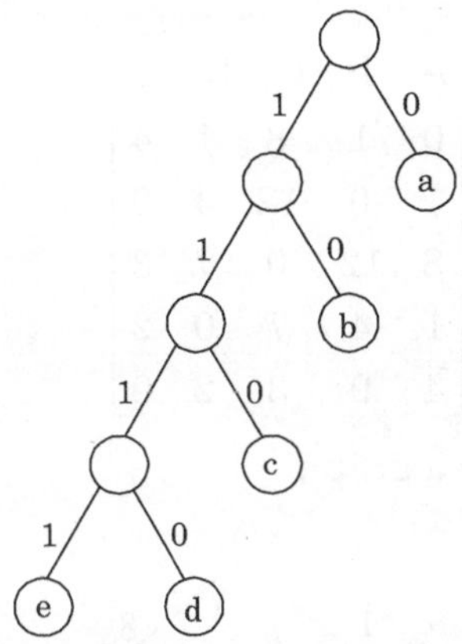

3.29 (d)

The average length of the tree is

$$= 1 \times \frac{1}{2} + 2 \times \frac{1}{4} + 3 \times \frac{1}{8} + 4 \times \frac{1}{16} + 5 \times \frac{1}{32}$$

$= 1.781$

3.30 (b)

After inserting an element at leaf in max heap we perform binary search on the path from the new leaf to the root to find the correct position for the newly inserted element. Comparisions will take θ(loglogn) but time will be θ(logn) because of swap operations.

3.31 (c)

Prim's algorith having special property which is, while finding minimum cost spanning tree, graph always be connected.

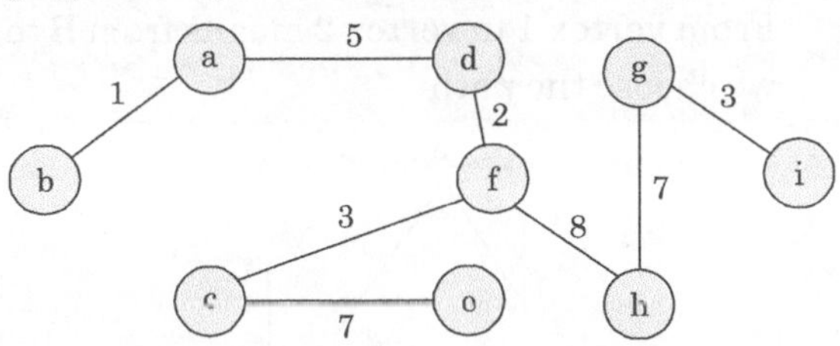

Minimum cost spanning tree using Prim's

Now for correct options, we need to check while entry for any edge except for 1st edge, only one vertex match from all left side (i.e. visited edges) edges.

(a) (a,b), (d,f) since (d,f) does not have any common vertex with (a,b) so not possible.

(b) (c,e), (c,f), (f,d), (d,a), (a,b), (g,h) since (g,h) edge does not have any common vertex from all left hand side edges, so not possible.

(c) (d,f), (f,c), (d,a), (a,b), (c,e), (f,h), (g,h), (g,i) is true since does not violate any rule.

3.32 (d)

Given graph G contains n vertices and $2n - 2 = 2(n - 1)$ edges. So spanning tree must contains $2n - 4$ edges. If G can be partitioned into two edge disjoint spanning tree. Then for any tree $2e = \sum_{i=1}^{n} d_i$ belongs to a tree if each d_i is positive and $e = n - 1$ but in given problem $e = 2n - 4 = 2(n - 2)$. So there is no two-vertex-disjoint paths between every pair of vertices.

3.33 (d)

Dijkstra's single source shortest path is not guaranteed to work for graphs with negative weight edges, but it works for the given graph.

Run the 1st pass: b1

b is minimum, so shortest distance to b is 1

After 1st pass, distances are: c3, e – 2

e is minimum, shortest distance to e is –2

After 2nd pass, distances are: c3, f0

f is minimum, so shortest distance to f is 0

After 3rd pass, distances are: c3, g3

Both are same, let us take g so shortest distance to g is 3

After 4th pass, distances are: c3, h5

c is minimum, so shortest distance to c is 3

After 5th pass, distance is: h – 2

h is minimum, so shortest distance to h is –2.

3.34 (b)

Bellman-Ford shortest path algorithm always finds any negative weighted cycle which is reachable from the source. If the negative weighted cycle is not reachable then, Bellman Ford algorithm not able to find that cycle.

3.35 (d)

In Kruskal's algorithm, edges are added in non decreasing order, starting from lowest edge weight, as long as they don't form a cycle with already added edges.

In choice (d): (b, e) edge weight is 2 is added first. This step is ok, since 2 is smallest in edge weight. Edge (e, f) is then added, whose edge weight of 3, is next highest, and hence this is also ok.

Then edge (b, c) is added, whose edge weight is 4, and this is not ok is Kruskal's algorithm, since there is an edge (a, c) of edge weight 3, which is less than the edge weight of (b, c). So (a, c) has to be added before (b, c). Notice also that adding (a, c) does not form a cycle with already added edges.

∴ Choice (d) is not the sequence that would be allowed by Kruskal's algorithm.

3.36 (c)

If i the node then $A[i] \geq A[2i]$ for $1 \leq i \leq n/2$ and $A[i] \geq A[2i + 1]$ for $1 \leq i \leq n/2$ where 2i is the left child and 2i + 1 right child of node i of a max heap.

Binary max-heap

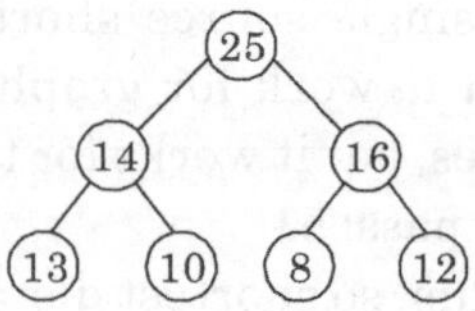

3.37 (d)

Deleter 25 : Replace it by 12

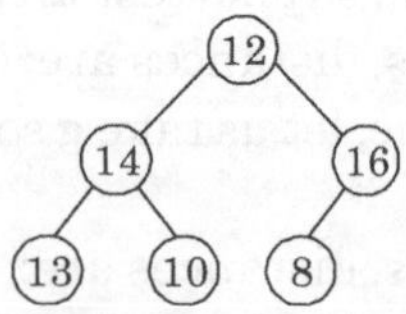

After heapfying :

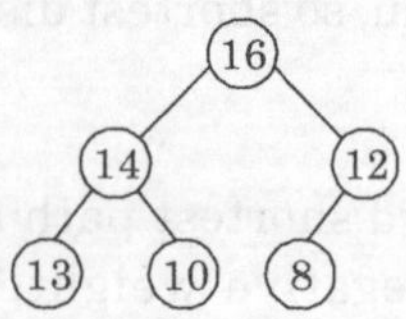

Deleter 16 : Replace it by 8

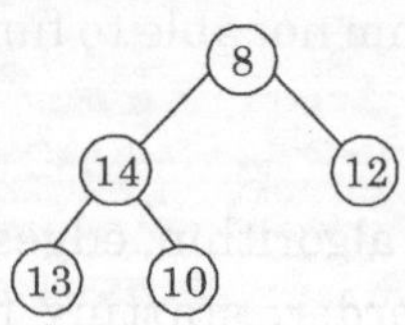

After heapfying :

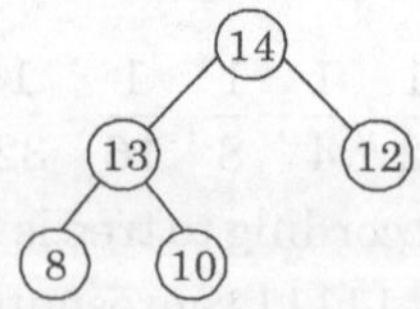

= {14, 13, 12, 8, 10}

3.39 (d)

$$W = \begin{array}{c} \\ A \\ B \\ C \\ D \\ E \end{array} \begin{array}{c} \begin{array}{ccccc} A & B & C & D & E \end{array} \\ \begin{bmatrix} 0 & 1 & 8 & 1 & 4 \\ 1 & 0 & 12 & 4 & 9 \\ 8 & 12 & 0 & 7 & 3 \\ 1 & 4 & 7 & 0 & 2 \\ 4 & 9 & 3 & 2 & 0 \end{bmatrix} \end{array}$$

We draw a graph

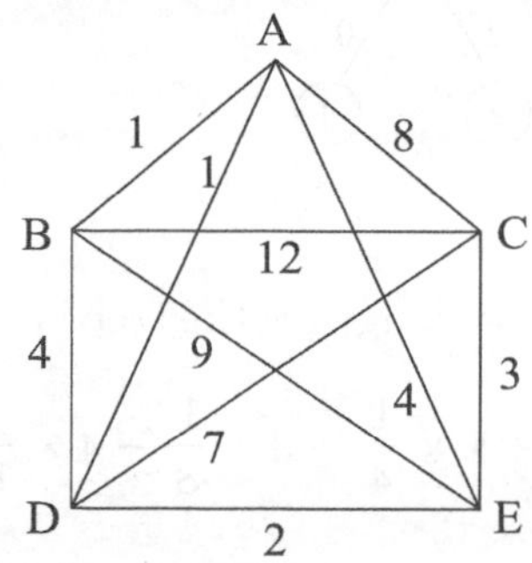

We have to make Minimum possible spanning tree having vertex A is a leaf node.

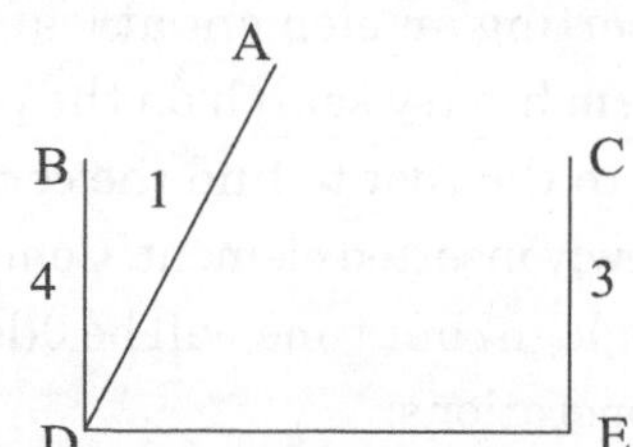

So Minimum possible weight

= 4 + 1 + 2 + 3 = 10

3.40 (b)

From vertex 1 to vertex 2 means from B to C. So we choose the path

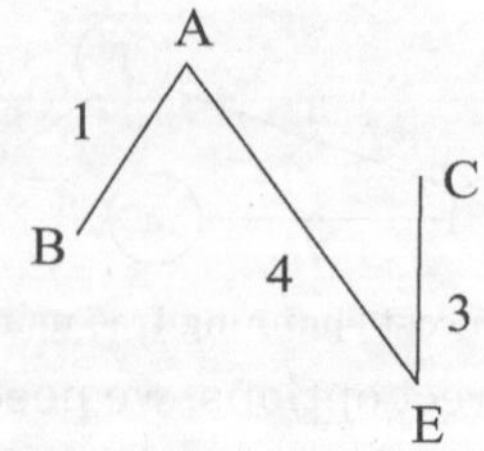

So weight = 1 + 4 + 3 = 8

3.41 (b)

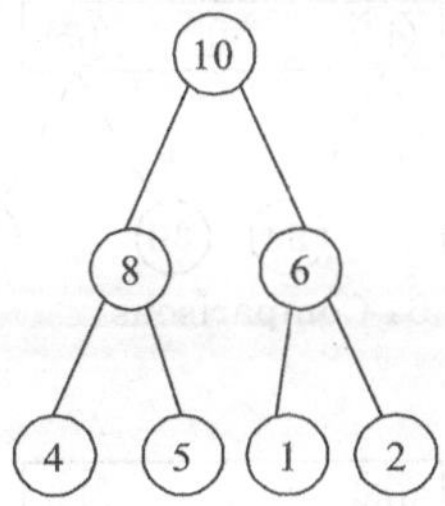

3.42 (b)

We observe a pattern in weight of MST being formed:

For n = 3 ⇒ (1 + 2 + 3) + (1)

For n = 4 ⇒ (1 + 2 + 3 + 4) + (1 + 2)

For n = 5 ⇒ (1 + 2 + 3 + 4 + 5) + (1 + 2 + 3)

For n = 6 ⇒ (1 + 2 + 3 + 4 + 5 + 6) + (1 + 2 + 3 + 4)

In general, total weight of MST for n:

$$= \sum_{i=1}^{n} i + \sum_{i=1}^{n-2} i = n^2 - n + 1$$

3.43 (c)

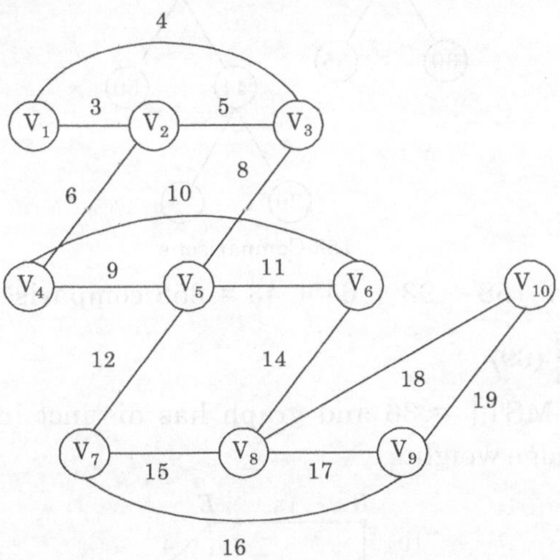

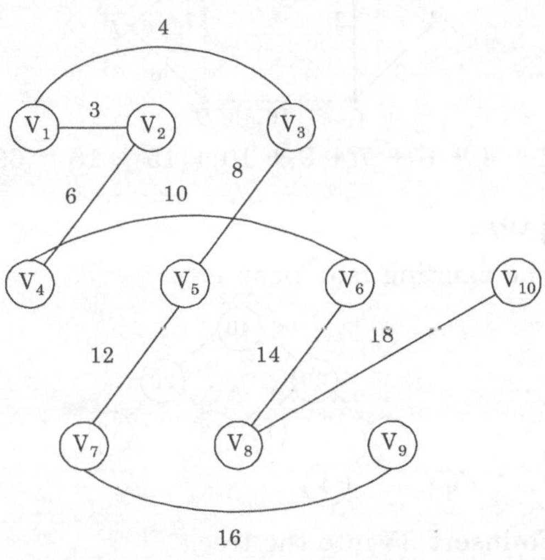

$V_5 \rightarrow V_6 : (V_5 - V_3 - V_1 - V_2 - V_4 - V_6)$

$\Downarrow$

31

$V_5 \rightarrow V_6 = 31$

3.44 (d)

E.g.

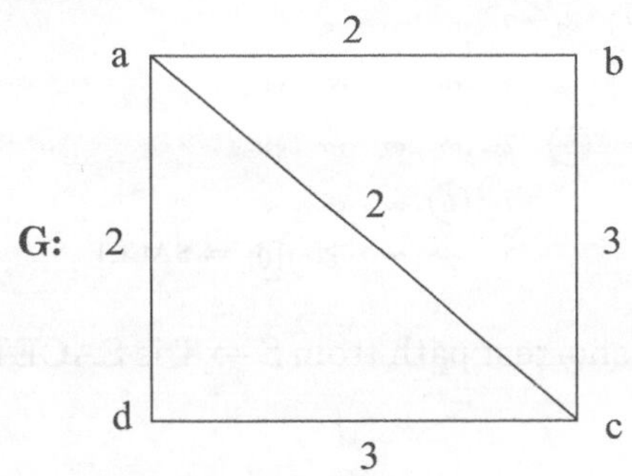

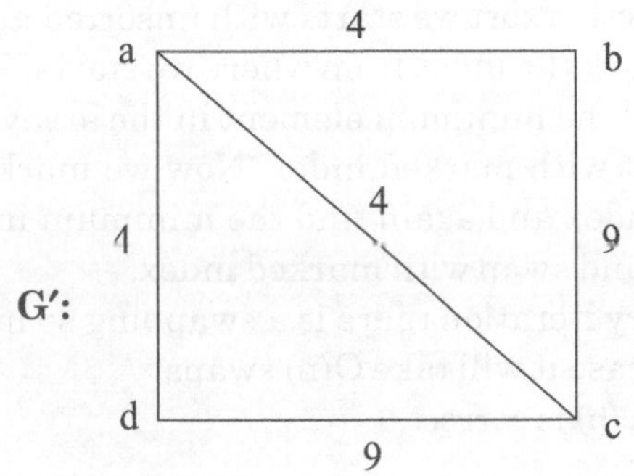

MST

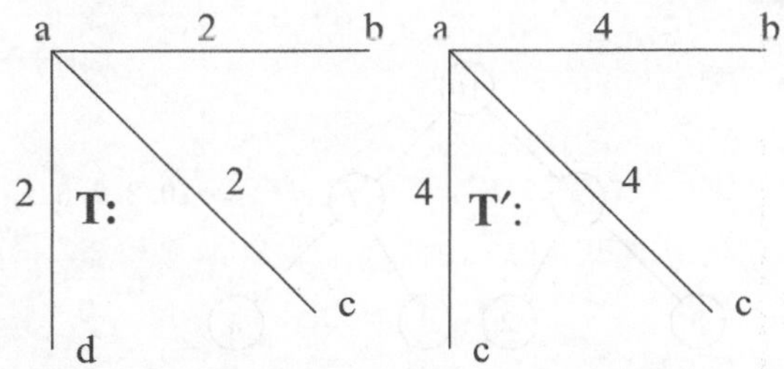

E.g.

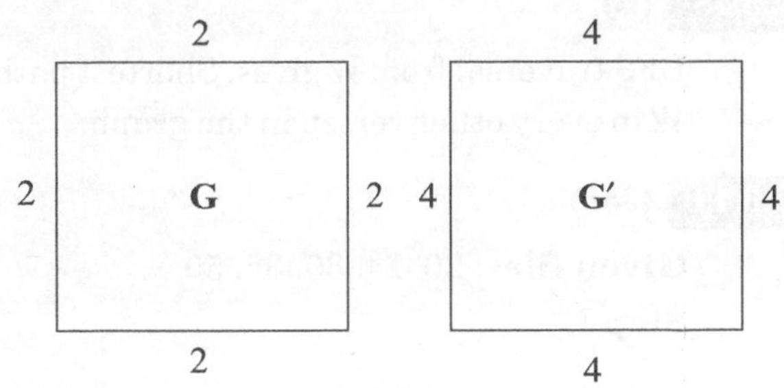

MST

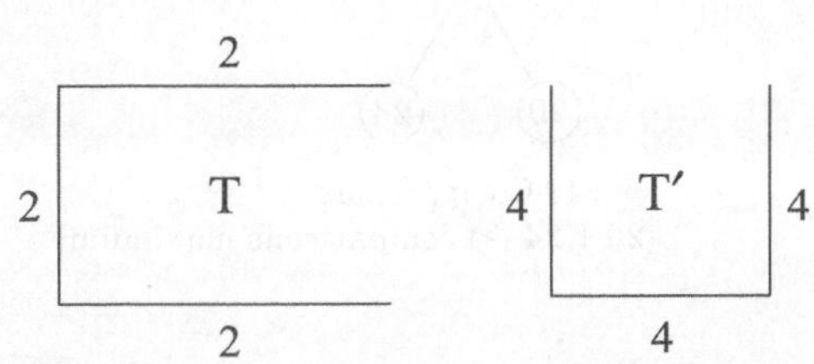

∴ we can make different as well as same MST also.

∴ option (d) is correct.

3.45 (d)

	A	B	C	D	E	F	G	T
S	4	(3)	∞	7	∞	∞	∞	∞
B	(4)		∞	7	∞	∞	∞	∞
A			(5)	7	∞	∞	∞	∞
C				7	(6)	∞	∞	∞
E				7		∞	8	(10) ⇒ SACET

∴ The shortest path from S → T is SACET.

3.46 (b)

In selection sort we starts with unsorted array, and mark the index from where we starts. Then we find the minimum element in the array and swap it with marked index. Now we mark the next index and again find the minimum in the array and swap with marked index.

In every iteration there is a swapping so in the worst case it will take O(n) swaps.

Option (b) is correct.

3.47 (a)

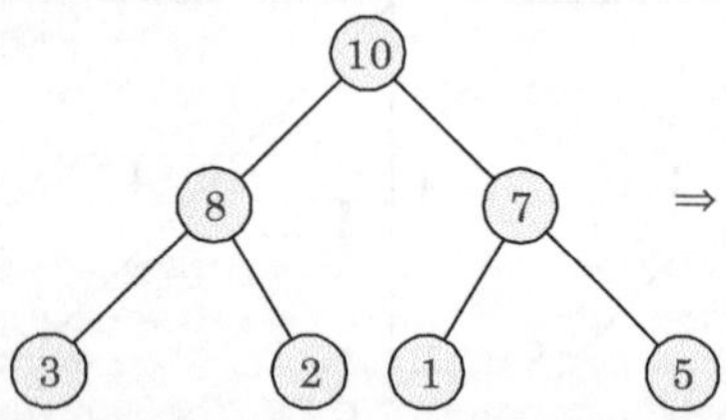

⇒ 10, 8, 7, 3, 2, 1, 5

3.48 (b)

BFS traversal from W gives. Shortest path from W to every other vertex in the graph.

3.49 (358)

Given files: 20, 24, 30, 35, 50

Step-I:

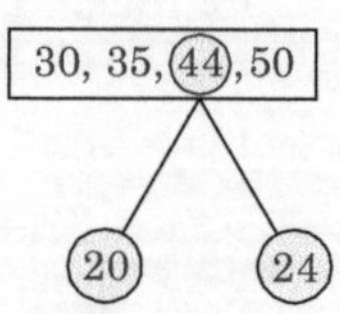

⇒ 43 Comparisons
[20 + 24 – 1 comparisons maximum]

Step-II:

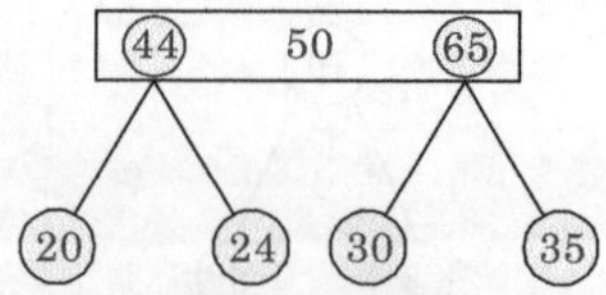

⇒ 64 Comparisons

Step-III:

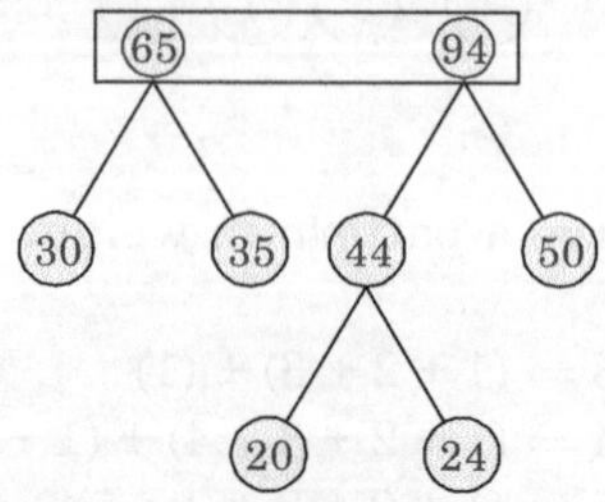

⇒ 93 Comparisons

Step-IV:

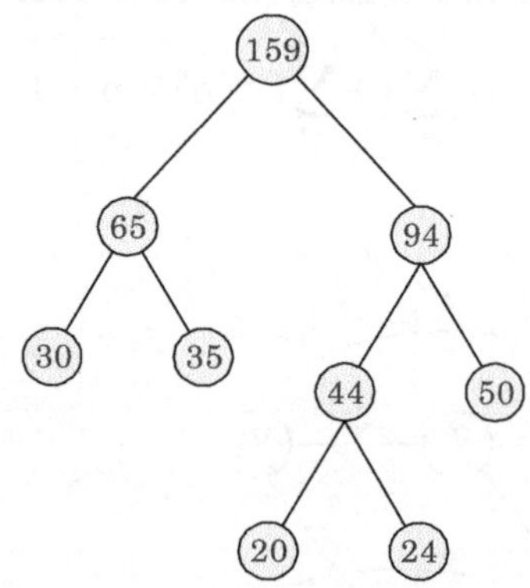

⇒ 158 Comparisons

∴ 158 + 93 + 64 + 43 = 358 comparisons

3.50 (69)

|MST| = 36 and graph has distinct integer edge weights.

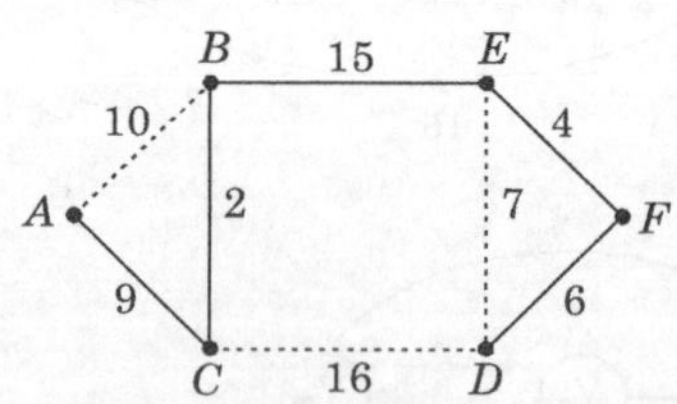

2 + 4 + 6 + 7 + 9 + 10 + 15 + 16 = 69

3.51 (b)

The existing max heap is

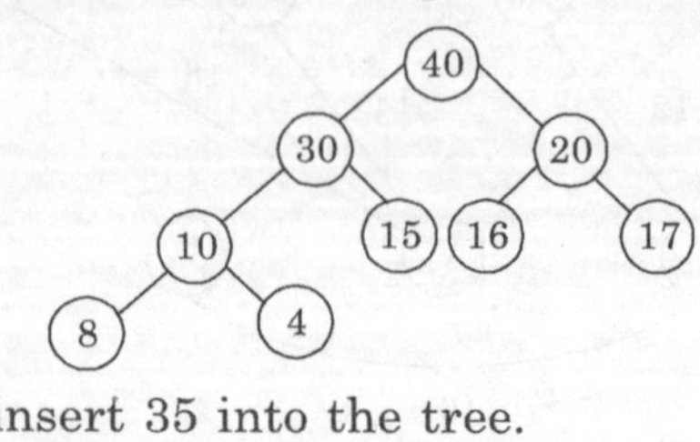

To insert 35 into the tree.

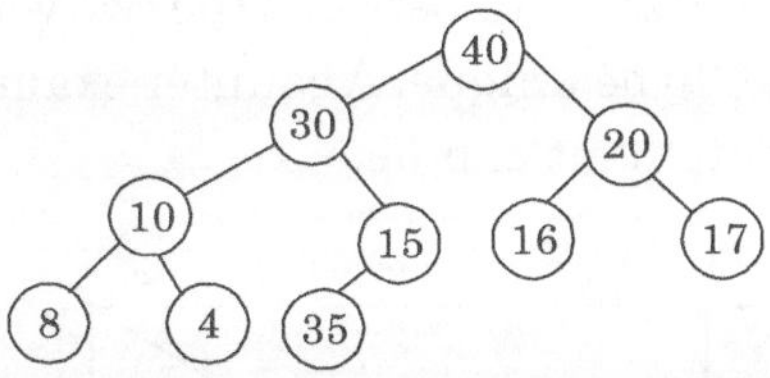

Now swap 15 and 35 to maintain max heap property.
Next swap 35 and 30 to maintain max heap property. Finally heap is

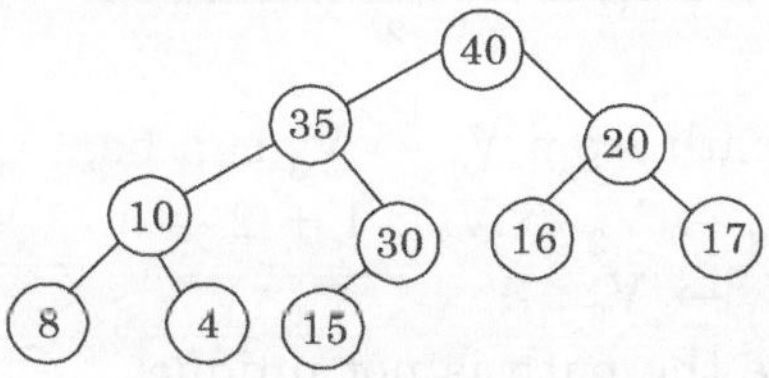

Array elements are : 40, 35, 20, 10, 30, 16, 17, 8, 4, 15.

3.52 (d)

Array:

0	1	2	3	4	5	6	7	8	9	10	11	12
89	19	50	17	12	15	2	5	7	11	6	9	100

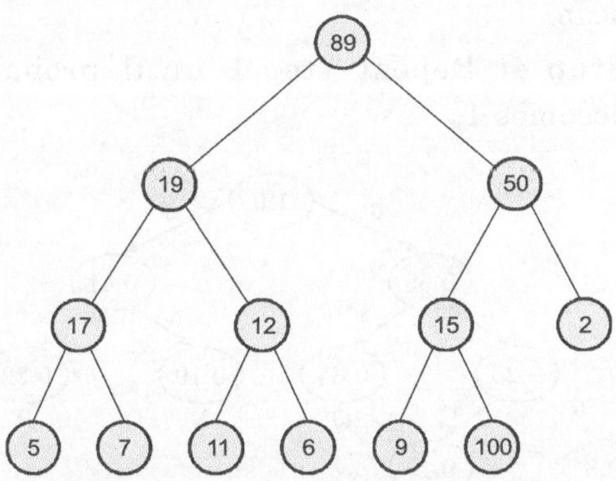

1st interchange : 100 & 15 (A[12] & A[5])
⇒ 2nd interchange : 100 & 50 (A[5] & A[27])
⇒ 3rd interchange : 100 & 89 (A[2] & A[0])

Maxheap Array:

100	19	89	17	12	50	2	5	7	11	6	9	15
0	1	2	3	4	5	6	7	8	9	10	11	12

3.53 (a)

Statement P: Since every edge weight is positive and we increase the value of every edge weight by same constant values. So minimum spanning tree of G does not change.

Statement Q: Taking an example:

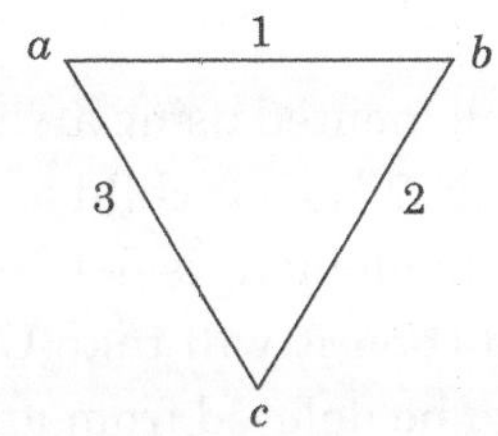

First path from 'a' to 'c' via 'b' have path value 3. But here path can be change 'a' to 'e' direct since paths value is same but path can be change.
So statement is wrong.

3.54 (7)

Since graph will be complete graph contain 4 vertex and 6 edges with weight 1, 2, 3, 4, 5, 6.

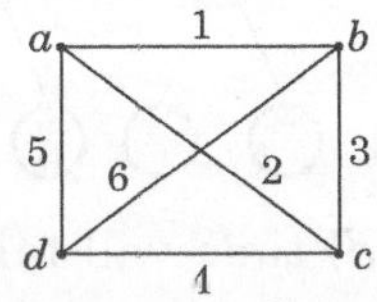

If we consider edge weights in such order, that form the cycle, then we have to choose two minimum we choose minimum value i.e., 4. So MST will be

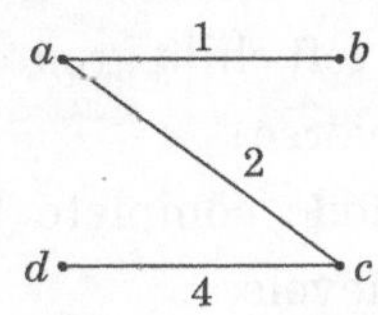

So max weight will be 1 + 2 + 4 = 7.

3.55 (b)

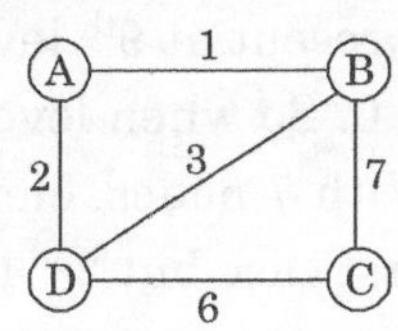

Here 'BD' is lightest edge in cycle 'BCD'. But here MST is not include edge 'BD' i.e.,

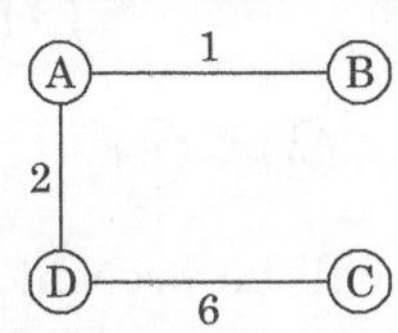

In second statement "if in any graph 'e' is the heaviest edge in then every MST must exclude it" is correct statement.

3.56 (b)

Heap is implemented using array. If 'i' is parent element then '$2i$' is left child and '$2i+1$' is right child. So if an element is delete from last level of the heap then it will take O(1) time. Since element can be deleted from any level of heap tree in worst case root element is deleted then at every level one element is exchange.

Example:

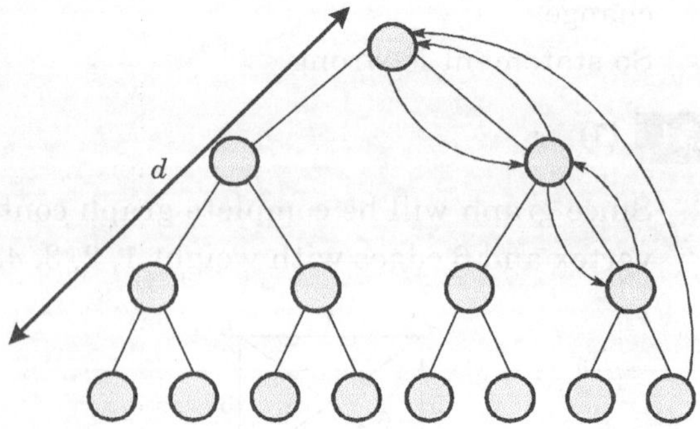

Minimum O(d) time will take if a element is deleted in heap tree but not O(1).

3.57 (8)

Min heap has property "root element is greater than from its left child and from its right child (don't matter left child greater or smaller than right or vice-versa)".

With 1023 node complete binary tree have maximum 9 levels.

1st minimum element can be present at 1st level. 2nd minimum element can be present at 2nd level ... and so on. So 9th minimum in worst case can be present at 9th level. When level is 1 = height is 0. So when level is 9 = height 8.

Example: With 7 nodes, element 3 can be at level 3 in worst case but 2nd level at best case.

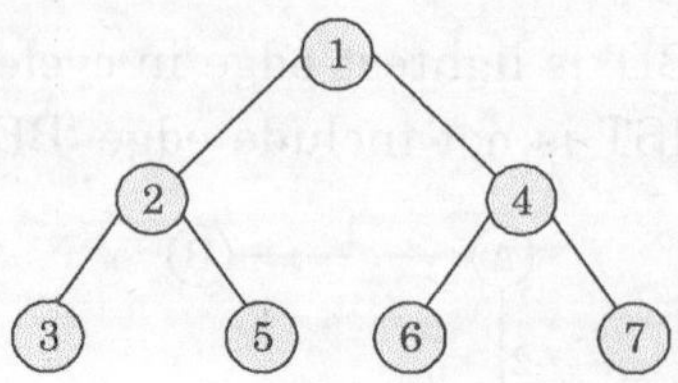

So integer 9 will be present at height 8.

3.58 (a)

- Since all the edge weights are unique, hence the minimum spanning tree of the graph will be unique.
- Shortest path between the two vertices need not to be unique. A counter example for the statement can be,

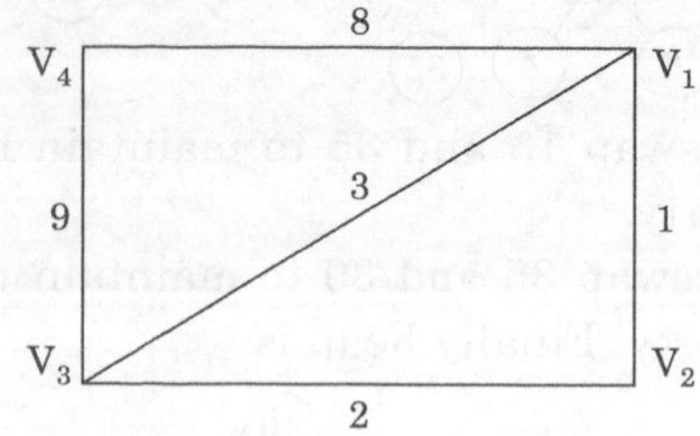

The path from $V_1 \rightarrow V_3$ can be,

I. $V_1 \rightarrow V_2 \rightarrow V_3 : 1 + 2 = 3$

II. $V_1 \rightarrow V_3 : 3$

Hence the path is not unique.

3.59 (225)

According to Huffman coding use Min heap data structure.

Step-1: Take two smallest probability.

Step-2: Add the probability taken in step-1.

Step-3: Put back previous sum in minimum heap.

Step-4: Repeat step-1 until probability becomes 1.

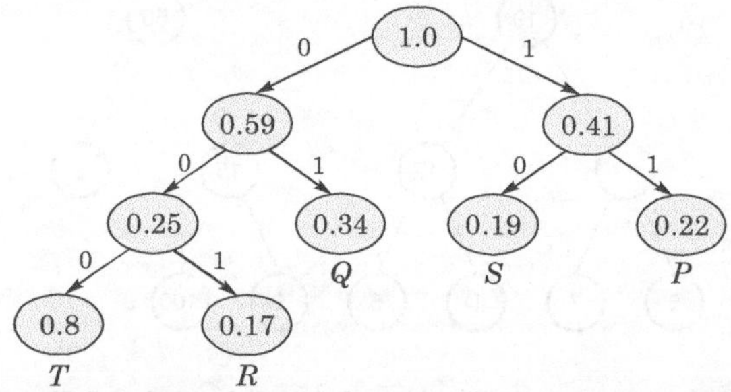

Number of bits required for each alphabet:

T = 3 bits (000)

S = 2 bits (10)

R = 3 bits (001)

P = 2 bits (11)

Q = 2 bits (01)

Average length/Char = Σ(Number of bits × Frequency of occurence of alphabet)

$= 3 \times 0.08 + 3 \times 0.17 + 2 \times 0.34 + 2 \times 0.19 + 2 \times 0.22 = 2.25$ bits

So, for 100 characters

$= 2.25$ bits $\times 100 = 225$ bits

3.60 (16)

This question is about Knapsack problem with greedy and dynamic solution with some constraints.

Total weight = 11 kgs

Constraints is: items should not split [either taken full are level it].

So, $V_{opt} = \text{Item}_1 (10) = 60 \text{ Rps}$

V_{greedy} :

Item Number	Weight in (Kgs)	Value in (Rupees)	Value/weight
1	10	60	60/10 = 6
2	7	28	28/7 = 4
3	4	20	20/4 = 5
4	2	24	24/2 = 12

Sorting in descending order:

Item	4	1	3	2
Value/weight	12	6	5	4

So, item picked $= \text{Item}_4 + \text{Item}_3$

$= 24 + 20 = 44 \text{ Rps}$

So, the value of $V_{opt} - V_{greedy}$

$= 60 \text{ Rps} - 44 \text{ Rps}$

$= 16 \text{ Rps}$

3.61 (4)

Given graph:

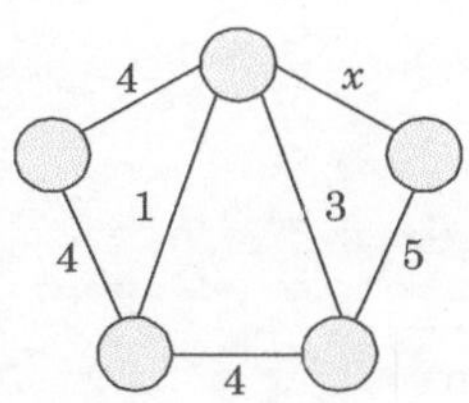

Minimum weight spanning tree according to given graph:

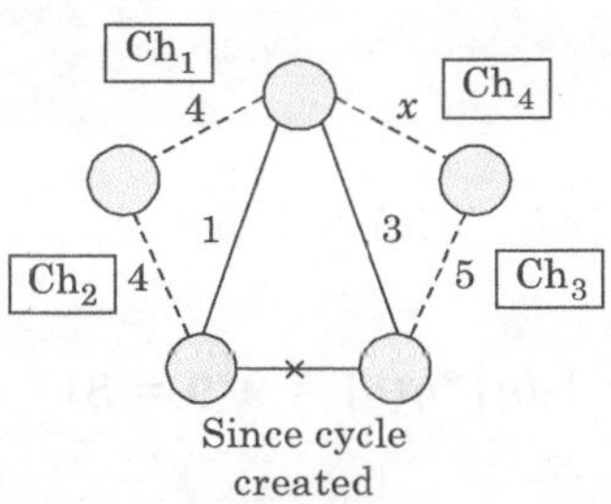

So, to get maximum number of MWST's we need to take $x = 5$. Hence number of MWST's are

$= 2 \times 2 = 4$

3.62 (d)

If no two edges of G have same weight surely G will have unique spanning tree is true. So I is true

Also if, for every cut of G, there is a unique minimum weight edge crossing the cut then G will have unique spanning tree is also true. So II is true

[**Note:** The converse of II is not true, but that is not relevant to this question]

So both I and II are true.

Option (d) is correct.

3.63 (511)

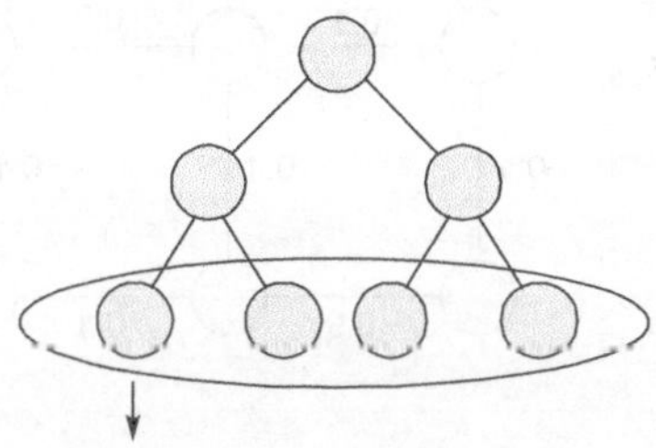

Maximum element at leaf level

Number of element $= \left\lceil \frac{1023}{2} \right\rceil = 512$

Now apply bubble sort requires 511 comparisons.

3.65 (99)

Consider small instance of 4-vertices

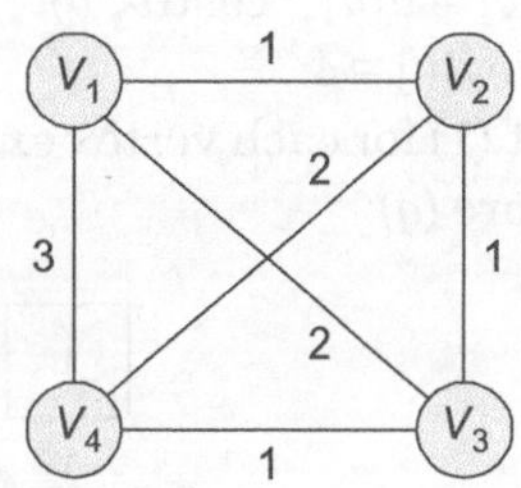

Clearly, 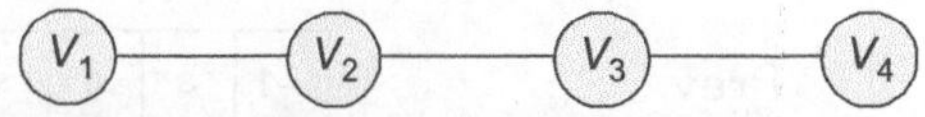

would be a MST with cost 3.

Hence for 100 vertices, 99 will the weight of minimum spanning tree.

3.67 (b, c)

R_7: Maximum amount earned by cutting rod of length '7' into 1, 2, 3, 4, 5, 6, 7 pieces (whichever way is maximum)

$p[7] = 18$

$p[6] + p[1] = 17 + 1 = 18$

(Also, R_7 is achieved by 3 diff. solution)

$p[5] + p[2] = 10 + 5 = 15$

$p[4] + p[3] = 9 + 8 = 17$

$p[5] + p[1] + p[1] = 10 + 1 + 1 = 12$

$p[4] + p[2] + p[1] = 9 + 5 + 1 = 15$

$p[4] + p[1] + p[1] + p[1] = 9 + 3(1) = 12$

$p[3] + p[4] = 17$

$p[3] + p[3] + p[1] = 8 + 8 + 1 = 17$

$p[3] + p[2] + p[2] = 8 + 5 + 5 = 18$

(R_7 is achieved by 3 diff. pieces)

$p[3] + 4\,p[1] = 8 + 4 = 12$

$p[2] + p[5] = 15$

$p[2] + p[4] + p[1] = 5 + 9 + 1 = 15$

$p[2] + p[3] + p[2] = 18$

$p[2] + 5\,p[15] = 5 + 5 = 10$

$p[2] + p[2] + 3\,p[1] = 5 + 5 + 3 = 13$

3.68 (3)

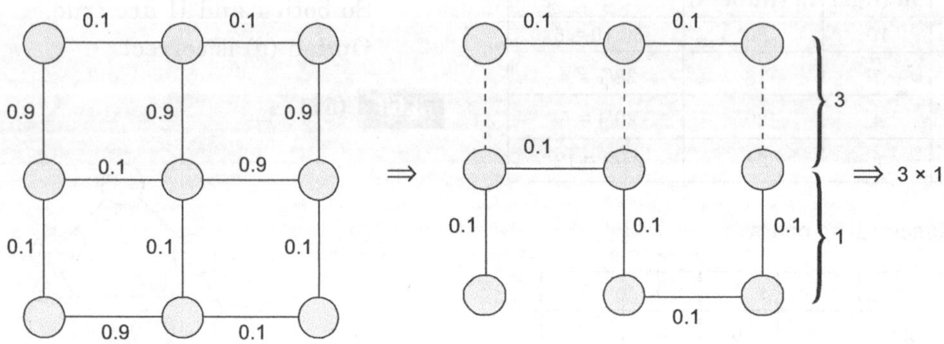

3.69 (929)

Modify Dijkstra's Algo to get longest path in terms of quality scores (Use Max heap)

Algo:

1. Choose vertex u which is maximum quality score value.
   V is set of adjacency of u.
   for (each vertex (v))
   { if ($q[v] < q[u]$*cost (u, v)]
   {
   $q[v] = q[u]$ * cost(u, v)
   prev[v] = 4
2. Repeat (1) for each vertex exactly once.

Quality score (q):

1	0	0	0	0	0	0	0	0
1	9	9	1	9	81	9	81	729

Initially, let $q[v] = 0$ or $-\infty$ (take smallest possible value)

Prev. :

–1	s	a	s	a	d	c	d	e

$S \Rightarrow \{a, c\}$ $\quad q(a) = 9,\ q(c) = 1$

$a \Rightarrow \{d, b\}$ $\quad q(d) = |ad| * q(a) = 1*9 = 9,\ q(b) = |ab| * q(a) = 1*9 = 9$

$b \Rightarrow \{e\}$ $\quad q(e) = |be| * q(b) = 1*9 = 9$

$d \Rightarrow \{e, g\}$ $\quad q(e) = |de| * q(d) = 9*9 = 81,\ q(g) = |dg| * q(d) = 9*9 = 81$

$e \Rightarrow \{t\}$ $\quad q(t) = |et| * q(e) = 9*81 = 729$

$g \Rightarrow \{t\}$ $\quad t$ already relaxed.

$c \Rightarrow \{f, d\}$ $\quad q(f) = |cf| * q(c) = 9*1 = 9$

$f \Rightarrow \{g\}$ $\quad g$ already relaxed

3.70 (d)

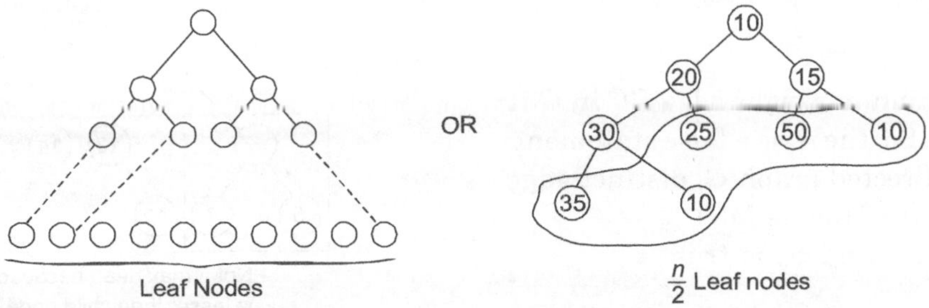

Maximum element is present somewhere in the leaf nodes.

$\therefore$ Find max in all leaf nodes from $\left(\left[\frac{n}{2}\right]+1\right)$.......$a[n]$

$\Rightarrow$ Number of comparisons $= \frac{n}{2} - 1 = \Theta(n)$

3.71 (a, b)

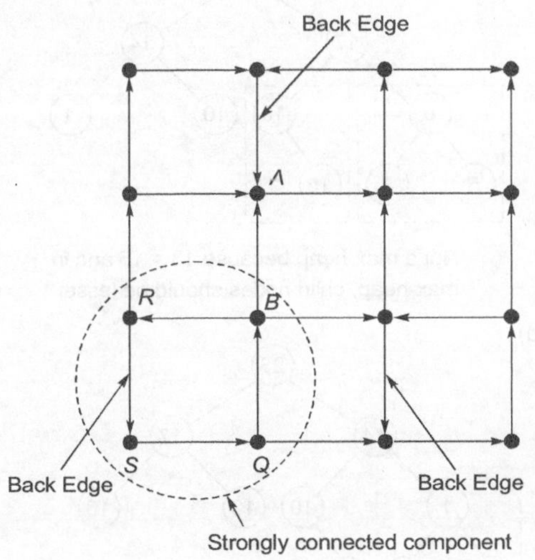

- As we can see there is cycle in given DAG. So, topological order is not possible.
- Statement (b) is also true.
- There is no path from A to B so statement (c) is false.
- In SQBR, it is strongly connected component. So, statement (d) is false.

3.72 (a)

Letters used in string :	a	b	c	d	e
Frequencies	1	2	2	2	3

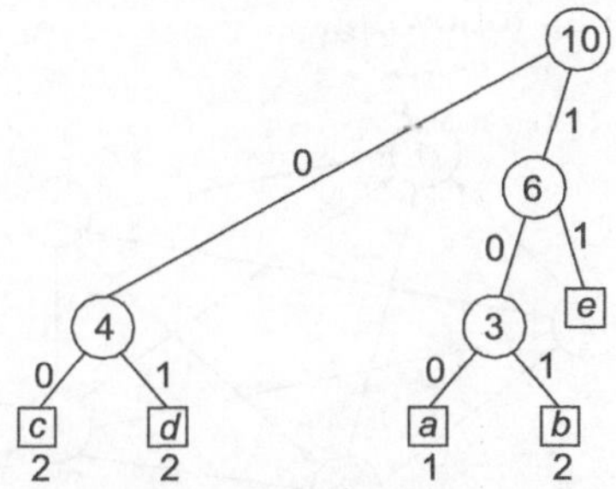

Letters used in string:

$\therefore$ Minimum length of encoded string = Sum frequency count using Huffman coding

$= 1 * 3 + 2 * 3 + 2 * 2 + 2 * 2 + 3 * 2$

$= 3 + 6 + 4 + 4 + 6 = 23$

3.73 (a)

S_1 : Consider the graph:

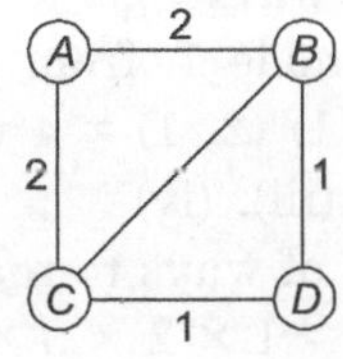

One of the possible MSTS can be:

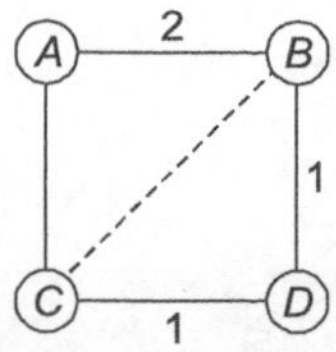

But the minimum weight edge BC in G is not present. So, the S_1 is a false statement.

S_2 : In any undirected graph G, distinct edge weights means Unique MST.

So, the S_1 is false and S_2 is true.

3.74 (a, b, c)

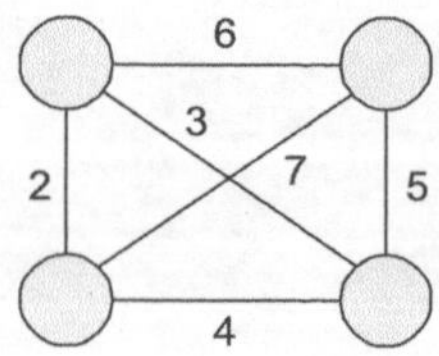

- 2nd minimum cost edge always in MST because 2nd minimum not forms cycle in MST.
- One of 3rd or 4th minimum in MST. If 3rd minimum forms cycle in MST the 4th minimum must be in MST.

3.75 (24)

According to given adjacency matrix, the graph will be as following:

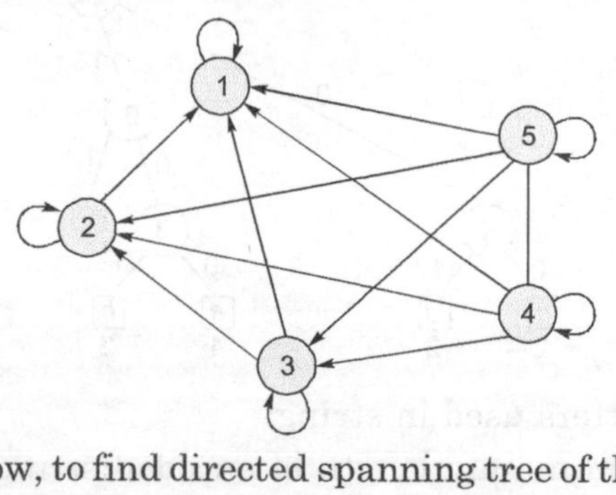

Now, to find directed spanning tree of the above graph as per definition given in question, $r = 5$ is given,

(i) To reach node 4, from 5 : (5, 4) = 1 way

(ii) To reach node 3, from {5, 4}: (5, 3), (4, 3) = 2 ways

(iii) To reach node 2, from {5, 4, 3}: (5, 2), (4, 2), (3, 2) = 3 ways

(iv) To reach node 1, from {5, 4, 3, 2}: (5, 1), (4, 1), (3, 1) (2, 1) = 4 ways

From (i), (ii), (iii), (iv),

Total number of ways to reach node 4, 5, 2, 1

From node 5 = 1 × 2 × 3 × 4 = 24 ways

So, there are 24 directed spanning trees for given graph.

3.76 (b)

(a)

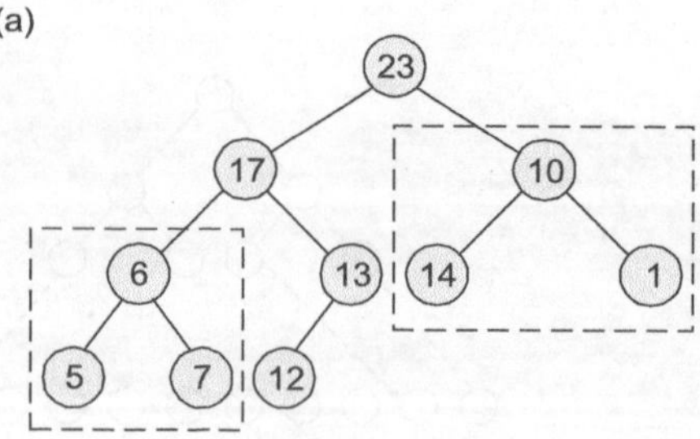

Not a max heap because parent node is lesser than child node (10 < 14, 6 < 7)

(b)

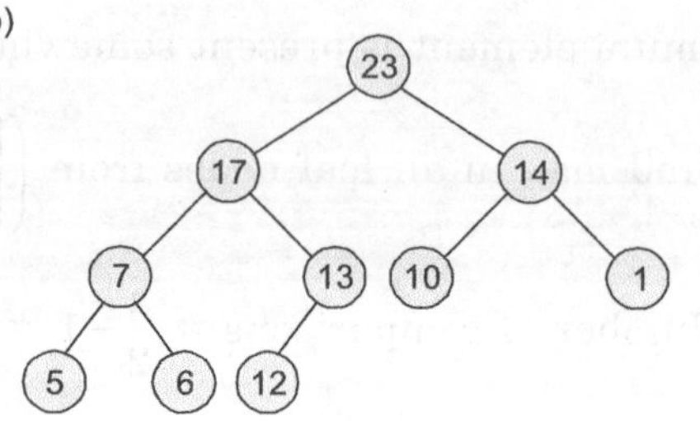

Max heap

(c)

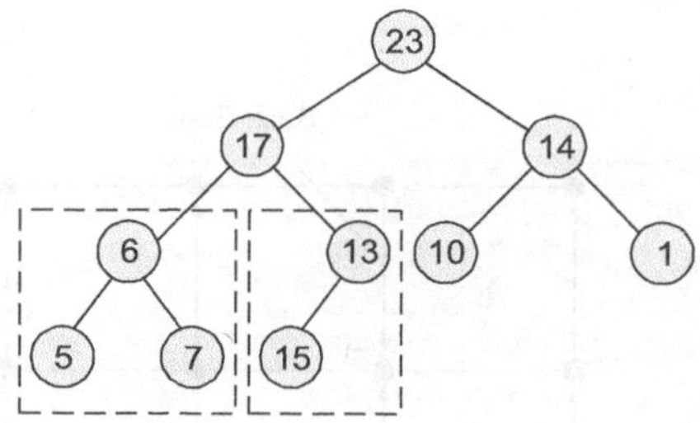

Not a max-heap because 13 < 15 and in max-heap, child nodes should be lesser

(d)

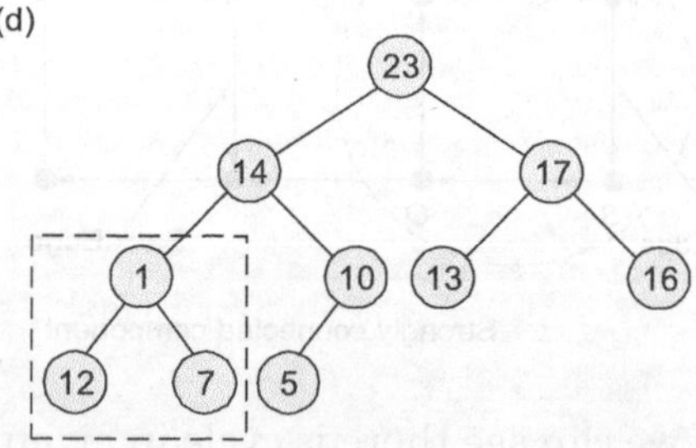

Not a max-heap because 1 < 12 and in max-heap, child nodes should be lesser

3.77 (b)

Insert element into max heap: required worst case $\log_2 n$ array shift operations or swaps

$$TC = O(\log n)$$

Delete max from max heap: required worst case $\log_2 n$ array shift operations

$$TC = O(\log n)$$

3.78 (16)

Complete connected graph of n vertices number of spanning trees : n^{n-2}
For 4 vertices complete graph number of spanning trees : $4^{4-2} = 16$

3.79 (60)

Number of vertices (V) = 100
{Number of edges present in forest of DFS traversal} = 40
[DFS spanning tee edges]
Number of connected components of graph (G) = {Number of vertices of graph (G)} – {Number of spanning tree edges of DFS traversal of graph G}
$$= 100 - 40 = 60$$

3.80 (d)

G : Directed graph
T : DFS spanning tree of G rooted at V and BFS spanning tree of G rooted at V.

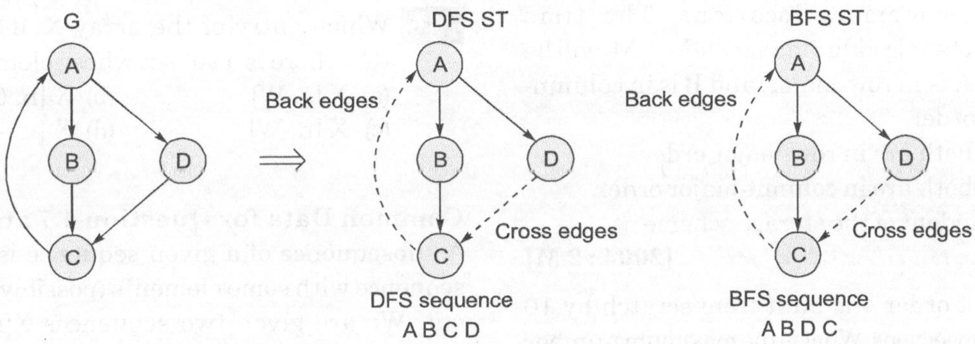

BFS spanning tree is also must be DFS spanning tree is given constraint so that no forward edges possible.

4 Dynamic Programming

4.1 Which one of the following algorithm design techniques is used in finding all pairs of shortest distances in a graph?
(a) Dynamic programming
(b) Backtracking
(c) Greedy
(d) Divide and Conquer **[1998 : 1 M]**

4.2 Two matrices M_1 and M_2 are to be stored in arrays A and B respectively. Each array can be stored either in row-major or column-major order in contiguous memory locations. The time complexity of an algorithm to compute $M_1 \times M_2$ will be
(a) best if A is in row-major, and B is in column-major order
(b) best if both are in row-major order
(c) best if both are in column-major order
(d) independent of the storage scheme

[2004 : 2 M]

4.3 A B-tree of order 4 is built from scratch by 10 successive insertions. What is the maximum number of node splitting operations that may take place?
(a) 3 (b) 4
(c) 5 (d) 6 **[2008 : 2 M]**

4.4 The subset-sum problem is defined as follows: Given a set S of n positive integers and a positive integer W; determine whether there is a subset of S whose elements sum to W.
An algorithm Q solves this problem in O(nW) time. Which of the following statements is false?
(a) Q solves the subset-sum problem in polynomial time when the input is encoded in unary
(b) Q solves the subset-sum problem is polynominal time when the input is encoded in binary
(c) The subset sum problem belongs to the class NP
(d) The subset sum problem in NP-hard

[2008 : 2 M]

Linked Answers Questions 4.5 and 4.6

The subset-sum problem is defined as follows. Given a set of n positive integers, $S = \{a_1, a_2, a_3, ..., a_n\}$ and a positive integer W is there a subset S whose elements sum of W? A dynamic program for solving this problem uses a 2-dimensional Boolean array, X with n rows and W-1 columns X [i, j], $1 \le i \le n$, $0 \le j \le W$, is TRUE if and only if there is a subset of $\{a_1, a_2....., a_i\}$ whose elements sum to j.

4.5 Which of the following is valid for $2 \le i \le n$ and $a_i \le j \le W$?
(a) $X[i, j] = X[i-1, j] \vee X[i, j-a_i]$
(b) $X[i, j] = X[i-1, j] \vee X[i-1, j-a_i]$
(c) $X[i, j] = X[i-1, j] \wedge X[i, j-a_i]$
(d) $X[i, j] = X[i-1, j] \wedge X[i-1, j-a_i]$

[2008 : 2 M]

4.6 Which entry of the array X, if TRUE, implies that there is a subset whose elements sum to W?
(a) X [1, W] (b) X [n, 0]
(c) X [n, W] (d) X [n –1, n]

[2008 : 2 M]

Common Data for Question 4.7 and 4.8

A subsequence of a given sequence is just the given sequence with some elements (possibly none or all) left out. We are given two sequences X[m] and Y[n] of lengths m and n, respectively, with indexes of X and Y starting from 0.

4.7 We wish to find the length of the longest common subsequence (LCS) of X[m] and Y[n] as $l(m, n)$, where an incomplete recursive definition for the function $l(i, j)$ to compute the length of the LCS of X[m] and Y[n] is given below:
$l(i, j) = 0$, if either i = 0 or j = 0
= expr1, if i, j > 0 and x[i –1] = Y [j –1]
= expr2, if i, j>0 and x[i–1] ≠ Y[j –1]
Which one of the following options is correct?
(a) expr1 ≡ $l(i-1, j) + 1$
(b) expr1 ≡ $l(i, j-1)$
(c) expr2 ≡ max ($l(i-1, j)$, $l(i, j-1)$)
(d) expr2 ≡ max ($l(i-1, j-1)$, $l(i, j)$)

[2009 : 2 M]

4.8 The values of $l(i,j)$ could be obtained by dynamic programming based on the correct recursive definition of $l(i, j)$ of the form given above, using an array L[M, N], where M = m + 1 and N = n + 1, such that L[i,j] = $l(i, j)$.

Which one of the following statements would be TRUE regarding the dynamic programming solution for the recursive definition of $l(i, j)$?

(a) All elements of L should be initialized to 0 for the values of l(i,j) to be properly computed
(b) The values of l(i, j) may be computed in a row major order or column major order of L [M, N]
(c) The values of l(i, j) cannot be computed in either row major order or column major order of L [M, N]
(d) L [p, q] needs to be computed before L [r, s] if either p < r or q < s **[2009 : 2 M]**

4.9 An algorithm to find the length of the longest monotonically increasing sequence of numbers in an array A[0: n – 1] is given below.

Let L_i denotes the length of the longest monotonically increasing sequence starting at index i in the array

Initialize $L_{n-1} = 1$

For all i such that $0 \le i \le n - 2$

$$L_i = \begin{cases} 1 + L_{i+1}, & \text{If } A[i] < A[i+1] \\ 1 & \text{Otherwise} \end{cases}$$

Finally the length of the longest monotonically increasing sequence is $\text{Max}(L_0, L_1, \ldots, L_{n+1})$.

Which of the following statements is TRUE?
(a) The algorithm uses dynamic programming paradigm
(b) The algorithm has a linear complexity and uses branch and bound paradigm
(c) The algorithm has a non-linear polynomial complexity and uses branch and bound paradigm
(d) The algorithm uses divide and conquer paradigm **[2011 : 1 M]**

4.10 What is the time complexity of Bellman-Ford single-source shortest path algorithm on a complete graph of n vertices?
(a) $\Theta(n^2)$ (b) $\Theta(n^2 \log n)$
(c) $\Theta(n^3)$ (d) $\Theta(n^3 \log n)$
[2013 : 1 M]

4.11 Consider two strings A = "$qpqrr$" and B = "$pqprqrp$". Let x be the length of the longest common subsequence (not necessarily contiguous) between A and B and let y be the number of such longest common subsequences between A and B. Then $x + 10y$ = ________.
[2014 (Set-2) : 2 M]

4.12 Consider the weighted undirected graph given by with 4 vertices, where the weight of edge $\{i, j\}$ is the entry W_{ij} in the matrix W.

$$W = \begin{bmatrix} 0 & 2 & 8 & 5 \\ 2 & 0 & 5 & 8 \\ 8 & 5 & 0 & x \\ 5 & 8 & x & 0 \end{bmatrix}$$

The largest possible integer value of x, for which at least one shortest path between some pair of vertices will contain the edge with weight x is ________.
[2016 (Set-1) : 2 M]

4.13 Let A_1, A_2, A_3 and A_4 be four matrices of dimensions 10×5, 5×20, 20×10, and 10×5, respectively. The minimum number of scalar multiplications required to find the product $A_1 A_2 A_3 A_4$ using the basic matrix multiplication method is ______.
[2016 (Set-2) : 2 M]

4.14 Assume that multiplying a matrix G_1 of dimension $p \times q$ with another matrix G_2 of dimension $q \times r$ requires pqr scalar multiplications. Computing the product of n matrices $G_1 G_2 G_3 \ldots G_n$ can be done by parenthesizing in different ways. Define $G_i G_{i+1}$ as an explicitly computed pair for a given paranthesization if they are directly multiplied. For example, in the matrix multiplication chain $G_1G_2G_3G_4G_5G_6$ using parenthesization $(G_1(G_2G_3))(G_4(G_5G_6))$. G_2G_3 and G_5G_6 are the only explicitly computed pairs.

Consider a matrix multiplication chain $F_1F_2F_3F_4F_5$, where matrices F_1, F_2, F_3, F_4 and F_5 are of dimensions 2×25, 25×3, 3×16, 16×1 and 1×1000, respectively. In the parenthesization of $F_1F_2F_3F_4F_5$ that minimizes the total number of scalar multiplications, the explicitly computed pairs is/are
(a) F_1F_2, F_3F_4 only (b) F_2F_3 only
(c) F_3F_4 only (d) F_1F_2, F_4F_5 only
[2018 : 2 M]

Answers Dynamic Programming

4.1	(a)	**4.2**	(d)	**4.3**	(a)	**4.4**	(d)	**4.5**	(b)	**4.6**	(c)	**4.7**	(c)	**4.8**	(b)	**4.9**	(a)
4.10	(c)	**4.11**	(34)	**4.12**	(12)	**4.13**	(1500)	**4.14**	(c)								

Explanations Dynamic Programming

4.1 (a)

Bellman-Ford algorithm is used to find all pairs shortest distances in a graph and it is dynamic programming technique.

4.2 (d)

Access time equal for all elements if it is stored in array. Accessing any element takes constant time.

4.3 (a)

A B-tree is similar to 2-3 tree. Consider a B-tree of order 4.

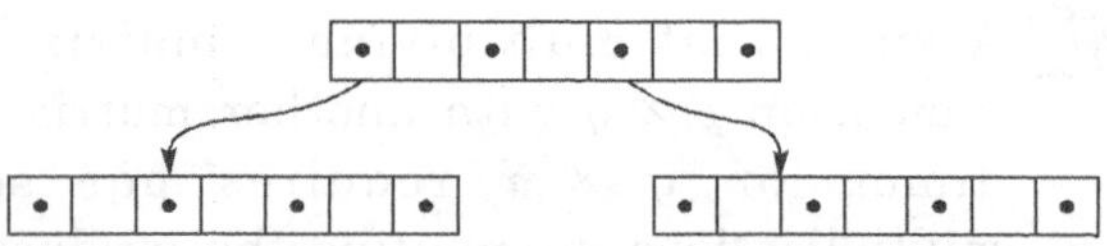

A B-tree of order m contains n records and if each contains b records on the average then the tree has about $\lceil n/b \rceil$ leaves. If we split k nodes along the path from leaves then

$$k \le 1 + \log_{m/2} \lceil n/b \rceil$$

In given problem n = 10, b = 3, m = 4 So

$$k \le 1 + \log_{4/2} \lceil 10/3 \rceil$$
$$k \le 1 + \log_2 4$$
$$k \le 1 + 2$$
$$k \le 3$$

4.4 (d)

If an algorithm Q solves subset-sum problem in O(nW) time where w is an integer which is constant so Q solves the subset-sum problem in O(n) time because W is a constant and input must encoded in binary.

4.5 (b)

The subset-sum problem is NP-complete problem.

For example given a set S = {–7, –3, –2, 5, 8} the algorithm determine whether the sum of the element of S′ is zero such that $S' \subseteq S$. For example if S′ = {– 3, – 2, 5} then return yes.

In given dynamic algorithm x is a two dimensional Boolean array contains n rows and W + 1 columns where W is a positive weight or sum.

$$x[i, j]\ 1 \le i \le n,\ 0 \le j \le W$$

The ith row determines the element of subset of S, and j^{th} column determines the corresponding weight.

If $2 \le i \le n$ and $a \le j \le W$

Let subset $S' = \{a_1, a_2, .., a_i\}$ the weight as $a_1, ..., a_2$ is W.

So $X[i, j] = X[i-1, j] \vee X[i-1, j-a_i]$

The row starts from 2nd position so i = i – 1. The column position starts from j but we will add upto a_i^{th} position so $j = j - a_i$.

4.6 (c)

The entry X[n, W] is true, because we find the weight W of subset of S which contains n elements.

4.7 (c)

$l(i, j)$

= 0; if either i = 0 or j = 0

= $l(i-1, j-1)$;

if i, j > 0 and $x[i-1]$=Y[j – 1]

= max ($l(i-1, j)$, $l(i, j-1)$);

if i, j > 0 and $x[i-1] \ne$ Y[j – 1]

$\therefore$ expr 2 $\equiv$ max($l(i-1, j)$, $l(i, j-1)$)

4.8 (b)

$l(i, j)$ values can be computed in a row major order or column major order of L[M, N].

4.9 (a)

The algorithm uses dynamic programming paradigm.

4.10 (c)

In a complete graph total no of edges is

$$= \frac{n(n-1)}{2}$$

Time complexity of Bellman-ford algo for a graph having n vertices and m edges = O(nm) for a complete graph time complexity

$$O\left(n \times \frac{n(n-1)}{2}\right) = O(n^3)$$

So, option (c) is correct.

4.11 (34)

$$A = qpqrr$$
$$B = pqprqrp$$
$$\text{LCS}(A, B) = qpqr$$
$$(\text{or}) = qprr$$
$$(\text{or}) = pqrr$$
$$x = \text{Length of LCS}(A, B) = 4$$
$$y = \text{Number of LCS}(A, B) = 3$$
$$\therefore \quad x + 10y = 4 + 10 \times 3 = 34$$

4.12 (12)

By drawing weighted undirected graph with 4 vertices from given matrix is

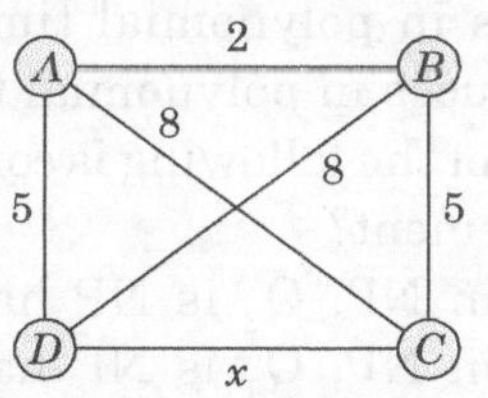

By analysis given graph we see their are two paths i.e., one is '*C B A D*' and '*C A D*' having weight 12 and 13 respectively to reach *C* to *D*. Since we have to find atleast 1 (means more than 1 or 1) sortest path between *C* to *D*. So we have to consider 12 as value of x, by doing this we have atleast 1 path between *C* to *D* i.e., 1 path *CBAD* and 2nd path *CD* having weight 12 each.

4.13 (1500)

$(A_1A_2A_3A_4)$

$(A_1A_2A_3)A_4$	$(A_1A_2)(A_3A_4)$	$A_1(A_2A_3A_4)$
1. 10 × 5 × 5 × 20 = 1000 (10 × 20)	1. 10 × 5 × 5 × 20 = 1000 (10 × 20)	1. 5 × 20 × 20 × 10 = 1000 (5 × 10)
2. 10 × 20 × 20 × 10 = 2000 (10 × 10)	2. 20 × 10 × 10 × 5 = 1000 (20 × 5)	2. 5 × 10 × 10 × 5 = 250 (5 × 10)
3. 10 × 10 × 10 × 5 = 500 (10 × 5)	3. 10 × 20 × 20 × 5 = 1000 (10 × 5)	3. 5 × 10 × 10 × 5 = 250 (5 × 10)
3500	3000	1500

The minimum number of multiplication required using basic matrix multiplication method will be 1500.

4.14 (c)

$$F_1 = 2 \times 25;$$
$$F_2 = 25 \times 3;$$
$$F_3 = 3 \times 16;$$
$$F_4 = 16 \times 1;$$
$$F_5 = 1 \times 1000$$

$F_1F_2F_3F_4F_5$

$(((F_1F_2)(F_3F_4))F_5)$	$((F_1(F_2(F_3F_4)))F_5)$	$(F_1((F_2F_3)F_4)F_5)$	$((F_1F_2)F_3(F_4F_5))$
1. 2×25×3 = 150	1. 3×16×1 = 48	1. 25×3×16 = 1200	1. 2×25×3 = 150
2. 3×16×1 = 48	2. 25×3×1 = 75	2. 25×16×1 = 400	2. 16×1×1000 = 16000
3. 2×3×1 = 6	3. 2×25×1 = 50	3. 2×25×1 = 50	3. 2×3×16 = 96
4. 2×1×1000 = 2000	4. 2×1×1000 = 2000	4. 2×1×1000 = 2000	4. 2×16×1000 = 32000
2204	2173	3650	48246

So, only (F_3F_4) must be necessary to calculate for minimum number of multiplication.

5 P and NP Concepts

5.1 A problem in NP is NP-complete if
(a) it can be reduced to the 3-SAT problem in polynomial time
(b) the 3-SAT problem can be reduced to it in polynomial time
(c) it can be reduced to any other problem in NP in polynomial time
(d) some problem in NP can be reduced to it in polynomial time **[2006 : 1 M]**

5.2 For problems X and Y, Y is NP-complete and X reduces to Y in polynomial time. Which of the following is TRUE?
(a) If X can be solved in polynomial time, then so can Y
(b) X is NP-complete
(c) X is NP-hard
(d) X is in NP, but not necessarily NP-complete

[2008 : 1 M]

5.3 Let π_A be a problem that belongs to the class NP. Then which one of the following is TRUE?
(a) There is no polynomial time algorithm for π_A
(b) If π_A can be solved deterministically in polynomial time, then P = NP
(c) If π_A is NP-hard, then it is NP-complete
(d) π_A may be undecidable **[2009 : 1 M]**

5.4 Which of the following statements are TRUE?
1. The problem of determining whether there exists a cycle in an undirected graph is in P.
2. The problem of determining whether there exists a cycle in an undirected graph is in NP.
3. If a problem A is NP-Complete, there exists a non-deterministic polynomial time algorithm to solve A.

(a) 1, 2 and 3 (b) 1 and 3 only
(c) 2 only (d) 3 only **[2013 : 1 M]**

5.5 Suppose a polynomial time algorithm is discovered that correctly computes the largest clique in a given graph. In this scenario, which one of the following represents the correct Venn diagram of the complexity classes P, NP and NP Complete (NPC)?

(a)
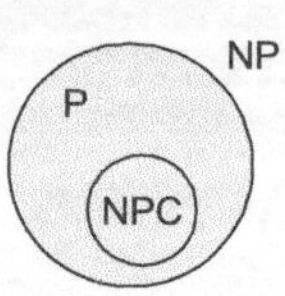

(b)
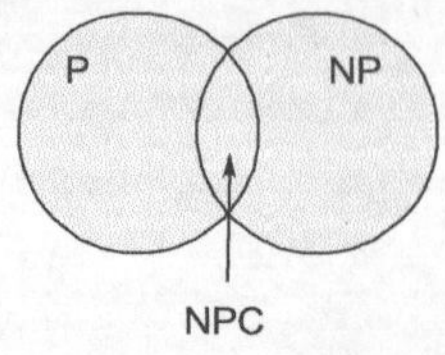

(c)
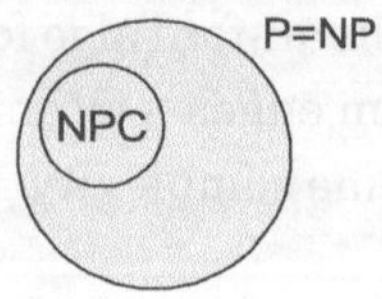

(d)
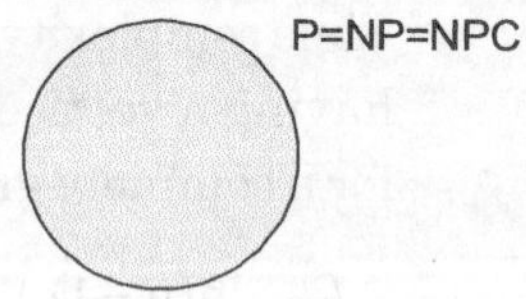

[2014 (Set-1) : 2 M]

5.6 Consider the decision problem 2CNFSAT defined as follows:
$\{\phi \mid \phi$ is a satisfiable propositional formula in CNF with at most two literals per clause$\}$
For example, $\phi = (x_1 \vee x_2) \wedge (x_1 \vee \overline{x}_3) \wedge (x_2 \vee x_4)$ is a Boolean formula and it is in 2CNFSAT. The decision problem 2CNFSAT is
(a) NP-Complete.
(b) solvable in polynomial time by reduction to directed graph reachability.
(c) solvable in constant time since any input instance is satisfiable.
(d) NP-hard, but not NP-complete.

[2014 (Set-3) : 2 M]

5.7 Consider two decision problems Q_1, Q_2 such that Q_1 reduces in polynomial time to 3-SAT and 3-SAT reduces in polynomial time to Q_2. Then which one of the following is consistent with the above statement?
(a) Q_1 is in NP, Q_2 is NP hard
(b) Q_2 is in NP, Q_1 is NP hard
(c) Both Q_1 and Q_2 are in NP
(d) Both Q_1 and Q_2 are NP hard

[2015 (Set-2) : 1 M]

5.8 Language L_1 is polynomial time reducible to language L_2. Language L_3 is polynomial time reducible to L_2, which in turn is polynomial time reducible to language L_4. Which of the following is/are True?
I. If $L_4 \in P$, $L_2 \in P$
II. If $L_1 \in P$ or $L_3 \in P$, then $L_2 \in P$
III. $L_1 \in P$, if and only if $L_3 \in P$
IV. If $L_4 \in P$, then $L_1 \in P$ and $L_3 \in P$
(a) II only
(b) III only
(c) I and IV only
(d) I only

[2015 (Set-3) : 2 M]

■■■■

Answers P & NP Concepts

5.1 (a) 5.2 (b) 5.3 (c) 5.4 (a) 5.5 (d) 5.6 (b) 5.7 (a) 5.8 (c)

Explanations P & NP Concepts

5.1 (a)

3-SAT being an NPC problem, reducing NP problem to 3-SAT would mean that NP problem is NPC.

5.2 (b)

X is reducible to NPC. Hence X is also NPC.

5.3 (c)

It is given that $\pi_A \in NP$

$\therefore$ If π_A is NP- hard, and since it is given that $\pi_A \in NP$, this means that π_A is NP-complete

$\therefore$ Choice (c) is correct.

Notice that choice (a) is incorrect since as P$\in$NP, some NP problems are actually P, and hence polynomial time algorithm can exist for these.

Choice (b) is incorrect since, If π_A can be solved deterministically in polynomial time, it does not generate that P = NP, unless of-course it is additionally known that π_A is NP-complete.

Choice (d) is incorrect since,

All problems belonging to P or NP have to be decidable.

5.4 (a)

1. By depth first search we can find whether there exits a cycle in an undirected graph in O(n^2) times so it is P problem.
2. As P $\subseteq$ NP so this problem will also be considered as NP problem.
3. This is the definition of NP-complete so it is trivially true.

So option (a) is correct

5.5 (d)

Clique is NP-complete problem

If clique is P then P = NPC

$\therefore$ P = NP = NPC

5.6 (b)

2CNF is P-problem.

Therefore it is solvable in polynomial time by reduction to directed graph reachability.

5.7 (a)

Given $Q_1 \leq$ 3-SAT and 3-SAT $\leq Q_2$

3-SAT is NP-complete.

$\therefore$ It is NP as well as NP-hard.

$Q_1 \leq$ 3-SAT and 3-SAT is NP $\Rightarrow$ Q_1 is NP-problem

3-SAT $\leq$ Q2 and 3-SAT is NP-hard. $\Rightarrow$ Q_2 is NP-hard problem.

So the strongest statements is Q_1 is in NP and Q_2 is NP-hard.

5.8 (c)

$L_1 \leq_P L_2$, $L_3 \leq_P L_2$, and $L_2 \leq_P L_4$

I. If $L_4 \in P$, $L_2 \in P$ is **true**

$L_2 \leq_P L_4$

II. If $L_1 \in P$ or $L_3 \in P$, then $L_2 \in P$ is **false**

$L_1 \leq_P L_2$, and $L_3 \leq_P L_2$

III. $L_1 \in P$, if and only if $L_3 \in P$ is **false**

If $L_1 \in P$ we cannot say anything about L_2 from $L_1 \leq_P L_2$

Since we cannot say anything about L_2 therefore we cannot say anything about L_3 either from $L_3 \leq_P L_2$.

So Statement III is **false**.

IV. If $L_4 \in P$, then $L_1 \in P$ and $L_3 \in P$ is **true**.

If $L_4 \in P$ then from $L_2 \leq_P L_4$ we can get that $L_2 \in P$.

and from $L_3 \leq_P L_2$ we can now get that $L_3 \in P$ also from $L_1 \leq_P L_2$ we can get that $L_1 \in P$

So Statement IV is **true**.

6 Miscellaneous Topics

6.1 The minimum number of interchanges needed to convert the array
89, 19, 40, 17, 12, 10, 2, 5, 7, 11, 6, 9, 70 into a heap with maximum element at the root is
(a) 0 (b) 1
(c) 2 (d) 3 **[1996 : 2 M]**

6.2 The average number of key comparisons done on a successful sequential search in list of length n is
(a) log n (b) $\frac{n-1}{2}$
(c) $\frac{n}{2}$ (d) $\frac{n+1}{2}$ **[1996 : 2 M]**

6.3 The correct matching for the following pairs is

List-I
A. All pairs shortest paths
B. Quick Sort
C. Minimum weight spanning tree
D. Connected Components

List-II
1. Greedy
2. Depth-First search
3. Dynamic Programming
4. Divide and Conquer

Codes:
(a) A – 2 B – 4 C – 1 D – 3
(b) A – 3 B – 4 C – 1 D – 2
(c) A – 3 B – 4 C – 2 D – 1
(d) A – 4 B – 1 C – 2 D – 3 **[1997 : 1 M]**

6.4 Give the correct matching for the following pairs:

List-I	List-II
A. O(log n)	**P.** Selection
B. O(n)	**Q.** Insertion sort
C. O(n log n)	**R.** Binary search
D. $O(n^2)$	**S.** Merge sort

Codes:
(a) A – R B – P C – Q D – S
(b) A – R B – P C – S D – Q
(c) A – P B – R C – S D – Q
(d) A – P B – S C – R D – Q **[1998 : 1 M]**

6.5 Suppose we want to arrange the n numbers stored in an array such that all negative values occur before all positive ones. Minimum number of exchanges required in the worst case is
(a) n – 1 (b) n
(c) n + 1 (d) None of these
[1999 : 1 M]

6.6 The number of articulation points of the following graph is

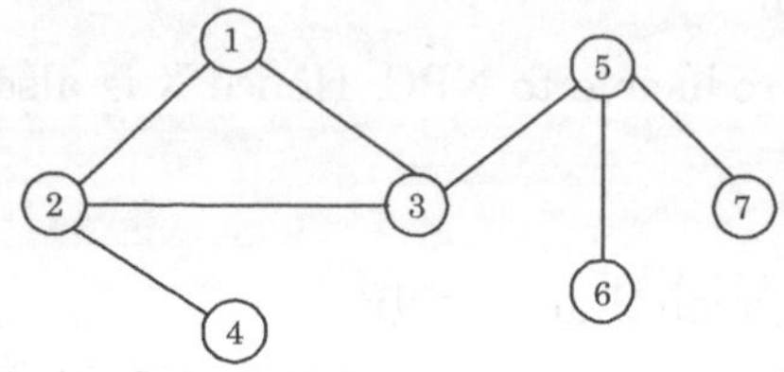

(a) 0 (b) 1
(c) 2 (d) 3 **[1999 : 1 M]**

6.7 If $T_1 = O(1)$, give the correct matching for the following pairs:
(m) $T_n = T_{n-1} + n$ (u) $T_n = O(n)$
(n) $T_n = T_{\frac{n}{2}} + n$ (v) $T_n = O(n \log n)$
(o) $T_n = T_{n/2} + n \log n$ (w) $T_n = O(n^2)$
(p) $T_n = T_{n-1} + \log n$ (x) $T_n = O(\log^2 n)$
(a) m – w, n – v, o – u, p – x
(b) m – w, n – u, o – x, p – v
(c) m – v, n – w, o – x, p – u
(d) m – w, n – u, o – v, p – v **[1999 : 2 M]**

Common Data for Q. 6.8 & Q. 6.9

In a permutation $a_1 \ldots a_n$ of n distinct integers, an inversion is a pair (a_i, a_j) such that $i < j$ and $a_i > a_j$.

6.8 If all permutations are equally likely, what is the expected number of inversions in a randomly chosen permutation of 1...n?
(a) $\frac{n(n-1)}{2}$ (b) $\frac{n(n-1)}{4}$
(c) $\frac{n(n+1)}{4}$ (d) $2n[\log_2 n]$
[2003 : 2 M]

6.9 What would be the worst case time complexity of the insertion sort algorithm, if the inputs are restricted to permutations of 1...n with at most n inversions?
(a) $\Theta(n^2)$ (b) $\Theta(n \log n)$
(c) $\Theta(n^{1.5})$ (d) $\Theta(n)$ **[2003 : 2 M]**

6.10 Suppose each set is represented as a linked list with elements in arbitrary order. Which of the operations among union, intersection, membership, cardinality will be the slowest?
(a) union only
(b) intersection, membership
(c) membership, cardinality
(d) union, intersection **[2004 : 2 M]**

6.11 A program takes as input a balanced binary search tree with n leaf nodes and computes the value of a function g(x) for each node x. If the cost of computing g(x) is min(number of leaf-nodes in left-subtree of x, number of leaf-nodes in right-subtree of x) then the worst case time complexity of the program is
(a) $\Theta(n)$ (b) $O(n \log n)$
(c) $O(n)^2$ (d) $O(2^n)$ **[2004 : 2 M]**

6.12 A scheme for storing binary trees in an array X is as follows. Indexing of X starts at 1 instead of 0. The root is stored at X[1]. For a node stored at X[i], the left child, if any, is stored in X[2i] and the right child, if any, in X[2i + 1]. To be able to store any binary tree on n vertices, the minimum size of X should be
(a) $\log_2 n$ (b) n
(c) 2n + 1 (d) 2n **[2006 : 1 M]**

6.13 Which one the following inplace sorting algorithms needs the minimum number of swaps?
(a) Quicksort (b) Insertion sort
(c) Selection sort (d) Heap sort
[2006 : 1 M]

6.14 An element in an array X is called a leader if it is greater than all elements to the right of it in X. The best algorithm to find all leaders in an array.
(a) Solves it in linear time using a left to right pass of the array
(b) Solves in linear time using a right to left pass
(c) Solves it is using divide and conquer in time $\Theta(n \log n)$
(d) Solves it in time $\Theta(n^2)$ **[2006 : 1 M]**

6.15 A set X can be represented by an array x[n] as follows:

$$x[i] = \begin{cases} 1; & \text{if } i \in X \\ 0; & \text{otherwise} \end{cases}$$

Consider the following algorithm in which x, y, and z are boolean arrays of size n:

```
algorithm zzz (x[ ], y[ ], z[ ])
{
  int i;
  for (i = 0; i < n; ++ i)
     z [i] = (x [i] ∧ ~ y[i]) ∨ (~x [i] ∧y[i]);
}
```

The set Z computed by the algorithm is
(a) $(X \cup Y)$ (b) $(X \cap Y)$
(c) $(X - Y) \cap (Y - X)$ (d) $(X - Y) \cup (Y - X)$
[2006 : 2 M]

6.16 Given two arrays of numbers $a_1, \ldots, a_n$ and $b_1, \ldots, b_n$ where each number is 0 or 1, the fastest algorithm to find the largest span (i, j) such that $a_i + a_{i+1} + \ldots + a_j = b_i + b_{i+1} + \ldots + b_j$, or report that there is no such span,
(a) Takes $O(3^n)$ and $\Omega(2^n)$ time if hashing is permitted
(b) Takes $O(n^3)$ and $\Omega(n^{2.5})$ time in the key comparison model
(c) Takes $\Theta(n)$ time and space
(d) Takes $O(\sqrt{n})$ time only if the sum of the 2n elements is an even number **[2006 : 2 M]**

6.17 Which of the following is the correct decomposition of the directed graph given below into its strongly connected components?

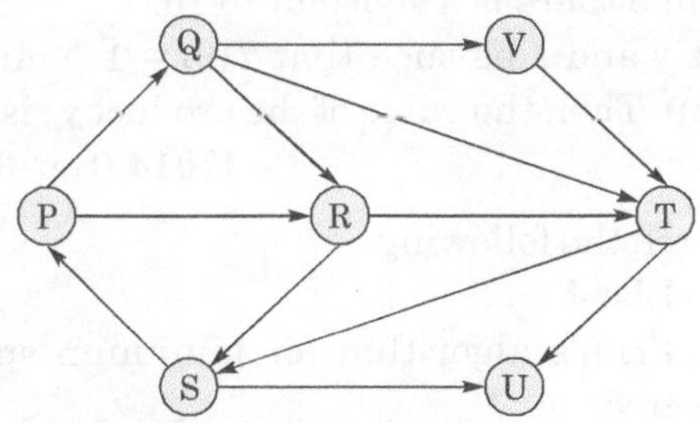

(a) {P, Q, R, S}, {T}, {U}, {V}
(b) {P,Q, R, S, T, V}, {U}
(c) {P, Q, S, T, V}, {R}, {U}
(d) {P, Q, R, S, T, U, V} **[2006 : 2 M]**

6.18 If we use Radix Sort to sort n integers in the range $(n^{k/12}, n^k]$, for some k > 0 which is independent of n, the time taken would be
(a) $\Theta(n)$ (b) $\Theta(kn)$
(c) $\Theta(n \log n)$ (d) $\Theta(n^2)$ **[2008 : 2 M]**

6.19 The number of elements that can be stored in $\Theta(\log n)$ time using heap sort is
(a) $\Theta(1)$ (b) $\Theta(\sqrt{\log n})$
(c) $\Theta\left(\frac{\log n}{\log \log n}\right)$ (d) $\Theta(\log n)$
[2013 : 2 M]

6.20 Consider the directed graph given below:

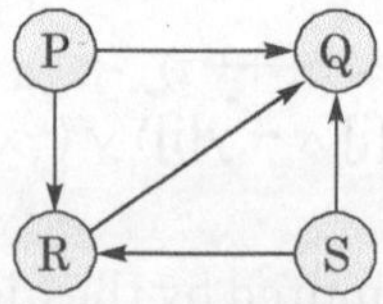

Which one of the following is TRUE?

(a) The graph does not have any topological ordering.
(b) Both PQRS and SRQP are topological orderings.
(c) Both PSRQ and SPRQ are topological orderings.
(d) PSRQ is the only topological ordering.

[2014 (Set-1) : 1 M]

6.21 Suppose you want to move from 0 to 100 on the number line. In each step, you either move right by a unit distance or you take a *shortcut*. A shortcut is simply a pre-specified pair of integers i, j with $i < j$. Given a shortcut i, j if you are at position i on the number line, you may directly move to j. Suppose $T(k)$ denotes the smallest number of steps needed to move from k to 100. Suppose further that there is at most 1 shortcut involving any number, and in particular from 9 there is a shortcut to 15.
Let y and z be such that $T(9) = 1 + \min(T(y), T(z))$. Then the value of the product yz is ______.

[2014 (Set-3) : 2 M]

6.22 Match the following:

List-I
A. Prim's algorithm for minimum spanning tree
B. Floyd-Warshall algorithm for all pairs shortest paths
C. Mergesort
D. Hamiltonian circuit

List-II
1. Backtracking
2. Greedy method
3. Dynamic programming
4. Divide and conquer

Codes:

	A	B	C	D
(a)	3	2	4	1
(b)	1	2	4	3
(c)	2	3	4	1
(d)	2	1	3	4

[2015 (Set-1) : 1 M]

6.23 Given below are some algorithms, and some algorithm design paradigms.

List-I
A. Dijkstra's Shortest Path
B. Floyd-Warshall algorithm to compute all pairs shortest path
C. Binary search on a sorted array
D. Backtracking search on a graph

List-II
1. Divide and Conquer
2. Dynamic Programming
3. Greedy design
4. Depth-first search
5. Breadth-first search

Match the above algorithms **(List-I)** to the corresponding design paradigm **(List-II)** they follow.

Codes:

	A	B	C	D
(a)	1	3	1	5
(b)	3	3	1	5
(c)	3	2	1	4
(d)	3	2	1	5

[2015 (Set-2) : 2 M]

6.24 Consider the following table:

Algorithms	Design Paradigms
(P) Kruskal	(i) Divide and Conquer
(Q) Quicksort	(ii) Greedy
(R) Floyd-Warshall	(iii) Dynamic Programming

Match the algorithms to the design paradigms they are based on.

(a) (P) ↔ (ii), (Q) ↔ (iii), (R) ↔ (i)
(b) (P) ↔ (iii), (Q) ↔ (i), (R) ↔ (ii)
(c) (P) ↔ (ii), (Q) ↔ (i), (R) ↔ (iii)
(d) (P) ↔ (i), (Q) ↔ (ii), (R) ↔ (iii)

[2017 (Set-1) : 1 M]

6.25 The Floyd-Warshall algorithm for all-pair shortest paths computation is based on

(a) Greedy paradigm
(b) Divide-and-conquer paradigm
(c) Dynamic Programming paradigm
(d) Neither Greedy nor Divide-and-Conquer nor Dynamic Programming paradigm.

[2016 (Set-2) : 1 M]

Answers Miscellaneous Topics

6.1 (c)	6.2 (d)	6.3 (b)	6.4 (b)	6.5 (d)	6.6 (d)	6.7 (d)	6.8 (b)	6.9 (d)
6.10 (d)	6.11 (a)	6.12 (b)	6.13 (c)	6.14 (b)	6.15 (d)	6.16 (c)	6.17 (b)	6.18 (b)
6.19 (c)	6.20 (c)	6.21 (150)	6.22 (c)	6.23 (c)	6.24 (c)	6.25 (c)		

Explanations Miscellaneous Topics

6.1 (c)

Tree representation of the given array is

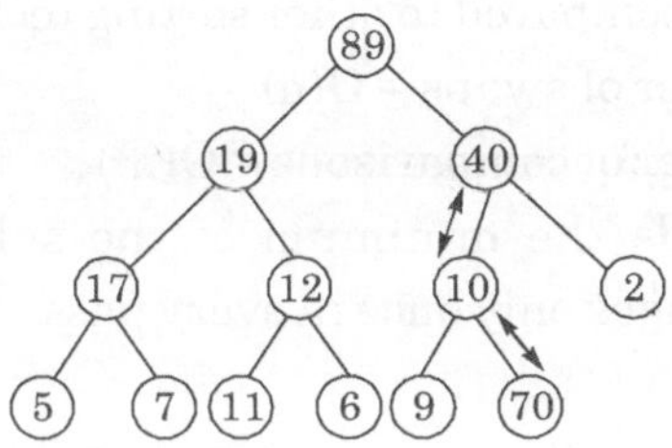

So two exchanges are required.

6.2 (d)

$$\text{Average case} = \frac{O(1)+n}{2} = \frac{n+1}{2}$$

6.3 (b)

All pairs shortest paths → Dynamic programming
Quick sort → Divide and conquer
Minimum weight spanining tree → Greedy technique
Connected components → Depth-First-Search

6.4 (b)

Binary ← O(logn)
[Linear search] Selection ← O(n)
Merge sort ← O(n log n)
Insertion sort ← O(n^2)

6.5 (d)

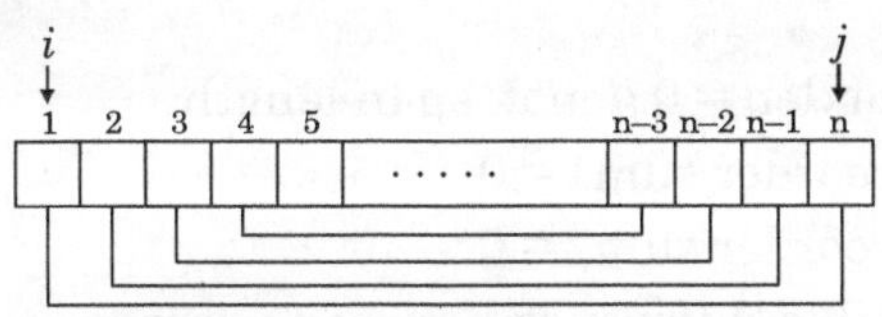

In worst case, minimum exchanges required

$$= \left\lfloor \frac{n}{2} \right\rfloor$$

(i) If (A[j] < A[j] && A[i] < 0) {i++;}
(ii) If (A[j] > A[i] && A[j] > 0) {j– –;}

Sorting is not needed, keep all negatives before positives. Worst case occurs when i and j reaches to middle position.

We require $\left\lfloor \frac{n}{2} \right\rfloor$ exchanges if array has first n/2 elements are positive and last n/2 elements are negative exactly.

6.6 (d)

Articulation point is that by removing that vertex, graph must be in disconnected state. 2, 3 and 5 are articulation points and also called as cut vertices.

6.7 (d)

T(n) = T(n–1)+n → T(n) = O(n^2)
T(n) = T(n/2)+n → T(n) = O(n)
T(n) = T(n/2)+n log n → T(n) = O(n log n)
T(n) = T(n–1)+ log n → T(n) = O(n log n)

6.8 (b)

Let $a_1, a_2, a_3, \ldots, a_n$ be a permutation of a set {1, 2, ..., n}. If i <j and $a_i > a_j$ the pair (a_i, a_j) is called an inversion of the permutation. For example the permutation 3142 has three inversions : (3, 1), (3, 2) and (4, 2). If all permutations are equally likely then according to "O. Terquen" the average number of inversions in a random permutation is

$$= \frac{1}{2}\binom{n}{2} = \frac{1}{2}({}^nC_2) = \frac{1}{2}\frac{\underline{|n}}{\underline{|2}\,\underline{|n-2}}$$

$$= \frac{1}{2}\frac{n(n-1)\,\underline{|n-2}}{2\cdot\underline{|n-2}} = \frac{n(n-1)}{4}$$

6.9 (d)

Insertion sort runs in O(n + f(n)) time where f(n) denotes number of inversions.
Therefore time complexity is Θ(n).

6.10 (d)

If each set is represented as a linked list with elements in arbitrary order. Membership cardinality takes O(1) time for an element in the set so the operation is very fast.

In the case of intersection, Let A and B are two sets. An operation such as $A \cap B$ requires time at least proportional to the sum of the sizes of the two sets. Since the list representing A and the list representant B must each be scanned at least once.

Similarly operation $A \cap B$ requires time at least proportional to the sum of the set sizes, since we must check for the some element appearing in both sets and delete one instance of each such element.

The well known algorithm using 2-3 tree (ordered or unordered) takes O(nlogn) if set contains n elements for union and intersection operation so it is slowest operation.

6.11 (b)

Balanced binary search tree with n leaf nodes [balanced binary search tree: almost full loaded binary tree]

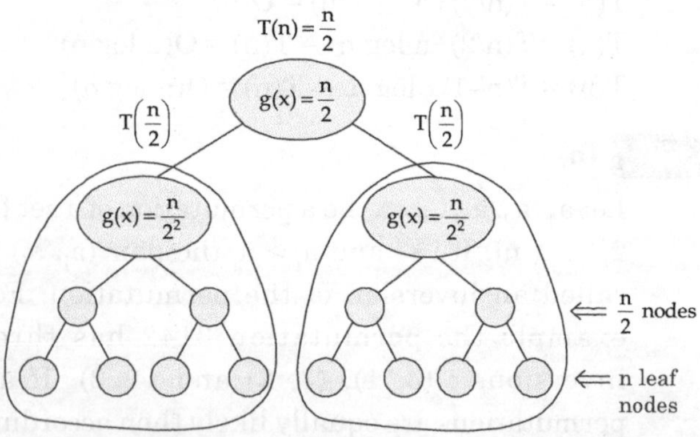

Compute g(x) in every node.

g(x) time complexity:

min (number of leaf nodes in left subtree of x, number of leaf nodes in right subtree of x)

Assume g(x) cost for all nodes with n leaf nodes are T(n).

$$T(n) = \underbrace{2T\left(\frac{n}{2}\right)}_{\substack{\text{g(x) cost for all nodes} \\ \text{tree with } \frac{n}{2} \text{ leaf nodes}}} + \underbrace{\frac{n}{2}}_{\text{g(x) cost for root}}$$

$\therefore \quad T(n) = \theta(n \log n)$

6.12 (b)

The root of the tree is stored at X[1],

Left of X[i] is stored at X[2i],

Right of X[i] is stored at X[2i + 1].

If tree is complete binary tree minimum it require n consecutive memory locations of x, x[1],...,x[n – 1], x[n].

6.13 (c)

In selection sort, minimum number of swaps taken compared to other sorting techniques.

Number of swaps = O(n)

Number of comparisons = $O(n^2)$

In finds the minimum of the subarray and exchanges only once in every pass.

6.14 (b)

Scan the array from right and select the leader. Now compare each element with the leader and find new leader. So it will take complete one scan of an array. So it is in linear time only.

6.15 (d)

Statement : $(x[i] \wedge \sim y[i]) \vee (\sim x[i] \wedge y[i])$ is equivalent to $(x[i] \cdot (y[i])') \vee ((x[i])' \cdot y[i])$ i.e. x[i] Ex-or y[i]. Which is represented by

$(x - y) \vee (y - x)$.

$\Rightarrow \quad (X - Y) \vee (Y - X)$

$\Rightarrow \quad (X \cdot Y') + (Y \cdot X')$

$\Rightarrow \quad (XY') + (X'Y) \equiv$ X Ex-or Y.

6.16 (c)

Two arrays $a_1 \ldots a_n$ and $b_1 \ldots b_n$ each number is 0 or 1.

Algo to find the largest span (i, j) such that

$a_i + a_{i+1} + \ldots + a_j = b_i + b_{i+1} + \ldots + b_j$

Algo largest span (a[n], b[n])

```
{
    maxlen = 0 // max span length
    preorder sum1 = 0
    preorder sum2 = 0
// possible difference between preorder sum of array a and b are –n to n.
(Total 2n + 1 possible differences)
for(i = 1; i ≤ 2n – 1; i++)
{
    diff[i] = –1 }
```

```
for(i = 1; i ≤ n; i++)
{
    preorder sum1 = preorder sum1 + a[i]
    preorder sum2 = preorder sum2 + b[i]
    currdiff = preorder sum1 – preorder sum2;
    diff index = n + currdiff
    if (currdiff == 0) // same number of 1's
        maxlen = i
    else if(diff [diffIndex] == – 1)
        diff [diffIndex] = i
        else
        {
            len = i – diff [diffIndex]
                if(len > maxlen)
                    maxlen = len;
        }
    }
}
```

TC : θ(n) and SC : θ(n)

∴ Option (c) is answer.

6.17 (b)

To find strongly connected component apply DFS on graph and find longest cycle. If you find a cycle the reverse the direction of every edge and again apply the DFS, if you found cycle this time also. Then this cycle is strongly connected component.

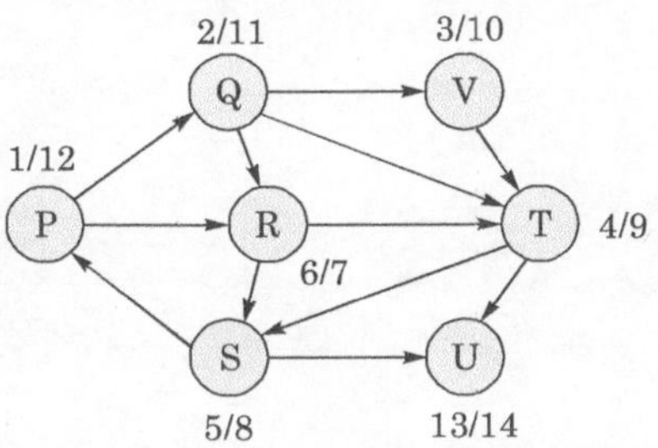

2 cycles
1. (PQVTSR)
2. (U)

Now change direction of every edge

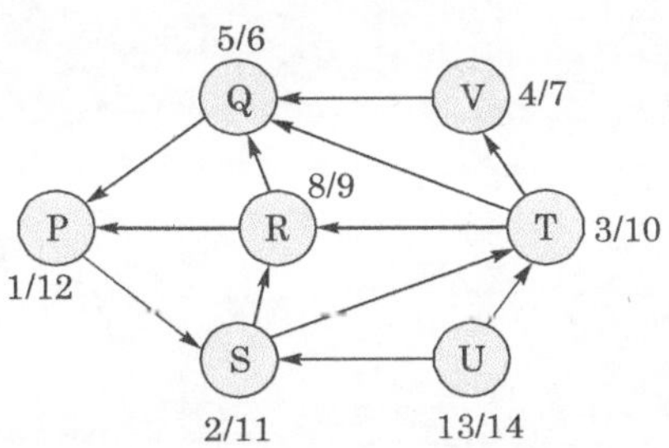

2 cycles
1. (PQVTSR)
2. (U)

So, 2 strongly connected component {P, Q, V, T, S, R} and {U} are present.

6.18 (b)

Convert element x whose range $[n^{k/12}, n^k]$ into radix n.

⇒ Number of digits of x is $\log_n n^k = k$ digits

(1 element conversion ⇒ atmost k divisions)

So, 1 conversion take k time and n element converting: θ(k) * n

Apply radix sort:

$$d = k$$

⇒ θ(k * n)

Convert radix n to radix 10 ⇒ θ(k) * n

∴ TC = θ(k * n)

6.19 (c)

If we have an array of x elements then time complexity to sort x elements using heap sort is Θ(x log x).

We need to find value of x so that time complexity to sort x elements gets Θ(log n) we can direct discard option (a) and (d).

Now check option (c)

$$= \Theta\left(\frac{\log n}{\log\log n}\log\left(\frac{\log n}{\log\log n}\right)\right)$$

$$= \Theta\left(\frac{\log n}{\log\log n}(\log\log n - \log\log\log n)\right)$$

$$= \Theta\left(\log n\left(\frac{\log\log n - \log\log\log n}{\log\log n}\right)\right)$$

$$= \Theta(\log n)$$

6.20 (c)

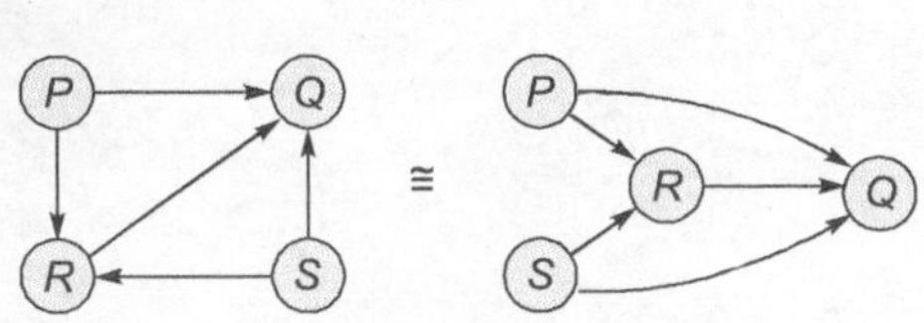

∴ Topological orderings: PSRQ and SPRQ

6.21 (150)

Given that there is shortcut from 9 to 15

OR

It can move right by a unit distance (9 to 10)

T(9) = 1 + Min (T(10), T(15))

⇒ y = 10, z = 15

∴ yz = 10 × 15 = 150

6.22 (c)

Prims algorithm for finding MST is a greedy algorithm. Floyd warshalls algorithm for finding all shortest paths is dynamic algorithm.
Merge sort is a Divide and Conquer algorithm.
Hamiltonian circuit is an NP-complete problem and done by using backtracking.

6.23 (c)

Dijktra's shortest path → Greedy design.
Floyd warshall algorithm to compute all pair shortest path → Dynamic programming.
Binary search on a sorted array → Divide and Conquer
Backtracking searching → Depth first search on a graph.

6.24 (c)

- Kruskal's algorithms which is used to find MST uses greedy approach.
- Quicksort uses divide and conquer approach by dividing the input array according to pivot element.
- Floyd Warshall which is used to find all pair shortest path uses dynamic programming.

6.25 (c)

Floyd Warshall algorithm for all pair shortest paths computation is based on dynamic programming paradigm.

■■■■

Compiler Design

UNIT VII

CONTENTS

Compiler Design

Syllabus

Lexical analysis, Parsing, Syntax directed translation, Runtime environments, Intermediate and target code generation, Local optimization, Data flow analyses: constant propagation, liveness analysis, common subexpression elimination.

Analysis of Previous GATE Papers

Exam Year	1 Mark Ques.	2 Marks Ques.	Total Marks
1995	3	1	**5**
1996	1	–	**1**
1997	1	–	**1**
1998	2	–	**7**
1999	–	1	**2**
2000	3	–	**3**
2001	2	–	**2**
2002	–	–	**–**
2003	3	5	**13**
2004	3	1	**5**
2005	1	4	**9**
2006	1	5	**11**
2007	1	5	**11**
2008	2	1	**4**
2009	1	–	**1**
2010	2	1	**4**
2011	1	1	**3**
2012	–	2	**4**
2013	1	3	**7**

Exam Year	1 Mark Ques.	2 Marks Ques.	Total Marks
2014 Set-1	1	1	**3**
2014 Set-2	2	2	**6**
2014 Set-3	2	1	**4**
2015 Set-1	1	2	**5**
2015 Set-2	2	1	**4**
2015 Set-3	1	1	**3**
2016 Set-1	1	2	**5**
2016 Set-2	1	1	**3**
2017 Set-1	2	1	**4**
2017 Set-2	2	1	**4**
2018	1	2	**5**
2019	2	2	**6**
2020	2	1	**4**
2021 Set-1	1	3	**7**
2021 Set-2	1	4	**9**
2022	2	1	**4**
2023	1	2	**5**
2024 Set-1	1	3	**7**
2024 Set-2	2	3	**8**

1 Lexical Analysis

1.1 Which of the following strings can definitely be said to be token without looking at the next input character while compiling a Pascal program?

I. begin **II.** Program **III.** <>

(a) I (b) II
(c) III (d) All of these

[1995 : 1 M]

1.2 In some programming languages an identifier is permitted to be a letter followed by any number of letters or digits. If L and D denotes the sets of letters and digits respectively. Which of the following expression defines an identifier?

(a) $(L + D)^+$ (b) $L(L + D)^*$
(c) $(L.D)^*$ (d) $L(L.D)^*$

[1995 : 1 M]

1.3 Type checking is normally done during
(a) Lexical analysis
(b) Syntax analysis
(c) Syntax directed translation
(d) Code optimization **[1998 : 1 M]**

1.4 The number of tokens in the Fortran statement DO 10 I = 1.25 is **[1999 : 2 M]**

1.5 The number of tokens in the following C statement printf("i = %d, &i = %x", i, &i); is **[2000 : 1 M]**

1.6 The number of tokens in the following C statement
printf("i = %d, &i = %x", i, &i); is
(a) 3 (b) 26
(c) 10 (d) 21 **[2000 : 1 M]**

1.7 Which of the following is NOT an advantage of using shared, dynamically linked libraries as opposed to using statically linked libraries?
(a) Smaller sizes of executable
(b) Lesser overall page fault rate in the system
(c) Faster program startup
(d) Existing programs need not be re-linked to take advantage of newer versions of libraries

[2003 : 2 M]

1.8 Consider a program P that consists of two source modules M_1 and M_2 contained in two different files. If M_1 contains a reference to a function defined in M_2, the reference will be resolved at
(a) Edit-time (b) Compile-time
(c) Link-time (d) Load-time

[2004 : 1 M]

1.9 Consider line number 3 of the following C-program.

```
int main ( ){                      /* Line 1 */
  int i, n;                        /* Line 2 */
  fro (i = 0, i < n, i + +);       /* Line 3 */
}
```

Identify the compiler's response about this line while creating the object - module
(a) No compilation error
(b) Only a lexical error
(c) Only syntactic errors
(d) Both lexical and syntactic errors

[2005 : 2 M]

1.10 Which of the following statements are TRUE?
I. There exist parsing algorithms for some programming languages whose complexities are less than $\theta(n^3)$.
II. A programming language which allows recursion can be implemented with static storage allocation.
III. No L-attributed definition can be evaluated in the framework of bottom-up parsing.
IV. Code improving transformations can be performed at both source language and intermediate code level.
(a) I and II (b) I and IV
(c) III and IV (d) I, III and IV

[2009 : 1 M]

1.11 Which data structure in a complier is used for managing information about variables and their attributes?
(a) Abstract syntax tree
(b) Symbol table
(c) Semantic stack
(d) Parse table **[2010 : 1 M]**

1.12 In a compiler, keywords of a language are recognized during
(a) parsing of the program
(b) the code generation
(c) the lexical analysis of the program
(d) dataflow analysis **[2011 : 1 M]**

1.13 Match the following:

List-I	List-II
P. Lexical analysis	**(i)** Leftmost derivation
Q. Top down parsing	**(ii)** Type checking
R. Semantic analysis	**(iii)** Regular expressions
S. Runtime environments	**(iv)** Activation records

Codes:
(a) P → (i), Q → (ii), R → (iv), S → (iii)
(b) P → (iii), Q → (i), R → (ii), S → (iv)
(c) P → (ii), Q → (iii), R → (i), S → (iv)
(d) P → (iv), Q → (i), R → (ii), S → (iii)

[2016 (Set-2) : 1 M]

1.14 Match the following according to input (from the left column) to the compiler phase (in the right column) that processes it:

List-I	List-II
P. Syntax tree	**(i)** Code generator
Q. Character stream	**(ii)** Syntax analyzer
R. Intermediate representation	**(iii)** Semantic analyzer
S. Token stream	**(iv)** Lexical analyzer

(a) P → (ii), Q → (iii), R → (iv), S → (i)
(b) P → (ii), Q → (i), R → (iii), S → (iv)
(c) P → (iii), Q → (iv), R → (i), S → (ii)
(d) P → (i), Q → (iv), R → (ii), S → (iii)

[2017 (Set-2) : 1 M]

1.15 A lexical analyzer uses the following patterns to recognize three tokens T_1, T_2 and T_3 over the alphabet {a, b, c}.

T_1 : a? (b | c)*a
T_2 : b? (a | c)*b
T_3 : c? (b | a)*c

Note that '*x*?' means 0 or 1 occurrence of the symbol *x*. Note also that the analyzer outputs the token that matches the longest possible prefix.

If the string bbaacabc is processed by the analyzer, which one of the following is the sequence of tokens it outputs?

(a) $T_1T_2T_3$
(b) $T_1T_1T_3$
(c) $T_2T_1T_3$
(d) T_3T_3

[2018 : 2 M]

1.16 Consider the following statements:

I. Symbol table is accessed only during lexical analysis and syntax analysis.

II. Compilers for programming languages that support recursion necessarily need heap storage for memory allocation in the run-time environment.

III. Errors violating the condition *'any variable must be declared before its use'* are detected during syntax analysis.

Which of the above statements is/are TRUE?

(a) I only
(b) I and III only
(c) I only
(d) None of I, II and III

[2020 : 1 M]

1.17 Consider the following ANSI C program:

```
int main() {
    Integer x;
    return 0;
}
```

Which one of the following phases in a seven-phase C compiler will throw an error?

(a) Syntax analyzer
(b) Machine dependent optimizer
(c) Semantic analyzer
(d) Lexical analyzer

[2021 (Set-2) : 2 M]

1.18 Consider the following statements regarding the front-end and back-end of a compiler.

S1: The front-end includes phases that are independent of the target hardware.

S2: The back-end includes phases that are specific to the target hardware.

S3: The back-end includes phases that are specific to the programming language used in the source code.

Identify the CORRECT option.

(a) Only S1 is TRUE.
(b) Only S1 and S2 are TRUE.
(c) S1, S2, and S3 are all TRUE.
(d) Only S1 and S3 are TRUE.

[2023 : 1 M]

1.19 Consider the following two sets:

Set X
P. Lexical Analyzer
Q. Syntax Analyzer
R. Intermediate Code Generator
S. Code Optimizer

Set Y
1. Abstract Syntax Tree
2. Token
3. Parse Tree
4. Constant Folding

Which one of the following options is the CORRECT match from **Set X** to **Set Y**?

(a) P - 4; Q - 1; R - 3; S - 2
(b) P - 2; Q - 1; R - 3; S - 4
(c) P - 4; Q - 3; R - 2; S - 1
(d) P - 2; Q - 3; R - 1; S - 4

[2024 (Set-2) : 1 M]

■■■■

Answers Lexical Analysis

1.1	(c)	**1.2**	(b)	**1.3**	(c)	**1.4**	(5)	**1.5**	(10)	**1.6**	(c)	**1.7**	(c)	**1.8**	(c)	**1.9**	(c)
1.10	(b)	**1.11**	(b)	**1.12**	(c)	**1.13**	(b)	**1.14**	(c)	**1.15.**	(d)	**1.16**	(d)	**1.17**	(c)	**1.18**	(b)
1.19	(d)																

Explanations Lexical Analysis

1.1 (c)

I. Begin can be followed by anything to form identifier, function name, etc.

II. Program can be followed by anything to form identifier, function name, etc.

III. <> can definitely be said to be token without looking at the next input character.

1.2 (b)

L : Set of letters
D: Set of digits
Identifier = letter (letter or digit)*
Identifier = L(L + D)*

1.3 (c)

Type checking verifies that a type of a construct matches that expected by its context.

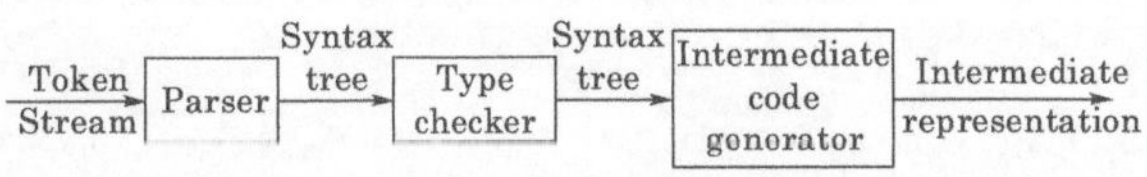

Position of type checker

1.4 (5)

In Fortran identifies can contain spaces.
'Do 10 I' is one token
'=' is another token and '1.25' is third token
∴ 3 tokens are identified in Fortran.
Whereas in C-language as following it counts:

DO	10	I	=	1.25
①	②	③	④	⑤

∴ The number of tokens in the C statement is 5.

1.5 (10)

Printf	(	"i - %d, & i = %x"	,	i	,	&	i	)	;
①	②	③	④	⑤	⑥	⑦	⑧	⑨	⑩

∴ The number of tokens in the C statement is 10.

1.6 (c)

Inside the printf (" ") the part between (" – ") count as '1'.

1.7 (c)

Statically linked libraries program run faster from startup since the linking has been performed during the compile time only and there is no need for the since at run time of program and program can run faster.

1.8 (c)

Modules may be compiled separately and all linking between references is made at link time.

1.9 (c)

Error is underlined function fro( )
Which is a syntactical error rather than lexical.

1.10 (b)

Recursion can not be implemented with static storage allocation.
L-attributed definition can be evaluated if all rules are at the end and all attributes are synthesized.
So I and IV are correct.

1.11 (b)

As we know symbol table is a Data structure containing record for each identifier, with fields for the attribute of the identifier.
The symbols entered into symbol table are nothing but identifier and operators. It contain information about variables and their attributes.

1.12 (c)

Lexical analysis of the program.

1.13 (b)

P. Lexical analysis → (iii) Regular expression
Q. Top down parsing → (i) LMD
R. Semantic Analysis → (ii) Type checking
S. Runtime Environments → (iv) Activation records

1.14 (c)

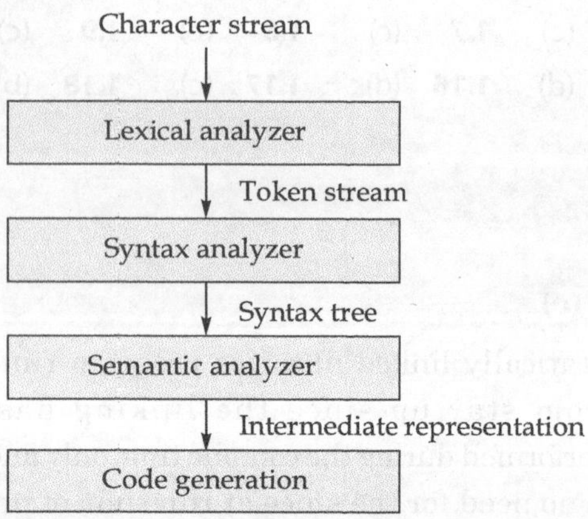

1.15 (d)

In T_3T_3, from first T_3 bbaac is taken, from second T_3 abc is taken, longest possible prefix.

Hence T_3T_3 token output.

1.16 (d)

None of the given statements are correct.

1.18 (b)

In compilers, the frontend converts source code for computer programmes into an intermediate representation, while the backend uses the intermediate representation to create code in an output language for computers. Usually, the backend performs optimization to create faster-running code. The parser part, which deals with source code, and the backend, which creates code and optimises, can be divided using the frontend/backend distinction. Some designs, like GCC, provide users the option to select from a variety of frontends (which parse several source languages) or backends (generating code for different target processors).

■■■■

2 Parsing Techniques

2.1 A shift reduce parser carries out the actions specified within braces immediately after reducing with the corresponding rule of grammar

$S \rightarrow xx\ W$ {print"1"}
$S \rightarrow y$ {print"2"}
$W \rightarrow Sz$ {print"3"}

What is the translation of xxxxyzz using the syntax directed translation scheme described by the above rules?

(a) 23131 (b) 11233
(c) 11231 (d) 33211 **[1995 : 2 M]**

2.2 The pass numbers for each of the following activities

(i) object code generation
(ii) literals added to literal table
(iii) listing printed
(iv) address resolution of local symbols that occur in a two pass assembler respectively are

(a) 1, 2, 1, 2 (b) 2, 1, 2, 1
(c) 2, 1, 1, 2 (d) 1, 2, 2, 2

[1996 : 1 M]

2.3 Which of the following statements is true?

(a) SLR parser is more powerful than LALR
(b) LALR parser is more powerful than Canonical LR parser
(c) Canonical LR parser is more powerful than LALR parser
(d) The parsers SLR, Canonical LR, and LALR have the same power **[1998 : 1 M]**

2.4 The process of assigning load addresses to the various parts of the program and adjusting the code and data in the program to reflect the assigned addresses is called

(a) Assembly (b) Parsing
(c) Relocation (d) Symbol resolution

[2001 : 1 M]

2.5 Which of the following statements is false?

(a) An unambiguous grammar has same leftmost and rightmost derivation
(b) An LL(1) parser is a top-down parser
(c) LALR is more powerful than SLR
(d) An ambiguous grammar can never be LR(k) for any k **[2001 : 1 M]**

2.6 Which of the following suffices to convert an arbitrary CFG to an LL(1) grammar?

(a) Removing left recursion alone
(b) Factoring the grammar alone
(c) Removing left recursion and factoring the grammar
(d) None of the above **[2003 : 1 M]**

2.7 Assume that the SLR parser for a grammar G has n_1 states and the LALR parser for G has n_2 states. The relationship between n_1 and n_2 is

(a) n_1 is necessarily less than n_2
(b) n_1 is necessarily equal to n_2
(c) n_1 is necessarily greater than n_2
(d) None of the above **[2003 : 1 M]**

2.8 Consider the grammar shown below:

$S \rightarrow i\ E\ t\ S\ S' \mid \alpha$
$S' \rightarrow e\ S \mid \varepsilon$
$E \rightarrow b$

In the predictive parse table M, of this grammar, the entries M[S′, e] and M[S′, $] respectively are

(a) $\{S' \rightarrow e\ S\}$ and $\{S' \rightarrow \varepsilon\}$
(b) $\{S' \rightarrow e\ S\}$ and { }
(c) $\{S' \rightarrow \varepsilon\}$ and $\{S' \rightarrow \varepsilon\}$
(d) $\{S' \rightarrow e\ S, S' \rightarrow \varepsilon\}$ and $\{S' \rightarrow \varepsilon\}$

[2003 : 2 M]

2.9 Consider the grammar shown below:

$S \rightarrow C\ C$
$C \rightarrow c\ C \mid d$

The grammar is

(a) LL (1)
(b) SLR (1) but not LL (1)
(c) LALR (1) but not SLR (1)
(d) LR (1) but not LALR (1) **[2003 : 2 M]**

2.10 Which of the following grammar rules violate the requirements of an operator grammar? P, Q, R are not terminals and r, s, t are terminals.

(i) $P \rightarrow Q\ R$ (ii) $P \rightarrow Q\ s\ R$
(iii) $P \rightarrow \varepsilon$ (iv) $P \rightarrow Q\ t\ R\ r$

(a) (i) only (b) (i) and (iii) only
(c) (ii) and (iii) only (d) (iii) and (iv) only

[2004 : 1 Mark]

2.11 The grammar $A \rightarrow AA \mid (A) \mid \varepsilon$ is not suitable for predictive-parsing because the grammar is

(a) ambiguous (b) left-recursive
(c) right-recursive (d) an operator-grammar

[2005 : 1 M]

2.12 Consider the grammar:

$S \rightarrow (S) \mid a$

Let the number of states in SLR (1), LR (1) and LALR(1) parsers for the grammar be n_1, n_2 and n_3 respectively. The following relationship holds good

(a) $n_1 < n_2 < n_3$ (b) $n_1 = n_3 < n_2$
(c) $n_1 = n_2 = n_3$ (d) $n_1 \geq n_3 \geq n_2$

[2005 : 2 M]

Linked Answer for Q. 2.13 & Q. 2.14:

Consider the context-free grammar:

$E \rightarrow E + E$
$E \rightarrow (E * E)$
$E \rightarrow id$

where E is the starting symbol, the set of terminals is { id, (, +,), * } and the set of non-terminals is {E}.

2.13 Which of the following terminal strings has more than one parse tree when parsed according to the above grammar?

(a) id + id + id + id
(b) id + (id* (id * id))
(c) (id* (id * id)) + id
(d) ((id * id + id) * id)

[2005 : 2 M]

2.14 For the terminal string with more than one parse tree obtained as solution to in above Question, how many parse trees are possible?

(a) 5 (b) 4
(c) 3 (d) 2 **[2005 : 2 M]**

2.15 Consider the following grammar:

$S \rightarrow S \times E$
$S \rightarrow E$
$E \rightarrow F + E$
$E \rightarrow F$
$F \rightarrow id$

Consider the following LR(0) items corresponding to the grammar above.

(i) $S \rightarrow S \times .E$
(ii) $E \rightarrow F. + E$
(iii) $E \rightarrow F + .E$

Given the items above, which two of them will appear in the same set in the canonical sets-of-items for the grammar?

(a) (i) and (ii) (b) (ii) and (iii)
(c) (i) and (iii) (d) None of these

[2006 : 1 M]

2.16 Consider the following grammar:

$S \rightarrow FR$
$R \rightarrow *S \mid \varepsilon$
$F \rightarrow id$

In the predictive parser table, M, of the grammar the entries M[S, id] and M[R, $] respectively

(a) $\{S \rightarrow FR\}$ and $\{R \rightarrow \varepsilon\}$
(b) $\{S \rightarrow FR\}$ and { }
(c) $\{S \rightarrow FR\}$ and $\{R \rightarrow *S\}$
(d) $\{F \rightarrow id\}$ and $\{R \rightarrow \varepsilon\}$ **[2006 : 2 M]**

Linked Data Questions 2.17 & 2.18:

2.17 Which one of the following grammars generates the language $L = \{a^i b^j \mid i \neq j\}$?

(a) $S \rightarrow AC \mid CB$
$C \rightarrow aCb \mid a \mid b$
$A \rightarrow aA \mid \varepsilon$
$B \rightarrow Bb \mid \varepsilon$

(b) $S \rightarrow aS \mid Sb \mid a \mid b$

(c) $S \rightarrow AC \mid CB$
$C \rightarrow aCb \mid \varepsilon$
$A \rightarrow aA \mid \varepsilon$
$B \rightarrow Bb \mid \varepsilon$

(d) $S \rightarrow AC \mid CB$
$C \rightarrow aCb \mid \varepsilon$
$A \rightarrow aA \mid a$
$B \rightarrow bB \mid b$

[2006 : 2 M]

2.18 In the correct grammar above, what is the length of the derivation (number of steps starting from S) to generate the string $a^l b^m$ with $l \neq m$?

(a) max $(l, m) + 2$ (b) $l + m + 2$
(c) $l + m + 3$ (d) max $(l, m) + 3$

[2006 : 2 M]

2.19 Which one of the following is a top-down parser?

(a) Recursive descent parser
(b) Operator precedence parser
(c) An LR(k) parser
(d) An LALR(k) parser **[2007 : 1 M]**

2.20 Consider the grammar with non-terminals $N = \{S, C, S_1\}$, terminals $T = \{a, b, i, t, e\}$, with S as the start symbol, and the following set of rules

$S \rightarrow iCtSS_1 \mid a$
$S_1 \rightarrow eS \mid \varepsilon$
$C \rightarrow b$

The grammar is NOT LL(1) because:

(a) It is left recursive
(b) It is right recursive
(c) It is ambiguous
(d) It is not context-free **[2007 : 2 M]**

2.21 Consider the following two statements:

P: Every regular grammar is LL (1)
Q: Every regular set has LR (1) grammar

Which of the following is TRUE?

(a) Both P and Q are true
(b) P is true and Q is false
(c) P is false and Q is true
(d) Both P and Q are false **[2007 : 2 M]**

Common Data for Q. 2.22 & Q. 2.23:

Consider the CFG with {S, A, B} as the non-terminal alphabet, {a, b} as the terminal alphabet, S as the start symbol and the following set of production rules

$S \rightarrow bA$ $\quad S \rightarrow aB$
$A \rightarrow a$ $\quad B \rightarrow b$
$A \rightarrow aS$ $\quad B \rightarrow bS$
$A \rightarrow bAA$ $\quad B \rightarrow aBB$

2.22 Which of the following strings is generated by the grammar?
(a) aaaabb (b) aabbbb
(c) aabbab (d) abbbba
[2007 : 2 M]

2.23 For the correct answer string to above question how many derivation trees are there?
(a) 1 (b) 2
(c) 3 (d) 4 **[2007 : 2 M]**

2.24 Which of the following describes a handle (as applicable to LR-parsing) appropriately?
(a) It is the position in a sentential form where the next shift or reduce operation will occur
(b) It is a non-terminal whose production will be used for reduction in the next step
(c) It is a production that may be used for reduction in a future step along with a position in the sentential form where the next shift or reduce operation will occur.
(d) It is the production p that will be used for reduction in the next step along with a position in the sentential form where the right hand side of the production may be found
[2008 : 1 M]

2.25 An LALR(1) parser for a grammar G can have shift-reduce (S-R) conflicts if and only if
(a) the SLR(1) parser for G has S-R conflicts
(b) the LR(1) parser for G has S-R conflicts
(c) the LR(0) parser for G has S-R conflicts
(d) the LALR(1) parser for G has reduce-reduce conflicts **[2008 : 2 M]**

2.26 The grammar $S \rightarrow aSa \mid bS \mid c$ is
(a) LL(1) but not LR(1)
(b) LR(1) but not LL(1)
(c) Both LL(1) and LR(1)
(d) Neither LL(1) nor LR(1)
[2010 : 2 M]

Linked Data Questions 2.27 and 2.28:

For the grammar below, a partial LL(1) parsing table is also presented along with the grammar. Entries that need to be filled are indicated as E1, E2, and E3. ε is the empty string, $ indicates end of input, and | separates alternate right hand sides of productions.

$S \rightarrow aAbB \mid bAaB \mid \varepsilon$
$A \rightarrow S$
$B \rightarrow S$

	a	b	$
S	E1	E2	$S \rightarrow \varepsilon$
A	$A \rightarrow S$	$A \rightarrow S$	error
B	$B \rightarrow S$	$B \rightarrow S$	E3

2.27 The FIRST and FOLLOW sets for the non-terminals A and B are
(a) FIRST (A) = {a, b, ε} = FIRST (B)
FOLLOW (A) = {a, b}
FOLLOW (B) = {a, b, $}
(b) FIRST (A) = {a, b, $}
FIRST (B) = {a, b, ε}
FOLLOW (A) = {a, b}
FOLLOW (B) = {$}
(c) FIRST (A) = {a, b, ε} = FIRST (B)
FOLLOW (A) = {a, b}
FOLLOW (B) = $
(d) FIRST (A) = {a, b} = FIRST (B)
FOLLOW (A) = {a, b}
FOLLOW (B) = {a, b} **[2012 : 2 M]**

2.28 The appropriate entries for E1, E2 and E3 are
(a) E1 : $S \rightarrow aAbB, A \rightarrow S$
E2 : $S \rightarrow bAaB, B \rightarrow S$
E3 : $B \rightarrow S$
(b) E1 : $S \rightarrow aAbB, S \rightarrow \varepsilon$
E2 : $S \rightarrow bAaB, B \rightarrow \varepsilon$
E3 : $S \rightarrow \varepsilon$
(c) E1 : $S \rightarrow aAbB, S \rightarrow \varepsilon$
E2 : $S \rightarrow bAaB, S \rightarrow \varepsilon$
E3 : $B \rightarrow S$
(d) E1 : $A \rightarrow S, S \rightarrow \varepsilon$
E2 : $B \rightarrow S, S \rightarrow \varepsilon$
E3 : $B \rightarrow S$ **[2012 : 2 M]**

2.29 What is the maximum number of reduces moves that can be taken by a bottom up parser for a grammar without epsilon and unit productions (i.e. of type $A \rightarrow \varepsilon$ and $A \rightarrow a$) to parse a string with n tokens?
(a) $\frac{n}{2}$ (b) $n - 1$
(c) $2n - 1$ (d) 2^n **[2013 : 1 M]**

2.30 Consider the following two sets of LR(1) items of LR(1) grammar.

$X \rightarrow c.X, c/d$ $\quad$ $X \rightarrow c.X, \$$
$X \rightarrow .cX, c/d$ $\quad$ $X \rightarrow .cX, \$$
$X \rightarrow .d, c/d$ $\quad$ $X \rightarrow .d, \$$

Which of the following statement related to merging of the two sets in the corresponding parser is/are FALSE?

1. Cannot be merged since look aheads are different.
2. Can be merged but will result in S-R conflict.
3. Can be merged but will result in R-R conflict.
4. Cannot be merged since goto on c will lead to two different sets.

(a) 1 only (b) 2 only
(c) 1 and 4 only (d) 1, 2, 3 and 4

[2013 : 2 M]

2.31 A canonical set of items is given below:

$S \rightarrow L \cdot > R$
$Q \rightarrow R \cdot$

On input symbol > the set has

(a) a shift-reduce conflict and a reduce-reduce conflict.
(b) a shift-reduce conflict but not a reduce-reduce conflict.
(c) a reduce-reduce conflict but not a shift-reduce conflict.
(d) neither a shift-reduce nor a reduce-reduce conflict.

[2014 (Set-1) : 2 M]

2.32 Consider the grammar defined by the following production rules, with two operators * and +

$S \rightarrow T * P$
$T \rightarrow U \mid T * U$
$P \rightarrow Q + P \mid Q$
$Q \rightarrow Id$
$U \rightarrow Id$

Which one of the following is TRUE?

(a) + is left associative, while * is right associative
(b) + is right associative, while * is left associative
(c) Both + and * are right associative
(d) Both + and * are left associative

[2014 (Set-2) : 1 M]

2.33 Which one of the following is TRUE at any valid state in shift-reduce parsing?

(a) Viable prefixes appear only at the bottom of the stack and not inside
(b) Viable prefixes appear only at the top of the stack and not inside
(c) The stack contains only a set of viable prefixes
(d) The stack never contains viable prefixes

[2015 (Set-1) : 1 M]

2.34 Match the following:

List-I
A. Lexical analysis
B. Parsing
C. Register allocation
D. Expression evaluation

List-II
1. Graph coloring
2. DFA minimization
3. Post-order traversal
4. Production tree

Codes:

	A	B	C	D
(a)	2	3	1	4
(b)	2	1	4	3
(c)	2	4	1	3
(d)	2	3	4	1

[2015 (Set-2) : 1 M]

2.35 Among simple LR (SLR), canonical LR, and look-ahead LR (LALR), which of the following pairs identify the method that is very easy to implement and the method that is the most powerful, in that order?

(a) SLR, LALR
(b) Canonical LR, LALR
(c) SLR, canonical LR
(d) LALR, canonical LR

[2015 (Set-3) : 1 M]

2.36 Consider the following grammar G:

$S \rightarrow F \mid H$
$F \rightarrow p \mid c$
$H \rightarrow d \mid c$

Where S, F and H are non-terminal symbols, *p*, *d* and *c* are terminal symbols. Which of the following statement(s) is/are correct?

S1: LL(1) can parse all strings that are generated using grammar G.
S2: LR(1) can parse all strings that are generated using grammar G.

(a) Only S1 (b) Only S2
(c) Both S1 and S2 (d) Neither S1 and S2

[2015 (Set-3) : 2 M]

2.37 Consider the following grammar:

$P \rightarrow xQRS$
$Q \rightarrow yz \mid z$
$R \rightarrow w \mid \epsilon$
$S \rightarrow y$

What is FOLLOW(Q)?

(a) {R} (b) {w}
(c) {w, y} (d) {w, \$}

[2017 (Set-1) : 1 M]

2.38 Which of the following statements about parser is/are CORRECT?

I. Canonical LR is more powerful than SLR.
II. SLR is more powerful than LALR.
III. SLR is more powerful than Canonical LR.

(a) I only (b) II only
(c) III only (d) II and III only

[2017 (Set-2) : 1 M]

2.39 Consider the following expression grammar G:

$E \rightarrow E - T \mid T$
$T \rightarrow T + F \mid F$
$F \rightarrow (E) \mid id$

Which of the following grammars is not left recursive, but is equivalent to G?

(a) $E \rightarrow E - T \mid T$
$T \rightarrow T + F \mid F$
$F \rightarrow (E) \mid id$

(b) $E \rightarrow TE'$
$E' \rightarrow -TE' \mid \in$
$T \rightarrow T + F \mid F$
$F \rightarrow (E) \mid id$

(c) $E \rightarrow TX$
$X \rightarrow -TX \mid \in$
$T \rightarrow FY$
$Y \rightarrow +FY \mid \in$
$F \rightarrow (E) \mid id$

(d) $E \rightarrow TX \mid (TX)$
$X \rightarrow -TX \mid +TX \mid \in$
$T \rightarrow id$

[2017 (Set-2) : 2 M]

2.40 Consider the grammar given below:

S → Aa
A → BD
B → b | ∈
D → d | ∈

Let a, b, d and \$ be indexed as follows:

a	b	d	\$
3	2	1	0

Compute the FOLLOW set of the non-terminal B and write the index values for the symbols in the FOLLOW set in the descending order. (For example, if the FOLLOW set is {a, b, d, \$}, then the answer should be 3210) **[2019 : 1 M]**

2.41 Which one of the following kinds of derivation is used by LR parsers?

(a) Rightmost
(b) Rightmost in reverse
(c) Leftmost
(d) Leftmost in reverse **[2019 : 1 M]**

2.42 Consider the augmented grammar given below:

S′ → S
S → <L> | id
L → L, S | S

Let I_0 = CLOSURE ({[S′ → S]}). The number of items in the set GOTO (I_0, <) is ______.

[2019 : 2 M]

2.43 Consider the following grammar:

S → aSB | d
B → b

The number of reduction steps taken by a bottom-up parser while accepting the string aaadbbb is ______. **[2020 : 1 M]**

2.44 Consider the following context-free grammar where the set of terminals is {a, b, c, d, f}

S → daT | Rf
T → aS | baT | ∈
R → caTR | ∈

The following is a partially-filled LL(1) parsing table.

	a	b	c	d	f	\$
S			①	S → daT	②	
T	T → aS	T → baT	③		T → ∈	④
R			R → caTR		R → ∈	

Which one of the following choices represents the correct combination for the numbered cells in the parsing table ("blank" denotes that the corresponding cell is empty)?

(a) ① S → Rf ② blank ③ blank ④ T → ∈
(b) ① blank ② S → Rf ③ T → ∈ ④ T → ∈
(c) ① blank ② S → Rf ③ blank ④ blank
(d) ① S → Rf ② S → Rf ③ T → ∈ ④ T → ∈

[2021 (Set-1) : 2 M]

2.45 Consider the following statements:

S_1: Every SLR(1) grammar is unambiguous but there are certain unambiguous grammars that are not SLR(1).

S_2: For any context-free grammar, there is a parser that takes at most $O(n^3)$ time to parse a string of length n.

Which one of the following options is correct?

(a) S_1 is true and S_2 is true
(b) S_1 is false and S_2 is false
(c) S_1 is true and S_2 is false
(d) S_1 is false and S_2 is true

[2021 (Set-1) : 1 M]

2.46 Consider the following augmented grammar with {#, @, <, >, a, b, c} as the set of terminals.

S′ → S
S → S # cS
S → SS
S → S@
S → <S>
S → a
S → b
S → c

Let I_0 = CLOSURE(S′ → •S}). The number of items in the set GOTO(GOTO(I_0, <), <) is ________. **[2021 (Set-2) : 2 M]**

2.47 Which one of the following statements is TRUE?
(a) The LALR(1) parser for a grammar G cannot have reduce-reduce conflict if the LR(1) parser for G does not have reduce-reduce conflict.
(b) Symbol table is accessed only during the lexical analysis phase.
(c) Data flow analysis is necessary for run-time memory management.
(d) LR(1) parsing is sufficient for deterministic context-free languages. **[2022 : 1 M]**

2.48 Consider the augmented grammar with {+, *, (,), id} as the set of terminals.

S′ → S
S → S + R | R
R → R * P | P
P → (S) | id

If I_0 is the set of two LR(0) items {[S′ → S.], [S → S. + R]}, then goto(closure(I_0), +) contains exactly ________ items. **[2022 : 1 M]**

2.49 Which of the following is/are Bottom-Up Parser(s)?
(a) LR Parser (b) Predictive Parser
(c) Shift-reduce Parser (d) LL(1) Parser **[2024 (Set-1) : 1 M]**

2.50 Consider the following grammar G, with S as the start symbol. The grammar G has three incomplete productions denoted by (1), (2), and (3).

$S \rightarrow daT \mid$ (1)
$T \rightarrow aS \mid bT \mid$ (2)
$R \rightarrow$ (3) $\mid \in$

The set of terminals is {a, b, c, d, f}. The FIRST and FOLLOW sets of the different non-terminals are as follows.

FIRST(S) = {c, d, f}, FIRST(T) = {$a, b, \in$}, FIRST(R) = {$c, \in$}
FOLLOW(S) = FOLLOW(T) = {$c, f, \$$}, FOLLOW(R) = {f}

Which one of the following options CORRECTLY fills in the incomplete productions?
(a) (1) $S \rightarrow fR$, (2) $T \rightarrow cT$, (3) $R \rightarrow cR$
(b) (1) $S \rightarrow fR$, (2) $T \rightarrow \in$, (3) $R \rightarrow cTR$
(c) (1) $S \rightarrow Rf$, (2) $T \rightarrow cT$, (3) $R \rightarrow cR$
(d) (1) $S \rightarrow Rf$, (2) $T \rightarrow \in$, (3) $R \rightarrow cTR$ **[2024 (Set-1) : 2 M]**

2.51 Consider the following context-free grammar where the start symbol is S and the set of terminals is {a, b, c, d}.

$S \rightarrow AaAb \mid BbBa$
$A \rightarrow cS \mid \in$
$B \rightarrow dS \mid \in$

The following is a partially-filled LL(1) parsing table.

	a	b	c	d	$
S	S → AaAb	S → BbBa	(1)	(2)	
A	A → ∈	(3)	A → cS		
B	(4)	B → ∈		B → dS	

Which one of the following options represents the CORRECT combination for the numbered cells in the parsing table?
Note: In the options, "blank" denotes that the corresponding cell is empty.
(a) (1) $S \rightarrow BbBa$ (2) $S \rightarrow AaAb$ (3) blank (4) blank
(b) (1) $S \rightarrow AaAb$ (2) $S \rightarrow BbBa$ (3) $A \rightarrow \in$ (4) $B \rightarrow \in$
(c) (1) $S \rightarrow AaAb$ (2) $S \rightarrow BbBa$ (3) blank (4) blank
(d) (1) $S \rightarrow BbBa$ (2) $S \rightarrow AaAb$ (3) $A \rightarrow \in$ (4) $B \rightarrow \in$ **[2024 (Set-2) : 2 M]**

2.52 Consider the following augmented grammar, which is to be parsed with a SLR parser. The set of terminals is {a, b, c, d, #, @}

S′ → S
S → SS | Aa | bAc | Bc | bBa
A → d#
B → @

Let I_0 = CLOSURE ({S′ → • S}). The number of items in the set GOTO(I_0, S) is ________. **[2024 (Set-2) : 2 M]**

Answers Parsing Techniques

2.1	(a)	**2.2**	(b)	**2.3**	(c)	**2.4**	(c)	**2.5**	(a)	**2.6**	(d)	**2.7**	(b)	**2.8**	(d)	**2.9**	(a)
2.10	(b)	**2.11**	(a)	**2.12**	(b)	**2.13**	(a)	**2.14**	(a)	**2.15**	(d)	**2.16**	(a)	**2.17**	(d)	**2.18**	(a)
2.19	(a)	**2.20**	(c)	**2.21**	(c)	**2.22**	(c)	**2.23**	(b)	**2.24**	(d)	**2.25**	(b)	**2.26**	(c)	**2.27**	(a)
2.28	(c)	**2.29**	(c)	**2.30**	(d)	**2.31**	(d)	**2.32**	(b)	**2.33**	(c)	**2.34**	(c)	**2.35**	(c)	**2.36**	(d)
2.37	(c)	**2.38**	(a)	**2.39**	(c)	**2.40**	(31)	**2.41**	(b)	**2.42**	(5)	**2.43**	(7)	**2.44**	(d)	**2.45**	(a)
2.46	(8)	**2.47**	(d)	**2.48**	(5)	**2.49**	(a, c)	**2.50**	(d)	**2.51**	(b)	**2.52**	(9)				

Explanations Parsing Techniques

2.1 (a)

Syntax directed tree of xxxxyzz is as follows:

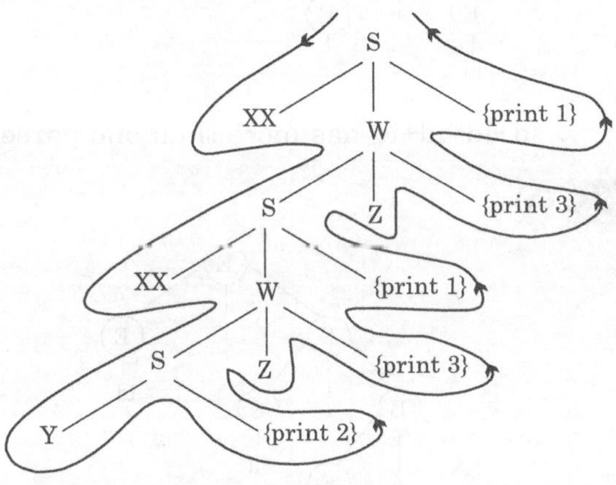

So translation by rules is 23131

[The code has translation rules at the end, so the result of top-down and bottom-up parsing will be same.]

2.2 (b)

A multi-pass compiler is a type of compiler that processes the source code or abstract syntax tree of a program several times. Two-pass compiler, on the first pass checks the syntax of statements and constructs a table of symbols, while on the second pass it actually translates program statements into machine language.

(i) Object code generation – 2

(ii) Literals added to literal table – 1

(iii) Listing printed – 2

(iv) Address resolution of local symbols that occur in a two pass assembler respectively – 1

2.3 (c)

LR parsers in term of power is

CLR > LALR > SLR > LR(O)

2.4 (c)

Relocation is the process of assigning load address to various parts of a program and adjusting the code and data in the program to reflect the assigned addresses.

2.5 (a)

(a) An unambiguous grammar can have different leftmost and rightmost derivation. However, an unambiguous grammar has only one derivation tree. So option (a) is false.

(b) LL(1) is a top-down parser.

(c) LALR is more powerful than SLR.

(d) For any parser, grammar should be unambiguous.

2.6 (d)

An ambiguous grammar can't be LL(1). Removing left recursion and factoring the grammar not always guarantees the unambiguity of grammar.

Therefore, option (d) is correct.

2.7 (b)

SLR parser has n_1 states for a grammar G LALR parser has n_2 states for a grammar G. The states of SLR and LALR parser are the states of corresponding states in a deterministic finite automata which recognizes the viable prefixes and both deterministic finite automata contains the equal number of states so $n_1 = n_2$.

2.8 (d)

The grammar is

$$S \rightarrow iEtSS' \mid a$$
$$S' \rightarrow eS \mid \varepsilon$$
$$E \rightarrow b$$

The predictive parser table M is

Non-terminal	a	b	e	i	t	$
S	$s \to a$			$S \to iEtSS'$		
S′			$S' \to \varepsilon$ $S' \to eS$			$S' \to \varepsilon$
E		$E \to b$				

So $M[S', e] = \{S' \to \varepsilon, S' \to eS\}$
$M[S', \$] = S \to \varepsilon$

2.9 (a)

Consider the grammar:

$S \to CC$
$C \to cC \mid d$

The above grammar is LL(1)
[it is unambiguous, not left recursive and left factored].
It is also LR(0) grammar, hence it is SLR(1), LALR(1) and CLR(1).
Therefore option (a) matches correctly.

2.10 (b)

In operator grammar the grammar has the property that no production side is ε or it doesn't contain adjacent nonterminals.
So $P \to QR$ and $P \to \varepsilon$ violate the requirements of an operator grammar.

2.11 (a)

The string ε has more than one parse tree for the given grammar.

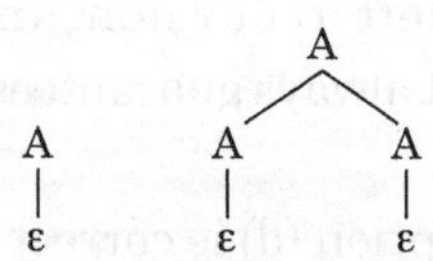

Given grammar is Ambiguous, that is the reason that it is not suitable for predictive parsing.

2.12 (b)

$S \to (S) \mid a$

Parser	Number of states
SLR(1)	n_1
LR(1)	n_2
LALR(1)	n_3

The number of states of deterministic finite automate in SLR(1) and LALR(1) parsers are equal, so $n_1 = n_3$. The number of states of deterministic finite automata in LR(1) is greater than number of states of deterministic finite automata of SLR(1) and LALR(1)
So, $n_1 = n_3 < n_2$.

2.13 (a)

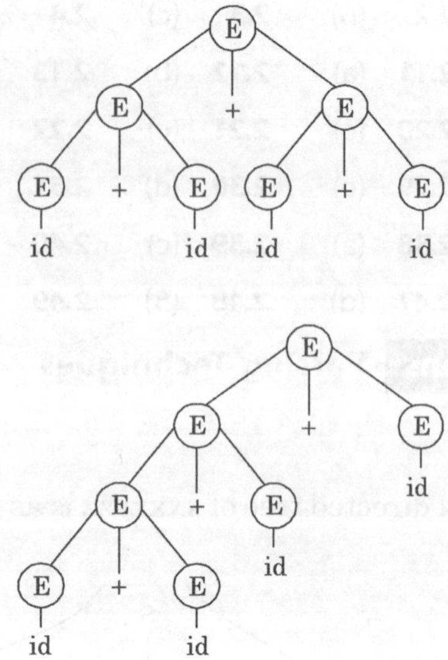

∴ id+id+id+id has more than one parse tree.

2.14 (a)

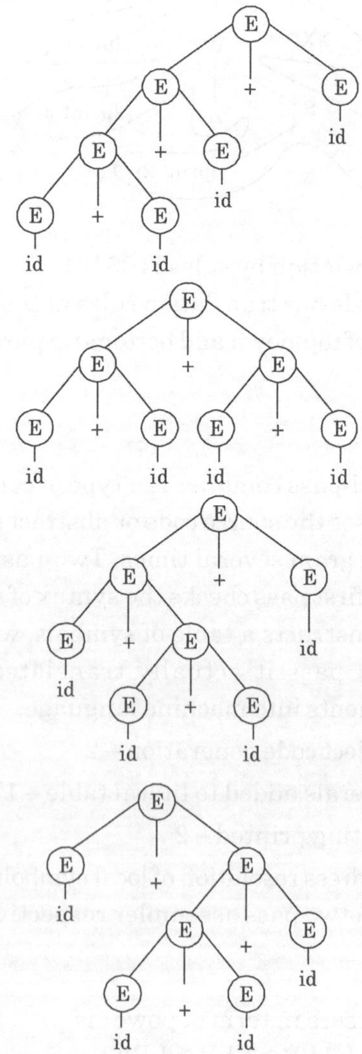

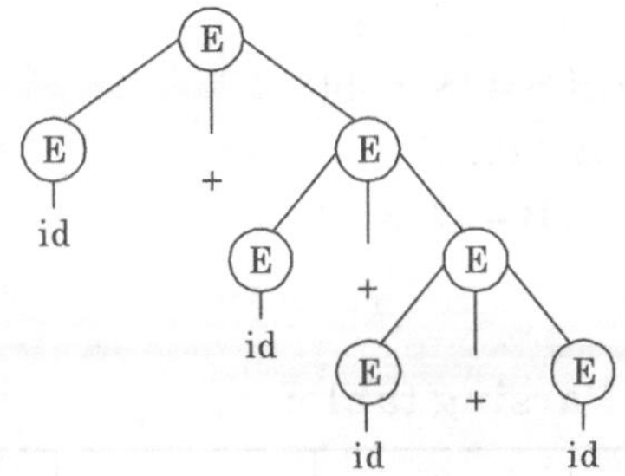

∴ 5 parse trees are exist for id+id+id+id

2.15 (d)

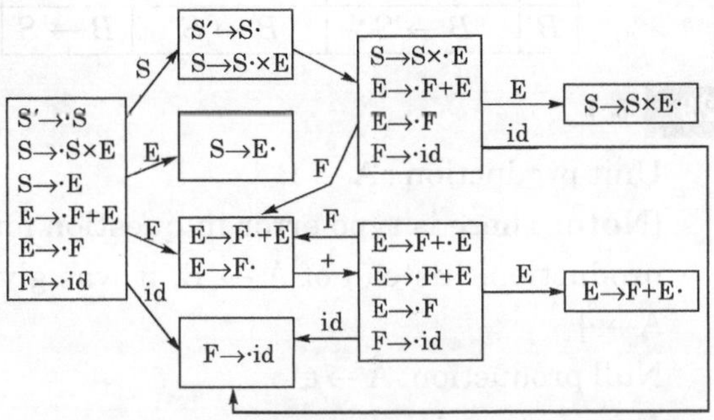

Given $S \rightarrow S\times\cdot E$, $E \rightarrow F\cdot+E$ and $E \rightarrow F+\cdot E$. Any of these two items not present in the same state of above DFA.

2.16 (a)

Construct Predictive parser table as follows:

Non-terminal	*id*	*	$
S	$S \rightarrow FR$		
R		$R \rightarrow *S$	$R \rightarrow \varepsilon$
F	$F \rightarrow id$		

So M[S, id] = {S → FR}
and M[R, $] = {R → ε}

2.17 (d)

$$S \rightarrow AC \mid CB$$
$$C \rightarrow aCb \mid \varepsilon$$
$$A \rightarrow aA \mid a$$
$$B \rightarrow bB \mid b$$

From the above grammar,

$C \rightarrow aCb \mid \varepsilon \Rightarrow L(C) = \{a^n b^n\}$

$A \rightarrow aA \mid a \Rightarrow L(A) = a^+$

$B \rightarrow bB \mid b \Rightarrow L(B) = b^+$

By substituting C, A and B in S.

$S \rightarrow AC \Rightarrow L(S) = \{a^m b^n \mid m > n\}$

$S \rightarrow CB \Rightarrow L(S) = \{a^m b^n \mid m < n\}$

$L(S) = \{a^m b^n \mid m > n \text{ or } m < n\}$

$= \{a^i b^j \mid i \neq j\}$

2.18 (a)

In the Q. 2.18 if grammar generates $a^n b^0$ or $a^0 b^n$ then the length is max(l, m) if it generates $a^l b^m$ initially we include two extra derivation so the total length is max(l, m) + 2.

2.19 (a)

Predictive parser and recursive descent parser are example of top-down parser. Whereas LR(k) and LALR(k) parser are bottom up parser.

2.20 (c)

$$S \rightarrow iCtSS_1 \mid a$$
$$S_1 \rightarrow eS \mid \varepsilon$$
$$C \rightarrow b$$

Consider the string = ibtibtaea. There are two parse trees to derive this string.

Parse Tree-1

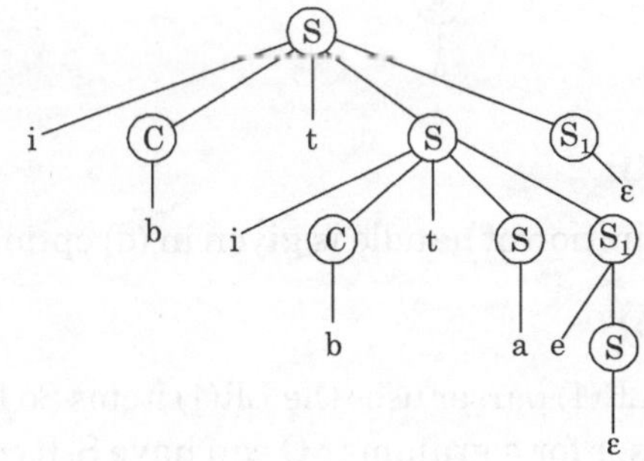

Parse Tree-2

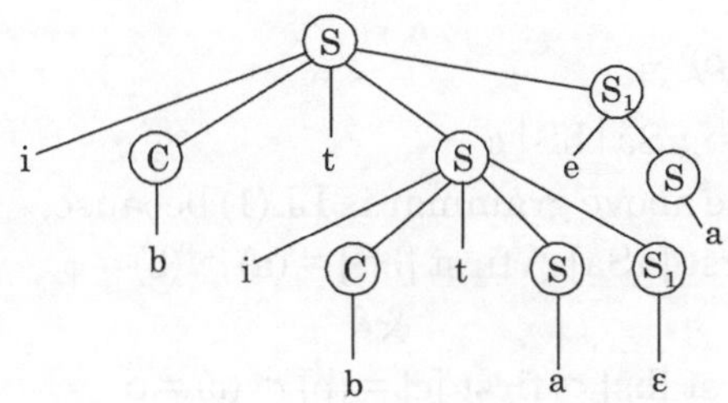

For the same string more than one parse tree is present to derive the string. Hence the given grammar is ambiguous.

2.21 (c)

- A regular grammar can be ambiguous. An ambiguous grammar can't be LL(1). Therefore, it is false that every regular grammar is LL(1)
- Every regular language is unambiguous i.e., for every regular language there exist atleast one grammar which is unambiguous.

Therefore, it is true that every regular set has LR(1) grammar.

∴ P is false and Q is true.

So option (c) is correct option.

2.22 (c)

$S \Rightarrow aB \Rightarrow aaBB \Rightarrow aabB \Rightarrow aabbS$

$\Rightarrow aabbaB \Rightarrow aabbab.$

So aaabbab is generated by the grammar.

2.23 (b)

There are two parse trees for the string aabbab.

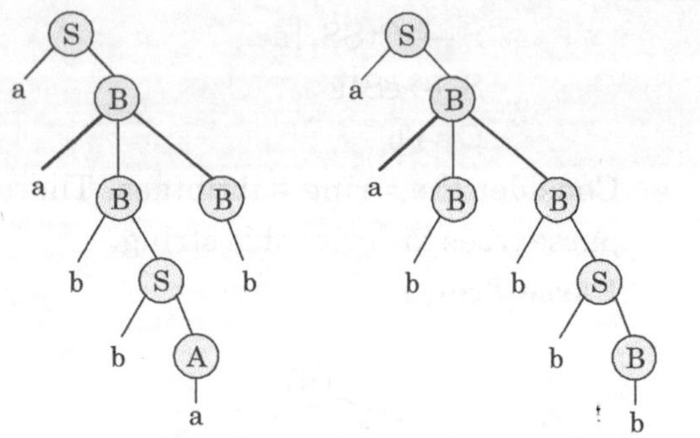

2.24 (d)

Definition of handle is given in (d) option.

2.25 (b)

LALR(1) parser uses the LR(1) items So LALR(1) parser for a grammar G can have S-R conflict if and only if the LR(1) Parser for G has S-R conflicts.

2.26 (c)

$S \rightarrow aSa \mid bS \mid c$

The above grammar is LL(1) because,

First [aSa] ∩ first [bS] = (a) ∩ (b) = ϕ

&&

First [bs] ∩ first [c] = (b) ∩ (c) = ϕ

&&

First [c] ∩ first [aSa] = (c) ∩ (a) = ϕ

As the above grammar is LL(1), also LR(1) because LL(1) grammar is always LR(1) grammar.

2.27 (a)

$S \rightarrow aAbB \mid bAaB \mid \varepsilon$

$A \rightarrow S$

$B \rightarrow S$

First of S = {a, b, ε}

First of A = {a, b, ε}

First of B = {a, b, ε}

Follow of S = {$, a, b}

Follow A = {b, a}

Follow of B = {$, a, b}

2.28 (c)

LL(1) Parsing table:

	a	*b*	$
S	*S* → *aAbB* *S* → ε	*S* → *aAbB* *S* → ε	*S* → ε
A	*A* → *S*	*A* → *S*	error
B	*B* → *S*	*B* → *S*	*B* → *S*

2.29 (c)

Unit production : A → B

[**Note:** There is typo error in question for unit production. Instead of A → B, it was given as A→a]

Null production: A → ε

Ex. S → AB

A → a

B → b

[2 length string] ab ⇒ Ab ⇒ AB ⇒ S, it requires three reductions.

∴ For n length string (2n – 1) reductions required.

2.30 (d)

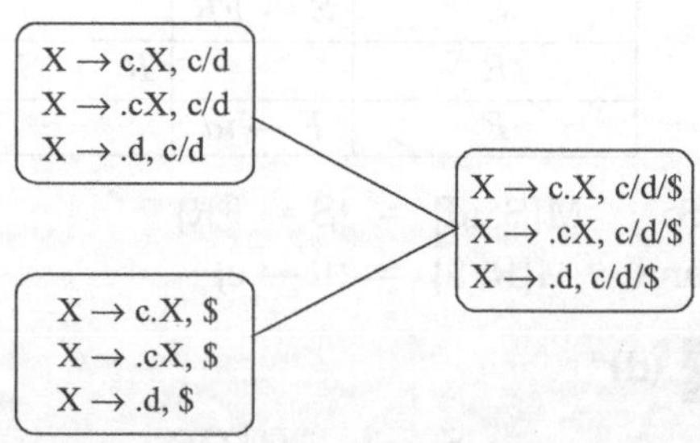

1. Cannot be merged since look-a-heads are different: FALSE, because merging does not depend on look ahead.
2. Can be merged but will results in S-R conflict: FALSE the two states are not containing reduce item, so after merging, merged state can not contain any S-R conflict.
3. Can be merged but will result in R-R conflict: FALSE, No RR conflict in merged state.
4. Cannot be merged since goto on c will lead to two different sets: FALSE, merging of states does not depends on goto.

2.31 (d)

$S \rightarrow L \cdot > R$

$Q \rightarrow R \cdot$

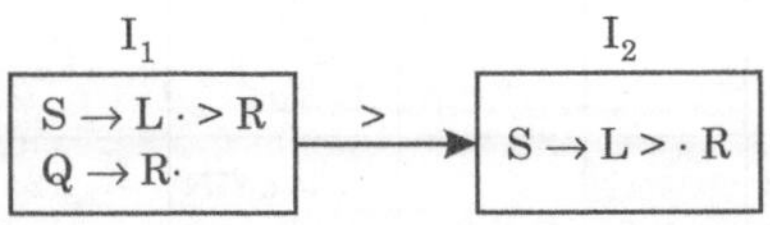

In above diagram, we see at I_2 state it has only one reduced item and there is no other item in the state. Therefore for input symbol '>', there is no conflict.

2.32 (b)

$S \rightarrow T * P$

$T \rightarrow U \mid T * U$

$P \rightarrow Q + P \mid Q$

$Q \rightarrow Id$

$U \rightarrow Id$

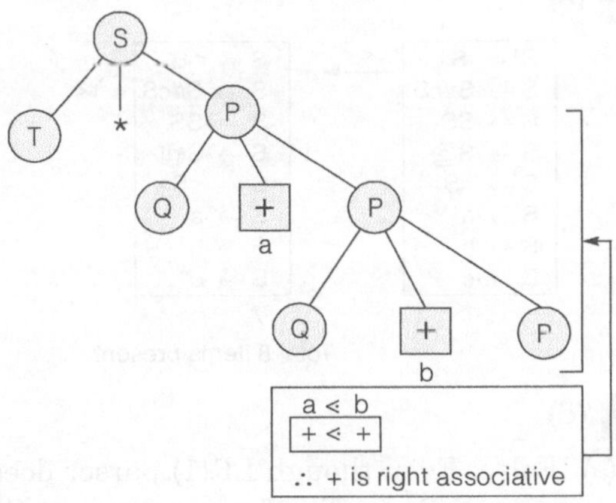

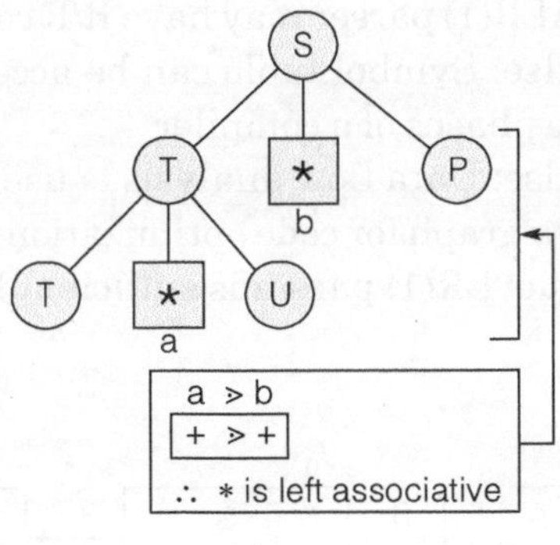

2.33 (c)

Stack contains only a set of viable prefixes.

2.34 (c)

A. Lexical Analysis → DFA minimization (2)

B. Parsing → Production Tree (3)

C. Register Allocation → Graph coloring (1)

D. Expression evaluation → Post-order traversal (4)

2.35 (c)

Among SLR, CLR and LALR parsers

(i) **SLR** parser is simple and very easy to implement compared to other parsers.

(ii) **CLR** parser is most powerful than other parsers because it can accepts more languages than other.

2.36 (d)

$S \rightarrow F \mid H$

$F \rightarrow p \mid c$

$H \rightarrow d \mid c$

Given grammar is ambiguous, since string "*c*" can be generated in two ways.

An ambiguous grammar can be neither LL nor LR since every LL and every LR grammar is unambiguous.

∴ Neither S1 nor S2 is correct.

2.37 (c)

$$\begin{aligned} \text{Follow (Q)} &= \text{First (RS)} \\ &= \{\{w\} - \in\} \cup \text{First (S)} \\ &= \{w, y\} \end{aligned}$$

2.38 (a)

Cononical LR is the most powerful parsers among all the LR(K) parsers.

2.39 (c)

$E \rightarrow E - T \mid T$

$T \rightarrow T + F \mid F$

$F \rightarrow (E) \mid id$

There are 2 left recurssion and both have to be removed. The following is the conversion

$E' \rightarrow -TE' \mid \in$

$E \rightarrow TE'$

$T' \rightarrow +FT' \mid \in$

$Y \rightarrow FT'$

$F \rightarrow (E) \mid id$

Now by putting E′ as X and T′ as Y we get option (c) which is

$E \rightarrow TX \qquad X \rightarrow -TX \mid \in$

$T \rightarrow FY \qquad Y \rightarrow +FY \mid \in$

$F \rightarrow (E) \mid id$

2.40 (31)

$S \rightarrow Aa$
$A \rightarrow BD$
$B \rightarrow b \mid \in$
$D \rightarrow d \mid \in$
Follow (B) = {d, a}
Hence their index in descending order is 31.

2.41 (b)

LR parser is a bottom up parser. Hence it uses right most derivation in reverse order.
So option (b) is the correct answer.

2.42 (5)

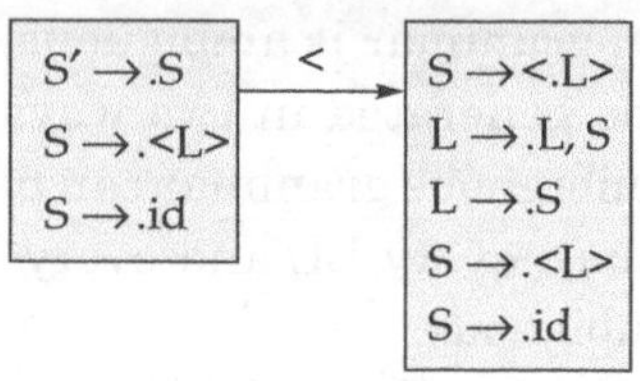

Total number of items in the set GOTO (I_0, ⟨) is 5.

2.43 (7)

$S \rightarrow aSB$
$\rightarrow aaSBB$ [$S \rightarrow aSB$]
$\rightarrow aaaSBBB$ [$S \rightarrow aSB$]
$\rightarrow aaadBBB$ [$S \rightarrow d$]
$\rightarrow aaadbBB$ [$B \rightarrow b$]
$\rightarrow aaadbbB$ [$B \rightarrow b$]
$\rightarrow aaadbbb$ [$B \rightarrow b$]

Total 7 steps required.

2.44 (d)

$S \rightarrow daT \mid Rf$
$T \rightarrow aS \mid baT \mid \in$
$R \rightarrow caTR \mid \in$

First (S) = $\{d, c, f\}$
Follow (S) = $\{c, f, \$\}$
First (T) = $\{a, b, \in\}$
Follow (T) = $\{c, f, \$\}$
First (R) = $\{c, \in\}$
Follow (R) = $\{f\}$

	a	b	c	d	f	$
S			① S → Rf	S → daT	② S → Rf	
T	T → aS	T → baT	③ T → ∈		T → ∈	④ T → ∈
R			R → caTR		R → ∈	

① S → Rf ② S → Rf ③ T → ∈ ④ T → ∈

Hence option (d) is answer.

2.45 (a)

S_1: Every SLR(1) is unambiguous but every unambiguous in not SLR(1). So S_1 is true.
S_2: Using CYK algorithm which takes at most $O(n^3)$ time to parse a string, where n is the length of the string. So S_2 is true.

Both S_1 is true and S_2 is true

2.46 (8)

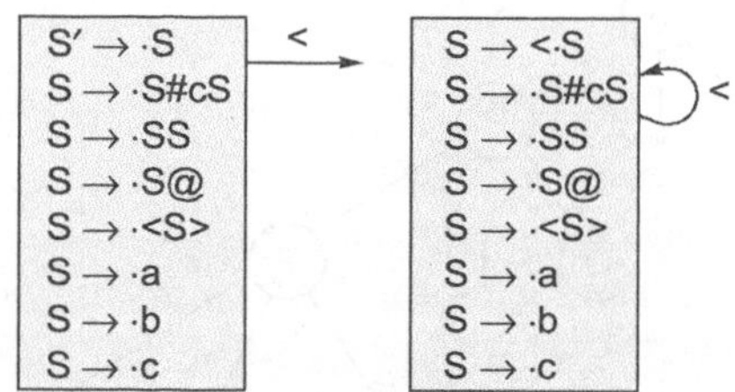

Total 8 items present

2.47 (d)

(a) False: Even though LR(1) parser does not have R/R conflict for a grammar G, the LALR(1) parser may have R/R conflict for G.
(b) False: Symbol table can be accessed by all the phases of a compiler
(c) False: Data flow analysis is used in control flow graph for code optimization
(d) True: LR(1) parser is sufficient for DCLF's.

2.48 (5)

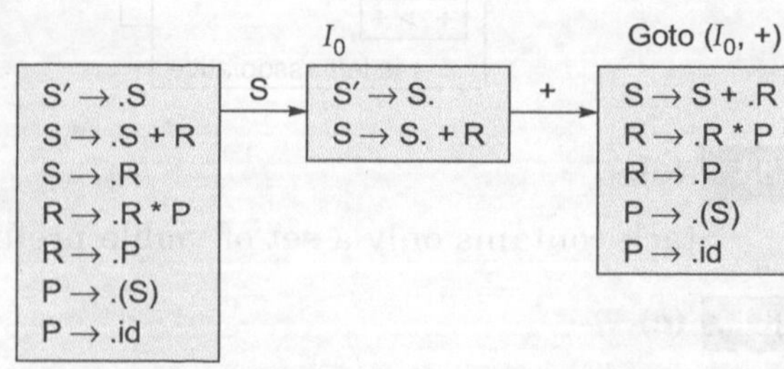

Goto (I_0, +) will have 5 elements.

2.49 (a, c)

LR parser and Shift-Reduce parsers are bottom up parsers.

Answer is (a) and (c).

2.50 (d)

By verification process the production will be

$S \rightarrow daT \mid Rf$

$T \rightarrow aS \mid bT \mid \in$

$R \rightarrow cTR \mid \in$

First(S) = $\{d, c, f\}$

First(T) = $\{a, b, \in\}$

First(R) = $\{c, \in\}$

2.51 (b)

$S \rightarrow AaAb \mid BbBa$

$A \rightarrow cS \mid \in$

$B \rightarrow dS \mid \in$

First (S) = $\{a, b, c, d\}$; Follow(S) = $\{a, b\}$

First (A) = $\{c, \in\}$; Follow(A) = $\{a, b\}$

First (B) = $\{d, \in\}$; Follow(B) = $\{a, b\}$

	a	b	c	d	$
S	S →AaAb	S →BbBa	S →AaAb	S →BbBb	
A	A →∈	A →∈	A →cS		
B	B →∈	B →∈		B →dS	

(1) $S \rightarrow AaAb$

(2) $S \rightarrow BbBa$

(3) $A \rightarrow \in$

(4) $B \rightarrow \in$

2.52 (9)

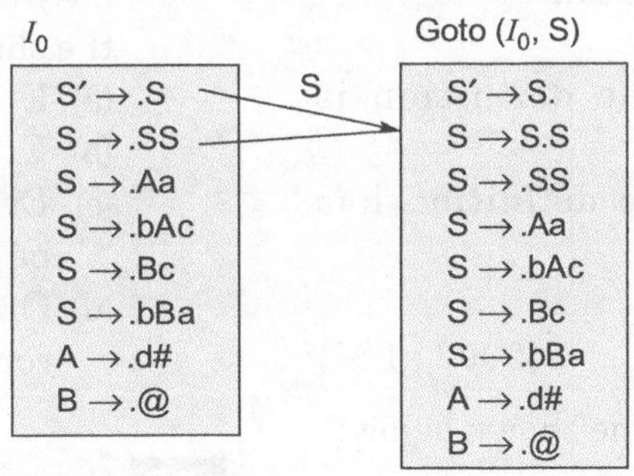

Total 9 items.

3 Syntax Directed Translation

3.1 A linker is given object modules for a set of programs that were compiled separately. What information need to be included in an object module?
(a) Object code
(b) Relocation bits
(c) Names and locations of all external symbols defined in the object module
(d) Absolute addresses of internal symbols

[1995 : 1 M]

3.2 In the following grammar:

X : : = X ⊕ Y/Y
Y : : = Z ⊙ Y/Z
Z : : = id

Which of the following is true?
(a) '⊕' is left associative while '⊙' is right associative
(b) Both '⊕' and '⊙' is left associative
(c) '⊕' is the right associative while '⊙' is left associative
(d) None of the above

[1997 : 1 M]

3.3 In a bottom-up evaluation of a syntax directed definition, inherited attributes can
(a) always be evaluated
(b) be evaluated only if the definition is L-attributed
(c) be evaluated only if the definition has synthesized attributes
(d) never be evaluated

[2003 : 1 M]

3.4 Consider the translation scheme shown below:
S → T R
R → + T {print ('+');} R | ε
T → num {print (num.val);}
Here num is a token that represents an integer and num.val represents the corresponding integer value. For an input string '9 + 5 + 2', this translation scheme will print
(a) 9 + 5 + 2
(b) 9 5 + 2 +
(c) 9 5 2 + +
(d) + + 9 5 2

[2003 : 2 M]

3.5 Consider the syntax directed definition shown below:
S → id : = E {gen (id.place = E.place;);}
$E \rightarrow E_1 + E_2$ {t = newtemp ( );
gen (t = E_1. place + E_2. place;);
E.place = t}
E → id {E.place = id.place;}
Here, gen is a function that generates the output code, and newtemp is a function that returns the name of a new temporary variable on every call. Assume that t_i's are the temporary variable names generated by newtemp.
For the statement 'X: = Y + Z', the 3-address code sequence generated by this definition is
(a) X = Y + Z
(b) t_1 = Y + Z; X = t_1
(c) t_1 = Y; t_2 = t_1 + Z; X = t_2
(d) t_1 = Y; t_2 = Z; t_3 = t_1 + t_2; X = t_3

[2003 : 2 M]

3.6 Consider the grammar rule $E \rightarrow E_1 - E_2$ for arithmetic expressions. The code generated is targeted to a CPU having a single user register. The subtraction operation requires the first operand to be in the register. If E_1 and E_2 do not have any common subexpression, in order to get the shortest possible code
(a) E_1 should be evaluated first
(b) E_2 should be evaluated first
(c) Evaluation of E_1 and E_2 should necessarily be interleaved
(d) Order of evaluation of E_1 and E_2 is of no consequence

[2004 : 1 M]

3.7 Consider the grammar with the following translation rules and E as the start symbol.

$E \rightarrow E_1$ # T	{E.value = E_1.value * T.value}
| T	{E.value = T.value}
$T \rightarrow T_1$ & F	{T.value = T_1.value + F.value}
| F	{T.value = F.value}
F → num	{F.value = num.value}

Compute E. value for the root of the parse tree for the expression: 2 # 3 & 5 # 6 & 4.
(a) 200 (b) 180
(c) 160 (d) 40

[2004 : 2 M]

3.8 Consider the grammar $E \rightarrow E + n \mid E \times n \mid n$
For a sentence $n + n \times n$, the handles in the right-sentential form of the reduction are
(a) n, E + n and E + n × n
(b) n, E + n and E + E × n
(c) n, n + n and n + n × n
(d) n, E + n and E × n **[2005 : 2 M]**

Common Data for Q. 3.9 & Q. 3.10:

Consider the following expression grammar. The semantic rules for expression calculation are stated next to each grammar production.

E → number E.val = number.val
  | E '+' E $E^{(1)}$.val = $E^{(2)}$.val + $E^{(3)}$.val
  | E '×' E $E^{(1)}$.val = $E^{(2)}$.val × $E^{(3)}$.val;

3.9 The above grammar and the semantic rules are fed to a yacc tool (which is an LALR(1) parser generator) for parsing and evaluating arithmetic expressions. Which one of the following is true about the action of yacc for the given grammar?
(a) It detects recursion and eliminates recursion
(b) It detects reduce-reduce conflict, and resolves
(c) It detects shift-reduce conflict, and resolves the conflict in favor of a shift over a reduce action
(d) It detects shift-reduce conflict, and resolves the conflict in favor of a reduce over a shift action **[2005 : 2 M]**

3.10 Assume the conflicts in Part (a) of this question are resolved and an LALR(1) parser is generated for parsing arithmetic expressions as per the given grammar. Consider an expression 3 × 2 + 1. What precedence and associativity properties does the generated parser realize?
(a) Equal precedence and left associativity; expression is evaluated to 7
(b) Equal precedence and right associativity; expression is evaluated to 9
(c) Precedence of '×' is higher than that of '+', and both operators are left associative; expression is evaluated to 7
(d) Precedence of '+' is higher than that of '×', and both operators are left associative; expression is evaluated to 9 **[2005 : 2 M]**

3.11 Consider the following translation scheme:

S → ER
R → *E {print("*"); R | ε
E → F + E {print ('+'); | F
F → (S) | id {print (id. value);}

Here id is a token that represents an integer and id.value represents the corresponding integer value. For an input '2 * 3 + 4', this translation scheme prints
(a) 2 * 3 + 4
(b) 2 * + 3 4
(c) 2 3 * 4 +
(d) 2 3 4 + *

[2006 : 2 M]

3.12 In a simplified computer the instructions are:
OP R_j, R_i – Performs R_j OP R_i and stores the result in register R_i
OP m, R_i – Performs val OP R_i and stores the result in R_i. val denotes the content of memory location m.
MOV m, R_i – Moves the content of memory location m to register R_i.
MOV Ri, m – Moves the content of register R_i to memory location m.
The computer has only two registers, and OP is either ADD or SUB.
Consider the following basic block:

$t_1 = a + b$
$t_2 = c + d$
$t_3 = e - t_2$
$t_4 = t_1 - t_3$

Assume that all operands are initially in memory. The final value of the computation should be in memory. What is the minimum number of MOV instructions in the code generated for this basic block?
(a) 2 (b) 3
(c) 5 (d) 6

[2007 : 2 M]

3.13 Consider the expression tree shown. Each leaf represents a numerical value, which can either be 0 or 1. Over all possible choices of the values at the leaves, the maximum possible value of the expression represented by the tree is _______.

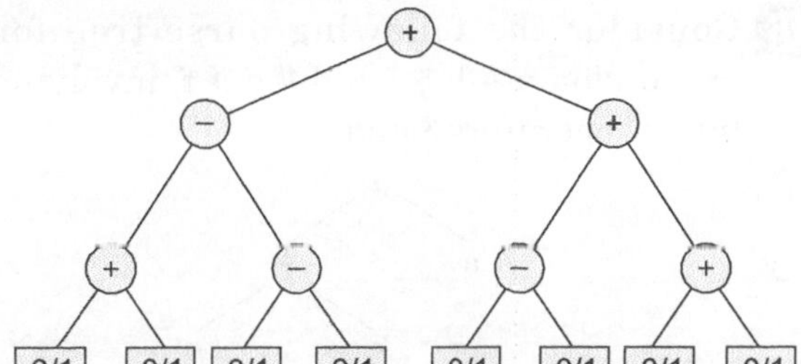

[2014 (Set-2) : 2 M]

3.14 One of the purposes of using intermediate code in compilers is to
(a) make parsing and semantic analysis simpler.
(b) improve error recovery and error reporting.
(c) increase the chances of reusing the machine-independent code optimizer in other compilers.
(d) improve the register allocation.
[2014 (Set-3) : 1 M]

3.15 The attributes of three arithmetic operators in some programming language are given below:

Operator	Precedence	Associativity	Arity
+	High	Left	Binary
–	Medium	Right	Binary
*	Low	Left	Binary

The value of the expression $2 - 5 + 1 - 7 * 3$ in this language is ______.
[2016 (Set-1) : 2 M]

3.16 Consider the following Syntax Directed Translation Scheme (SDTS), with non-terminals $\{S, A\}$ and terminals $\{a, b\}$.

$S \rightarrow aA$ {print 1}
$S \rightarrow a$ {print 2}
$A \rightarrow Sb$ {print 3}

Using the above SDTS, the output printed by a bottom-up parser, for the input *aab* is:
(a) 1 3 2 (b) 2 2 3
(c) 2 3 1 (d) Syntax error
[2016 (Set-1) : 2 M]

3.17 Which one of the following statements is FALSE?
(a) Context-free grammar can be used to specify both lexical and syntax rules.
(b) Type checking is done before parsing.
(c) High-level language programs can be translated to different Intermediate Representations.
(d) Arguments to a function can be passed using the program stack. **[2018 : 1 M]**

3.18 Consider the following parse tree for the expression a # b \$ c \$ d # e # f, involving two binary operators \$ and #.

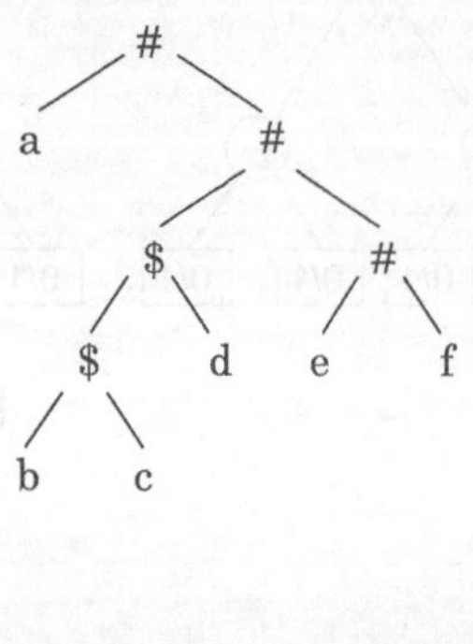

Which one of the following is correct for the given parse tree?
(a) \$ has higher precedence and is left associative; # is right associative
(b) # has higher precedence and is left associative; \$ is right associative
(c) \$ has higher precedence and is left associative; # is left associative
(d) # has higher precedence and is right associative; \$ is left associative
[2018 : 2 M]

3.19 Consider the following grammar and the semantic actions to support the inherited type declaration attributes. Let X_1, X_2, X_3, X_4, X_5 and X_6 be the place holders for the non-terminals D, T, L or L_1 in the following table:

Productions rule	Semantic action
$D \rightarrow TL$	X_1.type=X_2.type
$T \rightarrow$ int	T.type=int
$T \rightarrow$ float	T.type=float
$L \rightarrow L_1$, id	X_3.type=X_4.type addType(id.entry, X_5.type)
$L \rightarrow$ id	addType(id.entry, X_6.type)

Which one of the following are the appropriate choices for X_1, X_2, X_3 and X_4?
(a) $X_1 = L, X_2 = T, X_3 = L_1, X_4 = L$
(b) $X_1 = L, X_2 = L, X_3 = L_1, X_4 = T$
(c) $X_1 = T, X_2 = L, X_3 = L_1, X_4 = T$
(d) $X_1 = T, X_2 = L, X_3 = T, X_4 = L_1$
[2019 : 2 M]

3.20 Consider the productions $A \rightarrow PQ$ and $A \rightarrow XY$. Each of the five non-terminals A, P, Q, X and Y has two attributes: *s* is a synthesized attribute, and *i* is an inherited attribute. Consider the following rules.
Rule 1: $P.i = A.i + 2$, $Q.i = P.i + A.i$ and $A.s = P.s + Q.s$
Rule 2: $X.i = A.i + Y.s$ and $Y.i = X.s + A.i$
Which one of the following is TRUE?
(a) Neither Rule 1 nor Rule 2 is L-attributed.
(b) Both Rule 1 and Rule 2 are L-attributed.
(c) Only Rule 1 is L-attributed.
(d) Only Rule 2 is L-attributed.
[2020 : 2 M]

3.21 Consider the following grammar (that admits a series of declarations, followed by expressions) and the associated syntax directed translation (SDT) actions, given as pseudo-code:

$P \rightarrow D^* E^*$

$D \rightarrow$ int ID {record that ID.lexeme is of type int}

$D \rightarrow$ bool ID {record that ID.lexeme is of type bool}

$E \rightarrow E_1 + E_2$ {check that E_1.type = E_2.type = int; set E.type := int}

$E \rightarrow !E_1$ {check that E_1.type = bool; set E.type := bool}

$E \rightarrow$ ID {set E.type := int}

With respect to the above grammar, which one of the following choices is correct?

(a) The actions can be used to type-check syntactically correct integer variable declarations and integer expressions.

(a) The actions can be used to type-check syntactically correct boolean variable declarations and boolean expressions.

(c) The actions will lead to an infinite loop.

(d) The actions can be used to correctly type-check any syntactically correct program.

[2021 (Set-1) : 2 M]

3.22 Consider the following grammar along with translation rules.

$S \rightarrow S_1 \# T$	{S.val = S_1.val * T.val}
$S \rightarrow T$	{S.val = T.val}
$T \rightarrow T_1 \% R$	{T.val = T_1.val ÷ R.val}
$T \rightarrow R$	{T.val = R.val}
$R \rightarrow$ id	{R.val = id.val}

Here # and % are operators and id is a token that represents an integer and id.val represents the corresponding integer value. The set of non-terminals is {S, T, R, P} and a subscripted non-terminal indicates an instance of the non-terminal.

Using this translation scheme, the computed value of S.val for root of the parse tree for the expression 20 # 10 % 5 # 8 % 2 % 2 is ________.

[2022 : 2 M]

3.23 Consider the syntax directed translation given by the following grammar and semantic rules. Here N, I, F and B are non-terminals. N is the starting non-terminal, and #, 0 and 1 are lexical tokens corresponding to input letters "#", "0" and "1", respectively. X.val denotes the synthesized attribute (a numeric value) associated with a non-terminal X. I_1 and F_1 denote occurrences of I and F on the right hand side of a production, respectively. For the tokens 0 and 1, 0.val = 0 and 1.val = 1.

$N \rightarrow I \# F$	N.val = I.val + F.val
$I \rightarrow I_1 B$	I.val = (2 I_1.val) + B.val
$I \rightarrow B$	I.val = B.val
$F \rightarrow BF_1$	F.val = $\frac{1}{2}$ (B.val + F_1.val)
$F \rightarrow B$	F.val = $\frac{1}{2}$ B.val
$B \rightarrow 0$	B.val = 0.val
$B \rightarrow 1$	B.val = 1.val

The value computed by the translation scheme for the input string 10 # 011 is ________. (Rounded off to three decimal places)

[2023 : 2 M]

3.24 Consider the following syntax-directed definition (SDD).

$S \rightarrow DHTU$	{S.val = D.val + H.val + T.val + U.val; }
$D \rightarrow$ "M" D_1	{D.val = 5 + D_1.val; }
$D \rightarrow \epsilon$	{D.val = –5; }
$H \rightarrow$ "L" H_1	{H.val = 5 * 10 + H_1.val; }
$H \rightarrow \epsilon$	{H.val = –10; }
$T \rightarrow$ "C" T_1	{T.val = 5 * 100 + T_1.val; }
$T \rightarrow \epsilon$	{T.val = –5; }
$U \rightarrow$ "K"	{U.val = 5; }

Given "MMLK" as the input, which one of the following options is the CORRECT value computed by the SDD (in the attribute S.val)?

(a) 50 (b) 45
(c) 55 (d) 65

[2024 (Set-1) : 2 M]

3.25 Which of the following statements is/are FALSE?

(a) The attributes in a L-attributed definition cannot always be evaluated in a depth first order.

(b) An attribute grammar is a syntax-directed definition (SDD) in which the functions in the semantic rules have no side effects.

(c) All L-attributed definitions based on LR(1) grammar can be evaluated using a bottom-up parsing strategy.

(d) Synthesized attributes can be evaluated by a bottom-up parser as the input is parsed.

[2024 (Set-2) : 1 M]

■■■■

Answers Syntax Directed Translation

3.1	(d)	**3.2**	(a)	**3.3**	(c)	**3.4**	(b)	**3.5**	(b)	**3.6**	(b)	**3.7**	(c)	**3.8**	(d)	**3.9**	(c)
3.10	(b)	**3.11**	(d)	**3.12**	(b)	**3.13**	(6)	**3.14**	(c)	**3.15**	(9)	**3.16**	(c)	**3.17**	(b)	**3.18**	(a)
3.19	(a)	**3.20**	(c)	**3.21**	(a)	**3.22**	(80)	**3.23**	(2.375)	**3.24**	(b)	**3.25**	(a, c)				

Explanations Syntax Directed Translation

3.1 (d)

A linker is a computer program that takes one or more object files generated by a compiler and combines them into a single executable program. The linker also takes care of arranging the objects in a program's address space. Therefore absolute addresses of internal symbols need to be included in an object module.

3.2 (a)

Creating syntax tree for id ⊕ id ⊕ d :

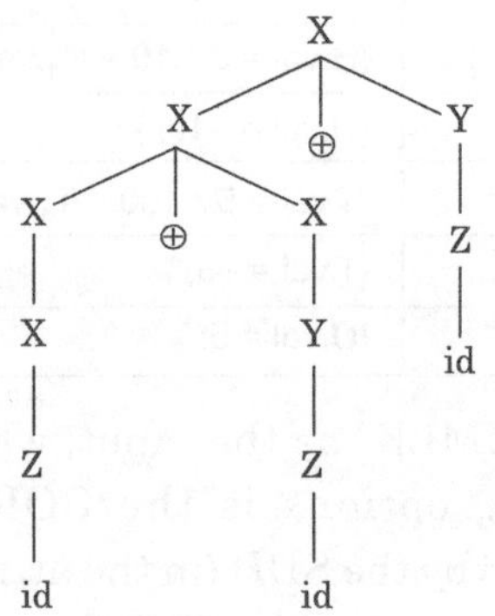

Therefore ⊕ is left associative.
Creating syntax tree for id ⊙ id ⊙ id :

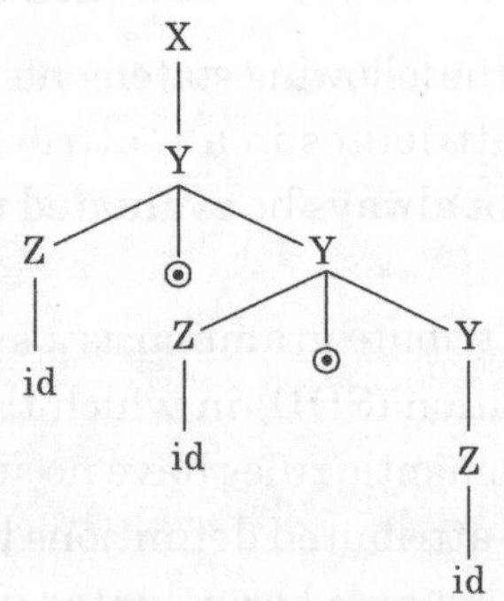

Therefore, ⊙ is right associative.

3.3 (c)

Every S(Synthesized) - attributed definition is L-attributed. For implementing inherited attributed during bottom-up parsing, extends to some, but not LR grammars. Consider the following example

Production	Semantic Rule
$S \rightarrow L$	L. count : = 0
$L \rightarrow L_1\ 1$	L. count : = L. count : + 1
$L \rightarrow E$	print (L.count)

In the example above the nonterminal L in $L \rightarrow E$ inherits the count of the number of 1's generated by S. Since the production $L \rightarrow E$ is the first that a bottom-up parser would reduce by, the translator at the time can't know the number of 1's in the input. So in a bottom -up evaluation of a syntax directed definition, inherits attributes can't be evaluated if the definition is L-attributed in the given example. So we can say that L-attributed definition is based on simple LR(1) grammar, but it can't be implemented always but inherit attributes can be evaluated only if the definition has synthesized attributes.

3.4 (b)

For the input '9 + 5 + 2' the translation scheme is 95 + 2 + shown below:

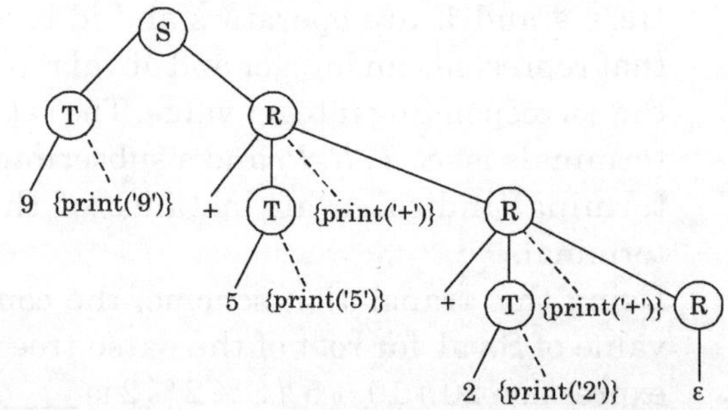

3.5 (b)

gen( ) function will be used only two times for X = Y + Z and only one temp variable is created with newtemp( ). ∴ t_1 = Y + Z; X = t_1

3.6 (b)

To optimize the solution evaluate the expression E_2. Then we can calculate E_1 and finally E_1 will be one of operands that will be in register and we can perform subtraction directly. But if we follow the opposite then we have to make move and store operations.

3.7 (c)

First we have to construct the parse tree.

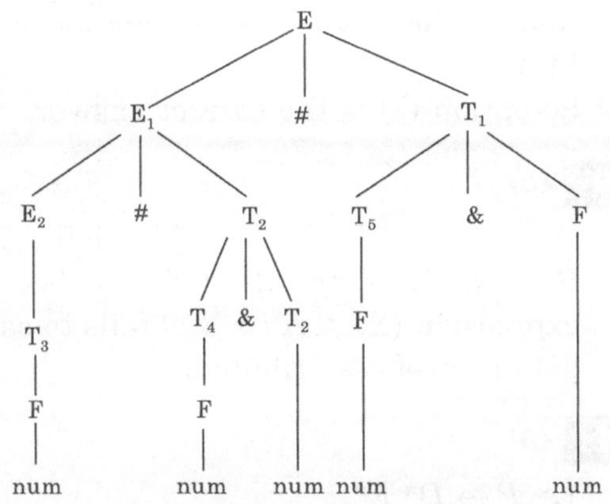

Then we construct the annonated parse tree or parse tree with value at the leaf node.

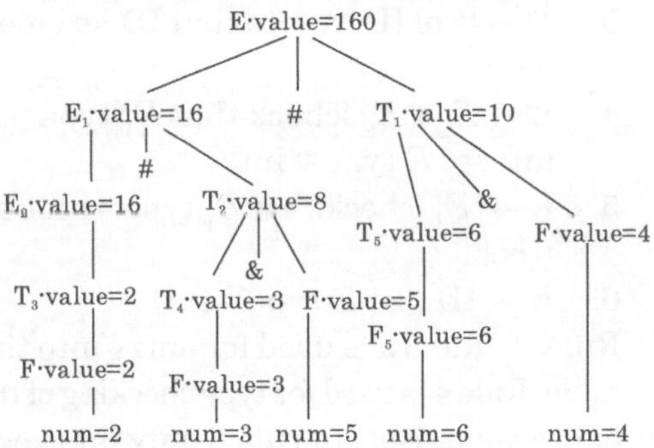

3.8 (d)

$E \rightarrow E + n \mid E \times n \mid n$

Input String $n + n \times n$

$\Rightarrow n + n \times n$

$\Rightarrow E + n \times n$ reduction $E \rightarrow n$

$\Rightarrow E \times n$ reduction $E \rightarrow E + n$

$\Rightarrow E$ reduction $E \rightarrow E \times n$

So the reductions are n, $E + n$, $E \times n$

3.9 (c)

$E \rightarrow$ number $\mid E' + E' \mid E' \times E'$

Then YACC compiler detects shift-reduce conflict, and resolves the conflict in favor of a shift over a reduce action.

Consider the following configuration where shift-reduce conflict occurs.

StackInput ... $E + E \times .E...\$$

3.10 (b)

SR conflict is resolved in favour of shift. All operators are shifted and evaluated from right end, because × and + have equal precedence. Hence, × and + follows right associatively.

$3 \times 2 + 1 = 3 \times 3 = 9$

3.11 (d)

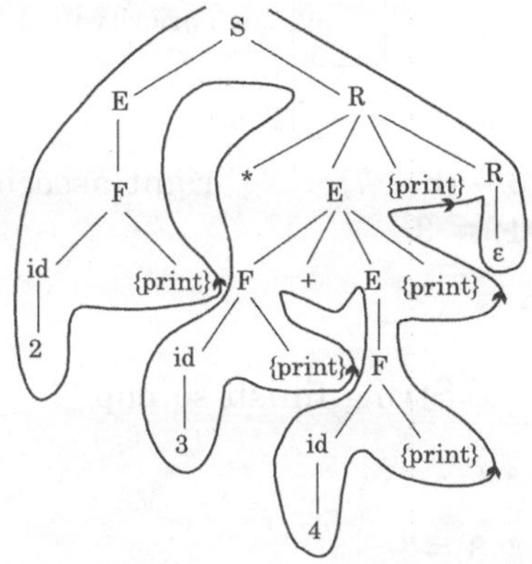

So an input 2 * 3 + 4, it prints from the above parse tree as 234 + *.

3.12 (b)

Checkit out using following code.

MOV	a, R_1	
opr	b, R_1	$t_1 = a + b$
MOV	d, R_2	
opr	c, R_2	$t_2 = c + d$
opr	e, R_2	$t_3 = e - t_2$
MOV	t_3, R_1	
opr	t_1, R_1	$t_4 = t_1 - t_3$

Minimum number of MOV instructions required = 3.

3.13 (6)

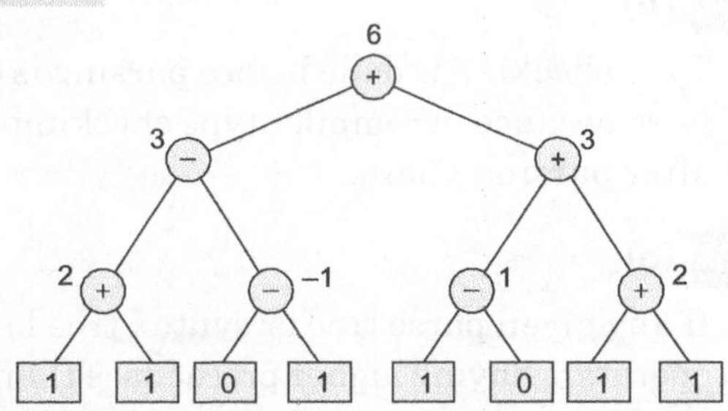

3.14 (c)

Intermediate code can be optimized using the machine-independent code optimizers. All compilers can use same machine independent code optimizers to optimize the intermediate code.

3.15 (9)

Since given expression in infix expression. So we use operator stack.

2 – 5 + 1 – 7 * 3

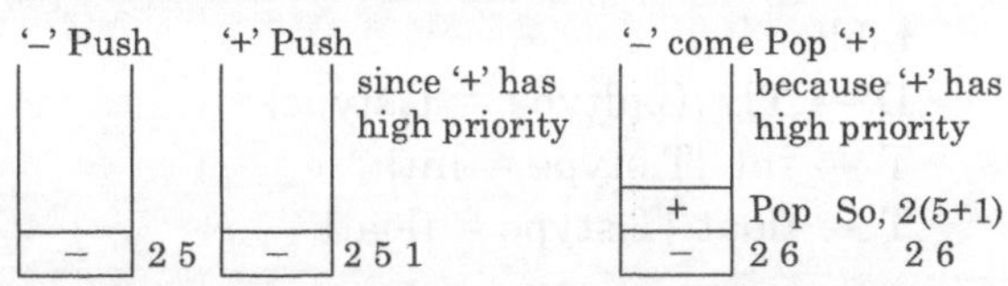

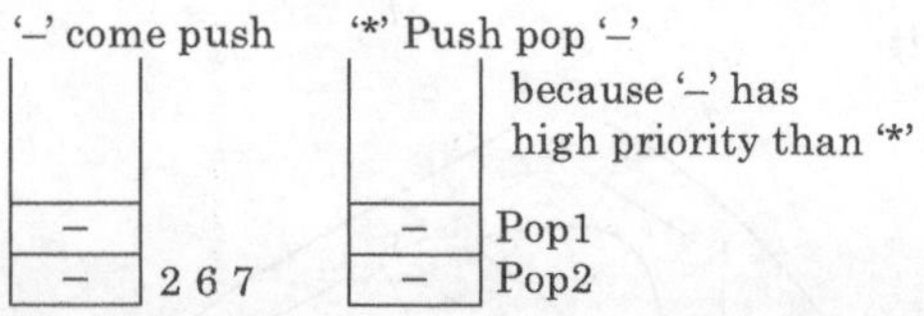

So, 2 – (6 – 7) ∴ '–' right associative.

2 – (–1) = 3

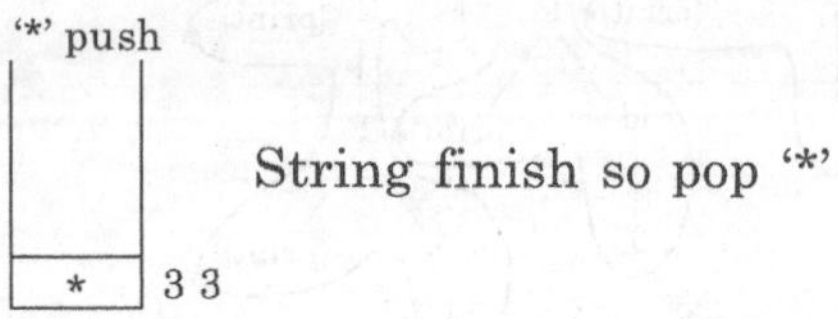

So, 3 * 3 = 9.

So, 2 – 5 + 1 – 7 * 3 evaluates to 9.

3.16 (c)

Input is '*aab*'.

So tree for given input.

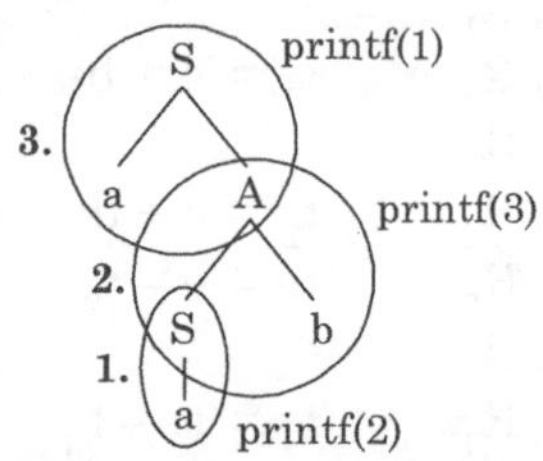

So output will be 2, 3 and 1 because printed order will be 1, 2, 3. So output: 2 3 1

3.17 (b)

Type checking is done before parsing is clearly false because in compiler type checking is done after parsing phase.

3.18 (a)

If any given parse tree or syntax tree low level operators having higher precedence than upper level operators.

Hence here $ is higher precedence than #. $ is left associative because in the sub expression b $ c $ d, b $ c will be evaluated first as per given tree.

As per the given tree structure right # is higher precedence than left #.

Hence it is right associative.

3.19 (a)

SDT for inserting type information in the symbol table:

D → TL {L.idtype = T.stype}

T → int {T.stype = int}

T → float {T.stype = float}

L → L_1, id {L_1.itype = L.itype}
addtype(id.entry, L.itype)

L → id addtype(id.entry, L.itype)

Replace these attribute information in the given SDT.

So option (a) is the correct answer.

3.20 (c)

Rule 1: It is L-attributed by definition.

Rule 2: It is not L-attributed because this expression (X.*i* = A.*i* + Y.*s)* fails to satisfy the definition of L-attributed.

3.21 (a)

1. $P \rightarrow D^* E^*$
2. $D \rightarrow$ int ID {record that ID.lexeme is of type int}
3. $D \rightarrow$ bool ID {record that ID.lexeme is of type bool}
4. $E \rightarrow E_1 + E_2$ {check that E_1.type = E_2.type = int; set E.type = int}
5. $E \rightarrow !E_1$ {check that E_1.type = bool; set E.type = bool}
6. $E \rightarrow$ ID {set E.type = int}

Rules 2 and 3 are used for entry into the symbol table. Rule 4 is used for type checking of the integer expression. But, in rule 6 only int type is set.

3.22 (80)

In the given SDT % has more precedence than # and % is left associative.

∴ The given expression 20#10%5#8%2%2 will be 20 # (10 % 5) # [(8 % 2) % 2].

Now replacing the operators # and % with * and ÷ respectively the expression will be

$$= 20 * (10 \div 5) * [(8 \div 2) \div 2]$$
$$= 20 * 2 * [4 \div 2] = 20 * 2 * 2 = 80$$

3.23 (2.375)

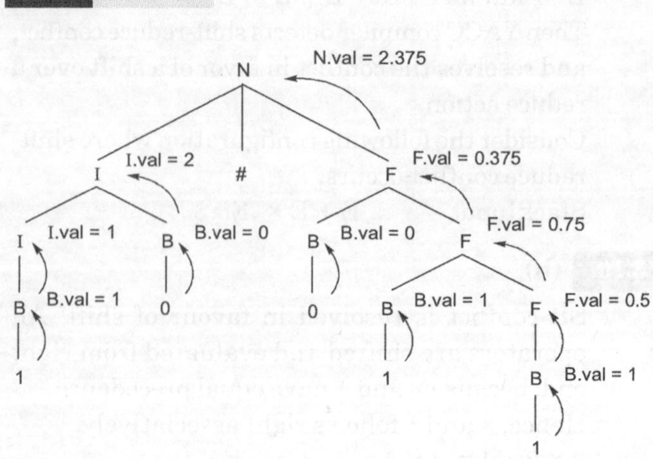

3.24 (b)

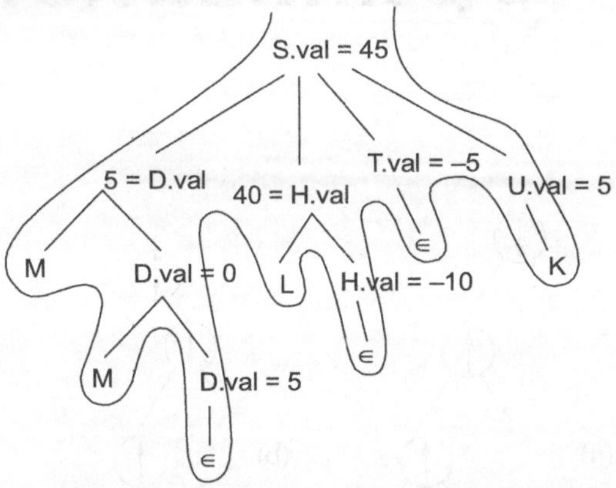

Answer is 45.

3.25 (a, c)

The attributes in a L-attributed definition can be evaluated in top-down and left to right order.

■■■■

4 Code Generation and Optimization

4.1 Consider the following C code segment:

```
for (i = 0; i < n; i ++)
{
    for (j = 0; j < n; j ++)
    {
        if (i % 2)
        {  x + = (4 * j + 5 * i);
           y + = (7 + 4 *j);
        }
    }
}
```

Which one of the following is false?

(a) The code contains loop invariant computation
(b) There is scope of common sub-expression elimination in this code
(c) There is scope strength reduction in this code
(d) There is scope of dead code elimination in this code

[2006 : 2 M]

4.2 Some code optimizations are carried out on the intermediate code because

(a) They enhance the portability of the complier to other target processors
(b) Program analysis is more accurate on intermediate code than on machine code
(c) The information from data flow analysis cannot otherwise be used for optimization
(d) The information from the front end cannot otherwise be used for optimization

[2008 : 1 M]

4.3 Which languages necessarily need heap allocation in the runtime environment?

(a) Those that support recursion
(b) Those that use dynamic scoping
(c) Those that allow dynamic data structure
(d) Those that use global variables

[2010 : 1 M]

4.4 Consider two binary operators '↑' and '↓' with the precedence of operator ↓ being lower than that of the operator ↑. Operator ↑ is right associative while operator ↓ is left associative. Which one of the following represents the parse tree for expression (7↓3↑4↑3↓'2)

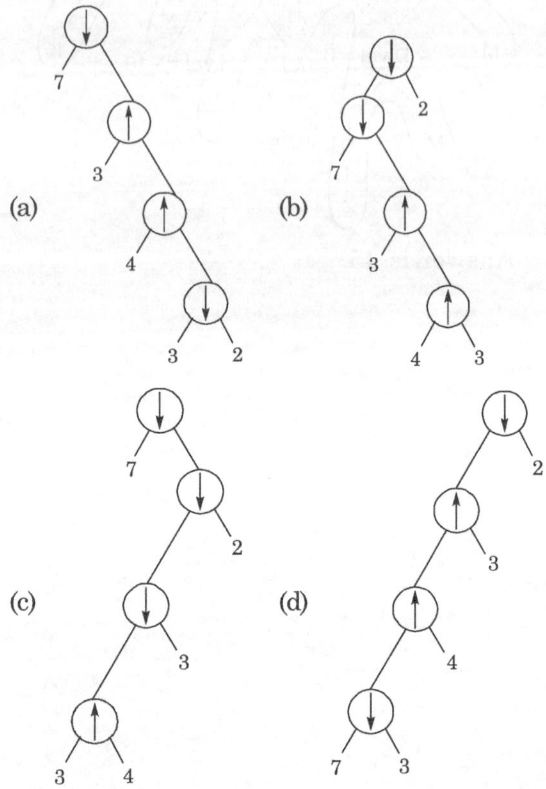

[2011 : 2 M]

4.5 Consider the program given below, in a block-structured pseudo-language with lexical scoping and nesting of procedures permitted.

```
Program main;
Var ...
Procedure A1;
   Var ...
   Call A2;
End A1
Procedure A2;
   Var ...
   Procedure A21;
            Var ...
            Call A1,
   End A21
   Call A21;
End A2
Call A1;
End main;
```

Consider the calling chain:

Main → A1 → A2 → A21 → A1

The correct set of activation records along with their access links is given by

(a)

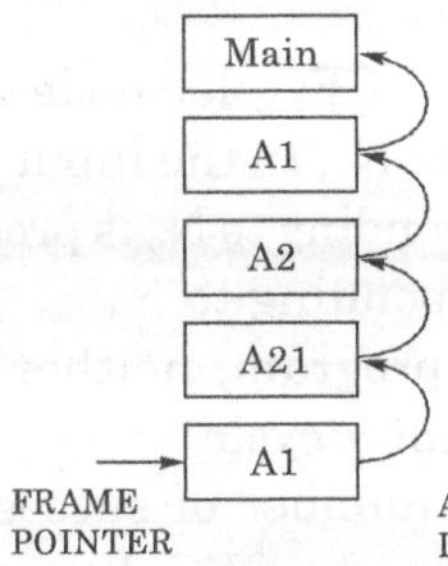

(b)

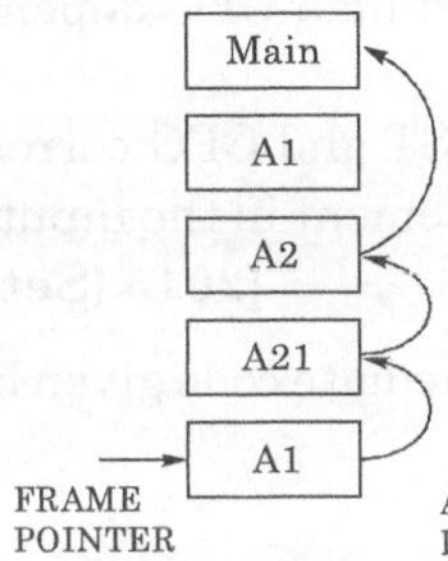

(c)

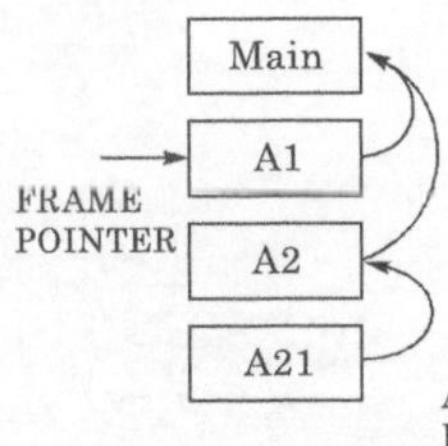

(d)

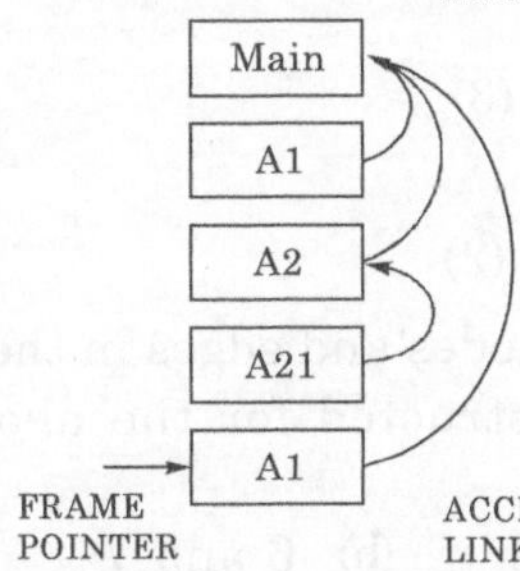

[2012 : 2 M]

Common Data for Questions 4.6 and 4.7:

The following code segment is executed on a processor which allows only register operands in its instructions. Each instruction can have almost two source operands and one destinations operand. Assume that all variables are dead after this code segment.

```
c = a + b;
d = c * a;
e = c + a;
x = c * c;
if (x > a) {
     y = a * a;
}
else {
     d = d * d;
     e = e * e;
}
```

4.6 Suppose the instruction set architecture of the processor has only two registers. The only allowed complier optimization is code motion, which moves statements from one place to another while preserving correctness. What is the minimum number of spills to memory in the compiled code?

(a) 0 (b) 1
(c) 2 (d) 3 **[2013 : 2 M]**

4.7 What is the minimum number of registers needed in the instruction set architecture of the processor to compile this code segment without any spill to memory? Do not apply any optimization other than optimizing register allocation?

(a) 3 (b) 4
(c) 5 (d) 6 **[2013 : 2 M]**

4.8 Which one of the following is **FALSE**?

(a) A basic block is a sequence of instructions where control enters the sequence at the beginning and exits at the end.
(b) Available expression analysis can be used for common subexpression elimination.
(c) Live variable analysis can be used for dead code elimination.
(d) $x = 4 \times 5 \Rightarrow 20$ is an example of common subexpression elimination.

[2014 (Set-1) : 1 M]

4.9 Which one of the following is NOT performed during compilation?

(a) Dynamic memory allocation
(b) Type checking
(c) Symbol table management
(d) Inline expansion **[2014 (Set-2) : 1 M]**

4.10 For a C program accessing X[i][j][k], the following intermediate code is generated by a compiler. Assume that the size of an integer is 32 bits and the size of a character is 8 bits.

```
t0 = i * 1024
t1 = j * 32
t2 = k * 4
t3 = t1 + t0
t4 = t3 + t2
t5 = X[t4]
```

Which one of the following statements about the source code for the C program is CORRECT?

(a) X is declared as "int X[32][32][8]".
(b) X is declared as "int X[4][1024][32]".
(c) X is declared as "char X[4][32][8]".
(d) X is declared as "char X[32][16][2]".

[2014 (Set-2) : 2 M]

4.11 Which of the following statements are CORRECT?

1. Static allocation of all data areas by a compiler makes it impossible to implement recursion.
2. Automatic garbage collection is essential to implement recursion.
3. Dynamic allocation of activation records is essential to implement recursion.
4. Both heap and stack are essential to implement recursion.

(a) 1 and 2 only (b) 2 and 3 only
(c) 3 and 4 only (d) 1 and 3 only

[2014 (Set-3) : 1 M]

4.12 Consider the basic block given below:

$a = b + c,\ c = a + d,\ d = b + c$
$e = d - b,\ a = e + b$

The minimum number of nodes and edges present in the DAG representation of the above basic block respectively are

(a) 6 and 6 (b) 8 and 10
(c) 9 and 12 (d) 4 and 4

[2014 (Set-3) : 2 M]

4.13 A variable x is said to be live at a statement S_i in a program if the following three conditions hold simultaneously:

1. There exists a statement S_j that uses x
2. There is a path from S_i to S_j in the flow graph corresponding to the program
3. The path has no intervening assignment to x including at S_i and S_j

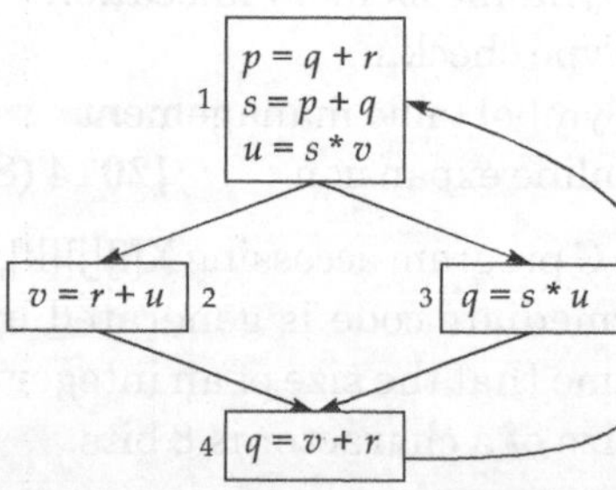

The variables which are live both at the statement in basic block 2 and at the statement in basic block 3 of the above control flow graph are

(a) p, s, u (b) r, s, u
(c) r, u (d) q, v

[2015 (Set-1) : 2 M]

4.14 The least number of temporary variables required to create a three-address code in static single assignment form for the expression q+r/3 + s – t * 5+u * v/w is _____.

[2015 (Set-1) : 2 M]

4.15 In the context of Abstract-Syntax-Tree (AST) and Control-Flow-Graph (CFG), which one of the following is True?

(a) In both AST and CFG, let node N_2 be the successor of node N_1. In the input program, the code corresponding to N_2 is present after the code corresponding to N_1
(b) For any input program, neither AST nor CFG will contain a cycle
(c) The maximum number of successors of a node in an AST and a CFG depends on the input program
(d) Each node in AST and CFG corresponds to at most one statement in the input program

[2015 (Set-2) : 1 M]

4.16 Consider the intermediate code given below:

1. i = 1
2. j = 1
3. t1 = 5 * i
4. t2 = t1 + j
5. t3 = 4 * t2
6. t4 = t3
7. a[t4] = –1
8. j = j + 1
9. if j <= 5 goto (3)
10. i = i + 1
11. if i < 5 goto (2)

The number of nodes and edges in the control-flow-graph constructed for the above code, respectively, are

(a) 5 and 7 (b) 6 and 7
(c) 5 and 5 (d) 7 and 8

[2015 (Set-2) : 2 M]

4.17 Consider the following code segment:

$x = u - t;$
$y = x * v;$
$x = y + w;$
$y = t - z;$
$y = x * y;$

The minimum number of *total* variables required to convert the above code segment to *static single assignment* form is _____.

[2016 (Set-1) : 1 M]

4.18 A student wrote two context-free grammars *G*1 and *G*2 for generating a single *C*-like array declaration. The dimension of the array is at least one. For example, int a [10] [3];

The grammars use D as the start symbol, and use six terminal symbols int; id [ ] num.

Grammar $G1$	**Grammar $G2$**
$D \rightarrow$ int L;	$D \rightarrow$ int L;
$L \rightarrow$ id [E	$L \rightarrow idE$
$E \rightarrow$ num]	$E \rightarrow E$ [num]
$E \rightarrow$ num] [E	$E \rightarrow$ [num]

Which of the grammars correctly generate the declaration mentioned above?

(a) Both $G1$ and $G2$
(b) Only $G1$
(c) Only $G2$
(d) Neither $G1$ nor $G2$

[2016 (Set-2) : 2 M]

4.19 Consider the following intermediate program in three address code

$p = a - b$
$q = p * c$
$p = u * v$
$q = p + q$

Which one of the following corresponds to a static single assignment form of the above code?

(a) $p_1 = a - b$
$q_1 = p_1 * c$
$p_1 = u * v$
$q_1 = p_1 + q_1$

(b) $p_3 = a - b$
$q_4 = p_3 * c$
$p_4 = u * v$
$q_5 = p_4 + q_4$

(c) $p_1 = a - b$
$q_1 = p_2 * c$
$p_3 = u * v$
$q_2 = p_4 + q_3$

(d) $p_1 = a - b$
$q_1 = p * c$
$q_2 = u * v$
$q_2 = p + q$

[2017 (Set-1) : 1 M]

4.20 Consider the expression $(a-1)*(((b+c)/3)+d))$. Let X be the minimum number of registers required by an optimal code generation (without any register spill) algorithm for a load/store architecture, in which (i) only load and store instructions can have memory operands and (ii) arithmetic instructions can have only register or immediate operands. The value of X is ________.

[2017 (Set-1) : 2 M]

4.21 Consider the following grammar:

stmt $\rightarrow$ **if** expr **then** expr **else** expr; stmt | 0
expr $\rightarrow$ term **relop** term | term
term $\rightarrow$ id | number
id $\rightarrow$ **a** | **b** | **c**
number $\rightarrow$ **[0-9]**

where **relop** is a relational operator (e.g., <, >, ...), 0 refers to the empty statement, and **if, then, else** are terminals.

Consider a program P following the above grammar containing ten **if** terminals. The number of control flow paths in P is ________.
For example, the program

if e_1 **then** e_2 **else** e_3

has 2 control flow paths, $e_1 \rightarrow e_2$ and $e_1 \rightarrow e_3$.

[2017 (Set-1) : 2 M]

4.22 Consider the following C code segment:

$a = b + c;$
$e = a + 1;$
$d = b + c;$
$f = d + 1;$
$g = e + f;$

In a compiler, this code segment is represented internally as a directed acyclic graph (DAG). The number of nodes in the DAG is ________.

[2021 (Set-1) : 2 M]

4.23 In the context of compilers, which of the following is/are NOT an intermediate representation of the source program?

(a) Symbol table
(b) Three address code
(c) Control Flow Graph (CFG)
(d) Abstract Syntax Tree (AST)

[2021 (Set-2) : 1 M]

4.24 For a statement S in a program, in the context of liveness analysis, the following sets are defined:

$USE(S)$: The set of variables used in S
$IN(S)$: The set of variables that are live at the entry of S
$OUT(S)$: The set of variables that are live at the exit of S

Consider a basic block that consists of two statements, S_1 followed by S_2.
Which one of the following statements is correct?

(a) $OUT(S_1) = USE(S_1) \cup IN(S_2)$
(b) $OUT(S_1) = IN(S_2) \cup OUT(S_2)$
(c) $OUT(S_1) = IN(S_2)$
(d) $OUT(S_1) = IN(S_1) \cup USE(S_1)$

[2021 (Set-2) : 2 M]

4.25 Consider the following ANSI C code segment:

```
z = x + 3 + y -> f1 + y -> f2;
for (i = 0; i < 200; i = i + 2) {
    if (z > i) {
```

```
        p = p + x + 3;
        q = q + y→f1;
    } else {
        p = p + y→f2;
        q = q + x + 3;
    }
}
```

Assume that the variable y points to a struct (allocated on the heap) containing two fields $f1$ and $f2$, and the local variables x, y, z, p, q, and i are allotted registers. Common sub-expression elimination (CSE) optimization is applied on the code. The number of addition and dereference operations (of the form $y \to f1$ or $y \to f2$) in the optimized code, respectively, are :

(a) 303 and 2 (b) 403 and 102
(c) 203 and 2 (d) 303 and 102

[2021 (Set-2) : 2 M]

4.26 Consider the control flow graph shown:

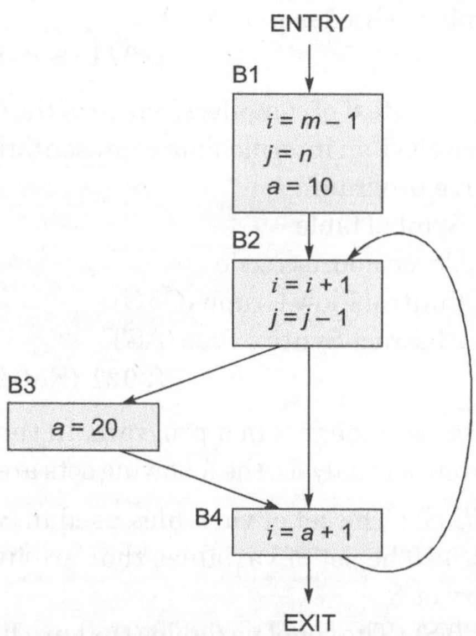

Which one of the following choices correctly lists the set of live variables at the exit point of each basic block?

(a) B1: { }, B2: {a}, B3: {a}, B4: {a}
(b) B1: {i, j}, B2: {a}, B3: {a}, B4: {i}
(c) B1: {a, i, j}, B2: {a, i, j}, B3: {a, i}, B4: {a}
(d) B1: {a, i, j}, B2: {a, j}, B3: {a, j}, B4: {a, i, j}

[2023 : 2 M]

4.27 Consider the following pseudo-code.

```
L1:   t1 = -1
L2:   t2 = 0
L3:   t3 = 0
L4:   t4 = 4 * t3
L5:   t5 = 4 * t2
L6:   t6 = t5 * M
L7:   t7 = t4 + t6
L8:   t8 = a[t7]
L9:   if t8 <= max goto L11
L10:  t1 = t8
L11:  t3 = t3 + 1
L12:  if t3 < M goto L4
L13:  t2 = t2 + 1
L14:  if t2 < N goto L3
L15:  max = t1
```

Which one of the following options CORRECTLY specifies the number of basic blocks and the number of instructions in the largest basic block, respectively?

(a) 6 and 6 (b) 7 and 6
(c) 6 and 7 (d) 7 and 7

[2024 (Set-1) : 2 M]

4.28 Consider the following expression: $x[i] = (p + r) * -s[i] + u/w$. The following sequence shows the list of triples representing the given expression, with entries missing for triples (1), (3), and (6).

(0)	+	p	r
(1)			
(2)	uminus	(1)	
(3)			
(4)	/	u	w
(5)	+	(3)	(4)
(6)			
(7)	=	(6)	(5)

Which of the following options fill in the missing entries CORRECTLY?

(a) (1) = [ ] s i (3) *(0) (2) (6) [ ] = x i
(b) (1) [ ] = s i(3) – (0) (2) (6) = [ ] x i
(c) (1) [ ] = s i(3) – (0) (2) (6) =[ ] x (5)
(d) (1) = [ ] s i (3) *(0) (2) (6) [ ] = x (5)

[2024 (Set-2) : 2 M]

■■■■

Answers Code Generation and Optimization

4.1	(d)	**4.2**	(b)	**4.3**	(c)	**4.4**	(b)	**4.5**	(d)	**4.6**	(b)	**4.7**	(b)	**4.8**	(d)	**4.9**	(a)
4.10	(a)	**4.11**	(d)	**4.12**	(a)	**4.13**	(*)	**4.14**	(8)	**4.15**	(c)	**4.16**	(b)	**4.17**	(10)	**4.18**	(a)
4.19	(b)	**4.20**	(2)	**4.21**	(1024)	**4.22**	(6)	**4.23**	(a)	**4.24**	(c)	**4.25**	(a)	**4.26**	(d)	**4.27**	(b)
4.28	(a)																

Explanations Code Generation and Optimization

4.1 (d)

(a) i%2 is inner loop invariant, it can be moved before inner loop.

(b) 4*j is common sub-expression appeared in two statements.

(c) 4*j can be reduced to j<<2 by strength reduction.

(d) There is no dead code in given code segment. So there is no scope of dead code elimination in this code.

Hence only option (d) is FALSE.

4.2 (b)

Some code optimizations are carried out on the intermediate code because program analysis is more accurate on intermediate code than on machine code.

4.3 (c)

Runtime environment means we deal with dynamic memory allocation and Heap is a dynamic data structure.

So it is clear that those languages that allow dynamic data structure necessarily need heap allocation in the runtime environment.

4.4 (b)

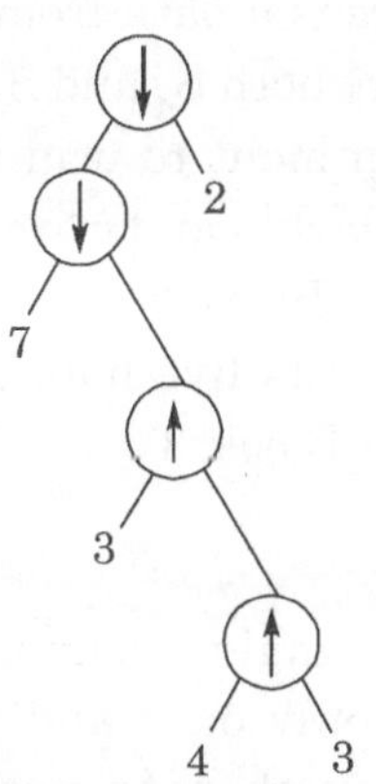

4.5 (d)

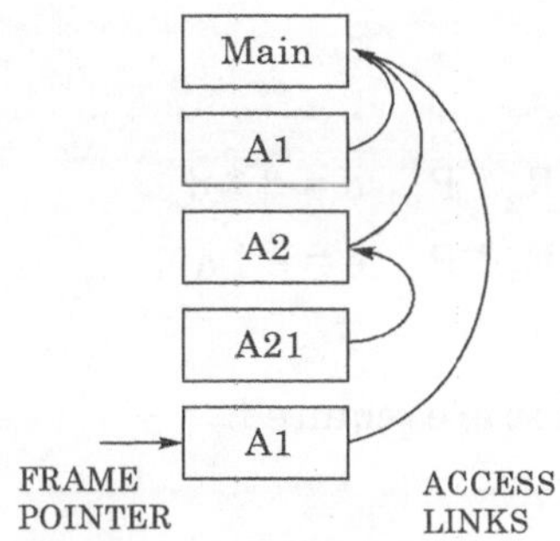

Given calling sequence from the program is:

Main → A1 → A2 → A21 → A1

A1 and A2 are defined in Main, so A1 and A2 access links are pointed to Main.

A21 definition is available in A2, hence A21 access link points to A2.

4.6 (b)

```
c = a + b;
x = c * c;                    ...(i)
if (x > a) {
    y = a * a;
}
else {
    d = c * a;                ...(ii)
    e = c + a;
    d = d * d;
    e = e * e;
}
```

(i) is to store c*c, it needs onememory spill.

(ii) is uses previous same (i) memory spill to store c*a.

Number of memory spills are used in above program is one. With the use of one memory cell and two registers the above optimized code can be executed.

4.7 (b)

$R_1 \leftarrow R_0 + R_1$ c = a + b
$R_2 \leftarrow R_0 * R_1$ d = c * a
$R_3 \leftarrow R_1 + R_0$ e = c + a
$R_1 \leftarrow R_1 * R_1$ x = c * c
if $(R_1 > R_0)$ if (x > a)
{
$R_1 \leftarrow R_0 * R_0$ y = a * a
}
else
{
$R_2 \leftarrow R_2 * R_1$ d = d * d
$R_3 \leftarrow R_3 * R_3$ e = e * a
}

4 registers are required.

4.8 (d)

$$x = 4 \times 5$$

$x = 20$ is an example for constant folding but not for common subexpression elimination.

4.9 (a)

Dynamic memory allocation performed during runtime whereas type checking, symbol table management and Inline expansion is performed during compilation.

4.10 (a)

Each integer element requires 4 bytes
int X [32] [32] [8];
To access X[i][j][k] :
X [(i ∗ 32 ∗ 8 + j ∗ 8 + k) ∗ 4] ...(1)
(integer size = 4)
From given intermediate code:
$X[t_4] = X[t_3 + t_2] = X[t_1 + t_0 + t_2]$
$= X[t_0 + t_1 + t_2]$
= X [(i ∗ 1024 + j ∗ 32 + k ∗ 4)] ...(2)
∴ (1) and (2) are equivalent.

4.11 (d)

Recursion can not be implemented using static allocation.
Recursion can be implemented using dynamic allocation.

4.12 (a)

$a = b + c$ ⇒ $a = e + b$
$c = a + d$ ⇒ $a = d - b - b = d$
$d = b + c$ ⇒ $a = b + c$
$e = d - b$ ⇒ $a = b + a + d$
$a = e + b$ ⇒ $a = b + b + c + d$

∴ $b + b + c + d$ is expression

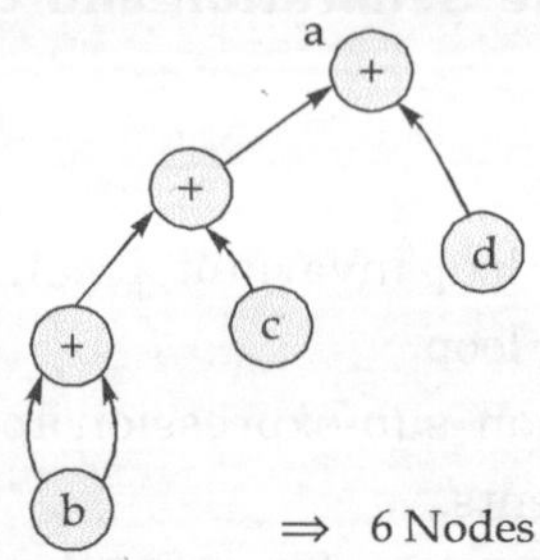

⇒ 6 Nodes and 6 edges

4.13 (*)

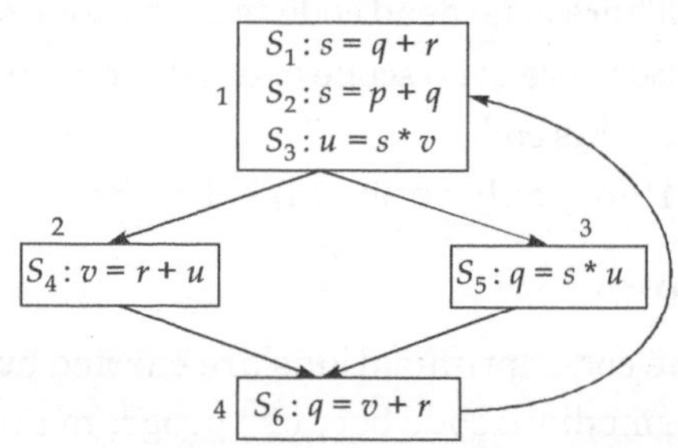

(i) p is not live both at Block 2 and Block 3. There is a path from (S_4 & S_5) to S_2, but S_1 has an assignment to p.
(ii) q is not live both at Block 2 and Block 3. There is a path from (S_4 & S_5) to S_1, but S_6 has an assignment to q in between the path.
(iii) r is live both at Block 2 and Block 3. There is a path from (S_4 & S_5) to S_6, S_6 uses r and no assignment to r in this path.
(iv) u is not live both at Block 2 and Block 3. There is a path from S_4 to S_5 and S_5 to S_4 where both S_4 and S_5 uses u but S_3 has an assignment to u in this path.
(v) Similarly s and v are not live both at Block 2 and Block 3

∴ Only r is live both at basic Block 2 and basic Block 3.

4.14 (8)

In static single assignment, every variable assigned only once and that variable can be used any number of times without assignment.

Expression: q + r/3 + s – t * 5 + u * v/w

$t_1 = r / 3$
$t_2 = q + t_1$
$t_3 = t_2 + s$
$t_4 = t * 5$
$t_5 = t_3 - t_4$
$t_6 = u * v$
$t_7 = t_6 / w$
$t_8 = t_5 + t_7$

∴ 8 temporary variables are required.

4.15 (c)

Maximum number of successors of a node in an AST and a CFG depends on input program. CFG can contain a cycle.
A CFG node may corresponds to more than one statement.

4.16 (b)

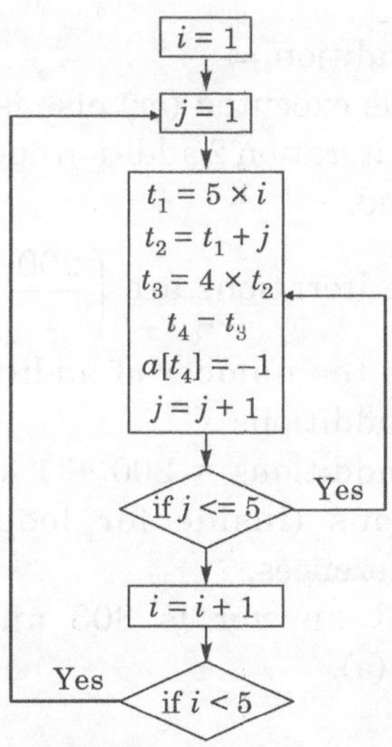

Number of nodes = 6
Number of edges = 7

4.17 (10)

$r_1 = r_2 - r_3;\quad r_4 = x * r_5;$
$x_6 = y + r_7;\quad r_8 = r_3 - r_9;$
$r_{10} = x * y;$

In static single assignment form 1 temporary register will be assign to only 1 variable and every variable is defined before it is used (we have to assign the register at a time of define the variable) throughout the code segments. So here 10 variables are needed.

4.18 (a)

Both *G*1 and *G*2 can generate int a[10][3]; as follows:

G1: D → int L; → int id[E ;
→ int id[num][E;
→ int id[num] [num];

G2: D → int L ; → int id E;
→ int id E [num];
→ int id [num][num];

4.19 (b)

Considering each option:

- In option (a) p_1, q_1 are used again for temporary storage which is not allowed under static single assignment.
- In option (c) in 2nd line it should be $q_1 = p_1 * c$.
- In option (d) in 2nd line it should be $q_1 = p_1 * c$.

The correct code is

$$p_3 = a - b,\ q_4 = p_3 * c$$
$$p_4 = u * v,\ q_5 = p_4 + q_4$$

4.20 (2)

Let T_1 and T_2 be two registers,

$T_1 \leftarrow b$
$T_2 \leftarrow c$
$T_1 \leftarrow (T_1 + T_2)$
$T_1 \leftarrow T_1 / 3$
$T_2 \leftarrow d$
$T_1 \leftarrow T_1 + T_2$
$T_2 \leftarrow a$
$T_2 \leftarrow T_2 - 1$
$T_1 \leftarrow T_1 * T_2$

Hence only 2 registers are required.

4.21 (1024)

The given problem, if e_1 then e_2 else e_3 has 2 paths $e_1 \rightarrow e_2$ and $e_1 \rightarrow e_3$.
So, for every statement, there will be 2 paths.
Since *P* contain 10 if terminals.

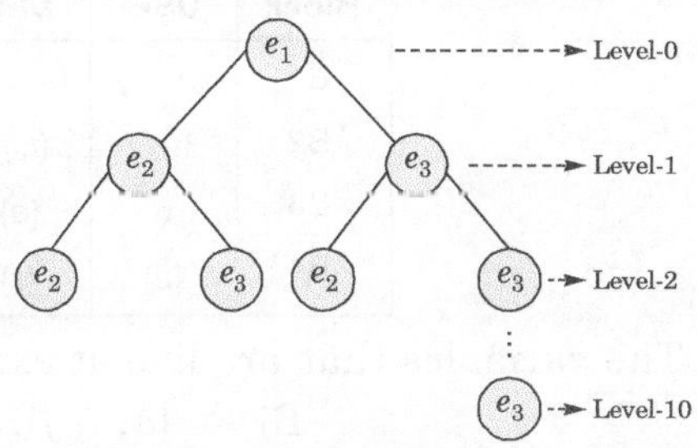

Hence, total control flow path = 2^{10} = 1024.

4.22 (6)

$a = b + c;\ e = a + 1;$
$d = b + c;\ f = d + 1;$
$g = e + f;$

Using common sub-expression elimination this code will become.

$a = b + c,\ e = a + 1$
$d = a,\ f = e$
$g = e + e$

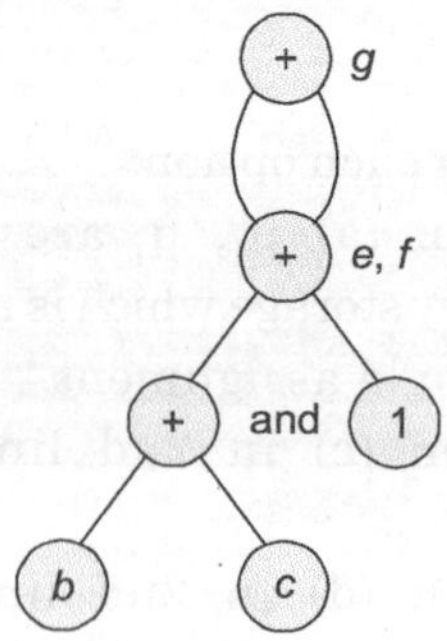

Number of nodes = 6

4.23 (a)

Symbol table is a data structure which is used for storing the information about the variables.
So, option (a) is correct.
There are three major categories of intermediate code representation.
Structural, linear and hybrid.
And CFG comes under the structural intermediate code representation.

4.24 (c)

In live variable analysis at any node, the set of variables live at just after the block are evaluated using the formula.

OUT = ∪ IN (Successor nodes)

So, the correct option is

$\text{OUT}(S_1) = \text{IN}(S_2)$

So, the correct option is (c).

4.25 (a)

$t_1 = y \rightarrow f_1$ (1 dereference)
$t_2 = y \rightarrow f_2$ (1 dereference)
$t_3 = x + 3$ (1 add)
$z = t_3 + t_1 + t_2$ (2 additions)

```
For (i = 0; i < 200; i+ = 2)
{
    if (z > i)
    {
        p = p + t3
        q = q + t1 (2 add)
    }
    else
    {
        p = p + p2
        q = q + t3 (2 add)
    }
}
```

If else condition ⇒
Either if is executed (or) else is executed.
⇒ At any iteration 2 addition operations will be executed.

So, in loop the iterations are $\left(\frac{200}{2}\right)$ = 100 times.

∴ In loop the number of additions = 100 × 2 = 200 additions

∴ Total additions = 200 + 1 + 2 + 100 loop additions (inside for loop) = 303 and 2 dereferences.

∴ Correct answer is 303 and 2 which is option (a).

4.26 (d)

IN = USE ∪ {OUT – DEF}
OUT = ∪ IN (successor)

			FIRST GO		SECOND GO		THIRD GO	
Block	USE	DEF	IN	OUT	IN	OUT	IN	OUT
B1	{m, n}	{a, i, j}	{m, n}	{i, j}	{m, n}	{a, i, j}	{m, n}	{a, i, j}
B2	{i, j}	{i, j}	{i, j}	{a}	{a, i, j}	{a, j}	{a, i, j}	{a, j}
B3	ϕ	{a}	ϕ	{a}	ϕ	{a, j}	{j}	{a, j}
B4	{a}	{i}	{a}	{i, j}	{a, j}	{a, i, j}	{a, j}	{a, i, j}

∴ The variables that are live at exit (i.e. live out) of each basic block are

B1 = {a, i, j}, B2 = {a, j}
B3 = {a, j}, B4 = {a, i, j}

4.27 (b)

L_1, L_3, L_4, L_{10}, L_{11}, L_{13} and L_{15} are leader statements

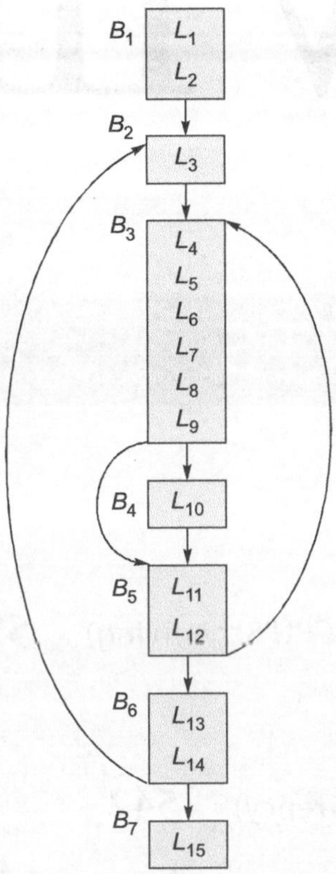

7 basic blocks. B_3 is the largest basic block and it contain 6 statements.

4.28 (a)

(1) = [ ] s i

(3) * (0) (2)

(6) [ ] = x i

■■■■

Operating System

UNIT

VIII

CONTENTS

Operating System

Syllabus

System calls, Processes, Threads, Inter-process communication, Concurrency, Synchronization, Deadlock, CPU and I/O scheduling, Memory management and virtual memory, File systems, I/O systems.

Analysis of Previous GATE Papers

Exam Year	1 Mark Ques.	2 Marks Ques.	Total Marks
1995	3	4	**11**
1996	4	3	**10**
1997	8	2	**12**
1998	8	2	**12**
1999	3	3	**9**
2000	1	2	**5**
2001	10	3	**16**
2002	2	4	**10**
2003	2	5	**12**
2004	3	4	**11**
2005	–	2	**4**
2006	1	8	**17**
2007	2	6	**14**
2008	2	5	**12**
2009	2	5	**12**
2010	3	2	**7**
2011	4	2	**8**
2012	1	4	**9**
2013	2	4	**10**

Exam Year	1 Mark Ques.	2 Marks Ques.	Total Marks
2014 Set-1	2	3	**8**
2014 Set-2	1	3	**7**
2014 Set-3	1	3	**7**
2015 Set-1	2	4	**10**
2015 Set-2	2	3	**8**
2015 Set-3	2	2	**6**
2016 Set-1	1	4	**9**
2016 Set-2	1	3	**7**
2017 Set-1	2	2	**6**
2017 Set-2	2	2	**6**
2018	2	3	**8**
2019	2	4	**10**
2020	2	4	**10**
2021 Set-1	4	1	**6**
2021 Set-2	1	3	**7**
2022	2	4	**10**
2023	3	3	**9**
2024 Set-1	2	4	**10**
2024 Set-2	2	4	**10**

1 Process Management-I

(Introduction, Process, Threads & CPU Scheduling)

1.1 Which scheduling policy is most suitable for a time-shared operating systems?
(a) Shortest Job First
(b) Round Robin
(c) First come first serve
(d) Elevator **[1995 : 1 M]**

1.2 The sequence ..... is an optimal non-preemptive scheduling sequence for the following jobs which leaves the CPU idle for ..... unit(s) of time.

Job	Arrival time	Burst time
1	0.0	9
2	0.6	5
3	1.0	1

(a) {3, 2, 1}, 1 (b) {2, 1, 3}, 0
(c) {3, 2, 1}, 0 (d) {1, 2, 3}, 5
[1995 : 2 M]

1.3 The process state transition diagram in figure is representative of

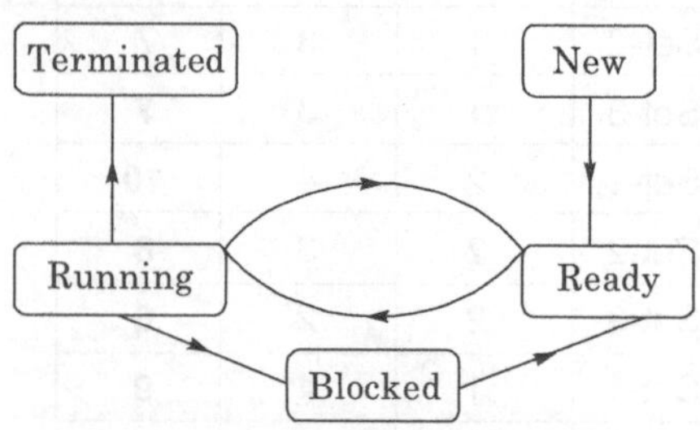

(a) a batch operating system
(b) an operating system with a preemptive scheduler
(c) an operating system with a non-preemptive scheduler
(d) a uni-programmed operating system
[1996 : 1 M]

1.4 Which of the following is an example of spooled device?
(a) A line printer used to print the output of a number of jobs
(b) A terminal used to enter input data to a running program
(c) A secondary storage device in a virtual memory system
(d) A graphic display device **[1996 : 1 M]**

1.5 Four jobs to be executed on a single processor system arrive at time 0^+ in the order A, B, C, D. Their burst CPU time requirements are 4, 1, 8, 1 time units respectively. The completion time of A under round robin scheduling with time slice of one time unit is
(a) 10 (b) 4
(c) 8 (d) 9 **[1996 : 2 M]**

1.6 Which of the following is an example of a spooled device?
(a) The terminal used to enter the input data for the C program being executed
(b) An output device used to print the output of a number of jobs.
(c) The secondary memory device in a virtual storage system
(d) The swapping area on a disk used by the swapper. **[1998 : 1 M]**

1.7 Consider n processes sharing the CPU in a round-robin fashion. Assuming that each process switch takes s seconds, what must be the quantum size q such that the overhead resulting from process switching is minimized but at the same time each process is guaranteed to get its turn at the CPU at least every t seconds?

(a) $q \le \frac{t-ns}{n-1}$ (b) $q \ge \frac{t-ns}{n-1}$

(c) $q \le \frac{t-ns}{n+1}$ (d) $q \ge \frac{t-ns}{n+1}$

[1998 : 1 M]

1.8 Four jobs are waiting to be run. Their expected run times are 6, 3, 5 and x in what order should they be run to minimize the average response time? **[1998 : 2 M]**

1.9 System calls are usually invoked by using
(a) a software interrupt
(b) polling
(c) an indirect jump
(d) a privileged instruction **[1999 : 1 M]**

1.10 A multi-user, multi-processing operating system cannot be implemented on hardware that does not support

(a) Address translation
(b) DMA for disk transfer
(c) At least two modes of CPU execution (privileged and non-privileged)
(d) Demand paging [1999 : 2 M]

1.11 Which of the following actions is/are typically not performed by the operating system when switching context from process A to process B?
(a) Saving current register values and restoring saved register values for process B.
(b) Changing address translation tables.
(c) Swapping out the memory image of process A to the disk.
(d) Invalidating the translation look-aside buffer. [1999 : 2 M]

1.12 A processor needs software interrupt to
(a) Test the interrupt system of the processor
(b) Implement co-routines
(c) Obtain system services which need execution of privileged instructions
(d) Return from subroutine [2001 : 1 M]

1.13 A CPU has two modes-privileged and non-privileged. In order to change the mode from privileged to non-privileged
(a) a hardware interrupt is needed
(b) a software interrupt is needed
(c) a privileged instruction (which does not generate an interrupt) is needed
(d) a non-privileged instruction (which does not generate an interrupt) is needed [2001 : 1 M]

1.14 Consider a set of n tasks with known runtimes $r_1, r_2, ..., r_n$ to be run on a uniprocessor machine. Which of the following processor scheduling algorithms will result in the maximum throughput?
(a) Round-Robin
(b) Shortest-Job-First
(c) Highest-Response-Ratio-Next
(d) First-Come-First-Served [2001 : 1 M]

1.15 Which of the following scheduling algorithms is non-preemptive?
(a) Round Robin
(b) First-in-First out
(c) Multilevel Queue Scheduling
(d) Multilevel Queue Scheduling with Feedback [2001 : 1 M]

1.16 Which of the following does not interrupt a running process?
(a) A device (b) Timer
(c) Scheduler process (d) Power failure [2001 : 2 M]

1.17 Which combination of the following features will suffice to characterize an OS as a multi-programmed OS? (i) more than one program may be loaded into main memory at the same time for execution. (ii) If a program waits for certain events such as I/O, another program is immediately scheduled for execution. (iii) If the execution of a program terminates, another program is immediately scheduled for execution
(a) (i) (b) (i) and (ii)
(c) (i) and (iii) (d) (i), (ii) and (iii) [2002 : 2 M]

1.18 Draw the process state transition diagram of an OS in which (i) each process is in one of the five states: created, ready, running, blocked (i.e. sleep or wait), or terminated, and (ii) only non-preemptive scheduling is used by the OS. Label the transitions appropriately. [2002 : 2 M]

1.19 A uni-processor computer system only has two processes, both of which alternate 10 ms CPU bursts with 90 ms I/O bursts. Both the processes were created at nearly the same time. The I/O of both processes can proceed in parallel. Which of the following scheduling strategies will result in the least CPU utilization (over a long period of time) for this system?
(a) First come first served scheduling
(b) Shortest remaining time first scheduling
(c) Static priority scheduling with different priorities for the two processes
(d) Round robin scheduling with a time quantum of 5 ms. [2003 : 2 M]

1.20 Consider the following statements with respect to user-level threads and kernel-supported threads
(i) Context switch is faster with Kernel-supported threads
(ii) For user-level threads, a system call can block the entire process
(iii) Kernel-supported threads can be scheduled independently
(iv) User-level threads are transparent to the Kernel
Which of the above statements are true?
(a) (ii), (iii) and (iv) only
(b) (ii) and (iii) only
(c) (i), and (iii) only
(d) (i) and (ii) only [2004 : 1 M]

1.21 Which one of the following is NOT shared by the threads of the same process?
(a) Stack
(b) Address Space
(c) File Descriptor Table
(d) Message Queue **[2004 : 1 M]**

1.22 A process executes the following segment of code:
for(i = 1; i < = n; i++)
fork();
The number of new processes created is
(a) n (b) ((n(n + 1))/2)
(c) $2^n - 1$ (d) $3^n - 1$
[2004 : 2 M]

1.23 Consider the following set of processes, with the arrival times and the CPU-burst times given in milliseconds

Process	Arrival time	Burst time
P1	0	5
P2	1	3
P3	2	3
P4	4	1

What is the average turnaround time for these processes with the preemptive Shortest Remaining Processing Time first (SRPT) algorithm?
(a) 5.50 (b) 5.75
(c) 6.00 (d) 6.25 **[2004 : 2 M]**

1.24 We wish to schedule three processes P1, P2 and P3 on a uniprocessor system. The priorities, CPU time requirements and arrival times of the processes are as shown below:

Process	Priority	CPU time required	Arrival time (hh:mm:ss)
P1	10 (highest)	20 sec	00:00:05
P2	9	10 sec	00:00:03
P3	8 (lowest)	15 sec	00:00:00

We have a choice of preemptive or non-preemptive scheduling. In preemptive scheduling, a late-arriving higher priority process can preempt a currently running process with lower priority. In non-preemptive scheduling, a late-arriving higher priority process must wait for the currently executing process to complete before it can be scheduled on the processor. What are the turnaround times (time from arrival till completion) of P2 using preemptive and non-preemptive scheduling respectively?
(a) 30 sec, 30 sec (b) 30 sec, 10 sec
(c) 42 sec, 42 sec (d) 30 sec, 42 sec
[2005 : 2 M]

1.25 Two shared resources R_1 and R_2 are used by processes P_1 and P_2. Each process has a certain priority for accessing each resource. Let T_{ij} denote the priority of P_i for accessing R_j. A process P_i can snatch a resource R_k from process P_j if T_{ik} is greater than T_{jk}.
Given the following :
I. $T_{11} > T_{21}$ II. $T_{12} > T_{22}$
III. $T_{11} < T_{21}$ IV. $T_{12} < T_{22}$
Which of the following conditions ensures that P_1 and P_2 can never deadlock?
(a) (I) and (IV) (b) (II) and (III)
(c) (I) and (II) (d) None of these
[2005 : 2 M]

1.26 Consider the following code fragment:
```
if (fork () = = 0)
{a = a + 5; printf("%d, %d\n", a, &a);}
else{a = a – 5; printf("%d, %d\n", a, &a);}
```
Let u, v be the values printed by the parent process, and x, y be the values printed by the child process. Which one of the following is TRUE?
(a) u = x + 10 and v = y
(b) u = x + 10 and v ≠ y
(c) u + 10 = x and v = y
(d) u + 10 = x and v ≠ y **[2005 : 2 M]**

1.27 Consider three CPU-intensive processes, which require 10, 20 and 30 time units and arrive at times 0, 2, and 6, respectively. How many context switches are needed if the operating system implements a shortest remaining time first scheduling algorithm? Do not count the context switches at time zero and at the end
(a) 1 (b) 2
(c) 3 (d) 4 **[2006 : 1 M]**

1.28 The process state transition diagram of an operating system is as given below.

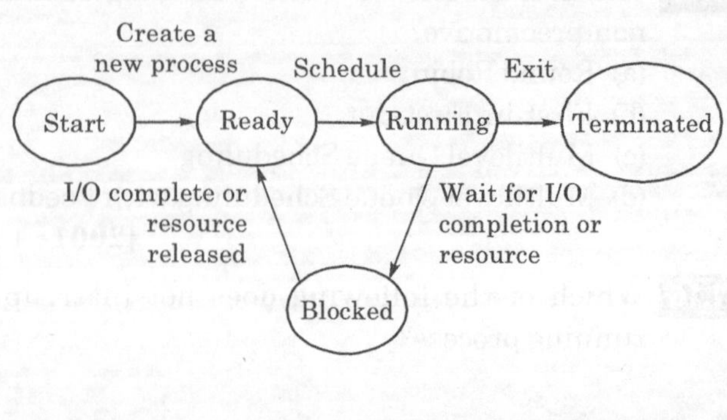

Which of the following must be FALSE about the above operating system?
(a) It is a multiprogrammed operating system
(b) It uses preemptive scheduling
(c) It uses non-preemptive scheduling
(d) It is a multi-user operating system
[2006 : 1 M]

1.29 The arrival time, priority, and durations of the CPU and I/O bursts for each of three processes P_1, P_2 and P_3 are given in the table below. Each process has a CPU burst followed by an I/O burst followed by another CPU burst. Assume that each process has its own I/O resource.

Process	Arrival time	Priority	Burst durations CPU, I/O CPU
P_1	0	2	1, 5, 3
P_2	2	3 (lowest)	3, 3, 1
P_3	3	1 (highest)	2, 3, 1

The multi-programmed operating system uses preemptive priority scheduling. What are the finish times of the processes P_1, P_2 and P_3?
(a) 11, 15, 9 (b) 10, 15, 9
(c) 11, 16, 10 (d) 12, 17, 11
[2006 : 2 M]

1.30 Consider three processes, all arriving at time zero, with total execution time of 10, 20 and 30 units, respectively. Each process spends the first 20% of execution time doing I/O, the next 70% of time doing computation, and the last 10% of time doing I/O again. The operating system uses a shortest remaining compute time first scheduling algorithm and schedules a new process either when the running process get blocked on I/O or when the running process finishes its compute burst. Assume that all I/O operations can be overlapped as much as possible. For what percentage of time does the CPU remain idle?
(a) 0% (b) 10.6%
(c) 30.0% (d) 89.4% **[2006 : 2 M]**

1.31 Consider three processes (process id 0, 1, 2, respectively) with compute time bursts 2, 4, and 8 time units. All processes arrive at time zero. Consider the longest remaining time first (LRTF) scheduling algorithm. In LRTF ties are broken by giving priority to the process with the lowest process id. The average turn around time is
(a) 13 units (b) 14 units
(c) 15 units (d) 16 units **[2006 : 2 M]**

1.32 **List-I** contains some CPU scheduling algorithms and **List-II** contains some applications. Match entries in **List-I** entries in **List-II**

List-I
A. Gang Scheduling
B. Rate Monotonic Scheduling
C. Fair Share Scheduling

List-II
1. Guaranteed Scheduling
2. Real-time Scheduling
3. Thread Scheduling

Code:

	A	B	C
(a)	3	2	1
(b)	1	2	3
(c)	2	3	1
(d)	1	3	2

[2007 : 1 M]

1.33 Consider the following statements about user level threads and kernel level threads. Which one of the following statements is FALSE?
(a) Context switch time is longer for Kernel level threads than for user level threads
(b) User level threads do not need any hardware support
(c) Related Kernel level threads can be scheduled on different processors in a multiprocessor system
(d) Blocking one Kernel level thread blocks all related threads **[2007 : 1 M]**

1.34 An operating system uses Shortest Remaining Time first (SRT) process scheduling algorithm. Consider the arrival times and execution times for the following processes

Process	Execution time	Arrival time
P1	20	0
P2	25	15
P3	10	30
P4	15	45

What is the total waiting time for process P2?
(a) 5 (b) 15
(c) 40 (d) 55 **[2007 : 2 M]**

1.35 If the time-slice used in the round-robin scheduling policy is more than the maximum time required to execute any process, then the policy will
(a) degenerate to shortest job first
(b) degenerate to priority scheduling
(c) degenerate to first come first serve
(d) None of the above **[2008 : 2 M]**

1.36 Which of the following statements about synchronous and asynchronous I/O is NOT true?
(a) An ISR is invoked on completion of I/O in synchronous I/O but not in asynchronous I/O.
(b) In both synchronous and asynchronous I/O an ISR (Interrupt Service Routine) is invoked after completion of the I/O.
(c) A process making a synchronous I/O call waits until I/O is complete, but a process making an asynchronous I/O call does not wait for completion of the I/O.
(d) In the case of synchronous I/O, the process waiting for the completion of I/O is woken up by the ISR that is invoked after the completion of I/O. **[2008 : 2 M]**

1.37 A process executes the following code for (i = 0; i < n; i ++) fork ( );
The total number of child processes created is
(a) n (b) $2^n - 1$
(c) 2^n (d) $2^{n+1} - 1$
[2008 : 2 M]

1.38 In the following process state transition diagram for a uniprocessor system, assume that there are always some processes in the ready state:

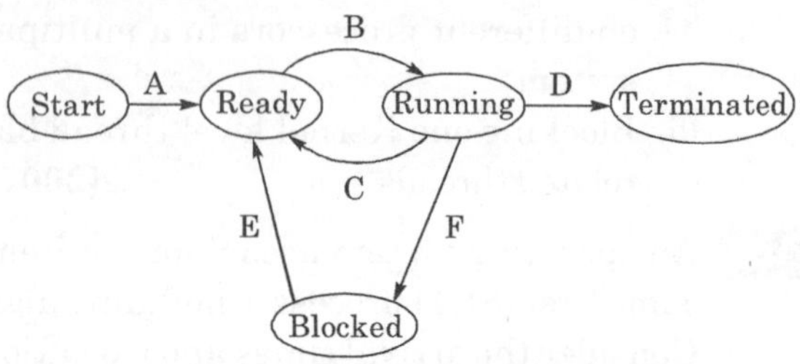

Now consider the following statements:
I. If a process makes a transition D, it would result in another process making transition A immediately.
II. A process P_2 in blocked state can make transition E while another process P_1 is in running state.
III. The OS uses preemptive scheduling.
IV. The OS uses non-preemptive scheduling.
Which of the above statements are TRUE?
(a) I and II (b) I and III
(c) II and III (d) II and IV
[2009 : 2 M]

1.39 Which of the following statements are true?
I. Shortest remaining time first scheduling may cause starvation.
II. Preemptive scheduling may cause starvation.
III. Round robin is better than FCFS in terms of response time.
(a) I only (b) I and III only
(c) II and III only (d) I, II and III
[2010 : 1 M]

1.40 Let the time taken to switch between user and Kernel modes of execution be t_1 while the time taken to switch between two processes be t_2. Which of the following is TRUE?
(a) $t_1 > t_2$
(b) $t_1 = t_2$
(c) $t_1 < t_2$
(d) nothing can be said about the relation between t_1 and t_2 **[2011 : 1 M]**

1.41 A computer handles several interrupt sources of which the following are relevant for this question.
- Interrupt from CPU temperature sensor (raises interrupt if CPU temperature is too high)
- Interrupt from Mouse (raises interrupt if the mouse is moved or a button is pressed)
- Interrupt from Keyboard (raises interrupt when a key is pressed or released)
- Interrupt from Hard Disk (raises interrupt when a disk read is completed)

Which one of these will be handled at the **HIGHEST** priority?
(a) Interrupt from Hard Disk
(b) Interrupt from Mouse
(c) Interrupt from Keyboard
(d) Interrupt from CPU temperature sensor
[2011 : 1 M]

1.42 A thread is usually defined as a "light weight process" because an Operating System (OS) maintains smaller data structures for a thread than for a process. In relation to this, which of the following is TRUE?
(a) On per-thread basis, the OS maintains only CPU register state
(b) The OS does not maintain a separate stack for each thread
(c) On per-thread basis, the OS does not maintain virtual memory state
(d) On per-thread basis, the OS maintains only scheduling and accounting information
[2011 : 1 M]

1.43 Consider the following table of arrival time and burst time for three processes P0, P1 and P2.

Process	Arrival time	Burst time
P0	0 ms	9 ms
P1	1 ms	4 ms
P2	2 ms	9 ms

The pre-emptive shortest job first scheduling algorithm is used. Scheduling is carried out only at arrival or completion of processes. What is the average waiting time for the three processes?

(a) 5.0 ms (b) 4.33 ms
(c) 6.33 ms (d) 7.33 ms

[2011 : 2 M]

1.44 A process executes the code:

```
fork();
fork();
fork();
```

The total number of child processes created is

(a) 3 (b) 4
(c) 7 (d) 8 **[2012 : 1 M]**

1.45 Consider the 3 processes, P1, P2 and P3 shown in the table:

Process Name	Arrival Time	Time Unit Required
P1	0	5
P2	1	7
P3	3	4

The completion order of the 3 processes under the policies FCFS and RR2 (round robin scheduling with CPU quantum of 2 time units) are:

(a) **FCFS:** P1, P2, P3 **RR2:** P1, P2, P3
(b) **FCFS:** P1, P3, P2 **RR2:** P1, P3, P2
(c) **FCFS:** P1, P2, P3 **RR2:** P1, P3, P2
(d) **FCFS:** P1, P3, P2 **RR2:** P1, P2, P3

[2012 : 2 M]

1.46 A scheduling algorithm assigns priority proportional to the waiting time of a process. Every process starts with priority zero (the lowest priority). The schedule reevaluates the process priorities every T time units and decides the next process to schedule. Which one of the following is TRUE if the processes have no I/O operations and all arrive at time zero?

(a) This algorithm is equivalent to the first come first serve algorithm
(b) This algorithm is equivalent to the round-round algorithm
(c) This algorithm is equivalent to the shortest-job-first algorithm
(d) This algorithm is equivalent to the shortest-remaining-time-first algorithm

[2013 : 1 M]

1.47 Which one of the following is **FALSE**?

(a) User level threads are not scheduled by the Kernel.
(b) When a user level thread is blocked, all other threads of its process are blocked.
(c) Context switching between user level threads is faster than context switching between Kernel level threads.
(d) Kernel level threads cannot share the code segment.

[2014 (Set-1) : 1 M]

1.48 Consider the following set of processes that need to be scheduled on a single CPU. All the times are given in milliseconds.

Process Name	Arrival Time	Execution Time
A	0	6
B	3	2
C	5	4
D	7	6
E	10	3

Using the shortest remaining time first scheduling algorithm, the average process turnaround time (in msec) is ________.

[2014 (Set-1) : 2 M]

1.49 Three processes A, B and C each execute a loop of 100 iterations. In each iteration of the loop, a process performs a single computation that requires t_c CPU milliseconds and then initiates a single I/O operation that lasts for t_{io} milliseconds. It is assumed that the computer where the processes execute has sufficient number of I/O devices and the OS of the computer assigns different I/O devices to each process. Also, the scheduling overhead of the OS is negligible. The processes have the following characteristics:

Process id	t_c	t_{io}
A	100 ms	500 ms
B	350 ms	500 ms
C	200 ms	500 ms

The processes A, B, and C are started at times 0, 5 and 10 milliseconds respectively, in a pure time sharing system (round robin scheduling) that uses a time slice of 50 milliseconds. The time in milliseconds at which process C would complete its first I/O operation is ________.

[2014 (Set-2) : 2 M]

1.50 An operating system uses shortest remaining time first scheduling algorithm for pre-emptive scheduling of processes. Consider the following set of processes with their arrival times and CPU burst times (in milliseconds):

Process	Arrival Time	Burst Time
P1	0	12
P2	2	4
P3	3	6
P4	8	5

The average waiting time (in milliseconds) of the processes is ________. **[2014 (Set-3) : 2 M]**

1.51 Consider a uniprocessor system executing three tasks T_1, T_2 and T_3, each of which is composed of an infinite sequence of jobs (or instances) which arrive periodically at intervals of 3, 7 and 20 milliseconds, respectively. The priority of each task is the inverse of its period, and the available tasks are scheduled in order of priority, with the highest priority task scheduled first. Each instance of T_1, T_2 and T_3 requires an execution time of 1, 2 and 4 milliseconds, respectively. Given that all tasks initially arrive at the beginning of the 1st millisecond and task preemptions are allowed, the first instance of T_3 completes its execution at the end of ________ milliseconds. **[2015 (Set-1) : 2 M]**

1.52 The maximum number of processes that can be in Ready state for a computer system with n CPUs is

(a) n (b) n^2
(c) 2^n (d) Independent of n

[2015 (Set-3) : 1 M]

1.53 For the processes listed in the following table, which of the following scheduling schemes will give the lowest average turnaround time?

Process	Arrival Time	Processing Time
A	0	3
B	1	6
C	4	4
D	6	2

(a) First Come First Serve
(b) Non-preemptive Shortest Job First
(c) Shortest Remaining Time
(d) Round-Robin with Quantum value two

[2015 (Set-3) : 2 M]

1.54 Consider an arbitrary set of CPU-bound processes with unequal CPU burst lengths submitted at the same time to a computer system. Which one of the following process scheduling algorithms would minimize the average waiting time in the ready queue?

(a) Shortest remaining time first
(b) Round-robin with time quantum less than the shortest CPU burst
(c) Uniform random
(d) Highest priority first with priority proportional to CPU burst length

[2016 (Set-1) : 1 M]

1.55 Consider the following processes, with the arrival time and the length of the CPU burst given in milliseconds. The scheduling algorithm used is preemptive shortest remaining-time first.

Process	Arrival Time	Burst Time
P_1	0	10
P_2	3	6
P_3	7	1
P_4	8	3

The average turn around time of these processes is ________ milliseconds.

[2016 (Set-2) : 2 M]

1.56 Threads of a process share

(a) global variables but not heap
(b) heap but not global variables
(c) neither global variables nor heap
(d) both heap and global variables

[2017 (Set-1) : 1 M]

1.57 Consider the following CPU processes with arrival times (in milliseconds) and length of CPU bursts (in milliseconds) as given below:

Process	Arrival time	Burst time
P_1	0	7
P_2	3	3
P_3	5	5
P_4	6	2

If the pre-emptive shortest remaining time first scheduling algorithm is used to schedule the processes, then the average waiting time across all processes is ________ milliseconds.

[2017 (Set-1) : 1 M]

1.58 Which of the following is/are shared by all the threads in a process?

I. Program counter II. Stack
III. Address space IV. Registers

(a) I and II only (b) III only
(c) IV only (d) III and IV only

[2017 (Set-2) : 1 M]

1.59 Consider the set of processes with arrival time (in milliseconds), CPU burst time (in milliseconds), and priority (0 is the highest priority) shown below. None of the processes have I/O burst time.

Process	Arrival Time	Burst Time	Priority
P_1	0	11	2
P_2	5	28	0
P_3	12	2	3
P_4	2	10	1
P_5	9	16	4

The average waiting time (in milliseconds) of all the processes using preemptive priority scheduling algorithm is ______.

[2017 (Set-2) : 2 M]

1.60 Consider the following C program is executed on a Unix/Linux system:

```
#include<unistd.h>
int main()
{
    int i;
    for (i = 0; i < 10; i++)
        if (i % 2 == 0) fork();
    return 0;
}
```

The total number of child processes created is ______.

[2019 : 1 M]

1.61 Consider the following four processes with arrival times (in milliseconds) and their length of CPU burst (in milliseconds) as shown below:

Process	P_1	P_2	P_3	P_4
Arrival time	0	1	3	4
CPU time	3	1	3	Z

These processes are run on a single processor using preemptive Shortest Remaining Time First scheduling algorithm. If the average waiting time of the processes is 1 millisecond, then the value of Z is ______.

[2019 : 2 M]

1.62 Consider the following statements about process state transitions for a system using preemptive scheduling.

I. A running process can move to ready state.
II. A ready process can move to running state.
III. A blocked process can move to running state.
IV. A blocked process can move to ready state.

Which of the above statements are TRUE?

(a) I, II, III and IV
(b) II and III only
(c) I, II and III only
(d) I, II and IV only

[2020 : 1 M]

1.63 Consider the following set of processes, assumed to have arrived at time 0. Consider the CPU scheduling algorithms Shortest Job First (SJF) and Round Robin (RR). For RR, assume that the processes are scheduled in the order P_1, P_2, P_3, P_4.

Processes	P_1	P_2	P_3	P_4
Burst time (in ms)	8	7	2	4

If the time quantum for RR is 4 ms, then the absolute value of the difference between the average turnaround times (in ms) of SJF and RR (round off to 2 decimal places) is ______.

[2020 : 2 M]

1.64 Three processes arrive at time zero with CPU bursts of 16, 20 and 10 milliseconds. If the scheduler has prior knowledge about the length of the CPU bursts, the minimum achievable average waiting time for these three processes in a non-preemptive scheduler (round to nearest integer) is ______ milliseconds.

[2021 (Set-1) : 1 M]

1.65 Which of the following standard C library functions will always invoke a system call when executed from a single-threaded process in a UNIX/Linux operating system?

(a) exit
(b) strlen
(c) sleep
(d) malloc

[2021 (Set-1) : 1 M]

1.66 Which of the following statement(s) is/are correct in the context of CPU scheduling?

(a) The goal is to only maximize CPU utilization and minimize throughput.
(b) Turnaround time includes waiting time.
(c) Round-robin policy can be used even when the CPU time required by each of the processes is not known apriori.
(d) Implementing preemptive scheduling needs hardware support.

[2021 (Set-2) : 1 M]

1.67 Consider four processes P, Q, R, and S scheduled on a CPU as per round robin algorithm with a time quantum of 4 units. The processes arrive in the order P, Q, R, S, all at time $t = 0$. There is exactly one context switch from S to Q, exactly one context switch from R to Q, and exactly two context switches from Q to R. There is no context switch from S to P. Switching to a ready process after the termination of another process is also considered a context switch. Which one of the following is NOT possible as CPU burst time (in time units) of these processes?

(a) P = 4, Q = 10, R = 6, S = 2
(b) P = 2, Q = 9, R = 5, S = 1
(c) P = 4, Q = 12, R = 5, S = 4
(d) P = 3, Q = 7, R = 7, S = 3

[2022 : 2 M]

1.68 Which one or more of the following need to be saved on a context switch from one thread (T1) of a process to another thread (T2) of the same process?

(a) Page table base register
(b) Stack pointer
(c) Program counter
(d) General purpose registers

[2023 : 1 M]

1.69 Which one or more of the following options guarantee that a computer system will transition from user mode to Kernel mode?

(a) Function call
(b) Malloc call
(c) Page fault
(d) System call

[2023 : 1 M]

1.70 Which one or more of the following CPU scheduling algorithms can potentially cause starvation?

(a) First-in First-Out
(b) Round Robin
(c) Priority Scheduling
(d) Shortest Job First

[2023 : 1 M]

1.71 Which of the following statements about threads is/are TRUE?

(a) Threads can only be implemented in Kernel space
(b) Each thread has its own file descriptor table for open files
(c) Threads belonging to a process are by default not protected from each other
(d) All the threads belonging to a process share a common stack

[2024 (Set-1) : 1 M]

1.72 Which of the following process state transitions is/are NOT possible?

(a) Waiting to Running
(b) Ready to Waiting
(c) Running to Terminated
(d) Running to Ready

[2024 (Set-1) : 1 M]

1.73 Consider the following two threads T1 and T2 that update two shared variables a and b. Assume that initially $a = b = 1$. Though context switching between threads can happen at any time, each statement of T1 or T2 is executed atomically without interruption.

T1	T2
$a = a + 1;$	$b = 2 * b;$
$b = b + 1;$	$a = 2 * a;$

Which one of the following options lists all the possible combinations of values of a and b after both T1 and T2 finish execution?

(a) ($a = 3$, $b = 4$); ($a = 4$, $b = 3$); ($a = 3$, $b = 3$)
(b) ($a = 4$, $b = 4$); ($a = 4$, $b = 3$); ($a = 3$, $b = 4$)
(c) ($a = 2$, $b = 2$); ($a = 2$, $b = 3$); ($a = 3$, $b = 4$)
(d) ($a = 4$, $b = 4$); ($a = 3$, $b = 3$); ($a = 4$, $b = 3$)

[2024 (Set-1) : 2 M]

1.74 Consider a process P running on a CPU. Which one or more of the following events will always trigger a context switch by the OS that results in process P moving to a non-running state (e.g., ready, blocked)?

(a) A timer interrupt is raised by the hardware.
(b) P tries to access a page that is in the swap space, triggering a page fault.
(c) An interrupt is raised by the disk to deliver data requested by some other process.
(d) P makes a blocking system call to read a block of data from the disk.

[2024 (Set-2) : 1 M]

1.75 Consider a multi-threaded program with two threads T1 and T2. The threads share two semaphores: s1 (initialized to 1) and s2 (initialized to 0). The threads also share a global variable x (initialized to 0). The threads execute the code shown below.

// code of T1	// code of T2
wait (s1);	wait (s1);
x = x + 1;	x = x + 1;
print(x);	print(x);
wait(s2);	signal(s2);
signal(s1);	signal(s1);

Which of the following outcomes is/are possible when threads T1 and T2 execute concurrently?

(a) T2 runs first and prints 1, T1 runs next and prints 2
(b) T2 runs first and prints 1, T1 does not print anything (deadlock)
(c) T1 runs first and prints 1, T2 runs next and prints 2
(d) T1 runs first and prints 1, T2 does not print anything (deadlock)

[2024 (Set-2) : 2 M]

1.76 Consider a single processor system with four processes A, B, C, and D, represented as given below, where for each process the first value is its arrival time, and the second value is its CPU burst time.

A(0, 10), B(2, 6), C(4, 3) and D(6, 7)

Which one of the following options gives the average waiting times when preemptive Shortest Remaining Time First (SRTF) and Non-Preemptive Shortest Job First (NP-SJF) CPU scheduling algorithms are applied to the processes?
(a) SRTF = 6, NP-SJF = 7
(b) SRTF = 7, NP-SJF = 7.5
(c) SRTF = 7, NP-SJF = 8.5
(d) SRTF = 6, NP-SJF = 7.5

[2024 (Set-2) : 2 M]

Answers Process Management-I

1.1	(b)	**1.2**	(a)	**1.3**	(b)	**1.4**	(a)	**1.5**	(d)	**1.6**	(b)	**1.7**	(a)	**1.8**	Sol.	**1.9**	(a)
1.10	(c)	**1.11**	(c)	**1.12**	(c)	**1.13**	(d)	**1.14**	(b)	**1.15**	(b)	**1.16**	(a)	**1.17**	(d)	**1.18**	Sol.
1.19	(d)	**1.20**	(a)	**1.21**	(a)	**1.22**	(c)	**1.23**	(a)	**1.24**	(d)	**1.25**	(c)	**1.26**	(d)	**1.27**	(b)
1.28	(b)	**1.29**	(b)	**1.30**	(b)	**1.31**	(a)	**1.32**	(a)	**1.33**	(d)	**1.34**	(b)	**1.35**	(c)	**1.36**	(b)
1.37	(b)	**1.38**	(c)	**1.39**	(d)	**1.40**	(c)	**1.41**	(d)	**1.42**	(c)	**1.43**	(a)	**1.44**	(c)	**1.45**	(c)
1.46	(b)	**1.47**	(d)	**1.48**	(7.2)	**1.49**	(1000)	**1.50**	(5.5)	**1.51**	(12)	**1.52**	(d)	**1.53**	(c)	**1.54**	(a)
1.55	(8.25)	**1.56**	(d)	**1.57**	(3)	**1.58**	(b)	**1.59**	(29)	**1.60**	(31)	**1.61**	(2)	**1.62**	(d)	**1.63**	(5.25)
1.64	(12)	**1.65**	(a, c)	**1.66**	(b, c, d)	**1.67**	(d)	**1.68**	(b, c, d)	**1.69**	(c, d)	**1.70**	(c, d)	**1.71**	(c)	**1.72**	(a, b)
1.73	(d)	**1.74**	(b, d)	**1.75**	(a, d)	**1.76**	(d)										

Explanations Process Management-I

1.1 (b)

Round-Robin scheduling working on time quantum, after certain time every process get back the CPU units for it's completion and the same phenomena used in time sharing system. So Round Robin is best for time sharing system.

1.2 (a)

In this sequence initially CPU ideal for '1' unit of time but after that, smaller jobs don't have to wait for so long time, so it is ideal.

1.3 (b)

Since there is an arrow from running to ready, so it is pre-emptive.

1.4 (a)

Spooled devices are those which fetches data from job waiting area one by one & spool means simultaneous peripheral operation on line, so printer is a spooling device.

1.5 (d)

Draw Gantt chart and place jobs according to round robin scheduling with time quantum '1' unit.

0–1	1–2	2–3	3–4	4–5	5–6	6–7	7–8	8–9	9–
A	B	C	D	A	C	A	C	A	C

So correct ans is '9'.

1.6 (b)

Spooled devices is output device used to print the output of jobs.

1.7 (a)

Let, $n = 4$

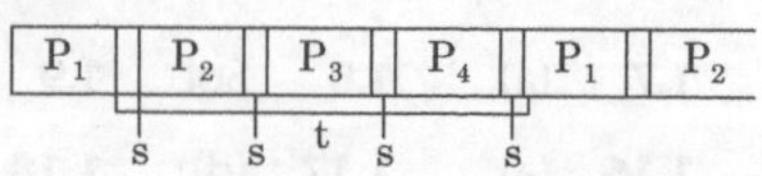

$$t \geq n \times s + q \times (n - 1)$$

So, $$q \leq \frac{t - ns}{n - 1}$$

1.8 Sol.

The average response time depends on the value of x,

If value of x < 3

then order of execution should be

x, 3, 5, 6

If value of x > 3 & x < 5

then order of execution should be

3, x, 5, 6

If value of x > 5 & x < 6

then order of execution should be

3, 5, x, 6

If value of x > 6

then order of execution should be

3, 5, 6, x

Note: Expected run times are in the increasing order.

1.9 (a)

System calls are invoked by using software interrupt.

1.10 (c)

The system has multi user and multi proccessing, the security is primary concern. Otherwise user program and operating system will be corrupted. That is the reason to ensure the security and protection hence two modes are required.

1.11 (c)

Swapping out the memory image of process A to the disk typically not performed by OS when switching context from process A to B.

1.12 (c)

To execute privileged instructions, system services can be obtained using software interrupt.

1.13 (d)

Because we want to change the mode from privileged to non-privileged, to the next instruction to be executed should be non-privileged instruction.

1.14 (b)

SJF executes shortest job very early so throughput increases in the case of SJF.

1.15 (b)

FCFS is non pre-emptive algorithm.

1.16 (a)

Device can not interrupt a running process but timer, scheduler and power failure can interrupt a running process.

1.17 (d)

(i) Virtual memory helps to load more than one program.
(ii) Multi programming, wait for an I/O of one process can execute other process.
(iii) After completion of one process, other process executes, immediately.

1.18 Sol.

Just draw the diagram by using all criteria as below.

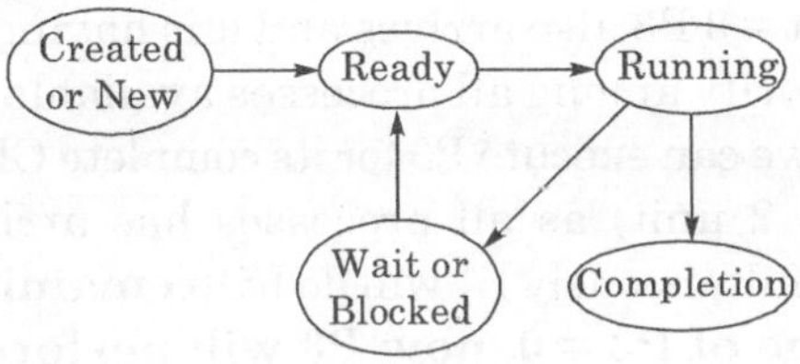

1.19 (d)

CPU = 10 ms
I/O = 90 ms

FCFS:

P1	P2	//////	P1	P2	//////	...
0	10	20	100	110	120	200

So CPU idle time is 80

Round Robin:

P1	P2	P1	P2	//////	P1	P2	P1	P2	//////	...
0	5	10	15	20	105	110	115	120	125	210

CPU idle time is 85.

Therefore, Round Robin scheduling with a time quantum of 5ms has least CPU utilization.

1.20 (a)

Consider the each statement:

(i) Statement is false because there is no connection between Kernel-supported threads and context switch.
(ii) Statement is true because it is the drawback of user-level threads blocking system call can block the entire process.
(iii) Statement is true because kernel-supported threads having own memory area and scheduled independently by the operating system.
(vi) Statement is true because kernel is unware about user level threads and there is no kernel support to user-level threads. So user-level threads are transparent to the Kernal.

1.21 (a)

Stack and registers are not shared by the threads of the same process while address space, message queue etc. are shared.

1.22 (c)

The number of process created will be $2^n - 1$.

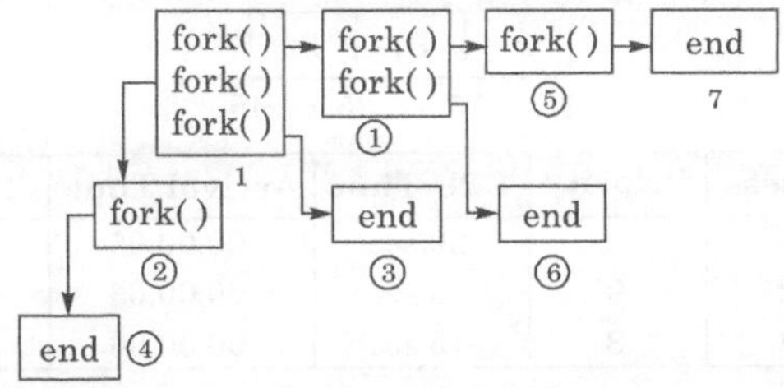

$$2^3 - 1 = 7$$

1.23 (a)

Process	Arrival Time	Burst Time
P_1	0	5
P_2	1	3
P_3	2	3
P_4	4	1

The Gantt chart for SRPT CPU scheduling algorithm is

P_1	P_2	P_2	P_2	P_4	P_3	P_3	P_3	P_1	P_1	P_1	P_1	
0	1	2	3	4	5	6	7	8	9	10	11	12

Process P_1 arrive at time 0, at time unit 1 the remaining time of P_1 is 4 and remaining time of P_2 is 3 so P_1 preempted. At the end of time unit 2 the remaining time of P_2 is 2 and P_3 is 3 so P_2 is not preempted At time unit 4 the remain time of P_2 is 0 and P_4 is scheduled. The same process repeated upto time unit 12.

The turnaround time is the interval between submission and completion of a given process. Calculate turnaround time for each process as

$P_1 = 12 - 0 = 12$
$P_2 = 4 - 1 = 3$
$P_3 = 6, P_4 = 1$

So Average Turn Around Time

$$= \frac{12 + 3 + 6 + 1}{4} = \frac{22}{4} = 5.50$$

1.24 (d)

Using preemptive priority scheduling.

Gantt chart:

P3	P2	P1	P2	P3	
0	3	5	25	33	45

Process	Priority	CPU Time	Arrival Time	TAT
P1	10	20 sec	00.00.05	20
P2	9	10 sec	00.00.03	30
P3	8	15 sec	00.00.00	45

So, Turn Around Time of P2 is 30.

Using non-preemptive priority scheduling.

Gantt chart:

P3	P1	P2
0 – 15	15 – 35	35 – 45

Process	Priority	CPU Time	Arrival Time	TAT
P1	10	20 sec	00.00.05	30
P2	9	10 sec	00.00.03	42
P3	8	15 sec	00.00.00	15

So, Turn Around Time of P2 is 42.

1.25 (c)

By following I and II conditions the process P_1 will get both the resouces R_1 and R_2.

If R_1 and R_2 are allocated to the Process P_1, then it will complete it's job and release it. After that process P_2 will get both the resources and complete it's Job.

1.26 (d)

Fork( ) returns 0 is child process and process ID of child process is parent process.

In child(x), $a = a + 5$

In paraent(u), $a = a - 5$. Therefore, $x = u + 10$

The physical addresses of parent and child processes will be different, $v \neq y$.

1.27 (b)

Let three processes are P_1, P_2 and P_3

Process	Arrival Time	Burst Time
P_1	0	10
P_2	2	20
P_3	6	30

The Gantt chart for SRTF scheduling algorithm is

P_1	P_2	P_3
0 – 10	10 – 30	30 – 60

So there is only two context switches, at time unit 10 context switch from P_1 to P_2 and at time unit 30 context switch from P_2 to P_3.

1.28 (b)

If it were a preemptive scheduling then there would have been a transition from Running state to Ready state. So it is non-preemptive scheduling.

1.29 (b)

At t = 0 only P1 is present so execute it for 1 unit (**remaining CPU time of P1 = 0, now it will perform I/O**).

At t = 1 no process is available for CPU so CPU will remain idle and P1 will perform I/O (**remaining I/O time of P1 = 4**)

At t = 2 P2 arrives and is available for CPU so execute it for 1 unit and simultaneously P1 will do I/O (**remaining CPU time of P2 = 2 & remaining I/O time of P1 = 3**).

At t = 3 P3 also arrives and it is having highest priority among all processes available for CPU so we can execute P3 for its complete CPU burst (i.e 2 unit) as all processes has arrived and simultaneously P1 will do I/O(**remaining CPU time of P3 = 0, now P3 will perform I/O & remaining I/O time of P1 = 1**).

At t = 5 only P2 is available for CPU so it will execute for 1 unit and P1 and P3 will perform I/O(**remaining CPU time of P2 = 1, remaining I/O time of P3 = 2, remaining I/O time of P1 = 0, now P1 will perform CPU**).

At t = 6 P2 and P1 are available for CPU but P1 will be selected because it is having highest priority then P2 so execute P1 for 1 unit (**remaining CPU time of P1 = 2 & remaining I/O time of P3 is 1**)

At t = 7 again P1 will be executed for 1 unit (**remaining CPU time of P1 = 1 & remaining I/O time of P3 = 0, now P3 will perform CPU**).

At t = 8 now P1, P2 & P3 are available for CPU so P3 will be selected based on highest priority and will be executed for 1 unit (**remaining CPU time of P3 = 0, so P3 completed at t = 9**).

At t = 9 P1 will be executed on CPU for 1 unit (**remaining CPU time of P1 = 0, so P1 completed at t = 10**).

At t = 10 only P2 is available for CPU so it will execute its remaining CPU burst (i.e 1 unit) (**remaining CPU time of P2 = 0, now P2 will perform I/O.**)

At t = 11 P2 will perform I/O for 3 units and CPU will remain IDLE.

At t = 14 P2 will perform CPU again and **P2 is completed at t = 15**

P1	IDLE	P2	P3	P2	P1	P1	P3	P1	P2	IDLE	P2
0	1	2	3	5	6	7	8	9	10	11	14 15

1.30 (b)

Process	Total burst time	I/O time	CPU time	I/O time
P1	10	2	7	1
P2	20	4	14	2
P3	30	6	21	3

The Gantt chart is

Idle	P1	P2	P3	Idle

0 2 9 23 44 47

Total time spent = 47

Idle time = 2 + 3 = 5

$$\% \text{ of idle time} = \frac{5}{47} \times 100 = 10.6\%$$

1.31 (a)

Process id	Arrival Time	CPU burst
P_0	0	2
P_1	0	4
P_2	0	8

The Gantt chart for LRTF CPU scheduling algorithm.

P_0	P_1	P_2	P_2	P_2	P_2	P_2	P_1	P_2	P_1	P_2	P_0	P_1	P_2

0 1 2 3 4 5 6 7 8 9 10 11 12 13 14

Turn around time for $P_0 = 12 - 0 = 12$

Turn around time for $P_1 = 13 - 0 = 13$

Turn around time for $P_2 = 14 - 0 = 14$

$$\text{Average Turn around Time} = \frac{39}{3} = 13$$

1.32 (a)

List-I

A. Gang Scheduling

B. Rate Monotonic Scheduling

C. Fair Show Scheduling

List-II

3. Thread Scheduling

2. Real-time Scheduling

1. Guaranteed Scheduling

1.33 (d)

In Kernel level threads, the blocking system call causes, the Kernel can schedule another thread in the application for execution. So statement (d) is false about Kernel level threads.

1.34 (b)

Process	Execution Time	Arrival Time
P_1	20	0
P_2	25	15
P_3	10	30
P_4	15	45

The Gantt chart for SRT Scheduling algorithm is

P_1	P_2	P_3	P_2	P_4

0 20 30 40 55 70

So the waiting time for

$$P_2 = (20 - 15) + (40 - 30) = 5 + 10 = 15$$

1.35 (c)

When time quantum used in round robin scheduling is more than maximum time required to execute any process then its behave like First Come First serve.

1.36 (b)

Statement (b) is not true because an ISR (Interrupt Service Routine) is invoked after completion of I/O in synchronous but not in asynchronous I/O.

1.37 (b)

fork( ) system call creates the child process initially number of processes is 0. After first fork( ), it creates a single process. After second fork( ), it creates one parent and two child processes. After n + 1 fork( ), the total number of processes is 2^n but we subtract the main process then total number of child processes is $2^n - 1$.

1.38 (c)

Transition C indicates that OS uses preemptive scheduling. II is also true.

1.39 (d)

In the shortest remaining time first scheduling, process will not get a chance if that one has the high remaining time so due to this, starvation take place.

Shortest remaining time first scheduling will cause starvation, is true because if a large CPU brust required process is there and if consider that all the processes which are coming have less CPU-brust then the former have to wait untill all finish execution gives starvation.

In preemptive scheduling, always one process preempt due to low priority then due to this starvation take place.

Response time = first response - process submission time

In the round robin every process first response gets after certain time quantum so due to this it give more better than FCFS in term of response time.

1.40 (c)

Process switching includes mode switching.
Context switching can occur only in Kernal mode.

1.41 (d)

Interrupt from CPU temperature sensor is given top priority to protect system resources. When CPU temperature is too high, the BIOS initiate an interrupt and informs the Operating System. OS gives top priority to this interrupt and immediately shuts down the system.

1.42 (c)

(a) **False:** On per thread basics, OS maintains both CPU register state and stack. Hence option (a) is false, because only CPU register state mentioned.

(b) **False:** OS maintains a separate stack for each thread.

(c) **True:** OS does not maintain virtual memory state.

(d) **False:** OS does not maintain scheduling and accounting information.

1.43 (a)

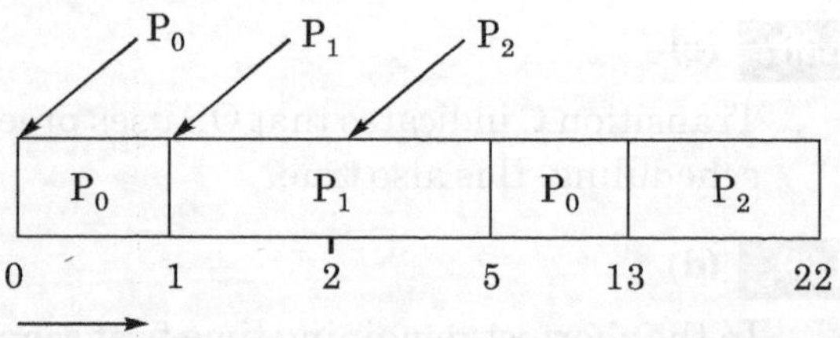

Time in milli seconds

Average waiting time

$$= \frac{w_1 + w_2 + w_3}{3}$$

$$= \frac{(5-1)+0+(13-2)}{3} = 5.0 \text{ ms}$$

1.44 (c)

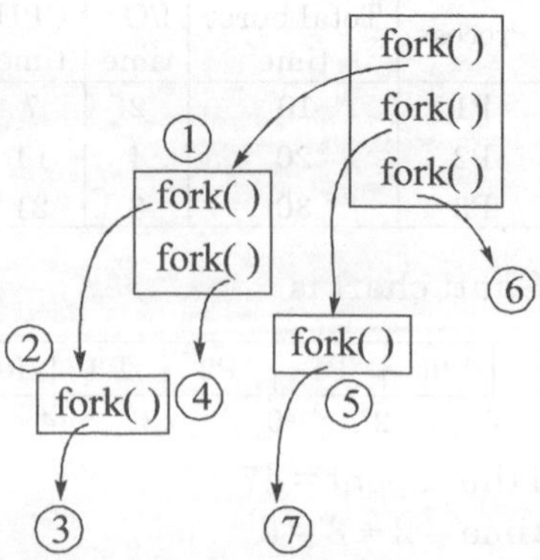

The number of child process created
$= 2^n - 1 = 2^3 - 1 = 7$
(where n is number of fork( ) statements)

1.45 (c)

RR Queue:

P1 P2 P1 P3 P2 P1 P3 P2
0 1 2 3 4 6 8 10

P1	P2	P1	P3	P2	P1	P3	P2

0 2 4 6 8 10 11 13 16

⇒ P1, P3, P2

FCFS: P1, P2, P3

Therefore option (c) is correct.

FCFS : P1, P2, P3

RR2 : P1, P3, P2

1.46 (b)

Lets take an example:

Process	Execution time	Arrival time
1	10	0
2	8	1
3	7	2
4	6	3

Consider scheduler schedule processes priority after S times units so order of execution will be $P_1\ P_2\ P_3\ P_4\ P_1\ P_2\ P_3\ P_4$ (S < execution time of any process) which is exactly same as round robin so answer is (b).

1.47 (d)

Kernel level threads shares the code segment.

1.48 (7.2)

	A.T	E.T
A	0	6
B	3	2
C	5	4
D	7	6
E	10	3

Using SRTF:

A	B	A	C	E	D

0 3 5 8 12 15 21

T.A.T (A) = 8 – 0 = 8
T.A.T (B) = 5 – 3 = 2
T.A.T (C) = 12 – 5 = 7
T.A.T (D) = 21 – 7 = 14
T.A.T (E) = 15 – 10 = 5

$$\text{Average T.A.T} = \frac{8+2+7+14+5}{5} = 7.2$$

1.49 (1000)

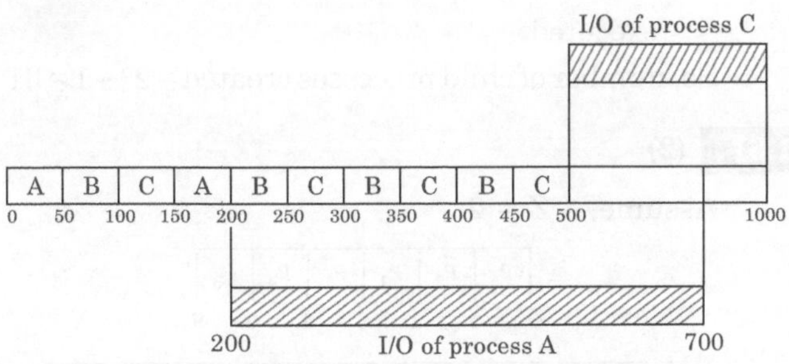

∴ At 1000 time units C completes its I/O.

1.50 (5.5)

SRTF:

P1	P2	P3	P4	P1

0 2 6 12 17 27

Waiting time of P1 = 17 – 2 = 15
Waiting time of P2 = 2 – 2 = 0
Waiting time of P3 = 6 – 3 = 3
Waiting time of P4 = 12 – 8 = 4

$$\text{Average waiting time} = \frac{15+0+3+4}{4} = 5.5$$

1.51 (12)

Periodic arrival times of T1: 0, 3, 6, 9, 12, 15, 18, 21,...

Priority of T1 = $\frac{1}{3}$, service time of $T_1 = 1$.

Periodic arrival times of T_2: 0, 7, 14, 21,...

Priority of $T_2 = \frac{1}{7}$, service time of $T_2 = 2$.

Periodic arrival times of T_3: 0, 20, 40,...

Priority of $T_3 = \frac{1}{20}$, service time of $T_3 = 4$.

T_1 has highest priority and T3 has lowest priority.

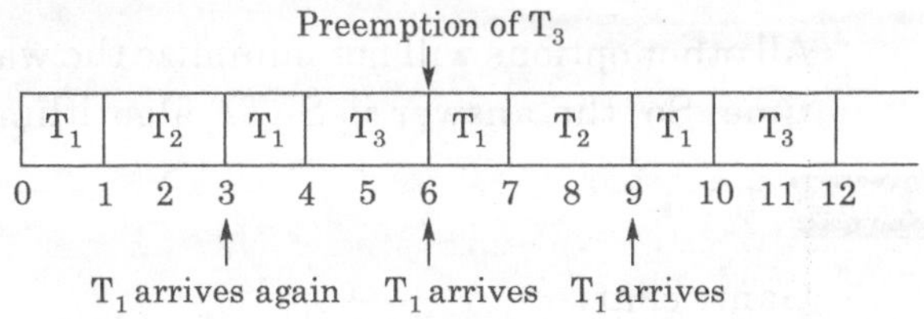

First instance of T_3 (4 units) completed at the end of 12 ms.

1.52 (d)

Maximum number of processes that can be in ready state is independent of number of processes (n).

1.53 (c)

Process	Arrival Time	B.T
A	0	3
B	1	6
C	4	4
D	6	2

(i) **FCFS:**

A	B	C	D

0 3 9 13 15

$$\text{Avg TAT} = \frac{3+8+9+9}{4} = \frac{29}{4} = 7.25$$

(ii) **Non-peemptive SJF:**

A	B	D	C

0 3 9 11 15

$$\text{Avg TAT} = \frac{3+8+5+11}{4} = 6.75$$

(iii) **SRTF:**

A	B	D	C	B

0 3 4 8 10 15

$$\text{Avg TAT} = \frac{3+14+4+4}{4} = \frac{25}{4} = 6.25$$

(iv) **RR:**

A	B	A	C	B	D	C	B

0 2 4 5 7 9 11 13 15

$$\text{Avg TAT} = \frac{5+14+9+5}{4} = \frac{33}{4} = 8.25$$

∴ SRTF has lowest turn around time.

1.54 (a)

To minimize the average waiting time, we need to select the shortest remaining time process first, because all are arriving at the same time, and they have unequal CPU burst times.

All other options will not minimize the waiting time. So, the answer is SRTF algorithm.

1.55 (8.25)

Gantt chart

P_1	P_2	P_3	P_2	P_4	P_1

0 3 7 8 10 13 20

Process Number	Arrival time	Burst time	Completion time	TAT
1	0	10	20	20
2	3	6	10	7
3	7	1	8	1
4	8	3	13	5
Average turn around time : 33/4 = 8.25				

Average turn around time is 8.25.

1.56 (d)

Generally every thread of a process have their own PC and stack. Both heap and global variables are shared by every thread of a process.

1.57 (3)

Using premptive SRTF algorithm Gantt chart will be,

P_1	P_2	P_4	P_1	P_3

0 3 6 8 12 17

(6 → P_2, 8 → P_4, 12 → P_1, 17 → P_3)

Turnaround time

$P_1 \rightarrow 12 - 0 = 12$

$P_2 \rightarrow 6 - 3 = 3$

$P_3 \rightarrow 17 - 5 = 12$

$P_4 \rightarrow 8 - 6 = 2$

Waiting time

$P_1 \rightarrow 12 - 7 = 5$

$P_2 \rightarrow 3 - 3 = 0$

$P_3 \rightarrow 12 - 5 = 7$

$P_4 \rightarrow 2 - 2 = 0$

$$\text{Average Waiting Time} = \frac{5+0+7+0}{4} = 3.0$$

1.58 (b)

All the threads share address space but other entities like, stack, PC, resisters are not shared and every thread will have its own.

1.59 (29)

Process	Arrival Time	Burst Time	Priority	C.T.	T.A.T.	W.T.
P_1	0	11	2	49	49	38
P_2	5	28	0 (high)	33	28	0
P_3	12	2	3	51	39	37
P_4	2	10	1	40	38	28
P_5	9	16	4	67	58	42
					Total	145

P_1	P_4	P_2	P_4	P_1	P_3	P_5

0 2 5 33 40 49 51 67

$$\text{Average waiting time} = \frac{145}{5} = 29$$

1.60 (31)

Number of child process created using n fork() calls $= 2^n - 1$

- fork(); statement inside the loop is executed when $i = 0, 2, 4, 6, 8$. So $n = 5$ times fork() is executed.

So, number of child processes created $= 2^5 - 1 = 31$

1.61 (2)

Assume, $Z = 2$

P_1	P_2	P_1	P_1	P_4	P_3

0 1 2 3 4 6 9

	Arrival time	CPU time	Completion time	Waiting time
P_1	0	3	4	1
P_2	1	1	2	0
P_3	3	3	9	3
P_4	4	Z = 2	6	0

$$\text{Average waiting time (WT)} = \frac{1+0+3+0}{4} = 1 \text{ ms}$$

Hence, $Z = 2$

1.62 (d)

Statement I, II and IV are correct.

1.63 (5.25)

SJF:

P_3	P_4	P_2	P_1

0 2 6 13 21

Turn Around Time (TAT) = (21 – 0) + (13 – 0) + (2 – 0) + (6 – 0)

$$\text{Average TAT} = \frac{42}{4} = 10.5$$

RR:

P_1	P_2	P_3	P_4	P_1	P_2

0 4 8 10 14 18 21

Turn Around Time (TAT) = (18 – 0) + (21 – 0) + (10 – 0) + (14 – 0)

= 18 + 21 + 10 + 14

$$\text{Average TAT} = \frac{63}{4} = 15.75$$

Hence, $|\text{SJF (TAT)} - \text{RR(TAT)}|$

$= |10.5 - 15.75| = 5.25$

1.64 **(12)**

WT	CT	P. No.	AT	BT
0	10	A	0	16
26	46	B	0	20
10	26	C	0	10

C	A	B

0 10 26 46

$$\text{Average writing time} = \frac{0+26+10}{3} = \frac{36}{3} = 12$$

1.66 **(b, c, d)**

- Goal is to maximize CPU utilization and maximize the throughput. So, statement (a) is false.
- Statement (b) is true, because turnaround time = completion time – arrival time and waiting time is included in this.
- Statement (c) is true because using time quantum, we can run the processes even if burst time is not known initially in round-robin.
- True for example, round robin scheduling requires hardware support which is timer.

1.67 **(d)**

P	Q	R	S	Q	R	Q

The above gantt chart satisfies all conditions given in the question the option (a), (b), (c) but not option (d). So, answer is option (d).

1.68 **(b, c, d)**

The status of a process or thread is saved via context switching. This makes it possible for the procedure to continue running later. During context switches, the registers (integer and floating-point), programme counter, condition registers (status register or flag register), and any other thread execution state must be stored. Registers, the stack pointer, and the programme counter must all be reloaded from the TCB of the new thread.

When the threads are from different processes, it is necessary to load the pointer for the top-level page table of the new address space in addition to saving and restoring the information mentioned above. (Notice that the process control block (PCB) already contains this top-level page table pointer from the previous process, so it is not necessary to preserve it.)

1.69 **(c, d)**

Page fault and system call required to change mode but others may not required.

1.70 **(c, d)**

In First Come First Serve (FCFS), if a process with a very large burst time comes before other processes, the other processes will have to wait for a very long time, but it is obvious that other processes will definitely get their chance to execute, so it is not likely that it will suffer from starvation.

There is no starvation in Round Robin because there is a defined time duration and every process will have a chance to run.

If higher priority processes keep coming in priority-based scheduling, low priority processes will starve.

If processes with short process times continue to arrive in Shortest Job First (SJF), processes with longer burst times will have to wait and starve.

■■■■

2 Process Management-II

(IPC, Synchronization and Concurrency)

2.1 A critical section is a program segment.
(a) Which should run in certain specified amount of time
(b) Which avoids deadlocks
(c) Where shared resources are accessed
(d) Which must be enclosed by a pair of semaphore operations, P and V

[1996 : 1 M]

2.2 A solution to the Dining Philosophers Problem which avoids deadlock is
(a) ensure that all philosophers pick up the left fork before the right fork
(b) ensure that all philosophers pick up the right fork before the left fork
(c) ensure that one particular philosophers picks up the left fork before the right fork, and that all other philosophers pick up the right fork before the left fork
(d) None of the above

[1996 : 2 M]

2.3 Each process Pi, i = 1 ... 9 is coded as follows

```
repeat
    P(mutex)
    {critical section}
    V(mutex)
forever
```

The code for P_{10} is identical except that it uses V(mutex) in place of P(mutex). What is the largest number of processes that can be inside the critical section at any moment?
(a) 1 (b) 2
(c) 3 (d) None of these

[1997 : 2 M]

2.4 When the result of a computation depends on the speed of the processes involved there is said to be
(a) cycle stealing (b) race condition
(c) a time lock (d) a deadlock

[1998 : 1 M]

2.5 A counting semaphore was initialized to 10. Then 6 P (wait) operations and 4V (signal) operations were completed on this semaphore. The resulting value of the semaphore is
(a) 0 (b) 8
(c) 10 (d) 12 **[1998 : 1 M]**

2.6 Let m[0]…m[4] be mutexes (binary semaphores) and P[0]...P[4] be processes. Suppose each process P[i] executes the following:

```
wait (m[i]);wait (m[(i+1) mode 4]);
............
release (m[i]); release (m[(i+1)mod 4]);
```

This could cause
(a) Thrashing
(b) Deadlock
(c) Starvation, but not deadlock
(d) None of the above **[2000 : 1 M]**

2.7 Suppose a processor does not have any stack pointer register. Which of the following statement is true?
(a) It cannot have subroutine call instruction
(b) It can have subroutine call instruction, but no nested subroutine calls
(c) Nested subroutine calls are possible, but interrupts are not
(d) All sequences of subroutine calls and also interrupts are possible **[2001 : 1 M]**

2.8 Where does the swap space reside?
(a) RAM (b) Disk
(c) ROM (d) On-chip cache

[2001 : 1 M]

2.9 Consider Peterson's algorithm for mutual exclusion between two concurrent processes i and j. The program executed by process is shown below:

```
repeat
    flag[i]=true;
    turn=j;
    while (P) do no-op;
    Enter critical section, perform actions, then
    exit critical section
    flag[i]=false;
    Perform other non-critical section actions.
Until false;
```

For the program to guarantee mutual exclusion, the predicate P in the while loop should be
(a) flag[j] = true and turn = i
(b) flag [j] = true and turn = j
(c) flag [i] = true and turn = j
(d) flag [i] = true and turn = i **[2001 : 2 M]**

Common Data for Q. 2.10 & 2.11:
Suppose we want to synchronize two concurrent processes P and Q using binary semaphores S and T. The code for the processes P and Q is shown below.

```
Process P:          Process Q:
while (1) {         while (1) {
  W:                  Y:
  print '0';          print '1';
  print '0';          print '1';
  X:                  Z:
}                   }
```

Synchronization statements can be inserted only at points W, X, Y and Z.

2.10 Which of the following will always lead to an output string with '001100110011'?
(a) P(S) at W, V(S) at X, P(T) at Y, V(T) at Z, S and T initially 1
(b) P(S) at W, V(T) at X, P(T) at Y, V(S) at Z, S initially 1, and T initially 0
(c) P(S) at W, V(T) at X, P(T) at Y, V(S) at Z, S and T initially 1
(d) P(S) at W, V(T) at X, P(T) at Y, V(T) at Z, S initially 1, and T initially 0 **[2003 : 2 M]**

2.11 Which of the following will ensure that the output string never contains a substring of the form 01^n0 or 10^n1 where n is odd?
(a) P(S) at W, V(S) at X, P(T) at Y, V(T) at Z, S and T initially 1
(b) P(S) at W, V(T) at X, P(T) at Y, V(S) at Z, S and T initially 1
(c) P(S) at W, V(S) at X, P(S) at Y, V(S) at Z, S initially 1
(d) V(S) at W, V(T) at X, P(S) at Y, P(T) at Z, S and T initially 1 **[2003 : 2 M]**

2.12 Consider two processes P_1 and P_2 accessing the shared variables X and Y protected by two binary semaphores S_X and S_Y respectively, both initialized to 1. P and V denote the usual semaphore operators, where P decrements the semaphore value, and V increments the semaphore value. The pseudocode of P_1 and P_2 is as follows:

```
P1:                  P2:
while true do {      while true do {
  L1 : ..........      L3 : ..........
  L2 : ..........      L4 : ..........
  X = X + 1;           Y = Y + 1;
  Y = Y – 1;           X = X – 1;
  V(SX);               V(SY);
  V(SY); }             V(SX); }
```

In order to avoid deadlock, the correct operators at L_1, L_2, L_3 and L_4 are respectively
(a) $P(S_Y)$, $P(S_X)$, $P(S_X)$, $P(S_Y)$
(b) $P(S_X)$, $P(S_Y)$, $P(S_Y)$, $P(S_X)$
(c) $P(S_X)$, $P(S_X)$, $P(S_Y)$, $P(S_Y)$
(d) $P(S_X)$, $P(S_Y)$, $P(S_X)$, $P(S_Y)$

[2004 : 2 M]

2.13 The semaphore variables full, empty and mutex are initialized to 0, n and 1, respectively. Process P_1 repeatedly adds one item at a time to a buffer of size n, and process P_2 repeatedly removes one item at a time from the same buffer using the programs given below. In the programs, K, L, M and N are unspecified statements.

```
P1 : while (1)
{
  K;
  P(mutex);
  Add an item to the buffer;
  V(mutex);
  L;
}
P2 : while (1)
{
  M;
  P(mutex);
  Remove an item from the buffer;
  V(mutex);
  N;
}
```

The statements K, L, M and N are respectively
(a) P(full), V(empty), P(full), V(empty)
(b) P(full), V(empty), P(empty), V(full)
(c) P(empty), V(full), P(empty), V(full)
(d) P(empty), V(full), P(full), V(empty)

[2004 : 2 M]

2.14 Given below is a program which when executed spawns two concurrent processes:

```
semaphore X : = 0 ;
/* Process now forks into concurrent processes
P1 & P2 */
P1: repeat forever      P2: repeat forever
V(X);                       P(X);
Compute;                    Compute;
P(X);                       V(X);
```

Consider the following statements about processes P1 and P2:
I. It is possible for process P1 to starve.
II. It is possible for process P2 to starve.
Which of the following holds?

(a) Both I and II are true.
(b) I is true but II is false.
(c) II is true but I is false
(d) Both I and II are false. **[2005 : 2 M]**

2.15 Two concurrent processes P1 and P2 use four shared resources R1, R2, R3 and R4, as shown below:

P1:	P2:
Compute:	Compute;
Use R1;	Use R1;
Use R2;	Use R2;
Use R3;	Use R3;
Use R4;	Use R4;

Both processes are started at the same time, and each resource can be accessed by only one process at a time. The following scheduling constraints exist between the access of resources by the processes:

- P2 must complete use of R1 before P1 gets access to R1.
- P1 must complete use of R2 before P2 gets access to R2.
- P2 must complete use of R3 before P1 gets access to R3.
- P1 must complete use of R4 before P2 gets access to R4.

There are no other scheduling constraints between the processes. If only binary semaphores are used to enforce the above scheduling constraints, what is the minimum number of binary semaphores needed?

(a) 1 (b) 2
(c) 3 (d) 4 **[2005 : 2 M]**

2.16 The atomic *fetch-and-set x, y* instruction unconditionally sets the memory location x to 1 and fetches the old value of x in y without allowing any intervening access to the memory location x. Consider the following implementation of P and V functions on a binary semaphore S.

```
void P(binary_semaphore *S) {
    unsigned y;
unsigned *x = & (S → value);
    do {
        fetch-and-set x, y;
    } while (y);
}
void V (binary_semphore *S) {
    S → value = 0;
}
```

Which one of the following is true?

(a) The implementation may not work if context switching is disabled in P
(b) Instead of using fetch-and-set, a pair of normal load/store can be used
(c) The implementation of V is wrong
(d) The code does not implement a binary semaphore **[2006 : 2 M]**

Common Data for Q. 2.17 & 2.18:

Barrier is a synchronization construct where a set of processes synchronizes globally i.e. each process in the set arrives at the barrier and waits for all others to arrive and then all processes leave the barrier. Let the number of processes in the set be three and S be a binary semaphore with the usual P and V functions. Consider the following C implementation of a barrier with line numbers shown on the left.

```
void barrier (void) {
 1 :  P (S) ;
 2 :  process_arrived + +;
 3 :  V (S) ;
 4 :  while (process_arrived ! = 3) ;
 5 :  P (S) ;
 6 :  process_left + + ;
 7 :  if (process_left = = 3){
 8 :      process_arrived = 0;
 9 :      process_left = 0;
10 :  }
11 :  V (S) ;
12 :  }
```

The variables process_arrived and process_left are shared among all processes and are initialized to zero. In a concurrent program all the three processes call the barrier function when they need to synchronize globally.

2.17 The above implementation of barrier is incorrect. Which one of the following is true?

(a) The barrier implementation is wrong due to the use of binary semaphore S
(b) The barrier implementation may lead to a deadlock if two barrier invocations are used in immediate succession
(c) Lines 6 to 10 need not be inside a critical section
(d) The barrier implementation is correct if there are only two processes instead of three

[2006 : 2 M]

2.18 Which one of the following rectifies the problem in the implementation?

(a) Lines 6 to 10 are simply replaced by process_arrived

(b) At the beginning of the barrier the first process to enter the barrier waits until process_arrived becomes zero before proceeding to execute P(S)
(c) Context switch is disabled at the beginning of the barrier and re-enabled at the end
(d) The variable process_left is made private instead of shared [2006 : 2 M]

2.19 Processes P1 and P2 use critical_flag in the following routine to achieve mutual exclusion. Assume that critical_flag is initialized to FALSE in the main program.

```
get_exclusive_access ( )
{
   if (critical _flag = = FALSE)
   {
      critical_flag = TRUE ;
      critical_region ( ) ;
      critical_flag = FALSE;
   }
}
```

Consider the following statements.
(i) It is possible for both P1 and P2 to access critical_region concurrently.
(ii) This may lead to a deadlock.
Which of the following holds?
(a) (i) is false and (ii) is true
(b) Both (i) and (ii) are false
(c) (i) is true and (ii) is false
(d) Both (i) and (ii) are true [2007 : 1 M]

2.20 Two processes, P1 and P2, need to access a critical section of code. Consider the following synchronization construct used by the processes:

```
/* P1 */
 while (true) {
  wantsl = true;
  while (wants2 = = true);
  /*    Critical
        Section */
  wants 1 = false;
}
/* Remainder section */
```

```
/* P2 */
 while (true) {
   wants2 = true;
   while (wantsl = = true);
   /*    Critical
         Section */
   wants 2 = false;
}
/* Remainder section */
```

Here, wants1 and wants2 are shared variables, Which are initialized to false. Which one of the following statements is TRUE about the above construct?
(a) It does not ensure mutual exclusion.
(b) It does not ensure bounded waiting.
(c) It requires that processes enter the critical section in strict alternation.
(d) It does not prevent deadlocks, but ensures mutual exclusion. [2007 : 2 M]

2.21 Synchronization in the classical readers and writers problem can be achieved through use of semaphores. In the following incomplete code for readers-writers problem, two binary semaphores mutex and wrt are used to obtain synchronization

```
wait (wrt)
writing is performed
signal (wrt)
wait (mutex)
readcount = readcount + 1
if readcount = 1 then S1
S2
reading is performed
S3
readcount = readcount – 1
if readcount = 0 then S4
Signal (mutex)
```

The values of S1, S2, S3, S4, (in that order) are
(a) signal (mutex), wait (wrt), signal (wrt), wait (mutex)
(b) signal (wrt), signal (mutex), wait (mutex), wait (wrt)
(c) wait (wrt), signal (mutex), wait (mutex), signal (wrt)
(d) signal (mutex), wait (mutex), signal (mutex), wait (mutex) [2007 : 2 M]

2.22 The P and V operations on counting semaphores, where s is a counting semaphore, are defined as follows:

P(s): $s = s - 1$;
  if $s < 0$ then wait;
V(s): $s = s + 1$;
  if $s \leq 0$ then wakeup a process waiting on s;

Assume that P_b and V_b, the wait and signal operations on binary semaphores are provided. Two binary semaphores X_b and Y_b are used to implement the semaphore operations P(s) and V(s) as follows:

```
P(s):   Pb(Xb);
        s = s – 1;
        if (s < 0) {
            Vb (Xb);
            Pb(Yb);
        }
        else Vb(Xb);
```

V(s): $P_b(X_b)$;
$s = s + 1$;
if ($s \leq 0$) V_b (Y_b);
$V_b(X_b)$;

The initial values of X_b and Y_b are respectively
(a) 0 and 0 (b) 0 and 1
(c) 1 and 0 (d) 1 and 1

[2008 : 2 M]

2.23 The enter_CS ( ) and leave_CS ( ) functions to implement critical section of a process are realized using test-and-set instruction as follows:

```
Void enter_CS (X)
{
    while (test-and-set (X));
}
Void leave_CS (X)
{
    X = 0 ;
}
```

In the above solution, X is a memory location associated with the CS and is initialized to 0. Now consider the following statements

I. The above solution to CS problem is deadlock-free.
II. The solution is starvation free.
III. The processes enter CS in FIFO order.
IV. More than one process can enter CS at the same time.

Which of the above statements are TRUE?
(a) I only (b) I and II
(c) II and III (d) IV only

[2009 : 2 M]

2.24 Consider the methods used by processes P1 and P2 for accessing their critical sections whenever needed, as given below. The initial values of shared boolean variables S1 and S2 are randomly assigned.

Method used by P1	Method used by P2
while (S1==S2); Critical Section S1=S2;	while (S1!=S2); Critical Section S2=not (S1);

Which one of the following statements describes the properties achieved?
(a) Mutual exclusion but not progress
(b) Progress but not mutual exclusion
(c) Neither mutual exclusion nor progress
(d) Both mutual exclusion and progress

[2010 : 1 M]

2.25 The following program consists of 3 concurrent processes and 3 binary semaphores. The semaphores are initialized as S0 = 1, S1 = 0, S2 = 0.

Process P0	Process P1	Process P2

while (true) {
wait (S0);
print '0'
release (S1);
release (S2);
}
	wait (S1);
	release(S0);
		wait (S2)
		release(S0);

How many times will process P0 print '0'?
(a) At least twice
(b) Exactly twice
(c) Exactly thrice
(d) Exactly once

[2010 : 2 M]

2.26 Fetch_And_Add(X, i) is an atomic Read-Modify-Write instruction that reads the value of memory location X, increments it by the value i, and returns the old value of X, it is used in the pseudocode shown below to implement a busy-wait lock. L is unsigned integer shared variable initialized to 0. The value of 0 corresponds to lock being available, while any non-zero value corresponds to the lock being not available.

```
AcquireLock (L){
    while ( Fetch_And_Add( L, 1 ) )
    L = 1;
}
ReleaseLock (L){
    L = 0;
}
```

This implementation
(a) fails as L can overflow
(b) fails as L can take on a non-zero value when the lock is actually available
(c) works correctly but may starve some processes
(d) works correctly without starvation

[2012 : 2 M]

2.27 Three concurrent processes X, Y and Z execute three different code segments that access and update ceratin shared variables. Process X execute the P operation (i.e., wait) on semaphores a, b and c; process Y executes the P operation on semaphores b, c and d; process Z executes the P operation on semaphores c, d and a before entering the respective code segments. After completing the execution of its code segments, each process invokes the V operation (i.e., signal) on its three semaphores. All semaphores are binary semaphores initialized to one. Which one of the following represents a deadlock-free order of invoking the P operations by the processes?

(a) X : P(a)P(b)P(c)
    Y : P(b)P(c)P(d)
    Z : P(c)P(d)P(a)

(b) X : P(b)P(a)P(c)
    Y : P(b)P(c)P(d)
    Z : P(a)P(c)P(d)

(c) X : P(b)P(a)P(c)
    Y : P(c)P(b)P(d)
    Z : P(a)P(c)P(d)

(d) X : P(a)P(b)P(c)
    Y : P(b)P(c)P(d)
    Z : P(c)P(d)P(a)

[2013 : 1 M]

2.28 A shared variable x, initialized to zero, is operated on by four concurrent process W, X, Y, Z as follows. Each of the processes W and X reads x from memory, increments by one, stores it to memory, and then terminates. Each of the processes Y and Z reads x from memory, decrements by two, stores it to memory and then terminates. Each process before reading x invokes the P operations (i.e., wait) on a counting semaphore S and invokes the V operation (i.e., signal) on the semaphore S after storing x to memory. Semaphore S is initialized to two. What is the maximum possible value of x after all processes complete execution?

(a) – 2 (b) – 1
(c) 1 (d) 2

[2013 : 2 M]

2.29 A certain computation generates two arrays a and b such that a[i] = f for $0 \le i \le n$ and b[i] = g(a[i]) for $0 \le i < n$. Suppose this computation is decomposed into two concurrent processes X and Y such that X computes the array a and Y computes the array b. The processes employ two binary semaphores R and S, both initialized to zero. The array a is shared by the two processes. The structure of the processes are shown below:

Process X	**Process Y**
private i;	private i;
for (i = 0; i < n; i++)	for (i = 0; i < n; i++)
{	{
a[i] = f(i);	EntryY(R, S);
ExitsX (R, S);	b[i] = g(a[i]);
}	}

Which one of the following represents the **CORRECT** implementations of ExitsX and EntryY?

(a)
```
ExitsX (R, S) {
    P(R);
    V(S)
}
EntryY(R, S) {
    P(S);
    V(R);
}
```

(b)
```
ExitsX (R, S) {
    V(R);
    V(S);
}
EntryY(R, S) {
    P(R);
    P(S);
}
```

(c)
```
ExitsX (R, S) {
    P(S);
    V(R)
}
EntryY(R, S) {
    V(S);
    P(R);
}
```

(d)
```
ExitsX (R, S) {
    V(R);
    P(S);
}
EntryY(R, S) {
    V(S);
    P(R);
}
```

[2013 : 2 M]

2.30 Consider the procedure below for the Producer-Consumer problem which uses semaphores:

```
semaphore n = 0;
semaphore s = 1;

void producer()              void consumer()
{                            {
  while(true)                  while (true)
  {                            {
     produce();                   semWait (s);
     semWait(s);                  semWait (n);
     addToBuffer();               removeFromBuffer ();
     semSignal(s);                semSignal (s);
     semSignal(n);                consume();
  }                            }
}                            }
```

Which one of the following is TRUE?

(a) The producer will be able to add an item to the buffer, but the consumer can never consume it.
(b) The consumer will remove no more than one item from the buffer.
(c) Deadlock occurs if the consumer succeeds in acquiring semaphore s when the buffer is empty.
(d) The starting value for the semaphore n must be 1 and not 0 for deadlock-free operation.

[2014 (Set-2) : 2 M]

2.31 The following two functions P1 and P2 that share a variable B with an initial value of 2 execute concurrently.

```
P1()              P2()
{                 {
   C = B – 1;        D = 2 * B;
   B = 2 * C;        B = D – 1;
}                 }
```

The number of distinct values that B can possibly take after the execution is ________.

[2015 (Set-1) : 1 M]

2.32 Two processes X and Y need to access a critical section. Consider the following synchronization construct used by both the processes

```
Process X
/* other code for process X */
while (true)
{  varP = true;
   while (varQ == true)
   { /* Critical Section */
     varP = false;
   }
}
/* other code for process X */
Process Y
/* other code for process Y */
while (true)
{  varQ = true;
   while (varP == true)
   { /* Critical Section */
     varQ = false;
   }
}
/* other code for process Y */
```

Here, varP and varQ are shared variables and both are initialized to false. Which one of the following statements is true?

(a) The proposed solution prevents deadlock but fails to guarantee mutual exclusion
(b) The proposed solution guarantees mutual exclusion but fails to prevent deadlock
(c) The proposed solution guarantees mutual exclusion and prevents deadlock
(d) The proposed solution fails to prevent deadlock and fails to guarantee mutual exclusion

[2015 (Set-3) : 1 M]

2.33 Consider the following proposed solution for the critical section problem. There are n processes: $P_0 \ldots P_{n-1}$. In the code, function *pmax* returns an integer not smaller than any of its arguments. For all i, $t[i]$ is initialized to zero.

Code for P_i :

```
do
{
    c[i]=1; t[i] = pmax (t[0],..., t[n–1]) +1; c[i] = 0;
    for every j ≠ i in {0,..., n – 1}
    {  while (c[j]);
       while (t[j] != 0 && t[j] <= t[i]);
    }
    Critical Section; t[i] = 0;
    Remainder Section;
}   while (true);
```

Which one of the following is TRUE about the above solution?

(a) At most one process can be in the critical section at any time
(b) The bounded wait condition is satisfied
(c) The progress condition is satisfied
(d) It cannot cause a deadlock

[2016 (Set-1) : 2 M]

2.34 Consider the following two-process synchronization solution:

```
Process 0
    Entry: loop while (turn == 1);
    (critical section)
    Exit: turn = 1;
Process 1
    Entry : loop while (turn == 0);
    (critical section)
    Exit: turn = 0;
```

The shared variable turn in initialized to zero. Which one of the following is **TRUE**?

(a) This is a correct two-process synchronization solution
(b) This solution violates mutual exclusion requirement
(c) This solution violates progress requirement
(d) This solution violates bounded wait requirement

[2016 (Set-2) : 2 M]

2.35 Consider a non-negative counting semaphore S. The operation P(S) decrements S, and V(S) increments S. During an execution, 20 P(S) operations and 12 V(S) operations are issued in some order. The largest initial value of S for which at lest one P(S) operation will remain blocked is ______. **[2016 (Set-2) : 2 M]**

2.36 A multithreaded program P executes with x number of threads and uses y number of locks for ensuring mutual exclusion while operating on shared memory locations. All locks in the program are non-reentrant, i.e., if a thread holds a lock l. then it cannot re-acquire lock l without releasing it. If a thread is unable to acquire a lock, it blocks until the lock becomes available. The minimum value of x and the minimum value of y together for which execution of P can result in a deadlock are:

(a) $x = 1, y = 2$
(b) $x = 2, y = 1$
(c) $x = 2, y = 2$
(d) $x = 1, y = 1$

[2017 (Set-1) : 2 M]

2.37 Consider the following solution to the producer-consumer synchronization problem. The shared buffer size is N. Three semaphores empty, full and mutex are defined with respective initial values of 0, N and 1. Semaphore empty denotes the number of available slots in the buffer, for the consumer to read from. Semaphore full denotes the number of available slots in the buffer, for the producer to write to. The placeholder variables, denoted by P, Q, R and S, in the code below can be assigned either empty or full. The valid semaphore operations are: wait () and signal ().

Producer:	Consumer

do {
wait (P);
wait (mutex);
//Add item to buffer
signal (mutex);
signal (Q);
} while (1);
	do {
	wait (R);
	wait (mutex);
	//Consume item from buffer
	signal (mutex);
	signal (S);
	} while (1);

Which one of the following assignment to P, Q, R and S will yield the correct solution?

(a) P: full, Q: full, R: empty, S; empty
(b) P: empty, Q: empty, R: full, S: full
(c) P: full, Q: empty, R: empty, S: full
(d) P: empty, Q: full, R: full, S: empty

[2018 : 2 M]

2.38 Consider three concurrent processes P1, P2 and P3 as shown below, which access a shared variable D that has been initialized to 100.

P1	P2	P3
: : D = D + 20 : :	: : D = D − 50 : :	: : D = D + 10 : :

The processes are executed on a uniprocessor system running a time-shared operating system. If the minimum and maximum possible values of D after the three processes have completed executed are X and Y respectively, then the value of Y − X is ______. **[2019 : 1 M]**

2.39 Consider the following snapshot of a system running n concurrent processes. Process i is holding X_i instances of a resource R, $1 \le i \le n$. Assume that all instances of R are currently in use. Further, for all i, process i can place a request for at most Y_i additional instances of R while holding the X_i instances it already has. Of the n processes, there are exactly two processes p and q such that $Y_p = Y_q = 0$. Which one of the following conditions guarantees that no other process apart from p and q can complete execution?

(a) $X_p + X_q < \text{Min}\ \{Y_k \mid 1 \le k \le n,\ k \ne p,\ k \ne q\}$
(b) $\text{Min}\ (X_p, X_q) \ge \text{Min}\ \{Y_k \mid 1 \le k \le n,\ k \ne p,\ k \ne q\}$
(c) $\text{Min}\ (X_p, X_q) \le \text{Max}\ \{Y_k \mid 1 \le k \le n,\ k \ne p,\ k \ne q\}$
(d) $X_p + X_q < \text{Max}\ \{Y_k \mid 1 \le k \le n,\ k \ne p,\ k \ne q\}$

[2019 : 2 M]

2.40 Each of a set of n processes executes the following code using two semaphores a and b initialized to 1 and 0, respectively. Assume that count is a shared variable initialized to 0 and not used in CODE SECTION P.

```
CODE SECTION P
```

wait(a); count = count + 1;
if (count == n) signal(b);
signal(a); wait(b); signal(b);

```
CODE SECTION Q
```

What does the code achieve?

(a) It ensures that at most two processes are in CODE SECTION Q at any time.
(b) It ensures that all processes execute CODE SECTION P mutually exclusively.
(c) It ensures that no process executes CODE SECTION Q before every process has finished CODE SECTION P.
(d) It ensures that at most $n - 1$ processes are in CODE SECTION P at any time.

[2020 : 2 M]

2.41 Consider the following pseudocode, where S is a semaphore initialized to 5 in line#2 and counter is a shared variable initialized to 0 in line#1. Assume that the increment operation in linc#7 is not atomic.

```
1.  int counter = 0;
2.  Semaphore S = init(5);
3.  void parop(void)
4.  {
5.        wait(S);
6.        wait(S);
7.        counter++;
8.        signal(S);
9.        signal(S);
10. }
```

If five threads execute the function parop concurrently, which of the following program behavior(s) is/are possible?

(a) The value of counter is 1 after all the threads successfully complete the execution of parop.
(b) The value of counter is 0 after all the threads successfully complete the execution of parop.
(c) There is a deadlock involving all the threads.

(d) The value of counter is 5 after all the threads successfully complete the execution of parop.

[2021 (Set-1) : 2 M]

2.42 Consider a computer system with multiple shared resource types, with one instance per resource type. Each instance can be owned by only one process at a time. Owning and freeing of resources are done by holding a global lock (L). The following scheme is used to own a resource instance:

```
function OwnResource (Resource R)
    Acquire lock L // a global lock
    if R is available then
        Acquire R
        Release lock L
    else
        if R is owned by another process P then
        Terminate P, after releasing all
        resources owned by P
            Acquire R
            Restart P
            Release lock L
        end if
    end if
end function
```

Which of the following choice(s) about the above scheme is/are correct?

(a) The scheme violates the mutual exclusion property.
(b) The scheme ensures that deadlocks will not occur.
(c) The scheme may lead to live-lock.
(d) The scheme may lead to starvation.

[2021 (Set-2) : 2 M]

2.43 Consider the following multi-threaded code segment (in a mix of C and pseudo-code), invoked by two processes P_1 and P_2, and each of the processes spawns two threads T_1 and T_2:

```
int x = 0;      // global
Lock L1;        // global
main() {
  create a thread to execute foo();  // Thread T1
  create a thread to execute foo();  // Thread T2
  wait for the two threads to finish execution;
  print (x);}
foo() {
  int y = 0;
  Acquire L1;
  x = x + 1;
  y = y + 1;
  Release L1;
  print (y); }
```

Which of the following statement(s) is/are correct?

(a) At least one of P_1 and P_2 will print the value of x as 4.
(b) Both P_1 and P_2 will print the value of x as 2.
(c) Both T_1 and T_2, in both the processes, will print the value of y as 1.
(d) At least one of the threads will print the value of y as 2.

[2021 (Set-2) : 2 M]

2.44 Consider the following threads, T_1, T_2 and T_3 executing on a single processor, synchronized using three binary semaphore variables, S_1, S_2 and S_2, operated upon using standard wait( ) and signal( ). The threads can be context switched in any order and at any time.

T_1	T_2	T_3
while (true) { wait (S_3); print ("C"); signal (S_2); }	while (true) { wait (S_1); print ("B"); signal (S_3); }	while (true) { wait (S_2); print ("A"); signal (S_1); }

Which initialization of the semaphores would print the sequence BCABCABCA.....?

(a) $S_1 = 1; S_2 = 1; S_3 = 1$
(b) $S_1 = 1; S_2 = 1; S_3 = 0$
(c) $S_1 = 1; S_2 = 0; S_3 = 0$
(d) $S_1 = 0; S_2 = 1; S_3 = 1$

[2022 : 1 M]

2.45 Consider the two functions incr and decr shown below:

```
incr() {                decr() {
    wait(s);                wait(s);
    X = X + 1;              X = X - 1;
    signal(s);              signal(s);
}                       }
```

There are 5 threads each invoking incr once, and 3 threads each invoking decr once, on the same shared variable X. The initial value of X is 10. Suppose there are two implementations of the semaphore s, as follows:

I-1: s is a binary semaphore initialized to 1.
I-2: s is a counting semaphore initialized to 2.

Let V1, V2 be the values of X at the end of execution of all the threads with implementations I-1, I-2, respectively.

Which one of the following choices corresponds to the minimum possible values of V1, V2, respectively?

(a) 15, 7
(b) 7, 7
(c) 12, 7
(d) 12, 8

[2023 : 2 M]

2.46 Consider the following code snippet using the fork( ) and wait( ) system calls. Assume that the code compiles and runs correctly, and that the system calls run successfully without any errors.

```
int x = 3;
while (x > 0) {
      fork( );
      printf("hello");
      wait(NULL);
      X--;
}
```

The total number of times the printf statement is executed is ________. **[2024 (Set-1) : 2 M]**

Answers Process Management-II

2.1	(c)	**2.2**	(c)	**2.3**	(d)	**2.4**	(b)	**2.5**	(b)	**2.6**	(b)	**2.7**	(a)	**2.8**	(b)	**2.9**	(b)
2.10	(b)	**2.11**	(c)	**2.12**	(d)	**2.13**	(d)	**2.14**	(d)	**2.15**	(b)	**2.16**	(a)	**2.17**	(b)	**2.18**	(b)
2.19	(c)	**2.20**	(d)	**2.21**	(c)	**2.22**	(c)	**2.23**	(a)	**2.24**	(a)	**2.25**	(a)	**2.26**	(b)	**2.27**	(b)
2.28	(d)	**2.29**	(c)	**2.30**	(c)	**2.31**	(3)	**2.32**	(a)	**2.33**	(a)	**2.34**	(c)	**2.35**	(7)	**2.36**	(d)
2.37	(c)	**2.38**	(80)	**2.39**	(a)	**2.40**	(c)	**2.41**	(a, c, d)			**2.42**	(b, c, d)	**2.43**	(b, c)	**2.44**	(c)
2.45	(c)	**2.46**	(14)														

Explanations Process Management-II

2.1 (c)

Critical section is an area in a program where the resources can be accessed.

2.2 (c)

Ensure that one particular philosophers picks up the left fork before the right fork, and that all other philosophers pick up the right fork before the left fork.

2.3 (d)

Everytime when process P_{10} enters it performs the 'UP' operation so after that one more process can enter into it, so this phenomena continue till end. So 10 processes at a time can enter into critical section.

2.4 (b)

Race condition.

2.5 (b)

Initially semaphore is 10, then 6 down operations are performed means (10 – 6) and 4 up operation means (4 + 4) = 8.
So option (b) is correct.

2.6 (b)

All processes are in deadlock condition i.e., not a single one process can enter into critical section.

2.7 (a)

If stack pointer register is not available then activation records in the stack can not be created. So it cannot have subroutine call instruction.

2.8 (b)

Disk has swap space.

2.9 (b)

To ensure the mutual exclusion, we have to take care that 'turn' value which is 'j' so we should not allow that process in CS which is having flag [j] = true.
Hence option (b) is correct.

2.10 (b)

To get the expected output , The sequence of processes are P, Q, P, Q, P, Q and so on we must initialize S = 1 and T = 0.

2.11 (c)

If the output string never contains a substring of the form 01^n0 or 10^n1 where n is odd then the procedures becomes.

S = 1

```
Process P:        Process Q:
While(1){         While(1){
W : P(S)          Y : P(S)
print '0';        print '1';
print '0';        print '1';
X : V(S)          Z : V(S)
}                 }
```

2.12 (d)

S_x and S_y are two binary semaphores if we assume P means wait and V means signal then for two process P_1 and P_2 we take alternate then there in no chance of dead lock.

2.13 (d)

Here full and empty are integer semaphore which denote number of full slots in buffer and number of empty slot in buffer at any time respectively. While mutex is binary semaphore to provide mutual exclusion.

Initially full = 0 empty = n and mutex = 1.

- When no item into buffer, since mutex = 1. So to restrict consumer to remove item P (full) must be there at M and when it leave the critical section made V (empty) at N i.e. one empty slot increase.
- When producer tries to add an item into buffer before entering to critical section, it must perform P (empty) i.e. one empty decrease and while leaving critical section make V (full) i.e. one new item added successfully.

So, correct order will be P (empty), V (full), P (full) and V (empty) to ensure consumer will be block until 1 item added into buffer.

2.14 (d)

None of the processes will ever starve. Starvation is said to be possible if in ALL the possible ways that a process want to enter CS, it is DENIED the access. But that can't happens here.

P1 is never starved because this is INFACT the first process that can enter. V(X) by P1 is always successful and hence it always enter the CS. So 1 is FALSE.

When P1 is in CS, X = 1. At this time P2 can make a successful DOWN operation P(X) and enter CS. So 2 is also FALSE.

But P2 must exit the CS before P1.

2.15 (b)

With one semaphore it is not possible, since binary semaphore is used.

So, try with 2 binary semaphore a and b assume initialize to 0.

P1:	P2:
P(a) R1	
	R1 V(a) P(b) R2
R2 V(b) P(a) R3	
	R3 V(a) P(b) R4
R4 V(b)	

So, minimum 2 semaphore needed.

2.16 (a)

The P and V function is as follows:

```
1.  void P (binary-semaphore * S){
2.     unsigned y;
3.     unsigned * x = & (S → value);
4.     do {
5.        fetch-and-set x, y;
6.     } while (y);
7.  }
8.  void V (binary-semaphore * S){
9.     S → value = 0;
10. }
```

fetch-and-set instruction always sets the memory location x = 1 and fetches the old value of x into y. The binary semphore * S takes only two values either 0 and 1. When we initiliaze S = 0 then in statement 3 this value will be start at location x and fetch-and-set instruction changes the value of x = 0 to x = 1 and y becomes 0. If there are more than two processes and context switching between processes is disabled in P then this implementation doesn't work properly and can't synchronize the processes.

2.17 (b)

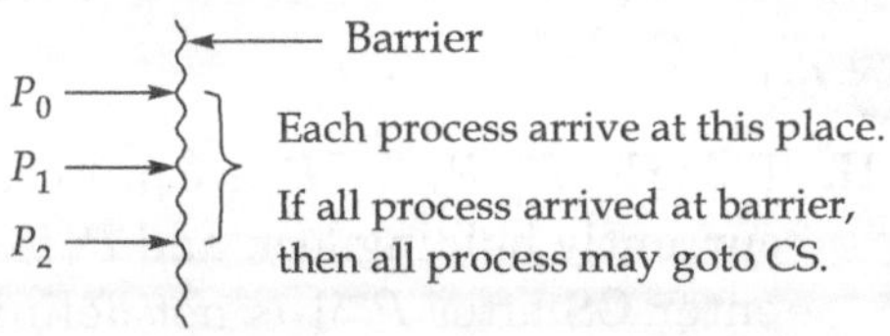

Given code:

```
1.  P(S);
2.  process_arrived++;
3.  V(S);
4.  while (process_arrived != 3);
5.  P(S);
6.  process_left++;
7.  if (process_left == 3)
8.  {  process_arrived = 0;
9.     process_left = 0;
10. }
11. V(S);
```

(Line 4: This time act as barrier)
(Lines 5–11: These code work like critical section)

Each process in the set arrive at barrier. We need to provide synchronization to process to enter barrier and wait for all other processes to arrive. Once they arrive, they start leaving the barrier and same thing goes on.

Process	Code
P_0	1. S = 0 2. P_A = 1 3. S = 1 4. (1! = 3); → condition true

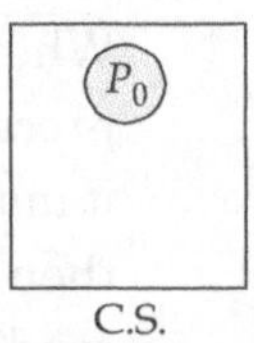

So, P_0 arrive at barrier and waiting for other process to arrive.

Process	Code
P_1	1. S = 0 2. P_A = 2 3. S = 1 4. (2! = 3); True

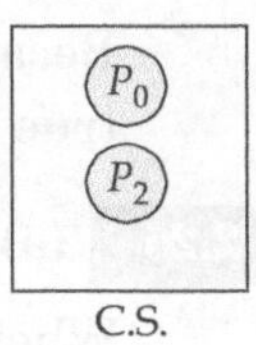

So, P_1 stop and waiting for other process.

Process	Code
P_2	1. S = 0 2. P_A = 3 3. S = 1 4. (3! = 3); False

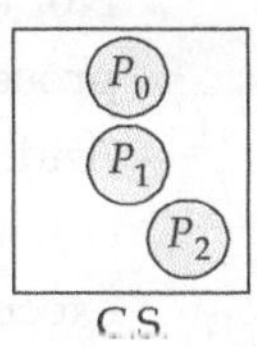

Now all 3-process arrived barrier, so while condition become false, to now the processes start leaving the barrier one by one.

Now suppose P_0 leaves the barrier, and not waiting for other processes (P_1 and P_2) to leave the barrier and immediately P_1 entering to barrier.

i.e.

Process	Code
P_0	5. S = 0 6. P_L = 1 7. (1 == 3) : false (so, 8 and 9 do not execute) 11. S = 1

Now S = 1

P_1 and P_2 still in loop.

So, we have two choices:

Choice 1: P_1 can leave the barrier

Choice 2: P_0 can enter into the barrier (Case for deadlock)

Process	Code
1.	S = 0
2.	P_A = 4
3.	S = 1
4.	(4! = 3); True

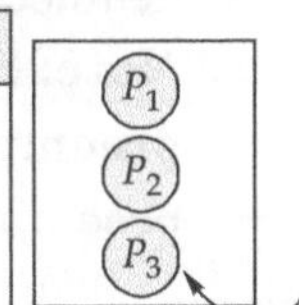

So, P_0 also enters the barrier and the while loop now, will be true forever because, we have 3 processes and all three processes stuck up at barrier.

Hence, no possibility of while loop to become false.

Hence, deadlock will occur.

Hence we conclude that, as soon as all the 3 processes arrive at barrier and until and unless, all the process leave the barrier, we will not call barrier again. So, two invocations if used in immediate succession, may lead to deadlock.

2.18 (b)

```
void barrier (void)
{
    while (process_left != 0);    ← Correction in implementation of code
    P(S);
    P_A++;
    V(S);
    while
    (P_A! = 3);
    P(S);
    P_L++;
    if (P_L == 3)
    {
        P_A = 0;
        P_L = 0;
    }
        V(S);
}
```

2.19 (c)

Initially critical_flag = false

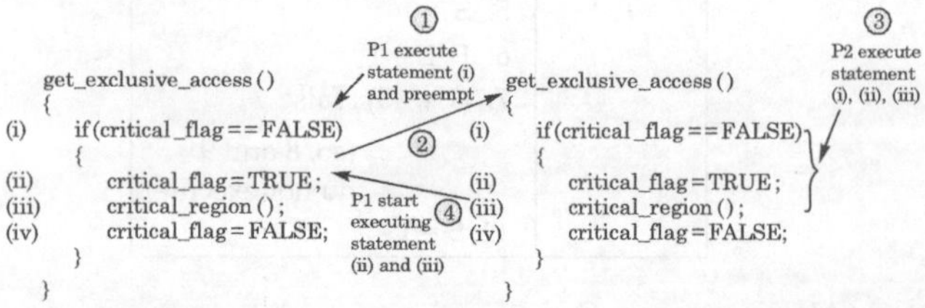

So, both process can enter in critical_region at same time.

So, no mutual exclusion.

We know that Dead lock → Mutual exclusion.

So, no Mutual exclusion → No deadlock. Here no deadlock possible.

Note: Here arrow shows order of execution and preemption is temporary, any process can preempt at any time.

2.20 (d)

In process P_1 if wants1 is true then wants2 entered in critical section and in process P_2 if wants2 is true then wants1 entered in critical section but in both cases a deadlock can occur but it ensures the mutual exclusion.

2.21 (c)

Writers code

wait (wrt)
writing is performed
signal (wrt)

- If read count = = 1 then it must perform wait (wrt)
  So, at same time reader and writer cannot access critical section.
- Since more than one reader an access the critical section simultaneously
  So, at S2 signal (mutex) must be done.

Readers code

wait (mutex)
read count = read count + 1
if (read count = =1)
then wait (wrt) S1
signal (mutex) S2
reading is performed
wait (mutex) S3
read count = read count = 1
If (read count = = 0)
then signal (wrt) S4
Signal (mutex)

Since by seeing S1 and S2 option (c) matches. Hence it is correct answer.

2.22 (c)

In given code, P(S) decrements the value of semaphore and V(S) increment the value of semaphore. P_b and V_b are wait and signal operations on binary semaphore. X_b and Y_b are binary semaphore. To avoid the mutual exclusion condition if the value of X_b is 1 and the value of $Y_b = 0$ then P(S) and V(S) works properly.

2.23 (a)

II. False let $P_1, P_2, \ldots, P_n$ are processes let P_1 is currently holding lock and P_2 is waiting to enter CS after P_1 it is not definite that P_2 will surely enter critical section and it may happen that each time P_2 try to enter CS it is blocked by some other process which may lead P_2 to starvation.

III. False as there is not fixed order whichever process get lock access can enter so not necessary to follow FIFO.

IV. Given code of test and set instruction satisfy mutual exclusion so atmost one process can enter CS

∴ By elimination only option (a) is true.

2.24 (a)

P1	P2
While (S1 == S2); Critical section S1 = S2;	While (S1! = S2); Critical section S2 = Not(S1);

We initially take $S_1 = 1$, and $S_2 = 1$.

When P1 execute S1 = S2 then it do nothing and process P1 go into critical section after executing it make the value of S1 = S2 and when P1 execute then P2 not execute because that checks different condition in while so both show mutual exclusion. But progress is not shown by these two because in this one process can be delayed infinitely.

Because again both P1 make its own condition true (S1 = S2 and S2 = not (S1);).

2.25 (a)

When $S_0 = 1$, $S_1 = 0$

So firstly the value of S_1 and $S_2 = 0$, P_1 and P_2 Not execute. The value of $S_0 = 1$, So it execute, process P_0 Now value of $S_0 = 0$

And it print zero one time when it release (S_1) call then value of S_1, S_2 increase by 1. So P_1 execute. It makes value of S_1 again zero and $S_0 = 1$, after it release (S_2) make value of $S_2 = 0$ by applying wait and $S_0 = 2$. So again P_0 process execute and wait (S_0) means it decrease by one so it print zero and this process continue. So P_0 print zero at least twice.

2.26 (b)

Assume P_1 executes until while condition and preempts before executing L = 1. Now P_2 executes all statements, hence L = 0. Then P_1 without checking L it makes L = 1 by executing the statement where it was preempted.

∴ It takes a non-zero value (L = 1) when the lock is actually available (L = 0)

2.27 (b)

Trick: To check deadlock in the given code proceed parallely i.e. first executed the first line of all the processes then executed the second line of all processes and so on.

(a)

X	Y	Z
1. P(a)	1. P(b)	1. P(c)
2. P(b)	2. P(c)	2. P(d)
3. P(c)	3. P(d)	3. P(a)

Value of a, b, c are 1 initially.

Execute first line of all processes i.e. $X_1 \rightarrow Y_1 \rightarrow Z_1$ after the executive of $X_1 \rightarrow Y_1 \rightarrow Z_1$ value of {a,b,c} = 0. Now no process is allowed to enter into the critical section. So deadlock.

(b)

X	Y	Z
1. P(b)	1. P(b)	1. P(a)
2. P(a)	2. P(c)	2. P(c)
3. P(c)	3. P(d)	3. P(d)

1. Execute X_1 so value of b gets 0.
2. As value of b = 0 when we execute Y_1 it get blocked (Y is blocked).
3. Now execute Z_1 so value of a get 0.
4. Now try to execute the second line of unblocked processes as value of a = 0 when we try to execute X_2 it get blocked (X is blocked)
5. Now execute Z_2, value of c gets 0.
6. As process X and Y are blocked so we can easily execute Z_3.

So no deadlock, like wise we can check option c and d, so option (b) is correct.

2.28 (d)

	X	W	Y	Z
1.	P(s)	P(s)	P(s)	P(s)
2.	R(x)	R(x)	R(x)	R(x)
3.	x = x +1	x = x + 1	x = x + 2	x = x + 2
4.	Store to MM	Store to MM	Store to MM	Store to MM
5.	V(s)	V(s)	V(s)	V(s)

1. Start with X. P(s) ⇒ s = 1 read x = 0, x = x+1 Now before store into the memory process X preempt to process Y.
2. Y will perform P(s) ⇒ s = 0, read x = 0 x = x – 2 then store x = – 2 and V(s) ⇒ s = 1
3. Z will perform P(s) ⇒ s = 0, read x = – 2 x = x – 2 then store x = – 4 and V(s) ⇒ s = 1
4. Now process x will store value of variable x into the MM so x = 1 and V(s) ⇒ s = 2
5. Now process W perform P(s) ⇒ s = 1 read x = 1 x = x + 1, store x = 2, V(s) ⇒ s = 2,

Option (d) is correct

2.29 (c)

The solution is using a binary semaphore. X and Y are two different processes whatever the values inserted by the processes in the array that will be used by process Y afterwards.

Option (a): It is wrong, because there exists a deadlock.

Option (b): It is wrong, because process X can insert multiple values in an array and terminate but process Y can consume only one value.

Option (c): It is correct, because each and every value inserted by the process X in the array will be immediately consumed by process Y.

Option (d): It is wrong, because process X can insert two values in the array but the process Y can consume only one value if the preemption occurs after V(S) operation in the entry seciton of process Y.

Process Y : - - - - - - -
- - - - - - -
Entry Y(R, S) {
V(S);

R = 0;
S = ~~0~~ 1;

⟶ Preemption

Process X : - - - - - - -
- - - - - - -
Exists (R, S) {
V(R);
P(S);
- - - - - - -
Exists (R, S) {
V(R);

R = ~~0~~ ~~1~~ 1;
S = ~~1~~ 0;

⟶ Preemption

Process Y : - - - - - - -
P(R);
- - - - - - -
Entry Y(R, S) {
V(S);

R = ~~1~~ 0;
S = ~~0~~ 1;

⟶ Preemption

Process X : P(S);

R = 0;
S = ~~1~~ 0;

2.30 (c)

Option (a): If producer want to produce an item, it will first perform wait(s); then add to buffer( ); then signal semaphore s, n

∴ Value of semaphore become

n = ~~0~~ 1;

s = ~~1~~ ~~0~~ 1;

If we now run consumer code, we have to perform wait option on semaphore s, n

∴ n = ~~0~~ ~~1~~ 0;

s = ~~1~~ ~~0~~ ~~1~~ 0;

after removing from buffer( ); it will signal(s) and consume( )

∴ Consumer can consume item produced by producer so option (a) is false.

Option (b): It is also false as after removing one item from buffer we can repeat above process any number of times.

Option (c): If consumer run before producer has produced i.e. if buffer is empty, it will perform wait option on 's' which is successful after that it perform an unsuccessful wait option on semaphore 'n'

So, s = ~~1~~ 0;

n = 0 (wait will lead to suspension)

Now if producer try to run and perform wait(s); it will also get suspended as both process are suspended and no other process to signal them so deadlock occurs.

∴ Option (c) is true.

Option (d): False, if we consume item in the buffer using consumer code, followed by process in option (c), deadlock will occur.

2.31 (3)

B = 2					
C = B − 1 B = 2 × C D = 2 × B B = D − 1 ⇓ B = 3	D = 2 × B B = D − 1 C = B − 1 B = 2 × C ⇓ B = 4	C = B − 1 D = 2 × B B = 2 × C B = D − 1 ⇓ B = 3	C = B − 1 D = 2 × B B = D − 1 B = 2 × C ⇓ B = 2	D = 2 × B C = B − 1 B = D − 1 B = 2 × C ⇓ B = 2	D = 2 × B C = B − 1 B = 2 × C B = D − 1 ⇓ B = 3

⇒ B = 2, 3 or 4

∴ Three distinct values are possible.

2.32 (a)

(i) Mutual exclusion fail in the given program. First, process *X* executes all statements till var*P* = true. Second, process *Y* executes all statements till var*Q* = true; Now, both process *X* and process *Y* both can execute while condition as true and then enters critical section simultaneously. Therefore mutual exclusion fails.

(ii) Deadlock is prevented by the given code.

2.33 (a)

Since *p*max returns as integer not smaller than any of its argument there may be a case when *t*[*j*] == *t*[*i*] and *t*[*j*] is not equal 0 for it.

∴ while (*t*[*j*] != 0 && *t*[*j*] <= *t*[*i*]);

will be true and process may get stuck in infinite loop which lead to deadlock and hence progress is also not satisfied.

As there is not proper definite waiting for process.

∴ Among process P_0 ..... P_{n-1} some process may not get chance to execute and may starve.

∴ Bounded wait is not satisfied.

2.34 (c)

Given problem is strict alteration problem it violates progress. As if initially when turn is initialized to zero and process 1 try to enter into critical section it will be stuck in while loop as a result condition for progress is violates because a process outside the critical section should not block another process from entering to critical section if it is free. Bounded wait is satisfied as after process 0 has executed process 1 will definitely get a chance to execute so there is bounded wait with respect to number of processes.

2.35 (7)

$$S - 20 + 12 = -1$$
$$S - 8 = -1$$
$$S = -1 + 8 = +7$$

So, the initial values of the semaphore should be '7'.

2.37 (c)

Full = N, empty = 0, mutex = 1

- Initially buffer will be empty, so consumer should not start first, so option b, D are eliminated.
- With option A consumer will never consume the item, so it is wrong.
- Option 'c' is correct answer which proper functionality of produce and consumer.

2.38 (80)

Initial value of D = 100

Minimum value occur when P1, P2, P3 read and P1, P3 update the value

P1 update D = 110

P3 update D = 120

and last P2 will update the D = 100 – 50 = 50 = X

Maximum value will occur when P2 and P3 read the initial value of D and update, first run P2 and update D = 50. P3 will run and update

$$D = 100 + 10 = 110$$

Now P1 will read the value of D = 110 and update

$$D = D + 20 = 110 + 20$$
$$D = 130 = Y$$
$$Y - X = 130 - 50 = 80$$

2.39 (a)

- Process 'p' is holding X_p resources and does not request any new resources. So 'p' can complete it's execution and releases X_p resources.
- Process 'q' is holding X_q resources and does not request any new resources. So 'q' can complete it's execution and release X_q resources.

  Now available resources = $X_p + X_q$
- Other than 'p' and 'q' no other processes completes it execution, means available resources ($X_p + X_q$) is less than the resources demanded or requested by all other processes. So, $X_p + X_q < \text{Min}\ \{Y_k \mid 1 \le k \le n, k \ne p, k \ne q\}$ holds true.

2.40 (c)

All the processes running the given code will remain block due to wait(b) until the value of count becomes n.

When the value of count will become equals to n, value of b changes to 1 which will subsequently allow a process to go into section Q.

Hence, no process can go to section Q until all the process executes code section P.

2.41 (a, c, d)

(a) True: (P_1) first process came executed wait(S); wait(S)

counter++; was partially executed and the process preempted. Other process came updated counter to 4.

Now P_1 came and wrote counter to 1. So (a) is correct.

(b) Not possible at least counter will be 1.

(c) True: All 5 processes executed first wait(S) and are blocked.

(d) True: Each process executed the code sequentially and counter was finally updated to 5.

Option (a), (c) and (d) are correct.

2.42 (b, c, d)

- Mutual exclusion is not violated.
- Also, there will be no deadlock because of forceful preemption of resources.
- This may lead to starvation if the process is keeps on coming and preempting each other like P_1 is preempted by P_2 and P_2 is preempted by P_3.
- Live-lock is also possible due to continuous preemption of resources.

For option (c) consider two processes P_1 and P_2 now P_1 enter the code acquires lock and resource. Now P_2 enters the else part kills P_1 and acquire R and restart P_1.

Now P_1 again acquire lock and kills the process P_2 this continues creating a live lock scenario but there is ambiguity in the code since "Release R" is not written any where so ambiguity is regarding how the process will release Resource R. According to the code, the only way to release the resource is by getting killed.

2.43 (b, c)

P_1 and P_2 can spawn two threads T_1 and T_2.

```
int x = 0; //global
Lock L1; //global
main( ) {
    foo( ); //Thread T1
    foo( ); //Thread T2
    print(x);
}
foo( ) {
a.  int y = 0;
b.  Acquire L1;
c.  x = x + 1;
d.  y = y + 1;
    Release L1;
    print(y);
}
```

- Let P_1 executed T_1 and in foo( ).

(a) $P_1 - T_1$ X: ~~0~~ 1

(b) $P_1 - T_1$

(c) $x = x + 1$ $(P_1 - T_1)$

.... Preempt T_1 of P_1.

Similarly, perform thread T_2 of P_1 then $x = 2$.

Now, if we similarly perform both threads of P_2 then x will be maximum 2.

Note: It is not mentioned about IPC or concurrent mechanism so the process will not shared the global variable. Each process will have its on own address space so the maximum value of x by each process can be 2 only.

Note: But as we know every foo call will have its own copy of variable y so y can not be more than 1 in any case.

2.44 (c)

Inorder to get the required output only semaphore S_1 should be initialized to 1, other semaphore should be initialized to 0.

So, $S_1 = 1;\ S_2 = 0;\ S_3 = 0$

Answer is option (c).

2.45 (c)

I-1: s is a binary semaphore initialized to 1. With binary semaphore, there can't be two threads that can simultaneously access the critical section, Thus, each thread with run incr( ) or decr( ) one by one.

Hence, only possible value of X at the end is 12.

I-2: With counting semaphore of value 2. Two processes at a time can access critical section.

Note:

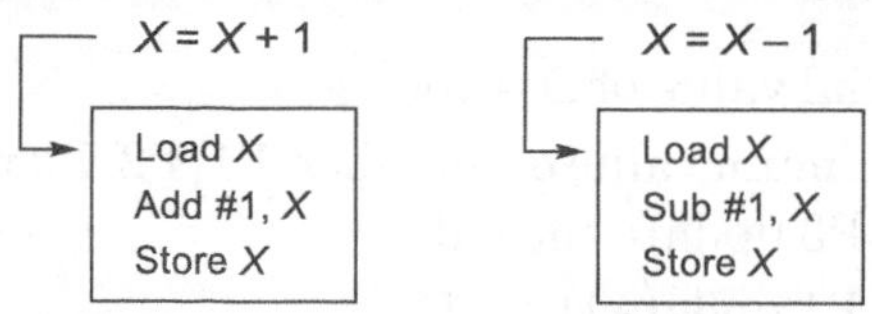

In order to minimize X value, we need to maximize the impact of decr( ) and minimize the impact of incr( ).

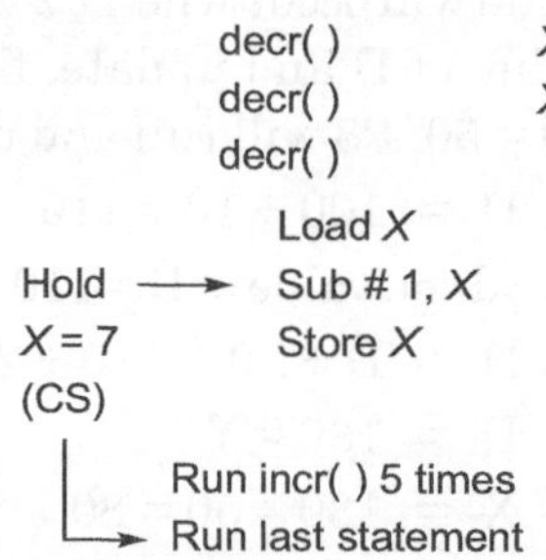

Thus, finally $\boxed{X = 7}$ will get stored.

3 Deadlock

3.1 An operating system contains 3 user processes each requiring 2 units of resource R. the minimum number of units of r such that no deadlocks will ever arise is

(a) 3 (b) 5

(c) 4 (d) 6 **[1997 : 2 M]**

3.2 A computer has six tape drives, with n processes competing for them. Each process may need two drives. What is the maximum value of n for the system to be deadlock free?

(a) 6 (b) 5

(c) 4 (d) 3 **[1998 : 1 M]**

3.3 Which of the following is NOT a valid deadlock prevention scheme?

(a) Release all resources before requesting a new resource.

(b) Number the resources uniquely and never request a lower numbered resource than the last one requested.

(c) Never request a resource after releasing any resource

(d) Request and all required resources be allocated before execution. **[2000 : 2 M]**

3.4 In a certain operating system, deadlock prevention is attemped using the following scheme. Each process is assigned a unique timestamp, and is restarted with the same timestamp if killed. Let P_h be the process holding a resource R, P_r be a process requesting for the same resource R, and $T(P_h)$ and $T(P_r)$ be their timestamps respectively. The decision to wait or preempt one of the processes is based on the following algorithm.

```
if T(Pr) < T(Ph) then
kill Pr
else
wait
```

Which one of the following is TRUE?

(a) The scheme is deadlock-free, but not starvation-free

(b) The scheme is not deadlock-free, but starvation-free

(c) The scheme is neither deadlock-free nor starvation-free

(d) The scheme is both deadlock-free and starvation-free **[2004 : 2 M]**

3.5 Suppose n processes, $P_1, ..., P_n$ share m identical resource units, which can be reserved and released one at a time. The maximum resource requirement of process P_i is s_i where $s_i > 0$. Which one of the following is a sufficient condition for ensuring that deadlock does not occur?

(a) $\forall i, s_i < m$ (b) $\forall i, s_i < n$

(c) $\sum_{i=1}^{n} s_i < (m + n)$ (d) $\sum_{i=1}^{n} s_i < (m * n)$

[2005 : 2 M]

3.6 Consider the solution to the bounded buffer producer/consumer problem by using general semaphores S, F, and E. The semaphore S is the mutual exclusion semaphore initialized to 1. The semaphore F corresponds to the number of free slots in the buffer and is initialized to N. The semaphore E corresponds to the number of elements in the buffer and is initialized to 0.

Producer Process	Consumer Process
Produce an item;	Wait (E);
Wait (F);	Wait (S);
Wait (S);	Remove an item from the buffer;
Append the item to the buffer;	Signal (S);
Signal (S);	Signal (F);
Signal (E);	Consumer the item;

Which of the following interchange operations may result in a deadlock?

I. Interchanging Wait (F) and Wait (S) in the Producer process.

II. Interchanging Signal (S) and Signal (F) in the Consumer process.

(a) I only (b) II only

(c) Neither I nor II (d) Both I and II

[2006 : 2 M]

3.7 Consider the following snapshot of a system running n processes. Process i is holding x_i instances of a resource R, for $1 \le i \le n$. Currently, all instances of R are occupied. Further, for all i, process i has placed a request for an additional y_i instances while holding the x_i instances it already has. There are exactly two processes p and q such that $y_p = y_q = 0$. Which one of the following can serve as a necessary condition to guarantee that the system is not approaching a deadlock?

(a) $\min(x_p, x_q) < \max_{k \neq p, q} y_k$
(b) $x_p + x_q \geq \max_{k \neq p, q} y_k$
(c) $\min(x_p, x_q) < 1$
(d) $\min(x_p, x_q) > 1$ **[2006 : 2 M]**

3.8 A single processor system has three resource types X, Y, and Z, which are shared by three processes. There are 5 units of each resource type. Consider the following scenario, where the column alloc denotes the number of units of each resource type allocated to each process, and the column request denotes the number of units of each resource type requested by a process in order to complete execution. Which of these processes will finish LAST?

	alloc XYZ	request XYZ
P0	121	103
P1	201	012
P2	221	120

(a) P0 (b) P1
(c) P2
(d) None of the above, since the system is in a deadlock **[2007 : 2 M]**

3.9 The following is a code with two threads, producer and consumer, that can run in parallel. Further, S and Q are binary semaphores equipped with the standard P and V operations.

```
semaphore S = 1, Q = 0;
integer x ;
producer :              consumer :
while (true) do         While (true) do
P(S) ;                  P(Q) ;
x = produce( ) ;        consume (x);
V(Q) ;                  V(S) ;
done                    done
```

Which of the following is TRUE about the program above?

(a) The process can deadlock
(b) One of the threads can starve
(c) Some of the items produced by the producer may be lost
(d) Values generated and stored in 'x' by the producer will always be consumed before the producer can generate a new value **[2008 : 2 M]**

3.10 An operating system implements a policy that requires a process to release all resources before making a request for another resource.
Select the TRUE statement from the following:

(a) Both starvation and deadlock can occur
(b) Starvation can occur but deadlock cannot occur
(c) Starvation cannot occur but deadlock can occur
(d) Neither starvation nor deadlock can occur **[2008 : 2 M]**

3.11 Which of the following is NOT true of deadlock prevention and deadlock avoidance schemes?

(a) In deadlock prevention, the request for resources is always granted if the resulting state is safe
(b) In deadlock avoidance, the request for resources is always granted if the resulting state is safe
(c) Deadlock avoidance is less restrictive than deadlock prevention
(d) Deadlock avoidance requires knowledge of resource requirements a priori **[2008 : 2 M]**

3.12 Consider a system with 4 types of resources R1 (3 units), R2 (2 units), R3 (3 units), R4 (2 units). A non-preemptive resource allocation policy is used. At any given instance, a request is not entertained if it cannot be completely satisfied. Three processes P1, P2, P3 request the resources as follows if executed independently.

Process P1 :	Process P2 :	Process P3 :
t = 0 : requests 2 units of R2	t = 0 : requests 2 units of R3	t = 0 : requests 1 unit of R4
t = 1 : requests 1 unit of R3	t = 2 : requests 1 unit of R4	t = 2 : requests 2 units of R1
t = 3 : requests 2 units of R1	t = 4 : requests 1 unit of R1	t = 5 : releases 2 units of R1
t = 5 : releases 1 unit of R2 and 1 unit of R1	t = 6 : releases 1 unit of R3	t = 7 : requests 1 unit of R2
t = 7 : releases 1 unit of R3	t = 8 : Finishes	t = 8 : requests 1 unit of R3
t = 8 : requests 2 units of R4		t = 9 : Finishes
t = 10 : Finishes		

Which one of the following statements is TRUE if all three processes run concurrently starting at time t = 0?

(a) All processes will finish without any deadlock
(b) Only P1 and P2 will be in deadlock
(c) Only P1 and P3 will be in deadlock
(d) All three processes will be in deadlock

[2009 : 2 M]

3.13 A system has n resources $R_0, R_1, ..., R_{n-1}$, and k processes $P_0, P_1, ..., P_{k-1}$. The implementation of the resource request logic of each process P_i, is as follows:

```
if (i%2 = = 0) {
    if (i < n) request R_i ;
    if (i + 2 < n) request R_(i+2);
}
else {
    if (i < n) request R_(n-i);
    if (i + 2 < n) request R_(n-i-2);
}
```

In which one of the following situations is a deadlock possible?

(a) n = 40, k = 26 (b) n = 21, k = 12
(c) n = 20, k = 10 (d) n = 41, k = 19

[2010 : 2 M]

3.14 An operating system uses the Banker's algorithm for deadlock avoidance when managing the allocation of three resource types *X*, *Y*, and *Z* to three processes P0, P1, and P2. The table given below presents the current system state. Here, the Allocation matrix shows the current number of resources of each type allocated to each process and the Max matrix shows the maximum number of resources of each type required by each process during its execution.

	Allocation			Max		
	X	*Y*	*Z*	*X*	*Y*	*Z*
P0	0	0	1	8	4	3
P1	3	2	0	6	2	0
P2	2	1	1	3	3	3

There are 3 units of type *X*, 2 units of type *Y* and 2 units of type *Z* still available. The system is currently in a **safe** state. Consider the following independent requests for additional resources in the current state:

REQ1: P0 requests 0 units of *X*, 0 units of *Y* and 2 units of *Z*

REQ2: P1 requests 2 units of *X*, 0 units of *Y* and 0 units of *Z*

Which one of the following is **TRUE**?

(a) Only REQ1 can be permitted.
(b) Only REQ2 can be permitted.
(c) Both REQ1 and REQ2 can be permitted.
(d) Neither REQ1 nor REQ2 can be permitted.

[2014 (Set-1) : 2 M]

3.15 A system contains three programs and each requires three tape units for its operation. The minimum number of tape units which the system must have such that deadlocks never arise is ___.

[2014 (Set-3) : 2 M]

3.16 A system has 6 identical resources and N processes competing for them. Each process can request atmost 2 resources. Which one of the following values of N could lead to a deadlock?

(a) 1 (b) 2
(c) 3 (d) 4

[2015 (Set-2) : 1 M]

3.17 Consider the following policies for preventing deadlock in a system with mutually exclusive resources.

I. Processes should acquire all their resources at the beginning of execution. If any resource is not available, all resources acquired so far are released.
II. The resources are numbered uniquely, and processes are allowed to request for resources only in increasing resource numbers.
III. The resources are numbered uniquely, and processes are allowed to request for resources only in decreasing resource numbers.
IV. The resources are numbered uniquely. A process is allowed to request only for a resource with resource number larger than its currently held resources.

Which of the above policies can be used for preventing deadlock?

(a) Any one of I and III but not II or IV
(b) Any one of I, III and IV but not II
(c) Any one of II and III but not I or IV
(d) Any one of I, II, III and IV

[2015 (Set-3) : 2 M]

3.18 A system shares 9 tape drives. The current allocation and maximum requirement of tape drives for three processes are shown below:

Process	Current Allocation	Maximum Requirement
P1	3	7
P2	1	6
P3	3	5

Which of the following best describes current state of the system?

(a) Safe, Deadlocked
(b) Safe, Not Deadlocked
(c) Not Safe, Deadlocked
(d) Not Safe, Not Deadlocked

[2017 (Set-2) : 2 M]

3.19 Consider a system with 3 processes that share 4 instances of the same resource type. Each process can request a maximum of K instances. Resource instances can be requested and released only one at a time. The largest value of K that will always avoid deadlock is ________. **[2018 : 1 M]**

3.20 In a system, there are three types of resources: E, F and G. Four processes P_0, P_1, P_2 and P_3 execute concurrently. At the outset, the processes have declared their maximum resource requirements using a matrix named Max as given below. For example Max[P_2, F] is the maximum number of instances of F that P_2 would require. The number of instances of the resources allocated to the various processes at any given state is given by a matrix named Allocation.

Consider a state of the system with the Allocation matrix as shown below and in which 3 instances of E and 3 instances of F are the only resources available.

Allocation			
	E	F	G
P_0	1	0	1
P_1	1	1	2
P_2	1	0	3
P_3	2	0	0

Max			
	E	F	G
P_0	4	3	1
P_1	2	1	4
P_2	1	3	3
P_3	5	4	1

From the perspective of deadlock avoidance, which one of the following is true?

(a) The system is in safe state.

(b) The system is not in safe state, but would be safe if one more instance of E were available.

(c) The system is not in safe state, but would be safe if one more instance of F were available.

(d) The system is not in safe state, but would be safe if one more instance of G were available.

[2018 : 2 M]

3.21 Which of the following statements is/are TRUE with respect to deadlocks?

(a) Circular wait is a necessary condition for the formation of deadlock.

(b) In a system where each resource has more than one instance, a cycle in its wait-for graph indicates the presence of a deadlock.

(c) If the current allocation of resources to processes leads the system to unsafe state, then deadlock will necessarily occur.

(d) In the resource-allocation graph of a system, if every edge is an assignment edge, then the system is not in deadlock state.

[2022 : 1 M]

■■■■

Answers Deadlock

3.1	(c)	**3.2**	(b)	**3.3**	(c)	**3.4**	(a)	**3.5**	(c)	**3.6**	(a)	**3.7**	(b)	**3.8**	(c)	**3.9**	(d)
3.10	(b)	**3.11**	(a)	**3.12**	(a)	**3.13**	(b)	**3.14**	(b)	**3.15**	(7)	**3.16**	(*)	**3.17**	(d)	**3.18**	(b)
3.19	(2)	**3.20**	(a)	**3.21**	(a, d)												

Explanations Deadlock

3.1 (c)

There are 3 processes, and demands '2' resources. If any one gets equal resources as it's demand, deadlock won't occur, and according to this question 4 resources are enough.

3.2 (b)

Every process needs 2 resources. With 5 resources atleast one can get whole 2 and executes. So after execution it frees 2 resources which can be used by other processes.

P_1	P_2	P_3	P_4	P_5
1	1	1	1	1

3.3 (c)

It is not a valid deadlock prevention scheme.

3.4 (a)

Given a deadlock prevention scheme, where each process is assigned a unique time stamp and if killed, it get restarted with same time stamp.

P_h : Process holding resource R.

P_r : Process requesting a resource R.

Now, consider the following scenario, where P_1, P_2, ... P_n are n-processes with given time stamp.

Process	P_1	P_2	P_3	P_4	P_5	....	P_n
Time stamp	10	20	30	40	50	....	....

Case-I: Let P_3 is holding resource R ($\therefore P_h = P_3$) and P_5 is requesting for resource R (so, $P_r = P_5$)

$\therefore$ Condition

if $T(P_r) < T(P_h)$

i.e. $T(P_5) < T(P_3)$

50 < 30 // False

Hence P_5 will wait for P_3.

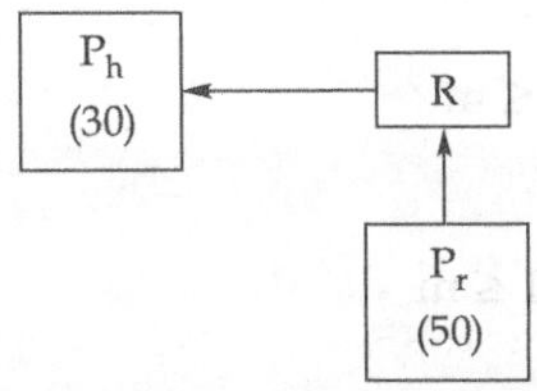

$\Rightarrow P_r$ will wait

Case-II: Now consider P_3 is requesting for resource R currently hold by process P_5, so checking condition.

if $T(P_r) < T(P_h)$

i.e. $T(P_3) < T(P_5)$

30 < 50 // True

$\therefore$ Kill P_r

So, P_3 will be killed and P_5 get resource R now suppose after sometime.

Process P_3 again requesting for resource R let now R is hold by P_6

Again checking

if $T(P_r) < T(P_h)$

30 < 60 // True

$\Rightarrow$ P_3 will get killed

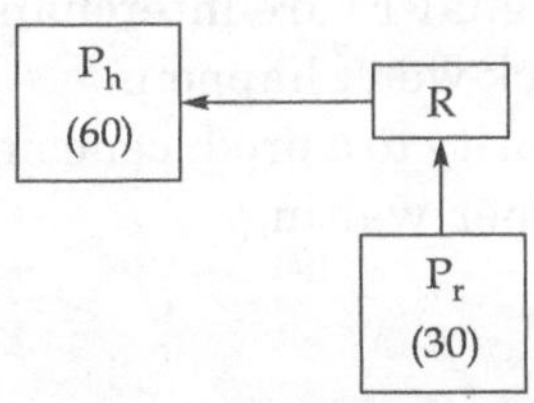

$\Rightarrow$ Kill P_r and restart with same time stamp, i.e. 30

P_3 will be killed again and similarly if P_7, P_8 ..... so on keep on arriving and holding resource R then innocent process P_3 may go to indefinite waiting which lead to starvation.

So, option (a) is answer.

3.5 (c)

Total number of resources shared by process P_1, ... P_n = m

Maximum resource requirement of process $P_i = S_i$

i.e.

Process	Requirement
P_1	S_1
P_2	S_2
⋮	
P_n	S_n

So the worst case allocation of resources for deadlock prevention. So, that deadlock do not occur is:

[Allot the resources to the processes, one less than the actual requirement (worst case deadlock will occur) + 1 (for no deadlock)]

So , we have

$(S_1 - 1) + (S_2 - 1) \ldots (S_n - 1) + 1 \le m$

$$\sum_{i=1}^{n} (S_i - 1) + 1 \le m$$

$$\sum_{i=1}^{n} S_i - \sum_{i=1}^{n} 1 + 1 \le m$$

$$\sum_{i=1}^{n} S_i - n + 1 \le m$$

$$\sum_{i=1}^{n} S_i \le m + n - 1$$

$$\sum_{i=1}^{n} S_i < (m + n)$$

3.6 (a)

Suppose the slots are full → F = 0.
Now, if Wait(F) and Wait(S) are interchanges and Wait(S) succeeds, The producer will wait for Wait(F) which is never going to succeed as Consumer would be waiting for Wait(S).
So, deadlock can happen.
If Signal(S) and Signal(F) are interchanged in Consumer, deadlock won't happen.
It will just give priority to a producer compared to the next consumer waiting.

3.7 (b)

Total number of processes = n
Process i is holding x_i instances of a resource R for $1 \le i \le n$.
Process i holding resource x_i and additional required y_i. There are two processes p and q such that $y_p = y_q = 0$.
For no dead lock can occur

$$x_p + x_q \ge (\max_{k \ne p,q} y_k)$$

3.8 (c)

Consider the matrix

	alloc XYZ	request XYZ
P_0	121	103
P_1	201	012
P_2	221	120

So the available matrix is (0, 1, 2) and the requirement of process P_1 is (0, 1, 2) so P_1 can use the available resources after releasing the resources the matrix becomes:

	alloc XYZ	request XYZ
P_0	121	103
P_1	000	000
P_2	221	120

In this time available matrix is (2 1 3) and required matrix of P_0 is (1 0 3) so P_0 uses the resources of the releasing the resource. Tthe available matrix is (3, 3, 4) so P_2 will use the resources, P_2 will finish last.

3.9 (d)

Consumer can consume only once the producer has produced the item, and producer can produce (except the first time) only once the consumer has consumed the item.

3.10 (b)

Releasing all resources before making a request for another resource is a technique of deadlock prevention. So, deadlock cannot occur.
However, we cannot say anything about starvation, it may occur or may not.

3.11 (a)

In deadlock prevention, the request for a resource may not be granted even if the resulting state is safe.

3.12 (a)

t	R1(3)	R2(2)	R3(3)	R1(2)
0	3	0	1	1
1	3	0	0	1
2	1	0	0	0
3	1	0	0	0
4	0	0	0	0
5	1	1	0	0
6	1	1	1	0
7	1	0	2	0
8	2	0	2	1
9	2	1	3	0
10				

3.13 (b)

Logic in question:
If both odd and even type of process need same resource, then deadlock may occur.

For even processes:

Request : R_i and R_{i+2}

For odd processes:

Request : R_{n-1} and R_{n-i-2}

(a)

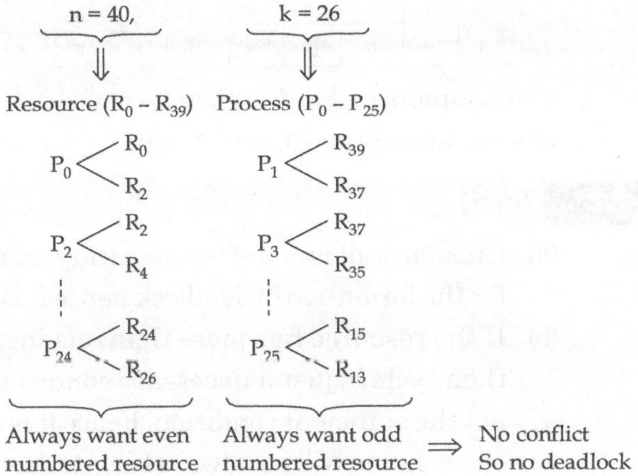

(b)

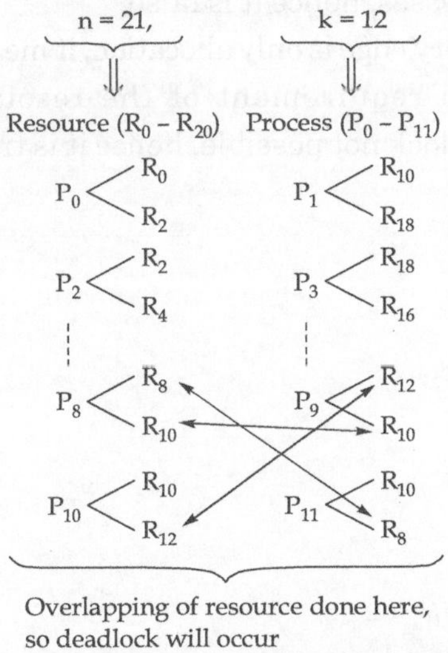

(c)

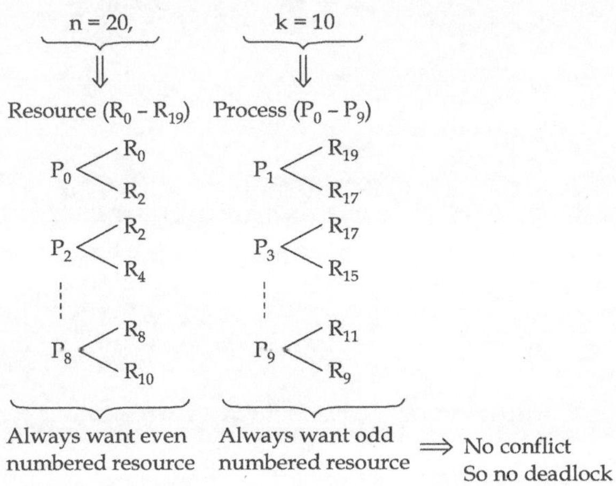

(d)

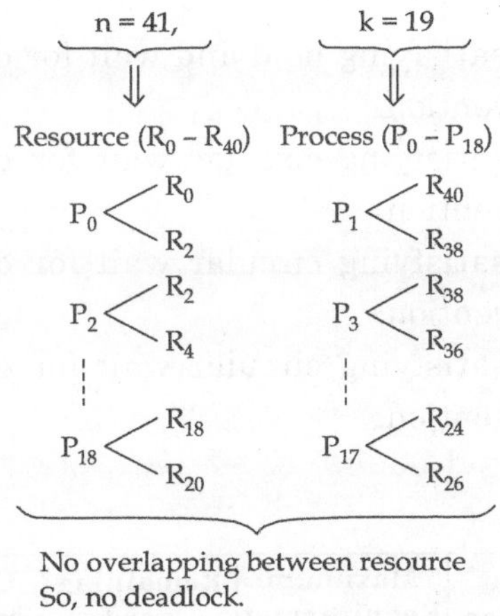

3.14 (b)

	Allocation	Max	Need	Need with additional requests
	X Y Z	X Y Z	X Y Z	X Y Z
P0	0 0 1	8 4 3	8 4 2	8 4 4 (REQ1)
P1	3 2 0	6 2 0	3 0 0	5 0 0 (REQ2)
P2	2 1 1	3 3 3	1 2 2	1 2 2

Available: X Y Z

3 2 2 → P2 can satisfy it needs

2 1 1 → P2 releases its allocation

5 3 3 → P1 satisfies it needs

3 2 0 → P1 releases its allocation

8 5 3 → P0 can not satisfy with additional requests

∴ Only REQ2 can be permitted.

3.15 (7)

Number of programs = 3

(P_1, P_2, P_3)

Tape required per program = 3

Allocating tape units, so that deadlock does not occur:

$$\underbrace{\overset{P_1}{(3-1)} + \overset{P_2}{(3-1)} + \overset{P_3}{(3-1)} + 1}_{\Downarrow \; 7}$$

Answer is 7.

3.16 (*)

N is number of processes, 6 identical resources

$N \geq 6 \times (2 - 1)$

If $N \geq 6 \Rightarrow$ It could lead to a deadlock

No option matches with $N \geq 6$

3.17 (d)

I. Dissatisfying hold and wait for deadlock prevention.
II. Dissatisfying circular wait for deadlock prevention.
III. Dissatisfying circular wait for deadlock prevention.
IV. Dissatisfying circular wait for deadlock prevention.

3.18 (b)

P.id	Current Allocation	Maximum Requirement	Remaining need	Current available
P1	3	7	4	9 − 7 = 2
P2	1	6	5	5
P3	3	5	2	8
				9

Safe sequence ⇒ P3 → P1 → P2
Safe and Not deadlocked

3.19 (2)

P_1 P_2 P_3

No deadlock

Maximum each process can request for 2 resources so, that there will not be any deadlock, because we, have only 4 resource available.
So, K value is '2'.

3.20 (a)

	Max need			Current allocation			Current available			Remaining need		
	E	F	G	E	F	G	E(3)	F(3)	G(0)	E	F	G
P_0	4	3	1	1	0	1	4	3	1	3	3	0
P_1	2	1	4	1	1	2	5	3	4	1	0	2
P_2	1	3	3	1	0	3	6	4	6	0	3	0
P_3	5	4	1	2	0	0	8	4	6	3	4	1

Safe sequence: P_0, P_2, P_1, P_3
Hence, system is in safe state.

3.21 (a, d)

(a) Circular wait is one of the necessary condition for the formation of deadlock hence it is true.
(b) If the resource has more than one instance then cycle is just a necessary condition but not the sufficient condition, hence it is false.
(c) Unsafe state will not always leads to deadlock, because it depends on behaviour of the processes, hence it is false.
(d) If every edge is only allocation, it means there is no requirement of the resources, so deadlock not possible, hence it is true.

■■■■

4 Memory Management and Virtual Memory

4.1 In a paged segmented scheme of memory management, the segment table itself must have a page table because
(a) the segment table is often too large to fit in one page
(b) each segment is spread over a number of pages
(c) segment tables point to page table and not to the physical locations of the segment
(d) the processor's description base register points to a page table **[1995 : 1 M]**

4.2 The capacity of a memory unit is defined by the number of words multiplied by the number of bits/word. How many separate address and data lines are needed for a memory of 4 K × 16?
(a) 10 address, 16 data lines
(b) 11 address, 8 data lines
(c) 12 address, 16 data lines
(d) 12 address, 12 data lines **[1995 : 2 M]**

4.3 The address sequence generated by tracing a particular program executing in a pure demand paging system with 100 records per page with 1 free main memory frame is recorded as follows. What is the number of page faults?
0100, 0200, 0430, 0499, 0510, 0530, 0560, 0120, 0220, 0240, 0260, 0320, 0370
(a) 13 (b) 8
(c) 7 (d) 10 **[1995 : 2 M]**

4.4 In a virtual memory system the address space specified by the address lines of the CPU must be ____ than the physical memory size and ____ than the secondary storage size.
(a) smaller, smaller (b) smaller, larger
(c) larger, smaller (d) larger, larger **[1995 : 2 M]**

4.5 A ROM is used to store the table for multiplication of two 8-bit unsigned integers. The size of ROM required is
(a) 256 × 16 (b) 64 K × 8
(c) 4K × 16 (d) 64 K × 16 **[1996 : 1 M]**

4.6 A 1000 Kbyte memory is managed using variable partitions but no compaction. It currently has two partitions of sizes 200 Kbytes and 260 Kbytes respectively. The smallest allocation request in Kbytes that could be denied is for
(a) 151 (b) 181
(c) 231 (d) 541 **[1996 : 2 M]**

4.7 Locality of reference implies that the page reference being made by a process
(a) will always be to the page used in the previous page reference
(b) is likely to be to one of the pages used in the last few page references
(c) will always be to one of the pages existing in memory
(d) will always lead to a page fault **[1997 : 1 M]**

4.8 Thrashing
(a) Reduces page I/O
(b) Decreases the degree of multipro-gramming
(c) Implies excessive page I/O
(d) Improve the system performance **[1997 : 1 M]**

4.9 Dirty bit for a page in a page table
(a) Helps avoid unnecessary writes on paging device
(b) Helps maintain LRU information
(c) Allows only read on a page
(d) None of the above **[1997 : 1 M**

4.10 In a resident–OS computer, which of the following system software must reside in the main memory under all situations?
(a) Assembler (b) Linker
(c) Loader (d) Compiler **[1998 : 1 M]**

4.11 The overlay tree for a program is as shown below:

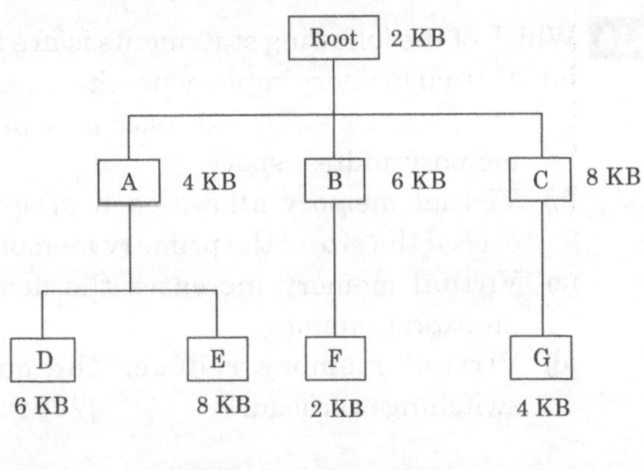

What will be the size of the partition (in physical memory) required to load (and run) this program?
(a) 12 KB (b) 14 KB
(c) 10 KB (d) 8 KB

[1998 : 2 M]

4.12 If an instruction takes i microseconds and a page fault takes an additional j microseconds, the effective instruction time if on the average a page fault occurs every k instruction is
(a) $i + \frac{j}{k}$ (b) $i + j * k$
(c) $\frac{i + j}{k}$ (d) $(i + j) * k$

[1998 : 2 M]

4.13 Which of the following is/are advantage of virtual memory?
(a) Faster access to memory on an average.
(b) Processes can be given protected address spaces.
(c) Linker can assign addresses independent of where the program will be loaded in physical memory.
(d) Programs larger than the physical memory size can be run. **[1999 : 2 M]**

4.14 Suppose the time to service a page fault is on the average 10 milliseconds, while a memory access takes 1 microsecond. Then a 99.99% hit ratio results in average memory access time of
(a) 1.9999 milliseconds
(b) 1 millisecond
(c) 9.999 microseconds
(d) 1.9999 microseconds **[2000 : 2 M]**

4.15 Consider a machine with 64 MB physical memory and a 32-bit virtual address space. If the page size of 4 KB, what is the approximate size of the page table?
(a) 16 MB (b) 8 MB
(c) 2 MB (d) 24 MB

[2001 : 2 M]

4.16 Which of the following statements is/are false?
(a) Virtual memory implements the translation of a program's address space in to physical memory address space.
(b) Virtual memory allows each program to exceed the size of the primary memory.
(c) Virtual memory increases the degree of multiprogramming.
(d) Virtual memory reduces the context switching overhead. **[2001 : 1 M]**

4.17 The process of assigning load addresses to the various parts of the program and adjusting the code and data in the program to reflect the assigned addresses is called
(a) assembly (b) parsing
(c) relocation (d) symbol resolution

[2001 : 1 M]

4.18 Consider a virtual memory system with FIFO page replacement policy. For an arbitrary page access pattern, increasing the number of page frames in main memory will
(a) always decrease the number of page faults
(b) always increase the number of page faults
(c) sometimes increase the number of page faults
(d) never affect the number of page faults

[2001 : 1 M]

4.19 Which of the following is not a form of memory?
(a) Instruction cache
(b) Instruction register
(c) Instruction opcode
(d) Translation look a side buffer

[2002 : 1 M]

4.20 The optimal page replacement algorithm will, select the page that
(a) Has not been used for the longest time in the past
(b) Will not be used for the longest time in the future
(c) Has been used least number of times
(d) Has been used most number of times

[2002 : 1 M]

4.21 Dynamic linking can cause security concerns because
(a) Security is dynamic
(b) The path for searching dynamic libraries is not known till runtime
(c) Linking is insecure
(d) Cryptographic procedures are not available for dynamic linking **[2002 : 2 M]**

4.22 In a system with 32 bit virtual addresses and 1 KB page size, use of one-level page tables for virtual to physical address translation is not practical because of
(a) the large amount of internal fragmentation
(b) the large amount of external fragmentation
(c) the large memory overhead in maintaining page tables
(d) the large computation overhead in the translation process **[2003 : 1 M]**

Common Data for Q. 4.23 & 4.24:

A processor uses 2-level page tables for virtual to physical address translation. Page tables for both levels are stored in the main memory. Virtual and physical addresses are both 32 bits wide. The memory is byte addressable. For virtual to physical address translation, the 10 most significant bits of the virtual address are used as index into the first level page table while the next 10 bits are used as index into the second level page table. The 12 least significant bits of the virtual address are used as offset within the page. Assume that the page table entries in both levels of page tables are 4 bytes wide. Further, the processor has a translation look-aside buffer (TLB), with a hit rate of 96%. The TLB caches recently used virtual page numbers and the corresponding physical page numbers. The processor also has a physically addressed cache with a hit ratio of 90%. Main memory access time is 10 ns, cache access time is 1 ns, and TLB access time is also 1 ns.

4.23 Assuming that no page faults occur, the average time taken to access a virtual address is approximately (to the nearest 0.5 ns)

(a) 1.5 ns (b) 2 ns
(c) 3 ns (d) 4 ns

[2003 : 2 M]

4.24 Suppose a process has only the following pages in its virtual address space: two contiguous code pages starting at virtual address 0 × 00000000, two contiguous data pages starting at virtual address 0 × 00400000, and a stack page starting at virtual address 0 × FFFFF000. The amount of memory required for storing the page tables of this process is

(a) 8 KB (b) 12 KB
(c) 16 KB (d) 20 KB

[2003 : 2 M]

4.25 Consider an operating system capable of loading and executing a single sequential user process at a time. The disk head scheduling algorithm used is First Come First Served (FCFS). If FCFS is replaced by Shortest Seek Time First (SSTF), claimed by the vendor to given 50% better benchmark results, what is the expected improvement in the I/O performance of user programs?

(a) 50% (b) 40%
(c) 25% (d) 0% **[2004 : 1 M]**

4.26 The minimum number of page frames that must be allocated to a running process in a virtual memory environment is determined by

(a) the instruction set architecture
(b) page size
(c) physical memory size
(d) number of processes in memory

[2004 : 1 M]

4.27 Consider a system with a two-level paging scheme in which a regular memory access takes 150 nanoseconds, and servicing a page fault takes 8 milliseconds. An average instruction takes 100 nanoseconds of CPU time, and two memory accesses. The TLB hit ratio is 90%, and the page fault rate is one in every 10,000 instructions. What is the effective average instruction execution time?

(a) 645 nanoseconds (b) 1050 nanoseconds
(c) 1215 nanoseconds (d) 1260 nanoseconds

[2004 : 2 M]

4.28 Consider a fully associative cache with 8 cache blocks (numbered 0-7) and the following sequence of memory block requests: 4, 3, 25, 8, 19, 6, 25, 8, 16, 35, 45, 22, 8, 3, 16, 25, 7.

If LRU replacement policy is used, which cache block will have memory block 7?

(a) 4 (b) 5
(c) 6 (d) 7 **[2004 : 2 M]**

4.29 In a virtual memory system, size of virtual address is 32-bit, size of physical address is 30-bit, page size is 4 Kbyte and size of each page table entry is 32-bit. The main memory is byte addressable. Which one of the following is the maximum number of bits that can be used for storing protection and other information in each page table entry?

(a) 2 (b) 10
(c) 12 (d) 14 **[2004 : 2 M]**

4.30 For each of the four processes P_1, P_2, P_3 and P_4. The total size in kilobytes (KB) and the number of segments are given below.

Process	Total size (in KB)	Number of segments
P_1	195	4
P_2	254	5
P_3	45	3
P_4	364	8

The page size is 1 KB. The size of an entry in the page table is 4 bytes. The size of an entry in the segment table is 8 bytes. The maximum size of a segment is 256 KB. The paging method for memory management uses two-level paging, and its storage overhead is P. The storage overhead for the segmentation method is S. The storage overhead for the segmentation and paging method is T. What is the relation among the overheads for the different methods of memory management in the concurrent execution of the above four processes?

(a) P < S < T (b) S < P < T
(c) S < T < P (d) T < S < P

[2006 : 2 M]

4.31 A CPU generates 32-bit virtual addresses. The page size is 4 KB. The processor has a translation look-aside buffer (TLB) which can hold a total of 128 page table entries and is 4-way set associative. The minimum size of the TLB tag is

(a) 11 bits (b) 13 bits
(c) 15 bits (d) 20 bits **[2006 : 2 M]**

4.32 A computer system supports 32-bit virtual addresses as well as 32-bit physical addresses. Since the virtual address space is of the same size as the physical address space, the operating system designers decide to get rid of the virtual memory entirely. Which one of the following is true?

(a) Efficient implementation of multi-user support is no longer possible
(b) The processor cache organization can be made more efficient now
(c) Hardware support for memory management is no longer needed
(d) CPU scheduling can be made more efficient now **[2006 : 2 M]**

4.33 Let a memory have four free blocks of sizes 4k, 8k, 20k, 2k. These blocks are allocated following the best-fit strategy. The allocation requests are stored in a queue as shown below:

Request No.	Request sizes	Usage Time
J1	2k	4
J2	14k	10
J3	3k	2
J4	6k	8
J5	6k	4
J6	10k	1
J7	7k	8
J8	20k	6

The time at which the request for J7 will be completed will be

(a) 16 (b) 19
(c) 20 (d) 37 **[2007 : 1 M]**

4.34 The address sequence generated by tracing a particular program executing in a pure demand paging system with 100 bytes per page is 0100, 0200, 0430, 0499, 0510, 0530, 0560, 0120, 0220, 0240, 0260, 0320, 0410.
Suppose that the memory can store only one page and if x is the address which causes a page fault then the bytes from addresses x to x + 99 are loaded on to the memory. How many page faults will occur?

(a) 0 (b) 4
(c) 7 (d) 8 **[2007 : 1 M]**

4.35 A demand paging system takes 100 time units to service a page fault and 300 time units to replace a dirty page. Memory access time is 1 time unit. The probability of a page fault is p. In case of a page fault, the probability of page being dirty is also p. It is observed that the average access time is 3 time units. Then the value of p is

(a) 0.194 (b) 0.233
(c) 0.514 (d) 0.981

[2007 : 2 M]

Common Data for Q. 4.36 & Q. 4.37:

A process has been allocated 3 page frames. Assume that none of the pages of the process are available in the memory initially. The process makes the following sequence of page references (reference string) : 1, 2, 1, 3, 7, 4, 5, 6, 3, 1.

4.36 If optimal page replacement policy is used, how many page faults occur for the above reference string?

(a) 7 (b) 8
(c) 9 (d) 10 **[2007 : 2 M]**

4.37 Least Recently Used (LRU) page replacement policy is a practical approximation to optimal page replacement. For the above reference string, how many more page faults occur with LRU than with the optimal page replacement policy?

(a) 0 (b) 1
(c) 2 (d) 3

[2007 : 2 M]

4.38 A paging scheme uses a Translation Look-aside Buffer (TLB). A TLB-access takes 10 ns and a main memory access takes 50 ns. What is the effective access time(in ns) if the TLB hit ratio is 90% and there is no page-fault?

(a) 54 (b) 60
(c) 65 (d) 75

[2008 : 1 M]

4.39 A processor uses 36 bit physical addresses and 32 bit virtual addresses, with a page frame size of 4 Kbytes. Each page table entry is of size 4 bytes. A three level page table is used for virtual-to-physical address translation, where the virtual address is used as follows:

- bits 30-31 are used to index into the first level page table,
- bits 21-29 are used to index into the second level page table
- bits 12-20 are used to index into the third level page table
- bits 0-11 are used as offset within the page

The number of bits required for addressing the next level page table (or page frame) in the page table entry of the first, second and third level page tables are respectively.

(a) 20, 20 and 20 (b) 24, 24, and 24
(c) 24, 24 and 20 (d) 25, 25 and 24

[2008 : 2 M]

4.40 Assume that a main memory with only 4 pages, each of 16 bytes, is initially empty. The CPU generates the following sequence of virtual addresses and uses the Least Recently Used (LRU) page replacement policy.

0, 4, 8, 20, 24, 36, 44, 12, 68,
72, 80, 84, 28, 32, 88, 92

How many page faults does this sequence cause? What are the page numbers of the pages present in the main memory at the end of the sequence?

(a) 6 and 1, 2, 3, 4 (b) 7 and 1, 2, 4, 5
(c) 8 and 1, 2, 4, 5 (d) 9 and 1, 2, 3, 5

[2008 : 2 M]

4.41 Match the following flag bits used in the context of virtual memory management on the **List-I (Name of the bit)** with the different purposes on the **List-II (Purpose)** of the table below:

List-I	List-II
I. Dirty	a. Page initialization
II. R/W	b. Write-back policy
III. Reference	c. Page protection
IV. Valid	d. Page replacement policy

Codes:

(a) I-d, II-a, III-b, IV-c (b) I-b, II-c, III-a, IV-d
(c) I-c, II-d, III-a, IV-b (d) I b, II-c, III-d, IV-a

[2008 : 2 M]

4.42 In which one of the following page replacement policies, Bleady's anomaly may occur?

(a) FIFO (b) Optimal
(c) LRU (d) MRU **[2009 : 1 M]**

4.43 The essential content(s) in each entry of a page table is/are

(a) virtual page number
(b) page frame number
(c) both virtual page number and page frame number
(d) access right information **[2009 : 1 M]**

4.44 A system uses FIFO policy for page replacement. It has 4 page frames with no pages loaded to begin with. The system first accesses 100 distinct pages in some order and then accesses the same 100 pages but now in the reverse order. How many page faults will occur?

(a) 196 (b) 192
(c) 197 (d) 195 **[2010 : 1 M]**

4.45 Let the page fault service time be 10 ms in a computer with average memory access time being 20 ns. If one page fault is generated for every 10^6 memory accesses, what is the effective access time for the memory?

(a) 21 ns (b) 30 ns
(c) 23 ns (d) 35 ns **[2011 : 1 M]**

4.46 Consider the virtual page reference string

1, 2, 3, 2, 4, 1, 3, 2, 4, 1

on a demand paged virtual memory system running on computer system that has main memory size of 3 page frames which are initially empty. Let LRU, FIFO and OPTIMAL denote the number of page faults under the corresponding page replacement policy. Then

(a) OPTIMAL < LRU < FIFO
(b) OPTIMAL < FIFO < LRU
(c) OPTIMAL = LRU
(d) OPTIMAL = FIFO **[2012 : 2 M]**

Linked Answer Questions 4.47 and 4.48:

A computer uses 46-bit virtual address, 32-bit physical address, and a three-level paged page table organization. The page table base register stores the base address of the first-level (T_1), which occupies exactly one page. Each entry of T_1 stores the base address of a page of the second-level table (T_2). Each entry of T_2 stores the base address of a page of the third-level table (T_3). Each entry of T_3 stores a page table entry (PTE). The PTE is 32 it in size. The processor used in the computer has a 1 MB 16-way-set associative virtually indexed physical tagged cache the cache block size is 64 bytes.

4.47 What is the size of a page in KB in this computer?

(a) 2 (b) 4
(c) 8 (d) 16 **[2013 : 2 M]**

4.48 What is the minimum number of page colours needed to guarantee that no two synonyms map to different sets in the processor cache of this computer?
(a) 2 (b) 4
(c) 8 (d) 16 **[2013 : 2 M]**

4.49 Assume that there are 3 page frames which are initially empty. If the page reference string is 1, 2, 3, 4, 2, 1, 5, 3, 2, 4, 6, the number of page faults using the *optimal replacement policy* is ________. **[2014 (Set-1) : 2 M]**

4.50 A computer has twenty physical page frames which contain pages numbered 101 through 120. Now a program accesses the pages numbered 1, 2, ... ,100 in that order, and repeats the access sequence THRICE.
Which one of the following page replacement policies experiences the same number of page faults as the optimal page replacement policy for this program?
(a) Least-recently-used (b) First-in-first-out
(c) Last-in-first-out (d) Most-recently-used
[2014 (Set-2) : 2 M]

4.51 A system uses 3 page frames for storing process pages in main memory. It uses the Least Recently Used (LRU) page replacement policy. Assume that all the page frames are initially empty. What is the total number of page faults that will occur while processing the page reference string given below?
4, 7, 6, 1, 7, 6, 1, 2, 7, 2
[2014 (Set-3) : 1 M]

4.52 Consider a paging hardware with a TLB. Assume that the entire page table and all the pages are in the physical memory. It takes 10 milliseconds to search the TLB and 80 milliseconds to access the physical memory. If the TLB hit ratio is 0.6, the effective memory access time (in milliseconds) is ________. **[2014 (Set-3) : 2 M]**

4.53 Consider a system with byte-addressable memory, 32 bit logical addresses, 4 kilobyte page size and page table entries of 4 bytes each. The size of the page table in the system in megabytes is ________. **[2015 (Set-1) : 1 M]**

4.54 Consider a main memory with five page frames and the following sequence of page references:
3, 8, 2, 3, 9, 1, 6, 3, 8, 9, 3, 6, 2, 1, 3
Which one of the following is true with respect to page replacement policies First-In-First Out (FIFO) and Least Recently Used (LRU)?
(a) Both incur the same number of page faults
(b) FIFO incurs 2 more page faults than LRU
(c) LRU incurs 2 more page faults than FIFO
(d) FIFO incurs 1 more page faults than LRU
[2015 (Set-1) : 2 M]

4.55 A computer system implements a 40 bit virtual address, page size of 8 kilobytes, and a 128-entry translation look-aside buffer (TLB) organized into 32 sets each having four ways. Assume that the TLB tag does not store any process id. The minimum length of the TLB tag in bits is ______.
[2015 (Set-2) : 1 M]

4.56 Consider six memory partitions of size 200 KB, 400 KB, 600 KB, 500 KB, 300 KB, and 250 KB, where KB refers to kilobyte. These partitions need to be allotted to four processes of sizes 357 KB, 210 KB, 468 KB and 491 KB in that order. If the best fit algorithm is used, which partitions are NOT allotted to any process?
(a) 200 KB and 300 KB
(b) 200 KB and 250 KB
(c) 250 KB and 300 KB
(d) 300 KB and 400 KB **[2015 (Set-2) : 2 M]**

4.57 A Computer system implements 8 kilobyte pages and a 32-bit physical address space. Each page table entry contains a valid bit, a dirty bit, three permission bits, and the translation. If the maximum size of the page table of a process is 24 megabytes, the length of the virtual address supported by the system is ________ bits.
[2015 (Set-2) : 2 M]

4.58 Consider a computer system with 40-bit virtual addressing and page size of sixteen kilobytes. If the computer system has a one-level page table per process and each page table entry requires 48 bits, then the size of the per-process page table is ________ megabytes. **[2016 (Set-1) : 2 M]**

4.59 Consider a computer system with ten physical page frames. The system is provided with an access sequence $(a_1, a_2, ..., a_{20}, a_1, a_2, ..., a_{20})$, where each a_i is a distinct virtual page number. The difference in the number of page faults between the last-in-first-out page replacement policy and the optimal page replacement policy is ________. **[2016 (Set-1) : 2 M]**

4.60 In which one of the following page replacement algorithms it is possible for the page fault rate to increase even when the number of allocated frames increases?

(a) LRU (Least Recently Used)
(b) OPT (Optimal Page Replacement)
(c) MRU (Most Recently Used)
(d) FIFO (First In First Out)

[2016 (Set-2) : 1 M]

4.61 Recall that Belady's anomaly is that the page-fault rate may increase as the number of allocated frames increases. Now, consider the following statements:

S1: Random page replacement algorithm (where a page chosen at random is replaced) suffers from Belady's anomaly

S2: LRU page replacement algorithm suffers from Belady's anomaly

Which of the following is CORRECT?
(a) S1 is true, S2 is true
(b) S1 is true, S2 is false
(c) S1 is false, S2 is true
(d) S1 is false, S2 is false

[2017 (Set-1) : 2 M]

4.62 Consider a process executing on an operating system that uses demand paging. The average time for a memory access in the system is M units if the corresponding memory page is available in memory and D units if the memory access causes a page fault. It has been experimentally measured that the average time taken for a memory access in the process is X units.

Which one of the following is the correct expression for the page fault rate experienced by the process?

(a) $\dfrac{(D-M)}{(X-M)}$ (b) $\dfrac{(X-M)}{(D-M)}$
(c) $\dfrac{(D-X)}{(D-M)}$ (d) $\dfrac{(X-M)}{(D-X)}$

[2018 : 1 M]

4.63 Assume that in a certain computer, the virtual addresses are 64 bits long and the physical addresses are 48 bits long. The memory is word addressable. The page size is 8 kB and the word size is 4 bytes. The Translation Look-aside Buffer (TLB) in the address translation path has 128 valid entries. At most how many distinct virtual addresses can be translated without any TLB miss?

(a) 16×2^{10} (b) 8×2^{20}
(c) 4×2^{20} (d) 256×2^{10}

[2019 : 2 M]

4.64 Consider allocation of memory to a new process. Assume that none of the existing holes in the memory will exactly fit the process's memory requirement. Hence, a new hole of smaller size will be created if allocation is made in any of the existing holes. Which one of the following statements is TRUE?
(a) The hole created by next fit is never larger than the hole created by best fit.
(b) The hole created by first fit is always larger than the hole created by next fit.
(c) The hole created by best fit is never larger than the hole created by first fit.
(d) The hole created by worst fit is always larger than the hole created by first fit.

[2020 : 1 M]

4.65 Consider a paging system that uses 1-level page table residing in main memory and a TLB for address translation. Each main memory access takes 100 ns and TLB lookup takes 20 ns. Each page transfer to/from the disk takes 5000 ns. Assume that the TLB hit ratio is 95%, page fault rate is 10%. Assume that for 20% of the total page faults, a dirty page has to be written back to disk before the required page is read in from disk. TLB update time is negligible. The average memory access time in ns (round off to 1 decimal places) is ______. **[2020 : 2 M]**

4.66 In the context of operating systems, which of the following statements is/are correct with respect to paging?
(a) Paging incurs memory overheads.
(b) Multi-level paging is necessary to support pages of different sizes.
(c) Page size has no impact on internal fragmentation.
(d) Paging helps solve the issue of external fragmentation. **[2021 (Set-1) : 1 M]**

4.67 Consider a three-level page table to translate a 39-bit virtual address to a physical address as shown below:

← 39-bit virtual address →

Level 1 offset	Level 2 offset	Level 3 offset	Page offset
9 bits	9 bits	9 bits	12 bits

The page size is 4 KB (1 KB = 2^{10} bytes) and page table entry size at every level is 8 bytes. A process *P* is currently using 2 GB (1 GB = 2^{30} bytes) virtual memory which is mapped to 2 GB of physical memory. The minimum amount of memory required for the page table of P across all levels is _______ KB.

[2021 (Set-2) : 2 M]

4.68 Which one of the following statements is FALSE?

(a) The TLB performs an associative search in parallel on all its valid entries using page number of incoming virtual address.

(b) If the virtual address of a word given by CPU has a TLB hit, but the subsequent search for the word results in a cache miss, then the word will always be present in the main memory.

(c) The memory access time using a given inverted page table is always same for all incoming virtual addresses.

(d) In a system that uses hashed page tables, if two distinct virtual addresses V1 and V2 map to the same value while hashing, then the memory access time of these addresses will not be the same.

[2022 : 2 M]

4.69 Consider a demand paging system with four page frames (initially empty) and LRU page replacement policy. For the following page reference string

7, 2, 7, 3, 2, 5, 3, 4, 6, 7, 7,1, 5, 6,1

the page fault rate, defined as the ratio of number of page faults to the number of memory accesses (rounded off to one decimal place) is _______.

[2022 : 2 M]

4.70 Consider the following two-dimensional array D in the C programming language, which is stored in row-major order:

int $D[128][128]$;

Demand paging is used for allocating memory and each physical page frame holds 512 elements of the array D. The Least Recently Used (LRU) page-replacement policy is used by the operating system. A total of 30 physical page frames are allocated to a process which executes the following code snippet:

```
for (int i = 0; i < 128; i++)
    for (int j = 0; j < 128; j++)
        D[j][i] *= 10;
```

The number of page faults generated during the execution of this code snippet is _______.

[2023 : 2 M]

4.71 Consider a computer system with 57-bit virtual addressing using multi-level tree-structured page tables with L levels for virtual to physical address translation. The page size is 4 KB (1 KB = 1024 B) and a page table entry at any of the levels occupies 8 bytes.

The value of L is _______.

[2023 : 2 M]

4.72 Consider a memory management system that uses a page size of 2 KB. Assume that both the physical and virtual addresses start from 0. Assume that the pages 0, 1, 2, and 3 are stored in the page frames 1, 3, 2, and 0, respectively. The physical address (in decimal format) corresponding to the virtual address 2500 (in decimal format) is _______.

[2024 (Set-1) : 2 M]

4.73 Which of the following tasks is/are the responsibility/responsibilities of the memory management unit (MMU) in a system with paging-based memory management?

(a) Allocate a new page table for a newly created process.

(b) Translate a virtual address to a physical address using the page table.

(c) Raise a trap when a process tries to write to a page marked with read-only permission in the page table.

(d) Raise a trap when a virtual address is not found in the page table.

[2024 (Set-2) : 1 M]

4.74 Consider a 32-bit system with 4 KB page size and page table entries of size 4 bytes each. Assume 1 KB = 2^{10} bytes. The OS uses a 2-level page table for memory management, with the page table containing an outer page directory and an inner page table. The OS allocates a page for the outer page directory upon process creation. The OS uses demand paging when allocating memory for the inner page table, i.e., a page of the inner page table is allocated only if it contains at least one valid page table entry.

An active process in this system accesses 2000 unique pages during its execution, and none of the pages are swapped out to disk. After it completes the page accesses, let X denote the minimum and Y denote the maximum number of pages across the two levels of the page table of the process.

The value of $X + Y$ is _______.

[2024 (Set-2) : 2 M]

■■■■

Answers Memory Management and Virtual Memory

4.1	(a)	**4.2**	(c)	**4.3**	(c)	**4.4**	(c)	**4.5**	(d)	**4.6**	(b)	**4.7**	(b)	**4.8**	(c)	**4.9**	(a)
4.10	(c)	**4.11**	(b)	**4.12**	(a)	**4.13**	(d)	**4.14**	(d)	**4.15**	(c)	**4.16**	(d)	**4.17**	(c)	**4.18**	(c)
4.19	(c)	**4.20**	(b)	**4.21**	(b)	**4.22**	(c)	**4.23**	(d)	**4.24**	(c)	**4.25**	(d)	**4.26**	(a)	**4.27**	(d)
4.28	(b)	**4.29**	(d)	**4.30**	(b)	**4.31**	(c)	**4.32**	(a)	**4.33**	(b)	**4.34**	(c)	**4.35**	(a)	**4.36**	(a)
4.37	(c)	**4.38**	(c)	**4.39**	(b)	**4.40**	(b)	**4.41**	(d)	**4.42**	(a)	**4.43**	(b)	**4.44**	(a)	**4.45**	(b)
4.46	(b)	**4.47**	(c)	**4.48**	(c)	**4.49**	(7)	**4.50**	(d)	**4.51**	(6)	**4.52**	(122)	**4.53**	(4)	**4.54**	(a)
4.55	(22)	**4.56**	(a)	**4.57**	(36)	**4.58**	(384)	**4.59**	(1)	**4.60**	(d)	**4.61**	(b)	**4.62**	(b)	**4.63**	(d)
4.64	(c)	**4.65**	(154.5)	**4.66**	(a, d)	**4.67**	(4108)	**4.68**	(c)	**4.69**	(0.6)	**4.70**	(4096)	**4.71**	(5)	**4.72**	(6596)
4.73	(b, c, d)	**4.74**	(1028)														

Explanations Memory Management and Virtual Memory

4.1 (a)

Segment table is often too large to fit in one page. So segment table itself must have a page table in a paged segmented scheme.

4.2 (c)

Size of memory $= 4\text{K} \times 16$ bits

$= 2^{12} \times 16$ bits

$\Rightarrow$ 12 address lines and 16 data lines are needed.

4.3 (c)

Given that only one frame is available which can hold 100 records at one time (0-99).

100	200	430	499	510	530
×	×	×	✓	✓	×
560	120	220	240	260	320
✓	×	×	✓	✓	×
370					
✓					

4.4 (c)

In virtual memory system, <u>Address Space</u> specified by the <u>address lines</u> of CPU must be <u>larger</u> than the <u>physical memory</u> size and <u>smaller</u> than the <u>secondary storage</u> size.

4.5 (d)

n-bit multiplication storage required

$= 2^{(n+n)} \times (n + n)$

For 8-bit data $= 2^{(8+8)} \times (8 + 8)$ bit

$= 2^{16} \times 16$ bit

$= 64\text{ K} \times 16$ bit

4.6 (b)

Initially memory available:

1000 K Byte

After allocation 200 K byte and 260 K byte in worst case.

1000 K Byte

180	200 KB	180	260 KB	180

So, in worst case partition left with size 180 KB that are not contiguous. So, request for allocation request 181 will be denied because it is asking smallest request.

4.7 (b)

Locality of reference is a phenomenon describing the same value, related storage location being frequently accessed.

4.8 (c)

Thrashing: After reaching the maximum number of jobs allowed in main memory. CPU efficiency drastically getting down this situation is called Trashing.

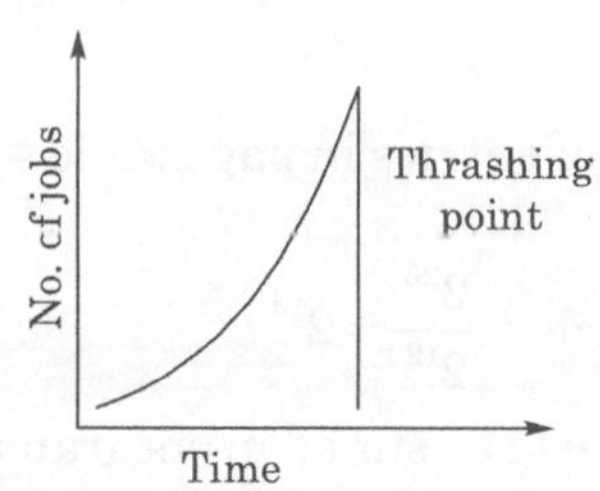

Thrashing implies excessive page I/O.

4.9 (a)

Which page has Dirty bit only those pages being written back in main memory, so it avoids unnecessary writing.

4.10 (c)

Loader is used to load the program into main memory.

4.11 (b)

Maximum space that can be needed at a time is 14 kB (using E). The maximum space to load the program is decided by the maximum size taken by rootine call sequence from the root. i.e., Root → A → E.

4.12 (a)

EMAT = p × s + (1 – p) × m

p: page fault rate

s: page fault service time

m: main memory access time

$$\Rightarrow \frac{(i+j)}{k} + \left(1 - \frac{1}{k}\right) \times i = i + \frac{j}{k}$$

4.13 (d)

Virtual memory allows the user can run the programs larger than the physical memory size.

4.14 (d)

Average memory access time = [(% of page miss) * (time of service a page fault) + (% of page hit) * (memory access time)]/100

So, average memory access time in microseconds is

$$\frac{(99.99 * 1 + 0.01 * 10 * 1000)}{100}$$

$$= \frac{(99.99 + 100)}{1000}$$

$$= \frac{199.99}{1000} = 1.9999\ \mu\text{sec}$$

4.15 (c)

$$\text{Number of entries in page table} = \frac{2^{32}}{2^{12}} = 2^{20}$$

$$\text{Frame size} = \frac{2^{26}}{2^{12}} = 2^{14}$$

∴ PT have to be stored in one frame so entry size must be 2 bytes, hence size of PT

$$= 2^{20} \times 2 = 2^{21} = 2\text{ MB}$$

4.16 (d)

In a system with virtual memory context switch includes extra overhead in switching of address spaces.

4.17 (c)

Relocation can change the assigned address of data and code in the program.

4.18 (c)

Bleady's Anamoly increasing number of frames = increase in page faults.

4.19 (c)

Instruction cache, instruction register and TLB are memories, but instruction opcode is the part of instruction that specify the operation to perform.

4.20 (b)

Optimal page replacement algorithm select a page that will not be used in the longest time in future.

4.21 (b)

In dynamic linking path for searching dynamic libraries is not known till runtime i.e. keep on changing, which can create security concern.

4.22 (c)

Given, Page Size = 1 KB

= 2^{10} bytes

Virtual address is 32 bit long

So required number of pages

$$= \frac{2^{32}}{2^{10}} = 2^{22}$$

2^{22} is a very large so the large memory overhead in maintaining page tables.

4.23 (d)

Average time taken to access a virtual address

$$= \left[\left(\frac{96}{100}\right)(1 + 0.9 \times 1 + 0.1 \times (1 + 10))\right]$$

$$+ [(0.04)(21 + 0.9 \times 1 + 0.1 \times (1 + 10))]$$

$$= [(.96)(1 + 0.9 + 0.1 \times 11)]$$

$$+ [(0.04)(21 + 0.9 + 0.1 \times 11)]$$

$$= 3.87\text{ ns} = 4\text{ ns}$$

4.24 (c)

Logical address = 32 bits

P_1	P_2	d
10 bits	10 bits	12 bits

P_1 = 1st level page table
P_2 = 2nd level page table
d = Page offset

First virtual address given

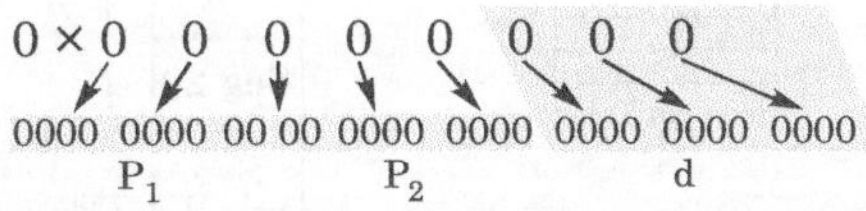

Address will be found in first level page table first block.

Second virtual address given

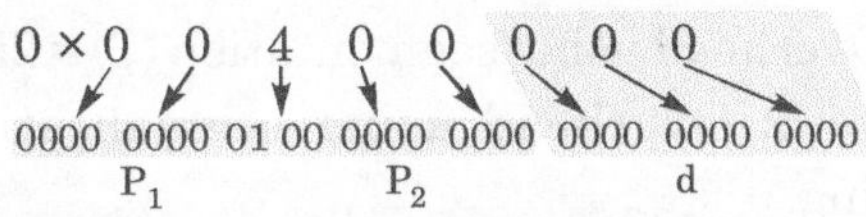

Address will be found in first level page table second block.

Third virtual address given

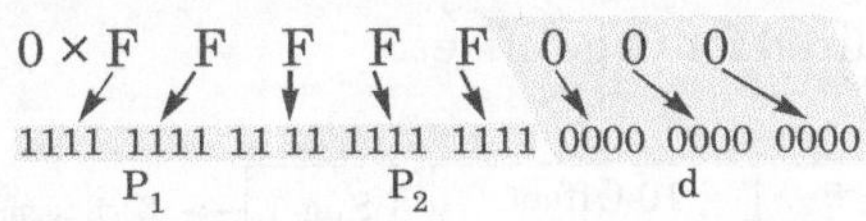

Address will be found in first level page table last block.

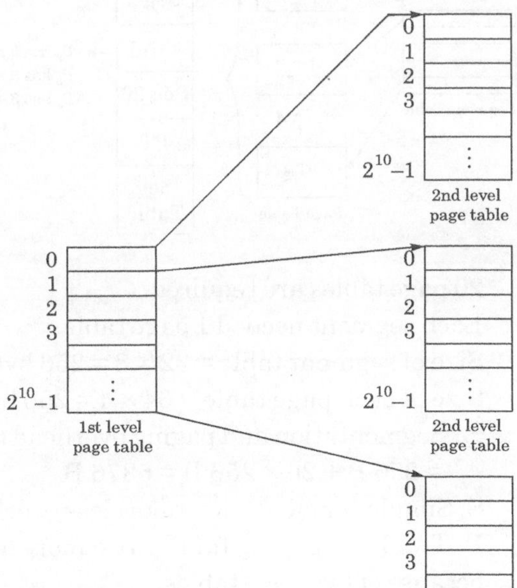

To execute the process we need to bring all the above 4 page tables in main memory.

$$4 \times 2^{10} \times 4\text{ B} = 16\text{ kB}$$

4.25 (d)

The I/O performance of a user program determined by many input and output devices not only by the disk. When we uses FCFS and replace it by SSTF it improves only disk driver performance not the entire I/O performance. So I/O improvement performance of user program is 0%.

4.26 (a)

Minimum number of page frames allocated to a running process is determined by the instruction set architecture.

4.27 (d)

Effective average instruction execution time = CPU time (100 ns) + Average time lost due to page fault per instruction $\left(= \dfrac{8\text{ ms}}{10000}\right)$ + Number of memory references × EMAT

$100 + 800 + 2 \times [0.9\,(0 + 150) + 0.1\,(0 + 3 \times 150)]$
– 1260

4.28 (b)

Fully associate cache with 8 block i.e. 1 set with 8 cache blocks.

Set	4	3	25	8	19	6	25	16
	B_0	B_1	B_2	B_3	B_4	B_5	B_6	B_7

Since set contain 8 blocks so, request for 45 replacement is needed. When 45 comes then

45 ~~4~~	3	25	8	19	6	25	16
B_0	B_1	B_2	B_3	B_4	B_5	B_6	B_7

For request 22:

45	22 ~~3~~	25	8	19	6	25	16
B_0	B_1	B_2	B_3	B_4	B_5	B_6	B_7

For request 8. Since already present, so no replacement.

For request 3:

45	22	25	8	3 ~~19~~	6	25	16
B_0	B_1	B_2	B_3	B_4	B_5	B_6	B_7

For request 16; since already present, so no replacement.

For request 25; since already present, so no replacement.

For request 7:

45	22	25	8	3	7 ~~6~~	25	16
B_0	B_1	B_2	B_3	B_4	B_5	B_6	B_7

So block 5 i.e. B_5 contain memory block 7.

4.29 (d)

In page table, number of entries is equal to number of pages in virtual memory and page table entry is equal to number of frames and other information i.e.

$$\text{So, number of frames} = \frac{\text{Physical memory}}{\text{Page size}}$$

$$= \frac{2^{30}}{2^{12}} = 2^{18}$$

So, number of bits for frames = $\log_2 (2^{18}) = 18$

We know that,

Page table entry = Number of bits for frames + Other information

$$32 = 18 + x$$

$$x = 32 - 18 = 14$$

So, 14 bits can be used for storing protection and other information.

Note: Each page table entry must contain frame number.

4.30 (b)

Overhead is considered as memory wasted by maintaining the page tables.

So, we need to find the overhead for every case in the given question.

Case-1: Overhead using two level paging.

Each page size is 1 KB.

Total pages = 195 + 254 + 45 + 364 = 858 KB pages.

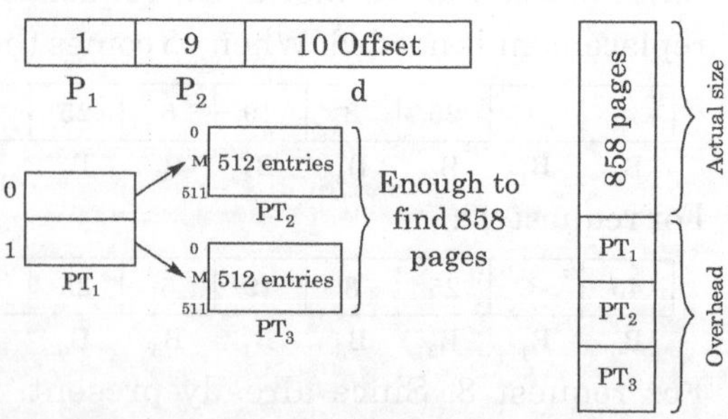

Each page table entry = 4 bytes.

Size of page table$_1$ = 2 × 4 = 8 bytes

Size of page table$_2$ = 2^9 × 4 = 2048 bytes

Size of page table$_3$ = 2^9 × 4 = 2048 bytes

∴ Overhead of paging (P) = 8 + 2048 + 2048 = 4104 bytes

Case-2: Overhead using segmentation.

Number of segments = 4 + 5 + 3 + 8 = 20 (5 bits for segment number)

Maximum segment size

$$= \max\left(\frac{195}{4}, \frac{254}{5}, \frac{45}{3}, \frac{364}{8}\right)$$

= max(49,51,15,46) = 51 bytes (helps to find segment offset)

Number of bits required to find the limit (segment offset) = 6

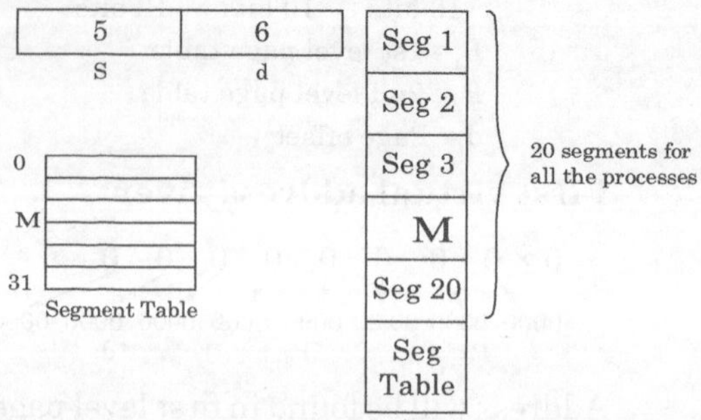

Size of segment table (aprox)

= 32 × 8 = 256 bytes

∴ Overhead using segmentation (S) = 256 bytes

Case-3: Overhead using segmentation and paging.

Number of segments = 20 ⇒ 5 bits required for segment number.

Maximum segment size = 51 bytes ⇒ 6 bits required for page offset.

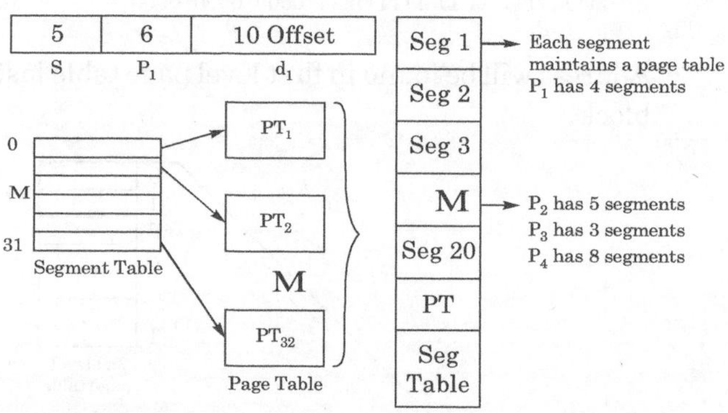

20 page tables are required.

Each segment needed 1 page table.

Size of segment table = 32 × 8 = 256 bytes.

Size of each page table = 64 × 4 = 256 bytes.

∴ Segmentation and paging overhead (T) = 256 B + 20 × 256 B = 5376 B

S: Simple segmentation takes less overhead.

P: Two level paging takes little more overhead because of two page tables.

T: Segmented paging takes more overhead because we have paging on both segment and segment table.

So, S < P < T.

4.31 (c)

Number of bits for virtual address = 32
Number of bits for pages = 4 KB
$= \log_2 (2^{12}) = 12$
So, number of bits for pages = 32 – 12 = 20

$$\text{Number of sets} = \frac{128}{4} = 32$$

Number of bits for sets = $\log_2 (32) = 5$
So, minimum tag size = 20 – 5 = 15

4.32 (a)

The main purpose of virtual memory is by allocating less memory to the process, we plan to increase degree of multiprogramming, by bringing more processes into memory. If virtual memory is removed then the efficient implementation of multiprogramming and multi user is no longer possible.

4.33 (b)

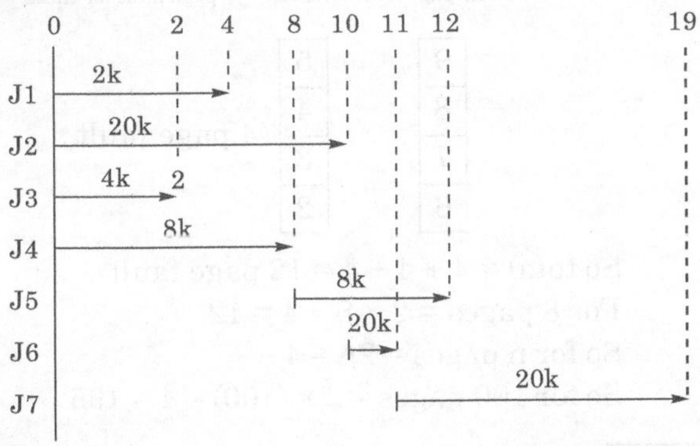

4.34 (c)

Size of page = 100
When page fault occur then from that page to next 99 byte are fetched into memory

Reference address	Address in memory	Hit/Fault
0100	0100 – 0199	Fault
0200	0200 – 0299	Fault
0430	0430 – 0529	Fault
0499	0430 – 0529	Hit
0510	0430 – 0529	Hit
0530	0530 – 0629	Fault
0560	0530 – 0629	Hit
0120	0120 – 0219	Fault
0220	0220 – 0319	Fault
0240	0220 – 0319	Hit
0260	0220 – 0319	Hit
0320	0320 – 0419	Fault
0410	0320 – 0419	Hit
		Total fault = 7

4.35 (a)

Average Access Time = $(1 - p)(T_{mem}) + p[p(\text{S.T. dirty}) + (1 - p)(\text{S.T. not dirty})]$

$3 = (1 - p)(1) + p(p \times 300) + (1 - p) \times 100)$
$3 = 1 - p + p^2\, 300 + p\, 100 - p^2\, 100$
$3 = 1 + 99\, p + 200\, p^2$
$200\, p^2 + 99\, p - 2 = 0$

$$p = \frac{b \pm \sqrt{b^2 - 4ac}}{2a} = 0.0194$$

4.36 (a)

Number of page frames = 3
Consider the optimal page replacement algorithm as follows:
1, 2, 1, 3, 7, 4, 5, 6, 3, 1

10 Page Frames										
	*	*	*	3	3	3	3	3	3	3
	*	2	2	2	7	4	5	6	6	6
	1	1	1	1	1	1	1	1	1	1
Referring String	1	2	1	3	7	4	5	6	3	1
Page	Page Fault	Page Fault		Page Fault	Page Fault	Page Fault	Page Fault	Page Fault		

So total number of page faults = 7

4.37 (c)

Consider the LRU page replacement policy

10 Page Frames										
				3	3	3	5	5	5	1
		2	2	2	7	7	7	6	6	6
	1	1	1	1	1	4	4	4	3	3
Referring String	1	2	1	3	7	4	5	6	3	1
Page	Page Fault	Page Fault		Page Fault	Page Fault	Page Fault	Page Fault	Page Fault	Page Fault	Page Fault

So total number of page fault = 9
Number of page fault in LRU – Number of page fault in optional = 9 – 7 = 2 more page fault.

4.38 (c)

Effective access time = hit ratio * (TLB access time + Memory access time) + miss ratio * (TLB access time + Page table access time + Memory access time)
= 0.9 * (10+50) + 0.1 *(10 + 50 + 50)
= 54 + 11 = 65

4.39 (b)

Page tables are stored in the main memory and main memory is divided into frames, it means addressing the page table is same as addressing the frames, so we just need to find, the number of bits required for addressing frames.

Number of bits required for first level = 24
Number of bits required for second level = 24
Number of bits required for third level = 24

4.40 (b)

Main memory is divided in page size = 16 bytes
We know that, page size = frame size
So in virtual address least 4 bit represent page size and remaining bits represents page number.
So, page number request are:

0 = 00 0000 i.e. Page number = 0

4 = 00 0100 i.e. Page number = 0

8 = 00 1000 i.e. Page number = 0

20 = 01 0100 i.e. Page number = 1

24 = 01 1000 i.e. Page number = 1

36 = 100 1000 i.e. Page number = 4

44 = 10 1100 i.e. Page number = 2

12 = 00 1100 i.e. Page number = 0

68 = 100 0100 i.e. Page number = 4

72 = 100 1000 i.e. Page number = 4

80 = 101 0000 i.e. Page number = 5

84 = 101 0100 i.e. Page number = 5

28 = 1 1100 i.e. Page number = 1

32 = 10 0000 i.e. Page number = 2

88 = 101 1000 i.e. Page number = 5

92 = 101 1100 i.e. Page number = 5

Request are:

0	0	0	1	1	4	2	0	4	4	5	5	1	2	5	5
						2	2	2	2	2	2	1	1	1	1
					4	4	4	4	4	4	4	4	4	4	4
			1	1	1	1	1	1	1	5	5	5	5	5	5
0	0	0	0	0	0	0	0	0	0	0	0	0	2	2	2
F	H	H	F	H	F	F	H	H	H	F	H	F	F	H	H

So, at the end pages 1, 2, 4, 5 are in memory and 7 page fault are present.

4.41 (d)

- **Dirty bits** used in write back policy.
- **R/W bits** used in page protection.
- **Reference bit** used in page replacement policy (second chance policy).
- **Valid bit** used when page is initialized.

4.42 (a)

In Bleady's Anomaly if number of frames are increased then number of page faults increases. This behaviour found only with FIFO.

4.43 (b)

Page frame numbers are most important, virtual page number maynot be stored entirely.

4.44 (a)

FIFO policy for page replacement used.
Access 100 distinct pages by taking some example: 2 3 4 5 6 7 8 9
So by loading it get

5
4
3
2

= 4 page fault,

9
8
7
6

= 4 page fault

and now access these page in reverse so

9 8 7 6 — So no page fault for these; 5 4 3 2 — page fault for these

9	5
8	4
7	3
6	2

4 page fault

So total = 4 + 4 + 4 = 12 page fault
For 8 pages = 2 × 8 – 4 = 12
So for n pages = 2n – 4
So for 100 pages = 2 × (100) – 4 = 196

4.45 (b)

Effective access time = [(1 – p)* access time when no page fault +p* access time during page fault]

$$= \left[1-\left(\frac{1}{10^6}\right)\right]*20\,\text{ns}+\left(\frac{1}{10^6}\right)*10\,\text{ms}$$

$$= \left(\frac{10^6-1}{10^6}\right)*20\,\text{ns}+\left(\frac{1}{10^6}\right)*10*10^6\,\text{ns} \approx 30\,\text{ns}$$

4.46 (b)

OPTIMAL

		3	3	3	3	3	2	2	2
	2	2	2	4	4	4	4	4	4
1	1	1	1	1	1	1	1	1	1
MISS	MISS	MISS	MISS	MISS	HIT	H	M	H	H

5 miss

FIFO

		3	3	3	3	3	2	2	2
	2	2	2	2	1	1	1	1	1
1	1	1	1	4	4	4	4	4	4
M	M	M	H	M	M	H	M	H	H

6 miss

LRU

		3	3	3	1	1	1	4	4
	2	2	2	2	2	3	3	3	1
1	1	1	1	4	4	4	2	2	2
M	M	M	H	M	M	M	M	M	M

⇒ 9 MISS

So optimal < FIFO < LRU

4.47 (c)

Assume Page Size = x bytes

We know that,

1st Page table size = Number of entries in page table × Page table entry size

$$= \frac{2^{46}}{x} \times 4\text{B} = \frac{2^{48}}{x}$$

2nd Page table size = Number of entries in 1st page table × Page table entry size

$$= \frac{2^{48}}{x^2} \times 4\text{B} = 2^{50}$$

3rd Page table size = Number of entries in 2nd page table × Page table entry size

$$= \frac{2^{50}}{x^3} \times 4\text{B} = \frac{2^{52}}{x^3}$$

Since 3rd level page table must fit into single page.

So, $$x = \frac{2^{52}}{x^3}$$

$$x^4 = 2^{52}$$

Take log both side

$$4(\log_2 x) = 52 (\log_2 2)$$

$$(\log_2 x) = 13$$

$$x = 2^{13} = 8 \text{ KB}$$

4.48 (c)

Since each cache line can contain 16,648 blocks, assuming best case 1024 bytes can accomodate in one line hence $\frac{2^{13}}{2^{10}} = 2^3$

⇒ Number of colours needed = 8.

4.49 (7)

Optimal page replacement policy:

✓	✓	✓	✓	×	×	✓	✓	×	×	✓
		3	4	4	4	4	4	4	4	6
	2	2	2	2	2	2	2	2	2	2
1	1	1	1	1	1	5	3	3	3	3
1	2	3	4	2	1	5	3	2	4	6

∴ 7 page faults

4.50 (d)

Pages:

1, 2,...20,...100, 1,...20,...100, 1,...20,...100,

Optimal page replacement:

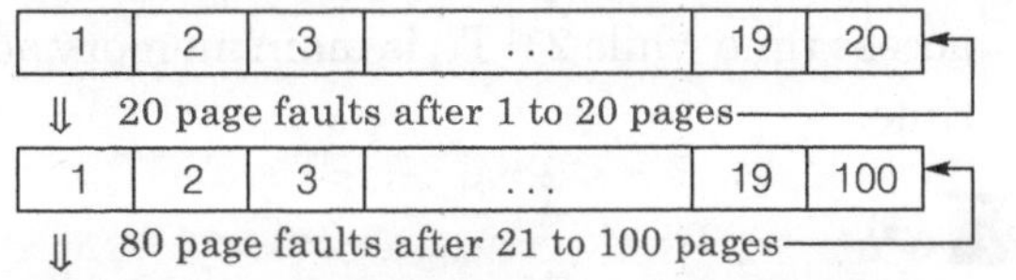

Therefore for first 100 pages there are 100 page faults

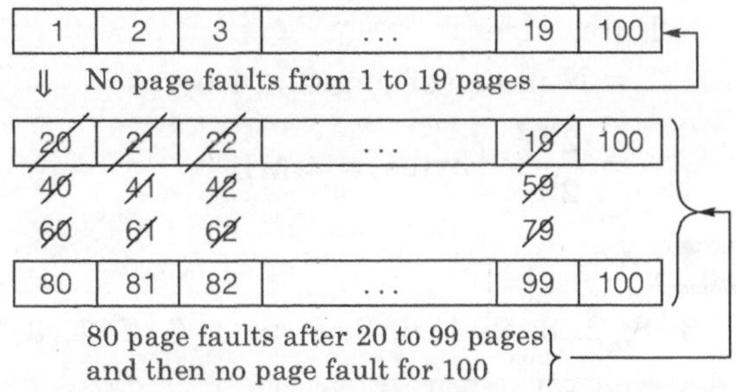

Therefore for second 100 pages there are 80 page faults

Similarly, for third 100 pages there will be 80 page faults.

∴ Total page faults for 300 pages = 100 + 80 + 80 = 260

Most recently used:

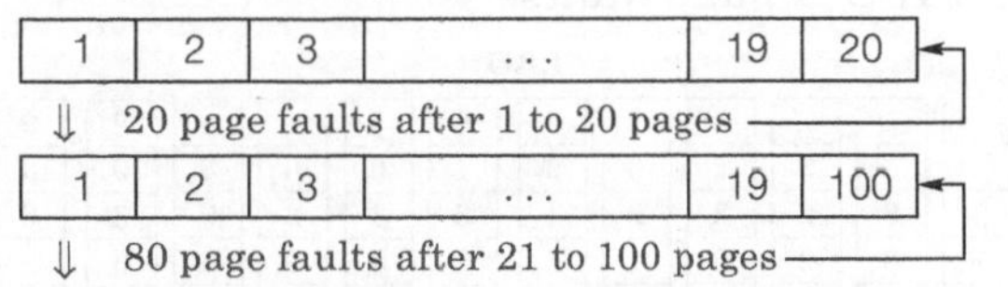

Therefore for first 100 pages there are 100 page faults

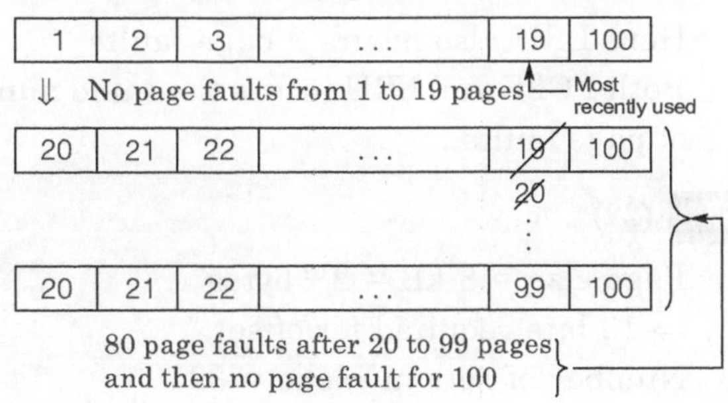

Therefore for second 100 pages there are 80 page faults

Similarly, for third 100 pages there will be 80 page faults.

∴ Total page faults for 300 pages = 100 + 80 + 80 = 260

4.51 (6)

✓	✓	✓	✓	×	×	×	✓	✓	×
		6	6	6	6	6	6	7	7
	7	7	7	7	7	7	2	2	2
4	4	4	1	1	1	1	1	1	1
4	7	6	1	7	6	1	2	7	2

∴ 6 page faults will occur using LRU

4.52 (122)

Effective Memory Access Time = H_{TLB} (T_{TLB} + T_M) + (1 – H_{TLB}) (T_{TLB} + T_M + T_M)
= 0.6 (10 + 80) + (1 – 0.6) (10 + 80 + 80)
= 0.6 (90) + 0.4 (170) = 122

Note: In case of TLB miss 1st T_M is page table access time while 2nd T_M is main memory access time.

4.53 (4)

Page table size = Number of page table entries × Entry size
= Number of pages × 4 bytes

$$= \frac{2^{32}}{2^{12}} \times 4 \text{ bytes} = 4 \text{ MB}$$

4.54 (a)

3, 8, 2, 3, 9, 1, 6, 3, 8, 9, 3, 6, 2, 1, 3

FIFO

	→		→		→		→	
1		1		1		1		1
9		9		9		9		2
2		2		2		8		8
8		8		3		3		3
3		6		6		6		6

FIFO 9 page faults.

LRU

				8	2	2	9	1	1	1	8	9	3
			8	2	3	9	1	6	6	8	9	3	9
		3	2	3	9	1	6	3	8	9	3	6	6
	3	8	3	9	1	6	3	8	9	3	6	2	2
3	8	2	9	1	6	3	8	9	3	6	2	1	1

Here LRU also incurs 9 page faults.
Both FCFS and LRU incurs the same number of page faults.

4.55 (22)

Page size = 8 kB = 2^{13} bytes
⇒ 13 bits required for offset.
Number of sets in cache = 32
⇒ 5 bits required for index (set offset).

40 bit

	5	13
22 bits for tag	Set offset	Page offset

∴ 22 tag bits are required.

4.56 (a)

The best fit algorithm finds the smallest sufficient partition.

	357	491	468		210
200 KB	400 KB	600 KB	500 KB	300 KB	250 KB

∴ The memory partitions 200 KB and 300 KB are not allotted.

4.57 (36)

Page size = 8 KB ⇒ 13 bit offset
Number of frame bits = 32 – 13 = 19 bits
Page table entry = Valid + Dirty + Permission bits + Translation (frame bits)
= 1 + 1 + 3 + Frame bits
= 5 + 19 = 24 bits
Page table size = 24 Mbytes
Number of pages = Number of page table entries

$$= \frac{24 \text{ Mbytes}}{24 \text{ Bits}} = 8 \text{ M} = 2^{23} \text{ pages}$$

∴ 23 bits needed for page and 13 bits offset
Length of virtual address = 23 + 13 = 36 bits.

4.58 (384)

Page table size = Number of entries in page table × Page table entry size

$$= \left(\frac{2^{40}}{2^{14}}\right) \times 48 \text{ bits} = 2^{26} \times 6 \text{ bytes}$$

$$= 64 \text{ M} \times 6 \text{ B} = 384 \text{ MB}$$

4.59 (1)

This question is little tricks, but not difficult. You can solve the problem by taking smaller example with smaller values.

Example: 1, 2, 3, 4, 1, 2, 3, 4 with '2' frames.

LIFO: 1, 2, 3, 4, 1, 2, 3, 4

	~~2~~	~~3~~	~~4~~	~~4~~	~~2~~	3	4
1	1	1	1	1	1	1	1
F	F	F	F		F	F	F

Total page faults = 7

Optimal: 1, 2, 3, 4, 1, 2, 3, 4

	~~2~~	~~3~~	4	4	4	4	4
1	1	1	1	~~1~~	~~2~~	3	3
F	F	F	F		F	F	

Total page faults = 6
7 – 6 = 1

4.60 (d)

Because of Belady's anamoly, it will happen in FIFO phase replacement algorithm.

4.61 **(b)**

Considering each statement:

- Random page replacement algorithm can behave like any algorithm probably FCFS too, hence it can suffer from Belady's anomaly.
- LRU page replacement algorithm doesn't suffer from Belady's anomaly.

4.62 **(b)**

$$\text{EMAT} = P \times S + (1 - P) \times M$$
$$X = P \times D + (1 - P) \times M$$
$$X = P \times D + M - P \times M$$
$$X - M = P(D - M)$$
$$P = \frac{X - M}{D - M}$$

4.63 **(d)**

1 word = 4 bytes, Page size = 8 kB = 2^{13} B

Number of words in 1 page = $\frac{2^{13}}{2^2} = 2^{11}$

TLB can hold 128 valid entries so, at most 128×2^{11} memory address can be addressed without TLB miss.

$$128 \times 2^{11} = 256 \times 2^{10}$$

4.64 **(c)**

The hole created by best fit is never larger than the hole created by first fit. This is correct option.

4.65 **(154.5)**

EMAT = 0.95 × (20 + 100) + 0.05 × (0.9 × (20 + 100 + 100) + 0.1 × [0.2 × (20 + 100 + 5000 + 5000) + 0.8 × (20 + 100 + 5000)]

= 154.5 ns

4.67 **(4108)**

39-bit virtual address

Level 1 offset	Level 2 offset	Level 3 offset	Page offset
9 bits	9 bits	9 bits	12 bits

Since process P is using 2 GB virtual memory and page size 2^{12} bytes.

So, Number of pages = $\frac{2^{31}}{2^{12}} = 2^{19}$

There are 2^9 entries per page table in 3rd level.

So, we need = $\frac{2^{19}}{2^9} = 2^{10}$ page tables in 3rd level

Now, it means 2^{10} entries in 2nd level so number of page tables in 2nd level

$$= \frac{2^{10}}{2^9} = 2 \text{ page table}$$

Now we have 2 entries in 1st level hence we need only 1 page table in 1st level.

Total number of page table = $2^{10} + 2 + 1$

= 1027 per table

Each page table has 2^9 entries and each entry size is 8 bytes.

So, total size of page table in bytes

= $1027 \times 2^9 \times 8$ bytes = 4108 KB

Alternative method:

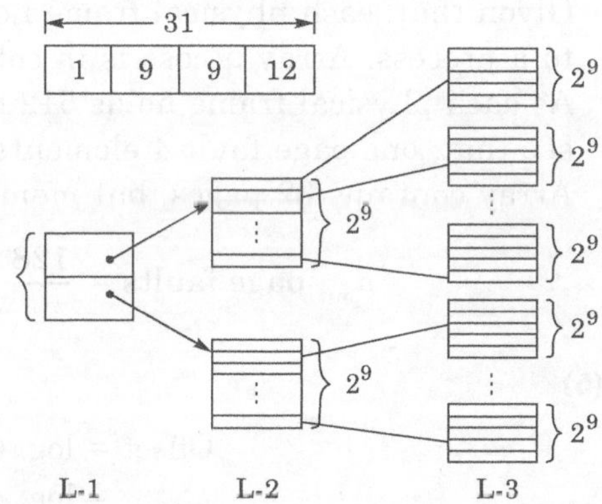

Total page table = $1 + 2^1 + 2^1 \times 2^9 = 1027$

Page table size = 1027×2^{12} = 4108 KB

4.68 **(c)**

(a) TLB performs parallel search. Hence true

(b) If the TLB hit means, word will surely present in main memory. Hence true

(c) Memory access time using inverted page table is always not same, because there is no indexing applied, we follows no equal linear search. Hence false

(d) If two distinct virtual address map to same value while hashing, they will be resolved using linked list, so memory access time will not be same. Hence true

4.69 **(0.6)**

9 page faults out of total 15 references $\frac{9}{15} = 0.6$

4.70 (4096)

Given array dimension is $D[128][128]$ which is stored in row major order.
It means that, 128 rows and 128 columns.

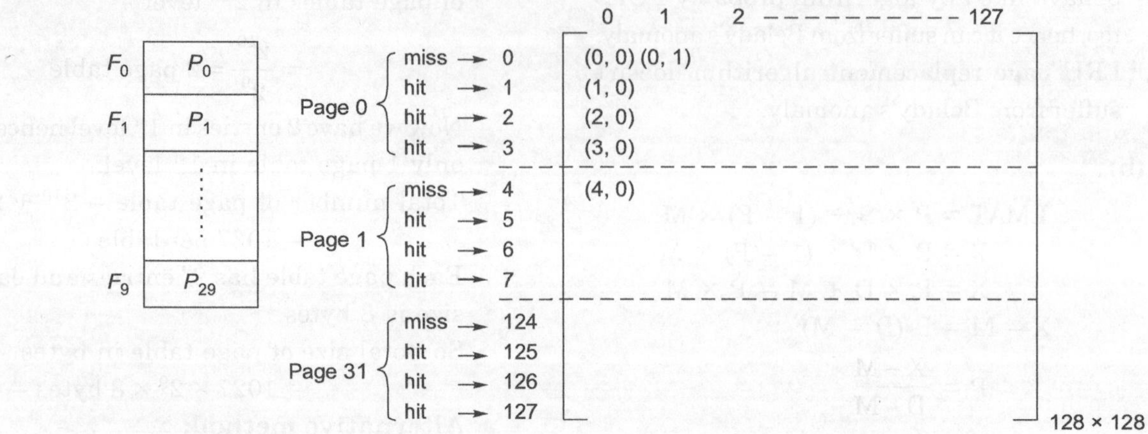

Given that, each physical frame holds 512 elements of the array D. 30 page frames are allocated to a process. Array access is in column major order $[D[j][i] \mathrel{*}= 10]$
As each physical frame holds 512 elements, the 1st 4 rows of the array belongs one page, we can see that, one page fault 4 elements.
Array contains 32 pages, but memory contains 30 page frames.

$$\therefore \quad n_{out} \text{ page faults} = \frac{128 \times 128}{4} = \frac{2^{14}}{2^2} = 4096$$

4.71 (5)

$$\text{Offset} = \log_2 (\text{Page size})$$
$$= \log_2 (4 \text{ KB}) = 12 \text{ bits}$$
$$\text{PTE} = 8 \text{ bytes}$$

For every level page table should fit into a page.

$$\text{So, number of entries in a page} = \frac{4 \text{ KB}}{8 \text{ B}} = 2^9$$

$$\text{Bits for a level} = \log_2 (2^9) = 9 \text{ bits}$$

Following is representation

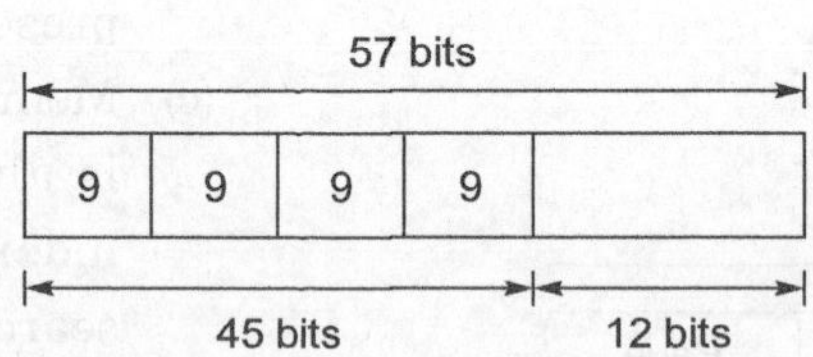

So, 5 levels required.

4.72 (6596)

$$\text{P.A.} \rightarrow (1100111000100)_2 \rightarrow (6596)_{10}$$

■■■■

5 File System and Device Management

5.1 The correct matching for the following pairs is

List-I	List-II
A. DMA I/O	1. High speed RAM
B. Cache	2. Disk
C. Interrupt I/O	3. Printer
D. Condition Code Register	4. ALU

Codes:
(a) A-4 B-3 C-1 D-2 (b) A-2 B-1 C-3 D-4
(c) A-4 B-3 C-2 D-1 (d) A-2 B-3 C-4 D-1
[1997 : 1 M]

5.2 The correct matching for the following pairs is

List-I	List-I
A. Disk scheduling	1. Round robin
B. Batch processing	2. SCAN
C. Time sharing	3. LIFO
D. Interrupt processing	4. FIFO

Codes:
(a) A-3, B-4, C-2, D-1
(b) A-4, B-3, C-2, D-1
(c) A-2, B-4, C-1, D-3
(d) A-3, B-4, C-3, D-2
[1997 : 1 M]

5.3 I/O redirection
(a) implies changing the name of a file
(b) can be employed to use an existing file as input file for a program
(c) implies connecting 2 programs through a pipe
(d) None of the above **[1997 : 1 M]**

5.4 When an interrupt occurs, an operating system
(a) ignores the interrupt
(b) always changes state of interrupted process after processing the interrupt
(c) always resumes execution of interrupted process after processing the interrupt
(d) may change state of interrupted process to 'blocked' and schedule another process
[1997 : 1 M]

5.5 Which of the following devices should get higher priority in assigning interrupts?
(a) Hard disk (b) Printer
(c) Keyboard (d) Floppy disk
[1998 : 1 M]

5.6 Which of the following is true?
(a) Unless enabled, a CPU will not be able to process interrupts.
(b) Loop instructions cannot be interrupted till they complete.
(c) A processor checks for interrupts before executing a new instruction.
(d) Only level triggered interrupts are possible on microprocessors **[1998 : 1 M]**

5.7 Listed below are some operating system abstractions (in the left column) and the hardware components (in the right column)?

A. Thread	1. Interrupt
B. Virtual address space	2. Memory
C. File system	3. CPU
D. Signal	4. Disk

(a) A-2, B-4, C-3, D-1
(b) A-1, B-2, C-3, D-4
(c) A-3, B-2, C-4, D-1
(d) A-4, B-1, C-2, D-3 **[1999 : 1 M]**

5.8 Which of the following disk scheduling strategies is likely to give the best throughput?
(a) Farthest cylinder next
(b) Nearest cylinder next
(c) First come first served
(d) Elevator algorithm **[1999 : 1 M]**

5.9 Which of the following requires a device driver?
(a) Register (b) Cache
(c) Main memory (d) Disk **[2001 : 1 M]**

5.10 In the index allocation scheme of blocks to a file, the maximum possible size of the file depends on
(a) the size of the blocks, and the size of the address of the blocks.
(b) the number of blocks used for the index, and the size of the blocks
(c) the size of the blocks, the number of blocks used for the index, and the size of the address of the blocks
(d) None of the above **[2002 : 2 M]**

5.11 Using a larger block size in a fixed block size file system leads to
(a) better disk throughput but poorer disk space utilization

(b) better disk throughput and better disk space utilization
(c) poorer disk throughput but better disk space utilization
(d) poorer disk throughput and poorer disk space utilization **[2003 : 1 M]**

5.12 A Unix-style I-node has 10 direct pointers and one single, one double and one triple indirect pointers. Disk block size is 1 Kbyte, disk block address is 32 bits, and 48-bit integers are used. What is the maximum possible file size?
(a) 2^{24} bytes (b) 2^{32} bytes
(c) 2^{34} bytes (d) 2^{48} bytes
[2004 : 2 M]

5.13 A disk has 200 tracks (numbered 0 through 199). At a given time, it was servicing the request of reading data from track 120, and at the previous request, service was for track 90. The pending requests (in order of their arrival) are for track numbers.

30 70 115 130 110 80 20 25

How many times will the head change its direction for the disk scheduling policies SSTF (Shortest Seek Time First) and FCFS (First Come Fist Serve)?
(a) 2 and 3 (b) 3 and 3
(c) 3 and 4 (d) 4 and 4
[2004 : 2 M]

5.14 In a particular Unix OS, each data block is of size 1024 bytes, each node has 10 direct data block addresses and three additional addresses: one for single indirect block, one for double indirect block and one for triple indirect block. Also, each block can contain addresses for 128 blocks. Which one of the following is approximately the maximum size of a file in the file system?
(a) 512 MB (b) 2 GB
(c) 8 GB (d) 16 GB
[2004 : 2 M]

5.15 In a computer system, four files of size 11050 bytes, 4990 bytes, 5170 bytes and 12640 bytes need to be stored. For storing these files on disk, we can use either 100 byte disk blocks or 200 byte disk blocks (but can't mix block sizes). For each block used to store a file, 4 bytes of bookkeeping information also needs to be stored on the disk. Thus, the total space used to store a file is the sum of the space taken to store the file and the space taken to store the book keeping information for the blocks allocated for storing the file. A disk block can store either bookkeeping information for a file or data from a file, but not both.
What is the total space required for storing the files using 100 byte disk blocks and 200 byte disk blocks respectively?
(a) 35400 and 35800 bytes
(b) 35800 and 35400 bytes
(c) 35600 and 35400 bytes
(d) 35400 and 35600 bytes **[2005 : 2 M]**

5.16 In the working-set strategy, which of the following is done by the operating system to prevent thrashing?
I. It initiates another process if there are enough extra frames.
II. It selects a process to suspend if the sum of the sizes of the working-sets.
III. Execeeds the total number of available frames.
(a) I only (b) II only
(c) Neither I nor II (d) Both I and II
[2006 : 1 M]

Directions for Question 5.17 to 5.18:
The head of a hard disk serves requests following the shortest seek time first (SSTF) policy. The head is initially positioned at track number 180.

5.17 Which of the request sets will cause the head to change its direction after servicing every request assuming that the head does not change direction if there is a tie in SSTF and all the requests arrive before the servicing starts?
(a) 11, 139, 170, 178, 181, 184, 201, 265
(b) 10, 138, 170, 178, 181, 185, 201, 265
(c) 10, 139, 169, 178, 181, 184, 201, 265
(d) 10, 138, 170, 178, 181, 185, 200, 265
[2007 : 2 M]

5.18 What is the maximum cardinality of the request set, so that the head changes its direction after servicing every request if the total number of tracks are 2048 and the head can start from any track?
(a) 9 (b) 10
(c) 11 (d) 12 **[2007 : 2 M]**

5.19 A virtual memory system uses First In First Out (FIFO) page replacement policy and allocates a fixed number of frames to a process. Consider the following statements:
P : Increasing the number of page frames allocated to a process sometimes increases the page fault rate.

Q : Some programs do not exhibit locality of reference.

Which one of the following is TRUE?

(a) Both P and Q are true, and Q is the reason for P

(b) Both P and Q are true, but Q is not the reason for P

(c) P is false, but Q is true

(d) Both P and Q are false **[2007 : 2 M]**

5.20 The data block of a very large file in the Unix file system are allocated using

(a) Contiguous allocation

(b) Linked allocation

(c) indexed allocation

(d) an extension of indexed allocation

[2008 : 1 M]

5.21 Consider a disk system with 100 cylinders. The requests to access the cylinders occur in following sequence:

4, 34, 10, 7, 19, 73, 2, 15, 6, 20

Assuming that the head is currently at cylinder 50, what is the time taken to satisfy all requests if it takes 1 ms to move from one cylinder to adjacent one and shortest seek time first policy is used?

(a) 95 ms (b) 119 ms

(c) 233 ms (d) 276 ms **[2009 : 2 M]**

5.22 A multilevel page table is preferred in comparison to a single level page table for translating virtual address to physical address because

(a) it reduces the memory access time to read or write a memory location

(b) it helps to reduce the size of page table needed to implement the virtual address space of a process

(c) it is required by the translation look a side buffer

(d) it helps to reduce the number of page faults in page replacement algorithms.

[2009 : 2 M]

5.23 A file system with 300 GByte disk uses a file descriptor with 8 direct block addresses, I indirect block address and 1 doubly indirect block address. The size of each disk block is 128 Bytes and the size of each disk block address is 8 Bytes. The maximum possible file size in this file system is

(a) 3 KBytes (b) 35 KBytes

(c) 280 KBytes

(d) dependent on the size of the disk

[2012 : 2 M]

5.24 Suppose a disk has 201 cylinders, numbered from 0 to 200. At some time the disk arm is at cylinder 100, and there is a queue of disk access requests for cylinders 30, 85, 90, 100, 105, 110, 135 and 145. If Shortest-Seek Time First (SSTF) is being used for scheduling the disk access, the request for cylinder 90 is serviced after servicing _______ number of requests. **[2014 (Set-1) : 1 M]**

5.25 A FAT (file allocation table) based file system is being used and the total overhead of each entry in the FAT is 4 bytes in size. Given a 100×10^6 bytes disk on which the file system is stored and data block size is 10^3 bytes, the maximum size of a file that can be stored on this disk in units of 10^6 bytes is _______. **[2014 (Set-2) : 1 M]**

5.26 Suppose the following disk request sequence (track numbers) for a disk with 100 tracks is given: 45, 20, 90, 10, 50, 60, 80, 25, 70. Assume that the initial position of the R/W head is on track 50. The additional distance that will be traversed by the R/W head when the Shortest Seek Time First (SSTF) algorithm is used compared to the SCAN (Elevator) algorithm (assuming that SCAN algorithm moves towards 100 when it starts execution) is _______ tracks.

[2015 (Set-1) : 2 M]

5.27 Consider a disk queue with requests for I/O to blocks on cylinders 47, 38, 121, 191, 87, 11, 92, 10. The C-LOOK scheduling algorithm is used. The head is initially at cylinder number 63, moving towards larger cylinder numbers on its servicing pass. The cylinders are numbered from 0 to 199. The total head movement (in number of cylinders) incurred while servicing these requests is _______. **[2016 (Set-1) : 2 M]**

5.28 In a file allocation system, which of the following allocation scheme(s) can be used if no external fragmentation is allowed?

I. Contiguous II. Linked

III. Indexed

(a) I and III only (b) II only

(c) III only (d) II and III only

[2017 (Set-2) : 1 M]

5.29 Consider a storage disk with 4 platters (numbered as 0, 1, 2 and 3), 200 cylinders (numbered as 0, 1, ..., 199) and 256 sectors per track (numbered as 0, 1, ..., 255). The following 6 disk requests of the form [sector number, cylinder number, platter number] are received by the disk controller at the same time:

[120, 72, 2], [180, 134, 1], [60, 20, 0],
[212, 86, 3], [56, 116, 2], [118, 16, 1]

Currently the head is positioned at sector number 100 of cylinder 80 and is moving towards higher cylinder numbers. The average power dissipation in moving the head over 100 cylinders is 20 milliwatts and for reversing the direction of the head movement once is 15 milliwatts. Power dissipation associated with rotational latency and switching of head between different platters is negligible.

The total power consumption in milliwatts to satisfy all of the above disk requests using the Shortest Seek Time First disk scheduling algorithm is ________. **[2018 : 2 M]**

5.30 The index node (inode) of a Unix-like file system has 12 direct, one single-indirect and one double-indirect pointers The disk block size is 4 kB and the disk block address is 32-bits long. The maximum possible file size is (rounded off to 1 decimal place) ________ GB. **[2019 : 2 M]**

5.31 Consider the following five disk access requests of the form (request id, cylinder number) that are present in the disk scheduler queue at a given time.

(P, 155), (Q, 85), (R, 110), (S, 30), (T, 115)

Assume the head is positioned at cylinder 100. The scheduler follows Shortest Seek Time First scheduling to service the requests,

Which one of the following statements is FALSE?

(a) R is serviced before P.
(b) T is serviced before P.
(c) Q is serviced after S but before T.
(d) The head reverses its direction of movement between servicing of Q and P.

[2020 : 2 M]

5.32 Consider a linear list based directory implementation in a file system. Each directory is a list of nodes, where each node contains the file name along with the file metadata, such as the list of pointers to the data blocks. Consider a given directory foo.

Which of the following operations will necessarily require a full scan of foo for successful completion?

(a) Opening of an existing file in foo
(b) Renaming of an existing file in foo
(c) Creation of a new file in foo
(d) Deletion of an existing file from foo

[2021 (Set-1) : 1 M]

5.33 Consider two files systems A and B, that use contiguous allocation and linked allocation, respectively. A file of size 100 blocks is already stored in A and also in B. Now, consider inserting a new block in the middle of the file (between 50th and 51st block), whose data is already available in the memory. Assume that there are enough free blocks at the end of the file and that the file control blocks are already in memory. Let the number of disk accesses required to insert a block in the middle of the file in A and B are n_A and n_B, respectively, then the value of $n_A + n_B$ is ________. **[2022 : 2 M]**

5.34 Consider a 512 GB hard disk with 32 storage surfaces. There are 4096 sectors per track and each sector holds 1024 bytes of data. The number of cylinders in the hard disk is ________. **[2024 (Set-1) : 2 M]**

5.35 Consider a disk with the following specifications: rotation speed of 6000 RPM, average seek time of 5 milliseconds, 500 sectors/track, 512 byte sectors. A file has content stored in 3000 sectors located randomly on the disk. Assuming average rotational latency, the total time (in seconds, rounded off to 2 decimal places) to read the entire file from the disk is ________. **[2024 (Set-2) : 2 M]**

■■■

Answers File System and Device Management

5.1	(b)	**5.2**	(c)	**5.3**	(b)	**5.4**	(d)	**5.5**	(a)	**5.6**	(a, c)	**5.7**	(c)	**5.8**	(b)	**5.9**	(d)
5.10	(c)	**5.11**	(a)	**5.12**	(c)	**5.13**	(c)	**5.14**	(b)	**5.15**	(c)	**5.16**	(b)	**5.17**	(b)	**5.18**	(c)
5.19	(b)	**5.20**	(d)	**5.21**	(b)	**5.22**	(b)	**5.23**	(b)	**5.24**	(3)	**5.25**	(99.6)	**5.26**	(10)	**5.27**	(346)
5.28	(d)	**5.29**	(85)	**5.30**	(4)	**5.31**	(c)	**5.32**	(b, c)	**5.33**	(153)	**5.34**	(4096)	**5.35**	(30.06)		

Explanations File System and Device Management

5.1 (b)

DMA I/O → Disk
Cache → High speed RAM
Interrupt I/O → Printer
Condition code register → ALU

5.2 (c)

Disk scheduling → SCAN
Batch processing → FIFO (Queue)
Time sharing → Round Robin (uses time slice)
Interrupt processing → LIFO (stack)

5.3 (b)

I/O redirection helps to use an existing file as input file for a program.

5.4 (d)

When an interrupt occurs, may change state of interrupted process to 'blocked' and schedule another process.

5.5 (a)

Hard disk should be higher priority compared to printer, keyboard, and floppy disk. Because all user programs and OS stored at hard disk.

5.6 (a, c)

(a) When interrupt is enabled, a CPU will be able to process the interrupt.

(b) CPU checks for interrupts before executing a new instruction.

5.7 (c)

Thread → CPU
Virtual address space → Memory
File system → Disk
Signal → Interrupt

5.8 (b)

Shortest seek time next or nearest cylinder next gives the best throughput.

5.9 (d)

Disk requires a device driver in computer system.

5.10 (c)

$$\frac{\text{DB size}}{\text{DBA}} = \text{Number of DBA's possible in one disk block.}$$

5.11 (a)

In a fixed block size file system, larger block size gives better disk throughput but poorer disk space utilization.

5.12 (c)

The number of disk block pointers that will be fit in one block

$$= \frac{2^{10}\text{ byte}}{32\text{ bit}} = \frac{2^{10}\text{ byte}}{4\text{ byte}} = 256$$

Maximum file size due to single indirection pointer

$$= 256 \times 1 = 256\text{ KB}$$

Maximum file size due to direct pointer

$$= 10 \times 1 = 10\text{ KB}$$

Maximum file size due to double indirection pointer

$$= 256 \times 256 \times 1\text{ KB}$$

Maximum file size due to triple indirection pointer

$$= 256 \times 256 \times 256 \times 1\text{ KB}$$

So the maximum file size

$$= 256 \times 256 \times 256 \times 1\text{ KB}$$
$$= 2^8 \times 2^8 \times 2^8 \times 2^{10}\text{ bytes}$$
$$= 2^{24} \times 2^{10}\text{ bytes} = 2^{34}\text{ bytes}$$

5.13 (c)

Using FCFS:

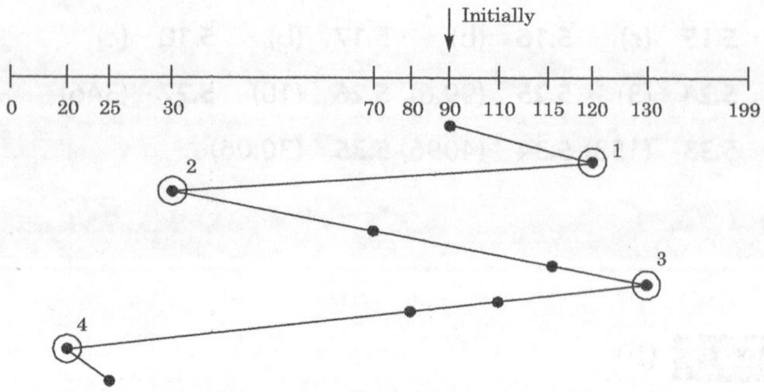

So, 4 head direction change.

Using SRTF:

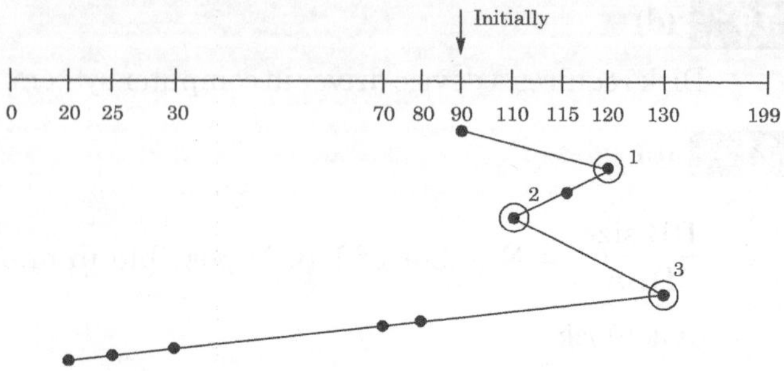

So, 3 head direction changed.

5.14 (b)

Maximum file size = [direct DBA + Number of (Data block size/DBA) + Number of (Data block size/DBA)^2 + Number of (Data block size/DBA)^3]* Data block size.

Data Block size = 1024 byte

(Disk Block size/DBA) = Number of Disk Block address Store inside One Block.

Maximum file size = [10+1*(128) + 1*(128*128*) + 1*(128*128*128)]*1024 Byte

= Approx 2 GB.

5.15 (c)

Using 100 bytes disk block:

100 B

File size	100 B block	Bytes needed for Book keeping	Block for Information
11050	111	111 × 4 = 444	5
4990	50	50 × 4 = 200	2
5170	52	52 × 4 = 208	3
12640	127	127 × 4 = 508	6
Total	**340**		**16**

Space = (340 + 16) × 100 = 35600

Using 200 bytes disk block:

200 B

File size	200 B block	Bytes needed for Book keeping	Block for Information
11050	56	56 × 4 = 224	2
4990	25	25 × 4 = 100	1
5170	26	26 × 4 = 104	1
12640	64	64 × 4 = 256	2
Total	**171**		**6**

Space = (171 + 6) × 200 = 35400

5.16 (b)

If there are enough extra frames, another process can be initiated. If the sum of the working-set sizes increases, exceeding the total number of available frames, the operating system selects a process to suspend. The process's pages are written out (swapped), and its frames are reallocated to other processes. The suspended process can be restarted later."

5.17 (b)

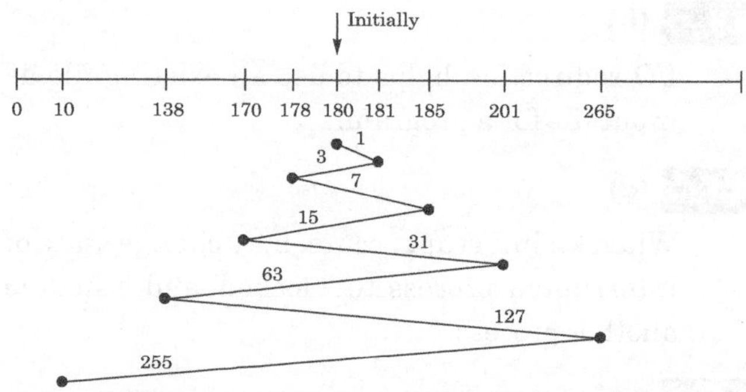

Initially at 180, so need to go at 181 since all option contain it.

Now 181 – 184 = 3 and 181 – 178 = 3 and 181 – 185 = 4.

Since on tie direction is not changed. So request service next will be 185. So option (b) and (d) can be answer.

Now after services 185, 185 – 200 = 15 and 185 – 170 = 15 and 185 – 201 = 16. So we need to choose request 201 since with 200 direction will not change.

5.18 (c)

If we see the pattern of distance of request, then 1, 3, 7, 15, 31, 63, 127, 255, 511, 1023, 2047 by using, new min distance = (old distance × 2) + 1. So, maximum 11 request can be satisfied.

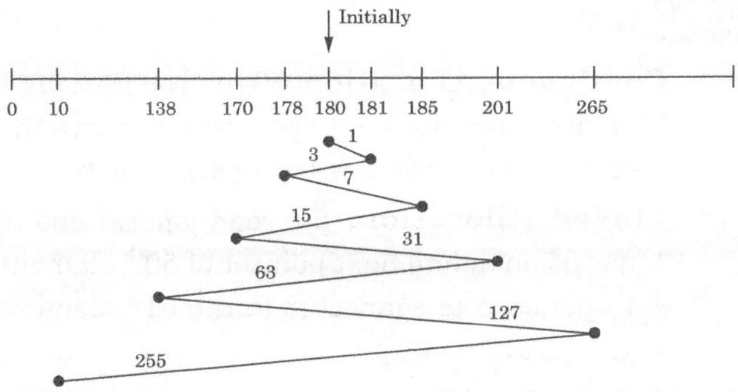

5.19 (b)

P and Q both are true but Q is not the reason for P because increasing the number of page frames allocated to a process sometimes increases the page fault rate or it is not concern with replacement policy.

5.20 (d)

An extension of index allocation is used to allocate the data block of a very large file.

5.21 (b)

Head is currently at cylinder 50,

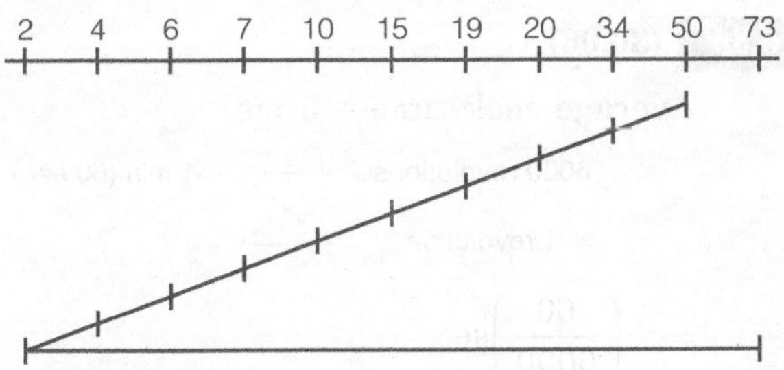

$$\begin{aligned} 50-34 &= 16, & 34-20 &= 14 \\ 20-19 &= 1, & 19-15 &= 4 \\ 15-10 &= 5, & 10-7 &= 3 \\ 7-6 &= 1, & 6-4 &= 2 \\ 4-2 &= 2, & 2-73 &= 71 \end{aligned}$$

$$\text{Total moves} = 16+14+1+4+5+3+1+2+2+71 = 119$$

5.22 (b)

Multilevel page table in preferred to reduce the size of page table needed to implement the virtual address space of a process.

5.23 (b)

Maximum possible size

$$= \left[\left(\begin{matrix}\text{address pointed by} \\ \text{doubly indrect block}\end{matrix}\right)^2 + \left(\begin{matrix}\text{addres pointed by} \\ \text{single address}\end{matrix}\right)\right.$$

+ Address points by single direct address] × [Block size]

$$= \left[\left(\frac{128}{8}\right)^2 + \left(\frac{128}{8}\right) + 8\right] \times 128 \text{ B}$$

$$= [2^8 + 2^4 + 2^3] \times 128 \text{ B}$$

$$= 32 \text{ kB} + 2 \text{ kB} + 1 \text{ kB} = 35 \text{ kB}$$

5.24 (3)

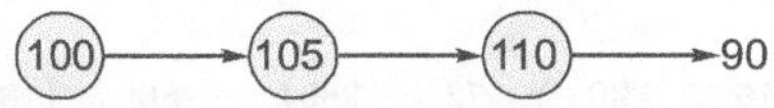

90 is serviced after servicing the 3 requests.

5.25 (99.6)

Overhead of each entry in FAT = 4 bytes

Block size = 10^3 bytes

Total size for each entry = 1004 bytes

$$\text{Number of entries in FAT} = \frac{100 \times 10^6}{1004} = 0.099601$$

Maximum size of a file

$$= 0.099601 \times 10^3 \text{ bytes} = 99.601 \times 10^6 \text{ bytes}$$

5.26 (10)

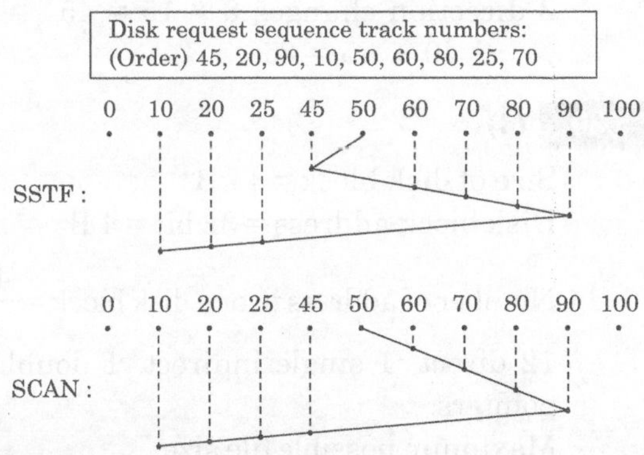

SSTF distance = (50 – 45)+(90 – 45)+(90 – 10)

= 5 + 45 + 80 = 130

SCAN distance = (90 – 50) + (90 – 10)

= 40 + 80 = 120

∴ SSTF distance – SCAN distance

= 130 – 120 = 10

5.27 (346)

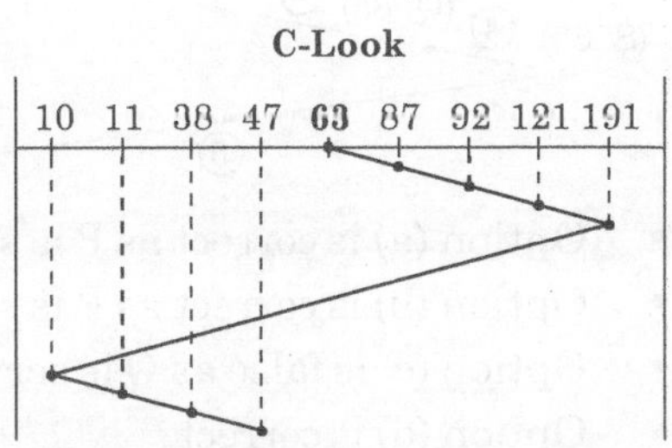

(84 – 63) + (92 – 87) + (121 – 92) + (191 – 121) + (191 – 10) + (11 – 10) + (38 – 11) + (47 – 38)
= 24 + 5 + 29 + 70 + 181 + 1 + 27 + 9 = 346

5.28 (d)

Linked and indexed allocations are non contigous so, they will not suffer from external fragmentation.

5.29 (85)

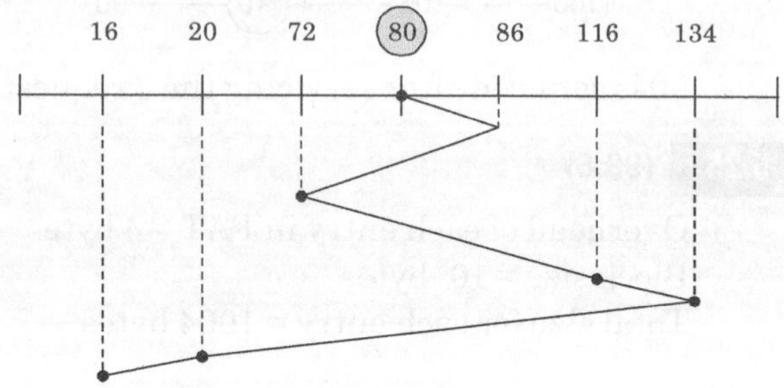

(86 – 80) + (86 – 72) + (134 – 72) + (134 – 16)
+ 62 + 118 + 14 = 200
100 → 20
200 → ?

$$\frac{200}{100} \times 20 = 40$$

3 direction changes 3 × 15 = 45
40 + 45 = 85

5.30 (4)

Size of disk block = 4 kB
Disk block address = 32 bit = 4 B

Number of address in one disk block = $\frac{4\text{ kB}}{4\text{ B}} = 2^{10}$

12 direct, 1 single indirect, 1 double indirect pointers.
Maximum possible file size

$$= (12 + 2^{10} + 2^{10} \times 2^{10}) \times 4\text{ kB}$$
$$= 4.00395\text{ GB} \cong 4\text{ GB}$$

5.31 (c)

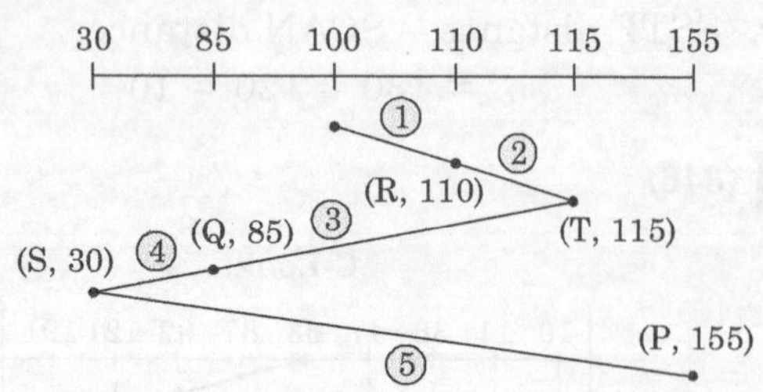

- Option (a) is correct as P is serviced last.
- Option (b) is correct as P is serviced last.
- Option (c) is false as Q is serviced before S.
- Option (d) is correct.

5.33 (153)

Contiguous Allocation: 50 read operations + 50 write operations + 1 operation to write the newly inserted block = 101 operations (n_A)

Linked Allocation: 50 read operations + 1 operation delete next pointer of 50th element + 1 operation to connect it to the 51st element = 52 operations (n_B)

Total operations:

$$\text{Required} = n_A + n_B$$
$$= 101 + 52 = 153$$

5.34 (4096)

Disk capacity = Number of surfaces × Number of tracks per surface × Number sectors per track × Number of bytes per sector

$$512\text{ GB} = 32 \times x \times 4096 \times 1024\text{ B}$$
$$2^{39}\text{ B} = 2^5 \times x \times 2^{12} \times 2^{10}\text{ B}$$
$$2^{39}\text{ B} = 2^{27}x\text{ B}$$
$$x = \frac{2^{39}\text{ B}}{2^{27}\text{ B}} = 2^{12}$$
$$x = 4096$$

5.35 (30.06)

Average seek time = 5 ms

6000 revolutions — 1 min (60 sec)
1 revolution — ?

$$\Rightarrow \left(\frac{60}{6000}\right)\text{sec}$$
$$\Rightarrow 10\text{ ms}$$

Average Rotational latency = $\frac{1}{2} \times 10\text{ ms} = 5\text{ ms}$

1 revolution time — 500 sectors
? — 1 sector

$$= \frac{10\text{ ms}}{500} = 0.02\text{ ms}$$

T_{avg} = Average seek time + Average Rotational latency + Transfer time + Over head

T_{avg} = (5 ms + 5 ms + 0.02 ms + 0) 3000
= 30060 ms = 30.06 sec

■■■■

6 Miscellaneous

6.1 A linker is given object modules for a set of programs that were compiled separately. What information need to be included in an object module?
(a) Object code
(b) Relocation bits
(c) Names and locations of all external symbols defined in the object module
(d) Absolute addresses of internal symbols

[1995 : 1 M]

6.2 When an interrupt occurs, an operating system
(a) ignores the interrupt
(b) always changes state of interrupted process after processing the interrupt
(c) always resumes execution of interrupted process after processing the interrupt
(d) may change state of interrupted process to 'blocked' and schedule another process

[1997 : 1 M]

6.3 A user level process in Unix traps the signal sent on a Ctrl-C input, and has a signal handling routine that saves appropriate files before terminating the process. When a Ctrl-C input is given to this process, what is the mode in which the signal handling routine executes?
(a) kernel mode (b) super user mode
(c) privileged mode (d) user mode

[2005 : 1 M]

6.4 A student wishes to create symbolic links in a computer system running Unix. Three text files named "file1", "file2" and "file3" exist in her current working directory, and the student has read and write permissions for all three files. Assume that file1 contains information about her hobbies, file2 contains information about her friends and file3 contains information about her courses. The student executes the following sequence of commands from her current working directory

```
ln -s file1 file2
ln -s file2 file3
```

Which of the following types of information would be lost from her file system?
I. Hobbies
II. Friends
III. Courses
(a) I and II only (b) II and III only
(c) II only (d) I and III only

[2005 : 1 M]

6.5 The shell command

```
find -name password -print
```

is executed in /etc directory of a computer system running Unix. Which of the following shell commands will give the same information as the above command when executed in the same directory?
(a) ls passwd
(b) cat passwd
(c) grep name passwd
(d) grep print passwd

[2005 : 1 M]

6.6 Consider n jobs $J_1, J_2, \ldots, J_n$ such that job J_i has execution time t_i and a non-negative integer weight w_i. The weighted mean completion time of the jobs is defined to be $\dfrac{\sum_{i=1}^{n} w_i T_i}{\sum_{i=1}^{n} w_i}$ where T_i is the completion time of job J_i. Assuming that there is only one processor available, in what order must the jobs be executed in order to minimize the weighted mean completion time of the jobs?
(a) Non-decreasing order of t_i
(b) Non-increasing order of w_i
(c) Non-increasing order of $w_i t_i$
(d) Non-increasing order of w_i / t_i

[2007 : 2 M]

Answers Miscellaneous

6.1 (c) **6.2** (d) **6.3** (a) **6.4** (b) **6.5** (a) **6.6** (d)

Explanations Miscellaneous

6.1 (c)

Linker definition: Linker is a program which links the compiled code with library files.

6.2 (d)

An interrupt is a signal from a device attached to a computer or from a program within the computer that causes the main program that operates the computer to stop and figure out what to do next. After the interrupt signal is sensed, it may change state of interrupted process to 'blocked' and schedule another process.

6.3 (a)

When user level process trapping the Ctrl+C signal then the trap signal is going through system call and that's why mode changed to Kernel mode from user mode and then the request is handling.

One more thing kernel mode and privilege mode are same, answer is Kernel mode (privilege mode).

6.4 (b)

In unix system: $ ln – s file 1, file 2 command replace existing file 2 by file 1. If file 2 exist then it is first removed then replace with file 1.

ln – s file 1 file 2
ln – s file 2 file 3

So, file 2 replace file 3 the file replace file 2 i.e. friends and courses are lost and hobbies are remaining.

6.6 (d)

To minimize the weighted mean completion time, jobs must be executed in increasing order of weight/time like FCFS scheduling algorithm. If we go with decreasing order of weight/time then it suffers from convoy effect.

■■■■

Databases

UNIT

IX

CONTENTS

Databases

Syllabus

ER-model, relational model (relational algebra, tuple calculus), Database design (integrity constraints, normal forms), Query languages (SQL). File structures (sequential files, indexing, B and B+ trees), Transactions and concurrency control.

Analysis of Previous GATE Papers

Exam Year	1 Mark Ques.	2 Marks Ques.	Total Marks
1995	–	1	**2**
1996	–	1	**2**
1997	–	2	**4**
1998	2	3	**8**
1999	4	3	**10**
2000	2	3	**8**
2001	2	2	**6**
2002	2	3	**8**
2003	2	3	**8**
2004	2	5	**12**
2005	3	4	**11**
2006	1	4	**9**
2007	–	6	**12**
2008	1	5	**11**
2009	–	5	**10**
2010	2	2	**6**
2011	–	2	**4**
2012	2	5	**12**
2013	–	3	**6**
2014 Set-1	2	3	**8**
2014 Set-2	2	3	**8**
2014 Set-3	2	3	**8**
2015 Set-1	2	2	**6**
2015 Set-2	2	2	**6**
2015 Set-3	2	2	**6**
2016 Set-1	4	1	**6**
2016 Set-2	2	2	**6**
2017 Set-1	2	3	**8**
2017 Set-2	2	3	**8**
2018	2	2	**6**
2019	2	3	**8**
2020	2	3	**8**
2021 Set-1	2	3	**8**
2021 Set-2	2	3	**8**
2022	3	2	**7**
2023	1	2	**5**
2024 Set-1	4	2	**8**
2024 Set-2	4	2	**8**

1 ER-Model

1.1 It is desired to design an object-oriented employee record system for a company. Each employee has a name, unique id and salary. Employees belong to different categories and their salary is determined by their category. The functions getName, getId and computeSalary are required. Given the class hierarchy below, possible locations for these functions are:
(i) getId is implemented in the superclass
(ii) getId is implemented in the subclass
(iii) getName is an abstract function in the superclass
(iv) getName is implemented in the superclass
(v) getName is implemented in the subclass
(vi) getSalary is an abstract function in the superclass
(vii) getSalary is implemented in the superclass
(viii) getSalary is implemented in the subclass

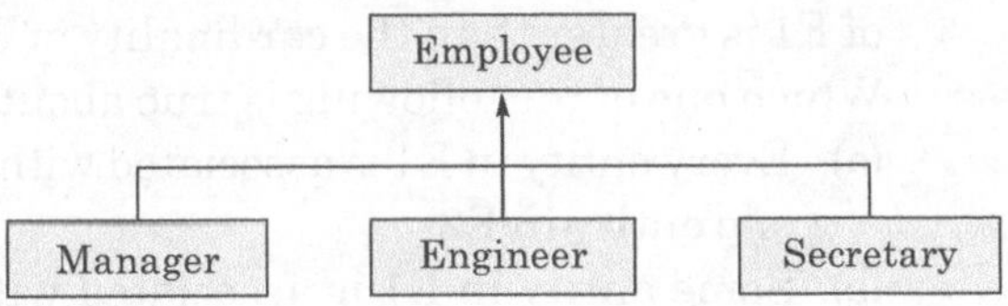

Choose the best design
(a) (i), (iv), (vi), (viii)
(b) (i), (iv), (vii)
(c) (i), (iii), (v), (vi), (viii)
(d) (ii), (v), (viii) **[2004 : 2 M]**

1.2 Consider the following entity relationship diagram (ERD), where two entities E1 and E2 have a relation R of cardinality 1 : m

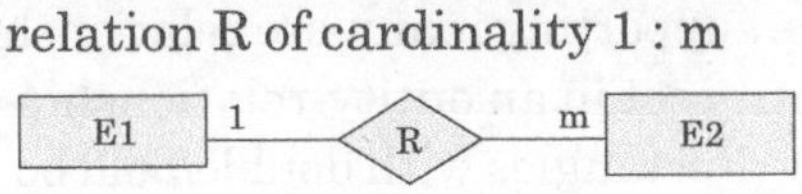

The attributes of E1 are A11, A12 and A13 where A11 is the key attribute. The attributes of E2 are A21, A22 and A23 where A21 is the key attribute and A23 is a multi-valued attribute. Relation R does not have any attribute. A relational database containing minimum number of tables with each table satisfying the requirements of the third normal form (3NF) is designed from the above ERD. The number of tables in the database is
(a) 2
(b) 3
(c) 5
(d) 4 **[2004 : 2 M]**

1.3 Consider the entities 'hotel room', and 'person' with a many to many relationship 'lodging' as shown below:

If we wish to store information about the rent payment to be made by person (s) occupying different hotel rooms, then this information should appear as an attribute of
(a) Person
(b) Hotel Room
(c) Lodging
(d) None of these

[2005 : 1 M]

1.4 Let E_1 and E_2 be two entities in an E/R diagram with simple single-valued attributes. R_1 and R_2 are two relationships between E_1 and E_2, where R_1 is one-to-many and R_2 is many-to-many. R_1 and R_2 do not have any attributes of their own. What is the minimum number of tables required to represent this situation in the relational model?
(a) 2
(b) 3
(c) 4
(d) 5

[2005 : 2 M]

1.5 The following table has two attributes A and C where A is the primary key and C is the foreign key referencing A with on-delete cascade.

A	C
2	4
3	4
4	3
5	2
7	2
9	5
6	4

The set of all tuples that must be additionally deleted to preserve referential integrity when the tuple (2, 4) is deleted is:
(a) (3, 4) and (6, 4)
(b) (5, 2) and (7, 2)
(c) (5, 2) (7, 2) and (9, 5)
(d) 1 **[2005 : 2 M]**

Common Data for Questions 1.6 & 1.7:

Consider the following ER diagram:

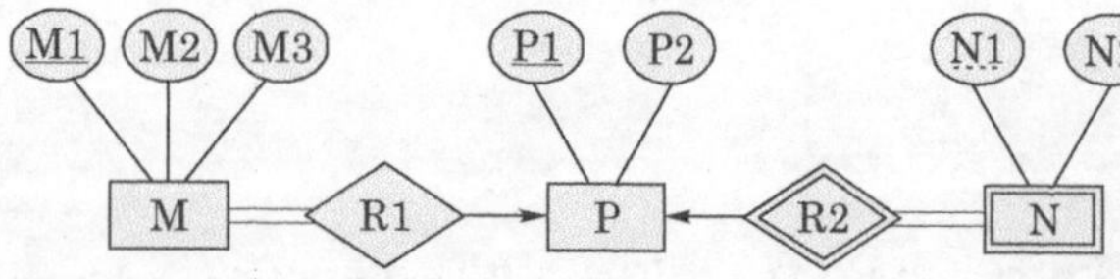

1.6 The minimum number of tables needed to represent M, N, P, R1, R2 is

(a) 2
(b) 3
(c) 4
(d) 5

[2008 : 2 M]

1.7 Which of the following is a correct attribute set for one of the tables for the correct answer to the above question?

(a) {M1, M2, M3, P1}
(b) {M1, P1, N1, N2}
(c) {M1, P1, N1}
(d) {M1, P1}

[2008 : 2 M]

1.8 Given the basic ER and relational models, which of the following is INCORRECT?

(a) An attribute of an entity can have more than one value
(b) An attribute of an entity can be composite
(c) In a row of a relational table, an attribute can have more than one value
(d) In a row of a relational table, an attribute can have exactly one value or a NULL value

[2012 : 1 M]

1.9 Consider an Entity-Relationship (ER) model in which entity sets E_1 and E_2 are connected by an m : n relationship R_{12}. E_1 and E_3 are connected by a 1 : n (1 on the side of E_1 and n on the side of E_3) relationship R_{13}.

E_1 has two single-valued attributes a_{11} and a_{12} of which a_{11} is the key attribute. E_2 has two single-valued attributes a_{21} and a_{22} of which a_{21} is the key attribute. E_3 has two single-valued attributes a_{31} and a_{32} of which a_{31} is the key attribute. The relationships do not have any attributes.

If a relational model is derived from the above ER model, then the minimum number of relations that would be generated if all the relations are in 3NF is ________.

[2015 (Set-1) : 2 M]

1.10 An ER model of a database consists of entity types A and B. These are connected by a relationship R which does not have its own attribute. Under which one of the following conditions, can the relational table for R be merged with that of A?

(a) Relationship R is one-to-many and the participation of A in R is total.
(b) Relationship R is one-to-many and the participation of A in R is partial.
(c) Relationship R is many-to-one and the participation of A in R is total.
(d) Relationship R is many-to-one and the participation of A in R is partial.

[2017 (Set-2) : 1 M]

1.11 In an Entity-Relationship (ER) model, suppose R is a many-to-one relationship from entity set E1 to entity set E2. Assume that E1 and E2 participate totally in R and that the cardinality of E1 is greater than the cardinality of E2. Which one of the following is true about R?

(a) Every entity in E1 is associated with exactly one entity in E2.
(b) Some entity in E1 is associated with more than one entity in E2.
(c) Every entity in E2 is associated with exactly one entity in E1.
(d) Every entity in E2 is associated with at most one entity in E1.

[2018 : 1 M]

1.12 Which one of the following is used to represent the supporting many-ore relationships of a weak entity set in an entity-relationship diagram?

(a) Rectangles with double/bold border
(b) Ovals with double/bold border
(c) Ovals that contain underlined identifiers
(d) Diamonds with double/bold border

[2020 : 1 M]

1.13 Let *S* be the specification: "Instructors teach courses. Students register for courses. Courses are allocated classrooms. Instructors guide students." Which one of the following ER diagrams CORRECTLY represents S?

(i)

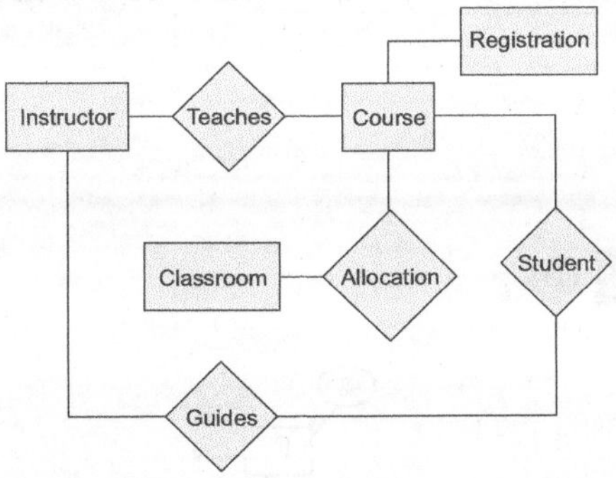

(ii)

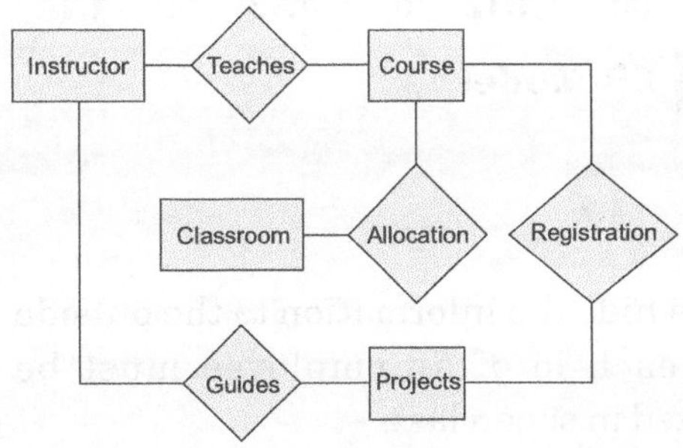

(iii)

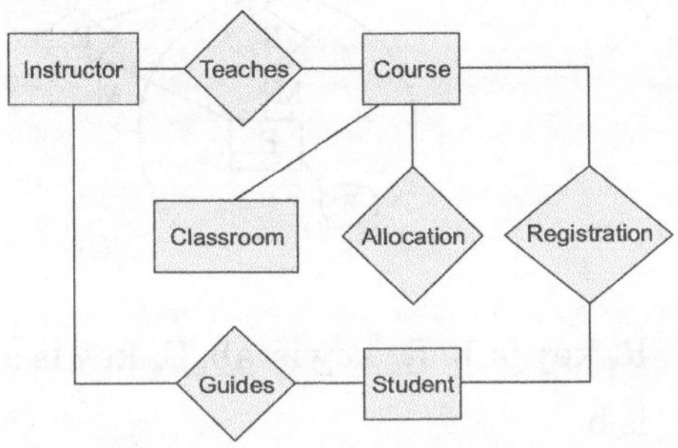

(iv)

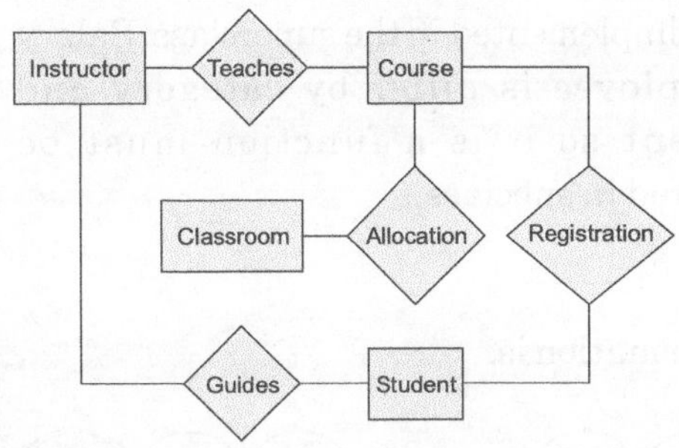

(a) (iii)
(b) (ii)
(c) (iv)
(d) (i)

[2024 (Set-1) : 1 M]

1.14 In the context of owner and weak entity sets in the ER (Entity-Relationship) data model, which one of the following statements is TRUE?

(a) The owner entity set MUST have total participation in the identifying relationship.
(b) Neither weak entity set nor owner entity set MUST have total participation in the identifying relationship.
(c) The weak entity set MUST have total participation in the identifying relationship.
(d) Both weak and owner entity sets MUST have total participation in the identifying relationship.

[2024 (Set-2) : 1 M]

■■■■

Answers ER-Model

1.1	(a)	**1.2**	(b)	**1.3**	(c)	**1.4**	(b)	**1.5**	(c)	**1.6**	(b)	**1.7**	(a)	**1.8**	(c)	**1.9**	(4)
1.10	(c)	**1.11**	(a)	**1.12**	(d)	**1.13**	(c)	**1.14**	(c)								

Explanations ER-Model

1.1 (a)

Superclass hide the information to the outside world so each id of an employee must be implemented in superclass.
Name is also an important attribute of employee so must be implemented in the superclass. Salary of an employee is differ by category and department so it is a function must be implemented in subclass.

1.2 (b)

Given information is:

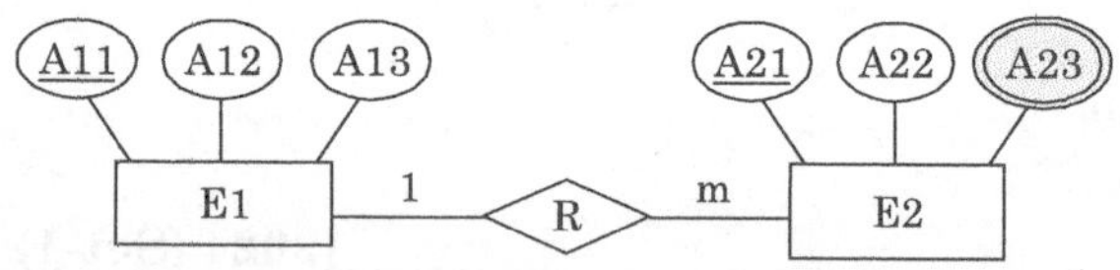

Relation table R combine with entity E2 (since 1-many relation. Relation combine with many side, in case of partial participation).

E1 (A11, A12, A13) with key <u>A11</u>

RE2 (A21, A22, A23, A11) with <u>A21, A23</u>

In RE2 partial dependence A21 → A22, A11 present. So, not in 2NF. To remove it we need 2 separate table i.e.

E1 (A11, A12, A13) with key <u>A11</u>

RE11 (A21, A23) with key <u>A21, A23</u>

RE22 (A21, A22, A11) with key <u>A21</u>

All the three tables are now in 3NF form.

1.3 (c)

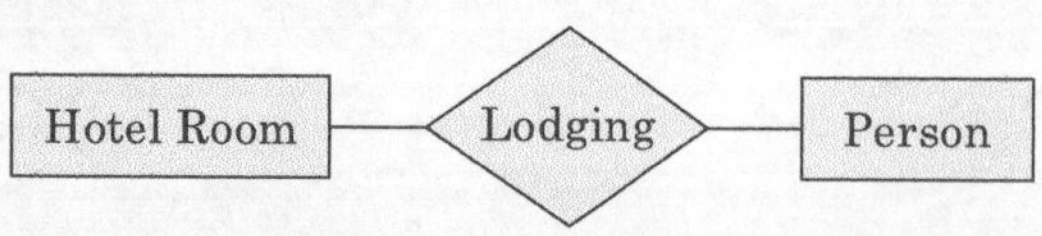

According to given E-R diagram, Hotel Room contain all the information about Hotel Room and Person contain all information related to Person. So, when any person do a payment for rent of room will be stored inside Lodging relation, since it is connected two entities.

1.4 (b)

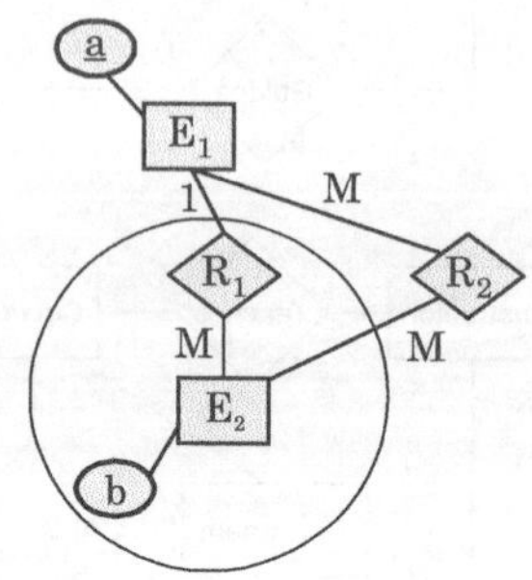

R_1 key is <u>b</u>, R_2 key is <u>ab</u>, E_1 key is a and E_2 key is b.

R_1 and E_2 can merge because keys are same.

∴ E_1, R_1E_2 and R_2 tables are required.

1.5 (c)

If (2, 4) is deleted then 2 is the primary key but in (5, 2) and (7, 2), 2 is the foreign key so these must be deleted. The primary key for (5, 2) and (7, 2) is 5 and 7 respectively but in (9, 5), 5 is the foreign key so it is also deleted.

1.6 (b)

R_1 relationship merges to entity set (M)
R_2 relationship is weak entity set N in one table
$M_R(\underline{M_1}\ M_2\ M_3\ P_1)\ P(\underline{P_1}\ P_2)\ NR_2\ (\underline{P_1 N_1}\ \ N_2)$
Total 3 relational tables required.

1.7 (a)

The correct attribute set is {M1, M2, M3, P1}

1.8 (c)

Option (a) is correct as there are multivalued attribute e.g. phone no (attribute)
Option (b) is also correct.
Option (d) 'Null' values are allowed in a row of relational database (Null value arc constraints only for primary keys).
As in 1NF also we remove multivalued attribute.
∴ Option (c) is incorrect.

1.9 (4)

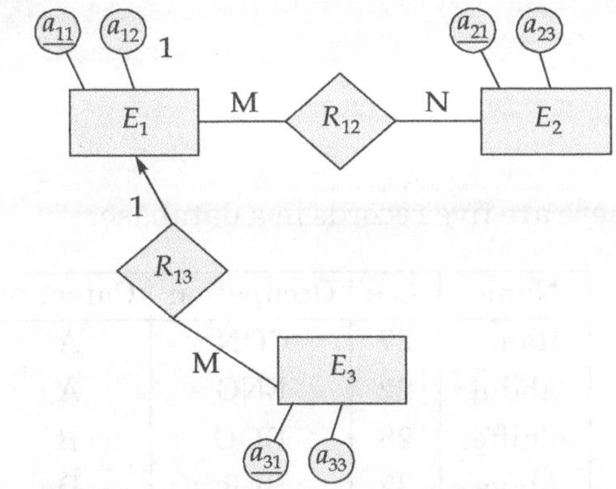

$E_3R_{13}(\underline{a_{31}}a_{33}a_{11})\ \{a_{31} \rightarrow a_{33}a_{11}\}$

$E_1(\underline{a_{11}}a_{12})\ \{a_{11} \rightarrow a_{12}\}$

$R_{12}(\underline{a_{11}a_{21}})$

$E_2(\underline{a_{21}}a_{23})\ \{a_{21} \rightarrow a_{23}\}$

Total minimum 4 tables required.

1.10 (c)

Entity sets A, B
Relationship set R
Relation R merges with that of A.

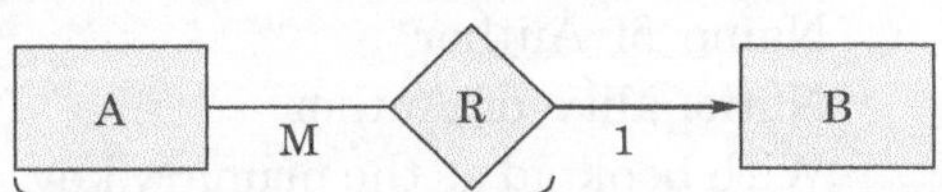

- Many to one relationship set can merge towards entity set 'A'.
- Partipation towards A side can be total / partial.

1.11 (a)

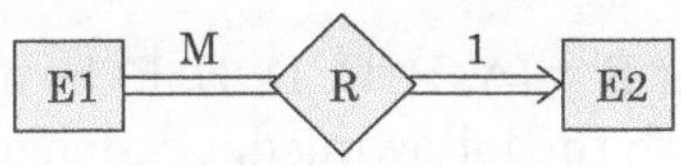

E1 entries > E2 entities

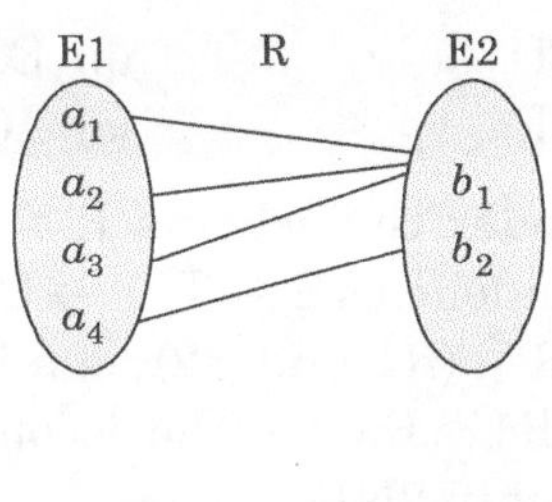

Every entity in E1 is associated with exactly one entity in E2.

Many: One cardinality and total participation both E1 E2 ends.

So that number of records of E1 = Number of records in R.

E1

<u>A</u>
a_1
a_2
a_3
a_4

R

<u>A</u>	B
a_1	b_1
a_2	b_1
a_3	b_1
a_4	b_2

E2

<u>B</u>
b_1
b_2

1.12 (d)

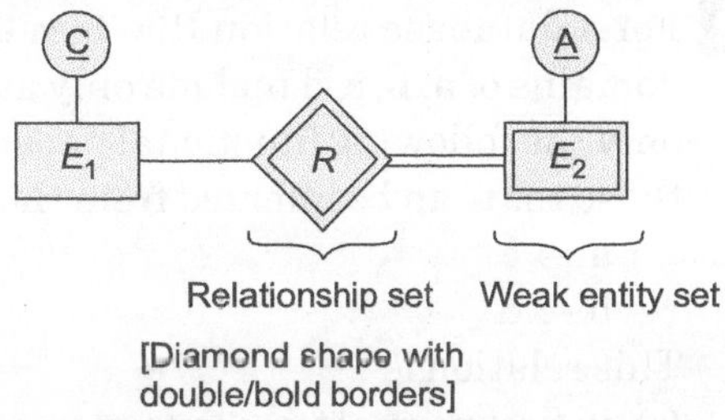

1.13 (c)

"Instructor", "Course", "Student", "Classroom" are entity sets.

"Teach" relationship set between Instructor & Course "Register" relationship set between Student & Course.

"Allocation" relationship set between course and Classroom.

"Guide" relationship set between Instructor & Student.

1.14 (c)

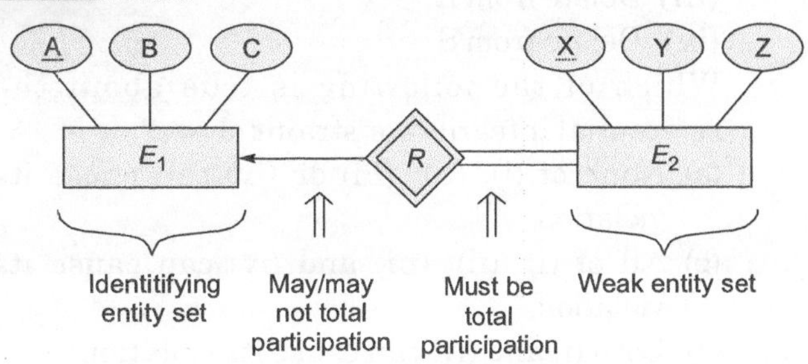

■■■■

2 Database Design: Functional Dependencies and Normalization

2.1 (a) Consider the relation scheme R(A, B, C) with the following functional dependencies:

$AB \rightarrow C$

$C \rightarrow A$

Show that the scheme R is the Third Normal Form (3NF) but not in Boyce- Code Normal Form (BCNF).

(b) Determine the minimal keys of relation R.

[1995 : 2 M]

2.2 For a database relation R(a, b, c, d), where the domains of a, b, c, d include only atomic values, only the following functional dependencies and those that can be inferred from them hold:

$a \rightarrow c$

$b \rightarrow d$

This relation is

(a) in first normal form but not in second normal form

(b) in second normal form but not in third normal form

(c) in third normal form

(d) None of the above

[1997 : 2 M]

2.3 Let R (a, b, c) and S(d, e, f) be two relations in which d is the foreign key of S that refers to the primary key of R. Consider the following four operations on R and S

(i) Insert into R

(ii) Insert into S

(iii) Delete from R

(iv) Delete from S

Which of the following is true about the referential integrity constraint above?

(a) None of (i), (ii), (iii) or (iv) can cause its violation

(b) All of (i), (ii), (iii) and (iv) can cause its violation

(c) Both (i) and (iv) can cause its violation

(d) Both (ii) and (iii) can cause its violation

[1997 : 2 M]

2.4 Which normal form is considered adequate for normal relational database design?

(a) 2 NF (b) 5 NF

(c) 4 NF (d) 3 NF **[1998 : 1 M]**

2.5 There are five records in a database:

Name	Age	Occupation	Category
Rama	27	CON	A
Abdul	22	ENG	A
Jeniffer	28	DOC	B
Maya	32	SER	D
Dev	24	MUS	C

There is an index file associated with this and it contains the values 1, 3, 2, 5 and 4. Which one of the fields is the index built from?

(a) Age (b) Name

(c) Occupation (d) Category

[1998 : 1 M]

2.6 Consider the following database relations containing the attributes:

Book_id

Subject_Category_of_book

Name_of_Author

Nationality_of_Author

With book_id as the primary key.

(a) What is the highest normal form satisfied by this relation?

(b) Suppose the attributes Book_title and Author_address are added to the relation, and the primary key is changed to {Name_of_Author, Book_title}, what will be the highest normal form satisfied by the relation? **[1998 : 2 M]**

2.7 Let R = (A, B, C, D, E, F) be a relation scheme with the following dependencies $C \rightarrow F$, $E \rightarrow A$, $EC \rightarrow D$, $A \rightarrow B$. Which of the following is a key for R?

(a) CD (b) EC

(c) AE (d) AC **[1999 : 1 M]**

2.8 Consider the schema R = (S T U V) and the dependencies $S \rightarrow T$, $T \rightarrow U$, $U \rightarrow V$ and $V \rightarrow S$. Let R = (R1 and R2) be a decomposition such that $R1 \cap R2 = \phi$. The decomposition is

(a) not in 2NF

(b) in 2NF but not 3NF

(c) in 3NF but not in 2NF

(d) in both 2NF and 3NF **[1999 : 2 M]**

2.9 Given the following relation instance:

X	Y	Z
1	4	2
1	5	3
1	6	3
3	2	2

Which of the following functional dependencies are satisfied by the instance?
(a) $XY \to Z$ and $Z \to Y$
(b) $YZ \to X$ and $Y \to Z$
(c) $YZ \to X$ and $X \to Z$
(d) $XZ \to Y$ and $Y \to X$ **[2000 : 2 M]**

2.10 Consider a schema R(A,B,C,D) and functional dependencies $A \to B$ and $C \to D$. Then the decomposition of R into R_1(AB) and R_2(CD) is
(a) dependency preserving and lossless join
(b) lossless join but not dependency preserving
(c) dependency preserving but not lossless join
(d) not dependency preserving and not lossless join
[2001 : 1 M]

2.11 R(A, B, C, D) is a relation. Which of the following does not have a lossless join, dependency preserving BCNF decomposition?
(a) $A \to B$, $B \to CD$ (b) $A \to B$, $B \to C$, $C \to D$
(c) $AB \to C$, $C \to AD$ (d) $A \to BCD$
[2001 : 2 M]

2.12 Relation R with an associated set of functional dependencies, F, is decomposed into BCNF. The redundancy (arising out of functional dependencies) in the resulting set of relation is
(a) Zero
(b) More than zero but less than that of an equivalent 3NF decomposition
(c) Proportional to the size of F^+
(d) Indeterminate
[2002 : 1 M]

2.13 Relation R is decomposed using a set of functional dependencies, F, and relation S is decomposed using another set of functional dependencies, G. One decomposition is definitely BCNF, the other is definitely 3 NF, but it is not known which is which. To make a guaranteed identification, which one of the following tests should be used on the decompositions? (Assume that the closures of F and G are available).
(a) Dependency-preservation
(b) Lossless-join
(c) BCNF definition
(d) 3 NF definition **[2002 : 2 M]**

2.14 From the following instance of a relation scheme R(A, B, C), we can conclude that:

A	B	C
1	1	1
1	1	0
2	3	2
2	3	2

(a) A functionally determines B and B functionally determines C
(b) A functionally determines B and B does not functionally determines C
(c) B does not functionally determines C
(d) A does not functionally determines B and B does not functionally determines C
[2002 : 2 M]

2.15 Consider the following functional dependencies in a database:
Data_of_Birth → Age
Age → Eligibility
Name → Roll_number
Roll_number → Name
Course_number → Course_name
Course_number → Instructor
(Roll_number; Course_number) → Grade
The relation (Roll_number; Name, Date_of_birth, Age) is
(a) in second normal form but not in third normal form
(b) in third normal form but not in BCNF
(c) in BCNF
(d) in none of the above **[2003 : 2 M]**

2.16 Consider the following relation schema pertaining to a students database:
Student (<u>rollno</u>, name, address)
Enroll (<u>rollno, courseno</u>, coursename)
where the primary keys are shown underlined. The number of tuples in the student and Enroll tables are 120 and 8 respectively. What are the maximum and minimum number of tuples that can be present in (Student * Enroll), where '*' denotes natural join?
(a) 8, 8 (b) 120, 8
(c) 960, 8 (d) 960, 120 **[2004 : 1 M]**

2.17 The relation scheme student Performance (name, courseNo, rollNo, grade) has the following functional dependencies:
name, courseNo → grade
RollNo, courseNo → grade
name → rollNo
rollNo → name

The highest normal form of this relation scheme is
(a) 2 NF (b) 3 NF
(c) BCNF (d) 4 NF

[2004 : 2 M]

2.18 A relation Empdtl is defined with attributes empcode (unique), name, street, city, state and pincode. For any pincode, there is only one city and state. Also, for any given street, city and state, there is just one pincode. In normalization terms, Empdtl is a relation in
(a) 1NF only
(b) 2NF and hence also in 1NF
(c) 3NF and hence also in 2NF and 1NF
(d) BCNF and hence also in 3NF, 2NF an 1NF

[2004 : 2 M]

2.19 A table has fields Fl, F2, F3, F4, F5 with the following functional dependencies:

$F1 \to F3$, $F2 \to F4$, $(F1, F2) \to F5$

In terms of Normalization, this table is in
(a) 1 NF
(b) 2 NF
(c) 3 NF
(d) None of these

[2005 : 1 M]

2.20 Which-one of the following statements about normal forms is FALSE?
(a) BCNF is stricter than 3 NF
(b) Loss less, dependency-preserving decomposition into 3 NF is always possible
(c) Loss less, dependency-preserving decomposition into BCNF is always possible
(d) Any relation with two attributes is BCNF

[2005 : 1 M]

2.21 Consider a relation scheme R = (A, B, C, D, E, H) on which the following functional dependencies hold:

$\{A \to B,\ BC \to D,\ E \to C,\ D \to A\}$

What are the candidate keys of R?
(a) AE, BE
(b) AE, BE, DE
(c) AEH, BEH, BCH
(d) AEH, BEH, DEH

[2005 : 2 M]

2.22 In a schema with attributes A, B, C, D and E following set of functional dependencies are given

$A \to B$
$A \to C$
$CD \to E$
$B \to D$
$E \to A$

Which of the following functional dependencies is NOT implied by the above set
(a) $CD \to AC$ (b) $BD \to CD$
(c) $BC \to CD$ (d) $AC \to BC$

[2005 : 2 M]

2.23 Consider the relations r_1(P,Q,R) and r_2 (R,S,T) with primary keys P and R respectively. The relation r_1 contains 2000 tuples and r_2 contains 2500 tuples. The maximum size of the join $r_1 \bowtie r_2$ is:
(a) 2000 (b) 2500
(c) 4500 (d) 5000 **[2006 : 1 M]**

2.24 Consider a relation R with five attributes V, W, X, Y, and Z. The following functional dependencies hold : $VY \to W$, $WX \to Z$, and $ZY \to V$. Which of the following is a candidate key for R?
(a) VXZ (b) VXY
(c) VWXY (d) VWXYZ

[2006 : 2 M]

2.25 The following functional dependencies are given: $AB \to CD$, $AF \to D$, $DE \to F$, $C \to G$, $F \to E$, $G \to A$.

Which one of the following options is false?
(a) $\{CF\}^+ = \{ACDEFG\}$
(b) $\{BG\}^+ = \{ABCDG\}$
(c) $\{AF\}^+ = \{ACDEFG\}$
(d) $\{AB\}^+ = \{ACDFG\}$

[2006 : 2 M]

2.26 Which one of the following statements is FALSE?
(a) Any relation with two attributes is in BCNF
(b) A relation in which every key has only one attribute is in 2NF
(c) A prime attribute can be transitively dependent on a key in 3NF relation
(d) A prime attribute can be transitively dependent on a key in a BCNF relation.

[2007 : 2 M]

2.27 Consider the following implications relating to functional and multivalued dependencies given below, which may or may not be correct.
(i) If $A \to\to B$ and $A \to\to C$ then $A \to BC$
(ii) If $A \to B$ and $A \to C$ then $A \to\to BC$
(iii) If $A \to\to BC$ and $A \to B$ then $A \to C$
(iv) If $A \to BC$ and $A \to B$ then $A \to\to C$

Exactly how many of the above implications are valid?
(a) 0 (b) 1
(c) 2 (d) 3 **[2007 : 2 M]**

2.28 Let R (A, B, C, D) be a relational schema with the following functional dependencies:

$A \rightarrow B$, $B \rightarrow C$, $C \rightarrow D$ and $D \rightarrow B$

The decomposition of R into (A, B), (B, C), (B, D)

(a) Gives a lossless join, and is dependency preserving
(b) Gives a lossless join, but is not dependency preserving
(c) Does not give a lossless join, but is dependency preserving
(d) Does not give a lossless join and is not dependency preserving **[2008 : 2 M]**

2.29 Let R (A, B, C, D, E, P, G) be a relational schema in which the following functional dependencies are known to hold:

$AB \rightarrow CD$, $DE \rightarrow P$, $C \rightarrow E$, $P \rightarrow C$ and $B \rightarrow G$

The relational schema R is

(a) in BCNF
(b) in 3NF, but not in BCNF
(c) in 2NF, but not in 3NF
(d) not in 2NF **[2008 : 2 M]**

2.30 Consider the following relational schemes for a library database:

Book (Title, Author, Catalog_no, Publisher, Year, price)

Collection (Title, Author, Catalog_no)

which the following functional dependencies:

I. Title Author $\rightarrow$ Catalog_no
II. Catalog_no $\rightarrow$ Title Author Publisher Year
III. Publisher Title Year $\rightarrow$ Price

Assume {Author, Title} is the key for both schemes: which of the following statements is true?

(a) Both Book and Collection are in BCNF
(b) Both Book and Collection are in 3NF only
(c) Book is in 2NF and Collection is in 3NF
(d) Both Book and Collection are in 2NF only **[2008 : 2 M]**

2.31 The following functional dependencies hold for relations R(A, B, C) and S(B, D, E):

$B \rightarrow A$
$A \rightarrow C$

The relation R contains 200 tuples and the relation S contains 100 tuples. What is the maximum number of tuples possible in the natural join $R \bowtie S$?

(a) 100 (b) 200
(c) 300 (d) 2000 **[2010 : 2 M]**

2.32 Consider a relational table with a single record from each registered student with the following attributes.

1. Registration_Num: Unique registration number of each registered student
2. UID: Unique identity number, unique at the national level for each citizen
3. BankAccount_Num: Unique account number at the bank. A student can have multiple accounts or joint accounts. This attribute stores the primary account number
4. Name: Name of the student
5. Hostel_Room: Room number of the hostel

Which of the following options is INCORRECT?

(a) BankAccount_Num is a candidate key
(b) Registration_Num can be a primary key
(c) UID is a candidate key if all students are from the same country
(d) If S is a superkey such that $S \cap UID$ is NULL then $S \cup UID$ is also a superkey **[2011 : 1 M]**

2.33 Which of the following is TRUE?

(a) Every relation in 3NF is also in BCNF
(b) A relation R is in 3NF if every non-prime attribute of R is fully functionally dependent on every key of R
(c) Every relation in BCNF is also in 3NF
(d) No relation can be in both BCNF and 3NF **[2012 : 1 M]**

Linked Answer Questions 2.34 and 2.35:

Relation R has eight attributes ABCDEFGH. Fields of R contain only atomic values.

$F = \{CH \rightarrow G, A \rightarrow BC, B \rightarrow CFH, E \rightarrow A, F \rightarrow EG\}$ is a set of functional dependencies (FDs) so that F is exactly the set of FDs that hold for R.

2.34 How many candidate keys does the relation R have?

(a) 3 (b) 4
(c) 5 (d) 6 **[2013 : 2 M]**

2.35 The relation R is

(a) in 1NF, but not in 2NF
(b) in 2NF, but not in 3NF
(c) in 3NF, but not in BCNF
(d) in BCNF **[2013 : 2 M]**

2.36 Consider the relation scheme $R = (E, F, G, H, I, J, K, L, M, N)$ and the set of functional dependencies $\{\{E,F\} \to \{G\}, \{F\} \to \{I,J\}, \{E,H\} \to \{K, L\}, \{K\} \to \{M\}, \{L\} \to \{N\}\}$ on R. What is the key for R?

(a) $\{E, F\}$ (b) $\{E, F, H\}$
(c) $\{E, F, H, K, L\}$ (d) $\{E\}$

[2014 (Set-1) : 1 M]

2.37 Given the following statements:

S1: A foreign key declaration can always be replaced by an equivalent check assertion in SQL.

S2: Given the table R(a,b,c) where a and b together form the primary key, the following is a valid table definition.

```
CREATE TABLE S (
a INTEGER,
d INTEGER,
e INTEGER,
PRIMARY KEY (d),
FOREIGN KEY (a) references R)
```

Which one of the following statements is **CORRECT**?

(a) S1 is TRUE and S2 is FALSE.
(b) Both S1 and S2 are TRUE.
(c) S1 is FALSE and S2 is TRUE.
(d) Both S1 and S2 are FALSE.

[2014 (Set-1) : 1 M]

2.38 Given the following two statements:

S1: Every table with two single-valued attributes is in 1NF, 2NF, 3NF and BCNF.

S2: $AB \to C, D \to E, E \to C$ is a minimal cover for the set of functional dependencies $AB \to C, D \to E, AB \to E, E \to C$.

Which one of the following is **CORRECT**?

(a) S1 is TRUE and S2 is FALSE.
(b) Both S1 and S2 are TRUE.
(c) S1 is FALSE and S2 is TRUE.
(d) Both S1 and S2 are FALSE.

[2014 (Set-1) : 2 M]

2.39 The maximum number of superkeys for the relation schema R(E,F,G,H) with E as the key is ________.

[2014 (Set-2) : 1 M]

2.40 Given an instance of the STUDENTS relation as shown below:

StudentID	StudentName	StudentEmail	StudentAge	CPI
2345	Shankar	shankar@math	X	9.4
1287	Swati	swati@ee	19	9.5
7853	Shankar	shankar@cse	19	9.4
9876	Swati	swati@mech	18	9.3
8765	Ganesh	ganesh@civil	19	8.7

For **(StudentName, StudentAge)** to be a key for this instance, the value X should NOT be equal to ________.

[2014 (Set-2) : 1 M]

2.41 A prime attribute of a relation scheme R is an attribute that appears

(a) in all candidate keys of R.
(b) in some candidate key of R.
(c) in a foreign key of R.
(d) only in the primary key of R.

[2014 (Set-3) : 1 M]

2.42 Consider the relation $X(P, Q, R, S, T, U)$ with the following set of functional dependencies

F = {
$\{P, R\} \to \{S, T\}$
$\{P, S, U\} \to \{Q, R\}$
}

Which of the following is the trivial functional dependency in F^+ is closure of F?

(a) $\{P, R\} \to \{S, T\}$ (b) $\{P, R\} \to \{R, T\}$
(c) $\{P, S\} \to \{S\}$ (d) $\{P, S, U\} \to \{Q\}$

[2015 (Set-3) : 1 M]

2.43 Which of the following is NOT a superkey in a relational schema with attributes V, W, X, Y, Z and primary key $V\,Y$?

(a) $V X Y Z$ (b) $V W X Z$
(c) $V W X Y$ (d) $V W X Y Z$

[2016 (Set-1) : 1 M]

2.44 A database of research articles in a journal uses the following schema:

(VOLUME, NUMBER, STARTPAGE, ENDPAGE, TITLE, YEAR, PRICE)

The primary key is (VOLUME, NUMBER, STARTPAGE, ENDPAGE) and the following functional dependencies exist in the schema.

(VOLUME, NUMBER, STARTPAGE, ENDPAGE) → TITLE
(VOLUME, NUMBER) → YEAR
(VOLUME, NUMBER, STARTPAGE, ENDPAGE) → PRICE

The database is redesigned to use the following schemas.

(VOLUME, NUMBER, STARTPAGE, ENDPAGE, TITLE, PRICE)
(VOLUME, NUMBER, YEAR)

Which is the weakest normal form that the new database satisfies, but the old one does not?

(a) 1NF (b) 2NF
(c) 3NF (d) BCNF

[2016 (Set-1) : 1 M]

2.45 The following functional dependencies hold true for the relational schema $R\{V, W, X, Y, Z\}$:

$V \to W$
$VW \to X$
$Y \to VX$
$Y \to Z$

Which of the following is irreducible equivalent for this set of functional dependencies?

(a) $V \to W$, $V \to X$, $Y \to V$, $Y \to Z$
(b) $V \to W$, $W \to X$, $Y \to V$, $Y \to Z$
(c) $V \to W$, $V \to X$, $Y \to V$, $Y \to X$, $Y \to Z$
(d) $V \to W$, $W \to X$, $Y \to V$, $Y \to X$, $Y \to Z$

[2017 (Set-1) : 1 M]

2.46 Consider the following tables T_1 and T_2:

T_1	
P	Q
2	2
3	8
7	3
5	8
6	9
8	5
9	8

T_2	
R	S
2	2
8	3
3	2
9	7
5	7
7	2

In table T_1, P is the primary key and Q is the foreign key referencing R in table T_2 with on-delete cascade and on-update cascade. In table T_2, R is the primary key and S is the foreign key referencing P in table T_1 with on-delete set NULL and on-update cascade. In order to delete record ⟨3, 8⟩ from table T_1, the number of additional records that need to be deleted from table T_1 is ______. **[2017 (Set-2) : 1 M]**

2.47 Consider the following four relational schemas. For each schema, all non-trivial functional dependencies are listed. The underlined attributes are the respective primary keys.

Schema I: Registration (<u>rollno</u>, courses)
Field 'courses' is a set-valued attribute containing the set of courses a student has registered for.
Non-trivial functions dependency:
rollno → courses

Schema II: Registration (<u>rollno, courseid</u>, email)
Non-trivial functional dependencies:
rollno, courseid → email
email → rollno

Schema III: Registration (<u>rollno, courseid</u>, marks, grade)
Non-trivial functional dependencies:
rollno, courseid → marks, grade
marks → grade

Schema IV: Registration (<u>rollno, courseid</u>, credit)
Non-trivial functional dependencies:
rollno, courseid → credit
courseid → credit

Which one of the relational schemas above is in 3NF but not in BCNF?

(a) Schema I (b) Schema II
(c) Schema III (d) Schema IV

[2018 : 2 M]

2.48 Let the set of functional dependencies F = {QR → S, R → P, S → Q} hold on a relation schema X = (PQRS). X is not in BCNF. Suppose X is decomposed into two schemas Y and Z where Y = (PR) and Z = (QRS).

Consider the two statements given below:

I. Both Y and Z are in BCNF
II. Decomposition of X into Y and Z is dependency preserving and lossless

Which of the above statements is/are correct?

(a) I only (b) Neither I nor II
(c) Both I and II (d) II only **[2019 : 2 M]**

2.49 Consider a relational table R that is in 3NF, but not in BCNF. Which one of the following statements is TRUE?

(a) R has a non-trivial functional dependency X → A, where X is not a superkey and A is a prime attribute.
(b) R has a non-trivial functional dependency X → A, where X is not a superkey and A is a non-prime attribute and X is not a proper subset of any key.
(c) R has a non-trivial functional dependency X → A, where X is not a superkey and A is a non-prime attribute and X is a proper subset of some key.
(d) A cell in R holds a set instead of an atomic value. **[2020 : 2 M]**

2.50 Consider the relation $R(P, Q, S, T, X, Y, Z, W)$ with the following functional dependencies

$$PQ \rightarrow X;\ P \rightarrow YX;\ Q \rightarrow Y;\ Y \rightarrow ZW$$

Consider the decomposition of the relation R into the constituent relations according to the following two decomposition schemes.

$D_1 : R = [(P, Q, S, T); (P, T, X); (Q, Y); (Y, Z, W)]$

$D_2 : R = [(P, Q, S); \{T, X); (Q, Y): (Y, Z, W)]$

Which one of the following options is correct?

(a) D_1 is a lossy decomposition, but D_2 is a lossless decomposition.

(b) Both D_1 and D_2 are lossless decompositions.

(c) Both D_1 and D_2 are lossy decompositions.

(d) D_1 is a lossless decomposition, but D_2 is a lossy decomposition.

[2021 (Set-1) : 2 M]

2.51 Consider the following statements S_1 and S_2 about the relational data model:

S_1: A relation scheme can have at most one foreign key.

S_2: A foreign key in a relation scheme R cannot be used to refer to tuples of R.

(a) Both S_1 and S_2 are false.

(b) S_1 is true and S_2 is false.

(c) S_1 is false and S_2 is true.

(d) Both S_1 and S_2 are true.

[2021 (Set-2) : 1 M]

2.52 Suppose the following functional dependencies hold on a relation U with attributes P, Q, R, S and T:

$$P \rightarrow QR,\ RS \rightarrow T$$

Which of the following functional dependencies can be inferred from the above functional dependencies?

(a) $PS \rightarrow T$

(b) $P \rightarrow R$

(c) $R \rightarrow T$

(d) $PS \rightarrow Q$

[2021 (Set-2) : 2 M]

2.53 In a relational data model, which one of the following statements is TRUE?

(a) A relation with only two attributes is always in BCNF.

(b) If all attributes of a relation are prime attributes, then the relation is in BCNF.

(c) Every relation has at least one non-prime attribute.

(d) BCNF decompositions preserve functional dependencies.

[2022 : 1 M]

2.54 Consider a relation $R(A, B, C, D, E)$ with the following three functional dependencies.

$$AB \rightarrow C;\ BC \rightarrow D;\ C \rightarrow E;$$

The number of superkeys in the relation R is ______.

[2022 : 1 M]

2.55 Which of the following statements about a relation R in first normal form (1NF) is/are TRUE?

(a) R cannot have a composite attribute

(b) R cannot have a foreign key

(c) R can have a multi-attribute key

(d) R cannot have more than one candidate key

[2024 (Set-1) : 1 M]

2.56 The symbol $\rightarrow$ indicates functional dependency in the context of a relational database. Which of the following options is/are TRUE?

(a) $(X \rightarrow Y$ and $Y \rightarrow Z)$ implies $X \rightarrow Z$

(b) $(X, Y) \rightarrow (Z, W)$ implies $X \rightarrow (Z, W)$

(c) $((X, Y) \rightarrow Z$ and $W \rightarrow Y)$ implies $(X, W) \rightarrow Z$

(d) $(X, Y) \rightarrow (Z, W)$ implies $(X, Y) \rightarrow Z$

[2024 (Set-1) : 2 M]

2.57 A functional dependency $F : X \rightarrow Y$ is termed as a useful functional dependency if and only if it satisfies all the following three conditions:

- X is not the empty set.
- Y is not the empty set.
- Intersection of X and Y is the empty set.

For a relation R with 4 attributes, the total number of possible useful functional dependencies is ______.

[2024 (Set-2) : 2 M]

■■■■

Answers Database Design: Functional Dependencies and Normalization

2.1	Sol.	2.2	(a)	2.3	(d)	2.4	(d)	2.5	(c)	2.6	Sol.	2.7	(b)	2.8	(a)	2.9	(b)
2.10	(c)	2.11	(c)	2.12	(a)	2.13	(c)	2.14	(c)	2.15	(d)	2.16	(a)	2.17	(b)	2.18	(b)
2.19	(a)	2.20	(c)	2.21	(d)	2.22	(b)	2.23	(a)	2.24	(b)	2.25	(c, d)	2.26	(d)	2.27	(c)
2.28	(a)	2.29	(d)	2.30	(c)	2.31	(a)	2.32	(a)	2.33	(c)	2.34	(b)	2.35	(a)	2.36	(b)
2.37	(d)	2.38	(a)	2.39	(8)	2.40	(19)	2.41	(b)	2.42	(c)	2.43	(b)	2.44	(b)	2.45	(a)
2.46	(0)	2.47	(b)	2.48	(d)	2.49	(a)	2.50	(d)	2.51	(a)	2.52	(a,b,d)	2.53	(a)	2.54	(8)
2.55	(a, c)	2.56	(a, c, d)	2.57	(50)												

Explanations Database Design: Functional Dependencies and Normalization

2.1 Sol.

(a) Relation scheme R(A, B, C) has following functional dependencies:

$AB \rightarrow C$

$C \rightarrow A$

The candidate keys of relation scheme R(A, B, C) is {AB, BC}

Functional dependency

$AB \rightarrow C$

is allowed in both BCNF & 3NF

Functional dependency

$C \rightarrow A$

Here both C & A are proper subset of candidate key. Therefore functional dependency $C \rightarrow A$ is allowed in 3NF however, not allowed in BCNF.

Therefore R is in 3NF but not in BCNF.

(b) The minimal keys of relation R are {AB, BC}

2.2 (a)

Relation R(a, b, c, d) has following functional dependencies:

$a \rightarrow c$

$b \rightarrow d$

The candidate key of relation R is {ab}

Attribute a & b are prime attribute whereas c & d are non-prime attribute.

Both the FD's of R is of kind

"Proper subset of candidate key → Non-prime attribute"

Above type of FD's are not allowed in 2NF therefore, relation R is in 1NF but not in 2NF.

2.3 (d)

Insertion into S can cause inconsistency, since it has foreign key which refers to the primary key of R.

Deletion from R can cause inconsistency because it's primary key is the foreign key for S.

2.4 (d)

There is not always a decomposition into BCNF that is lossless and dependency preserving. So 3NF is considered adequate for normal relational database design.

2.5 (c)

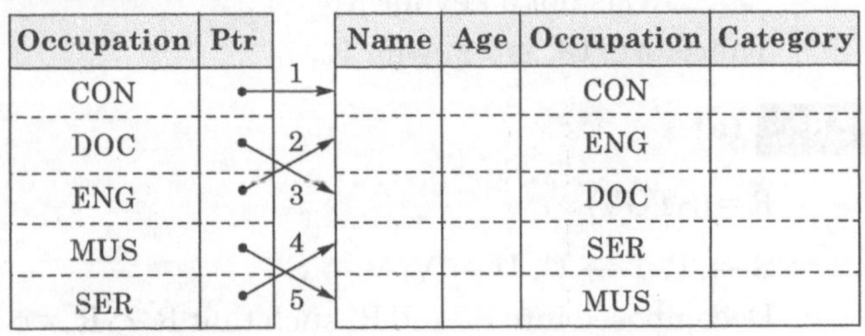

Pointers of index pointed to [1, 3, 2, 5, 4] records index key's of index file always sorting sequence order [CON, DOC, ENG, SER, MUS] index entries pointed to [1, 3, 2, 5, 4] records respectively.

2.6 Sol.

(a) R(bid, Catalog, NameAuthor, NationalityAuthor)

Primary key <u>bid</u>

Because bid primary key

{bid → Catalog NameAuthor, NationalityAuthor}

Given relation in <u>BCNF</u>.

(b) Two attributes BookTitle AuthorAddress added to relation

R(bid, Catalog, NameAuthor, NationalityAuthor, BookTitle, AuthorAddress)

{bid → Catalog NameAuthor, NationalityAuthor, NameAuthor BookTitle → bid catalog NationalityAuthor AuthorAddress}

[First FD because of previous relation is 2nd FD because of new primary key]

Candidate keys

[NameAuthor, BookTitle, bid, booktitle]

Partial dependencies exists (Not in 2NF)

Rel R in 1NF.

2.7 (b)

We can choose the correct option by finding the closure of each option

(a) CD

$(CD)^+ = \{C, D, F\}$

∴ CD is not a key for R.

(b) $(EC)^+ = \{A, B, C, D, E, F\}$

∴ EC is a key for R.

(c) $(AE)^+ = \{A, B, E\}$

∴ AE is not a key for R

(d) $(AC)^+ = \{A, B, C, F\}$

∴ AC is not a key for R

Therefore, EC is a key for R.

2.8 (a)

R = (STUV)

$S \rightarrow T$, $T \rightarrow U$, $U \rightarrow V$, $V \rightarrow S$

Decomposed into R_1 and R_2 such that $R_1 \cap R_2 = \phi$ is lossy join decomposition which is not allowed in 2NF, 3NF, BCNF.

2.9 (b)

(a) $XY \rightarrow Z$ is satisfied whereas $Z \rightarrow Y$ is not satisfied.

(b) $YZ \rightarrow X$ & $Y \rightarrow Z$ both are satisfied

(c) $YZ \rightarrow X$ is satisfied whereas $X \rightarrow Z$ is not satisfied

(d) $XZ \rightarrow Y$ is not satisfied whereas $Y \rightarrow X$ is satisfied.

2.10 (c)

R(A, B, C, D)

R_1(AB) and R_2(CD)

$F_1 = (A \rightarrow B)$ and $F_2 = \{C \rightarrow D\}$

All the original functional dependency can be derived from F_1 and F_2, hence the decomposition is dependency preserving. The attribute B doesnot functionally determine either 'A' or 'C' or 'D'. The attribute D doesnot functionally determine either 'A' or 'B' or 'C'.

∴ It is not lossless join (or it is lossy).

2.11 (c)

(a) {AB, BCD} is BCNF decomposition is loseless join and dependency preserving.

(b) {AB, BC, CD} is BCNF decomposition is loseless join and dependency preserving.

(c) {CAD, BC} is BCNF decomposition is loseless join but not dependency preserving.

(d) Given functional dependece is in BCNF.

2.12 (a)

Degree of redundancy over functional dependencies (single valued FD's) is zero percent.

2.13 (c)

The relation which is in BCNF is also in 3NF whereas the relation which is in 3NF may be or may not be in BCNF. Therefore, to make a guaranteed identification, BCNF definition should be used on the decomposition.

2.14 (c)

Since given snapshot is only an instance of the database. Just by checking the snapshot, we can't determine whether something is functional determined or not but on the contrary to this, we can correctly determine something which is not functionally determined.

Hence B does not functionally determine C is the correct answer.

2.15 (d)

Relation (Roll_number, Name, Date-of-birth, Age)

Fd's which are the member of the relation (other FD's can discard)

Date-of-birth → Age

Name → Roll_number

Roll_number → Name

Candidate keys {Name, date-of-birth}, {Rollnumber, date-of-birth}

Date-of-birth → Age is partial dependency which is not allowed in 2NF. Highest NF of given relation 1 NF.

2.16 (a)

Student (rno, name, address) {rno} unique.

Enroll (rno, cno, Cname) {rnocno} unique.

Each record of Enroll can be mapped one record of student based on natural join condition. So maximum 8 tuples in natural join and rno of Enroll foreign key references to student. So minimum also 8 tuples in natural join.

2.17 (b)

Student (name, courseNo, rollno, grade)
Rollno, CourseNo → grade
Rollno, CourseNo → grade
Name → rollno
rollno → name
candidate keys {name, courseNo}, {rollno, CourseNo}
Highest NF of given relation 3NF but not BCNF.

2.18 (b)

Relation Empdtl (empcode, name, street, city, state, pincode) and functional dependency:
empcode → name, street, city, state, pincode
pincode → state, city
enter and street, city, state → pincode.
Here candidate key is empcode. So, relation is in 2NF but pincode → state form non-prime to non prime which is not allowed in 3NF. Hence relation is in 2NF but not in 3NF.

2.19 (a)

Candidate key is F1, F2 in given relation.

So, F1 → F3 and F2 → F4 forms partial dependency which is not allowed in 2NF. Hence relation is in 1NF only.

2.20 (c)

Lossless join decomposition into BCNF always possible. But dependency preserving BCNF decomposition may not possible for all relations.

2.21 (d)

R = (A, B, C, D, E, H)

FD = {A → B, BC → D, E → C, D → A,}

The candidate keys of R is AEH, BEH, DEH because both of these generates the {A, B, C, D, E, H}

2.22 (b)

Apply membership test for all the given Functional Dependencies.

1. CD → AC
   CD^+ = CDEAB
2. BD → CD
   BD^+ = BD

i.e. BD cannot derive CD and hence is not implied. Similarly for rest two can be done.

2.23 (a)

Relation r_1 contains 2000 tuples, and r_2 contains 2500 tuples.
Maximum common R values are possible for join of r_1 and r_2 is 2000 i.e. equal to number of tuples in foreign key relation.

2.24 (b)

$(VY)^+$ = VYW
$(VYX)^+$ = VYWXZ
We have a dependencies ZY → V and WX → Z.
∴ VYX, ZYX and WXY are candidate keys.
⇒ VXY is candidate key for R.

2.25 (c, d)

(c) $(AF)^+$ = {AFED} but given
$(AF)^+$ = {ACDEFG} which is wrong.

(d) $(AB)^+$ = {ABCDG} but given
$(AB)^+$ = {ACDFG} which is wrong.

2.26 (d)

Statement (d) is false because in BCNF relation a prime attribute can't be transitively dependent on a key.

2.27 (c)

(i) If A→→B and A→→C then A→BC not valid

(ii) If A→B and A→C then A→BC, therefore A→→BC is also valid.

(iii) If A→→BC and A→B, then A→C not valid

(iv) If A→BC and A→B then A→C, therefore A→→C also valid.

Every FD can replace as multi-valued dependency and multi valued dependency not allowed to split.

2.28 (a)

R(A, B, C, D) $\{A \to B, B \to C, C \to D, D \to B\}$ [AB, BC, BD] $R_1(AB)$, $R_2(BC)$, $R_3(BD)$ ⋈ B is key for R_2 $R_1 \bowtie R_2$ (ABC) ⋈ B is key for R_3 [R_1, R_2, R_3] Lossless join decomposition.	R(A, B, C, D) $\{A \to B, B \to C, C \to D, D \to B\}$ $R_1(AB)$ | $R_2(BC)$ | $R_3(BD)$ $A \to B$ | $B \to C$, $C \to B$ | $B \to D$, $D \to B$ are FDs of subrelations Every FD and R implied in sub relations. Dependency preserving decomposition.

2.29 (d)

Not in 2NF because here candidate key is AB and in FD's proper subset of C.K. determine the non prime attribute i.e. $B \to G$.

2.30 (c)

book (title, author, catalogno, publisher, year, price)

{file author → catalogno}

catalogno → title author publisher year

publisher title year price}

Candidate keys {<u>title author</u>, <u>catalogno</u>}

book is in <u>2NF</u> between not 3NF

Collection (title author catalogno)

{title author → catalogno

catalog → title author}

candidate keys {<u>title author</u>, <u>catalogno</u>}

collection is in <u>BCNF</u> so also in 3NF.

2.31 (a)

B → A where R(A, B, C) and S(B, D, E)

A → C

R contains 200 tuples

S contains 100 tuples

Natural join $\bowtie = R \bowtie S = \pi[\sigma(R \times S)]$

R = 200 tuples and S = 100 tuples.

$R \bowtie S = 100$

[common in both or we can say Distinct equality between all common Attribute]

2.32 (a)

BankAccount_Num can not be a key, because if students has only joint accounts then it can not identify uniquely.

2.33 (c)

BCNF is more restrictive than 3NF. A relation is in BCNF iff X → Y where 'X' is super key and in 3NF iff other 'X' is super key or 'Y' is prime attribute.

∴ Every relation in BCNF is also in 3NF.

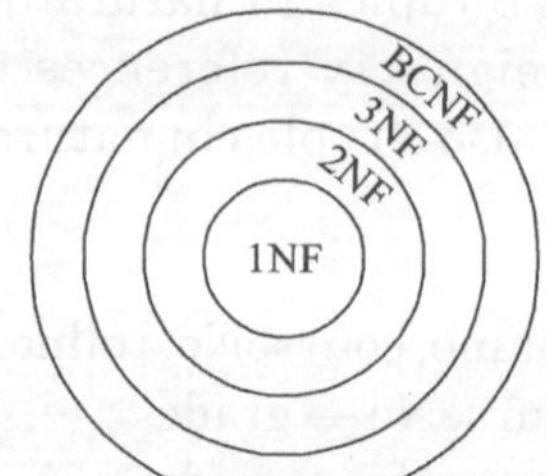

2.34 (b)

Relation contain 8 attributes ABCDEFGH

F = {CH → G, A → BC, B → CFH, E → A, F → EG}

Candidate key : AD, ED, BD, FD

2.35 (a)

Option (a) is correct because partial dependencies exist in functional dependency.

2.36 (b)

$\{E, F, H\} \to \{E, F, H, G, I, J, K, L, M, N\}$

∴ EFH is a key for the relation R.

2.37 (d)

S1: False because check assertion meant for to specify given attribute values in specific range and not allowed null. Foreign key constraint meant for attribute values should belongs to referenced key values and allowed null's.

S2: False

R(<u>AB</u>C) <u>AB</u> primary key

S(A<u>D</u>E) <u>D</u> primary key

'A' foreign key references to R not allowed because referenced key two attributes <u>AB</u> so that foreign key also must be two attributes.

2.38 (a)

S1: Every two attribute relation is BCNF

∴ S1 is true

S2: $FD = \{AB \to C, D \to E, AB \to E, E \to C\}$

$\{AB \to C, D \to E, E \to C\}$ is not a minimal cover.

AB can not determine E.

∴ S1 is TRUE and S2 is FALSE

2.39 (8)

Super keys:
= {E, EF, EG, EH, EFG, EFH, EGH, EFGH}

2.40 (19)

(StudentName, StudentAge) is a key

Shankar X
Shankar 19 } here X should not be 19

2.41 (b)

A prime attribute is an attribute that appears in some candidate key of given relation *R*.

2.42 (c)

$PS \rightarrow S$ is trivial FD, because $\{P, S\} \supseteq \{S\}$.

2.43 (b)

In *V W X Z*, no complete candidate key exists. So, it is not super key.

2.44 (b)

Journal (Volume, Number, Startpage, Endpage, Title, Year, Price)
Primary key: Volume, Number, Startpage, Endpage
FD's: Volume Number Startpage Endpage → Title
Volume number → Year
Volume Number, Startpage Endpage → Price
Given relation 1NF but not 2NF. This DB is redesigned following schemas
R_1(Volume, Number Startpage Endpage Title Price) which has FD's
Volume Number, Startpage Endpage → Title
Volume Number Startpage Endpage → Price
Which is in BCNF.
R_2(Volume, Number, Year)
Volume Number → Year
Which is also in BCNF.
Journal in 1NF.
R_1 R_2 in BCNF.
Weakest NF which satisfy R_1 and R_2 and fails for journal is 2NF.

2.45 (a)

$\{V \rightarrow W, VW \rightarrow X, Y \rightarrow V, Y \rightarrow X, Y \rightarrow Z\}$
W extraneous from $VW \rightarrow X$
$\{V \rightarrow W, V \rightarrow X, Y \rightarrow V, Y \rightarrow X, Y \rightarrow Z\}$
$Y \rightarrow X$ is redundant FD from above set
$\{V \rightarrow W, V \rightarrow X, Y \rightarrow V, Y \rightarrow Z\}$ is minimal cover.

2.46 (0)

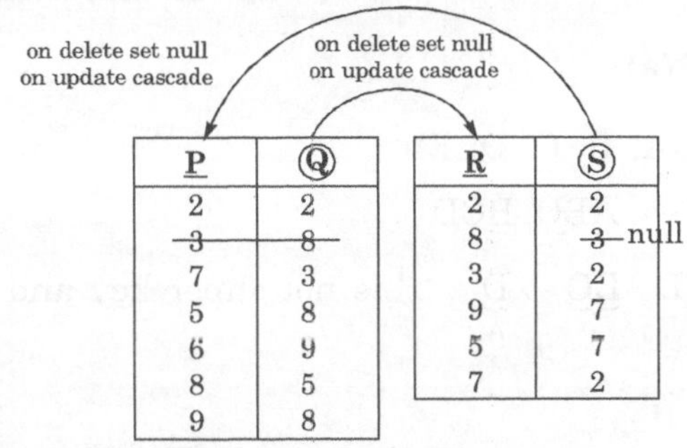

No other record need to delete because of deletion of (3, 8) record from T_1.

2.47 (b)

Schema II: Registration (rollno, courseid, email)
Primary key [rollno, courseid]
Non-trivial functional dependencies:
{ rollno, courseid → email
email → rollno}
candidate keys {rollno, courseid},
{email courseid}

Given relation is in 3NF but not in BCNF.

2.48 (d)

X(PQRS) {QR → S, R → P, S → Q} decomposed into

Y(PR)	Z(QRS)
{R → P}	{QR → S, S → Q}
Candidate key : R	Candidate key : QR, RS
Relation Y in BCNF	Relation Z in 3NF but not BCNF

Common attribute between Y and Z relations is R which is key for relation Y.

So that given decomposition is lossless join decomposition.

R → P in Y

$\left.\begin{matrix} QR \to S \\ S \to Q \end{matrix}\right\}$ are in Z

and dependency preserving decomposition.

2.49 (a)

R(A, B, C, D, E)

Key: ABC, BCD

FD: $\underbrace{BC}_{x} \to \underbrace{D}_{A}$: 'x' is not super key and 'A' is prime attribute.

2.50 (d)

- D_1: R_1(P, Q, S, T) R_2(P, T, X) R_3(Q, Y) R_4(Y, Z, W)

R_1R_2(P, Q, S, T, X) — PT → X [common attributes is a key of R_2 so lossless R_{12}]

R_{34}(Q, Y, Z, W) — Y → ZW [common attributes is a key of R_4 so lossless R_{34}]

R_{12}(P, Q, S, T, X) R_{34}(Q, Y, Z, W)

Q^+ = {Q, Y, Z, W}

R_{1234}(P, Q, S, T, X, Y, Z, W)

Q → YZW

[common attributes is a key of R_{34} overall lossless decomposition]

- D_2: R_1(P, Q, S) R_2(T, X) R_3(Q, Y) R_4(Y, Z, W)

R_1(T, X) R_1(P, Q, S) R_{34}(Q, Y, Z, W) Y → ZW [lossless R_{34}]

R_2(T, X) R_{134}(P, Q, S, Y, Z, W) Q → YZW [lossless R_{134}]

No common attribute ⇒ Lossy

D_1 is a lossless decomposition, but D_2 is a lossy decomposition.

2.51 (a)

S_1 : A relation scheme can have at most one foreign key.

e.g. $E_1(\underline{AB}) \xleftarrow{f_k} R(\underline{AC}) \xrightarrow{f_k} E_2(\underline{CD})$

More than 1 f_k also possible.

S_2 : A foreign key in a relational scheme R_1 cannot be used to refer to tuples of R.

f_k : Sup → Eid

Emp (Eid	ename	Sup)
e_1	A	E_1
e_2	A	E_1
e_3	A	E_2

e.g. Self referential relationship.

So, both S_1 and S_2 are false.

2.52 (a, b, d)

1. $PS \rightarrow T$	$(PS)^+ = \{P, S, Q, R, T\}$
2. $P \rightarrow R$	$(P)^+ = \{P, Q, R\}$
3. $R \rightarrow T$	$(R)^+ = \{R\}$
4. $PS \rightarrow Q$	$(PS)^+ = \{P, S, Q, R, T\}$

2.53 (a)

Relation with 2 attributes always in BCNF

R(AB) {A → B} ⇒ BCNF

{B → A} ⇒ BCNF

{A → B, B → A} ⇒ BCNF

{No non-trivial FD's} ⇒ BCNF

- If all prime attributes than relation always in 3NF but may not BCNF.
- Not mandatory of atleast one non-prime attribute in RDBMS table.
- Not every relation can decompose into BCNF with dependency preserving.

2.54 (8)

AB : Candidate key

Number of superkeys 8.

2.55 (a, c)

1NF relation:

- Every attribute of R must be atomic.
- Not allowed multivalued attributes.
- Not allowed composite attributes.
- Must have atleast one candidate key.
- May consist foreign keys.
- Can have composite key [multi-attribute key].

2.56 (a, c, d)

- $X \rightarrow Y, Y \rightarrow Z \Rightarrow X \rightarrow Z$
- $XY \rightarrow ZW \nRightarrow X \rightarrow ZW$
- $XY \rightarrow Z, W \rightarrow Y \Rightarrow XW \rightarrow Z$
- $XY \rightarrow ZW \Rightarrow XY \rightarrow Z$

2.57 (50)

$$R(ABCD)\begin{Bmatrix} X \rightarrow Y \text{ FD set} \\ X, Y \text{ non empty sets and } X \cap Y = \phi \end{Bmatrix}$$

Any non-empty sub set

$A \rightarrow \boxed{\{B, C, D\}} : 2^3 - 1$	$AB \rightarrow \boxed{\{C, D\}} : 2^2 - 1$	$ABC \rightarrow D$
$B \rightarrow \{A, C, D\} : 2^3 - 1$	$AC \rightarrow \{B, D\} : 2^2 - 1$	$BCD \rightarrow A$
$C \rightarrow \{A, B, D\} : 2^3 - 1$	$AD \rightarrow \{B, C\} : 2^2 - 1$	$ACD \rightarrow B$
$D \rightarrow \{A, B, C\} : 2^3 - 1$	$BC \rightarrow \{A, D\} : 2^2 - 1$	$ABD \rightarrow C$
	$BD \rightarrow \{A, C\} : 2^2 - 1$	
	$CD \rightarrow \{A, B\} : 2^2 - 1$	

$\Rightarrow \quad 4 \times 7 + 6 \times 3 + 4 = 50$

■■■■

3 Structured Query Language (SQL)

3.1 Suppose we have a database consisting of the following three relations.
FREQUENTS (student, parlor) giving the parlors each student visits.
SERVES (parlor, ice-cream) indicating what kind of ice-creams each parlor serves.
LIKES (student, ice-cream) indicating what ice-creams each student likes.
(Assume that each student likes at least one ice-cream and frequents at least one parlor)
Express the following in SQL:
Print the students that frequent at least one parlor that serves some icecream that they like.
[1998 : 2 M]

3.2 Which of the following is/are correct?
(a) An SQL query automatically eliminates duplicates
(b) An SQL query will not work if there are no indexes on the relations
(c) SQL permits attribute names to be repeated in the same relation
(d) None of the above
[1999 : 2 M]

3.3 Consider the set of relations:
EMP (Employee-no, Dept-no, Employee-name, Salary)
DEPT (Dept-no, Dept-name, Location)
Write an SQL query to:
(a) Find all employee names who work in departments located at 'Calcutta' and whose salary is greater than Rs.50,000.
(b) Calculate, for each department number, the number of employees with a salary greater than Rs.1,00,000. **[1999 : 2 M]**

3.4 In SQL, relations can contain null values, and comparisons with null values are treated as unknown. Suppose all comparisons with a null value are treated as false. Which of the following pairs is not equivalent?
(a) x = 5, not (not (x = 5))
(b) x = 5, x > 4 and x < 6, where x is an integer
(c) $x \neq 5$, not (x = 5)
(d) None of the above **[2000 : 2 M]**

3.5 Given relations r(w, x) and s(y, z), the result of
select distinct w, x
from r, s
is guaranteed to be same as r, provided
(a) r has no duplicates and s is non-empty
(b) r and s have no duplicates
(c) s has no duplicates and r is non-empty
(d) r and s have the same number of tuples
[2000 : 2 M]

3.6 Consider the following SQL query:
select distinct $a_1, a_2, ..., a_n$
from $r_1, r_2 ..., r_m$
where ρ
For an arbitrary predicate ρ, this query is equivalent to which of the following relational algebra expressions?
(a) $\Pi_{a_1,a_2,...,a_n} \sigma_\rho(r_1 \times r_2 \times ... \times r_m)$
(b) $\Pi_{a_1,a_2,...,a_n} \sigma_\rho(r_1 \bowtie r_2 \bowtie ... \bowtie r_m)$
(c) $\Pi_{a_1,a_2,...,a_n} \sigma_\rho(r_1 \cup r_2 \cup ... \cup r_m)$
(d) $\Pi_{a_1,a_2,...,a_n} \sigma_\rho(r_1 \cap r_2 \cap ... \cap r_m)$ **[2003 : 1 M]**

3.7 Consider the set of relations shown below and the SQL query that follow:
Students:
(Roll_number, Name, Date_of_birth)
Courses:
(Course_number, Course_name, Instructor)
Grades:
(Roll_number, Course_number, Grade)
select distinct Name
from Students, Courses, Grades
where
Students.Roll_number= Grades.Roll_number
and
Courses.Instructor=Korth and
Courses.Course_number=
Grades.Course_number and
Grades. grade=A
Which of the following sets is computed by the above query?
(a) Names of students who have got an A grade in all courses taught by Korth
(b) Names of students who have got an A grade in all courses

(c) Name of students who have got an A grade in at least one of the courses taught by Korth
(d) None of the above **[2003 : 2 M]**

3.8 The employee information in a company is stored in the relation:
Employee (name, sex, salary, deptName)
Consider the following SQL query

```
select deptName
from Employee
where sex = 'M'
group by deptName
having avg (salary)>
    (select avg (salary) from Employee)
```

It returns the names of the department in which
(a) the average salary is more than the average salary in the company
(b) the average salary of male employees is more than the average salary of all male employees in the company
(c) the average salary of male employees is more than the average salary of employees in the same department
(d) the average salary of male employees is more than the average salary in the company
[2004 : 2 M]

3.9 A relational database contains two tables student and department in which student table has columns roll_no, name and dept_id and department table has columns dept_id and dept_name. The following insert statements were executed successfully to populate the empty tables:
Insert into department values (1, 'Mathematics')
Insert into department values (2, 'Physics')
Insert into student values (l, 'Navin', 1)
Insert into student values (2, 'Mukesh', 2)
Insert into student values (3, 'Gita', 1)
How many rows and columns will be retrieved by the following SQL statement? Select * from student, department
(a) 0 row and 4 columns
(b) 3 rows and 4 columns
(c) 3 rows and 5 columns
(d) 6 rows and 5 columns **[2004 : 2 M]**

3.10 A table T1 in a relational database has the following rows and columns:

roll no.	marks
1	10
2	20
3	30
4	Null

The following sequence of SQL statements was successfully executed on table T1.
Update T1 set marks = marks +5
Select avg(marks) from T1
What is the output of the select statement?
(a) 18.75 (b) 20
(c) 25 (d) Null
[2004 : 2 M]

3.11 Consider two tables in a relational database with columns and rows as follows:

Table : Student

Roll_no	Name	Dept_id
1	ABC	1
2	DEF	1
3	GHI	2
4	JKL	3

Table : Department

Dept_id	Dept_name
1	A
2	B
3	C

Roll_no is the primary key of the Student table, Dept_id is the primary key of the Department table and Student.Dept_id is a foreign key from Department.Dept_id
What will happen if we try to execute the following two SQL statements?
(i) update Student set Dept_id = Null where Roll_no = 1
(ii) update Department set Dept_id = Null where Dept_id = 1
(a) Both (i) and (ii) will fail
(b) (i) will fail but (ii) will succeed
(c) (i) will succeed but (ii) will fail
(d) Both (i) and (ii) will succeed
[2004 : 2 M]

3.12 A company maintains records of sales made by its salespersons and pays them commission based on each individual's total sales made in a year. This data is maintained in a table with following schema:
salesinfo = (salespersonid, totalsales, commission)
In a certain year, due to better business results, the company decides to further reward its salespersons by enhancing the commission paid to them as per the following formula:
If commission < = 50000, enhance it by 2%
If 50000 < commission < = 100000, enhance it by 4%

If commission > 100000, enhance it by 6%

The IT staff has written three different SQL scripts to calculate enhancement for each slab, each of these scripts is to run as a separate transaction as follows:

T1 Update salesinfo
Set commission = commission * 1.02
Where commission < = 50000;

T2 Update salesinfo
Set commission = commission * 1.04
Where commission > 50000 and commission is < = 100000;

T3 Update salesinfo
Set commission = commission * 1.06
Where commission > 100000;

Which of the following options of running these transactions will update the commission of all salespersons correctly?

(a) Execute T1 followed by T2 followed by T3
(b) Execute T2, followed by T3; T1 running concurrently throughout
(c) Execute T3 followed by T2; T1 running concurrently throughout
(d) Execute T3 followed by T2 followed by T1

[2005 : 2 M]

3.13 In an inventory management system implemented at a trading corporation, there are several tables designed to hold all the information. Amongst these, the following two tables hold information on which items are supplied by which suppliers, and which warehouse keeps which items along with the stock-level of these items.

Supply = (supplierid, itemcode)
Inventory = (itemcode, warehouse, stocklevel)

For a specific information required by the management, following SQL query has been written

```
Select distinct STMP.supplierid
From Supply as STMP
Where not unique (Select ITMP.supplierid
From Inventory, Supply as ITMP
Where STMP.supplierid = ITMP.supplierid
And ITMP.itemcode = Inventory.itemcode
And Inventory.warehouse = 'Nagpur');
```

For the warehouse at Nagpur, this query will find all suppliers who

(a) do not supply any item
(b) supply exactly one item
(c) supply one or more items
(d) supply two or more items

[2005 : 2 M]

3.14 The relation book (title, price) contains the titles and prices of different books. Assuming that no two books have the same price, what does the following SQL

```
select title
from book as B
where (select count (*)
  from book as T
  where T. price > B. Price) < 5
```

(a) Titles of the four most expensive books
(b) Title of the fifth most inexpensive book
(c) Title of the fifth most expensive book
(d) Titles of the five most expensive books

[2005 : 2 M]

Directions for Question 3.15 to 3.16:

Consider a database with three relation instances shown below. The primary keys for the Drivers and Cars relation are *did* and *cid* respectively and the records are stored in ascending order of these primary keys as given in the tables. No indexing is available in the database.

D : Drivers relation

did	dname	rating	age
22	Karthikeyan	7	25
29	Salman	1	33
31	Boris	8	55
32	Amoldt	8	25
58	Schumacher	10	35
64	Sachin	7	35
71	Senna	10	16
74	Sachin	9	35
85	Rahul	3	25
95	Ralph	3	53

R : Reserves relation

did	cid	day
22	101	10/10/06
22	102	10/10/06
22	103	08/10/06
22	104	07/10/06
31	102	10/11/06
31	103	06/11/06
31	104	12/11/06
64	101	05/09/06
64	102	08/09/06
74	103	08/09/06

C : Cars relation

cid	cname	colour
101	Renault	blue
102	Renault	red
103	Ferrari	green
104	Jaguar	red

3.15 What is the output of the following SQL query?

```
select D.dname
from Drivers D
where D.did in (
    select R.did
    from Cars C, Reserves R
    where R.cid=C.cid and C.colour = 'red'
    intersect
    select R.did
    from Cars C, Reserves R
    where R.cid=C.cid and C.colour= 'green'
    )
```

(a) Karthikeyan, Boris
(b) Sachin, Salman
(c) Karthikeyan, Boris, Sachin
(d) Schumacher, Senna **[2006 : 2 M]**

3.16 Let n be the number of comparisons performed when the above SQL query is optimally executed. If linear search is used to locate a tuple in a relation using primary key, then n lies in the range

(a) 36 – 40 (b) 44 – 48
(c) 60 – 64 (d) 100 – 104 **[2006 : 2 M]**

3.17 Consider the relation account (customer, balance) where customer is a primary key and there are no null values. We would like to rank customers according to decreasing balance. The customer with the largest balance gets rank 1. Ties are not broken but ranks are skipped: if exactly two customers have the largest balance they each get rank 1 and rank 2 is not assigned.

Query 1: Select A.customer, count (B.customer) from account A, account B where A.balance < = B.balance Group by A.customer

Query 2: Select A.customer, 1 + count (B.customer) from account A, account B where A.balance < B.balance group by A.customer

Consider these statements about Query 1 and Query 2.

1. Query 1 will produce the same row set as Query 2 for some but not all databases.
2. Both Query 1 Query 2 are correct implementations of the specification.
3. Query 1 is a correct implementation of the specification but Query 2 is not.
4. Neither Query 1 nor Query 2 is a correct implementation of the specification.
5. Assigning rank with a pure relational Query takes less time than scanning in decreasing balance order and assigning ranks using ODBC.

Which two of the above statements are correct?

(a) 2 and 5 (b) 1 and 3
(c) 1 and 4 (d) 3 and 5 **[2006 : 2 M]**

3.18 Consider the relation enrolled (student, course) in which (student, course) is the primary key, and the relation paid (student, amount) where student is the primary key. Assume no null values and no foreign keys or integrity constraints. Given the following four queries:

Query 1: Select student from enrolled where student in (select student form paid)

Query 2: Select student from paid where student in (select student from enrolled)

Query 3: Select E.student from enrolled E, paid P where E.student = P.student

Query 4: Select student from paid where exists (select * from enrolled where enrolled.student = paid.student)

Which one of the following statements is correct?

(a) All queries return identical row sets for any database
(b) Query 2 and Query 4 return identical row sets for all databases but there exist databases for which Query 1 and Query 2 return different row sets
(c) There exist databases for which Query 3 returns strictly fewer rows than Query 2
(d) There exist databases for which Query 4 will encounter an intergrity violation at runtime **[2006 : 2 M]**

3.19 Consider the table employee (<u>empId</u>, name, department, salary) and the two queries Q_1, Q_2 below. Assuming that department 5 has more than one employee, and we want to find the employees who get higher salary than anyone in the department 5, which one of the statements is TRUE for any arbitrary employee table?

Q_1 :
```
Select e.empId
From employee e
Where not exists
(Select * From employee s Where
s.department = "5" and s.salary >= e.salary)
```

Q_2 :
```
Select e. empId
From employee e
Where e.salary > Any
(Select distinct salary From employee s
Where s.department = "5")
```

(a) Q_1 is the correct query
(b) Q_2 is the correct query
(c) Both Q_1 and Q_2 produce the same answer
(d) Neither Q_1 nor Q_2 is the correct query **[2007 : 2 M]**

Directions for Question 3.20 to 3.21:

Consider the following relational schema:

Student (school-id, sch-roll-no, sname, saddress)

School (school-id, sch-name, sch-address, sch-phone)

Enrolment(school-id sch-roll-no, erollno, examname)

ExamResult (erollno, examname, marks)

3.20 What does the following SQL query output?

```
SELECT   sch-name, COUNT (*)
FROM     School C, Enrolment E, ExamResult R
WHERE    E.school-id C.school-id
         AND
         E.examname = R.examname AND
         E.Erollno = R.erollno
         AND
         R.marks = 100 AND S.school-id IN
         (SELECT school-id
         FROM student
         GROUP BY school-id
         HAVING COUNT (*) > 200)
GROUP By school-id
```

(a) for each school with more than 200 students appearing in exams, the name of the school and the number of 100s scored by its students

(b) for each school with more than 200 students in it, the name of the school and the number of 100s scored by its students

(c) for each school with more than 200 students in it, the name of the school and the number of its students scoring 100 in at least one exam

(d) nothing; the query has a syntax error

[2008 : 2 M]

3.21 Consider the following tuple relational calculus query.

$\{t \mid \exists E \in \text{Enrolment } t = E.\text{school-id} \wedge \mid\{x \mid x \text{ Enrolment} \wedge x.\text{school-id} = t \wedge (\exists B \text{ ExamResult } B.\text{erollno} = x.\text{erollno} \wedge B.\text{examname} = x.\text{examname} \wedge B.\text{marks} > 35)\}\mid \div \mid\{x \mid x \text{ Enrolment} \wedge x.\text{school-id} = t\}\mid * 100 > 35\}$

If a student needs to score more than 35 marks to pass an exam, what does the query return?

(a) The empty set

(b) schools with more than 35% of its students enrolled in some exam or the other

(c) schools with a pass percentage above 35% over all exams taken together

(d) schools with a pass percentage above 35% over each exam

[2008 : 2 M]

3.22 Consider a database table T containing two columns X and Y each of type integer. After the creation of the table, one record (X = 1, Y = 1) is inserted in the table.

Let MX and MY denote the respective maximum values of X and Y among all records in the table at any point in time. Using MX and MY, new records are inserted in the table 128 times with X and Y values being MX + 1, 2*MY + 1 respectively. It may be noted that each time after the insertion, values of MX and MY change.

What will be the output of the following SQL query after the steps mentioned above are carried out?

SELECT Y FROM T WHERE X = 7;

(a) 127 (b) 255

(c) 129 (d) 257 **[2011 : 2 M]**

3.23 Database table by name Loan_Records is given below:

Borrower	Bank_Manager	Loan_Amount
Ramesh	Sunderajan	10000.00
Suresh	Ramgopal	5000.00
Mahesh	Sunderajan	7000.00

What is the output of the following SQL query?

```
SELECT count(*)
FROM (
    (SELECT Borrower, Bank_Manager FROM
    Loan_Records) AS S NATURAL JOIN
    (SELECT Bank-Manager, Loan_Amount
    FROM Loan_Records) AS T
    );
```

(a) 3 (b) 9

(c) 5 (d) 6 **[2011 : 2 M]**

3.24 Which of the following statements are TRUE about an SQL query?

P: An SQL query can contain a HAVING clause even if it does not have a GROUP BY clause

Q: An SQL query can contain a HAVING clause only if it has a GROUP BY clause

R: All attributes used in the GROUP BY clause must appear in the SELECT clause

S: Not all attributes used in the GROUP BY clause need to appear in the SELECT clause

(a) P and R (b) P and S

(c) Q and R (d) Q and S

[2012 : 1 M]

Common Data for Questions 3.25 and 3.26:

Consider the following relations A, B and C:

A.

Id	Name	Age
12	Arun	60
15	Shreya	24
99	Rohit	11

B.

Id	Name	Age
15	Shreya	24
25	Hari	40
98	Rohit	20
99	Rohit	11

C.

Id	Phone	Age
10	2200	02
99	2100	01

3.25 How many tuples does the result of the following relational algebra expression contain? Assume that the schema of A ∪ B is the same as that of A.

$$(A \cup B) \bowtie_{A.Id > 40 \vee C.Id < 15} C$$

(a) 7 (b) 4
(c) 5 (d) 9 **[2012 : 2 M]**

3.26 How many tuples does the result of the following SQL query contain?

```
SELECT A.Id
FROM A
WHERE A.Age > All (SELECT B.Age
                   FROM B
                   WHERE B.Name='Arun')
```

(a) 4 (b) 3
(c) 0 (d) 1 **[2012 : 2 M]**

3.27 Given the following schema:

employees(emp-id, first-name, last-name, hire-date, dept-id, salary)
departments(dept-id, dept-name, manager-id, location-id)

You want to display the last names and hire dates of all latest hires in their respective departments in the location ID 1700. You issue the following query:

```
SQL > SELECT last-name, hire-date
      FROM employees
      WHERE (dept-id, hire-date) IN
      (SELECT dept-id, MAX(hire-date) FROM
      employees JOIN departments
      USING(dept-id)
      WHERE location-id = 1700
      GROUP BY dept-id);
```

What is the outcome?

(a) It executes but does not give the correct result.
(b) It executes and gives the correct result.
(c) It generates an error because of pairwise comparison.
(d) It generates an error because the GROUP BY clause cannot be used with table joins in a subquery.

[2014 (Set-1) : 2 M]

3.28 SQL allows duplicate tuples in relations, and correspondingly defines the multiplicity of tuples in result of joins. Which one of the following queries always gives the same answer as the nested query shown below:

select * from R where a in (select S.a from S)

(a) select R.* from R, S where R.a = S.a
(b) select distinct R.* from R,S where R.a = S.a
(c) select R.* from R,(select distinct a from S) as S1 where R.a = S1.a
(d) select R.* from R,S where R.a = S.a and is unique R

[2014 (Set-2) : 2 M]

3.29 What is the optimized version of the relation algebra expression $\pi_{A1}(\pi_{A2}(\sigma_{F1}(\sigma_{F2}(r))))$, where $A1, A2$ are sets of attributes in r with $A1 \subset A2$ and $F1, F2$ are Boolean expressions based on the attributes in r?

(a) $\pi_{A1}(\sigma_{(F1 \wedge F2)}(r))$ (b) $\pi_{A1}(\sigma_{(F1 \vee F2)}(r))$
(c) $\pi_{A2}(\sigma_{(F1 \wedge F2)}(r))$ (d) $\pi_{A2}(\sigma_{(F1 \vee F2)}(r))$

[2014 (Set-3) : 1 M]

3.30 Consider the following relational schema:

employee(<u>**empId**</u>, empName,empDept)

customer(<u>**custId**</u>, custName,salesRepId,rating)

salesRepId is a foreign key referring to **empId** of the employee relation. Assume that each employee makes a sale to at least one customer. What does the following query return?

```
SELECT empName
FROM employee E
WHERE NOT EXISTS (SELECT custId
                  FROM customer C
                  WHERE C.salesRepId = E.empId
                  AND C.rating <> 'GOOD');
```

(a) Names of all the employees with at least one of their customers having a 'GOOD' rating.
(b) Names of all the employees with at most one of their customers having a 'GOOD' rating.

(c) Names of all the employees with none of their customers having a 'GOOD' rating.

(d) Names of all the employees with all their customers having a 'GOOD' rating.

[2014 (Set-3) : 2 M]

3.31 Consider the following relations:

Student

Roll_No	Student_Name
1	Raj
2	Rohit
3	Raj

Performance

Roll_No	Course	Marks
1	Math	80
1	English	70
2	Math	75
3	English	80
2	Physics	65
3	Math	80

Consider the following SQL query.

```
SELECT S. Student_Name, sum (P.Marks)
FROM Student S, Performance P
WHERE S. Roll_No =P.Roll_No
GROUP BY S.Student_Name
```

The number of rows that will be returned by the SQL query is ________. **[2015 (Set-1) : 2 M]**

3.32 SELECT operation in SQL is equivalent to

(a) the selection operation in relational algebra

(b) the selection operation in relational algebra, except that SELECT in SQL retains duplicates

(c) the projection operation in relational algebra

(d) the projection operation in relational algebra, except that SELECT in SQL retains duplicates **[2015 (Set-1) : 1 M]**

3.33 Consider the following relation:

Cinema (*theater, address, capacity*)

Which of the following options will be needed at the end of the SQL query

```
SELECT P1.address
FROM Cinema P1
```

such that it always finds the addresses of theaters with maximum capacity?

(a) WHERE P1.*capacity* > = All (select P2. *capacity* from Cinema P2)

(b) WHERE P1.*capacity* > = Any (select P2. *capacity* from Cinema P2)

(c) WHERE P1.*capacity* > All (select max (P2. *capacity*) from Cinema P2)

(d) WHERE P1.*capacity* > Any (select max (P2. *capacity*) from Cinema P2)

[2015 (Set-3) : 1 M]

3.34 Consider the following database table named water_schemes:

water_schemes		
scheme_no	**district_name**	**capacity**
1	Ajmer	20
1	Bikaner	10
2	Bikaner	10
3	Bikaner	20
1	Churu	10
2	Churu	20
1	Dungargarh	10

The number of tuples returned by the following SQL query is:

```
with total (name, capacity) as
    select district_name, sum (capacity)
    from water_schemes
    group by district_name
with total_avg (capacity) as
    select avg (capacity)
    from total
select name
    from total, total_avg
    where total.capacity ≥ total_avg. capacity
```

[2016 (Set-2) : 2 M]

3.35 Consider a database that has the relation schema EMP (EmpId, EmpName, and DeptName). An instance of the schema EMP and a SQL query on it are given below:

EMP

EmpId	EmpName	DeptName
1	XYA	AA
2	XYB	AA
3	XYC	AA
4	XYD	AA
5	XYE	AB
6	XYF	AB
7	XYG	AB
8	XYH	AC
9	XYI	AC
10	XYJ	AC
11	XYK	AD
12	XYL	AD
13	XYM	AE

```
SELECT AVG(EC.Num)
FROM EC
WHERE (DeptName, Num) IN
    (SELECT DeptName, COUNT(EmpId) AS
                          EC( DeptName, Num)
     FROM EMP
     GROUP BY DeptName)
```

The output of executing the SQL query is ________. **[2017 (Set-1) : 1 M]**

3.36 Consider the following database table named *top_scorer*.

top_scorer

player	country	goals
Klose	Germany	16
Ronaldo	Brazil	15
G Muller	Germany	14
Fontaine	France	13
Pele	Brazil	12
Klinsmann	Germany	11
Kocsis	Hungary	11
Batistuta	Argentina	10
Cubillas	Peru	10
Lato	Poland	10
Lineker	England	10
T Muller	Germany	10
Rahn	Germany	10

Consider the following SQL query:

```
SELECT ta.player FROM top_scorer AS ta
WHERE ta.goals > ALL (SELECT tb.goals
                      FROM top_scorer AS tb
                      WHERE tb.country = 'Spain')
AND ta.goals >   ANY (SELECT tc.goals
                      FROM top_scorer AS tc
                      WHERE tc.country
                      = 'Germany')
```

The number of tuples returned by the above SQL query is ________.

[2017 (Set-2) : 2 M]

3.37 Consider the following two tables and four queries in SQL.

Book (<u>isbn</u>, bname), Stock (<u>isbn</u>, copies)

Query 1: SELECT B.isbn, S.copies
FROM Book B INNER JOIN Stock S
ON B.isbn = S.isbn;

Query 2: SELECT B.isbn, S.copies
FROM Book B LEFT OUTER JOIN Stock S
ON B.isbn = S.isbn;

Query 3: SELECT B.isbn, S.copies
FROM Book B RIGHT OUTER JOIN Stock S
ON B.isbn = S.isbn;

Query 4: SELECT B.isbn, S.copies
FROM Book B FULL OUTER JOIN Stock S
ON B.isbn = S.isbn;

Which one of the queries above is certain to have an output that is superset of the outputs of the other three queries?

(a) Query 1
(b) Query 2
(c) Query 3
(d) Query 4 **[2018 : 1 M]**

3.38 A relational database contains two tables Student and Performance as shown below:

Student

Roll_no.	Student_name
1	Amit
2	Priya
3	Vinit
4	Rohan
5	Smita

Performance

Roll_no.	Student_code	Marks
1	A	86
1	B	95
1	C	90
2	A	89
2	C	92
3	C	80

The primary key of the Student table is Roll_no. For the Performance table, the columns Roll_no. and Subject_code together from the primary key. Consider the SQL query given below:

```
SELECT S.Student_name, sum(P.Marks)
FROM Student S, Performance P
WHERE P.Marks > 84
GROUP BY S.Student_name;
```

The number of rows returned by the above SQL query is ________. **[2019 : 2 M]**

3.39 Consider a relational database containing the following schemas.

Catalog

<u>sno</u>	<u>pno</u>	cost
S1	P1	150
S1	P2	50
S1	P3	100
S2	P4	200
S2	P5	250
S3	P1	250
S3	P2	150
S3	P5	300
S3	P4	250

Supplies

<u>sno</u>	sname	location
S1	M/s Royal furniture	Delhi
S2	M/s Balaji furniture	Bangalore
S3	M/s Premium furniture	Chennai

Parts

<u>pno</u>	sname	part_spec
P1	Table	Wood
P2	Chair	Wood
P3	Table	Steel
P4	Almirah	Steel
P5	Almirah	Wood

The primary key of each table is indicated by underlining the constituent fields.

```
SELECT  s.sno, s.name
FROM    Suppliers s, Catalogue c
WHERE   s.no = c.sno AND
        cost > (SELECT AVG (cost)
```

FROM Catalogue
WHERE pno = 'P4'
GROUP BY pno);

The number of rows returned by the above SQL query is

(a) 2 (b) 5
(c) 4 (d) 0 **[2020 : 1 M]**

3.40 The relation scheme given below is used to store information about the employees of a company, where empId Iis the key and deptId indicates the department to which the employee is assigned. Each employee is assigned to exactly one department.

emp(empId, name, gender, salary, deptId)

Consider the following SQL query:

select deptId, count(*)
from emp
where gender = "female" and salary > (select avg(salary) from emp)
group by deptId;

The above query gives, for each department in the company, the number of female employees whose salary is greater than the average salary of

(a) Female employees in the department.
(b) Employees in the department.
(c) Female employees in the company.
(d) Employees in the company.

[2021 (Set-2) : 2 M]

3.41 Consider the relational database with the following four schemas and their respective instances.

Student(sNo, sName, dNo) Dept(dNo, dName)

Course(cNo, cName, dNo) Register(sNo, cNo)

Student

sNo	sName	dNo
S01	James	D01
S02	Rocky	D02
S03	Jackson	D02
S04	Jane	D01
S05	Milli	D02

Dept

dNo	dName
D01	CSE
D02	EEE

Course

cNo	sName	dNo
C11	DS	D01
C12	OS	D01
C21	DE	D02
C22	PT	D02
C23	CV	D03

Register

sNo	cNo
S01	C11
S01	C12
S02	C11
S03	C21
S03	C22
S03	C23
S04	C11
S04	C12
S05	C11
S05	C21

SQL Query:

SELECT * FROM Student AS S WHERE NOT EXIST
(SELECT cNo FROM Course WHERE dNo = "D01" EXCEPT
SELECT cNo FROM Register WHERE sNo = S.sNo)

The number of rows returned by the above SQL query is ______. **[2022 : 2 M]**

3.42 Consider the following table named Student in a relational database. The primary key of this table is rollNum.

Student

rollNum	name	gender	marks
1	Naman	M	62
2	Aliya	F	70
3	Aliya	F	80
4	James	M	82
5	Swati	F	65

The SQL query below is executed on this database.

SELECT *
FROM Student
WHERE gender = 'F' AND
marks > 65;

The number of rows returned by the query is ______. **[2023 : 2 M]**

■■■■

Answers Structured Query Language (SQL)

3.1	Sol.	3.2	(d)	3.3	Sol.	3.4	(c)	3.5	(a)	3.6	(a)	3.7	(c)	3.8	(d)	3.9	(d)
3.10	(c)	3.11	(c)	3.12	(d)	3.13	(d)	3.14	(d)	3.15	(a)	3.16	(c)	3.17	(c)	3.18	(b)
3.19	(b)	3.20	(d)	3.21	(c)	3.22	(a)	3.23	(c)	3.24	(c)	3.25	(a)	3.26	(b)	3.27	(b)
3.28	(c)	3.29	(a)	3.30	(d)	3.31	(2)	3.32	(d)	3.33	(a)	3.34	(2)	3.35	(2.6)	3.36	(7)
3.37	(d)	3.38	(5)	3.39	(c)	3.40	(d)	3.41	(2)	3.42	(2)						

Explanations Structured Query Language (SQL)

3.1 Sol.

```
Select  Student
From    FREQUENTS
Where   parlor in (select parlor from SERVES
        where ice-cream in (select ice-cream
        from LIKES where LIKES.student =
        FREQEUNTS.student))
```

3.2 (d)

To eliminate duplicate rows, we need to use "DISTNICT" keyword with SELECT in SQL query.

SQL not permits attribute names to be repeated in the same relation. Therefore option (d) is correct.

3.3 Sol.

(a)
```
Select  Employee-name
From    EMP
Where   Salary > 50,000
and     Dept-no in  (Select Dept-no from
                    Dept where Location
                    = 'Calcutta')
```

or

```
Select  EMP. Employee-name
From    EMP, DEPT
Where   EMP. Dept-no=DEPT.Dept-no
and     DEPT.Location= 'Calcutta' and
        EMP.Salary > 50,000
```

(b)
```
Select  Dept-no, count (Employee-no)
From    EMP
Where   Salary > 1,00,000
group by Dept-no.
```

3.4 (c)

Comparison with Null treated as false i.e.

null = 5 false

null ≠ 5 false

null < 5 false

Given expression:

(a) X = 5   not(not(X = 5))
(false of X is null)   (false of X is null)
both are equal even X is null.

(b) X = 5   X > 4 and X < 6
(false of X is null)   (false of X is null)
both are equal even X is null.

(c) X ≠ 5   not(X = 5))
(false of X is null)   (true of X is null)
both are equal even X is null.
So that these expressions are not equal of X is null and comparison with X treated as false.

3.5 (a)

If any of the relation is empty in cross-join then the overall result is empty. Therefore S need to be non-empty. To have same rows of 'r' then 'r' should not have duplicates.

3.6 (a)

The given SQL is

Select distinct $a_1, a_2, ..., a_n$

from $r_1, r_2, ..., r_m$

where P

All possible combination of tuples from r_1, r_2, r_m is denoted by $r_1 \times r_2 \times ... \times r_m$

If P is a predicate then to select the all, condition is denoted by $\sigma_P (r_1 \times r_2 \times ... \times r_m)$

If we wants to select only some tuples in the relation then composite expression for above SQL is

$$\Pi_{a_1, a_2, ..., a_n} \sigma_\rho(r_1 \times r_2 \times ... \times r_m)$$

3.7 (c)

There are three relations

Select distinct name, select the name of students and then there are three predicates

Courses. instructor = korth specify the courses taught by korth. The other two predicates specify that student can earn a grade at least A from courses so the SQL query compute.
Name of students who have got an A grade in at least one of the courses taught by korth.

3.8 (d)

Select deptName
from employee
where sex = 'M'
grouply deptName
having avg (salary) > (select avg (salary) from employees)

[Avg salary of male employees in each department] [Avg salary in the company]

It returns the names of departments in which the average salary of male employees is more than the average salary in the company.

3.9 (d)

There is no specific joining condition specified, so it will retrieve Cartesian product of the tables.
Number of rows = Product of number of rows in each relation = 3 * 2 = 6
Number of columns = Sum of number of columns = 3 + 2 = 5.

3.10 (c)

Update on null gives null.
Average function ignores null values.
So, Average will produce (15 + 25 + 35)/3 = 25.

3.11 (c)

Foreign key

Student

Roll_no	Name	Dept_id
1	ABC	1
2	DEF	1
3	GHI	2
4	JKL	3

Department

Dept_id	Dept_name
1	A
2	B
3	C

1. On execute 1st SQL statement we try to put Null where Roll_no = 1. Since foreign key can contain Null. So allowed to update.
2. On execute 2nd SQL statement we try to put Null where Roll_no = 1, since primary key of department table. Hence cannot be Null. So it will fail to execute.

3.12 (d)

Commission enhancement formula is:
If commission ≤ 50000, enhance it by 2%.
If 50000 < commission ≤ 100000, enhance it by 4%.
If commission > 100000, enhance it by 6%.
If we execute T1 first, then there may be case that amount less than equal to 50000 now comes to range 50000 to 100000, now if we execute T2 we will get wrong output.
So, first we need to evaluate T3 since all are increasing after calculation, then T2 then T1. Hence order will be T3 followed by T2 followed by T1.

3.13 (d)

Select ITMP.supplierid
From Inventory.Supply as ITMP
Where STMP.supplierid = ITMP.suppliered
And ITMP.itemcode = Inventory.itemcode
And Inventory.warehouse = 'Nagpur';

Inner query will give supplier id of supplier who supply atleast one item at 'Nagpur'. Here if supplier supply more than 1 item then supplier id will be duplicated in result.
When outer query execute, then it will return distinct supplier id of those supplier who supplier atleast two item at Nagpur.
Note: Not unique returns true for repeated values.

3.14 (d)

Select title
FROM book as B
Where (Select count (*)
FROM book as T
Where T.price > B.price) < 5

Inner query counts number of books more price than B.price of count < 5 then B.title in result if count 0 means no other book more price ⇒ most expensive
If count 1, 2, 3, 4 more one, two three, four other books more price ⇒ 2nd, 3rd, 4th 5th most expensive books.
Query retrieves titles of fire most expensive books.

3.15 (a)

For color = "Red", did = {22, 22, 31, 64}
For color = "Green", did = {22, 31, 74}
Intersection of Red and Green will give = {22, 31}
which is Karthikeyan and Boris.

3.16 (c)

(i) Sub Query
(Select R.did
From Cars C, Reserves R
Where R.cid = C.cid and C.color = red)

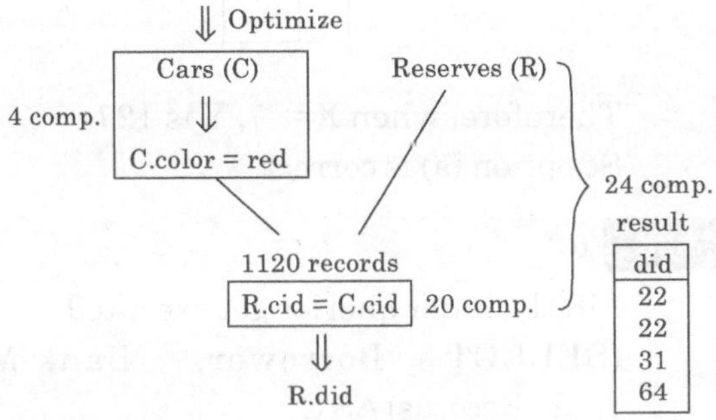

(ii) Sub Query
(Select R.did
From Cars C, Reserves R
Where R.cid = C.cid and C.color = green)

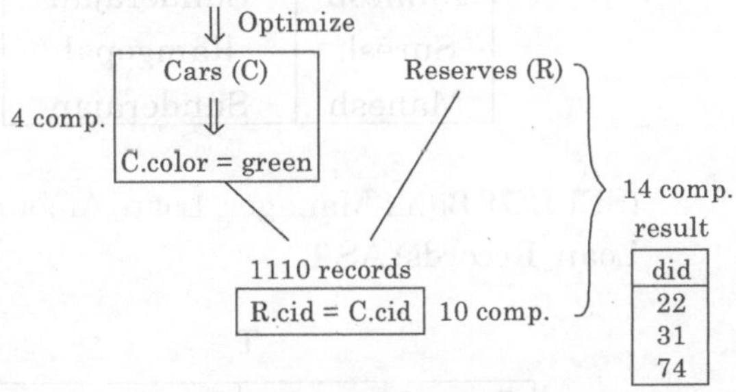

(iii) Intersection of sub query result:
⇒ Intersection can translate into inner join.
⇒ Minimum '6' comparisons to get result of intersection result [22, 31]

(iv) Retrieve driver name whose id in [22, 31] approximately. In also works as inner join 4 comparisons.

Total cost 24 + 14 + 6 + 4 = 48 (approximately)

3.17 (c)

Query 1:

(1) FROM Acc A, Acc B

cid	bal
C1	30
C2	30
C3	25
C4	25
C5	20
C6	15

×

cid	bal
C1	30
C2	30
C3	25
C4	25
C5	20
C6	15

(2) Where A.bal < = B.bal → Group by A.cid

{C1 30 C1 30
C1 30 C2 30}

{C2 30 C1 30
C2 30 C2 30}

{C3 25 C1 30
C3 25 C2 30
C3 25 C3 25
C3 25 C4 25}
⋮

(4) Select A.cid count (B.cid)

C1	2
C2	2
C3	4
C4	4
C5	5
C6	6

result

Query 1 wrong as per given specification.

Query 2:

(1) From acc A, Acc B
(2) Where A.bal < B.bal
(3) Group by A.cid
(4) Select A.cid, count + 1

cid	bal	cid	bal
C3	25	C1	30
C3	25	C2	30
C4	25	C1	30
C4	25	C2	30
C5	20	C1	30
C5	20	C2	30
C5	20	C3	20
C5	20	C4	20
C6	15	C1	30
C6	15	C2	30
C6	15	C3	25
C6	15	C4	25
C6	15	C5	20

C3	3
C4	3
C5	5
C6	6

result

Wrong because 1st rank cust not in result.
Both Query 1 Query 2 wrong for given specific between some records of Query 1 Query 2 can same for some database.

3.18 (b)

Enroll (<u>student, course</u>) paid (<u>student</u>, amount)

student column not unique for enroll and unique for paid relation.

Query 1: Retrieving student from enroll for which student in paid. [May result duplicate student values].

Query 2: Retrieving student from paid for which students in enroll. [result distance student values]

Query 3: Retrieving student by joining enroll and paid which results same as query 1.

Query 4: Retrieving students from paid if exists in enroll, which results same as query 2.

3.19 (b)

Query 1: Retrieves employees who gets more salary than every employee of dept 5.

Query 2: Retrieves employees who gets more salary than some employee of dept 5.

3.20 (d)

SQL standard not allowed to select any other attribute in SELECT clause other than attributes and group by clause if group by clause exists.

In query outer query group's based on Schoo_id and select clause selects Schname with is not allowed.

So that query has a syntax error.

3.22 (a)

Initially the table is

X	Y
1	1

Then MX = 1, MY = 1

Using MX and MY new records inserted is $(1 + 1, 2 \times 1 + 1) = (2, 3)$

X	Y
1	1
2	3

Now MX = 2, MY = 3

New record is $(2 + 1, 3 \times 2 + 1) = (3, 7)$

X	Y
1	1
2	3
3	7

Records are inserted in the table 128 times and looks like as follows.

X	Y
1	1
2	3
3	7
4	15
5	31
6	63
7	127
8	255
⋮	

Therefore, when X = 7, Y is 127.

So option (a) is correct.

3.23 (c)

Firstly, inner queries get executed

(SELECT Borrower, Bank-Manager Loan_Records) AS S

S

Borrower	Bank-Manager
Ramesh	Sunderajan
Suresh	Ramgopal
Mahesh	Sunderajan

(SELECT Bank-Manager, Loan_Amount FROM Loan_Records) AS T

T

Bank-Manager	Loan_Amount
Sunderajan	10000.00
Ramgopal	5000.00
Sunderajan	7000.00

S Natural Join T

Borrower	Bank-Manager	Loan_Amount
Ramesh	Sunderajan	10000.00
Ramesh	Sunderajan	7000.00
Suresh	Ramgopal	5000.00
Mahesh	Sunderajan	7000.00
Mahesh	Sunderajan	10000.00

Therefore, the output of the SQL query contains 5 records.

3.24 (c)

As per SQL standard HAVING clause allowed only if GROUP BY clause exists.
Like basic clauses:
SELECT [DISTINCT] Atributes
FROM relations
[WHERE condition]
[GROUP BY attributes
[HAVING condition]]
All attributes used in the GROUP BY clause must appear in the SELECT clause.

3.25 (a)

$A \cup B$ =

ID	Name	Age
12	Arun	60
15	Shreya	24
99	Rohit	11
25	Hari	40
98	Rohit	20

$(A \cup B) \bowtie_{A.Id > 40 \vee C.Id < 15} C$

ID	Name	Age	ID	Phone	Area
12	Arun	60	10	2200	02
15	Shreya	24	10	2200	02
99	Rohit	11	10	2200	02
25	Hari	40	10	2200	02
98	Rohit	20	10	2200	02
99	Rohit	11	99	2100	01
98	Rohit	20	99	2100	01

3.26 (b)

In this (select B. Age
From B
where B. Name = Arun)
⇓
ϕ

So all A. Age will be selected, so Ans 3.

3.27 (b)

Given SQL query will correctly display the last names and hire dates of all latest hires in their respective departments in the location ID 1700.

3.28 (c)

Select * from R where a in (select s.a from S)

$\cong$

Select R.* from R, (select distinct a from S) as $S1$ where $R.a = S1.a$.

3.29 (a)

(*i*) $\pi_{A1}(\pi_{A2}(X)) = \pi_{A1}(X)$, since $A1 \subset A2$

(*ii*) $\sigma_{F1}(\sigma_{F2}(X)) = \sigma_{F1 \wedge F2}(X)$

$\therefore \pi_{A1}(\pi_{A2}(\sigma_{F1}(\sigma_{F2}(r)))) = \pi_{A1}(\sigma_{F1 \wedge F2}(r))$

3.30 (d)

SELECT empName
FROM employee E
WHERE NOT EXISTS

```
(SELECT CustId
FROM Customer C
WHERE C.SalesRepId = E.empId
AND C.rating <> 'GOOD'
```

where, the SQL query in the box represents all customers having other than 'good' while the complete query represents name of all employees with all their customers having a 'good rating'.

3.31 (2)

Student_Name	Marks
Raj	310
Rohit	140

3.32 (d)

Π in relation algebra is similar to SELECT in SQL but the only difference is that 'Π' gives distinct rows by eliminating duplicates but SELECT does not remove duplicates.

3.33 (a)

The following query finds the addresses of theaters with maximum capacity:

SELECT P1.address
FROM Cinema P1
WHERE P1.*capacity* > = All (select P2. *capacity* from Cinema P2)

3.34 (2)

Total

Name	Capacity
Ajmer	20
Bikaner	40
Churu	30
Dungargarth	10

is result of first query.

Total average

Capacity
25

is result of second subquery.

Select name
from Total, Total_Avg
Where total capacity ≥ Total_Avg capcity
Query results 2 records.

3.35 (2.6)

EMP

EmpId	EmpName	DeptName
1	XYA	AA
2	XYB	AA
3	XYC	AA
4	XYD	AA
5	XYE	AB
6	XYF	AB
7	XYG	AB
8	XYH	AC
9	XYI	AC
10	XYJ	AC
11	XYK	AD
12	XYL	AD
13	XYM	AE

```
SELECT AVG(EC.Num)
FROM EC
WHERE (DeptName, Num) IN
    (SELECT DeptName, COUNT(EmpId) AS
                              EC( DeptName, Num)
    FROM EMP
    GROUP BY DeptName)
```

Result of inner query

EC

Dept name	Number
AA	4
AB	3
AC	3
AD	2
AE	1

Result of outer query : $\frac{13}{5} = 2.6$

3.36 (7)

Select ta.player
FROM top_scorer as ta
WHERE ta.goals > ALL (SELECT tb.goals
    FROM top_scorer as tb
    WHERE tb.country = 'Spain')
AND ta.goals > ANY (SELECT tc.goals
    FROM top_scorer as tc
    WHERE tc.country =
    'Germany')

Number of tuples in result 7.

3.37 (d)

Book

isbn	bname
2	A
4	B
6	C
8	D
10	E

Stock

isbn	copies
4	100
6	200
10	200
12	400

Query 1:

isbn	copies
4	100
6	200
10	200

Query 2:

isbn	copies
4	100
6	200
10	200
2	Null
8	Null

Query 3:

isbn	copies
4	100
6	200
10	200
12	400

Query 4:

isbn	copies
4	100
6	200
10	200
2	Null
8	Null
12	400

Query 4 is full outer join so that full order join record set superset of records compare to inner join, left outer join and right outer join.

3.38 (5)

Student

Roll_no.	Student_name
1	Amit
2	Priya
3	Vinit
4	Rohan
5	Smita

×

Performance

Roll_no.	Student_code	Marks
1	A	86
1	B	95
1	C	90
2	A	89
2	C	92
3	C	80

In where condition no condition over Roll_no so query produces all groups.

Total 5 different student names all 5 group records in result.

3.39 (c)

```
SELECT S.sno, S.sname
    FROM Supplier S, catalog C
    WHERE S.sno = C.sno AND
        cost > (SELECT AVG (cost)
            FROM Catalog
            WHERE pno = 'P4' GroupBy pno);
```

Catalog

sno	pno	cost
S1	P1	150
S1	P2	50
S1	P3	100
S2	P4	200
S2	P5	250
S3	P1	250
S3	P2	150
S3	P5	300
S3	P4	250

AVG
225

sno	pno	cost
S2	P4	200
S4	P4	250

Results

sno	sname
S1	RF
S2	BF
S3	PF
S3	PF

∴ 4 rows in table.

3.40 (d)

emp	empId	Name	Gender	Salary	deptid
21K>20K	e_1	X	Female	210K	CS
	e_2	Y	Male	19K	CS
25K>20K	e_3	XZ	Female	25K	EC
	e_4	YZ	Male	14K	EC
21K>20K	e_5	a	Female	21K	CS

Average salary of all employees in the company.

O/P

deptid	count(*)
CS	2
EC	1

For each department, number of female employees whose salary is greater than average salary of employees in the company.

3.41 (2)

- Given SQL query retrieves student records who register for all courses of dno = D01.
- Course id's and dept D01 are $\{C_{11}C_{12}\}$.
- Result students who register and all $C_{11}C_{12}$ courses which are SO1, S04 students records.

3.42 (2)

rollNum	name	gender	marks	
1	Naman	M	62	×
2	Aliya	F	70	✓
3	Aliya	F	80	✓
4	James	M	82	×
5	Swati	F	65	×

```
SELECT *
FROM Student
WHERE gender = 'F' AND
    marks > 65;
```

2 tuples in result.

4 Relational Model: Relational Algebra and Tuple Calculus

4.1 A library relational database system uses the following schema
USERS (User #, UserName, HomeTown)
BOOK (Book #, BookTitle, AuthorName)
ISSUED (Book #, User #, Date)
Explain in one English sentence, what each of the following relational algebra queries is designed to determine.
(a) $\sigma_{User\# = 6} (\pi_{User\#, BookTitle} ((USERS \bowtie ISSUED) \bowtie BOOK))$
(b) $\sigma_{AuthorName} (BOOK \bowtie (\sigma_{HomeTown=Delhi} (USERS \bowtie ISSUED)))$

[1996 : 2 M]

4.2 Which of the following query transformations (i.e. replacing the LHS expression by the RHS expression) is incorrect? R_1 and R_2 are relations, C_1, C_2 are selection conditions and A_1, A_2 are attributes of R_1?
(a) $\sigma_{c1}(\sigma_{c2} (R_1)) \rightarrow \sigma_{c2}(\sigma_{c1}(R_1))$
(b) $\sigma_{c1}(\pi_{A1} (R_1)) \rightarrow \pi_{A1}(\sigma_{c1}(R_1))$
(c) $\sigma_{c1}(R_1 \cup R_2) \rightarrow \sigma_{c1}(R_1) \cup \sigma_{c1}(R_2)$
(d) $\pi_{A1}(\sigma_{c1}(R_1)) \rightarrow \sigma_{c1}(\pi_{A1}(R_1))$

[1998 : 2 M]

4.3 Consider the join of a relation R with a relation S. If R has m tuples and S has n tuples then the maximum and minimum sizes of the join respectively are
(a) m + n and 0
(b) mn and 0
(c) m + n and |m − n|
(d) mn and m + n **[1999 : 1 M]**

4.4 The relational algebra expression equivalent to the following tuple calculus expression:
$\{t \mid t \in r \wedge (t [A] = 10 \wedge t [B] = 20)\}$ is
(a) $\sigma_{(A = 10 \vee B = 20)} (r)$
(b) $\sigma_{(A = 10)} (r) \cup \sigma_{(B = 20)} (r)$
(c) $\sigma_{(A = 10)} (r) \cap \sigma_{(B = 20)} (r)$
(d) $\sigma_{(A = 10)} (r) - \sigma_{(B = 20)} (r)$ **[1999 : 1 M]**

4.5 Given the relations:
employee (name, salary, deptno), and
department (deptno, deptname, address)
Which of the following queries cannot be expressed using the basic relational algebra operations ($\sigma, \pi, \times, \bowtie, \cup, \cap, -$)?
(a) Department address of every employee
(b) Employees whose name is the same as their department name
(c) The sum of all employees salaries
(d) All employees of a given department

[2000 : 1 M]

4.6 Which of the following relational calculus expressions is not safe?
(a) $\{t \mid \exists u \in R_1 (t[A] = u[A]) \wedge \neg \exists s \in R_2 (t[A] = s[A])\}$
(b) $\{t \mid \forall u \in R_1 (u[A] = \text{“x”} \Rightarrow \exists s \in R_2 (t[A] = s[A] \wedge s[A] = u[A]))\}$
(c) $\{t \mid \neg (t \in R_1)\}$
(d) $\{t \mid \exists u \in R_1 (t[A] = u[A]) \wedge \exists s \in R_2 (t[A] = s[A])\}$

[2001 : 2 M]

4.7 With regard to the expressive power of the formal relational query languages, which of the following statements is true?
(a) Relational algebra is more powerful than relational calculus
(b) Relational algebra has the same power as relational calculus
(c) Relational algebra has the same power as safe relational calculus
(d) None of the above **[2002 : 1 M]**

4.8 Let $R_1(\underline{A}, B, C)$ and $R_2(\underline{D}, E)$ be two relation schema, where the primary keys are shown underlined, and let C be a foreign key in R_1 referring to R_2. Suppose there is no violation of the above referential integrity constraint in the corresponding relation instances r_1 and r_2. Which one of the following relational algebra expressions would necessarily produce an empty relation?
(a) $\Pi_D(r_2) - \Pi_C(r_1)$
(b) $\Pi_C(r_1) - \Pi_D(r_2)$
(c) $\Pi_D(r_1 \bowtie_{C \neq D} R_2) - \Pi_C(r_1)$
(d) $\Pi_C(r_1 \bowtie_{C = D} R_2)$ **[2004 : 1 M]**

4.9 Consider the relation Student (name, sex, marks), where the primary key is shown underlined, pertaining to students in a class that has at least one boy and one girl. What does the following relational algebra expression produce?
(**Note:** ρ is rename operator)

$$\Pi_{name}(\sigma_{sex=female}(Student)) - \Pi_{name}(Student \bowtie_{(sex = female \wedge x = male \wedge marks \le m)} \rho_{n, x, m}(Student))$$

(a) names of girl students with the highest marks
(b) names of girl students with more marks than some boy student
(c) names of girl students with marks not less than some boy student
(d) names of girl students with more marks than all the boy students **[2004 : 2 M]**

4.10 Let r be a relation instance with schema R = (A, B, C, D). We define $r_1 = \Pi_{A,B,C}(r)$ and $r_2 = \Pi_{A,D}(r)$. let $S = r_1 * r_2$ where * denotes natural join. Given that the decomposition of r into r_1 and r_2 is lossy, which one of the following is TRUE?

(a) $s \subset r$ (b) $r \cup s = r$
(c) $r \subset s$ (d) $r * s = s$

[2005 : 1 M]

4.11 A table 'student' with schema (roll, name, hostel, marks), and another table 'hobby' with schema (roll, hobbyname) contains records as shown below:

Table Student

Roll	Name	Hostel	Marks
1798	Manoj Rathod	7	95
2154	Soumic Banerjee	5	68
2369	Gumma Reddy	7	86
2581	Pradeep Pendse	6	92
2643	Suhas Kulkarni	5	78
2711	Nitin Kadam	8	72
2872	Kiran Vora	5	92
2926	Manoj Kunkalikar	5	94
2959	Hemant Karkhanis	7	88
3125	Rajesh Doshi	5	82

Table Hobby

Roll	Hobbyname
1798	chess
1798	music
2154	music
2369	swimming
2581	cricket
2643	chess
2643	hockey
2711	volleyball
2872	football
2926	cricket
2959	photography
3125	music
3125	chess

The following SQL query is executed on the above tables:

```
select hostel
from student natural join hobby
where marks > = 75 and roll between 2000 and 3000;
```

Relations S and H with the same schema as those of these two tables respectively contain the same information as tuples. A new relation S′ is obtained by the following relational algebra operation:

$$S' = \Pi_{hostel}((\sigma_{s.roll = H.roll}(\sigma_{marks > 75 \text{ and } roll > 2000 \text{ and } roll < 3000}(S)) \times (H))$$

The difference between the number of rows output by the SQL statement and the number of tuples in S′ is

(a) 6 (b) 4
(c) 2 (d) 0

[2005 : 2 M]

4.12 Which of the following relational query languages have the same expressive power?

I. Relational algebra.
II. Tuple relational calculus restricted to safe expressions.
III. Domain relational calculus restricted to safe expressions.

(a) II and III only (b) I and II only
(c) I and III only (d) I, II and III

[2006 : 1 M]

4.13 Consider the relation enrolled (student, course) in which (student, course) is the primary key, and the relation paid (student, amount) where student is the primary key. Assume no null values and no foreign keys or integrity constraints. Assume that amounts 6000, 7000, 8000, 9000 and 10000 were each paid by 20% of the students. Consider these query plans (Plan 1 on left, Plan 2 on right) to "list all courses taken by students who have paid more than x."

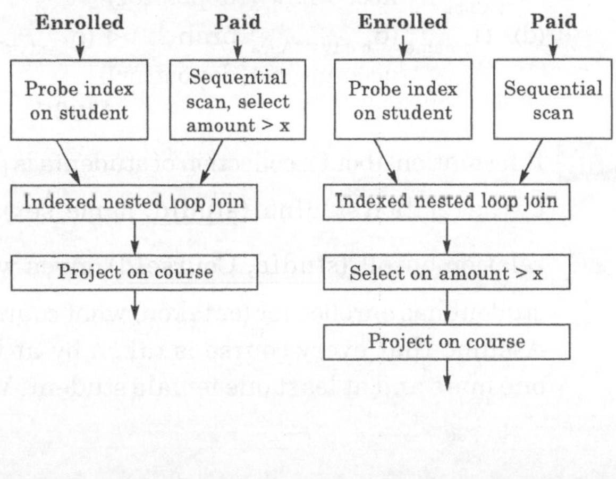

A disk seek takes 4 ms.disk data transfer bandwidth is 300 MB/s and checking a tuple to see if amount is greater than x takes 10 μs. Which of the following statements is correct?

(a) Plan 1 and Plan 2 will not output identical row sets for all databases
(b) A course may be listed more than once in the output of Plan 1 for some databases
(c) For x = 5000, Plan 1 executes faster than Plan 2 for all databases
(d) For x = 9000, Plan 1 executes slower than Plan 2 for all databases **[2006 : 2 M]**

4.14 Consider a selection of the form $\sigma_{A \leq 100}(r)$, where r is a relation with 1000 tuples. Assume that the attribute values for A among the tuples are uniformly distributed in the interval [0, 500]. Which one of the following options is the best estimate of the number of tuples returned by the given selection query?

(a) 50
(b) 100
(c) 150
(d) 200 **[2007 : 2 M]**

4.15 Consider the following relation schemas:

b-Schema = (b-name, b-city, assets)
a-Schema = (a-num, b-name, bal)
d-Schema = (c-name, a-number)

Let branch, account and depositor be respectively instances of the above schemas. Assume that account and depositor relations are much bigger than the branch relation.

Consider the following query:

$\Pi_{c\text{-name}} (\sigma_{b\text{-city}=\text{"Agra"} \wedge bal<0} (branch \bowtie (account \bowtie depositor))$

Which one of the following queries is the most efficient version of the above query?

(a) $\Pi_{c\text{-name}} (\sigma_{bal < 0} (\sigma_{b\text{-city} = \text{"Agra"}}\ branch \bowtie account) \bowtie depositor)$
(b) $\Pi_{c\text{-name}} (\sigma_{b\text{-city} = \text{"Agra"}}\ branch \bowtie (\sigma_{bal < 0}\ account \bowtie depositor))$
(c) $\Pi_{c\text{-name}} (\sigma_{b\text{-city} = \text{"Agra"}}\ branch \bowtie \sigma_{b\text{-city}= \text{"Agra"} \wedge bal < 0}\ account \bowtie depositor)$
(d) $\Pi_{c\text{-name}} (\sigma_{b\text{-city} = \text{"Agra"}}\ branch \bowtie (\sigma_{b\text{-city}=\text{"Agra"} \wedge bal < 0}\ account \bowtie depositor))$ **[2007 : 2 M]**

4.16 Information about a collection of students is given by the relation studInfo (<u>studId</u>, name, sex). The relation enroll (<u>studId, CourseId</u>) gives which student has enrolled for (or taken) what course(s). Assume that every course is taken by at least one male and at least one female student. What does the following relational algebra expression represent?

$\Pi_{courseId} ((\Pi_{studId} (\sigma_{sex = \text{"female"}}(studInfo) \times \Pi_{courseId}(enroll)) - enroll)$

(a) Courses in which all the female students are enrolled
(b) Courses in which a proper subset of female students are enrolled
(c) Courses in which only male students are enrolled
(d) None of the above **[2007 : 2 M]**

4.17 Consider the relation employee (<u>name</u>, sex, supervisorName) with name as the key. supervisorName gives the name of the supervisor of the employee under consideration. What does the following Tuple Relational Calculus query produce?

$\{e.name \mid employee(e) \wedge (\forall x) [\neg employee(x) \vee x.supervisorName \neq e.name \vee x.sex = \text{"male"}]\}$

(a) Names of employees with a male supervisor
(b) Names of employees with no immediate male subordinates
(c) Names of employees with no immediate female subordinates
(d) Names of employees with a female supervisor **[2007 : 2 M]**

4.18 Let R and S be two relations with the following schema:

R(<u>P, Q</u>, R1, R2, R3)
S(<u>P, Q</u>, S1, S2)

where {P, Q} is the key for both schemas. Which of the following queries are equivalent?

I. $\Pi_P(R \bowtie S)$
II. $\Pi_P(R) \bowtie \Pi_P(S)$
III. $\Pi_P(\Pi_{P,Q}(R) \cap \Pi_{P,Q}(S))$
IV. $\Pi_P(\Pi_{P,Q}(R) - (\Pi_{P,Q}(R) - \Pi_{P,Q}(S)))$

(a) Only I and II
(b) Only I and III
(c) Only I, II and III
(d) Only I, III and IV **[2008 : 2 M]**

4.19 Let R and S be relational schemes such that R = {a, b, c} and S = {c}. Now consider the following queries on the database :

I. $\pi_{R-S}(r) - \pi_{R-S}(\pi_{R-S}(r) \times S - \pi_{R-S,S}(r))$
II. $\{t \mid t \in \pi_{R-S}(r) \wedge \forall u \in s(\exists v \in r(u = v[s] \wedge t = v[R-S]))\}$

III. $\{t \mid t \in \pi_{R-S}(r) \wedge \forall v \in r(\exists u \in s(u = v[s] \wedge t = v[R-S]))\}$

IV. Select R.a, R.b
from R, S
where R.c = S.c

Which of the above queries are equivalent?

(a) I and II
(b) I and III
(c) II and IV
(d) III and IV

[2009 : 2 M]

Common Data for Questions 4.20 and 4.21:

Consider the following relational schema:

Suppliers(<u>sid: integer,</u> sname: string, city: string, street: string)

Parts(<u>pid: integer,</u> pname:string, color:string)

Catalog(<u>sid: integer, pid: integer,</u> cost:real)

4.20 Consider the following relational query on the above database:

```
SELECT S.sname
FROM  Suppliers S
WHERE S. sid NOT IN (SELECT C.sid
    FROM  Catalog C
    WHERE C.pid NOT IN (SELECT P.pid
    FROM Parts P
    WHERE P.color <> 'blue'))
```

Assume that relations corresponding to the above schema are not empty. Which one of the following is the correct interpretation of the above query?

(a) Find the names of all suppliers who have supplied a non-blue part
(b) Find the names of all suppliers who have not supplied a non-blue part
(c) Find the names of all suppliers who have supplied only blue parts
(d) Find the names of all suppliers who have not supplied only blue parts

[2009 : 2 M]

4.21 Assume that, in the suppliers relation above, each supplier and each street within a city has a unique name, and (sname, city) forms a candidate key. No other functional dependencies are implied other than those implied by primary and candidate keys. Which one of the following is TRUE about the above schema?

(a) The schema is in BCNF
(b) The schema is in 3NF but not in BCNF
(c) The schema is in 2NF but not in 3 NF
(d) The schema is not in 2NF

[2009 : 2 M]

4.22 Suppose $R_1(A, B)$ and $R_2(C, D)$ are two relation schemes. Let r_1 and r_2 be the corresponding relation instances. B is a foreign key that refers to C in R_2. If data in r_1 and r_2 satisfy referential integrity constraints, which of the following is ALWAYS TRUE?

(a) $\Pi_B(r_1) - \Pi_C(r_2) = \phi$
(b) $\Pi_C(r_2) - \Pi_B(r_1) = \phi$
(c) $\Pi_B(r_1) = \Pi_C(r_2)$
(d) $\Pi_B(r_1) - \Pi_C(r_2) \neq \phi$

[2012 : 2 M]

4.23 Consider the following relational schema:

Student (<u>rollno: integer:</u> sname string)

Courses (<u>courseno: integer,</u> cname: string)

Registration (<u>rollno: integer, courseno: integer,</u> percent: real)

Which of the following queries are equivalent to this query in English?

"Find the distinct names of all students who score more than 90% in the course numbered 107"

I. SELECT DISTINCT S.sname
FROM Students as S, Registration as R
WHERE R.rollno=S.roll.no AND R.courseno=107 and R.percent>90

II. $\Pi_{sname}(\sigma_{courseno=107 \wedge percent>90}$ (Regis-tration $\bowtie$ Student))

III. $\{T \mid \exists S \in$ Students, $\exists R \in$ Registration (S.rollno=R.rollno $\wedge$ R.courseno=107 $\wedge$ R.percent>90$\wedge$T.sname=S.sname)$\}$

IV. $\{<S_N> \mid \exists S_R \exists R_P (<S_R, S_N> \in \text{Student} \wedge <S_R, 107, R_P> \in \text{Registration} \wedge R_P > 90)\}$

(a) I, II, III and IV
(b) I, II and III only
(c) I, II and IV only
(d) I, III and IV only

[2013 : 2 M]

4.24 Consider a join (relation algebra) between relations r(R) and s(S) using the nested loop method. There are 3 buffers each of size equal to disk block size, out of which one buffer is reserved for intermediate results.

Assuming size(r(R)) < size(s(S)), the join will have fewer number of disk block accesses if

(a) relation r(R) is in the outer loop.
(b) relation s(S) is in the outer loop.
(c) join selection factor between r(R) and s(S) is more than 0.5.
(d) join selection factor between r(R) and s(S) is less than 0.5.

[2014 (Set-2) : 2 M]

4.25 Consider the relational schema given below, where **eId** of the **dependent** is a foreign key referring to **empId** of the relation **employee**. Assume that every employee has at least one associated dependent in the **dependent** relation.

employee(empId, empName, empAge)

dependent(depId, eId, depName, depAge)

Consider the following relational algebra query:

$\Pi_{empId}(employee) - \Pi_{empId}(employee \bowtie_{(empId = eId) \wedge (empAge \leq depAge)} dependent)$

The above query evaluates to the set of empIds of employees whose age is greater than that of

(a) some dependent.
(b) all dependents.
(c) some of his/her dependents.
(d) all of his/her dependents.

[2014 (Set-3) : 2 M]

4.26 Consider two relations R_1 (A, B) with the tuples (1, 5), (3, 7) and R_2(A, C) = (1, 7), (4, 9). Assume that R(A,B,C) is the full natural outer join of R_1 and R_2. Consider the following tuples of the form (A,B,C): a = (1, 5, null), b = (1, null, 7), c = (3, null, 9), d = (4, 7, null), e = (1, 5, 7), f = (3, 7, null), g = (4, null, 9). Which one of the following statements is correct?

(a) R contains a, b, e, f, g but not c, d
(b) R contains all of a, b, c, d, e, f, g
(c) R contains e, f, g but not a, b
(d) R contains e but not f, g

[2015 (Set-2) : 1 M]

4.27 Consider a database that has the relation schemas EMP(EmpId, EmpName, Deptid) and DEPT(DeptName, Deptid). Note that the Deptid can be permitted to be NULL in the relation EMP. Consider the following queries on the database expressed in tuple relational calculus.

I. $\{t \mid \exists u \in EMP(t[EmpName] = u[EmpName] \wedge \forall v \in DEPT(t[DeptId] \neq v[DeptId]))\}$

II. $\{t \mid \exists u \in EMP(t[EmpName] = u[EmpName] \wedge \exists v \in DEPT(t[DeptId] \neq v[DeptId]))\}$

III. $\{t \mid \exists u \in EMP(t[EmpName] = u[EmpName] \wedge \exists v \in DEPT(t[DeptId] = v[DeptId]))\}$

Which of the above queries are safe?

(a) I and II only
(b) I and III only
(c) II and III only
(d) I, II and III

[2017 (Set-1) : 1 M]

4.28 Consider a database that has the relation schema CR(StudentName, CourseName). An instance of the schema CR is as given below:

CR

Student Name	Course Name
SA	CA
SA	CB
SA	CC
SB	CB
SB	CC
SC	CA
SC	CB
SC	CC
SD	CA
SD	CB
SD	CC
SD	CD
SE	CD
SE	CA
SE	CB
SF	CA
SF	CB
SF	CC

The following query is made on the database.

$T_1 \leftarrow \pi_{CourseName}(\sigma_{StudentName = 'SA'}(CR))$

$T_2 \leftarrow CR \div T_1$

The number of rows in T_2 is ________.

[2017 (Set-1) : 2 M]

4.29 Consider the relations $r(A, B)$ and $s(B, C)$, where $s.B$ is a primary key and $r.B$ is a foreign key referencing $s.B$. Consider the query

$$Q : r \bowtie (\sigma_{B<5}(s))$$

Let LOJ denote the natural left outer-join operation. Assume that r and s contain no null values.

Which one of the following queries is NOT equivalent to Q?

(a) $\sigma_{B<5}(r \bowtie s)$
(b) $\sigma_{B<5}(r \text{ LOJ } s)$
(c) $r \text{ LOJ } (\sigma_{B<5}(s))$
(d) $\sigma_{B<5}(r) \text{ LOJ } s$

[2018 : 2 M]

4.30 Consider the following relations P(X, Y. Z), Q(X, Y, T) and R(Y, V).

P

X	Y	Z
X1	Y1	Z1
X1	Y1	Z2
X2	Y2	Z2
X2	Y4	Z4

Q

X	Y	T
X2	Y1	2
X1	Y2	5
X1	Y1	6
X3	Y3	1

R

Y	V
Y1	V1
Y3	V2
Y2	V3
Y2	V2

How many tuples will be returned by the following relational algebra query?
$\Pi_x(\sigma_{(P.Y = R.Y \wedge R.V = V2)}(P \times R)) - \Pi_x(\sigma_{(Q.Y = R.Y \wedge Q.T > 2)}(Q \times R))$

[2019 : 2 M]

4.31 A relation $r(A, B)$ in a relational database has 1200 tuples. The attribute A has integer values ranging from 6 to 20, and the attribute B has integer values ranging from 1 to 20. Assume that the attributes A and B are independently distributed.
The estimated number of tuples in the output of $\sigma_{(A > 10) \vee (B = 18)}(r)$ is ________.

[2021 (Set-1) : 1 M]

4.32 The following relation records the age of 500 employees of a company, where empNo (indicating the employee number) is the key:

empAge(<u>empNo</u>, age)

Consider the following relational algebra expression:

$\Pi_{empNo}(\text{empAge} \bowtie_{(age > age1)} \rho_{empNo1,\, age1}(\text{empAge}))$

What does the above expression generate?
(a) Employee numbers of all employees whose age is not the minimum.
(b) Employee numbers of all employees whose age is the minimum.
(c) Employee numbers of only those employees whose age is the maximum.
(d) Employee numbers of only those employees whose age is more than the age of exactly one other employee.

[2021 (Set-1) : 2 M]

4.33 Consider the following three relations in a relational database.

Employee (<u>eId</u>, Name), Brand (<u>bId</u>, bName), Own (<u>eId, bId</u>)

Which of the following relational algebra expressions return the set of eIds who own all the brands?
(a) $\pi_{eId}(\pi_{eId,\, bId}(\text{Own}) / \pi_{bId}(\text{Brand}))$
(b) $\pi_{eId}(\text{Own}) - \pi_{eId}((\pi_{eId}(\text{Own}) \times \pi_{bId}(\text{Brand})) - \pi_{eId,\, bId}(\text{Own}))$
(c) $\pi_{eId}(\pi_{eId,\, bId}(\text{Own}) / \pi_{bId}(\text{Own}))$
(d) $\pi_{eId}((\pi_{eId}(\text{Own}) \times \pi_{bId}(\text{Own}) / \pi_{bId}(\text{Brand}))$

[2022 : 1 M]

4.34 Which one of the options given below refers to the degree (or arity) of a relation in relational database systems?
(a) Number of attributes of its relation schema.
(b) Number of tuples stored in the relation.
(c) Number of entries in the relation.
(d) Number of distinct domains of its relation schema.

[2023 : 1 Mark]

4.35 Consider the following two relations, $R(A, B)$ and $S(A, C)$:

R	
A	B
10	20
20	30
30	40
30	50
50	95

S	
A	B
10	90
30	45
40	80

The total number of tuples obtained by evaluating the following expression

$$\sigma_{B < C}(R \bowtie_{R.A = S.A} S)$$

is ________.

[2024 (Set-1) : 1 M]

4.36 The relation schema, Person (<u>pid</u>, city), describes the city of residence for every person uniquely identified by pid. The following relational algebra operators are available: selection, projection, cross product, and rename.
To find the list of cities where at least 3 persons reside, using the above operators, the minimum number of cross product operations that must be used is
(a) 1
(b) 2
(c) 4
(d) 3

[2024 (Set-2) : 2 M]

■■■■

Answers Relational Model: Relational Algebra and Tuple Calculus

4.1	Sol.	4.2	(d)	4.3	(b)	4.4	(c)	4.5	(c)	4.6	(c)	4.7	(c)	4.8	(b)	4.9	(d)
4.10	(c)	4.11	(b)	4.12	(d)	4.13	(c)	4.14	(d)	4.15	(b)	4.16	(b)	4.17	(c)	4.18	(d)
4.19	(c)	4.20	(a)	4.21	(b)	4.22	(a)	4.23	(a)	4.24	(a)	4.25	(d)	4.26	(c)	4.27	(d)
4.28	(4)	4.29	(c)	4.30	(1)	4.31	(205)	4.32	(a)	4.33	(a, b)	4.34	(a)	4.35	(2)	4.36	(b)

Explanations Relational Model: Relational Algebra and Tuple Calculus

4.1 Sol.

(a) Determines titles of the books issued to user with user # 6

(b) Determines name of authors of books that are issued to users who have home town 'Delhi'.

4.2 (d)

$$\pi_{A1}(\sigma_{c1}(R_1)) \rightarrow \sigma_{c1}(\pi_{A1}(R_1))$$

If the selection condition (c_1) is on attribute over A_2, then we can not replace LHS expression by RHS expression. In RHS first projection is on A_1 then selection condition c_1 will fail because no A_2.

4.3 (b)

Whenever there is no foreign key constraint between two tables then the max. & min. no. of tuples in their join is mn and 0 respectively.

4.4 (c)

The given tuple calculus expression give tuples where A = 10 & B = 20 of relation r.
Therefore, relational algebra expression
$\sigma_{A=10}(r) \cap \sigma_{B=20}(r)$ is equivalent to given tuple calculus expression.

4.5 (c)

Aggregate operations like sum, average, count can't be expressed in terms of basic relational algebra operations. Aggregate function requires extended relational algebra. (Min, Max can be expressed using basic RA operations)

4.6 (c)

The query $\{t \mid \neg(t \in R_1)\}$ is syntactically correct. However, it ask for all tuples t such that t is not in R_1.

That set of such t tuples is obviously infinite, in the context of such as the infinite domain set of all integers. Therefore, this is unsafe query.

4.7 (c)

Every query that can be expressed using a safe relational calculus query can also be expressed as a relational algebra query.
Therefore, relational algebra has the same power as safe relational calculus.

4.8 (b)

$R_1(A, B, C)$
$R_2(D,E)$
C is the foreign key in R_1 and referring to primary key of R_2.
D is the primary key of R_2.
In relation R_2 column D values are superset or equal of R_1 column C. Because of integrity constraints. So $\pi_C(R_1) - \pi_D(R_2)$ is empty.

4.9 (d)

The given query computes the names of girl students with more marks than all the boy students.

4.10 (c)

Given, $R = (A, B, C, D)$

$$r_1 = \pi_{A,B,C}(r)$$
$$r_2 = \pi_{A,D}(r)$$
$$s = r_1 * r_2$$

On applying Natural Join all the tupples which are there already, will be present in S apart from this is some tupples have duplicate values of 'A', then the resultant tupples will increases.
Hence, $r \subset s$ is the correct answer.

4.11 (b)

SQL query will return following tupples:

Hostel
7
6
5
5
5
5
7

output contain duplicate since distinct hostel are not selected

So number of tupples are 7.
Select operation in Relation algebra by default return distinct attributes i.e.

Hostel
7
6
5

So number of tupples are 3.
The difference between the number count = Number of (SQL) – Number (RA) = 7 – 3 = 4

4.12 (d)

Given three relational query languages have same expressive power.

4.13 (c)

Enroll (Student, Course) paid (Student, amount)

Assume that amounts 6000, 7000, 8000, 9000, 10000 were each paid by 20% of the students

$$\text{Plan1: } \pi_{course}\left(\text{Enroll} \bowtie \underset{amount > X}{\sigma(\text{paid})}\right)$$

$$\text{Plan2: } \pi_{course}\left(\underset{amount > X}{\sigma(\text{ Enroll}} \bowtie \text{paid})\right)$$

If X = 5000 Plan1 Plan2 join cost equal because amounts of paid relation are greater than 5000. between Plan1 required (amount > 5000) comparisons only to paid relation and Plan2 requried (amount > 5000) comparisons to the result of Enroll ⋈ paid.
So that plan 1 executes faster than Plan2 if X = 9000 join cost of Plan1 less than Plan2 because plan1 joins enroll records with 20% of paid records. So that plan1 executes faster than Plan2.

4.14 (d)

Number of tuples in relation r = 1000
Each value of A is appeared twice, because uniformly distributed in [0, 500]
$\sigma_{A \le 100}$ (r) returns 200 tuples.

4.15 (b)

As b is very small compared to a and d which is also taken in account.
$\sigma_{b\text{-city}=\text{“Agra”}}$ branch which already filter city as agra making it small, and before that $\sigma_{bal<0}$ account ⋈ depositor filter and give selected tables so ⋈ between them will give same result and better one.

4.16 (b)

Let the tables are like
Stud Info

StudId	Name	Sex
1	Sita	F
2	Ram	M
3	Geeta	F
4	Shyam	M
5	Radha	F

Enroll

StudId	CourseId
1	A
3	A
1	B
2	B
3	B
5	B
1	C
4	C
2	D
4	D

Now the query π_{studId} ($\sigma_{sex = \text{“females”}}$ (Stud Info)) results in

S

StudId
1
3
5

$\pi_{courseId}$ (Enroll)

C

CourseId
A
B
C
D

$((S \times C)\_enroll)$

StudentId	CourseId
1	A
1	B
1	C
1	D
3	A
3	B
3	C
3	D
5	A
5	B
5	C
5	D

T

StudentId	CourseId
1	D
3	C
3	D
5	A
5	C
5	D

Selecting course-id from above

CourseId
D
C
A

A, C, D are courses which is enrolled by proper subset of female students.

4.17 (c)

Emp(<u>name</u>, Sex, Supervisor_name)

{e.name | emplyee(e) λ

$(\forall_X)$ (¬ employee(X) ∨

X.SupervisorName ≠ e.name ∨

X.sex = "male")}

Retrieves e.name of employee(e) and every "X" if X is employer then [X is not supervisor (r) subordinate of e.name is male] Which is equal to retrieves employees who has no immediate subordinate.

4.18 (d)

In I, Ps from natural join of R & S are selected. In III, all Ps from intersection of (P, Q) pairs present in R and S. IV is also equivalent to III because (R-(R-S)) = R ⋈ S.

II is not equivalent as it may also include Ps where Qs are not same in R and S.

4.19 (c)

R(A, B, C) S(C)

I. $\pi_{R-S}(R) - \pi_{R-S}(\pi_{R-S}(R) \times S - \pi_{R-S,\,S}(R)) \equiv$
$\pi_{AB}(R) - \pi_{AB}(\pi_{AB}(R) \times \pi_C(S) - \pi_{ABC}(R) \equiv R/S$

II. $\{t \mid t \in \underbrace{\pi(R)}_{R-S} \wedge \underbrace{\forall U \in S(\exists V \in R(U = V[S] \wedge t = V[R-S]))}\}$

"AB" of R and for every "C" of S there must be some C of R which equal to A, B and C.

Which also equal to R/S.

III. TRC query which equal to S/R.

IV. SQL query equal to R ⋉ S.

So query I and II are equal.

4.20 (a)

Suppliers (<u>sid</u>)	Catalog (sid pid)	Parts (<u>pid</u> color)
S_1	S_1 P_1	P_1 red
S_2	S_1 P_2	P_2 red
S_2	S_2 P_2	P_3 blue
S_4	S_2 P_5	P_4 blue
	S_3 P_3	P_5 green
	S_3 P_4	

sid's supplied sids not in

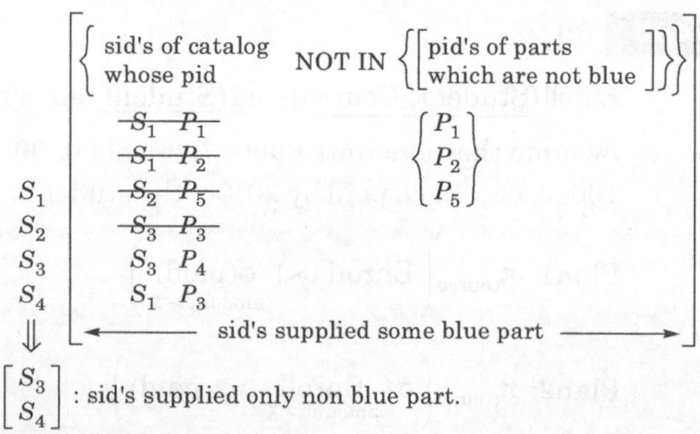

4.21 (b)

Suppliers (Sid, Sname, City, Street)

1. Each supplier and each street within a city has a unique name
   Sid Street City → Sname
2. (Sname, city) forms candidate key so that
   (Sname city) → Sid Street
3. <u>Sid</u> primary key so that
   Sid → Sname City Street

Each FD of above satisfy BCNF.

4.22 (a)

R_1 (A, B)
R_2 (C, D)
B is a foreign key and referring to C and C is candidate key

So, $\Pi_B(r_1) - \Pi_C(r_2) = \phi$

E.g.,

R_1

A	B
1	5
2	5
3	7
4	7

R_2

C	D
5	1
6	2
7	3
8	4

Now, (B) $\Pi_C(r_2) - \Pi_B(r_1) = \{6, 8\}$

(C) $\Pi_B(r_1) \neq \Pi_C(r_2)$

$\{5, 7\}$ $\{5, 6, 7, 8\}$

(D) $\Pi_B(r_1) - \Pi_C(r_2) = \phi$

4.23 (a)

i SQL query
ii RA query
iii TRC query
iv DRC query

} All four queries Correct for Given specification

4.24 (a)

Join will have fewer number of disk block accesses if outer loop has smaller relation ($r(R)$).

4.25 (d)

$\pi_{\text{empId}}(\text{employee}) - \pi_{\text{empId}}(\text{employee} \bowtie_{(\text{empId=eId}) \wedge (\text{empAge} \leq \text{depAge})} \text{dependent})$

Where,

(*i*) $\pi_{\text{empId}}(\text{employee} \bowtie_{(\text{empId=eId}) \wedge (\text{empAge} \leq \text{depAge})} \text{dependent})$

Means

All employees whose age is less than or equal to that of all of his dependents.

(*ii*) $\pi_{\text{empId}}(\text{employee}) - \pi_{\text{empId}}(\text{employee} \bowtie_{(\text{empId=eId}) \wedge (\text{empAge} \leq \text{depAge})} \text{dependent})$.

Means

All employees whose age is greater than that of all of his dependents.

4.26 (c)

R_1

A	B
1	5
3	7

R_2

A	C
1	7
4	9

$R = R_1 ⟗ R_2$

R

A	B	C	
1	5	7	→ e
3	7	Null	→ f
4	Null	9	→ g

R contains e, f, g but not a, b.

4.27 (d)

I. $\{t \mid \exists u \in \text{EMP}(t[\text{EmpName}] = u[\text{EmpName}] \wedge \forall v \in \text{DEPT}(t[\text{DeptId}] \neq v[\text{DeptId}]))\}$
results empname who does not belongs to any departments (safe query).

II. $\{t \mid \exists u \in \text{EMP}(t[\text{EmpName}] = u[\text{EmpName}] \wedge \exists v \in \text{DEPT}(t[\text{DeptId}] \neq v[\text{DeptId}]))\}$
results empname who does not belongs to some departments (safe query).

III. $\{t \mid \exists u \in \text{EMP}(t[\text{EmpName}] = u[\text{EmpName}] \wedge \exists v \in \text{DEPT}(t[\text{DeptId}] = v[\text{DeptId}]))\}$
results empname who belongs to same departments (safe query).

$\therefore$ All are safe queries.

4.28 (4)

T_1 result :

Course Name
CA
CB
CC

T_2 result : CR $\div T_1 \Rightarrow$ student name for which every course name of CA, CB, CC is

SA
SC
SD
SF

4 tuples in result

4.29 (c)

R

A	(B)
a_1	2
a_2	4
a_3	4
a_4	6
a_5	6

S

<u>B</u>	C
2	c_1
4	c_2
4	c_2
8	c_2

Foreign key ⇑ (R.B) → Primary key ⇑ (S.B)

Given query: $r \bowtie \left(\sigma_{B<5}(s)\right)$

A	B	C
a_1	2	c_1
a_2	4	c_1
a_3	4	c_1

$A: \sigma_{B<5}(r \bowtie s)$

A	B	C
a_1	2	c_1
a_2	4	c_1
a_3	4	c_1

B: $\sigma_{B<5}(r ⟕ s)$

A	B	C
a_1	2	c_1
a_2	4	c_1
a_3	4	c_1

C: $r ⟕ (\sigma_{B<5}(s))$

A	B	C
a_1	2	c_1
a_2	4	c_1
a_3	4	c_1
a_4	6	Null
a_5	6	Null

D: $\sigma_{B<5}(r) ⟕ s$

A	B	C
a_1	2	c_1
a_2	4	c_1
a_3	4	c_1

Option "c" query result not equal to given query.

4.30 (1)

P

X	Y	Z
X1	Y1	Z1
X1	Y1	Z2
X2	Y2	Z2
X2	Y4	Z4

R

Y	V
Y1	V1
Y3	V2
Y2	V3
Y2	V2

$$\Pi_x(\sigma_{(P.Y = R.Y \wedge R.V = V2)}(P \times R)) \Rightarrow \begin{array}{|c|}\hline X \\ \hline X2 \\ \hline\end{array} \quad ...(I)$$

Q

X	Y	T
X2	Y1	2
X1	Y2	5
X1	Y1	6
X3	Y3	1

R

Y	V
Y1	V1
Y3	V2
Y2	V3
Y2	V2

$$\Pi_x(\sigma_{(Q.Y = R.Y \wedge Q.T > 2)}(Q \times R)) \Rightarrow \begin{array}{|c|}\hline X \\ \hline X1 \\ \hline\end{array} \quad ...(II)$$

$$I - II \Rightarrow \begin{array}{|c|}\hline X \\ \hline X2 \\ \hline\end{array}$$ one record in result.

4.31 (205)

- There are 10 distinct integer for $(A > 10)$ out of 15.
- There are 1 distinct integer for $(B = 18)$ out of 20.

$$P(A > 10) = \frac{10}{15} = \frac{2}{3},\ P(B = 18) = \frac{1}{20}$$

$$P((A > 10) \wedge (B = 18)) = \frac{2}{3} \times \frac{1}{20} = \frac{1}{30}$$

Now, $P((A > 10) \vee (B = 18)) = P(A > 10) + P(B = 18) - P((A > 10) \wedge (B = 18))$

$$= \frac{2}{3} + \frac{1}{20} - \frac{1}{30} = \frac{40 + 3 - 2}{60} = \frac{41}{60}$$

$$\text{Estimated number of tuples} = \frac{41}{60} \times 1200 = 820 \text{ tuples}$$

or

Using method exclusion principle: for each (A > 10) i.e. [11 – 20] there are 20 distinct B i.e. (1 – 20). So current total = 20 × 10 = 200.
For each B = 18 there are 15 distinct A.
So current total = 200 + 15 = 215
But here we counted some tuples twice i.e. B = 18 and A > 10, there are 10 such tuples.
So, final number of tuples are = 215 – 10 = 205.
Therefore both the answers are correct.

4.32 (a)

empAge(<u>empNo</u>, age)

Π_{empNo} (empAge $\bowtie_{(\text{age} > \text{age1})}$ $\rho_{\text{empNo1, age1}}$ (empAge))

Retrieve empNo values of emptAge those are having 'age' greater than some age.

⇓

Retrieve Employee Number (empNo) of all employee whose age is greater than some employee' age.

⇓

Employee number of all employees whose age is not the minimum. [Since it is greater than atleast 1 age].

4.33 (a, b)

- $\pi_{\text{eid bid}}$ (Own) / π_{bid} (Brand) ⇒ Results eid's which owns every brand of brand relation.
- Option (b) expansion and division using basic operators.

4.34 (a)

The cardinality of a relation is the number of tuples, whereas the degree of a relation is the number of attributes.

4.35 (2)

$R \bowtie_{R.A = S.A} S$ result ⇒

A	B	A	C
10	20	10	90
30	40	30	45
30	50	30	45

$\sigma_{B<C}(R \bowtie_{R.A = S.A} S)$ result ⇒

A	B	A	C
10	20	10	90
30	40	30	45

Results 2 tuples

4.36 (b)

$$\pi_{T_1.\text{city}}\left(\sigma_{T_1.\text{city} = T_2.\text{city} \wedge T_2.\text{city} = T_3.\text{city} \wedge T_1.\text{pid} \neq T_2.\text{pid} \wedge T_2.\text{pid} \neq T_3.\text{pid} \wedge T_1.\text{pid} \neq T_3.\text{pid}}\left(\rho_{(T_1,\ \text{person})} \times \rho_{(T_2,\ \text{person})} \times \rho_{(T_3,\ \text{person})}\right)\right)$$

Retrieves city with atleast three persons required three instances of 'person' relations with two cross products.

■■■■

5 Transactions and Concurrency Control

5.1 For the schedule given below, which of the following is correct?

1	Read A	
2		Read B
3	Write A	
4		Read A
5		Write A
6		Write B
7	Read B	
8	Write B	

(a) This schedule is serializable and can occur in a scheme using 2PL protocol.
(b) This schedule is serializable but cannot occur in a scheme using 2PL protocol.
(c) This schedule is not serializable but can occur in a scheme using 2PL protocol.
(d) This schedule is not serializable and cannot occur in a scheme using 2PL protocol.

[1999 : 2 M]

5.2 Which of the following scenarios may lead to an irrecoverable error in a database system?
(a) A transaction writes a data item after it is read by an uncommitted transaction
(b) A transaction reads a data item after it is read by an uncommitted transaction
(c) A transaction reads a data item after it is written by a committed transaction
(d) A transaction reads a data item after it is written by an uncommitted transaction

[2003 : 1 M]

5.3 Consider three data items D1, D2, and D3, and the following execution schedule of transactions T1, T2 and T3. In the diagram, R(D) and W(D) denote the actions reading and writing the data item D respectively.

T1	T2	T3
	R(D3);	
	R(D2);	
	W(D2);	
		R(D2);
		R(D3);
R(D1);		
W(D1);		
		W(D2);
		W(D3);
	R(D1);	
R(D2);		
W(D2);		
	W(D1);	

(a) The schedule is serializable as T2; T3; T1;
(b) The schedule is serializable as T2; T1; T3;
(c) The schedule is serializable as T3; T2; T1;
(d) The schedule is not serializable

[2003 : 2 M]

5.4 Which level of locking provides the highest degree of concurrency in a relational database?
(a) Page
(b) Table
(c) Row
(d) Page, table and row level locking allow the same degree of concurrency

[2004 : 1 M]

5.5 Consider the following schedule S of transactions T1 and T2:

T1	T2
Read(A)	
A = A–10	
	Read(A)
	Temp = 0.2 * A
	Write(A)
	Read(B)
Write(A)	
Read(B)	
B = B + 10	
Write(B)	
	B = B + Temp
	Write(B)

Which of the following is TRUE about the schedule S ?
(a) S is serializable only as T1, T2
(b) S is serializable only as T2, T1
(c) S is serializable both as T1, T2 and T2, T1
(d) S is not serializable either as T1 or as T2

[2004 : 2 M]

5.6 Amongst the ACID properties of a transaction, the 'Durability' property requires that the changes made to the database by a successful transaction persist
(a) Except in case of an Operating System crash
(b) Except in case of a Disk crash
(c) Except in case of a power failure
(d) Always, even if there is a failure of any kind

[2005 : 1 M]

5.7 Consider the following log sequence of two transactions on a bank account, with initial balance 12000, that transfer 2000 to a mortgage payment and, then apply a 5% interest.

1. T1: start
2. T1: B old = 12000, new = 10000
3. T1: M old = 0, new = 2000
4. T1: commit
5. T2: start
6. T2: B old = 10000, new = 10500
7. T2: commit

Suppose the database system crashed just before log record 7 is written. When the system is restarted, which one statement is true of the recovery procedure?

(a) We must redo log record 6 to set B to 10500
(b) We must undo log record 6 to set B to 10000 and then redo log records 2 and 3
(c) We need not redo log records 2 and 3 because transaction T1 has committed
(d) We can apply redo and undo operations in arbitrary order because they are idempotent

[2006 : 1 M]

5.8 Consider the following schedules involving two transactions. Which one of the following statements is TRUE?

S_1 : $r_1(X); r_1(Y); r_2(X); r_2(Y); w_2(Y); w_1(X)$
S_2 : $r_1(X); r_2(X); r_2(Y); W_2(Y); r_1(Y); w_1(X)$

(a) Both S_1 and S_2 are conflict serializable
(b) S_1 is conflict serializable and S_2 is not conflict serializable
(c) S_1 is not conflict serializable and S_2 is conflict serializable
(d) Both S_1 and S_2 are not conflict serializable

[2007 : 2 M]

5.9 Consider the following two transactions : T_1 and T_2.

T_1:	T_2:
read(A);	read(B);
read(B);	read(A);
If A = 0 then B ← B + 1;	If B ≠ 0 then A ← A – 1;
write(B);	write(A);

Which of the following schemes, using shared and exclusive locks, satisfy the requirements for strict two phase locking for the above transactions?

(a)

S1 :	S2 :
lock S(A);	lock S(B);
read (A);	read (B);
lock S(B);	lock S(A);
read (B);	read (A);
if A = 0	if B ≠ 0
then B ← B+1;	then A ← A–1;
write (B);	write (A);
commit;	commit;
unlock (A);	unlock (B);
unlock (B);	unlock (A);

(b)

S1 :	S2 :
lock X(A);	lock X(B);
read (A);	read (B);
lock X(B);	lock X(A);
read (B);	read (A);
if A = 0	if B ≠ 0
then B ← B+1;	then A ← A–1;
write (B);	write (A);
unlock (A);	unlock (A);
commit;	commit;
unlock (B);	unlock (A);

(c)

S1 :	S2 :
lock S(A);	lock S(B);
read (A);	read (B);
lock X(B);	lock X(A);
read (B);	read (A);
if A = 0	if B ≠ 0
then B ← B+1;	then A ← A– 1;
write (B);	write (A);
unlock (A);	unlock (A);
commit;	commit;
unlock (B);	unlock (A);

(d)

S1 :	S2 :
lock S(A);	lock S(B);
read (A);	read (B);
lock X(B);	lock X(A);
read (B);	read (A);
if A = 0	if B ≠ 0
then B ← B+1;	then A ← A–1;
write (B);	write (A);
unlock (A);	unlock (A);
unlock (B);	unlock (B);
commit;	commit;

[2007 : 2 M]

5.10 Consider the following three schedules of transactions T1, T2 and T3. [Notation: In the following NYO represents the action Y (R for read, W for write) performed by transaction N on object O].

S1: 2RA 2WA 3RC 2WB 3WA 3WC 1RA 1RB 1WA 1WB

S2: 3RC 2RA 2WA 2WB 3WA 1RA
1RB 1WA 1WB 3WC
S3: 2RA 3RC 3WA 2WA 2WB 3WC
1RA 1RB 1WA 1WB

Which of the following statements is TRUE?

(a) S1, S2 and S3 are all conflict equivalent to each other
(b) No two of S1, S2 and S3 are conflict equivalent to each other
(c) S2 is conflict equivalent to S3, but not to S1
(d) S1 is conflict equivalent to S2, but not to S3

[2008 : 2 M]

5.11 Consider two transactions T_1 and T_2, and four schedules S_1, S_2, S_3, S_4 of T_1 and T_2 as given below:

$T_1 : R_1[x]\ W_1[x]\ W_1[y]$
$T_2 : R_2[x]\ R_2[y]\ W_2[y]$
$S_1 : R_1[x]\ R_2[x]\ R_2[y]\ W_1[x]\ W_1[y]\ W_2[y]$
$S_2 : R_1[x]\ R_2[x]\ R_2[y]\ W_1[x]\ W_2[y]\ W_1[y]$
$S_3 : R_1[x]\ W_1[x]\ R_2[x]\ W_1[y]\ R_2[y]\ W_2[y]$
$S_4 : R_2[x]\ R_2[y]\ R_1[x]\ W_1[x]\ W_1[y]\ W_2[y]$

Which of the above schedules are conflict-serializable?

(a) S_1 and S_2 (b) S_2 and S_3
(c) S_3 only (d) S_4 only

[2009 : 2 M]

5.12 Which of the following concurrency control protocols ensure both conflict serializability and freedom from deadlock?

I. 2-phase locking
II. Time-stamp ordering

(a) I only (b) II only
(c) Both I and II (d) Neither I nor II

[2010 : 1 M]

5.13 Consider the following schedule for transactions T1, T2 and T3:

T1	T2	T3
Read (X)		
	Read (Y)	
		Read (Y)
	Write (Y)	
Write (X)		
		Write (X)
	Read (X)	
	Write (X)	

Which one of the schedules below is the correct serialization of the above?

(a) T1 → T3 → T2 (b) T2 → T1 → T3
(c) T2 → T3 → T1 (d) T3 → T1 → T2

[2010 : 2 M]

5.14 Consider the following transactions with data items P and Q initialized to zero:

T_1 : read (P);
read (Q);
if P = 0 then Q: = Q + 1;
write (Q);
T_2 : read (Q);
read (P);
if Q = 0 then P: = P + 1;
write (P);

Any non-serial interleaving of T_1 and T_2 for concurrent execution leads to

(a) a serializable schedule
(b) a schedule that is not conflict serializable
(c) a conflict serializable schedule
(d) a schedule for which a precedence graph cannot be drawn

[2012 : 2 M]

5.15 An index is clustered, if

(a) it is on a set of fields that form a candidate key.
(b) it is on a set of fields that include the primary key.
(c) the data record of the file are organized in the same order as the date entries of the index
(d) the data records of the file are organized not in the same order as the data entries of the index.

[2013 : 1 M]

5.16 Consider the following four schedules due to three transactions (indicated by the subscript) using *read* and *write* on a data item *x*, denoted by *r*(*x*) and *w*(*x*) respectively. Which one of them is conflict serializable?

(a) $r_1(x); r_2(x); w_1(x); r_3(x); w_2(x)$
(b) $r_2(x); r_1(x); w_2(x); r_3(x); w_1(x)$
(c) $r_3(x); r_2(x); r_1(x); w_2(x); w_1(x)$
(d) $r_2(x); w_2(x); r_3(x); r_1(x); w_1(x)$

[2014 (Set-1) : 2 M]

5.17 Consider the following schedule S of transactions T1, T2, T3 and T4:

T1	T2	T3	T4
	Reads(*X*)		
		Writes(*X*) Commit	
Writes(*X*) Commit			
	Writes(*Y*) Reads(*Z*) Commit		
			Reads(*X*) Reads(*Y*) Commit

Which one of the following statements is CORRECT?
(a) S is conflict-serializable but not recoverable
(b) S is not conflict-serializable but is recoverable
(c) S is both conflict-serializable and recoverable
(d) S is neither conflict-serializable nor is it recoverable

[2014 (Set-2) : 2 M]

5.18 Consider the transactions T1, T2, and T3 and the schedules S1 and S2 given below:
T1: r1 (X); r1 (Z); w1 (X); w1(Z)
T2: r2 (Y); r2 (Z); w2 (Z)
T3: r3 (Y); r3 (X); w3 (Y)
S1 : r1 (X); r3 (Y); r3 (X); r2 (Y); r2 (Z); w3(Y); w2 (Z); r1(Z); w1(X); w1(Z)
S2 : r1 (X); r3 (Y); r2 (Y); r3 (X); r1(Z); r2(Z); w3 (Y); w1 (X); w2 (Z); w1(Z)

Which one of the following statements about the schedules is **TRUE**?
(a) Only S1 is conflict-serializable.
(b) Only S2 is conflict-serializable.
(c) Both S1 and S2 are conflict-serializable.
(d) Neither S1 nor S2 is conflict-serializable.

[2014 (Set-3) : 2 M]

5.19 Consider the following transaction involving two bank accounts x and y.
read (x) ; $x := x - 50$; write (x); read (y); $y := y + 50$; write (y)
The constraint that the sum of the accounts x and y should remain constant is that of
(a) Atomicity (b) Consistency
(c) Isolation (d) Durability

[2015 (Set-2) : 1 M]

5.20 Consider a simple checkpointing protocol and the following set of operations in the log.
(start, T4); (write, T4, y, 2, 3); (start, T1); (commit, T4); (write, T1, z, 5, 7);
(checkpoint);
(start, T2); (write, T2, x, 1, 9); (commit, T2); (start, T3); (write, T3, z, 7, 2);
If a crash happens now and the system tries to recover using both undo and redo operations, what are the contents of the undo list and the redo list
(a) Undo: T3, T1; Redo: T2
(b) Undo: T3, T1; Redo: T2, T4
(c) Undo: none; Redo: T2, T4, T3, T1
(d) Undo: T3, T1, T4; Redo: T2

[2015 (Set-2) : 1 M]

5.21 Consider the following partial Schedule S involving two transactions $T1$ and $T2$. Only the read and the write operations have been shown. The read operation on data item P is denoted by read(P) and the write operation on data item P is denoted by write(P).

Time instance	Transaction-id	
	T1	*T2*
1	read(A)	
2	write(A)	
3		read(C)
4		write(C)
5		read(B)
6		write(B)
7		read(A)
8		commit
9	read(B)	

Suppose that the transaction $T1$ fails immediately after time instance 9. Which one of the following statements is correct?
(a) $T2$ must be aborted and then both $T1$ and $T2$ must be restarted to ensure transaction atomicity
(b) Schedule S is nonrecoverable and cannot ensure transaction atomicity
(c) Only $T2$ must be aborted and then restarted to ensure transaction atomicity
(d) Schedule S is recoverable and can ensure atomicity and nothing else needs to be done

[2015 (Set-3) : 2 M]

5.22 Consider the following directed graph:

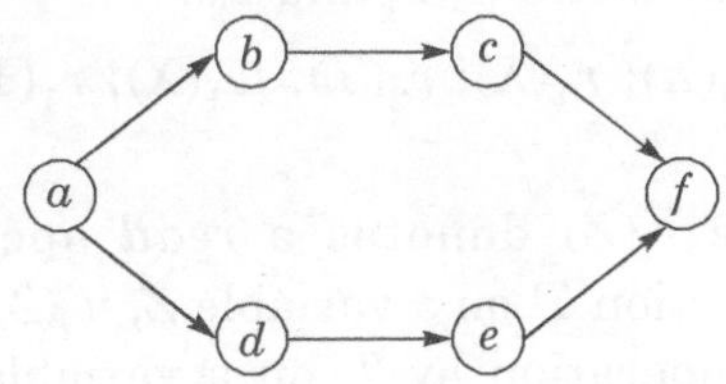

The number of different topological ordering of the vertices of the graph is ____.

[2016 (Set-1) : 1 M]

5.23 Which one of the following is NOT a part of the ACID properties of database transactions?
(a) Atomicity
(b) Consistency
(c) Isolation
(d) Deadlock-freedom

[2016 (Set-1) : 1 M]

5.24 Consider the following two phase locking protocol. Suppose a transaction T accesses (for read or write operations), a certain set of objects $\{O_1,...,O_k\}$. This is done in the following manner:
Step 1. T acquires exclusive locks to $O_1,...,O_k$ in increasing order of their addresses.
Step 2. The required operations are performed.
Step 3. All locks are released.
This protocol will
(a) guarantee serializability and deadlock-freedom
(b) guarantee neither serializability nor deadlock-freedom
(c) guarantee serializability but not deadlock-freedom
(d) guarantee deadlock-freedom but not serializability

[2016 (Set-1) : 2 M]

5.25 Suppose a database schedule S involves transactions $T_1,..., T_n$. Construct the precedence graph of S with vertices representing the transactions and edges representing the conflicts. If S is serializable, which one of the following orderings of the vertices of the precedence graph is guaranteed to yield a serial schedule?
(a) Topological order
(b) Depth-first order
(c) Breadth-first order
(d) Ascending order of transaction indices

[2016 (Set-2) : 1 M]

5.26 Consider the following database schedule with two transactions, T_1 and T_2.
$S = r_2(X); r_1(X); r_2(Y); w_1(X); r_1(Y); w_2(X); a_1; a_2$
where $r_i(Z)$ denotes a *read* operation by transaction T_i on a variable Z, $w_i(Z)$ denotes a *write* operation by T_i on a variable Z and a_i denotes an *abort* by transation T_i.
Which one of the following statements about the above schedule is **TRUE**?
(a) S is non-recoverable
(b) S is recoverable, but has a cascading abort
(c) S does not have a cascading abort
(d) S is strict

[2016 (Set-2) : 2 M]

5.27 In a database system, unique timestamps are assigned to each transaction using Lamport's logical clock. Let $TS(T_1)$ and $TS(T_2)$ be the timestamps of transactions T_1 and T_2 respectively. Besides, T_1 holds a lock on the resource R and T_2 has requested a conflicting lock on the same resource R. The following algorithm is used to prevent deadlocks in the database system assuming that a killed transaction is restarted with the same timestamp.
if $TS(T_2) < TS(T_1)$ then
  T_1 is killed
else T_2 waits
Assume any transaction that is not killed terminates eventually. Which of the following is TRUE about the database system that uses the above algorithm to prevent deadlocks?
(a) The database system is both deadlock-free and starvation-free.
(b) The database system is deadlock-free, but not starvation-free.
(c) The database system is starvation-free, but not deadlock-free.
(d) The database system is neither deadlock-free nor starvation-free.

[2017 (Set-1) : 2 M]

5.28 Two transactions T_1 and T_2 are given as
T_1: $r_1(X)\ w_1(X)\ r_1(Y)\ w_1(Y)$
T_2: $r_2(Y)\ w_2(Y)\ r_2(Z)\ w_2(Z)$
where $r_i(V)$ denotes a read operation by transaction T_i on a variable V and $w_i(V)$ denotes a write operation by transaction T_i on a variable V. The total number of conflict serializable schedules that can be formed by T_1 and T_2 is _______.

[2017 (Set-2) : 2 M]

5.29 Consider the following two statements about database transaction schedules:
I. Strict two-phase locking protocol generates conflict serializable schedules that are also recoverable.
II. Timestamp-ordering concurrency control protocol with Thomas' Write Rule can generate view serializable schedules that are not conflict serializable.
Which of the above statements is/are TRUE?
(a) Neither I nor II
(b) I only
(c) II only
(d) Both I and II

[2019 : 1 M]

5.30 Consider a schedule of transactions T_1 and T_2:

T_1	RA			RC		WD		WD	Commit	
T_2		RB	WB		RD		WC			Commit

Here, RX stands for "Read(X)" and WX stands for "Write(X)". Which one of the following schedules is conflict equivalent to the above schedule?

(a)

T_1					RA	RC	WD	WB	Commit	
T_2	RB	WB	RD	WC						Commit

(b)

T_1				RA	RC	WD	WB		Commit	
T_2	RB	WB	RD					WC		Commit

(c)

T_1	RA	RC	WD	WB					Commit	
T_2					RB	WB	RD	WC		Commit

(d)

T_1	RA	RC	WD				WB		Commit	
T_2				RB	WB	RD		WC		Commit

[2020 : 2 M]

5.31 Suppose a database system crashes again while recovering from a previous crash. Assume checkpointing is not done by the database either during the transactions or during recovery.

Which of the following statements is/are correct?

(a) The same undo and redo list will be used while recovering again.
(b) The system cannot recover any further.
(c) The database will become inconsistent.
(d) All the transactions that are already undone and redone will not be recovered again.

[2021 (Set-1) : 1 M]

5.32 Let $r_i(z)$ and $w_i(z)$ denote read and write operations respectively on a data item z by a transaction T_i. Consider the following two schedules:

$S_1: r_1(x), r_1(y), r_2(x), r_2(y), w_2(y), w_1(x)$

$S_2: r_1(x), r_2(x), r_2(y), w_2(y), r_1(y), w_1(x)$

Which one of the following options is correct?

(a) S_1 is not conflict serializable, and S_2 is conflict serializable.
(b) Neither S_1 nor S_2 is conflict serializable.
(c) Both S_1 and S_2 are conflict serializable.
(d) S_1 is conflict serializable, and S_2 is not conflict serializable.

[2021 (Set-1) : 2 M]

5.33 Let S be the following schedule of operations of three transactions T_1, T_2 and T_3 in a relational database system:

$R_2(Y), R_1(X), R_3(Z), R_1(Y), W_1(X), R_2(Z), W_2(Y), R_3(X), W_3(Z)$

Consider the statements P and Q below:

P : S is conflict-serializable.

Q : If T_3 commits before T_1 finishes, then S is recoverable.

Which one of the following choices is correct?

(a) P is true and Q is false
(b) Both P and Q are false
(c) Both P and Q are true
(d) P is false and Q is true

[2021 (Set-2) : 2 M]

5.34 Let $R_i(z)$ and $W_i(z)$ denote read and write operations on a data element z by a transaction T_i, respectively. Consider the schedule S with four transactions.

$$S: R_4(x), R_2(x), R_3(x), R_1(y), W_1(y), W_2(x), W_3(y), R_4(y)$$

Which one of the following serial schedules is conflict equivalent to S?

(a) $T_1 \rightarrow T_3 \rightarrow T_4 \rightarrow T_2$
(b) $T_1 \rightarrow T_4 \rightarrow T_3 \rightarrow T_2$
(c) $T_4 \rightarrow T_1 \rightarrow T_3 \rightarrow T_2$
(d) $T_3 \rightarrow T_1 \rightarrow T_4 \rightarrow T_2$

[2022 : 2 M]

5.35 Consider the following read-write schedule S over three transactions T_1, T_2 and T_3, where the subscripts in the schedule indicate transaction IDs:

$$S : r_1(z); w_1(z); r_2(x) ; r_3(y); w_3(y); r_2(y); w_2(x); w_2(y);$$

Which of the following transaction schedules is/are conflict equivalent to S?

(a) $T_3 T_1 T_2$ (b) $T_1 T_2 T_3$
(c) $T_3 T_2 T_1$ (d) $T_1 T_3 T_2$

[2024 (Set-1) : 2 M]

5.36 Once the DBMS informs the user that a transaction has been successfully completed, its effect should persist even if the system crashes before all its changes are reflected on disk. This property is called

(a) durability (b) atomicity
(c) consistency (d) isolation

[2024 (Set-2) : 1 M]

5.37 Which of the following statements about the Two Phase Locking (2PL) protocol is/are TRUE?

(a) 2PL permits only serializable schedules
(b) With 2PL, a transaction always locks the data item being read or written just before every operation and always releases the lock just after the operation.
(c) With 2PL, once a lock is released on any data item inside a transaction, no more locks on any data item can be obtained inside that transaction.
(d) A deadlock is possible with 2PL.

[2024 (Set-2) : 1 M]

■■■■

Answers Transactions & Concurrency Control

5.1 (d)	**5.2** (d)	**5.3** (d)	**5.4** (c)	**5.5** (d)	**5.6** (d)	**5.7** (b)	**5.8** (c)	**5.9** (c)
5.10 (c)	**5.11** (b)	**5.12** (b)	**5.13** (a)	**5.14** (b)	**5.15** (c)	**5.16** (d)	**5.17** (c)	**5.18** (a)
5.19 (b)	**5.20** (a)	**5.21** (b)	**5.22** (6)	**5.23** (d)	**5.24** (a)	**5.25** (a)	**5.26** (c)	**5.27** (a)
5.28 (54)	**5.29** (d)	**5.30** (b)	**5.31** (a)	**5.32** (a)	**5.33** (a)	**5.34** (a)	**5.35** (a, c, d)	
5.36 (a)	**5.37** (a, c, d)							

Explanations Transactions & Concurrency Control

5.1 (d)

Initial read of A is done by T_1 whereas final write of B is also done by T_1. Therefore $W_1(A)$, $R_2(A)$ and $W_2(B)$, $R_1(B)$ are conflicting pairs. The schedule is neither $T_1 \rightarrow T_2$ nor $T_2 \rightarrow T_1$ serializable. Since schedule is not serializable, it can't occur in a scheme using 2PL protocol.

5.2 (d)

(a)

T_1	T_2
R(X)	
	W(X)

Not forms irrecoverable

(b)

T_1	T_2
R(X)	
	R(X)

Not forms irrecoverable

(c)

T_1	T_2
W(X)	
Count	
	R(X)

Committed read not forms irrecoverable

(d)

T_1	T_2
W(X)	
	R(X)
Count	

Dirty read may cause irrecoverable of T_2 commit before C/R of T_1

5.3 (d)

We have sequence W2(D2) → W1(D1). Which means T2 → T1 whereas we also have W1(D1) → R2(D1) which means T1 → T2. Both are not possible. Therefore, schedule is not serializable.

5.4 (c)

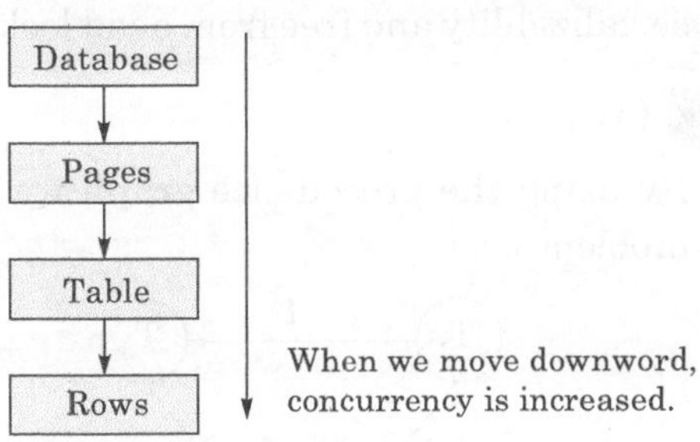

So degree of concurrency is highest when lock is applied on rows.

5.5 (d)

S is not serializable because T1 and T2 forms a cycle.

5.6 (d)

"Durability" property requires that the changes made to the database by a successive transaction persist always even in every kind of failure.

5.7 (b)

As per the recovery shcemes used, in database there is a need to undo all the write_item operations of the active transaction from the log and redo the write_item operations of the committed transactions from the log.
Hence, we must undo log record 6 and redo log records 2 and 3.

5.8 (c)

Consider the table for transaction

time ↓	S_1	S_2
	$r_1(x)$;	$r_1(x)$;
	$r_1(y)$;	$r_2(x)$;
	$r_2(x)$;	$r_2(y)$;
	$r_2(y)$;	$w_2(y)$;
	$w_2(y)$;	$r_1(y)$;
	$w_1(x)$;	$w_1(x)$;

There is no serializable conflict in transaction S_1 but for statement w_1 (x); there is a conflict in S_2 but it is also serializable.

5.9 (c)

According to 2-phase locking protocol no Locking can be done after the first Unlock.
According to Strict 2PL all exclusive locks should be released after the commit.
In option (c) no exclusive lock release until commit operation. Hence a strict 2 PL property satisfied.

5.10 (c)

(i) Schedule S1: 2RA, 2WA, 3RC, 2WB, 3WA, 3WC, 1RA, 1RB, 1WA, 1WB.

T1	T2	T3
	R(A)	
	W(A)	
		R(C)
	W(B)	
		W(A)
		W(C)
R(A)		
R(B)		
W(A)		
W(B)		

S1 is conflict serializable.

T2 → T3 → T1

(ii) Schedule S2: 3RC, 2RA, 2WA, 2WB, 3WA, 1RA, 1RB, 1WA, 1WB, 3WC.

T1	T2	T3
		R(C)
	R(A)	
	W(A)	
	W(B)	
		W(A)
R(A)		
R(B)		
W(A)		
W(B)		
		W(C)

S2 is conflict serializable.

T2 → T3 → T1

S1 is conflict equivalent to S2.

(iii) Schedule S3: 2RA, 3RC, 3WA, 2WA, 3WB, 3WC,1RA, 1RB,1WA, 1WB.

T1	T2	T3
	R(A)	
		R(C)
		W(A)
	W(A)	
	W(B)	
		W(C)
R(A)		
R(B)		
W(A)		
W(B)		

S3 is not conflict serializable.

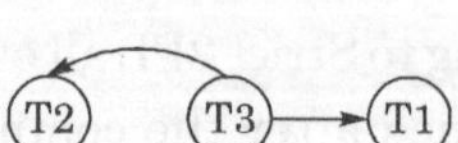

Therefore S1 and S2 are conflict cycle equivalent schedules but S3 is not equivalent to S1 and S3.

5.11 (b)

Schedule S2

T1	T2
R[X]	
	R[X]
	R[Y]
W[X]	
	W[Y]
W[Y]	

dependency graph 

S2 have no cycles.

Schedule S3

T1	T2
R[X]	
W[X]	
	R[X]
W[Y]	
	R[Y]
	W[Y]

dependency graph 

S3 also have no cycles.

So S2 & S3 are conflict-serializable.

5.12 (b)

In 2-phase locking concurrency control protocol it ensures the conflict serializable schedule but it may not free from deadlock. Ex.

T_1	T_2
l(A)	l(A)
W(A)	
	W(B)
u(A)	
Deny → l(B)	l(A) ← Denied
T_1 in waiting T_2 for unlock W(B)	W(A)

In time stamp ordering protocol it ensure conflict serializablity and free from dead lock.

5.13 (a)

By using the precedence graph we solve thus problem

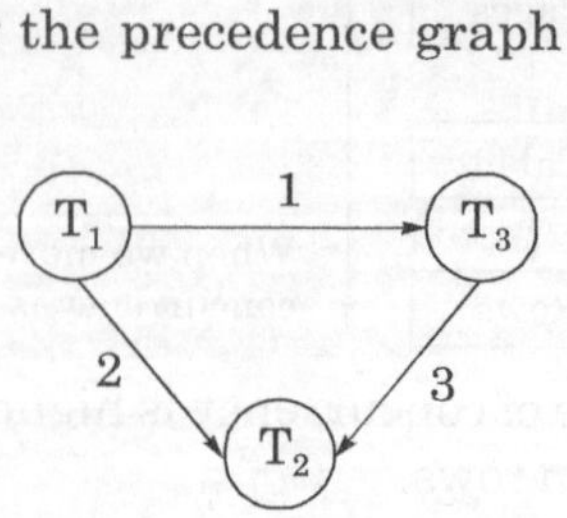

1. Read (x) in T_1 is followed by write (x) in T_2
2. Read (x) in T_1 is followed by write (x) in T_3
3. Read (y) in T_3 is followed by write (y) in T_2

So it in clear from precedence graph

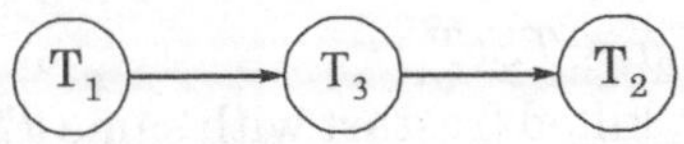

5.14 (b)

T_1	T_2
$r_1(P)$	$r_2(Q)$
$r_4(Q)$	$r_2(P)$
$w_1(Q)$	$w_2(P)$

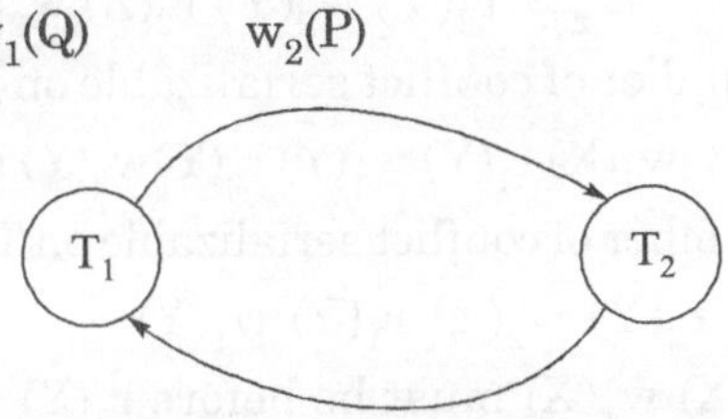

Cycle present so not conflict serializable.

5.15 (c)

Clustered index sort the data in the table based on their key values of the column on which clustered index is created.
So option (c) is correct.

5.16 (d)

(a) $r_1(x); r_2(x); w_1(x); r_3(x); w_2(x)$

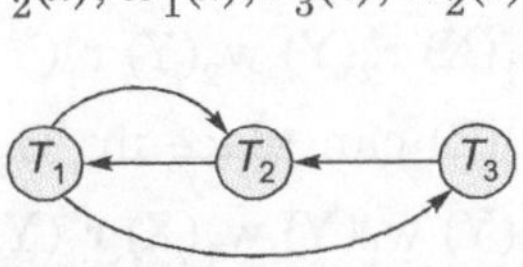

Contains cycle; Not conflict serializable

(b) $r_2(x); r_1(x); w_2(x); r_3(x); w_1(x)$

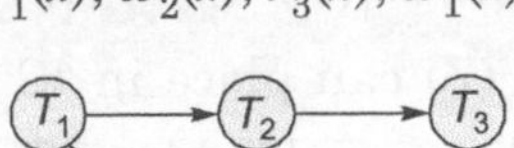

Contains cycle; Not conflict serializable

(c) $r_3(x); r_2(x); r_1(x); w_2(x); w_1(x)$

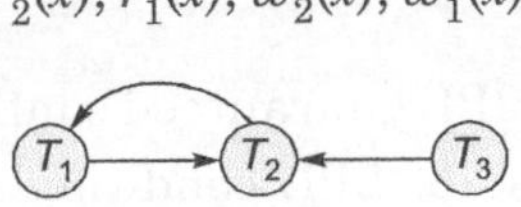

Contains cycle; Not conflict serializable

(d) $r_2(x); w_2(x); r_3(x); r_1(x); w_1(x)$

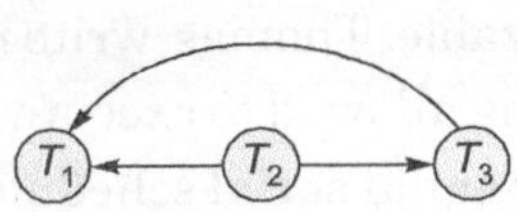

Not contains cycle; conflict serializable.

5.17 (c)

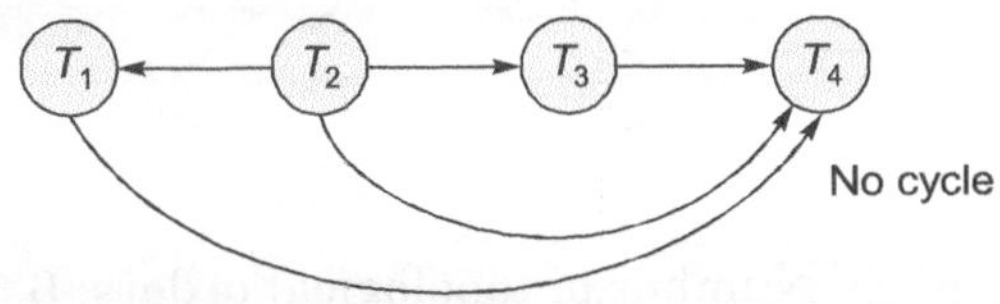

∴ Conflict serializable and it is also recoverable.

5.18 (a)

S1: $r1(X); r3(Y); r3(X); r2(Y); r2(Z); w3(Y); w2(Z); r1(Z); w1(X); w1(Z)$

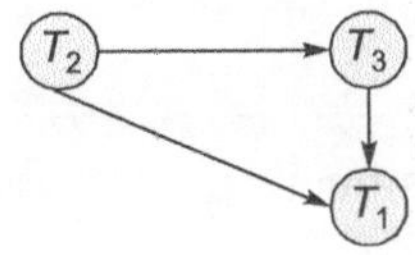

No cycle ⇒ S1 is conflict serializable.

S2: $r1(X); r3(Y); r2(Y); r3(X); r1(Z); r2(Z); w3(Y); w1(X); w2(Z); w1(Z)$

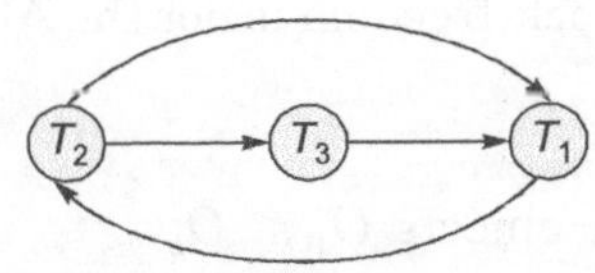

Cycle ⇒ S2 is not conflict serializable.

5.19 (b)

Consistency ensures the given constraint.

5.20 (a)

The transactions that committed before checkpoint can be ignored as their updations are retained in data base.
Those transactions that are committed after checkpoint can be repeated (REDO). The transactions that are not committed after checkpoint have to be UNDONE UNDO : T3, T1 // T1 and T3 are not committed after checkpoint
REDO : T2 // T2 committed after checkpoint

5.21 (b)

T_1 writes on A.
T_2 is performing read operation on A, and T_2 commits before T_1.
Hence it's DIRTY Read or Uncommitted Read.
Hence it's non-recoverable as T_2 is committed before T_1 commit.
It can not ensure transaction atomicity.

5.22 (6)

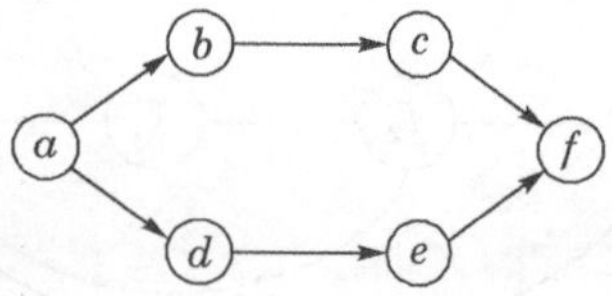

Number of topological orders: 6

a b c d e f
a b d e c f
a d b c e f
a d b c e f
a d e b c f
a d b e c f

5.23 (d)

ACID stands for
A: Atomicity
C: Consistency
I: Isolation
D: Durability
So, deadlock-freedom is not the ACID property.

5.24 (a)

2PL over objects $O_1 \ldots O_k$
Step-1: T acquires exclusive lock to $O_1 \ldots O_k$ in increasing order of their address.
Step-2: The required operations are performed.
Step-3: All locks are released.
Because of 2PL it guarantee serializability and objects locks in increasing order of address and all objects locks before read/write which avoids deadlock.

5.25 (a)

For acyclic precedence graph of conflict serializable schedule equal serial schedule is topological order of acyclic precedence graph.

5.26 (c)

No uncommitted reads so that its casecadeless rollback recoverable because $T_1w_1(x)$ before T_1 commit / Rollback $T_2w_2(x)$.
So not strict recoverable.

T_1	T_2
	$r_2(x)$
$r_1(x)$	
	$r_2(y)$
$w_1(x)$	
$r_1(y)$	
	$w_2(x)$
a_1	
	a_2

5.27 (a)

T_1 holds lock on resource R,
T_2 requires conflict lock on same resource R

Wait for graph 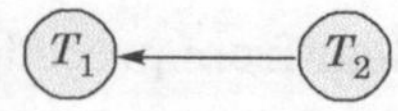

if $TS(T_2) < TS(T_1)$
Then T_1 killed [restart with same TS value] else T_2 waits.
Avoids both deadlock and starvation.

5.28 (54)

$$T_1 : r_1(X)\ w_1(X)\ r_1(Y)\ w_1(Y)$$
$$T_2 : r_2(Y)\ w_2(Y)\ r_2(Z)\ w_2(Z)$$

(i) Number of conflict serializable on $T_1 \rightarrow T_2 : 1$
$r_1(X)w_1(X)\ r_1(Y)\ w_1(Y)\ r_2(Y)w_2(Y)\ r_2(Z)\ w_2(Z)$

(ii) Number of conflict serializable on $T_2 \rightarrow T_1 : 53$
$S : r_2(Y)\ w_2(Y)\ r_1(Y)\ w_1(Y)$
$r_1(X)\ w_1(X)$ must be before $r_1(Y)$

So that $(r_2(Y)\ w_2(Y))\ (r_1(X)\ w_1(X))$ can place.
${}^4C_2 = 6$ ways.

1. $r_2(Y)\ w_2(Y)\ r_1(X)\ w_1(X)\ r_1(Y)\ w_1(Y)$
   $r_2(Z)\ w_2(Z)$ can place in ${}^6C_2 = 15$ ways.
2. $r_2(Y)\ r_1(X)\ w_1(X)\ w_2(Y)\ r_1(Y)\ w_1(Y)$
   $r_2(Z)\ w_2(Z)$ can place in ${}^4C_2 = 6$ ways.
3. $r_2(Y)\ r_1(X)\ w_2(Y)\ w_1(X)\ r_1(Y)\ w_1(Y)$
   $r_2(Z)\ w_2(Z)$ can place in ${}^5C_2 = 10$ ways.
4. $r_1(X)\ w_1(X)\ r_2(Y)\ w_2(Y)\ r_1(Y)\ w_1(Y)$
   $r_2(Z)\ w_2(Z)$ can place in ${}^4C_2 = 6$ ways.
5. $r_1(X)\ r_2(Y)\ w_2(Y)\ w_1(X)\ r_1(Y)\ w_1(Y)$
   $r_2(Z)\ w_2(Z)$ can place in ${}^5C_2 = 10$ ways.
6. $r_1(X)\ r_2(Y)\ w_1(X)\ w_2(Y)\ r_1(Y)\ w_1(Y)$
   $r_2(Z)\ w_2(Z)$ can place in ${}^4C_2 = 6$ ways.

Total conflict serializable of T_1 and T_2 = 53+1 = 54 ways.

5.29 (d)

I. Strict 2PL guaranteed conflict serializable because of 2PL condition and also strict recoverable.

II. Thomas Write time stamp ordering ensures serializable. Thomas write rule time stamp ordering allowed to execute schedule which is view equal serial schedule based on time stamp ordering.

5.30 (b)

T_1	T_2
$r_1(A)$	
	$r_2(B)$
	$w_2(B)$
$r_1(C)$	
	$r_2(D)$
$w_1(D)$	
	$w_2(c)$
$w_1(B)$	
Commit	
	Commit

Conflict equal

$\Longrightarrow$ Conflict equal

T_1	T_2
	$r_2(B)$
	$w_2(B)$
	$r_2(D)$
$r_1(A)$	
$r_1(C)$	
$w_1(D)$	
$w_1(B)$	
	$w_2(c)$
Commit	
	Commit

5.32 (a)

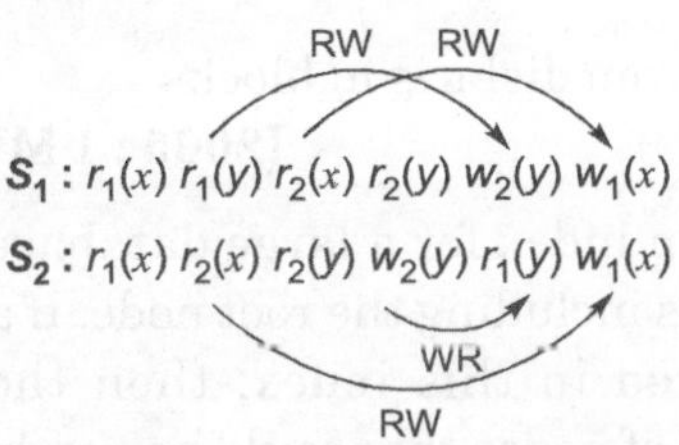

S_1 :
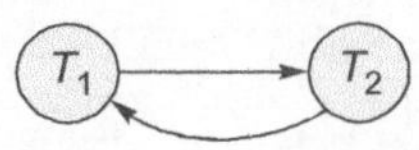

Cycle exists in precedence graph
∴ Not conflict serializable.

S_2 :
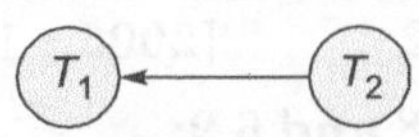

No cycle
∴ Conflict serializable.

S_1 is not conflict serializable, and S_2 is conflict serializable.

5.33 (a)

No cycle in precedence graph.
∴ Conflict serializable.

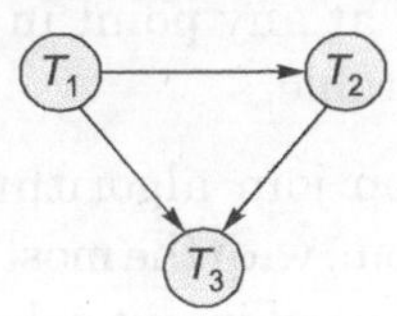

T_1	T_2	T_3
⋮	⋮	⋮
$W_1(X)$		
	$R_2(Z)$	
	$W_2(Y)$	
		$R_3(X)$
		$W_3(Z)$

T_2 is doing dirty read of updated X by T_1.
So, recoverability only possible if T_3 commits after commit/RB of T_1.
So, option (a) is correct.

5.34 (a)

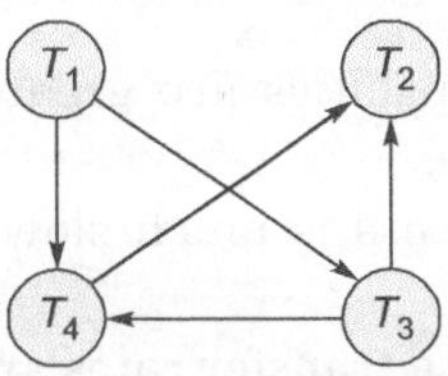

Precedence graph of schedule S.
Topological order = Conflict equal serial order
$\Rightarrow T_1\ T_3\ T_4\ T_2$

5.35 (a, c, d)

Precedence graph of schedule(s):

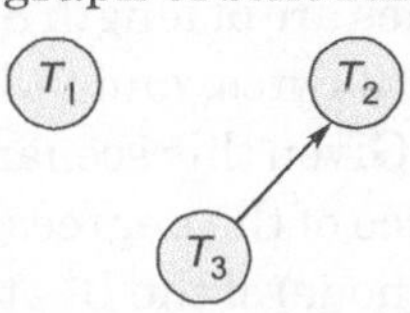

Serial schedules conflict equal to S are
$T_1 T_3 T_2$
$T_3 T_1 T_2$
$T_3 T_2 T_1$

5.36 (a)

Durability: Transaction should able to recover under any case of failure.
After transaction completes successfully, the changes it has made it to be DB persist, even if there are system failure.
RAID design uses recover from disk crash.

5.37 (a, c, d)

1. Transaction (T) not allowed to request lock on any data item during unlocking phase of T.
2. Always guaranteed for serializability.
3. Not free from deadlock.

■■■■

6 File Structures

6.1 Which of the following is correct?
(a) B-trees are for storing data on disk and B^+ trees are for main memory.
(b) Range queries are faster on B^+ trees.
(c) B-trees are for primary indexes and B^+ trees are for secondary indexes.
(d) The height of a B^+ tree is independent of the number of records. **[1999 : 1 M]**

6.2 B^+ trees are preferred to binary trees in databases because
(a) Disk capacities are greater than memory capacities
(b) Disk access is much slower than memory access
(c) Disk data transfer rates are much less than memory data transfer rates
(d) Disks are more reliable than memory
[2000 : 1 M]

6.3 A B^+ tree index is to be built on the Name attribute of the relation STUDENT. Assume that all student names are of length 8 bytes, disk blocks are of size 512 bytes, and index pointers are of size 4 bytes. Given this scenario, what would be the best choice of the degree (i.e. the number of pointers per node) of the B^+ - tree?
(a) 16 (b) 42
(c) 43 (d) 44 **[2002 : 2 M]**

6.4 The order of an internal node in a B^+ tree index is the maximum number of children it can have. Suppose that a child pointer takes 6 bytes, the search field value takes 14 bytes, and the block size is 512 bytes. What is the order of the internal node?
(a) 24 (b) 25
(c) 26 (d) 27 **[2004 : 2 M]**

6.5 Consider a table T in a relational database with a key field K. A B-tree of order p is used as an access structure on K, where p denotes the max imum number of tree pointers in a B-tree index node. Assume that K is 10 bytes long; disk block size is 512 bytes; each data pointer P_D is 8 bytes long and each block pointer P_B is 5 bytes long. In order for each B-tree node to fit in a single disk block, the maximum value of p is
(a) 20 (b) 22
(c) 23 (d) 32 **[2004 : 2 M]**

6.6 Which one of the following is a key factor for preferring B^+ trees to binary search trees for indexing database relations?
(a) Database relations have a large number of record
(b) Database relations are sorted on the primary key
(c) B^+ trees require less memory than binary search trees
(d) Data transfer from disks is in blocks
[2005 : 1 M]

6.7 A B-tree used as an index for a large database table has four levels including the root node. If a new key is inserted in this index, then the maximum number of nodes that could be newly created in the process are
(a) 5 (b) 4
(c) 3 (d) 2 **[2005 : 1 M]**

Linked Answer Questions 6.8 and 6.9:

A database table T1 has 2000 records and occupies 80 disk blocks. Another table T2 has 400 records and occupies 20 disk blocks. These two tables have to be joined as per a specified join condition that needs to be evaluated for every pair of records from these two tables. The memory buffer space available can hold exactly one block of records for T1 and one block of records for T2 simultaneously at any point in time. No index is available on either table.

6.8 If Nested-loop join algorithm is employed to perform the join, with the most appropriate choice of table to be used in outer loop, the number of block accesses required for reading the data are
(a) 800000 (b) 40080
(c) 32020 (d) 100 **[2005 : 2 M]**

6.9 If, instead of Nested-loop join, Block nested-loop join is used, again with the most appropriate choice of table in the outer loop, the reduction in number of block accesses required for reading the data will be
(a) 0 (b) 30400
(c) 38400 (d) 798400
[2005 : 2 M]

6.10 In a database file structure, the search key field is 9 bytes long, the block size is 512 bytes, a record pointer is 7 bytes and a block pointer is 6 bytes. The largest possible order of a non-leaf node in a B+ tree implementing this file structure is
(a) 23 (b) 24
(c) 34 (d) 44 **[2006 : 2 M]**

6.11 The order of a leaf node in a B^+ tree is the maximum number of (value, data record pointer) pairs it can hold. Given that the block size is 1K bytes, data record pointer is 7 bytes long, the value field is 9 bytes long and a block pointer is 6 bytes long, what is the order of the leaf node?
(a) 63 (b) 64
(c) 67 (d) 68 **[2007 : 2 M]**

Directions for question 6.12 to 6.13:

Consider the B^+ tree in the adjoining figure, where each node has at most two keys and three links.

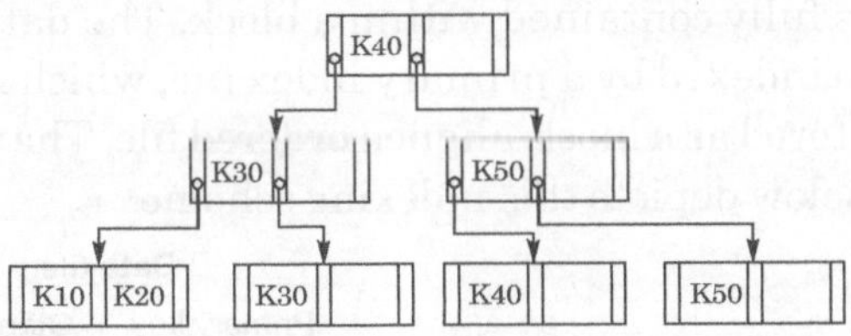

6.12 Keys K15 and then K25 are inserted into this tree in that order. Exactly how many of the following nodes (disregarding the links) will be present in the tree after the two insertions?

[K30 | K50] [K25 | K30] [K20 | K25] [K50 | K20]

(a) 1 (b) 2
(c) 3 (d) 4 **[2007 : 2 M]**

6.13 Now the key K50 is deleted from the B^+ tree resulting after the two insertions made earlier. Consider the following statements about the B^+ tree resulting after this deletion.
(i) The height of the tree remains the same.
(ii) The node [k20] (disregarding the links) is present in the tree.
(iii) The root node remains unchanged (disregarding the links).

Which one of the following options is true ?
(a) Statements (i) and (ii) are true
(b) Statements (ii) and (iii) are truc
(c) Statements (iii) and (i) are true
(d) All the statements are false
[2007 : 2 M]

6.14 A clustering index is defined on the fields which are of type
(a) Non-key and ordering
(b) Non-key and non-ordering
(c) Key and ordering
(d) Key and non-ordering **[2008 : 1 M]**

6.15 Consider a file of 16384 records. Each record is 32 bytes long and its key field is of size 6 bytes. The file is ordered on a non-key field, and the file organization is unspanned. The file is stored in a file system with block size 1024 bytes, and the size of a block pointer is 10 bytes. If the secondary index is built on the key field of the file, and a multilevel index scheme is used to store the secondary index, the number of first-level and second-level blocks in the multilevel index are respectively
(a) 8 and 0 (b) 128 and 6
(c) 256 and 4 (d) 512 and 5
[2008 : 2 M]

6.16 The following key values are inserted into a B^+ tree in which order of the internal nodes is 3, and that of the leaf nodes is 2, in the sequence given below. The order of internal nodes is the maximum number of tree pointers in each node, and the order of leaf nodes is the maximum number of data items that can be stored in it. The B^+ tree is initially empty.

10, 3, 6, 8, 4, 2, 1

The maximum number of times leaf nodes would get split up as a result of these insertions is
(a) 2 (b) 3
(c) 4 (d) 5 **[2009 : 2 M]**

6.17 Consider a B^+ tree in which the maximum number of keys in a node is 5. What is the minimum number of keys in any non-root node?
(a) 1 (b) 2
(c) 3 (d) 4 **[2010 : 1 M]**

6.18 A file is organized so that the ordering of data records is the same as or close to the ordering of data entries in some index. Then that index is called
(a) Dense (b) Sparse
(c) Clustered (d) Unclustered
[2015 (Set-1) : 1 M]

6.19 With reference to the B^+ tree index of order 1 shown below, the minimum number of nodes (including the Root node) that must be fetched in order to satisfy the following query: "Get all records with a search key greater than or equal to 7 and less than 15" is ________.

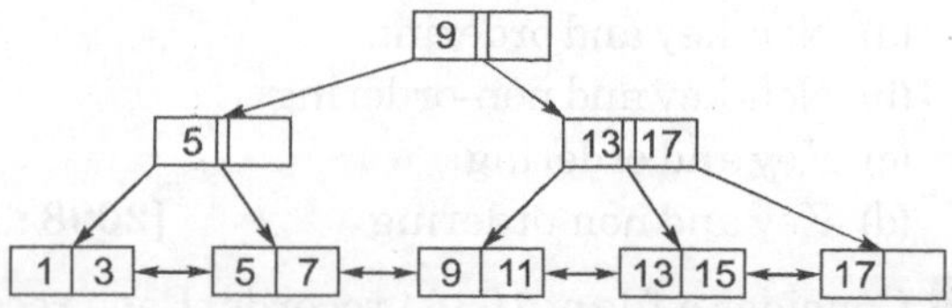

[2015 (Set-2) : 1 M]

6.20 Consider a B^+ tree in which the search key is 12 bytes long, block size is 1024 bytes, record pointer is 10 bytes long and block pointer is 8 bytes long. The maximum number of keys that can be accommodated in each non-leaf node of the tree is ________.

[2015 (Set-3) : 2 M]

6.21 B^+ trees are considered BALANCED because
(a) the lengths of the paths from the root to all leaf nodes are all equal.
(b) the lengths of the paths from the root to all leaf nodes differ from each other by at most 1.
(c) the number of children of any two non-leaf sibling nodes differ by at most 1.
(d) the number of records in any two leaf nodes differ by at most 1. **[2016 (Set-2) : 1 M]**

6.22 In a B^+ tree, if the search-key value is 8 bytes long, the block size is 512 bytes and the block pointer size is 2 bytes, then the maximum order of the B^+ tree is________.

[2017 (Set-2) : 2 M]

6.23 Which one of the following statements is NOT correct about the B^+ tree data structure used for creating an index of a relational database table?
(a) Key values in each node are kept in sorted order
(b) Each leaf node has a pointer to the next leaf node
(c) B^+ tree is a height balanced tree
(d) Non-leaf nodes have pointers to data records

[2019 : 1 M]

6.24 Consider a database implemented using B^+ tree for file indexing and installed on a disk drive with block size of 4 KB. The size of search key is 12 bytes and the size of tree/disk pointer is 8 bytes. Assume that the database has one million records. Also assume that no node of the B+ tree and no records are present initially in main memory. Consider that each record fits into one disk block. The minimum number of disk accesses required to retrieve any record in the database is ________. **[2020 : 2 M]**

6.25 A data file consisting of 1,50,000 student-records is stored on a hard disk with block size of 4096 bytes. The data file is sorted on the primary key RollNo. The size of a record pointer for this disk is 7 bytes. Each student-record has a candidate key attributed called ANum of size 12 bytes. Suppose an index file with records consisting of two fields, ANum value and the record pointer to the corresponding student record, is built and stored on the same disk. Assume that the records of data file and index file are not split across disk blocks. The number of blocks in the index file is ________.

[2021 (Set-2) : 1 M]

6.26 Consider a database of fixed-length records, stored as an ordered file. The database has 25,000 records, with each record being 100 bytes, of which the primary key occupies 15 bytes. The data file is block-aligned in that each data record is fully contained within a block. The database is indexed by a primary index file, which is also stored as a block-aligned ordered file. The figure below depicts this indexing scheme.

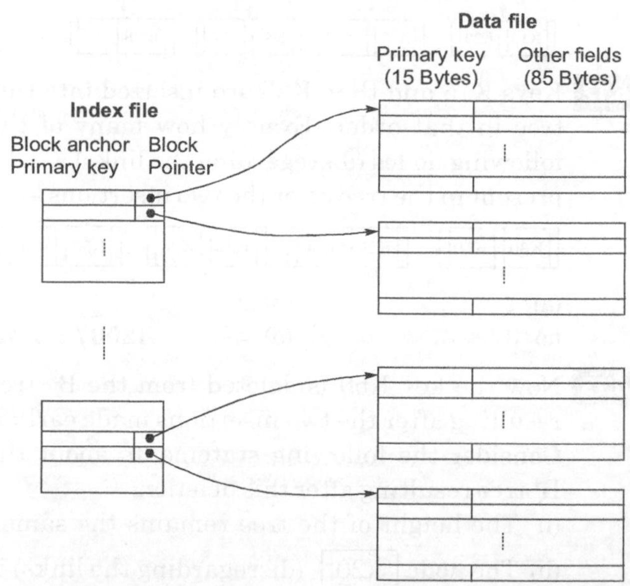

Suppose the block size of the file system is 1024 bytes, and a pointer to *a* block occupies 5 bytes. The system uses binary search on the index file to search for a record with a given key. You may assume that a binary search on an index file of b blocks takes $\lceil \log_2 b \rceil$ block accesses in the worst case. Given a key, the number of block accesses required to identify the block in the data file that may contain a record with the key, in the worst case, is________. **[2023 : 2 M]**

6.27 In a B+ tree, the requirement of at least half-full (50%) node occupancy is relaxed for which one of the following cases?
(a) Only the root node
(b) Only the leftmost leaf node
(c) All leaf nodes
(d) All internal nodes **[2024 (Set-1) : 1 M]**

6.28 Which of the following file organizations is/are I/O efficient for the scan operation in DBMS?
(a) Heap
(b) Sorted
(c) Unclustered hash index
(d) Unclustered tree index
[2024 (Set-2) : 1 M]

Answers File Structures

6.1	(b)	6.2	(b)	6.3	(c)	6.4	(c)	6.5	(c)	6.6	(d)	6.7	(a)	6.8	(c)	6.9	(b)
6.10	(c)	6.11	(a)	6.12	(a)	6.13	(a)	6.14	(a)	6.15	(c)	6.16	(c)	6.17	(b)	6.18	(c)
6.19	(5)	6.20	(50)	6.21	(a)	6.22	(52)	6.23	(d)	6.24	(4)	6.25	(698)	6.26	(6)	6.27	(a)
6.28	(a, b)																

Explanations File Structures

6.1 (b)

Most database systems use indexes built on some form of a B^+-tree due to its many advantages, in particular its support for range queries. Leaf nodes are linked together in B^+ trees hence range queries are faster.

6.2 (b)

Disk access is slow and B^+ tree provide search in less number of disk hits. This is primarily because unlike binary search trees, B^+ trees have very high fanout (typically on the order of 100 or more), which reduces the number of I/O operations required to find an element in the tree.

6.3 (c)

Let n be the degree
Given, k, key size (length of the name = 8 byte attribute of student)
Disk block size, B = 512 bytes
Index pointer size, b = 4 bytes
Degree of B^+ tree can be calculated if we know the maximum number of key a internal node can have the formula for that is

$$(n-1)\,k + n * b = \text{Block size}$$
$$(n-1)\,8 + n * 4 = 512$$
$$12n = 520$$
$$n = \frac{520}{12} = 43$$

So, the answer is option (c).

6.4 (c)

Size of child pointer = 6 byte
Size of search field value takes 14 bytes
Block size = 512
The order of internal node = P
$\Rightarrow (P-1)14 + P * 6 \leq 512$
$P = 26$

6.5 (c)

Format of B-tree node with order P:
$P(P_B) + (P-1)(K + P_D) \leq$ Block size
By putting value of P_B, P_D and K.
$P(5) + (P-1)(10+8) \leq 512$ B
$5P + 18P - 18 \leq 512$
$23P \leq 512 + 18$

$$P = \left\lfloor \frac{530}{23} \right\rfloor = 23$$

6.6 (d)

The transfer of data from disk to primary memory is in the form of data blocks, if a data block is larger than indexing is easy due to this B^+ tree is better than binary search tree data structure if large amount of data can be access.

6.7 (a)

Assume in B-tree all node contain its full capacity i.e. if order is P then P – 1 key + data pointer. Since it has 4 level and if new element is inserted than at each level overflow results in one new node and extra node comes to new level 5.
So total 5 new nodes are created.

6.8 (c)

Here condition given only 1 block of T1 and 1 block of T2 can present simultaneously inside the memory buffer space.

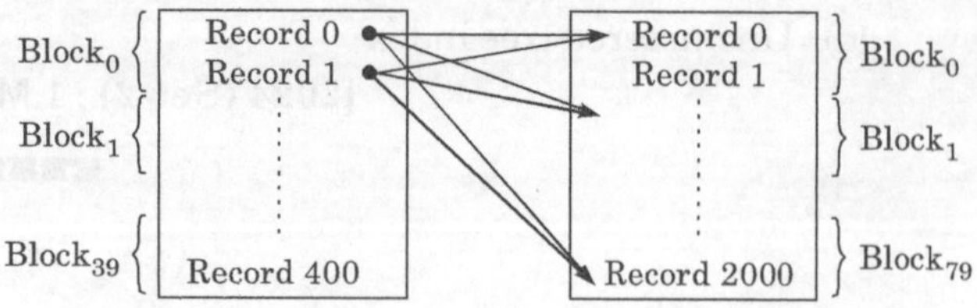

In Nested loop join every record of one table every block of second table is loaded.

So, If T1 is outer = 2000 × 20 + 80 = 40080

If T2 is outer = 400 × 80 + 20 = 32020

So we go with T2 as outer table = 32020

6.9 (b)

In Nested block loop join, in place of for every record of table 1, fetch every block of table 2, we go for every block of table 1, fetch every block of table 2.

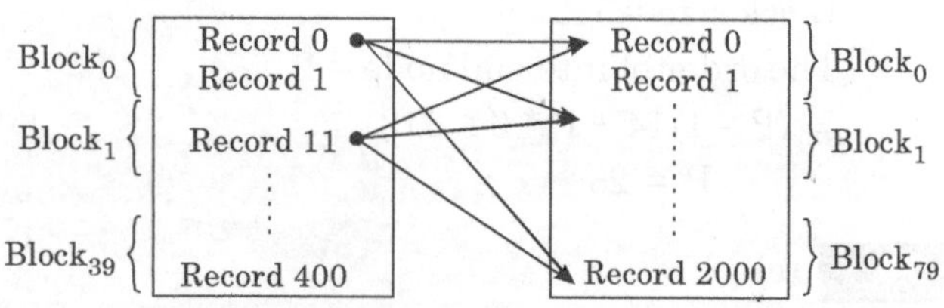

So, If T1 is outer = 80 × 20 + 80 = 1680

If T2 is outer = 80 × 20 + 20 = 1620

So we go with T2 as outer table = 1620

Reduction in number of block access

= 32020 – 1620 = 30400

6.10 (c)

Format of B+ tree non leaf node with order p:

(p – 1) (key) + p block pointer ≤ block size

Putting value of key, data pointer and block pointer.

$$(p-1)(9) + p(6) \le 512$$

$$9p - 9 + 6p \le 512$$

$$15p \le 521$$

$$p = \left\lfloor \frac{521}{15} \right\rfloor$$

$$p = 34$$

6.11 (a)

Let the order of leaf node is n.

Block size = 1K = 1024 bytes

So, 6 + 7n + (n) 9 ≤ 1024

n = 63

6.12 (a)

It can be seen that B+ tree is right biased

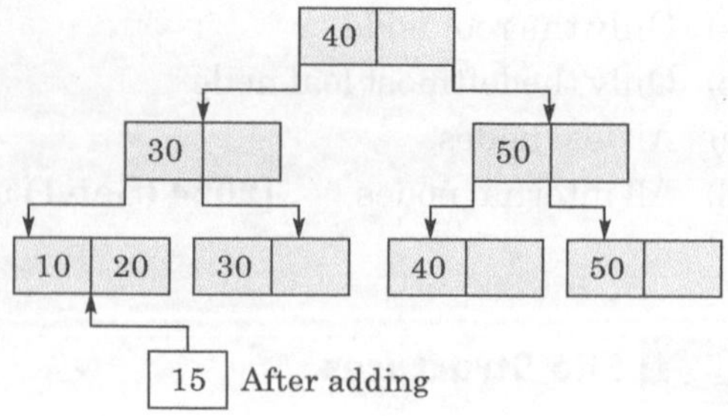

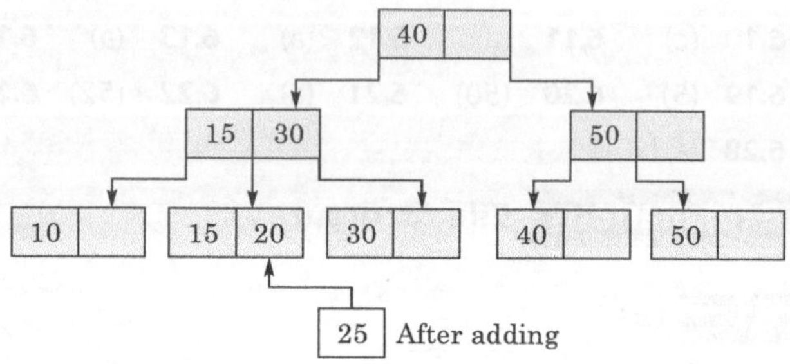

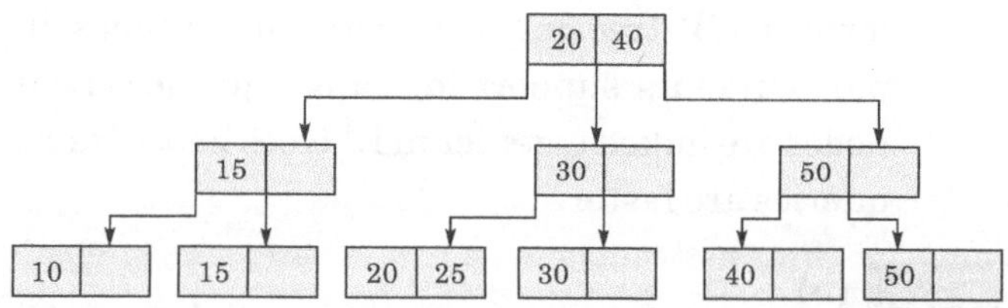

Matching node is [20 | 25] which is one only.

6.13 (a)

After deleting key 50 from the above part:

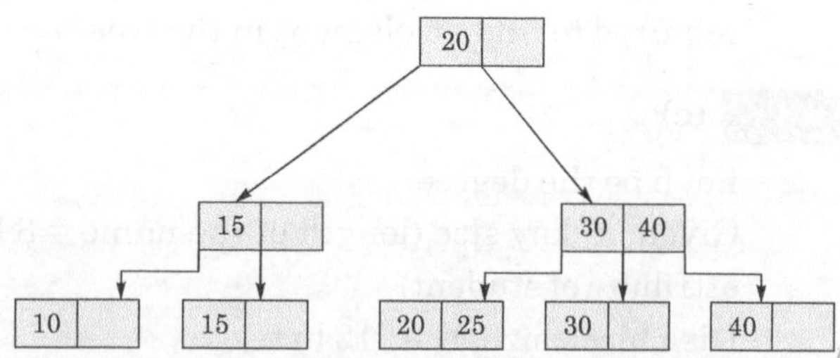

Height of tree is remain same. So true

The key 20 is remain in the tree. So true

The root node contain key 20, previously it was key 20 and 40. So false

6.14 (a)

If records of a file are physically ordered on a nonkey field which doesn't have a distinct value for each record that field is called the clustering field.

6.15 (c)

Total number of records = 16384
Record size = 32 B, Block size = 1024 B

Number of records inside block $= \dfrac{1024\text{B}}{32\text{B}} = 32$

In index block number of entries $= \dfrac{1024}{6+10} = 64$

Since secondary indexing is used:

So, at 1st level index block $= \dfrac{16384}{64} = 256$

at 2nd level index block $= \dfrac{256}{64} = 4$

6.16 (c)

B⁺ tree is initially empty

insert 10 : 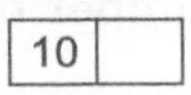

insert 3 : 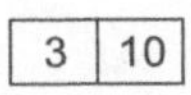

insert 6 : 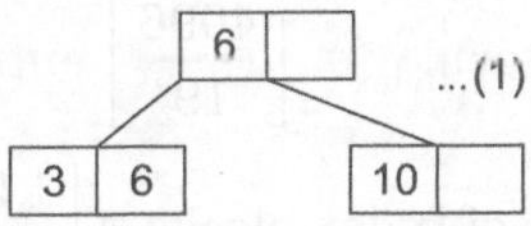

insert 8 : 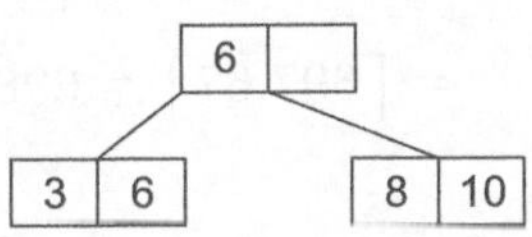

insert 4 : 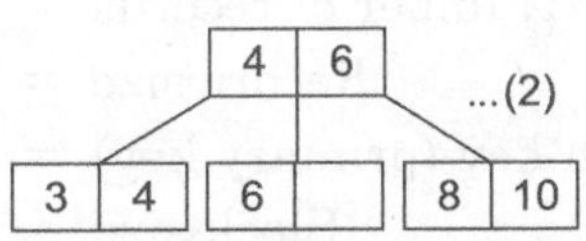

insert 2 : 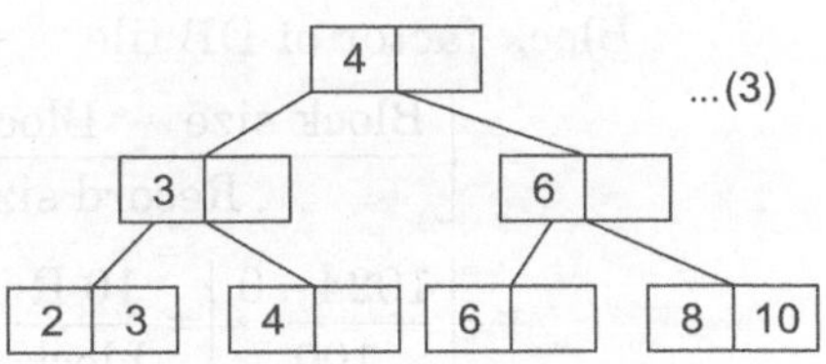

insert 1 : 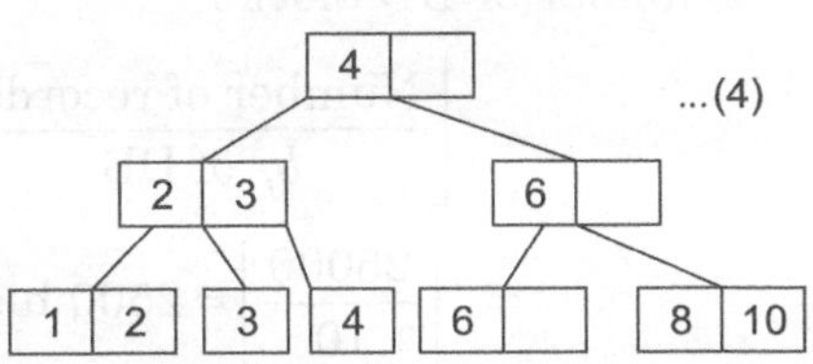

So total splits = 4.

6.17 (b)

In a B⁺ tree the root node have minimum two block pointer and maximum P block pointer where P = order, key = order – 1 and in the non-root node the minimum number of keys $= \left\lceil \dfrac{P}{2} \right\rceil - 1$

So in the question key = 5, order = 6
So minimum number of keys in non-root node $= \left\lceil \dfrac{6}{2} \right\rceil - 1 = 2$

6.18 (c)

File is organized so that the ordering of data records is the same as or close to the ordering of data entries in some index. Then that index is called clustered index.

6.19 (5)

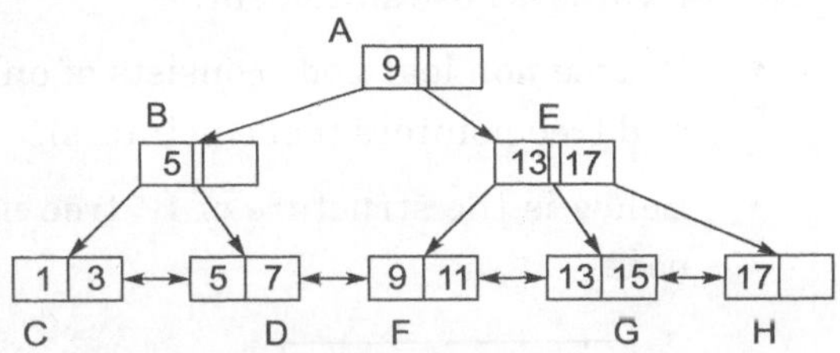

We need three node accesses to search for key 7.

In B⁺ tree all neighbouring leaf nodes are connected (Via block pointers) and it is possible to access remaining records after 7 using linear search for records. Hence two more nodes needed to fetch records less than 15.

Nodes A, B, D, F and G are needed to get all records greater than or equal to 7 and less than 15.

6.20 (50)

For Non-leaf node:

$$p \times \text{key size} + (p+1)\, B_p \le \text{block size}$$
$$p \times 12 + (p+1)8 \le 1024$$
$$20p \le 1016$$
$$p \le \lceil 50.8 \rceil$$
$$p = 51$$
$$\text{Number of keys} = p - 1$$
$$= 51 - 1 = 50$$

6.21 (a)

B⁺ tree balanced because the length of the paths from the root to all leaf nodes are all equal and every internal node must filled by.

6.22 (52)

B^+ tree

Search key : 8 bytes

Block size : 512 bytes

Block pointer : 2 bytes

Maximum order of B^+ tree node??

$P \times B_P + (P - 1) \times \text{Key} \le \text{Block size}$

$P \times 2 + (P - 1)8 \le 512$

$10P \le 520$

$$P = \left\lfloor \frac{520}{10} \right\rfloor = 52$$

6.23 (d)

- B^+ tree non leaf node have pointer to data records is false statement.
- B^+ tree non leaf node consists of only keys and tree pointers (node pointers).
- Below is the structure of B^+ tree non leaf node

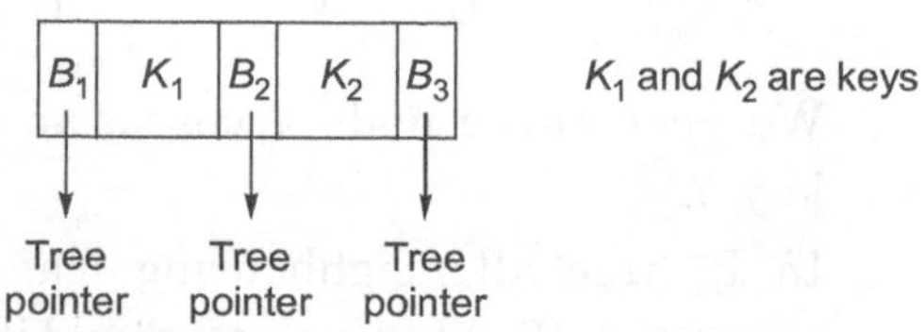

6.24 (4)

Block size = 4 KB

Search key = 12 bytes

Tree pointer = 8 bytes

DB records1 million = 1000000

B+ tree index each record fits into one block

⇒ Keys given to find levels of B+ tree : Bulk loading B+ tree design we can use

⇒ Order of B tree nodes $P \times B_p + (P - 1)$ Keys $\le$ Block size

$P \times 8 + (P - 1)12 \le 4096$

$20P \le 4108$

$$P = \left\lfloor \frac{4108}{20} \right\rfloor = 205$$

⇒ Minimum levels and index: Max fill factor of each node.

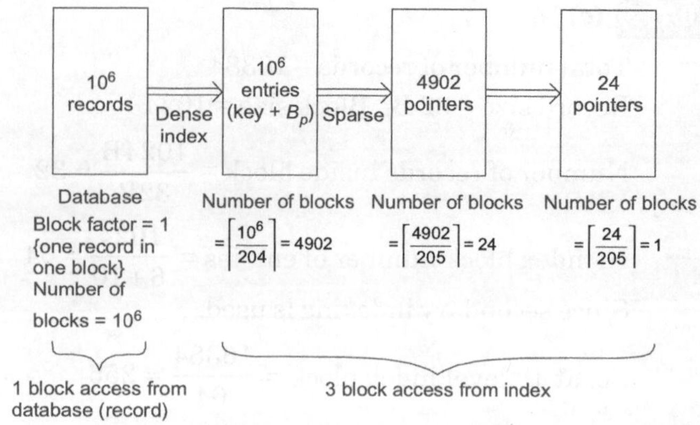

⇒ Minimum 4 block access required to access record.

6.25 (698)

Index entries = Number of DB file record (∵ Dense index)

Block factor for Index,

$$(\text{Block factor})_{\text{Index}} = \left\lfloor \frac{4096}{19} \right\rfloor \text{ entries/block} = 215$$

$$\therefore \text{Number of Index blocks} = \left\lceil \frac{1,50,000}{215} \right\rceil \text{ blocks}$$

$$= \lceil 697.67 \rceil = 698 \text{ index blocks}$$

6.26 (6)

Number of records = 25000

Record size = 100 byte

Search key (primary key) = 15 bytes

Block size = 1024 bytes

Pointer size = 5 bytes

Block factor of DB file

$$= \left\lfloor \frac{\text{Block size} - \text{Block header}}{\text{Record size}} \right\rfloor$$

$$= \left\lfloor \frac{1024 - 0}{100} \right\rfloor = \frac{10\text{ R}}{\text{block}}$$

Number of DB blocks

$$= \left\lceil \frac{\text{Number of records}}{b_f \text{ of DB}} \right\rceil$$

$$= \left\lceil \frac{25000}{10} \right\rceil = 2500 \text{ blocks}$$

$$\text{Block factor of index file} = \left\lfloor \frac{\text{B} - \text{H}}{\text{K} + \text{P}} \right\rfloor$$

$$= \left\lfloor \frac{1024 - 0}{15 + 5} \right\rfloor = 50 \text{ entries/block}$$

Sparse index used.
[Number of index entries = Number of block of DB]
Number of index blocks at I[st] level

$$= \left\lfloor \frac{2500}{51} \right\rfloor = 49$$

Number of index block's access required to identify block in DB file $= \lceil \log_2 49 \rceil = 6$

6.27 (a)

- In B+ tree every node must occupy at least 50% except root node.
- Root node can be min 2 block pointers with one key allowed.

6.28 (a, b)

File type	Scan Cost
Heap	P.D
Unclustered hash index	P.D.R + read index
Sorted	P.D
Unclustered tree index	P.D.R + read index

P : Number of pages in the file.
D : Amount of time required to read or write in page.
R : Number of records in a particular page.

■■■■

Computer Networks

UNIT

X

CONTENTS

Computer Networks Syllabus

Concept of layering: OSI and TCP/IP Protocol Stacks; Basics of packet, circuit and virtual circuit-switching; Data link layer: framing, error detection, Medium Access Control, Ethernet bridging; Routing protocols: shortest path, flooding, distance vector and link state routing; Fragmentation and IP addressing, IPv4, CIDR notation, Basics of IP support protocols (ARP, DHCP, ICMP), Network Address Translation (NAT); Transport layer: flow control and congestion control, UDP, TCP, sockets; Application layer protocols: DNS, SMTP, HTTP, FTP, Email.

Analysis of Previous GATE Papers

Exam Year	1 Mark Ques.	2 Marks Ques.	Total Marks
2003	2	3	**8**
2004	3	4	**11**
2005	5	2	**15**
2006	1	5	**11**
2007	2	6	**14**
2008	1	4	**9**
2009	–	5	**10**
2010	2	3	**8**
2011	2	2	**6**
2012	3	3	**9**
2013	3	2	**7**
2014 Set-1	2	3	**8**
2014 Set-2	3	2	**7**
2014 Set-3	3	3	**9**
2015 Set-1	4	2	**8**

Exam Year	1 Mark Ques.	2 Marks Ques.	Total Marks
2015 Set-2	2	3	**8**
2015 Set-3	2	3	**8**
2016 Set-1	2	4	**10**
2016 Set-2	3	4	**11**
2017 Set-1	2	3	**8**
2017 Set-2	3	2	**7**
2018	3	2	**7**
2019	2	4	**10**
2020	2	2	**6**
2021 Set-1	1	4	**9**
2021 Set-2	1	3	**7**
2022	2	4	**10**
2023	2	3	**8**
2024 Set-1	3	3	**9**
2024 Set-2	3	3	**9**

1 ISO/OSI Stack and SWP

1.1 Host A is sending data to host B over a full duplex link. A and B are using the sliding window protocol for flow control. The send and receive window sizes are 5 packets each. Data packets (sent only from A to B) are all 1000 bytes long and the transmission time for such a packet is 50 μs. Acknowledgment packets (sent only from B to A) are very small and require negligible transmission time. The propagation delay over the link is 200 μs. What is the maximum achievable throughput in this communication?
(a) 7.69×10^6 Bps (b) 11.11×10^6 Bps
(c) 12.33×10^6 Bps (d) 15.00×10^6 Bps
[2003 : 2 M]

1.2 Choose the best matching between **List-I** and **List-II**.

List-I
P. Data link layer
Q. Network layer
R. Transport layer

List-II
1. Ensures reliable transport of data over a physical point-to-point link
2. Encodes / decodes data for physical transmission
3. Allow end-to-end communication between two processes
4. Routes data from one network node to the next

	P	Q	R
(a)	1	4	3
(b)	2	4	1
(c)	2	3	1
(d)	1	3	2

[2004 : 1 M]

1.3 How many 8-bit characters can be transmitted per second over a 9600 baud serial communication link using asynchronous mode of transmission with one start bit, eight data bits, and one parity bit and two stop bits?
(a) 600 (b) 800
(c) 876 (d) 1200 **[2004 : 1 M]**

1.4 A serial transmission T1 uses 8 information bits, 2 start bits, 1 stop bit and 1 parity bit for each character. A synchronous transmission T2 uses 3 eight-bit sync characters followed by 30 eight-bit information characters. If the bit rate is 1200 bits/second in both cases, what are the transfer rates of T1 and T2?
(a) 100 characters/sec, 153 characters/sec
(b) 80 characters/sec, 136 characters/sec
(c) 100 characters/sec, 136 characters/sec
(d) 80 characters/sec, 153 characters/sec
[2004 : 2 M]

1.5 In a data link protocol, the frame delimiter flag is given by 0111. Assuming that bit stuffing is employed, the transmitter sends the data sequence 01110110 as
(a) 01101011 (b) 011010110
(c) 011101100 (d) 0110101100
[2004 : 2 M]

1.6 In a sliding window ARQ scheme, the transmitter's window size is N and the receiver's window size is M. The minimum number of distinct sequence numbers required to ensure correct operation of the ARQ scheme is
(a) min (M, N) (b) max (M, N)
(c) M + N (d) MN
[2004 : 2 M]

1.7 A 20 Kbps satellite link has a propagation delay of 400 ms. The transmitter employs the "go back n ARQ" scheme with n set to 10. Assuming that each frame is 100 bytes long, what is the maximum data rate possible?
(a) 5 Kbps (b) 10 Kbps
(c) 15 Kbps (d) 20 Kbps
[2004 : 2 M]

1.8 Consider a parity check code with three data bits and four parity check bits. Three of the code words are 0101011, 1001101 and 1110001.
Which of the following are also code words?
I. 0010111 **II.** 0110110
III. 1011010 **IV.** 0111010
(a) I and III (b) I, II and III
(c) II and IV (d) I, II, III and IV
[2004 : 2 M]

1.9 The maximum window size for data transmission using the selective reject protocol with n-bit frame sequence numbers is

(a) 2^n (b) 2^{n-1}
(c) $2^n - 1$ (d) 2^{n-2} **[2005 : 1 M]**

1.10 Consider the following message M = 1010001101. The cyclic redundancy check (CRC) for this message using the divisor polynomial $x^5 + x^4 + x^2 + 1$ is :
(a) 01110 (b) 01011
(c) 10101 (d) 10110 **[2005 : 2 M]**

1.11 A channel has a bit rate of 4 kbps and one-way propagation delay of 20 ms. The channel uses stop and wait protocol. The transmission time of the acknowledgement frame is negligible. To get a channel efficiency of at least 50%, the minimum frame size should be
(a) 80 bytes (b) 80 bits
(c) 160 bytes (d) 160 bits
[2005 : 2 M]

1.12 In the 4B/5B encoding scheme, every 4 bits of data are encoded in a 5-bit codeword. It is required that the codewords have at most 1 leading and at most 1 trailing zero. How many such codewords are possible?
(a) 14 (b) 16
(c) 18 (d) 20 **[2006 : 2 M]**

1.13 On a wireless link, the probability of packet error is 0.2. A stop-and-wait protocol is used to transfer data across the link. The channel condition is assumed to be independent from transmission to transmission. What is the average number of transmission attempts required to transfer 100 packets?
(a) 100 (b) 125
(c) 150 (d) 200 **[2006 : 2 M]**

1.14 Station A uses 32 byte packets to transmit messages to Station B using a sliding window protocol. The round trip delay between A and B is 80 milliseconds and the bottleneck bandwidth on the path between A and B is 128 kbps. What is the optimal window size that A should use?
(a) 20 (b) 40
(c) 160 (d) 320 **[2006 : 2 M]**

1.15 Station A needs to send a message consisting of 9 packets to Station B using a siding window (window size 3) and go-back-n error control strategy. All packets are ready and immediately available for transmission. If every 5th packet that A transmits gets lost (but no acks from B ever get lost), then what is the number of packets that A will transmit for sending the message to B?
(a) 12 (b) 14
(c) 16 (d) 18 **[2006 : 2 M]**

1.16 The message 11001001 is to be transmitted using the CRC polynomial $x^3 + 1$ to protect it from errors. The message that should be transmitted is:
(a) 11001001000
(b) 11001001011
(c) 11001010
(d) 110010010011
[2007 : 2 M]

1.17 An error correcting code has the following code words: 00000000, 00001111, 01010101, 10101010, 11110000. What is the maximum number of bit errors that can be corrected ?
(a) 0 (b) 1
(c) 2 (d) 3 **[2007 : 2 M]**

1.18 In the waveform (a) given below, a bit stream is encoded by Manchester encoding scheme. The same bit stream is encoded in a different coding scheme in wave form (b). The bit stream and the coding scheme are

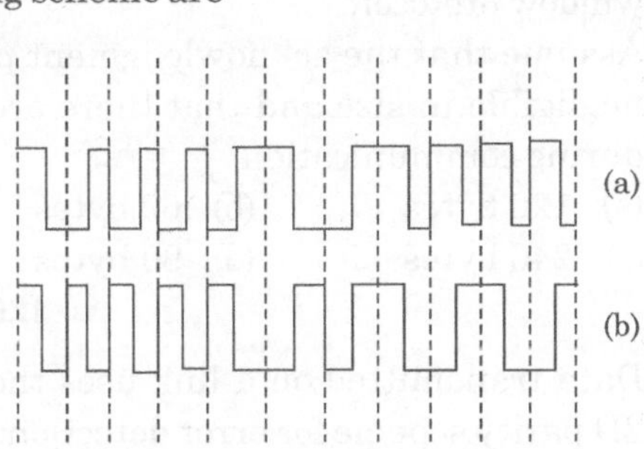

(a) 1000010111 and Differential Manchester respectively
(b) 0111101000 and Differential Manchester respectively
(c) 1000010111 and Integral Manchester respectively
(d) 0111101000 and Integral Manchester respectively **[2007 : 2 M]**

1.19 Let us consider a statistical time division multiplexing of packets. The number of sources is 10. In a time unit, a source transmits a packet of 1000 bits. The number of sources sending data for the first 20 time units is 6, 9, 3, 7, 2, 2, 2, 3, 4, 6, 1, 10, 7, 5, 8, 3, 6, 2, 9, 5 respectively. The output capacity of multiplexer is 5000 bits per time unit. Then the average number of backlogged of packets per time unit during the given period is
(a) 5 (b) 4.45
(c) 3.45 (d) 0 **[2007 : 2 M]**

1.20 Your are given the following four bytes:
10100011 00110111 11101001 10101011
Which of the following are substrings of the base 64 encoding of the above four bytes?
(a) zdp (b) fpq
(c) qwA (d) oze **[2007 : 2 M]**

1.21 How many bytes of data can be sent in 15 seconds over a serial link with baud rate of 9600 in asynchronous mode with odd parity and two stop bits in the frame?
(a) 10,000 bytes
(b) 12,000 bytes
(c) 15,000 bytes
(d) 27,000 bytes
[2008 : 1 M]

1.22 A 1 Mbps satellite link connects two ground stations. The altitude of the satellite is 36,504 km and speed of the signal is 3×10^8 m/s. What should be the packet size for a channel utilization of 25% for a satellite link using go-back-127 sliding window protocol?
Assume that the acknowledgment packets are negligible in size and that there are no errors during communication.
(a) 120 bytes (b) 60 bytes
(c) 240 bytes (d) 90 bytes
[2008 : 2 M]

1.23 Data transmitted on a link uses the following 2D parity scheme for error detection:
Each sequence of 28 bits is arranged in a 4 × 7 matrix (rows r_0 through r_3, and columns d_7 through d_1) and is padded with a column d_0 and row r_4 of parity bits computed using the Even parity scheme. Each bit of column d_0 (respectively, row r_4) gives the parity of the corresponding row (respectively, column). These 40 bits are transmitted over the data link.

	d_7	d_6	d_5	d_4	d_3	d_2	d_1	d_0
r_0	0	1	0	1	0	0	1	1
r_1	1	1	0	0	1	1	1	0
r_2	0	0	0	1	0	1	0	0
r_3	0	1	1	0	1	0	1	0
r_4	1	1	0	0	0	1	1	0

The table shows data received by a receiver and has n corrupted bits. What is the minimum possible value of n?
(a) 1 (b) 2
(c) 3 (d) 4 **[2008 : 2 M]**

1.24 Let G(x) be the generator polynomial used for CRC checking. What is the condition that should be satisfied by G(x) to detect odd number of bits in error?
(a) G(x) contains more than two terms
(b) G(x) does not divide $1 + x^k$, for any k not exceeding the frame length
(c) 1 + x is a factor of G(x)
(d) G(x) has an odd number of terms
[2009 : 2 M]

Linked Answer for Q.1.25 and Q. 1.26:

Frames of 1000 bits are sent over a 10^6 bps duplex link between two hosts. The propagation time is 25 ms. Frames are to be transmitted into this link to maximally pack them in transit (within the link).

1.25 What is the minimum number of bits (l) that will be required to represent the sequence numbers distinctly? Assume that no time gap needs to be given between transmission of two frames.
(a) $l = 2$ (b) $l = 3$
(c) $l = 4$ (d) $l = 5$ **[2009 : 2 M]**

1.26 Suppose that the sliding window protocol is used with the sender window size of 2^l, where l is the number of bits identified in the earlier part and acknowledgments are always piggy backed. After sending 2^l frames, what is the minimum time the sender will have to wait before starting transmission of the next frame? (Identify the closest choice ignoring the frame processing time).
(a) 16 ms (b) 18 ms
(c) 20 ms (d) 22 ms **[2009 : 2 M]**

1.27 In the following pairs of OSI protocol layer/sub-layer and its functionality, the **INCORRECT** pair is
(a) Network layer and Routing
(b) Data Link Layer and Bit synchronization
(c) Transport layer and End-to-end process communication
(d) Medium Access Control sub-layer and Channel sharing **[2014 (Set-3) : 1 M]**

1.28 A bit-stuffing based framing protocol uses an 8-bit delimiter pattern of 01111110. If the output bit-string after stuffing is 01111100101, then the input bit-string is
(a) 0111110100 (b) 0111110101
(c) 0111111101 (d) 0111111111
[2014 (Set-3) : 1 M]

1.29 Suppose that the stop-and-wait protocol is used on a link with a bit rate of 64 kilobits per second and 20 milliseconds propagation delay. Assume that the transmission time for the acknowledgment and the processing time at nodes are negligible. Then the minimum frame size in bytes to achieve a link utilization of at least 50% is ________. **[2015 (Set-1) : 2 M]**

1.30 A link has a transmission speed of 10^6 bits/sec. It uses data packets of size 1000 bytes each. Assume that the acknowledgment has negligible transmission delay, and that its propagation delay is the same as the data propagation delay. Also assume that the processing delays at nodes are negligible. The efficiency of the stop-and-wait protocol in this setup is exactly 25%. The value of the one-way propagation delay (in milliseconds) is ________. **[2015 (Set-2) : 1 M]**

1.31 Consider a network connecting two systems located 8000 kilometers apart. The bandwidth of the network is 500×10^6 bits per second. The propagation speed of the media is 4×10^6 meters per second. It is needed to design a Go-Back-N sliding window protocol for this network. The average packet size is 10^7 bits. The network is to be used to its full capacity.
Assume that processing delays at nodes are negligible. Then, the minimum size in bits of the sequence number field has to be ______.

[2015 (Set-3) : 2 M]

1.32 Two hosts are connected via a packet switch with 10^7 bits per second links. Each link has a propagation delay of 20 microseconds. The switch begins forwarding a packet 35 microseconds after it receives the same. If 10000 bits of data are to be transmitted between the two hosts using a packet size of 5000 bits, the time elapsed between the transmission of the first bit of data and the reception of the last bit of the data in microseconds is ________.

[2015 (Set-3) : 2 M]

1.33 Consider two hosts *X* and *Y* connected by a single direct link of rate 10^6 bits/sec. The distance between the two hosts is 10,000 km and the propagation speed along the link is 2×10^8 m/sec. Host *X* sends a file of 50,000 bytes as one large message to host *Y* continuously. Let the transmission and propagation delays be *p* milliseconds and *q* milliseconds, respectively.
Then the values of *p* and *q* are
(a) $p = 50$ and $q = 100$
(b) $p = 50$ and $q = 400$
(c) $p = 100$ and $q = 50$
(d) $p = 400$ and $q = 50$

[2017 (Set-2) : 2 M]

1.34 Consider the following two statements:
S_1: Destination MAC address of an ARP reply is a broadcast address.
S_2: Destination MAC address of an ARP request is a broadcast address.
Which one of the following choices is correct?
(a) S_1 is false and S_2 is true.
(b) Both S_1 and S_2 are false.
(c) Both S_1 and S_2 are true,
(d) S_1 is true and S_2 is false.

[2021 (Set-1) : 1 M]

1.35 Consider the sliding window flow-control protocol operating between a sender and a receiver over a full-duplex error-free link. Assume the following:
- The time taken for processing the data frame by the receiver is negligible.
- The time taken for processing the acknowledgement frame by the sender is negligible.
- The sender has infinite number of frames available for transmission.
- The size of the data frame is 2000 bits and the size of the acknowledgement frame is 10 bits.
- The link data rate in each direction is 1 Mbps (= 10^6 bits per second).
- One way propagation delay of the link is 100 milliseconds.

The minimum value of the sender's window size in terms of the number of frames, (rounded to the nearest integer) needed to achieve a link utilization of 50% is ________.

[2021 (Set-1) : 2 M]

1.36 Assume that a 12-bit Hamming codeword consisting of 8-bit data and 4 check bits is $d_8\, d_7\, d_6\, d_5\, c_8\, d_4\, d_4\, d_3\, d_2\, c_4\, d_1\, c_2\, c_1$, where the data bits and the check bits are given in the following tables:

Data bits							
d_8	d_7	d_6	d_5	d_4	d_3	d_2	d_1
1	1	0	x	0	1	0	1

Check bits			
c_8	c_4	c_2	c_1
y	0	1	0

Which one of the following choices gives the correct values of *x* and *y*?

(a) x is 0 and y is 0 (b) x is 0 and y is 1
(c) x is 1 and y is 0 (d) x is 1 and y is 1

[2021 (Set-1) : 2 M]

1.37 Consider the cyclic redundancy check (CRC) based error detecting scheme having the generator polynomial $X^3 + X + 1$. Suppose the message $m_4m_3m_2m_1m_0 = 11000$ is to be transmitted. Check bits $c_2c_1c_0$ are appended at the end of the message by the transmitter using the above CRC scheme, The transmitted bit string is denoted by $m_4m_3m_2m_1m_0c_2c_1c_0$. The value of the checkbit sequence $c_2c_1c_0$ is

(a) 110 (b) 111
(c) 100 (d) 101

[2021 (Set-2) : 2 M]

1.38 Consider a 100 Mbps link between an earth station (sender) and a satellite (receiver) at an altitude of 2100 km. The signal propagates at a speed of 3×10^8 m/s. The time taken (in milliseconds, rounded off to two decimal places) for the receiver to completely receive a packet of 1000 bytes transmitted by the sender is ________.

[2022 : 2 M]

1.39 Suppose two hosts are connected by a point-to-point link and they are configured to use Stop-and-Wait protocol for reliable data transfer. Identify in which one of the following scenarios, the utilization of the link is the lowest.

(a) Longer link length and lower transmission rate
(b) Longer link length and higher transmission rate
(c) Shorter link length and lower transmission rate
(d) Shorter link length and higher transmission rate

[2023 : 1 M]

■■■

Answers ISO/OSI Stack and SWP

1.1	(b)	**1.2**	(a)	**1.3**	(b)	**1.4**	(c)	**1.5**	(d)	**1.6**	(c)	**1.7**	(b)	**1.8**	(a)	**1.9**	(b)
1.10	(a)	**1.11**	(d)	**1.12**	(c)	**1.13**	(b)	**1.14**	(b)	**1.15**	(c)	**1.16**	(b)	**1.17**	(b)	**1.18**	(a)
1.19	(b)	**1.20**	(b)	**1.21**	(b)	**1.22**	(a)	**1.23**	(c)	**1.24**	(c)	**1.25**	(d)	**1.26**	(c)	**1.27**	(b)
1.28	(b)	**1.29**	(320)	**1.30**	(12)	**1.31**	(8)	**1.32**	(1575)	**1.33**	(d)	**1.34**	(a)	**1.35**	(51)	**1.36**	(a)
1.37	(c)	**1.38**	(7.08)	**1.39**	(b)												

Explanations ISO/OSI Stack and SWP

1.1 (b)

Given, Window size

$n = 5$ packets

Packet size = 1000 byte

Total packet size = 5 × 1000 = 5000 bytes

Total time = Transmission Time + Propagation Time

$= 5 \times 50 + 200\ \mu s = 250 + 200\ \mu s$

$= 450\ \mu s = 450 \times 10^{-6}$ s

Maximum Achievable throughput

$$= \frac{\text{Total Size}}{\text{Total Time}} = \frac{5000}{450 \times 10^{-6}}$$

$$= \frac{5000 \times 10^{6}}{450} = 11.11 \times 10^{6} \text{ Bps}$$

1.2 (a)

P. Datal link layer:

1. Ensures reliable transport of data over a physical point-to point link- Reliable mens error correction & detection done by Data link layer.

Q. Network layer:

4. Routes data from one network node to the next according to routing algorithm.

R. Transport layer:

3. Allows end-to- end communication between two processes with the help of TCP and UDP protocol.

1.3 (b)

Total number of bits = 12

modulation Rate = 9600 baud

Number of characters (8 bit character) are

$$\text{transmitted} = \frac{9600}{12 \text{ bits}} = 800$$

1.4 (c)

In serial transmission T_1:

To send 1 character = 8 bit data + 2 + 1 + 1

= 12 bits

$$\text{So transfer rate} = \frac{1200 \text{ bits/sec}}{12 \text{ bits}}$$

= 100 character/sec

In synchronous transmission T_2:

To send 30 eight bit = 30 × 8 = 240 bits

Synchronize character = 3 × 8 = 24 bits

So, to send 30 eight bit data = 240 + 24 = 264 bits

$$\text{So, Transfer rate (for 30 char)} = \frac{1200 \text{ bits/sec}}{264 \text{ bits}}$$

= 4.545 (30 char/sec)

So, In per character = 4.545 × 30

= 136.2 character/sec

8 bit data + 2 + 1 + 1 = 12 bit

1.5 (d)

Frame delimiter flag = 0111

- If we put 0 after first 1 i.e. 01011, there is very high probability that in message 1 followed by 0, so need, two many bits to transmit data.
- So we need to put 0 after two 1 followed by 0 i.e., 01101

Hence data to be send = 0 1 1 0 1 0 1 1 0 0

1.6 (c)

In sliding window protocol :

Minimum number of sequence number = M + N

Since at any time number of unacknowledged frames are M at receiver and same time N packets are transmitted by sender.

So, M + N ≤ minimum sequence number.

1.7 (b)

$$T_r = \frac{100 \times 8 \text{ bits}}{20 \text{ Kbps}} = 40 \text{ ms}, \quad T_p = 400 \text{ ms}$$

$$a = \frac{T_p}{T_x} = \frac{400}{40} = 10$$

$$\text{Efficiency of GBN} = \frac{W}{(1+2a)}$$

where, w = Window size

$$= 10 = \frac{10}{(1+20)} = \frac{10}{21}$$

BW utilization or throughput or max data rate

$$= \text{Efficiency} \times \text{BW} = \left(\frac{10}{21}\right) \times 20$$

It is nearly 10 Kbps

1.8 (a)

Consider first 3 bits x_1, x_2 and x_3 are data bits and last 4 bits c_1, c_2, c_3 and c_4 are parity check bits i.e., 0101011 will be $x_1(0)$, $x_2(1)$, $x_3(0)$, $c_1(1)$, $c_2(0)$, $c_3(1)$ and $c_4(1)$.

By analysing pattern used to create parity bits i.e., here

	x_1	x_2	x_3	$c_1=x_1 \oplus x_2$	$c_2=x_1 \oplus x_3$	$c_3=x_2 \oplus x_3$	$c_4=x_1 \oplus x_2 \oplus x_3$
1.	0	1	0	1	0	1	1
2.	1	0	0	1	1	0	1
3.	1	1	1	0	0	0	1

So by applying same rule to find other code words.

	x_1	x_2	x_3	$c_1=x_1 \oplus x_2$	$c_2=x_1 \oplus x_3$	$c_3=x_2 \oplus x_3$	$c_4=x_1 \oplus x_2 \oplus x_3$
4.	0	0	1	0	1	1	1
5.	0	1	1	1≠0 so fail			
6.	1	0	1	1	0	1	0
7.	0	1	1	1	1≠0 so fail		

So only I[st] and IIIrd are code word.

1.9 (b)

In case of the selective Reject Protocol, the

$$\text{maximum window size} = \frac{2^n}{2} = 2^{n-1}$$

In the case of Selective Reject the window size will be half.

1.10 (a)

Generator polynomial is of degree 5 so append 5 0's to the end of data and then divide new data by generator polynomial $x^5 + x^4 + x^2 + 1$ = 110101.

```
110101 ) 101000110100000 (
         110101
         ------
          0111011
           110101
          -------
           0111011
            110101
           --------
            00111110
              110101
            --------
              00101100
                110101
              --------
                0110010
                 110101
                -------
                 0001110
```

Remainder is 01110

1.11 (d)

Propagation time = 20 m sec

Bandwidth = 4 Kbps

When efficiency is atleast 50%, then

$$\text{T.T.} \geq 2 \text{ P.T.}$$

$$\frac{x}{4 \text{ kbps}} \geq 2 \times 20 \text{ msec}$$

$$x \geq 2 \times 20 \times 10^{-3} + 4 \times 10^3 \text{ bits}$$

$$x \geq 40 \times 4 \text{ bits}$$

$$x \geq 160 \text{ bits}$$

1.12 (c)

With 5 bits total number of encoding = $2^5 = 32$

Condition :

Case : 1 When [0] 1 0/1 0/1 [1] = 2^2 = 4 encoding

Case : 2 When [1] 0/1 0/1 0/1 [1] = 2^3 = 8 encoding

Case : 3 When [0] 1 0/1 1 [0] = 2 encoding

Case : 4 When [1] 0/1 0/1 1 [0] = 2^2 = 4 encoding

So total encoding = 8 + 4 + 2 + 4 = 18 encoding

1.13 (b)

In sliding window protocol (stop and wait) :

Average number of transmission of single packet

$$= \left(\frac{1}{1-P}\right)$$

(where P = probability of packet error).

$$\text{So for 100 packets} = 100 \times \left(\frac{1}{1-0.2}\right)$$

$$= 100 \times \frac{1}{0.8} = 125$$

1.14 (b)

Given round trip delay t = 80 ms

$$= 80 \times 10^{-3} \text{ sec}$$

$$R = 128 \text{ kbps} = 128 \times 10^3 \text{ bps}$$

$$L = Rt$$

$$= 128 \times 10^3 \times 80 \times 10^{-3}$$

$$= 128 \times 80$$

So, optional window size = n

$$= \frac{128 \times 80}{32 \times 8} = \frac{10240}{256}$$

$$n = 40$$

1.15 (c)

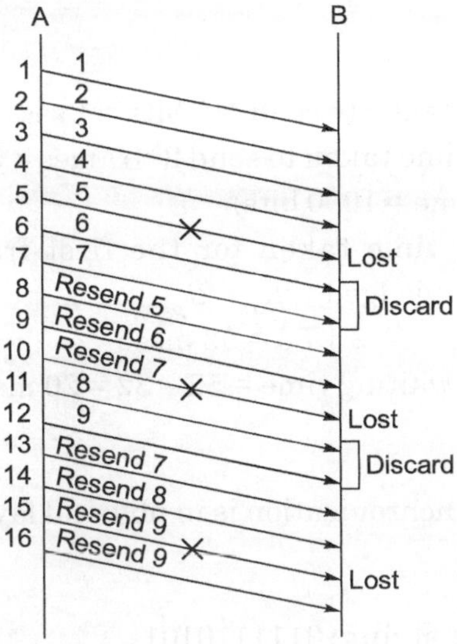

1.16 (b)

P(x) = 11001001
divisor D(x) = 1001 and CRC remainder is 011.
So the transmitted message is 11001001011.

1.17 (b)

The maximum hamming distance among the gives code words is 8. i.e.

$$\begin{array}{r} 01010101 \\ +\ 10101010 \\ \hline 11111111 \\ \hline \end{array}$$

Minimum of all hamming distance is 4.
So, the hamming distance to correct the 'd' errors should be 2d + 1.

$\therefore \quad 2d + 1 = 4$

$d \simeq 1$

∴ Maximum 3 bit errors can be corrected.

1.18 (a)

'0' – Low to high, '1' – High to low.
There is a transition at the middle of each bit period.
∴ The bit pattern is 100 00 10111.

1.19 (b)

Output capacity of multiplexer = 5000 b per unit time.
Each packet contain number of bits = 1000 b
So, number of packets send in one time unit

$$= \frac{5000}{1000} = 5.$$

If number of transmitted packet at any unit of time more than 5, then extra packet will be backlock of present packet.

S.no.	Number of Packets	Backlock
1.	6	1 (backlock to next pkt
2.	9+1 (backlock from previous)	5
3.	3+5 (backlock from previous)	3
4.	7+3 (backlock from previous)	5
5.	2+5 (backlock from previous)	2
6.	2+2 (backlock from previous)	0
7.	2	0
8.	3	0
9.	4	0
10.	6	1
11.	1+1 (backlock from previous)	0
12.	10	5
13.	7+5 (backlock from previous)	7
14.	5+7	7
15.	8+7	10
16.	3+10	8
17.	6+8	9
18.	2+9	6
19.	9+6	10
20.	5+10	10

$$\therefore \text{ Average number of backlock} = \frac{89}{20} = 4.45$$

1.20 (b)

In base 64 encoding :
Number of bits in data = Multiple of 6 bits i.e. log (64).
Since data given is 4 B = 32 bits.
So number of padding bits = 36 – 32 = 4 bits (minimum).
Padding bits are always adding at LSB side.
So,

101000	110011	011111	101001	101010	110000
40	51	31	41	42	48
(o)	(z)	(f)	(p)	(q)	(w)

So, substring which match is fpq.

1.21 (b)

Number of bits needed to send 1 byte of data = 8 + 1 + 1 + 2 = 12 bits.
Baud rate = 9600 bps = 800 char/sec
Hence, 800 bytes are sent in 1 sec.
In 15 sec = 15 × 800 = 12,000 bytes

Note: In given Question 1 start bit is miss which is always present in asynchronous transmission mode.

1.22 (a)

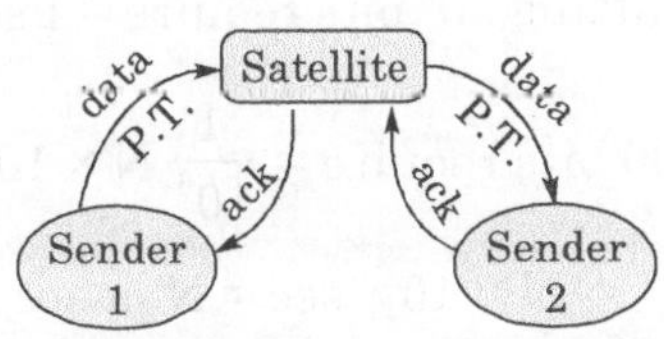

So propagation delay = Propagation from S_1 to S + Propagation from S + S_2 = P.T. + P.T. = 2 P.T.

$$\text{Propagation delay } (T_p) = \frac{\text{Distance}}{\text{Speed}}$$

$$= \frac{2 \times 36504 \times 10^3 \text{m}}{3 \times 10^8 \text{m}} \text{sec}$$

$$= 24336 \times 10^{-5} \text{ sec} = 243360 \text{ μsec}$$

$$\text{Transmission delay } (T_t) = x \text{ μsec}$$

$$a = \frac{T_p}{T_t} = \frac{243360}{x}$$

$$\text{Efficiency } (\eta) = \frac{1}{1+2a}$$

$$\Rightarrow \quad 0.25 = \frac{127x}{x + 2 \times 243360}$$

$$\Rightarrow \quad x = 960 \text{ bits} = 120 \text{ Bytes}$$

1.23 (c)

To find number of corrupted bits recievd by reciever we need to calculate hamming distance of every row and every column.

	d_7	d_6	d_5	d_4	d_3	d_2	d_1	d_0	Hamming distance
r_0	0	1	0	1	0	0	1	1	0 (no error)
r_1	1	1	~~0~~ 1	0	1	1	1	0	1 (error)
r_2	0	0	0	1	0	1	0	0	0 (no error)
r_3	0	1	1	0	1	0	1	0	0 (no error)
r_4	1	1	0	0	0	~~1~~ 0	1	~~0~~ 1	0 (no error)
	0 no error	**0** no error	**1** error	**0** no error	**0** no error	**1** error	**0** no error	**1** error	

So, error bits are (r_1, d_5), (r_4, d_2), and (r_4, d_0). After change these three bits we have no corrupted bits.

1.24 (c)

The polynomial generator used for CRC checking must satisfy at least two condition to detect odd numbers of errors:

1. It should be not divisible by x.
2. It should be divisible by 1 + x.

Therefore (1 + x) is a factor of G(x).

1.25 (d)

The link is a Duplex hence we need not wait for twice the propagation time for sending the frame belonging to next window. If the sender window is of size N.

Transmitting 10^6 bits require = 1 sec

$$\text{N} \times 1000 \text{ bits require} = \frac{1}{10^6} \text{N} \times 10^3$$

$$= \text{N} \times 10^{-3} \text{ sec} = \text{N m sec}$$

$$\text{Nm sec} = 25 \text{ m sec}, \text{N} = 25 < 2^5$$

∴ Minimum number of bits required is 5 to represent sequence numbers disinctly.

1.26 (c)

Time taken to send 10^6 bits = 1 sec

∴ Time taken to send 2^5 frames = 32 msec

(1 frame = 1000 bits)

Total time taken for the first frame to be acknowledged = $2t_x + 2t_p$

= 2 + 50 = 52 msec

Then waiting time = 52 – 32 = 20 msec

1.27 (b)

Bit synchronization is in physical layer.

1.28 (b)

Input string: 0111110101

Output string: 01111100101

After five consecutive 1's in the input, bit 0 is inserted.

1.29 (320)

For atleast 50% efficiency: $t_d = 2\ p_d$

∴ Minimum frame size = 64 kbps × 40 ms

= 2560 bits = 320 bytes

1.30 (12)

$$\text{Efficiency of Stop \& Wait} = \frac{1}{1+2a}$$

$$\Rightarrow \quad 0.25 = \frac{1}{1+2a}$$

$$a = \frac{p_d}{t_d}, \; t_d = 8 \text{ ms}$$

$$\therefore \; 0.25 = \frac{1}{1 + 2 \times \frac{p_d}{8\text{ms}}} \Rightarrow p_d = 12 \text{ ms}$$

1.31 (8)

$$d = 8000 \text{ km}$$

$$\text{Band width} = 500 \times 10^6 \text{ bps}$$

Propagation delay:

$$P_d = \frac{8000 \times 10^3}{4 \times 10^6} = 2 \text{ sec}$$

$$\text{Packet size} = 10^7 \text{ bits}$$

Transmission delay:

$$t_d = \frac{10^7}{500 \times 10^6} = \frac{1}{50} \text{ sec}$$

$$n = 100\%$$

$$\frac{100}{100} = \frac{N}{1+2a}$$

$$a = \frac{P_d}{t_d} = \frac{2}{\frac{1}{50}} \text{ sec} = 100$$

$$\frac{100}{100} = \frac{N}{201}$$

$$N = 201$$

⇒ Number of packets = 201

∴ 8 bits are required for sequence number.

1.32 (1575)

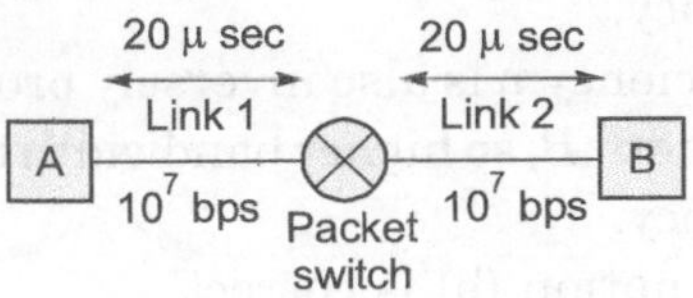

Extra delay at switch = 35 μsec for each packet.
Data = 10000 bits

$$\text{Number of packets} = \frac{10000}{5000} = 2 \text{ packet}$$

Transmission delay for one packet = 500 μsec.
At t = **500** μsec, last bit of packet 1 is placed on link 1 by A and Transmission of packet begins.
At t = **520**, last bit of packet 1 reaches switch.
At t = **555**, first bit of packet 1 is placed on link 2 by switch.
At t = **1000**, last bit of packet 2 is placed on link 1 by A.
At t = **1020**, last bit of packet 2 reaches switch.
Note: pkt 2 need to wait upto 1055 μsec before switch transfers it.
Last bit of packet1 will be placed on link 2 by switch at 1055 μsec.
Hence No additional delay for packet 2.
At t = **1055** packet two first bit is placed on link 2.
At t = **1575** last bit of packet 2 reaches B.
∴ 1575 μsec is required.

1.33 (d)

$$\text{Transmission time} = \frac{\text{Data size}}{\text{Bandwidth}}$$

$$= \frac{50000 \times 8 \text{ bits}}{10^6 \text{ bits/sec}} = 400 \text{ msec}$$

$$\text{Propagation time} = \frac{\text{Distance}}{\text{Velocity}} = \frac{10000 \times 10^3 \text{ m}}{2 \times 10^8 \text{ m/sec}}$$

$$= 50 \times 10^{5-8} \text{ sec} = 50 \text{ msec}$$

1.34 (a)

ARP request is broadcasting.
ARP reply is unicasting.

1.35 (51)

$$T_t(\text{packet}) = \frac{L}{\text{B.W}}$$

$$\Rightarrow \frac{2000 \text{ bits}}{10^6 \text{ bps}} = 2 \times 10^{-3} \text{ sec} = 2 \text{ millisec}$$

$$T_t(\text{Ack}) = \frac{L}{\text{B.W}}$$

$$\Rightarrow \frac{10 \text{ bits}}{10^6 \text{ bps}} = 10^{-5} \text{ sec}$$

$$= 10^{-2} \text{ millisec} = 0.01 \text{ millisec}$$

$$T_p = 100 \text{ millisec}$$

$$\text{Total time} = T_t(\text{packet}) + T_p + T_t(\text{Ack})$$

$$= 2 + 100 + 0.01$$

$$= 102.01 \text{ millisec}$$

$$\text{Efficiency} = 50\% = \frac{1}{2}$$

$$\text{Efficiency} = \frac{\text{Useful time}}{\text{Total time}}$$

$$\frac{1}{2} = \frac{n \times T_t}{\text{Total time}}$$

$$\Rightarrow 2 \times n = 102.01$$

$$\Rightarrow n = \frac{102.01}{2} \Rightarrow 51.005$$

For minimum we have to take ceil, hence size of window = 51.

1.36 (a)

1	2	3	4	5	6	7	8
1	0	1	0	x	0	1	1

c_1	c_2	c_4	c_8
0	1	0	y

c_1 = 1 3 5 7 9 11
0 1 0 0 x 1 ⟷ $x = 0$

1	2	3	4	5	6	7	8	9	10	11	12
0	1	1	0	0	1	0	y	x	0	1	1

c_8 = 8 9 10 11 12
0 0 0 1 1 ⟷ $y = 0$

1.37 (c)

$$x^3 + x + 1 = 1011$$

```
1011 ) 11000000
       1011
        1110
        1011
         1010
         1011
          100
```

1.38 (7.08)

$$T.T = \frac{\text{Data size}}{\text{Bandwidth}}$$

$$= \frac{1000 \times 8 \text{ bits}}{10^8 \text{ bits/sec}} = 8 \times 10^{-5} \text{ sec}$$

$$T.T = 0.08 \text{ millisec}$$

$$P.T = \frac{\text{Length}}{\text{Velocity}}$$

$$= \frac{2100 \times 10^3 \text{ m}}{3 \times 10^8 \text{ m/sec}}$$

$$= 7 \times 10^{-3} \text{ sec} = 7 \text{ millisec}$$

$$\text{Total time} = 7 + 0.08 = 7.08 \text{ millisec}$$

1.39 (b)

Utilization of link = Efficiency

Efficiency of stop and wait protocol is

$$\eta = \left(\frac{1}{1+\left(\frac{2T_p}{T_t}\right)}\right) = \left(\frac{1}{1+\left(\frac{2d \times B}{v \times L}\right)}\right)$$

where, d = Link length (Distance)

v = Speed

B = Bandwidth (Transmission rate)

and L = Packet length

As efficiency η is inversely proportional to distance d, so longer distance means lower efficiency.

As efficiency η is also inversely proportional to bandwidth B, so higher bandwidth means lower efficiency.

Hence option (b) is correct.

■■■■

2 LAN

2.1 A 2 km long broadcast LAN has 10^7 bps bandwidth and uses CSMA/CD. The signal travels along the wire at 2×10^8 m/s. What is the minimum packet size that can be used on this network?
(a) 50 bytes (b) 100 bytes
(c) 200 bytes (d) None of these
[2003 : 2 M]

2.2 A host is connected to a Department network which is part of a University network. The University network, in turn, is part of the Internet. The largest network in which the Ethernet address of the host is unique is
(a) the subnet to which the host belongs
(b) the Department network
(c) the University network
(d) the Internet **[2004 : 1 M]**

2.3 Consider a simplified time slotted MAC protocol, where each host always has data to send and transmits with probability p = 0.2 in every slot. There is no backoff and one frame can be transmitted in one slot. If more than one host transmits in the same slot, then the transmissions are unsuccessful due to collision. What is the maximum number of hosts which this protocol can support, if each host has to be provided a minimum through put of 0.16 frames per time slot?
(a) 1 (b) 2
(c) 3 (d) 4 **[2004 : 2 M]**

2.4 A and B are the only two stations on an Ethernet. Each has a steady queue of frames to send. Both A and B attempt to transmit a frame, collide, and A wins the first backoff race. At the end of this successful transmission by A, both A and B attempt to transmit and collide. The probability that A wins the second backoff race is
(a) 0.5 (b) 0.625
(c) 0.75 (d) 1.0 **[2004 : 2 M]**

2.5 In a network of LANs connected by bridges, packets are sent from one LAN to another through intermediate bridges. Since more than one path may exist between two LANs, packets may have to be routed through multiple bridges. Why is the spanning tree algorithm used for bridge-routing?
(a) For shortest path routing between LANs
(b) For avoiding loops in the routing paths
(c) For fault tolerance
(d) For minimizing collisions **[2005 : 1 M]**

2.6 Which of the following statements is TRUE about CSMA/CD
(a) IEEE 802.11 wireless LAN runs CSMA/CD protocol
(b) Ethernet is not based on CSMA/CD protocol
(c) CSMA/CD is not suitable for a high propagation delay network like satellite network
(d) There is no contention in a CSMA/CD network **[2005 : 1 M]**

2.7 Which of the following statements is FALSE regarding a bridge
(a) Bridge is a layer 2 device
(b) Bridge reduces collision domain
(c) Bridge is used to connect two or more LAN segments
(d) Bridge reduces broadcast domain
[2005 : 1 M]

2.8 A network with CSMA/CD protocol in the MAC layer is running at 1 Gbps over a 1 km cable with no repeaters. The signal speed in the cable is 2×10^8 m/sec. The minimum frame size for this network should be
(a) 10000 bits (b) 10000 bytes
(c) 5000 bits (d) 5000 bytes
[2005 : 2 M]

2.9 In a TDM medium access control bus LAN, each station is assigned one time slot per cycle for transmission. Assume that the length of each time slot is the time to transmit 100 bits plus the end-to-end propagation delay. Assume a propagation speed of 2×10^8 m/sec. The length of the LAN is 1 km with a bandwidth of 10 Mbps. The maximum number of stations that can be allowed in the LAN so that the throughput of each station can be 2/3 Mbps is
(a) 3 (b) 5
(c) 10 (d) 20 **[2005 : 2 M]**

2.10 A router has two full-duplex Ethernet interfaces each operating at 100 Mb/s. Ethernet frames are at least 84 bytes long (including the Preamble and the Inter-Packet-Gap).
The maximum packet processing time at the router for wirespeed forwarding to be possible is (in micro-seconds)
(a) 0.01 (b) 3.36
(c) 6.72 (d) 8 **[2006 : 2 M]**

2.11 In Ehernet when Manchester encoding is used, the bit rate is
(a) Half the baud rate
(b) Twice the baud rate
(c) Same as the baud rate
(d) None of these **[2007 : 1 M]**

2.12 A broadcast channel has 10 nodes and total capacity of 10 Mbps. It uses polling for medium access. Once a node finishes transmission, there is a polling delay of 80 μs to poll the next node. Whenever a node is polled, it is allowed to transmit a maximum of 1000 bytes. The maximum throughput of the broadcast channel is
(a) 1 Mbps (b) 100/11 Mbps
(c) 10 Mbps (d) 100 Mbps
[2007 : 2 M]

2.13 There are n stations in a slotted LAN. Each station attempts to transmit with a probability p in each time slot. What is the probability that ONLY one station transmits in a given time slot?
(a) $np(1-p)^{n-1}$ (b) $(1-p)^{n-1}$
(c) $p(1-p)^{n-1}$ (d) $1-(1-p)^{n-1}$
[2007 : 2 M]

2.14 The distance between two stations M and N is L kilometres. All frames are K bits long. The propagation delay per kilometre is t seconds. Let R bits/second be the channel capacity. Assuming that processing delay is negligible, the minimum number of bits for the sequence number field in a frame for maximum utilization, when the sliding window protocol is used, is:
(a) $\left\lceil \log_2 \frac{2LtR+2K}{K} \right\rceil$ (b) $\left\lceil \log_2 \frac{2LtR}{K} \right\rceil$
(c) $\left\lceil \log_2 \frac{2LtR+K}{K} \right\rceil$ (d) $\left\lceil \log_2 \frac{2LtR+K}{2K} \right\rceil$
[2007 : 2 M]

2.15 The minimum frame size required for a CSMA/CD based computer network running at 1 Gbps on a 200 m cable with a link speed of 2×10^8m/s is
(a) 125 bytes (b) 250 bytes
(c) 500 bytes (d) None of these
[2008 : 2 M]

2.16 Consider a source computer (S) transmitting a file of size 10^6 bits to a destination computer (D) over a network of two routers (R_1 and R_2) and three links (L_1, L_2 and L_3). L_1 connects S to R_1; L_2 connects R_1 to R_2; and L_3 connects R_2 to D. Let each link be of length 100 km. Assume signals travel over each link at a speed of 10^8 meters per second. Assume that the link bandwidth on each link is 1 Mbps. Let the file be broken down into 1000 packets each of size 1000 bits. Find the total sum of transmission and propagation delays in transmitting the file from S to D?
(a) 1005 ms (b) 1010 ms
(c) 3000 ms (d) 3003 ms
[2012 : 2 M]

2.17 Determine the maximum length of the cable (in km) for transmitting data at a rate of 500 Mbps in an Ethernet LAN with frames of size 10,000 bits. Assume the signal speed in the cable to be 2,00,000 km/s.
(a) 1 (b) 2
(c) 2.5 (d) 5 **[2013 : 2 M]**

2.18 Consider a LAN with four nodes S_1, S_2, S_3 and S_4. Time is divided into fixed-size slots, and a node can begin its transmission only at the beginning of a slot. A collision is said to have occurred if more than one node transmit in the same slot. The probabilities of generation of a frame in a time slot by S_1, S_2, S_3 and S_4 are 0.1, 0.2, 0.3 and 0.4, respectively. The probability of sending a frame in the first slot without any collision by any of these four stations is ______.
[2015 (Set-1) : 2 M]

2.19 Consider a CSMA/CD network that transmits data at a rate of 100 Mbps (10^8 bits per second) over a 1 km (kilometer) cable with no repeaters. If the minimum frame size required for this network is 1250 bytes, what is the signal speed (km/sec) in the cable?
(a) 8000 (b) 10000
(c) 16000 (d) 20000
[2015 (Set-3) : 1]

2.20 Which one of the following protocols is NOT used to resolve one form of address to another one?
(a) DNS (b) ARP
(c) DHCP (d) RARP
[2016 (Set-1) : 2 M]

2.21 Which of the following is/are example(s) of statefull application layer protocols?
(i) HTTP (ii) FTP
(iii) TCP (iv) POP3
(a) (i) and (ii) only (b) (ii) and (iii) only
(c) (ii) and (iv) only (d) (iv) only
[2016 (Set-1) : 2 M]

2.22 A sender uses the Stop-and-Wait ARQ protocol for reliable transmission of frames. Frames are of size 1000 bytes and the transmission rate at the sender is 80 Kbps (1 Kbps = 1000 bits/second). Size of an acknowledgment is 100 bytes and the transmission rate at the receiver is 8 Kbps. The one-way propagation delay is 100 milliseconds. Assuming no frame is lost, the sender throughput is ______ bytes/second.
[2016 (Set-1) : 2 M]

2.23 In an Ethernet local area network, which one of the following statements is **TRUE**?
(a) A station stops to sense the channel once it starts transmitting a frame.
(b) The purpose of the jamming signal is to pad the frames that are smaller than the minimum frame size.
(c) A station continues to transmit the packet even after the collision is detected.
(d) The exponential backoff mechanism reduces the probability of collision on retransmissions.
[2016 (Set-2) : 2 M]

2.24 A network has a data transmission bandwidth of 20×10^6 bits per second. It uses CSMA/CD in the MAC layer. The maximum signal propagation time from one node to another node is 40 microseconds. The minimum size of a frame in the network is ______ bytes.
[2016 (Set-2) : 2 M]

2.25 Consider a 128×10^3 bits/second satellite communication link with one way propagation delay of 150 milliseconds. Selective retransmission (repeat) protocol is used on this link to send data with a frame size of 1 kilobyte. Neglect the transmission time of acknowledgement. The minimum number of bits required for the sequence number field to achieve 100% utilization is ______.
[2016 (Set-2) : 2 M]

2.26 For the IEEE 802.11 MAC protocol for wireless communication, which of th following statements is/are **TRUE**?
I. At least three non-overlapping channels are available for transmission.
II. The RTS-CTS mechanism is used for collision detection.
III. Unicast frames are ACKed
(a) All I, II, and III (b) I and III only
(c) II and III only (d) II only
[2016 (Set-2) : 2 M]

2.27 The values of parameters for the Stop-and-Wait ARQ protocol are as given below:
- Bit rate of the transmission channel = 1 Mbps.
- Propagation delay from sender to receiver = 0.75 ms.
- Time to process a frame = 0.25 ms.
- Number of bytes in the information frame = 1980.
- Number of bytes in the acknowledge frame = 20.
- Number of overhead bytes in the information frame = 20.

Assume that there are no transmission errors. Then, the transmission efficiency (expressed in percentage) of the Stop-and-Wait ARQ protocol for the above parameters is ______ (correct to 2 decimal places). **[2017 (Set-1) : 2 M]**

2.28 A computer network uses polynomials over GF(2) for error checking with 8 bits as information bits and uses $x^3 + x + 1$ as the generator polynomial to generate the check bits. In this network, the message 01011011 is transmitted as
(a) 01011011010 (b) 01011011011
(c) 01011011101 (d) 01011011100
[2017 (Set-1) : 2 M]

2.29 Consider a binary code that consists of only four valid codewords as given below:
00000, 01011, 10101, 11110
Let the minimum Hamming distance of the code be p and the maximum number of erroneous bits that can be corrected by the code be q. Then the values of p and q are
(a) $p = 3$ and $q = 1$ (b) $p = 3$ and $q = 2$
(c) $p = 4$ and $q = 1$ (d) $p = 4$ and $q = 2$
[2017 (Set-2) : 2 M]

2.30 Consider a simple communication system where multiple nodes are connected by a shared broadcast medium (like Ethernet or wireless). The nodes in the system use the following carrier-sense based medium access protocol. A node that receives a packet to transmit will carrier-sense the medium for 5 units of time. If the node does not detect any other transmission in this duration, it starts transmitting its packet in the next time unit. If the node detects another transmission, it waits until this other transmission finishes, and then begins to carrier-sense for 5 time units again. Once they start to transmit, nodes do not perform any collision detection and continue transmission even if a collision occurs. All transmissions last for 20 units of time. Assume that the transmission signal travels at the speed of 10 meters per unit time in the medium.

Assume that the system has two nodes P and Q, located at a distance d meters from each other. P starts transmitting a packet at time $t = 0$ after successfully completing its carrier-sense phase. Node Q has a packet to transmit at time $t = 0$ and begins to carrier-sense the medium.

The maximum distance d (in meters, rounded to the closest integer) that allows Q to successfully avoid a collision between its proposed transmission and P's ongoing transmission is ________.

[2018 : 2 M]

2.31 Consider that 15 machines need to be connected in a LAN using 8-port Ethernet switches. Assume that these switches do not have any separate uplink ports. The minimum number of switches needed is ________.

[2019 : 2 M]

2.32 Suppose that in an IP-over-Ethernet network, a machine X wishes to find the MAC address of another machine Y in its subnet. Which one of the following techniques can be used for this?

(a) X sends an ARP request packet with broadcast IP address in its local subnet
(b) X sends an ARP request packet to the local gateway's MAC address which then finds the MAC address of Y and sends to X
(c) X sends an ARP request packet with broadcast MAC address in its local subnet
(d) X sends an ARP request packet to the local gateway's IP address which then finds the MAC address of Y and sends to X

[2019 : 2 M]

2.33 Consider a network using the pure ALOHA medium access control protocol, where each frame is of length 1,000 bits. The channel transmission rate is 1 Mbps (= 10^6 bits per second). The aggregate number of transmissions across all the nodes (including new frame transmissions and retransmitted frames due to collisions) is modelled as a Poisson process with a rate of 1,000 frames per second, Throughput is defined as the average number of frames successfully transmitted per second. The throughput of the network (rounded to the nearest integer) is ________.

[2021 (Set-2) : 2 M]

2.34 Node X has a TCP connection open to node Y. The packets from X to Y go through an intermediate IP router R. Ethernet switch S is the first switch on the network path between X and R. Consider a packet sent from X to Y over this connection.

Which of the following statements is/are TRUE about the destination IP and MAC addresses on this packet at the time it leaves X?

(a) The destination IP address is the IP address of Y.
(b) The destination MAC address is the MAC address of Y.
(c) The destination MAC address is the MAC address of S.
(d) The destination IP address is the IP address of R.

[2024 (Set-2) : 1 M]

2.35 Consider an Ethernet segment with a transmission speed of 10^8 bits/sec and a maximum segment length of 500 meters. If the speed of propagation of the signal in the medium is 2×10^8 meters/sec, then the minimum frame size (in bits) required for collision detection is ________.

[2024 (Set-2) : 2 M]

■■■■

Answers LAN

2.1 (d)	2.2 (d)	2.3 (b)	2.4 (b)	2.5 (b)	2.6 (c)	2.7 (d)	2.8 (a)	2.9 (c)
2.10 (b)	2.11 (a)	2.12 (b)	2.13 (a)	2.14 (c)	2.15 (b)	2.16 (a)	2.17 (b)	2.18 (0.4404)
2.19 (d)	2.20 (a)	2.21 (c)	2.22 (2500)	2.23 (d)	2.24 (200)	2.25 (4)	2.26 (b)	2.27 (89.33)
2.28 (c)	2.29 (a)	2.30 (50)	2.31 (3)	2.32 (c)	2.33 (135)	2.34 (a)	2.35 (500)	

Explanations LAN

2.1 (d)

In CSMA/CD the minimum frame size

$= 2 \times z \times$ data rate (Bandwidth)

$\because \quad z = \frac{d}{v} = \frac{\text{Distance}}{\text{Velocity}}$

$d = 2\text{ km} = 2 \times 10^3\text{ m}$

$v = 2 \times 10^8\text{ m/sec.}$

Data rate $= 10^7$ bps

$$z = \frac{d}{v} = \frac{2 \times 10^3 \text{m}}{2 \times 10^8 \text{m/sec}} = 10^{-5}\text{ sec}$$

$\therefore$ Minimum packet size $= 2 * 10^{-5} * 10^7$

$= 2 * 10^2$

$= 200 \text{ bits} = \frac{200}{8}$

$= 25$ bytes

2.2 (d)

Address used in ethernet is MAC address, also known as ethernet address. MAC address of 2 host in any network cannot be same, since MAC address assigned by manufacture in such a way that no two system has same MAC. So largest network will be internet, because MAC of two system always unique.

2.3 (b)

Consider 'n' host are possible.

In slotted aloha, only one sender send data.

So, probability of successful transmission

$= (1 - 0.2)^{n-1}(0.2)$

Through put $= 0.16$ frame

So, for what value of n, through put = probability of successful transmission.

$0.16 = (1 - 0.2)^{n-1}(0.2)$

$(0.8)^1 = (0.8)^{n-1}$

Comparing powers $\Rightarrow n - 1 = 1$

$n = 2$

2.4 (b)

When both station A and B transmit and collied:

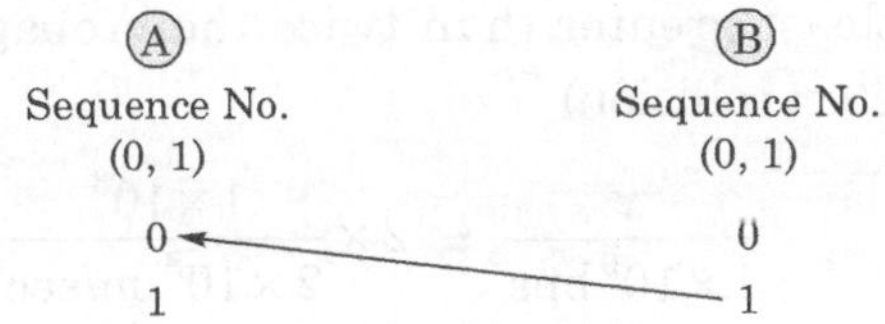

Probability of A win 1st backoff race $= \frac{1}{4} = 0.25$

When again A and B transmit and collied

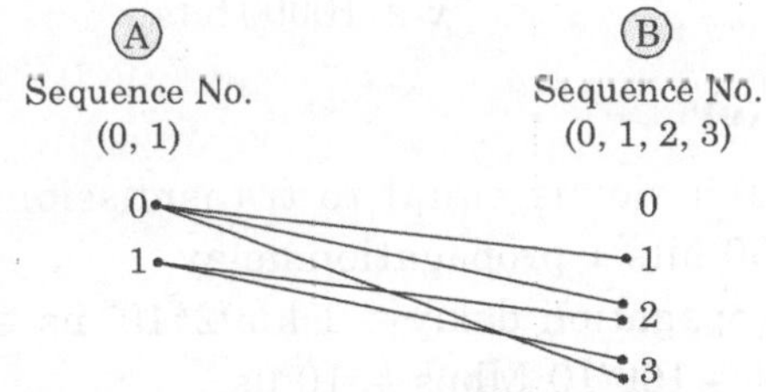

Probability of A win 2nd backoff race $= \frac{5}{8} = 0.625$

2.5 (b)

The spanning tree approach is a mechanism in which bridges automatically develop a routing table and update that table in response to changing topology. The algorithm consists of three mechanisms

1. Frame Forwarding
2. Address Learning
3. Loop resolution.

2.6 (c)

- IEEE 802.11 wireless LAN run CSMA/CA instead of CSMA/CD, while ethernet uses CSMA/CD.
- In CSMA/CD network more one station can transmit data at a same time, hance contention possible.
- CSMA/CD is not used for network which has high propagation delay since, efficiency will be very less when collision is happen.

2.7 (d)

Bridge is a 2 layer device i.e. work at physical and data link layer. It can reduce collision domain but cannot reduce broadcast domain. Bridge can connect more than 1 LAN.
Note : Hub is neither collision nor broadcast domain separator.

2.8 (a)

In CSMA/CD minimum frame size calculated as: T.T ≥ 2 P.T. (Since transmission time of frame atleast greater than twice the propagation to detect collision)

$$\frac{x}{1\times10^9\text{ bps}} = 2\times\frac{1\times10^3}{2\times10^8\text{ m/sec}}$$

$$x \geq \frac{2\times10^3\times10^9\text{ bpsec}}{2\times10^8\text{ m/sec}}$$

$$x \geq 10^4\text{ bits}$$

$$x \geq 10000\text{ bits}$$

2.9 (c)

Each slot is equal to transmission time of 100 bits + propagation delay.
Propagation delay = 1 km/2*10⁸ms = 5 μs.
T_x = 100/10 Mbps = 10 μs.
Let there are maximum N number of station then Length of cycle is = N*(10+5) = 15N μs.
In a whole cycle each user transmit for only 10 μs.
Therefore efficiency is (10/15N).
Throughput of each station is (10/15N)* 10 Mbps which is given as 2/3 Mbps
N = (10*10*3)/(15*2) = 10

2.10 (b)

It is given that 2 full duplex ethernet interfaces, each operating at 100 Mbps. So, we are needed that packets are comes at speeds of 200 Mbps. Since, transmission of 84 byte long packet takes 84 B/100 Mbps = 6.72 μsec on single interface. Now if we get 2 packets of processing times 6.72 μsec, we cannot manage both packets in given time, which results overflow of incomming queue.
So, 6.72/2 = 3.36 μsec processing time possible to forward packet from router interface and till 6.72 μsec we send 2 packets one at 3.36 μsec and second at 6.72 μsec.

2.11 (a)

In Manchester encoding, two signal changes to represent a bit. Therefore baud rate (number of signals/sec) = 2*bit rate (number of bits/sec).
Hence, bit rate is half the baud rate.

2.12 (b)

Transmission time for one frame

$$= \frac{1000\text{ B}}{10\text{ Mbps}} = \frac{8000}{10\times10^6}$$

In polling mechanism, each station involve in polling and one station will be elected to transmit data.
So, total time to transmit frame

$$= 800\ \mu\text{ sec} + 80\ \mu\text{ sec}$$
$$= 880\ \mu\text{ sec}$$

$$\text{Through put} = \frac{10\times800\ \mu\text{sec}}{880\ \mu\text{sec}}$$

$$= \frac{800}{88} = \frac{100}{11}\text{ Mbps}$$

2.13 (a)

If there are n stations in a slotted LAN. Each station attempts to transmit with a probability p in each time slot and therefore acquires the medium. This is just the binomial probability distribution that any one station attempts to transmit and the others do not. Let the event is denoted by P(E)

$$P(E) = {}^nC_1\, p^1(1-p)^{n-1}$$

$$= \frac{\lfloor n}{\lfloor 1\,\lfloor n-1}\, p(1-p)^{n-1}$$

$$= \frac{n\,\lfloor n-1}{\lfloor n-1}\, p(1-p)^{n-1}$$

$$= np(1-p)^{n-1}$$

2.14 (c)

Frame size K bit long
Propagation delay t sec/km
Channel capacity = R bits/sec

$$U = \frac{w\cdot\frac{K}{R}\text{sec}}{\frac{K}{R}\text{sec} + 2\text{ Lt}}$$

$$1 = \frac{\frac{wK}{R}\sec}{\frac{K + 2\,LtR}{R}},\ w = \frac{K + 2\,LtR}{K}$$

$$2^n = \frac{K + 2\,LtR}{K},\ n = \left\lceil \log_2 \frac{K + 2\,LtR}{K} \right\rceil$$

2.15 (b)

In CSMA/CD :

T.T. ≥ 2.P.T. (To detect the collision, frame should be transmitted atleast 2PT)

$$\frac{x}{1\times10^9\text{ bps}} \geq 2\times\left(\frac{200\text{ m}}{2\times10^8\text{ m/sec}}\right)$$

$$x \geq 2\times\left(\frac{200\text{ m}}{2\times10^8\text{ m/sec}}\times1\times10^9\text{ bps}\right)$$

$$x \geq \frac{400\times10^9}{2\times10^8}\text{bits}$$

$$x \geq 200\times10\text{ bits}$$

$$x \geq \frac{2000}{8}\text{ bytes}$$

$$x \geq 250\text{ bytes}$$

2.16 (a)

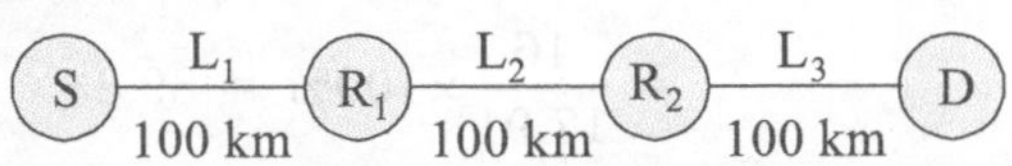

Propagation delay to travel from S to R_1

$$= \frac{\text{Distance}}{\text{Link Speed}} = \frac{10^5}{10^8} = 1\text{ ms}$$

Total propagation delay to travel from S to D = 3 × 1 ms = 3 ms

Total transmission delay for 1 packet

$$= 3\times\frac{(\text{Number of bits})}{\text{Bandwidth}}$$

$$= 3\times\frac{(1000)}{10^6} = 3\text{ ms}$$

So the first packet will take 6 ms to reach D. While first packet was reaching D, other packets must have been processing in parallel. So D will receive remaining packets 1 packet per 1 ms from R_2. So remaining 999 packets will take 999 ms and total time will be 999 + 6 = 1005 ms.

2.17 (b)

$$\frac{10000\text{ bit}}{500\times10^6\text{ bits/sec}} = \frac{2\times L}{2\times10^5\text{ km/sec}}$$

$$L = 2\text{ km}$$

2.18 (0.4404)

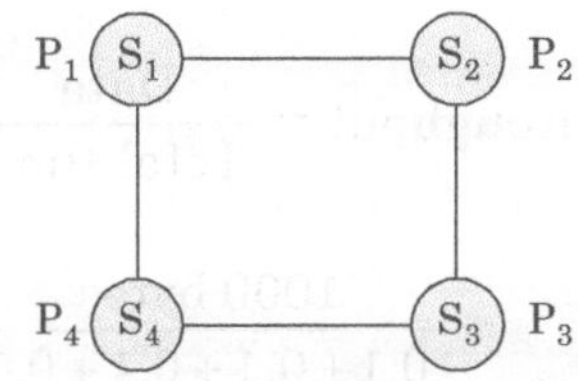

P_i is the probability of frame being generated by station S_i

Required probability = Only one station generation frames

$$= (P_1\cdot\bar{P}_2\cdot\bar{P}_3\cdot\bar{P}_4) + (\bar{P}_1\cdot P_2\cdot\bar{P}_3\cdot\bar{P}_4) + (\bar{P}_1\cdot\bar{P}_2\cdot P_3\cdot\bar{P}_4) + (\bar{P}_1\cdot\bar{P}_2\cdot\bar{P}_3\cdot P_4)$$

= (0.1 × 0.8 × 0.7 × 0.6) + (0.9 × 0.2 × 0.7 × 0.6) + (0.9 × 0.8 × 0.3 × 0.6) + (0.9 × 0.8 × 0.7 × 0.4)

= 0.0336 + 0.0756 + 0.1296 + 0.2016 = 0.4404

2.19 (d)

Band width = 100 × 10^6 bps

d = 1000 m

For CSMA/CD: $t_d \geq 2P_d$

$$\frac{1250\times8}{100\times10^6} = 2\times\frac{1000}{x\text{ m/s}}$$

$$x = \frac{2\times1000\times100\times10^6}{1250\times8}$$

= 2 × 10^7 = 20000 (km/sec)

2.20 (a)

- DNS is used for mapping host name to IP address.
- ARP is address resulution protocol used to map IP address with MAC address.
- RARP is reverse address resolution protocol used to map MAC address with IP address.
- DHCP is also used to map MAC address with IP address.

Since all options are used to find one address to another address but option (c) is most appropriate answer.

2.21 (c)

FTP has control and data connection it requires authorization. HTTP is stateless protocol. TCP is not application layer but it is statefull. POP3 is application protocol and it gets state with help of TCP.

2.22 (2500)

$$\text{Sender throughput} = \frac{\text{Data}}{\text{Total time}}$$

$$= \frac{1000 \text{ bytes}}{0.1+0.1+0.1+0.1}$$

$$= 2500 \text{ bytes/sec}$$

2.23 (d)

Exponential back of algorithm reduce the possibility of collisions in next iteration.

2.24 (200)

Bandwidth = 20×16^6 bps
Propagation time = 40 μsec
For ethernet,
Transmission time = 2 × Propagation time

$$\frac{\text{Frame size}}{20 \times 10^6 \text{bits/sec}} = 2 \times 40 \ \mu\text{sec}$$

$$\text{Frame size} = 20 \times 10^6 \times 2 \times 40 \times 10^{-6}$$

$$= 1600 \text{ bits}$$

$$= \frac{1600}{8} = 200 \text{ bytes}$$

2.25 (4)

$$\text{Transmission time} = \frac{\text{Frame size}}{\text{Bandwidth}}$$

$$\text{T.T.} = \frac{1024 \times 8}{128 \times 10^3} = \frac{1028}{16} \text{msec}$$

$$= 64 \text{ msec } [\because 1 \text{ K} = 2^{10}]$$

$$\text{Efficiency} = \text{w.s.} \times \frac{\text{T.T.}}{\text{T.T.}+2\text{P.T.}}$$

$$100\% = \text{w.s.} \times \frac{1}{1+2\dfrac{\text{P.T.}}{\text{T.T.}}}$$

$$1+2\frac{\text{P.T.}}{\text{T.T.}} = \text{w.s.}$$

$$1+2\left[\frac{150}{64}\right] = \text{w.s.}$$

$$1 + 2[2.34] = \text{w.s.}$$

$$\text{w.s.} = \lceil 5.68 \rceil = 6$$

In selective repeate ARQ.

Total window size ≥ $\log_2$ (S.w.s. + R.w.s.) and Sender w.s. = Reciever w.s.
So, total w.s. ≥ $\log_2 [6 + 6] = \log_2[12] = 4$

2.26 (b)

RTS-CTS is used for collision avoidance but not for collision detection.

2.27 (89.33)

Processing time of frame = 0.25 msec

$$\text{Transmission time of data} = \frac{\text{Data size}}{\text{Bandwidth}}$$

$$= \frac{(1980 + 20) \text{ bytes}}{10^6 \text{ bits/sec}} = 16 \text{ msec}$$

Propagation time = 0.75 msec
Transmission time of ACK

$$= \frac{20 \times 8 \text{ bits}}{10^6 \text{ bits/sec}} = 0.16 \text{ msec}$$

Transmission efficiency of stop and wait ARQ

$$= \frac{16}{0.25+16+1.5+0.16}$$

$$= \frac{16}{17.91} \times 100\% = 89.33\%$$

2.28 (c)

CRC generator = $x^3 + x + 1$

$$= 1.x^3 + 0.x^2 + 1.x^1 + 1.x^0$$

$$= 1011$$

Data = 01011011

```
1011 ) 01011011000 ( 01000011
       0000↓
        1011
        1011↓
         0000
         0000↓
          0001
          0000↓
           0011
           0000↓
            0110
            0000↓
             1100
             1011↓
              1110
              1011
               101
```

So the message 01011011 is transmitted as 01011011101.

2.29 (a)

Given codewords

00000, 01011, 10101, 11110

Minimum hamming distance = Min(hamming distance between every 2 pair) = 3

To correct errors, the condition is

$2d + 1 = 3$

$\Rightarrow \quad d = 1$

2.30 (50)

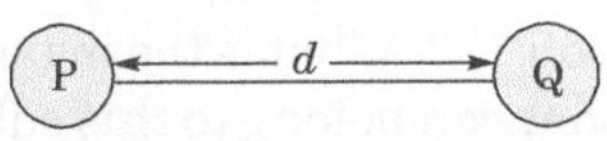

- At time $t = 0$, station P start transmitting packet (since it already sense the channel free). At the same time station Q sensing the channel to check channel is free or not.
- **To avoid collision:** Sensing of node Q ≥ Transmitting a packet
  i.e. if channel is found free at 5 unit time or more than 5 unit time, station Q data will be collied with station P any where in middle.
  So, distance between P and Q = sensing of node Q ≥ transmitting a packet.
  $d = 5$ time unit
  $d = 5$ unit × 10 meter/unit
  $d = 50$ meter

Note: Idea is simple, 1st bit of node P data reach to node Q with in 5 time units or at exact 5 time unit, then node Q understand that someone is transmitting the data and wait for next time unit to sense the channel is free or not.

2.31 (3)

3 switches of ethernet are required to connect 15 computers.

2.33 (135)

1 frames takes = $Tt = \frac{L}{\text{B.W.}}$

$\Rightarrow \quad \frac{1000}{10^6} = 1$ millisec

1000 frame $Tt = 1000 \times 1$ millisec = 1 sec

In 1 sec, 1000 frames sends, which is 1 millisec per frame.

So, $G = 1$

Efficiency of Pure Aloha (η) = $G \times e^{-2G}$

where G = Number of requests per time slot willing to transmit

e = Mathematical constant approximately equal to 2.718

So, $\eta = 1 \times 2.718^{(-2 \times 1)} = 0.1353$

Therefore, in 1 sec 1000 frames = 0.1353 × 1000 = 135.3 (closest integer) ⇒ 135

Throughput ⇒ 135

■■■■

3 TCP, UDP and IP

3.1 Which of the following assertions is FALSE about the Internet Protocol (IP)?
(a) It is possible for a computer to have multiple IP addresses
(b) IP packets from the same source to the same destination can take different routes in the network
(c) IP ensures that a packet is farwarded if it is unable to reach its destination within a given number of hopes
(d) The packet source cannot set the route of an outgoing packets; the route is determined only by the routing tables in the routers on the way **[2003 : 1 M]**

3.2 Which of the following functionalities must be implemented by a transport protocol over and above the network protocol?
(a) Recovery from packet losses
(b) Detection of duplicate packets
(c) Packet delivery in the correct order
(d) End to end connectivity
[2003 : 1 M]

3.3 The subnet mask for a particular network is 255.255.31.0 Which of the following pairs of IP addresses could belong to this network?
(a) 172.57.88.62 and 172.56.87.23.2
(b) 10.35.28.2 and 10.35.29.4
(c) 191.203.31.87 and 191.234.31.88
(d) 128.8.129.43 and 128.8.161.55
[2003 : 2 M]

3.4 Which of the following is NOT true with respect to a transparent bridge and a router?
(a) Both bridge and router selectively farward data packets
(b) A bridge uses IP addresses while a router uses MAC addresses
(c) A bridge builds up its routing table by inspecting incoming packets
(d) A router can connect between a LAN and a WAN **[2004 : 1 M]**

3.5 Which one of the following statements is FALSE?
(a) TCP guarantees a minimum communication rate
(b) TCP ensures in-order delivery
(c) TCP reacts to congestion by reducing sender window size
(d) TCP employs retransmission to compensate for packet loss **[2004 : 1 M]**

3.6 A subnet has been assigned a subnet mask of 255.255.255.192. What is the maximum number of hosts that can belong to this subnet?
(a) 14 (b) 30
(c) 62 (d) 126 **[2004 : 1 M]**

3.7 In TCP, a unique sequence number assigned to each
(a) byte (b) word
(c) segment (d) message
[2004 : 1 M]

3.8 In the TCP/IP protocol suite, which one of the following is NOT part of the IP header?
(a) Fragment Offset
(b) Source IP address
(c) Destination IP address
(d) Destination port number **[2004 : 2 M]**

3.9 A TCP message consisting of 2100 bytes is passed to IP for delivery across two networks. The first network can carry a maximum payload of 1200 bytes per frame and the second network can carry a maximum payload of 400 bytes per frame, excluding network overhead. Assume that IP overhead per packet is 20 bytes. What is the total IP overhead in the second network for this transmission?
(a) 40 bytes (b) 80 bytes
(c) 120 bytes (d) 160 bytes
[2004 : 2 M]

3.10 Suppose that the maximum transmit window size for a TCP connection is 12000 bytes. Each packet consists of 2000 bytes. At some point of time, the connection is in slow-start phase with a current transmit window of 4000 bytes. Subsequently, the transmitter receives two acknowledgement. Assume that no packets are lost and there are no time-outs. What is the maximum possible value of the current transmit window?
(a) 4000 bytes (b) 8000 bytes
(c) 10000 bytes (d) 12000 bytes
[2004 : 2 M]

3.11 The routing table of a router is shown below:

Destination	Subnet Mask	Interface
128.75.43.0	255.255.255.0	Eth0
128.75.43.0	255.255.255.128	Eth1
192.12.17.5	255.255.255.255	Eth3
default		Eth2

On which interfaces will the router farward packets addressed to destinations 128.75.43.16 and 192.12.17.10 respectively?

(a) Eth1 and Eth2
(b) Eth0 and Eth2
(c) Eth0 and Eth3
(d) Eth1 and Eth3 **[2004 : 2 M]**

Common Data Questions Q. 3.12 & Q. 3.13:

Consider three IP networks A, B and C. Host H_A in network A sends messages each containing 180 bytes of application data to a host H_C in network C. The TCP layer prefixes a 20 byte header to the message. This passes through an intermediate network B. The maximum packet size, including 20 byte IP header, in each network is

A: 1000 bytes
B: 100 bytes
C: 1000 bytes

The network A and B are connected through a 1 Mbps link, while B and C are connected by a 512 Kbps link (bps = bits per second).

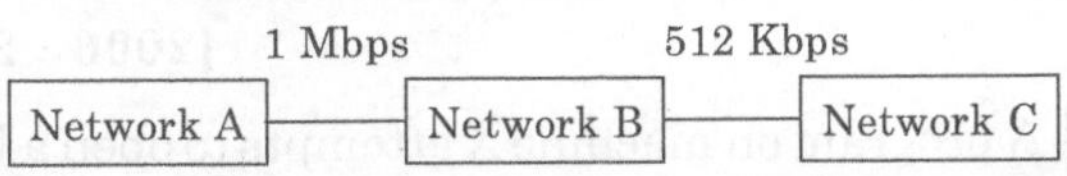

3.12 Assuming that the packets are correctly delivered, how many bytes, including headers, are delivered to the IP layer at the destination for one application message, in the best case? Consider only data packets.

(a) 200 (b) 220
(a) 240 (d) 260 **[2004 : 2 M]**

3.13 What is the rate at which application data is transferred to host H_c? Ignore errors, acknowledgments, and other overheads.

(a) 325.5 Kbps (b) 354.5 Kbps
(c) 409.6 Kbps (d) 512.0 Kbps **[2004 : 2 M]**

3.14 Packets of the same session may be routed through different paths in

(a) TCP, but not UDP
(b) TCP and UDP
(c) UDP, but not TCP
(d) Neither TCP, nor UDP **[2005 : 1 M]**

3.15 The address resolution protocol (ARP) is used for

(a) Finding the IP address from the DNS
(b) Finding the IP address of the default gateway
(c) Finding the IP address that corresponds to a MAC address
(d) Finding the MAC address that corresponds to an IP address **[2005 : 1 M]**

3.16 An organization has a class B network and wishes to form subnets for 64 departments. The subnet mask would be

(a) 255.255.0.0 (b) 255.255.64.0
(c) 255.255.128.0 (d) 255.255.252.0 **[2005 : 1 M]**

3.17 In a packet switching network, packets are routed from source to destination along a single path having two intermediate nodes. If the message size is 24 bytes and each packet contains a header of 3 bytes, then the optimum packet size is

(a) 4 (b) 6
(c) 7 (d) 9 **[2005 : 2 M]**

3.18 Suppose the round trip propagation delay for a 10 Mbps Ethernet having 48-bit jamming signal is 46.4 μs. The minimum frame size is:

(a) 94 (b) 416
(c) 464 (d) 512 **[2005 : 2 M]**

3.19 On a TCP connection, current congestion window size is Congestion Window = 4 KB. The window size advertised by the receiver is Advertise Window = 6 KB. The last byte sent by the sender is LastByteSent = 10240 and the last byte acknowledged by the receiver is LastByteAcked = 8192. The current window size at the sender is

(a) 2048 bytes (b) 4096 bytes
(c) 6144 bytes (d) 8192 bytes **[2005 : 2 M]**

3.20 In a communication network, a packet of length L bits takes link L1 with a probability of p_1 or link L2 with a probability of p_2. Link L1 and L2 have bit error probability of b_1 and b_2 respectively. The probability that the packet will be received without error via either L1 or L2 is

(a) $(1 - b_1)^L p_1 + (1 - b_2)^L p_2$
(b) $[1 - (b_1 + b_2)^L]p_1p_2$
(c) $(1 - b_1)^L (1 - b_2)^L p_1p_2$
(d) $1 - (b_1^L p_1 + b_2^L p_2)$ **[2005 : 2 M]**

3.21 A company has a class C network address of 204.204.204.0. It wishes to have three subnets, one with 100 hosts and two with 50 hosts each.

Which one of the following options represents a feasible set of subnet address/subnet mask pairs?

(a) 204.204.204.128/255.255.255.192
204.204.204.0/255.255.255.128
204.204.204.64/255.255.255.128

(b) 204.204.204.0/255.255.255.192
204.204.204.192/255.255.255.128
204.204.204.64/255.255.255.128

(c) 204.204.204.128/255.255.255.128
204.204.204.192/255.255.255.192
204.204.204.224/255.255.255.192

(d) 204.204.204.128/255.255.255.128
204.204.204.64/255.255.255.192
204.204.204.0/255.255.255.192

[2005 : 2 M]

3.22 Which of the following statements is TRUE?

(a) Both Ethernet frame and IP packet include checksum fields
(b) Ethernet frame includes a checksum field and IP packet includes a CRC field
(c) Ethernet frame includes a CRC field and IP packet includes a checksum field
(d) Both Ethernet frame and IP packet include CRC fields **[2006 : 1 M]**

3.23 For which one of the following reasons does Internet Protocol (IP) use the time-to-live (TTL) field in the IP datagram header?

(a) Ensure packets reach destination within that time
(b) Discard packets that reach later than that time
(c) Prevent packets from looping indefinitely
(d) Limit the time for which a packet gets queued in intermediate routers **[2006 : 1 M]**

3.24 A router uses the following routing table:

Destination	Mask	Interface
144.16.0.0	255.255.0.0	eth0
144.16.64.0	255.255.224.0	eth1
144.16.68.0	255.255.255.0	eth2
144.16.68.64	255.255.255.224	eth3

A packet bearing a destination address 144.16.68.117 arrives at the router. On which interface will it be forwarded?

(a) eth0 (b) eth1
(c) eth2 (d) eth3 **[2006 : 2 M]**

3.25 Suppose that it takes 1 unit of time to transmit a packet (of fixed size) on a communication link. The link layer uses a window flow control protocol with a window size of N packets. Each packet causes an ack or a nak to be generated by the receiver, and ack/nak transmission times are negligible. Further, the round trip time on the link is equal to N units. Consider time i > N. If only acks have been received till time i(no naks), then the throughput evaluated at the transmitter at time i(in packets per unit time) is

(a) $\frac{1-N}{i}$ (b) $\frac{i}{(N+i)}$
(c) 1 (d) $1 - e^{(i/N)}$

[2006 : 2 M]

3.26 A link of capacity 100 Mbps is carrying traffic from a number of sources. Each source generates an on-off traffic stream; when the source is on, the rate of traffic is 10 Mbps, and when the source is off, the rate of traffic is zero. The duty cycle, which is the ratio of on-time to off-time, is 1 : 2. When there is no buffer at the link, the minimum number of sources that can be multiplexed on the link so that link capacity is not wasted and no data loss occurs is S_1. Assuming that all sources are synchronized and that the link is provided with a large buffer, the maximum number of sources that can be multiplexed so that no data loss occurs is S_2. The values of S_1 and S_2 are, respectively,

(a) 10 and 30 (b) 12 and 25
(c) 5 and 33 (d) 15 and 22

[2006 : 2 M]

3.27 A program on machine X attempts to open a UDP connection to port 5376 on a machine Y, and a TCP connection to port 8632 on machine Z. However, there are no applications listening at the corresponding ports on Y and Z. An ICMP Port Unreachable error will be generated by

(a) Y but not Z
(b) Z but not Y
(c) Neither Y nor Z
(d) Both Y and Z

[2006 : 2 M]

3.28 A subnetted Class B network has the following broadcast address : 144.16.95.255. Its subnet mask

(a) is necessarily 255.255.224.0
(b) is necessarily 255.255.240.0
(c) is necessarily 255.255.248.0.
(d) could be any one of 255.255.224.0, 255.255.240.0, 255.255.248.0

[2006 : 2 M]

3.29 Two computers C1 and C2 are configured as follows. C1 has IP address 203. 197.2.53 and netmask 255.255.128.0. C2 has IP address 203.197.75.201 and netmask 255.255.192.0. Which one of the following statements is true?

(a) C1 and C2 both assume they are on the same network
(b) C2 assumes C1 is on same network, but C1 assumes C2 is on a different network
(c) C1 assumes C2 is on same network, but C2 assumes C1 is on a different network
(d) C1 and C2 both assume they are on different networks **[2006 : 2 M]**

3.30 Consider the following statements about the timeout value used in TCP.

(i) The timeout value is set to the RTT (Round Trip Time) measured during TCP connection establishment for the entire duration of the connection.
(ii) Appropriate RTT estimation algorithm is used to set the timeout value of a TCP connection.
(iii) Timeout value is set to twice the propagation delay from the sender to the receiver.

Which of the following choices hold?

(a) (i) is false, but (ii) and (iii) are true
(b) (i) and (iii) are false, but (ii) is true
(c) (i) and (ii) are false, but (iii) is true
(d) (i), (ii) and (iii) are false **[2007 : 1 M]**

3.31 Consider a TCP connection in a state where there are no outstanding ACKs. The sender sends two segments back to back. The sequence numbers of the first and second segments are 230 and 290 respectively. The first segment was lost, but the second segment was received correctly by the receiver.

Let X be the amount of data carried in the first segment (in bytes), and Y be the ACK number sent by the receiver. The values of X and Y (in that order) are

(a) 60 and 290 (b) 230 and 291
(c) 60 and 231 (d) 60 and 230 **[2007 : 1 M]**

3.32 The address of a class B host is to be split into subnets with a 6-bit subnet number. What is the maximum number of subnets and the maximum number of hosts in each subnet?

(a) 62 subnets and 262142 hosts
(b) 64 subnets and 262142 hosts
(c) 62 subnets and 1022 hosts
(d) 64 subnets and 1024 hosts **[2007 : 2 M]**

3.33 What is the maximum size of data that the application layer can pass on to the TCP layer below?

(a) Any size
(b) 2^{16} bytes – size of TCP header
(c) 2^{16} bytes
(d) 1500 bytes **[2008 : 1 M]**

3.34 Which of the following system calls results in the sending of SYN packets?

(a) socket (b) bind
(c) listen (d) connect **[2008 : 1 M]**

3.35 Which of the following statements are TRUE?

S1: TCP handles both congestion and flow control
S2: UDP handles congestion but not flow control
S3: Fast retransmit deals with congestion but not flow control
S4: Slow start mechanism deals with both congestion and flow control

(a) S1, S2 and S3 only
(b) S1 and S3 only
(c) S3 and S4 only
(d) S1, S3 and S4 only **[2008 : 2 M]**

3.36 The three way handshake for TCP connection establishment is shown below:

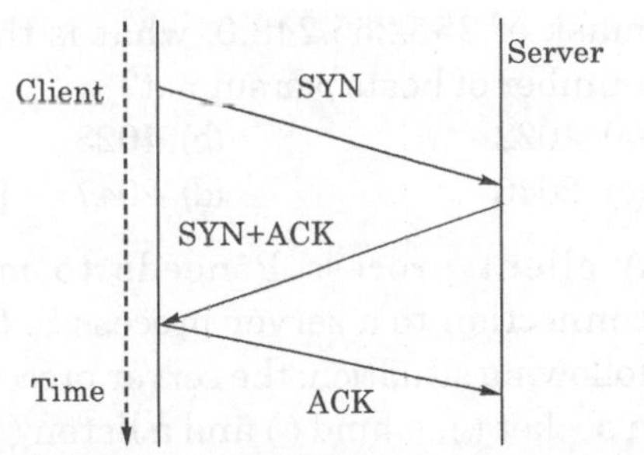

Which of the following statements are TRUE?

S1: Loss of SYN + ACK from the server will not establish a connection
S2: Loss of ACK from the client cannot establish the connection
S3: The server moves LISTEN → SYN_RCVD → SYN_SENT → ESTABLISHED in the state machine on no packet loss
S4: The server moves LISTEN → SYN_RCVD → ESTABLISHED in the state machine on no packet loss.

(a) S2 and S3 only
(b) S1 and S4 only
(c) S1 and S3 only
(d) S2 and S4 only

[2008 : 2 M]

Directions for Question 3.37 to 3.38:

Host X has IP address 192.168.1.97 and is connected through two routers R1 and R2 to another host Y with IP address 192.168.1.80. Router R1 has IP addresses 192.168.1.135 and 192.168.1.110. R2 has IP addresses 192.168.1.67 and 192.168.1.155. The netmask used in the network is 255.255.255.224.

3.37 Given the information above, how many distinct subnets are guaranteed to already exist in the network?
(a) 1 (b) 2
(c) 3 (d) 6 **[2008 : 2 M]**

3.38 Which IP address should X configure its gateway as?
(a) 192.168.1.67 (b) 192.168.1.110
(c) 192.168.1.135 (d) 192.168.1.155
[2008 : 2 M]

3.39 In the slow start phase of the TCP congesting control algorithm, the size of the congestion window
(a) Does not increase
(b) Increases linearly
(c) Increases quadratically
(d) Increases exponentially **[2008 : 2 M]**

3.40 If a class B network on the Internet has a subnet mask of 255.255.248.0, what is the maximum number of hosts per subnet?
(a) 1022 (b) 1023
(c) 2046 (d) 2047 **[2008 : 2 M]**

3.41 A client process P needs to make a TCP connection to a server process S. Consider the following situation: the server process S executes a socket ( ), a bind ( ) and a listen ( ) system call in that order, following which it is preempted. Subsequently, the client process P executes a socket ( ) system call followed by connect ( ) system call to connect to the server process S. The server process has not executed any accept ( ) system call. Which one of the following events could take place?
(a) connect( ) system call returns successfully
(b) connect( ) system call blocks
(c) connect( ) system call returns an error
(d) connect( ) system call results in a core dump
[2008 : 2 M]

3.42 While opening a TCP connection, the initial sequence number is to be derived using a time-of-day (ToD) clock that keeps running even when the host is down. The low order 32 bits of the counter of the ToD clock is to be used for the initial sequence numbers. The clock counter increments once per millisecond. The maximum packet lifetime is given to be 64s.
Which one of the choices given below is closest to the minimum permissible rate at which sequence numbers used for packets of a connection can increase?
(a) 0.015/s (b) 0.064/s
(c) 0.135/s (d) 0.327/s
[2009 : 2 M]

3.43 One of the header fields in an IP datagram is the Time-to-Live (TTL) field. Which of the following statements best explains the need for this field?
(a) It can be used to prioritize packets
(b) It can be used to reduce delays
(c) It can be used to optimize throughput
(d) It can be used to prevent packet looping
[2010 : 1 M]

3.44 Suppose computers A and B have IP addresses 10.105.1.113 and 10.105.1.91 respectively and they both use the same netmask N. Which of the values of N given below should not be used if A and B should belong to the same network?
(a) 255.255.255.0
(b) 255.255.255.128
(c) 255.255.255.192
(d) 255.255.255.224
[2010 : 2 M]

3.45 A layer-4 firewall (a device that can look at all protocol headers up to the transport layer) CANNOT
(a) block entire HTTP traffic during 9:00 pm and 5:00 am
(b) block all ICMP traffic
(c) stop incoming traffic from a specific IP address but allow outgoing traffic to the same IP address
(d) block TCP traffic from a specific user on a multi-user system during 9:00 pm and 5:00 am **[2011 : 1 M]**

3.46 Which of the following transport layer protocols is used to support electronic mail?
(a) SMTP (b) IP
(c) TCP (d) UDP **[2012 : 1 M]**

3.47 In the IPv4 addressing format, the number of networks allowed under Class C addresses is
(a) 2^{14} (b) 2^{7}
(c) 2^{21} (d) 2^{24} **[2012 : 1 M]**

3.48 An Internet Service Provider (ISP) has the following chunk of CIDR-based IP addresses available with it : 245.248.128.0/20. The ISP wants to give half of this chunk of addresses to Organization A, and a quarter of Organization B, while retaining the remaining with itself. Which of the following is a valid allocation of addresses to A and B?
(a) 245.248.136.0/21 and 245.248.128.0/22
(b) 245.248.128.0/21 and 245.248.128.0/22
(c) 245.248.132.0/22 and 245.248.132.0/21
(d) 245.248.136.0/24 and 245.248.132.0/21

[2012 : 2 M]

3.49 Consider an instance of TCP's Additive Increase Multiplicative Decrease (AIMD) algorithm where the windows size at the start of the slow start phase is 2 MSS and the threshold at the start of the first transmission is 8 MSS. Assume that a time-out occurs during the fifth transmission. Find the congestion window size at the end of the tenth transmission.
(a) 8 MSS (b) 14 MSS
(c) 7 MSS (d) 12 MSS

[2012 : 2 M]

3.50 In an IPv4 datagram, the M bit is 0, the value of HLEN is 10, the value of total length is 400 and the fragment offset value if 300. The position of the datagram, the sequence numbers of the first and the last bytes of the payload, respectively are
(a) Last fragment, 2400 and 2789
(b) First fragment, 2400 and 2759
(c) Last fragment, 2400 and 2759
(d) Middle fragment, 300 and 689

[2013 : 2 M]

3.51 Let the size of congestion window of a TCP connection be 32 KB when a timeout occurs. The round trip time of the connection is 100 msec and the maximum segment size used is 2 KB. The time taken (in msec) by the TCP connection to get back to 32 KB congestion window is ________.

[2014 (Set-1) : 2 M]

3.52 Consider a selective repeat sliding window protocol that uses a frame size of 1 KB to send data on a 1.5 Mbps link with a one-way latency of 50 msec. To achieve a link utilization of 60%, the minimum number of bits required to represent the sequence number field is ________.

[2014 (Set-1) : 2 M]

3.53 Which one of the following socket API functions converts an unconnected active TCP socket into a passive socket?
(a) connect (b) bind
(c) listen (d) accept

[2014 (Set-2) : 1 M]

3.54 In the diagram shown below, L1 is an Ethernet LAN and L2 is a Token-Ring LAN. An IP packet originates from sender S and traverses to R, as shown. The links within each ISP and across the two ISPs, are all point-to-point optical links. The initial value of the TTL field is 32. The maximum possible value of the TTL field when R receives the datagram is ________.

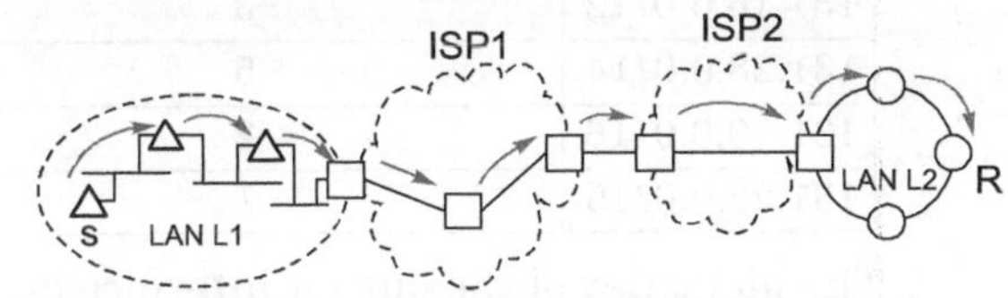

[2014 (Set-2) : 1 M]

3.55 Consider the store and forward packet switched network given below. Assume that the bandwidth of each link is 10^6 bytes/sec. A user on host *A* sends a file of size 10^3 bytes to host *B* through routers *R*1 and *R*2 in three different ways. In the first case a single packet containing the complete file is transmitted from *A* to *B*. In the second case, the file is split into 10 equal parts, and these packets are transmitted from *A* to *B*. In the third case, the file is split into 20 equal parts and these packets are sent from *A* to *B*. Each packet contains 100 bytes of header information along with the user data. Consider only transmission time and ignore processing, queuing and propagation delays. Also assume that there are no errors during transmission. Let *T*1, *T*2 and *T*3 be the times taken to transmit the file in the first, second and third case respectively. Which one of the following is CORRECT?

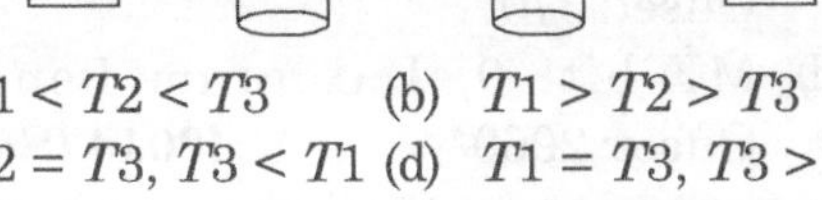

(a) $T1 < T2 < T3$ (b) $T1 > T2 > T3$
(c) $T2 = T3$, $T3 < T1$ (d) $T1 = T3$, $T3 > T2$

[2014 (Set-2) : 2 M]

3.56 Host A (on TCP/IP v4 network A) sends an IP datagram D to host B (also on TCP/IP v4 network B). Assume that no error occurred during the transmission of D. When D reaches B, which of the following IP header field(s) may be different from that of the original datagram D?

(i) TTL
(ii) Checksum
(iii) Fragment Offset
(a) (i) only
(b) (i) and (ii) only
(c) (ii) and (iii) only
(d) (i), (ii) and (iii)

[2014 (Set-3) : 1 M]

3.57 An IP router implementing Classless Inter-domain Routing (CIDR) receives a packet with address 131.23.151.76. The router's routing table has the following entries:

Prefix	Output Interface Identifier
131.16.0.0/12	3
131.28.0.0/14	5
131.19.0.0/16	2
131.22.0.0/15	1

The identifier of the output interface on which this packet will be forwarded is _______.

[2014 (Set-3) : 2 M]

3.58 Every host in an IPv4 network has a 1-second resolution real-time clock with battery backup. Each host needs to generate up to 1000 unique identifiers per second. Assume that each host has a globally unique IPv4 address. Design a 50-bit globally unique ID for this purpose. After what period (in seconds) will the identifiers generated by a host wrap around? **[2014 (Set-3) : 2 M]**

3.59 An IP router with a Maximum Transmission Unit (MTU) of 1500 bytes has received an IP packet of size 4404 bytes with an IP header of length 20 bytes. The values of the relevant fields in the header of the third IP fragment generated by the router for this packet are
(a) MF bit: 0, Datagram Length: 1444; Offset: 370
(b) MF bit: 1, Datagram Length: 1424; Offset: 185
(c) MF bit: 1, Datagram Length: 1500; Offset: 370
(d) MF bit: 0, Datagram Length: 1424; Offset: 2960 **[2014 (Set-3) : 2 M]**

3.60 Suppose two hosts use a TCP connection to transfer a large file. Which of the following statements is/are FALSE with respect to the TCP connection?

1. If the sequence number of a segment is m, then the sequence number of the subsequent segment is always m+1.
2. If the estimated round trip time at any given point of time is t sec, the value of the retransmission timeout is always set to greater than or equal to t sec.
3. The size of the advertised window never changes during the course of the TCP connection.
4. The number of unacknowledged bytes at the sender is always less than or equal to the advertised window.

(a) 3 only (b) 1 and 3 only
(c) 1 and 4 only (d) 2 and 4 only

[2015 (Set-1) : 1 M]

3.61 Which one of the following fields of an IP header is NOT modified by a typical IP router?
(a) Checksum (b) Source address
(c) Time to Live (TTL) (d) Length

[2015 (Set-1) : 1 M]

3.62 Identify the correct order in which a server process must invoke the function calls **accept**, **bind**, **listen**, and **recv** according to UNIX socket API.
(a) listen, accept, bind recv
(b) bind, listen, accept, recv
(c) bind, accept, listen, recv
(d) accept, listen, bind, recv

[2015 (Set-2) : 1 M]

3.63 Assume that the bandwidth for a TCP connection is 1048560 bits/sec. Let α be the value of RTT in milliseconds (rounded off to the nearest integer) after which the TCP window scale option is needed. Let β be the maximum possible window size with window scale option. Then the values of α and β are
(a) 63 milliseconds 65535×2^{14}
(b) 63 milliseconds 65535×2^{16}
(c) 500 milliseconds 65535×2^{14}
(d) 500 milliseconds 65535×2^{16}

[2015 (Set-2) : 2 M]

3.64 Consider the following routing table at an IP router:

Network No.	Net Mask	Next Hop
128.96.170.0	255.255.254.0	Interface 0
128.96.168.0	255.255.254.0	Interface 1
128.96.166.0	255.255.254.0	R2
128.96.164.0	255.255.252.0	R3
0.0.0.0	Default	R4

For each IP address in Group-I identify the correct choice of the next hop from Group-II using the entries from the routing table above.

List-I	List-II
A. 128.96.171.92	1. Interface 0
B. 128.96.167.151	2. Interface 1
C. 128.96.163.121	3. R2
D. 128.96.165.121	4. R3
	5. R4

Codes:

	A	B	C	D
(a)	1	3	5	4
(b)	1	4	2	5
(c)	2	3	4	5
(d)	2	3	5	4

[2015 (Set-2) : 2 M]

3.65 Host A sends a UDP datagram containing 8880 bytes of user data to host B over an Ethernet LAN. Ethernet frames may carry data up to 1500 bytes (i.e. MTU = 1500 bytes). Size of UDP header is 8 bytes and size of IP header is 20 bytes. There is no option field in IP header. How may total number of IP fragments will be transmitted and what will be the contents of offset field in the last fragment?

(a) 6 and 925 (b) 6 and 7400
(c) 7 and 1110 (d) 7 and 8880

[2015 (Set-2) : 2 M]

3.66 Consider the following statements:

I. TCP connections are full duplex.
II. TCP has no option for selective acknowledgment
III. TCP connections are message streams.

(a) Only I is correct
(b) Only I and II are correct
(c) Only II and III are correct
(d) All of I, II and III are correct

[2015 (Set-3) : 1 M]

3.67 In the network 200.10.11.144/27, the fourth octet (in decimal) of the last IP address of the network which can be assigned to a host is ______.

[2015 (Set-3) : 2 M]

3.68 An IP datagram of size 1000 bytes arrives at a router. The router has to forward this packet on a link whose MTU (maximum transmission unit) is 100 bytes. Assume that the size of the IP header is 20 bytes. The number of fragments that the IP datagram will be divided into for transmission is ______. **[2016 (Set-1) : 2 M]**

3.69 For a host machine that uses the token bucket algorithm for congestion control, the token bucket has a capacity of 1 megabyte and the maximum output rate is 20 megabytes per second. Tokens arrive at a rate to sustain output at a rate of 10 megabytes per second. The token bucket is currently full and the machine needs to send 12 megabytes of data.

The minimum time required to transmit the data is ______ seconds. **[2016 (Set-1) : 2 M]**

3.70 Consider socket API on a Linux machine that supports connected UDP sockets. A connected UDP socket is a UDP socket on which **connect** function has already been called. Which of the following statements is/are CORRECT?

I. A connected UDP socket can be used to communicate with multiple peers simultaneously.
II. A process can successfully call **connect** function again for an already connected UDP socket.

(a) I only (b) II only
(c) Both I and II (d) Neither I nor II

[2017 (Set-2) : 1 M]

3.71 The maximum number of IPv4 router addresses that can be listed in the record route (RR) option field of an IPv4 header is ______.

[2017 (Set-2) : 1 M]

3.72 Consider the following statements about the routing protocols, Routing Information Protocol (RIP) and Open Shortest Path First (OSPF) in an IPv4 network.

I. RIP uses distance vector routing
II. RIP packets are sent using UDP
III. OSPF packets are sent using TCP
IV. OSPF operation is based on link-state routing

Which of the statements above are CORRECT?

(a) I and IV only (b) I, II and III only
(c) I, II and IV only (d) II, III and IV only

[2017 (Set-2) : 1 M]

3.73 Consider a TCP client and a TCP server running on two different machines. After completing data transfer, the TCP client calls **close** to terminate the connection and a FIN segment is sent to the TCP server. Server-side TCP responds by sending an ACK. which is received by the client-side TCP. As per the TCP connection state diagram (RFC 793). In which state does the client-side TCP connection wait for the FIN from the server-side TCP?

(a) LAST-ACK (b) TIME-WAIT
(c) FIN-WAIT-1 (d) FIN-WAIT-2

[2017 (Set-1) : 1 M]

3.74 Consider a long-lived TCP session with an end-to-end bandwidth of 1 Gbps (= 10^9 bits-per-second). The session starts with a sequence number of 1234. The minimum time (in seconds, rounded to the closest integer) before this sequence number can be used again is ________.

[2018 : 1 M]

3.75 Consider the following statements regarding the slow start phase of the TCP congestion control algorithm. Note the *cwnd* stands for the TCP congestion window and MSS denotes the Maximum Segment Size.

(i) The *cwnd* increases by 2 MSS on every successful acknowledgment.
(ii) The *cwnd* approximately doubles on every successful acknowledgment.
(iii) The *cwnd* increases by 1 MSS every round trip time.
(iv) The *cwnd* approximately doubles every round trip time.

Which one of the following is correct?

(a) Only (ii) and (iii) are true
(b) Only (i) and (iii) are true
(c) Only (iv) is true
(d) Only (i) and (iv) are true **[2018 : 1 M]**

3.76 Match the following:

Field
P. UDP Header's Port Number
Q. Ethernet MAC Address
R. IPv6 Next Header
S. TCP Header's Sequence Number

Length in bits
I. 48
II. 8
III. 32
IV. 16

(a) P-III, Q-IV, R-II, S-I
(b) P-II, Q-I, R-IV, S-III
(c) P-IV, Q-I, R-II, S-III
(d) P-IV, Q-I, R-III, S-II **[2018 : 1 M]**

3.77 Consider three machines M, N and P with IP addresses 100.10.5.2, 100.10.5.5 and 100.10.5.6 respectively. The subnet mask is set to 255.255.255.252 for all the three machines. Which one of the following is true?

(a) M, N and P all belong to the same subnet
(b) Only N and P belong to the same subnet
(c) M, N, and P belong to three different subnets
(d) Only M and N belong to the same subnet

[2019 : 2 M]

3.78 Consider a TCP connection between a client and a server with the following specifications: the round trip time is 6 ns, the size of the receiver advertised window is 50 KB, slow-start threshold at the client is 32 KB, and the maximum segment size is 2 KB. The connection is established at time $t = 0$. Assume that there are no timeouts and errors during transmission. Then the size of the congestion window (in KB) at time $t + 60$ ms after all acknowledgments are processed is ________.

[2020 : 2 M]

3.79 An organization requires a range of IP addresses to assign one to each of its 1500 computers. The organization has approached an Internet Service Provider (ISP) for this task. The ISP uses CIDR and serves the requests from the available IP address space 202.61.0.0/17. The ISP wants to assign an address space to the organization which will minimize the number of routing entries in the ISP's router using route aggregation. Which of the following address spaces are potential candidates from which the ISP can allot any one to the organization?

I. 202.61.84.0/21 **II.** 202.61.104.0/21
III. 202.61.64.0/21 **IV.** 202.61.144.0/21

(a) I and IV only (b) III and IV only
(c) I and II only (d) II and III only

[2020 : 2 M]

3.80 A TCP server application is programmed to listen on port number P on host S. A TCP client is connected to the TCP server over the network. Consider that while the TCP connection was active, the server machine S crashed and rebooted. Assume that the client does not use the TCP keepalive timer.

Which of the following behaviors is/are possible?

(a) If the client sends a packet after the server reboot, it will receive a FIN segment.
(b) If the client was waiting to receive a packet, it may wait indefinitely.
(c) The TCP server application on S can listen on P after reboot.
(d) If the client sends a packet after the server reboot, it will receive a RST segment.

[2021 (Set-1) : 2 M]

3.81 Consider two hosts P and Q connected through a router R. The maximum transfer unit (MTU) value of the link between P and R is 1500 bytes, and between R and Q is 820 bytes.

A TCP segment of size 1400 bytes was transferred from P to Q through R, with IP identification value as 0x1234. Assume that the IP header size is 20 bytes. Further, the packet is allowed to be fragmented, i.e., Don't Fragment (DF) flag in the IP header is not set by P.

Which of the following statements is/are correct?

(a) Two fragments are created at R and the IP datagram size carrying the second fragment is 620 bytes.

(b) If the second fragment is lost, P is required to resend the whole TCP segment.

(c) TCP destination port can be determined by analysing only the second fragment.

(d) If the second fragment is lost, R will resend the fragment with the IP identification value 0x1234

[2021 (Set-1) : 2 M]

3.82 Consider the three-way handshake mechanism followed during TCP connection establishment between hosts P and Q. Let X and Y be two random 32 bit starting sequence numbers chosen by P and Q respectively. Suppose P sends a TCP connection request message to Q with a TCP segment having SYN bit = 1. SEQ number = X, and ACK bit = 0. Suppose Q accepts the connection request. Which one of the following choices represents the information present in the TCP segment header that is sent by Q to P?

(a) SYN bit = 1, SEQ number = Y, ACK bit = 1, ACK number = $X + 1$, FIN bit = 0

(b) SYN bit = 1, SEQ number = Y, ACK bit = 1, ACK number = X, FIN bit = 0

(c) SYN bit = 0, SEQ number = $X + 1$, ACK bit = 0, ACK number = Y, FIN bit = 1

(d) SYN bit = 1, SEQ number = $X + 1$, ACK bit = 0, ACK number = Y, FIN bit = 0

[2021 (Set-2) : 1 M]

3.83 Consider the data transfer using TCP over a 1 Gbps link. Assuming that the maximum segment lifetime (MSL) is set to 60 seconds, the minimum number of bits required for the sequence number field of the TCP header, to prevent the sequence number space from wrapping around during the MSL is ________.

[2022 : 2 M]

3.84 Suppose you are asked to design a new reliable byte-stream transport protocol like TCP. This protocol, named myTCP, runs over a 100 Mbps network with Round Trip Time of 150 milliseconds and the maximum segment lifetime of 2 minutes.

Which of the following is/are valid lengths of the Sequence Number field in the myTCP header?

(a) 30 bits (b) 32 bits

(c) 34 bits (d) 36 bits

[2023 : 2 M]

3.85 Suppose in a web browser, you click on the www.gate-2023.in URL. The browser cache is empty. The IP address for this URL is not cached in your local host, so a DNS lookup is triggered (by the local DNS server deployed on your local host) over the 3-tier DNS hierarchy in an iterative mode. No resource records are cached anywhere across all DNS servers.

Let RTT denote the round trip time between your local host and DNS servers in the DNS hierarchy. The round trip time between the local host and the web server hosting www.gate-2023.in is also equal to RTT. The HTML file associated with the URL is small enough to have negligible transmission time and negligible rendering time by your web browser, which references 10 equally small objects on the same web server.

Which of the following statements is/are CORRECT about the minimum elapsed time between clicking on the URL and your browser fully rendering it?

(a) 7 RTTs, in case of non-persistent HTTP with 5 parallel TCP connections.

(b) 5 RTTs, in case of persistent HTTP with pipelining.

(c) 9 RTTs, in case of non-persistent HTTP with 5 parallel TCP connections.

(d) 6 RTTs, in case of persistent HTTP with pipelining.

[2023 : 2 M]

3.86 The forwarding table of a router is shown below:

Subnet Number	Subnet mask	Interface ID
200.150.0.0	255.250.0.0	1
200.150.64.0	255.255.224.0	2
200.150.68.0	255.255.255.0	3
200.150.68.64	255.255.255.224	4
Default		0

A packet addressed to a destination address 200.150.68.118 arrives at the router. It will be forwarded to the interface with ID ________.

[2023 : 2 M]

3.87 TCP client P successfully establishes a connection to TCP server Q. Let N_P denote the sequence number in the SYN sent from P to Q. Let N_Q denote the acknowledgement number in the SYN ACK from Q to P. Which of the following statements is/are CORRECT?

(a) The acknowledgement number N_Q is equal to N_P
(b) The sequence number N_Q is always 0 for a new connection
(c) The acknowledgement number N_Q is equal to $N_P + 1$
(d) The sequence number N_P is chosen randomly by P

[2024 (Set-1) : 1 M]

3.88 Which of the following fields is/are modified in the IP header of a packet going out of a network address translation (NAT) device from an internal network to an external network?

(a) Total Length (b) Source IP
(c) Destination IP (d) Header Checksum

[2024 (Set-1) : 1 M]

3.89 Consider sending an IP datagram of size 1420 bytes (including 20 bytes of IP header) from a sender to a receiver over a path of two links with a router between them. The first link (sender to router) has an MTU (Maximum Transmission Unit) size of 542 bytes, while the second link (router to receiver) has an MTU size of 360 bytes. The number of fragments that would be delivered at the receiver is ________.

[2024 (Set-1) : 2 M]

3.90 Consider the entries shown below in the forwarding table of an IP router. Each entry consists of an IP prefix and the corresponding next hop router for packets whose destination IP address matches the prefix. The notation "/N" in a prefix indicates a subnet mask with the most significant N bits set to 1.

Prefix	Next hop router
10.1.1.0/24	R1
10.1.1.128/25	R2
10.1.1.64/26	R3
10.1.1.192/26	R4

This router forwards 20 packets each to 5 hosts. The IP addresses of the hosts are 10.1.1.16, 10.1.1.72, 10.1.1.132, 10.1.1.191, and 10.1.1.205. The number of packets forwarded via the next hop router R2 is ________.

[2024 (Set-1) : 2 M]

3.91 Which of the following fields of an IP header is/are always modified by any router before it forwards the IP packet?

(a) Time to Live (TTL) (b) Header Checksum
(c) Protocol (d) Source IP Address

[2024 (Set-2) : 1 M]

3.92 Which of the following statements about IPv4 fragmentation is/are TRUE?

(a) The fragmentation of an IP datagram is performed only at the source of the datagram.
(b) The reassembly of fragments is performed at all intermediate routers along the path from the source to the destination.
(c) The reassembly of fragments is performed only at the destination of the datagram.
(d) The fragmentation of an IP datagram is performed at any IP router which finds that the size of the datagram to be transmitted exceeds the MTU.

[2024 (Set-2) : 1 M]

3.93 Which one of the following CIDR prefixes exactly represents the range of IP address 10.12.2.0 to 10.12.3.255?

(a) 10.12.0.0/22 (b) 10.12.2.0/22
(c) 10.12.2.0/24 (d) 10.12.2.0/23

[2024 (Set-2) : 2 M]

3.94 Consider a TCP connection operating at a point of time with the congestion window of size 12 MSS (Maximum Segment Size), when a timeout occurs due to packet loss. Assuming that all the segments transmitted in the next two RTTs (Round Trip Time) are acknowledged correctly, the congestion window size (in MSS) during the third RTT will be ________.

[2024 (Set-2) : 2 M]

Answers TCP, UDP and IP

3.1	(d)	**3.2**	(d)	**3.3**	(d)	**3.4**	(b)	**3.5**	(a)	**3.6**	(c)	**3.7**	(a)	**3.8**	(d)	**3.9**	(c)
3.10	(b)	**3.11**	(a)	**3.12**	(d)	**3.13**	(b)	**3.14**	(b)	**3.15**	(d)	**3.16**	(d)	**3.17**	(d)	**3.18**	(c)
3.19	(b)	**3.20**	(a)	**3.21**	(d)	**3.22**	(c)	**3.23**	(c)	**3.24**	(c)	**3.25**	(a)	**3.26**	(a)	**3.27**	(c)
3.28	(d)	**3.29**	(c)	**3.30**	(b)	**3.31**	(d)	**3.32**	(c)	**3.33**	(a)	**3.34**	(d)	**3.35**	(b)	**3.36**	(b)
3.37	(c)	**3.38**	(b)	**3.39**	(d)	**3.40**	(c)	**3.41**	(c)	**3.42**	(b)	**3.43**	(d)	**3.44**	(d)	**3.45**	(a)
3.46	(c)	**3.47**	(c)	**3.48**	(a)	**3.49**	(c)	**3.50**	(c)	**3.51**	(1100)	**3.52**	(5)	**3.53**	(c)	**3.54**	(26)
3.55	(d)	**3.56**	(d)	**3.57**	(1)	**3.58**	(256)	**3.59**	(a)	**3.60**	(b)	**3.61**	(b)	**3.62**	(b)	**3.63**	(c)
3.64	(a)	**3.65**	(c)	**3.66**	(a)	**3.67**	(158)	**3.68**	(13)	**3.69**	(1.1)	**3.70**	(b)	**3.71**	(36)	**3.72**	(c)
3.73	(d)	**3.74**	(34 or 35)			**3.75**	(c)	**3.76**	(c)	**3.77**	(b)	**3.78**	(44)	**3.79**	(d)	**3.80**	(b, c, d)
3.81	(a, b)	**3.82**	(a)	**3.83**	(33)	**3.84**	(b, c, d)	**3.85**	(c, d)	**3.86**	(3)	**3.87**	(c, d)	**3.88**	(b, d)	**3.89**	(6)
3.90	(40)	**3.91**	(a, b)	**3.92**	(c, d)	**3.93**	(d)	**3.94**	(4)								

Explanations TCP, UDP and IP

3.1 (d)

Consider each choice separately

Choicc (a): It is possible for a computer to have multiple IP addresses-IP addresses specify the network connection not to a host computer so if a host computer moves from one network to another, its IP address must change. In the network the IP address for a computer is unique but when we move the host computes from one network to another network, its IP address must be changed.

Choice (b): IP packets from the same source to the same destination can take different routes in the network. In packet switching network the routes are determined by routing algorithms. It may be possible that different networks follow different routing algorithms so the statement is true.

Choice (c): IP ensures that a packet is discarded if it is unable to reach its destination within a given number of hops, so statement is true.

Choice (d): The packet source cannot set the route of an outgoing packet, the route is determined only by the routing tables in the routers on the way. The usual IP routing algorithm employs an internet routing table on each machine (computer) that stores informations about possible destination and how to reach them. Because both hosts (computer) and routers route datagrams both have IP routing tables so the statement is false.

3.2 (d)

The transport protocol provides an end-to end connectivity that shields network layer protocol from the details of the intervening network or networks. A transport protocol can be either connection oriented such as TCP, or connectionless such as UDP.

3.3 (d)

Subnet mask for a particular network, 255.255.31.0.

OR

we can say that subnet mask is,
11111111.11111111.00011111.0.
So, only the set of IP addresses i.e., 128.8.129.43 and 128.8.161.55 belongs to this particular network.

3.4 (b)

Choice (b) is not true.
A bridge operates at layer 2 (Data Link Layer) so it use MAC address while routes operates at layer 3 (Network Layer) so it uses IP addresses.

3.5 (a)

Some of the services that TCP *does not* provide: it does not guarantee a minimum transmission rate, a sending process is not permitted to transmit at any rate it wishes, the sending rate is regulated by TCP congestion control, or it may force the sender to send at a low average rate. Hence 1st statement is false. All other statements are true.

3.6 (c)

Since you have 6 subnet bits so we can make (64 – 2) = 62 hosts.

3.7 (a)

TCP sequences each byte in the packet. Assigning a sequence number to indicate the first byte in a multi-byte packet does this. The second packet will have a sequence number equal to the first sequence number plus the number of bytes in the first packet.

3.8 (d)

Destination Port number is NOT present in IP header.
Because the IP header has nothing to do with the port number.
Port numbers are used by the transport layer to ensure process to process delivery.

3.9 (c)

In the question they have directly given the payload so 2100 will be divided into 1200 and 904 (900 is not a multiple of 8 so we have to pad 4 bits in order to make it a multiple of 8).
Now in second network payload is 400B.
So 1200 bytes packet will be divided in 400, 400, 400 with each having 20B header, and 900 will be divided into 400, 400 and 104 (4 bits padded) each having 20 B header.
So total overhead is 20*6 = 120 B

3.10 (b)

In slow start phase, the sender increase the size of current transmission window by MSS for each ACK.

Here Packet size = 2000 B,
Data to be send = 4000 B

$$\text{So, number of packet 2 be sent} = \frac{4000\text{ B}}{2000\text{ B}} = 2$$

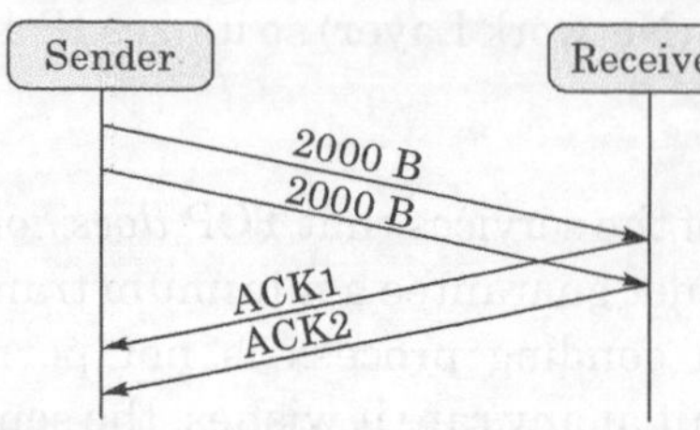

So, Maximum value of transmit window
= 4000 × (2 × 2000)
= 4000 + 4000 = 8000 B

3.11 (a)

Solution for IP address: 128.75.43.16

$$\begin{array}{r} 10000000.01010011.00101011.00010000 \\ 11111111.11111111.11111111.00000000 \\ \hline 10000000.01010011.00101011.00000000 \end{array}$$

i.e. 128.75.43.0 i.e. Eth0
128.75.43.0
255.255.255.128
128.75.43.0 Eth1
But number of 1's in Eth1 is more than Eth0
So answer will (a).

3.12 (d)

Network A sends a message of size 180 byte. Network B having the maximum packet size limit is 100 byte including 20 byte header. So the possible combination of sending packet from network A to C is

Header		Header		Header	
20	80	20	80	20	40

So the packet size = 100 + 100 + 60 = 260 byte

3.13 (b)

Overall B received by Network C = 260 B
(TCP header + IP header included)
Effective data transfer from network B to network C = 180 B
Hence effective efficiency:

$$\eta = \frac{180}{260} \times 512 = 354.46 \text{ Kbps}$$

3.14 (b)

TCP and UDP are not network layer protocols. TCP and UDP both are transport layer protocols. Routing packets from source to destination done by network layer and depends on routing algorithm so there is no interface of TCP and UDP at network layer.

3.15 (d)

The Address Resolution Protocol (ARP), allows a host to find the MAC (Physical) address of a target host on the same physical network, given only the targets IP address.

3.16 (d)

In a class B network initial two octets are all 1's but the third octet specifies the physical network for subnet of 64 department or 2^6 so initial 6 bits of third octets are 1's.

11111111 · 11111111 · 11111100 · 00000000
255 · 255 · 252 · 0

3.17 (d)

Let S denotes the source station and D denotes the destination station. x and y are two intermediate nodes between S and D.

S x y D

Message size = 24 bytes, Header (control information) = 3 bytes consider the first choice (a). Packet size = 4 then message size = 4 – 3 = 1 byte so it required 24 messages each containtaing 3 byte header so the transmission time for header overhead increases. Consider the second, third and fourth choices.

(b) Packet size = 6 then message size = 6 – 3 = 3 bytes (required 8 packets)

(c) Packet size = 7 then message size = 7 – 3 = 4 bytes (Required 6 packets)

(d) Packet size = 9 then message size = 9 – 3 = 6 bytes (Required 4 packets)

So 4 packet is the optional message size & 9 is the optional packet size.

3.18 (c)

Round trip propagation delay is $2*T_p$ minimum frame size of ethernet can be found by using formula

Frame size = $(2T_p)$ * (Bandwidth)
= 46.4 ms * 10 Mbps = 464 k bits

3.19 (b)

In TCP congestion control

Current window size at sender = min(congestion window, Advertise window)
= min(4 KB, 6 KB) = 4 KB

3.20 (a)

Required probability = Probability of selecting a link L1 × (Probability of number of bit errors in L1)^L + Probability of selecting a link L2 × (Probability of number of bit errors in L2)^L

$$= p_1(1 - b_1)^L + p_2(1 - b_2)^L$$

3.21 (d)

In Class C network first three octet are reserved for net id, so we have total 8 bits for host and subnets.

If we want to distribute addresses in subnets so first we should consider the subnet with maximum host, here which is subnet with 100 hosts.

For 100 hosts we require 7 bits for host id, one bit remains for subnet, which we fix to 1 for this subnet.

Subnet mask for this subnet is 255.255.255.10000000 = 255.255.255.128 and subnet address is 204.204.204.128.

For first subnet we have fixed 17th bit to 1 now for second and third subnet 17th bit will be 0.

50 host require 6 bit for host id , two bit remain for subnet id in which one bit is already fixed to zero.

We can configure 18th bit only for these subnets; for second subnet we fix 18th bit to 1 and for third subnet to 0.

Subnet mask for **2nd subnet** and **3rd subnet** is 255.255.255.11000000 = 255.255.255.192.

Subnet address for **2nd subnet** is 204.204.204.64 and **3rd subnet** is 204.204.204.0.

3.22 (c)

Ethernet uses a **Cyclic Redundancy Check** (CRC) algorithm to detect transmission errors. The Internet Protocol (IP) and most higher-layer protocols of the Internet Protocol Suite (ICMP, IGMP, UDP, UDP-Lite, TCP) use a common checksum algorithm to validate the integrity of the packets that they exchange.

3.23 (c)

TTL is a mechanism that limits the lifespan or lifetime of data in a network. Once the prescribed event count or timespan has elapsed, data is discarded. TTL prevents a data packet from circulating indefinitely.

3.24 (c)

When a packet comes to a router, it matches destination network with those available in routing tables to route the packet.

The route which has longest network address match will be used to route the packet.

The longest matching mask is 255.255.255.0 and hence forwarded to eth2.

3.25 (a)

Transmission time for a packet = 1 unit

So, T.T. for N packets = N units

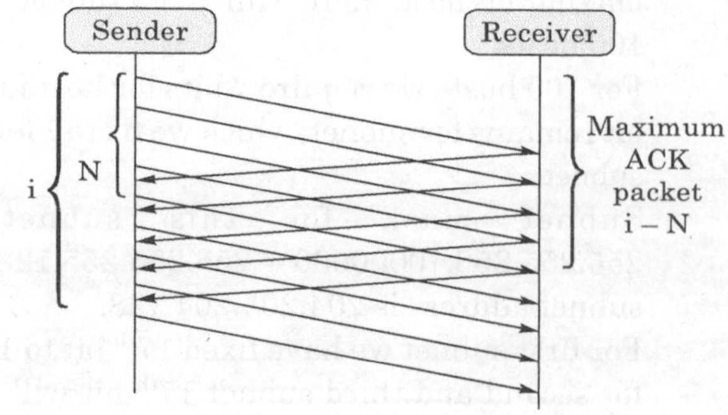

Since in round trip time all packets must be ack = N

Number of ack getting till time i = i – N

So, throughput = $\frac{i-N}{i} = \left(\frac{1-N}{i}\right)$

3.26 (a)

Since there is no buffer, and constraint given, there should not be any data loss and no wastage of capacity as well.

Calculate for the extreme case when all sources are on time (that in transmitting).

10 Mbps * n_station = 100 Mbps

n_station = 10

The link is provided with large buffer, so calculate expected value of bandwidth usage:

$E = \frac{1}{3}\times 10 + \frac{1}{3}\times 10 + \ldots$ n_station times

$E \le 100$ Mbps

$\frac{1}{3}\times 10 \times$ n_station ≤ 100 Mbps

Therefore n_station = 30

3.27 (c)

An ICMP port unreachable error will be generated for both UDP as well as TCP.

It is independent on the transport layer protocol.

3.28 (d)

Broadcast address in class B :

144 . 16 . 95 . 255 = 144 . 16 . 01011111 . 11111111

We know that in broadcast address all the host bit will be 1. So any subnet mask tells how many bits we borrowed from hosts bits in class B network.

So,
```
    144.16  . 01011111 . 11111111
    255.255 . 11100000 . 00000000
    -----------------------------
    144.16  . 01000000 . 00000000
              Subnet     Host
              bits       bits
```

Since all Host bits are 1 in given IP address. So, 255.255.224.0 can be used as subnet mask.

```
    144.16  . 01011111 . 11111111
    255.255 . 11110000 . 00000000
    -----------------------------
    144.16  . 01010000 . 00000000
              Subnet     Host
              bits       bits
```

Since all Host bits are 1 in given IP address. So, 255.255.240 can be used as subnet mask.

```
    144.16  . 01011111 . 11111111
    255.255 . 11111000 . 00000000
    -----------------------------
    144.16  . 01011000 . 00000000
              Subnet     Host
              bits       bits
```

Since all Host bits are 1 in given IP address, so 255.255.248 can be used as subnet mask.

3.29 (c)

The two computer C_1 and C_2 are configured as follows:

	C_1	C_2
IP Address	203.197.2.53	203.197.75.201
AND		AND
Netmask	255.255.128.0	255.255.192.0
	203.197.0.0	203.197.64.0
	Network Address	Network Address

203.197.2.53 → 203.197.0000010.00111001

255.255.192.0 → 255.255.11000000.00000000

203.197.0.0   203.197.00000000.00000000

203.197.75.201

255.255.128.0

203.192.0.0

∴ C_1 assumes C_2 is on same network but C_2 assumes C_1 is on different network.

3.30 (b)

Appropriate RTT estimation algorithm is used to set the timeout value of a TCP connection. Hence, the other statements regarding timeout values are incorrect.

3.31 (d)

TCP connection:

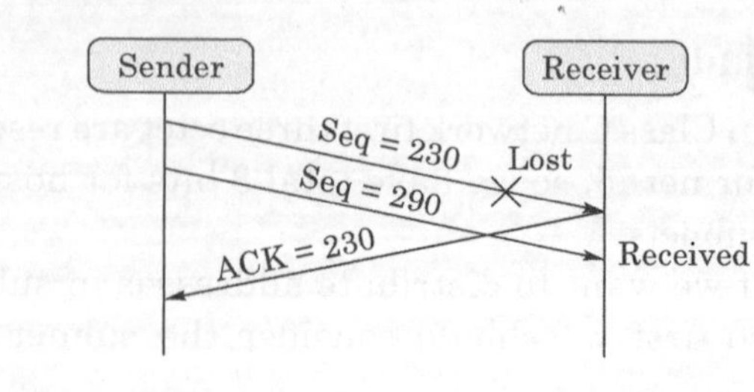

First packet size = 290 – 230 = 60 byte

So, X = Size of packet = 60 byte and

ACK = Y = 230

3.32 (c)

The class B is defined as follows:

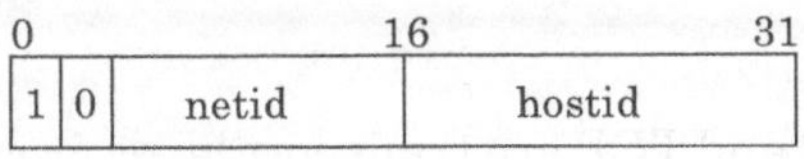

Maximum number of subnets

$= 2^6 - 2 = 64 - 2 = 62$

Maximum number of hosts in each subnet

$= 2^{16-6} - 2 = 2^{10} - 2 = 1022$

3.33 (a)

Since there is no maximum limit of data that application can send, so option (a) is correct.

3.34 (d)

Connect system call is responsible for synchronize the packets.

3.35 (b)

S1 is true because TCP has its congestion control and flow control mechanisms.

UDP itself does note have any flow control or congestion control mechanism. Therefore S2 is false.

Fast retransmit and fast recovery are part of TCP congestion control algorithms. Therefore S3 is True.

In slow start algorithm we use congestion and advertised windows and these act as flow control windows.

The congestion windows flow control imposed by the sender while the advertised window is flow control imposed by receiver.

∴ S4 is True.

3.36 (b)

Initially server is in LISTEN mode when a SYN is received, the server sends SYN+Ack and goes to SYN_RCVD state. When ack from client is received it moves to ESTABLISHED state.

∴ S3 is false and S4 is true.

The loss of SYN+Ack from server will not allow the client to move to ESTABLISHED state. Hence connection can not be established.

3.37 (c)

By performing bitwise AND between given IP addresses and subnetmask such as

192.168. 1 .97	192.168. 1 .80
255.255.255.224	255.255.255.224
192.168. 1 .01100001	192.168. 1 .01010000
192.168. 1 .135	192.168. 1 .110
255.255.255.224	255.255.255.224
192.168. 1 .10000111	192.168. 1 .01101110
192.168. 1 .67	192.168. 1 .155
255.255.255.224	255.255.255.224
192.168. 1 .01000011	192.168. 1 .10011011

Their are 3 distinct subnet present i.e., 011, 010, 100.

So answer should be option (c).

3.38 (b)

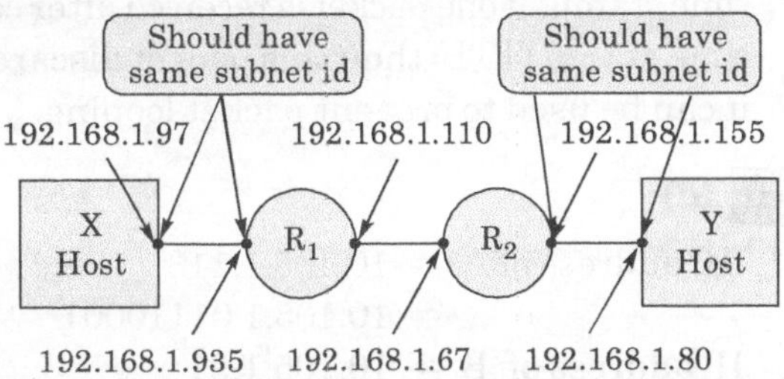

To connect host X to router R_1 subnet id of both should be same.

So,

192.168. 1 . 97	192.168. 1 .110	192.168. 1 .135
255.255.255.224	255.255.255.224	255.255.255.224
192.168. 1 . 96	192.168. 1 . 96	192.168. 1 .128

matched

So, IP address of X should configured as 192.168.1.110.

3.39 (d)

In the slow-start (additive) phase of the TCP congestion control algorithm, the size of congestion window increases exponentially TCP acts the window size as follows allowed-window = min (receiver_advertisment, congestion-window)

In slow-start (additive) phase whenever starting traffic on a new connection or increasing traffic after a period of congestion, start the congestion window at the size of a single segment and increases the congestion window by one segment each time an acknowledgment arrives.

3.40 (c)

11111111 11111111 11111000 00000000

So the number of hosts per subnet

$= 2^{11} - 2$

$= 2048 - 2 = 2046$

3.41 (c)

Connect( ) system call returns an error.

3.42 (b)

Maximum packet lifetime = 64 seconds
Sequence of one packet change after 64 sec (otherwise packet duplication possible) and clock counter increment per millisec. therefore min. possible rate = 0.064/sec.

3.43 (d)

In the IP datagram there is time to live (TTL) field. It is mainly used to prevent packet looping. Means every packet is associated with certain time stamp. If one packet is received after certain time stamp (TTL) then this packet discarded so it can be used to prevent packet looping.

3.44 (d)

IP address of A = 10.105.1.113
= 10.105.1.01110001
IP address of B = 10.105.1.91
= 10.105.1.01011011

For finding network address we take AND of IP address with the netmask N. So we check one by one

(a) (i) 255.255.255.0 = N } Taking AND
10.105.1.01110001 → IP address of A
10.105.1.00000000

(ii) By taking and with B
255.255.255.00000000
10.105.1.01011011
10.105.1.00000000

So both belong to same network

(b) (i) 255.255.255.128
255.255.255.10000000
10.105.1.01110001 IP of A
10.105.1.00000000

(ii) 255.255.255.10000000
10.105.1.01011011 IP of B
10.105.1.00000000

So both belong to same N/W

(d) (i) 255.255.255.11100000
10.105.1.01110001 IP of A
10.105.1.01100000

(ii) 255.255.255.11100000
10.105.1.01011011 IP of B
10.105.1.01100000

Both are not same so it not belong to same network.

3.45 (a)

To block TCP traffic from specific user, requires information of specific user which is done by application layer. To block HTTP data, layer-4, i.e., transport layer can be used becuase it can block port used by HTTP.

3.46 (c)

UDP and TCP are transport layer protocol. TCP supports electronic mail.
∴ Option (c) is correct.

3.47 (c)

Number of Networks in class 'C' are 2^{21} as in class 'C' there

24 bits Net id | 8 bits host id

out of 24 bits 3-bits are used for representation class 'C' i.e. 110
∴ 21 bits with 21 bits we can make 2^{21} networks.

3.48 (a)

It has 20 bit mask, so total number of hosts are 2^{12}.
It wants half of the chunk to A, therefore mask of A should be 21 bits, so it will contain 2^{11} hosts.
∴ Two choices for A
⇒ 245.248.1000 0000.0/21
245.248.1000 1000.0/21
∴ A can be assigned
⇒ 245.248.136.0/21
245.248.128.0/22

Now, one fourth needs to be assigned to B, hence its mask should be of 22 bits. But if we look at choice (b), the host in A and B are conflicting, therefore option (a) is correct.

3.49 (c)

Threshold = 8 MSS
Window size for
1st transmission = 2 MSS
Window size for
2nd transmission = 4 MSS
Windows size for
3rd transmission = 8 MSS
} Slow start

Threshold reached, increase linearly (according to AIMD)
Window size for 4th transmission = 9 MSS
Window size for 5th transmission = 10 MSS

Time out occur so new threshold $= \frac{10}{2} = 5\,\text{MSS}$
Window size for 6th transmission = 2 MSS
Windows size for 7th transmission = 4 MSS
Threshold reached
Window size for 8th transmission = 5 MSS
Window size for 9th transmission = 6 MSS
Window size for 10th transmission = 7 MSS

3.50 (c)

M = 0 : Means no more fragmentation so it represent the last fragment.
HLEN = 10 :
Header length $= 10 \times 4 = 40$ bytes
Payload : $400 - 40 = 360$ bytes [0 to 359]
Fragment offset: = 300 means 300×8
= 2400 bytes
Sequence number of first fragment = 2400
Sequence number of last fragment
$= 2400 + 359 = 2759$
Option (c) is correct.

3.51 (1100)

Congestion window = 32 KB
Threshold = 16 KB
2 | 4 | 8 | 16 | 18 | 20 | 22 | 24 | 26 | 28 | 30 | 32
So, time taken to reach 32 KB = 11 segment × 100 ms = 1100 ms

3.52 (5)

Given, SR protocol
and $L = 1\text{ KB} = 1024 \times 8$ bits
$BW = 1.5$ Mbps, $PT = 50$ ms
$\eta = 60\%$

Now, $$TT = \frac{L}{BW} = \frac{1024 \times 8}{1.5 \times 10^6} = 5.46 \text{ ms}$$

We know, $$\eta = \frac{N \times TT}{TT \times 2PT} \times 100$$

$$60 = \frac{N \times 5.46}{5.46 \times 2 \times 50} \times 100$$

$$N = 11.58$$

{N is the window size}
As we know in SR:
$$2^{m-1} = N$$
$$m = \log_2 N + 1 = 5$$
{m: number of bits to represent sequence number}

3.53 (c)

Listen( ) converts an unconnected active TCP socket into a passive socket.

3.54 (26)

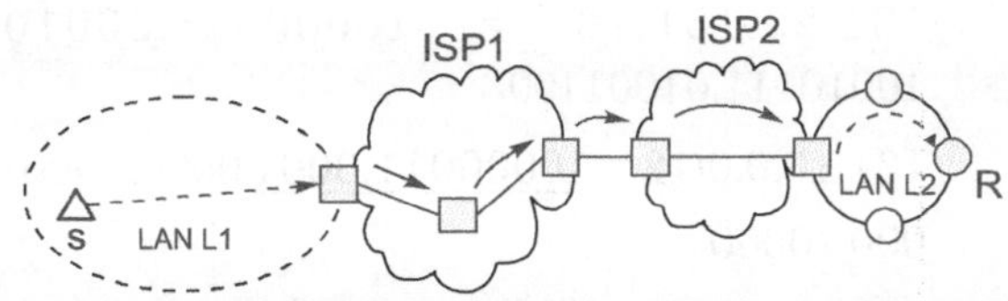

$32 - 6 = 26$ is maximum possible value of TTL.

3.55 (d)

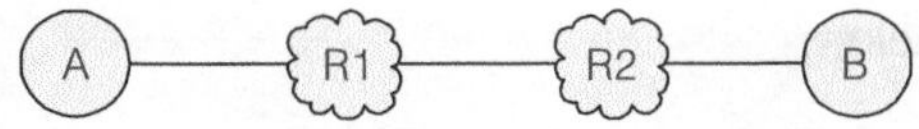

Bandwidth : 10^6 bps
Packet size: 1000 bytes
1st transmission: 1000 +100 = 1100 bytes transmitted at a time from A

$$T_1: \frac{1000 + 100\,(\text{overhead})\,B}{10^6 \text{ bps}} = \frac{1100 \times 8}{10^6 \text{bps}}$$

$$T_1 = \frac{8800}{10^6 \text{bps}} = 8.8 \text{ msec}$$

$$T_1 = 8.8 \times 3 = 26.4 \text{ msec}$$

2nd transmission: 100 +100 = 200 bytes transmitted 10 times from A

$$T_2: \frac{100 + 100\,(\text{overhead})\,B}{10^6 \text{ bps}} = \frac{200 \times 8}{10^6 \text{bps}}$$

$$T_2 = \frac{1600}{10^6 \text{bps}} = 1.6 \text{ msec}$$

$$T_2 = (3 \times 1.6) + (9 \times 1.6) = 19.2 \text{ msec}$$

3rd transmission: 50 +100 = 150 bytes transmitted 20 times from A

$$T_3: \frac{50 + 100\,(\text{overhead})\,B}{10^6 \text{ bps}} = \frac{150 \times 8}{10^6 \text{bps}}$$

$$T_3 = \frac{1200}{10^6 \text{bps}} = 1.2 \text{ msec}$$

$$T_3 = (3 \times 1.2) + (19 \times 1.2) = 26.4 \text{ msec}$$

$T_1 = T_3 = 26.4$ msec
$T_2 = 19.2$ msec, $T_2 < T_3 = T_1$

3.56 (d)

TTL changes from one hop to the next
Check sum changes on each hop due to TTL change.
Fragment offset may be computed again.
∴ TTL, check sum and fragment offset may be changed in the transmission.

3.57 (1)

131.23.151.76 = 10000011.00010111.10010111.01001100

131.16.0.0/12 = 10000011.0001 0000.0000 0000.0000 0000

131.22.0.0/15 = 10000011.00010110.0000 0000.0000 0000

Longest prefix match is: 131.22.0.0/15

∴ Interface 1 will be selected to forward the packet.

3.58 (256)

Each host needs to generate 1000 unique identifiers per second which requires

$\lceil \lg 1000 \rceil = 10$ bits

Now, these 10 bits along with 32 bit globally unique IP address will give a globally unique 42 bit ID's which stays constant.

Since we are allowed 50 bits we can use the next 8 bits using the clock which changes every second. Thus our Id's will wrap around once in $2^8 = 256$ seconds.

3.59 (a)

MTU = 1500 bytes

Header length = 20 bytes

Packet size = 4404 bytes

Fragment 1: 1480 + 20

Fragment 2: 1480 + 20

Fragment 3: 1424 + 20

Datagram length for third IP fragment = 1444

MF bit = 0

$$\text{Offset} = \frac{1480}{4} = 370$$

3.60 (b)

Statement 1 is false because, the sequence number of the subsequent segment depends on the number of byte characters in the current segment.

Statement 2 True, because RTO timer is always ≥ Estimated RTT.

Statement 3 False. Receiver window is used to give the sender an idea of how much free buffer space is available at the receiver.

Statement 4 is true, TCP is not permitted to overflow the allocated receiver buffer.

Hence when the sender cannot send any more data receiver window would be 0 and hence. The buffer would have unacknowledged data.

3.61 (b)

Only source address field of IP header is NOT modified by a router.

3.62 (b)

Order of function calls invoked: Bind(), Listen(), Accept() and Recv().

3.63 (c)

Maximum windows size is the amount of data that can be transmitted in an RTT.

$$\therefore \text{RTT} = \frac{65535 \times 8}{1048560 \text{ bps}} = 500 \text{ ms}$$

For scaling factor 14 bits are used

∴ Option (c) is correct.

3.64 (a)

IP Address	Net Mask	Network No.	Next Hop
128.96.171.92	255.255.254.0	128.96.170.0	Interface 0
128.96.167.151	255.255.254.0	128.96.166.0	R2
128.96.163.151	default	0.0.0.0	R4
128.96.165.121	255.255.252.0	128.96.164.0	R3

3.65 (c)

User data in UDP datagram = 8880 bytes

UDP header = 8 bytes

IP header = 20 bytes

We need to break (8880+8) bytes into small fragments.

Ethernet frame can hold (1500 – IP header) = 1500 – 20 = 1480 payload of data

1st fragment 0 to 184 (1480 bytes)

2nd fragment 185 to 369 (1480 bytes)

3rd fragment 370 to 554 (1480 bytes)

4th fragment 555 to 739 (1480 bytes)

5th fragment 740 to 924 (1480 bytes)

6th fragment 925 to 1109 (1480 bytes)

7th fragment 1110 to 1117 (8 bytes)

∴ 7 fragments with 1110 as the offset of last fragment.

3.66 (a)

TCP is a byte stream protocol. Hence III is false. TCP can use both selective ACK and Cumulative Acknowledgment. Hence II is false. TCP connection are full duplex.

∴ Statement I is correct.

3.67 (158)

200.10.11.144/27

255.255.255.224

$$224 \rightarrow \underbrace{111}\,00000$$
$$144 \rightarrow \underbrace{100}_{\text{N/w ID.}}\,10000$$
$$\text{First IP} \rightarrow \underbrace{100}\,00001$$
$$\vdots$$
$$\text{Last IP} \rightarrow \underbrace{100}\,11110$$

i.e. last IP is 200.10.11.158

∴ 158 is the fourth octet of the last IP address.

3.68 (13)

MTU is 100 bytes, IP header is 20 bytes, IP datagram is 1000 bytes.

So, number of fragments are 13.

3.69 (1.1)

Time taken to transmit 1 MB when output rate is 20 MBps, capacity is 1 MB and token arrival rate is 10 MBps is

$$C + \rho S = MS$$
$$1\text{ MB} + (10\text{ MBps}) \times S \leftarrow (20\text{ MBps}) \times S$$
$$S = \frac{1\text{ MB}}{(20-10)\text{MBps}} = \frac{1\text{ MB}}{10\text{ MBps}} = 0.1\text{sec}$$

In 0.1 sec data transmit = 0.1 × Output rate
= 0.1 × 20 MBps = 2 MB

Remaining data = (12 MB – 2 MB) = 10 MB

So to transmit 1 MB takes 0.1 sec

then for 10 MB = 10 × 0.1 sec = 1 sec

Total time = (0.1 + 1) sec = 1.1 sec

3.70 (b)

Bind( ) function creates local address.

Connect( ) function is specfying remote address.

An unconnected UDP socket is just a Bind( ) function.

A connected UDP socket is one more step above i.e. connect( ) function [Just behaves like TCP].

3.71 (36)

In IPv4, options and padding 40 bytes are alloted.

Maximum nine routers addresses are allowed.

Each IPv4 address is 32 bits or 4 bytes

So 4 × 9 = 36 bytes

Extra bytes are used for the option.

3.72 (c)

RIP uses distance vector routing. OSPF uses link-state routing protocols.

RIP uses UDP as transport protocol.

OSPF neither uses TCP nor UDP.

[Link state packet should be given to all routers in subnet so it's not possible with TCP. These link state packets should be reliable at that same time which is not posible with UDP].

3.73 (d)

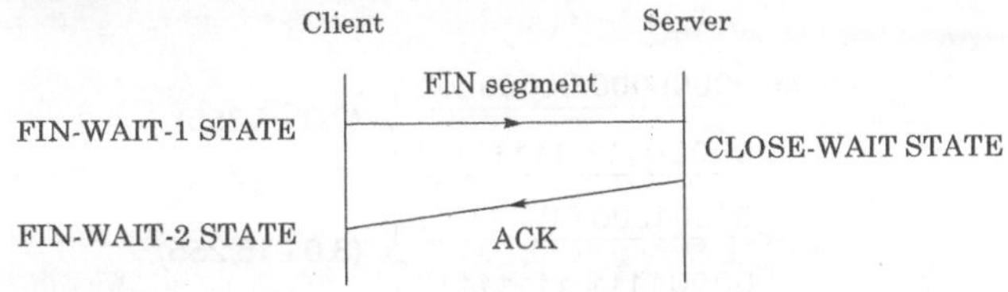

3.74 (34 or 35)

Total number of sequence number possible in TCP = 2^{32}

Given that,

In 10^9 bits = 1 sec

So, for 2^{32} bytes, $\frac{2^{32} \times 8\text{ Bits}}{10^9\text{ Bits}} = 34.35$

Answer could be 34 or 35, since integer value is asked.

3.75 (c)

cwnd approximately doubles every round trip time.

3.76 (c)

UDP Header's Port Number → 16 bit

Ethernet MAC Address ⇒ 48 bit

IPv6 Next Header ⇒ 8 bit

TCP Header's Sequence Number ⇒ 32 bot

3.77 (b)

	100.10.5.2	2	00000010
M →	255.255.255.252	252	11111100
	194.56.10.0	0	00000000
	100.10.5.5	5	00000101
N →	255.255.255.252	252	11111100
	194.56.10.4	4	00000100
	100.10.5.6	6	00000110
P →	255.255.255.252	252	11111100
	194.56.10.4	4	00000100

N and P belongs to same subnet.

3.78 (44)

At time t = 0 there are not timeouts

Threshold = 32 KB

MSS = 2 KB

At time t + 60, window size = 44

3.79 (d)

/17 ⇒ 11111111 11111111 10000000 00000000
⇒ 255.255.128.0

$$1500 \simeq 2048 = 2^{11} = 2^3 \times 2^8$$
$$= 2^{32-21} = 2^{11} = /21$$
$$/17 \Rightarrow 202.61.0.0/17$$

0.0 ⇒ 00000000.00000000
00000111.11111111 } (0.0 - 7.255)

00001000.00000000
00001111.11111111 } (8.0 - 15.255)

00010000.00000000
00010111.11111111 } (16.0 - 23.255)

Sequence is 0, 8, 16, 24, 32, 40, 48, 56, 64, 72, 80, 96, 104, 112, 120.
Hence 64 and 104 is present in sequence so it is the possible IP addresses.
So option (d) is correct.

3.80 (b, c, d)

(a) **False:** The situation resolves itself when client tries to send data to server over the dead connection, and server replies with an RST packet (not FIN), causing client to finally to close the connection forcibly.

FIN is used to close TCP connections gracefully in each direction (normal close of connection), while TCP RST is used in a scenario where TCP connections cannot recover from errors and the connection needs to reset forcibly.

(b) **True:** Since broken connections can only be detected by sending data, the receiving side will wait forever. This scenario is called a "half-open connection" because one side realizes the connection was lost but the other side believes it is still active.

(c) **True:** Yes, a TCP Server can listen to same port number even after reboot. For example, the SMTP service application usually listens on TCP port 25 for incoming requests. So, even after reboot the port 25 is assigned to SMTP.

(d) **True:** The situation resolves itself when client tries to send data to server over the dead connection, and server replies with an RST packet (not FIN).

3.81 (a, b)

Data = 1400 B

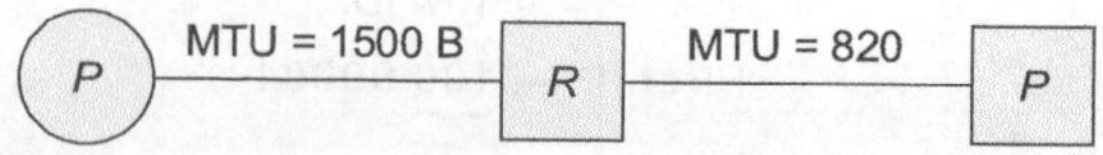

For First Link – PR – There will be no fragment
For Second Link – RQ

First Fragment = 800(Data) + 20(Header) = 820 B

Second Fragment = 1400 – 800 = 600 + 20(H) = 620 B

3.82 (a)

Q will send the SYN bit = 1 to the connection establishment.
Q seq number will be Y different from X.
ACK bit = 1 because sending the ACK.
ACK number = X + 1 (Next seq number id)
FIN bit = 0 (Because establishing the connection)

3.83 (33)

$$\text{B.W} = 1 \text{ Gbps} = \frac{2^{30}}{8} \text{ bytes/sec}$$

$$1 \text{ sec} = \frac{2^{30}}{8} \text{ bytes}$$

$$\Rightarrow 1 \text{ sec} = \frac{2^{30}}{8} \text{ sequence number}$$

$$\Rightarrow 60 \text{ sec} = \frac{2^{30} \times 60}{8} \text{ sequence number}$$

Number of sequence bits required

$$= \log_2 \left[\frac{2^{30} \times 60}{8}\right]$$

$$= \log_2 2^{30} + \log_2^{60} - \log_2^{8}$$

$$= 30 + 5.9 - 3 = 32.9 \simeq 33$$

3.84 (b, c, d)

Given, Bandwidth = 100 Mbps
i.e. in 1 second = 100×2^{20} bits can be transferred.
Since MSL is 2 minutes than, number of bits that can be transferred in 2 minutes

$$= 120 \times 100 \times 2^{20} \text{ bits}$$

Note: Sequence number is assigned to each byte in TCP.
Total bytes = 1500×2^{20} or 31 bits atleast

3.85 (c, d)

Default mode of http pipelining will take 3 RTTs (one for connection + one for sending 10 objects + one for closing connection) objects.

And in iterative mode for DNS is 3 RTTs (Root + Top domain + Authoritative).

So total 6 RTTs in case of persistent HTTP with pipeline.

For non-persistent with 5 parallel TCP:

1. DNS lookup will take 3 RTTs.
2. TCP connection: 2 RTTs
3. HTTP request / response: Since we are using non-persistent HTTP, each TCP connection can be used to request one object at a time. Since there are 10 objects in total, and we have 5 parallel TCP connections, it will take 2 rounds of HTTP requests / responses to fetch all the objects. Therefore, this step will take a minimum of 4 RTTs (2 RTTs for the first round and 2 RTTs for the second round).

$\therefore$ 3 + 2 + 4 RTTs i.e. 9 RTTs

$\therefore$ option (c)

3.86 (3)

The first step is to find matching table entries for incoming packets with the address 200.150.68.118. This can be done by taking bitwise AND with subnet mask. The addresses are matched by three entries, 255.255.0.0 and 255.255.224.0, and 255.255.255.0 So, which interface – 1, 2 or 3 – should accept the packet? We employ longest prefix matching to select one of the three. The interface is the matching table entry with the highest level of specificity. 255.255.255.0 is the address that is the most specific, so interface 3 receives the packet.

Note: Longest prefix matching is the rule to find the entry in a table which has the longest prefix matching with the incoming packet's destination IP and forward the packet to the corresponding next hope.

250.150.68.118	200.150.68.118
255.255.0.0	255.255.224.0
255.150.0.0 (matching)	200.150.64.0 (matching)

250.150.68.118	200.150.68.118
255.255.255.0	255.255.255.224
200.150.68.0 (matching)	200.150.68.96 (mismatch)

3.87 (c, d)

The acknowledgment number will always be sequence number of the next expected data. The initial sequence number will always be random number within range of 0 to $2^{32} - 1$.

■■■■

4 Routing and Application Layer

4.1 Which one of the following statements is FALSE?
(a) Packet switching leads to better utilization of bandwidth resources than circuit switching
(b) Packet switching results in less variation in delay than circuit switching
(c) Packet switching requires more per-packet processing than circuit switching
(d) Packet switching can lead to reordering unlike in circuit switching

[2004 : 1 M]

4.2 Consider the three commands : PROMPT, HEAD and RCPT. Which of the following options indicate a correct association of these commands with protocols where these are used?
(a) HTTP, SMTP, FTP
(b) FTP, HTTP, SMTP
(c) HTTP, FTP, SMTP
(d) SMTP, HTTP, FTP **[2005 : 1 M]**

4.3 Traceroute reports a possible route that is taken by packets moving from some host A to some other host B. Which of the following options represents the technique used by traceroute to identify these hosts
(a) By progressively querying routers about the next router on the path to B using ICMP packets, starting with the first router
(b) By requiring each router to append the address to the ICMP packet as it is forwarded to B. The list of all routers en-route to B is returned by B in an ICMP reply packet
(c) By ensuring that an ICMP reply packet is returned to A by each router en-route to B, in the ascending order of their hop distance from A
(d) By locally computing the shortest path from A to B **[2005 : 1 M]**

4.4 Count to infinity is a problem associated with
(a) Link state routing protocol.
(b) Distance vector routing protocol.
(c) DNS while resolving host name.
(d) TCP for congestion control.

[2005 : 1 M]

4.5 Assume that "host1.mydomain.dom" has an IP address of 145.128.16.8. Which of the following options would be most appropriate as a subsequence of steps in performing the reverse lookup of 145.128.16.8? In the following options "NS" is an abbreviation of "nameserver".
(a) Query a NS for the root domain and then NS for the "dom" domains
(b) Directly query a NS for "dom" and then a NS for "mydomain.dom" domains
(c) Query a NS for in-addr.arpa and then a NS for 128.145.in-addr.arpa domains
(d) Directly query a NS for 145.in-addr.arpa and then a NS for 128.145.in-addr.arpa domains **[2005 : 2 M]**

4.6 HELO and PORT, respectively, are commands from the protocols
(a) FTP and HTTP
(b) TELNET and POP3
(c) HTTP and TELNET
(d) SMTP and FTP **[2006 : 2 M]**

Linked Data for Q.4.7 and Q.4.8:

Consider the diagram shown below where a number of LANs are connected by (transparent) bridges. In order to avoid packets looping through circuits in the graph, the bridges organize themselves in a spanning tree. First, the root bridge is identified as the bridge with the least serial number. Next, the root sends out (one or more) data units to enable the setting up of the spanning tree of shortest paths from the root bridge to each bridge.

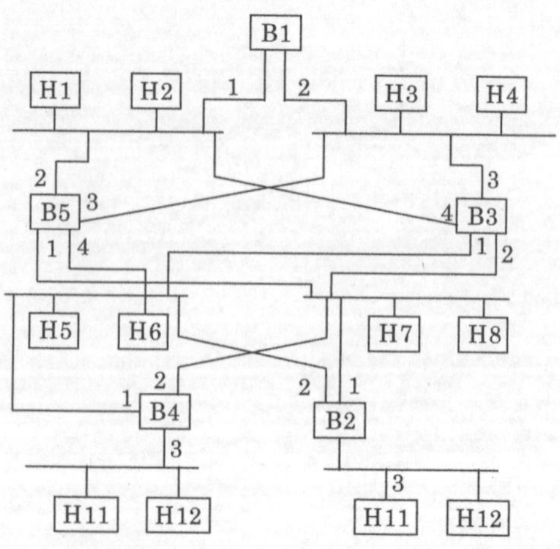

Each bridge identifies a port (the root port) through which it will forward frames to the root bridge. Port conflicts are always resolved in favour of the port with the lower index value. When there is a possibility of multiple bridges forwarding to the same LAN (but not through the root port), ties are broken as follows : bridges closest to the root get preference and between such bridges, the one with the lowest serial number is preferred.

4.7 For the given connection of LANs by bridges, which one of the following choices represents the depth first traversal of the spanning tree of bridges?
(a) B1, B5, B3, B4, B2 (b) B1, B3, B5, B2, B4
(c) B1, B5, B2, B3, B4 (d) B1, B3, B4, B5, B2
[2006 : 2 M]

4.8 Consider the correct spanning tree for the previous question. Let host H1 send out a broadcast ping packet. Which of the following options represents the correct forwarding table on B3?

(a)

Hosts	Port
H1, H2, H3, H4	3
H5, H6, H9, H10	1
H7, H8, H11, H12	2

(b)

Hosts	Port
H1, H2	4
H3, H4	3
H5, H6	1
H7, H8, H9, H10	2
H11, H12	

(c)

Hosts	Port
H3, H4	3
H5, H6, H9, H10	1
H1, H2	4
H7, H8, H11, H12	2

(d)

Hosts	Port
H1, H2, H3, H4	3
H5, H7, H9, H10	1
H7, H8, H11, H12	4

[2006 : 2 M]

4.9 Which one of the following uses UDP as the transport protocol?1
(a) HTTP (b) Telnet
(c) DNS (d) SMTP **[2007 : 1 M]**

4.10 For the network given in the figure below, the routing tables of the four nodes A, E, D and G are shown. Suppose that F has estimated its delay to its neighbors, A, E, D and G as 8, 10, 12 and 6 msecs respectively and updates its routing table using distance vector routing technique.

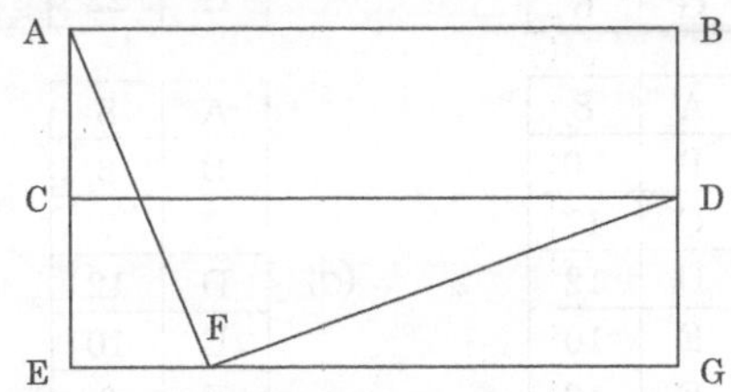

Routing Table of A	
A	0
B	40
C	14
D	17
E	21
F	9
G	24

Routing Table of E	
A	24
B	27
C	7
D	20
E	0
F	11
G	22

Routing Table of D	
A	20
B	8
C	30
D	0
E	14
F	7
G	22

Routing Table of G	
A	21
B	24
C	22
D	19
E	22
F	10
G	0

(a)

A	8
B	20
C	17
D	12
E	10
F	0
G	6

(b)

A	21
B	8
C	7
D	19
E	14
F	0
G	22

(c)

A	8
B	20
C	17
D	12
E	10
F	16
G	6

(d)

A	8
B	8
C	7
D	12
E	10
F	0
G	6

[2007 : 2 M]

4.11 A group of 15 routers are interconnected in a centralized complete binary tree with a router at each tree node. Router i communicates with router j by sending a message to the root of the tree. The root then sends the message back down to router j. The mean number of hops per message, assuming all possible router pairs are equally likely is

(a) 3 (b) 4.26
(c) 4.53 (d) 5.26 **[2007 : 2 M]**

4.12 Consider the following clauses:

(i) Not inherently suitable for client authentication.
(ii) Not a state sensitive protocol.
(iii) Must be operated with more than one server.
(iv) Suitable for structured message organization.
(v) May need two ports on the serve side for proper operation.

The option that has the maximum number of correct matches is

(a) IMAP-(i), FTP-(ii), HTTP-(iii), DNS-(iv), POP3-(v)
(b) FTP-(i), POP3-(ii), SMTP-(iii), HTTP-(iv), IMAP-(v)
(c) POP3-(i), SMTP-(ii), DNS-(iii), IMAP-(iv), HTTP-(v)
(d) SMTP-(i), HTTP-(ii), IMAP-(iii), DNS-(iv), FTP-(v)

[2007 : 2 M]

4.13 Match the following:

P. SMTP	**1.** Application layer
Q. BGP	**2.** Transport layer
R. TCP	**3.** Data link layer
S. PPP	**4.** Network layer
	5. Physical layer

Codes:
(a) P-2, Q-1, R-3, S-5
(b) P-1, Q-4, R-2, S-3
(c) P-1, Q-4, R-2, S-5
(d) P-2, Q-4, R-1, S-3 **[2007 : 2 M]**

4.14 Two popular routing algorithms are Distance Vector(DV) and Link State (LS) routing. Which of the following are true?

S1: Count to infinity is a problem only with DV and not LS routing
S2: In LS, the shortest path algorithm is run only at one node
S3: In DV, the shortest path algorithm is run only at one node
S4: DV requires lesser number of network messages than LS

(a) S1, S2 and S4 only
(b) S2 and S3 only
(c) S1, S2 and S3 only
(d) S1 and S4 only **[2008 : 2 M]**

Linked Answer Questions 4.15 and 4.16:

Consider a simple graph with unit edge costs. Each node in the graph represents a router. Each node maintains a routing table indicating the next hop router to be used to relay a packet to its destination and the cost of the path to the destination through that router. Initially, the routing table is empty. The routing table is synchronously updated as follows. In each updation interval, three tasks are performed.

(i) A node determines whether its neighbours in the graph are accessible. If so, it sets the tentative cost to each accessible neighbour as 1. Otherwise, the cost is set to ∞.
(ii) From each accessible neighbour, it gets the costs to relay to other nodes via that neighbour (as the next hop).
(iii) Each node updates its routing table based on the information received in the previous two steps by choosing the minimum cost.

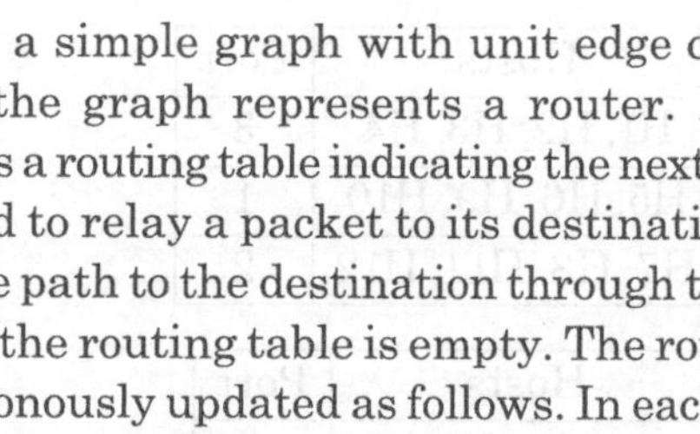

4.15 For the graph given above, possible routing tables for various nodes after they have stabilized, are shown in the following options. Identify the correct table.

(a)

Table for node A		
A	–	–
B	B	1
C	C	1
D	B	3
E	C	3
F	C	4

(b)

Table for node C		
A	A	1
B	B	1
C	–	–
D	D	1
E	E	1
F	E	3

(c)

Table for node B		
A	A	1
B	–	–
C	C	1
D	D	1
E	C	2
F	D	2

(d)

Table for node D		
A	B	3
B	B	1
C	C	1
D	–	–
E	E	1
F	F	1

[2005 : 2 M]

4.16 Continuing from the earlier problem, suppose at some time t, when the costs have stabilized, node A goes down. The cost from node F to node A at time (t + 100) is :

(a) > 100 but finite (b) ∞
(c) 3 (d) > 3 and ≤ 100

[2005 : 2 M]

4.17 A computer on a 10 Mbps network is regulated by a token bucket. The token bucket is filled at a rate of 2 Mbps. It is initially filled to capacity with 16 Megabits. What is the maximum duration for which the computer can transmit at the full 10 Mbps?

(a) 1.6 seconds (b) 2 seconds
(c) 5 seconds (d) 8 seconds

[2008 : 2 M]

4.18 Which one of the following is not a client-server application?

(a) Internet chat
(b) Web browsing
(c) E-mail
(d) Ping

[2010 : 1 M]

Linked Data Questions Q. 4.19 and Q. 4.20:

Consider a network with 6 routers R1 to R6 connected with links having weights as shown in the following diagram.

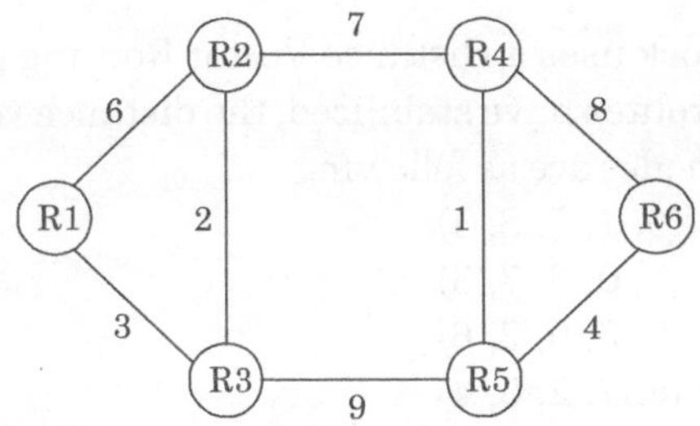

4.19 All the routers use the distance vector based routing algorithm to update their routing tables. Each router starts with its routing table initialized to contain an entry for each neighbour with the weight of the respective connecting link. After all the routing tables stabilize, how many links in the network will never be used for carrying any data?

(a) 4 (b) 3
(c) 2 (d) 1 **[2010 : 2 M]**

4.20 Suppose the weights of all unused links in the previous question are changed to 2 and the distance vector algorithm is used again until all routing tables stabilize. How many links will now remain unused?

(a) 0 (b) 1
(c) 2 (d) 3 **[2010 : 2 M]**

4.21 Consider different activities related to email

m1: Send an email from a mail client to a mail server

m2: Download an email from mailbox server to a mail client

m3: Checking email in a web browser

(a) m1:HTTP m2:SMTP m3:POP
(b) m1:SMTP m2:FTP m3:HTTP
(c) m1:SMTP m2:POP m3:HTTP
(d) m1:POP m2:SMTP m3:IMAP

[2011 : 1 M]

Linked Data Question Q.4.22 and Q.4.23:

Consider a network with five nodes, N1 to N5, as shown below:

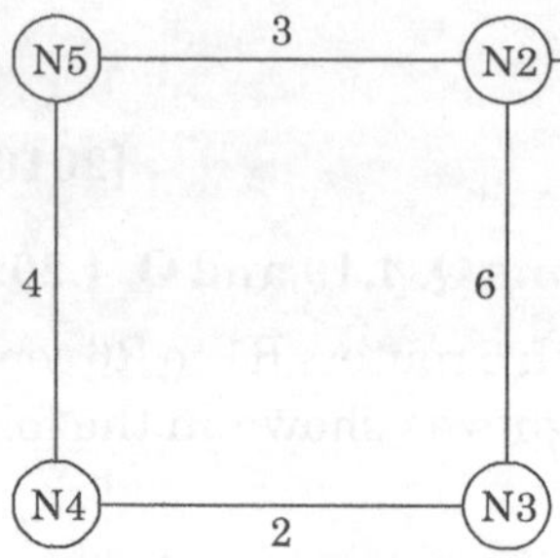

The network uses a Distance Vector Routing protocol. Once the routes have stabilized, the distance vectors at different nodes are as following.

N1: (0, 1, 7, 8, 4)
N2: (1, 0, 6, 7, 3)
N3: (7, 6, 0, 2, 6)
N4: (8, 7, 2, 0, 4)
B5: (4, 3, 6, 4, 0)

Each distance vector is the distance of the best known path at that instance to nodes, N1 to N5, where the distance to itself is 0. Also, all links are symmetric and the cost is identical in both directions. In each round, all nodes exchange their distance vectors with their respective neighbors. Then all nodes update their distance vectors. In between two rounds, any change in cost of a link will cause the two incident nodes to change only that entry in their distance vectors.

4.22 The cost of link N2-N3 reduces to 2(in both directions). After the next round of updates, what will be the new distance vector at node, N3?

(a) (3, 2, 0, 2, 5) (b) (3, 2, 0, 2, 6)
(c) (7, 2, 0, 2, 5) (d) (7, 2, 0, 2, 6)

[2011 : 2 M]

4.23 After the update in the previous question, the link N1-N2 goes down. N2 will reflect this change immediately in its distance vector as cost ∞. After the NEXT ROUND of update, what will be the cost to N1 in the distance vector of N3?

(a) 3 (b) 9
(c) 10 (d) ∞ **[2011 : 2 M]**

4.24 The Protocol Data Unit (PDU) for the application layer in the Internet stack is

(a) Segment (b) Datagram
(c) Message (d) Frame **[2012 : 1 M]**

4.25 The transport layer protocols used for real time multimedia, file transfer, DNS and email respectively are

(a) TCP, UDP, UDP and TCP
(b) UDP, TCP, TCP and UDP
(c) UDP, TCP, UDP and TCP
(d) TCP, UDP, TCP and UDP **[2013 : 1 M**

4.26 Assume that source S and destination D are connected through two intermediate routers labels R. Determine how many times each packet has to visit the network layer and the data link layer during a transmission form S to D.

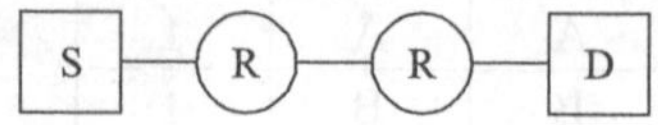

(a) Network layer - 4 times and Data link layer - 4 times
(b) Network layer - 4 times and Data link layer - 3 times
(c) Network layer - 4 times and Data link layer - 6 times
(d) Network layer - 2 times and Data link layer - 6 times **[2013 : 1 M]**

4.27 Consider the following three statements about link state and distance vector routing protocols, for a large network with 500 network nodes and 4000 links.

S1: The computational overhead in link state protocols is higher than in distance vector protocols.
S2: A distance vector protocol (with split horizon) avoids persistent routing loops, but not a link state protocol.
S3: After a topology change, a link state protocol will converge faster than a distance vector protocol.

Which one of the following is correct about S1, S2, and S3?

(a) S1, S2, and S3 are all true.
(b) S1, S2, and S3 are all false.
(c) S1 and S2 are true, but S3 is false.
(d) S1 and S3 are true, but S2 is false.

[2014 (Set-1) : 1 M]

4.28 Which one of the following is TRUE about the interior gateway routing protocols - Routing Information Protocol (RIP) and Open Shortest Path First (OSPF)

(a) RIP uses distance vector routing and OSPF uses link state routing
(b) OSPF uses distance vector routing and RIP uses link state routing
(c) Both RIP and OSPF use link state routing
(d) Both RIP and OSPF use distance vector routing **[2014 (Set-2) : 1 M]**

4.29 An IP machine Q has a path to another IP machine H via three IP routers $R1$, $R2$, and $R3$.

$Q — R1— R2 — R3 — H$

H acts as an HTTP server, and Q connects to H via HTTP and downloads a file. Session layer encryption is used, with DES as the shared key encryption protocol. Consider the following four pieces of information:

[I1] The URL of the file downloaded by Q
[I2] The TCP port numbers at Q and H
[I3] The IP addresses of Q and H
[I4] The link layer addresses of Q and H

Which of I1, I2, I3, and I4 can an intruder learn through sniffing at $R2$ alone?

(a) Only I1 and I2 (b) Only I1
(c) Only I2 and I3 (d) Only I3 and I4

[2014 (Set-2) : 2 M]

4.30 In one of the pairs of protocols given below, both the protocols can use multiple TCP connections between the same client and the server. Which one is that?

(a) HTTP, FTP (b) HTTP, TELNET
(c) FTP, SMTP (d) HTTP, SMTP

[2015 (Set-1) : 1 M]

4.31 Identify the correct sequence in which the following packets are transmitted on the network by a host when a browser requests a webpage from a remote server, assuming that the host has just been restarted.

(a) HTTP GET request, DNS query, TCP SYN
(b) DNS query, HTTP GET request, TCP SYN
(c) DNS query, TCP SYN, HTTP GET request
(d) TCP SYN, DNS query, HTTP GET request

[2016 (Set-2) : 1 M]

4.32 Consider an IP packet with a length of 4500 bytes that includes a 20-byte IPv4 header and a 40-byte TCP header. The packet is forwarded to an IPv4 router that supports a Maximum Transmission Unit (MTU) of 600 bytes. Assume that the length of the IP header in all the outgoing fragments of this packet is 20 bytes. Assume that the fragmentation offset value stored in the first fragment is 0.

The fragmentation offset value stored in the third fragment is ________. **[2018 : 2 M]**

4.33 Which of the following protocol pairs can be used to send and retrieve e-mails (in that order)?

(a) IMAP, SMTP (d) SMTP, MIME
(c) IMAP, POP3 (d) SMTP, POP3

[2019 : 1 M]

4.34 Consider the following statements about the functionality of an IP based router.

I. A router does not modify the IP packets during forwarding.
II. It is not necessary for a router to implement any routing protocol.
III. A router should reassemble IP fragments if the MTU of the outgoing link is larger than the size of the incoming IP packet.

Which of the above statements is/are TRUE?

(a) I only
(b) I and II only
(c) II and III only
(d) II only

[2020 : 1 M]

4.35 Assume that you have made a request for a web page through your web browser to a web server. Initially the browser cache is empty. Further, the browser is configured to send HTTP requests in non-persistent mode. The web page contains text and five very small images. The minimum number of TCP connections required to display the web page completely in your browser is ________. **[2020 : 1 M]**

4.36 Consider a computer network using the distance vector routing algorithm in its network layer. The partial topology of the network is as shown below.

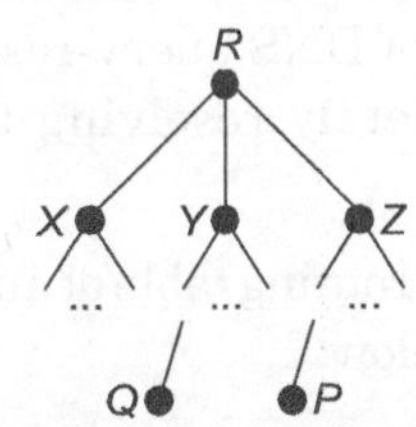

The objective is to find the shortest-cost path from the router R to routers P and Q. Assume that R does not initially know the shortest routes to P and Q. Assume that R has three neighbouring routers denoted as X, Y and Z. During one iteration, R measures its distance to its neighbours X, Y and Z as 3, 2 and 5, respectively. Router R gets routing vectors from its neighbours that indicate that the distance to router P from routers X, Y and Z are 7, 6 and 5, respectively. The routing vector also indicates that the distance to router Q from routers X, Y and Z are 4, 6 and 8, respectively. Which of the following statement(s) is/are correct with respect to the new routing table of R, after updation during this iteration?

(a) The distance from R to Q will be stored as 7.
(b) The distance from R to P will be stored as 10.
(c) The next hop router for a packet from R to Q is Z.
(d) The next hop router for a packet from R to P is Y. **[2021 (Set-2) : 2 M]**

4.37 Consider an enterprise network with two Ethernet segments, a web server and a firewall, connected via three routers as shown below:

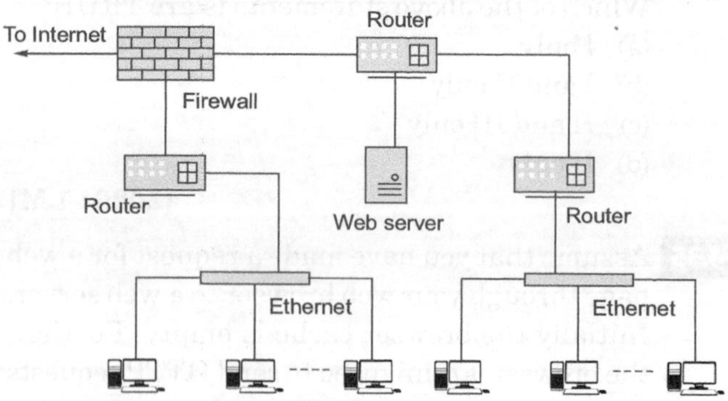

What is the number of subnets inside the enterprise network?
(a) 3 (b) 12
(c) 6 (d) 8 **[2022 : 1 M]**

4.38 Consider the resolution of the domain name www.gate.org.in by a DNS resolver. Assume that no resource records are cached anywhere across the DNS servers and that iterative query mechanism is used in the resolution. The number of DNS query-response pairs involved in completely resolving the domain name is ________. **[2022 : 1 M]**

4.39 Consider routing table of an organization's router shown below:

Subnet number	Subnet mask	Next hop
12.20.164.0	255.255.252.0	R1
12.20.170.0	255.255.254.0	R2
12.20.168.0	255.255.254.0	Interface 0
12.20.166.0	255.255.254.0	Interface 1
default		R3

Which of the following prefixes in CIDR notation can be collectively used to correctly aggregate all of the subnets in the routing table?
(a) 12.20.164.0/20
(b) 12.20.164.0/22
(c) 12.20.164.0/21
(d) 12.20.168.0/22

[2022 : 2 M]

4.40 Consider a network with three routers P, Q, R shown in the figure below. All the links have cost of unity.

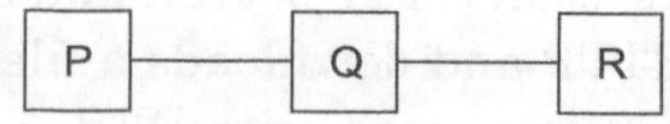

The routers exchange distance vector routing information and have converged on the routing tables, after which the link Q-R fails. Assume that P and Q send out routing updates at random times, each at the same average rate. The probability of a routing loop formation (rounded off to one decimal place) between P and Q, leading to count-to-infinity problem, is ________. **[2022 : 2 M]**

4.41 Which of the following statements is/are INCORRECT about the OSPF (Open Shortest Path First) routing protocol used in the Internet?
(a) OSPF implements Bellman-Ford algorithm to find shortest paths.
(b) OSPF uses Dijkstra's shortest path algorithm to implement least-cost path routing.
(c) OSPF is used as an inter-domain routing protocol.
(d) OSPF implements hierarchical routing.

[2023 : 1 M]

4.42 A user starts browsing a webpage hosted at a remote server. The browser opens a single connection to fetch the entire webpage from the server. The webpage consists of a top-level index page with multiple embedded image objects. Assume that all caches (e.g., DNS cache, browser cache) are all initially empty. The following packets leave the user's computer in some order.
(i) HTTP GET request for the index page.
(ii) DNS request to resolve the web server's name to its IP address.
(iii) HTTP GET request for an image object.
(iv) TCP SYN to open a connection to the web server.
Which one of the following is the CORRECT chronological order (earliest in time to latest) of the packets leaving the computer?
(a) (iv), (ii), (iii), (i)
(b) (iv), (ii), (i), (iii)
(c) (ii), (iv), (i), (iii)
(d) (ii), (iv), (iii), (i)

[2024 (Set-1) : 1 M]

4.43 Consider a network path $P-Q-R$ between nodes P and R via router Q. Node P sends a file of size 10^6 bytes to R via this path by splitting the file into chunks of 10^3 bytes each. Node P sends these chunks one after the other without any wait time between the successive chunk transmissions. Assume that the size of extra headers added to these chunks is negligible, and that the chunk size is less than the MTU.

Each of the links $P-Q$ and $Q-R$ has a bandwidth of 10^6 bits/sec, and negligible propagation latency. Router Q immediately transmits every packet it receives from P to R, with negligible processing and queueing delays. Router Q can simultaneously receive on link $P-Q$ and transmit on link $Q-R$.

Assume P starts transmitting the chunks at time $t = 0$.

Which one of the following options gives the time (in seconds, rounded off to 3 decimal places) at which R receives all the chunks of the file?

(a) 16.000

(b) 8.000

(c) 8.008

(d) 15.992

[2024 (Set-1) : 2 M]

Answers Routing and Application Layer

4.1	(b)	**4.2**	(b)	**4.3**	(c)	**4.4**	(b)	**4.5**	(c)	**4.6**	(d)	**4.7**	(a)	**4.8**	(a)	**4.9**	(c)
4.10	(a)	**4.11**	(d)	**4.12**	(d)	**4.13**	(b)	**4.14**	(d)	**4.15**	(c)	**4.16**	(b)	**4.17**	(b)	**4.18**	(d)
4.19	(c)	**4.20**	(b)	**4.21**	(c)	**4.22**	(a)	**4.23**	(c)	**4.24**	(c)	**4.25**	(c)	**4.26**	(c)	**4.27**	(d)
4.28	(a)	**4.29**	(c)	**4.30**	(a)	**4.31**	(c)	**4.32**	(144)	**4.33**	(d)	**4.34**	(d)	**4.35**	(6)	**4.36**	(a, d)
4.37	(c)	**4.38**	(4)	**4.39**	(b, d)	**4.40**	(0.5)	**4.41**	(a, c)	**4.42**	(c)	**4.43**	(c)				

Explanations Routing and Application Layer

4.1 (b)

- Circuit switching has 3 phase : connection establishment, Data transfer and tear down, while packet switching has only one phase data transfer based on routing.
- Circuit switching has less variation in packets than packet switching.
- Circuit switching send packet in order, while packet switching send out of order, so reordering needs in packet switching.
- In packet switching no resource reservation needed while circuit switching has resource reservation so packet switching leads to better utilization of resources i.e. bandwidth.
- In circuit switching no packet processing needed at intermediate path routers while in packet switching needed. So, packet switching require more packet processing time than circuit switching.

4.2 (b)

RCPT: Recipient to, As the name suggest it is used in SMTP(Simple Mail Transfer protocol).
HEAD: This is used in HTTP to get the meta-information, to decide the category of packet.
Prompt: Turns off prompting for individual files when using the mget or mput commands.

4.3 (c)

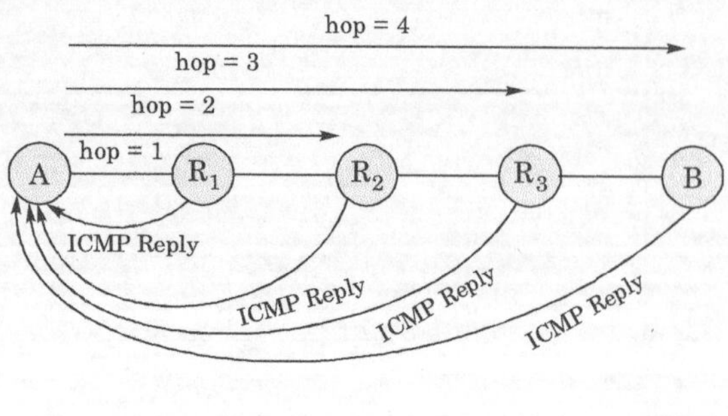

Trace route reports a possible route that is taken by packet moving from host A to host B by ensuring that an ICMP reply packet returned to A by each router enroute to B, in the ascending order of their hop distance.

4.4 (b)

In distance vector routing the count-to-infinity problem happens when a router is unable to reach an adjoining network. A second router, 1 hop away from the first router, thinks that the unreachable network is 2 hops away. Meanwhile, the first router then updates its records to say it is 3 hops away from the unreachable network based on the fact it is 1 hop from the second router, which says it is 2 hops from the unreachable network.
The routers continue incrementing their hop count until the maximum (15), "infinity", is reached.

4.5 (c)

The appropriate sequence of steps performing the reverse lookup of 145.128.16.8 are:
Step-1: Query a Name server for in-addr.arpa. Because information regarding the IP address 145.128.16.8 is located at domain name. 145.128.16.8 in-addr.arpa.
Step-2: Second setup is request Name server for 128.45.in-addr.arpa domain.
This is because the query is for PTR records. The resolver reverses the address and appends the in-addr.arpa to the end of the reversed address. This forms the fully qualified domain name for which to be searched is reverse lookup zone.

4.6 (d)

HELO: Initiates a conversation with the mail server. When using this command you can specify your domain name so that the mail server knows who you are.

The PORT command is sent by an FTP client to establish a secondary connection (address and port) for data to travel over.

4.7 (a)

B1-B5-B2-B3-B4

4.8 (a)

Use spanning tree generated in previous question.

4.9 (c)

Domain Name System (DNS) maps a name onto an IP address, an application program calls a library procedure called the resolver, passing name as a parameter. The resolver sends a UDP packet to a load DNS server, which then looks up the name and returns the IP address to the resolver, which then returns it to caller.

So DNS uses transport layer protocol UDP.

4.10 (a)

Routing table of F:

A	(8) (34) (32) (27) min(F→A, F→E→A, F→D→A, F→G→A)	= (8)
B	(48) (37) (20) (30) min(F→A→B, F→E→B, F→D→B, F→G→B)	= (20)
C	(22) (17) (42) (28) min(F→A→C, F→E→C, F→D→C, F→G→C)	= (17)
D	(25) (30) (12) (25) min(F→A→D, F→E→D, F→D→D, F→G→D)	= (12)
E	(29) (10) (12) (28) min(F→A→E, F→E→E, F→D→E, F→G→E)	= (10)
F	F→F	= (0)
G	(32) (32) (34) (6) min(F→A→G, F→E→G, F→D→G, F→G)	= (6)

4.11 (d)

Consider Complete tree:

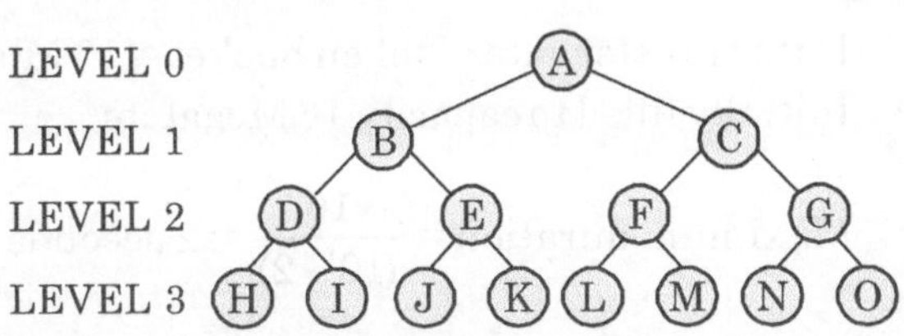

Forward packet to the router at Level 3: Total 6 hops are required if A wants to communicate with any level 3 node. Similarly, 5 hops are required if H wants to communicate with any level 2 node, 4 hops are required if H wants to communicate with any level 1 node and 3 hops are required if H wants to communicate with any level 0 node.

Hops required if H wants to communicates with all other nodes

$= (8 - 1)*6 + 4*5 + 2*4 + 1*3 = 73.$

If all 8 nodes of level 3 communicates with all other nodes then hops required = 73 * 8 = 584.

Similarly, Hops required if D wants to communicates with all other nodes = 8*5 + (4 – 1)*4 + 2*3 + 1*2 = 60.

If all 4 nodes of level 2 communicates with all other nodes then hops required = 60*4 = 240.

Hops required if B wants to communicates with all other nodes

$= 8*4 + 4*3 + (2 - 1)*2 + 1*1 = 47.$

If all 2 nodes of level 1 communicates with all other nodes then hops required = 47*2 = 94.

Hops required if A wants to communicates with all other nodes = 8*3+4*2+2*1 = 34.

Total hops required when all nodes communicate with all other nodes

$= 584 + 240 + 94 + 34 = 952.$

Total number of messages

$= 2*{}^{15}C_2 = 2*(15*14/2) = 2*105 = 210.$

(Here 2 is multiplied with ${}^{15}C_2$, because in communication between A and B, A sends message to B and B sends message to A)

Mean number of hops per message

= 952/210 = 4.53

4.12 (d)

IMAP : It distributes mail boxes across multiple servers.

FTP: Requires two ports: 20 and 21 for FTP-data and FTP-control respectively.

HTTP: HTTP is a stateless protocol. Hence not a state sensitive protocol.

DNS: This protocol maintains it's database is a structured and hierarchial manner.

SMTP: Intersect standard for e-mail transmission and not suitable for client server communication.

4.13 (b)

P. SMTP = Application Layer protocol
Q. BGP = Network Layer protocol
R. TCP = Transport Layer protocol
S. PPP = Data Link Layer protocol

4.14 (d)

- Count to infinity problem occur only in distance vector router since each router dependent only on neighbours but link state routing using complete information of network.
- Link state routing stabilize very fast than distance vector routing because it uses flooding.
- Distance vector and link state routing apply on each router i.e. shortest path calculate at each router.
- Distance vector require less number of messages than link state routing.

4.15 (c)

For node A routing table will be:

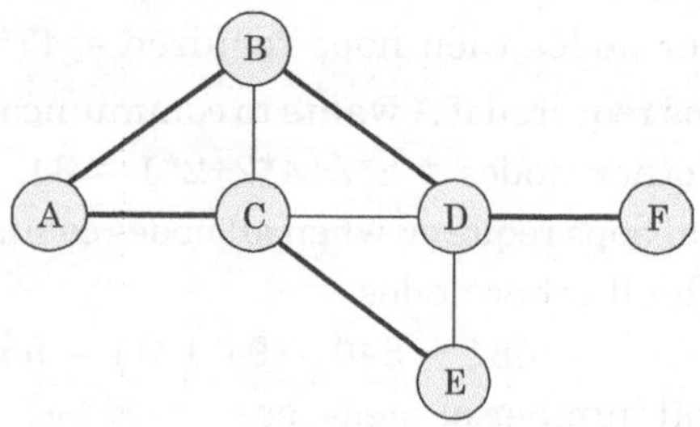

Bold edges showing the path:

Table for node A		
A	–	–
B	B	1
C	C	1
D	B	2
E	C	2
F	C	3

For node B routing table will be:

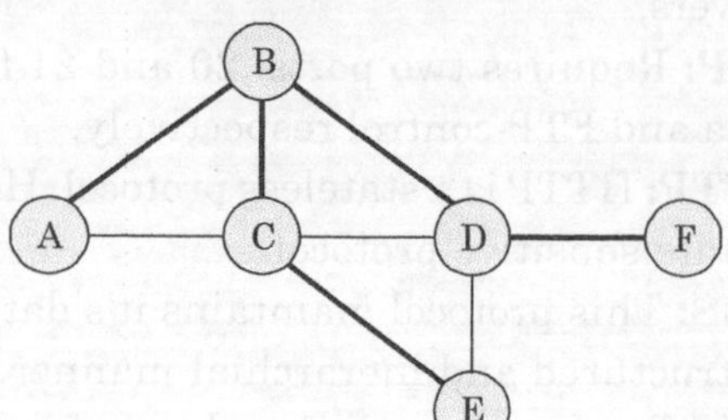

Bold edges showing the path:

Table for node B		
A	A	1
B	–	–
C	C	1
D	D	1
E	C	2
F	D	2

For node C routing table will be:

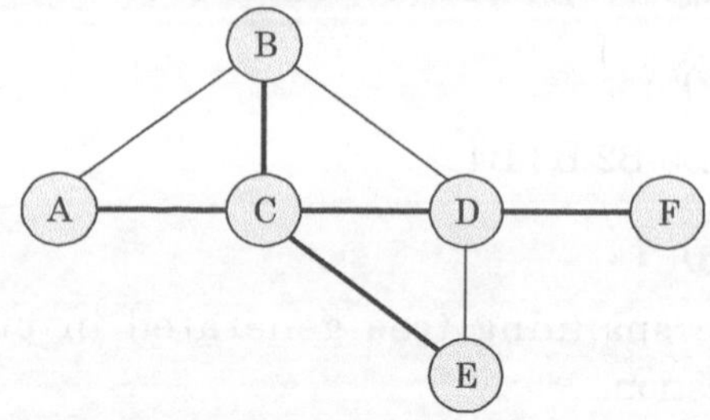

Bold edges showing the path:

Table for node C		
A	A	1
B	B	1
C	–	–
D	D	1
E	E	1
F	D	2

For node D routing table will be:

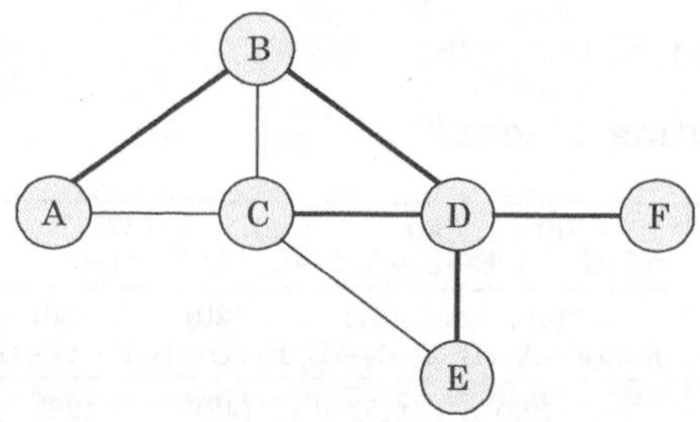

Bold edges showing the path:

Table for node D		
A	B	2
B	B	1
C	C	1
D	–	–
E	E	1
F	F	1

4.17 (b)

Data transfer rate of token bucket = 10 Mbps
Initially filled to capacity 16 Megabits

$$\text{Maximum duration} = \frac{16}{(10-2)} = 2 \text{ seconds}$$

4.18 (d)

Internet chat, web browsing and E-mail all are client-server application. Ping is a utility. It is mainly used to check the connection between two computers, there is a chance both are client or one is client and another server. In chat system first user authentication required and it requires server.

4.19 (c)

Routing table

for R1

		Next hop
R2	5	R3
R3	3	R3
R4	12	R3
R5	12	R3
R6	16	R3

for R2

		Next hop
R1	5	R3
R3	2	R3
R4	7	R4
R5	8	R4
R6	12	R4

for R3

		Next hop
R1	3	R1
R2	2	R2
R4	9	R4
R5	9	R5
R6	13	R5

for R4

		Next hop
R1	12	R2
R2	7	R2
R3	9	R2
R5	1	R5
R6	5	R5

for R5

		Next hop
R1	12	R3
R2	8	R4
R3	9	R3
R4	1	R4
R6	4	R6

for R6

		Next hop
R1	16	R5
R2	12	R5
R3	13	R5
R4	5	R5
R5	4	R5

So it is clear visualize from the all routing table construction that we never use the direct path between R1 —6— R2 and R4 —8— R6.

So two links never be used for carrying data.

4.20 (b)

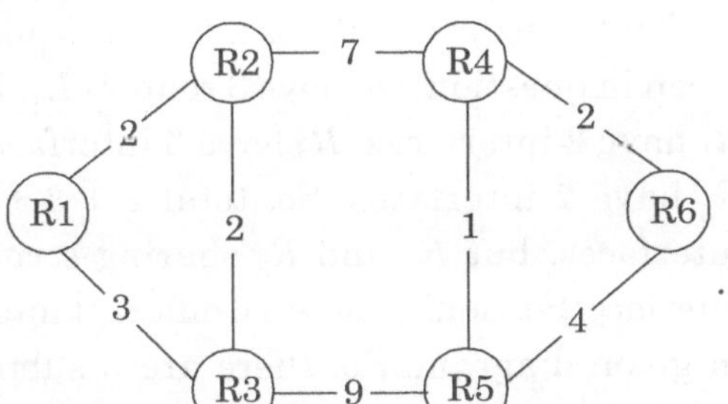

As in the previous question

R1 —— R2 and

R4 —— R6 not used.

So we give them weight = 2 by Apply.

Same method only one link R5-R6 not used.

4.21 (c)

SMTP is typically used by user clients for sending mails. POP is used by clients for receiving mails. Checking mails in web browser is a simple HTTP process.

4.22 (a)

In the next round, every node will send and receive distance vectors to and from neighbours, and update its distance vector. N3 will receive (1, 0, 2, 7, 3) from N2 and it will update distances to N1 and N5 as 3 and 5 respectively.

4.23 (c)

In the next round, N3 will receive distance from N2 to N1 as infinite. It will receive distance from N4 to N1 as 8.

So it will update distance to N1 as 8 + 2 = 10.

4.24 (c)

Message	→	Application layer
Segment	→	Transport layer
Datagram	→	Network layer
Frame	→	Data link layer

4.25 (c)

UDP: Transport layer protocol which is unreliable but fast.

TCP: Transport layer connection oriented protocol which is secure and reliable but comparatively slow.

So for real time multimedia we need fast processing so UDP is suitable for it.

For file transfer we need security so TCP is suitable for file transfer.

DNS always use UDP.

For email we need security so uses TCP.

Option (c) is correct.

4.26 (c)

In the layered architecture sender and receiver uses all the 5 layers and router do processing upto the network layer.

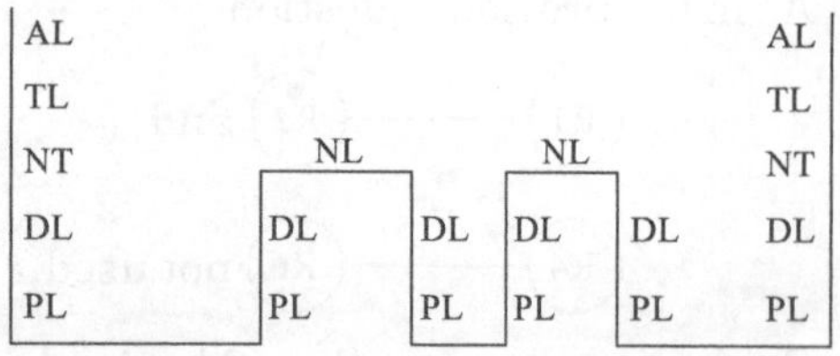

So network layer is visited 4 times and data link layer is visited 6 times.

4.27 (d)

S1 is TRUE: Processing time in link state protocol is high because of link state database.

S2 is FALSE: Link state protocol is a loop free protocol. Link state protocol avoids persistent routing loops.

S3 is TRUE: Link state protocol is fast convergence because of link state database i.e., it contains complete information of the network. This information is given to the router by the link state packets.

4.28 (a)

RIP uses distance vector routing
OSPF uses link state routing

4.29 (c)

Given R_1, R_2 and R_3 are routers
At R_2 intruder can learn the TCP port numbers and IP address of Q and H.

4.30 (a)

HTTP and FTP protocols can use multiple TCP connections between the same client and the server. FTP uses data and control connections used with two separate TCP connections.

4.31 (c)

DNS query, TCP SYN, HTTP GET request.

4.32 (144)

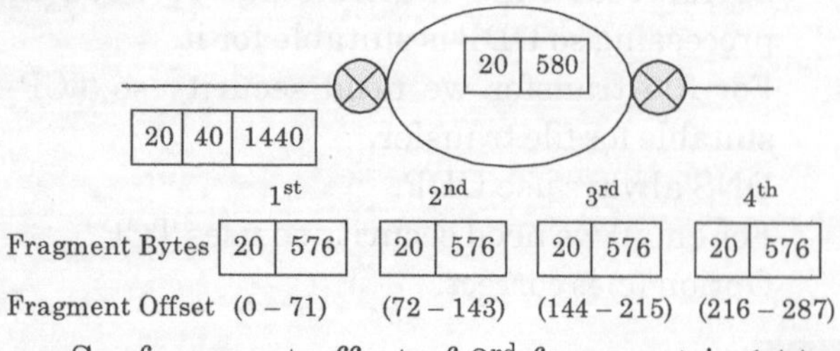

So, fragment offset of 3rd fragment is 144.

4.33 (d)

SMTP is push protocol and to send e-mail and POP3 is pull protocol i.e. to retrieve e-mail.

4.34 (d)

I. A router modifies the IP packets during forwarding because TTL is changing.

II. A router can also be used in LAN network it does not require routing protocol at that time.

III. Packet fragmentation is done if packet size is more than MTU.

4.35 (6)

In non persistent HTTP for every objects there is a TCP connection required.

Hence 1 TCP connection for text and 5 TCP connections for images required.

So total 6 connections are required.

4.36 (a, d)

Given R gets the distance vector (3, 2, 5)
After the one iteration distance vector from X to P, Y to P, and Z to P is (7, 6, 5) respectively.
The distance vector from R to P via XYZ is $(3 + 7, 2 + 6, 5 + 5) = (10, 8, 10)$
So, take minimum distance from R to P which is 8 via Y.
After the iteration distance vector from X to Q, Y to Q, Z to Q is (4, 6, 8) respectively.
The distance vector from R to Q via XYZ is $(3 + 4, 2 + 6, 5 + 8) = (7, 8, 13)$.
So, take minimum distance from R to Q which is 7 via X.

4.37 (c)

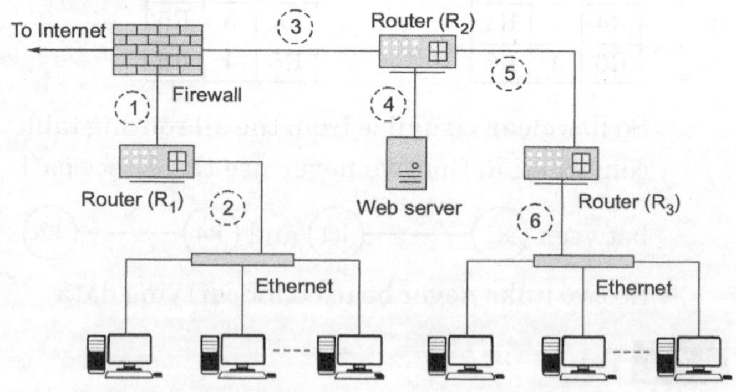

Each interface of a router is a subnet. In diagram given in question, we have 3 routers R_1, R_2, R_3. R_1 have 2 interfaces, R_2 have 3 interfaces and R_3 have 2 interfaces. So, total 2 + 3 + 2 = 7 interfaces, but R_2 and R_3 sharing a common interface. So, finally there a 6 different interfaces in given diagram. So, there are 6 subnets in given enterprises.

4.38 **(4)**

In the iterative query, the DNS resolver goes to root server and then forwards to top level domain and forwards to secondary to level domain and gets IP address from authoritative DNS server.

4.39 **(b, d)**

12.20.164.0 255.255.252.0 R_1

164.0 = | 101001 | 00.00000000

12.20.170.0 255.255.254.0 R_2

170.0 = | 101010 | 1 | 0.00000000

12.20.168.0 255.255.254.0 Interface 0

168.0 = | 101010 | 0 | 0.00000000

12.20.166.0 255.255.254.0 Interface 1

166.0 = | 101001 | 1 | 0.00000000

Default R_3

164.0	**168.0**
165.0	169.0
166.0	**170.0**
167.0	171.0

Multiples of 4 $\Rightarrow 4 \times 256 \Rightarrow 2^{10} \Rightarrow 2^{32-22}$

0, 4, 8, 12, 16, 20, 24, 28, 32, 36, 40, 44, 48, 52, 56, 60, 64, 68, 72, 76, 80, 84......... 156, 160, **[164]**, **[168]**, 172, ...

Multiples of 8 $\Rightarrow 8 \times 256$

$\Rightarrow 2^3 \times 2^8 \Rightarrow 2^{11} \Rightarrow 2^{32-21}$

152, 160, 168, 176, 184, ...

Multiples of 16 $\Rightarrow 16 \times 256$

$\Rightarrow 2^4 \times 2^8 \Rightarrow 2^{12} \Rightarrow 2^{32-21}$

154, 160, 176, 192, ...

4.40 **(0.5)**

It is given that P and Q are sending routing update at same average rate, so in a particular time period of sending update, there are two possibilities, one is P sends before Q and other is Q sends before P.

Now, We have 2 conditions:

1. Q sends update to P first, then P finds the value of R as ∞ and count-to-infinity will not occur.
2. P sends update to Q first, then Q found that P has a way to reach R, so it is count-to-infinity problem.

As Q - R is broken , So R cannot send update to anyone.

From above two conditions (total possibilities) only one condition leads to count-to-infinity problem and each possibility have equal probability of occurrence, so probability is $\frac{1}{2} = 0.5$ for count-to-infinity.

4.41 **(a, c)**

The protocol minimizes the amount of routing protocol traffic it produces by employing the Dijkstra algorithm to recalculate routes whenever the network topology changes.

It offers support for numerous, equally expensive approaches.

Area routing offers a multi-level hierarchy (two levels for OSPF) that conceals the topology inside a specified area of the AS from routers outside of that area. As a result, routing protocol traffic is reduced and an additional degree of routing protection is made possible.

To ensure that only reputable routers participate in the AS's routing exchanges, all protocol exchanges can be verified.

4.42 **(c)**

Sequence from client side:

(i) DNS request to get IP address of server.

(ii) Make a TCP connection with sever.

(iii) HTTP GET request for the index page.

(iv) HTTP GET request for an image object as image is inside index page.

■■■■

UNIT

XI

General Aptitude

Syllabus

Verbal Aptitude : Basic English grammar: tenses, articles, adjectives, prepositions, conjunctions, verb-noun agreement, and other parts of speech Basic vocabulary: words, idioms, and phrases in context Reading and comprehension Narrative sequencing.

Quantitative Aptitude : Data interpretation: data graphs (bar graphs, pie charts, and other graphs representing data), 2- and 3-dimensional plots, maps, and tables Numerical computation and estimation: ratios, percentages, powers, exponents and logarithms, permutations and combinations, and series Mensuration and geometry Elementary statistics and probability

Analytical Aptitude : Logic: deduction and induction ; Analogy : Numerical relations and reasoning

Spatial Aptitude : Transformation of shapes: translation, rotation, scaling, mirroring, assembling, and grouping Paper folding, cutting, and patterns in 2 and 3 dimensions

Analysis of Previous GATE Papers

Exam Year	1 Mark Ques.	2 Mark Ques.	Total Marks
2010	5	5	**15**
2011	5	5	**15**
2012	5	5	**15**
2013	5	5	**15**
2014 Set-1	5	5	**15**
2014 Set-2	5	5	**15**
2014 Set-3	5	5	**15**
2015 Set-1	5	5	**15**
2015 Set-2	5	5	**15**
2015 Set-3	5	5	**15**
2016 Set-1	5	5	**15**

Exam Year	1 Mark Ques.	2 Mark Ques.	Total Marks
2016 Set-2	5	5	**15**
2017 Set-1	5	5	**15**
2017 Set-2	5	5	**15**
2018	5	5	**15**
2019	5	5	**15**
2020	5	5	**15**
2021 Set-1	5	5	**15**
2021 Set-2	5	5	**15**
2022	5	5	**15**
2023	5	5	**15**
2024 Set-1	5	5	**15**
2024 Set-2	5	5	**15**

General Aptitude

1.1 Which of the following options is the closest in meaning to the word below?

Circuitous

(a) Cyclic (b) Indirect
(c) Confusing (d) Crooked

[2010 : 1 M]

1.2 The question below consists of a pair of related words followed by four pairs of words. Select the pair that best expresses the relation in the original pair.

Unemployed : Worker

(a) Fallow : Land (b) Unaware : Sleeper
(c) Wit : Jester (d) Renovated : House

[2010 : 1 M]

1.3 Choose the most appropriate word from the options given below to complete the following sentence:

If we manage to .... our natural resources, we would leave a better planet for our children.

(a) uphold (b) restrain
(c) cherish (d) conserve **[2010 : 1 M]**

1.4 Choose the most appropriate word from the options given below to complete the following sentence:

His rather casual remarks on politics .... his lack of seriousness about the subject.

(a) masked (b) belied
(c) betrayed (d) suppressed

[2010 : 1 M]

1.5 25 persons are in a room. 15 of them play hockey, 17 of them play football and 10 of them play both hockey and football. Then the number of persons playing neither hockey nor football is

(a) 2 (b) 17
(c) 13 (d) 3 **[2010 : 1 M]**

1.6 Modern warfare has changed from large scale clashes of armies to suppression of civilian populations. Chemical agents that do their work silently appear to be suited to such warfare; and regretfully, there exist people in military establishments who think that chemical agents are usefull fools for their cause.

Which of the following statements best sums up the meaning of the above passage?

(a) Modern warfare has resulted in civil strife.
(b) Chemical agents are useful in modern warfare.
(c) Use of chemical agents in warfare would be undesirable.
(d) People in military establishments like to use chemical agents in war. **[2010 : 2 M]**

1.7 If 137 + 276 = 435 how much is 731 + 672?

(a) 534 (b) 1403
(c) 1623 (d) 1531 **[2010 : 2 M]**

1.8 5 skilled workers can build a wall in 20 days; 8 semi-skilled workers can build a wall in 25 days; 10 unskilled workers can build a wall in 30 days. If a team has 2 skilled, 6 semi-skilled and 5 unskilled workers, how long will it take to build the wall?

(a) 20 days (b) 18 days
(c) 16 days (d) 15 days **[2010 : 2 M]**

1.9 Given digits 2, 2, 3, 3, 3, 4, 4, 4, 4 how many distinct 4 digit numbers greater than 3000 can be formed?

(a) 50 (b) 51
(c) 52 (d) 54 **[2010 : 2 M]**

1.10 Hari (H), Gita (G), Irfan (I) and Saira (S) are siblings (i.e. brothers and sisters). All were born on Ist January. The age difference between any two successive siblings (that is born one after another) is less than 3 years. Given the following facts:

1. Hari's age + Gita's age > Irfan's age + Saira's age.
2. The age difference between Gita and Saira is 1 year. However, Gita is not the oldest and Saira is not the youngest.
3. There are no twins.

In what order were they born (oldest first)?

(a) HSIG (b) SGHI
(c) IGSH (d) IHSG **[2010 : 2 M]**

1.11 Which of the following options is the closest in the meaning to the word below:

Inexplicable

(a) Incomprehensible (b) Indelible
(c) Inextricable (d) Infallible

[2011 : 1 M]

1.12 If $\text{Log}(P) = \left(\frac{1}{2}\right)\text{Log}(Q) = \left(\frac{1}{3}\right)\text{Log}(R)$, then which of the following options is TRUE?

(a) $P^2 = Q^3R^2$ (b) $Q^2 = PR$
(c) $Q^2 = R^3P$ (d) $R = P^2Q^2$

[2011 : 1 M]

1.13 Choose the most appropriate word(s) from the options given below to complete the following sentence.

In contemplated __________ Singapore for my vacation but decided against it.

(a) to visit (b) having to visit
(c) visiting (d) for a visit

[2011 : 1 M]

1.14 Choose the most appropriate word from the options given below to complete the following sentence.

If you are trying to make a strong impression on your audience, you cannot do so by being understated, tentative or

(a) hyperbolic (b) restrained
(c) argumentative (d) indifferent

[2011 : 1 M]

1.15 Choose the word from the options given below that is most nearly opposite in meaning to the given word:

Amalgamate

(a) merge (b) split
(c) collect (d) separate **[2011 : 1 M]**

1.16 Few school curricula include a unit on how to deal with bereavement and grief, and yet all students at some point in their lives suffer from losses through death and parting.

Based on the above passage which topic would not be included in a unit on bereavement?

(a) how to write a letter of condolence
(b) what emotional stages are passed through in the healing process
(c) what the leading causes of death are
(d) how to give support to a grieving friend

[2011 : 2 M]

1.17 P, Q, R and S are four types of dangerous microbes recently found in a human habitat. The area of each circle with its diameter printed in brackets represents the growth of a single microbe surviving human immunity system within 24 hours of entering the body. The danger to human beings varies proportionately with the toxicity, potency and growth attributed to a microbe shown in the figure below:

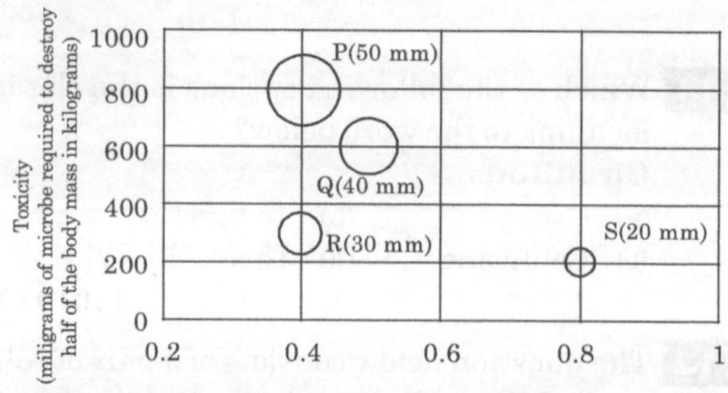

A pharmaceutical company is contemplating the development of a vaccine against the most dangerous microbe. Which microbe should the company target in its first attempt?

(a) P (b) Q
(c) R (d) S **[2011 : 2 M]**

1.18 The variable cost (V) of manufacturing a product varies according to the equation V = 4q, where q is the quantity produced. The fixed cost (F) of production of same product reduces with q according to the equation F = 100/q. How many units should be produced to minimize the total cost (V + F)?

(a) 5 (b) 4
(c) 7 (d) 6 **[2011 : 2 M]**

1.19 A transporter receives the same number of orders each day. Currently, he has some pending orders (backlog) to be shipped. If he uses 7 trucks, then at the end of the 4th day he can clear all the orders. Alternatively, if he uses only 3 trucks, then all the orders are cleared at the end of the 10th day. What is the minimum number of trucks required so that there will be no pending order at the end of the 5th day?

(a) 4 (b) 5
(c) 6 (d) 7 **[2011 : 2 M]**

1.20 A container originally contains 10 litres of pure spirit. From this container 1 litre of spirit is replaced with 1 litre of water. Subsequently, 1 litre of the mixture is again replaced with 1 litre of water and this processes is repeated one more time. How much spirit is now left in the container?

(a) 7.58 litres (b) 7.84 litres
(c) 7 litres (d) 7.29 litres

[2011 : 2 M]

1.21 The cost function for a product in a firm is given by $5q^2$, where q is the amount of production. The firm can sell the product at a market price of ₹ 50 per unit. The number of units to be produced by the firm such that the profit is maximized is
(a) 5 (b) 10
(c) 15 (d) 25 **[2012 : 1 M]**

1.22 Choose the most appropriate alternative from the options given below to complete the following sentence:
Despite several ____ the mission succeeded in its attempt to resolve the conflict.
(a) attempts (b) setbacks
(c) meetings (d) delegations
[2012 : 1 M]

1.23 Which one of the following options is the closest in meaning to the word given below?
Mitigate
(a) Diminish (b) Divulge
(c) Dedicate (d) Denote **[2012 : 1 M]**

1.24 Choose the grammatically INCORRECT sentence:
(a) They gave us the money back less the service charges to Three Hundred rupees.
(b) This country's expenditure is not less than that of Bangladesh.
(c) The committee initially asked for a funding of Fifty Lakh rupees, but later settled for a lesser sum.
(d) This country's expenditure on educational reforms is very less. **[2012 : 1 M]**

1.25 Choose the most appropriate alternative from the options given below to complete the following sentence:
Suresh's dog is the one ____ was hurt in the stampede.
(a) that (b) which
(c) who (d) whom **[2012 : 1 M]**

1.26 Wanted Temporary, Part-time persons for the post of Field Interviewer to conduct personal interviews to collect and collate economic data. Requirements: High School-pass, must be available for Day, Evening and Saturday work. Transportation paid, expenses reimbursed.
Which one of the following is the best inference from the above advertisement?
(a) Gender-discriminatory
(b) Xenophobic
(c) Not designed to make the post attractive
(d) Not gender-discriminatory **[2012 : 2 M]**

1.27 A political party orders an arch for the entrance to the ground in which the annual conventions is being held. The profile of the arch follows the equation $y = 2x - 0.1\,x^2$ where y is the height of the arch in meters. The maximum possible height of the arch is
(a) 8 meters (b) 10 meters
(c) 12 meters (d) 14 meters
[2012 : 2 M]

1.28 An automobile plant contracted to buy shock absorbers from two supplies X and Y. X supplies 60% and Y supplies 40% of the shock absorbers. All shock absorbers are subjected to a quality test. The ones that pass the quality test are considered reliable. Of X's shock absorbers, 96% are reliable. OF Y's shock absorbers, 72% are reliable. The probability that a randomly chosen shock absorber, which is found to be reliable, is made by Y is
(a) 0.288 (b) 0.334
(c) 0.667 (d) 0.720
[2012 : 2 M]

1.29 Which of the following assertions are CORRECT?
P: Adding 7 to each entry in a list adds 7 to the mean of the list
Q: Adding 7 to each entry in a list adds 7 to the standard deviation of the list
R: Doubling each entry n a list doubles the mean of the list
S: Doubling each entry in a list leaves the standard deviation of the list unchanged
(a) P, Q (b) Q, R
(c) P, R (d) R, S **[2012 : 2 M]**

1.30 Given the sequence of terms, AD CG FK JP, the next time is
(a) OV (b) OW
(c) PV (d) PW
[2012 : 2 M]

1.31 Which one of the following options is the closest in meaning to the word given below?
Nadir
(a) Highest (b) Lowest
(c) Medium (d) Integration
[2013 : 1 M]

1.32 Complete the sentence:
Universalism is to particularism as diffuseness is to ____.
(a) specificity (b) neutrality
(c) generality (d) adaptation
[2013 : 1 M]

1.33 What will be the maximum sum of 44, 42, 40, ...?

(a) 502 (b) 504
(c) 506 (d) 500 **[2013 : 1 M]**

1.34 Were you a bird, you ______ in the sky.

(a) would fly (b) shall fly
(c) should fly (d) shall have flown **[2013 : 1 M]**

1.35 Choose the grammatically INCORRECT sentence:

(a) He is of Asian origin.
(b) They belonged to Africa
(c) She is an European.
(d) They migrated from India to Australia. **[2013 : 1 M]**

1.36 Find the sum of the expression

$$\frac{1}{\sqrt{1}+\sqrt{2}}+\frac{1}{\sqrt{2}+\sqrt{3}}+\frac{1}{\sqrt{3}+\sqrt{4}}+.....+\frac{1}{\sqrt{80}+\sqrt{81}}$$

(a) 7 (b) 8
(c) 9 (d) 10 **[2013 : 2 M]**

1.37 Out of all the 2-digit integers between 1 and 100, a 2-digit number has to be selected at random. What is the probability that the selected number is not divisible by 7?

(a) $\frac{13}{90}$ (b) $\frac{12}{90}$
(c) $\frac{78}{90}$ (d) $\frac{77}{90}$ **[2013 : 2 M]**

1.38 After several defeats in wars, Robert Bruce went in exile and wanted to commit suicide. Just before committing suicide, the came across a spider attempting tirelessly to have its net. Time and again, the spider failed but the did not deter it to refrain from making attempts. Such attempts by the spider made Bruce curious. Thus, Bruce started observing the near-impossible goal of the spider to have the net. Ultimately, the spider succeeded in having its net despite several failures. Such act of the spider encouraged Bruce not to commit suicide. And then, Bruce went back again and won many a battle, and the rest is history.
Which one of the following assertions is best supported by the above information?

(a) Failure is the pillar of success
(b) Honesty is the best policy
(c) Life begins and ends with adventures
(d) No adversity justifies giving up hope **[2013 : 2 M]**

1.39 A tourist covers half of this journey by train at 60 km/h, half of the remainder by bus at 30 km/h and the rest by cycle at 10 km/h. The average speed of the tourist in km/h during his entire journey is

(a) 36 (b) 30
(c) 24 (d) 18 **[2013 : 2 M]**

1.40 The current erection cost of a structure is Rs. 13,200. If the labour wages per day increase by $\frac{1}{5}$ of the current wages and the working hours decrease by $\frac{1}{24}$ of the current period, then the new cost of erection in Rs., is

(a) 16, 500 (b) 15, 180
(c) 11, 000 (d) 10, 120 **[2013 : 2 M]**

1.41 Which of the following options is the closest in meaning to the phrase underlined in the sentence below?

It is fascinating to see life forms <u>cope with</u> varied environmental conditions.

(a) adopt to (b) adapt to
(c) adept in (d) accept with **[2014 (Set-1) : 1 M]**

1.42 Choose the most appropriate word from the options given below to complete the following sentence.

He could not understand the judges awarding her the first prize, because he thought that her performance was quite ________.

(a) superb (b) medium
(c) mediocre (d) exhilarating **[2014 (Set-1) : 1 M]**

1.43 In a press meet on the recent scam, the minister said, "The buck stops here". What did the minister convey by the statement?

(a) He wants all the money
(b) He will return the money
(c) He will assume final responsibility
(d) He will resist all enquiries **[2014 (Set-1) : 1 M]**

1.44 If $\left(\frac{z+1}{z}\right)^2 = 98$, compute $\left(\frac{z^2+1}{z^2}\right)$. **[2014 (Set-1) : 1 M]**

1.45 The roots of $ax^2 + bx + c = 0$ are real and positive. a, b and c are real. Then $ax^2 + b\,|x| + c = 0$ has

(a) no roots (b) 2 real roots
(c) 3 real roots (d) 4 real roots

[2014 (Set-1) : 1 M]

1.46 The Palghat Gap (or Palakkad Gap), a region about 30 km wide in the southern part of the Western Ghats in India, is lower than the hilly terrain to its north and south. The exact reasons for the formation of this gap are not clear. It results in the neighbouring regions of Tamil Nadu getting more rainfall from the South West monsoon and the neighbouring regions of Kerala having higher summer temperatures.
What can be inferred from this passage?
(a) The Palghat gap is caused by high rainfall and high temperatures in southern Tamil Nadu and Kerala
(b) The regions in Tamil Nadu and Kerala that are near the Palghat Gap are low-lying
(c) The low terrain of the Palghat Gap has a significant impact on weather patterns in neighbouring parts of Tamil Nadu and Kerala
(d) Higher summer temperatures result in higher rainfall near the Palghat Gap area

[2014 (Set-1) : 2 M]

1.47 Geneticists say that they are very close to confirming the genetic roots of psychiatric illnesses such as depression and schizophrenia, and consequently, that doctors will be able to eradicate these diseases through early identification and gene therapy.
On which of the following assumptions does the statement above rely?
(a) Strategies are now available for eliminating psychiatric illnesses
(b) Certain psychiatric illnesses have a genetic basis
(c) All human diseases can be traced back to genes and how they are expressed
(d) In the future, genetics will become the only relevant field for identifying psychiatric illnesses

[2014 (Set-1) : 2 M]

1.48 Round-trip tickets to a tourist destination are eligible for a discount of 10% on the total fare. In addition, groups of 4 or more get a discount of 5% on the total fare. If the one way single person fare is Rs 100, a group of 5 tourists purchasing round-trip tickets will be charged Rs ______.

[2014 (Set-1) : 2 M]

1.49 In a survey, 300 respondents were asked whether they own a vehicle or not. If yes, they were further asked to mention whether they own a car or scooter or both. Their responses are tabulated below. What percent of respondents do not own a scooter?

		Men	Women
Own vehicle	Car	40	34
	Scooter	30	20
	Both	60	46
Do not own vehicle		20	50

[2014 (Set-1) : 2 M]

1.50 When a point inside of a tetrahedron (a solid with four triangular surfaces) is connected by straight lines to its corners, how many (new) internal planes are created with these lines? ______.

[2014 (Set-1) : 2 M]

1.51 Choose the most appropriate phrase from the options given below to complete the following sentence. India is a post-colonial country because
(a) it was a former British colony
(b) Indian Information Technology professionals have colonized the world
(c) India does not follow any colonial practices
(d) India has helped other countries gain freedom

[2014 (Set-2) : 1 M]

1.52 Who ______ was coming to see us this evening?
(a) you said (b) did you say
(c) did you say that (d) had you said

[2014 (Set-2) : 1 M]

1.53 Match the columns:

Column 1	Column 2
1. eradicate	P. misrepresent
2. distort	Q. soak completely
3. saturate	R. use
4. utilize	S. destroy utterly

Codes:
(a) 1:S, 2:P, 3:Q, 4:R
(b) 1:P, 2:Q, 3:R, 4:S
(c) 1:Q, 2:R, 3:S, 4:P
(d) 1:S, 2:P, 3:R, 4:Q

[2014 (Set-2) : 1 M]

1.54 What is the average of all multiples of 10 from 2 to 198?
(a) 90 (b) 100
(c) 110 (d) 120

[2014 (Set-2) : 1 M]

1.55 The value of $\sqrt{12+\sqrt{12+\sqrt{12+...}}}$ is
(a) 3.464 (b) 3.932
(c) 4.000 (d) 4.444
[2014 (Set-2) : 1 M]

1.56 The old city of Koenigsberg, which had a German majority population before World War 2, is now called Kaliningrad. After the events of the war, Kaliningrad is now a Russian territory and has a predominantly Russian population. It is bordered by the Baltic Sea on the north and the countries of Poland to the south and west and Lithuania to the east respectively. Which of the statements below can be inferred from this passage?
(a) Kaliningrad was historically Russian in its ethnic make up
(b) Kaliningrad is a part of Russia despite it not being contiguous with the rest of Russia
(c) Koenigsberg was renamed Kaliningrad, as that was its original Russian name
(d) Poland and Lithuania are on the route from Kaliningrad to the rest of Russia
[2014 (Set-2) : 2 M]

1.57 The number of people diagnosed with dengue fever (contracted from the bite of a mosquito) in north India is twice the number diagnosed last year. Municipal authorities have concluded that measures to control the mosquito population have failed in this region.
Which one of the following statements, if true, does not contradict this conclusion?
(a) A high proportion of the affected population has returned from neighbouring countries where dengue is prevalent
(b) More cases of dengue are now reported because of an increase in the Municipal Office's administrative efficiency
(c) Many more cases of dengue are being diagnosed this year since the introduction of a new and effective diagnostic test
(d) The number of people with malarial fever (also contracted from mosquito bites) has increased this year
[2014 (Set-2) : 2 M]

1.58 If x is real and $|x^2 - 2x + 3| = 11$, then possible values of $|-x^3 + x^2 - x|$ include
(a) 2, 4 (b) 2, 14
(c) 4, 52 (d) 14, 52
[2014 (Set-2) : 2 M]

1.59 The ratio of male to female students in a college for five years is plotted in the following line graph. If the number of female students doubled in 2009, by what percent did the number of male students increase in 2009?

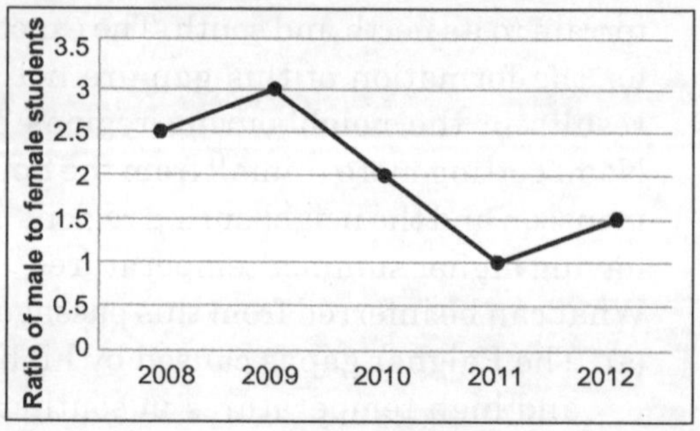

[2014 (Set-2) : 2 M]

1.60 At what time between 6 a.m. and 7 a.m. will the minute hand and hour hand of a clock make an angle closest to 60° ?
(a) 6 : 22 a.m. (b) 6 : 27 a.m.
(c) 6 : 38 a.m. (d) 6 : 45 a.m.
[2014 (Set-2) : 2 M]

1.61 <u>While trying to collect</u> (I) an envelope <u>from under the table</u> (II), <u>Mr. X fell down</u> (III) and <u>was losing consciousness</u> (IV).
Which one of the above underlined parts of the sentence is NOT appropriate?
(a) I (b) II
(c) III (d) IV
[2014 (Set-3) : 1 M]

1.62 If she _____ how to calibrate the instrument, she _____ done the experiment.
(a) knows, will have
(b) knew, had
(c) had known, could have
(d) should have known, would have
[2014 (Set-3) : 1 M]

1.63 Choose the word that is opposite in meaning to the word "coherent".
(a) sticky
(b) well-connected
(c) rambling
(d) friendly
[2014 (Set-3) : 1 M]

1.64 Which number does not belong in the series below?

2, 5, 10, 17, 26, 37, 50, 64

(a) 17 (b) 37
(c) 64 (d) 26

[2014 (Set-3) : 1 M]

1.65 The table below has question-wise data on the performance of students in an examination. The marks for each question are also listed. There is no negative or partial marking in the examination.

Q. No.	Marks	Answered Correctly	Answered Wrongly	Not Attempted
1	2	21	17	6
2	3	15	27	2
3	2	23	18	3

What is the average of the marks obtained by the class in the examination?

(a) 1.34 (b) 1.74
(c) 3.02 (d) 3.91

[2014 (Set-3) : 1 M]

1.66 A dance programme is scheduled for 10.00 a.m. Some students are participating in the programme and they need to come an hour earlier than the start of the event. These students should be accompanied by a parent. Other students and parents should come in time for the programme. The instruction you think that is appropriate for this is

(a) Students should come at 9.00 a.m. and parents should come at 10.00 a.m.
(b) Participating students should come at 9.00 a.m. accompanied by a parent, and other parents and students should come by 10.00 a.m.
(c) Students who are not participating should come by 10.00 a.m. and they should not bring their parents. Participating students should come at 9.00 a.m.
(d) Participating students should come before 9.00 a.m. Parents who accompany them should come at 9.00 a.m. All others should come at 10.00 a.m.

[2014 (Set-3) : 2 M]

1.67 By the beginning of the 20th century, several hypotheses were being proposed, suggesting a paradigm shift in our understanding of the universe. However, the clinching evidence was provided by experimental measurements of the position of a star which was directly behind our sun. Which of the following inference(s) may be drawn from the above passage?

(i) Our understanding of the universe changes based on the positions of stars
(ii) Paradigm shifts usually occur at the beginning of centuries
(iii) Stars are important objects in the universe
(iv) Experimental evidence was important in confirming this paradigm shift

(a) (i), (ii) and (iv)
(b) (iii) only
(c) (i) and (iv)
(d) (iv) only

[2014 (Set-3) : 2 M]

1.68 The Gross Domestic Product (GDP) in Rupees grew at 7% during 2012-2013. For international comparison, the GDP is compared in US Dollars (USD) after conversion based on the market exchange rate. During the period 2012-2013 the exchange rate for the USD increased from Rs. 50/ USD to Rs. 60/ USD. India's GDP in USD during the period 2012-2013

(a) increased by 5 %
(b) decreased by 13%
(c) decreased by 20%
(d) decreased by 11%

[2014 (Set-3) : 2 M]

1.69 The ratio of male to female students in a college for five years is plotted in the following line graph. If the number of female students in 2011 and 2012 is equal, what is the ratio of male students in 2012 to male students in 2011?

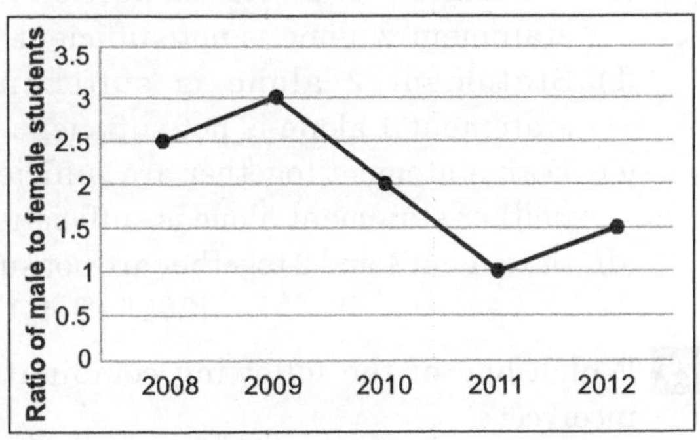

(a) 1 : 1 (b) 2 : 1
(c) 1.5 : 1 (d) 2.5 : 1

[2014 (Set-3) : 2 M]

1.70 Consider the equation: $(7526)_8 - (Y)_8 = (4364)_8$, where $(X)_N$ stands for X to the base N. Find Y.

(a) 1634 (b) 1737
(c) 3142 (d) 3162

[2014 (Set-3) : 2 M]

1.71 Didn't you buy _____ when you went shopping?
(a) any paper
(b) much paper
(c) no paper
(d) a few paper

[2015 (Set-1) : 1 M]

1.72 Which of the following options is the closet in meaning to the sentence below?
She enjoyed herself immensely at the party.
(a) She had a terrible time at the party
(b) She had a horrible time at the party
(c) She had a terrific time at the party
(d) She had a terrifying time at the party

[2015 (Set-1) : 1 M]

1.73 Given Set A = {2, 3, 4, 5} and Set B = {11, 12, 13, 14, 15}, two numbers are randomly selected, one from each set. What is the probability that the sum of the two numbers equals 16?
(a) 0.20 (b) 0.25
(c) 0.30 (d) 0.33

[2015 (Set-1) : 1 M]

1.74 Based on the given statements, select the most appropriate option to solve the given question. If two floors in a certain building are 9 feet apart, how many steps are there in a set of stairs that extends from the first floor to the second floor of the building?
Statements:
1. Each step is $\frac{3}{4}$ foot high.
2. Each step is 1 foot wide.
(a) Statement 1 alone is sufficient, but statement 2 alone is not sufficient
(b) Statement 2 alone is sufficient, but statement 1 alone is not sufficient
(c) Both statement together are sufficient, but neither statement alone is sufficient
(d) Statement 1 and 2 together are not sufficient

[2015 (Set-1) : 1 M]

1.75 Which one of the following combinations is incorrect?
(a) Acquiescence - Submission
(b) Wheedle - Roundabout
(c) Flippancy - Lightness
(d) Profligate - Extravagant

[2015 (Set-1) : 1 M]

1.76 The number of students in a class who have answered correctly, wrongly, or not attempted each question in an exam, are listed in the table below. The marks for each question are also listed. There is no negative or partial marking.

Q.No.	Marks	Answered Correctly	Answered Wrongly	Not Attempted
1	2	21	17	6
2	3	15	27	2
3	1	11	29	4
4	2	23	18	3
5	5	31	12	1

What is the average of the marks obtained by the class in the examination?
(a) 2.290 (b) 2.970
(c) 6.795 (d) 8.795

[2015 (Set-1) : 2 M]

1.77 Select the alternative meaning of the underlined part of the sentence.
The chain snatchers **<u>took to their heels</u>** when the police party arrived.
(a) took shelter in a thick jungle
(b) open indiscriminate fire
(c) took to flight
(d) unconditionally surrendered

[2015 (Set-1) : 2 M]

1.78 The given statement is following by some courses of action. Assuming the statement to be true, decide the correct option.
Statement: There has been a significant drop in the water level in the lakes supplying water to the city.
Course of action:
1. The water supply authority should impose a partial cut in supply to tackle the situation.
2. The government should appeal to all the residents through mass media for minimal use of water.
3. The government should ban the water supply in lower areas.
(a) Statement 1 and 2 follow
(b) Statement 1 and 3 follow
(c) Statement 2 and 3 follow
(d) All statements follow

[2015 (Set-1) : 2 M]

1.79 The pie chart below has the breakup of the number of students, from different departments in an engineering college for the year 2012. The proportion of male to female students in each department is 5 : 4. There are 40 males in Electrical Engineering. What is the different between the numbers of female students in the Civil department and the female students in the Mechanical department?

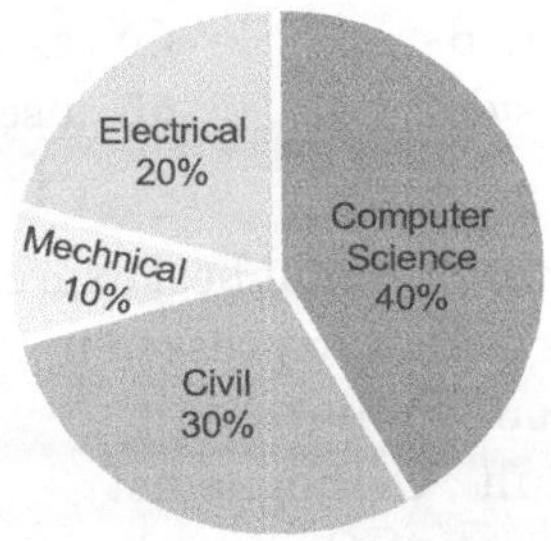

[2015 (Set-1) : 2 M]

1.80 The probabilities that a student passes in Mathematics, Physics and Chemistry are m, p, and c respectively. Of these subjects, the student has 75% chance of passing in at least one, a 50% chance of passing in at least two and a 40% chance of passing in exactly two. Following relations are drawn in m, p and c:

1. $p + m + c = \frac{27}{20}$
2. $p + m + c = \frac{13}{20}$
3. $(p) \times (m) \times (c) = \frac{1}{10}$

(a) Only relation 1 is true
(b) Only relation 2 is true
(c) Relations 2 and 3 are true
(d) Relations 1 and 3 are true.

[2015 (Set-1) : 2 M]

1.81 We ______ our friend's birthday and we ______ how to make it up to him.

(a) completely forgot --- don't just know
(b) forgot completely --- don't just know
(c) completely forgot --- just don't know
(d) forgot completely --- just don't know

[2015 (Set-2) : 1 M]

1.82 Choose the statement where underlined word is used correctly.

(a) The industrialist had a personnel jet
(b) I write my experience in my personnel diary
(c) All personnel are being given the day off
(d) Being religious is a personnel aspect

[2015 (Set-2) : 1 M]

1.83 Consider a function $f(x) = 1 - |x|$ on $-1 \le x \le 1$. The value of x at which the function attains a maximum, and the maximum value of the function are:

(a) 0, –1
(b) –1, 0
(c) 0, 1
(d) –1, 2

[2015 (Set-2) : 1 M]

1.84 A generic term that includes various items of clothing such as a skirt, a pair of trousers and a shirt is

(a) fabric
(b) textile
(c) fibre
(d) apparel

[2015 (Set-2) : 1 M]

1.85 Based on the given statements, select the most appropriate option to solve the given question. What will be the total weight of 10 poles each of same weight?

Statements:

1. One fourth of the weight of a pole is 5 Kg.
2. The total weight of these poles is 160 Kg more than the total weight of two poles.

(a) Statement 1 alone is sufficient, but Statement 2 alone is not sufficient
(b) Statement 2 alone is sufficient, but Statement 1 alone is not sufficient
(c) Both statement together are sufficient, but neither statement alone is sufficient
(d) Both statements 1 and 2 are not sufficient

[2015 (Set-2) : 1 M]

1.86 If the list of letters, P, R, S, T, U is an arithmetic sequence, which of the following are also in arithmetic sequence?

1. 2P, 2R, 2S, 2T, 2U
2. P–3, R–3, S–3, T–3, U–3
3. P^2, R^2, S^2, T^2, U^2

(a) 1 only
(b) 1 and 2
(c) 2 and 3
(d) 1 and 3

[2015 (Set-2) : 2 M]

1.87 Four branches of a company are located at M, N, O and P. M is north of N at a distance of 4 km; P is south of O at a distance of 2 km; N is southeast of O by 1 km. What is the distance between M and P in km?

(a) 5.34
(b) 6.74
(c) 28.5
(d) 45.49

[2015 (Set-2) : 2 M]

1.88 In a triangle PQR, PS is the angle bisector of $\angle QPR$ and $\angle QPS = 60°$. What is the length of PS?

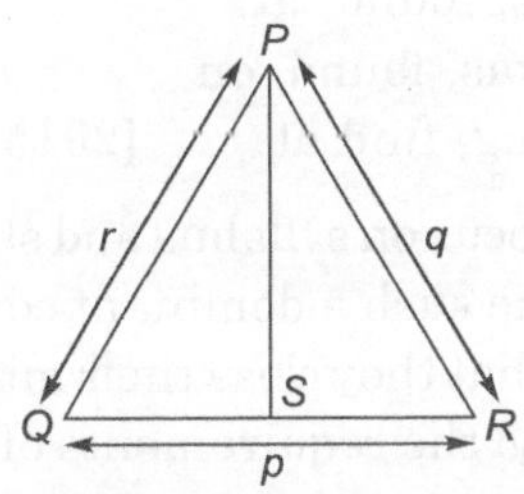

(a) $\dfrac{(q+r)}{qr}$ (b) $\dfrac{qr}{(q+r)}$

(c) $\sqrt{(q^2+r^2)}$ (d) $\dfrac{(q+r)^2}{qr}$

[2015 (Set-2) : 2 M]

1.89 If p, q, r, s are distinct integers such that:
f(p, q, r, s) = max (p, q, r, s)
g(p, q, r, s) = min (p, q, r, s)
h(p, q, r, s) = remainder of (p × q) /(r × s) if (p × q) > (r × s) or remainder of (r × s)/ (p × q) if (r × s) > (p × q)
Also a function fgh (p, q, r, s) = f(p, q, r, s) × g(p, q, r, s) × h(p, q, r, s).
Also the same operation are valid with two variable functions of the form f(p, q).
What is the value of fg(h(2, 5, 7, 3), 4, 6, 8)?

[2015 (Set-2) : 2 M]

1.90 Out of the following four sentences, select the most suitable sentence with respect to grammar and usage:
(a) Since the report lacked needed information, it was of no use to them
(b) The report was useless to them because there were no needed information in it
(c) Since the report did not contain the needed information, it was not real useful to them
(d) Since the report lacked needed information, it would not had been useful to them

[2015 (Set-2) : 2 M]

1.91 If ROAD is written as URDG, then SWAN should be written as:
(a) VXDQ (b) VZDQ
(c) VZDP (d) UXDQ

[2015 (Set-3) : 1 M]

1.92 The Tamil version of ______ John Abraham-starrer Madras Cafe ______ cleared by the Censor Board with no cuts last week, but the film's distributors ______ no takers among the exhibitors for a release in Tamil Nadu ______ this Friday.
(a) Mr., was, found, on
(b) a, was, found, at
(c) the, was, found, on
(d) a, being, find at **[2015 (Set-3) : 1 M]**

1.93 Extreme focus on syllabus and studying for tests has become such a dominant concern of Indian students that they close their minds to anything ______ to the requirements of the exam.
(a) related (b) extraneous
(c) outside (d) useful

[2015 (Set-3) : 1 M]

1.94 Select the pair that best expresses a relationship similar to that expressed in the pair:
Children : Pediatrician
(a) Adult : Orthopaedist
(b) Females : Gynaecologist
(c) Kidney : Nephrologist
(d) Skin : Dermatologist **[2015 (Set-3) : 1 M]**

1.95 A function $f(x)$ is linear and has a value of 29 at $x = -2$ and 39 at $x = 3$. Find its value at $x = 5$.
(a) 59 (b) 45
(c) 43 (d) 35

[2015 (Set-3) : 1 M]

1.96 Alexander turned his attention towards India, since he had conquered Persia. Which one of the statements below is logically valid and can be inferred from the above sentence?
(a) Alexander would not have turned his attention towards India had he not conquered Persia
(b) Alexander was not ready to rest on his laurels, and wanted to march to India
(c) Alexander was completely in control of his army and could command it to move towards India.
(d) Since Alexander's kingdom extended to Indian borders after the conquest of Persia, he was keen to move further.

[2015 (Set-3) : 2 M]

1.97 The head of a newly formed government desires to appoint five of the six selected members P, Q, R, S, T and U to portfolios of Home, Power, Defense, Telecom, and Finance. U does not want any portfolio if S gets one of the five. R wants either Home or Finance or no portfolio. Q says that if S gets either Power of Telecom, then she must get the other one. T insists on a portfolio if P gets one.
Which is the valid distribution of portfolios?
(a) P-Home, Q-Power, R-Defense, S-Telecom, T-Finance
(b) R-Home, S-Power, P-Defense, Q-Telecom, T-Finance
(c) P-Home, Q-Power, T-Defense, S-Telecom, U-Finance
(d) Q-Home, U-Power, T-Defense, R-Telecom, P-Finance

[2015 (Set-3) : 2 M]

1.98 Choose the most appropriate equation for the function drawn as a thick line, in the plot below.

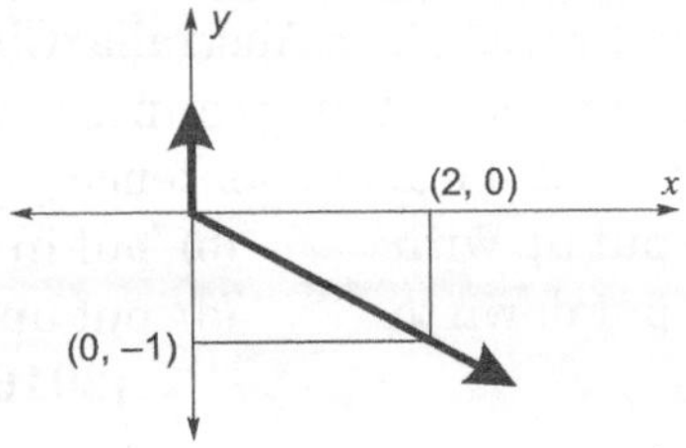

(a) $x = y - |y|$ (b) $x = -(y - |y|)$
(c) $x = y + |y|$ (d) $x = -(y + |y|)$

[2015 (Set-3) : 2 M]

1.99 Most experts feel that in spite of possessing all the technical skills required to be a batsman of the highest order, he is unlikely to be so due to lack of requisite temperament.
He was guilty of throwing away his wicket several times after working hard to lay a strong foundation. His critics pointed out that until he addressed this problem, success at the highest level will continue to elude him.
Which of the statement(s) below is/are logically valid and can be inferred from the above passage?
(i) He was already a successful batsman at the highest level.
(ii) He has to improve his temperament in order to become a great batsman.
(iii) He failed to make many of his good starts count.
(iv) Improving his technical skills will guarantee success.
(a) (iii) and (iv) (b) (ii) and (iii)
(c) (i), (ii) and (iii) (d) (ii) only

[2015 (Set-3) : 2 M]

1.100 The exports and imports (in crores of Rs.) of a country from the year 2000 to 2007 are given in the following bar chart. In which year is the combined percentage increase in imports and exports the highest?

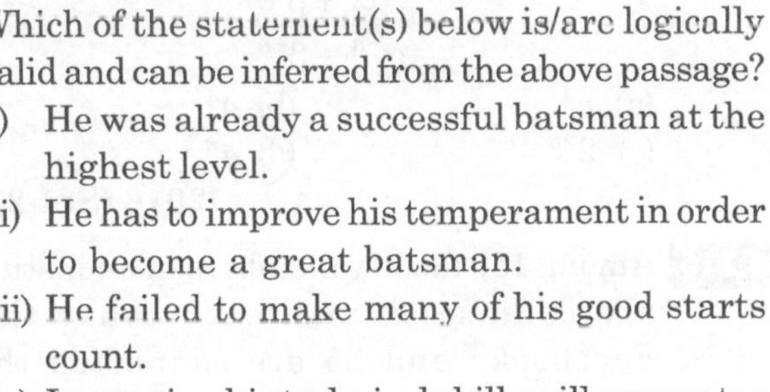

[2015 (Set-3) : 2 M]

1.101 Out of the following four sentences, select the most suitable sentence with respect to grammar and usage.
(a) I will not leave the place until the minister does not meet me.
(b) I will not leave the place until the minister doesn't meet me.
(c) I will not leave the place until the minister meet me.
(d) I will not leave the place until the minister meets me.

[2016 (Set-1) : 1 M]

1.102 A rewording of something written or spoken is a ________.
(a) paraphrase (b) paradox
(c) paradigm (d) paraffin

[2016 (Set-1) : 1 M]

1.103 Archimedes said, "Give me a lever long enough and a fulcrum on which to place it, and I will move the world." The sentence above is an example of a ________ statement.
(a) figurative (b) collateral
(c) literal (d) figurine

[2016 (Set-1) : 1 M]

1.104 If 'relftaga' means carefree, 'otaga' means careful and 'fertaga' means careless, which of the following could mean 'aftercare'?
(a) zentaga (b) tagafer
(c) tagazen (d) relffer

[2016 (Set-1) : 1 M]

1.105 A cube is built using 64 cubic blocks of side one unit. After it is built, one cubic block is removed from every corner of the cube. The resulting surface area of the body (in square units) after the removal is ________.
(a) 56 (b) 64
(c) 72 (d) 96

[2016 (Set-1) : 1 M]

1.106 A shaving set company sells 4 different types of razors, Elegance, Smooth, Soft and Executive. Elegance sells at Rs. 48, Smooth at Rs. 63, Soft at Rs. 78 and Executive at Rs. 173 per piece. The table below shows the numbers of each razor sold in each quarter of a year.

Quarter/ Product	Elegance	Smooth	Soft	Executive
Q1	27300	20009	17602	9999
Q2	25222	19392	18445	8942
Q3	28976	22429	19544	10234
Q4	21012	18229	16595	10109

Which product contributes the greatest fraction to the revenue of the company in that year?
(a) Elegance (b) Executive
(c) Smooth (d) Soft
[2016 (Set-1) : 2 M]

1.107 Indian currency notes show the denomination indicated in at least seventeen languages. If this is not an indication of the nation's diversity, nothing else is. Which of the following can be logically inferred from the above sentences?
(a) India is a country of exactly seventeen languages.
(b) Linguistic pluralism is the only indicator of a nation's diversity.
(c) Indian currency notes have sufficient space for all the Indian languages.
(d) Linguistic pluralism is strong evidence of India's diversity. **[2016 (Set-1) : 2 M]**

1.108 Consider the following statements relating to the level of poker play of four players *P*, *Q*, *R* and S. I. *P* always beats *Q* II. *R* always beats *S* III. *S* loses to *P* only sometimes IV. R always loses to *Q*. Which of the following can be logically inferred from the above statements? (i) *P* is likely to beat all the three other players (ii) S is the absolute worst player in the set
(a) (i) only (b) (ii) only
(c) (i) and (ii) (d) Neither (i) nor (ii)
[2016 (Set-1) : 2 M]

1.109 If $f(x) = 2x^7 + 3x - 5$, which of the following is a factor $f(x)$?
(a) $(x^3 + 8)$ (b) $(x - 1)$
(c) $(2x - 5)$ (d) $(x + 1)$
[2016 (Set-1) : 2 M]

1.110 In a process, the number of cycles to failure decreases exponentially with an increase in load. At a load of 80 units, it takes 100 cycles for failure. When the load is halved, it takes 10000 cycles for failure. The load for which the failure will happen in 5000 cycles is ______.
(a) 40.00 (b) 46.02
(c) 60.01 (d) 92.02
[2016 (Set-1) : 2 M]

1.111 The man who is now Municipal Commissioner worked as ______.
(a) the security guard at a university
(b) a security guard at the university
(c) a security guard at university
(d) the security guard at the university
[2016 (Set-2) : 1 M]

1.112 Nobody knows how the Indian cricket team is going to cope with the difficult and seamer-friendly wickets in Australia. Choose the option which is closest in meaning to the underlined phrase in the above sentence.
(a) put up with (b) put in with
(c) put down to (d) put up against
[2016 (Set-2) : 1 M]

1.113 Find the odd one in the following group of words.
mock, deride, praise, jeer
(a) mock (b) deride
(c) praise (d) jeer
[2016 (Set-2) : 1 M]

1.114 Pick the odd one from the following options.
(a) CADBE (b) JHKIL
(c) XVYWZ (d) ONPMQ
[2016 (Set-2) : 1 M]

1.115 In a quadratic function, the value of the product of the roots (α, β) is 4. Find the value of

$$\frac{\alpha^n + \beta^n}{\alpha^{-n} + \beta^{-n}}$$

(a) n^4 (b) 4^n
(c) 2^{2n-1} (d) 4^{n-1}
[2016 (Set-2) : 1 M]

1.116 Among 150 faculty members in an institute, 55 are connected with each other through Facebook® and 85 are connected through WhatsApp®. 30 faculty members do not have Facebook® or WhatsApp® accounts. The number of faculty members connected only through Facebook® accounts is ______.
(a) 35 (b) 45
(c) 65 (d) 90
[2016 (Set-2) : 2 M]

1.117 Computers were invented for performing only high-end useful computations. However, it is no understatement that they have taken over our world today. The internet, for example, is ubiquitous. Many believe that the internet itself is an unintended consequence of the original invention. With the advent of mobile computing on our phones, a whole new dimension is now enabled. One is left wondering if all these developments are good or, more importantly, required.
Which of the statement(s) below is/are logically valid and can be inferred from the above paragraph?

(i) The author believes that computers are not good for us.
(ii) Mobile computers and the internet are both intended inventions

(a) (i) only (b) (ii) only
(c) Both (i) and (ii) (d) Neither (i) nor (ii)

[2016 (Set-2) : 2 M]

1.118 All hill-stations have a lake. Ooty has two lakes. Which of the statement(s) below is/are logically valid and can be inferred from the above sentences?

(i) Ooty is not a hill-station.
(ii) No hill-station can have more than one lake.

(a) (i) only (b) (ii) only
(c) both (i) and (ii) (d) neither (i) nor (ii)

[2016 (Set-2) : 2 M]

1.119 In a 2×4 rectangle grid shown below, each cell is a rectangle. How many rectangles can be observed in the grid?

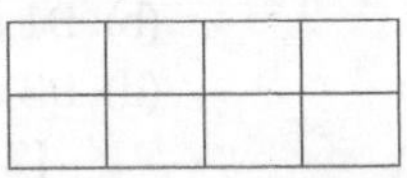

(a) 21 (b) 27
(c) 30 (d) 36

[2016 (Set-2) : 2 M]

1.120 Choose the correct expression for $f(x)$ given in the graph.

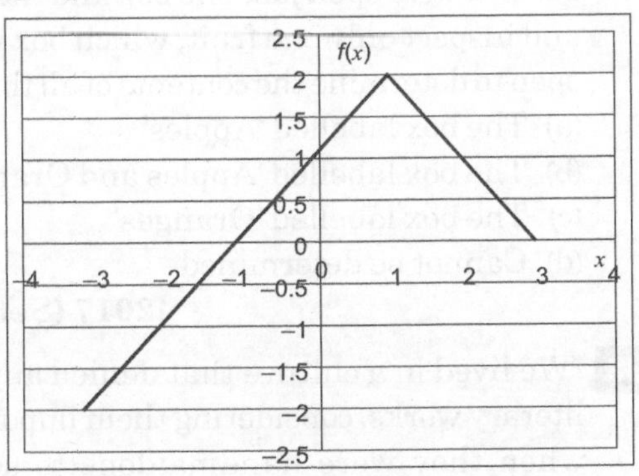

(a) $f(x) = 1 - |x - 1|$ (b) $f(x) = 1 + |x - 1|$
(c) $f(x) = 2 - |x - 1|$ (d) $f(x) = 2 + |x - 1|$

[2016 (Set-2) : 2 M]

1.121 The probability that a k-digit number does NOT contain the digits 0, 5 or 9 is

(a) 0.3^k (b) 0.6^k
(c) 0.7^k (d) 0.9^k

[2017 (Sct-1) : 1 M]

1.122 Find the smallest number y such that that $y \times 162$ is a perfect cube.

(a) 24 (b) 27
(c) 32 (d) 36

[2017 (Set-1) : 1 M]

1.123 After Rajendra Chola returned from his voyage to Indonesia, he ________ to visit the temple in Thanjavur.

(a) was wishing (b) is wishing
(c) wished (d) had wished

[2017 (Set-1) : 1 M]

1.124 Research in the workplace reveal that people work for many reasons ________.

(a) money beside (b) beside money
(c) money besides (d) besides money

[2017 (Set-1) : 1 M]

1.125 Rahul, Murali, Srinivas and Arul are seated around a square table. Rahul is sitting to the left of Murali. Srinivas is sitting to the right of Arul. Which of the following pairs are seated opposite each other?

(a) Rahul and Murali
(b) Srinivas and Arul
(c) Srinivas and Murali
(d) Srinivas and Rahul

[2017 (Set-1) : 1 M]

1.126 "The hold of the nationalist imagination on our colonial past is such that anything inadequately or improperly nationalist is just not history." Which of the following statements best reflects the author's opinion?

(a) Nationalists are highly imaginative
(b) History is viewed through the filter of nationalism
(c) Our colonial past never happened
(d) Nationalism has to be both adequately and properly imagined. **[2017 (Set-1) : 2 M]**

1.127 A contour line joins locations having the same height above the mean sea level. The following is a contour plot of a geographical region. Contour lines are shown at 25 m intervals in this plot. If in a flood, the water level rises to 525 m, which of the villages P, Q, R, S, T get submerged?

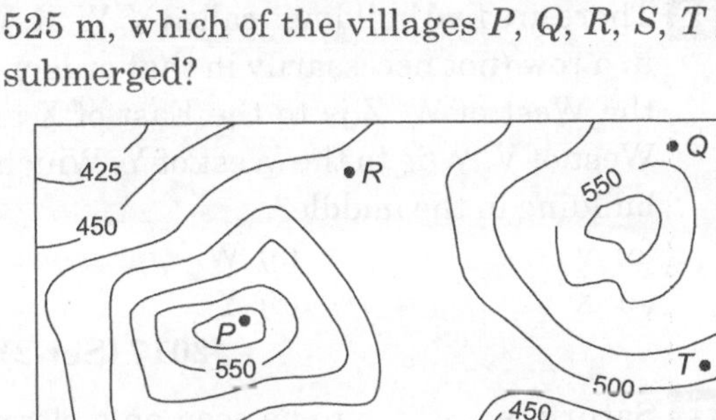

(a) P, Q (b) P, Q, T
(c) R, S, T (d) Q, R, S

[2017 (Set-1) : 2 M]

1.128 The expression $\frac{(x+y)-|x-y|}{2}$ is equal to

(a) the maximum of x and y
(b) the minimum of x and y
(c) 1
(d) None of the above

[2017 (Set-1) : 2 M]

1.129 Arun, Gulab, Neel and Shweta must choose one shirt each from a pile of four shirts coloured red, pink, blue and white respectively. Arun dislikes the colour red and Shweta dislikes the colour white. Gulab and Neel like all the colours. In how many different ways can they choose the shirts so that no one has a shirt with a colour he or she dislikes?

(a) 21 (b) 18
(c) 16 (d) 14

[2017 (Set-1) : 2 M]

1.130 Six people are seated around a circular table. There are at least two men and two women. There are at least three right-handed persons. Every woman has a left-handed person to her immediate right. None of the women are right-handed. The number of women at the table is

(a) 2 (b) 3
(c) 4 (d) Cannot be determined

[2017 (Set-2) : 1 M]

1.131 A test has twenty questions worth 100 marks in total. There are two types of questions. Multiple choice questions are worth 3 marks each and essay questions are worth 11 marks each. How many multiple choice questions does the exam have?

(a) 12 (b) 15
(c) 18 (d) 19

[2017 (Set-2) : 1 M]

1.132 There are five buildings called V, W. X, Y and Z in a row (not necessarily in that order). V is to the West of W. Z is to the East of X and the West of V. W is to the West of Y. Which is the building in the middle?

(a) V (b) W
(c) X (d) Y

[2017 (Set-2) : 1 M]

1.133 Saturn is ________ to be seen on a clear night with the naked eye.

(a) enough bright (b) bright enough
(c) as enough bright (d) bright as enough

[2017 (Set-2) : 1 M]

1.134 Choose the option with words that are not synonyms.

(a) aversion, dislike
(b) luminous, radiant
(c) plunder, loot
(d) yielding, resistant

[2017 (Set-2) : 1 M]

1.135 There are 3 red socks. 4 green socks and 3 blue socks. You choose 2 socks. The probability that they are of the same colour is

(a) $\frac{1}{5}$ (b) $\frac{7}{30}$
(c) $\frac{1}{4}$ (d) $\frac{4}{15}$

[2017 (Set-2) : 1 M]

1.136 X is a 30 digit number starting with the digit 4 followed by the digit 7. Then the number X^3 will have

(a) 90 digits (b) 91 digits
(c) 92 digits (d) 93 digits

[2017 (Set-2) : 2 M]

1.137 There are three boxes. One contains apples, another contains oranges and the last one contains both apples and oranges. All three are known to be incorrectly labelled. If you are permitted to open just one box and then pull out and inspect only one fruit, which box would you open to determine the contents of all three boxes?

(a) The box labelled 'Apples'
(b) The box labelled 'Apples and Oranges'
(c) The box labelled 'Oranges'
(d) Cannot be determined

[2017 (Set-2) : 2 M]

1.138 "We lived in a culture that denied any merit to literary works, considering them important only when they were handmaidens to something seemingly more urgent-namely ideology. This was a country where all gestures, even the most private, were interpreted in political terms,". The author's belief that ideology is not as important as literature is revealed by the word:

(a) 'culture' (b) 'seemingly'
(c) 'urgent' (d) 'political'

[2017 (Set-2) : 2 M]

1.139 The number of roots of $e^x + 0.5x^2 - 2 = 0$ in the range [–5, 5] is

(a) 0 (b) 1
(c) 2 (d) 3

[2017 (Set-2) : 2 M]

1.140 An air pressure contour line joins locations in a region having the same atmospheric pressure. The following is an air pressure contour plot is a geographical region. Contour lines are shown at 0.05 bar intervals in this plot.

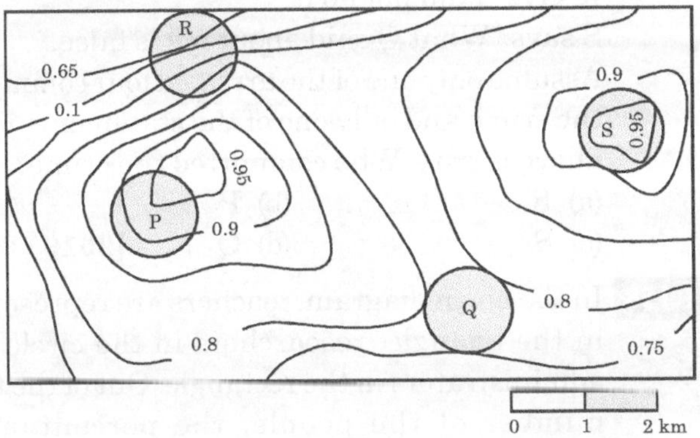

If the possibility of a thunderstorm is given by how fast air pressure rises or drops over a region, which of the following regions is most likely to have a thunderstorm?

(a) P (b) Q
(c) R (d) S

[2017 (Set-2) : 2 M]

1.141 What is the missing number in the following sequence?

2, 12, 60, 240, 720, 1440, _______, 0

(a) 2880 (b) 1440
(c) 720 (d) 0 [2018 : 1 M]

1.142 "From where are they bringing their books? _____ bringing _____ books from _____."

The words that best fill the blanks in the above sentence are

(a) Their, they're, there
(b) They're, their, there
(c) There, their, they're
(d) They're, there, there [2018 : 1 M]

1.143 What would be the smallest natural number which when divided either by 20 or by 42 or by 76 leaves a remainder of 7 in each case?

(a) 3047 (b) 6047
(c) 7987 (d) 63847 [2018 : 1 M]

1.144 "A _______ investigation can sometimes yield new facts, but typically organized ones are more successful."

The word that best fills the blank in the above sentence is

(a) meandering (b) timely
(c) consistent (d) systematic

[2018 : 1 M]

1.145 The area of a square is d. What is the area of the circle which has the diagonal of the square as its diameter?

(a) πd (b) πd^2
(c) $\frac{1}{4}\pi d^2$ (d) $\frac{1}{2}\pi d$ [2018 : 1 M]

1.146 In $pqr \neq 0$ and $p^{-x} = \frac{1}{q}$, $q^{-y} = \frac{1}{r}$, $r^{-z} = \frac{1}{p}$, what is the value of the product xyz?

(a) –1 (b) $\frac{1}{pqr}$
(c) 1 (d) pqr [2018 : 2 M]

1.147 In appreciation of the social improvements completed in a town, a wealthy philanthropist decided to gift Rs 750 to each male senior citizen in the town and Rs 1000 to each female senior citizen. Altogether, there were 300 senior citizens eligible for this gift. However, only $\frac{8}{9^{th}}$ of the eligible men and $\frac{2}{3^{rd}}$ of the eligible women claimed the gift. How much money (in Rupees) did the philanthropist give away in total?

(a) 1,50,000 (b) 2,00,000
(c) 1,75,000 (d) 1,51,000

[2018 : 2 M]

1.148 In the figure below, $\angle DEC + \angle BFC$ is equal to _______.

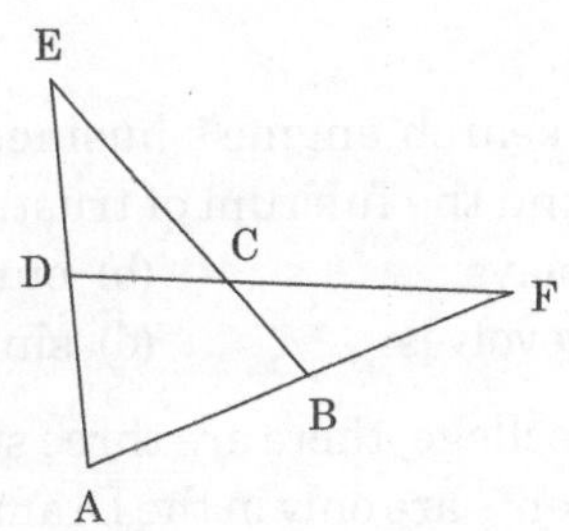

(a) $\angle BCD - \angle BAD$ (b) $\angle BAD + \angle BCF$
(c) $\angle BAD + \angle BCD$ (d) $\angle CBA + \angle ADC$

[2018 : 2 M]

1.149 A six sided unbiased die with four green faces and two red faces is rolled seven times. Which of the following combinations is the most likely outcome of the experiment?

(a) Three green faces and four red faces.
(b) Four green faces and three red faces.
(c) Five green faces and two red faces.
(d) Six green faces and one red face.

[2018 : 2 M]

1.150 In a party, 60% of the invited guests are male and 40% are female. If 80% of the invited guests attended the party and if all the invited female guests attended, what would be the ratio of males to females among the attendees in the party?
(a) 2 : 3 (b) 1 : 1
(c) 3 : 2 (d) 2 : 1 **[2018 : 2 M]**

1.151 Two cars start at the same time from the same location and go in the same direction. The speed of the first car is 50 km/h and the speed of the second car is 60 km/h. The number of hours it takes for the distance between the two cars to be 20 km is ________.
(a) 1 (b) 2
(c) 3 (d) 6 **[2019 : 1 M]**

1.152 The expenditure on the project ________ as follows; equipments Rs. 20 lakhs, salaries Rs. 12 lakhs, and contingency Rs.3 lakhs.
(a) break (b) breaks
(c) breaks down (d) break down
[2019 : 1 M]

1.153 Ten friends planned to share equally the cost of buying a gift for their teacher. When two of them decided not to contribute, each of the other friends had to pay Rs. 150 more. The cost of the gift was Rs. ________.
(a) 12000 (b) 6000
(c) 3000 (d) 666 **[2019 : 1 M]**

1.154 A court is to a judge as ________ is to a teacher.
(a) a syllabus (b) a school
(c) a student (d) a punishment
[2019 : 1 M]

1.155 The search engine's business model ________ around the fulcrum of trust.
(a) plays (b) bursts
(c) revolves (d) sinks **[2019 : 1 M]**

1.156 In a college, there are three student clubs. Sixty students are only in the Drama club. 80 students are only in the Dance club, 30 students are only in the Maths club. 40 students are in both Drama and Dance clubs. 12 students are in both Dance and Maths clubs, 7 students are in both Drama and Maths clubs, and 2 students are in all the clubs. If 75% of the students in the college are not in any of these clubs, then the total number of students in the college is ________.
(a) 225 (b) 1000
(c) 975 (d) 900
[2019 : 2 M]

1.157 The police arrested four criminals – P, Q, R and S. The criminals knew each other. They made the following statements:
P says "Q committed the crime."
Q says "S committed the crime."
R says "I did not do it."
S says "What Q said about me is false."
Assume only one of the arrested four committed the crime and only one of the statements made above is true. Who committed the crime?
(a) R (b) P
(c) S (d) Q **[2019 : 2 M]**

1.158 In the given diagram, teachers are represented in the triangle, researchers in the circle and administrators in the rectangle. Out of the total number of the people, the percentage of administrators shall be in the range of ________.

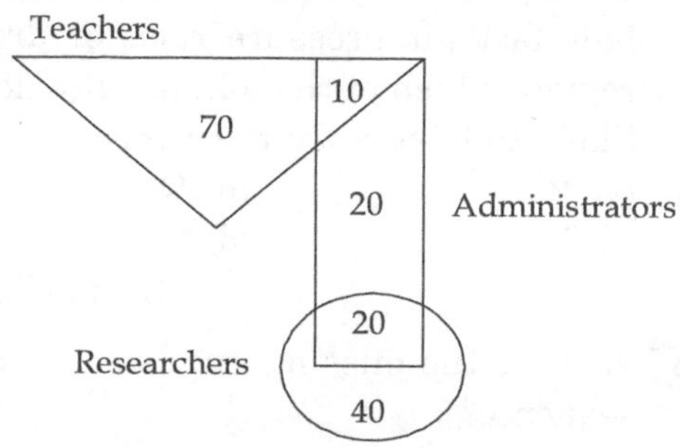

(a) 46 to 60 (b) 0 to 15
(c) 31 to 45 (d) 16 to 30
[2019 : 2 M]

1.159 Three of the five students allocated to a hostel put in special requests to the warden. Given the floor plan of the vacant rooms, select the allocation plan that will accommodate all their requests.
Request by X : Due to pollen allergy. I want to avoid a wing next to the garden.
Request by Y : 1 want to live as far from the washrooms as possible, since I am very sensitive to smell.
Request by Z : I believe in Vaastu and so want to stay in the South-West wing.
The shaded rooms are already occupied. WR is washroom.

(a) 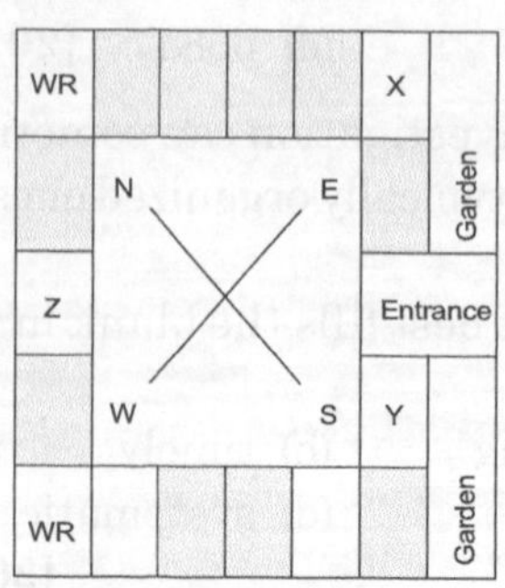

(b)

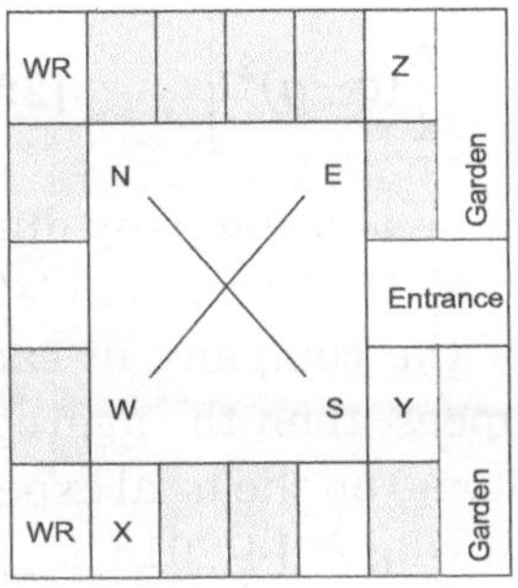

(c)

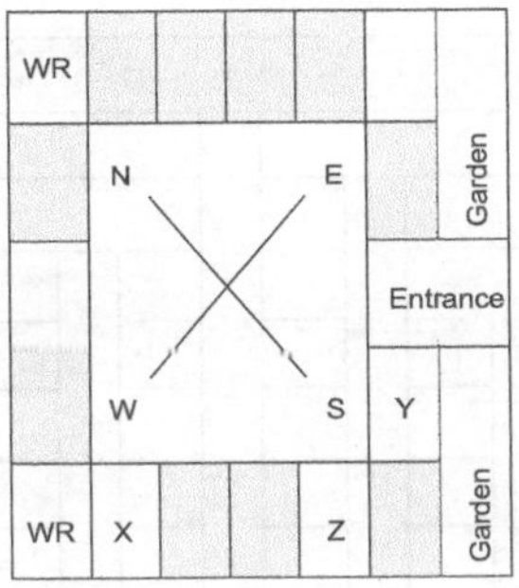

(d)

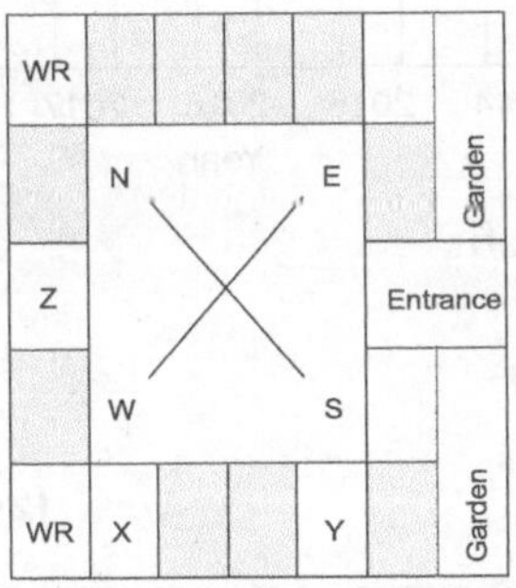

[2019 : 2 M]

1.160 "A recent High Court judgement has sought to dispel the idea of begging as a disease — which leads to its stigmatization and criminalization — and to regard it as a symptom. The underlying disease is the failure of the state to protect citizens who fall through the social security net". Which one of the following statements can be inferred from the given passage?

(a) Begging is an offence that has to be dealt with firmly

(b) Beggars are created because of the lack of social welfare schemes

(c) Begging has to be banned because it adversely affects the welfare of the state

(d) Beggars are lazy people who beg because they are unwilling to work

[2019 : 2 M]

1.161 Raman is confident of speaking English ________ six months as he has been practising regularly ________ the last three weeks.

(a) for, since (b) within, for

(c) during, for (d) for, in **[2020 : 1 M]**

1.162 There are multiple routes to reach from node 1 to node 2, as shown in the network.

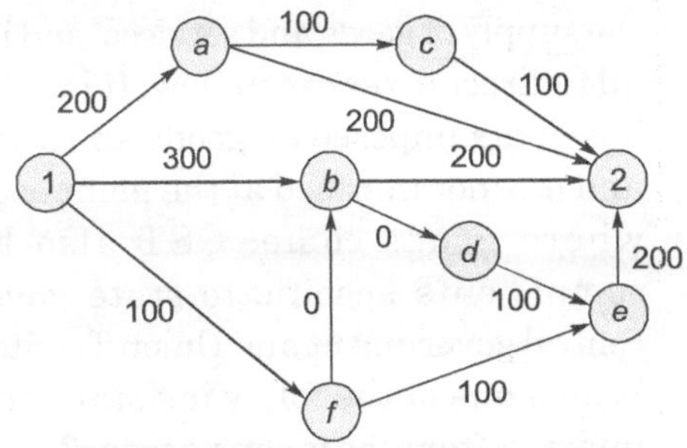

The cost of travel on an edge between two nodes is given in rupees. Nodes 'a' 'b', 'c', 'd', 'e' and 'f' are toll booths. The toll price at toll booths marked 'a' and 'c' is Rs. 200. and is Rs. 100 for the other toll booths. Which is the cheapest route from node 1 to node 2?

(a) 1-a-C-2 (b) 1-b-2

(c) 1-f-b-2 (d) 1-f-e-2 **[2020 : 1 M]**

1.163 His knowledge of the subject was excellent but his classroom performance was ________.

(a) extremely poor (b) praiseworthy

(c) desirable (d) good **[2020 : 1 M]**

1.164 Select the word that fits the analogy:

Cook : Cook :: Fly : ________

(a) Flying (b) Flyer

(c) Flew (d) Flighter

[2020 : 1 M]

1.165 The dawn of the 21st century witnessed the melting glaciers oscillating between giving too much and too little to billions of people who depend on their for fresh water. The UN climate report estimates that without deep cuts to man-made emissions, at least 30% of the northern hemisphere's surface permafrost could melt by the end of the century. Given this situation of imminent global exodus of billions of people displaced by rising seas, nation-states need to rethink their carbon footprint for political concerns, if not for environmental ones.

Which one of the following statements can be inferred from the given passage?

(a) Billions of people are affected by melting glaciers.

(b) Billions of people are responsible for man-made emissions.

(c) Nation-states do not have environmental concerns.

(d) Nation-states are responsible for providing fresh water to billions of people.

[2020 : 1 M]

1.166 Goods and Services Tax (GST) is an indirect tax introduced in India in 2017 that is imposed on the supply of goods and services, and it subsumes all indirect taxes except few. It is a destination-based tax imposed on goods and services used, and it is not imposed at the point of origin from where goods come. GST also has a few components specific to state governments, central government and Union Territories (UTs). Which one of the following statements can be inferred from the given passage?

(a) GST is imposed on the production of goods and services.
(b) GST does not have a component specific to UT.
(c) GST is imposed at the point of usage of goods and services.
(d) GST includes all indirect taxes.

[2020 : 2 M]

1.167 If P = 3, R = 27, T = 243, then Q + S = ________.

(a) 110 (b) 80
(c) 90 (d) 40 **[2020 : 2 M]**

1.168 Two straight lines are drawn perpendicular to each other in X-Y plane. If α and β are the acute angles the straight lines make with the X-axis, then $\alpha + \beta$ is ________.

(a) 180° (b) 120°
(c) 90° (d) 60° **[2020 : 2 M]**

1.169 The figure below shows an annular ring with outer and inner radii as b and a, respectively. The annular space has been painted in the form of blue colour circles touching the outer and inner periphery of annular space. If maximum n number of circles can be painted, then the unpainted are available in annular space is ________.

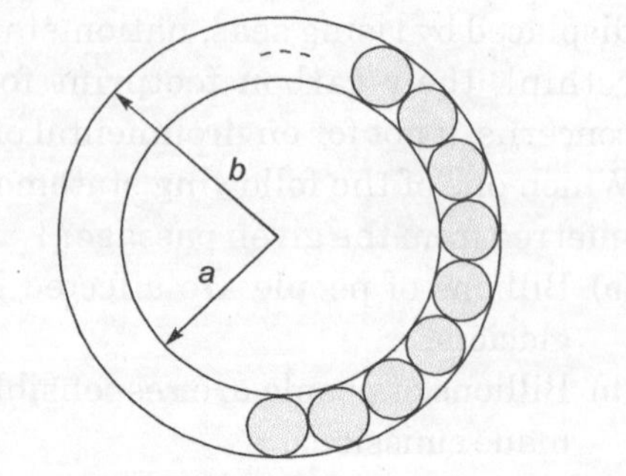

(a) $\pi[(b^2 - a^2) + n(b - a)^2]$
(b) $\pi\left[(b^2 - a^2) + \frac{\pi}{4}(b - a)^2\right]$
(c) $\pi[(b^2 - a^2) - n(b - a)^2]$
(d) $\pi\left[(b^2 - a^2) - \frac{\pi}{4}(b - a)^2\right]$ **[2020 : 2 M]**

1.170 The total revenue of a company during 2014-2018 is shown in the bar graph. If the total expenditure of the company in each year is 500 million rupees, then the aggregate profit loss (in percentage) on the total expenditure of the company during 2014-2018 is ________.

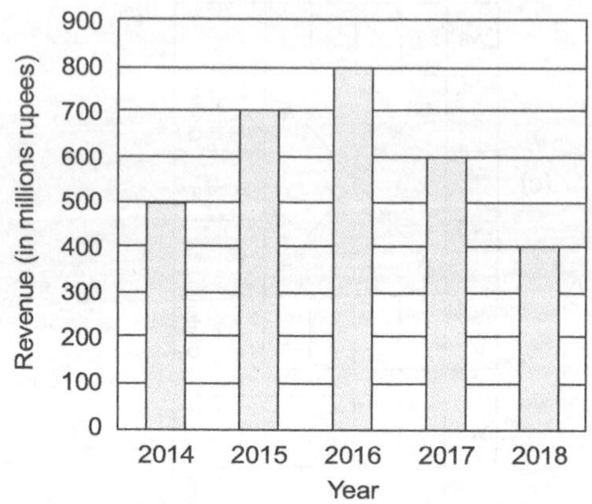

(a) 16.67% profit
(b) 20% loss
(c) 16.67% loss
(d) 20% profit

[2020 : 2 M]

1.171 Given below are two statements I and II and two conclusions I and II:

Statement:
I. All bacteria are microorganisms.
II. All pathogens are microorganisms.

Conclusions:
I. Some pathogens are bacteria.
II. All pathogens are not bacteria.

Based on the above statements and conclusions, which one of the following options is logically CORRECT?

(a) Only conclusion I is correct
(b) Either conclusion I or II is correct.
(c) Only conclusion II is correct
(d) Neither conclusion I nor II is correct.

[2021 (Set-1) : 2 M]

1.172 There are five bags each containing identical sets of ten distinct chocolates. One chocolate is picked from each bag.
The probability that at least two chocolates are identical is ________.

(a) 0.6976 (b) 0.3024
(c) 0.4235 (d) 0.8125

[2021 (Set-1) : 2 M]

1.173 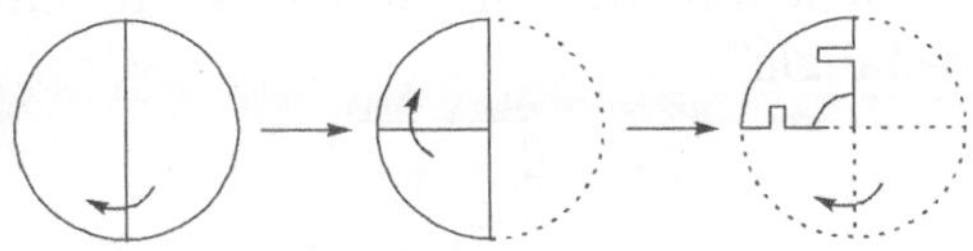

A circular sheet of paper is folded along the lines in the directions shown. The paper, after being punched in the filial folded state as shown and unfolded in the reverse order of folding, will look like ________.

(a) 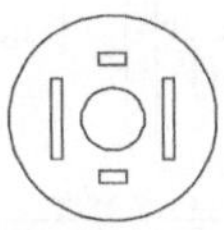 (b) 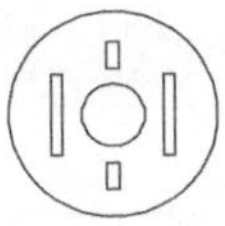

(c) 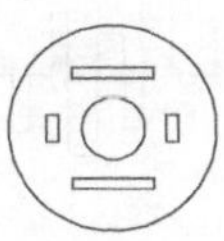 (d) 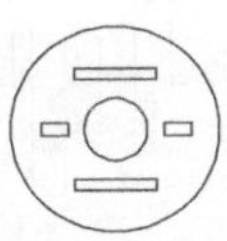

[2021 (Set-1) : 1 M]

1.174 We have 2 rectangular sheets of paper, M and N, of dimensions 6 cm × 1 cm each. Sheet M is rolled to form an open cylinder by bringing the short edges of the sheet together. Sheet N is cut into equal square patches and assembled to form the largest possible closed cube. Assuming the ends of the cylinder are closed, the ratio of the volume of the cylinder to that of the cube is ________.

(a) $\frac{9}{\pi}$ (b) 3π

(c) $\frac{\pi}{2}$ (d) $\frac{3}{\pi}$

[2021 (Set-1) : 2 M]

1.175 Consider the following sentences:

(i) Everybody in the class is prepared for the exam.

(ii) Babu invited Danish to his home because he enjoys playing chess.

Which of the following is the CORRECT observation about the above two sentences?

(a) (i) is grammatically incorrect and (ii) is unambiguous

(b) (i) is grammatically correct and (ii) is ambiguous

(c) (i) is grammatically incorrect and (ii) is ambiguous

(d) (i) is grammatically correct and (ii) is unambiguous **[2021 (Set-1) : 1 M]**

1.176 Some people suggest anti-obesity measures (AOM) such as displaying calorie information in restaurant menus. Such measures sidestep addressing the core problems that cause obesity: poverty and income inequality.

Which one of the following statements summarizes the passage?

(a) The proposed AOM addresses the care problems that cause obesity.

(b) If obesity reduces, poverty will naturally reduce, since obesity causes poverty.

(c) AOM are addressing the problem superficially.

(d) AOM are addressing the core problems and are likely to succeed.

[2021 (Set-1) : 2 M]

1.177 ________ is to *surgery* as *writer* is to ________

Which one of the following options maintains a similar logical relation in the above sentence?

(a) Plan, outline (b) Hospital, library

(c) Doctor, book (d) Medicine, grammar

[2021 (Set-1) : 1 M]

1.178 A polygon is convex if, for every pair of points. P and Q belonging to the polygon, the line segment PQ lies completely inside or on the polygon.

Which one of the following is NOT a convex polygon?

(a) 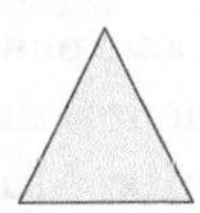 (b) 

(c) 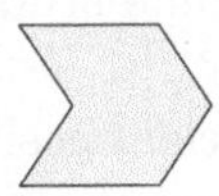 (d) 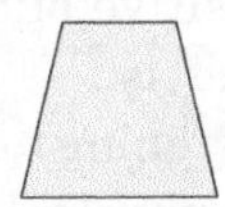

[2021 (Set-1) : 1 Mark]

1.179

Items	Cost (₹)	Profit%	Marked price (₹)
P	5400	–	5860
Q	–	25	10000

Details of prices of two items P and Q are presented in the above table. The ratio of cost of item P to cost of item Q is 3 : 4. Discount is calculated as the difference between the marked price and the selling price. The profit percentage is calculated as the ratio of the difference between selling price and cost, to the cost

$$\left(\text{Profit\%} = \frac{\text{Selling price} - \text{Cost}}{\text{Cost}} \times 100\right).$$

The discount on item Q, as a percentage of its marked price, is ________.

(a) 25 (b) 10
(c) 5 (d) 12.5

[2021 (Set-1) : 2 M]

1.180 The ratio of boys to girls in a class is 7 to 3. Among the options below, an acceptable value for the total number of students in the class is:

(a) 37 (b) 50
(c) 21 (d) 73

[2021 (Set-1) : 1 M]

1.181 If $\left(x-\frac{1}{2}\right)^2 - \left(x-\frac{3}{2}\right)^2 = x+2$, then the value of x is:

(a) 2 (b) 8
(c) 4 (d) 6

[2021 (Set-2) : 1 M]

1.182 Listening to music during exercise improves exercise performance and reduces discomfort. Scientists researched whether listening to music while studying can help students learn better and the results were inconclusive. Students who needed external stimulation for studying fared worse while students who did not need any external stimulation benefited from music.

Which one of the following statements is the CORRECT inference of the above passage?

(a) Listening to music has a clear positive effect on learning in all students. Music has a positive effect only in some students who exercise.
(b) Listening to music has a clear positive effect on physical exercise. Music has a positive effect on learning only in some students.
(c) Listening to music has a clear positive effect both on physical exercise and on learning.
(d) Listening to music has no effect on learning and a positive effect on physical exercise.

[2021 (Set-2) : 2 M]

1.183 Pen : Write : : Knife : ______

Which one of the following options maintains a similar logical relation in the above?

(a) Sharp (b) Cut
(c) Vegetables (d) Blunt

[2021 (Set-2) : 1 M]

1.184 Gauri said that she can play keyboard ______ her sister.

(a) as worse as (b) as better as
(c) as well as (d) as nicest as

[2021 (Set-2) : 1 M]

1.185 What will be the remaining part to make it a rectangle?

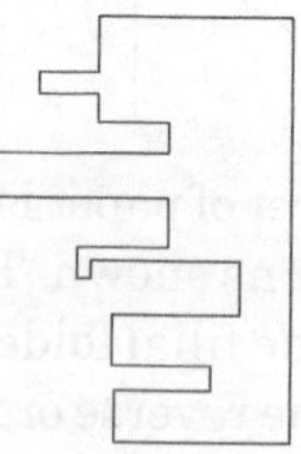

(a) 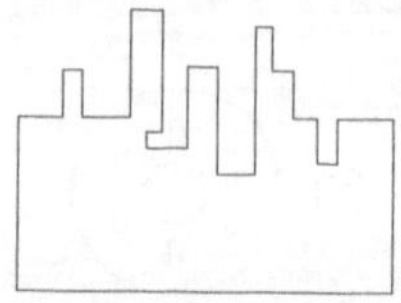 (b) 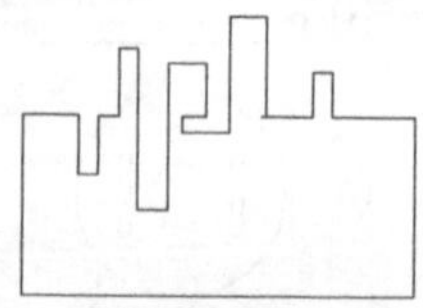

(c) 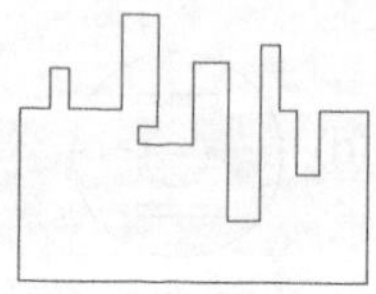 (d) 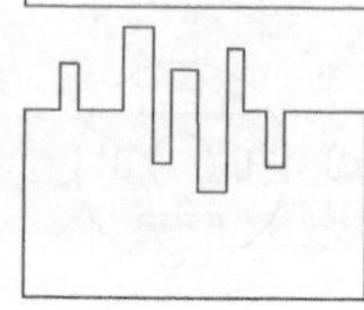

[2021 (Set-2) : 2 M]

1.186 If θ is the angle, in degrees, between the longest diagonal of the cube and any one of the edges of the cube, then cos θ = ______.

(a) $\frac{1}{\sqrt{3}}$ (b) $\frac{1}{2}$
(c) $\frac{1}{\sqrt{2}}$ (d) $\frac{\sqrt{3}}{2}$

[2021 (Set-2) : 1 M]

1.187 Six students P, Q, R, S, T and U, with distinct heights, compare their heights and make the following observations :

Observation I : S is taller than R.
Observation II : Q is the shortest of all.
Observation III : U is taller than only one student.
Observation IV : T is taller than S but is not the tallest.

The number of students that are taller than R is the same as the number of students shorter than ______.

(a) R (b) P
(c) S (d) T

[2021 (Set-2) : 2 M]

1.188 The number of student in three classes is in the ratio 3 : 13 : 6. If 18 students are added to each class, the ratio changes to 15 : 35 : 21.

The total number of students in all the three classes in the beginning was:

(a) 88 (b) 110
(c) 66 (d) 22

[2021 (Set-2) : 2 M]

1.189 The number of units of a product sold in three different years and the respective net profits are presented in the figure above. The cost/unit in year 3 was Rs. 1, which was half the cost/unit in year 2. The cost/unit in year 3 was one-third of the cost/unit in year 1. Taxes were paid on the selling price at 10%, 13% and 15% respectively for the three years. Net profit is calculated as the difference between the selling price and the sum of cost and taxes paid in that year.

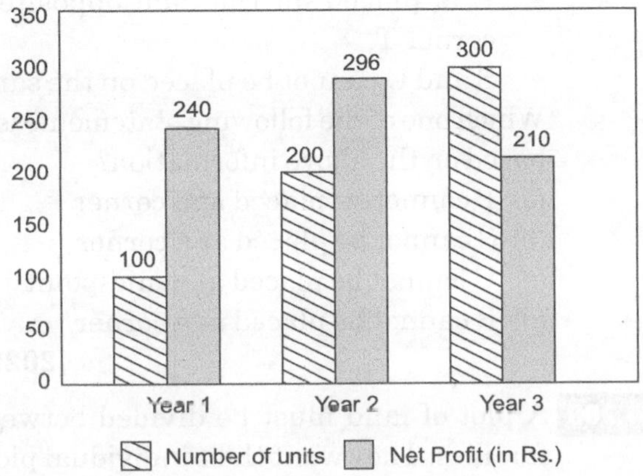

The ratio of the selling price in Year 2 to the selling price in Year 3 is ________.

(a) 3 : 4 (b) 1 : 2
(c) 1 : 1 (d) 4 : 3

[2021 (Set-2) : 2 M]

1.190 A transparent square sheet shown above is folded along the dotted line. The folded sheet will look like ________.

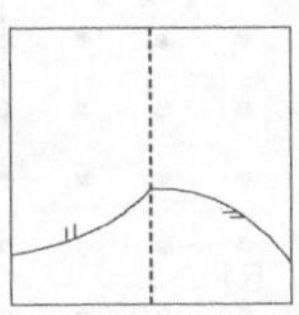

(a) 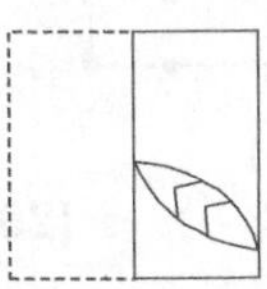 (b) 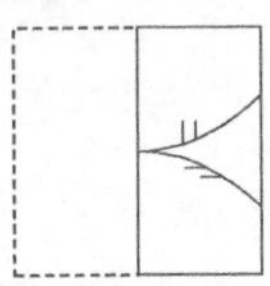

(c) 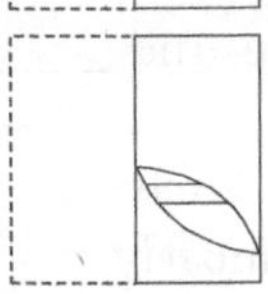 (d) 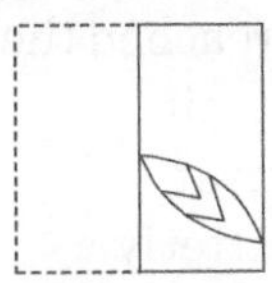

[2021 (Set-2) : 1 M]

1.191 The ________ is too high for it to be considered ________.

(a) fair / fare (b) faer / fair
(c) fare / fare (d) fare / fair

[2022 : 1 M]

1.192 A function $y(x)$ is defined in the interval [0, 1] on the x-axis as

$$y(x) = \begin{cases} 2 \text{ if } 0 \le x < \frac{1}{3} \\ 3 \text{ if } \frac{1}{3} \le x < \frac{3}{4} \\ 1 \text{ if } \frac{3}{4} \le x < 1 \end{cases}$$

Which one of the following is the area under the curve for the interval [0, 1] on the x-axis?

(a) $\frac{5}{6}$ (b) $\frac{6}{5}$
(c) $\frac{13}{6}$ (d) $\frac{6}{13}$

[2022 : 1 M]

1.193 Let r be a root of the equation $x^2 + 2x + 6 = 0$. Then the value of the expression $(r + 2)(r + 3)(r + 4)(r + 5)$ is

(a) 51 (b) –51
(c) 126 (d) –126

[2022 : 1 M]

1.194 Given below are four statements:

Statement 1: All students are inquisitive.
Statement 2: Some students are inquisitive.
Statement 3: No student is inquisitive.
Statement 4: Some students are not inquisitive.

From the given four statements, find the two statements that **CANNOT BE TRUE** simultaneously, assuming that there is at least one student in the class.

(a) Statement 1 and Statement 3
(b) Statement 1 and Statement 2
(c) Statement 2 and Statement 4
(d) Statement 3 and Statement 4

[2022 : 1 M]

1.195 A palindrome is a word that reads the same forwards and backwards. In a game of words, a player has the following two plates painted with letters.

A	D

From the additional plates given in the options, which one of the combinations of additional plates would allow the player to construct a five-letter palindrome. The player should use all the five plates exactly once. The plates can be rotated in their plane.

(a) | D | ᗡ | J |
(b) | R | ∀ | ᴚ |

(c) 

(d) 

[2022 : 1 M]

1.196 Some people believe that "what gets measured, improves". Some others believe that "what gets measured, gets gamed". One possible reason for the difference in the beliefs is the work culture in organizations. In organizations with good work culture, metrics help improve outcomes. However, the same metrics are counterproductive in organizations with poor work culture.

Which one of the following is the CORRECT logical inference based on the information in the above passage?

(a) Metrics are useful in organizations with poor work culture.
(b) Metrics are useful in organizations with good work culture.
(c) Metrics are always counterproductive in organizations with good work culture.
(d) Metrics are never useful in organizations with good work culture.

[2022 : 2 M]

1.197 In a recently conducted national entrance test, boys constituted 65% of those who appeared for the test. Girls constituted the remaining candidates and they accounted for 60% of the qualified candidates.

Which one of the following is the correct logical inference based on the information provided in the above passage?

(a) Equal number of boys and girls qualified
(b) Equal number of boys and girls appeared for the test
(c) The number of boys who appeared for the test is less than the number of girls who appeared
(d) The number of boys who qualified the test is less than the number of girls who qualified

[2022 : 2 M]

1.198 A box contains five balls of same size and shape. Three of them are green coloured balls and two of them are orange coloured balls. Balls are drawn from the box one at a time. If a green ball is drawn, it is not replaced. If an orange ball is drawn, it is replaced with another orange ball.

First ball is drawn. What is the probability of getting an orange ball in the next draw?

(a) $\frac{1}{2}$ (b) $\frac{8}{25}$
(c) $\frac{19}{50}$ (d) $\frac{23}{50}$

[2022 : 2 M]

1.199 The corners and mid-points of the sides of a triangle are named using the distinct letters P, Q, R, S, T and U, but not necessarily in the same order. Consider the following statements:

- The line joining P and R is parallel to the line joining Q and S.
- P is placed on the side opposite to the corner T.
- S and U cannot be placed on the same side.

Which one of the following statements is correct based on the above information?

(a) P cannot be placed at a corner
(b) S cannot be placed at a corner
(c) U cannot be placed at a mid-point
(d) R cannot be placed at a corner

[2022 : 2 M]

1.200 A plot of land must be divided between four families. They want their individual plots to be similar in shape, not necessarily equal in area. The land has equally spaced poles, marked as dots in the below figure. Two ropes, R1 and R2, are already present and cannot be moved.

What is the least number of additional straight ropes needed to create the desired plots? A single rope can pass through three poles that are aligned in a straight line.

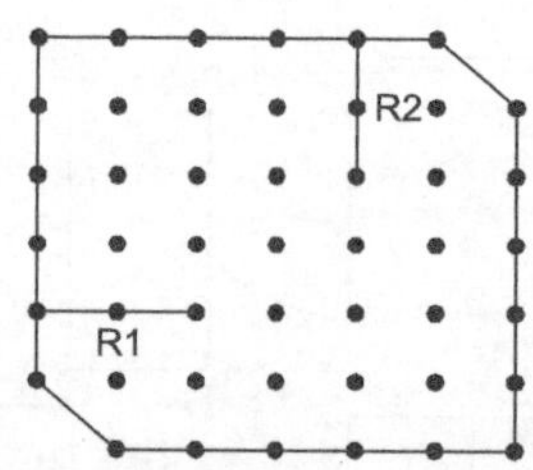

(a) 2 (b) 4
(c) 5 (d) 3

[2022 : 2 M]

1.201 We reached the station late, and _______ missed the train.

(a) near (b) nearly
(c) utterly (d) mostly

[2023 : 1 M]

1.202 Kind : _______ : : Often : Frequently
(By word meaning)

(a) Mean (b) Type
(c) Cruel (d) Kindly

[2023 : 1 M]

1.203 A series of natural numbers $F_1, F_2, F_3, F_4, F_5, F_6, F_7, \ldots$ beys $F_{n+1} = F_n + F_{n+1}$ for all integers $n \geq 2$.

If $F_6 = 37$, and $F_7 = 60$, then what is F_1?

(a) 4 (b) 5

(c) 8 (d) 9

[2023 : 1 M]

1.204 A survey for a certain year found that 90% of pregnant women received medical care at least once before giving birth. Of these women, 60% received medical care from doctors, while 40% received medical care from other healthcare providers.

Given this information, which one of the following statements can be inferred with certainty?

(a) More than half of the pregnant women received medical care at least once from a doctor.

(b) Less than half of the pregnant women received medical care at least once from a doctor.

(c) More than half of the pregnant women received medical care at most once from a doctor.

(d) Less than half of the pregnant women received medical care at most once from a doctor. [2023 : 1 M]

1.205 Looking at the surface of a smooth 3-dimensional object from the outside, which one of the following options is TRUE?

(a) The surface of the object must be concave everywhere.

(b) The surface of the object must be convex everywhere.

(c) The surface of the object may be concave in some places and convex in other places.

(d) The object can have edges, but no corners.

[2023 : 1 M]

1.206 The country of Zombieland is in distress since more than 75% of its working population is suffering from serious health issues. Studies conducted by competent health experts concluded that a complete lack of physical exercise among its working population was one of the leading causes of their health issues. As one of the measures to address the problem, the Government of Zombieland has decided to provide monetary incentives to those who ride bicycles to work.

Based only on the information provided above, which one of the following statements can be logically inferred with certainty?

(a) All the working population of Zombieland will henceforth ride bicycles to work.

(b) Riding bicycles will ensure that all of the working population of Zombieland is free of health issues.

(c) The health experts suggested to the Government of Zombieland to declare riding bicycles as mandatory.

(d) The Government of Zombieland believes that riding bicycles is a form of physical exercise.

[2023 : 2 M]

1.207 Consider two functions of time (t),

$$f(t) = 0.01t^2$$
$$g(t) = 4t$$

where $0 < t < \infty$.

Now consider the following two statements:

(i) For some $t > 0$, $g(t) > f(t)$

(ii) There exists a T, such that $f(t) > g(t)$ for all $t > T$.

Which one of the following options is TRUE?

(a) Only (i) is correct

(b) Only (ii) is correct

(c) Both (i) and (ii) are correct

(d) Neither (i) nor (ii) is correct [2023 : 2 M]

1.208 Which one of the following sentence sequences creates a coherent narrative?

(i) Once on the terrace, on her way to her small room in the corner, she notices the man right away.

(ii) She begins to pant by the time she has climbed all the stairs.

(iii) Mina has bought vegetables and rice at the market, so her bags are heavy.

(iv) He was leaning against the parapet, watching the traffic below.

(a) (i), (ii), (iv), (iii)

(b) (ii), (iii), (i), (iv)

(c) (iv), (ii), (i), (iii)

(d) (iii), (ii), (i), (iv)

[2023 : 2 M]

1.209 $f(x)$ and $g(y)$ are functions of x and y, respectively, and $f(x) = g(y)$ for all real values of x and y Which one of the following options is necessarily TRUE for all x and y?

(a) $f(x) = 0$ and $g(y) = 0$

(b) $f(x) = g(y) = \text{constant}$

(c) $f(x) \neq \text{constant}$ and $g(y) \neq \text{constant}$

(d) $f(x) + g(y) = f(x) - g(y)$ [2023 : 2 M]

1.210 Which one of the options best describes the transformation of the 2-dimensional figure P to Q, and then to R, as shown?

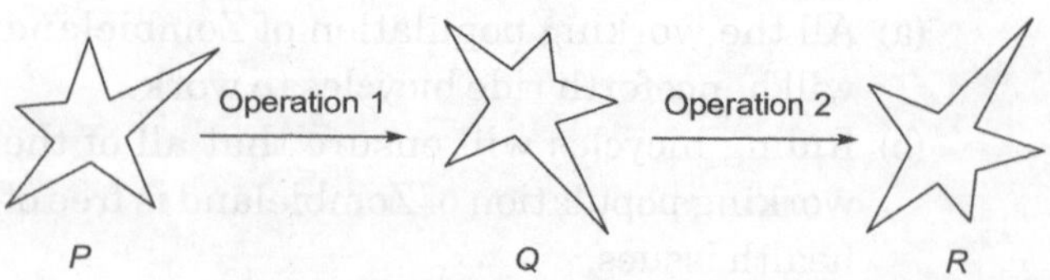

(a) Operation 1: A clockwise rotation by 90° about an axis perpendicular to the plane of the figure
Operation 2: A reflection along a horizontal line
(b) Operation 1: A counter clockwise rotation by 90° about an axis perpendicular to the plane of the figure
Operation 2: A reflection along a horizontal line
(c) Operation 1: A clockwise rotation by 90° about an axis perpendicular to the plane of the figure
Operation 2: A reflection along a vertical line
(d) Operation 1: A counter clockwise rotation by 180° about an axis perpendicular to the plane of the figure
Operation 2: A reflection along a vertical line

[2023 : 2 M]

1.211 For positive non-zero real variables p and q, if $\log(p^2 + q^2) = \log p + \log q + 2 \log 3$, then, the value of $\frac{p^4+q^4}{p^2q^2}$ is

(a) 9 (b) 79
(c) 81 (d) 83

[2024 (Set-1) : 1 M]

1.212 A rectangular paper sheet of dimensions 54 cm × 4 cm is taken. The two longer edges of the sheet are joined together to create a cylindrical tube. A cube whose surface area is equal to the area of the sheet is also taken.
Then, the ratio of the volume of the cylindrical tube to the volume of the cube is

(a) $\frac{2}{\pi}$ (b) $\frac{1}{\pi}$
(c) $\frac{4}{\pi}$ (d) $\frac{3}{\pi}$

[2024 (Set-1) : 2 M]

1.213 In the given text, the blanks are numbered (i) - (iv). Select the best match for all the blanks.

Steve was advised to keep his head ___(i)___ before heading ___(ii)___ to bat; for, while he had a head ___(iii)___ batting, he could only do so with a cool head ___(iv)___ his shoulders.

(a) (i) on (ii) down (iii) for (iv) on
(b) (i) down (ii) down (iii) on (iv) for
(c) (i) down (ii) out (iii) for (iv) on
(d) (i) on (ii) out (iii) on (iv) for

[2024 (Set-1) : 2 M]

1.214 A rectangular paper of 20 cm × 8 cm is folded 3 times. Each fold is made along the line of symmetry, which is perpendicular to its long edge. The perimeter of the final folded sheet (in cm) is

(a) 18 (b) 20
(c) 24 (d) 21

[2024 (Set-1) : 2 M]

1.215 The pie chart present the percentage contribution of different macronutrients to a typical 2,000 kcal diet of a person.

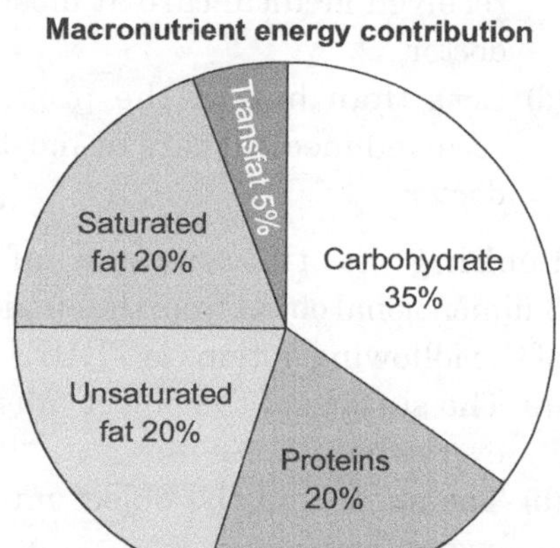

The typical energy density (kcal/g) of these macronutrients is given in the table.

Macronutrient	Energy density (kcal/g)
Carbohydrates	4
Proteins	4
Unsaturared fat	9
Saturared fat	9
Trans fat	9

The total fat (all three types), in grams, this person consumes is

(a) 3,600 (b) 100
(c) 44.4 (d) 77.8

[2024 (Set-1) : 2 M]

1.216 If '→' denotes increasing order of intensity, then the meaning of the words [dry → arid → parched] is analogous to [diet → fast → ______].
Which one of the given options is appropriate to fill the blank?
(a) reject
(b) feast
(c) deny
(d) starve

[2024 (Set-1) : 1 M]

1.217 The least number of squares to be added in the figure to make AB a line of symmetry is

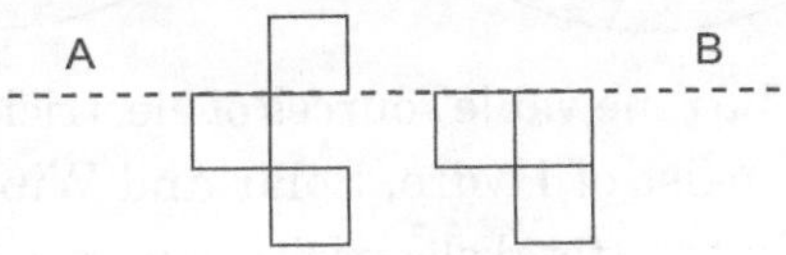

(a) 4 (b) 5
(c) 7 (d) 6

[2024 (Set-1) : 2 M]

1.218 The number of coins of ₹1, ₹5, and ₹10 denominations that a person has are in the ratio 5 : 3 : 13. Of the total amount, the percentage of money in ₹5 coins is
(a) $14\frac{2}{7}\%$ (b) 10%
(c) 21% (d) 30%

[2024 (Set-1) : 1 M]

1.219 If two distinct non-zero real variables x and y are such that $(x + y)$ is proportional to $(x - y)$ then the value of $\frac{x}{y}$
(a) is a constant
(b) depends on xy
(c) depends only on x and not on y
(d) depends only on y and not on x

[2024 (Set-1) : 1 M]

1.220 Consider the following sample of numbers:
9, 18, 11, 14, 15, 17, 10, 69, 11, 13
The median of the sample is
(a) 14 (b) 18.7
(c) 13.5 (d) 11

[2024 (Set-1) : 1 M]

1.221 If '→' denotes increasing order of intensity, then the meaning of the words [walk → jog → sprint] is analogous to [bothered → ______ → daunted].
Which one of the given options is appropriate to fill the blank?
(a) phased (b) phrased
(c) fazed (d) fused

[2024 (Set-2) : 1 M]

1.222 In the 4 × 4 array shown below, each cell of the first three rows has either a cross (X) or a number.

1	X	4	3
X	5	5	4
3	X	6	X

The number in a cell represents the count of the immediate neighboring cells (left, right, top, bottom, diagonals) NOT having a cross (X). Given that the last row has no crosses (X), the sum of the four numbers to be filled in the last row is
(a) 9 (b) 11
(c) 12 (d) 10

[2024 (Set-2) : 2 M]

1.223 In the sequence 6, 9, 14, x, 30, 41, a possible value of x is
(a) 18 (b) 21
(c) 25 (d) 20

[2024 (Set-2) : 1 M]

1.224 Two wizards try to create a spell using all the four elements, *water*, *air*, *fire*, and *earth*. For this, they decide to mix all these elements in all possible orders. They also decide to work independently. After trying all possible combination of elements, they conclude that the spell does not work. How many attempts does each wizard make before coming to this conclusion, independently?
(a) 12 (b) 24
(c) 16 (d) 48

[2024 (Set-2) : 1 M]

1.225 A person sold two different items at the same price. He made 10% profit in one item, and 10% loss in the other item. In selling these two items, the person made a total of
(a) 1% profit (b) 1% loss
(c) 2% profit (d) 2% loss

[2024 (Set-2) : 2 M]

1.226 A cube is to be cut into 8 pieces of equal size and shape. Here, each cut should be straight and it should not stop till it reaches the other end of the cube.

The minimum number of such cuts required is

(a) 8 (b) 3
(c) 7 (d) 4

[2024 (Set-2) : 2 M]

1.227 Sequence the following sentences in a coherent passage.

P : This fortuitous geological event generated a colossal amount of energy and heat that resulted in the rocks rising to an average height of 4 km across the contact zone.

Q : Thus, the geophysicists tend to think of the Himalayas as an active geological event rather than as a static geological feature.

R : The natural process of the cooling of this massive edifice absorbed large quantities of atmospheric carbon dioxide, altering the earth's atmosphere and making it better suited for life.

S : Many millennia ago, a breakaway chunk of bedrock from the Antarctic Plate collided with the massive Eurasian Plate.

(a) SRPQ (b) SPRQ
(c) QPSR (d) QSPR

[2024 (Set-2) : 2 M]

1.228 In an engineering college of 10,000 students, 1,500 like neither their core branches nor other branches. The number of students who like their core branches is $\frac{1}{4}^{\text{th}}$ of the number of students who like other branches. The number of students who like both their core and other branches is 500.

The number of students who like their core branches is

(a) 1800 (b) 1500
(c) 3500 (d) 1600

[2024 (Set-2) : 1 M]

1.229 The pie charts depict the shares of various power generation technologies in the total electricity generation of a country for the years 2007 and 2023.

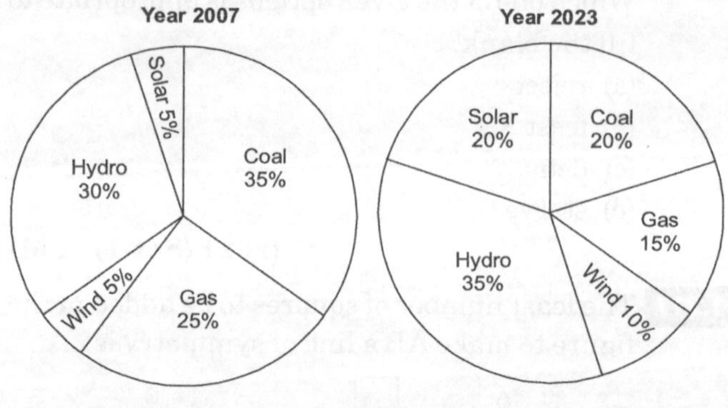

The renewable sources of electricity generation consist of Hydro, Solar and Wind. Assuming that the total electricity generated remains the same from 2007 to 2023, what is the percentage increase in the share of the renewable sources of electricity generation over this period?

(a) 77.5% (b) 50%
(c) 62.5% (d) 25%

[2024 (Set-2) : 2 M]

1.230 For positive non-zero real variables x and y, if

$\ln\left(\frac{x+y}{2}\right) = \frac{1}{2}[\ln(x) + \ln(y)]$ then, the value of

$\frac{x}{y} + \frac{y}{x}$ is

(a) 4 (b) $\frac{1}{2}$
(c) 2 (d) 1

[2024 (Set-2) : 1 M]

■■■■

Answers General Aptitude

1.1	(b)	1.2	(a)	1.3	(d)	1.4	(c)	1.5	(d)	1.6	(d)	1.7	(c)	1.8	(d)	1.9	(b)
1.10	(b)	1.11	(a)	1.12	(b)	1.13	(c)	1.14	(b)	1.15	(d)	1.16	(c)	1.17	(d)	1.18	(a)
1.19	(c)	1.20	(d)	1.21	(a)	1.22	(b)	1.23	(a)	1.24	(b)	1.25	(a)	1.26	(d)	1.27	(b)
1.28	(b)	1.29	(c)	1.30	(a)	1.31	(b)	1.32	(a)	1.33	(c)	1.34	(a)	1.35	(c)	1.36	(b)
1.37	(d)	1.38	(d)	1.39	(c)	1.40	(b)	1.41	(b)	1.42	(c)	1.43	(c)	1.44	(96)	1.45	(d)
1.46	(c)	1.47	(b)	1.48	(850)	1.49	(48)	1.50	(6)	1.51	(a)	1.52	(b)	1.53	(a)	1.54	(b)
1.55	(c)	1.56	(b)	1.57	(d)	1.58	(d)	1.59	(140)	1.60.	(a)	1.61	(d)	1.62	(c)	1.63	(c)
1.64	(c)	1.65	(c)	1.66	(b)	1.67	(d)	1.68	(d)	1.69	(c)	1.70	(c)	1.71	(a)	1.72	(c)
1.73	(a)	1.74	(a)	1.75	(b)	1.76	(c)	1.77	(c)	1.78	(a)	1.79	(32)	1.80	(d)	1.81	(c)
1.82	(c)	1.83	(c)	1.84	(d)	1.85	(a)	1.86	(b)	1.87	(a)	1.88	(b)	1.89	(8)	1.90	(a)
1.91	(b)	1.92	(c)	1.93	(b)	1.94	(b)	1.95	(c)	1.96	(d)	1.97	(b)	1.98	(b)	1.99	(b)
1.100	(2006)	1.101	(d)	1.102	(a)	1.103	(a)	1.104	(c)	1.105	(d)	1.106	(b)	1.107	(d)	1.108	(d)
1.109	(b)	1.110	(b)	1.111	(b)	1.112	(a)	1.113	(c)	1.114	(d)	1.115	(a)	1.116	(a)	1.117	(d)
1.118	(d)	1.119	(c)	1.120	(c)	1.121	(c)	1.122	(d)	1.123	(c)	1.124	(d)	1.125	(c)	1.126	(b)
1.127	(c)	1.128	(b)	1.129	(d)	1.130	(a)	1.131	(b)	1.132	(a)	1.133	(b)	1.134	(d)	1.135	(d)
1.136	(a)	1.137	(b)	1.138	(c)	1.139	(c)	1.140	(c)	1.141	(b)	1.142	(b)	1.143	(c)	1.144	(a)
1.145	(d)	1.146	(c)	1.147	(b)	1.148	(a)	1.149	(c)	1.150	(b)	1.151	(b)	1.152	(c)	1.153	(b)
1.154	(b)	1.155	(c)	1.156	(d)	1.157	(a)	1.158	(c)	1.159	(c)	1.160	(b)	1.161	(b)	1.162	(c)
1.163	(a)	1.164	(b)	1.165	(a)	1.166	(c)	1.167	(c)	1.168	(c)	1.169	(d)	1.170	(d)	1.171	(b, d)
1.172	(a)	1.173	(c)	1.174	(a)	1.175	(b)	1.176	(c)	1.177	(c)	1.178	(c)	1.179	(b)	1.180	(b)
1.181	(c)	1.182	(b)	1.183	(b)	1.184	(c)	1.185	(c)	1.186	(a)	1.187	(c)	1.188	(a)	1.189	(d)
1.190	(a)	1.191	(d)	1.192	(c)	1.193	(d)	1.194	(a)	1.195	(b)	1.196	(b)	1.197	(d)	1.198	(d)
1.199	(b)	1.200	(d)	1.201	(b)	1.202	(b, d)	1.203	(a)	1.204	(a)	1.205	(c)	1.206	(d)	1.207	(c)
1.208	(d)	1.209	(b)	1.210	(a)	1.211	(b)	1.212	(b)	1.213	(c)	1.214	(a)	1.215	(b)	1.216	(d)
1.217	(d)	1.218	(b)	1.219	(a)	1.220	(c)	1.221	(c)	1.222	(b)	1.223	(b)	1.224	(b)	1.225	(b)
1.226	(b)	1.227	(b)	1.228	(a)	1.229	(c)	1.230	(c)								

Explanations General Aptitude

1.1 (b)

Circuitous: Deviating from a straight course ⇒ Indirect

(a) Cyclic: Recurring in cycle
(b) Indirect: Not leading by straight line
(c) Confusing: Lacking clarity
(d) Crooked: For shapes (irregular in shape)

1.2 (a)

Unemployed: Worker ⇒ Here one is opposite to other.

(a) Fallow: Land ⇒ Fallow means undeveloped land.
(b) Unaware: sleeper ⇒ Both are same unaware or asleep.
(c) Wit: Jester ⇒ Wit means ability to make jokes and jester is a joker.
(d) Renovated: House ⇒ Renovate means to make better and house can be renovated.

1.3 (d)

(a) Uphold: cause to remain ⇒ not appropriate
(b) Restrain: keep under control ⇒ not appropriate
(c) Cherish: be fond of ⇒ not related
(d) Conserve: Keep in safety and protect from harm, decay, loss, or destruction ⇒ most appropriate.

1.4 (c)

(a) Masked: Hide under a false appearance ⇒ opposite
(b) Belied: Be in contradiction with ⇒ not appropriate
(c) Betrayed: Reveal unintentionally⇒ most appropriate
(d) Suppressed: To put down by force or authority ⇒ irrelevant

1.5 (d)

Using the set theory formula
n(A) : Number of people who play hockey = 15
n(B) : Number of people who play football = 17
$n(A \cap B)$: Persons who play both hockey and football = 10
$n(A \cup B)$: Persons who play either hockey or football or both
Using the formula

$n(A \cup B) = n(A) + n(B) - n(A \cap B)$

$n(A \cup B) = 15 + 17 - 10 = 22$

Thus people who play neither hockey nor football = 25 − 22 = 3

Alternative Method:

Refer to Venn diagram given below:

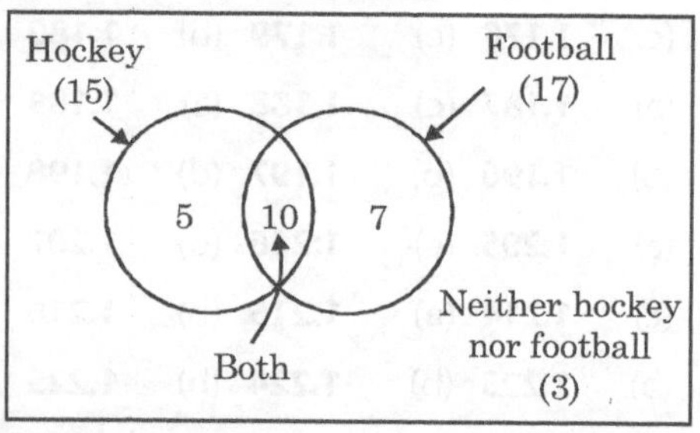

1.6 (d)

(a) Modern warfare has resulted in civil strife: There is no direct consequence of warfare given, so it is not appropriate.
(b) Chemical agents are useful in modern warfare: Passage does not say whether chemical agents are useful or not, so not appropriate.
(c) Use of chemical agents in warfare would be undesirable: Given that people in military think these are useful, undesirable is wrong
(d) People in military establishments like to use chemical agents in war; Correct choice as last statement tells that military people think that chemical agents are useful tools for their cause (work silently in warfare).

1.7 (c)

137 + 276 = 435
This is a octahedral addition thus
731 + 672(8) = 1623

Alternative Method

7 and 6 added is becoming five means the given two numbers are added on base 8.

$$\begin{array}{r} (137)_8 \\ +(276)_8 \\ \hline (435)_8 \end{array}$$

Hence we have to add the another two given set of numbers also on base 8.

$$\begin{array}{r} (731)_8 \\ +(672)_8 \\ \hline (1623)_8 \end{array}$$

Hence the overall problem was based on identyfying base which was 8 and adding number on base 8.

1.8 (d)

Per day work or rate of 5 skilled workers = $\frac{1}{20}$

⇒ Per day work or rate of one skill worker

$= \frac{1}{5 \times 20} = \frac{1}{100}$

Similarly Per day work or rate of 8 semiskilled workers = $\frac{1}{25}$

⇒ Per day work or rate of one semi-skill worker

$= \frac{1}{8 \times 25} = \frac{1}{200}$

And per day work or rate of 10 unskilled workers

$= \frac{1}{30}$

⇒ Per day work or rate of one semi-skill worker

$= \frac{1}{10 \times 30} = \frac{1}{300}$

Thus total per day work of 2 skilled, 6 semiskilled and 5 unskilled workers

$$= \frac{2}{100} + \frac{6}{200} + \frac{5}{300} = \frac{12+18+10}{600} = \frac{40}{600} = \frac{1}{15}$$

Thus time to complete the work is 15 days.

Alternative Method:

Let one day work of skilled semi-skilled and unskilled worker be a, b, c units respectively.

$5a \times 20 = 8b + 25 = 10c \times 30 =$ Total unit of work

$100a = 200b = 300c$

$a = 2b = 3c$

$$\Rightarrow \quad b = \frac{a}{2} \text{ and } c = \frac{a}{3}$$

Given that 2 skilled, 6 semi-skilled and 5 unskilled workers are working. Let they finish the work in 'x' days.

(2a + 6b + 5c)x = 5a + 20 = Total units of work

$$\left(2a + 3a + \frac{5}{3}a\right)x = 5a \times 20$$

$$\frac{20a}{3}x = 5a \times 20$$

$$x = 15 \text{ days}$$

1.9 (b)

We have to make 4 digit numbers, so the number should be start with 3 or 4, two cases possible;

Case (1) thousands digit is 3

Now other three digits may be any of 2, 2, 3, 3, 4, 4, 4, 4.

(a) Using 2, 2, 3
$\Rightarrow$ 223, 232, 322 -------

$\left(\frac{3!}{2!} = 3 \text{ numbers are possible}\right)$

(b) Using 2, 2, 4 $\Rightarrow$ 224, 242, 422

$\left(\frac{3!}{2!} = 3 \text{ numbers are possible}\right)$

(c) Using 2, 3, 3 $\Rightarrow$ 233, 323, 332

$\left(\frac{3!}{2!} = 3 \text{ numbers are possible}\right)$

(d) Using 2, 3, 4 $\Rightarrow$ 234, 243, 324, 342, 423, 432
(3! = 6 numbers are possible)

(e) Using 2, 4, 4 $\Rightarrow$ 244, 424, 442

$\left(\frac{3!}{2!} = 3 \text{ numbers are possible}\right)$

(f) Using 3, 3, 4 $\Rightarrow$ 334, 343, 433

$\left(\frac{3!}{2!} = 3 \text{ numbers are possible}\right)$

(g) Using 3, 4, 4 $\Rightarrow$ 344, 434, 443

$\left(\frac{3!}{2!} = 3 \text{ numbers are possible}\right)$

(h) Using 4, 4, 4 $\Rightarrow$ 444

$\left(\frac{3!}{3!} = 1 \text{ numbers are possible}\right)$

Total 4 digit numbers in case 1 = 3 + 3 + 3 + 6 + 3 + 3 + 3 + 1 = 25

Case (2): thousands digit is 4 ; Now other three digits may be any of 2, 2, 3, 3, 3, 4, 4, 4.

(a) Using 2, 2, 3 $\Rightarrow$ 223, 232, 322

$\left(\frac{3!}{2!} = 3 \text{ numbers are possible}\right)$

(b) Using 2, 2, 4 $\Rightarrow$ 224, 242, 422

$\left(\frac{3!}{2!} = 3 \text{ numbers are possible}\right)$

(c) Using 2, 3, 3 $\Rightarrow$ 233, 323, 332

$\left(\frac{3!}{2!} = 3 \text{ numbers are possible}\right)$

(d) Using 2, 3, 4 $\Rightarrow$ 234, 243, 324, 342, 423, 432

$\left(\frac{3!}{2!} = 3 \text{ numbers are possible}\right)$

(e) Using 2, 4, 4 $\Rightarrow$ 244, 424, 442

$\left(\frac{3!}{2!} = 3 \text{ numbers are possible}\right)$

(f) Using 3, 3, 3 $\Rightarrow$ 333

$\left(\frac{3!}{3!} - 1 \text{ numbers are possible}\right)$

(g) Using 3, 3, 4 $\Rightarrow$ 334, 343, 433

$\left(\frac{3!}{2!} = 3 \text{ numbers are possible}\right)$

(h) Using 3, 4, 4 $\Rightarrow$ 344, 434, 443

$\left(\frac{3!}{2!} = 3 \text{ numbers are possible}\right)$

(i) Using 4, 4, 4 $\Rightarrow$ 444

$\left(\frac{3!}{3!} = 1 \text{ numbers are possible}\right)$

Total 4 digit numbers in case (2)
= 3 + 3 + 3 + 6 + 3 + 3 = 1 + 3 + 1 = 26

Thus total 4 digits numbers using case (1) and case (2)
= 25 + 26 = 51

***Alternative Method / Short-cut method**

As the number is greater than 3000. So thousand's place can be tiehr 3 or 4.

Let's consider the following two cases

Case (I) When thousand's place is 3.

3 a b c

If there is no restricton on number of two's, three's and four's. Then each of a, b, c can be filled with 2 or 3 or 4 each in 3 ways.

So 3 × 3 × 3 = 27 numbers are there. Out of which 3222, 3333 are invalid as 2 can be used twice & three thrice only so number of such valid numbers begining with 3 are 27 – 2 = 25.

Case (II) When thousand's place is 4

4 a b c

Without restriction on number of 2's, 3's and 4's a, b, c (as explained in case I) can be filled in 27 ways.

Out of these 27 numbers, 4 2 2 2 is only invalid as two have to be used twice only.

So valid numbers are 27 – 1 = 26.

Total numbers from Case (I) & Case (II) 25 + 26 = 51.

1.10 (b)

Suppose: Hari's age : H, Gita's age : G,
Saira's age : S, Irfan's age : I

- H + G > I + S
- Using Statement (2) both G – S = 1 or S – G=1; G can't be oldest and S can't be youngest.
- There are no twins thus using statement (2) either GS or SG possible.

(a) HSIG: not possible as there is I between S and G which is not possible using statement (3)

(b) SGHI: SG order is possible, S > G > H > I and G + H > S + I (possible)
Because if { S = G + 1; and G = H+1 and H = I +2 then G + (I+2) > (G+1) +I}

(c) IGSH: according to this I > G and S > H thus adding these both inequalities we get I + S > G + H which is opposite of statement (2) thus not possible.

(d) IHSG: according to this I > H and S > G thus adding both inequalities I + S > H + G which is opposite of statement (2) thus not possible.

1.12 (b)

$$\text{Log}(P) = \left(\frac{1}{2}\right)\text{Log}(Q) = \left(\frac{1}{3}\right)\text{Log}(R)$$

$$Q = P^2,\ R = P^3,\ Q = R^{2/3}$$

$$\Rightarrow \quad Q^2 = Q.Q = P^2.P^2 = P.P^3 = P.R$$

1.17 (d)

As we can understand that the danger of a microbe to human being will be directly proportional to potency and growth. At the same time, it will be inversely proportional to toxicity defined (more dangerous will a microbe be if lesser of its milligram is required). So level of dangerous

$\alpha \dfrac{P\uparrow \times G\uparrow}{T\downarrow}$ where P, G and T are the potency, growth and toxicity as defined in question.

So $\quad D_i = \dfrac{KPG}{T}$ ...(i)

Where K is constant of proportionality.

So level of dangerous of S will be maximum given by $D_S = \dfrac{0.8\times\pi(10\text{ mm})^2}{200}$.

Similar calculation for D_P, D_Q, D_R can be done based on (i) to find out that D_S is maximum and so most dangerous among them.

1.18 (a)

(T.C.) Total cost = V + f

$$\text{T.C} = 4q + \frac{100}{q}$$

As we have to minimize total cost.

Using options

(a) $q = 5$, T.C. $= 4\times 5 + \dfrac{100}{5} = 40$

(b) $q = 4$, T.C. $= 4\times 4 + \dfrac{100}{4} = 41$

(c) $q = 7$, T.C. $= 4\times 7 + \dfrac{100}{7} = 42.285$

(d) $q = 6$, T.C. $= 4\times 6 + \dfrac{100}{6} = 40.\overline{66}$

Hence, T.C. is minimum at q = 5, **(a) ans.**

* always put options to get answer fast.

Alternate Solution:

or $\quad \text{T.C.} = 4q + \dfrac{100}{q}$

for minimum $\quad \dfrac{d}{dq}\left(4q + \dfrac{100}{q}\right) = 0$

$\Rightarrow \quad q^2 = 25$ or $q = 5$

$$\frac{d}{dq}\left(4q + \frac{100}{q}\right)_{q=5} > 0$$

Hence, TC (Total cost) is minimum at q = 5 **(a) ans.**

1.19 (c)

Let 'y' we the backlog with transporter and 'x' be the number of orders each day. So, as per conditions given in question

$$4x + y = 28 \quad \text{...(i)}$$

$$10x + y = 30 \quad \text{...(ii)}$$

Solving (i) and (ii),

$$x = \frac{1}{3} \text{ and } y = \frac{80}{3}$$

Since, we need to find out number of trucks so that no pending order will be there at the end of 5th day.

$$5x + y = n \times 5$$

So we need to find 'n', where 'n' is the number of trucks required.

$$n = \frac{5x + y}{5} = \frac{5\times\frac{1}{3}+\frac{80}{3}}{5} = \frac{\frac{85}{3}}{5}$$

Hence, $5.\overline{66}$ truck will be required. As number of trucks have to be natural number. Hence, 6 trucks will be required.

1.20 (d)

Short-cut Method:

Every time if we take 1 litre of mixture out and replace with water, content of pure spirit will keep on reducing by 10%.

So, final quantity of spirit after 3 such operations are

$$10 \times 0.9 \times 0.9 \times 0.9 = 7.29 \text{ litres}$$

Alternate Solution:

$$\frac{\text{Quantity of spirit left after } n^{th} \text{ operation}}{\text{Intial quantity of spirit}} = \left(\frac{a-b}{a}\right)^n = \left(1-\frac{b}{a}\right)^n$$

where 'a' is initial quantity of pure spirit and 'b' is quantity taken out and replaced every time.

Hence, quantity of spirit left after 3rd operation

$$= \text{Initial quantity} \times \left(1-\frac{1}{10}\right)^3$$

$$= 10 \times 0.9 \times 0.9 \times 0.9 = 7.29 \text{ litres}$$

1.21 (a)

Total Cost $C = 5q^2$

Total Sales Revenue,

$S = 50q$

Profit (P)= $(S - C) = 50q - 5q^2$

As the profit have to be maximized

$P = 50q - 5q^2$

Using options

(a) $q = 5$

$P = 50 \times 5 - 5 \times 5^2 = 125$

(b) $q = 10$

$P = 50 \times 10 - 5 \times 10^2 = 0$

(c) $q = 15$

$P = 50 \times 15 - 5 \times 15^2 = $ –ive

(d) $q = 25$

$P = 50 \times 25 - 5 \times 25^2 = $ –ive

Hence the maximum profit will happen at $q = 5$

Alternative solution:

$$P = 50q - 5q^2$$

For max/min $\frac{dp}{dq} = 0,$

$$\Rightarrow \quad 50 - 10q = 0$$

$$q = 10$$

$$\left.\frac{d^2p}{dq^2}\right]_{q=5} = -10 \text{ which is negative}$$

Hence maximum profit will happen at

$$q = 5$$

1.22 (b)

Setbacks: Despite several setbacks the mission succeeded in its attempt to resolve the conflict. The word 'Despite' indicates that there has to be a contrast in the sentence, use of the word 'Setbacks' in the blank indicates that despite many problems the mission was successful.

1.23 (a)

Diminish: Mitigate means to reduce, to lessen etc. So only the word Diminish is close. Rest all choices have no link with the given word.

Divulge means to disclose or reveal which has no link with the given word.

1.24 (b)

The country's expenditure is not less than that of Bangladesh.

The correct statement should be: The country's expenditure is not less than that of Bangladesh's. Because the country's expenditure cannot be compared with Bangladesh. It should be compared with Bangladesh's expenditure.

1.25 (a)

Suresh's dog is the one that was hurt in the stampede.

That is used with restrictive clauses.

1.26 (d)

Not gender-discriminatory.

Choice (A) cannot be considered since there is no gender discrimination mentioned in the argument.

Choice (B) Xenophobic is one who has fear of foreigners, no link with the given argument.

Choice (C) It is wrong to say that the profile has not been designed to make the post attractive, since there are certain features which have been added to make the profile lucrative (which are given towards the end of the advertisement, like Transportation paid, expenses reimbursed).

1.27 (b)

$$y = 2x - 0.1x^2$$

For y (height) to be maximum $\frac{dy}{dx} = 0$

$$\frac{dy}{dx} = 2x - 2x = 0$$

$$\Rightarrow \quad x = 10$$

$$\left.\frac{d^2y}{dx^2}\right]_{x=10} = -0.2\ (\text{–ive}), \quad \frac{d^2y}{dx^2} < 0$$

Hence maximum height will be at x = 10

$$y = 2 \times 10 - 0.1 \times (10)^2 = 10$$

Maximum height = 10 meters.

1.28 (b)

Let total 100 shock absorbes are supplied

so X supplies = 60 (40% of 100)

Y supplies = 40 (60% of 100)

Reliable supply by X = 96% of 60 = 57.6

Reliable supply by Y = 72% of 40 = 28.8

So the required probability that random chosen reliable shock absorber is made by Y is

Required probability

$$= \frac{28.8}{57.6 + 28.8} = 0.333 \approx 0.334$$

1.29 (c)

Adding each and every observation by constant increases the arithmetic mean by same constant. Also multiplying each and every observation by constant mutiplies the arithmetic mean by same constant. While the standard deviation will not show similar effect.

Alternative Solution:

$$\frac{x_1 + x_2 + x_3 \cdots x_N}{N} = \overline{x}$$

$$\frac{(x_1 + 7) + (x_2 + 7) + \cdots + (x_n + 7)}{N} = \overline{x} + 7$$

and $$\frac{2(x_1 + x_2 \cdots x_N)}{N} = 2\overline{x}$$

Hence only P, R are correct.

1.30 (a)

AD, CG, FK, JP are given terms. The first letter in each term A, C, F, J are respectively 1, 3, 6, 10 letter of alphabetical sequence.

A = 1; 1st letter of alphabetical sequence

C = 1 + 2 = 3 ; 3rd letter of alphabetical sequence

F = 1 + 2 + 3 = 6 ; 6th letter of alphabetical sequence

G = 1 + 2 + 3 + 4 = 10; 10th letter of alphabetical sequence

The next term should begin with 1 + 2 + 3 + 4 + 5 = 15th letter of alphabetical sequence. So next term should begin with O.

Also in the series AD, CG, FK, JP, two letters in each term are separated by 2, 3, 4, 5 letters in between them respectively. Hence next term should be OV as there should be a gap of 6 alphabets between O and V.

Hence answer is OV.

1.31 (b)

Lowest

Nadir which is opposite to the word Zenith (the highest point) means the lowest point.

1.32 (a)

Specificity

Universalism and Particularism are opposite words. Similarly, we require a word that indicates an opposite to diffuseness.

Diffuseness refers to spreading or causing to spread in all directions. Specificity is an appropriate opposite.

1.33 (c)

For sum to be maximum, we will consider only positive terms

$$44 + 42 + \ldots\ldots 2$$

or $$2 + 4 + 6 \ldots\ldots 42 + 44$$

$$2[1 + 2 + 3 \ldots\ldots 22]$$

$$= \frac{2[22 \times 23]}{2} = 506$$

Alternate:

Sum of even Natural Numbers

$$2 + 4 + 6 \ldots 44 = n(n + 1)$$

$$22 \times 23 = 506$$

1.34 (a)

Would fly

This is a sentence based on Conditional Sentences. In the first clause imagination (Were) has been used, so the main clause should comprise of 'would + I form of verb' usage.

1.35 (c)

She is an European.

The usage of articles a and an is not decided by the first alphabet of the word but the sound / pronunciation of the word. European though starts with 'E' which is a vowel but it does not yield a vowel sound. When read aloud, it leaves the sound of 'Y' which is a consonant sound.

So the correct sentence should be, "She is a European".

1.36 (b)

The question can be written as..

$$\frac{1}{\sqrt{2}+\sqrt{1}}+\frac{1}{\sqrt{3}+\sqrt{2}}+\frac{1}{\sqrt{4}+\sqrt{3}}\cdots\frac{1}{\sqrt{8}+\sqrt{80}}$$

Rationalize each term by multiplying it with its congugate in both numerator and denominator.

$$\left[\frac{1}{\sqrt{2}+\sqrt{1}}\times\frac{\sqrt{2}-\sqrt{1}}{\sqrt{2}-\sqrt{1}}\right]\times\left[\frac{1}{\sqrt{3}+\sqrt{2}}\times\frac{\sqrt{3}-\sqrt{2}}{\sqrt{3}-\sqrt{2}}\right]\times\ldots\ldots$$

$$\ldots\ldots\left[\frac{1}{\sqrt{8}\times\sqrt{80}}\times\frac{\sqrt{81}-\sqrt{80}}{\sqrt{81}-\sqrt{80}}\right]$$

Which can be resolved as

$$\left[\sqrt{2}-\sqrt{1}\right]+\left[\sqrt{3}-\sqrt{2}\right]\ldots\ldots\ldots\left[\sqrt{81}-\sqrt{80}\right]$$

(as denominator in each term will become 1)

Now cancelling like terms we will be left with

$$\sqrt{81}-\sqrt{1} = (9-1) = 8$$

1.37 (d)

There are Ninety (90) two digited numbers between 1 and 100 (typically 10 to 99). Among these numbers which are divisible by 7 are (14, 21, ...98) only 13 numbers.

$$\left[\because\left(\frac{100}{7}\right)-1=13\right]$$

So, the required probability that the number is not divisible by $1-\frac{13}{90}=\frac{77}{90}$.

1.38 (d)

No adversity justifies giving up hope.

Options (b) and (c) can be clearly discarded since honesty and adventure making have no relationship with the given argument.

Option (a) may appear close. But remember, every failure may not result into success as it happens in the case of spider.

1.39 (c)

$$\text{Average speed} = \frac{\text{Total distance}}{\text{Total time}}$$

$$= \frac{D}{\frac{\frac{D}{2}}{60}+\frac{\frac{D}{4}}{30}+\frac{\frac{D}{4}}{10}} = 24 \text{ kmph}$$

1.40 (b)

$$\text{Now cost} = 13200\left[1+\frac{1}{5}\right]\left[1-\frac{1}{24}\right]$$

$$= 13200\left[\frac{6}{5}\right]\times\left[\frac{23}{24}\right] = 15,180$$

1.44 (96)

$$\left(\frac{z+1}{z}\right)^2 = 98$$

$$\Rightarrow \frac{z^2+1}{z^2+2} = 98 \Rightarrow \frac{z^2+1}{z^2} = 96$$

1.45 (d)

Given $ax^2 + bx + c = 0$

roots are real and positive

For e.g. : $x^2 - 5x + 6 = 0$ roots are 2, 3

Then roots of $ax^2 + b|x| + c = 0$

then roots of $x^2 - 5|x| + 6 = 0$

$\therefore |x|^2 - 5|x| + 6 = 0$

$\left.\begin{array}{l}\text{If } x > 0\\ |x| = x\end{array}\right\}\quad x^2 - 5x + 6 = 0$

$(x-2)(x-3) = 0 \qquad x = 2, 3$ are roots

$\left.\begin{array}{l}\text{If } x < 0\\ |x| = -x\end{array}\right\}\quad x^2 + 5x + 6 = 0$

$(x+2)(x+3) = 0 \qquad x = -2, -3$ are roots

So if $x^2 - 5x + 6 = 0$ has two real roots

then $|x|^2 - 5|x| + 6 = 0$ has four real roots

$ax^2 + bx + c = 0$ has real roots

then $ax^2 + b|x| + c = 0$ has 4 real roots.

1.48 (850)

One way fare of single person = 100

Undiscounted round trip fare of singe person = 200

Undiscounted round trip fare of five persons = 1000

Round trip will result in a discount of 10%. Also additional discount of 5% on total fare for 4 or more person.

So a term of 5 person will be eligible for overall discount of (10% + 5%) = 15% discount

Discounted round trip for 5 persons

$1000 \times 0.85 = 850$ Rs.

1.49 (48)

Following table can be redrawn:

		Men	Women	Total
Own vehicle	Car	40	34	74
	Scooter	30	20	50
	Both	60	46	106
Do not own vehicle		20	50	70

Now following venn-diagram can be drawn

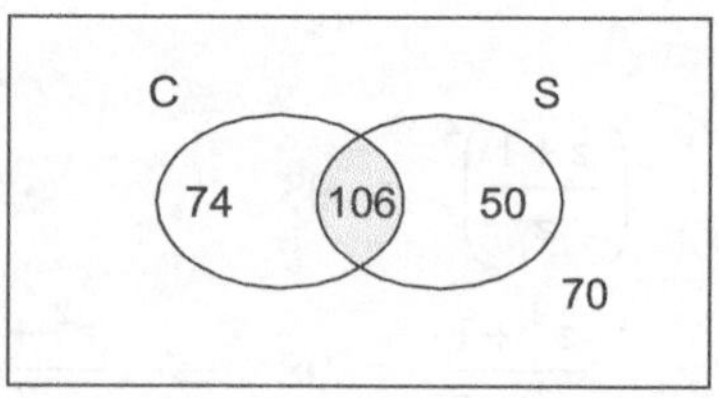

So % of people not owing a scooter

$$\left[\left(\frac{74+70}{300}\right)\times 100\right] = \frac{144}{300}\times 100 = 48\%$$

1.50 (6)

By visualization we can see 6 new internal planes will be created through these lines.

1.54 (b)

There are 19 multiples of 10 from 2 to 198 typically 10, 20, 30 ··· 190.

Average of these number will be

$$= \frac{10+20+30+\cdots+190}{19}$$

$$= 10\frac{[1+2+3+\cdots+19]}{19}$$

$$= 10\frac{[19\times 20]}{19\times 2} = 100$$

1.55 (c)

Let, $y = \sqrt{12+\sqrt{12+\sqrt{12+...}}}$

So, $y = \sqrt{12+y}$

Squaring $y^2 = 12 + y$

$y^2 - y - 12 = 0$

$\Rightarrow (y-4)(y+3) = 0$

$y = 4$

Option (c) $\rightarrow$ 4.000

1.58 (d)

$x^2 - 2x + 3 = 11$

$\Rightarrow x^2 - 2x - 8 = 0$

$\Rightarrow x^2 - 4x + 2x - 8 = 0$

$\Rightarrow x(x-4) + 2(x-4) = 0$

$\Rightarrow x = 4, -2$

(i) put $x = 4$ in $|-x^3 + x^2 - x|$

$\Rightarrow |-(4)^3 + (4)^2 - (4)| = 52$

(ii) put $x = -2$ in $|-x^3 + x^2 - x|$

$\Rightarrow |-(-2)^3 + (-2)^2 - (-2)| = 14$

Possible values = 14, 52

1.59 (140)

Given ratio of male to female is 2.5 in year 2008.

So let us assume male as 500 and female as 200.

So ratio $\frac{M}{F}$ in 2008 = 2.5

Now female doubles up in 2009

Number of female in 2009 = $2 \times 200 = 400$

$\frac{M}{F}$ ratio in 2009 is 3

$$\frac{M}{400} = 3 \Rightarrow M = 1200$$

So number of males in 2009 is 1200.

% increase in number of males from 2008 to 2009

$$\% \text{ increase} = \frac{1200-500}{500}\times 100 = 140\%$$

1.60 (a)

Any angle and corresponding time can be driven by

$$\left[5x - \left(\frac{D^0}{6}\right)\right]\times\frac{12}{11}$$

Between 6 and 7 pm angle have to be 60°

Put x = 6 and D = 60°

$$\left[5\times6-\frac{60°}{6}\right]\times\frac{12}{11} = 20\times\frac{12}{11} = 21\frac{9}{11}$$

So at $4:21\frac{9}{11}$ angle between minutes and hour hand will be 60°.

In given question angle have to close to 60° nearest option is 6 : 22.

1.64 (c)

Given series: 2, 5, 10, 17, 26, 37, 50, 64

$1^2+1, 2^2+1, 3^2+1, 4^2+1, 5^2+1, 6^2+1, 7^2+1, 8^2+1$

$\Rightarrow$ 2, 5, 10, 17, 26, 37, 50, 65

$\therefore$ 64 is not in the series.

1.65 (c)

Q.No.	Marks	Answered Correctly	Answered Wrongly	Not Attemped	Total Students
1	2	21	17	6	44
2	3	15	27	2	44
3	2	23	18	3	44

$$\text{Average} = \frac{\text{Total marks scored}}{\text{Total number of students}}$$

$$= \frac{21\times2+15\times3+23\times2}{44} = 3.022727 \simeq 3.02$$

1.68 (d)

Let GDP before year 2012-2013 be 100 Rs.

During 2012-2013 GDP increased by 7%.

So, GDP in year 2013-2013 = 107 Rs.

Now exchange rate for USD change from Rs. 50/USD to Rs. 60/USD.

i.e. value of Rupee reduced to 5/6 of original value.

So, new GDP (in USD) = $107\times\frac{5}{6}$ = 89.16

Effective change is from 100 to 89.16

= 10.83% decrease $\simeq$ 11% decrease

1.69 (c)

Ratio of male to female 2001 is given.

Let there be 100 males and 100 female in 2011.

Now number of female in 2012 is same as number of females in 2012 as given in question.

So female in 2012 is also 100.

Ratio of male of female in 2012 is 1.5

$$\frac{M}{F} = 1.5$$

$$\Rightarrow \quad \frac{M}{F} = 1.5 \Rightarrow M = 150$$

So there are 150 males n 2012.

Ratio of male in 2012 to male in 2011

$$= \frac{150}{100} = \frac{1.5}{1}$$

1.70 (c)

$(7526) - (Y)_8 = (4364)_8$

So, $Y = (7526)_8 - (4364)_8$

$= (7526 - 4364)_8 = (31420)_8$

1.71 (a)

Didn't you buy <u>any paper</u> when you went shopping?

1.72 (c)

She enjoyed herself immensely at the party. The sentence with closest meaning is " she had a terrific time at the party".

1.73 (a)

Total number of ways of selecting two numbers

$= {}^4C_1 \times {}^5C_1 = 4 \times 5 = 20$

Favourable outcomes are {2, 14}, {3, 13}, {4, 12}, {5, 11}

$$\therefore \text{ Probability} = \frac{4}{20} = \frac{1}{5} = 0.20$$

1.74 (a)

Statement 1:

$$\text{Number of steps} = \frac{9}{\frac{3}{4}} = 12 \text{ steps}$$

$\therefore$ Statement 1 is alone sufficient.

Statement 2:

Only width of step is given which is not sufficient to calculate number of steps.

$\therefore$ Option (a) is correct.

1.75 (b)

The incorrect combination is

Wheedle – Roundabout

1.76 (c)

Total number of students = 44

Marks obtained by answering 1st question correctly $\Rightarrow 2 \times 12 = 42$

From 2nd question answered correctly
$\Rightarrow 15 \times 3 = 45$
From 3rd question answered correctly
$\Rightarrow 11 \times 1 = 11$
From 4th question answered correctly
$\Rightarrow 2 \times 223 = 46$
From 5th question answered correctly
$\Rightarrow 5 \times 31 = 155$
Total = 299

$\therefore$ Average marks $= \dfrac{299}{44} = 6.795$

1.77 (c)

The chain snatchers **took to their heels** when the police party arrived.
The alternative meaning for "took to their heels" is **"took to flight"**.

1.78 (a)

Only statement 1 and 2 follows.

1.79 (32)

Let total number of students in Electrical be T_e.
Male students in Electrical is 40.

So $\dfrac{5}{9} \times T_e = 40 \Rightarrow T_e = 72$

20% are Electrical students.

Therefore $\dfrac{20}{100} \times \text{Total} = 72 \Rightarrow \text{total} = 360$

Number of females in Civil

$= 30\% \text{ of } 360 \times \dfrac{4}{9} = 48$

Number of females in Mechanical

$= 10\% \text{ of } 360 \times \dfrac{4}{9} = 16$

Different between the numbers of female students in the Civil department and the female students in the Mechanical department = 48 – 16 = 32.

1.80 (d)

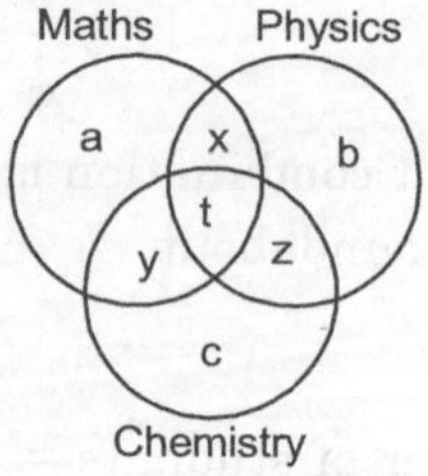

$75\% = a + b + c + (x + y + z) + t$
$(x + y + z) + t = 50\%$
$x + y + z = 40\% \Rightarrow t = 10\%$
$\therefore$ Statement 3 is correct.
$a + b + c = 25$
Also Maths + Physics + Chemistry
$= (a + x + y + t)+(b + x + z + t)+(c + y + z + t)$
$= (a + b + c) + 2(x + y + z) + 3t$
$= 25 + 2 \times 40 + 3 \times 10 = 135\% = \dfrac{27}{20}$
$\therefore$ Statement 1 is correct.

1.81 (c)

We **completely forgot** our friend's birthday and we **just don't know** how to make it up to him.

1.82 (c)

All **personnel** are being given the day off.

1.83 (c)

$f(x) = 1 - |x|$ on $-1 \le x \le 1$
$|x| = -x$ on $x < 0 = x$ on $x \ge 0$
Substituting in the given function we get
$f(x) = 1 - (-x)$ on $-1 \le x \le 0$
$= 1 - x$ on $0 \le x \le 1$
Which is same as
$f(x) = 1 + x$ on $-1 \le x \le 0$
$= 1 - x$ on $0 \le x \le 1$
The graph of this function is shown below

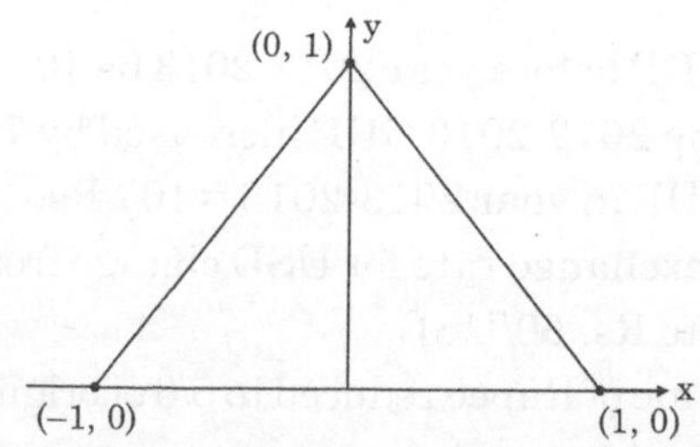

From the graph we can easily see that the function attains the maximum at x = 0 and the maximum value of the function is 1.

1.84 (d)

Apparel include cloth items such as skirt, trousers, shirts, etc.

1.85 (a)

Statements 1: $\dfrac{1}{4}^{th}$ of weight of 1 pole

= 5 kg $\Rightarrow$ Weight of 1 pole = 20 kg
$\therefore$ Statement 1 is alone sufficient but statement 2 alone is not sufficient.

1.86 (b)

Adding by a constant, subtracting by a constant, and multiplying by a constant to the Arithmatic sequence do not destroy the property of Arithmatic progression.

1.87 (a)

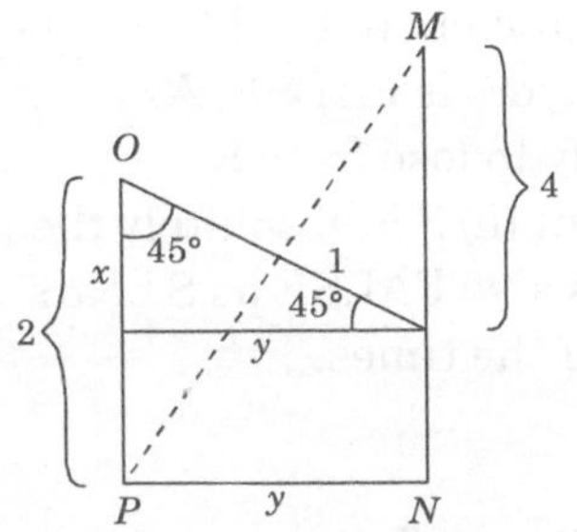

$$\frac{1}{x} = \sin 45°, \frac{1}{y} = \sin 45°$$

$$x = \frac{1}{\sqrt{2}} \quad y = \frac{1}{\sqrt{2}}$$

$$PM^2 = y^2 + \left(4+2-\frac{1}{\sqrt{2}}\right)^2 = \left(\frac{1}{\sqrt{2}}\right)^2 + \left(6-\frac{1}{\sqrt{2}}\right)^2$$

$$PM = \sqrt{\frac{1}{2}+28.01} = 5.34$$

1.88 (b)

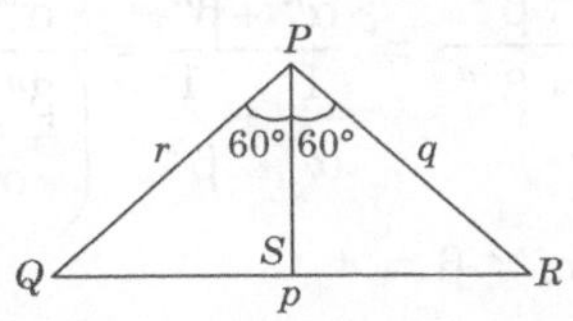

$$\text{Area of } \Delta PQS = \frac{1}{2} \times PS \times r \times \sin 60°$$

$$\text{Area of } \Delta PRS = \frac{1}{2} \times PS \times q \times \sin 60°$$

$$\text{Area of } \Delta PQR = \frac{1}{2} PS \frac{\sqrt{3}}{2}(q+r) \quad ...(1)$$

$$\text{Also } \Delta PQR = \frac{1}{2} \times r \times q \times \sin 120° \quad ...(2)$$

$$\text{Equating (1) and (2)} PS = \frac{qr}{r+q}$$

1.89 (8)

h(2, 5, 7, 8) = h(p, q, r, s)

(r × s) > (p × q)

Therefore $h(2, 5, 7, 3)$ returns remainder of

$$\frac{7 \times 3}{2 \times 5} = 1$$

Now $f_g(1, 4, 6, 8) = f(1, 4, 6, 8) \times g(1, 4, 6, 8) = 8 \times 1 = 8$

1.90 (a)

Since the report lacked needed information, it was of no use to them.

1.91 (b)

R O A D

↓ ↓ ↓ ↓

U R D G

Replace with 3rd letter in alphabetical order.

Therefore SWAN → VZDQ

1.92 (c)

The Tamil version of **the** John Abraham-starrer Madras Cafe **was** cleared by the Censor Board with no cuts last week, but the film's distributors **found** no takers among the exhibitors for a release in Tamil Nadu **on** this Friday.

1.93 (b)

Extreme focus on syllabus and studying for tests has become such a dominant concern of Indian students that they close their minds to anything **extraneous** to the requirements of the exam.

1.94 (b)

Females and Gynaecologist relationship is similar.

1.95 (c)

$f(x) = ax + b$ is the linear form

$f(-2) = -2a + b = 29$

$f(3) = 3a + b = 39$

$\Rightarrow \quad 5a = 10$

$\Rightarrow \quad a = 2, \quad b = 33$

$f(x) = ax + b \Rightarrow f(5) = 2 \times 5 + 33 = 43$

1.96 (d)

Alexander turned his attention towards India, since he had conquered Persia.

The following statement is logically valid and can be inferred from the above sentence is: "Since Alexander's kingdom extended to Indian borders after the conquest of Persia, he was keen to move further".

1.97 (b)

R-Home, S-Power, P-Defense, Q-Telecom, T-Finance.

1.98 (b)

From the given graph at x = 2, y = –1

Put this in each of the four options and test

Option (a) $2 = -1 - |-1| = -2$ (False)

Option (b) $2 = -(-1 - |-1|) = 2$ (True)

Option (c) $2 = -1 + |-1| = 0$ (False)

Option (d) $2 = -(-1 + |-1|) = 0$ (False)

So option (b) is the right answer.

1.99 (b)

The following statements are logically valid and can be inferred from the given passage.

(ii) He has to improve his temperament in order to become a great batsman.

(iii) He failed to make many of his good starts count.

1.100 (2006)

Year	2000	2001	2002	2003	2004	2005	2006	2007
Sum of Imports and Exports	40 + 50 = 90	60 + 50 = 110	70 + 60 = 130	70 + 60 = 130	70 + 80 = 150	70 + 90 = 160	100 + 120 = 220	120 + 110 = 230
Change		20	20	0	20	30	60	10

Therefore combined percentage increases is highest in 2006.

1.105 (d)

Original surface area = $6(4)^2 = 96$

If corner cubes are removed, three exposed surfaces are removed which will create 3 new surfaces in original large cube. So surface area will remain unchanged, i.e. 96.

1.106 (b)

Elegance:

(27300 + 25222 + 28976 + 21012) × 48 = A

Executive:

(999 + 8942 + 10234 + 10234 + 10109) ×173 = B

Smooth:

(20009 + 9392 + 22429 + 18229) × 63 = C

Soft

(17602 +18445 + 19544 + 16595) × 78 = D

Which is highest for B (executive).

1.108 (d)

From conditions I, II and IV, following inference can be drawn

P always beats Q

Q always beats R

and R always beats S

Thus, arranging players from best performing to worst performing is

P – Q – R – S

Now, condition III states that S loses to P only sometimes. That means S most of the times beats the otherwise best performing player P.

Thus, statement (i) P is likely to beat all the other three players is FALSE, As P is likely to lose from S.

Statement (ii) S is absolutely the worst player in the set is also FALSE as S beats P most of the times.

1.109 (b)

$$f(x) = 2x^7 + 3x - 5$$

Use option (b), if $(x - 1)$ will be a factor then on putting $x - 1 = 0$ i.e. $(x = 1)$ in $f(x)$.

$$f(1) = 2(1)^7 + 3 - 5 = 5 - 5 = 0$$

1.114 (d)

Same pattern follows in all options A, B, C only (d) doesn't follow this pattern.

Hence odd number out (d).

1.115 (a)

$$\frac{\alpha^n + \beta^n}{\alpha^{-n} + \beta^{-n}} = \frac{\alpha^n + \beta^n}{\frac{1}{\alpha^n} + \frac{1}{\beta^n}} = \frac{\alpha^n + \beta^n}{\left(\frac{\beta^n + \alpha^n}{\alpha^n \beta^n}\right)} = (\alpha\beta)^n$$

Given , $\alpha \times \beta = 4$

Hence, $(\alpha\beta)^n = 4^n$

1.116 (a)

Following venn diagram can be drawn as

$$n(A \cup B) = n(A) + n(B) - n(A \cap B)$$

$$120 = 55 + 85 - n(A \cap B)$$

$$n(A \cap B) = 20$$

So only facebook 35.

1.119 (c)

Number of rectangles will be

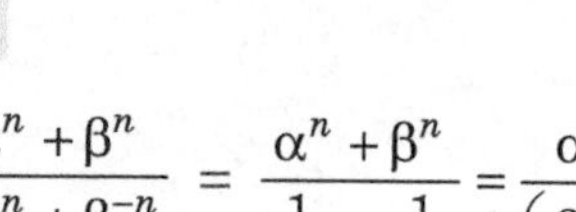

$${}^5C_2 \times {}^3C_2$$

$$10 \times 3 = 30$$

As whenever any two horizontal and any two vertical lines are choosen their intersection will produce a rectangle.

1.121 (c)

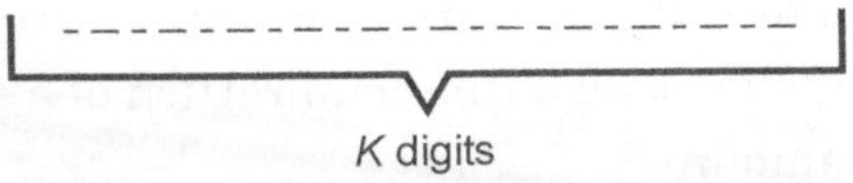

Each digit can be filled in 7 ways as 0, 5 and 9 is not allowed, so each of these places can be filled by 1, 2, 3, 4, 6, 7, 8 so required probability is $\left(\frac{7}{10}\right)^k$ or 0.7^k

$$\frac{fC}{TC} = \frac{7^k}{10^k} = 0.7^k$$

1.122 (d)

$$y \times 162 = \text{Perfect cube}$$

Option, $y = 24$

$\Rightarrow 2^3 \times 3\ (2 \times 81) \neq$ Not perfect cube

$y = 27$

$\Rightarrow 3^3 \times (2 \times 3^4) \neq$ Not perfect cube

$y = 32$

$\Rightarrow 2^5 \times 2 \times 3^4 \neq$ Not perfect cube

$y = 36$

$\Rightarrow 2^2 \times 3^2 \times 2 \times 3^4 = 2^3 \times 3^6 = (2 \times 3^2)^3$ is a perfect cube.

Hence the answer is, $y = 36$

1.123 (c)

Correct option is; Wished

After Rajender Chola **retured** from his voyage to Indonesia, he **wished** to visit the temple in Thanjavur.

Both are events of past. Use of past perfect form is unwarranted as it reflects part of past.

1.124 (d)

Besides money:

Research in the work place reveals the people works for many reasons **besides money**.

Besides conveys the meaning of 'in addition' Beside means 'next to'.

1.125 (c)

Following seating arrangement can be drawn

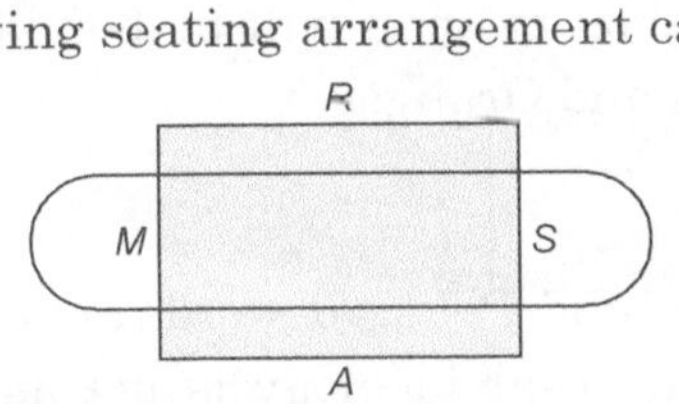

Therefore correct option is (c).

1.126 (b)

History is viewed through the filter of rationalism.

1.127 (c)

P will not get submerged as it is @ 550. So P is not present in correct options. Hence (i) and (ii) options are incorrect. Now compare Q and T. As T is between 500 and 525. T will get submerged. Hence ans (c) as among option (c) and (d). T is present in only option (c).

1.128 (b)

$$\frac{(x+y)-|x-y|}{2} \quad \text{...(i)}$$

If $x > y$ $\quad |x-y| = x - y$;

If $x < y$ $\quad |x-y| = y - x$

Now if $x > y$, above expression (i) becomes

$$\frac{(x+y)-(x-y)}{2} = \frac{2y}{2}$$

$= y =$ minimum of (x, y) as $x > y$

Now if $x < y$;

$$\frac{x+y-(y-x)}{2} = \frac{2x}{2} = x = \text{minimum of } (x, y)$$

as $x < y$

Therefore correct answer is option (b).

Alternate solution:

Use easy values,

$x = 1$ and $y = -2$

Now, $\dfrac{(1-2)-|1-(-2)|}{2} = -2$

or $x = 2$ and $y = -1$

$$\frac{(2-1)-|2-(-1)|}{2} = \frac{1-3}{2} = -1$$

Which is minimum of (x, y).

1.129 (d)

As there are 4 people A, G, N, S and four colours so without any restriction total ways have to be $4 \times 4 = 16$.

Now, Arun $\rightarrow$ dislikes Red and Shweta dislikes white so $16 - 2 = 14$ ways.

Therefore correct answer should be option (d).

Alternate solution:

Only one option is less than 16.

Therefore correct answer should be option (d).

1.130 (a)

Out of six people. 3 place definitely occupied by right handed people as at least 2 women are there so these two will sit adjacently. Now as only on seat is left it will be occupied by a left handed man because on right side of this seat is sitting an right handed man.

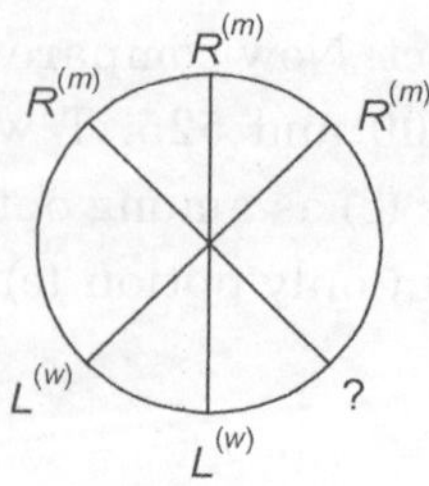

? have to be a left handed man.

Here, $R^{(m)}$ indicates right handed man and $L^{(w)}$ indicates left handed women.

Therefore answer should be only 2 women.

1.131 (b)

Total marks of questions = 100
Let x be the number of Essay questions = $11x$
Let y be the number of MCQ questions = $3y$
Total number of question given

$$x + y = 20 \quad ...(1)$$

and

$$11x + 3y = 100 \quad ...(2)$$

using equation (1) and (2), we get

$$11(20 - y) + 3y = 100$$
$$220 - 11y + 3y = 100$$
$$220 - 8y = 100$$
$$120 = 8y$$
$$y = 15$$

So, number of questions of MCQ types are 15.

1.132 (a)

According to the condition given:

1. V is to the West of W, so order will be VW.
2. Z is to the East of X and West of V, so order will be XZ and ZV.
3. W is the West of Y, so order will be WY.

Combining all conditions will gives:
(i) XZ and ZV will gives XZV ...(1)
(ii) VW and WY will gives VWY ...(2)
(iii) XZV and VWY will gives XZVWY from (1) and (2).

So, V is the middle building.

1.133 (b)

Bright is an adjective that is to express a bold object.
For *e.g.*: She is bright enough to know that this is a very good offer.
Similarly, the statement will be saturn is bright enough to be seen on a clear night with the naked eye.

1.134 (d)

- Yielding means "giving a product or generating a financial return of a specified amount."
- Resistant means "to be immune or imprevious to something."

By the meaning, it can be observed that both are not synonyms to each other.

1.135 (d)

Number of socks of red colour = 3
Number of socks of green colour = 4
Number of socks of blue colour = 3
Sample space for choosing 2 socks = $^{10}C_2$
2 socks of sames colours equals to $^3C_2 + {}^4C_2 + {}^3C_2$

So, required probability $= \dfrac{^3C_2 + {}^4C_2 + {}^3C_2}{^{10}C_2}$

$$= \frac{3+6+2}{45} = \frac{12}{45} = \frac{4}{15}$$

1.136 (a)

According to the question 30 digit number will be X = 47777 ... (29 times) in worst case.
It can be written as in power of 10 i.e., $X = 4.7777... \times 10^{29}$
Now, we have to find X^3.
So, $(4.777... \times 10^{29})^3 = (4.777...)^3 \times 10^{87}$
if we round up 4.777... will give 5, and $(5)^3$ is maximum 125, so 3 more digits.
So, X^3 requires 3 + 87 = 90 digits.

1.137 (b)

One box contain only Apples.
One box contain only Oranges.
One box contain both Apples and Oranges.
Given that all labelling on boxes are incorrect.
So, opening box with labelling apples and oranges will gives box with only Oranges or only Apples and we can guess rest two boxes content. So, correct box to open will be box labelled with "Apples and Oranges."

1.138 (c)

'Culture' and 'Political' words are not related to the belief that ideology is not as important.
Similarly generally means 'as far as one known' and is the correct phrase that has been used by the author.

1.139 (c)

Given equation: $e^x + 0.5x^2 - 2 = 0$

To find roots of equation, we can do

$$e^x = -0.5x^2 + 2$$

Graph will look likes:

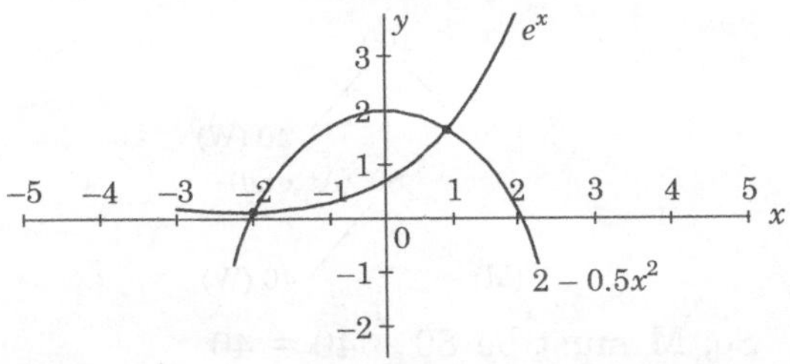

So from graph, two graphs meating at only two points which are roots of the equation. So 2 roots in range [–5, 5].

1.140 (c)

It is given that, possiblity of thunderstorm is given by how fast air pressure rises or drops over a region. Hence the probabilities will be higher where the variance of air pressure is higher which is there is region 'R'.

Hence 'R' is the answer.

1.141 (b)

2, 12, 60, 240, 720, 1440, 1440, 0

(2 × 6) (12 × 5) (60 × 4) (240 × 3) (720 × 2) (1440 × 1) (1440 × 0)

So, 1440 must be answer.

1.142 (b)

They're used for pointing group.

Their is pointing people.

There is used for place.

1.143 (c)

Number is divided by either by 20 or 42 or by 76

K × LCM(20, 42, 76) + constant difference

= 7980 K + 7 (K is natural number)

Least number will be 7980 + 7 = 7987.

1.144 (a)

Meandering : wandering aimlessly/indirect.

1.145 (d)

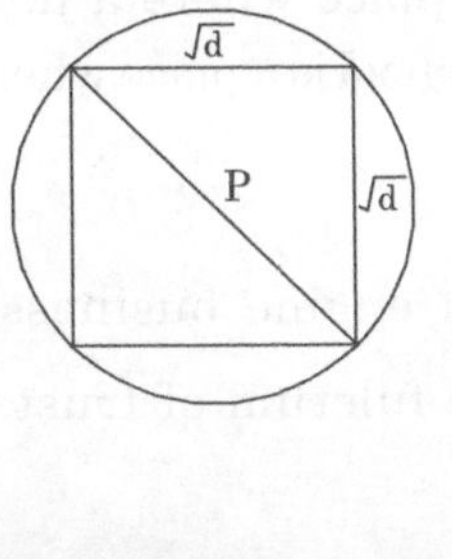

$$\text{Area of square} = d$$

$$\text{Side one side of square} = \sqrt{d}$$

$$\text{Diagonal of square} = \sqrt{d+d} = \sqrt{2d}$$

$$\text{Area of circle} = \pi r^2$$

$$= \pi\left(\frac{\sqrt{2d}}{2}\right)^2 = \pi\frac{d}{2} = \frac{1}{2}\pi d$$

1.146 (c)

$$p^{-x} = \frac{1}{q},\ q^{-y} = \frac{1}{r},\ r^{-z} = \frac{1}{p}$$

Put $x = 1,\ y = 1,\ z = 1$

$$p^{-1} = \frac{1}{q},\ q^{-1} = \frac{1}{r},\ r^{-1} = \frac{1}{p}$$

i.e. $\frac{1}{p} = \frac{1}{q},\ \frac{1}{q} = \frac{1}{r},\ \frac{1}{r} = \frac{1}{p}$

Which is true i.e. $\frac{1}{p} = \frac{1}{q} = \frac{1}{r}$

So, $xyz = 1$

Alternate Solution:

$$p^x = q,\ q^y = r,\ r^z = p$$

$$x \log p = \log q$$

$$y \log q = \log r$$

$$z \log r = \log p$$

$$x = \frac{\log q}{yp}$$

$$y = \frac{yr}{\log q}$$

$$z = \frac{\log p}{yr}$$

$$x \times y \times z = 1$$

1.147 (b)

Male + Female = 300 ...(i)

$$\text{Total money} = \frac{8}{9}\text{M} \times 750 + \frac{2}{3}\text{F} \times 1000$$

$$= \frac{6000}{9}\times\text{M} + \frac{6000}{9}\times\text{F}$$

$$= \frac{6000}{9}\times(\text{M}+\text{F})$$

From equation (i)

$$\text{Total money} = \frac{6000}{9}\times(300) = 2{,}00{,}000$$

1.148 (a)

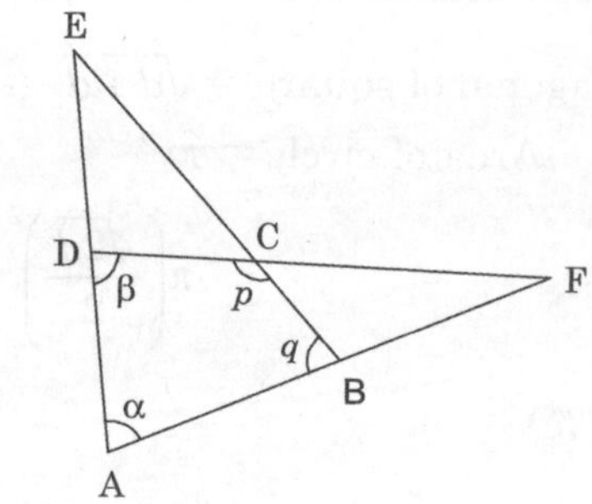

$$\angle E + \angle F = ?$$

$$\alpha + q + E = 180 \quad ...(i)$$

$$\alpha + \beta + F = 180 \quad ...(ii)$$

$$\alpha + \beta + p + q = 360 \quad ...(iii)$$

Equation (i) + (ii) = (iii)

$$\alpha + q + E + \alpha + \beta + F = \alpha + \beta + p + q$$

$$E + F = p - \alpha$$

1.149 (c)

Four green, two red face

$$P(G) = \frac{4}{6} = \frac{2}{3}$$

$$P(R) = \frac{1}{3}$$

$$n = 7$$

Option (1),

$$P(G = 3) = {}^7C_3 \times \left(\frac{2}{3}\right)^3 \times \left(\frac{1}{3}\right)^4$$

$$= \frac{35 \times 2^3}{(3)^7} = \frac{35 \times 2^3}{(3)^7}$$

(2),

$$P(G = 4) = {}^7C_4 \times \left(\frac{2}{3}\right)^4 \times \left(\frac{1}{3}\right)^3$$

$$= \frac{35 \times 2^4}{(3)^7} = \frac{35 \times 2^4}{(3)^7}$$

(3),

$$P(G = 5) = {}^7C_5 \times \left(\frac{2}{3}\right)^5 \times \left(\frac{1}{3}\right)^2$$

$$= \frac{21 \times 2^5}{(3)^7} = \frac{42 \times 2^4}{(3)^7}$$

(4),

$$P(G = 6) = {}^7C_6 \times \left(\frac{2}{3}\right)^6 \times \left(\frac{1}{3}\right)$$

$$= \frac{7 \times 2^6}{(3)^7} = \frac{28 \times 2^4}{(3)^7}$$

Option 3 is maximum value.

So, five green faces and two red faces.

1.150 (b)

Let total number of people are 100.

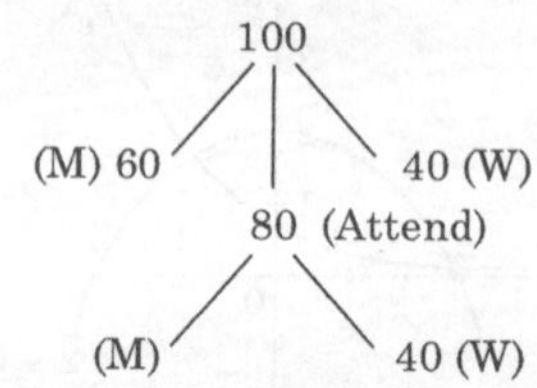

So, M must be 80 – 40 = 40

Ratio of male to female

40 : 40

1 : 1

1.151 (b)

Speed of car A = 50 km/hr

Speed of car B = 60 km/hr

Since, both cars A and B are moving in same direction, the relative speed

= 60 – 50 = 10 km/hr

Distance required between them = 20 km

$$\therefore \quad \text{Time} = \frac{\text{Distance}}{\text{Speed}} = \frac{20}{10} = 2 \text{ hrs}$$

1.152 (c)

'Breaks down' is a transitive phrasal verb which means to divide something such as a total amount into separate parts.

Option (c) is most appropriate.

1.153 (b)

Let share of each student = x

Total cost of gift = $10 \times x$

$$x = 8(x + 150)$$

$$x = 600$$

Total cost = 10 × 600 = 6000

1.154 (b)

A 'court' is for a 'judge' as a 'school' is for a 'teacher'.

Court is a place where a judge works.

Similarly, school is a place where a teacher works.

1.155 (c)

The search engine business model <u>resolves</u> around the fulcrum of trust.

Fulcrum is any thing that plays a central or essential role in an activity, event, or situation. Thus, out of the given options, 'revolves' is the most appropriate.

1.156 (d)

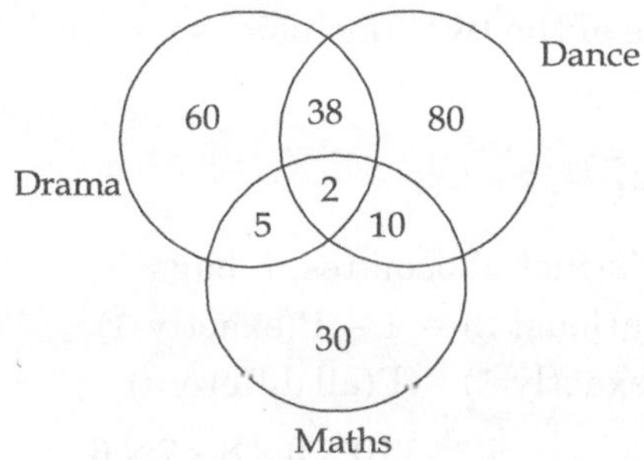

Total number of students = 60 + 80 + 30 + 38 + 5 + 10 + 2 = 225

$$25\% = 225$$

$$\Rightarrow \quad 100\% = \frac{225}{25}\times 100 = 900$$

1.157 (a)

Suffix C as criminal, NC as not criminal.

Case I:

Criminals	P	Q	R	S
Assumption	T	F	F	F
Result	Q_C	S_{NC}	R_C	S_C

Q and R are criminal in the result is impossible because only one person committed the crime.

Case II:

Criminals	P	Q	R	S
Assumption	F	T	F	F
Result	Q_{NC}	S_C	R_C	S_C

S and R are criminal in the result is impossible because only one person committed the crime.

Case III:

Criminals	P	Q	R	S
Assumption	F	F	T	F
Result	Q_{NC}	S_{NC}	R_{NC}	S_C

S_{NC} and S_C in the result which is contradiction. [S committed crime and same time not committed crime which is contradiction]

Case IV:

Criminals	P	Q	R	S
Assumption	F	F	F	T
Result	Q_{NC}	S_{NC}	R_C	S_{NC}

R is criminal in the result.

Hence this case satisfies only one person committed the crime.

So option (a) is correct.

1.158 (c)

Percentage of Administrators

$$= \frac{\text{Administrators}}{\text{Total}}\times 100 = \frac{50}{160}\times 100 = 31.25$$

1.160 (b)

The passage states that the underlying disease behind begging is the failure of the state to protect citizens who fall through the social security net.

Only option (b) can be inferred from this.

1.161 (b)

- 'within' is a preposition that is used to express something that occurs inside a particular period of time.
- 'for' is used here because
  - (i) Sentence is in 'present perfect continuous tense'.
  - (ii) For is used when we talk about a period of time.

1.162 (c)

- Cost of 1-a-C-2 : 200 + 200 + 100 + 100 + 100 = 700
- Cost of 1-b-2 : 300 + 100 + 200 = 600
- Cost of 1-f-b-2 : 100 + 100 + 100 + 200 = 500
- Cost of 1-f-e-2 :100 + 100 + 100 + 200 + 200 = 700

Hence, 1-f-b-2 is having minimum cost.

1.163 (a)

'But' is used for introducing an idea which contrasts with the statement that has been already said.

1.164 (b)

(i) Relation is verb : noun

(ii) One who cooks is a cook, similarly one who flies any aircraft is a flyer.

1.166 (c)

According to passage GST is imposed on the supply of goods and services hence option (a) is incorrect which states that GST is imposed on production of goods and services.
In the passage it is mentioned that GST has a two components specific to UTs. So option (b) is incorrect.
Passage states that GST is a destination based tax which means it is imposed at the point of usage of goods and services. Hence (c) is correct.
Option (d) says GST includes all indirect taxes but as per passage it subsumes all indirect taxes **except few**.

1.167 (c)

P = 3 R = 27 T = 243

P	Q	R	S	T
3^1	3^2	3^3	3^4	3^5

Hence, Q + S = 90

1.168 (d)

Unpainted area = Area of annular ring – Area of n blue color circles
Area of annular ring = $\pi b^2 - \pi a^2$
i.e. $\pi(b^2 - a^2)$

$$\text{Area of 1 blue circle} = \pi\left(\frac{b-a}{2}\right)^2$$

$$\text{Hence, area of } n \text{ blue circles} = n\pi\left(\frac{b-a}{2}\right)^2$$

$$\Rightarrow \text{Unpainted area} = \pi\left[(b^2-a^2) - \frac{\pi}{4}(b-a)^2\right]$$

1.170 (d)

Total expenditure = 2500 million
Total revenue = 3000 million

$$\text{So, profit \%} = \frac{500}{2500} \times 100 = 20\%$$

1.171 (b, d)

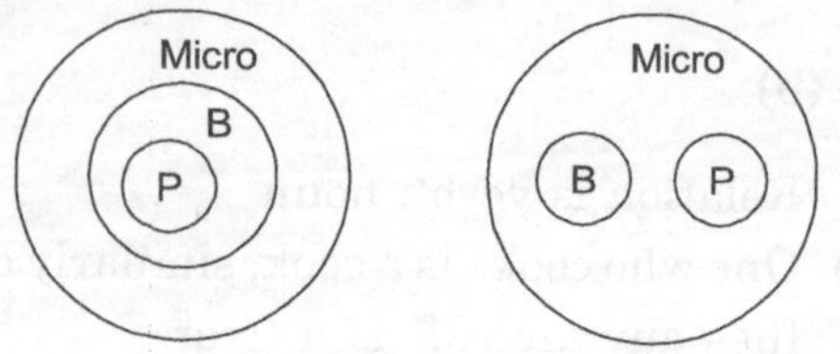

Case 1 Case 2

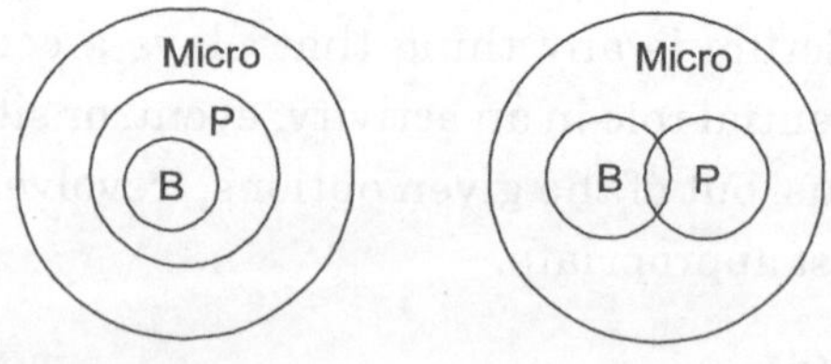

Case 3 Case 4

None of the two conclusions will satisfy all the 4 cases.

1.172 (a)

10 distinct chocolates, 5 bags
P(atleast 2) = 1 – P(exactly 1)
P(exactly 1) = P(all different)

$$= \frac{10 \times 9 \times 8 \times 7 \times 6}{10^5}$$

$$\text{P(atleast 2)} = 1 - \frac{10 \times 9 \times 8 \times 7 \times 6}{10^5}$$

$$= 1 - 0.3024 = 0.6976$$

1.174 (a)

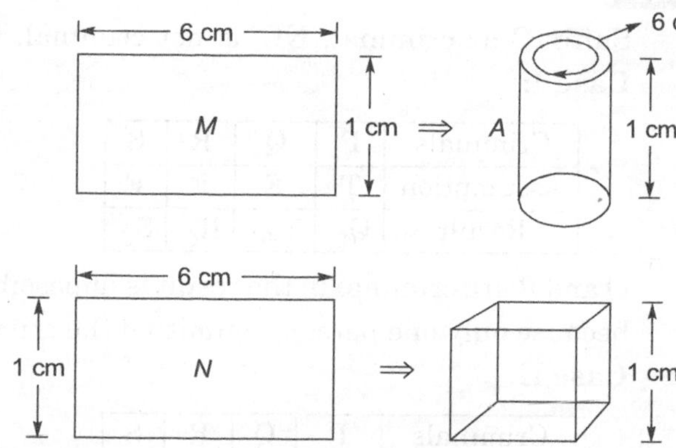

Volume of cylinder = $\pi r^2 h$
Now, $2\pi r = 6$ (figure A)

$$r = \frac{3}{\pi}$$

$$\text{Volume of cylinder} = \pi \times \frac{3}{\pi} \times \frac{3}{\pi} \times 1 = \frac{9}{\pi}$$

$$\text{Volume of cube} = (1)^3$$

$$\text{Ratio} = \frac{\frac{9}{\pi}}{1} = \frac{9}{\pi}$$

1.175 (b)

Everybody is singular and takes singular verb is, first statement is correct and the purpose of invitation is unclear so second statement is vague and that sounds ambiguous.

1.176 (c)

Superficially is the deciding key word which means apparently/seemingly.

1.177 (c)

Doctor performs surgery as writer writes book.

1.179 (b)

$$\Rightarrow \quad CP_P = 5400, \quad MP_P = 5860$$

$$\frac{5400 \times 4}{5} = CP_Q$$

$$CP_Q = 7200, \; SP_Q = ?$$

$$\text{Profit \%} = \frac{\text{Profit}}{\text{CP}} \times 100 \text{ [for Q]}$$

$$25 = \left(\frac{SP - 7200}{7200}\right)100$$

$$SP_Q = 9000$$

$$\text{Discount Q} = MP_Q - SP_Q = 1000$$

$$\text{Discount \%} = \frac{1000}{10000} \times 100 = 10\%$$

1.180 (b)

Boys = $7x$, Girls = $3x$

Total number of students = $10x$

Now, $7x$ and $3x$ should be integers.

Hence, x should be integer.

So $10x$ should be an integer.

Only 50 satisfies the above constraints.

1.181 (c)

$$\left(x-\frac{1}{2}\right)^2 - \left(x-\frac{3}{2}\right)^2 = x + 2$$

Using $a^2 - b^2 = (a + b)(a - b)$

$$\left(x-\frac{1}{2}+x-\frac{3}{2}\right)\left(x-\frac{1}{2}-x+\frac{3}{2}\right) = x + 2$$

$$(2x - 2) = x + 2$$

$$2x - x = 2 + 2$$

$$x = 4$$

1.182 (b)

"Only in some students" is the key in option (b) and that matches well with the given informations in the passage.

1.183 (b)

Pen is used to write and knife is used to cut.

1.184 (c)

(as – as) and (so – as) are used in positive degree of comparison only. Worse, better are comparative degree and nicest is superlative degree, which are not fit between as – as and so – as.

1.185 (c)

Option (c) is flipped.

1.186 (a)

Angle of longest diagonal of cube with an edge of cube.

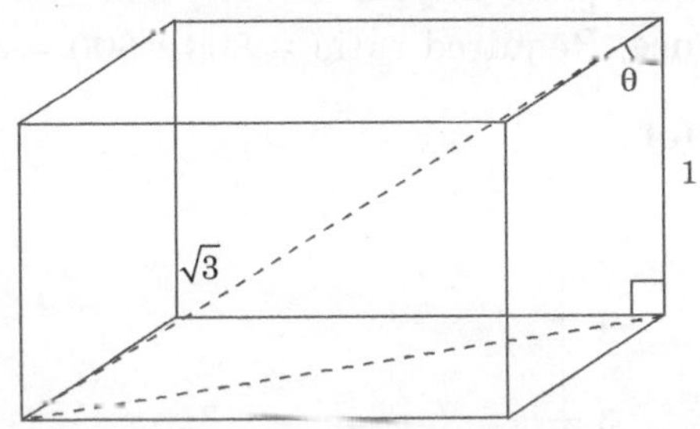

$$\cos\theta = \frac{\text{Base}}{\text{Hypotenuse}} = \frac{1}{\sqrt{3}}$$

1.187 (c)

S > R ...(i)

Q is shortest and U is taller than only one. ...(ii)

T > S ...(iii)

Hence, possible order is:

P > T > S > R > U > Q.

$\because$ Number of students taller than R = 3.

$\therefore$ Number of students shorter than S = 3.

Hence, option (c) is the correct answer.

1.188 (a)

3 : 13 : 6

Let $3k + 13k + 6k = n$

Now
$$\frac{+18 \;+\; 18 \;+\; 18}{15 : 35 : 21} \quad ...(1)$$

$$15y + 35y + 21y = 22k + 54$$

$$71y = 22k + 54$$

Put value of k and satisfy

Here for $k = 4$

On putting $k = 4$ in equation (1)

$$n = 88$$

1.189 (d)

Cost/unit in year 3 = Rs. 1
Cost/unit in year 2 = Rs. 2
Cost/unit in year 1 = Rs. 3

Net Profit = S.P. – (Cost + Taxes)

In year 2, 296 = S.P. – (2 × 200 + 0.13 S.P.)
S.P. = 800

Selling price in year 2 = Rs. 800

In year 3, 210 = S.P. – (300 × 1 + 0.15 S.P.)
210 = S.P. – 300 – 0.15 S.P.

$$\text{S.P.} = \frac{210+300}{1-0.15}$$

Selling price in year 3 = Rs. 600

Hence, Required ratio = 800 : 600 = 4 : 3

1.192 (c)

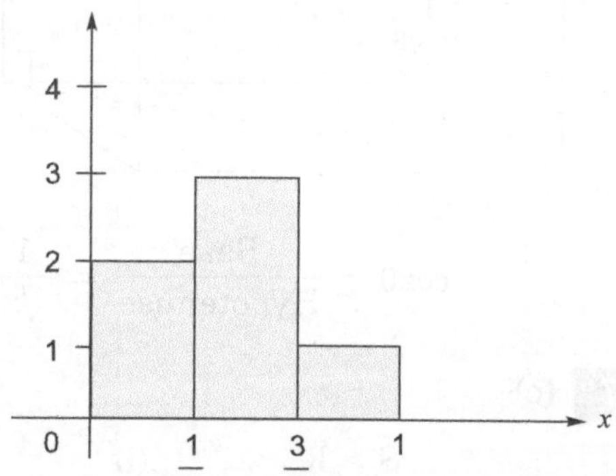

$$A = \frac{1}{3}\times 2+\left(\frac{3}{4}-\frac{1}{3}\right)\times 3+\left(1-\frac{3}{4}\right)\times 1=\frac{13}{6}$$

1.193 (d)

$= (r + 2)(r + 3)(r + 4)(r + 5)$
$= (r^2 + 5r + 6)(r^2 + 9r + 20)$

Now, $r^2 + 2r = -6$
$= (3r + 6 - 6)(7r + 20 - 6)$
$= 3r(7r + 14)$
$= 21r(r + 2)$
$= 21(r^2 + 2r) = 21 \times (-6) = -126$

1.194 (a)

- If all children are inquisitive is true. Then no children are inquisitive is false.
- There is a possibility both (i) and (iii) can be false.
  Assume that some children are inquisitive in that case both (i) and (iii) are false.

1.197 (d)

Appeared = x

$$\text{Boys} = \frac{65x}{100}$$

$$\text{Girls} = \frac{35x}{100}$$

Qualified = y

$$\text{Boys} = \frac{40y}{100}$$

$$\text{Girls} = \frac{60y}{100}$$

1.198 (d)

3 green balls and 2 orange balls are there.
Second ball is orange,

$$P = P(G\cap O)+P(O\cap O)$$

$$= \frac{3}{5}\times\frac{2}{4}+\frac{2}{5}\times\frac{2}{5}=\frac{46}{100}=0.46$$

1.199 (b)

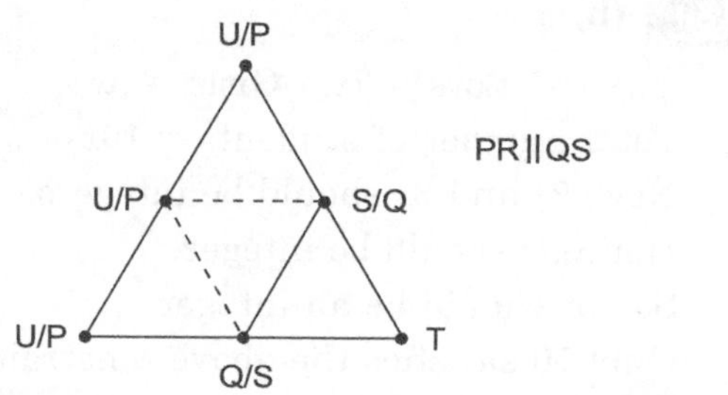

1.200 (d)

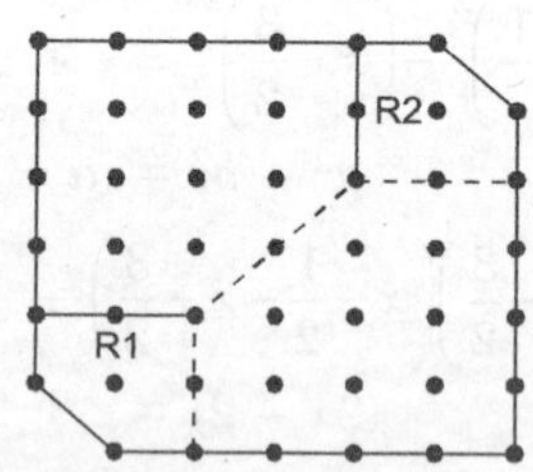

1.203 (a)

$F_{n+1} = F_n + F_{n+1} \quad \forall n \geq 2$
$F_{n-1} = F_{n+1} - F_n$

For $n = 6$, $F_5 = F_7 - F_6$
$= 60 - 37$
($\because$ Given that $F_7 = 60$, $F_6 = 37$)
$= 23$

For $n = 5$, $F_4 = F_6 - F_5$
$= 37 - 23 = 14$

For $n = 4$, $F_3 = F_5 - F_4$
$= 23 - 14 = 9$

For $n = 3$, $F_2 = F_4 - F_3$
$= 14 - 9 = 5$

For $n = 2$, $F_1 = F_3 - F_2 = 9 - 5 = 4$

$\therefore$ $F_1 = 4$

1.207 (c)

$f(t) = 0.01t^2$
$g(t) = 4t$

(i) For some $t > 0$, $g(t) > f(t)$ is true
For example if $t = 1$, $g(t) = 4$, $f(t) = 0.01$
$\therefore$ $g(t) > f(t)$ for some $t = 1$

(ii) There exist $T = 400$ such that
$f(t) > g(t)\ \forall t > 400$ so it is true

1.209 (b)

Given that $f(x) = g(y)$ for all real value of x and y it is necessary that image of 'x' under f is same as image of 'y' is same as image of 'y' using 'g' for all real values of x and y

i.e. $f(x) = g(y) = \text{Constant}$

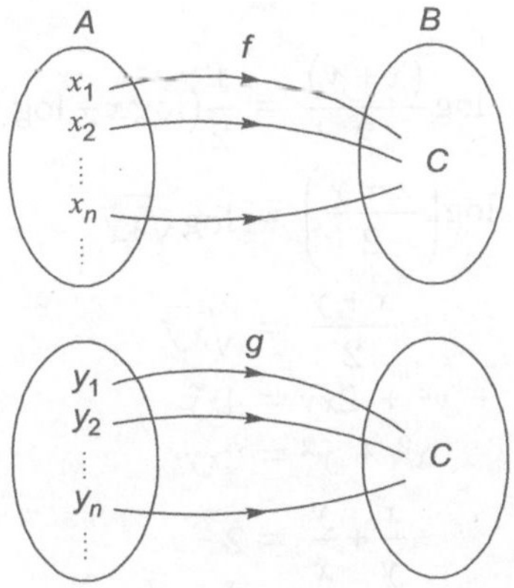

$\therefore$ $f(x) = g(y)\ \forall x, y \in R$

1.211 (b)

$$\log(p^2 + q^2) = \log p + \log q + 2 \log 3$$
$$= \log p + \log q + \log 9$$
$$\log (p^2 + q^2) = \log(9pq)$$
$$p^2 + q^2 = 9pq$$
$$p^4 + q^4 + 2p^2q^2 = 81p^2q^2$$
$$p^4 + q^4 = 79p^2q^2$$

The value of $\dfrac{p^4 + q^4}{p^2q^2} = \dfrac{79p^2q^2}{p^2q^2} = 79$

1.212 (b)

Area of rectangular sheet $= 54 \times 4 = 216\ \text{cm}^2$

Surface area of cube $= 6\ (\text{side})^2$

$216 = 6\ (\text{side})^2$

Side $= 6$

Volume of cube $= 6^3 = 216$

Radius of cylinder $= \dfrac{4}{2\pi}$

Volume of cylinder $= \pi r^2 h$

$$= \pi\left(\frac{4}{2\pi}\right)^2 \times 54 = \frac{216}{\pi}$$

$$\frac{\text{Volume of cylinder}}{\text{Volume of cube}} = \frac{\frac{216}{\pi}}{216} = \frac{1}{\pi}$$

1.214 (a)

Given dimension 20×8 folded 3 times symmetrically along perpendicular to its longest edge.

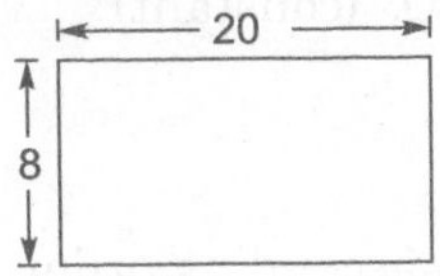

After folding first times

We get
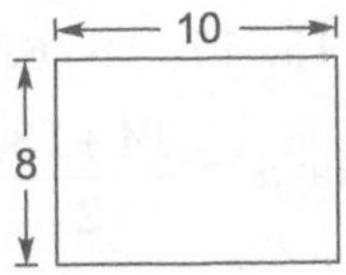

After folding second times

We get
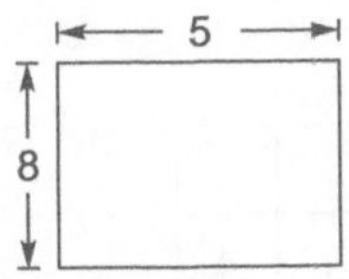

After folding third times

We get
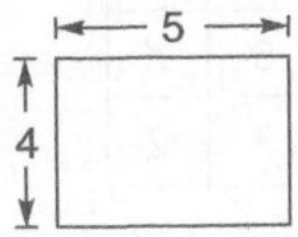

So, the perimeter of final figure $= 2(4 + 5)$
$= 18$ unit

1.215 (b)

Total fat = Saturated fat + unsaturated fat
$= 20 + 20 + 5 = 45\%$

Total fat in Kcal $= 2000 \times \dfrac{45}{100} = 900$ Kcal

Given all fat energy density $= 9$ Kcal/g

There of total fat in gram $= \dfrac{900}{9} = 100$ g

1.217 (d)

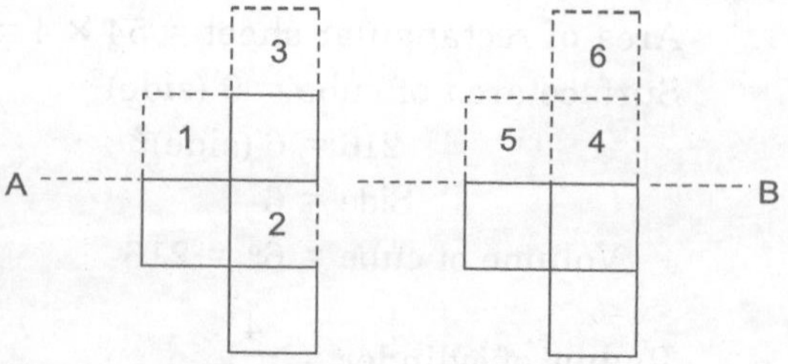

1.218 (b)

$$\text{Percentage of ₹5} = \frac{5\times3}{1\times5+5\times3+10\times13}\times100$$

$$= \frac{15}{150}\times100 = 10\%$$

1.219 (a)

Given $(x + y) \propto (x - y)$

$\therefore \quad (x + y) = k(x - y)$

$$x + y = kx - ky$$

$$(k - 1)x = (k + 1)y$$

$$\frac{x}{y} = \frac{k+1}{k-1} = k' \text{ (constant)}$$

1.220 (c)

Arrange given series in ascending order

9, 10, 11, 11, 13, 14, 15, 17, 18, 69

$$\text{Median of the given data} = \frac{14+13}{2} = 13.5$$

1.222 (b)

1	X	4	3
X	5	5	4
3	X	6	X
2	4	3	2

Sum of fourth row = 2 + 4 + 3 + 2 = 11

1.223 (b)

$$6 + 3 = 9$$
$$9 + 5 = 14$$
$$14 + 7 = \boxed{21}$$
$$21 + 9 = 30$$
$$30 + 11 = 41$$

1.224 (b)

Combination of $4 \times 3 \times 2 \times 1 = 24$

1.225 (b)

If S.P. is same than overall loss $= \dfrac{x^2}{100} = 1\%$

1.228 (a)

Number of students = 10000

Number of students not interested in both branches = 1500

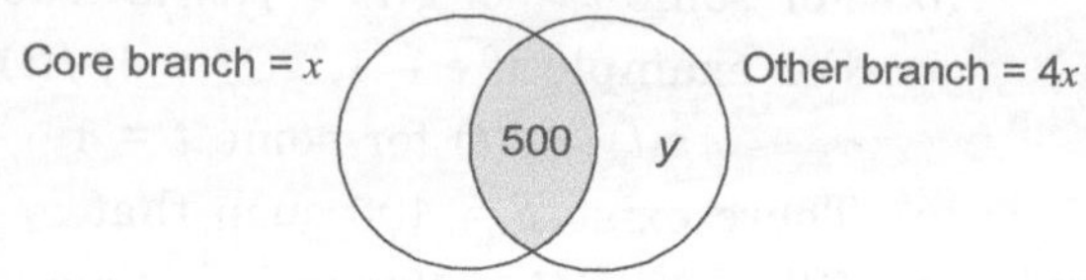

$\therefore$ When diagram represent $10000 - 1500 = 8500$

x = Core

y = Other branch (core branch not include) = $4x$

$$x + 4x = 8500 + 500$$
$$5x = 9000$$
$$x = 1800$$

1.230 (c)

$$\log\frac{(x+y)}{2} = \frac{1}{2}(\log x + \log y)$$

$$\log\left(\frac{x+y}{2}\right) = \log\sqrt{xy}$$

$$\frac{x+y}{2} = \sqrt{xy}$$

$$x^2 + y^2 + 2xy = 4xy$$

$$x^2 + y^2 = 2xy$$

$$\frac{x}{y} + \frac{y}{x} = 2$$

■■■■